Handbook of Polymer Science and Technology

Volume 1:
SYNTHESIS AND PROPERTIES

Handbook of Polymer Science and Technology

edited by
Nicholas P. Cheremisinoff

Volume 1   Synthesis and Properties
Volume 2   Performance Properties of Plastics and Elastomers
Volume 3   Applications and Processing Operations
Volume 4   Composites and Specialty Applications

Other Volumes in Preparation

Handbook of Polymer Science and Technology

Volume 1:
SYNTHESIS AND PROPERTIES

edited by

Nicholas P. Cheremisinoff

MARCEL DEKKER, INC.　　New York and Basel

ISBN 0-8247-8173-2

Copyright © 1989 by MARCEL DEKKER, INC. All Rights Reserved

Neither this book nor any part may be reproduced or transmitted in any form or by any means, electronic or mechanical, including photocopying, microfilming, and recording, or by any information storage and retrieval system, without permission in writing from the publisher.

MARCEL DEKKER, INC.
270 Madison Avenue, New York, New York 10016

Current printing (last digit):
10 9 8 7 6 5 4 3 2

PRINTED IN THE UNITED STATES OF AMERICA

Preface

The *Handbook of Polymer Science and Technology* is a series of volumes aimed at providing a comprehensive, authoritative guide to the field of applied polymer science as well as the technologies related to its engineering applications. Its purpose is to help provide unification between the theoretical foundations of polymer science and such practical manufacturing concepts as synthesis, property characterization, fabrication and processing, and product design and development. The handbook is organized by topic into a multivolume series so that each subject area can be given in-depth treatment in a single volume.

Volume 1, *Synthesis and Properties*, covers both applied and theoretical principles of polymer chemistry. Material coverage includes polymerization kinetics, reactor design and operation, and the molecular characterization of polymeric products. Section I is organized into twelve chapters providing in-depth discussions and reviews of polymerization kinetics. The synthesis and the structural and engineering properties of important polymer classes such as PVC, nylon, polystyrene derivatives, and polyesters are highlighted. Included in this section are detailed discussions on catalysis as related to the production of these various polymers. Section II contains seven chapters covering analytical techniques for characterizing molecular and viscoelastic properties. An understanding of the principles behind key characterization techniques and the proper interpretation of data derived are essential to product development. These particular discussions are heavily cross-referenced and are referred to in subsequent volumes dealing with material performance properties.

This first volume of the *Handbook of Polymer Science and Technology* series represents the efforts of twenty-three experts. Each author is to be viewed responsible for the statements, recommendations and data within his or her respective chapter. In addition, the work embodies the opinions and suggestions of numerous engineers, chemists, and researchers, all of whom have helped to shape the organization of not only this first volume but the entire series. Heartfelt gratitude is extended to all these individuals for their time and effort.

Nicholas P. Cheremisinoff

Contents

Preface iii

Contributors vii

I SYNTHESIS AND REACTION KINETICS

1. Overview of Polymerization Kinetics 1
   Zhou Pu

2. Overview of Polymerization Reactor Technology 67
   K. Y. Choi

3. Principles of Polymerizations with Ziegler–Natta Catalysts 103
   Y. V. Kissin

4. Synthesis and Characterization of Block Copolymers 133
   Baki Hazer

5. Synthesis and Characterization of Aromatic Polyesters 177
   Yoshio Imai and Masa-aki Kakimoto

6. Nylon Polymerization 211
   Santosh K. Gupta

7. The Radical Polymerization of Vinyl Esters 249
   Mikiharu Kamachi

8. Electroinitiated Cationic Polymerization 271
   Levant Toppare

9. Radiation-Induced Reactions of Polystyrene Derivatives 307
   Katsumi Tanigaki and Kazuo Tateishi

10.	Advances in PVC Polymerization Bertil Törnell	347
11.	Kinetic Modeling of Polymerization Reactions Anil Kumar and Pankaj K. Khandelwal	375
12.	Optimization of Polymerization Reactors Jorge N. Farber	429

II POLYMER CHARACTERIZATION AND MOLECULAR STRUCTURE

13.	Techniques for Polymer Property Characterization Nicholas P. Cheremisinoff	471
14.	Optical Microscopy for Studying Molecular Ordering in Polymers Christopher Viney	505
15.	Microscopy Techniques for Polymer Characterization L. Bartosiewicz and C. J. Kelly	537
16.	Viscoelastic and Molecular Characterization of Commercial Polymers I. P. Briedis	605
17.	Thermally Stimulated Depolarization Techniques for Studying Polymer Relaxation Alexander L. Kovarskii	643
18.	Structural Characterization of Styrene–Butadiene Copolymers by Ozonolysis—GPC and HPLC Methods Yasuyuki Tanaka	677
19.	Gelation of Polymer Solutions Jean-Michel Guenet	711

Index	763
About the Author	785

Contributors

L. Bartosiewicz   Physical Analysis & Technical Services Department, Ford Motor Company, Dearborn, Michigan

I. P. Briedis   Institute of Polymers Mechanics, Latvian SSR Academy of Sciences, Riga, USSR

Nicholas P. Cheremisinoff   Polymers Technology Division, Exxon Chemical Company, Linden, New Jersey

K. Y. Choi   Department of Chemical and Nuclear Engineering, University of Maryland, College Park, Maryland

Jorge N. Farber* Planta Piloto de Ingenieria Quimica, UNS–CONICET, Bahía Blanca, Argentina

Jean-Michel Guenet   Institute Charles Sadron (CNRS), Strasbourg, France

Santosh K. Gupta†   Department of Chemical Engineering, University of Notre Dame, Notre Dame, Indiana

Baki Hazer   Department of Chemistry, Karadeniz Technical University, Trabzon, Turkey

Yoshio Imai   Department of Textile and Polymeric Materials, Tokyo Institute of Technology, Meguro-ku, Tokyo, Japan

Masa-aki Kakimoto   Department of Textile and Polymeric Materials, Tokyo Institute of Technology, Meguro-ku, Tokyo, Japan

Mikiharu Kamachi   Department of Macromolecular Science, Faculty of Science, Osaka University, Toyonaka, Osaka, Japan

---

Current affiliations:
*Tremco Ltd., Toronto, Ontario, Canada
†On leave from Indian Institute of Technology, Kanpur, India

C. J. Kelly   Central Laboratory, Ford Motor Company, Dearborn, Michigan

Pankaj K. Khandelwal   Department of Chemical Engineering, Indian Institute of Technology–Kanpur, Kanpur, India

Y. V. Kissin   Mobil Chemical Company, Edison, New Jersey

Alexander L. Kovarskii   Institute of Chemical Physics, Academy of Sciences of the USSR, Moscow, USSR

Anil Kumar   Department of Chemical Engineering, Indian Institute of Technology–Kanpur, Kanpur, India

Zhou Pu   Department of Organo–Chemical Engineering, Jiangsu Institute of Chemical Technology, Changzhou, Jiangsu, People's Republic of China

Yasuyuki Tanaka   Department of Material Systems Engineering, Faculty of Technology, Tokyo University of Agriculture and Technology, Konganei, Tokyo, Japan

Katsumi Tanigaki   Fundamental Research Laboratories, NEC Corporation, Miyamae-ku, Kawasaki City, Japan

Kazuo Tateishi   Environment Research Laboratories, NEC Corporation, Miyamae-ku, Kawasaki City, Japan

Levant Toppare   Macromolecular Research Division, Department of Chemistry, Faculty of Arts and Sciences, Middle East Technical University, Ankara, Turkey

Bertil Törnell   Department of Chemical Engineering, Chemical Center, University of Lund, Lund, Sweden

Christopher Viney*   IBM Almaden Research Center, San Jose, California

---

Current affiliation: Department of Materials Science and Engineering, University of Washington, Seattle, Washington

I
SYNTHESIS AND REACTION KINETICS

1
Overview of Polymerization Kinetics

Zhou Pu
Jiangsu Institute of Chemical Technology
Changzhou, Jiangsu, People's Republic of China

INTRODUCTION	1
KINETICS OF CHAIN REACTION POLYMERIZATION	2
Radical Chain Polymerization	3
Cationic Polymerization	10
Anionic Polymerization	14
Ring-Opening Polymerization	19
Chain Copolymerization	23
Ziegler–Natta Polymerization	29
KINETICS OF STEP POLYMERIZATION	34
KINETICS OF GROUP TRANSFER POLYMERIZATION	47
THE CALCULATION OF MOLECULAR WEIGHT DISTRIBUTIONS FROM KINETIC SCHEMES	51
REFERENCES	62

INTRODUCTION

There are three major facets in the study of polymerization kinetics. The first is the experimental side, including the measurement of polymerization rates. The second is the reaction mechanism. The third, the theoretical study of polymerization, now is highly developed. In addition, the molecular weight distribution (MWD) is closely related to polymerization kinetics. Polymerization generally is classified into chain and step polymerization based on the polymerization mechanism [1]. Chain polymerizations require an initiator, from which an initiator species with a reactive center is produced. The reactive center may be a free radical, a cation, or an anion. The full-sized polymer molecules are produced almost immediately after the start of the reaction. Step polymerizations proceed by the stepwise reaction between the functional groups of reactants as in reactions. The size of the polymer molecule increases at a relatively slow rate in such polymerization. Recently a new method of polymerization, group transfer polymerization (GTP), has received a good deal of attention [2–4]. Here the reactive group is transferred from the initiator to the incoming monomer. However, few studies have appeared on kinetic data and reaction mechanisms based on this new method [5–8].

KINETICS OF CHAIN REACTION POLYMERIZATION

The kinetics of chain reaction polymerization have been studied theoretically and experimentally in great detail; extensive reviews of this topic have appeared [1, 9, 10].

Chain polymerization is initiated by a reactive species R^* produced from some compound I termed an initiator:

$$I \rightarrow R^* \tag{1}$$

The reactive species, which may be a free radical, cation, or anion, adds to a monomer molecule by opening the π-bond to form a new radical cation or anion center as the case may be. The process is repeated as many more monomer molecules are successively added to continuously propagate the reactive center:

$$R^* \xrightarrow{CH_2=CHY} R-CH_2-\underset{Y}{\overset{H}{C^*}} \xrightarrow{CH_2=CHY} R-CH_2-\underset{Y}{\overset{H}{C}}-CH_2-\underset{Y}{\overset{H}{C^*}} \dashrightarrow$$

$$R-\left[-CH_2-\underset{Y}{\overset{H}{C}}-\right]_m-CH_2-\underset{Y}{\overset{H}{C^*}} \tag{2}$$

Polymer growth is terminated at some point by destruction of the reactive center by an appropriate reaction depending on the type of reactive center and the particular reaction conditions. This type of system is treated mathematically as the steady-state approximation [11]. The essential criterion for the existence of a steady state is that the overall rate of change of concentration of chain carriers is very much less than both their rate of production and their rate of destruction. It must be emphasized that this treatment is, after all, an approximate method. Although steady-state treatment is not strictly accurate, it is nevertheless of wide utility, as we shall see in the following discussion.

Whether a particular monomer can be converted into a polymer depends on thermodynamic and kinetic considerations. Polymerization is possible only if the free energy difference ΔG between monomer and polymer is negative [12]. A negative ΔG does not, however, mean that polymerization will be observed under a particular set of reaction conditions (type of initiation, temperature, etc.). The ability to carry out a thermodynamically feasible polymerization depends on its kinetic feasibility: on whether the process proceeds at a reasonable rate under a proposed set of reaction conditions.

Although radical, cationic, and anionic initiators are used in chain polymerizations, they cannot be used indiscriminately, since all three types of initiation do not work for all monomers [13]. Monomers showing varying degrees of selectivity with regard to the type of behaviors can be seen in Table 1. Thus, although the polymerization of all the monomers in Table 1 is thermodynamically feasible, kinetic feasibility is achieved in many cases only with a specific type of initiation.

Table 1 Type of Chain Polymerization Undergone by Various Unsaturated Monomers

	Type of initiation		
Monomer	Radical	Cationic	Anionic
Ethylene	+	−	+
1-Alkyl olefins (α-olefins)	−	+	−
1,1-Dialkyl olefins	−	+	−
1,3-Dienes	+	+	+
Styrene, α-methyl styrene	+	+	+
Halogenated olefins	+	−	−
Vinyl esters (CH_2=CHOCOR)	+	−	−
Acrylates, methacrylates	+	−	+
Acrylonitrile, methacrylonitrile	+	−	+
Acrylamide, methacrylamide	+	−	+
Vinyl ethers	−	+	−
N-Vinyl carbazole	+	+	−
N-Vinyl pyrrolidone	+	+	−
Aldehydes, ketones	−	+	+

Radical Chain Polymerization

Radical chain polymerization is a chain reaction consisting of a sequence of three steps: initiation, propagation, and termination. The initiation step is considered to involve two reactions. The first is the production of free radicals by any one of a number of reactions. The usual case is the homolytic dissociation of an initiator or catalyst species I to yield a pair of radicals R:

$$I \xrightarrow{k_d} 2R \cdot \tag{3}$$

where k_d is the rate constant for the catalyst dissociation. The second part of the initiation involves the addition of this radical to the first monomer molecule to produce the chain initiating species M_1:

$$R \cdot + M \xrightarrow{k_i} M_1 \cdot \tag{4}$$

where M represents a monomer molecule and k_i is the rate constant for the initiation step. For the polymerization of CH_2=CHY, Eq. 4 takes the form

$$R \cdot + CH_2=CHY \rightarrow R-CH_2-\underset{\underset{Y}{|}}{\overset{\overset{H}{|}}{C}} \cdot \tag{5}$$

Propagation consists of the growth of $M_1 \cdot$ by the successive additions of large numbers of monomer molecules according to Eq. 2. The successive additions may be represented by

$$M_1\cdot + M \xrightarrow{k_{p1}} M_2\cdot \tag{6}$$

$$M_2\cdot + M \xrightarrow{k_{p2}} M_3\cdot \tag{7}$$

$$M_3\cdot + M \xrightarrow{k_{p3}} M_4\cdot \tag{8}$$

etc., or in general terms

$$M_n\cdot + M \xrightarrow{k_{pn}} M_{n+1}\cdot \tag{9}$$

where k_{p1}, k_{p2}, k_{p3} ... k_{pn} are the rate constants for propagation. Propagation with the growth of a chain to high polymer proportions takes place very rapidly. The value of k_p for radical chain polymerization is much larger than the values usually encountered in step polymerization [13].

At some point, the propagating polymer chain stops growing and terminates. Termination with the annihilation of the radical centers occurs by bimolecular reaction between radicals. Two radicals react with each other by combination (coupling):

$$\sim\sim CH_2-\underset{\underset{Y}{|}}{\overset{\overset{H}{|}}{C}}\cdot + \cdot\underset{\underset{Y}{|}}{\overset{\overset{H}{|}}{C}}-CH_2\sim\sim \xrightarrow{k_{tc}} \sim\sim CH_2-\underset{\underset{Y}{|}}{\overset{\overset{H}{|}}{C}}-\underset{\underset{Y}{|}}{\overset{\overset{H}{|}}{C}}-CH_2\sim\sim \tag{10}$$

or, rarely, by disproportionation in which a hydrogen radical that is beta to one radical center is transferred to another radical center. This results in the formation of two polymer molecules, one saturated and one unsaturated:

$$\sim\sim CH_2-\underset{\underset{Y}{|}}{\overset{\overset{H}{|}}{C}}\cdot + \cdot\underset{\underset{Y}{|}}{\overset{\overset{H}{|}}{C}}-CH_2\sim\sim \xrightarrow{k_{td}} \sim\sim CH_2-\underset{\underset{Y}{|}}{\overset{\overset{H}{|}}{CH}} + \underset{\underset{Y}{|}}{\overset{\overset{H}{|}}{C}}=\overset{\overset{H}{|}}{C}\sim\sim \tag{11}$$

Termination can also occur by a combination of coupling and disproportionation. The two different modes of termination can be represented in general terms by

$$M_n\cdot + M_m\cdot \xrightarrow{k_{tc}} M_{n+m} \tag{12}$$

$$M_n\cdot + M_m\cdot \xrightarrow{k_{td}} M_n + M_m \tag{13}$$

where k_{tc} and k_{td} are the rate constants for termination by coupling and disproportionation, respectively. It can also be expressed by

$$M_n\cdot + M_m\cdot \xrightarrow{k_t} \text{dead polymer} \tag{14}$$

where the particular mode of termination is not specified and

$$k_t = k_{tc} + k_{td} \tag{15}$$

Besides these three reactions, chain transfer also takes place in many polymerizations. It often leads to a decrease in the molecular weight of the polymer. This effect results from the premature termination of a growing polymer by the transfer of a hydrogen or other atom or species to the polymer from some compound present in the system: the monomer, initiator, or solvent, as the case may be. It may be depicted as

$$M_n\cdot + XA \xrightarrow{k_{tr}} M_n - X + A\cdot \qquad (16)$$

where XA may be monomer, initiator, solvent, or other substance, X is the atom or species transferred, and A is a new radical which then reinitiates polymerization.

The Simple Kinetic Equation of Radical Chain Polymerization

In deriving the simple kinetic relations pertaining to the preceding reaction scheme, four simplifying assumptions are normally made.

1. The length of the propagating chain is large, so that the total rate of monomer consumption may be equated to the rate of consumption in the propagation alone.

2. The rate constants for both propagation and termination are independent of chain length, i.e., independent of the size of the propagating polymer molecule, and of conversion, so that $k_{p1} = k_{p2} = k_{p2} = k_{pn}$, and similarly for individual values of k_{tc} and k_{td}. It is reasonable [14], especially for long chains, since the reactivity of a radical is determined by the molecular structure in the vicinity of the unpaired electron and in a homopolymerization all propagating radicals have the same structure, differing only in the lengths of the polymer chains of which they form the terminal unit.

3. A steady state in radical concentration is established, allowing the rate of change of concentration of any radical intermediate to be equated to zero. Such a situation is closely approximate in many systems, as the radical concentration builds up (rapidly in the initial stages) at a constant rate of radical formation, an increase in termination follows the increase in radical concentration, and the rates of the two processes are equal at infinite time.

4. Termination involving primary radicals, i.e., bimolecular termination between $R\cdot$ and $M_n\cdot$, is negligible. This assumption is valid for most systems under conditions of long chain length, since the concentration of $R\cdot$ is small compared with the total radical concentration.

With the aid of these four assumptions, the rate expression may be derived. The monomer disappears by the initiation reaction as well as by the propagation reactions. The rate of monomer disappearance, which is synonymous with the rate of polymerization, is given by

$$\frac{-d(M)}{dt} = R_i + R_p \qquad (17)$$

where R_i and R_p are the rates of initiation and propagation, respectively. However, the number of monomer molecules reacting in the initiation step is far less than the number in the propagation step for a process producing a high polymer. To a very close approximation the former can be neglected and the polymerization rate is given simply by the rate of propagation:

$$\frac{-d(M)}{dt} = R_p \qquad (18)$$

The rate of propagation, and therefore the rate of polymerization, is the sum of many individual propagation steps. Since the rate constants for the propagation steps are the same, one can express the polymerization rate by

$$R_p = k_p[\text{M·}][\text{M}] \tag{19}$$

where [M] is the monomer concentration and [M·] the total concentration of all chain radicals, i.e., $[\text{M·}] = \Sigma_n[\text{M}_n\text{·}]$.

Equation 19 for the polymerization rate is not directly usable because the radical concentrations are too low to measure quantitatively. It is therefore desirable to eliminate [M·] from Eq. 19. To do this, the steady-state assumption is made. The rate of change of the concentration of radicals quickly becomes and remains zero during the course of polymerization. This is equivalent to stating that the rates of initiation R_i and termination R_t of radicals are equal, or

$$R_i = 2k_t[\text{M·}]^2 \tag{20}$$

The right side of Eq. 20 represents the rate of termination. There is no specification as to whether termination is by coupling or disproportionation since both follow the same kinetic expression. The use of factor 2 in the termination rate equation follows the generally accepted convention for reactions destroying radicals in pairs. It is also generally employed for reactions creating radicals in pairs, as in Eq. 3. (In using the polymer literature one should be aware that the factor 2 is not universally employed.) Rearranging Eq. 20 yields

$$[\text{M·}] = \left(\frac{R_i}{2k_t}\right)^{1/2} \tag{21}$$

and substituting Eq. 21 into Eq. 19 yields

$$R_p = k_p[\text{M}]\left(\frac{R_i}{2k_t}\right)^{1/2} \tag{22}$$

for the rate of polymerization. From Eq. 22 it can be concluded that the polymerization rate is dependent on the square root of the initiation rate.

The initiation rate is simply shown as R_i. A variety of initiator systems can be used to bring about the polymerization. Radicals can be produced by a variety of methods, thermal, photochemical, redox, etc.

The thermal homolytic dissociation of initiators is the most widely used mode of generating radicals to initiate polymerization—for both commercial polymerizations and theoretical studies. Several different types of peroxides are widely used, for example,

$$\text{Ph-C(=O)-O-O-C(=O)-Ph} \longrightarrow 2\ \text{Ph-C(=O)-O·} \tag{23}$$

The rate of producing primary radicals by thermal homolysis of an initiator R_d (Eq. 3) is given by

$$R_d = 2fk_d[\text{I}] \tag{24}$$

where [I] is the concentration of the initiator and f the initiator efficiency. The value of f is usually less than unity.

The initiation reaction in polymerization is composed of two steps (Eqs. 3 and 4) as discussed previously. In most polymerizations, the second step (the addition of the primary radical to the monomer) is much faster than the first step. The rate of initiation is then given by

$$R_i = 2fk_d[I] \tag{25}$$

Substitution of Eq. 25 into Eq. 22 yields

$$R_p = k_p[M]\left(\frac{fk_d[I]}{k_t}\right)^{1/2} \tag{26}$$

This expression has been abundantly confirmed for many different monomer–initiator combinations over wide ranges of monomer and initiator concentrations [14–19].

Many oxidation–reduction reactions produce radicals that can be used to initiate polymerization [20–22]. This type of initiation is referred to as redox initiation, redox catalysis, or redox activation. For example, the reduction of hydrogen peroxide with ferrous ion may be used to initiate polymerization.

$$H_2O_2 + Fe^{2+} \rightarrow HO^- + HO\cdot + Fe^{3+} \tag{27}$$

The initiation and polymerization rates will be given by appropriate expressions that are very similar to those developed previously:

$$R_i = k_d[\text{reductant}][\text{oxidant}] \tag{28}$$

$$R_p = k_p[M]\left(\frac{k_d[\text{reductant}][\text{oxidant}]}{2k_t}\right)^{1/2} \tag{29}$$

Photochemical or photoinitiated polymerizations occur when radicals are produced by ultraviolet and visible light irradiation of a reaction system [23, 24]. In general, light absorption results in radical production by either of two pathways:

1. Some compound or compounds in the system undergo excitation by energy absorption and subsequent decomposition into radicals.
2. Some compound undergoes excitation and the excited species interacts with a second compound (by either energy transfer or redox reaction) to form radicals derived from the latter and/or former compound(s).

The irradiation of some monomers results in the formation of an excited state M* by the absorption of light photons (quanta):

$$M + h\nu \rightarrow M^* \tag{30}$$

The excited species undergoes homolysis to produce radicals

$$M^* \rightarrow R\cdot + R'\cdot \tag{31}$$

capable of initiating the polymerization of the monomer. The rate of photochemical initiation is given by

$$R_i = 2\phi I_a \tag{32}$$

where I_a is the intensity of absorbed light in moles of light quanta (called Einsteins in photochemistry) per liter-second and ϕ is the number of propagating chains initiated per light photon absorbed.

An expression for the polymerization rate is obtained by combining Eq. 32 with Eq. 22 to yield

$$R_p = k_p[\text{M}]\left(\frac{\phi I_a}{k_t}\right)^{1/2} \tag{33}$$

It is often convenient to express the absorbed light intensity by

$$I_a = I_0\epsilon[\text{A}]b \tag{34}$$

where I_0 is the incident light intensity, A the species which undergoes photoexcitation, ϵ the molar absorptivity (extinction coefficient) of A at the particular frequency of radiation absorbed, and b the thickness of the reaction system being irradiated. Combination of Eq. 34 and Eqs. 32 and 22 allows R_i and R_p to be expressed as

$$R_i = 2\phi\epsilon I_0[\text{A}]b \tag{35}$$

$$R_p = k_p[\text{M}]\left(\frac{\phi\epsilon I_0[\text{A}]b}{k_t}\right)^{1/2} \tag{36}$$

The use of Eqs. 35 and 36 assumes that the incident light intensity does not vary appreciably throughout the thickness of the reaction vessel. This will be true only when the absorption of light is low or very thin reaction vessels are employed. For most polymerizations, an expression for I can be obtained from the Lambert–Beer law

$$I = I_0 e^{-\epsilon[\text{A}]b} \tag{37}$$

where I is the incident light at a distance b into the reaction vessel. The light intensity absorbed by the reaction system is then given by

$$I_a = I_0(1 - e^{-\epsilon[\text{A}]b}) \tag{38}$$

The polymerization rate can then be analyzed by combining Eqs. 33 and 38 to yield

$$R_p = k_p[\text{M}]\left\{\frac{\phi I_0(1 - e^{-\epsilon[\text{A}]b})}{k_t}\right\}^{1/2} \tag{39}$$

For a more comprehensive discussion the reader is referred to standard textbooks on the theory of polymerization kinetics [1, 11, 13, 14, 20].

The rest of this section describes the kinetics of inhibition or retardation [25–31]. Inhibitors prevent the formation of measurable amounts of polymer under conditions that otherwise would permit such polymerization. Retarders reduce the rate at which polymer is formed. Polymerization is completely stopped by benzoquinone, a typical inhibitor, during an induction or inhibition period. Nitrobenzene, a retarder, lowers the polymerization rate without an inhibition period.

The kinetics of retarded or inhibited polymerizations can be analyzed using a scheme consisting of the usual initiation, propagation, and termination reactions in addition to the inhibition reaction

$$\text{M}_n\cdot + \text{Z} \xrightarrow{k_z} \text{M}_n + \text{Z}\cdot \tag{40}$$

where Z is the inhibitor or retarder. The kinetics are relatively simplified if one assumes that the inhibitor Z· does not reinitiate polymerization and that it terminates without regeneration of the original inhibitor molecule.

The steady-state assumption for the radical concentration leads to

$$\frac{d[M\cdot]}{dt} = R_i - 2k_t[M\cdot]^2 - k_z[Z][M\cdot] = 0 \tag{41}$$

which can be combined with Eq. 19 to yield

$$\frac{2R_p^2 k_t}{k_p^2[M]^2} + \frac{R_p[Z]k_z}{k_p[M]} - R_i = 0 \tag{42}$$

A consideration of Eq. 42 shows that R is inversely proportional to the ratio k_z/k_p of the rate constants for inhibition and propagation. This ratio is often referred to as the inhibition constant z:

$$z = \frac{k_z}{k_p} \tag{43}$$

Two limiting cases of Eq. 42 exist. When the second term is negligible compared to the first, the polymerization is not retarded, and Eq. 42 simplifies to Eq. 25. For the case where the retardation is strong ($k_z/k_p \gg 1$), normal bimolecular termination will be negligible. Under these conditions the first term in Eq. 42 is negligible and one has

$$\frac{R_p[Z]k_z}{k_p[M]} - R_i = 0 \tag{44}$$

or

$$R_p = \frac{k_p[M]R_i}{k_z[Z]} = \frac{-d[M]}{dt} \tag{45}$$

Equations 44 and 45 show the rate of retarded polymerization to be dependent on the first power of the initiation rate. Further, R_p varies inversely as the inhibition concentration.

Ionic Chain Polymerization

Almost all monomers containing the carbon–carbon double bond undergo radical polymerization, whereas ionic polymerization is highly selective (see Table 1). Cationic polymerization is essentially limited to those monomers with electron-releasing substituents such as alkoxy, phenyl, vinyl, and 1,1-dialkyl [32–38]. Anionic polymerization takes place with monomers possessing electron-withdrawing groups such as nitrile, carboxyl, phenyl, and vinyl [39–43]. The selectivity of ionic polymerization is due to the very strict requirements for stabilization of anionic and cationic propagating species. The rates of ionic polymerizations are usually faster than those of radical polymerizations. The highly polar hydroxylic solvents (water, alcohols) react with and destroy most ionic initiators. Other polar solvents such as ketones prevent initiation of polymerization by forming highly stable complexes with the initiators. Ionic polymerizations are therefore usually carried out in solvents of low or moderate polarity such as methyl chloride, ethylene dichloride, and pentane, although moderately high-polarity solvents such as nitrobenzene are also used. In such solvents one usually does not have only a single type of propagating species. For any propagating species such as $\sim\!\!\sim\!\!BA$ in cationic polymerization, one can visualize the range of behaviors from one extreme of a completely covalent species (I) to the other of a completely free (and highly solvated) ion (IV)

$\sim\!\!\sim\!\!BA$	$\sim\!\!\sim\!\!B^+(A)^-$	$\sim\!\!\sim\!\!B^+ \parallel A^-$	$\sim\!\!\sim\!\!B^+ \; A^-$
(I)	(II)	(III)	(IV)

The intermediate species include the tightly bound or contact ion pair (II) (also referred to as the intimate ion pair) and the solvent-separated or loose external ion pair (III). The intimate ion pair has a counterion or gegenion, of opposite charge close to the propagating center (unseparated by solvent) through its lifetime. A propagating cationic chain end has a negative counterion; a propagating anionic chain end has a positive counterion. It is often useful to consider that two types of propagating species are involved, an ion pair and a free ion (IV), coexisting in equilibrium with each other. That is the reason why the cationic polymerization is very complicated. Ionic polymerizations are characterized by a wide variety of initiation and termination. Termination of a propagating chain occurs by its reaction with the gegenion, solvent, or other species present in the ionic polymerization system.

Cationic Polymerization

Various initiators can be used to bring about the polymerization of a monomer with electron-releasing substituents. Protonic acids can be used to some extent to initiate cationic polymerization by protonation of the olefin. The method depends on the use of an acid that is strong enough to produce a reasonable concentration of the protonated species:

$$HA + RR'C=CH_2 \rightarrow RR'C^+(A)^- \quad \text{(46)}$$
$$\qquad\qquad\qquad\qquad\quad |$$
$$\qquad\qquad\qquad\qquad\; CH_3$$

but the anion of the acid should not be highly nucleophilic, or it will terminate the protonated olefin by combination (i.e., by covalent bond formation):

$$RR'C^+(A)^- \rightarrow RR'C-CH_3 \quad \text{(47)}$$
$$\quad |\qquad\qquad\qquad |$$
$$\;CH_3\qquad\qquad\; A$$

A variety of Lewis acids used to initiate cationic polymerization, generally at low temperatures, may yield high polymer molecular weights [44]. These include metal halide (e.g., BF_3 [45a], $AlCl_3$, $SnCl_5$, $SbCl_5$, $ZnCl_2$, $TiCl_4$, PCl_5) [45b], organometallic derivatives (e.g., $RAlCl_2$, R_2AlCl, R_3Al) [46, 47], and oxyhalides (e.g., $POCl_3$, CrO_2Cl, $SOCl_2$, $VOCl_3$) [48, 49]. Lewis acids are by far the most important means of initiating cationic polymerization from the industrial viewpoint. Initiation by Lewis acids either requires or proceeds faster in the presence of either a proton donor (protogen) such as water, alcohol, and organic acids or a cation donor (cationogen) such as t-butyl chloride or triphenylmethyl fluoride. Thus dry isobutylene is unaffected by dry boron trifluoride but polymerization occurs immediately when trace amounts of water are added [50]. The initiator and coinitiator, comprising an initiating system, react to form an initiator–coinitiator complex (or syncatalyst system), which then proceeds to donate a proton or other cation (usually a carbenium ion) to the monomer and thus to initiate propagation. The coinitiation process for boron trifluoride and water is

$$BF_3 + H_2O \rightleftharpoons BF_3 \cdot OH_2$$
$$BF_3 \cdot OH_2 + (CH_3)_2C=CH_2 \rightarrow (CH_3)_3C^+(BF_3OH)^- \quad \text{(48)}$$

Initiation by aluminum chloride and t-butyl chloride is described by

$$AlCl_3 + (CH_3)_3CCl \rightleftharpoons (CH_3)_3C^+(AlCl_4)^-$$
$$(CH_3)_3C^+(AlCl_4)^- + CH(C_6H_5){=}CH_2 \rightarrow (CH_3)_3CCH_2C^+H(C_6H_5)(AlCl_4)^- \quad (49)$$

The initiation process can be generalized as

$$I + ZY \underset{}{\overset{k}{\rightleftharpoons}} Y^+(IZ)^-$$
$$Y^+(IZ)^- + M \overset{k_i}{\rightarrow} YM^+(IZ)^- \quad (50)$$

where I, ZY, and M represent the coinitiator, initiator, and monomer, respectively, and k represents the rate constant of the complex reaction. The stronger Lewis acids such as $AlCl_3$, $AlBr_3$, and $TiCl_4$ may be able to initiate polymerization by a self-ionization process in addition to the coinitiation process, for example, the bimolecular ionization

$$2AlBr_3 \rightleftharpoons AlBr_2^+(AlBr_4)^- \quad (51)$$

followed by reaction with the monomer

$$AlBr_2^+(AlBr_4)^- + M \rightarrow AlBr_2M^+(AlBr_4)^- \quad (52)$$

The Lewis acid acts as both the initiator and coinitiator in this process. An alternate self-ionization mechanism is the direct addition of initiator to monomer,

$$TiCl_4 + M \rightarrow TiCl_3M^+Cl^- \quad (53)$$

Other cationic initiators that have been studied include acetyl perchlorate [51, 52], iodine [53], electrolytic initiation [54], and ionizing radiation [55, 56]. Acetyl perchlorate initiates polymerization probably by addition of the acylium ion to monomer:

$$CH_3\underset{\underset{O}{\|}}{C}{}^+(ClO_4)^- + M \rightarrow CH_3\underset{\underset{O}{\|}}{C}M^+(ClO_4)^- \quad (54)$$

Initiation may be accomplished by addition of iodine followed by ionization:

$$I_2 + CH_2{=}CH(OR) \rightarrow ICH_2{-}CHI(OR) \overset{I_2}{\rightarrow} ICH_2{-}C^+H(OR)(I_3)^- \quad (55)$$

Electrolytic or electroinitiated polymerization involves initiation by cations formed via electrolysis of some component (either deliberately added, perhaps as the electrolyte, or adventitiously present) in the reaction system. Thus initiation in the presence of perchlorate ion proceeds by oxidation of perchlorate followed by hydrogen abstraction,

$$ClO_4^- \overset{-e}{\rightarrow} ClO_4\cdot \overset{HA}{\rightarrow} HClO_4 \quad (56)$$

where HA is monomer, solvent, or other species in the system. Perchloric acid is the actual initiator.

Ionizing radiation also initiates cationic polymerization, probably through monomer radical cations,

$$M \xrightarrow{\text{radiation}} \cdot M^+ + e \tag{57}$$

This initiation is of interest in that it is the only method of initiation that yields the free ion species devoid of a gegenion. The initiator ion pair produced in the initiation step proceeds to grow by the successive addition of monomer molecules:

$$H\text{-}[CH_2C(CH_3)_2\text{-}]_n^+ (BF_3OH)^- + (CH_3)_2C=CH_2 \rightarrow$$
$$H\text{-}[CH_2C(CH_3)_2\text{-}]_n CH_2C^+(CH_3)_2(BF_3OH)^- \tag{58}$$

or

$$HM_n^+(IZ)^- + M \xrightarrow{k_p} HM_nM^+(IZ)^- \tag{59}$$

This addition can be thought of as occurring by an insertion of monomer between the carbonium ion and its negative gegenion.

Various reactions lead to termination of chain growth in cationic polymerization. Chain transfer to a monomer is one of the most common chain-breaking reactions for many monomers. It usually involves transfer of a proton to a monomer molecule with the formation of terminal unsaturation in the polymer molecule:

$$H\text{-}[CH_2C(CH_3)_2\text{-}]_n CH_2C^+(CH_3)_2(BF_3OH)^-$$
$$+ CH_2=C(CH_3)_2 \rightarrow (CH_3)_3C^+(BF_3OH)^-$$
$$+ H\text{-}[CH_2C(CH_3)_2\text{-}]_n CH_2C(CH_3)=CH_2 \tag{60}$$

or

$$HM_nM^+(IZ)^- + M \xrightarrow{k_{tr}\cdot M} M_{n+1} + HM^+(IZ)^- \tag{61}$$

Termination can also take place by rearrangement of the propagating ion pair. Spontaneous termination involves regeneration of the initiator–coinitiator complex by expulsion from the propagating ion pair with the polymer molecule left with terminal unsaturation:

$$H\text{-}[CH_2C(CH_3)_2\text{-}]_n CH_2C^+(CH_3)_2(BF_3OH)^- \rightarrow BF_3OH_2$$
$$+ H\text{-}[CH_2C(CH_3)_2\text{-}]_n CH_2C(CH_3)=CH_2 \tag{62}$$

or, in general terms,

$$HM_nM^+(IZ)^- \xrightarrow{k_{tr}} M_{n+1} + H^+(IZ)^- \tag{63}$$

Termination by combination of the propagating carbonium ion wih the counterion,

$$HM_nM^+(IZ)^- \xrightarrow{k_t} HM_nMIZ \tag{64}$$

occurs, for example, in the trifluoroacetic acid–catalyzed polymerization of styrene:

$$H\text{-}[CH_2CH(Ph)\text{-}]_n CH_2\overset{+}{C}H(Ph)\ (OCOCF_3)^- \longrightarrow H\text{-}[CH_2CH(Ph)\text{-}]_n CH_2CH(Ph)\text{-}OCOCF_3 \tag{65}$$

Intramolecular electrophilic aromatic substitution by a backbiting mechanism occurs in the polymerization of a styrene and most other aromatic monomers

$$\sim\sim CH_2-CH-CH_2-\overset{+}{C}H-\ (IZ)^- \longrightarrow \sim\sim CH_2\overset{H}{\underset{H}{\diamond}} + \overset{+}{H}(IZ)^- \quad (66)$$

with the formation of a terminal indanyl structure and regeneration of the initiator–coinitiator complex [57, 58].

Various transfer agents (denoted by XA), present as a solvent, impurity, or substances deliberately added to the system, can terminate the growing polymer chain by transfer of a negative fragment A:

$$HM_nM^+(IZ)^- + XA \xrightarrow{k_{tr(S)}} HM_nMA + X^+(IZ)^- \quad (67)$$

Certain substances such as amines, trialkylphosphines, and thiophene act as inhibitors or retarders of cationic polymerization. Termination by amines involves formation of stable quaternary ions which are unreactive to propagation [59], for example,

$$HM_nM^+(IZ)^- + :NR_3 \rightarrow HM_nMN^+R_3(IZ)^- \quad (68)$$

The overall kinetics of cationic polymerizations vary considerably, depending largely on the mode of termination in a particular system. Consider the case of termination by combination of the propagating ion pair with the counterion. The derivation of the rate expression for this polymerization under steady-state conditions ($R_i = R_t$) follows in a manner analogous to that used in radical polymerization. The rates of initiation, propagation, and termination are given by

$$R_i = Kk_i[I][ZY][M] \quad (69)$$
$$R_p = k_p[YM^+(IZ)^-][M] \quad (70)$$
$$R_t = k_t[YM^+(IZ)^-] \quad (71)$$

where $[YM^+(IZ)^-]$ is the total concentration of all sized propagating centers

$$[YM^+(IZ)^-] = \frac{Kk_i[I][ZY][M]}{k_t} \quad (72)$$

Combining Eq. 70 with Eq. 72 yields the rate of polymerization as

$$R_p = \frac{R_i k_p [M]}{k_t} = \frac{Kk_i k_p [I][ZY][M]^2}{k_t} \quad (73)$$

and the rate of chain transfer to monomer as

$$R_{tr(M)} = k_{tr(M)}[YM^+(IZ)^-][M] \quad (74)$$

For the case where chain transfer terminates the kinetic chain, the polymerization rate is decreased and R_p is given by

$$R_p = \frac{Kk_i k_p [I][ZY][M]^2}{k_t + k_{tr(S)}[S]} \quad (75)$$

where S and $k_{tr(S)}$ are the concentration of the chain transfer agent and the chain transfer rate constant for the transfer agent, respectively.

Initiation, which consists of the formation of radical cations from the monomer, is followed by their addition to the monomer:

$$\cdot M^+ + M \xrightarrow{k_i} \cdot MM^+ \tag{76}$$

Propagation follows by successive additions of monomer at the cationic center with radical propagation not being favored at low temperatures in superpure systems. In the usual situation Eq. 57 is rate-determining and R_i is given by

$$R_i = IG[M] \tag{77}$$

where I is the radiation intensity and G the number of cationic species formed per 100 eV of energy absorbed. For polymerizations where termination occurs primarily by a reaction such as Eq. 68, R_p is given by

$$R_p = \frac{k_p IG[M]^2}{k_t + k_{tr(S)}[S]} \tag{78}$$

which is equivalent to the expression for initiation by chemical means. However, in sufficiently pure systems (concentrations of water and other terminating agents $< 10^{-7} \sim 10^{-10}M$), termination of a propagating carbonium ion occurs by combination with a negative fragment Y^-,

$$\text{\ensuremath{\sim\!\!\sim\!\!\sim}} M^+ + Y^- \xrightarrow{k_t'} \text{\ensuremath{\sim\!\!\sim\!\!\sim}} M\text{—}Y \tag{79}$$

Y^- is either a solvated electron or the product of the displaced electron formed during initiation reacting with some component of the reaction system [60–63]. In this case the termination rate is

$$R_t' = k_t'[M][Y^-] = k_t'[M]^2 \tag{80}$$

if other terminations are absent. The polymerization rate becomes

$$R_p = k_p[M]^{3/2}\left(\frac{GI}{k_t'}\right)^{1/2} = k_p[M]\left(\frac{R_i}{k_t'}\right)^{1/2} \tag{81}$$

which is different from the usual case in cationic polymerization in that R_p is one-half order in R_i. This is then the same as observed in radical chain polymerization.

Anionic Polymerization

Anionic polymerizations comprise those systems in which the growing polymer chain has a terminal reactive carbon atom with a partial or full negative charge. Although it has been known for many years that the alkali metals or their alkyls would initiate vinyl polymerization [64–67], systematic kinetic studies started only in the mid-1960s [68].

Anionic chain polymerizations are different from cationic polymerizations, although they share some characteristics. Anionic chain polymerizations are usually not as temperature-sensitive as cationic polymerizations. Most anionic polymerizations proceed well at ambient temperatures and higher, since the identities of the initiating species and counterions are much better established. Termination occurs by transfer of a positive fragment, usually a proton, from the solvent or some transfer agent, although other modes of termination are also known. Many anionic chain polymerizations are, however, devoid of any termination reaction.

Several basic initiators have been used to initiate anionic polymerization. These include covalent or ionic metal amides such as $NaNH_2$ and $LiN(C_2H_5)_2$ [69, 70]; alkoxides, hydroxides, cyanides, phosphines, and amines [71, 72]; and organometallic compounds such as $n\text{-}C_4H_9Li$, ⌬MgBr [73]. Initiation usually involves the addition of a nucleophile (base) B^-: to the monomer. Alkyllithium initiators are among the most useful, being employed commercially in the polymerization of butadiene and isoprene. Initiation proceeds by addition of the initiator to the monomer:

$$C_4H_9Li + CH_2{=}CHY \rightarrow C_4H_9{-}CH_2{-}\underset{H}{\overset{Y}{C^-}}{:}(Li)^+ \tag{82}$$

followed by propagation:

$$C_4H_9{-}CH_2{-}\underset{H}{\overset{Y}{C^-}}{:}(Li)^+ + n\text{-}CH_2{=}CHY \rightarrow C_4H_9{-}(CH_2CHY)_n{-}CH_2{-}\underset{H}{\overset{Y}{C^-}}{:}(Li)^+ \tag{83}$$

In the relatively few anionic polymerizations initiated by neutral nucleophiles such as an amine,

$$R_3N{:} + CH_2{=}\underset{H}{\overset{Y}{C}} \rightarrow R_3N^+{-}CH_2{-}\underset{H}{\overset{Y}{C^-}}{:} \dashrightarrow R_3N^+{-}(CH_2CHY)_n CH_2{-}\underset{H}{\overset{Y}{C^-}}{:} \tag{84}$$

the propagating species is proposed to be a zwitterion.

Szwarc and co-workers [35] studied the interesting polymerization initiated by aromatic radical anions such as sodium naphthalene and sodium biphenyl. Initiation proceeds by the prior formation of the active initiator, the naphthalene radical anion

$$Na + \text{(naphthalene)} \longrightarrow [\text{(naphthalene)}^{\bullet -}]\ Na^+ \tag{85}$$

The reaction involves the transfer of an electron from the alkali metal to naphthalene. The naphthalene anion radical (which is colored greenish-blue) transfers an electron to a monomer such as styrene to form the styryl radical anion,

$$[\text{(naphthalene)}^{\bullet -}]\ Na^+ + CH{=}CH_2\text{(Ph)} \longrightarrow \text{(naphthalene)} + [\overset{..}{C}H{-}\overset{..}{C}H_2\text{(Ph)} \leftarrow \overset{..}{C}H{-}\overset{..}{C}H_2\text{(Ph)}]\ Na^+ \tag{86}$$

Besides the foregoing initiators, an alkali metal—for example, lithium, sodium, and potassium—can directly transfer an electron to a monomer such as styrene to form a styryl radical anion,

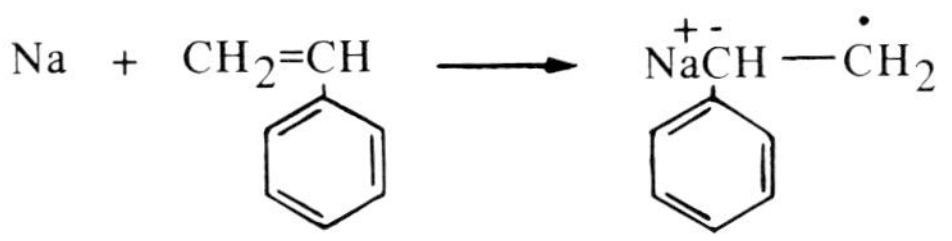

$$Na + CH_2=CH\text{-Ph} \longrightarrow Na^+ \bar{C}H\text{-Ph}\text{—}\dot{C}H_2 \qquad (87)$$

but this reaction is often accompanied by chain propagation and then the molecular weight distribution is wide. This reaction is a heterogeneous one and the utilization ratio of sodium is low.

Termination of a propagating anion by combination with the counterion occurs only in a few instances. Termination by combination of the anion with a metal counterion does not take place. Many anionic polymerizations, especially of nonpolar monomers such as styrene and 1,3-butadiene, take place under conditions in which there are no effective termination reactions. Propagating anionic centers remain intact because transfer of a proton (or other positive species) from the solvent does not occur. Such nonterminated polymeric anions are referred to as living polymers. Living polymers are produced as long as one employs solvents (e.g., tetrahydrofuran, 1,2-dimethoxyethane, dioxane), which are inactive in terms of terminating the propagating anion by chain transfer. The polymerization rate for an anionic system where termination occurs simultaneously with propagation follows in exactly the manner followed for cationic polymerizations. For potassium amide–initiated polymerization in liquid ammonia, initiation involves the dissociation of potassium amide followed by addition of amide ion to the first monomer unit,

$$KHN \overset{k}{\rightleftharpoons} K^+ + H_2N^-: \qquad (88)$$

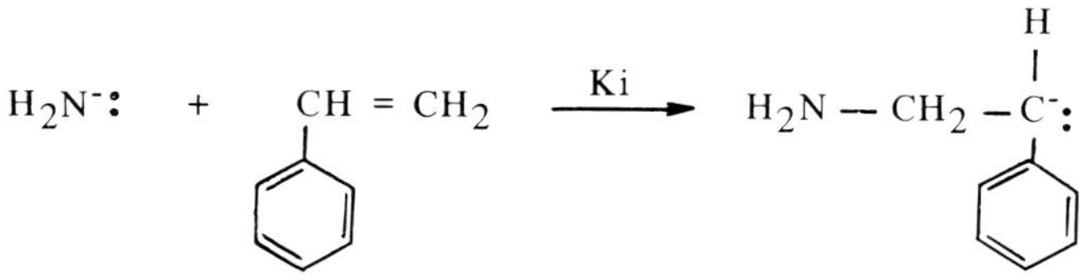

$$H_2N^-: + CH=CH_2\text{(Ph)} \overset{K_i}{\longrightarrow} H_2N-CH_2-\overset{H}{\underset{Ph}{C^-}}: \qquad (89)$$

The rate of initiation is given by

$$R_i = k_i[H_2N:][M] \qquad (90)$$

or

$$R_i = \frac{Kk_i[M][KNH_2]}{[K^+]} \qquad (91)$$

Propagation proceeds according to

$$H_2N\text{—}M_n^- + M \overset{k_p}{\to} H_2N\text{—}M_nM^- \qquad (92)$$

with a rate given by

$$R_p = k_p[M^-][M] \qquad (93)$$

where [M$^-$] represents the total concentration of the propagating anionic centers.

Chain transfer to solvent

$$H_2N\text{—}M_n^- + NH_3 \xrightarrow{k_{tr(NH_3)}} H_2N\text{—}M_n\text{—}H + NH_2^- \tag{94}$$

is extensive but does not terminate the kinetic chain since amide ion is regenerated. Termination occurs by transfer to adventitious water,

$$H_2N\text{—}M_n^- + H_2O \xrightarrow{k_{tr(H_2O)}} H_2N\text{—}M_n\text{—}H + HO^- \tag{95}$$

or other impurity present. The rates of Eqs. 94 and 95 are given by

$$R_{tr(NH_3)} = k_{tr(NH_3)}[M^-][NH_3] \tag{96}$$
$$R_{tr(H_2O)} = k_{tr(H_2O)}[M^-][H_2O] \tag{97}$$

The polymerization rate, derived in the usual manner by combining Eqs. 90, 91, 93, and 97 with the assumption of a steady state for [M$^-$], is obtained as

$$R_p = \frac{Kk_ik_p[M]^2[KNH_2]}{k_{tr(H_2O)}[K^+][H_2O]} \tag{98}$$

There is little interest, theoretical or practical, in the above mentioned systems. The interest, generally speaking, is in living polymerizations. The rate of polymerization in nonterminating systems is expressed simply as the rate of propagation

$$R_p = k_p[M^-][M] \tag{99}$$

where [M$^-$] is the total concentration of living anionic propagating centers. The concentration of living ends can be determined spectrophotometrically as propagating carbanions have strong absorption peaks in the visible or near ultraviolet region.

The rate of polymerization is appropriately expressed as the sum of the rates for the free propagating anion P$^-$ and the ion pair P$^-$(C$^+$)

$$R_p = k_p^-[P^-][M] + k_p^\mp[P^-(C^+)][M] \tag{100}$$

where k_p^- and $k_p^\mp$ are the propagation rate constants for the free ion and ion pair, and [M] is the monomer concentration. C$^+$ in Eq. 100 is the positive gegenion. Comparison of Eqs. 99 and 100 yields the apparent k_p^{app} as

$$k_p^{app} = \frac{k_p^-[P^-] + k_p^\mp[P^-(C^+)]}{[M^-]} \tag{101}$$

The two propagating species are in equilibrium according to

$$P^-(C^+) \overset{K}{\rightleftharpoons} P^- + C^+ \tag{102}$$

governed by the dissociation constant K given by

$$K = \frac{[P^-][C^+]}{[P^-(C^+)]} \tag{103}$$

For the case where [P$^-$] = [C$^+$], that is, there is no source of any ion other than P$^-$(C$^+$), the concentration of free ions is

$$[P^-] = (K[P^-(C^+)])^{1/2} \tag{104}$$

The extent of dissociation is small under most conditions. The concentration of ion pairs is close to the total concentration of living ends and Eq. 104 can be rewritten as

$$[P^-] = (K[M^-])^{1/2} \tag{105}$$

The concentration of ion pairs is given by

$$[P^-(C^+)] = [M^-] - (K[M^-])^{1/2} \tag{106}$$

Combination of Eqs. 101, 105, and 106 yields k_p^{app} as a function of $[M^-]$:

$$k_p^{app} = k_p^{\mp} + \frac{(k_p^- - k_p^{\mp})K^{1/2}}{[M^-]^{1/2}} \tag{107}$$

Polymerizations can also be carried out in the presence of excess gegenion by adding a strongly dissociating salt (e.g., NaBPh$_4$ to supply excess Na$^+$). The concentration of free ions, depressed by the common ion effect, is given by

$$[P^-] = \frac{K[M^-]}{[C^+]} \tag{108}$$

When the added salt is strongly dissociated and the ion pairs slightly dissociated, the gegenion concentration is very close to that of the added salt [CZ]:

$$[C^+] \simeq [CZ] \tag{109}$$

The concentrations of free anions and ion pairs are given by

$$P = \frac{K[M^-]}{[CZ]} \tag{110}$$

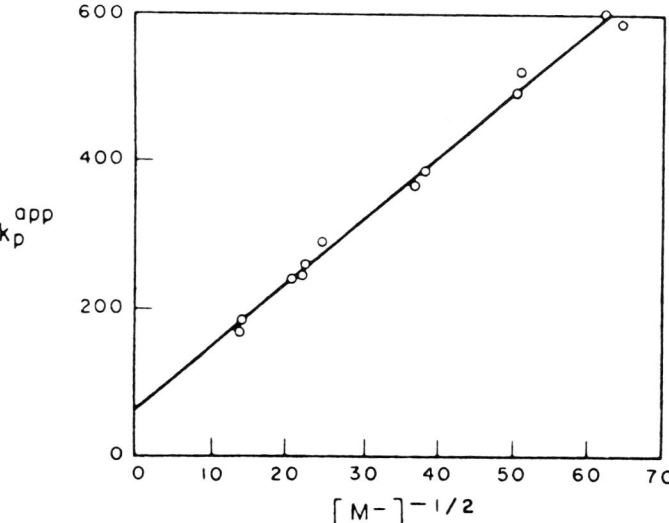

Figure 1 Polymerization of styrene by sodium naphthalene in 3-methyltetrahydrofuran at 20°C. (After Ref. 74.)

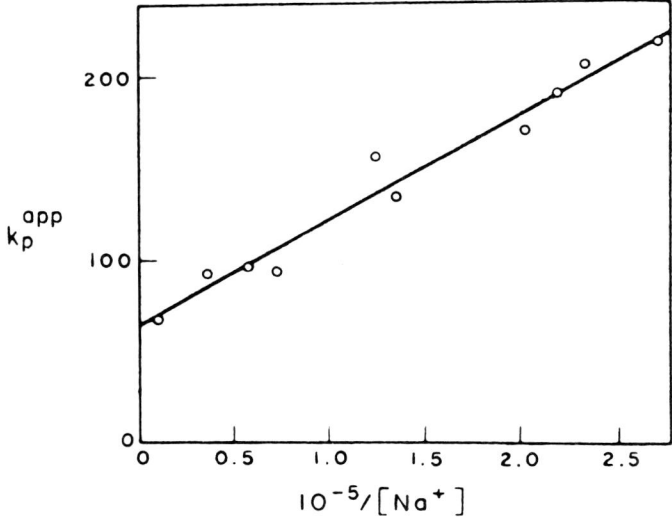

Figure 2 Polymerization of styrene by sodium naphthalene in 3-methyltetrahydrofuran at 20°C in the presence of sodium tetraphenylboride. (After Ref. 74.)

$$[P^-(C^+)] = [M^-] - \frac{K[M^-]}{[CZ]} \tag{111}$$

which are combined with Eq. 101 to yield

$$k_p^{app} = k_p^{\mp} + \frac{(k_p^- - k_p^{\mp})K}{[CZ]} \tag{112}$$

Equations 107 and 112 allow one to obtain k_p^-, $k_p^{\mp}$, and K from k_p^{app} values obtained in the absence and presence of added common ion. A plot of k_p^{app} obtained in the absence of added common ion versus $[M^-]^{1/2}$ yields a straight line whose slope and intercept are $k_p^{\mp}$ and $(k_p^- - k_p^{\mp})$, respectively. A plot of k_p^{app} obtained in the presence of added common ion versus $[CZ]^{-1}$ yields a straight line whose slope and intercept are $k_p^{\mp}$ and $(k_p^- - k_p^{\mp})K$, respectively. Figures 1 and 2 show these plots for polystyryl sodium in 3-methyltetrahydrofuran at 20°C.

Ring-Opening Polymerization

The ring-opening polymerization of cyclic monomer such as cyclic ethers, acetals, esters, amides, and siloxanes is of commercial interest in a number of cases [75–77]. It also includes the polymerizations of ethylene and propylene oxides (R = —H and —CH$_3$, respectively),

$$n\mathrm{H_2C}\overset{O}{\underset{}{\diagup\!\diagdown}}\mathrm{CHR} \rightarrow \{\!\!-\mathrm{O}\!-\!\underset{\underset{R}{|}}{\mathrm{CH}}\!-\!\mathrm{CH_2}\!-\!\}_n \tag{113}$$

ε-caprolactam,

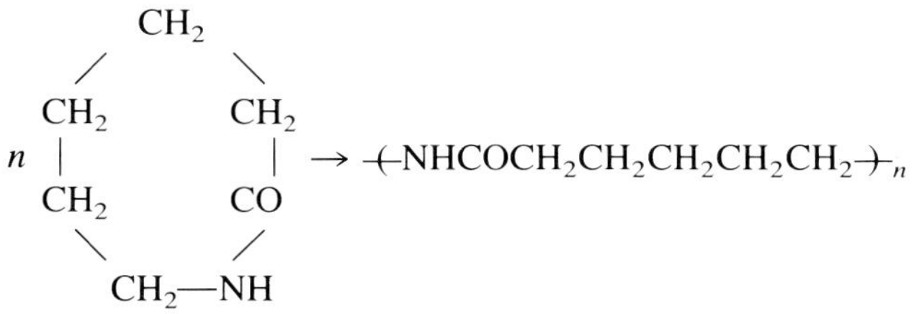

$$n \begin{array}{c} \diagup CH_2 \diagdown \\ CH_2 CH_2 \\ | | \\ CH_2 CO \\ \diagdown CH_2-NH \diagup \end{array} \rightarrow +\!\!(\text{NHCOCH}_2\text{CH}_2\text{CH}_2\text{CH}_2\text{CH}_2)\!\!+_n \qquad (114)$$

and 3,3-bis(chloromethyl)oxetane,

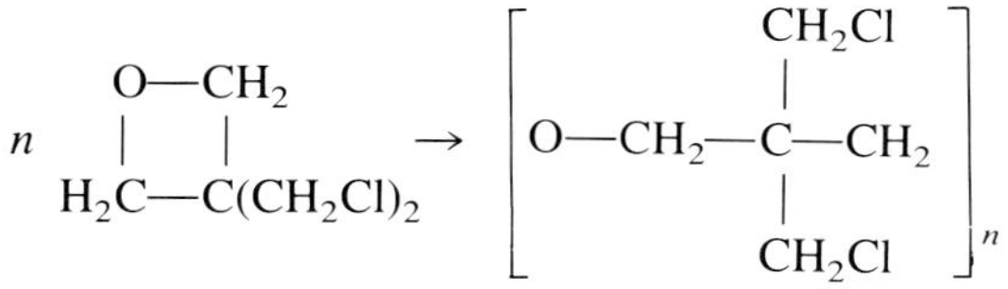

$$n \begin{array}{c} O-CH_2 \\ || \\ H_2C-C(CH_2Cl)_2 \end{array} \rightarrow \left[\begin{array}{c} CH_2Cl \\ | \\ O-CH_2-C-CH_2 \\ | \\ CH_2Cl \end{array} \right]_n \qquad (115)$$

Ring-opening polymerizations have been initiated by ionic initiators as well as initiators that are molecular species. Initiation results in opening of the ring to form an initiator species M*, which may be either an ion or neutral molecule depending on the initiator. This can be generalized as

$$R\!-\!Z + C \rightarrow M^* \qquad (116)$$

where Z is the functional group in the monomer and C the ionic or molecular initiator. Ionic ring-opening polymerizations include those initiated by species such as Na, Ro$^-$, HO$^-$, H$^+$, and BF$_3$. The prime initiator of the molecular type is water. Ionic initiators are usually more reactive than the molecular ones. Most monomers require the use of stronger ionic initiators. Ionic ring-opening polymerizations show most of the general characteristics for ionic polymerization.

The initiator species M* grows by successive ring-opening additions of many monomer molecules,

$$M^* + nR\!-\!Z \rightarrow M(R\!-\!Z)_n^* \qquad (117)$$

The anionic polymerization of epoxides such as ethylene and propylene oxides can be initiated by hydroxides, alkoxides, metal oxides, organometallic compounds, and other bases [78–80]. Thus the polymerization of ethylene oxide by the initiator M$^+$A$^-$ involves initiation,

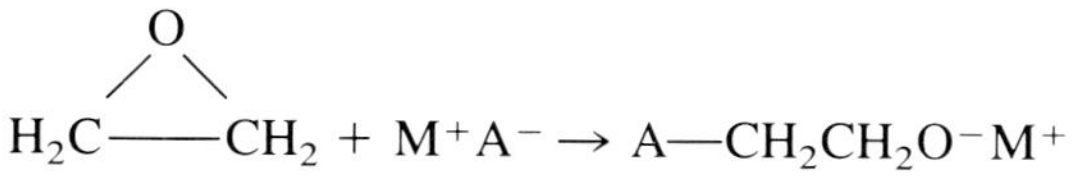

$$\underset{H_2C\!-\!-\!-\!CH_2}{\overset{O}{\diagup\!\diagdown}} + M^+A^- \rightarrow A\!-\!CH_2CH_2O^-M^+ \qquad (118)$$

followed by propagation,

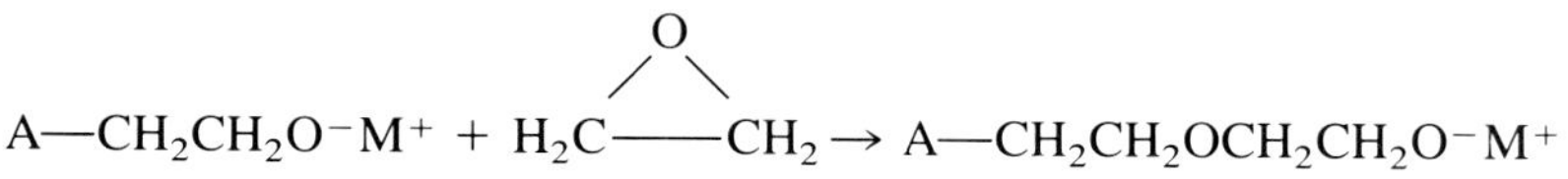

$$A\!-\!CH_2CH_2O^-M^+ + \underset{H_2C\!-\!-\!-\!CH_2}{\overset{O}{\diagup\!\diagdown}} \rightarrow A\!-\!CH_2CH_2OCH_2CH_2O^-M^+ \qquad (119)$$

which may be generalized as

$$A(CH_2CH_2O)_nCH_2CH_2O^-M^+ + H_2C\overset{O}{\overset{\diagup\diagdown}{\text{———}}}CH_2 \rightarrow$$
$$A(CH_2CH_2O)_{n+1}CH_2CH_2O^-M^+ \qquad (120)$$

The polymer molecular weights obtained in epoxide polymerizations are low because of the presence of a chain transfer to monomer reaction in the system. The transfer reaction involves hydrogen abstraction from the alkyl substituent on the epoxide ring followed by very rapid ring cleavage to form an allyl ether anion,

$$\sim\!\!\sim\!\!CH_2\!-\!\underset{\underset{CH_3}{|}}{CH}\!-\!O^-Na^+ + CH_3CH\overset{O}{\overset{\diagup\diagdown}{\text{———}}}CH_2 \xrightarrow{k_{tr(M)}} CH_2\!-\!\underset{\underset{CH_3}{|}}{CH}\!-\!OH$$

$$+ H_2C\overset{O}{\overset{\diagup\diagdown}{\text{———}}}CH\!-\!CH_2^-Na^+ \qquad (121)$$

$$H_2C\overset{O}{\overset{\diagup\diagdown}{\text{———}}}CH\!-\!CH_2^-Na^+ \rightarrow CH_2\!=\!CH\!-\!CH_2O^-Na^+ \qquad (122)$$

The rate of monomer disappearance may be given by the sum of the rates of propagation and transfer reactions,

$$\frac{-d[M]}{dt} = [k_p + k_{tr(M)}][M][I]_0 \qquad (123)$$

The increase in the concentration of polymer chains [N] is given by the rate of the transfer reaction,

$$\frac{d[N]}{dt} = k_{tr(M)}[M][I]_0 \qquad (124)$$

After dividing Eq. 124 by Eq. 123, combining the result with $C_M = k_{tr(M)}/k_p$ and integrating, one obtains

$$[N] = [N]_0 + \frac{C_M}{1 + C_M}([M]_0 - [M]) \qquad (125)$$

where $[N]_0$ is the concentration of polymer chains in the absence of chain transfer to monomer and $[M]_0$ is the initial value of monomer.

A variety of initiator systems of the types used in the cationic polymerization of alkenes can be used to generate the tertiary oxonium ion–propagating species. Very strong protonic acids such as concentrated sulfuric, trifluoroacetic, fluorosulfuric, and trifluoromethane sulfonic acids initiate polymerization via the initial formation of a secondary oxonium ion,

$$H^+A^- + \underset{R}{\overset{R}{\bigotimes}}O \longrightarrow HO^+\underset{R}{\overset{R}{\bigotimes}} \quad A^- \qquad (126)$$

which reacts with a second monomer molecule to form the tertiary oxonium ion,

$$\text{HO}^+\underset{A^-}{\overset{R}{\diamondsuit}}_R + \text{O}\underset{R}{\overset{R}{\diamondsuit}}_R \longrightarrow \text{HOCH}_2\text{CH}_2\text{CH}_2 - \overset{+}{\text{O}}\underset{A^-}{\overset{R}{\diamondsuit}}_R \quad (127)$$

Lewis acids such as BF_3 and $SnCl_4$, in conjunction with water or some other protogen or cationogen, almost always initiate polymerization of cyclic ethers. Interaction between the coinitiator and initiator yields a coinitiator–initiator complex, for example,

$$BF_3 + H_2O \rightarrow BF_3 \cdot OH_2 \qquad (128)$$

which initiates polymerization by acting as a proton donor in an initiation sequence similar to Eqs. 126 and 127. Some Lewis acids may initiate polymerization alone, probably via self-ionization:

$$2PF_5 \rightarrow PF_4^+(PF_6)^- \qquad (129)$$

Carbonium ions have also been used to initiate the polymerization of cyclic ethers. The initiation reaction generally consists of the same sequence as in Eqs. 126 and 127 except that the initiating species is R^+A^- instead of H^+A^-. Carbonium ions formed from

$$F_3CSO_3CH_3 \rightarrow CH_3^+(F_3CSO_3)^- \qquad (130)$$

$$\text{Ph}-\text{CHCl}-\text{Ph} + AgSbF_6 \longrightarrow AgCl + \text{Ph}-CH^+(SbF_6)^-\text{Ph} \qquad (131)$$

and a variety of other reactions have been used [81–83].

Under certain conditions, cationic cyclic ether polymerizations have the characteristics of living polymerizations in that the propagating species are long-lived and narrow molecular weight distributions are observed. Living polymerizations occur when initiation is fast relative to propagation, for example, when superacids and their esters or acylium and 1,3-dioxolan-2-ylium salts

$$\underset{R-C^+\ (A^-)}{\overset{O}{\|}} \qquad \underset{H\ (A^-)}{\overset{O\overset{+}{\underset{}{\diagup}}O}{\diagdown\diagup}}$$

are used in conjunction with stable counterions such as AsF_6^-, PF_6^-, and SbF_6^-.

Many ether polymerizations can be described by kinetic expressions very similar to those used in alkene polymerizations. The polymerization rate may be given by

$$R_p = k_p[M^*][M] \qquad (132)$$

where $[M^*]$ is the concentration of the propagating oxonium ion.

Ring-opening polymerizations that take place without termination and with a propagation–depropagation equilibrium are described in a different manner [84–86]. (The following treatment for reversible ring-opening polymerizations is also applicable to other reversible polymerizations such as those of alkenes or carbonyl monomers.) The propagation–depropagation equilibrium can be expressed by

$$M_n^* + M \underset{k_{dp}}{\overset{k_p}{\rightleftharpoons}} M_{n+1}^* \tag{133}$$

The polymerization rate is given by the difference between the rates of the propagation and depropagation reactions:

$$R_p = \frac{-d[M]}{dt} = k_p[M^*][M] - k_{dp}[M^*] \tag{134}$$

At equilibrium, the polymerization rate is zero and Eq. 134 becomes

$$k_p[M]_C = k_{dp} \tag{135}$$

where $[M]_C$ is the equilibrium monomer concentration. Combination of Eqs. 134 and 135 gives the polymerization rate as

$$\frac{-d[M]}{dt} = k_p[M^*]([M] - [M]_C) \tag{136}$$

which can be integrated to yield

$$\ln\left(\frac{[M]_0 - [M]_C}{[M] - [M]_C}\right) = k_p[M^*]t \tag{137}$$

where $[M]_0$ is the initial monomer concentration.

Chain Copolymerization

Copolymerization may be defined as any process whereby two or more monomers are incorporated as integral parts of a polymer. A copolymer is the product resulting from such a process. It is not necessary that the relative numbers of the different types of unit be the same in different molecules of the copolymer or even in different portions of a single molecule. Several excellent treatments of copolymerization are available [87–90]. Yamada [89], in particular, offers an exhaustive exposition.

For purposes of discussion the kinetic aspects may be divided into two parts, factors affecting copolymer composition and structure and those controlling rates of copolymerization [91]. Consider the case for the copolymerization of the two monomers M_1 and M_2. Although radical copolymerization has been more extensively studied and is more important than ionic copolymerizations, we consider the general case without specifying whether the mode of initiation is by a radical, anionic, or cationic species. Copolymerization of the two monomers leads to two types of propagating species: one with M_1 at the propagating end and the other with M_2 there. These can be represented by M_1^* and M_2^*, where the asterisk indicates that as a propagating species, the monomer may be a radical, a carbonium ion, or a carbanion depending on the particular case. If it is assumed that the reactivity of the propagating species is dependent only on the monomer unit at the end of the chain (referred to as the end or ultimate unit), four propagation reactions are then possible. Monomers M_1 and M_2 can each add either to a propagating chain ending in M or to one ending in M_2:

$$M_1^* + M_1 \overset{k_{11}}{\rightarrow} M_1^* \tag{138}$$

$$M_1^* + M_2 \xrightarrow{k_{12}} M_2^* \tag{139}$$

$$M_2^* + M_1 \xrightarrow{k_{21}} M_1^* \tag{140}$$

$$M_2^* + M_2 \xrightarrow{k_{22}} M_2^* \tag{141}$$

where k_{11} is the rate constant for a propagating chain ending in M_1 adding to monomer M_1, k_{12} that for a propagating chain ending in M_1 adding to monomer M_2, and so on. The propagation of a reactive center with the addition of the same monomer (i.e., reactions 138 and 141) is often referred to as homopropagation or self-propagation; propagation of a reactive center with the addition of the other monomer (reactions 139 and 140) is referred to as cross-propagation or a crossover reaction. All propagation reactions are assumed to be irreversible. Monomer M_1 disappears by reactions 138 and 140, while monomer M_2 disappears by reactions 139 and 141. The rates of disappearance of the two monomers, which are synonymous with their rates of entry into the copolymer, are given by

$$\frac{-d[M_1]}{dt} = k_{11}[M_1^*][M_1] + k_{21}[M_2^*][M_1] \tag{142}$$

$$\frac{-d[M_2]}{dt} = k_{12}[M_1^*][M_2] + k_{22}[M_2^*][M_2] \tag{143}$$

Dividing Eq. 142 by Eq. 143 yields the ratio of the rates at which the two monomers enter the copolymer, that is, the copolymer composition, as

$$\frac{d[M_1]}{d[M_2]} = \frac{k_{11}[M_1^*][M_1] + k_{21}[M_2^*][M_1]}{k_{12}[M_1^*][M_2] + k_{22}[M_2^*][M_2]} \tag{144}$$

To remove the concentration terms in M_1^* and M_2^* from Eq. 144, a steady-state concentration is assumed for each of the reactive species M_1^* and M_2^* separately. For the concentrations of M_1^* and M_2^* to remain constant their rates of interconversion must be equal. In other words, the rates of reactions 139 and 140 are equal:

$$k_{21}[M_2^*][M_1] = k_{12}[M_1^*][M_2] \tag{145}$$

Equation 145 can be rearranged and combined with Eq. 144 to yield

$$\frac{d[M_1]}{d[M_2]} = \frac{k_{11}k_{21}[M_2^*][M_1]^2/k_{12}[M_2] + k_{21}[M_2^*][M_1]}{k_{22}[M_2^*][M_2] + k_{21}[M_2^*][M_1]} \tag{146}$$

Dividing the numerator and denominator of the right side of Eq. 146 by $k_{21}[M_2^*][M_2]$ and combining the result with the parameters r_1 and r_2, which are defined by $r_1 = k_{11}/k_{12}$ and $r_2 = k_{22}/k_{21}$, one finally obtains

$$\frac{d[M_1]}{d[M_2]} = \frac{[M_1](r_1[M_1] + [M_2])}{[M_2](r_2[M_2] + [M_1])} \tag{147}$$

Equation 147 is known as the copolymerization equation or the Mayo–Lewis equation [91–94]. The copolymer composition, $d[M_1]/d[M_2]$, is the molar ratio of the two monomer units in the copolymer. $d[M_1]/d[M_2]$ is expressed by Eq. 147 as being related to the concentrations of the two monomers in the feed $[M_1]$ and $[M_2]$, and the parameters r_1 and r_2. The parameters r_1 and r_2 are termed the monomer reactivity ratios. Each r is the ratio of the rate constant for a reactive propagating species adding its own type of monomer to the rate constant for its addition of the other monomer. The tendency of two monomers to copolymerize is noted by r values between zero and unity. An r_1 value greater than unity means that M_1^* preferentially adds M_1 instead of M_2, while an r_1 value less than unity means that M_1^* preferentially adds M_2. An r_1 value of zero would mean that M_1 is incapable of undergoing homopolymerization.

The Mayo–Lewis equation can also be expressed in terms of mole fractions instead of concentrations. If f_1 and f_2 are the mole fractions of monomers M_1 and M_2 in the feed, and F_1 and F_2 are the mole fractions of M_1 and M_2 in the copolymer, then

$$F_1 = \frac{r_1 f_1^2 + f_1 f_2}{r_1 f_1^2 + 2 f_1 f_2 + r_2 f_2^2} \tag{148}$$

where

$$f_1 = 1 - f_2 = \frac{[M_1]}{[M_1] + [M_2]} \quad \text{and} \quad F_1 = 1 - F_2 = \frac{d[M_1]}{d[M_1] + d[M_2]}$$

The preceding copolymerization equation is equally applicable to radical, cationic, and anionic chain copolymerizations, although the r_1 and r_2 values for any particular comonomer pair can be different depending on the mode of initiation. The r_1 and r_2 values, for example, for the comonomer pairs of styrene (M_1) and methyl methacrylate (M_2) are 0.52 and 0.46 in radical copolymerization, 10 and 0.1 in cationic copolymerization, and 0.1 and 6 in anionic copolymerization [95, 96]. Figure 3 shows that these different r_1 and r_2 values give rise to large differences in the copolymer composition depending on the mode of initiation. The copolymer composition as a function of the feed composition in the various cases are shown in Figure 4 [97].

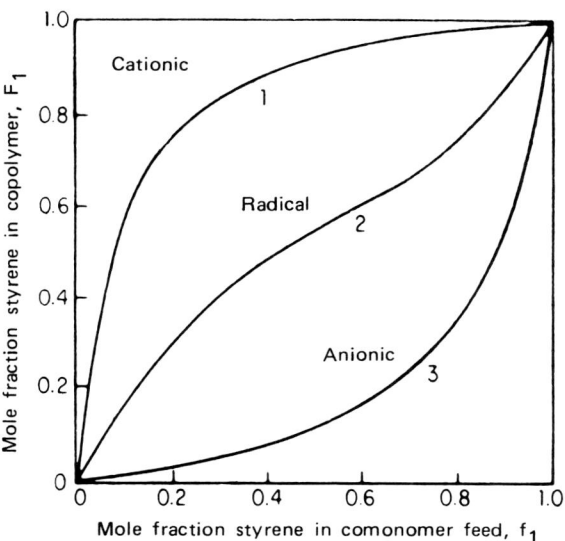

Figure 3 Dependence of the instantaneous copolymer composition F_1 on the comonomer feed composition f_1 for styrene-methacrylate in cationic (curve 1), radical (curve 2), and anionic (curve 3) copolymerizations initiated by $SnCl_4$, benzoyl peroxide, and Na-liquid NH_3, respectively. (After Ref. 96.)

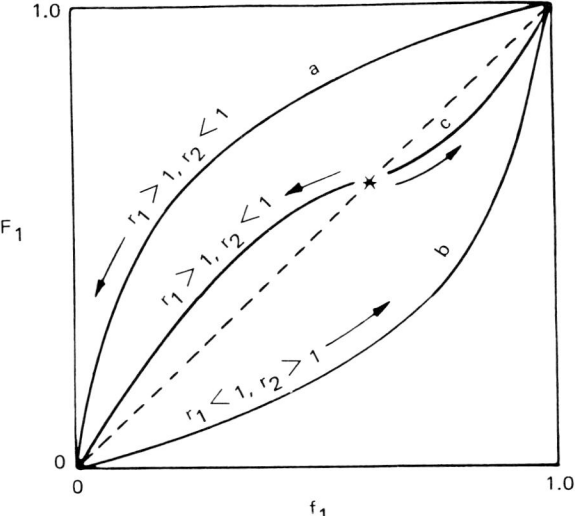

Figure 4 The mole fraction of monomer F_1 in copolymer as a function of f_1 in the comonomer feed for an ideal copolymer (dashed line), $r_1 = r_2 = 1$; (a) $r_1 > 1$; $r_2 < 1$; (b) $r_1 < 1$; $r_2 > 1$; (c) $r_1 < 1$; $r_2 < 1$. At the crossover point (*) $F_1 = f_1 = (1 - r_2)/(2 - r_1 - r_2)$. (After Ref. 97.)

Thus far it has been assumed that the reactivity of a propagating species is determined solely by the nature of the terminal monomer from which that species was derived. The possibility that the penultimate monomer unit may influence the reactivity of the species (i.e., the rate constants of the addition of, say, M_2 to $M_1M_1^*$ and $M_2M_1^*$ are not equal) was envisaged by Merz et al. [98]. To derive the copolymer composition equation in such circumstances, for the binary copolymerization of M_1, M_2, it is necessary to consider the eight propagation reactions:

$$M_1M_1^* + M_1 \xrightarrow{k_{111}} M_1M_1M_1^*$$

$$M_1M_1^* + M_2 \xrightarrow{k_{112}} M_1M_1M_2^*$$

$$M_2M_1^* + M_1 \xrightarrow{k_{211}} M_2M_1M_1^*$$

$$M_2M_1^* + M_2 \xrightarrow{k_{212}} M_2M_1M_2^*$$

$$M_2M_2^* + M_1 \xrightarrow{k_{221}} M_2M_2M_1^*$$

$$M_2M_2^* + M_2 \xrightarrow{k_{222}} M_2M_2M_2^*$$

$$M_1M_2^* + M_1 \xrightarrow{k_{121}} M_1M_2M_1^*$$

$$M_1M_2^* + M_2 \xrightarrow{k_{122}} M_1M_2M_2^* \tag{149}$$

The composition equation for Eq. 149 is

$$\frac{d[M_1]}{d[M_2]} = \left(1 + r_1' \frac{[M_1] \cdot r_1[M_1] + [M_2]}{[M_2] \cdot r_1'[M_1] + [M_2]}\right) \bigg/ \left(1 + r_2' \frac{[M_2] \cdot r_2[M_2] + [M_1]}{[M_1] \cdot r_2'[M_2] + [M_1]}\right) \quad (150)$$

where

$$r = \frac{k_{111}}{k_{112}}, \quad r_1' = \frac{k_{211}}{k_{212}}, \quad r_2 = \frac{k_{222}}{k_{221}}, \quad r_2' = \frac{k_{122}}{k_{121}}$$

If $r_1 = r_1'$ and $r_2 = r_2'$, Eq. 150 reduces to the Mayo–Lewis equation.

Near the so-called "ceiling temperature" [12] of one component, in addition to propagation, depropagation reactions are present, and these affect copolymer compositions. Lowry has derived copolymer composition for the foregoing effect [99]. He assumed that monomer M_2 has tendency to depropagate if it is attached to another M_2 unit, while monomer M_1 has absolutely no tendency to depropagate. His scheme is as follows:

$$M_1^* + M_1 \xrightarrow{k_{11}} M_1 M_1^*$$

$$M_1^* + M_2 \xrightarrow{k_{12}} M_1 M_2^*$$

$$M_2^* + M_1 \xrightarrow{k_{21}} M_2 M_1^*$$

$$M_2^* + M_2 \underset{k_{22'}}{\overset{k_{22}}{\rightleftarrows}} M_2 M_2^* \quad (151)$$

The steady-state assumptions are given by

$$K_{12}[M_1^*][M_2] + K_{22}'[(M_2)_2^*] = K_{22}[(M_2)_1^*][M_2] + K_{21}[(M_2)_1^*][M_1] \quad (152)$$

$$K_{22}[(M_2)_1^*][M_2] + K_{22}'[(M_2)_3^*] = K_{22}[(M_2)_2^*][M_2] + K_{21}[(M_2)_2^*][M_1] + K_{22}'[(M_2)_2^*] \quad (153)$$

$$K_{21}[M_1] \sum_{i=1}^{\infty} [(M_2)_i^*] = K_{12}[M_1^*][M_2] \quad (154)$$

where $[(M_2)_i^*]$ denotes an active species containing i units of monomer 2, immediately preceded by one or more units of monomer 1, at the active end of the growing polymer molecule. One may finally obtain

$$\frac{d[M_2]}{d[M_1]} = \frac{[M_2]}{(r_1[M_1] + [M_2])(1 - \alpha)} \quad (155)$$

where

$$\alpha = \frac{1}{2}([1 + r_3[M_2] + (r_3/r_2)[M_1]] - \{[1 + r_3[M_2] + (r_3/r_2)[M_1]]^2 - 4r_3[M_2]\}^{1/2})$$

$$r_1 = \frac{k_{11}}{k_{12}}, \quad r_2 = \frac{k_{22}}{k_{21}}, \quad r_3 = \frac{k_{22}}{k'_{22}}$$

While a study of copolymer compositions provides a wealth of data on reactivities, it gives no information on the overall rates of reaction. To obtain an expression describing rates of copolymerization it is necessary to include, in addition to the various propagation reactions considered previously, one initiation reaction for each monomer and all possible bimolecular termination reactions. Two different approaches have been used to derive expressions for the rate of copolymerization.

One approach assumes the termination reaction to be chemically controlled. Its reaction scheme consists of four propagation reactions (Eqs. 138 to 141) and three termination steps:

$$M_1\cdot + M_1\cdot \xrightarrow{k_{t11}} \text{polymer}$$

$$M_2\cdot + M_2\cdot \xrightarrow{k_{t22}} \text{polymer} \tag{156}$$

$$M_1\cdot + M_2\cdot \xrightarrow{k_{t12}} \text{polymer}$$

The overall rate of copolymerization is given by the sum of the four propagation rates:

$$R_p = \frac{d[M_1] + d[M_2]}{dt}$$
$$= k_{11}[M_1\cdot][M_1] + k_{12}[M_1\cdot][M_2] + k_{22}[M_2\cdot][M_2] + k_{21}[M_2\cdot][M_1] \tag{157}$$

Two steady-state assumptions are made. One is given by

$$k_{21}[M_2\cdot][M_1] = k_{12}[M_1\cdot][M_2] \tag{158}$$

The other is given by Eq. 145. The rate of initiation may be expressed as

$$R_i = 2k_{t11}[M_1\cdot]^2 + 2k_{t12}[M_1\cdot][M_2\cdot] + 2k_{t22}[M_2\cdot]^2 \tag{159}$$

By combining the preceding equations and using the notations r_1 and r_2, one obtains for the copolymerization rate

$$R_p = \frac{(r_1[M_1]^2 + 2[M_1][M_2] + r_2[M_2]^2)R_i^{1/2}}{(r_1^2\delta_1^2[M_1]^2 + 2\phi r_1 r_2 \delta_1 \delta_2 [M_1][M_2] + r_2^2\delta_2^2[M_2]^2)^{1/2}} \tag{160}$$

where

$$\delta = \left(\frac{2k_{t11}}{k_{11}^2}\right)^{1/2}, \quad \delta_2 = \left(\frac{2k_{t22}}{k_{22}^2}\right)^{1/2}, \quad \phi = \frac{k_{t12}}{2(k_{t11}k_{t22})^{1/2}}$$

Equation 160 is known as the Walling equation [100].

North and co-workers [101–103] have pointed out that the usual treatment of copolymerization with the ϕ factor can be ambiguous, since it is fairly well established that termination in radical polymerization is generally diffusion-controlled. Thus ϕ cannot be interpreted primarily in terms of the chemical effects of the radical ends. A more applicable kinetic expression for the rate of diffusion-controlled copolymerization is obtained by considering the termination reaction as the following scheme:

$$M_1\cdot + M_1\cdot \xrightarrow{k_{t(12)}} \text{dead copolymer}$$

$$M_1\cdot + M_2\cdot \xrightarrow{k_{t(12)}} \text{dead copolymer} \tag{161}$$

$$M_2\cdot + M_2\cdot \xrightarrow{k_{t(12)}} \text{dead copolymer}$$

where the termination rate constant $k_{t(12)}$ is a function of the copolymer composition. The condition for the steady state for the total concentration of radicals then takes the form

$$R_i = 2k_{t(12)}([M_1\cdot] + [M_2\cdot])^2 \tag{162}$$

Combination of the preceding equations with notation r_1 and r_2 yields the rate of copolymerization as

$$R_p = \frac{(r_1[M_1]^2 + 2[M_1][M_2] + r_2[M_2]^2)R_i^{1/2}}{k_{t(12)}^{1/2}\{(r_1[M]/k_{11}) + (r_2[M_2]/k_{22})\}} \tag{163}$$

where

$$k_{t(12)} = F_1 k_{t11} + F_2 k_{t22}$$

and the other symbols are as defined earlier.

Ziegler–Natta Polymerization

Ziegler–Natta catalysts have been defined as the products of reactions between two different groups of metal compounds, compounds of the transition elements of groups IV to VIII, known as the catalysts, and compounds such as the hydrides, alkyls, or aryls of the elements of groups I to IV, termed the cocatalysts [104]. The important characteristics of Ziegler–Natta catalysts lie in their ability to produce stereoregular polymers, i.e., nonbranching polymers from ethylene, and isotactic polymers from α-olefins such as propylene. This stereospecificity has given Ziegler–Natta catalysts considerable scientific and industrial importance. Because of the complex behavior of Ziegler–Natta catalysts, almost all studies, with the exception of the pioneer work of Natta and Pasquon [105], have been directed toward the pure organic chemistry of these compounds with little consideration of kinetic effects. Kinetic studies become essential for heterogeneous catalysts such as many Ziegler–Natta systems. From kinetic data the rate expression, number of polymerization centers, and other information can be determined, all of which are important in the formulation of mechanisms for the reactions. The kinetic data also provide important basic information for process engineering (process design, reactor design, etc.).

Ziegler–Natta polymerization kinetics are, as indicated, difficult to study experimentally, particularly those based on insoluble transition metal compounds. They are also complicated by the interaction of concurrent chemical reactions and by physical processes that may have either accelerating or retarding influences on the polymerization. The following points should be taken into account in considering polymerization mechanisms.

1. *Surface activity.* Some of the active sites on crystal edges and other lattice defects will be poisoned, some reversibly, others irreversibly. The initial concentration of sites will depend on the composition of the catalyst, i.e., the incorporation of defects during its preparation, and on the size of the crystallites.

2. *New site formation.* Breakup of crystallites as a result of the growth of polymer chains will expose fresh sites. Stirring breaks up the agglomeration of polymer coated with catalyst particles and, if sufficiently vigorous, may result in cleavage of crystals.

3. *Multiplicity of catalyst species.* More than one type of initiating species may be formed in both homogeneous and heterogeneous catalysts, but in the latter identification and determination are rendered more difficult since they cannot be studied directly by physical techniques such as electron-spin resonance (ESR) and nuclear magnetic resonance (NMR).

4. *Catalyst decay.* Catalyst sites can undergo decomposition with loss of activity and in some cases the activity can be restored, e.g., by oxidation. There may be great differences in catalyst stability even with those of allied structure obtained from the same reactions.

5. *Mass transfer of monomer.* Solution of gaseous monomers may become rate-controlling and will be affected by the speed and efficiency of stirring.

6. *Encapsulation of catalyst sites.* Propagation, metal–alkyl transfer, and monomer transfer may become diffusion-controlled when the catalyst particles become encapsulated with insoluble polymer.

Following the precedent of α-olefins, we first present the kinetic mechanism of anionic coordination polymerization. The process of initiation can be depicted as

$$[\text{Cat}]^+\text{---}R^- + CH_2\text{=}\underset{\underset{CH_3}{|}}{CH} \xrightarrow{k_i} [\text{Cat}]^+\text{---}C^-H_2\text{---}\underset{\underset{CH_3}{|}}{CH}\text{---}R \tag{164}$$

where $[\text{Cat}]^+$ is the cationic counterparts of the Ziegler–Natta catalyst. The chain propagation may be represented by

$$[\text{Cat}]^+\text{---}CH_2\text{---}\underset{\underset{CH_3}{|}}{CH}\text{---}R + nCH_2\text{=}\underset{\underset{CH_3}{|}}{CH} \xrightarrow{k_p}$$

$$[\text{Cat}]^+\text{---}C^-H_2\text{---}\underset{\underset{CH_3}{|}}{CH}\text{---}(CH_2\text{---}\underset{\underset{CH_3}{|}}{CH})_n\text{---}R \tag{165}$$

The chain terminations are as follows:

1. Self-termination

$$[Cat]^+\text{---}C^-H_2\text{---}CH\text{---}(CH_2\text{---}CH)_n\text{---}R \xrightarrow{k_t}$$
$$\qquad\qquad\quad |\qquad\quad\;\; |$$
$$\qquad\qquad\; CH_3\qquad CH_3$$

$$[Cat]\text{---}H + CH_2\text{=}C\text{---}(CH_2\text{---}CH)_n\text{---}R \qquad (166)$$
$$\qquad\qquad\qquad\quad |\qquad\quad\;\; |$$
$$\qquad\qquad\qquad CH_3\qquad CH_3$$

This reaction never proceeds until the temperature has been more than 50°C.

2. Transfer to monomer:

$$[Cat]^+\text{---}C^-H_2\text{---}CH\text{---}(CH_2\text{---}CH)_n\text{---}R + CH_2\text{=}CH \xrightarrow{k_{tr(n)}} [Cat]^+\text{---}CH_2\text{---}CH_2$$
$$\qquad\qquad\quad |\qquad\quad\;\; |\qquad\qquad\qquad |\qquad\qquad\qquad\qquad\quad |$$
$$\qquad\qquad\; CH_3\qquad CH_3\qquad\qquad\quad CH_3\qquad\qquad\qquad\qquad CH_3$$

$$+ CH_2\text{=}C\text{---}(CH_2\text{---}CH)_n\text{---}R \qquad (167)$$
$$\qquad\quad |\qquad\quad\;\; |$$
$$\quad\; CH_3\qquad CH_3$$

3. Transfer to aluminum alkyl:

$$[Cat]^+\text{---}C^-H_2\text{---}CH\text{---}(CH_2\text{---}CH)_n\text{---}R + AlR_3 \xrightarrow{k_{tr(A)}} [Cat]^+\text{---}R^-$$
$$\qquad\qquad\quad |\qquad\quad\;\; |$$
$$\qquad\qquad\; CH_3\qquad CH_3$$

$$+ R_2Al\text{---}CH_2\text{---}CH\text{---}(CH_2\text{---}CH)_n\text{---}R \qquad (168)$$
$$\qquad\qquad\qquad\quad |\qquad\quad\;\; |$$
$$\qquad\qquad\qquad CH_3\qquad CH_3$$

where

$$R_2Al\text{---}CH_2\text{---}CH\text{---}(CH_2\text{---}CH)_n R \xrightarrow{regeneration} R_2AlH + CH\text{=}C\text{---}(CH_2\text{---}CH)_n\text{---}R$$
$$\qquad\qquad\qquad |\qquad\quad\;\; |\qquad\qquad\qquad\qquad\qquad\qquad\qquad |\qquad\quad\;\; |$$
$$\qquad\qquad\; CH_3\qquad CH_3\qquad\qquad\qquad\qquad\qquad\qquad CH_3\qquad CH_3 \quad (169)$$

4. Transfer to H_2, where H_2 is used to regulate the molecular weight of polymer in the process of polymerization:

$$[Cat]^+\text{---}C^-H_2\text{---}CH\text{---}(CH_2\text{---}CH)_n\text{---}R + H_2 \xrightarrow{k_{tr(H)}} [Cat]\text{---}H$$
$$\qquad\qquad\quad |\qquad\quad\;\; |$$
$$\qquad\qquad\; CH_3\qquad CH_3$$

$$+ CH_3\text{---}CH\text{---}(CH_2\text{---}CH)_n\text{---}R \qquad (170)$$
$$\qquad\qquad |\qquad\quad\;\; |$$
$$\qquad\quad CH_3\qquad CH_3$$

It is necessary to note that Eqs. 166 to 170 express the chemical reaction process only in the case of heterogeneous catalystic system without any reference to macroscopic factors. In practice all these factors must be taken into account.

We next consider the rate expression for Ziegler–Natta polymerization. The concentration of active centers ([C*]) and polymerization rate R_p are given by

$$[C^*] = f[T]^a[A]^b \tag{171}$$

and

$$R_p = k_p[C^*][M]^c \tag{172}$$

where [T], [A], and [M] are the concentrations of a transition metal compound, metal alkyl, and monomer and the coefficients a, b, and c depend on the polymerization mechanism; f is the efficiency of utilization of the transition metal compound. [C*] is the concentration of active centers under steady-state conditions. The effects of initiation, transfer, and termination are considered elsewhere. With heterogeneous catalysts where concentrations of active species result from adsorption equilibria of components on to the surface, the rate is more properly expressed in the form

$$R_p = k_p \theta_M \theta_A S \tag{173}$$

where θ_M and θ_A are the fractions of catalyst surface covered by monomer and metal alkyl and S is the surface area of the transition metal compound [106]. However, most polymerizations are first order in monomer; hence θ_M will be proportional to [M], S clearly will be related to [T] at the steady state, and at a fixed ratio of [A]/[T] the effect of the metal alkyl can be included in the rate constant. It is thus convenient to express the rate constants using molar concentrations as in Eq. 172.

The coefficient a will be unity unless the active center is formed from the transition metal complex in an aggregative or dissociative step, and, in general, only a fraction of the transition metal will participate in the polymerization. When the [A]/[T] ratio is increased the rate usually increases either to a steady value or to a maximum and then declines, and, dependent on the range of values of [A]/[T] and the type of catalyst, b will be positive, zero, or negative. As one or two monomer molecules may coordinate with the catalyst, the exponent c will have a value between 0 and 2, depending on the interaction between monomer and catalyst and the rate of propagation.

If active sites are produced by reversible reaction of the metal alkyl with the transition metal compound, i.e.,

$$T + A \underset{}{\overset{k_A}{\rightleftharpoons}} C^* \tag{174}$$

their concentration at equilibrium is given by

$$[C^*] = \frac{fk_A[A][T]}{1 + k_A[A]} \quad \text{(provided } [A] \gg f[T]) \tag{175}$$

If the monomer does not coordinate with the catalyst complex, the rate of polymerization will be given by

$$R_p = k_p[C^*][M] = \frac{fk_pk_A[A][T][M]}{1 + k_A[A]} \tag{176}$$

As the concentration of metal alkyl is increased the rate will rise to the maximum value $(fk_p[T][M])$, and at constant [M] and [T] there will be a linear relationship between $[A]/R_p$ and [A]. As the lower aluminum alkyls are dimeric in solution but coordinate on the catalyst surface in the monomeric form, the concentration term will be $[A]^{1/2}$. If the monomer is coordinated prior to addition, the rate is not necessarily directly proportional to the concentration in solution. The reaction scheme may be written

$$M_nC^* + M \underset{k_{-1}}{\overset{k_1}{\rightleftharpoons}} M_nC^*M \overset{k_p}{\rightarrow} M_{n+1}C^* \tag{177}$$

The total concentration of active centers, Eq. 175, will be the sum of uncoordinated and coordinated sites, i.e.,

$$[C^*] = [\Sigma M_iC^*] + [\Sigma M_iC^*M] \tag{178}$$

and the rate of polymerization is given by

$$R_p = k_p[\Sigma M_iC^*M] = \frac{k_p k_1 [C^*][M]}{k_1[M] + (k_{-1} + k_p)} \tag{179}$$

When the complex between monomer and catalyst is weak ($k_1 \ll k_{-1}$), the polymerization is first order in monomer and

$$R_p \simeq \frac{k_1 k_p [C^*][M]}{(k_{-1} + k_p)} \tag{180}$$

If the monomer is strongly coordinated, $k_1 \gg k_{-1}$ but is not rate controlling in respect of propagation, i.e., $k_1 \gg k_p$, the rate will be independent of monomer concentration ($R_p \simeq k_p[C^*]$). Second-order dependence of rate on monomer will result from insertion of weakly coordinated monomer molecules in pairs or if the coordination of a second monomer molecule facilitates the insertion of a weakly complexed molecule, i.e.,

$$M_nC^* + M \rightleftharpoons M_nC^*M + M \rightleftharpoons M_nC^*M_2 \rightarrow M_{n+2}C^* \tag{181}$$

or

$$M_nC^* + M \rightleftharpoons M_nC^*M + M \rightarrow M_{n+1}C^*M \tag{182}$$

In the second case if the transition metal–olefin complex is stable, the polymerization will be first order in monomer [107].

If monomer and metal alkyl are adsorbed at the transition metal to produce the growing chains, an excess of one or another of the reagents will occupy all the sites. Hence with increasing [A]/[T] the polymerization rate should rise to a maximum and then decline. The concentration of active centers is given by

$$[C^*] = \frac{fk_A k_M [A][M][T]}{(1 + k_A[A] + k_M[M])^2} \simeq \frac{fk_A k_M [A][M][T]}{(1 + k_A[M])^2} \quad \text{for } k_A \gg k_M \tag{183}$$

The maximum rate with increase in [A] at constant [M] and [T] occurs at

$$[A]_{max} = \frac{k_M[M] + 1}{k_A} \tag{184}$$

with

$$[C^*]_{max} = \frac{f[M][T]}{4(k_M[M] + 1)} \tag{185}$$

There have been some books on the kinetics of Ziegler–Natta polymerization [108–110], and the reader is referred to the original literature for more detailed information.

KINETICS OF STEP POLYMERIZATION

A number of different chemical reactions may be used to synthesize polymeric materials by step polymerization [111]. These include esterification, amidation, the formation of urethanes, and aromatic substitution. Step polymerization usually proceeds by the reactions between two different functional groups, for example, hydroxyl and carboxyl groups, or isocyanate and hydroxyl groups.

All step polymerizations fall into two groups depending on the type of monomer(s) employed. The first involves two different bifunctional and/or polyfunctional monomers possessing only one type of functional group. The second involves a single monomer containing both types of functional group. These two groups of reactions can be represented in a general manner by the equations

$$nA-A + nB-B \rightarrow +(A-AB-B)_n \tag{186}$$

and

$$nA-B \rightarrow +(A-B)_n \tag{187}$$

where A and B are the two different types of functional group. For example, the synthesis of polyamides can be obtained from the reaction of diamines with diacids,

$$nH_2N-R-NH_2 + nHO_2C-R'-CO_2H \rightarrow H+(NH-R-NHCO-R'-CO)_n OH + (2n-1)H_2O \tag{188}$$

or from the reaction of amino acids with themselves,

$$nH_2N-R-CO_2H \rightarrow H+(NH-R-CO)_n OH + (n-1)H_2O \tag{189}$$

Step polymerization proceeds by a relatively slow increase in molecular weight of the polymer. Consider the synthesis of a polycondensate. It proceeds from monomer to dimer, trimer, tetramer, pentamer, and so on:

$$\begin{aligned}
&\text{Monomer} + \text{monomer} \rightarrow \text{dimer} \\
&\text{Dimer} + \text{monomer} \rightarrow \text{trimer} \\
&\text{Dimer} + \text{dimer} \rightarrow \text{tetramer} \\
&\text{Trimer} + \text{monomer} \rightarrow \text{tetramer} \\
&\text{Trimer} + \text{dimer} \rightarrow \text{pentamer} \\
&\text{Trimer} + \text{trimer} \rightarrow \text{hexamer} \\
&\text{Tetramer} + \text{monomer} \rightarrow \text{pentamer} \\
&\text{Tetramer} + \text{dimer} \rightarrow \text{hexamer} \\
&\text{Tetramer} + \text{trimer} \rightarrow \text{heptamer} \\
&\text{Tetramer} + \text{tetramer} \rightarrow \text{octamer}
\end{aligned} \tag{190}$$

etc., which can be expressed as the general reaction

$$n\text{-mer} + m\text{-mer} \rightarrow (n+m)\text{-mer} \tag{191}$$

The kinetics of such a situation with innumerable separate reactions would normally be difficult to analyze. However, the kinetic analysis is greatly simplified if one assumes that the reactivities of both functional groups' monomers (e.g., both hydroxyls of a diol) are the same, the reactivity of one functional group of a bifunctional reactant is the same

Table 2 Rate Constant for Esterification (25°C) in Homologous Compounds[a]

Molecular size (X)	$k \times 10^4$ for $H(CH_2)_x CO_2 H$	$k \times 10^4$ for $(CH_2)_x(CO_2H)_2$
1	22.1	
2	15.3	6.0
3	7.5	8.7
4	7.5	8.4
5	7.4	7.8
6		7.3
8	7.5	
9	7.4	
11	7.6	
13	7.5	
15	7.7	
17	7.7	

[a]All rate constants are in units of (moles carboxyl groups per liter) per second.
Source: Data from Ref. 112.

irrespective of whether or not the other functional group has reacted, and the reactivity of a functional group is independent of the size of the molecule to which it is attached (i.e., independent of the values of n and m). These simplifying assumptions, often referred to as the concept of equal reactivity of functional groups, make the kinetics of step polymerization identical to those for the analogous small-molecule reaction. After many long-term and heated debates these questions have recently been answered. These assumptions are justified on the basis that many step polymerizations have reaction rate constants that are independent of the reaction time or polymer molecular weight. It is, however, useful to examine in detail the experimental and theoretical justifications for these assumptions. Consider, for example, the rate constant data in the first column in Table 2 for the esterification of a series of homologous carboxylic acids [112]:

$$H(CH_2)_x CO_2 H + C_2 H_5 OH \xrightarrow{HCl} H(CH_2)_x CO_2 C_2 H_5 + H_2 O \qquad (192)$$

It is evident that, although there is a decrease in reactivity with increased molecular size, the effect is significant only at a very small size. The reaction rate constant very quickly (at $X = 3$) reaches a limiting value, which remains constant and independent of molecular size. Analogous results are found for the polyesterification of sebacoyl chloride with α,ω-alkane diol:

$$HO(CH_2)_x OH + Cl-OC(CH_2)_8 CO-Cl \xrightarrow{-HCl} +O(CH_2)_x OCO(CH_2)_8 CO+_n \qquad (193)$$

The rate constant for esterification is independent of X for the compounds studied (Table 3) [113]. The results for sebacoyl chloride offer direct evidence of the concept of functional group reactivity being independent of molecular size, since the rate constant is

Table 3 Rate Constants for Polyesterification (26.9°C) of Sebacoyl Chloride with α,ω-Alkane Diols in Dioxane[a]

Molecular size (X)	$k \times 10^3$ for $HO(CH_2)_xOH$
5	0.60
6	0.63
7	0.65
8	0.62
9	0.65
10	0.62

[a]Rate constants are in units of (moles per liter) per second.
Source: Data from Ref. 113.

independent of n as well as X. There is also ample theoretical justification for the independence of the reactivity of a functional group of molecular size [114]. According to the collision theory of reaction rates [115, 116], the reactivity of functional group, i.e., rate constant, can be written as

$$K = A \exp\left(\frac{-E}{RT}\right) \tag{194}$$

where A is the frequency factor, $\exp(-E/RT)$ is the familiar Boltzmann constant, where E is the activated energy, R the gas constant, and T the temperature, respectively. Here the activated energy E is independent of viscosity as well as molecular size. Thus the reactivity of the functional group is dependent only on the frequency factor A. The observed reactivity of a functional group is dependent on the collision frequency of group, not on the diffusion rate of the whole molecule. The collision frequency is the number of collisions one functional group makes with other functional groups per unit of time. A terminal functional group attached to a growing polymer has much greater mobility than would be expected from the mobility of the polymer molecule as a whole. The functional group has appreciable mobility due to the rearrangements that occur in nearby segments of the polymer chain. The collision rate of such a functional group with neighboring groups will be about the same as small molecules.

The so-called transition state rate theory [117] can offer more strict evidence that the functional group reactivity is independent of molecular size. In light of this theory, the reaction process may be represented as

$$\text{Reactant} \rightleftharpoons \text{Activated complex} \rightleftharpoons \text{Product} \tag{195}$$
$$\quad (a) \qquad\qquad (ab)^* \qquad\qquad (b)$$

The rate constant is

$$K = k^*\left(\frac{k'T}{h}\right) \tag{196}$$

and for a bimolecular reaction

$$k^* = \left(\frac{Q^*_{ab}}{Q_a Q_b}\right) \exp\left(\frac{-E_0}{RT}\right) \tag{197}$$

where K is the rate constant, k' the Boltzmann constant, h the Planck constant, and k^* the constant for the equilibrium (Eq. 196), Q^*_{ab}, Q_a, and Q_b the respective partition functions for activated complex (ab)*, reactant (a), and product (b), E_0 the energy of activation for the reaction when all the substances concerned are in their lowest energy levels, and the other symbols are as defined earlier. As is well known, the values of Q^*_{ab}, Q_a, and Q_b have no tendency to change unless the structure of a terminal functional group is changed. Although the molecular weight of a polycondensate continuously increases with reaction time, the structure of the terminal functional group does not change. Of course, the equilibrium constant k^* will be kept constant. In other words, the activated complex (ab)* is in equilibrium with the reactant (a) and the product (b) under such circumstances. The concentration of activated complex is also not affected by molecular size. Therefore, the reactivity of the functional group and the rate constant are independent of molecular size.

To illustrate the general form of the kinetics of a typical step polymerization consider the polyesterification of a diacid and a diol. Simple esterification is a well-known acid-catalyzed reaction and polyesterification follows the same course [118]. The reaction involves protonation of the carboxylic acid,

$$\underset{}{\text{\textasciitilde\textasciitilde C}}\overset{O}{\overset{\|}{-}}\text{OH} + \text{HA} \underset{k_2}{\overset{k_1}{\rightleftharpoons}} \underset{}{\text{\textasciitilde\textasciitilde C}}\overset{OH}{\overset{|}{-}}\text{OH} + \text{A}^- \tag{198}$$

$$+$$
$$\text{(I)}$$

followed by the reaction of protonated species with alcohol to yield the ester,

$$\underset{+}{\overset{OH}{\overset{|}{\text{\textasciitilde\textasciitilde C}-\text{OH}}}} + \text{\textasciitilde\textasciitilde OH} \underset{k_4}{\overset{k_3}{\rightleftharpoons}} \underset{\underset{+}{\overset{|}{\text{\textasciitilde\textasciitilde OH}}}}{\overset{OH}{\overset{|}{\text{\textasciitilde\textasciitilde C}-\text{OH}}}} \tag{199}$$

$$\text{(II)}$$

$$\underset{\underset{+}{\overset{|}{\text{\textasciitilde\textasciitilde OH}}}}{\overset{OH}{\overset{|}{\text{\textasciitilde\textasciitilde C}-\text{OH}}}} \overset{k_5}{\rightleftharpoons} \text{\textasciitilde\textasciitilde C}\overset{O}{\overset{\|}{-}}\text{O\textasciitilde\textasciitilde} + \text{H}_2\text{O} + \text{H}^+ \tag{200}$$

Polyesterifications, like many other step polymerizations, are equilibrium reactions. However, from the practical viewpoint of obtaining high yields on high molecular weight product such polymerizations are run in a manner so as to continuously shift the equilibrium in the direction of the polymer. In the case of polyesterification this is easily accomplished by the removal of water, which is a product of the reaction. The rate of polymerization R_p can then be expressed as the rate of disappearance of carboxyl groups, $-d[\text{CO}_2\text{H}]/dt$. For the usual polyesterification, the polymerization rate is synonymous with the rate of formation of species (II), that is, k_4 is nonexistent, and k_1, k_2, and k_5 are large compared to k_3. An expression for the reaction rate can be obtained following general procedures [119]. The rate of polyesterification is given by

$$R_p = \frac{-d[\text{COOH}]}{dt} = k_3[\text{C}^+(\text{OH})_2][\text{OH}] \tag{201}$$

where [COOH], [OH], and [C$^+$(OH)$_2$] represent the concentrations of carboxyl, hydroxyl, and protonated carboxyl groups, respectively.

Equation 201 is inconvenient in that the concentration of protonated carboxyl groups is not easily determined experimentally. One can obtain a more convenient expression for the polymerization rate by substituting for [C$^+$(OH)$_2$] from the equilibrium expression

$$K = \frac{k_1}{k_2} = \frac{[\text{C}^+(\text{OH})_2][\text{A}^-]}{[\text{COOH}][\text{HA}]} \tag{202}$$

for the protonation reaction. Combination of Eqs. 201 and 202 yields

$$\frac{-d[\text{COOH}]}{dt} = \frac{k_1 k_3 [\text{COOH}][\text{OH}][\text{HA}]}{k_2 [\text{A}^-]} \tag{203}$$

This equation can also be written in the form

$$\frac{-d[\text{COOH}]}{dt} = \frac{k_1 k_3 [\text{COOH}][\text{OH}][\text{H}^+]}{k_2 K_{\text{HA}}} \tag{204}$$

where K_{HA} is the acid dissociation constant for HA. Two distinct kinetic situations arise from Eq. 204 depending on the identity of HA, that is, on whether or not a strong acid such as sulfuric acid is added as an external catalyst.

In the absence of an externally added strong acid the diacid monomer acts as its own catalyst for the esterification reaction. If [HA] is replaced by [COOH], Eq. 202 can be written in the usual form [1]

$$\frac{-d[\text{COOH}]}{dt} = k[\text{COOH}]^2[\text{OH}] \tag{205}$$

where the three rate constants k_1, k_2, and k_3 and the concentration term [A$^-$] have been collected into the experimentally determined rate constant K. This equation shows the reaction is third order overall with a second-order dependence on the carboxyl concentration.

For most polymerizations the concentrations of the two functional groups are very nearly stoichiometric, and Eq. 205 can be written as

$$\frac{-d[\text{M}]}{dt} = k[\text{M}]^3 \tag{206}$$

or

$$\frac{-d[\text{M}]}{[\text{M}]^3} = k\, dt \tag{207}$$

where [M] is the concentration of hydroxyl groups or carboxyl groups. Integration of Eq. 207 yields

$$2kt = \frac{1}{[M]^2} - \frac{1}{[M]_0^2} \tag{208}$$

where $[M]_0$ is the initial (at $t = 0$) concentration of hydroxyl or carboxyl groups. It is convenient at this point to write Eq. 208 in terms of the extent or fraction of reaction p defined as the fraction of the hydroxyl or carboxyl functional groups that has reacted at time t. p is also referred to as the extent or fraction of conversion. The concentration $[M]$ at time t of either hydroxyl or carboxyl groups is then given by

$$[M] = [M]_0 - [M]_0 p = [M]_0 (1 - p) \tag{209}$$

Combination of Eqs. 208 and 209 yields

$$2[M]_0^2 kt = \frac{1}{(1-p)^2} - 1 \tag{210}$$

Equation 210 indicates that a plot of $1/(1 - p)$ versus t should be linear. This behavior has been generally observed in polyesterifications. Figure 5 shows the results for the polymerization of diethylene glycol, $(HOCH_2CH_2)_2O$, and adipic acid.

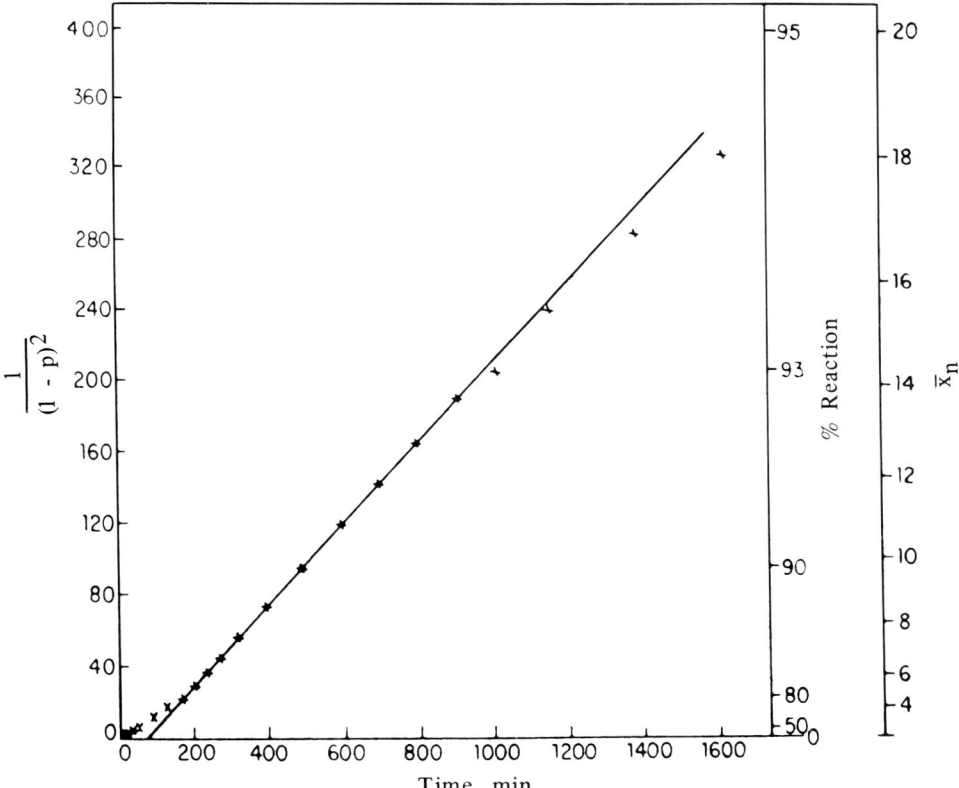

Figure 5 Third-order plot of the self-catalyzed polyesterification of adipic acid with diethylene glycol at 166°C. (After Ref. 120.)

The experimental points deviate from the third-order plot in the initial region below 80% conversion and in the later stages above 93% conversion. The rest agree with Eq. 219. These deviations have led various workers to suggest alternative kinetic expressions, that is, second-order and two and one-half-order overall dependencies according to

$$\frac{-d[\text{COOH}]}{dt} = k[\text{COOH}][\text{OH}] \tag{211}$$

and

$$\frac{-d[\text{COOH}]}{dt} = k[\text{COOH}]^{3/2}[\text{OH}] \tag{212}$$

A plot of the experimental rate data according to Eq. 211 fits the experimental data well only in the region between 50 and 86% conversion with an excessively poor fit above 86% conversion. On the other hand, a plot according to Eq. 212 fits reasonably well up to

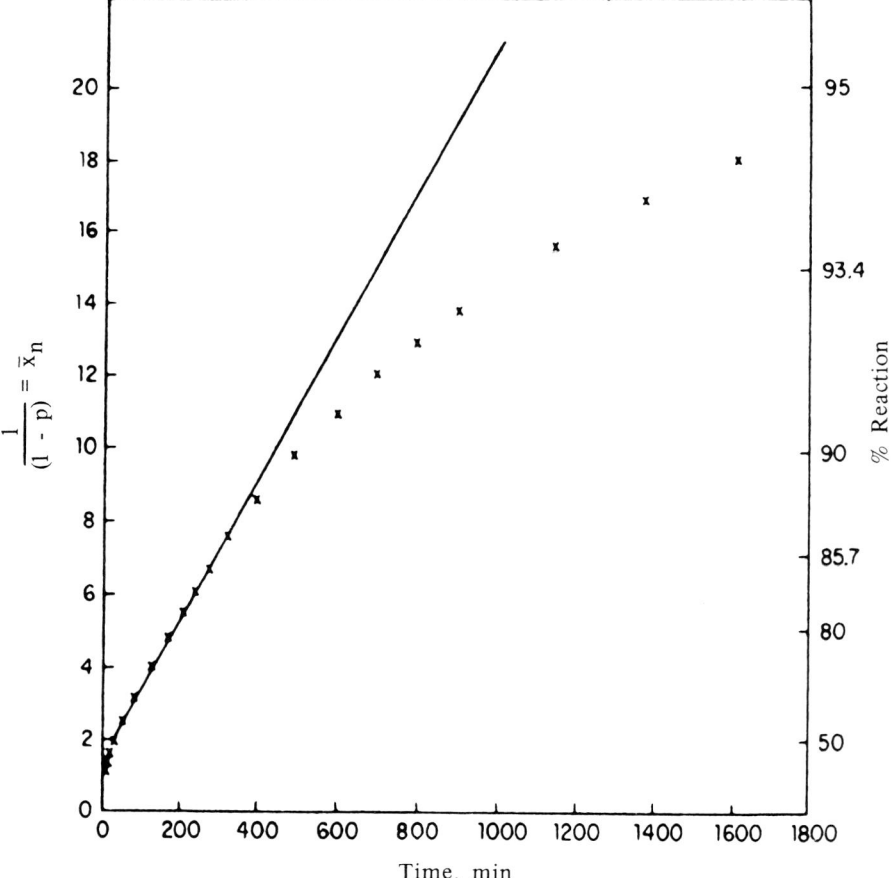

Figure 6 Second-order plot of the self-catalyzed polyesterification of adipic acid with diethylene glycol at 166°C. (After Ref. 120.)

about 70% conversion but deviates badly above that point. The results of the two alternative kinetic plots are shown in Figs. 6 and 7, respectively [120, 121].

The third-order plot fits the experimental data much better than does either of the others at the higher conversions. The fit of the data to the third-order plot is reasonably good over a much greater range of the higher conversion region. It is of prime importance, since high molecular weight polymer is obtained only at high conversions. The nonlinearity in the low conversion region was observed by Flory [120] for the esterifications between

$CH_3(CH_2)_4CO_2H$ and $HOCH_2CH_2OCH_2CH_2OH$
(caproic acid) (diethylene glycol)

$CH_3(CH_2)_{10}CO_2H$ and $CH_3(CH_2)_{10}CH_2OH$
(lauric acid) (lauryl alcohol)

$HO_2C(CH_2)_4CO_2H$ and $CH_3(CH_2)_{10}CH_2OH$
(adipic acid) (lauryl alcohol)

and ascribed the deviation from linearity to the large changes that take place in the reaction medium.

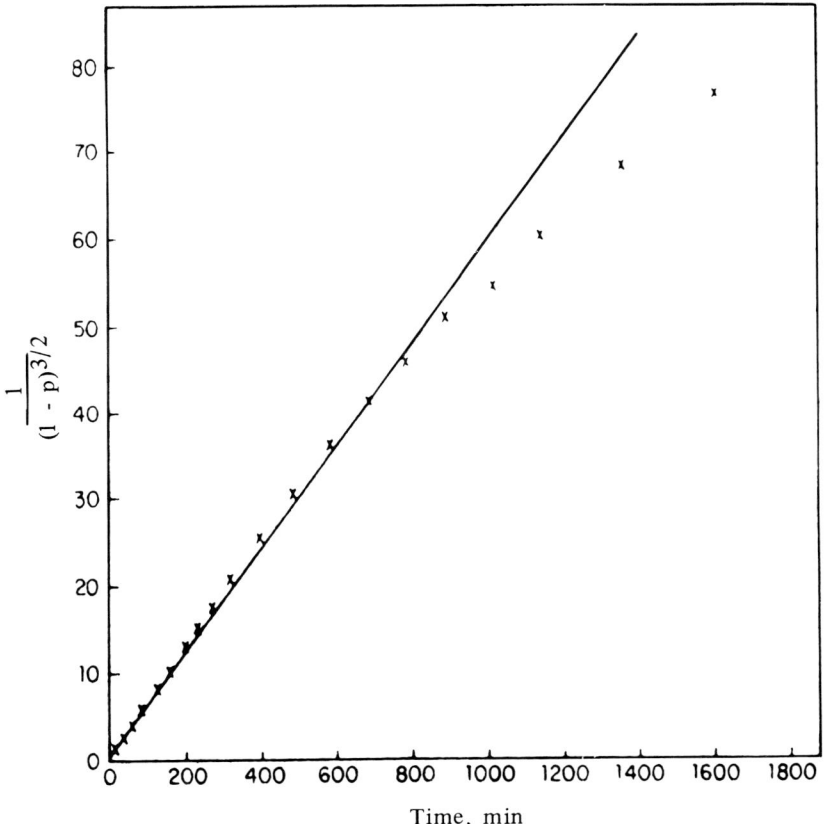

Figure 7 Two and one-half order plot of the self-catalyzed polyesterification of adipic acid with diethylene glycol at 166°C. (After Ref. 120.)

Odian [13] has pointed out that [HA] or [H$^+$] is not an accurate measure of the acidity of a nonaqueous system or a system containing high concentrations of an acid. In such systems the acidity function h_0 may be the more applicable indicator of the acidity of the reaction system. The h_0 function for a system [122] is defined by

$$h_0 = \frac{a_{H^+} r_B}{r_{BH^+}} = \frac{r_{H^+}[H^+] r_B}{r_{BH^+}} \tag{213}$$

where a_{H^+} and r_{H^+} are the activity and activity coefficient, respectively, of H$^+$ in the system, and r_B/r_{BH^+} is the ratio of the activity coefficients of a base B and its conjugate acid BH$^+$. For the esterification reaction the carboxyl group and its protonated form would be considered as B and BH$^+$, respectively. The h_0 function has been found to be the applicable measure of acidity in many reaction systems that are highly concentrated or nonaqueous solutions. The use of the h_0 function would involve the substitution of the appropriate expression from Eq. 213 for [H$^+$] in Eq. 204 to yield

$$\frac{-d[\text{COOH}]}{dt} = \frac{k_1 k_3 [\text{COOH}][\text{OH}] h_0 r_{C^+(OH)_2}}{k_2 k_{HA} r_{H^+} r_{COOH}} \tag{214}$$

The h_0 function has been successfully used in the kinetic analysis of the acid-catalyzed polymerization reaction of formaldehyde with phenol and other aromatic compounds [123].

The nonlinearity observed in the third-order plot in the final stages of the polyesterification is probably not due to any of the foregoing reasons, since the reaction system is fairly dilute and of relatively low polarity. It is more likely that several other factors are responsible for the nonlinear region above 93% conversion. Polyesterifications, like many step polymerizations, are carried out at moderate to high temperatures not only to achieve fast reaction rates but also to aid in removal of the small molecule byproduct. Partial vacuum is usually also employed to drive the system toward high molecular weight. Under these conditions small amounts of one or the other or both reactants may be lost by degradation or volatilization. Although such losses may not be important initially, they can become very significant during the late stages of the reaction. Thus a loss of only 0.3% of one reactant can lead to an error of almost 5% in the concentration of that reactant at 93% conversion. Another possible reason for the observed nonlinearity is an increase in the rate of the reverse reaction. It often becomes progressively more difficult to displace the equilibrium to the right as the conversion increases. This is caused in large part by the greatly increased viscosity of the reaction medium at high conversions. This large viscosity increase decreases the efficiency of water removal and may lead to the observed decrease in the reaction rate with increasing conversion.

We next consider the effect of added strong acids (such as sulfuric acid or p-toluenesulfonic acid) as catalysts upon the kinetic behavior of polyesterification. The slow increase in molecular weight was simply a consequence of the third-order kinetics of the direct polyesterification reaction. Under these conditions, [HA] in Eq. 202 or [H$^+$] in Eq. 204 is the concentration of the catalyst. Since this remains constant throughout the course of the polymerization, Eq. 204 can be written as

$$\frac{-d[M]}{dt} = k'[M]^2 \tag{215}$$

where the various constant terms in Eqs. 202 and 203 have been collected into the experimentally determinable ratio constant k'.

Integration of Eq. 215 yields

$$k't = \frac{1}{[M]} - \frac{1}{[M]_0} \qquad (216)$$

Combining Eq. 216 with Eq. 209 yields the dependence of the degree of polymerization on reaction time as

$$[M]_0 kt = \frac{1}{1-p} - 1 \qquad (217)$$

Data for the polymerization of diethylene glycol with adipic acid catalyzed by p-toluenesulfonic acid are shown in Fig. 8. The plot follows Eq. 217 with the degree of polymerization increasing linearly with action time. The kinetics of step polymerizations other than polyesterification are considered later.

First we discuss the kinetics of polyamidation. Nylon-6,6,poly(hexamethylene adipamide), prepared by reacting substantially equimolar quantities of adipic acid and hexamethylene diamine, has for many years been the world's major wholly synthetic fiber.

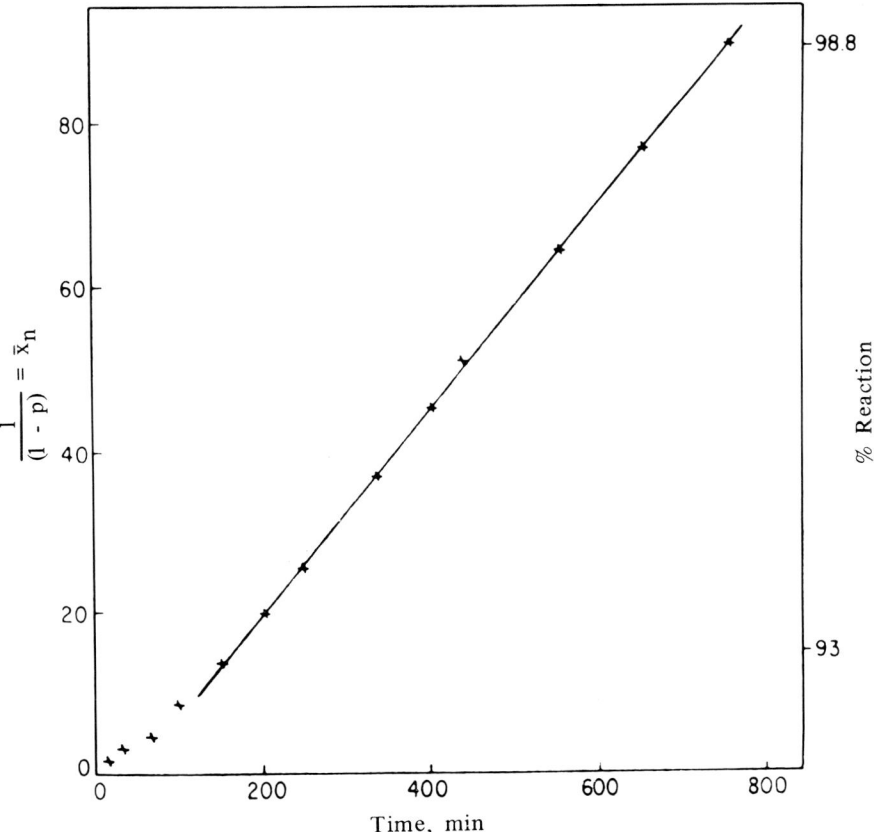

Figure 8 Polyesterification of adipic acid with diethylene glycol at 109°C catalyzed by 0.4 mole percent p-toluenesulfonic acid. (After Ref. 120.)

Nylon-6,6 is made commercially from adipic acid and hexamethylene diamine with the intermediate formation of a salt. Isolation of this salt provides a means of further purification to the stringent levels required for polymerization to high molecular weight:

$$n\text{HOOC(CH}_2)_4\text{COOH} + n\text{H}_2\text{N(CH}_2)_6\text{NH}_2$$
$$= n(\text{O}^-\text{OC(CH}_2)_4\text{COO}^-)(\text{H}_3\text{N}^+(\text{CH}_2)_6\text{N}^+\text{H}_3) \qquad (218)$$
$$\rightleftharpoons n\text{HO}\text{+}\text{CO(CH}_2)_4\text{CONH(CH}_2)_6\text{NH}\text{+}_n\text{H} + (2n-1)\text{H}_2\text{O}$$

The polyamidation reaction when carried out in the melt appears to follow second-order kinetics [1, 124–126]. Thus for equal concentrations of —NH$_2$ and —COOH groups, the second-order equation applies:

$$\frac{-d[\text{M}]}{dt} = k[\text{M}]^2 \qquad (219)$$

where [M] is the concentration of amino or carboxyl groups. The integrated form of this equation is

$$[\text{M}] = \frac{[\text{M}]_0}{1 - [\text{M}]_0 kt} \qquad (220)$$

where $[\text{M}]_0$ is the initial concentration (at time $t = 0$). When end group concentrations are relatively low, as at conversions greater than roughly 90%, a third-order reaction becomes increasingly important. Under these conditions the formation of the amide group seems to be catalyzed by carboxyl, although there is still a strong second-order contribution to the rate of reaction in some cases [127]. For the third-order reaction

$$\frac{-d[\text{COOH}]}{dt} = k_1[\text{COOH}]^2[\text{NH}_2] \qquad (221)$$

For equimolar quantities of diamine and dicarboxylic acid, this equation simplifies to

$$\frac{-d[\text{COOH}]}{dt} = k_1[\text{COOH}]^3 \qquad (222)$$

Integrating, we have

$$\frac{1}{[\text{COOH}]^2} - \frac{1}{[\text{COOH}]_0^2} = 2k_1 t \qquad (223)$$

where $[\text{COOH}]_0$ is the concentration of the carboxyl group at time $t = 0$. It is analogous to the Eq. 208, as mentioned previously. The carboxyl-catalyzed reverse reaction (hydrolysis) must be taken into account. If k' is the rate constant for hydrolysis, then

$$\frac{-d[\text{COOH}]}{dt} = k[\text{COOH}]^2[\text{NH}_2] - k'[\text{COOH}][\text{CONH}][\text{H}_2\text{O}] \qquad (224)$$

For kinetic studies at high conversion, when the concentration of amide groups may be regarded as constant, the substitution

$$k'[\text{CONH}][\text{H}_2\text{O}] = k[\text{COOH}]_{\text{equ}}[\text{NH}_2]_{\text{equ}} \qquad (225)$$

can be introduced. Using this substitution, the third-order equation for the carboxyl-catalyzed reaction is

$$\frac{-d[\text{COOH}]}{dt} = k[\text{COOH}]([\text{COOH}][\text{NH}_2] - [\text{COOH}]_{\text{equ}}[\text{NH}_2]_{\text{equ}}) \qquad (226)$$

where the subscript equ denotes equilibrium concentrations. For balanced systems, where equimolar quantities of diamine and diacid were used, substitution and integration gives

$$\ln \frac{[COOH]^2}{[COOH]^{-2} - [COOH]^2_{equ}} = 2k[COOH]^2_{equ}t + \text{constant} \qquad (227)$$

Typical values of rate constants reported [128] for polyamidation in m-cresol include 7.7×10^{-4} g mmol^{-1} sec^{-1} for the preparation of poly(m-xylylene adipamide) and 4.2×10^{-4} g mmol^{-1} sec^{-1} for poly(p-xylylene sebacamide), as noted in Table 4.

Polyurethanes vary widely in their chemical composition, the starting materials which are used in their synthesis, the final polymer structure, and their uses. The one chemical feature all types have in common, and hence their name, is an abundance of urethane groups, NHCOO. A polyurethane may be formed from a simple glycol and a diisocyanate,

$$R(NCO)_2 + G(OH)_2 \rightarrow \{O-\underset{\underset{O}{\|}}{C}NH-R-NH\underset{\underset{O}{\|}}{C}O-G\}_n \qquad (228)$$

where G(OH)$_2$ is glycol. The glycol may be replaced by many other raw materials which terminate in hydroxyl groups, for example, polyethers, polyesters, and castor oil. The preparation may also utilize a reaction between a bis(carbamoyl chloride) and a glycol,

$$R(R'NCOCl)_2 + G(OH)_2 \xrightarrow{-HCl} \{O-\underset{\underset{O}{\|}}{C}NR'-R-NR'\underset{\underset{O}{\|}}{C}O-G\}_n \qquad (229)$$

a bis(chloroformate) and a diamine,

$$R(NH_2)_2 + G(OCOCl)_2 \xrightarrow{-HCl} \{O-\underset{\underset{O}{\|}}{C}NH-R-NH\underset{\underset{O}{\|}}{C}O-G\}_n \qquad (230)$$

a bis(urethane) and a glycol,

$$R(NHCOOR'')_2 + G(OH)_2 \xrightarrow{-R''OH} \{O-\underset{\underset{O}{\|}}{C}NH-R-NH\underset{\underset{O}{\|}}{C}O-G\}_n \qquad (231)$$

Table 4 Rate Constant for Polyamide Formation in m-Cresol

Salt or amino acid[a]	Temperature (°C)	Rate constant[b]
m-Xylylene diammonium adipate (I)	175	7.7×10^{-4}
Aminoundecanoic acid	176	1.8×10^{-4}
(I) and aminoenanthic acid	170	2.5×10^{-4}
(I) and aminoundecanoic acid	170	3.2×10^{-4}
Aminoenanthic acid	165	0.8×10^{-4}
p-Xylylene diammonium sebacate (II)	174	4.2×10^{-4}
(II) and aminoundecanoic acid	172	3.8×10^{-4}

[a]Equimolar quantities.
[b]In grams per millimole per second.
Source: Data from Ref. 128 for 30% solutions in m-cresol.

and similar reactions. The procedure most frequently used is the reaction between a diisocyanate and a polyhydroxy compound.

A number of reviews of the reactions of isocyanates with hydroxy compounds, including a thorough survey of the catalysis and kinetics of the reactions, have been published [129–133].

The isocyanate group is of such a structure that it may be expected to reflect the resonance forms

$$(R\text{—}\ddot{\underset{..}{N}}^-\text{—}C^+\text{=}\ddot{\underset{..}{O}} \leftrightarrow R\text{—}\ddot{N}\text{=}C\text{=}\ddot{\underset{..}{O}} \leftrightarrow R\text{—}\ddot{N}\text{=}C^+\text{—}\ddot{\underset{..}{O}}\text{:}^-) \tag{232}$$

These resonance structures suggest the probability of ionic reactions, with electron donors attacking the carbonyl carbon and electron acceptors attacking the oxygen or nitrogen. Catalysis by Lewis acids and bases should be common. One may visualize the reaction of an isocyanate with an alcohol as involving the following steps, when a basic catalyst is present:

$$\text{RNCO} + \text{:B} \rightleftharpoons \text{RN=C—O} \overset{\text{ROH}}{\rightleftharpoons} \begin{matrix} \text{H} \\ \diagdown \diagup \\ \text{O} \\ \downarrow \\ \text{RN—C—O} \\ \uparrow \\ \text{B} \end{matrix} \overset{\text{H}}{\rightarrow} \begin{matrix} \text{H} \\ | \\ \text{RN—COR} \\ \| \\ \text{O} \end{matrix} + \text{:B} \tag{233}$$

An acid catalyst could function in the following way:

$$\text{RNCO} + \text{HA} \rightleftharpoons (\text{R—N=C—}\ddot{\underset{..}{O}}\text{:} \rightarrow \text{HA}) \overset{\text{R'OH}}{\longrightarrow} \text{RNHCOOR'} + \text{HA} \tag{234}$$

Baker et al. published the first detailed kinetic treatment of the isocyanate–alcohol reaction [134–136]. They proposed the existence of both a catalyzed reaction,

$$(\text{R—N=C=O} \leftrightarrow \text{RN=C}^+\text{—O}^-) + \text{:B} \underset{k_2}{\overset{k_1}{\rightleftharpoons}} \text{RN=C—O}^- \overset{\text{R'OH}}{\underset{k_3}{\longrightarrow}} \begin{matrix} \text{RNCOOR'} \\ \text{H} \end{matrix} + \text{:B} \tag{235}$$

and an uncatalyzed one,

$$\text{RNCO} + \text{R'OH} \overset{k_0}{\rightarrow} \text{RNHCOOR'} \tag{236}$$

Assuming the steady-state condition, the concentration of the complex RNCO:B was given by the equation

$$(\text{complex}) = \frac{k_1(\text{RNCO})(\text{B})}{k_2 + k_3(\text{R'OH})} \tag{237}$$

and the rate of disappearance of isocyanate and rate of formation of the product was

$$\frac{-d[\text{RNCO}]}{dt} = k_0[\text{R'OH}][\text{RNCO}] + \frac{k_1 k_3 [\text{RNCO}][\text{R'OH}][\text{B}]}{k_2 + k_3[\text{R'OH}]} \tag{238}$$

From this, the experimentally observed rate constant for any one set of conditions would be

$$k_{\text{exp}} = k_0 + \frac{k_1 k_3 [\text{B}]}{k_2 + k_3 [\text{R'OH}]} \tag{239}$$

If the value of k_2 is much greater than $k_3[R'OH]$, or if $[R'OH]$ is constant in a set of experiments, this may be further simplified to

$$k_{exp} = k_0 + k_c[B] \tag{240}$$

where k_c, the catalytic rate constant, is given by

$$k_c = \frac{k_1 k_3}{k_2 + k_3[R'OH]} \tag{241}$$

KINETICS OF GROUP TRANSFER POLYMERIZATION

Group transfer polymerization (GTP) is a fundamentally new method for polymer formation [137–144]. It was discovered and claimed by DuPont in 1983. It involves the repeated addition of monomer to a growing polymer chain end which carries a reactive silyl ketene acetal group. During the addition, the silyl group transfers to incoming monomer regenerating a new ketene acetal function ready for reaction with more monomer, hence the name group transfer polymerization. GTP works best with the methacrylate family of monomers, to which most of the published data refer. GTP requires a catalyst. Nucleophilic catalysts such as soluble fluorides, bifluorides, azides and cyanides activate the initiator. For ease of synthesis, tris(dimethylamino) sulfonium salts are preferable. GTP provides "living" polymer rapidly at room temperature and offers new dimensions in the construction and design of polymer chains. The process is illustrated by Eqs. 242 and 243, for methalmethacrylate monomer:

$$\underset{\substack{|\\CH_3}}{CH_3-C=C}-O-Si(CH_3)_3 + CH_2=\underset{\substack{\|\\CH_3O}}{C}-C-O-CH_3 \xrightarrow{\text{catalyst}}$$

$$CH_3-\underset{\substack{|\\CH_3}}{\overset{\substack{C-O-CH_3\\\|\\O}}{C}}-CH_2-\underset{\substack{|\\CH_3}}{\overset{\substack{CH_3\\|\\O}}{C=C}}-O-Si(CH_3)_3 \tag{242}$$

$$CH_3-\underset{\substack{|\\CH_3}}{\overset{\substack{C-O-CH_3\\\|\\O}}{C}}-CH_2-\underset{\substack{|\\CH_3}}{\overset{\substack{CH_3\\|\\O}}{C=C}}-O-Si(CH_3)_3 + CH_2=\underset{\substack{\|\\CH_3O}}{C}-C-O-CH_3 \dashrightarrow$$

$$CH_3-\underset{\substack{|\\CH_3}}{\overset{\substack{C-O-CH_3\\\|\\O}}{C}}\left(CH_2-\underset{\substack{|\\CH_3}}{\overset{\substack{C-O-CH_3\\\|\\O}}{C}}\right)_n CH_2-\underset{\substack{|\\CH_3CH_3}}{\overset{\substack{CH_3\\|\\O}}{C=C}}-O-Si(CH_3)_3 \tag{243}$$

Only a few studies have been done on the mechanisms of GTP. An associative mechanism was proposed for nucleophilic catalysis. It is outlined in Scheme 1 [138, 145, 146]:

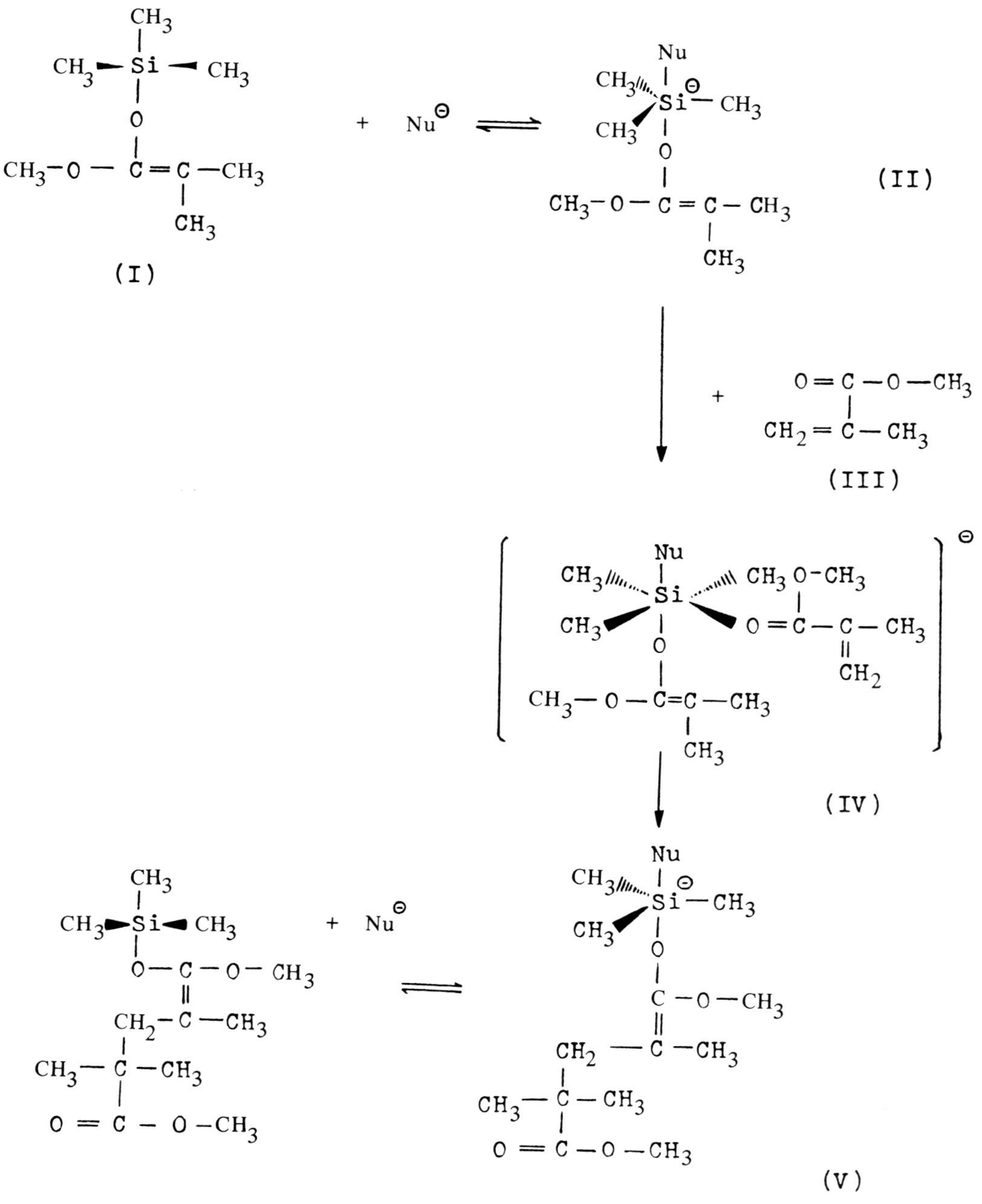

Scheme 1

The nucleophilic catalyst (e.g., HF_2^-) coordinates to the silicon atom of the initiator [1-methoxyl-1-(trimethylsiloxy)-2-methyl propene] (I) to provide a pentacoordinate species (II). The activated initiator and monomer (III) are proposed for a hypervalent silicon intermediate (IV). A new carbon–carbon bond is created between initiator and monomer and the trimethylsilyl group is transferred to the carbonyl oxygen of the monomer, forming a new silicon–oxygen bond, cleaving the old silicon–oxygen bond. This transfer regenerates a structure similar to that of the initiator. The propagation reaction [(II) to (V)] should proceed as long as monomer is present. It has been shown that dissociation of (II) to free enolate is unlikely. However, there is no direct evidence, so far, for a "concerted" mechanism symbolized by structure (IV).

The following kinetic scheme is consistent with the mechanism shown above:

$$C + I \underset{}{\overset{k_i^*}{\rightleftharpoons}} I^*$$
$$+M \downarrow k_i \quad \text{Initiation}$$
$$C + N \underset{}{\overset{k_p^*}{\rightleftharpoons}} N_2^* \quad \quad \quad \quad \quad \quad (245)$$
$$+M \downarrow k_p \quad \text{Propagation}$$
$$C + N \underset{}{\overset{k_p^*}{\rightleftharpoons}} N_3^*$$
etc.

where C is the catalyst, I the initiator, M the monomer, N_n the polymer with degree of polymerization n, N_n^* the activated n-mer, and k's are the relevant rate constants.

If initiator and polymer have the same chemical structure, then

$$I = N_1 \quad \text{and} \quad I^* = N_1^*$$

To a first approximation we assume that

$$k_i^* = k_p^* = k^* \quad \text{and} \quad k_i = k_p$$

The scheme then simplifies to

$$C + N_i \underset{}{\overset{k^*}{\rightleftharpoons}} N_i^* \quad (i \geq 1)$$
$$+M \downarrow k_p$$
$$C + N_{i+1} \underset{}{\overset{k^*}{\rightleftharpoons}} N_{i+1}^* \quad \quad \quad \quad \quad \quad (246)$$

The rate of polymerization is given by

$$R_p = \frac{-d[M]}{dt} = k_p[M]\sum_{i=1}^{\infty}[N_i^*] = k_p[M][N^*] \quad (247)$$

For a living system $[N^*]$ = constant and integration leads to

$$\ln \frac{[M]_0}{[M]} = k_p[N^*]t = k_{app}t \quad (248)$$

The concentration of activated polymers is given by the mass action law:

$$[N^*] = k^*[N][C] = k^*([I]_0 - [N^*])([C]_0 - [N^*]) \quad (249)$$

Usually catalyst is added only in very low amounts, thus $[I]_0 \gg [N^*]$. This leads to

$$[N^*] = k^*[I]_0([C]_0 - [N^*]) \tag{250}$$

$$[N^*] = \frac{k^*[I]_0}{1 + k^*[I]_0}[C]_0 \tag{251}$$

and

$$k_{app} = k_p \frac{k^*[I]_0}{1 + k^*[I]_0}[C]_0 \tag{252}$$

Two limiting cases can be discussed as follows. First, $k^*[I]_0 \gg 1$, i.e., the activation equilibrium is shifted to the right. Then

$$[N^*] = [C]_0 \quad \text{and} \quad k_{app} = k_p[C] \tag{253}$$

Second, $k^*[I]_0 \ll 1$, i.e., the activation equilibrium is shifted to the left. This leads to

$$[N^*] = k^*[I]_0[C]_0 \quad \text{and} \quad k_{app} = k_p k^*[I]_0[C]_0 \tag{254}$$

Thus it follows from the proposed mechanism that the kinetic orders of the reaction, defined by

$$R_p = \text{constant } [M]^\alpha [I]^\beta [C]^\tau \tag{255}$$

where $\alpha = 1$, $0 \leq \beta \leq 1$, and $\tau = 1$.

Müller and co-workers researched this question in detail. Their results are as follows [147]. For high catalyst concentration the reaction proceeds in an ideal manner, initiation is fast compared to propagation, and the propagation reaction follows first-order kinetics (indicating the absence of termination reactions). For low catalyst concentration low reaction rates are observed initially ("induction period") followed by increased rates in a linear manner. Finally, the rates decrease again, indicating the presence of terminations. The rate constants are smaller than expected. It is conceivable that during the induction period some deactivation of active centers occurs, leading to lower apparent rate constants.

The results of the influence of the initiator concentration on the reaction rate have shown that the time-conversion curves are characterized by a slight induction period at the beginning of the reaction followed by a linear phase in the case of higher initiator concentrations. There are indications of the presence of termination reactions at higher conversions and the inhibition due to the initiator. This may be explained by the following extended kinetic scheme:

$$\begin{array}{c} C + I \underset{}{\overset{k_i^*}{\rightleftharpoons}} N_1^* \xrightarrow[+I]{k_{t,i}} X^* \underset{}{\overset{k_x^*}{\rightleftharpoons}} X + C \\ +M \downarrow k_p \\ C + N \underset{}{\overset{k_p^*}{\rightleftharpoons}} N_2^* \xrightarrow[+N_i]{k_{t,p}} X^* \underset{}{\overset{k_x^*}{\rightleftharpoons}} X + C \end{array} \tag{256}$$

Initiator or polymer may react with activated initiator or polymer leading to an inactive product X capable of binding catalyst ($k_x^* \geq k_i^* \simeq k_p^*$). The reduced actual catalyst concentration will be reflected in lower apparent rate constants. It is likely that the induction periods observed relate to the same phenomenon.

THE CALCULATION OF MOLECULAR WEIGHT DISTRIBUTIONS FROM KINETIC SCHEMES

When we speak of the molecular weight of a polymer, it means something quite different from that which applies to small-sized compounds. Polymers differ from small-sized compounds in that they are polydisperse and heterogeneous in molecular weight. Polymers, in their purest form, are mixtures of molecules of different molecular weights. The control of molecular weight distribution (MWD) is often used to obtain and improve certain desired physical properties in a polymeric product.

However, only in a few very simple cases can the molecular weight distribution of a polymer resulting from a polymerization reaction be obtained by the steady-state assumption. In a common case, for example, vinyl polymerization with termination by disproportionation, the MWD has to be evaluated without the assumption of steady state. This assumption, in fact, is tenable only for questions of reaction rate, but it is inconsistent with most of the results of MWD. In all other cases the MWD must be obtained by solving the set of simultaneous differential equations by which the rates of different reactions are described. These differential equations can be solved as such in some cases to give analytical results; in some cases the solutions can be obtained by the appropriate classical methods; in others they can be solved successively and occasionally the solution can be guessed and then verified. These methods have been described for a number of kinetic schemes in several textbooks [11, 148, 149] and we use them only by way of illustration; it is the purpose of this section to describe some general methods, including probability, linear operator, and graphic methods and to show how to use them from the easy to the difficult. It is, of course, impractical, if not impossible, to describe them in detail in such limited space. Another method, consisting of transforming the kinetic equations into a partial differential equation, has been proposed and used with success, but it is unnecessary to give details here and we refer the interested reader to the articles for further information [150, 151].

We begin by illustrating the use of the probability method to obtain the MWD when the initiator is only supplied at the beginning and without termination [152]. An example is the case of caprolactam initiated by organic acid or amine. The chemical reactions we consider are

$$N_n^* + M \xrightarrow{k_p} N_{n+1}^*, \qquad n = 0,1,2,\ldots \tag{257}$$

where M is the monomer, N_n^* the active n-mer, and k_p the rate constant for propagation.

In terms of this mechanism we can write a set of simultaneous differential equations as follows:

$$\frac{dN_0^*}{dt} = -k_p N_0^* M \tag{258}$$

$$\frac{dN_n^*}{dt} = -k_p N_n^* M + k_p N_{n-1}^* M, \qquad n = 1,2,\ldots \tag{259}$$

Assuming one initiator to be present in a system consisting of M_0 monomers, the probability that monomers have reacted is defined as q. It follows that $(1 - q)$ is the probability of finding monomers unreacted. When n monomers have been linked into the initiator, the probability of finding the other $(M_0 - n)$ unreacted monomers is then $q^n(1 - q)^{M_0 - q}$.

These n reacted monomers, in fact, are unable to appoint, that is, the probability of finding the n-mer should be multiplied by a combinatorial factor:

$$W = \frac{M_0!}{n!(M_0 - n)!} q^n (1 - q)^{M_0-q} \tag{260}$$

In actual fact, q is small and $M_0 \gg n$. If putting $\alpha = M_0 q$, then†

$$W = \frac{\alpha^n e^{-\alpha}}{n!} \tag{261}$$

As we know this is a Poisson distribution.

For a system containing I_0 initiators and M_0 monomers, the probability of finding the n-mer is

$$W' = I_0 W' = I_0 \frac{\alpha^n e^{-\alpha}}{n!} \tag{262}$$

That is the MWD function of the preceding polymerization reaction. This MWD function contains only one parameter α which can be determined from the following condition:

$$M_0 - M' = \Sigma n W' = I_0 \alpha \tag{263}$$

Thus

$$\alpha = \frac{M_0 - M'}{I_0} \tag{264}$$

where M' is the number of the unreacted monomers. Putting $n = \alpha + X$ and substituting into Eq. 262 and using Stirling approximation for n,

$$n! = \sqrt{2\pi n}\left(\frac{n}{e}\right)^n \left(1 + \frac{1}{12n} + \frac{1}{288n^2} + \cdots\right) \tag{265}$$

Thus

$$n! \simeq \sqrt{2\pi n}\left(\frac{n}{e}\right)^n \tag{266}$$

Assuming $X \ll \alpha$ (this assumption is generally satisfied since the shape of this distribution curve is steep), we obtain

$$W(X) = \frac{I_0}{\sqrt{2\pi\alpha}} e^{-X/2\alpha} \tag{267}$$

That is the well-known Gaussian distribution function. The proportion of the polymers with $|X| \gg X_0$ can be calculated:

---

†$1 - q \simeq e^{-q}$, $(1 - q)^{M_0 - q} \simeq e^{-M_0 q} = e^{-\alpha}$

$\frac{M_0!}{(M_0 - n)!} = (M_0 - n + 1)(M_0 - n + 2) \cdots (M_0) \simeq M_0^n$

$\therefore W = \frac{M_0^n q^n e^{-\alpha}}{n!} = \frac{\alpha^n e^{-\alpha}}{n!}$

$$\frac{1}{I_0} \sum_{|X| \geq X_0} W(X) = 2 \frac{1}{\sqrt{2\pi\alpha}} \int_{x_0}^{\infty} e^{-X/2\alpha} dX = \frac{2}{\sqrt{\pi}} \int_{x_0/\sqrt{2\alpha}}^{\infty} e^{-y^2} dy \tag{268}$$

The integral value of the right side in Eq. 268 can be obtained directly from the table of Gaussian functions. Here some data are excerpted:

X	0.6745	1.000	1.500	2.000	2.500	3.000
Integral value	0.5000	0.317	0.134	0.0455	0.0124	0.0027

It is thus clear that the greater part of the polymers are within the range of $\alpha - 3\sqrt{\alpha} \leq n \leq \alpha + 3\sqrt{\alpha}$. This is therefore a perfectly monodispersed polymer. According to Eq. 262 and the definitions of average molecular weights, the number-, weight-, and z-average molecular weights can be calculated as follows:

$$\overline{M}_n = \frac{w \Sigma n W'}{\Sigma W'} = w\alpha \tag{269}$$

$$\overline{M}_w = \frac{w \Sigma n^2 W'}{\Sigma n W'} = w(\alpha + 1) \tag{270}$$

$$\overline{M}_z = \frac{w \Sigma n^3 W'}{\Sigma n^2 W'} = w \frac{(\alpha^2 + 3\alpha + 1)}{(\alpha + 1)} \tag{271}$$

where w is the molecular weight of the monomer. The ratios of these three average molecular weights are

$$\overline{M}_n : \overline{M}_w : \overline{M}_z = 1 : \left(1 + \frac{1}{\alpha}\right) : \left(1 + \frac{2\alpha + 1}{\alpha^2 + \alpha}\right) \tag{272}$$

If α is large, Eq. 272 tends to $1:1:1$. It indicates once again that the polymer has a narrow MWD.

Next we illustrate how to solve this kind of equation by kinetic methods. A typical method is the so-called linear operator notation [153]. It is sometimes convenient to adopt the notation $Dy, D^2y, \ldots, D^n y$ to denote $dy/dx, d^2y/dx^2, \ldots, d^n y/dx^n$. The symbols D, $D^2, \ldots, D^n$ are called differential operators and have properties analogous to those of algebraic quantities. The general linear differential equation of order n has the form

$$a_0(x) \frac{d^n y}{dx^n} + a_1(x) \frac{d^{n-1} y}{dx^{n-1}} + \cdots + a_{n-1}(x) \frac{dy}{dx} + a_n(x) y = R(x) \tag{273}$$

Using this notation, we write Eq. 273 as

$$[a_0(x) D^n + a_1(x) D^{n-1} + \cdots + a_{n-1}(x) D + a_n(x)] y = R(x) \tag{274}$$

or, briefly,

$$\phi(D) y = R(x) \tag{275}$$

where $\phi(D) = a_0(x) D^n + a_1(x) D^{n-1} + \cdots + a_{n-1}(x) D + a_n(x)$ is called an operator polynomial in D.

An operator L is called a linear operator if for any constants A, B and functions u, v to which L can be applied, we have

$$L(Au + Bv) = AL(u) + BL(v) \tag{276}$$

The operators D, D^2, ... and $\phi(D)$ are linear operators. Let $1/\phi(D) \cdot R(x)$ be defined as a particular solution y_p such that $\phi(D)y_p = R(x)$. We call $1/\phi(D)$ an inverse operator. By referring to the entries in the table of inverse operator techniques, the labor involved in finding solutions of $\phi(D)y = R(x)$ is often diminished considerably.

The problem of monomer transfer in living polymerization may be treated by the linear operator method [154]. This process may be considered to involve three elementary reactions:

Initiation:

$$I + M \xrightarrow{k_i} N_1^* \tag{277}$$

Propagation:

$$N_{n-1}^* + M \xrightarrow{k_p} N_n^* \quad (n > 1) \tag{278}$$

Transfer:

$$N_n^* + M \xrightarrow{k_t} N_n' + N_1^* \tag{279}$$

where I, M, N_n^*, and N_n' denote the initiator, monomer, active n-mer, and dead n-mer, respectively; and k_i, k_p, and k_t represent the rate constants for initiation, propagation, and transfer.

The differential rate equations corresponding to foregoing kinetic scheme are

$$\frac{dI}{dt} = -k_i IM \tag{280}$$

$$\frac{dN_1^*}{dt} = k_i IM + k_t \sum_{n \geq 1} N_n^* M - k_p N_1^* M - k_t N_1^* M \tag{281}$$

$$\frac{dN_n^*}{dt} = k_p N_{n-1}^* M - k_p N_n^* M - k_t N_n^* M \tag{282}$$

$$\frac{dN_n'}{dt} = k_t N_n^* M \tag{283}$$

It is sometimes convenient to solve the equations with the aid of the linear operator method. Letting $X = \int_0^t k_p M \, dt$, we obtain $dX/dt = k_p M$. For convenience, we adopt the following notation:

$$\frac{k_i}{k_p} = \alpha, \quad \frac{k_t}{k_p} = \beta$$

Substituting into Eqs. 280 to 283, and using linear operator, we obtain the following equations:

$$[D + \alpha]I = 0 \tag{284}$$

$$[D + (1 + \beta)]N_1^* = (\alpha - \beta)I + \beta I_0 \tag{285}$$

$$[D + (1 + \beta)]N_n^* = N_{n-1}^* \tag{286}$$

$$DN_n' = \beta N_n^* \tag{287}$$

Using the operator technique, we can write successively the solutions:

$$I = I_0 c^{-\alpha x} \tag{288}$$

$$N_1^* = I_0 \left[\frac{\beta}{1+\beta}(1 - e^{-(1+\beta)x}) + \frac{(\alpha - \beta)}{1+\beta-\alpha} e^{-\alpha x}(1 - e^{-(1+\beta-\alpha)x}) \right] \tag{289}$$

$$N_n^* = \frac{I_0 \beta}{(1+\beta)^n} \Gamma_{(1+\beta)x(n)} + \frac{I_0(\alpha - \beta)}{(1+\beta-\alpha)^n} e^{-\alpha x} \Gamma_{(1+\beta-\alpha)x(n)} \tag{290}$$

$$N_n' = I_0 \beta \left\{ \left[\beta \left(X - \frac{n}{1+\beta} \right) + 1 - \frac{\beta}{\alpha} \right] (1+\beta)^{-n} \Gamma_{(1+\beta)x(n)} \right.$$

$$\left. - \frac{(\alpha - \beta)}{\alpha(1+\beta-\alpha)^n} e^{-\alpha x} \Gamma_{(1+\beta-\alpha)x(n)} + \frac{n\beta x^n}{(1+\beta)n!} e^{-(1+\beta)x} \right\} \tag{291}$$

where I_0 is the initial value of the initiator at time $t = 0$, and $\Gamma_{(1+\beta)x(n)}$ and $\Gamma_{(1+\beta-\alpha)x(n)}$ are incomplete gamma functions. This is defined as

$$\Gamma_{(1+\beta)x(n)} = \frac{1}{(n-1)!} \int_0^x [(1+\beta)x]^{n-1} e^{-(1+\beta)x} d[(1+\beta)x]$$

Finally we illustrate the use of the graph method to treat a more complicated problem. Graph theory is an important branch of combinatory mathematics [155]. It has undergone extensive development since the late 1950s and still accounts for a major portion of all research efforts. Its applications continue to grow rapidly. The subject of chemistry was introduced to graph theory much earlier [156]. In recent years, graph theory has expanded into polymer science. Some graphs are used to study random condensation processes by analogy with the f-functional group [157, 158]. The graphs may also be applied to solve polymerization kinetic systems, especially in non–steady-state cases [159].

The graph consists of a vertex and an edge. Graph theory provides simple techniques for constructing models of systems, and powerful methods for their analysis and optimization. For example, a matrix of order $n \times n$, $A = (a_{ij})_{n \times n}$, may be represented as a graph consisting of n vertices. Each nonzero element a_{ij} in matrix A is an oriented edge which flows from i to j,

Thus, the matrix

$$A = \begin{bmatrix} 1 & 4 & 7 \\ 2 & 5 & 8 \\ 3 & 6 & 9 \end{bmatrix} \tag{292}$$

can be shown as

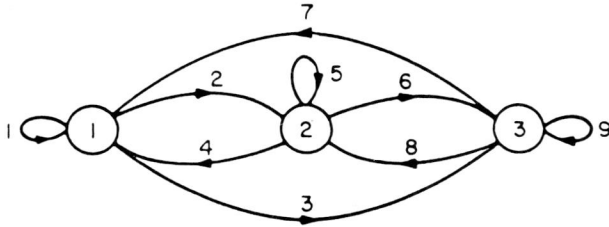

These graphs which consist of n vertices corresponding to $n \times n$ matrices are called Coates graphs. We introduce the calculating rule for a Coates graph as follows.

A set of equations having the form

$$\begin{bmatrix} a_{11} & & & \\ a_{12} & a_{22} & & \\ a_{13} & a_{23} & a_{33} & \\ a_{14} & a_{24} & a_{34} & a_{44} \end{bmatrix} \begin{bmatrix} X_1 \\ X_2 \\ X_3 \\ X_4 \end{bmatrix} = \begin{bmatrix} 1 \\ 0 \\ 0 \\ 0 \end{bmatrix} \tag{293}$$

which is four linear equations with four unknowns, x_1, x_2, x_3, and x_4, can be written more briefly as

$$A \cdot X = B \tag{294}$$

where A, X, and B represent the corresponding matrices in Eq. 293. Its coefficient matrix A may be shown as

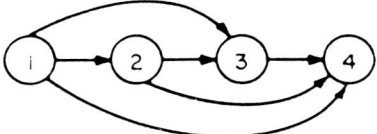

The notations a_{ii} and a_{ij} are denoted as the weight of the vertex i and the edge which flows from i to j, respectively. Thus each unknown x_i in Eq. 293 can be attained directly. The value of x_i equals the sum of all paths' contributions and the contribution of each path may then be calculated. The weight product of all edges along the path is taken as the numerator and that of the vertices the denominator. Multiplying it by $(-1)^k$, we can obtain the contribution of one path where k is the number of edges. We take 1 as the numerator for a source vertex. According to these rules, we can write the unknown of Eq. 293 in the following way.

X_1: There is only a source vertex:

① $\dfrac{1}{a_{11}}$

$$X_1 = \frac{1}{a_{11}}$$

X_2: One path:

①→② $(-1)\dfrac{a_{12}}{a_{11}a_{22}}$

$$X_2 = -\frac{a_{12}}{a_{11}a_{22}}$$

X_3: Two paths:

①→②→③ $(-1)^2 \dfrac{a_{12}a_{23}}{a_{11}a_{22}a_{33}}$

①→③ $(-1)\dfrac{a_{13}}{a_{11}a_{33}}$

$$X_3 = -\frac{a_{13}}{a_{11}a_{33}} + \frac{a_{12}a_{23}}{a_{11}a_{22}a_{33}}$$

X_4: Four paths:

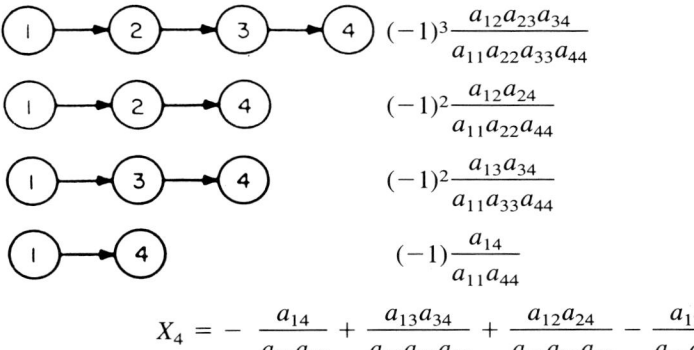

$$X_4 = -\frac{a_{14}}{a_{11}a_{44}} + \frac{a_{13}a_{34}}{a_{11}a_{33}a_{44}} + \frac{a_{12}a_{24}}{a_{11}a_{22}a_{44}} - \frac{a_{12}a_{23}a_{34}}{a_{11}a_{22}a_{33}a_{44}}$$

There are m source vertices, if the constant matrix B contains m nonzero elements and if these elements are not equal to 1 but $a, b, c, \ldots$, then correspondingly we multiply by $a, b, c, \ldots$.

We now consider the influence of the penultimate chain element on the monomer transfer in living polymerization as an instance to show the use of graph theory [160]. The penultimate chain element will influence the chain propagation in living polymerization. If we consider this effect, the kinetic mechanism may be represented as follows:

Initiation:

$$1 + M \xrightarrow{k_i} N_1^* \tag{295}$$

Propagation:

$$N_1^* + M \xrightarrow{k_p'} N_2^* \tag{296}$$

$$N_1\ddagger + M \xrightarrow{k_p''} N_2^* \tag{297}$$

$$N_{n-1}^* + M \xrightarrow{k_p} N_n^* \quad (n \geq 3) \tag{298}$$

Transfer:

$$N_n^* + M \xrightarrow{k_t} N_n' + N_1\ddagger \tag{299}$$

where $N_1\ddagger$ is the active monomer due to transfer, which is different from that due to initiation, N_1^*.

The kinetic equations are

$$\frac{dI}{dt} = -k_i IM \tag{300}$$

$$\frac{dN_1^*}{dt} = k_i IM - k_p' N_1^* M - k_t N_1^* M \tag{301}$$

$$\frac{dN_1^+}{dt} = k_t \sum_{n \geq 1} N_n^* M - k_p'' N_1^{\pm} M \tag{302}$$

$$\frac{dN_2^*}{dt} = k_p' N_1^* M + k_p'' N_1^{\pm} M - k_p N_2^* M - k_t N_2^* M \tag{303}$$

$$\frac{dN_n^*}{dt} = k_p N_{n-1}^* M - k_p N_n^* M - k_t N_n^* M \tag{304}$$

$$\frac{dN_n'}{dt} = k_t N_n^* M \tag{305}$$

Using the Laplace transformation we can convert the set of differential equations into a set of algebraic equations. Its matrix form can be written as follows:

$$\begin{bmatrix} a_{11} & & & & & & \\ a_{12} & a_{22} & & & & & \\ a_{13} & & a_{33} & & & & \\ & a_{24} & a_{34} & a_{44} & & & \\ & & & a_{45} & a_{55} & & \\ & & & & \cdots & & \\ & & & & & a_{m-1,m} & a_{mm} \end{bmatrix} \begin{bmatrix} X_1 \\ X_2 \\ X_3 \\ X_4 \\ \vdots \\ \vdots \\ X_m \end{bmatrix} = \begin{bmatrix} A \\ 0 \\ B \\ 0 \\ \vdots \\ \vdots \\ 0 \end{bmatrix} \tag{306}$$

where $m = n + 3$.

The graph corresponding to the preceding matrix consists of two source vertices and m edges. It is as follows:

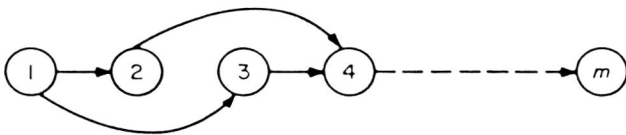

The respective solutions to Eq. 306 are as follows:

$$X_1 = \frac{A}{a_{11}}$$

$$X_2 = -\frac{Aa_{12}}{a_{11}a_{22}}$$

$$X_3 = -\frac{Aa_{13}}{a_{11}a_3} + \frac{B}{a_{33}}$$

$$X_m = (-1)^{m-2} \frac{Aa_{12}a_{24} \cdots a_{m-1,m}}{a_{11}a_{22}a_{44} \cdots a_{mm}} + (-1)^{m-2} \frac{Aa_{13}a_{34} \cdots a_{m-1,m}}{a_{11}a_{33}a_{44} \cdots a_{mm}}$$

$$+ (-1)^{m-3} \frac{Ba_{34}a_{45} \cdots a_{m-1,m}}{a_{33}a_{44}a_{55} \cdots a_{mm}}$$

Using Heaviside expansion or the table of Laplace transforms, we obtain:

$$N_n^* = I_0 e^{-(1+\beta)X} \left\{ \frac{(\beta - \alpha)(\gamma_1 \alpha - \gamma_2 \beta - \gamma_1 \gamma_2)}{(\gamma_1 + \beta - \alpha)(\gamma_2 + \beta - \alpha)(1 + \beta - \alpha)^{n-1}} \sum_{i=n-1}^{\infty} \frac{[(1 + \beta - \alpha)X]^i}{i!} \right.$$

$$\left. - \frac{\gamma_1 \alpha}{(\gamma_1 + \beta - \alpha)(1 - \gamma_1)^{n-1}} \sum_{i=n-1}^{\infty} \frac{[(1 - \gamma_1)X]^i}{i!} \right.$$

$$+ \frac{\gamma_2\alpha\beta}{(\gamma_2 + \beta - \alpha)(\gamma_2 + \beta)(1 - \gamma_2)^{n-1}} \sum_{i=n-1}^{\infty} \frac{[(1 - \gamma_2)X]^i}{i!} \qquad (307)$$

$$+ \frac{\gamma_2\beta}{(\gamma_2 + \beta)(1 + \beta)^{n-1}} \sum_{i=n-1}^{\infty} \frac{[(1 + \beta)X]^i}{i!} \bigg\}$$

$$N'_n = I_0 e^{-(1+\beta)X} \bigg\{ \frac{\beta}{(1 + \beta)^{n-1}} \bigg[\frac{(\beta - \alpha)(\gamma_1\alpha - \gamma_2\beta - \gamma_1\gamma_2)}{\alpha(\gamma_1 + \beta - \alpha)(\gamma_2 + \beta - \alpha)}$$

$$- \frac{\gamma_1\alpha}{(\gamma_1 + \beta - \alpha)(\gamma_1 + \beta)} + \frac{\gamma_2\alpha\beta}{(\gamma_2 + \beta - \alpha)(\gamma_2 + \beta)^2}$$

$$+ \frac{\gamma_2\beta}{(\gamma_2 + \beta)}\bigg(X - \frac{n-1}{1 + \beta}\bigg)\bigg] \sum_{i=n-1}^{\infty} \frac{[(1 + \beta)X]^i}{i!}$$

$$- \frac{\beta(\beta - \alpha)(\gamma_1\alpha - \gamma_2\beta - \gamma_1\gamma_2)}{\alpha(\gamma_1 + \beta - \alpha)(\gamma_2 + \beta - \alpha)(1 + \beta - \alpha)^{n-1}} \sum_{i=n-1}^{\infty} \frac{[(1 + \beta - \alpha)X]^i}{i!} \qquad (308)$$

$$+ \frac{\gamma_2\beta^2 X^{n-1}}{(\gamma_2 + \beta)(1 + \beta)(n - 2)!}$$

$$+ \frac{\gamma_1\alpha\beta}{(\gamma_1 + \beta - \alpha)(\gamma_1 + \beta)(1 - \gamma_1)^{n-1}} \sum_{i=n-1}^{\infty} \frac{[(1 - \gamma_1)X]^i}{i!}$$

$$- \frac{\gamma_2\alpha\beta^2}{(\gamma_2 + \beta - \alpha)(\gamma_2 + \beta)^2(1 - \gamma_2)^{n-1}} \sum_{i=n-1}^{\infty} \frac{[(1 - \gamma_2)X]^i}{i!} \bigg\}$$

Since $N_n = N_n^* + N'_n$, we have

$$N_n = I_0 e^{-(1+\beta)X} \bigg\{ \frac{\beta}{(1 + \beta)^{n-1}} \bigg[\frac{(\beta - \alpha)(\gamma_1\alpha - \gamma_2\beta - \gamma_1\gamma_2)}{\alpha(\gamma_1 + \beta - \alpha)(\gamma_2 + \beta - \alpha)}$$

$$- \frac{\gamma_1\alpha}{(\gamma_1 + \beta - \alpha)(\gamma_1 + \beta)} + \frac{\gamma_2\alpha\beta}{(\gamma_2 + \beta - \alpha)(\gamma_2 + \beta)^2}$$

$$+ \frac{\gamma_2\beta}{(\gamma_2 + \beta)}\bigg(X + \frac{1}{\beta} - \frac{n-1}{1 + \beta}\bigg)\bigg] \sum_{i=n-1}^{\infty} \frac{[(1 + \beta)X]^i}{i!} \qquad (309)$$

$$- \frac{(\beta - \alpha)^2(\gamma_1\alpha - \gamma_2\beta - \gamma_1\gamma_2)}{\alpha(\gamma_1 + \beta - \alpha)(\gamma_2 + \beta - \alpha)(1 + \beta - \alpha)^{n-1}} \sum_{i=n-1}^{\infty} \frac{[(1 + \beta - \alpha)X]^i}{i!}$$

$$- \frac{\gamma_1^2\alpha}{(\gamma_1 + \beta - \alpha)(\gamma_1 + \beta)(1 - \gamma_1)^{n-1}} \sum_{i=n-1}^{\infty} \frac{[(1 - \gamma_1)X]^i}{i!}$$

$$+ \frac{\gamma_2^2\alpha\beta}{(\gamma_2 + \beta - \alpha)(\gamma_2 + \beta)^2(1 - \gamma_2)^{n-1}} \sum_{i=n-1}^{\infty} \frac{[(1 - \gamma_2)X]^i}{i!}$$

$$+ \frac{\gamma_2\beta^2}{(\gamma_2 + \beta)(1 + \beta)(n - 2)!} X^{n-1} \bigg\}$$

where $\alpha = k_i/k_p$, $\beta = k_t/k_p$, $\gamma_1 = k'_p/k_p$ and $\gamma_2 = k''_p/k_p$, respectively, and the other variables are as defined earlier.

Next we obtain the molecular weight distribution of the step polymerization. For A-B type monomers such as hydroxy and amino acids the step polymerization takes place with external supplied stoichiometry. Here A and B are denoted as two different types of functional groups, respectively. They can be reacted with each other and converted into a high polymer. The MWD has been derived by Flory by a probability approach based on the concept of equal reactivity of functional groups. Consider the step polymerization of an ω-hydrox acid consisting of n structural units as an example [161]:

$$H{-}ORCO{-}ORCO{-} \ldots {-}ORCO{-}OH$$
$$\phantom{H{-}OR}1 \phantom{CO{-}OR}2 \phantom{CO{-}\ldots{-}ORC}n{-}1 \phantom{O{-}}n$$

In the statistical analysis of such polymerization the extent of reaction, p, is defined as the probability that a carboxyl group has reacted at time t. (In kinetic treatments p is often used as the equivalent expression, the fraction of carboxyl groups which has reacted.) It follows that $(1 - p)$ is the probability of finding a carboxyl group unreacted. A molecule containing n repeating units may be designated as an n-mer. Such a molecule contains $(n - 1)$ reacted carboxyl groups. The probability of finding a reacted carboxyl group in the molecule is p, and that of finding $(n - 1)$ of them in the same molecule is p^{n-1}. The probability of finding the complete molecule, then, is $p^{n-1}(1 - p)$, where $(1 - p)$ is the probability of finding the unreacted group. The fraction of all the molecules which are n-mers is the same as the probability of finding the complete molecule. If the total number of carboxyls is N_0, then the number of polymer molecules (N) is

$$N = N_0(1 - p) \tag{310}$$

The number of n-mers (N_n) is

$$\begin{aligned}N_n &= Np^{n-1}(1 - p) \\ &= N_0 p^{n-1}(1 - p)^2\end{aligned} \tag{311}$$

Neglecting the weights of the end groups, the weight fraction (W_n) of n-mers, is

$$\begin{aligned}W_n &= \frac{nN_n}{N_0} \\ &= n(1 - p)^2 p^{n-1}\end{aligned} \tag{312}$$

Equations 311 and 312 give the number and weight distribution functions, respectively, for A-B type step polymerizations at the extent of reaction p. These distributions are usually referred to as the most probable or Flory or Flory–Schulz distributions. This MWD can also be obtained by solving kinetical equations.

The kinetic scheme may be written as follows:

$$N_n + N_i \underset{k_r}{\overset{k}{\rightleftharpoons}} N_{n+i} + H_2O \tag{313}$$

$$N_i + N_{n-i} \underset{k_r}{\overset{k}{\rightleftharpoons}} N_n + H_2O \tag{314}$$

Neglecting the reverse reactions, the kinetic equations corresponding to the foregoing scheme are

$$\frac{dN_n}{dt} = k \sum_{i=1}^{n-1} N_i N_{n-i} - 2kN_n \sum_{i=1}^{\infty} N_i, \quad n = 1, 2, 3, \ldots \tag{315}$$

where the first term represents the formation of N_n and the second term the consumption of N_n. The use of the factor 2 in the second term means that each of A group and B group in the N_n has one chance of reaction.

To add up all terms in Eq. 315, we have

$$\frac{d\Sigma N_n}{dt} = -k(\Sigma N_n)^2 \tag{316}$$

According to the definition of the extent of reaction p, we obtain

$$p = \frac{N_0 - \Sigma N_n}{N_0} \tag{317}$$

and

$$\Sigma N_n = N_0(1 - p) \tag{318}$$

Substituting Eq. 318 into Eq. 316, we have

$$\frac{dP}{dt} = kN_0(1 - p)^2 \tag{319}$$

Equation 315 divided by Eq. 319 gives

$$\frac{-dN_n}{dp} = \frac{2N_n}{(1-p)} - \frac{\sum_{i=1}^{n-1} N_i N_{n-i}}{N_0(1-p)^2} \tag{320}$$

When $n = 1$, Eq. 320 becomes

$$\frac{dN_1}{dp} = \frac{-2N_1}{1-p} \tag{321}$$

Using the initial condition $N_1|_{p=0} = N_0$, we have

$$N_1 = N_0(1 - p)^2 \tag{322}$$

When $n = 2$, Eq. 320 becomes

$$\frac{dN_2}{dp} = \frac{N_1^2}{N_0(1-p)^2} - \frac{2N_2}{1-p} \tag{323}$$

Substituting Eq. 322 into Eq. 323 and solving the latter, we obtain

$$N_2 = N_0 p(1 - p)^2 \tag{324}$$

Proceeding along this line, we finally obtain the following expression by the aid of mathematical induction:

$$N_n = N_0 p^{n-1}(1 - p)^2 \tag{325}$$

It is thus obvious that the same goal has been reached by different routes. The same derivations hold for the cases of A_2-B_2 type and A_2-$B(B')$ type monomers. It is unnecessary to go into detail here.

REFERENCES

1. P. J. Flory, *Principles of Polymer Chemistry*, Cornell University Press, Ithaca, N.Y., 1953.
2. O. W. Webster et al., *J. Am. Chem. Soc.*, *105*, 5706 (1983).
3. H. M. Thomas, *Science*, *222*, 39 (1983).
4. D. S. Yu, *Hua Xue Tong Bao*, *10*, 41 (1984).
5. F. Bandermann et al., *Makromol. Chem. Rapid Commun.*, *6*, 335 (1985).
6. D. Y. Sogah and W. B. Farnham, in *Organosilicon and Bioorganosilicon Chemistry* (H. Sakurai, ed.), Wiley, New York, p. 219 (1986).
7. W. B. Farnham and D. Y. Sogah, *Polymer Preprints. ACS Div. Polym. Chem.*, *27*(1), 167 (1986).
8. H. D. Sitz et al., *Polymer Preprints. ACS Div. Polym. Chem.*, *27*(1), 169 (1986).
9. C. H. Bamford et al., *Kinetics of Vinyl Polymerization by Radical Mechanism*, Butterworths, London (1958).
10. F. W. Billmeyer, *Textbook of Polymer Science*, Interscience, New York (1962).
11. V. N. Kondratiev, in *Comprehensive Chemical Kinetics*, Vol. 2 (C. H. Bamford and C. F. H. Tipper, eds.), American Elsevier, New York, Chap. 2 (1969).
12. H. Sawada, *Thermodynamics of Polymerization*, Marcel Dekker, New York, Chaps 1, 2, 5 (1976).
13. G. Odian, *Principles of Polymerization*, 2nd ed., Wiley, New York (1981).
14. G. C. Eastmond, in *Comprehensive Chemical Kinetics*, Vol. 14A (C. H. Bamford and C. F. H. Tipper, eds.), American Elsevier, New York, Chap. 1 (1976).
15. C. V. Schulz and F. Blaschke, *Z. Phys. Chem. (Leipzig)*, *B51*, 15 (1942).
16. G. F. Santee et al., *Makromol. Chem.*, *73*, 177 (1964).
17. A. Vrancken and G. Smets, *Makromol. Chem.*, *30*, 197 (1959).
18. P. E. M. Allen and C. R. Patrick, *Kinetics and Mechanisms of Polymerization Reaction*, Wiley, New York, Chaps. 2–4, 7 (1974).
19. M. Kamachi et al., *J. Polym. Sci. Polym. Chem. Ed.*, *16*, 1789 (1978); *Polym. J.*, *9*, 307 (1977).
20. E. S. Huyser, *Free Radical Chain Reactions*, Wiley, New York, pp. 314–330 (1970).
21. K. F. O'Driscoll et al., *J. Polym. Sci.*, *A3*, 1567 (1965).
22. S. E. Morsi et al., *Eur. Polym. J.*, *13*, 851 (1977).
23. A. Ledwith, *Pure Appl. Chem.*, *49*, 431 (1977).
24. G. Oster and N. L. Yang, *Chem. Rev.*, *68*, 125 (1968).
25. S. G. Foord, *J. Chem. Soc.*, *1940*, 48.
26. G. Goldfinger et al., *J. Phys. Chem.*, *47*, 578 (1943).
27. G. V. Schulz and S. Gertrud, *Chem. Ber.*, *80*, 232 (1947).
28. F. R. Mayo et al., *J. Am. Chem. Soc.*, *72*, 1051 (1950).
29. H. W. Melville and W. F. Watson, *Trans. Faraday Soc.*, *44*, 886 (1948).
30. L. F. Fieser and A. E. Oxford, *J. Am. Chem. Soc.*, *64*, 2060 (1942).
31. J. W. Breitenbach and K. Horeischy, *Chem. Ber.*, *74*, 1386 (1941).
32. C. M. Fontana, in *The Chemistry of Cationic Polymerization* (P. H. Plesch, ed.), Macmillan, New York, p. 96 (1963).
33. C. M. Fontana, in *Cationic Polymerization and Related Complexes* (P. H. Plesch, ed.), Heffer and Sons, Cambridge, pp. 2, 74 (1953).
34. H. Güterbock, *Polyisobutylen*, Springer, Berlin, p. 65 (1959).
35. (a) M. Szwarc, *Carbanions, Living Polymers and Electron Transfer Processes*, Interscience, New York, p. 18 (1968); (b) D. N. Bhattacharyya et al., *J. Phys. Chem.*, *69*, 624 (1965).
36. Z. Zlamal, in *Kinetics and Mechanism of Polymerization Series*, Vol. 1 (G. E. Ham, ed.), Marcel Dekker, New York, p. 108 (1969).

37. D. J. Dunn, in *Developments in Polymerization*, Vol. 1 (R. N. Haward, ed.), Applied Science Publishers, London (1979).
38. J. P. Kennedy, *Cationic Polymerization of Olefins: A Critical Inventory*, Wiley, New York, p. 28 (1975).
39. J. E. McGrath, *Anionic Polymerization Kinetics, Mechanisms, and Synthesis*, ACS Symposium Series 166, Washington, D.C., p. 112 (1981).
40. S. Bywater, in *Anionic Polymerization of Olefins*, Vol. 15 (C. H. Bamford and C. F. H. Tipper, eds.), American Elsevier, New York, p. 83 (1976).
41. S. Bywater, in *Progress in Polymer Science*, Vol. 4 (A. D. Jenkins, ed.), Pergamon, New York, p. 124 (1975).
42. D. H. Richards, in *Developments in Polymerization*, Vol. 1 (R. N. Haward, ed.), Applied Science Publishers, Essex, England, p. 15 (1979).
43. A. Parry, in *Reactivity, Mechanism and Structure in Polymer Chemistry* (A. D. Jenkins and A. Ledwith, eds.), Wiley-Interscience, New York, p. 77 (1974).
44. J. P. Kennedy and M. Ernest, *Carbocationic Polymerization*, Wiley-Interscience, New York, p. 58 (1981).
45. (a) J. P. Kennedy and S. C. Feinberg, *J. Polym. Sci. Polym. Chem. Ed.*, 16, 2191 (1978); (b) J. P. Kennedy et al., *J. Polym. Sci. Polym. Chem. Ed.*, 15, 2801, 2869 (1977).
46. M. DiMains et al., *Makromol. Chem.*, 178, 2223 (1977).
47. L. Reibel et al., *J. Polym. Sci. Polym. Chem. Ed.*, 17, 2757 (1979).
48. M. Biswas and M. A. Kabir, *J. Polym. Sci. Polym. Chem. Ed.*, 17, 673 (1979); *Eur. Polym. J.*, 14, 861 (1978).
49. T. Taninaka et al., *Eur. Polym. J.*, 14, 199 (1978).
50. A. G. Evans and G. W. Meadows, *Trans. Faraday Soc.*, 46, 327 (1950).
51. M. Sawamoto et al., *J. Polym. Sci. Polym. Chem. Ed.*, 16, 2675 (1978); *Makromol. Chem.*, 178, 1497 (1977).
52. A. F. Johnson and R. N. Young, *J. Polym. Sci. Symp.*, 56, 211 (1976).
53. A. Gandini and H. Cheradame, *Adv. Polym. Sci.*, 34/35, 1 (1980).
54. P. Cerrai et al., *Eur. Polym. J.*, 15, 153 (1979); *Eur. Polym. J.*, 12, 247 (1976).
55. S. Egusa et al., *J. Polym. Sci. Polym. Chem. Ed.*, 16, 729 (1978).
56. (a) A. M. Goineau et al., *J. Macromol. Sci. Chem.*, A11, 99 (1979); (b) A. Deffieux et al., *Polym. Bull.*, 2, 469 (1980).
57. J. M. Rooney, *J. Polym. Sci. Symp.*, 56, 47 (1976); *Makromol. Chem.*, 179, 165, 2419 (1978).
58. K. Hatada et al., *Polym. Bull.*, 2, 15 (1980).
59. M. Biswas and P. Kamannarayana, *J. Polym. Sci. Polym. Chem. Ed.*, 14, 2071 (1976).
60. F. Williams et al., *Trans. Faraday Soc.*, 63, 1501 (1967).
61. K. Hayashi et al., *Trans. Faraday Soc.*, 63, 1489 (1967).
62. C. C. Allen et al., *J. Appl. Polym. Sci.*, 18, 709 (1974).
63. H. Kubota et al., *J. Macromol. Sci. Chem.*, A12, 1299 (1978).
64. F. E. Matthews and E. H. Strange, Brit. Pat. 24,790 (1910).
65. C. Harries, U.S. Pat. 1,058,056 (1913).
66. K. Ziegler et al., *Ann. Chem.*, 473, 1 (1929).
67. K. Ziegler and J. Smid, *Ann. Chem.*, 511, 13 (1934).
68. M. Szwarc et al., *Prog. React. Kinet.*, 2, 219 (1964).
69. C. G. Oyerberger et al., *J. Polym. Sci.*, 45, 127 (1960).
70. S. A. Hurley and P. J. T. Tait, *J. Polym. Sci. Polym. Chem. Ed.*, 14, 1565 (1976).
71. P. Hrdlovic et al., *Eur. Polym. J.*, 15, 229 (1979).
72. E. F. Donnelly et al., *J. Polym. Sci. Polym. Lett. Ed.*, 15, 399 (1977).
73. Ch. B. Tsvetanov, *Eur. Polym. J.*, 15, 503 (1979).
74. B. J. Schmitt and G. V. Schultz, *Eur. Polym. J.*, 11, 119 (1975).

75. N. C. Billingham, in *Developments in Polymerization*, Vol. 1 (R. N. Haward, ed.), Applied Science Publishers, London, Chap. 4 (1979).
76. K. C. Frisch and R. L. Reegan, eds., *Ring-Opening Polymerization*, Marcel Dekker, New York (1969).
77. T. Saegusa and E. Goethals, eds., *Ring-Opening Polymerization*, American Chemical Society, Washington, D.C. (1977).
78. P. Dreyfuss and M. P. Dreyfuss, in *Comprehensive Chemical Kinetics*, Vol. 15 (C. H. Bamford and C. F. H. Tipper, eds.), American Elsevier, New York, Chap. 4 (1976).
79. Y. Ishii and S. Sakai, in *Ring-Opening Polymerization* (K. C. Frisch and S. L. Reegan, eds.), Marcel Dekker, New York, Chap. 2 (1969).
80. T. Saegusa and S. Kobayashi, in *Progress in Polymer Science (Japan)*, Vol. 6 (S. Onogi and K. Uno, eds.), Halsted Press (Wiley), New York, pp. 107–151 (1973).
81. T. Saegusa et al., *Macromolecules*, 9, 231 (1976).
82. G. Pruckmayr and T. K. Wu, *Macromolecules*, 11, 265, 662 (1978).
83. K. Matyjaszewski, *J. Polym. Sci. Polym. Chem. Ed.*, 15, 247 (1977).
84. F. Afshar-Taromi et al., *Makromol. Chem.*, 179, 849 (1978).
85. (a) M. Szwarc, *Makromol. Chem. Suppl.*, 3, 327 (1979); (b) S. Kobayashi et al., *Macromolecules*, 7, 415 (1974).
86. L. F. Beste and H. K. Hall, Jr., *J. Phys. Chem.*, 68, 269 (1964).
87. T. Alfrey, J. J. Bohrer, and H. F. Mark, *Copolymerization*, Interscience, New York (1952).
88. G. E. Ham, in *Copolymerization* (G. E. Ham, ed.), Wiley-Interscience, New York, Chap. 1 (1964).
89. A. Yamada, *Copolymerization*, Vol. 1: *Reaction Analytics* (1975); Vol. 2: *Reaction Regulation* (1976). SPSJ (Bai Fu Kan, Tokyo).
90. G. C. Eastman, in *Comprehensive Chemical Kinetics*, Vol. 14A (C. H. Bamford and C. F. H. Tipper, eds.), American Elsevier, New York, Chap. 4 (1976).
91. F. R. Mayo and F. M. Lewis, *J. Am. Chem. Soc.*, 66, 1594 (1944).
92. F. T. Wall, *J. Am. Chem. Soc.*, 66, 2050 (1944).
93. T. Alfrey and G. Goldfinger, *J. Chem. Phys.*, 12, 205 (1944).
94. R. Simha and H. Branson, *J. Chem. Phys.*, 12, 253 (1944).
95. Y. Landler, *Compt. Rend.*, 230, 539 (1950).
96. D. C. Pepper, *Q. Rev. (London)*, 8, 88 (1954).
97. V. Meyer and G. G. Lowry, *J. Polym. Sci.*, 3A, 2843 (1965).
98. E. Merz et al., *J. Polym. Sci.*, 1, 75 (1946).
99. G. C. Lowry, *J. Polym. Sci.*, 42, 463 (1960).
100. C. Walling, *J. Am. Chem. Soc.*, 71, 1930 (1949); P. Wittmer, *Makromol. Chem. Suppl.*, 3, 129 (1979).
101. A. M. North, *Polymer*, 4, 134 (1963).
102. J. N. Atherton and A. M. North, *Trans. Faraday Soc.*, 58, 2049 (1962).
103. P. Wittmer, *Makromol. Chem. Suppl.*, 3, 129 (1979).
104. D. O. Jordan, *The Stereochemistry of Macromolecules*, Marcel Dekker, New York (1967).
105. G. Natta and I. Pasquon, *Adv. Catalysis*, 11, 1 (1959).
106. W. M. Saltman et al., *J. Am. Chem. Soc.*, 80, 5615 (1958).
107. Yu. V. Kissin and N. M. Chirkov, *Eur. Polym. J.*, 6, 525 (1970).
108. T. Keii, *Kinetics of Ziegler–Natta Polymerization*, Kodansha, Tokyo (1972).
109. W. Cooper, in *Comprehensive Chemical Kinetics*, Vol. 15 (C. H. Bamford and C. F. H. Tipper, eds.), American Elsevier, New York, Chap. 3 (1976).
110. R. W. Lenz and F. Ciardelli, in *NATO Advanced Study Institutes Series*, D. Reidel, Dordrecht, pp. 85–112 (1980).
111. R. W. Lenz, *Organic Chemistry of Synthetic High Polymers*, Interscience, New York (1967).
112. B. V. Bhide and J. J. Sudborough, *J. Indian Inst. Sci.*, 8A, 89 (1925).

113. K. Ueberreiter and M. Engel, *Makromol. Chem., 178*, 2257 (1977).
114. E. Robinowitch, *Trans. Faraday Soc., 33*, 1225 (1937).
115. W. C. McLewis, *J. Chem. Soc., 113*, 471 (1918).
116. C. N. Hinshelwood and E. A. Moelwyn-Hughes, *Proc. Roy. Soc., 131A*, 177 (1931).
117. H. Eyring, *J. Chem. Phys., 3*, 107, 492 (1935); H. Eyring, *Chem. Rev., 17*, 65 (1935).
118. I. Vancso-Szmercsanyi and Vancso-Szmercsayi, *Eur. Polym. J., 5*, 145, 155 (1969).
119. A. A. Frost and R. G. Pearson, *Kinetics and Mechanism*, 2nd ed., Wiley, New York, Chaps. 2, 3, 9 (1961).
120. P. J. Flory, *J. Am. Chem. Soc., 61*, 3334 (1939).
121. (a) D. H. Solomon, *J. Macromol. Sci. Rev. Macromol. Chem., C1*(1), 179 (1967); (b) D. H. Solomon, in *Step-Growth Polymerizations* (D. H. Solomon, ed.), Marcel Dekker, New York, Chap. 1 (1972).
122. J. Hine, *Physical Organic Chemistry*, McGraw-Hill, New York, Chap. 2 (1962).
123. M. Imoto and T. Tanigaki, *J. Chem. Soc. Ind.* (Kogyo Kagaku Zasshi), *66*, 517 (1963).
124. G. Champetier and R. Vergoz, *Rec. Trav. Chim. Pays-Bas, 69*, 85 (1950).
125. P. J. Flory, U.S. Pat. 2,244,192 (1941).
126. V. V. Korshak and T. M. Frunze, *Synthetic Heterochain Polyamides*, Daniel Davey & Co., New York (1964).
127. J. Charles et al., *C. R. Acad. Sci., 256*, 3107 (1963).
128. B. A. Zhubanov et al., *Nauk. Kaz. SSR, Ser. Khim., 17*(4), 69 (1967); *Chem. Abstr., 69*, 10763 (1968).
129. J. H. Saunders and K. C. Frisch, *Polyurethanes, Chemistry and Technology*, Vol. 1: *Chemistry*, Interscience, New York (1962).
130. A. Farkas and G. A. Mills, in *Advances in Catalysis*, Vol. 13 (D. D. Eley, P. W. Selwood, and P. B. Weisz, eds.), Academic Press, New York, p. 393 (1962).
131. D. J. Lyman, *Macromol. Chem., 1*, 191 (1966).
132. P. C. Johnson, in *Advances in Polyurethane Technology* (J. M. Buist and H. Gudgeon, eds.), Elsevier, Essex, London, Chap. 1 (1968).
133. J. H. Saunders and K. C. Frisch, in *Polyurethanes, Chemistry and Technology*, Vol. 1: *Chemistry*, Interscience, New York, pp. 211–215 (1962); Vol. 2: *Technology*, Interscience, New York (1964).
134. J. W. Baker and J. Gaunt, *J. Chem. Soc.*, 24 (1949).
135. J. W. Baker and J. Gaunt, *J. Chem. Soc., 713*, (1947).
136. J. W. Baker and J. Gaunt, *J. Chem. Soc., 9*, 19, 27 (1949).
137. O. W. Webster, et al., *J. Am. Chem. Soc., 105*, 5706 (1983).
138. O. W. Webster, Eur. Pat. Appl. E.P. 68887 (1983).
139. W. B. Farham and D. Y. Sogah, U.S. Pat. 4,414,372 (1983).
140. D. Y. Sogah and O. W. Webster, *J. Polym. Soc. Lett. Ed., 21*, 927 (1983).
141. O. W. Webster et al., *J. Macromol. Sci., Chem., A21*(8,9), 943 (1984).
142. W. R. Hertler et al., *Macromolecules, 17*, 1417 (1984).
143. D. Y. Sogah and W. B. Farnham, *Seventh International Symposium on Organosilicon Chemistry*, Kyoto, Japan, September 9, 1984, Ellis Horwood, London.
144. D. Y. Sogah et al., *ACS Polymer Preprints, 25*(2), 3 (1984).
145. W. R. Hertler et al., *Macromolecules, 17* 1415 (1984).
146. D. Y. Sogah and O. W. Webster, *Macromolecules, 19* 1775 (1986).
147. Axel. H. E. Müller et al., *Makromol. Chem.* (in press).
148. L. Küchler, *Polymerisation Kinetik*, Springer, Berlin (1951).
149. G. M. Burnett, *Mechanism of Polymer Reactions*, Interscience, London (1954).
150. C. H. Bamford and H. Tompa, *Trans. Faraday Soc., 50*, 1097 (1954).
151. O. Saito et al., *J. Polym. Sci., Part A-2, 7*, 1937 (1969).
152. Tang Ao Qin, *Statistical Theory of Polymerization Reaction*, Scientific Publishing House, Beijing (1985). (Chinese)

153. M. R. Spiegel, *Advanced Mathematics for Engineers and Scientists,* McGraw-Hill, New York (1971).
154. De Yue Yan, *Gaofenzi Tongxun, 10,* 450 (1979). (Chinese)
155. F. Harary, *Graph Theory,* Addison-Wesley, Reading, Mass. (1969).
156. A. Cayley, *Phil. Mag., 13,* 10 (1857).
157. D. S. Butler et al., *Proc. Roy. Soc. London A, 259,* 29 (1966).
158. M. Gordon and T. G. Parker, *Proc. Roy. Soc. London A, 69,* 181 (1970).
159. Zhou Pu et al., *Polymer, 27,* 275 (1986).
160. Zhou Pu et al., *Makromol. Chem., 186,* 159 (1985).
161. P. J. Flory, in *High Molecular Weight Organic Compounds* (R. E. Burk and O. Grummitt, eds.), Interscience, New York, p. 27 (1949).

2
Overview of Polymerization Reactor Technology

K. Y. Choi
University of Maryland
College Park, Maryland

INTRODUCTION	67
REACTORS FOR INDUSTRIAL POLYMERIZATION PROCESSES	70
Reactors for Radical and Ionic Polymerizations	70
Reactors for Coordination Polymerization of Olefins	77
Reactors for Step Growth Polymerization	87
Polymerizations in Screw Reactors	95
CONTROL OF POLYMERIZATION REACTORS	98
REFERENCES	100

INTRODUCTION

Polymers are chemical aggregates assembled by the combination of a large number of small molecules (monomers); the action of the chemical reactions through which the monomers are transformed to polymers is termed polymerization. More than one monomer can be polymerized (e.g., copolymers) to yield polymers of improved properties over homopolymers or their mixtures. Currently, synthetic polymers are of enormous industrial importance and numerous polymers are manufactured by a variety of polymerization processes.

One convenient way of classifying polymerization reactions is based on their reaction kinetics as shown in Table 1. In *addition polymerization* single monomer molecules are attached successively to the growing polymer molecules. For example, in free radical polymerization, monomer linkage occurs with chain termination either by combination or disproportionation. In anionic polymerization the monomer linkage occurs without termination. In *polycondensation or step growth polymerization* polymer linkage occurs between two macromolecules containing reactive functional end groups. There are many excellent references on the detailed discussion of polymerization kinetics and mechanisms for various polymerization systems [1–6].

Polymerization processes can also be classified according to the types of reaction medium and kinetic mechanisms as shown in Table 2. Note that a variety of combinations

Table 1 Kinetics of Polymerization Reactions

Addition polymerization
  Monomer linkage with termination (e.g., free radical polymerization)

$$I \xrightarrow{k_d} 2R\cdot \qquad \text{Initiation}$$

$$R\cdot + M \xrightarrow{k_i} P_1$$

$$P_n + M \xrightarrow{k_p} P_{n+1} \ (n \geq 1) \qquad \text{Propagation}$$

$$P_n + A \xrightarrow{k_f} P_1 + M_n \qquad \text{Chain transfer}$$

$$P_n + P_m \xrightarrow{k_{tc}} M_{n+m} \qquad \text{Termination}$$
$$\qquad \searrow^{k_{td}} M_n + M_m$$

  Monomer linkage without termination (e.g., anionic living polymerization)

$$I + M \xrightarrow{k_i} P_1 \qquad \text{Initiation}$$

$$P_n + M \xrightarrow{k_p} P_{n+1} \qquad \text{Propagation}$$

Step growth polymerization
  Polymer linkage (e.g., polycondensation)

$$P_n + P_m \xrightarrow{k_p} P_{n+m} \qquad \text{Propagation}$$

Table 2 Kinetic Mechanisms and Reaction Media Employed in Polymerization Reactors

Reaction medium	Kinetic mechanism			
	Free radical	Ionic	Condensation	Coordination catalysis
Homogeneous (bulk and solution)	Vinyl polymers (styrene, MMA, LDPE)	Polyethers (ethylene oxide)	Polyesters, polyamides (PET, nylon)	Polyolefins with soluble catalyst (ethylene–propylene copolymers)
Heterogeneous				
Emulsion	Vinyl polymers (styrene, PVC)			
Suspension	Vinyl polymers (styrene, PVC)			
Precipitation (bulk and solution)	Vinyl polymers (PVC)	Polyacetals (formaldehyde) Vinyls (isobutylene–butyl rubber)	Polyamides (nylon interfacial polymerization)	Polyolefins with insoluble catalyst (liquid and gas phase processes for HDPE, LLDPE, polypropylene)

Source: Ref. 7.

of polymerization modes is possible. Each class of polymerization system has its own peculiar design and control problems. Some of the issues that must be faced in designing and operating the polymerization reactors are as follows [7]:

1. *Kinetics and Phase Behavior.* Some polymerizations go through several phase changes in the course of the reaction. For example, in the bulk polymerization of vinyl chloride, the reaction medium begins as a low-viscosity liquid, progresses to a slurry (the PVC polymer, which is insoluble in the monomer, precipitates), becomes a paste as the monomer disappears, and finishes as a solid powder. An understanding of such polymerization kinetics is essential for reactor design.

2. *Material Mixing and Conveying.* With phase changes during polymerization, there can be serious nonidealities in micromixing and macromixing, particularly at high conversions. Often a large amount of mechanical energy is required for mixing and conveying. Sometimes, sticking and fouling of the reactor surfaces by polymer is a difficult design problem.

3. *Heat Removal.* Most polymerizations are characterized by large release of reaction heat as monomer is converted to polymer. Table 3 lists heats of polymerization values for some common monomers. In addition, the mechanical energy required for mixing may be converted to heat under highly viscous conditions. Removal of this reaction heat is often difficult for high-conversion polymerization because of high viscosity, heat transfer area fouling, and change of phase during reaction. In many industrial polymerization situations, disastrous reactor runaway is an ever-present potential hazard because of these heat removal difficulties.

4. *Quality of Polymer Product.* Product quality is a much more complex issue in polymerization than in more conventional short-chain reactions. Because the molecular architecture of the polymer is so sensitive to reactor operating conditions, upsets in feed conditions, mixing, reactor temperature, etc., can alter critical molecular properties such as molecular weight distribution, polymer composition distribution, chain sequence distribution, degree of chain branching, and stereoregularity. In addition, the morphological

Table 3 Heats of Polymerization for Some Common Monomers

Monomer	Heat of polymerization (cal mol^{-1} at 25°C)
Ethylene	−21.2
Propylene	−19.5
Butadiene	−17.6
Styrene	−16.7
Vinyl chloride	−22.9
Vinylidene chloride	−18.0
Vinyl acetate	−21.2
Methyl acrylate	−18.5
Methyl methacrylate	−13.2
Acrylonitrile	−18.4
Formaldehyde	−7.4

form of the polymer is often a key quality parameter to be controlled. One of the greatest difficulties in achieving quality control of the polymer product is that the actual customer specifications may be in terms of nonmolecular parameters such as tensile strength, crack resistance, temperature stability, color or clarity, or absorption capacity for plasticizer. The quantitative relationship between these product quality parameters and reactor operating conditions may be the least understood area of polymerization reaction engineering. The lack of on-line sensors to measure polymer properties makes the control of polymerization reactors a special challenge.

REACTORS FOR INDUSTRIAL POLYMERIZATION PROCESSES

Reactors for Radical and Ionic Polymerizations

A variety of polymerization reactors are used in industry to polymerize ethylenically unsaturated monomers such as ethylene, styrene, vinyl chloride, acrylonitrile, and methyl methacrylate by free radical or ionic polymerization mechanisms. These polymers are manufactured by batch, semibatch, and continuous processes depending on the characteristics of monomers, catalysts, polymerization conditions, and process economics.

Commercial processes are subdivided into bulk, solution, suspension, and emulsion polymerization. In bulk polymerization, polymers may be soluble in their own monomers [e.g., polystyrene (PS), polymethyl methacrylate (PMMA), polyvinyl acetate (PVA)] or insoluble in their monomers [e.g., polyvinyl chloride (PVC), polyacrylonitrile (PA)]. In solution polymerization, solvent miscible with monomer dissolves polymer (e.g., styrene in ethylbenzene). Some polymers are swollen by their own monomers at specific reaction conditions (e.g., polyethylene, polyacrylamide). In suspension polymerization, organic monomer phase in small droplets dispersed by stirring in aqueous phase is polymerized into hard solid particles (e.g., PS, PMMA). Emulsion polymerization differs from suspension polymerization in two important respects: the initiator is located in the aqueous phase, and the polymer particles produced are typically of the order of 0.1 μm in diameter, some 10 times smaller than the smallest encountered in suspension or dispersion polymerization. Emulsion polymerization can also be carried out in systems using an aqueous monomer solution emulsified in a continuous oil phase using an appropriate water-in-oil emulsifier (inverse emulsion polymerization). In both suspension and emulsion polymerizations good temperature controls can be obtained.

Ionic polymerizations are not as well understood as free radical polymerizations because of the involvement of heterogeneous inorganic catalysts, which are fast and strong in sensitivity of reaction rates to very minor impurities. Polydiene and diene–styrene copolymers are typical examples of industrial ionic polymerization using alkyllithium catalysts. These polymerization systems are unique in that they have precise control over such polymer properties as composition, microstructure, molecular weight, MWD, choice of functional end groups, and even copolymer sequence distribution [8].

A selection of polymerization reactor configurations for a given polymerization reaction depends on many factors such as production rate, reaction conditions (e.g., temperature, pressure, viscosity), product properties, and investment and operating cost. Table 4 illustrates several reactor designs for styrene polymerization on the basis of reactor function. Note that various reactor configurations are available to meet specific polymerization conditions.

In what follows, various types of polymerization reactor systems for industrial free

Table 4 Examples of Styrene Polymerization Reactors

	Process type	
Reactor function	Batch	Continuous
Bulk polymerization		
Polymer < 20% concentration	Conventional kettle with: Turbine agitator	CSTR with: Turbine agitator Tubular reactors Agitated towers
Polymer 20–50% concentration	Large turbine, anchor or helical agitator	Turbine, anchor or helical agitator
Polymer 30–80% concentration	Anchor or helical agitator proprietary and patented stirred reactors	Anchor, helical agitators, or special designs
Polymer > 80% concentration	Press, unagitated tower	Tubular reactors Unagitated towers
Suspension	Conventional kettle with turbine agitators	No commercial application

Source: Ref. 9.

radical polymerization are discussed. Table 5 lists the types of reactors used for the different polymerization reactors [10]. Among the various types of polymerization reactors, batch and continuous stirred tank reactors are the most widely used. The batch polymerization reactors are used extensively in the polymer industry to manufacture a variety of polymers of numerous grades. In general, the processing procedure of batch polymerization can be broken down into several steps as seen in Table 6. Clearly, the turnaround time can be reduced considerably by reducing the polymerization time through advanced reactor operating technology. The batch polymerization reactors are mostly stirred reactors with large turbine agitators and jacket cooling. Anchor or helical agitators are also used to overcome heat transfer and mixing problems at high viscosities. When one polymerizes styrene in bulk in a stirred tank reactor, monomer conversion of only 30–40% is obtainable due to extremely high viscosity of the polymerization mixture above this conversion level. Polymer–monomer mixtures are then separated in extruder-type devolatilizers. Monomer vapor recovered is condensed and recycled to the polymerization reactor, whereas polymers are extruded and pelletized.

Static reactors such as the polymerization press (modified plate-and-frame filter press) have been used for highly viscous bulk styrene polymerization [11]. Here, styrene monomers are polymerized in frames alternating between cooling plates through which water (or steam) can be circulated. The polymer blocks formed inside the frames are cooled, removed, crushed, and granulated. The product polymer generally has a moderate amount of residual styrene as well as a broad molecular weight distribution, largely caused by the various temperature histories seen by the reaction mass during the course of polymerization.

In operating batch polymerization reactors adequate temperature control is extremely

Table 5 Polymerization Processes and Industrially Employed Reactors

	Polymerization reactions	Monomer linkage — with termination				Monomer linkage — without termination		Polymer linkage		
		Solution polymerization	Precipitation polymerization	Bead polymerization	Emulsion polymerization	Solution polymerization	Precipitation polymerization	Solution or melt polycondensation	Interfacial polycondensation	Solid phase polycondensation
BR	Batch stirred tank	○	○	○	○	○	○	○	○	
BR	Semibatch	○				○	○			
BR	Piston	○								
BR	Tumbler									○
BR	Batch stirred tank + filter press	○								
BR	Batch stirred tank + autoclave with gate paddle nuscer				○					
BR	2-Component mixer + mold						○	○		
CPFR	Continuous tube	○						○		
CPFR	Continuous tube with distributed feed	○								
CPFR	Mixer + conveyor						○	○		
CPFR	Extruder						○	○		
Cascade	Tower or tower cascade	○						○		○
Cascade	Stirred tank + tower	○						○		
Cascade	Stirred tank cascade				○	○	○	○		
Cascade	Rotating ring disc							○		
CSTR	(Double) Loop						○			
CSTR	Continuous stirred tank	○	○		○		○			
CSTR	Fluidized bed						○			

Source: Ref. 10.

Table 6 Batch Polymerization Cycle

Example reactor: 3700 U.S. gal, 5300 pounds per batch

Cycle	Standard hours
Materials charging and purging	½
Heating	1
Polymerization	8
Unreacted monomer stripping	1
Cooling	1
Discharge	½
Reactor flushing and cleaning	1
Maintenance	½
Inefficiency	½
Total turnaround time	14
Pounds/operating hour	380

Source: Ref. 11.

important to maximize the polymer productivity and reactor safety and to optimize the product properties. Removal of polymerization heat becomes more difficult as the reactor volume increases because of reduced heat transfer area to reactor volume ratio. Large polymerization reactors require, in addition to a cooling jacket, the installation of internal cooling baffles or bundles of cooling coils and frequently of reflux condensers. In most batch-free radical polymerization processes, special time-varying temperature and reactant (e.g., initiators, monomers) feed programs are required to produce the polymer properties desired.

Continuous reactors have some advantages over batch reactors in that polymers can be produced in larger quantities and polymer quality control through process automation can be carried out more effectively than in batch polymerization processes. However, a deep understanding of polymerization reactor behavior is essential in designing and operating the continuous polymerization reactors. Continuous reactors used in industrial polymerization processes vary from stirred tank reactor(s) to tubular reactors (e.g., ethylene polymerization) and numerous modifications of such basic reactor configurations exist. Figure 1 illustrates two types of continuous tower reactors for bulk styrene polymerization. The tower reactor system shown in Figure 1a was developed by I. G. Farben in the 1930s. This employs batch prepolymerization reactors from which the prepolymers are transferred to a tower reactor whose temperature profile is controlled from 100 to 200°C. The product polymer melt is discharged from the bottom of the tower by an extruder, cooled, pelletized, and bagged. Figure 1b shows a continuous bulk styrene polymerization reactor system which consists of a series of towers using slow agitation and grids of pipes through which a mixture of dipehnyl oxide is circulated for temperature control.

Figure 2 illustrates reactor cascades used for high impact polystyrene (HIPS) polymerization. In the polymerization process shown in Figure 2a, a fresh feed of 8% polybutadiene rubber in styrene is added with antioxidant and recycled monomer to the first reactor operating at 124°C and about 18% conversion at about 40% fillage. The agitator is a horizontal shaft on which a set of paddles is mounted. Since the temperature in each

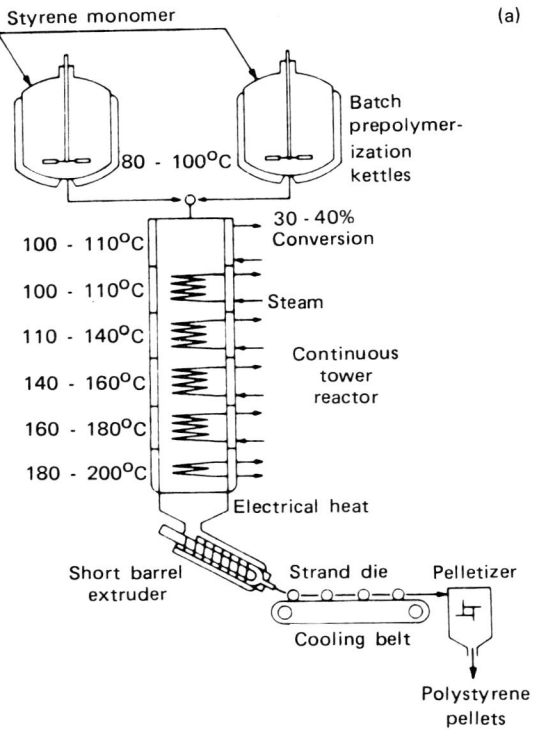
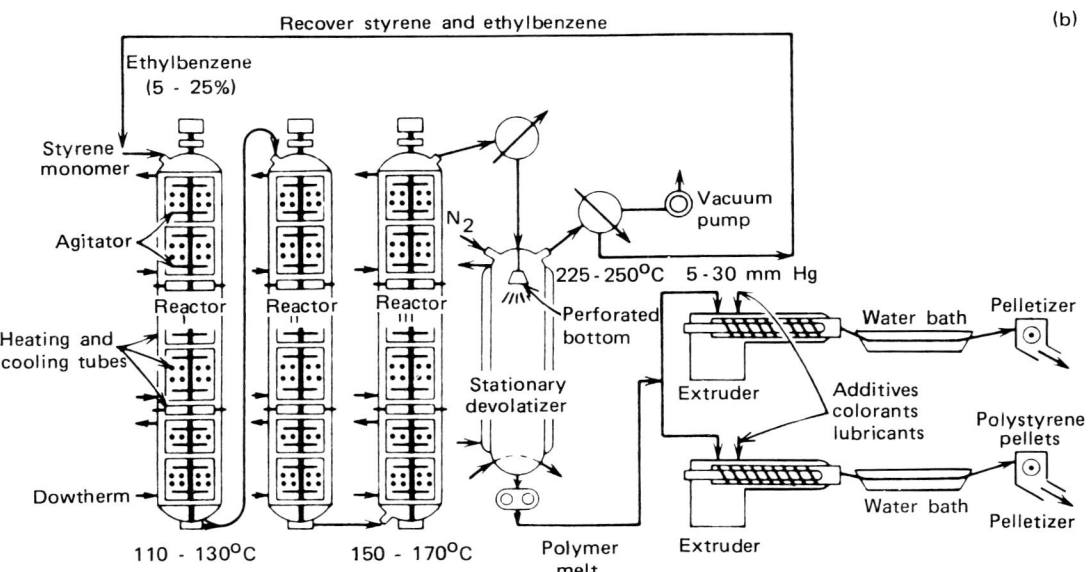

Figure 1 (*a*) Continuous tower reactor for bulk styrene polymerization. (*b*) Continuous bulk polymerization of styrene in a series of towers. (From Ref. 11.)

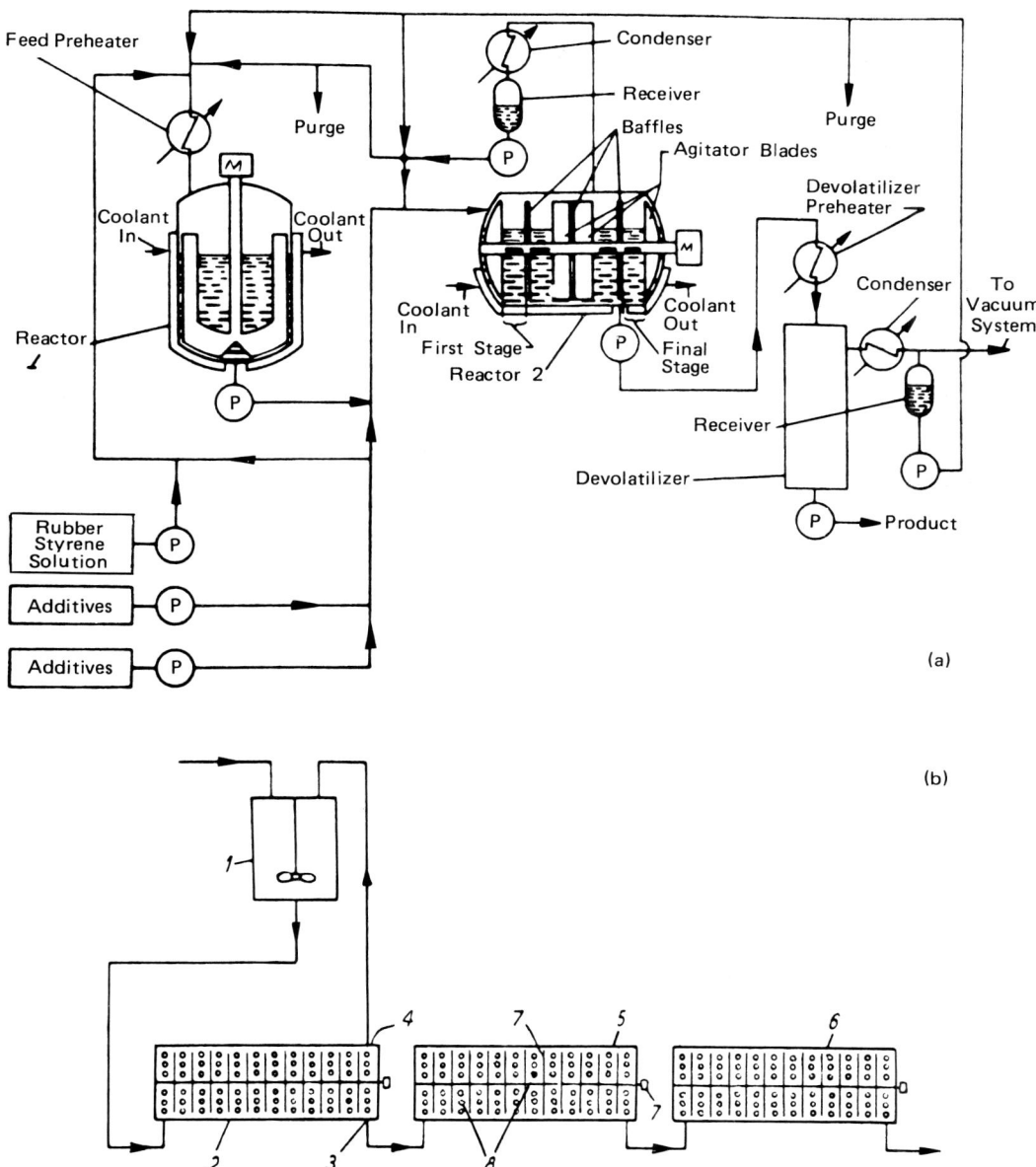

Figure 2 (a) Schematic of stirred horizontal reactor for HIPS process. (b) Schematic of multiple horizontal reactor system for continuous bulk HIPS polymerization. (From Ref. 9.)

compartment can be varied, it is claimed that the linear flow behavior provided by the reactor staging results in more favorable rubber phase morphology than would be the case if the second reactor were operated as a single continuous stirred tank reactor [12]. The similar reaction configuration shown in Figure 2b is also used for efficient control of styrene–rubber morphology. Here the first stage reactor is maintained just beyond the phase inversion point and the dissolved styrene reacts to form either a graft copolymer with the rubber or a homopolymer in the linear flow reactor train. Many other types of styrene polymerization reactor system are discussed in Refs. 9 and 11. Figure 3 shows a classical high-pressure, high-temperature tubular ethylene polymerization reactor system operating at 30,000–50,000 psi and 200–300°C using free radical initiators such as peroxides or oxygen.

Although bulk and solution polymerizations have been carried out in continuous reactors for many monomers, suspension polymerization has been exclusively conducted in batch reactors. However, there is extensive literature describing continuous suspension polymerization in different kinds of reactors [13–16]. For continuous polymerization, the reactor has to accomplish certain requirements [15]: (a) narrow residence time distribution to achieve high conversion; (b) good mixing of the two phases to obtain polymers with proper particle size distribution; (c) no dead space and gas phase within the reactor to avoid reactor fouling; and (d) large heat transfer surface area for heat removal.

Figure 4 is a schematic of a pilot plant scale continuous suspension polymerization reactor system [15]. The reactor consists of a tube with a blade stirrer. Vinyl acetate, which contains the initiator, and water, which contains the dispersion agent, are pumped in parallel flow through the tube reactor from top to bottom. The conversion was above 90%, and good particle size distribution was obtained.

Emulsion polymerization is used extensively to produce many polymers used for coating and adhesive applications, especially for those products that can be used in latex

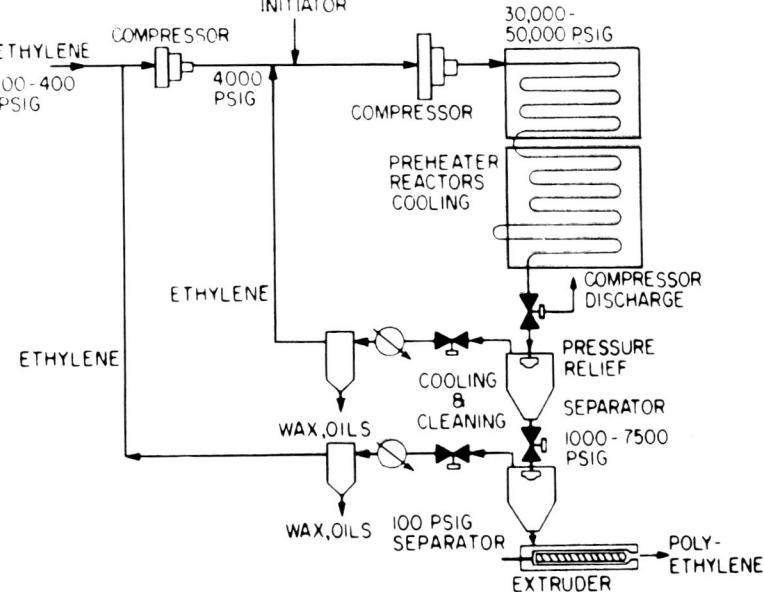

Figure 3 Tubular reactor for high-pressure ethylene polymerization.

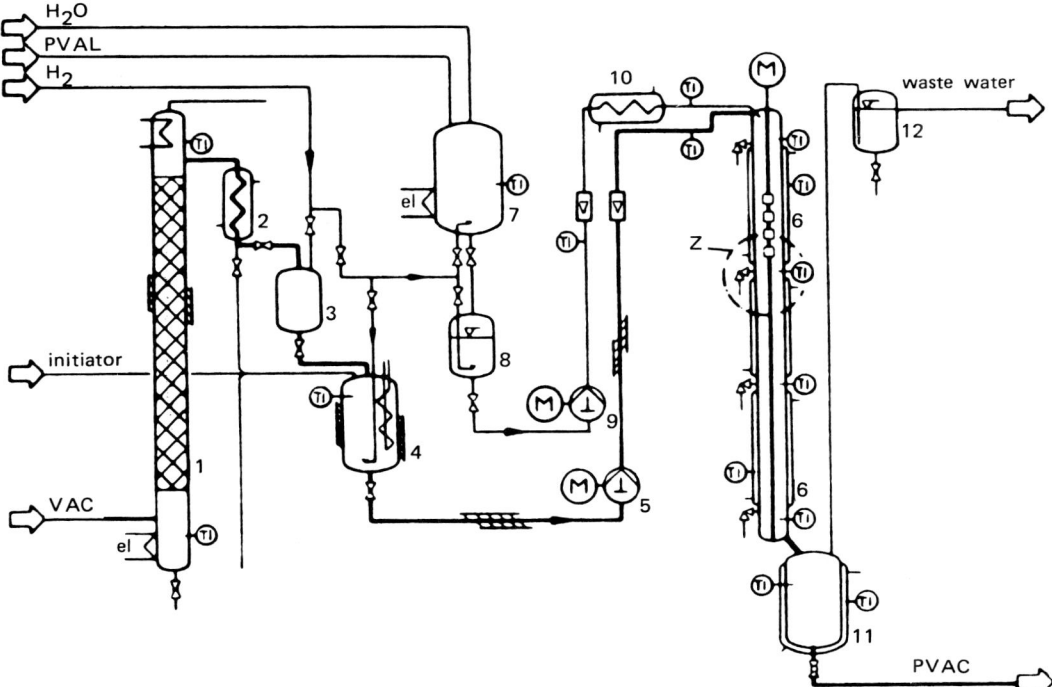

Figure 4 Flow sheet of the continuous suspension polymerization of vinyl acetate. (From Ref. 15.)

form. Emulsion polymerization uses free radical polymerization mechanisms with unsaturated monomers. The heterogeneous nature of the reaction mixture, however, has a significant influence on the chemical and physical reaction mechanisms and on the nature of the final product [17, 20].

Reactors for Coordination Polymerization of Olefins

Coordination polymerization is an important industrial process to produce olefin polymers such as polyethylene and polypropylene with Ziegler-type catalysts. When heterogeneous catalysts are used, olefin monomers are polymerized by an anionic coordination insertion of monomer into a transition metal–carbon bond. Since the 1970s, there has been remarkable progress in olefin polymerization technology due to the development of extremely efficient organotransition metallic catalysts and innovative reactor technology. Extensive reviews of polymerization kinetics and mechanisms for various olefin polymerization systems can be found in Refs. 21–23.

Ethylene Polymerization Reactor Technology

High-density polyethylene (HDPE) is a major thermoplastic used for a variety of applications. Polyethylene has been manufactured since the early 1950s, but because of its industrial importance, the production of HDPE continues to be a very active area of technology. New improved catalysts and processes have been developed to produce HDPE with tailored properties at lower manufacturing cost.

Polyethylene of density ranging from 0.91 to 0.94 g cm^{-3} is classified broadly as low-density polyethylene (LDPE), and this polymer can be manufactured by high- or low-pressure processes. After nearly a half century of LDPE production with high-pressure tubular or autoclave reactors, radical new technology capable of operating at less than 300 psi and near 100°C emerged in the mid-1970s. This low-pressure LDPE technology has rapidly established itself as a low-cost route to polyethylene having many process and property advantages over conventional high-pressure LDPE.

A key to the new low-pressure LDPE technology is the family of transition metal catalysts that triggers the polymerization reaction at very low pressure and temperature. Moreover, most of these new low-pressure processes utilize higher α-olefins such as butene-1 or hexene-1 as a comonomer to regulate the density of polyethylene. The α-olefin comonomers add side groups which spread the plates of the polyethylene crystal apart sufficiently to reduce the density of the polyethylene to 0.920 g cm^{-3} or even lower.

The LDPE manufactured by copolymerizing ethylene with butene-1 has a substantial degree of ethyl group branching and vinyl unsaturation, whereas the LDPE prepared by conventional high-pressure technology has longer chain branches and essentially no vinyl unsaturation. The observable differences in polymer properties between high-pressure LDPEs and low-pressure LDPEs are caused by the linearity of the main polymer chains, the molecular weight distribution, and the type of chain branching. Figure 5 illustrates schematically the polymer chain structure of various polyethylenes.

Three types of polymerization processes are used today for low-pressure ethylene polymerization: (a) liquid slurry polymerization, (b) solution polymerization, and (c) gas phase polymerization. Table 7 summarizes the features of various polyethylene processes.

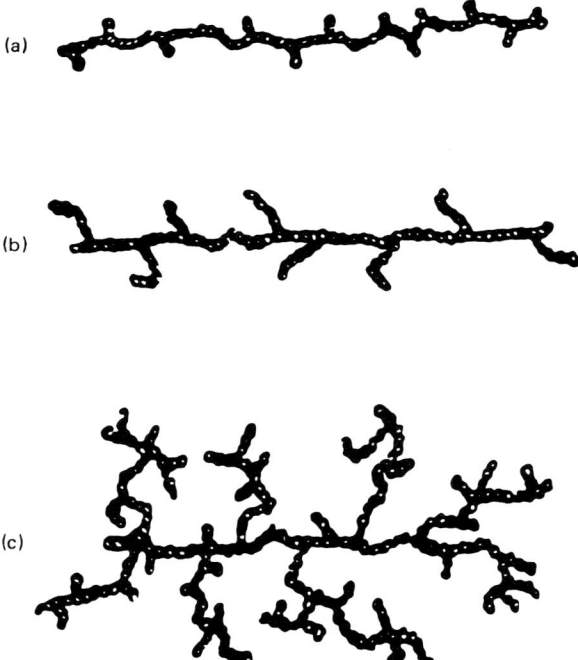

Figure 5 Schematic chain structure of various polyethylenes. (*a*) HDPE (by low-pressure process), (*b*) LLDPE (by low-pressure process), and (*c*) LDPE (by high-pressure process).

Table 7 Low-Pressure Ethylene Polymerization Processes

	Process								
	Liquid slurry			Solution		Gas		
Company:	Phillips	Solvay	Hoechst, Montedison, Dow, Mitsubishi	Stamicarbon	DuPont	Union Carbide	BASF	Amoco
Reactor	Loop reactor	Loop reactor	CSTR	CSTR	CSTR	Fluidized bed	Stirred bed	Compartmented stirred bed
Diluent (or solvent)	i-Butane	n-Hexane	n-Hexane	n-Hexane	Cyclohexane			
Catalyst	Supported Cr-catalyst	Mg-supported Z-N catalyst	Mg-supported Z-N catalyst	Solution form Ti-Mg-Al component	$TiCl_4/VOCl_3/Al(i\text{-}Bu)_3$	Supported Cr-catalyst/Ti-Mg catalyst	Supported Cr-catalyst	Supported Z-N catalyst
Catalyst productivity (g/g cat)	3,000–10,000	11,000	3,000			9,000	8,500	2,000–40,000
Temperature (°C)	85–110	80	80–90	130	140–150	85–100	110	70–95
Pressure (atm)	30–35	30	7.8–35	30	80	20–30	35	20–30
Residence time (hr)	1.5	2.5	2.0–2.7	0.17	0.08–0.17	3–5	4	
Conversion (%)	97–98	97	95–98	95		99		
Comonomer	1-Hexene				1-Octene	Butene-1	Butene-1	Butene-1 or propylene
MW control	Temperature	H_2	H_2	Narrow (M_w/M_n = 3.5–5.0)	H_2	H_2	H_2	H_2
MWD		Narrow-broad					Narrow-broad	

Source: Ref. 23.

Liquid slurry polymerization. The liquid slurry polymerization process encompasses by far the largest group of HDPE technologies. In most cases this process utilizes a catalyst of activity such that catalyst deashing is not required. Excellent temperature control is a major attraction of the liquid slurry process. However, when linear low-density polyethylene (LLDPE) is made by copolymerizing ethylene with a higher α-olefin comonomer in a liquid slurry process, the swelling of the polymer in the slurry medium is major problem. The swelling severely lowers the polymer production rate and polymer density to a minimum of about 0.925–0.935 g cm^{-3}. Currently, long-jacketed loop reactors and continuous stirred tank reactors (CSTRs) are most widely used for slurry polymerization.

The loop reactors, which are recycled tubular reactors, are used by the Phillips Petroleum Co. and Solvay et Cie [24]. The Phillips process is characterized by the use of a light hydrocarbon diluent such as isopentane or isobutane in loop reactors which consist of four jacketed vertical pipes. Figure 6 shows the schematic flow diagram for the loop reactor polyethylene process. The use of high-activity supported chromium oxide catalyst eliminates the need to deash the product. This reactor is operated at about 35 atm and 85–110°C with an average polymer residence time of 1.5 hr. Solid concentrations in the reactor and effluent are reported as 18 and 50 wt %, respectively. The reactor diameter is 30 in. (O.D.) and the length of the reactor loop is about 450 ft. Although reactor fouling was more severe at high temperatures with increased solubility of the polymer in the diluent, higher operating temperatures have become possible in the latest Phillips process because of reduced reactor fouling. This is accomplished by injecting a small quantity of

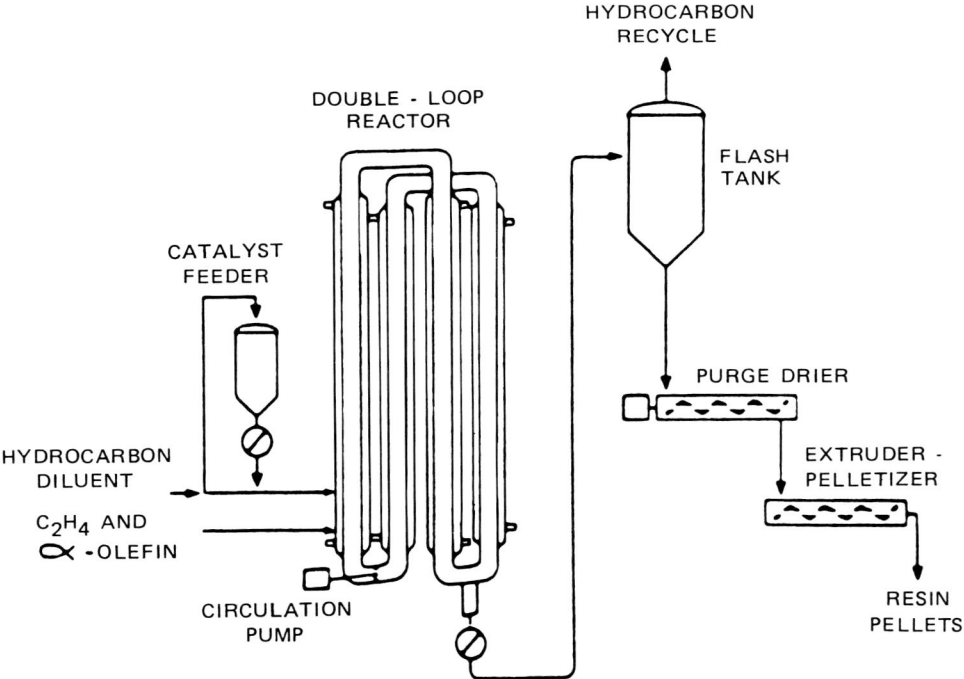

Figure 6 Schematic of loop reactor system for ethylene polymerization. (From Ref. 24.)

additives such as sodium dioctyl sulfosuccinate and aluminum mono- and dihexadecyl-salicylic acid. The contents of the reactor are discharged continuously to a flash tank, utilizing the sensible heat to evaporate most of the diluent and the small amount of unreacted ethylene. The dry polyethylene powder is discharged from the flash tank through a series of licks to a drier to remove additional diluent dissolved in the polymer particles. Then the powder is purged with nitrogen and transported pneumatically to the finishing area for stabilization and pelletizing. It has been reported that ethylene conversion in the reactor is around 98%. This process relies on polymerization temperature for average molecular weight control, while the MWD is controlled by the type of catalyst employed and certain proprietary operational adjustments which alter MWD.

Continuous stirred tank reactors are also widely used for hexane slurry ethylene polymerization by many manufacturers. In the Hoechst process, the reaction is carried out in four CSTRs arranged in series such that the slurry phase and the vapor phase move in concurrent flow. Polymerization occurs at 100 psig and 85°C with 98% conversion of ethylene. The residence time in the reactor is about 2.7 hr. The product slurry is pumped into centrifuges, which separate the bulk of the hydrocarbon diluent liquid from the polymer fluff.

Solution polymerization. Solution processes have some unique advantages over slurry processes in that the MWD can be controlled better, and the process variables are also more easily controlled because the polymerization occurs in a homogeneous phase. The high polymerization temperature (130–150°C) also leads to high reaction rates and high polymer throughputs from the reactor. However, very high molecular weight polymers cannot be produced easily at these high temperatures, and since the solid content is relatively low compared with the slurry process, greater diluent recovery may be required. Figure 7 shows the DuPont solution polymerization process. The catalyst components,

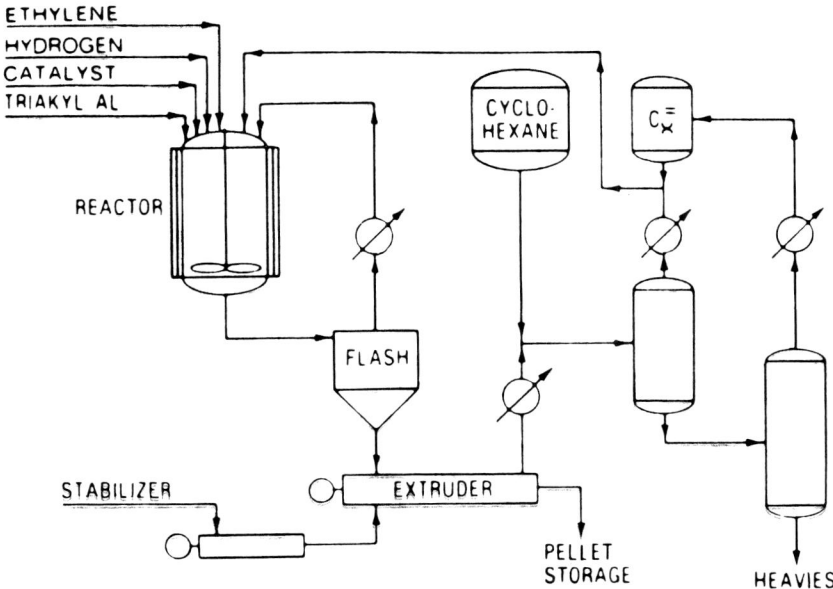

Figure 7 DuPont solution polymerization process. (From Ref. 24.)

cyclohexane, ethylene, octene-1, and hydrogen, are charged continuously to a CSTR operating at a temperature in excess of 150°C and a pressure of about 80 atm. Because of the short residence time (5–10 min) and high polymer concentration (~35%), relatively small reactors may be used either in series or parallel to alter the MWD of the product.

Gas phase polymerization. Polymer separation, diluent recovery, and polymer drying steps may be completely eliminated in gas phase polymerization. The first commercial gas phase HDPE plant using a fluidized bed reactor was constructed by Union Carbide in 1968. Other gas phase polymerization processes using different types of reactors have also been developed by BASF, Naphthachimie, and Amoco. In the fluidized bed polymerization process shown in Figure 8 [25, 26], ethylene and butene-1 are copolymerized over a chromium- or titanium-based high-activity catalyst. Dry catalyst particles less than 250 μm are injected to the fluidized bed reactor, which is maintained by a large volume of circulating gas. Reaction occurs at 300 psig and 85–100°C for HDPE or 75–100°C for LLDPE. The heat of polymerization is adequately removed by the circulating cooled monomer gas at high velocity. The catalyst activity in this process is more than 600 kg g^{-1} Cr (or Ti). The polymerization rate is controlled by the rate of catalyst injection to the bed. The gas velocity in the fluidized bed is maintained at about 5–10 times the weight of the fresh feed. The polymer particles are withdrawn intermittently or continuously near the distributor plate through a gas lock chamber into discharge tanks, where the ethylene gas is separated by adjusting the product withdrawal rate.

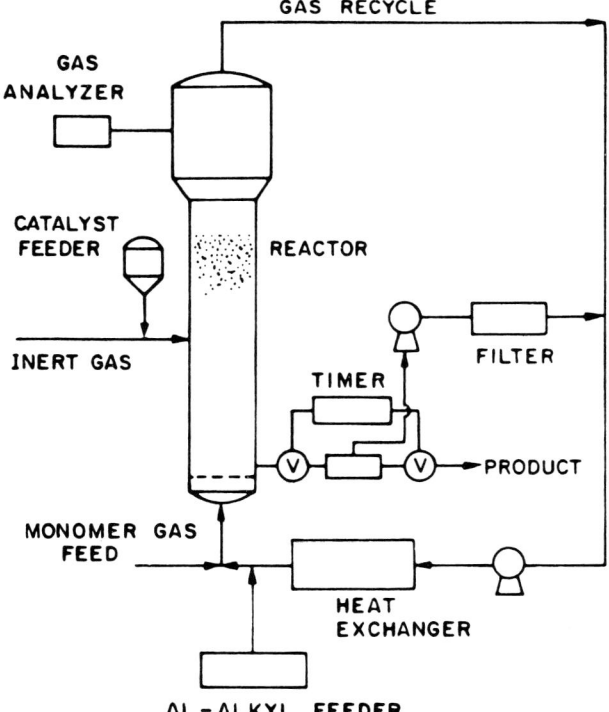

Figure 8 Gas phase polymerization in a fluidized bed reactor. (From Ref. 23.)

In a newer version of the Union Carbide fluidized bed polymerization process described in Ref. 27, a smaller reactor containing an internal cooler is used for more efficient temperature control. Although the conversion per pass is low (2–5%), the overall conversion of ethylene and butene to polymer is about 97%. Polymer particles grow to an average size of about 1000 μm in diameter during their 3–5 hr residence in the fluidized bed reactor. When high-activity catalyst is used, the average particle size of the polymer becomes about 15–20 times larger than the size of the original catalyst particle. Average molecular weight is controlled principally by manipulating the chain transfer agent concentration (hydrogen) and the polymerization temperature, while the MWD is strongly affected by the specific catalyst type and, to a much lower extent, by reactor operating conditions. The gas phase LLDPE reactor is operated at a temperature close to the polymer's softening point. Two potential difficulties in operating the gas phase fluidized bed reactors are (a) the possibility of sintering and agglomeration of the polymer in the bed due to poor fluidization and/or poor heat removal, and (b) the possible inability to achieve control of polymer particle size and shape under a broad range of operating conditions.

In Naphthachimie's gas phase process [28], highly active Ziegler-type catalysts supported on magnesium compound are used to polymerize ethylene in a similar fluidized bed reactor. BASF uses a continuous stirred bed reactor (CSBR) for gas phase ethylene polymerization as shown in Figure 9 [29]. The reactor is operated at a higher pressure and temperature (500 psig, 100–110°C) than employed in Union Carbide's fluidized bed process. The makeup feed and ethylene recycle enter at the bottom of the reactor and rise through the bed at a very slow velocity of about 0.1 ft/sec. The polymer bed is agitated

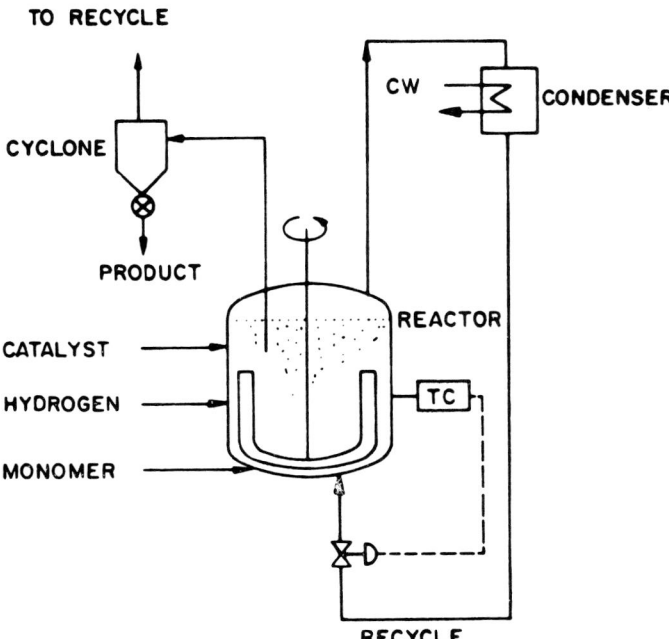

Figure 9 Continuous stirred bed reactor for gas phase olefin polymerization. (From Ref. 29.)

rather than fluidized. A fairly uniform temperature of about 110°C is maintained in the bed, and the unreacted ethylene gas leaves the top of the reactor at 105°C. About 9% of the ethylene recycle leaves the reactor with effluent polyethylene powder and is separated from the polymer at 40 psig. The recycle ethylene is compressed to 1500 psig and cooled to 32°C before being reintroduced to the reactor.

In the Amoco (Standard Oil of Indiana) process, a compartmented horizontal reactor as depicted in Figure 10 is used. The compartments permit variation in temperature and hydrogen pressure in the reactor as a means of controlling polymerization rate and MWD.

Propylene Polymerization Reactors

Since its discovery in 1954, polypropylene has become one of the most important commodity polymers. Among the outstanding properties of polypropylene are low density, high melting point, high tensile strength, great rigidity, better stress crack resistance than HDPE, and high resistance to acid, alkali, solvent, and other chemical attack. The introduction of highly active and highly stereospecific Ziegler–Natta catalysts has allowed low-cost manufacture of polypropylenes with desired properties. Currently, polypropylene is produced by liquid slurry, bulk, and gas phase processes using various types of transition metallic catalysts.

Recent progress in propylene polymerization reactor technology is summarized in Table 8. Brockmeier [30] surveyed these processes, and Table 9 provides a comparison of

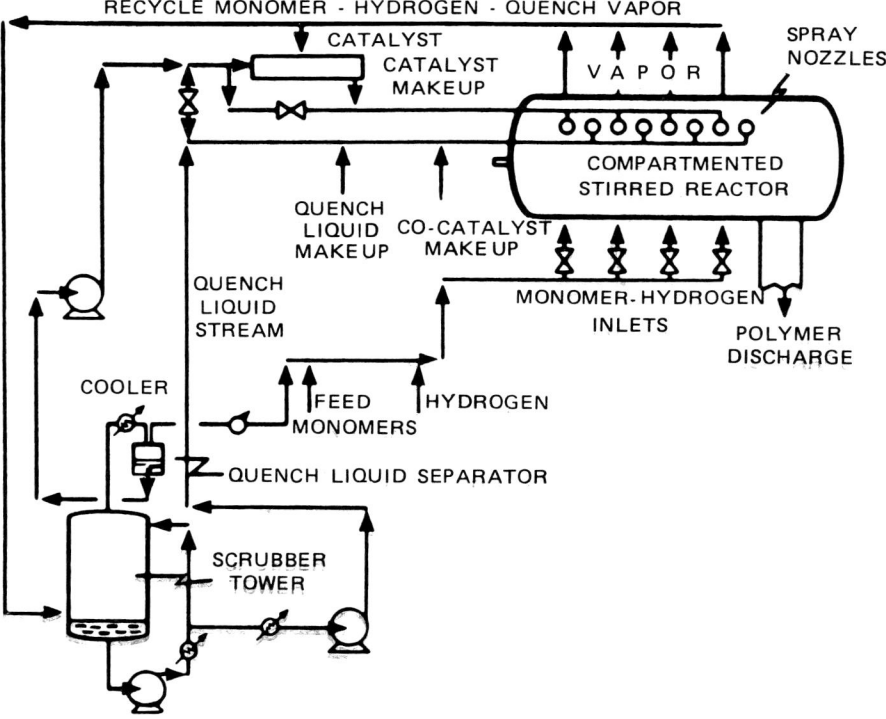

Figure 10 Compartmented stirred reactor for gas phase olefin polymerization. (From Ref. 30.)

Table 8 Progress in Polypropylene Processes

Process	Polymerization	Solvent	Catalyst deashing	Product extraction	Examples
First generation	Slurry	Yes	Yes	Yes	Sumitomo, Mitsubishi, Hercules
	Bulk	No	Yes	Yes	Dart, ShowaDenko
Second generation	Slurry	Yes	No	No	Montedison/Mitsui
	Bulk	No	No	Yes	Sumitomo
Third generation	Gas phase	No	No	No	BASF, Amoco

process features. In this section, characteristics of various polymerization reactor systems are described.

Liquid slurry polymerization. Among current industrial propylene polymerization processes, the liquid slurry process is still the most widely used. Propylene is polymerized in a continuous stirred tank reactor at a temperature below 90°C and at a pressure high enough to maintain the propylene in the liquid phase. The slurry contains 10–20% solids of polymer particles. These particles are concentrated in and removed from the reactor as a 20–40% solids slurry, which is then flashed to remove the remaining propylene and hydrocarbon diluent. In a modern simplified slurry polymerization process, highly active and stereospecific catalysts are used and deashing and extraction steps are completely eliminated from the process (see Table 9). The major advantages of the simplified slurry processes are high content of solids in the slurry, good temperature control, flexibility of reactor operations, simplified overall process, high productivity of isotactic polypropylene, and a very small waste stream from solvent recovery.

Bulk phase (liquid pool) polymerization. The bulk process is similar to the liquid slurry process except that liquefied monomer becomes the reaction medium. Insoluble polymer formed on the suspended catalysts precipitates out to form a slurry. An obvious advantage of bulk polymerization is the high monomer concentration due to the absence of inert solvent, which, if present, must be recovered and repurified. Recently the bulk polymerization process has been improved significantly by the use of extremely efficient catalyst systems producing highly isotactic polypropylene. The new "second generation" bulk phase processes have been developed on a commercial production scale by the Sumitomo Co. [31], and jointly by Montedison (Italy) and Mitsui Petrochemicals (Japan) [32, 33]. These new bulk processes have the following advantages: (a) the space–time yield of the reactor is high and (b) the reactor can be operated at reduced temperature and pressure because of the increased reaction rate. A broad range of propylene homopolymer, random copolymer, and block copolymer grades is manufactured by this process.

Gas phase polymerization. Gas phase polymerization is also used for propylene polymerization. Here, gaseous monomers are polymerized over solid Ziegler-type catalyst in the presence of aluminum alkyl cocatalyst. In an commercial gas phase polymerization processes using different reactors, catalyst deashing and product purification steps are not required because highly efficient catalyst systems are used [34, 35]. Continuous stirred

Table 9 Polypropylene Process Comparison

Process type	Reactor		Operating conditions					Catalyst			¢/lb PP	Energy cost factor[b]
	Type	Heat removal	Temperature (°C)	Pressure (psig)	C_3H_6 (mole fraction)	Slurry concentration (wt%)	Residence time (hr)	Yield (lb/lb)	I.I. (%)[a]	Use of support		
Conventional slurry	CSTR	50% reflux, 50% jacket	77	155	0.20	35	3.0	1,000 (est)		No	28.9	1.0
Simplified slurry	CSTR	100% jacket	65	200	0.39		3.3	70,000	93–96	Yes	27.7	0.67
Simplified bulk	CSTR	60% reflux, 40% jacket	54	315	0.85	60	2.0	90,000	89–95	Yes	27.7	0.55
Gas phase	Stirred bed	100% evaporation	80	425	0.90	Powder	4.2	4,500	75–95 (est)	No	27.7	0.41

[a]I.I. = Isotactic index.
[b]Energy cost factor is set to 1 for conventional slurry process for comparison.
Source: Ref. 30.

bed reactors, horizontal compartmented reactors, and fluidized bed reactors are used for propylene polymerization in the gas phase.

The continuous stirred bed reactor shown in Figure 9 is used by BASF at about 30–35 atm and 80–95°C. Into it, via separate streams, highly purified or partially liquefied propylene monomer, high-activity $TiCl_3$ catalyst, and aluminum alkyl cocatalyst are injected. A fraction of unreacted monomer evaporates during the reaction while absorbing the heat of polymerization. The propylene vapor from the reactor is condensed and returned continuously to maintain the desired polymerization temperature. The reactor pressure is high enough that condensation or partial condensation can be achieved with the available plant cooling water. The polymer is taken as a dense phase to a separator from which unreacted propylene is removed, recompressed, distilled to remove any by-product materials, and recycled back to the reactor. The gas velocity of the vaporized monomer is about 1–2 cm sec^{-1}. At this flow rate the polymer powder in the polymerization zone is not fluidized and the polymer is thus prevented from entering the cooling zone with the gaseous monomer. Reactor pressure is controlled by manipulating the feed rate of makeup monomer. The liquid monomer recycle ratio (i.e., propylene recirculated/makeup propylene) is about 4–5 and the catalyst yield is about 25 kg g^{-1} catalyst.

The gas phase polymerization reactor of Amoco is characterized by its unique design and operation. As shown in Figure 10, the reactor is a horizontal, cylindrical vessel, stirred by paddles mounted on an axial shaft with the lower section of the reactor divided into several compartments. The compartments permit variation in temperature and gas phase composition. Specially prepared high-activity unsupported or supported titanium chloride catalyst, which is temporarily inactivated by ethanol, is fed to the reactor with inert quench liquid and reactivated in the reaction zone by an aluminum alkyl cocatalyst which is sprayed onto the polymer bed. The heat of reaction is removed by evaporating liquid propylene or a quench liquid such as isobutane or isopentane. The temperature is controlled by manipulating the quench liquid flows and/or amount of catalyst injected to the reactor. It has been claimed that a narrow or broad MWD ($M_w/M_n = 6-12$) is obtainable by varying hydrogen concentration, temperature, and catalyst composition in each compartment. According to an example in Ref. 36, an active $TiCl_3$ catalyst with diethylaluminum chloride cocatalyst is charged to the reactor every 30 min. The reactor temperature is maintained at 71°C by continuously sprayed isopentane at the appropriate rate onto the 30 rpm stirred polymer bed. The reactor pressure is controlled at 300 psig by controlling temperature in the condenser at about 50°C. The polymer yield obtained is 10 kg g^{-1} catalyst with tacticity of 96% and the polymer is removed from the reactor as a melt. The product is then pelletized into a desired size.

The fluidized bed reactor illustrated in Figure 8 is also used for propylene polymerization using superhigh-activity catalyst [35]. When a two-component catalyst system is used, the titanium chloride catalyst and aluminum alkyl cocatalyst are added separately, one (titanium) at the top, and one (aluminum alkyl) at the bottom [37–39]. The fluidized bed reactors are also used to produce ethylene–propylene block copolymers.

Reactors for Step Growth Polymerization

Polycondensation is important to industry in synthesizing various polymers such as polyesters (unsaturated polyesters, alkyd resins, polyethylene terephthalate), polyamides, polyurethans, polycarbonates, polyarylates, and certain silicon- and sulfur-containing polymers. Many of the engineering and specialty polymers are produced by polycondensation reactions.

In step growth polymerization or polycondensation, polymer linkage occurs between two macromolecules containing reactive functional end groups. The primary characteristics of this polymerization is that any two species containing functional end groups in a reaction mixture can react each other (see Table 1). Detailed reviews of various polycondensation kinetics and mechanisms as well as reactor modeling techniques can be found in Refs. 1–6 and 40–42.

Polycondensation can be carried out by various polymerization techniques including melt polymerization, solution polymerization, interfacial polymerization, emulsion polymerization, and solid-state polymerization. In melt polymerization, condensation reactions occur in homogeneous polymer melt at high temperature and low pressure. The polymer products obtained by this process are generally pure and no additional product purification steps are required. Since the viscosity of the polymerizing mass increases dramatically as the conversion increases, the removal of small molecules (e.g., byproduct) from the viscous mass is often the rate-controlling process. High temperature employed in many melt polycondensation processes may cause unwanted side product formation. Therefore, optimum polymerization reactor operating conditions must be chosen to produce high-quality polymers.

Solution polycondensation is used in industry to produce polyurethans, polycarbonates, and certain types of polyamides and polyesters. The polycondensation in solution is most frequently used when melt polymerization is impossible or too difficult because of the high melting point of the resulting polymer. In solution polymerization, polymer formed may be readily soluble in the solvent employed or the solution may contain only the monomer in which the polymer is soluble only slightly or not at all. Solution polycondensation takes place at lower temperature than melt polymerization and enables an efficient heat transfer to be maintained. However, solution polycondensation requires polymer separation from solution, repurification of solvent, and polymer washing and drying.

In interfacial polymerization, the formation of polymer takes place at the interface of two nonmiscible liquids in which the starting reactants are dissolved separately, and it takes place almost exclusively by a diffusion mechanism. The system is usually stirred to ensure better contact of two liquids. The polymers formed at the interface are filtered off, washed, and dried. Polymers which can be prepared by interfacial polymerization are polyamides, polyureas, polyurethans, polyesters, polysulfon amides, phenol–formaldehyde polymers, etc. In interfacial polymerization, the materials employed need not be of the highest degree of purity and the polycondensation takes place rapidly at low temperature under atmospheric pressure. Polymers obtained by interfacial polycondensation have a high melting point. Highly reactive monomers, which constitute a large volume of the reaction mixture, and the regeneration of the organic phase are required in this process. Interfacial polymerization is conveniently used to produce polymers which are difficult or impossible to prepare by other methods.

In emulsion polycondensation the polymer formation reaction occurs in the bulk of one of the liquid phases. This method is employed for irreversible, exothermic polycondensations, accompanied by the liberation of a side product. Highly reactive monomers—dichlorides of dicarboxylic acids, diamines, etc.—are most suitable for emulsion polycondensation. To date, the number of polymers preparable by emulsion polymerization is small and the technological aspects of this technique are still largely unknown.

Solid-state polymerization is conducted to produce very high molecular weight polymers at a temperature above the glass transition temperature but below the melting point of the polymers so that reactive functional end groups are activated. The polycondensation by-product is removed by applying high vacuum or by conducting the polymerization in an inert gas stream. Since the polymerization rate is very slow, this method is used mostly to prepare special grade polymers.

In most polycondensation reactions, the reaction equilibrium, which is affected by the eliminated compound (condensation by-product), often determines the extent of the reaction. In the initial stage of the polycondensation, the degree of polymerization increases very slowly with the fractional conversion of functional end groups. To prepare polymers of high molecular weight, the reactions must be driven to very high conversion ($> 99\%$), which also accompanies high viscosity of the polymerizing mass. Thus the mass transfer is often the controlling factor. The polycondensation reactor design must provide for large surface areas and a short diffusion path, and high vacuum is frequently applied. This implies that the central reactor design problems reside in overcoming severe mass transfer resistance to the removal of volatiles from the reactor and in minimizing unwanted side product formation. In the following discussion, characteristics of several industrially very important polycondensation reactor systems are described in order to illustrate unique reactor technology.

Reactors for Polyethylene Terephthalate (PET)

PET polymers are manufactured by a stagewise melt polymerization process which consists of transesterification, prepolymerization, and finishing polymerization steps. In the transesterification stage, dimethyl terephthalate (DMT) or terephthalic acid (TPA) is converted into bishydroxyethyl terephthalate in the presence of metal acetate catalyst. Ethylene glycol (EG), DMT, and catalyst are fed at a molar rate of 1.7 to 2.0:1 (with TPA, this ratio is 1.3 to 1.5:1) and a temperature of 160–180°C. This stage takes 3–4 hr. The reaction by-product methanol (water when TPA is used) is separated from EG vapors in a reflux column. After adding stabilizer and additives, the prepolymer is forced through a superfine filter to the second (prepolycondensation) stage. The polycondensation reaction is performed under vacuum (15–25 torr) and elevated temperatures (280–300°C). The resulting EG is removed by a vacuum pump while the prepolycondensation product—a low molecular weight [degree of polymerization (DP) of about 30], relatively nonviscous material (about 40 poise)—is pumped to the finishing polycondensation stage after a residence time of about 2 hr. The finishing polymerization stage, which is operated at 0.5–1 torr, requires special wiped film or extruder-type reactors in order to handle the high viscosity of the polymer (DP of about 100 at a few thousand poise). Many manufacturers have devised their own proprietary agitation systems to minimize the buildup of solidifying polymer layers. The EG vapors contaminated with oligomers are drawn off through a special condensing system and sent to a recovery unit. The polymer melt can either be sent to a direct spinning plant or transformed into pellets.

PET polymers are produced by either a semibatch or continuous process. Figure 11 shows a schematic diagram of the Hitachi continuous PET process using TPA as a raw material [43, 44]. TPA and EG are fed directly into the slurry mixing tank. The mole ratio of EG and TPA is approximately equal to the stoichiometric ratio for the reaction. The slurry is fed to the esterification stage, which consists of two reactors arranged in series. As shown, the second reactor is of the cascade type. For the polycondensation, three reactors are used. The degree of vacuum in the reactors increases toward downstream so the vacuum degree of each reactor is moderate for the degree of polymerization. The

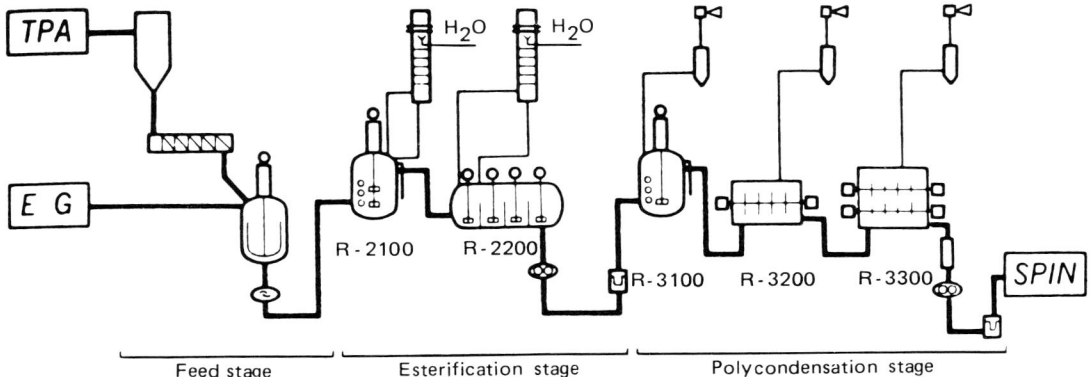

Figure 11 Continuous melt polycondensation process for PET manufacture. (From Ref. 43.)

second reactor is a special type of horizontal single-shaft reactor and the finishing polymerization reactor is a twin-shaft processor with spectacle-shaped blades.

The final stages of polycondensation are characterized by a rapid rise in molecular weight as the condensation product is formed and diffuses out of the polymer. The process occurring is one of desorption accompanied by chemical reaction. Beyond a 90% conversion or a DP of 30, the viscosity of the reaction mixtures becomes very high and the overall reaction becomes diffusion controlled. In addition to the usual problems of difficult heat transfer, mixing, and material conveyance encountered in polymerization reactions conducted on an industrial scale, condensation polymerization presents a serious mass transfer problem: a volatile by-product (EG or water) has to be continuously removed from the reaction mixture in order to maintain favorable equilibrium and reaction rates. To resolve these difficulties, various types of reactors have been developed over the years for use in the finishing stage polymerization [45–51]. In a wiped film reactor, a small portion of the mixture is exposed to vacuum in the form of a film (or several films) with large surface area, and the by-product is removed across this surface. To prevent by-product depletion and thus maintain a maximum driving force for diffusion at any axial position along the film, each point across it is periodically remixed with the bulk at that axial position by means of the rotating action of the screw or rotor. A complete analysis of this reactor would require not only a careful enumeration of various polymerization reactions in the bulk as well as the film, but also a modeling of the transport processes within the film. Ravindranath and Mashelkar present an excellent review of PET polymerization kinetics and reactor modeling and design problems [52, 53].

Reactors for Polyamide Polymerization

Various polyamides can be produced by several different methods: batch polymerization from aqueous salt (nylon salt) solution, direct polymerization from molten intermediates, solid-state polymerization of dry salt, and interfacial polymerization. Among many reactor systems, continuous reactors are the most widely used.

Schematic diagrams of the continuous polymerization of ϵ-caprolactam to produce polyamide 6 and 66 are illustrated in Figure 12 [10]. The so called VK tube is used in the polyamide 6 process. Reactive end groups are formed by hydrolyzing the caprolactam to

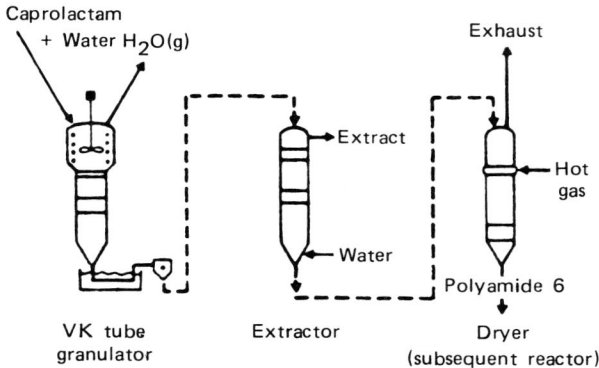

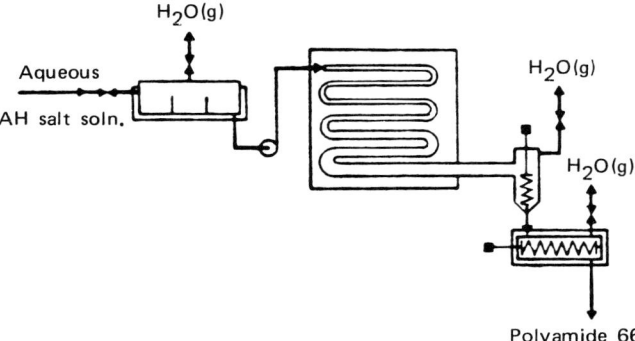

Figure 12 Flow diagrams of continuous production of polyamide 6 and 66. (From Ref. 10.)

ε-aminocaproic acid. A lactam melt with a relatively high water content (~15%) is fed to the top of the VK tube and equipped with a stirrer and a heating coil. The water vaporizes at the top, when viscosity is still low, to give a residue of the desired composition. In the lower part of the tube, the equilibrium degree of polymerization is reached with an increasing viscosity of the melt. The polymer is drawn off at the bottom and granulated. Its equilibrium content of caprolactam and oligomers is about 10% at a final temperature of 270°C. The monomer and oligomers are extracted from the chips with hot water, and the polymer is subsequently dried with hot gas in a vertical cylinder dryer. Intensive drying can produce a further reaction in the solid state and according to the polycondensation equilibrium a higher degree of polymerization is reached.

The polymerization of polyamide 66 is carried out in several different reactors connected in series as shown in Figure 12. The starting material is an aqueous solution of AH salt, containing equivalent quantities of hexamethylene diamine and adipic acid. The solution with about 60% solid content is fed into the first horizontal cylindrical reactor then divided into several compartments where the water is drawn off as vapor and precondensate of low molar weight is formed. This is pumped into the second reactor, which is a heated tube reactor with a gradually increasing diameter. Polycondensation proceeds here and water vapor forms at falling pressure. The next step is the removal of wastes in a steam separator followed by feeding the polymer melt by means of a screw conveyor into the last reactor, which consists of a heated screw conveyor where water vapor is again withdrawn and the final polycondensation equilibrium is attained.

Reactors for Polycarbonates

Polycarbonates (PC) are important as versatile engineering thermoplastics because of their outstanding mechanical properties, electrical resistance, heat resistance, dimensional stability, and transparency. The self-extinguishing feature and adaptability to a host of colors and pigments are especially attractive. All commercial polycarbonates are based on bisphenol A (BPA), although chloro- and methyl-substituted BPA have been reported to produce polycarbonates with interesting properties. At present, polycarbonates are commercially produced by either interfacial polymerization or melt polymerization. The solution process, which was once a major commercial process, is no longer used in industry because of its inferior economics. In what follows, the interfacial and the melt polymerization processes for the manufacture of polycarbonates are described [54–57].

Interfacial polymerization. The interfacial polymerization of polycarbonates involves a reaction of BPA with phosgene at the interface between an inert organic methylene chloride solution and an aqueous caustic solution. The reaction takes place in two steps. First, phosgene reacts with BPA to form monochloroformates. Then the polycondensation takes place between BPA hydroxyl groups and chloroformates in the presence of triethylamine as a catalyst, yielding the polymer that remains dissolved in the organic phase. During the polymerization, the by-product (hydrochloric acid) reacts with the aqueous caustic phase to form sodium chloride. After the polymerization, the organic polymer solution is separated from the aqueous phase and purified. The polymer is recovered from the purified polymer solution by precipitation or evaporation. The solid polymer is crushed and extruded into pellets.

The interfacial polymerization can be conducted by batch or continuous reactors. Figure 13 is a schematic of a typical batch polycarbonate polymerization process. Gaseous or liquid phosgene is pumped into a well-stirred reactor containing BPA in an organic–aqueous dispersion. Aqueous sodium hydroxide is added during reaction to maintain an alkaline pH. A molecular weight regulator, most often *para-tert*-butylphenol or phenol is added to the reactor to inactivate some of the end groups in the oligomer carbonates. At the end of polymerization, the polymer-containing organic phase is separated and washed to remove the catalyst and impurities such as sodium chloride. The washed solution is then concentrated by flashing and is passed through a devolatilizing extruder to produce a molten ribbon, which is then cooled and chopped into pellets.

In the continuous process, multiple tubular reactors in series or a cascade of stirred tank reactors is used. In this process, oligomeric chloroformates are produced by the phosgenation of BPA prior to polycondensation in the presence of a catalyst. Polymerization conditions for batch and continuous processes are summarized as follows [57]:

Process	Phosgenation		Polycondensation	
	Temperature (°C)	Time (min)	Temperature (°C)	Time (min)
Batch process	25–30	20–30	25–30	15–45
Continuous process	34	12	34–36	24
Continuous process alternative	71–75	1.3	74–85	4–21

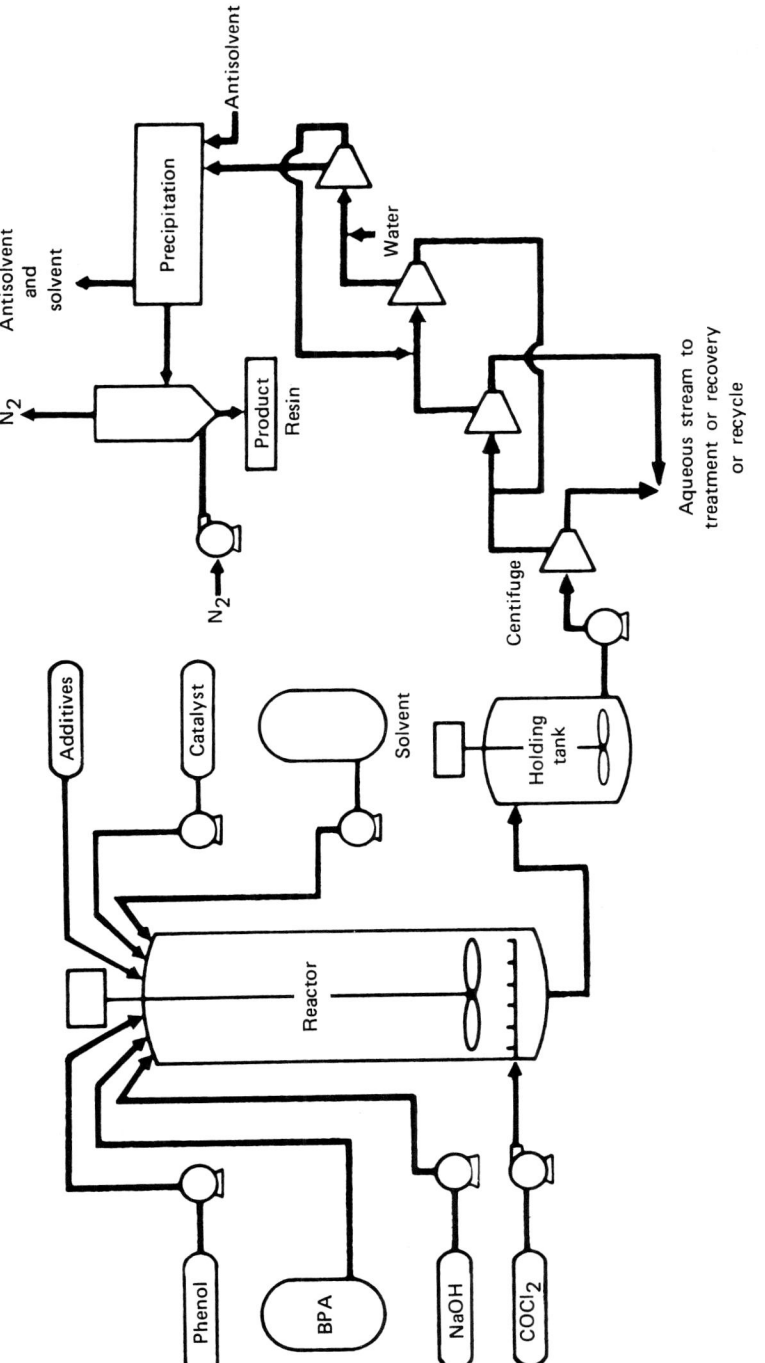

Figure 13 Batch polycarbonate polymerization process. (From Ref. 54.)

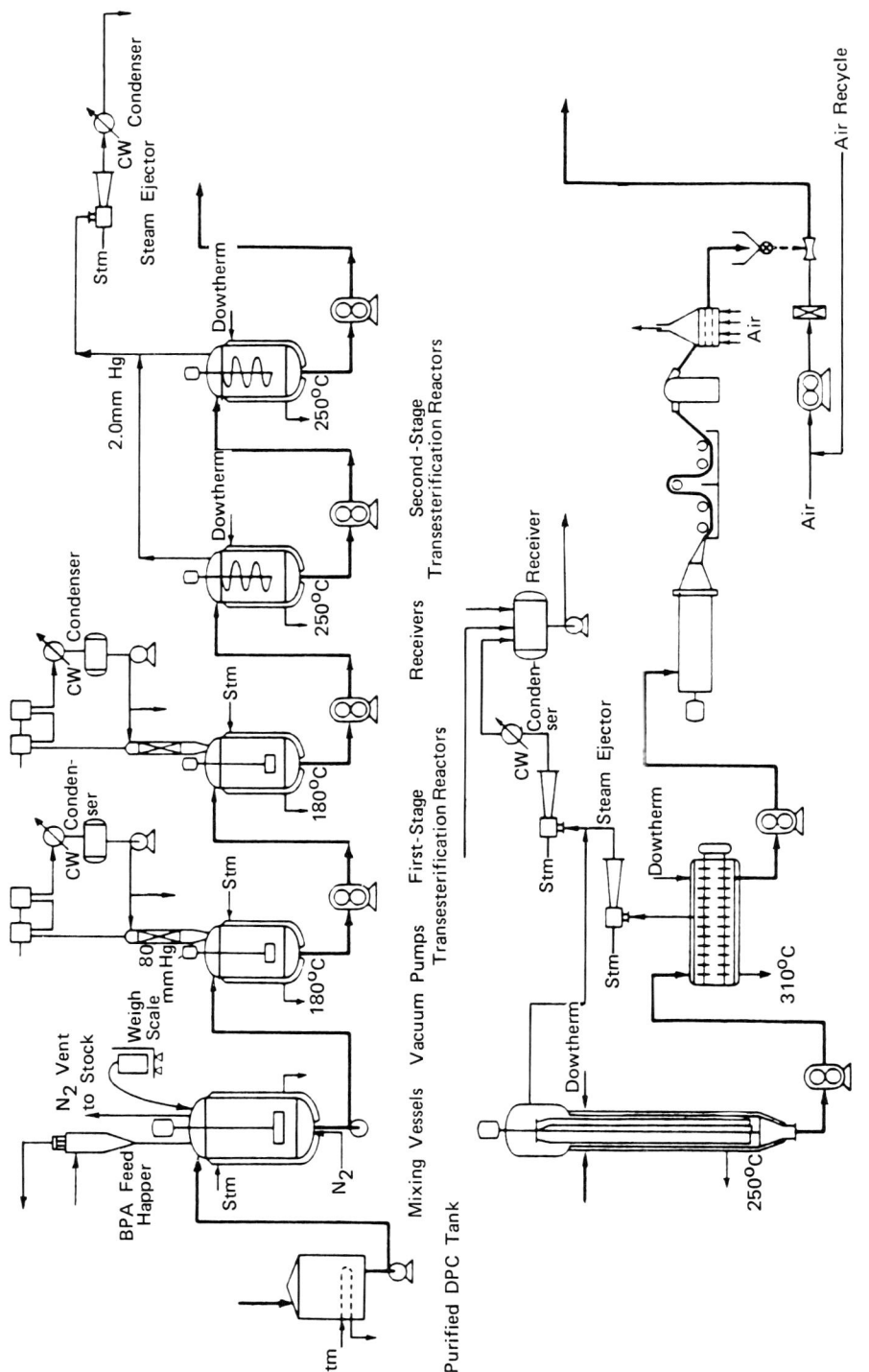

Figure 14 Continuous melt polycarbonate polymerization process. (From Ref. 57.)

Melt polymerization processes. The melt polymerization process or the interesterification process is another important commercial method for manufacturing polycarbonates widely used in Western Europe and Japan. There are several ways to carry out melt polymerization. Conventionally, the polymer is produced by transesterifying diphenyl carbonate with BPA in the presence of a catalyst such as a sodium salt of BPA. Monophenyl carbonate of BPA is formed first. It reacts further to produce oligomers having phenoxy and hydroxy end groups. The oligomers are then polycondensed to give polycarbonate. During the polymerization, the reaction temperature is kept above the melting point of the reaction mass, initially about 150°C. It is increased in steps to a final temperature of 300°C. Meanwhile, the pressure is reduced, also in steps, from an initial 760 torr to less than 1 torr at the final stage of the polymerization. During the high-temperature, low-pressure polymerization, the viscosity of polycarbonates increases dramatically and specially designed extruder-type reactors are required to finish the polycondensation. Figure 14 is a schematic process flow diagram of a continuous melt polycarbonate polymerization process.

An alternative process is to first prepare an oligomer, with an average molecular weight less than 3000, from BPA and phosgene by an interfacial process. The oligomer is then transesterified with additional BPA to form the desired polycarbonate. Another process is to prepare a prepolymer from BPA and phosgene, also by an interfacial polymerization. The prepolymer with a molecular weight greater than 10,000 is polycondensed by a melt polymerization to the finished polymer.

Polyarylates, which are aromatic polyesters from diphenols, and dicarboxylic acids are prepared by a process similar to polycarbonate polymerization. In a semicontinuous process illustrated in Figure 15, a prepolymer with a molecular weight of 2000–3000 is obtained in a stirred batch reactor. This prepolymer is drained off, cooled, crushed, and subjected to a continuous reaction in an extruder-type high-temperature, low-pressure reactor.

Polymerizations in Screw Reactors

Twin-screw, corotating extruders have long been used by the compounding industry for mixing and dispersing of one or more components into a polymer matrix. In recent years, there has been interest in using twin-screw extruders to produce high-performance spe-

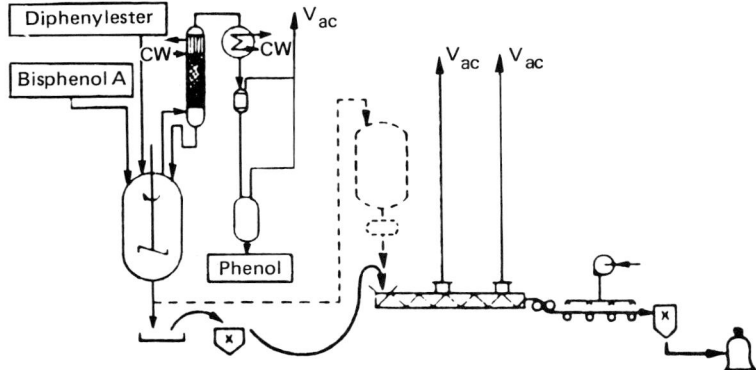

Figure 15 Polyarylate polymerization process. (From Ref. 58.)

cialty polymers has grown rapidly because of the excellent control of mixing characteristics and the ability of the extruders to accept gas, liquid, or solid feed at any point along the extruder length.

The batch-size versatility of such extruders compared to full-scale reactor polymerization has made it financially attractive for polymer or resin manufacturers to go beyond the property limitations of blends and alloys in specialty materials. In the reactive extrusion process, the appropriate monomer(s) or prepolymers and initiator(s) are fed to an extruder where the polymerization takes place and produces polymer which can be forced into a mold or through a die to give a finished article. This is a bulk polymerization with the extruder providing the mixing and heat control. Table 10 presents examples of polymers which can be produced by twin-screw reactors. Note that the reactor system can be used for almost any type of polymerization mechanisms.

Figure 16 illustrates typical extruder polymerization reactor systems. In both reactors, vents located at the upstream end of the barrel can be used to remove volatile gas byproduct as quickly as it forms, promoting the completion of the reaction. In a study on methyl methacrylate polymerization in a twin-screw extruder [60] it was found that reactor residence time distribution and molecular weight distribution showed strong interaction with reaction kinetics and fluid dynamics. The sharper residence time distribution is caused by faster reaction building viscosity near the extruder entrance and the molecular weight distribution was broader when the residence time distribution was broad.

Table 10 Examples of Reactions in Screw Reactors [59]

Final product	Feed product	Type of reaction
Polyurethane	Polyol + diisocyanate + aromatic diamine	Polyaddition
Polyethylene terephthalate	Bis(hydroxyethyl) terephthalate	Polycondensation
Polybutylene terephthalate	Bis(hydroxybutyl) terephthalate	Polycondensation
Polyamide	Precondensate	Polycondensation
Polyarylate	Bisphenol A + phthalic acids	Polycondensation
Polyoximethylene	Trioxane + comonomer	Ionic polymerization
Block copolymer	Isoprene, 1,3-butadiene + styrene	Ionic block copolymerization
Polyamide 6	Caprolactam	Ionic polymerization
(SAN)	Styrene + acrylonitrile prepolymer	Free radical copolymerization
Poly(ethylene + vinyl acetate)	Polyethylene + vinylacetate	Radical grafting
Polystyrene maleic anhydride adduct	Polystyrene + maleic anhydride	Radical grafting
Poly(alkyl methacrylate)	Methacrylate ester	Radical polymerization
Polyol, amines	Polyurethane scrap	Hydrolysis
Glucose	Wood, flour, straw, acid	Hydrolysis
Indoxyl	Phenylglycine	Cyclization
Terephthalic acid	Isophthalic acid	Isomerization
Potassium aluminate	Clay + alkali solution	Salt formation

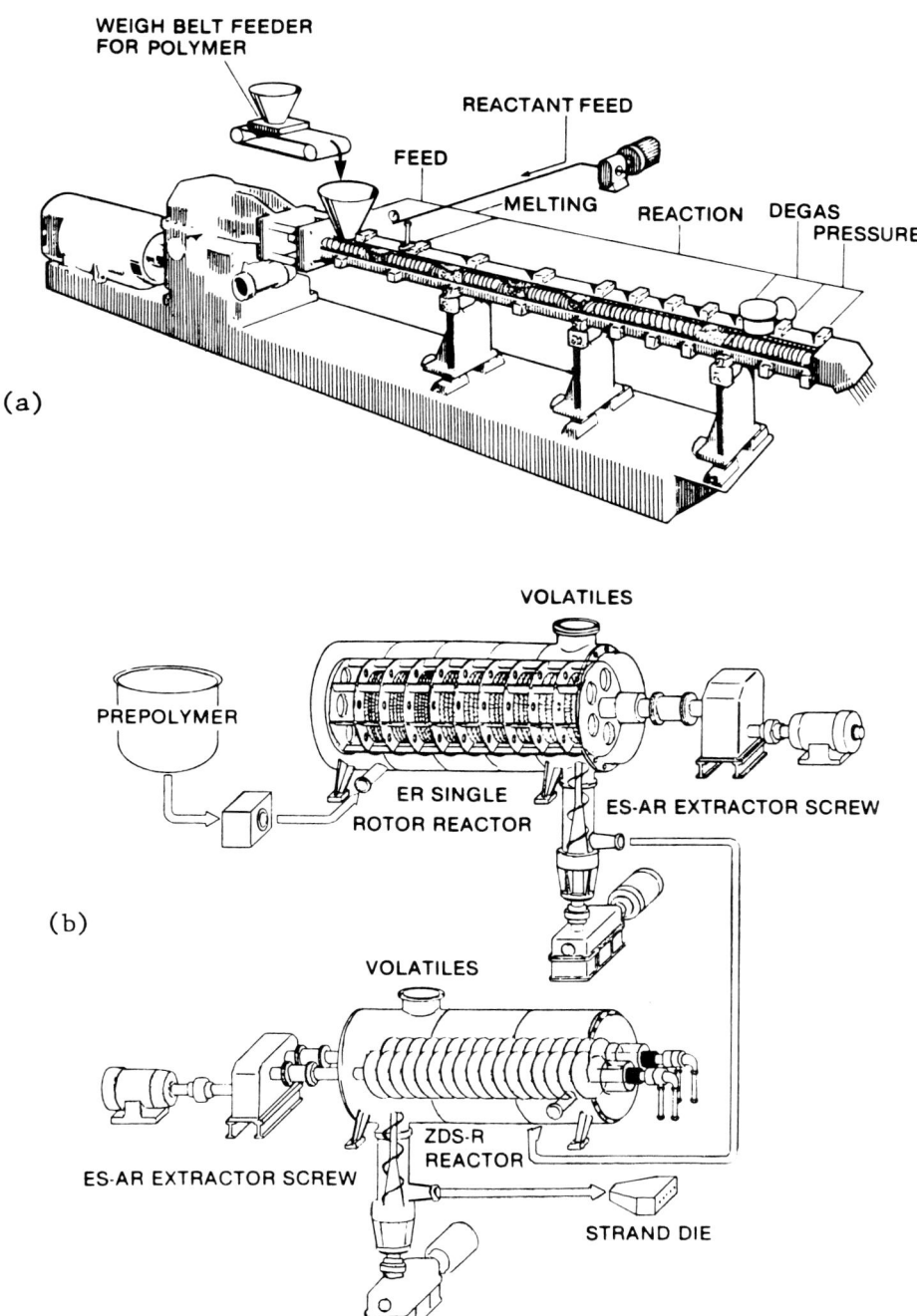

Figure 16 Typical extruder polymerization processes. (From Ref. 59.)

Some unique advantages of this polymerizing extrusion process are as follows [61]:

1. Due to the extruder's ability to handle ultrahigh-viscosity monomer and polymer components without the need for solvents, dramatic raw material and solvent recovery cost reductions are possible.
2. An optimum reactor residence time can be achieved by balancing screw rpm and feed metering. Since materials removal and reaction occur simultaneously, there is no postreaction to degrade batch quality.
3. Temperature runaway situation is not likely to take place in the extruder.
4. Continuous exposure of solid, viscous, or gaseous surfaces to each other in an extruder promotes more complete reaction throughout the entire reacted mass within the specified residence time. Such improved component distribution, for example, has made it possible to chlorinate polymers in an extruder by reacting them with gaseous halogen.
5. Since the hotter, more reactive grafting monomers and oligomers (e.g., maleic anhydride and polyamines) can be added in noncritical quantities through sequential injection ports, new types of polymers can be produced. A key example is maleated tie layer resins, which contain high concentrations of maleic anhydride that could only have been added incrementally in an extruder.

CONTROL OF POLYMERIZATION REACTORS

Precise control of polymerization reactors is crucial in producing polymers with desired properties at maximum productivity. Being characterized by large heat release, nonlinear steady-state behavior, strong parametric sensitivity, nonlinear dynamics, and a lack of adequate on-line sensors for quality monitoring, polymerization reactors present many challenging design and control problems. In this section, fundamental polymerization reactor control problems and practical control techniques are described. For more comprehensive discussion of various polymerization reactor controls, the reader should see Refs. 62–65.

In batch processes for exothermic polymerization reactions, achieving desired polymer control and maximizing polymer yield are the two most important control objectives. These objectives can be achieved by manipulating some reactor operating variables such as temperature, pressure, and reactant and catalyst addition rates according to predetermined polymerization recipes. The exothermic nature of the reaction, poor heat transfer in large reactors, and poor mixing and autocatalytic reactions combine to make batch polymerization reactor temperature control difficult. In general, batch polymerization reactor controllers are designed by the following procedure:

1. Synthesis of open-loop time-varying control policies which minimize properly defined performance index (task level control). Inevitably, this step requires a detailed process model and deep physical insight into the process behavior as a prerequisite.

2. Design of control systems which will drive the plant to follow the open-loop control policies obtained in (1) as closely as possible (execution level control).

Modeling of various polymerization systems has been reported in the literature and the references [5, 66–71] will be particularly useful for comprehensive review of current polymerization reactor modeling techniques.

For task level control, optimal open-loop control techniques are used with detailed reactor models to derive reactor control policies. Classical maximum principle and approximation methods such as control vector parameterization techniques are commonly used to synthesize such optimal strategies [72–81]. Since the jacket coolant flow rate is mostly manipulated to control the polymerization temperature, a priori programing of reactor temperature set point and coolant temperature set point are required. When isothermal polymerization is desired, the control objective is to minimize the heating period with minimal temperature overshoot or oscillations. Typical control policy is a bang-bang control in which switching times for cooling and hot water valves must be chosen properly. Figure 17 illustrates various batch polymerization reactor policies with jacket coolant temperature as a control variable [82].

For the control of polymer properties such as MWD and copolymer composition distribution, the semibatch mode of operation is often employed and the feed rates of reactants or initiators are varied [70, 83–86]. Since many polymer property control

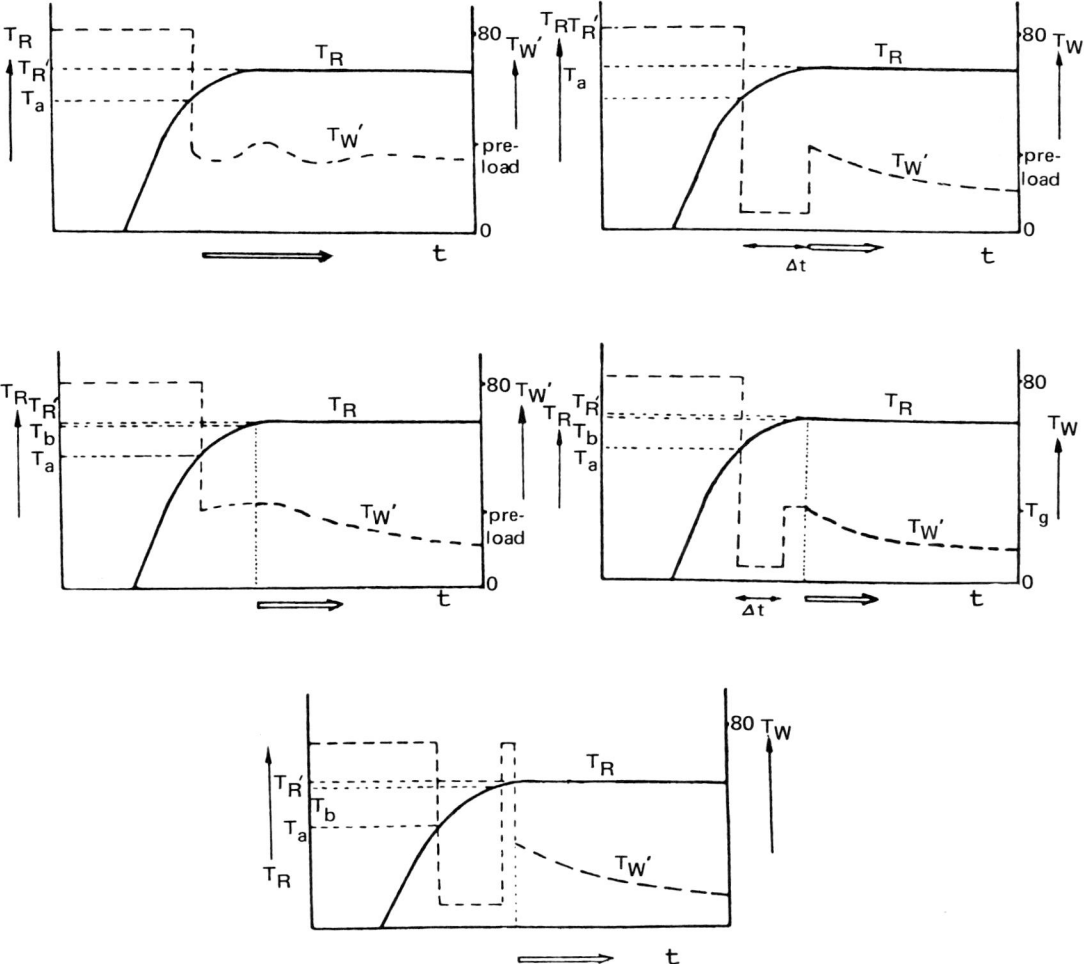

Figure 17 Various batch polymerization reactor temperature control policies. (From Ref. 82.)

problems in batch or continuous polymerization processes are due to a lack of on-line sensors for property measurements, application of optimal state estimation technique (e.g., extended Kalman filtering) will be an attractive alternative with a few on-line measurements (e.g., temperature, pressure, viscosity, conversion) and a limited number of delayed off-line laboratory measurements. The idea is to use the estimated polymer properties in the feedback control scheme to reduce the variation in polymer product quality from the desired value. Recent work reported in the literature suggests that nonlinear state estimation techniques can be successfully implemented in polymerization reactor systems [62, 87–91]. Stochastic control techniques such as adaptive control or self-tuning control are also used for polymerization reactor control. In a self-tuning adaptive control scheme, the parameters of an input–output process model are identified by on-line parameter estimation and the controller parameters are directly estimated to minimize the variance of the output signal deviation [92, 93].

Continuous polymerization reactors often exhibit strongly nonlinear steady-state and dynamic behavior, which makes reactor control very difficult. In many polymerization systems including free radical polymerization and solid catalyzed coordination polymerization, multiple steady states, sustained oscillations, and runaway phenomena may take place for some polymerization conditions. Thus continuous reactors must be designed and controlled in such a way that unwanted dynamic behavior can be avoided [94–100]. Certainly more research is needed to develop improved polymer property control algorithms for complex and nonlinear polymerization reactor systems. Development of accurate, easy to use, and inexpensive on-line sensors is urgently needed for precise control of polymer properties.

REFERENCES

1. F. J. Odian, *Principles of Polymerization*, 2nd ed., Wiley, New York (1981).
2. F. W. Billmeyer, Jr., *Texbook of Polymer Science*, 3rd ed., Wiley, New York (1984).
3. P. J. Flory, *Principles of Polymer Chemistry*, Cornell University Press, Ithaca, N. Y. (1953).
4. H. G. Elias, *Macromolecules*, Vol. 2, Plenum, New York (1977).
5. W. H. Ray and R. L. Laurence, in *Chemical Reactor Theory—A Review* (L. Lapidus and N. R. Amundson, eds.), Prentice-Hall, New York, p. 532 (1977).
6. W. H. Ray, *J. Macromol. Sci.-Revs. Macromol. Chem.*, C8(1): 1 (1972).
7. W. H. Ray, Plenary lecture, 7th Int. Sym. Chem. React. Eng., October 1982, Boston.
8. H. L. Hsieh and R. C. Farrar, *ACS Symp. Ser.*, 166: 389 (1981).
9. R. H. M. Simon and D. C. Chappelear, *ACS Symp. Ser.*, 104: 71 (1979).
10. H. Gerrens, *Ger. Chem. Eng.*, 4: 1 (1981); *Chemtech*, June, 380 (1982).
11. N. Platzer, *Ind. Eng. Chem.*, 62(1): 6 (1970).
12. D. E. Carter and R. H. M. Simon, U.S. Pat. 3,903,202 (1975); Monsanto.
13. A. H. Stark, U.S. Pat. 3,007,903 (1961); Dow Chemical.
14. S. Kato, K. Uku, and H. Morikawa, U.S. Pat. 4,487,898 (1984); Kanegafuchi Chemical.
15. K. H. Reichert and H. U. Moritz, *J. Appl. Poly. Sci. Appl. Poly. Symp.*, 36: 151 (1981).
16. K. H. Reichert, H. U. Moritz, C. Gabel, and G. Deiringer, in *Polymer Reaction Engineering* (K. H. Reichert and W. Geiseler, eds.), Hanser Publishers, Munich, p. 153 (1983).
17. I. Piirma, *Emulsion Polymerization*, Academic Press, New York (1982).
18. G. W. Poehlin, *ACS Symp. Ser.*, 285; 131 (1985).
19. A. L. Rollin, W. I. Patterson, J. Archambault, and B. Bataille, *ACS Symp. Ser.*, 104: 113 (1979).
20. C. E. Schildknecht and I. Skeist, *Polymerization Processes*, Wiley, New York (1977).

21. J. Boor, Jr., *Ziegler–Natta Catalysts and Polymerizations*, Academic Press, New York (1979).
22. T. Keii, *Kinetics of Ziegler–Natta Polymerization*, Kodansha, Tokyo (1972).
23. K. Y. Choi and W. H. Ray, *J. Macromol. Sci. Rev. Macromol. Chem. Phys.*, $C25(1)$: 1 (1985), 56 (1985).
24. J. N. Short, *MMI Symp. Ser.*, 4: 651 (1983).
25. *Chem. Eng.*, 84 (Apr. 23, 1979).
26. *Chem. Eng.*, 79 (Oct. 22, 1979).
27. G. L. Brown and D. F. Warner, U.S. Pat. 4,255,542 (1981); Union Carbide.
28. I. R. Dormenval, L. Havas, and P. Mangin, U.S. Pat. 3,922,322 (1975); Naphthachimie.
29. K. Wisseroth, U.S. Pat. 4,012,573 (1977); BASF.
30. N. F. Brockmeier, *MMI Symp. Ser.*, 4: Harwood Academic Pub., 671 (1983).
31. K. Matsuyama, A. Shiga, M. Kakugo, and H. Hashimoto, *Hydrocarbon Proc.*, 31 (Nov. 1980).
32. N. Kashiwa, Paper presented at the 5th Int. Symp. on Cationic and Other Ionic Poly., Kyoto, 1980.
33. *Chem. Eng.*, 80 (Apr. 20, 1982).
34. *Oil and Gas J.*, 64 (Nov. 23, 1970).
35. *Chem. Eng. News*, 8 (Nov. 14, 1983).
36. J. L. Jezl, E. F. Peters, R. D. Hall, and J. W. Shepard, U.S. Pat. 3,985,083 (1976); Amoco.
37. W. A. Fraser and G. S. Cioloszyk, U.S. Pat. 4,243,619 (1981); Union Carbide.
38. R. J. Jorgensen, G. L. Goeke, and F. J. Karol, U.S. Pat. 4,349,648 (1982); Union Carbide.
39. G. L. Goeke, B. E. Wagner, and F. J. Karol, U.S. Pat. 4,354,009 (1982); Union Carbide.
40. L. B. Sokolov, *Synthesis of Polymers by Polycondensation*, Israel Program for Scientific Translation, Jerusalem (1968).
41. D. H. Solomon, *Step-Growth Polymerization*, Marcel Dekker, New York (1972).
42. P. W. Morgan, *Condensation Polymers: By Interfacial and Solution Methods*, Interscience, New York (1965).
43. H. Yokoyama and T. Sano, *Chem. Econ. Eng. Rev.*, $11(10/11)$: 31 (1979).
44. T. Sano and H. Yokoyama, *Hitachi Rev.*, $28(2)$: 83 (1979).
45. L. L. Kilpatrick, U.S. Pat. 3,248,180 (1966); DuPont.
46. J. E. Crawford, R. W. Edward, E. D. Henze, and W. C. L. Wu, U.S. Pat. 3,619,145 (1971); Mobil Oil.
47. J. W. Crawford, R. W. Edwards, E. D. Henze, and W. C. L. Wu, U.S. Pat. 3,522,214 (1970); Mobil Oil.
48. K. Ogata, K. Kazama, S. Suzuki, and Y. Morimatsu, U.S. Pat. 3,442,868 (1969); Teijin Ltd.
49. J. M. Gerbel, U.S. Pat. 3,633,645 (1972); Luwa A. G.
50. M. Hachiya, K. Hiratsuka, and H. Fukumori, U.S. Pat. 3,532,151 (1970); Hitachi.
51. P. Ellwood, *Chem. Eng.*, 98 (Nov. 20, 1967).
52. K. Ravindranath and R. A. Mashelkar, *Chem. Eng. Sci.*, $41(9)$: 2197 (1986).
53. K. Ravindranath and R. A. Mashelkar, *Chem. Eng. Sci.*, $4(12)$: 2969 (1986).
54. S. K. Sikdar, *Chemtech*, 112 (Feb. 1987).
55. H. Schnell, *Chemistry and Physics of Polycarbonates*, Interscience, New York (1964).
56. E. D. Oliver, Polycarbonates, *SRI Report*, 50 (1969).
57. Y. R. Chin, Polycarbonates, *SRI Report*, 50B (1982).
58. G. Bier, *Polymer*, $15(8)$: 527 (1974).
59. *Polym. Proc. News*, $13(2)$. Werner & Pfleiderer Corp., N.J.
60. N. P. Stuber and M. Tirrell, *Polym. Proc. Eng.*, $3(1,2)$: 71 (1985).
61. *Modern Plastics*, 56 (July 1985).
62. W. H. Ray, *Proc. Am. Contr. Conf.*, 842, June 1985, Boston.
63. J. F. MacGregor, A. Penlidis, and A. E. Hamielec, *Polym. Proc. Eng.*, $2(2,3)$: 179 (1984).

64. H. Amrehn, *Automatica, 13*: 533 (1977).
65. K. Hoogendoorn and R. Shaw, *IFAC PRP 4, Automation,* Ghent, Belgium (1980).
66. W. H. Ray, *J. Macromol. Sci. Revs. Macromol. Chem., C8*(1): 1 (1972).
67. L. H. Garcia-Rubio, J. F. MacGregor, and A. E. Hamielec, *ACS Symp. Ser., 197*: 87 (1982).
68. W. H. Ray, *IFAC PRP 4 Automation,* Ghent, Belgium (1980).
69. S. A. Volfson, G. G. Aleksanyan, E. I. Maksimov, in *Advances in Polymer Science* (Z. A. Rogovin, ed.), Wiley, p. 305 (1974).
70. A. E. Hamielec and J. F. MacGregor, in *Polymer Reaction Engineering* (K. H. Reichert and W. Geiseler, eds.), Hanser Publishers, Munich, p. 22 (1983).
71. W. H. Ray, T. L. Douglas, and E. W. Godsalve, *Macromolecules, 4*: 166 (1971).
72. G. Z. A. Wu, L. A. Denton, and R. L. Laurence, *Polym. Eng. Sci., 22*(1), 1 (192).
73. S. A. Chen and W. F. Jeng, *Chem. Eng. Sci., 33*: 735 (1978).
74. M. Tirrell and K. Gromley, *Chem. Eng. Sci., 36*: 367 (1981).
75. S. A. Chen and N. W. Huang, *Chem. Eng. Sci., 36*: 1295 (1981).
76. S. A. Chen and K. F. Lin, *Chem. Eng. Sci., 35*: 2325 (1980).
77. Y. Yoshimoto, H. Yanagawa, T. Suzuki, T. Araki, and Y. Inaba, *Int. Chem. Eng., 11*(1): 147 (1971).
78. K. Osakada and L. T. Fan, *J. Appl. Polym. Sci., 14*: 3065 (1970).
79. J. Hicks, A. Mohan, and W. H. Ray, *Can. J. Chem. Eng., 47*(12): 590 (1969).
80. G. A. Hicks and W. H. Ray, *Can. J. Chem. Eng., 49*(8) (1971).
81. A. Tsoukas, M. Tirrell, and G. Stephanopoulos, *Chem. Eng. Sci., 37*(12): 1785 (1982).
82. K. Hoogendoorn and R. Shaw, *IFAC PRP 4 Automation,* Ghent, Belgium (1980).
83. A. F. Johnson, B. Khali, and J. Ramsay, *IFAC,* Antwerp, Belgium (1983).
84. A. Guyot, J. Guillot, C. Pichot, and L. R. Guerrero, *ACS Symp. Ser., 165*: 415 (1981).
85. A. W. Hanson and R. L. Zimmerman, *Ind. Eng. Chem., 49*(11): 1803 (1957).
86. R. J. Hanna, *Ind. Eng. Chem., 49*(2): 208 (1957).
87. G. A. Ardell and B. Gumowski, *Chem. Eng. Prog., 77* (June 1983).
88. H. Schuler, *IFAC PRP 4 Automation,* Ghent, Belgium (1980).
89. J. C. Hyun and S. G. Bankoff, *Chem. Eng. Sci., 31*: 953 (1976).
90. S. Papadopoulou and E. D. Gilles, in *Polymer Reaction Engineering* (K. H. Reichert and W. Geiseler, eds.), Huthig & Wepf Verlag, Heidelberg, p. 243 (1986).
91. T. W. Taylor, V. Gonzalez, and K. F. Jensen, in *Polymer Reaction Engineering* (K. H. Reichert and W. Geiseler, eds.), Huthig & Wepf Verlag, Heidelberg, p. 261 (1986).
92. C. Kiparissides and S. L. Shah, *Automatica, 19*(3): 225 (1983).
93. K. M. Kwalik and F. J. Shork, *Proc. Am. Cont. Conf.,* 872, Boston, June 1985, p. 872.
94. R. Jaisinghani and W. H. Ray, *Chem. Eng. Sci., 32*: 811 (1977).
95. A. D. Schmidt, A. B. Clinch, and W. H. Ray, *Chem. Eng. Sci., 39*(3): 419 (1984).
96. J. Hamer, T. A. Akramov, and W. H. Ray, *Chem. Eng. Sci., 32*: 811 (1977).
97. K. Y. Choi, *Polym. Eng. Sci., 26*(4): 975 (1986).
98. K. Y. Choi and W. H. Ray, *Chem. Eng. Sci., 40*(12): 2261 (1985).
99. M. Nomura and M. Harada, *ACS Symp. Ser., 165*: 121 (1981).
100. L. S. Henderson, *Chem. Eng. Prog.,* 42 (March 1987).

3
Principles of Polymerizations with Ziegler–Natta Catalysts

Y. V. Kissin
Mobil Chemical Company
Edison, New Jersey

INTRODUCTION	103
CLASSIFICATION OF ZIEGLER–NATTA CATALYSTS	104
COMPOSITION AND STRUCTURE OF ZIEGLER–NATTA CATALYSTS	105
Organometallic Compounds	105
Transition Metal Compounds	106
POLYMERIZATION MECHANISM OF ZIEGLER–NATTA CATALYSTS	109
COPOLYMERIZATION REACTIONS WITH ZIEGLER–NATTA CATALYSTS	113
POLYMERIZATION OF DIENES	115
STEREOSPECIFIC POLYMERIZATION REACTIONS	119
INHOMOGENEITY OF ZIEGLER–NATTA CATALYSTS	124
COMMERCIAL PROCESSES UTILIZING ZIEGLER–NATTA CATALYSTS	126
Gas Phase Processes	126
Suspension (Slurry) Processes	128
Solution Processes	129
REFERENCES	130

INTRODUCTION

Ziegler–Natta catalysts is a generic term that describes a large variety of transition metal-based catalysts capable of polymerizing and copolymerizing alkenes (olefins) and dienes. The products of these polymerization reactions, polyolefins (alkene polymers), polydienes, and alkene copolymers, are manufactured commercially in a very large volume and have numerous applications as general-purpose and engineering plastics, synthetic rubbers, and elastomers. The list of polymers produced with the application of Ziegler–Natta catalysts includes such widely known products as high-density polyethylene; linear low-density polyethylene; crystalline isotactic polyolefins (polypropylene, polybutene, and polymethylpentene); ethylene–propylene elastomers; and synthetic rubbers based on polybutadiene and polyisoprene.

    The catalysts used for synthesis of these polymeric products are named after German chemist Karl Ziegler and Italian chemist Giulio Natta. In early 1950s these scientists discovered catalytically active compositions, determined principles of their action, and described the structures and properties of polymers produced with the catalysts [1–3]. The monumental contributions of Ziegler and Natta received general recognition when these scientists were jointly awarded the Nobel Prize in Chemistry in 1963.

Ziegler–Natta catalysts have been used for commercial manufacture of various polymeric materials since 1956. Today, the total volume of plastics, elastomers, and rubbers produced with them worldwide exceeds 10 million metric tons. Together these polymers represent the largest commodity plastics as well as the largest commodity chemicals in the world.

This chapter describes the composition and structure of Ziegler–Natta catalysts, the polymerization chemistry of alkenes and dienes, and the structure of polymers produced with the catalysts. Physical and mechanical properties of commercially produced polyolefins and polydienes as well as the technologies of their processing into various articles are discussed in the corresponding articles (see POLYETHYLENE, POLYPROPYLENE, ELASTOMERS, SYNTHETIC RUBBERS).

CLASSIFICATION OF ZIEGLER–NATTA CATALYSTS

An overwhelming majority of Ziegler–Natta catalysts consist of two components. One of the components is a derivative of a transition metal, such as titanium, vanadium, or zirconium. The second component is an organometallic compound, usually an organoaluminum compound [4, pp. 33–35; 5, pp. 94–104]. Typical transition metal compounds are $TiCl_4$, $TiCl_3$, $TiCl_2$, $Ti(OR)_4$, TiI_4, $(C_5H_5)_2TiCl_2$, VCl_4, $VOCl_3$, VCl_3, V acetylacetonate, $ZrCl_4$, Zr tetrabenzyl, and $(C_5H_5)_2ZrCl_2$. Typical organoaluminum compounds are $Al(C_2H_5)_3$, $Al(i-C_4H_9)_3$, $Al(n-C_6H_{13})_3$, $Al(C_2H_5)_2Cl$, $Al(i-C_4H_9)_2Cl$, $Al(C_2H_5)Cl_2$, and $Al_2(C_2H_5)_3Cl_3$.

Neither of the principal catalyst components, if used alone, can polymerize alkenes. (A few exceptions from this rule include $TiCl_2$ and tetrabenzyl titanium, which, by themselves, can act as catalysts [4, pp. 285–305]. However, when the two components of Ziegler–Natta catalysts are mixed, a series of chemical reactions takes place, and some of the products of these reactions (called active centers of Ziegler–Natta catalysts) readily polymerize various alkenes and dienes. Although Ziegler–Natta catalysts have been known for more than 35 years, the exact chemical structure of the active centers still usually remains unknown. Only recent, detailed spectroscopic studies allow identification of some complexes of transition metals as true catalytic species (discussed later).

In addition to two principal components, modern commercial Ziegler–Natta catalysts also contain supports, inert carriers, and modifiers. The most widely used supports and carriers include $MgCl_2$, silica, alumina, and various polymers [5, pp. 105–109; 6]. Supports and carriers differ in the way they affect the catalysts. Supports, although inactive by themselves, have a significant influence on the performance of the catalysts: they usually increase catalytic activity or change the properties of the produced polymers. Carriers do not affect the catalyst performance in any noticeable degree, and their use is warranted by technological factors. For instance, carriers dilute very active solid catalysts, make them more easily transportable, agglomerate catalysts in particles of a specific shape, etc.

Until 1970s, all polymerization reactions with application of Ziegler–Natta catalysts were carried out in inert hydrocarbon media, such as hexane, heptane, or toluene. The media readily dissolve monomers (alkenes and dienes) and organometallic compounds,

Table 1 Classification of Ziegler–Natta Catalysts

Catalyst for ethylene polymerization	Catalyst for polymerization of α-olefins
Homogeneous catalysts	
$(C_5H_5)_2TiCl_2$–$Al(C_2H_5)_2Cl$,	VCl_4–$Al(C_2H_5)_2Cl$ (at low temperatures),
$V(acac)_3$–$Al(C_2H_5)_2Cl$,	ethylene(tetrahydroindenyl)$_2$ZrCl$_2$–
$(C_5H_5)_2ZrCl_2$–$(CH_3AlO)_n$,	–$(CH_3AlO)_n$
VCl_4–$Al(C_2H_5)_2Cl$ (at low temperatures),	
$Ti(OR)_4$–$Al(C_2H_5)_3$ (dimerization catalysts)	
Pseudo-homogeneous catalysts	
$TiCl_4$–$Al(C_2H_5)_2Cl$,	$TiCl_4$–$Al(C_2H_5)_3$,
VCl_4–$Al(C_2H_5)_2Cl$,	$VOCl_3$–$Al(C_2H_5)_3$
$VOCl_3$–$Al_2(C_2H_5)_3Cl_3$	
Heterogeneous catalysts	
δ-$TiCl_3$–$Al(C_2H_5)_3$,	α-$TiCl_3$–$Al(C_2H_5)_3$,
$TiCl_4$–silica–$Al(C_2H_5)_3$,	δ-$TiCl_3$–$Al(C_2H_5)_3$,
$TiCl_4/MgCl_2$/silica–$Al(C_2H_5)_3$,	VCl_3–$Al(i$-$C_4H_9)_3$,
$VOCl_3/MgCl_2$/silica–$Al(C_2H_5)_3$	$TiCl_4/MgCl_2$/donor–$Al(C_2H_5)_3$/donor,
	$TiCl_4/TiCl_3$–$Al(C_2H_5)_2Cl$

the second components of the catalysts. Ziegler–Natta catalysts are traditionally classified with respect to their solubility in the polymerization medium (see examples in Table 1):

1. *Soluble (homogeneous) catalysts.* The catalysts in which both the starting transition metal compounds and all products of their interaction with organometallic compounds, including active centers, are soluble in the reaction medium. Homogeneous catalysts listed in Table 1 are widely used in laboratories and some of them have industrial significance.

2. *Pseudo-homogeneous catalysts.* The catalysts in which the starting transition metal compounds are soluble in the reaction medium but interact with organometallic compounds to form insoluble products. Some pseudo-homogeneous systems are important industrial catalysts for polymerization and copolymerization of alkenes and dienes (Table 1). The first industrial catalysts for ethylene polymerization discovered by Ziegler in 1953, $TiCl_4$–$Al(C_2H_5)_2Cl$ and $TiCl_4$–$Al(C_2H_5)_3$, belong to this class [1].

3. *Heterogeneous catalysts.* The catalysts in which both the starting transition metal derivatives and the products of their interaction with organometallic compounds are insoluble in the polymerization medium. This class includes the most important catalysts used in industry for alkene polymerization. All supported catalysts belong to this class.

COMPOSITION AND STRUCTURE OF ZIEGLER–NATTA CATALYSTS

Organometallic Compounds

All organometallic compounds used as components of Ziegler–Natta catalysts, including organoaluminum compounds, are liquids with high boiling points. They readily dissolve

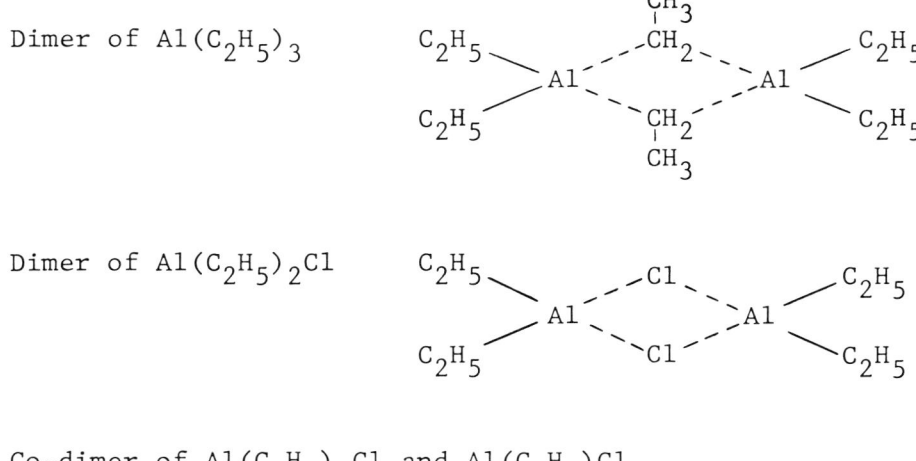

Dimeric structure of organoaluminum compounds

Scheme 1

in all saturated and aromatic hydrocarbons. Most of the organometallic compounds form stable dimers [7, pp. 93, 94, 101–104], as in Scheme 1. Only few such compounds, e.g., Al(i-C$_4$H$_9$)$_3$ and Zn(C$_2$H$_5$)$_2$, exist in the monomeric form.

All these organometallic compounds exhibit very high reactivity toward various organic and inorganic compounds [5, pp. 153–177; 7, pp. 205–213]. For example, they all react violently with water and alcohols, burn when exposed to air, etc. For this reason, the organometallic compounds are always handled with extreme caution, in inert atmosphere. As a rule, they are used not as pure substances but as dilute (~25 wt %) solutions in inert hydrocarbons such as hexane, heptane, or toluene.

Transition Metal Compounds

Some of transition metal compounds used as the components of Ziegler–Natta catalysts are heavy liquids or solids readily soluble in aliphatic and aromatic hydrocarbons. Their properties are listed in Table 2. Most of these compounds are very reactive with water and many other polar and hydroxyl group–containing reagents. Their handling also requires caution and is carried out in inert atmosphere.

Another group of transition metal derivatives used as Ziegler–Natta catalysts includes crystalline solids insoluble in hydrocarbons. Table 3 lists some parameters of their crystal structure. Four crystalline forms of TiCl$_3$ listed in the table were widely used for commercial synthesis of various polyolefins and alkene copolymers during the first 15 years after

Table 2 Properties of Some Soluble Transition Metal Compounds Used as Components of Ziegler–Natta Catalysts

Compound	Melting point (°C)	Boiling point (°C)	Application[a]
$TiCl_4$	−25	136.4	1,2
$Ti(OC_2H_5)_4$		122°/1 mm Hg	1,3
$Ti(Oi\text{-}C_3H_7)_4$	20	58°/1 mm Hg	1,3
$(C_5H_5)_2TiCl_2$	291		1
VCl_4	−28	148.5	1,4,5
$VOCl_3$	−77	126.7	1,4
V acetylacetonate	250 (dec.)		1,5

[a]Applications are as follows: (1) in combination with $Al(C_2H_5)_2Cl$, ethylene polymerization catalyst; (2) in combination with $Al(C_2H_5)_3$, catalyst for polymerization of various olefins and dienes; (3) in combination with $Al(C_2H_5)_3$, ethylene dimerization catalyst; (4) in combination with $Al(C_2H_5)_2Cl$ or $Al_2(C_2H_5)_3Cl_3$, catalyst for ethylene–propylene copolymerization; (5) in combination with $Al(C_2H_5)_2Cl$ at low temperatures, catalyst for syndiospecific propylene polymerization.

Table 3 Crystalline Structure of Solid Transition Metal Compounds Used in Ziegler–Natta Catalysis

Compound	Hexagonal cell parameter		Interatomic distance	
	a (nm)	b (nm)	M–Cl (nm)	M–M (nm)
$\alpha\text{-}TiCl_3$	0.612	1.750	0.250	0.354
$\gamma\text{-}TiCl_3$	0.613	1.740	0.250	0.354
$\delta\text{-}TiCl_3$	0.613	1.740	0.250	0.354
$\beta\text{-}TiCl_3$	0.627	0.582	0.245	0.291
VCl_3	0.601	1.734	0.245	0.347

Source: Refs. 5, 8.

the discovery of the catalysts and they still remain important in laboratory practice. Three of the $TiCl_3$ modifications, the α-form, the γ-form, and the δ-form, as well as $TiBr_3$, VCl_3, and $CrCl_3$, have the same elementary pattern shown in Figure 1: a sandwichlike, three-layered, flat sheet containing a layer of the transition metal cations surrounded by two layers of chlorine anions. The three $TiCl_3$ modifications differ only in the relative stacking arrangements of these layered aggregates. Very small $TiCl_3$ crystallites of the same structure are also present on the surface of many modern supported catalysts. All hydrocarbon-insoluble components of Ziegler–Natta catalysts easily react with water, alcohols, and oxygen.

The area of commercial Ziegler–Natta catalysis has greatly expanded since the late 1960s. Initially only a few industrially used catalysts existed, the three most important be-

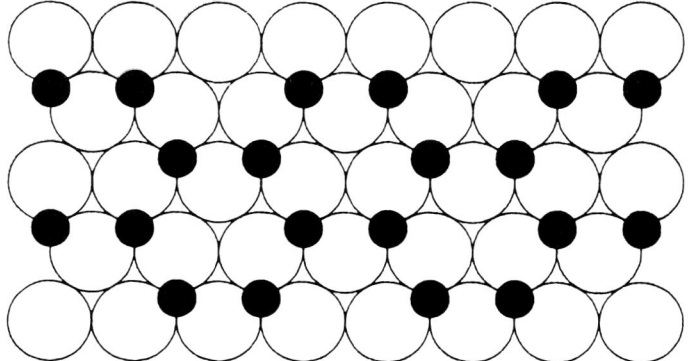

Figure 1 Elementary pattern of the MX_3 crystals (M = Ti, V, Cr; X = Cl, Br). The top layer of halogen atoms (open circles) in the three-layered structure is removed to expose the arrangement of the transition metal atoms (closed circles). (From Ref. 5.)

ing the $TiCl_4$–$Al(C_2H_5)_2Cl$ catalyst for polyethylene production, the δ-$TiCl_3 \cdot 0.33AlCl_3$–$Al(C_2H_5)_2Cl$ catalyst for production of polypropylene and other crystalline polyolefins, and the $VOCl_3$–$Al_2(C_2H_5)_3Cl_3$ catalyst for production of ethylene–propylene copolymers. However, continuous efforts in the development of new catalytic compositions brought about drastic improvements in catalyst performance. These efforts were directed toward two principal goals: increase of catalyst productivity and improvement of polymer quality. To achieve these goals, a large variety of catalytic modifiers were evaluated and special supports were developed. Together these improvements increased polymerization activity 100- to 200-fold. Today the majority of commercial catalysts for synthesis of polyolefins are supported. The most widely used support is $MgCl_2$ [5, pp. 105–109; 9]. Crystalline $MgCl_2$ has the same elementary structure as $TiCl_3$ (seen in Fig. 1) and, when titanium derivatives are deposited on the $MgCl_2$ surface, they can cocrystallize with $MgCl_2$. Companies that manufacture Ziegler–Natta catalysts and use them for polymer production have developed numerous proprietary catalytic compositions for alkene polymerization. The two following examples demonstrate the procedures usually involved in the synthesis of industrial-supported Ziegler–Natta catalysts.

Catalyst for Ethylene Polymerization [10]

Anhydrous $MgCl_2$ and $TiCl_4$ (taken in the 3:1 molar ratio) are dissolved in tetrahydrofuran. This solution is mixed with silica powder (catalyst carrier) that has been preliminarily dehydrated and treated with $Al(C_2H_5)_3$. Tetrahydrofuran is dried in a nitrogen flow, leaving silica impregnated with the mixture of $MgCl_2$ and $TiCl_4$. This product is subsequently treated with the solutions of $Al(C_2H_5)Cl_2$ and $Al(C_2H_5)_3$ in hexane, and the solvent is again removed to produce the final catalytic composition containing derivatives of three metals, titanium, magnesium, and aluminum. This catalyst is used in combination with $Al(C_2H_5)_3$ for polymerization of ethylene and for its copolymerization with other alkenes.

Catalyst for Propylene Polymerization [11]

Anhydrous $MgCl_2$ is ball-milled for 70 hr in inert atmosphere in the presence of an organic diester (e.g., the ester of maleic or malonic acid) and $TiCl_4$, at the Mg:Ti:ester molar ratio of 12:1:1. The ground mixture is treated with dichloroethane at 80°C for 4 hr, filtered, carefully washed with heptane, and dried. This catalyst, when used for propylene polymerization in combination with $Al(C_2H_5)_3$, exhibits very high activity but produces polymer of very poor quality, containing only 20–50% of the desirable crystalline fraction. However, its performance can be greatly improved if, in addition to $Al(C_2H_5)_3$, it is treated with a small quantity of a modifier, such as dimethoxydiphenylsilane, $(C_6H_5)_2Si(OCH_3)_2$, or diethoxydiphenylsilane, $(C_6H_5)_2Si(OC_2H_5)_2$, at the Al:Si molar ratio of 5:20. The introduction of such modifiers in the process of propylene polymerization brings about an increase of the yield of crystalline polypropylene to 95–97%. The same catalytic compositions are used for synthesis of other crystalline polyolefins, such as polybutene and polymethylpentene.

Ethylene–propylene elastomers are produced in copolymerization reactions performed in the presence of vanadium-based catalysts, such as $VOCl_3$–$Al(i$-$C_4H_9)_2Cl$ or $VOCl_3$–$Al_2(C_2H_5)_3Cl_3$, which are modified by chloroorganic compounds [3; 4, pp. 566–572; 12; 13]. The reaction is usually carried out in a hydrocarbon solution at 35–50°C (the amorphous polymer remains dissolved in the reaction medium). The same technique is used for tertiary polymerization of ethylene, propylene, and unconjugated dienes, such as ethylidenenorbornene or dicyclopentadiene. These polymer products contain double bonds in their chains and they can be vulcanized with sulfur to produce crosslinked synthetic rubbers [14].

Polymerization of the dienes butadiene and isoprene requires the use of special Ziegler–Natta catalytic systems [3; 4, pp. 130–148; 15; 16]. Crystalline *trans*-1,4-polybutadiene is produced by polymerization of butadiene with the pseudo-homogeneous $TiCl_4$–$Al(C_2H_5)_3$ catalyst at 20–30°C. Introduction of iodine derivatives in this catalytic system drastically alters its behavior: after the modification the catalyst polymerizes butadiene to the amorphous *cis*-1,4 polymer, which finds wide applications for tire manufacture. Several methods for introduction of iodine into these catalytic systems exist, e.g., the use of such catalytic systems as TiI_4–$Al(i$-$C_4H_9)_3$, $TiCl_4$–$Al(C_2H_5)_2I$, and the ternary, $TiCl_4$–$Al(i$-$C_4H_9)_3$–I_2 system. The last catalytic system was chosen for commercial manufacture of *cis*-1,4-polybutadiene. *Cis*-1,4-polyisoprene has structure and properties comparable to those of natural rubber. This polymer can be produced in the hexane medium with the $TiCl_4$–$Al(i$-$C_4H_9)_3$ catalyst, at the Al:Ti ratio of ~1.0, in the 20–50°C range.

POLYMERIZATION MECHANISM OF ZIEGLER–NATTA CATALYSTS

Active centers of Ziegler–Natta catalysts are formed in the process of interaction between two components of the catalysts, transition metal compounds and organometallic compounds. This interaction usually proceeds at a very high rate and includes several consecutive chemical reactions. The most important of these reactions is the exchange of halogen atoms in the transition metal compounds and alkyl groups in the organometallic compounds [4, 325–339; 5, 109–118]:

$$\text{Ti–Cl} + \text{C}_2\text{H}_5\text{–Al} \Big< \rightarrow \text{Ti–C}_2\text{H}_5 + \text{Cl–Al} \Big< \tag{1}$$

The chemical bond formed in Reaction 1 between the transition metal, such as titanium or vanadium, and carbon is usually unstable. It breaks either spontaneously or in reactions with an excess of organoaluminum compounds. For example, the interaction between

$TiCl_4$ and $Al(C_2H_5)_2Cl$ in solution at moderate temperatures can be described as a sequence of two reactions [4, pp. 325–339]:

Alkylation reaction:

$$TiCl_4 + Al(C_2H_5)_2Cl \rightarrow Cl_3TiC_2H_5 + Al(C_2H_5)Cl_2 \tag{2}$$

Decomposition of the Ti–C bond:

$$2Cl_3TiC_2H_5 \rightarrow 2TiCl_3 + C_2H_6 + C_2H_4 \tag{3}$$

Chemists have identified all final products of these reactions: solid $TiCl_3$ (the β modification), $Al(C_2H_5)Cl_2$, ethane, and ethylene. Similarly, interaction between $(C_2H_5)_2TiCl_2$ and $Al(C_2H_5)_2Cl$ in aromatic solvents proceeds in three stages [5, pp. 373–379; 17]:

Complex formation:

$$(C_5H_5)_2TiCl_2 + Al(C_2H_5)_2Cl \rightarrow (C_4H_5)_2TiCl_2 \cdot Al(C_2H_5)_2Cl \tag{4}$$

Alkylation reaction:

$$(C_5H_5)_2TiCl_2 \cdot Al(C_2H_5)_2Cl \rightarrow (C_5H_5)_2TiCl(C_2H_5) \cdot Al(C_2H_5)Cl_2 \tag{5}$$

Decomposition of the Ti–C bond:

$$2(C_5H_5)_2TiCl(C_2H_5) \cdot Al(C_2H_5)Cl_2 \rightarrow 2(C_5H_5)_2TiCl \cdot Al(C_2H_5)Cl_2$$
$$+ C_2H_6 + C_2H_4 \tag{6}$$

Both sequences of reactions, Reactions 2–3 and 4–6, result in reduction of the transition metal: its valence state decreases from 4+ to 3+.

Similar reactions occur when organoaluminum compounds contact solid transition metal components of Ziegler–Natta catalysts. These reactions also include formation of chemical bonds between transition metal atoms and carbon atoms. For example, interaction between solid $TiCl_3$ and $Al(C_2H_5)_3$ can be described as [4, pp. 325–339; 5, pp. 109–118]

$$TiCl_3 + Al(C_2H_5)_3 \rightarrow Cl_2TiC_2H_5 + Al(C_2H_5)_2Cl \tag{7}$$

However, in such cases the reaction is usually confined to the surface of the $TiCl_3$ crystals. Experiments with radioactive-labeled $Al(^{14}C_2H_5)_3$ confirmed the existence of the Ti–C bonds on the surface of the catalysts [3; 5, pp. 109–118].

Although the transition metal–carbon bonds are not very stable and can disappear in several reactions (such as Reaction 3 or 6), a significant fraction of these bonds survives under conditions typical for polymerization of alkenes and dienes: at temperatures in the 30–100°C range and reaction times from 0.5 to 5 hr. These transition metal–carbon bonds are the principal constituents of active centers in all Ziegler–Natta catalysts. Instability of the transition metal–carbon bonds strongly affects performance of Ziegler–Natta catalysts: activity of most catalysts decreases with time.

From the point of view of Ziegler–Natta catalysis the most important feature of the transition atom–carbon bonds is their ability to react with double bonds of alkene molecules. The reaction proceeds in two stages [3; 4, pp. 325–339; 5, pp. 373–379]:

1. Formation of a complex between an alkene molecule and the transition metal atom M (alkene coordination):

$$M\text{—}C_2H_5 + CH_2\text{=}CH\text{—}R \rightarrow \underset{\underset{CH_2\text{=}CHR}{\uparrow}}{M\text{—}C_2H_5} \quad (8)$$

2. Insertion of the coordinated alkene molecule into the metal–carbon bond M—C:

$$\underset{\underset{CH_2\text{=}CHR}{\uparrow}}{M\text{—}C_2H_5} \rightarrow M\text{—}CH_2\text{—}CHR\text{—}C_2H_5 \quad (9)$$

This alkene insertion reaction is the principal reaction responsible for the activity of all Ziegler–Natta catalysts. Repetitive insertions of alkene molecules into the metal–carbon bonds result in lengthening of the alkyl groups attached to the transition metal atoms. The alkyl groups can grow to include many thousands of alkene units, i.e., they become polymer molecules.

Experimental proofs of alkene insertion reactions were presented in several specialized experiments. For example, the soluble complex formed in Reaction 5 contains the Ti–C_2H_5 bond. The ^{13}C spectroscopic study revealed that when this complex contacts the ^{13}C-labeled ethylene at low temperatures, the ethylene insertion into the Ti–C_2H_5 bond indeed takes place [17, 18]:

$$\begin{array}{c} C_5H_5 \quad C_2H_5 \\ \diagdown \quad \diagup \\ Ti \\ \diagup \quad \diagdown \\ C_5H_5 \quad Cl \cdot Al(C_2H_5)Cl_2 \end{array} + {}^{13}CH_2\text{=}{}^{13}CH_2 \rightarrow \begin{array}{c} C_5H_5 \quad {}^{13}CH_2\text{—}{}^{13}CH_2\text{—}C_2H_5 \\ \diagdown \quad \diagup \\ Ti \\ \diagup \quad \diagdown \\ C_5H_5 \quad Cl \cdot Al(C_2H_5)Cl_2 \end{array} \quad (10)$$

Several spectroscopic studies in which the original alkyl group attached to the transition metal atom was labeled with ^{13}C provide further proof of the alkene insertion reaction [5, pp. 373–379; 19]. When $TiCl_3$ reacts with $Al(^{13}CH_3)_3$ or with $Al(^{13}CH_2\text{–}CH_3)_3$, the Ti–$^{13}CH_3$ or Ti–$^{13}CH_2$–CH_3-labeled bonds are formed (in Reaction 7). Polymerization of propylene with such catalysts results in the formation of polymer chains with specifically labeled chain ends:

$$\text{Ti–}{}^{13}CH_3 + CH_2\text{=}CH\text{—}CH_3 \rightarrow \text{Ti—}CH_2\text{—}CH(CH_3)\text{—}{}^{13}CH_3 \quad (11)$$

$$\text{Ti–}{}^{13}CH_2\text{—}CH_3 + CH_2\text{=}CH\text{—}CH_3 \rightarrow \text{Ti—}CH_2\text{—}CH(CH_3)\text{—}{}^{13}CH_2\text{—}CH_3 \quad (12)$$

The ^{13}C NMR spectroscopic technique not only allowed identification of the labels in the polymer chains but also confirmed the detailed structure of the chain ends shown in Reactions 11 and 12.

Active centers of the heterogeneous Ziegler–Natta catalysts are situated on the surfaces of the catalyst particles, at distances of ~2–10 nm [4, pp. 180–211; 5, 200–205]. Figure 2 shows schematically the structure of such a center with the growing polypropylene chain attached to the titanium atom and with the coordinated propylene molecule positioned for insertion into the Ti–C bond. The closest neighbors of the transition metal atoms in these centers are halogen atoms, other transition metal atoms, adsorbed organoaluminum compounds, and, in the case of supported catalysts, atoms belonging to the support.

The kinetic mechanism of alkene polymerization consists of several steps. The first step is the formation of active centers: a series of reactions between catalyst components, including the formation of the transition metal–carbon bonds (as in Reactions 2, 5, and 7).

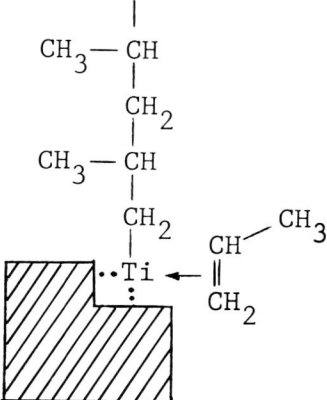

Figure 2 The scheme of the active center of a heterogeneous Ziegler–Natta catalyst.

The next reaction step consists of the insertion of the first alkene molecule in the transition metal–carbon bond. This is the chain initiation reaction (Reactions 8 and 9). It is followed by thousands of consecutive insertion reactions of alkene molecules (chain growth reactions):

$$M-(CH_2-CHR)_n-H + CH_2=CH-R \rightarrow M-(CH_2-CHR)_{n+1}-H \quad (13)$$

With titanium-based catalysts, all of these insertion reactions proceed in such a way that the CH_2 group of the alkene molecule faces, after insertion, the transition metal atom (primary insertion reaction) [4, pp. 325–339; 5, 373–379; 19]. This regularity of the insertion reactions in the process of polymer chain growth ensures the regular, head-to-tail structure of polymer chains. However, some vanadium-based polymerization catalysts reverse the pattern of the insertion reactions, especially at low temperatures. In such cases the CHR groups of the inserted alkene molecules face the transition metal atoms (secondary alkene insertion) [19].

Periodically, the growing polymer chains disengage from the transition metal atoms. Several reactions are responsible for the separation of polymer chains from the active centers [3; 4, pp. 244–258; 5, pp. 16–25]. One of them is the chain transfer to monomer:

$$M-CH_2-CHR-polymer + CH_2=CH-R \rightarrow M-CH_2-CH_2R$$
$$+ CH_2=CR-polymer \quad (14)$$

Another reaction limiting the polymer chain length is the exchange reaction between the chain-growing center and an organometallic compound:

$$M-CH_2-CHR-polymer + Al(C_2H_5)_3 \rightarrow M-C_2H_5$$
$$+ (C_2H_5)_2Al-CH_2-CHR-polymer \quad (15)$$

The polymer chain attached to the aluminum atom cannot further increase its length under typical polymerization conditions. The third reaction resulting in polymer chain disengagement from the active center is the β-elimination reaction of the growing polymer center:

$$M-CH_2-CHR-polymer \rightarrow M-H + CH_2=CR-polymer \quad (16)$$

Usually all three reactions limiting chain growth (Reactions 14, 15, and 16) occur very infrequently compared to the chain growth reaction (Reaction 13) and, as a result, polymer molecules produced by most modern Ziegler–Natta catalysts consist of a very large number of monomer units (50,000–100,000), i.e., they have very high molecular weight. Such polymers cannot be effectively processed with the existing equipment. The problem of excessively long polymer chains was solved when chemists discovered that hydrogen, if introduced into the polymerization system, reacts with growing polymer chains:

$$M—CH_2—CHR—polymer + H_2 \rightarrow M—H + CH_3—CHR—polymer \qquad (17)$$

Because the rate of Reaction 17 is proportional to the hydrogen pressure in a reactor, the use of hydrogen provides an effective independent means of molecular weight control. Today hydrogen is the principal chain-terminating agent used in industrial processes utilizing Ziegler–Natta catalysts [3; 4, pp. 244–258; 5, pp. 16–25; 10; 11].

All Ziegler–Natta catalysts are very sensitive to the overwhelming majority of organic and inorganic substances. Only few of the compounds, as discussed earlier, activate or modify the catalysts; all others poison them [8]. However, the catalysts have a built-in protective mechanism due to the presence of organoaluminum compounds in the reaction medium. The organoaluminum compounds very rapidly react with most common catalytic poisons, such as water, oxygen, and alcohols, and convert them into other compounds, usually aluminum alkoxides R_2AlOR' [7, 8] that are less detrimental to catalyst performance. Some other poisons, e.g., carbon monoxide, carbon dioxide, acetylene, and allene, do not interact with organoaluminum compounds. If small quantities of these poisons are deliberately added to the reactor, they immediately halt the polymerization reaction. The mechanism of this poisoning has been well established [8]. All these compounds form complexes with transition metal atoms. The complexes have a structure similar to that of alkene–transition metal complexes (Reaction 8, Fig. 2), but the complexes with these poisons are more stable. As a result, such poisons interfere with the formation of the alkene complexes and with the alkene insertion reactions (Reaction 9). This type of poisoning is nearly completely reversible: when the poisons are removed from the system, the active centers resume the polymerization reaction [8].

COPOLYMERIZATION REACTIONS WITH ZIEGLER–NATTA CATALYSTS

Ziegler–Natta catalysts easily copolymerize various alkenes [3; 4, pp. 563–584; 5, pp. 61–87; 12; 13; 20]. In copolymerization, two or more different alkene molecules are competitively inserted into the transition metal–carbon bonds of the growing centers. In the simplest case, when two alkenes, $CH_2=CH—R'$ and $CH_2=CH—R''$, copolymerize, four chain growth reactions should be considered instead of a single chain growth reaction in homopolymerization processes (Reaction 13):

$$M—CH_2—CHR'—polymer + CH_2=CH—R' \xrightarrow{k_{11}}$$
$$M—CH_2—CHR'—CH_2—CHR'—polymer \qquad (18)$$

$$M—CH_2—CHR'—polymer + CH_2=CH—R'' \xrightarrow{k_{12}}$$
$$M—CH_2—CHR''—CH_2—CHR'—polymer \qquad (19)$$

$$M—CH_2—CHR''—polymer + CH_2=CH—R'' \xrightarrow{k_{22}}$$
$$M—CH_2—CHR''—CH_2—CHR''—polymer \qquad (20)$$

$$M—CH_2—CHR''—polymer + CH_2=CH—R' \xrightarrow{k_{21}}$$
$$M—CH_2—CHR'—CH_2—CHR''—polymer \qquad (21)$$

Two of the reactions represent homopolymerization growth processes: those of the $CH_2{=}CH{-}R'$ monomer (Reaction 18, reaction rate constant k_{11}) and of $CH_2{=}CH{-}R''$ (Reaction 20, k_{22}); two other reactions represent cross-growth processes (Reactions 19 and 21, k_{12} and k_{21}). The values of all four rate constants in this scheme are different. This means that the rate of the insertion reaction of a particular alkene depends not only on the structure of the alkene molecule (e.g., on the type of the alkyl group R in alkene $CH_2{=}CH{-}R$) but also on the structure of the last monomer unit attached to the transition metal atom. The four rate constants in Reactions 18–21 are traditionally grouped into two ratios (reactivity ratios): $r_1 = k_{11}/k_{12}$ and $r_2 = k_{22}/k_{21}$ [4, pp. 563–584; 5, pp. 61–87; 20]. The reactivity ratios for different alkene pairs vary widely. For example, in ethylene–propylene copolymerization with vanadium-based catalysts, the r_1 value (the ethylene reactivity ratio) is ~10–15 and the r_2 value (propylene reactivity ratio) is ~0.04–0.06 [4, pp. 563–584; 12; 13].

The reactivity ratios represent a convenient measure of alkene reactivities in polymerization reactions with Ziegler–Natta catalysts. Alkene reactivities mostly depend on the

Table 4 Relative Reactivities of Olefins in Polymerization Reactions with Ziegler–Natta Catalysts, Compared to Propylene Reactivity

Olefin	Structure	Relative reactivity
Ethylene	$CH_2{=}CH_2$	20
Propylene	$CH_2{=}CH{-}CH_3$	1 (standard)
1-Butene	$CH_2{=}CH{-}C_2H_5$	0.40
1-Pentene	$CH_2{=}CH{-}C_3H_7$	0.35
1-Hexene	$CH_2{=}CH{-}C_4H_9$	0.25
1-Decene	$CH_2{=}CH{-}C_8H_{17}$	0.15
1-Octadecene	$CH_2{=}CH{-}C_{16}H_{33}$	0.12
3-Methyl-1-butene	$CH_2{=}CH{-}CH(CH_3){-}CH_3$	0.042
3-Methyl-1-pentene	$CH_2{=}CH{-}CH(CH_3){-}C_2H_5$	0.045
4-Methyl-1-pentene	$CH_2{=}CH{-}CH_2{-}CH(CH_3){-}CH_3$	0.15
4-Methyl-1-hexene	$CH_2{=}CH{-}CH_2{-}CH(CH_3){-}CH_2{-}CH_3$	0.13
5-Methyl-1-hexene	$CH_2{=}CH{-}CH_2{-}CH_2{-}CH(CH_3){-}CH_3$	0.42
Vinylcyclohexane	$CH_2{=}CH{-}cyclo{-}C_6H_{11}$	0.012
Styrene	$CH_2{=}CH{-}C_6H_5$	~0.2
Vinylnaphthalene	$CH_2{=}CH{-}C_{10}H_7$	~0.1

Source: Data from Refs. 5, 21.

structure of alkyl groups attached to the double bond in alkene molecules [5, pp. 61–87; 20; 21]. Table 4 lists relative reactivities of various alkenes, as compared with propylene reactivity. Analysis of the table shows that for alkenes with linear alkyl groups, from ethylene to 1-octadecene, the reactivity decreases with the increase in the length of an alkyl group. Branched alkyl groups reduce reactivity of the double bond in the polymerization reaction much more effectively than linear alkyl groups of the same size. Especially low reactivity was found for those alkenes that have specifically branched alkyl groups, with the branching point positioned next to the double bond, as in 3-methyl-1-butene and vinylcyclohexane. In addition to alkenes and dienes, Ziegler–Natta catalysts can polymerize vinylaromatic compounds such as styrene. Styrene is a very reactive monomer in radical and cationic polymerization reactions, but it exhibits very modest reactivity with Ziegler–Natta catalysts compared to reactivities of such alkenes as ethylene or propylene.

Alkene copolymerization reactions have great commercial significance. The reactions provide the basis for the manufacture of linear low-density polyethylene (copolymers of ethylene containing 3–5 mol % of 1-butene, 1-hexene, 1-octene, or 4-methyl-1-pentene); impact-resistant polypropylene (propylene–ethylene block copolymer); commercial polymethylpentene (copolymer of 4-methyl-1-pentene containing 3–5% of a higher linear alkene); ethylene–propylene elastomers; and ethylene–propylene synthetic rubbers (copolymers of ethylene, propylene, and unconjugated dienes). The choice of catalysts used for alkene copolymerization depends on the type of copolymers. Copolymers of ethylene containing low concentrations of higher alkenes are produced with the same catalysts as those used for ethylene homopolymerization. Block copolymers of propylene and copolymers of 4-methyl-1-pentene are synthesized with the same catalysts as those used for production of homopolymers of propylene and methylpentene. Ethylene–propylene elastomers with the ethylene content of 50–70 wt % are produced with catalysts based on vanadium compounds, as described earlier.

POLYMERIZATION OF DIENES

Conjugated dienes, such as butadiene and isoprene, can be easily polymerized with Ziegler–Natta catalysts. Diene polymers contain several different structures of monomeric units [3; 4, pp. 366–380; 15; 16]. For example, five different structures can be formed in polymerization of isoprene, CH_2=$C(CH_3)$—CH=CH_2, as seen in Scheme 2. Two of the polymer structures duplicate natural products: cis-1,4-polyisoprene is identical to natural rubber (Hevea) and trans-1,4-polyisoprene is identical to natural gutta-percha (Balata). Different types of Ziegler–Natta catalysts are capable of producing four structures, cis-1,4-; trans-1,4-; 1,2-; and cyclized polymers, as noted in Table 5 [10, 11]. Butadiene, CH_2=CH—CH=CH_2, can also be polymerized to similar structures (in this case the 1,2- and 3,4-structures shown in Scheme 2 are identical), as seen in Table 6 [3; 4, pp. 366–380; 15; 16].

The mechanism of diene polymerization with Ziegler–Natta catalysts has several features in common with the mechanism of alkene polymerization. In both cases, the driving force of the polymerization reaction is monomer insertion into the transition metal–carbon bond. However, the reaction of diene polymerization is more mechanistically complex. In the case of 1,4- polymerization, the growing chain end formally has the following structure (called η^1 allylic form):

M—CH_2—CH=CH—CH_2—polymer

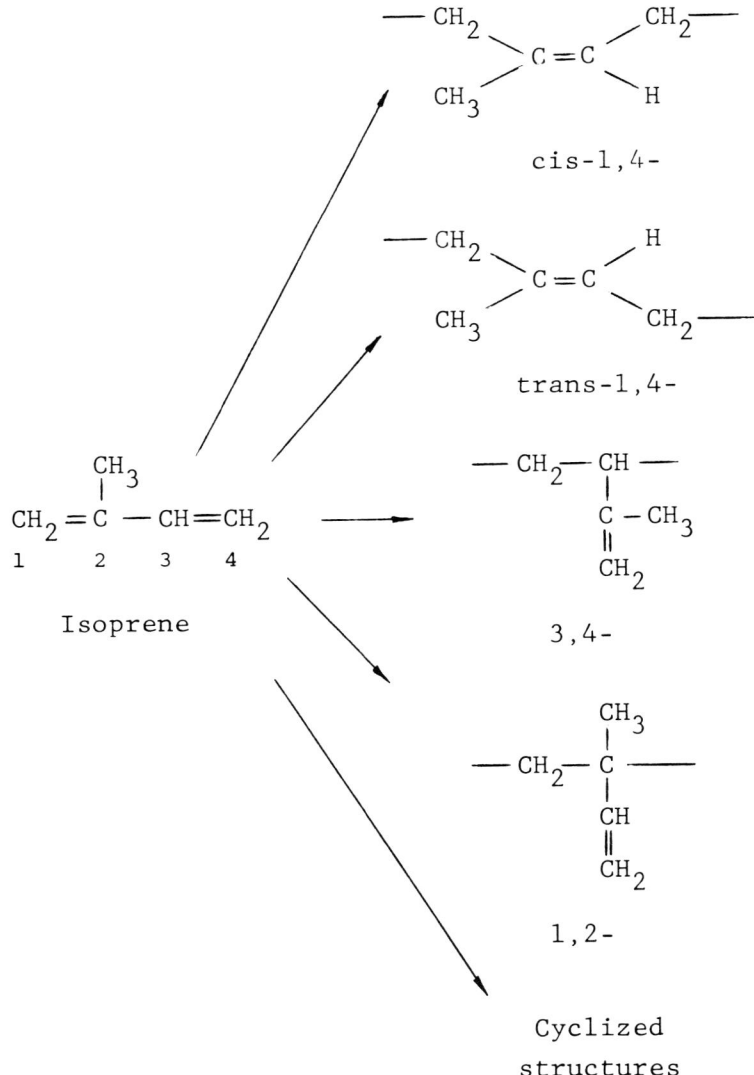

Structures of monomer units formed in polymerization of isoprene

Scheme 2

Table 5 Isoprene Polymerization

Monomer unit structure	Catalyst
trans-1,4-	$TiCl_4$–$A(C_2H_5)_3$; Al:Ti ~ 2
cis-1,4-	$TiCl_4$–$A(C_2H_5)_3$; Al:Ti = 0.5–1.5
3,4-	$Ti(Oi$-$C_3H_7)_4$–$Al(C_2H_5)_3$
Cyclized	$TiCl_4$–$A(C_2H_5)_3$; Al:Ti > 3

Table 6 Butadiene Polymerization

Monomer unit structure	Catalyst
trans-1,4-	$TiCl_3$–$A(C_2H_5)_3$; VCl_4–$Al(i$-$C_4H_9)_3$
cis-1,4-	TiI_4–$A(i$-$C_4H_9)_3$; $CoCl_2$–pyridine–$Al(i$-$C_4H_9)_3$
1,2-	Cr acetylacetonate–$Al(C_2H_5)_3$
	$Cr(CO)_5$–pyridine–$Al(C_2H_5)_3$

However, according to numerous spectroscopic studies, interaction of the transition metal atom M with the double bond of the attached monomer unit in the η^1 form results in the rapid equilibrium isomerization of the growing chain end to the conjugated η^3 allylic form [22]:

$$M-CH_2-CH=CH-CH_2-\text{polymer} \quad \rightarrow \quad M \leftarrow \begin{array}{c} CH_2 \\ \diagdown \\ CH \\ \diagup \\ CH-CH_2-\text{polymer} \end{array} \quad (22)$$

η^1 allylic form $\qquad\qquad\qquad\qquad\qquad$ η^3 allylic form

The η^3 allylic form is more thermodynamically stable and its isomerization to the η^1 allylic form occurs only if a diene molecule coordinates with the transition metal atom. The η^3 allylic form can exist in two isomeric states, anti and syn. Scheme 2 shows the structures of these isomers (**b** and **b′**) and the reactions that lead to the formation of different monomeric units in the case of butadiene polymerization [22]. When a diene molecule approaches a polymerization center, it can form two different complexes with it, with the participation of either both double bonds or only one (**a** and **a′**). The two-bond coordination **a′** produces the *anti*-η^3 form **b′**, which, after the following monomer coordination, forms the *cis*-CH=CH bond **c′**, i.e., the *cis*-1,4- polymerization step occurs. Two routes can account for the formation of the *trans*-1,4 units **c**. In the first, the monomer molecule forms the one-bond coordination complex **a** that, after monomer insertion, produces the *syn*-η^3 form **b** and eventually the *trans*-CH=CH bond **c**. The second route, **a′** → **b′** → **b** → **c**, includes the *anti* → *syn* isomerization of the η^3 form, which is

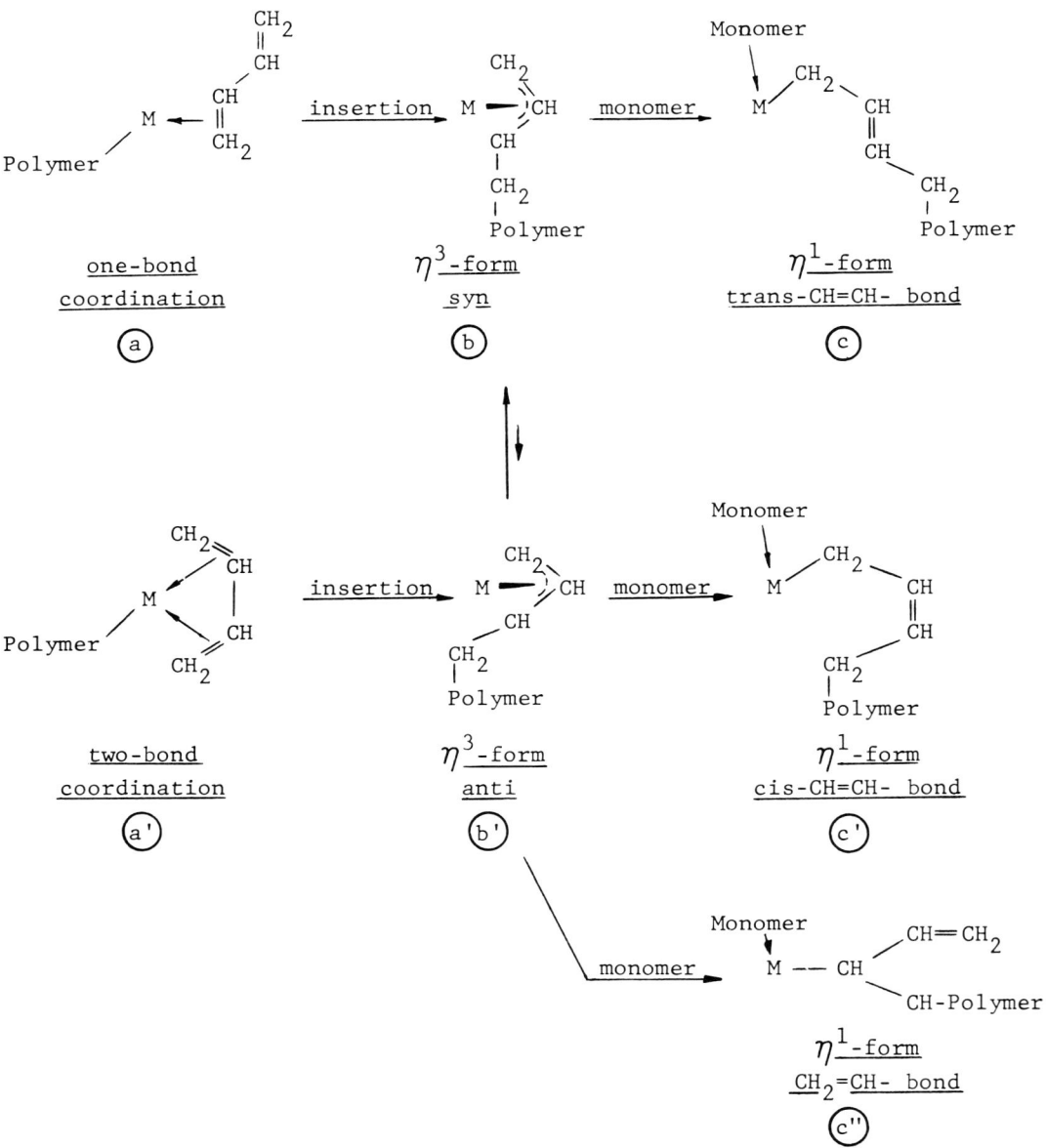

Mechanism of diene polymerization with Ziegler-Natta catalysts

Scheme 3

thermodynamically favorable. Finally, the 1,2- polymerization of dienes **c″** occurs when, in the process of the η^3- → η^1- isomerization under the influence of monomer, the metal atom M in the η^3 form, **b** or **b′**, attaches itself to the CH group rather than to the CH$_2$ group.

STEREOSPECIFIC POLYMERIZATION REACTIONS

Giulio Natta and his co-workers discovered in their early studies that alkene polymers produced in reactions involving heterogeneous catalysts are highly crystalline [2, 3]. This discovery had very important consequences. Crystalline alkene polymers (polyolefins) have very attractive mechanical properties and some of them (polypropylene, polybutene, and polymethylpentene) are widely used as general-purpose and engineering materials [3, 4]. All crystalline polymers of alkenes are also excellent dielectrics and find numerous applications as insulators.

Detailed crystallographic studies of crystalline polyolefins revealed that their crystallinity stems from a specific feature of their molecular structure called stereoregularity [2; 3; 4, pp. 33–35; 5, pp. 221–226]. Stereoregularity is the existence of a special order in the spatial structure of a polymer chain. This spatial order is twofold. First, during polymer synthesis monomer units are linked with each other in some definite regular manner which depends on the properties of the catalyst and on polymerization conditions. Second, these regularly linked polymer chains spontaneously arrange themselves in a specific, spatially regular form, usually in a helix form.

Regularity of the first type, that of monomer unit linking, arises in the course of polymerization. It cannot be altered or destroyed in any subsequent physical transformation of a polymer such as melting or stretching. A polymer chain of any alkene polymer has a backbone with the chemical structure

$$-CH_2-\underset{\underset{R}{|}}{CH}-CH_2-\underset{\underset{R}{|}}{CH}-CH_2-\underset{\underset{R}{|}}{CH}-CH_2-\underset{\underset{R}{|}}{CH}-CH_2-\underset{\underset{R}{|}}{CH}-CH_2-\underset{\underset{R}{|}}{CH}-$$

Here R are various side groups attached to the CH groups: CH$_3$ in polypropylene, C$_2$H$_5$ in polybutene, etc. All carbon atoms in these polymer molecules have the tetrahedral spatial arrangement of their four neighbors, atoms C or H. If one draws the polymer backbone in a flat zigzag form (Fig. 3), two regular patterns of the side group positions are easily envisaged [3; 4, pp. 33–35; 5, pp. 221–226]:

1. All side groups R can lie on one side of the zigzag plane (above or below it), as in Figure 3*a*. Such polymers are called *isotactic*. All crystalline polymers of alkenes produced with heterogeneous Ziegler–Natta catalysts are isotactic, as was concluded after numerous x-ray and NMR studies of the polymers.
2. The side groups R can alternatively occupy positions above and below the backbone plane (Fig. 3*b*). Such polymers are called *syndiotactic*. Two syndiotactic crystalline alkene polymers, syndiotactic polypropylene and syndiotactic polystyrene, are known to exist.

If no regular pattern could be found in the arrangement of the side groups the polymer is said to be *atactic* (Fig. 3*c*). Atactic polyolefins are usually produced as by-products in the synthesis of isotactic polymers. All atactic polyolefins are amorphous rubberlike or solid materials.

The phenomenon of stereoregularity is not limited to alkene polymers. Many chemi-

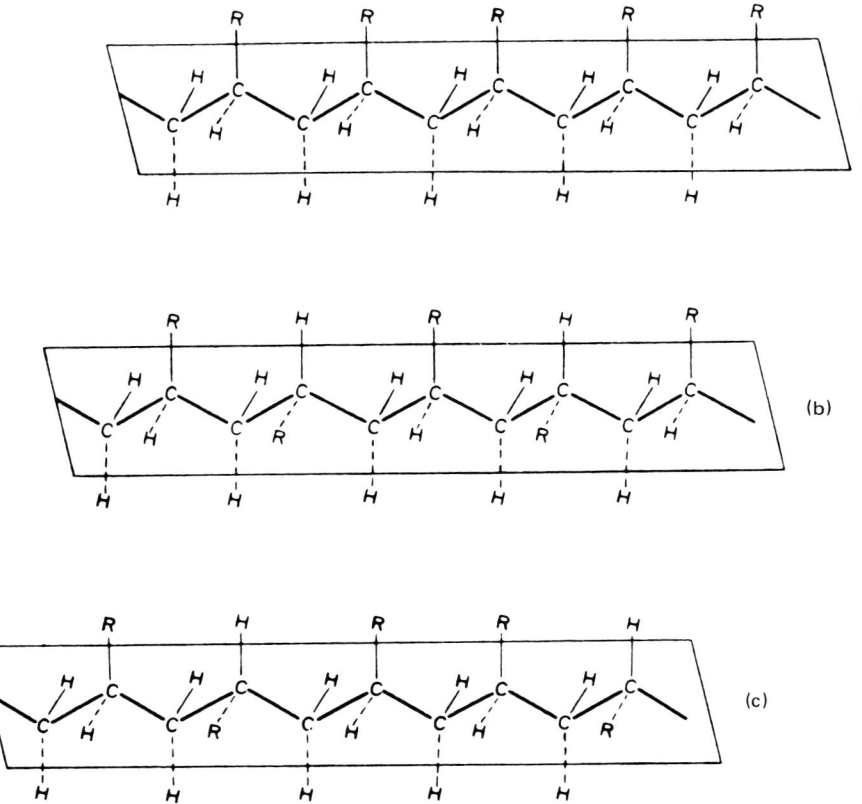

Figure 3 Representation of polyolefin chains in the flat zigzag form: (*a*) isotactic polymer; (*b*) syndiotactic polymer; (*c*) atactic polymer.

cally regularly built polymers—as polystyrene, polyvinyl chloride, polyacrylates, polymethacrylates, polypropylene oxide, etc.—can also exist in the stereoregular isotactic or syndiotactic forms. The same considerations apply to 1,2-polybutadiene and 3,4-polyisoprene. From the structural point of view, these diene polymers can be regarded as alkene polymers with side groups —CH=CH$_2$ and —CH(CH$_3$)=CH$_2$. They too, in principle, can exist in isotactic and syndiotactic forms.

Regularity of the second type refers to the real spatial structure of polymer chains. The flat structure of isotactic polyolefins shown in Figure 3a is entirely imaginary. In reality, such chains are always coiled in a helix [3; 5, pp. 221–226]. The form of the helix depends mainly on the size and the shape of the side group R. Usually, for a given R, only one helical type is possible.

Isotactic polyolefins form three types of helixes, as seen in Figure 4. The helixes differ in the number of monomer units per one turn of the helix: three units (found in polypropylene, poly-1-butene, poly-1,2-butadiene, and polystyrene); 3.5 units per turn (seven monomer units per two turns of the helix, in poly-4-methyl-1-pentene); and four units per turn (in poly-3-methyl-1-butene and polyvinylcyclohexane).

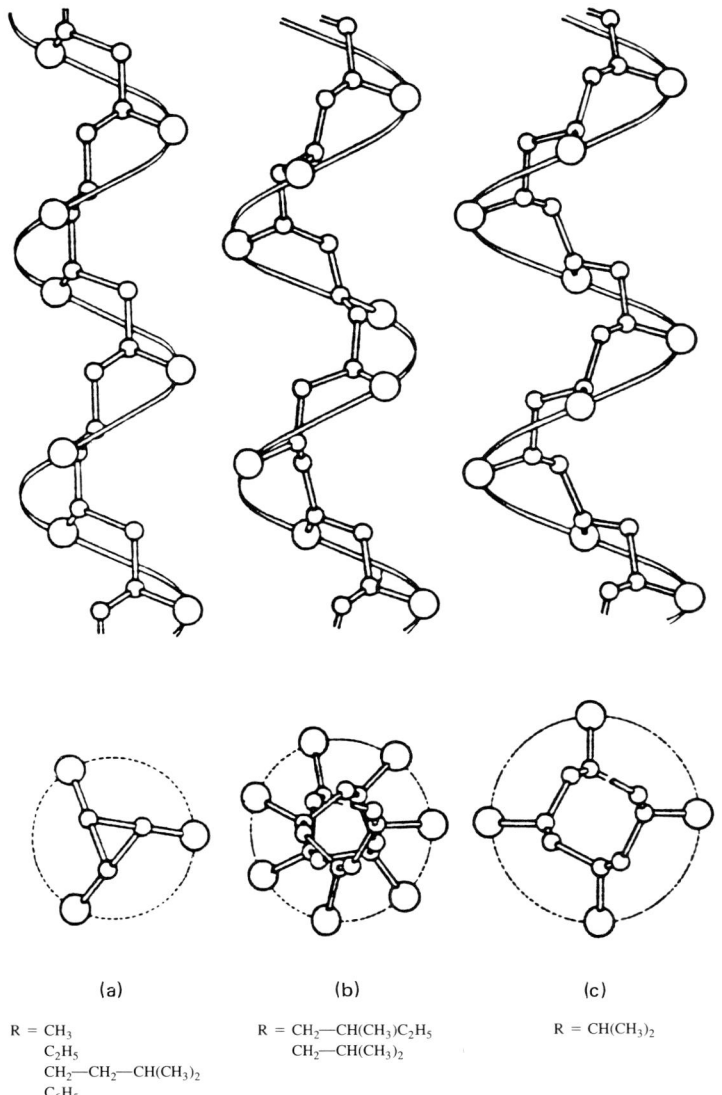

(a)　　　　　　　(b)　　　　　　　(c)

R = CH$_3$　　　　R = CH$_2$—CH(CH$_3$)C$_2$H$_5$　　R = CH(CH$_3$)$_2$
C$_2$H$_5$　　　　　CH$_2$—CH(CH$_3$)$_2$
CH$_2$—CH$_2$—CH(CH$_3$)$_2$
C$_6$H$_5$

Figure 4 Models of chain helixes of various isotactic polyolefins. Open circle = R.

All polyolefin helixes easily aggregate in three-dimensional regular structures—crystals. Sometimes the same helixes can, depending on crystallization conditions, form different types of crystals, but usually only one of the crystal structures is thermodynamically stable. Table 7 lists parameters of polyolefin crystal structures and their melting points. Included in the table are linear polyethylene and isotactic polystyrenes. Polyethylene has no regularly positioned side groups in its chains and the definition of stereoregularity does not apply to this polymer. Polystyrene and its derivatives, such as poly(methylstyrenes) are not traditionally regarded as polyolefins. However, these poly-

Table 7 Crystalline Structures and Properties of Polyolefins

Polymer	Structural unit	Crystal system	Density Crystal	Density Polymer	Melting point (°C)
Polyethylene	—CH_2—CH_2—	Orthrombic $a = 0.742$, $b = 0.494$, $c = 0.255$ nm (stable form)	1.00	0.96	138
		Monoclinic $a = 0.809$, $b = 0.253$, $c = 0.479$ nm, $\beta = 108°$ (metastable form)	0.988		
Isotactic polypropylene	—CH_2—CH— | CH_3	Monoclinic $a = 0.665$, $b = 2.096$, $c = 0.65$ nm, $\beta = 99°$ (stable form)	0.936	0.92	176
		Hexagonal $a = 0.191$, $c = 0.65$ nm (metastable form)	0.92		
Syndiotactic polypropylene	—CH_2—CH— | CH_3	Orthorhombic $a = 1.45$, $b = 0.56$, $c = 0.74$ nm	0.93	0.91	156
Isotactic poly-1-butene	—CH_2—CH— | C_2H_5	Trigonal $a = 1.77$, $c = 0.65$ nm (stable form)	0.95	0.915	136
		Tetragonal $a = 1.45$, $b = 2.11$ nm (unstable form)	0.92		124
Isotactic poly-1-pentene	—CH_2—CH— | C_3H_7	Monoclinic $a = 1.135$, $b = 2.085$, $c = 0.65$ nm, $\beta = 100°$	0.92	0.87	80
Isotactic poly-3-methyl-1-butene	—CH_2—CH— | CH / \ CH_3 CH_3	Monoclinic $a = 0.96$, $b = 1.71$, $c = 0.68$ nm, $\gamma = 116°$	0.93	0.90	350
Isotactic poly-4-methyl-1-pentene	—CH_2—CH— | CH_2 | CH / \ CH_3 CH_3	Tetragonal $a = 1.86$, $c = 1.385$ nm	0.812	0.83	245

Table 7 (Continued)

Polymer	Structural unit	Crystal system	Density Crystal	Density Polymer	Melting point (°C)
Isotactic poly-4-methyl-1-hexene	—CH$_2$—CH— | CH$_2$ | CH—CH$_3$ | C$_2$H$_5$	Tetragonal $a = 1.98$, $c = 1.35$ nm	0.845	0.86	200
Isotactic polyvinyl cyclohexane	CH$_2$—CH—(cyclohexyl)	Tetragonal $a = 2.20$, $c = 0.64$ nm	0.94	0.93	385
Isotactic polystyrene	—CH$_2$—CH—(phenyl)	Trigonal $a = 2.19$, $c = 0.81$ nm	1.13	1.08	240

Source: Refs. 23, 24.

mers are produced with the same Ziegler–Natta catalysts as polyolefins and they have many features similar to those of polyolefins: the same types of the polymer helixes, high melting points, etc. [3].

Most isotactic polyolefins are 70–90% crytalline products. Melting points of crystalline polyolefins depend mostly on the structure of their side groups (R in Fig. 3). In several cases the melting points exceed 350°C. (Melting points listed in Table 7 characterize highly isotactic, carefully annealed samples. Melting points of commercial resins are usually ~10°C lower.) The differences in properties of isotactic and sterically irregular (atactic) polymers can be seen clearly from the following comparisons:

1. Isotactic polypropylene is a highly crystalline material with a melting point of 176°C, whereas atactic polypropylene is a soft, amorphous polymer.
2. Isotactic polystyrene is a highly crystalline material with a melting point of ~240°C, whereas atactic polystyrene (the polymer widely used in numerous applications) is a completely amorphous, glassy polymer with a softening point of ~80°C.

Only one syndiotactic polyolefin, polypropylene, as well as syndiotactic polystyrene and one syndiotactic polydiene, poly-1,2-butadiene, have been synthesized. Chains of syndiotactic polypropylene can exist both in the flat zigzag form (Fig. 3ib) and in the helical form that contains four monomer units per one chain turn. This polymer is also highly crystalline. Its principal structural parameters are given in Table 7.

Table 8 lists the data on the crystal structures of all known forms of polybutadiene and polyisoprene, including isotactic and syndiotactic poly-1,2-butadienes. The densities and the melting points listed in the table characterize polymers with the highest attainable

Table 8 Crystalline Structure and Properties of Polydienes

Polymer	Structural unit	Crystal form	Crystal density	Melting point (°C)
trans-1,4-Polybutadiene	—CH$_2$\CH=CH\CH$_2$—	Monoclinic $a = 0.86$, $b = 0.91$, $c = 0.48$ nm, $\beta = 114°$	1.04	148
cis-1,4-Polybutadiene	—CH$_2$\CH=CH/CH$_2$—	Monoclinic $a = 0.46$, $b = 0.95$, $c = 0.86$ nm, $\beta = 109°$	1.01	~2
Isotactic poly-1,2-butadiene	—CH$_2$—CH— | CH ‖ CH$_2$	Trigonal $a = 1.73$, $b = 0.65$ nm	0.96	126
Syndiotactic poly-1,2-butadiene	—CH$_2$—CH— | CH ‖ CH$_2$	Orthorhombic $a = 1.01$, $b = 0.66$, $c = 0.51$ nm	0.964	156
trans-1,4-Polyisoprene	—CH$_2$\CH=C(CH$_3$)\CH$_2$—	Orthorhombic $a = 0.78$, $b = 1.18$, $c = 0.47$ nm (stable form)	1.05	64
		Monoclinic $a = 0.80$, $b = 0.63$, $c = 0.88$ nm, $\beta = 102°$ (metastable form)	1.05	74
cis-1,4-Polyisoprene	—CH$_2$\CH=C(CH$_3$)/CH$_2$—	Monoclinic $a = 1.25$, $b = 0.89$, $c = 0.81$ nm, $\beta = 92°$	1.02	<0

Source: Refs. 16, 23.

regularity. Polymers produced with industrial catalysts usually have steric purity of ~95%. As a consequence, their melting points and densities are lower.

INHOMOGENEITY OF ZIEGLER–NATTA CATALYSTS

All alkene polymerization products obtained with heterogeneous Ziegler–Natta catalysts are mixtures containing polymer molecules of different stereoregularity [3; 5, pp. 322–346]. These polymers can be efficiently separated by the solvent fractionation technique.

Atactic, amorphous polymers easily dissolve in many aliphatic solvents such as heptane, whereas isotactic, highly crystalline polymers have much lower solubility or are completely insoluble, even at high temperatures. Polypropylene represents the best studied example. Any propylene polymer produced with a Ziegler–Natta catalyst contains the highly crytalline, isotactic fraction that is insoluble in boiling heptane, an amorphous atactic fraction soluble in cold heptane or in boiling ether, and an intermediate fraction of reduced crystallinity that is soluble in boiling heptane (named stereoblock polymer) [3]. Relative amounts of these fractions depend mostly on the catalyst used for polymer synthesis. Several examples of polypropylene fractionations are given in Table 9. Crystalline isotactic polymer has a much higher commercial value, and polymerization catalysts are specially designed or modified to maximize the yield of the isotactic material. For example, modern supported catalysts for propylene polymerization (see above), if used without any modification, produce propylene polymers of very poor quality, containing merely 20–50% of the isotactic fraction. However, when such catalysts are modified by addition of aromatic-acid esters or alkoxyarylsilanes, the yield of the crystalline isotactic fraction increases to 95–97% [11]. Fractionation techniques were applied to other polyolefins and produced similar results—all these polymers contain both isotactic, highly crystalline material and amorphous, atactic fractions [3; 5, pp. 322–346]. The fractionation data indicate that the active centers in heterogeneous Ziegler–Natta catalysts are not identical with respect to isospecificity: the ability to produce isotactic polymers. The catalysts contain centers ranging from highly isospecific to completely aspecific, which produce atactic polymers [5, pp. 322–346]. When various modifiers are added to such catalysts they usually have very little effect on the isospecific centers, but they poison or alter the aspecific centers and, in this way, greatly improve the quality of polymer products [5, pp. 322–346; 11].

Existence of different types of active centers in heterogeneous catalysts is a very general phenomenon, common for many other catalytic processes. The situation is op-

Table 9 Fractionation Data for Polypropylene Produced with Several Unsupported Ziegler–Natta Catalysts at 70°C

Catalyst	Fraction	Content (%)
$TiCl_4$–$Al(C_2H_5)_3$	Isotactic	35–50
	Stereoblock	10–20
	Atactic	40–55
α-$TiCl_3$–$Al(C_2H_5)_3$	Isotactic	77–80
	Stereoblock	8–12
	Atactic	8–13
δ-$TiCl_3$–$Al(C_2H_5)_3$	Isotactic	70–75
	Stereoblock	10–15
	Atactic	15–20
δ-$TiCl_3$–$Al(C_2H_5)_2Cl$	Isotactic	90–95
(first commercial catalyst	Stereoblock	2–4
for propylene polymerization)	Atactic	3–5

Source: Ref. 5.

posite for homogeneous polymerization catalysts. Such catalysts often have only one type of active center and they produce sterically uniform products. For example, the homogeneous $(C_5H_5)_2ZrCl_2$-$(CH_3AlO)_n$ catalyst (designed for ethylene polymerization; see Table 1) polymerizes propylene to a completely atactic material that does not contain any crystalline fraction [25]. On the other hand, another homogeneous catalyst, ethylene (tetrahydroindenyl)$_2$ZrCl$_2$-$(CH_3AlO)_n$ (also listed in Table 1) polymerizes propylene to a nearly completely isotactic material [26]. This behavior is different from that of heterogeneous catalysts, which always produce mixtures of polymer molecules exhibiting various degrees of isotacticity.

The existence of different types of active centers in heterogeneous Ziegler–Natta catalysts has several other manifestations. For example, alkene copolymers produced with heterogeneous catalysts always contain copolymer molecules of different compositions. This occurs because different types of active centers have different reactivities in copolymerization reactions, i.e., different reactivity ratios r_1 and r_2. On the other hand, homogeneous catalysts produce compositionally uniform copolymers, i.e., the products in which all copolymer molecules have approximately the same composition. This distinction is very important in the case of ethylene–propylene copolymers that are used as elastomers and synthetic rubbers. Such copolymers usually contain approximately equal amounts of ethylene and propylene units in their molecules. However, when these copolymers are produced with heterogeneous Ziegler–Natta catalysts, they often contain fractions having both high ethylene and high propylene content. These fractions crystallize, which is detrimental to the elastic properties of the copolymers. For this reason, although copolymerization of ethylene and propylene can be carried out in laboratory with virtually any Ziegler–Natta catalyst, the ethylene–propylene elastomers of high quality are produced in industry with homogeneous catalysts, such as VCl_4–$Al_2(C_2H_5)_3Cl_3$ or $VOCl_3$–$Al_2(C_2H_5)_3Cl_3$ [3; 4, pp. 563–584; 12; 13].

COMMERCIAL PROCESSES UTILIZING ZIEGLER–NATTA CATALYSTS

Design of commercial processes for polymerization and copolymerization of alkenes and dienes with Ziegler–Natta catalysts depends mostly on the physical states of monomers and polymers at the reaction temperatures. Industrial firms developed numerous proprietary technological schemes for polymerization processes. These processes can be, in general terms, divided into three groups: gas phase processes, suspension processes, and solution processes. Most of the technological schemes are continuous: monomers and catalysts (and, optionally, solvents) are continuously fed into polymerization reactors, and polymers are continuously removed from the reactors.

Gas Phase Processes

In these processes [11, 27, 28], the reactor contains only a gaseous monomer and a large volume of polymer powder. Particles of the fresh catalyst are continuously fed into the reactor and immediately start polymerizing the monomer. As a result, each catalyst particle rapidly becomes encapsulated in a polymer envelope and slowly increases in size due to polymer growth. The bed of polymer particles is thoroughly mixed to ensure uniformity of the process. The volume of the polymer bed increases with polymerization, and small fractions of the bed are periodically removed from the reactor in short pulses.

Several existing types of the gas phase reactors differ mostly in the design of the polymer-stirring method and in the way the heat of polymerization is removed from the reactor. Alkene polymerization reactions generate much heat, ~25 kcal per mole of alkene, and, due to poor heat transfer properties of the polymer powder, the monomer is the only viable agent for heat removal. The monomer continuously circulates through the reactor and an adjacent cooling device where the monomer either cools or, in the case of propylene, condenses into a liquid. The liquid returns to the polymer bed, immediately evaporates, and cools the bed. Agitation of the polymer bed in such reactors can be achieved with various anchor-type stirrers. Union Carbide Company designed a fluid bed method [29] for polymerization of ethylene and propylene in the gas phase (Fig. 5). In this method, the polymer–particle bed in a large, tower-type reactor is both stirred and cooled by gaseous monomer that circulates at a high velocity between the reactor and the cooler. The volume of the catalyst feed into the reactor and the volume of the removed polymer are matched to maintain a constant level of the polymer bed. An average catalyst residence time, from the moment a catalyst particle enters the reactor till the moment when the same, polymer-covered particle leaves it, is 4–5 hr.

The gas phase polymerization method is limited to synthesis of polymers with a high to medium degree of crystallinity. It is used for manufacture of three large-volume products:

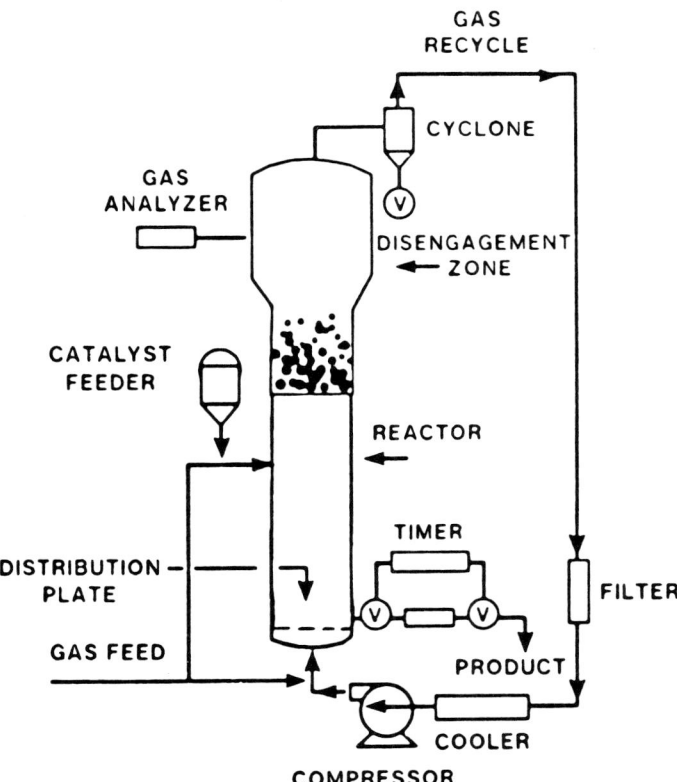

Figure 5 Gas phase polymerization scheme utilizing a fluid bed reactor. (From Ref. 29.)

polypropylene, high-density polyethylene (ethylene homopolymer), and linear low-density polyethylenes—the copolymers of ethylene containing 3–4 mol % of 1-butene, 1-hexene, or 4-methyl-1-pentene. This technology has significant economical advantages and is widely used throughout the world.

Suspension (Slurry) Processes

In these processes [9, 27, 28], the polymerization reaction is carried out in a liquid medium that suspends catalysts and polymer particles. The liquid in such processes can be either an inert solvent or a liquefied monomer (e.g., liquid propylene, liquid 1-butene). The polymer growing on the catalyst particles remains insoluble in the liquid during the whole polymerization process. The liquid absorbs the heat of polymerization and releases it to an external cooling jacket. Slurry polymerization is the earliest technology for polyolefin synthesis with Ziegler–Natta catalysts. The initial design utilized standard cylinder-shaped reactors equipped with impeller stirrers. An example of such a technological scheme for synthesis of polypropylene is shown in Figure 6. The suspension of polymer particles leaves the reactor and unreacted monomer is flash evaporated and returns to the monomer stream. The polymer is separated and the suspending liquid, after purification, returns to the reactor. The method of polymer separation depends on the type of liquid. When it has a sufficiently high boiling point (e.g., heptane), the separation is usually achieved by centrifugation. In polypropylene synthesis, the liquid leaving the centrifuge usually contains a small quantity of dissolved atactic polymer that is recovered by solvent steaming. If the liquid has a low boiling point (e.g., isobutane), it is removed by flash evaporation.

Phillips Petroleum Company developed an important modification of the suspension polymerization process [30]. In this method, the suspension of polymer particles in a

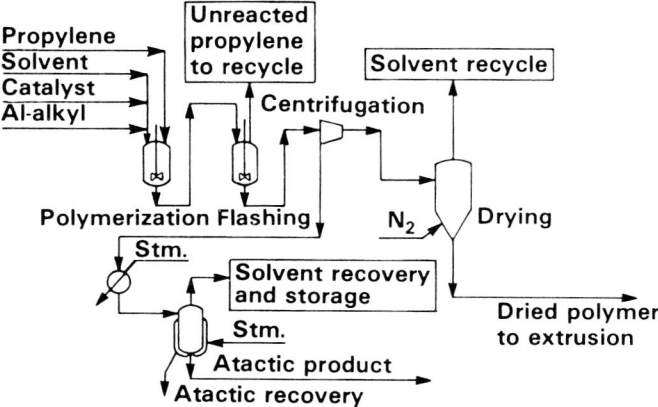

Figure 6 Suspension (slurry) polymerization scheme with a stirred-tank reactor. (From Ref. 28.)

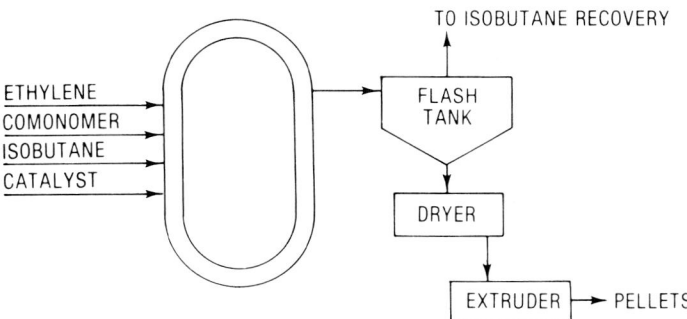

Figure 7 Suspension (slurry) polymerization scheme with a loop reactor. (From Ref. 30.)

liquid medium circulates at a high speed in a closed loop (Fig. 7). Rapid circulation ensures effective removal of the polymerization heat through the reactor walls. Such loop reactors usually use light suspension liquids, which are separated from the polymer by flash evaporation after leaving the reactor.

Solution Processes

The third technological principle also uses organic liquids as the medium for polymerization reactions [9, 27, 28]. However, in this case the polymer is dissolved in the liquid. Polymer solutions have high viscosity, which represents an obvious disadvantage of the solution method. Additionally, separation of the dissolved polymer from the solvent is usually a difficult task. Nevertheless, solution technology enjoys significant popularity. In many cases, formation of polymer solutions cannot be avoided. For example, amorphous ethylene–propylene elastomers are prepared by copolymerization of ethylene and propylene with homogeneous catalysts based on vanadium compounds [3, 12–14]. Both the catalysts and the copolymers are soluble in the reaction medium (usually hexane). The viscosity of the solution increases with increasing copolymer concentration and when polymer concentration reaches 8–10% the solution leaves the reactor and the copolymer is separated from the solvent by hot-water stripping.

In several technological schemes formation of polymer solutions is a result of specific polymerization conditions. For example, semicrystalline ethylene–alkene copolymers can be produced at high temperatures, 150–180°C, with the application of very active, short-lived Ziegler–Natta catalysts [9, 27, 31]. Polyethylene melt has very high viscosity at this temperature, therefore a diluent is added to the reactor to decrease viscosity and to facilitate mixing and temperature control. When, after a residence time of 5–10 min, the hot mixture of the polymer and the diluent leaves the reactor, the low-boiling solvent evaporates very rapidly and the polymer melt is immediately processed into pellets. An example of this reaction scheme used for synthesis of linear low-density polyethylene (ethylene–octene copolymer) is seen in Figure 8.

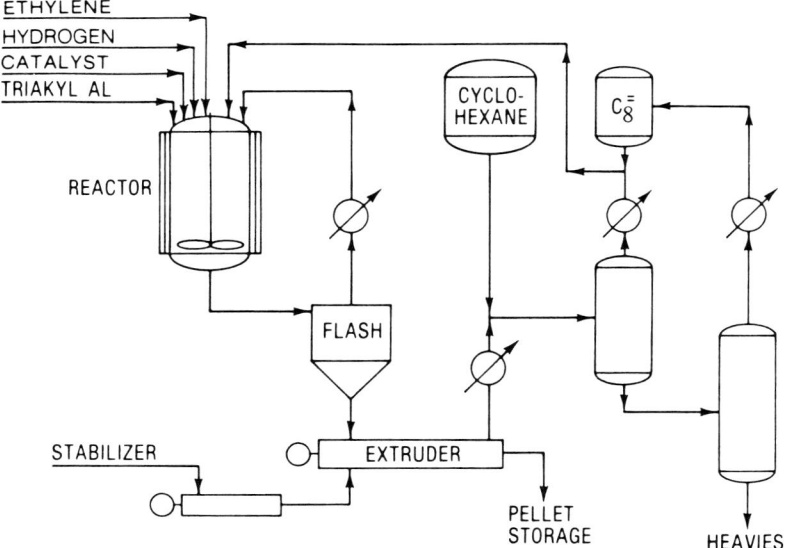

Figure 8 High-temperature solution polymerization scheme. (From Ref. 27.)

REFERENCES

1. K. Ziegler, H. Breil, H. Martin, and E. Holzkamp, Germ. Pat. 973,626 (Apr. 14, 1960, prior. Nov. 18, 1953).
2. G. Natta, P. Pino, P. Corradini, F. Danusso, E. Mantica, G. Mazzanti, and G. Moraglio, *J. Am. Chem. Soc.*, 77: 1708 (1955).
3. G. Natta and F. Danusso, eds., *Stereoregular Polymers and Stereospecific Polymerizations*, Pergamon Press, Oxford (1967).
4. J. Boor, Jr., *Ziegler–Natta Catalysts and Polymerizations*, Academic Press, New York (1979).
5. Y. V. Kissin, *Isospecific Polymerization of Olefins with Heterogeneous Ziegler–Natta Catalysts*, Springer-Verlag, New York (1985).
6. J. C. W. Chien, in *Preparation and Properties of Stereoregular Polymers* (R. W. Lenz and F. Ciardelli, eds.), D. Reidel, Dordrecht, p. 113 (1980).
7. T. Mole and E. Jeffrey, *Organoaluminum Compounds*, Elsevier, Amsterdam (1972).
8. G. Natta, P. Corradini, and G. Allegra, *J. Polym. Sci.*, 51: 399 (1961).
9. T. E. Nowlin, *Prog. Polym. Sci.*, 11: 29 (1985).
10. G. L. Goeke, B. E. Wagner, and F. J. Karol, U.S. Pat. 4,354,009 (1982).
11. E. Albizzati, Eur. Pat. Appl. 83101332.1 (filed Feb. 11, 1983).
12. C. Cozewith and G. Ver Strate, *Macromolecules*, 4: 482 (1971).
13. H. N. Friedlander, in *Encyclopedia of Polymer Science and Technology*, Vol. 6 (H. F. Mark, N. G. Gaylord, and N. M. Bikales, eds.), Wiley-Interscience, New York, p. 338 (1967).
14. G. G. Evans, in *Ziegler-Natta and Metathesis Polymerization* (R. P. Quirk, ed.), Cambridge University Press, New York (1988).
15. S. E. Horne, Jr., in *Transition Metal Catalyzed Polymerizations: Alkenes and Dienes*, Part B (R. P. Quirk, ed.), Harwood, New York, p. 527 (1985).
16. D. P. Tate and T. W. Bethea, in *Encyclopedia of Polymer Science and Engineering*, Vol. 2 (H. F. Mark, N. M. Bikales, C. G. Overberger, G. Wenes, and J. I. Kroschwitz, eds.), Wiley-Interscience, New York, p. 537 (1984).

17. G. Henrici-Olive and S. Olive, *Adv. Polym. Sci.*, 6: 421 (1969).
18. G. Fink, in *Transition Metal Catalyzed Polymerizations: Alkenes and Dienes*, Part B (R. P. Quirk, ed.), Harwood, New York, p. 495 (1985).
19. A. Zambelli, M. C. Sacchi, and P. Locatelli, in *Transition Metal Catalyzed Polymerizations: Alkenes and Dienes*, Part A (R. P. Quirk, ed.), Harwood, New York, p. 83 (1985).
20. Y. V. Kissin, *Adv. Polym. Sci.*, 15: 91 (1974).
21. Y. V. Kissin, in *Transition Metal Catalyzed Polymerizations: Alkenes and Dienes*, Part B (R. P. Quirk, ed.), Harwood, New York, p. 597 (1985).
22. L. Porri, in *Structural Order in Polymers* (F. Ciardelli and P. Giusti, eds.), Pergamon Press, Oxford, p. 51 (1981).
23. H. Tadokoro, *Structure of Crystalline Polymers*, Wiley, New York, pp. 355–358 (1979).
24. Y. V. Kissin and D. L. Beach, *Encyclopedia of Polymer Science and Engineering*, Vol. 9 (H. F. Mark, N. M. Bikales, C. G. Overberger, G. Wenes, and J. K. Kroschwitz, eds.), Wiley-Interscience, New York (1984).
25. W. Kaminsky, in *Transition Metal Catalyzed Polymerizations: Alkenes and Dienes*, Part A (R. P. Quirk, ed.), Harwood, New York, p. 225 (1985).
26. W. Kaminsky, K. Kulper, H. H. Brintzinger, and F. R. Wald, *Angew. Chem. Int. Ed. Engl.*, 24: 507 (1985).
27. J. N. Short, in *Transition Metal Catalyzed Polymerizations: Alkenes and Dienes*, Part B (R. P. Quirk, ed.), Harwood, New York, p. 651 (1985).
28. N. N. Brockmeier, in *Transition Metal Catalyzed Polymerizations: Alkenes and Dienes*, Part B (R. P. Quirk, ed.), Harwood, New York, p. 671 (1985).
29. A. R. Miller, U.S. Pat. 4,003,712 (1977).
30. J. P. Hogan, in *Kirk-Othmer Encyclopedia of Chemical Technology*, 3rd ed., vol. 16 (M. Grayson, ed.), Wiley-Interscience, New York, p. 421 (1981).
31. J. P. Machon, in *Transition Metal Catalyzed Polymerizations: Alkenes and Dienes*, Part B (R. P. Quirk, ed.), Harwood, New York, p. 639 (1985).

4

Synthesis and Characterization of Block Copolymers

Baki Hazer
Karadeniz Technical University
Trabzon, Turkey

INTRODUCTION	133
SYNTHESIS OF BLOCK COPOLYMERS	135
Block Copolymerization via the Free Radical Mechanism	135
Block Copolymerization via the Step-Growth Procedure	148
Block Copolymerization via the Anionic Mechanism	153
Block Copolymers by Cationic Polymerization	156
Block Copolymers by Coordination Polymerizations	159
Block Copolymers by Coupling Reactions	159
CHARACTERIZATION BY BLOCK COPOLYMERS	165
CONCLUSION	168
REFERENCES	170

INTRODUCTION

The science and technology of block copolymers, a new branch of polymer chemistry, has developed gradually over the last four decades. A tremendous amount of research has been performed on the synthesis and characterization of block copolymers. Many books [1–3], proceedings of symposia [4, 5], and papers [6–10] offer excellent reviews of research on block copolymers.

Block copolymers are polymers in which chemically different blocks (or sequences) bind each other in macromolecular chains. They have also been considered as materials formed by different homopolymers chemically bonded end to end. Thus the effects of the blocks are asserted individually in the block copolymer.

Although the terms block copolymers and block polymers are used synonymously in the literature to describe polymers with block structure, block copolymers is used in this chapter. Block copolymers can be divided into four main classes: diblock, triblock, multiblock, and star (or radial) block copolymers (Fig. 1).

Except in rare instances, polymer blocks in copolymer are immiscible. Because of the immiscibility of the polymer sequences, block copolymers show phase segregation in

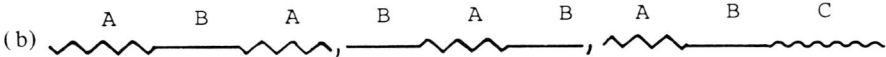
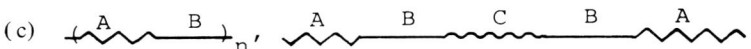
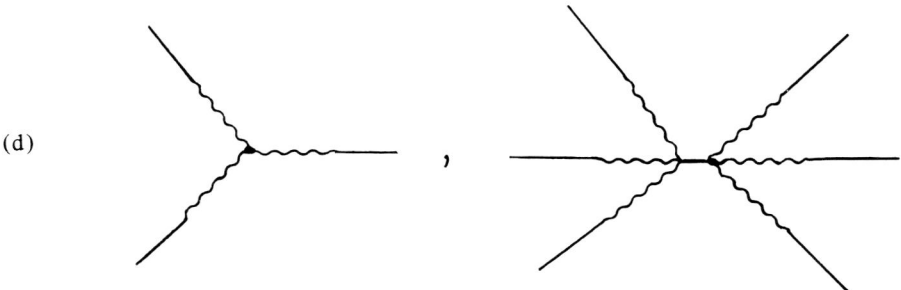

Figure 1 Schematic representation of block copolymer structures. (*a*) diblock, AB; (*b*) triblock, ABA, BAB, ABC; (*c*) multiblock, $+AB+_n$, ABCBA; and (*d*) star or radial block, $(AB)_x$ with three arms (left) and six arms (right), where $n > 1$, $n = 2,3,4, \ldots$; A,B,C, ... denote dissimilar polymer blocks.

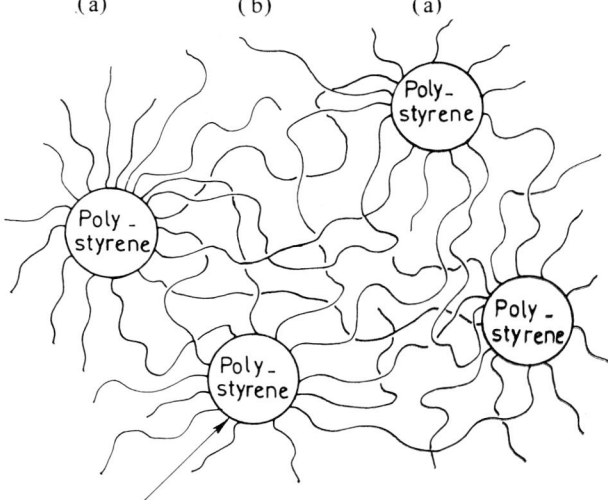

Figure 2 Schematic of the domain structure of styrene/butadiene/styrene triblock copolymers (thermoplastic elastomer). (*a*) polystyrene, (*b*) polybutadiene, and (*a*) polystyrene. (From Ref. 9.)

accordance with the phase rule and phase segregation causes domains. The domains composed of one type blocks were dispersed in the other block matrix [10]. The sizes of the domains in block copolymers are much smaller than aggregates found in a mixture of homopolymers exhibiting phase separation. Therefore, homopolymer mixtures (or blends) are opaque if the homopolymers have different refractive index, whereas block copolymers, which contain very small domains (~200 Å), are transparent. Because of this minute domain size, phase segregation in block copolymers is called microphase separation. For instance, in a Kraton thermoplastic elastomer—an ABA block copolymer of styrene (A) and butadiene or isoprene (B)—polystyrene blocks aggregate to form a domain in the continuous polybutadiene matrix as shown in Figure 2.

Graft copolymers are also formed by bonding chemically different polymer blocks and are very similar to block copolymers in many ways. Whereas block copolymers mainly have linear macromolecular chains, graft copolymers are branched and have a main chain of polymer A to which a number of B sequences are grafted (B blocks are grafted on A main chain):

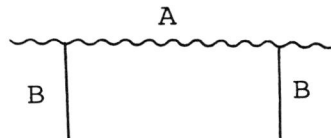

The domain formation imparts valuable features to block copolymers. First, phase separation in block copolymers results in distinct transition temperatures that correspond closely to each phase. Thus block copolymers can be used up to the glass transition temperature, T_g, of hard segments and their range extends from the T_g of the soft segment to the T_g of the hard segment. By contrast, if both blocks are soft segments or hard segments, they will have a narrow T_g range. These narrow T_g ranges will be either low temperatures (in case both are soft segments) or high temperatures (in case both are hard segments). Second, block copolymers that include two or more glassy segments such as ABA, $+AB+_n$, and $(AB)_x$ types have features of vulcanized rubber. Domains in these materials serve as physical crosslinking and reinforcement sites. Although these polymers are perfectly linear chains without chemical crosslinks between them, they behave as though they were vulcanized rubbers at ambient temperature. They are thermoplastic, can be easily molded, and are soluble in common solvents. Some commercial block copolymers are tabulated in Table 1.

SYNTHESIS OF BLOCK COPOLYMERS

The first synthesis of block copolymers was carried out by Bolland and Melville [11] via living macroradicals in 1938. But block copolymers assumed great industrial importance after Szwarc et al. [12] synthesized well-defined block copolymers via living macroanions in 1956. To date, synthesis has been effected by general polymerization methods such as free radical, step growth, anionic, cationic, coordination, and coupling. In this chapter these methods are summarized by considering general papers which discussed block copolymerization up to the end of 1986.

Block Copolymerization via the Free Radical Mechanism

In the synthesis of block copolymers via the free radical mechanism, homopolymer formation with block copolymers is generally unavoidable and purification of block copolymers is usually necessary.

Table 1 Some Commercial Block Copolymers

Trade name	Type of block copolymer	Segment, A	Segment, B	Property
Kraton 1101	ABA	Polystyrene	Polybutadiene	Thermoplastic elastomer
Kraton 1107	ABA	Polystyrene	Polyisoprene	Thermoplastic elastomer
Pluronic	ABA	Polyoxyethylene	Polyoxypropylene	Nonionic surfactant
Tetronic	$(AB)_4$ Starblock copolymer with four arms	Polyoxyethylene	Polyoxypropylene	Nonionic surfactant
Estane	$-(AB)_n-$	Polyurethane	Polybutylene adipate	Elastomeric fiber
Lycra	$-(AB)_n-$	Polyurethane	Polyoxymethylene glycol	Elastomeric fiber
Hytrel	$-(AB)_n-$	Polybutylene-terephthalate	Polytetramethylene glycol	Elastomeric fiber

As in the general polymerization procedure, block copolymerization is carried out via free radicals which are produced by irradiation or thermally from disulfides, peroxy-, and/or azoinitiators [13–15]. A mechanochemical method is also used to produce free radicals from macromolecules. Block copolymerization can occur via the free radical mechanism in six ways: long-lived macroradicals, mechanochemical synthesis, oligoperoxides, macrobisperoxides, macroazoinitiators, and azoperoxidic initiators.

Long-Lived Macroradicals

Monomer radicals can be directly produced by irradiation of the vinyl molecules, and polymer radicals formed in this medium continue adding monomer even after removal of the light source [11, 15–19]. By adding a second monomer on the polymer radicals a polymer of the second monomer will be formed on top of the first polymer and block copolymer will be obtained with low yield [20]. Block copolymers of some vinyl monomers were also prepared by γ-irradiation of vinyl acetate emulsions [21] and acrylamide crystals [22].

Otsu and co-workers synthesized block copolymers by using organic disulfides such as tetraethylthiuram disulfide (TETD) in the photopolymerization of some vinyl monomers under UV light [23–25]. This method, recently named the iniferter technique [24], includes the reactions of photochemical initiation, chain transfer reaction, and primary radical termination:

Initiation:

$$Et_2N-\underset{\underset{TETD}{}}{\overset{\overset{S}{\|}}{C}}-S-S-\overset{\overset{S}{\|}}{C}-NEt_2 \xrightarrow{h\nu} 2\,Et_2N-\underset{R\cdot}{\overset{\overset{S}{\|}}{C}}-S\cdot$$

$$R\cdot + M_1 \xrightarrow{h\nu} RM_1\cdot$$
$$RM_1\cdot + (n-1)M_1 \rightarrow R(M_1)_n\cdot \quad \text{(propagation)}$$

Chain transfer:

$$R(M_1)_n\cdot + TETD \rightarrow R(M_1)_n R$$

Termination:

$$2R(M_1)_n\cdot \rightarrow R(M_1)_{2n}R$$

and

$$R(M_1)_n\cdot + R\cdot \rightarrow R(M_1)_n R$$

Block copolymerization:

$$(M_1)_n R \rightleftharpoons (M_1)_n\cdot + R\cdot$$
$$(M_1)_n\cdot + mM_2 \rightarrow (M_1)_n(M_2)_m R$$

By a similar technique, the polymers having dye (e.g., eosin, safranin, acridin orange) are also photosensitive and dissociate polymer radicals under visible light [26].

Long-lived macroradicals can also be formed by UV irradiation of some vinyl monomers such as N-phenyl methacrylamide, N-methyl acrylamide, and acryloyl-L-valine in the presence of di-*tert*-butyl peroxide [27, 28] or di-*tert*-butyl peroxyfumarate [29] as sensitizer. The formation of long-lived macroradicals of these vinyl monomers in a block copolymerization matrix was investigated by ESR spectroscopy, as was the synthesis of block copolymer by the reaction of living polymer radicals with vinyl monomers. Recently it was shown that block copolymers can also be obtained by long-lived polymer radicals formed in viscous poor solvents such as silicone oil or phosphoric acid under UV irradiation [30, 31].

Bamford and Mullik [32] showed that some transition metal carbonyls, particularly $Mn_2(CO)_{10}$ and $Re_2(CO)_{10}$ in the presence of an additive such as tetrafluoroethylene or acetylene dicarboxylic acid dimethylester (MeOOC—C≡C—COOMe), can photoinitiate the polymerization of common vinyl monomers at ambient temperature. The resulting polymers are active thermal initiators at higher temperatures (e.g., 100°C):

$$nM_1 + C_2F_2 \xrightarrow[Mn_2(CO)_{10}]{h\nu} \sim\!\!M_1\!\!\sim\!\!CF_2CF_2Mn(CO)_5$$

$$\sim\!\!M_1\!\!\sim\!\!CF_2CF_2Mn(CO)_5 \xrightarrow[(100°C)]{\Delta} \sim\!\!M_1\!\!\sim\!\!\dot{C}F_2 + \dot{M}n(CO)_5$$
$$\downarrow + nM_2$$
$$\sim\!\!M_1\!\!\sim\!\!CF_2CF_2\!\!\sim\!\!\sim\!\!M_2\!\!\sim$$

These transition metal carbonyls react easily with reactive organic halides and give rise to free radicals, which can initiate the polymerization:

$$Mn_2(CO)_{10} + RCCl_3 \xrightarrow{h\nu} Mn_2(CO)_{10}Cl + R\dot{C}Cl_2$$

$$R\dot{C}Cl_2 \xrightarrow{nM_1} RCCl_2\!\!\sim\!\!M_1\!\!\sim$$

The polymer with -Br and -CBr$_3$ terminal groups obtained by polymerizing monomer in the presence of CBr_4 as transfer agent leads to alternating block copolymers by a similar reaction [33]:

$$\sim\sim M_1 \sim\sim CBr_3 \xrightarrow[Mn_2(CO)_{10}]{h\nu} \sim\sim M_1 \sim\sim \dot{C}Br_2$$

$$\sim\sim M_1 \sim\sim \dot{C}Br_2 + nM_2 \rightarrow \sim\sim M_1 \sim\sim CBr_2 \sim\sim\sim M_2 \sim\sim$$

Mechanochemical Synthesis

When polymer chains are broken by the application of large stresses, macroradicals are produced and can be used in block copolymer synthesis. These mechanical processes include high-speed stirring [34–37] and ultrasonic waves [38, 39] to produce macroradicals. Polystyrene–monomer and polymethyl methacrylate–monomer mixtures are subjected to grind by high-speed stirring (e.g., 30,000 rpm) at around 0–12°C under nitrogen atmosphere. Block copolymers are thus obtained with high conversions. Minoura and coworkers synthesized block copolymers by high-speed stirring of polyethylene oxide (PEO) glycol and methyl methacrylate in solution [36, 37]. They found that the rate of polymerization is proportional to (a) the square root of the shear rate of the PEO molecule, (b) monomer concentration, and (c) PEO polymer concentration.

When ultrasonic waves are passed through the polymer solutions, degradation later occurs. During ultrasonic degradation two macroradicals will be formed each time by chain rupture [38, 39]. The degradation increases with an increase of sonic intensity and with a decrease of solution concentration. From PEO, hydroxyethyl cellulose, polysiloxanes, and polyvinyl chloride, macroradicals can be produced this way. High-voltage electrical discharge and freezing or shaking of polymer solutions may result in breaking of polymer molecules [40].

Oligoperoxides

Compounds having more than one peroxygen group can be used in block copolymer synthesis in two polymerization steps. In the first step, by using the oligoperoxide in the polymerization of a vinyl monomer, an active polymer which has undecomposed peroxygen groups in the polymer backbone is obtained. When another monomer is polymerized in the presence of this active polymer as free radical initiator in the second step, block copolymer with high yield is obtained. Two polymerization steps can be shown as follows:

$$nM_1 + \text{oligoperoxide} \rightarrow \underset{\text{Active polymer}}{M_1 \sim\sim OO \sim\sim OO \sim\sim}$$

and

$$M_1 \sim\sim OO \sim\sim OO \sim\sim + nM_2 \rightarrow \underset{\text{Block copolymer}}{M_1 \sim\sim M_2 \sim\sim}$$

The first oligoperoxide was prepared by the reaction of phtaloyl dichloride with sodium peroxide at 0–5°C by Pechmann and Vanino [41], who could not understand that the product was a polymeric structure because of its insolubility in common solvents. Woodward and Smets [42] suggested that it was a polymeric structure and represented its formula as follows:

$$\{O-\underset{\underset{O}{\|}}{C}-\underset{}{\bigcirc}-\underset{\underset{O}{\|}}{C}-OO-\underset{\underset{O}{\|}}{C}-\underset{}{\bigcirc}-\underset{\underset{O}{\|}}{C}-O\}_n$$

Phtaloyl peroxide oligomer

They used this oligoperoxide as initiator in the polymerization of some vinyl monomers at 70–75°C and active polymers and other vinyl monomers gave the block copolymers of styrene and vinyl acetate with styrene and methyl methacrylate at 100°C.

The series of dicarboxylic acids has been used in the synthesis of oligoperoxides, which usually are soluble in solvents. Oligoperoxides can be synthesized by the reaction of equimolar amounts of diacid chlorides or mono-α-substituted dichlorides of diacids with peroxides having difunctionality such as sodium peroxide [43, 44] (or hydrogene peroxide), dipercarboxylic acids [45–47], and dihydroperoxides [48–50]. A typical example of a condensation reaction to prepare oligoperoxide can be shown as follows [49]:

$$Cl-\underset{\underset{O}{\|}}{C}(CH_2)_m\underset{\underset{O}{\|}}{C}-Cl + HOO-\underset{\underset{CH_3}{|}}{\overset{\overset{CH_3}{|}}{C}}(CH_2)_2\underset{\underset{CH_3}{|}}{\overset{\overset{CH_3}{|}}{C}}-OOH \xrightarrow[+NaOH]{0-5°C}$$

$$\left[-\underset{\underset{O}{\|}}{C}(CH_2)_4\underset{\underset{O}{\|}}{C}-OO-\underset{\underset{CH_3}{|}}{\overset{\overset{CH_3}{|}}{C}}(CH_2)_2\underset{\underset{CH_3}{|}}{\overset{\overset{CH_3}{|}}{C}}-OO-\right]_p$$

Oligoperoxide, $p = 3-4$

In case of monofunctional hydroperoxides (e.g., butylhydroperoxide, perbenzoic acid), mono-, di-, and triperoxides are prepared by a similar method [51–53]. Synthesis of oligoperoxide has been performed by the interfacial condensation reaction [41, 43, 46, 49] of aqueous phase containing NaOH and peroxide with organic phase containing diacid chloride at 0–5°C. Recently a different synthesis method of polymeric peroxides was reported by Hazer and Baysal [54]. They prepared a polymeric peroxycarbamate from the reaction of equimolar amounts of an aliphatic diisocyanate and dihydroperoxide. Some oligoperoxides are tabulated in Table 2.

As the number of peroxy groups increases in oligoperoxide compounds, the amount of inactive polymer in the block copolymer decreases [47]. For instance, when using diperoxide compound (two peroxy groups per molecule), the gravimetric proportion of the inactive polymer reaches 25%; for oligoperoxide with $p = 10$ (10 peroxide groups per molecule) it is less than 1%. Therefore, to reduce homopolymer formation in the synthesis of block copolymers, it is important to prepare oligoperoxide having as many peroxy groups as possible.

Oligomeric peroxy compounds containing peroxy groups of different activities exhibit superiority as polymerization initiators. It was found that the acylperoxy group decomposes earlier than the peroxyester group (e.g., $E_a = 30$ kcal mol^{-1} and 36 kcal mol^{-1}, respectively) [52]. At the same time the peroxy group more remote from the substituent in the mono-α-substituted oligoperoxides also decomposes earlier than the peroxy group closest to the substituent (e.g., $E_a = 24–27$ kcal mol^{-1} and 30–32 kcal mol^{-1}, respectively) [46].

Table 2 Oligoperoxides Used in Block Polymerization

Formula and name	Notes	Reference
Oligoterephtalylperoxide $+O-C(=O)-C_6H_4-C(=O)-OO-C_6H_4-C(=O)-O+_n$		41, 42
Oligoacylperoxide $+C(=O)(CH_2)_xC(=O)-OO+_n$	$x = 8, 5, 7$	43, 44
Oligo(mono-α-substituted acyl peroxide) $+C(=O)(CH_2)_nC(=O)-OO-C(=O)\,CH(X)(CH_2)_mCOO+_p$	$p = 9-21$ $n = 4-8$ $m = 1-4, 6, 7$ $x = CH_3, CH_2, Cl, Br$	46, 47
Oligo(acyl-ester peroxide) $+OO-C(=O)-R-C(=O)-OO-C(=O)-R-C(=O)-OO-C(CH_3)_2(CH_2)_2-$ ... $-C(CH_3)_2-OO-C(=O)-R-C+_p$	$R = o\text{-}C_6H_4$ $(CH_2)_8$ $(CH_2)_4$ $(CH_2)_7$ $p = 3$	48
Oligoperester $[OO-C(CH_3)_2(CH_2)_2C(CH_3)_2-OO-C(=O)(CH_2)_4C(=O)]_p$	$p = 3, 4$	49
Diperester (diperoxide) $ROO-C(=O)(CH_2)_nC(H)(X)-C(=O)-OOR$	$R = C_6H_5-C(=O)-$ $CH_3(CH_2)_6C(=O)-$ $(CH_3)_3C-$ $C_{10}H_{21}C(=O)-$ $x = CH_3, CH_2, Cl, Br, H$	50, 52
Diacylperoxyperester (diperoxide) $(CH_3)_3COO-C(=O)(CH_2)_xC(=O)-OO-C(=O)-R$	$R = C_6H_5-$ $C_7H_{15}-$ $C_9H_{19}-$ $x = 2, 3$	51

Table 2 Continued

Formula and name	Notes	Reference
Diacylperoxyperester (triperoxide) $(CH_3)_3COOC(CH_2)_xC(=O)-OO-C(=O)(CH_2)_x-$ $-C(=O)-OO-C(CH_3)_3$	$x = 2, 3$	51
Oligoperoxycarbamate $\begin{array}{c} CH_3 \quad O \quad H \quad\quad H \quad O \\ \mid \quad\; \parallel \quad \mid \quad\quad \mid \quad \parallel \\ HC-O(C-N-R_1-N-C-OO-R_2-OO)_n \\ \mid \\ CH_3 \end{array}$ $\begin{array}{c} O \quad H \quad\quad H \quad O \quad CH_3 \\ \parallel \quad \mid \quad\quad \mid \quad \parallel \quad \mid \\ -C-N-R_1-N-C-O-CH \\ \mid \\ CH_3 \end{array}$	$R_1 = -\langle H \rangle-CH_2-\langle H \rangle-$ with CH_3 groups $R_2 = -C(CH_2)_2C-$ with CH_3 groups $n = 6$	53

The other important point in oligoperoxide copolymerization is the number of methylene groups in the initial diperacids and dibasic acid chlorides. Lengthening the methylene groups between the peroxy groups reduces the rigidity of the oligomeric chain, which promotes the solubility of the peroxides [43, 46, 49].

Macrobisperoxides

When polymers having peroxide groups were used in the polymerization of some vinyl monomers, block copolymers were obtained in one or two steps. Ceresa reported block copolymer synthesis by peroxidized polymers [55]. During the bulk polymerization of methyl methacrylate, styrene, and vinyl acetate in the presence of dissolved oxygen, peroxy linkages are introduced into the polymer backbone:

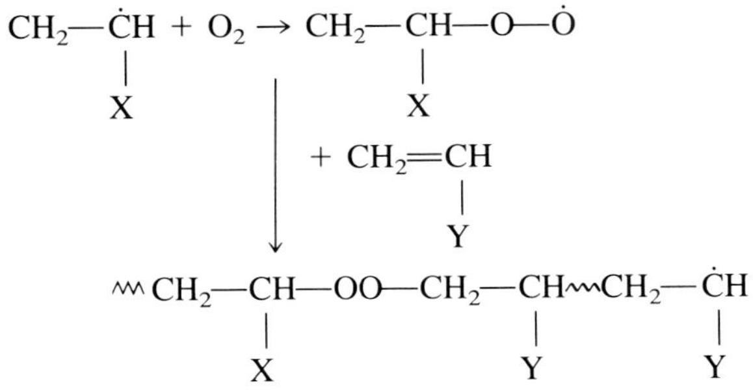

$$CH_2-\overset{\cdot}{C}H + O_2 \rightarrow CH_2-CH-O-\overset{\cdot}{O}$$
$$\underset{X}{\mid} \qquad\qquad\qquad \underset{X}{\mid}$$

$$+ \; CH_2=CH$$
$$\underset{Y}{\mid}$$

$$\sim\!\!\sim CH_2-CH-OO-CH_2-CH\!\sim\!\!\sim\!CH_2-\overset{\cdot}{C}H$$
$$\;\;\underset{X}{\mid} \qquad\qquad \underset{Y}{\mid} \qquad\quad\; \underset{Y}{\mid}$$

$-(AB)_n$-Type Block Copolymer

Smets and co-workers also described the synthesis and characterization of vinyl chloride block copolymers [56, 57]. The emulsion polymerizations of styrene and methacrylic and acrylic esters were initiated with peroxidized polypropylene obtained by the reaction of isotactic polypropylene with ozone and oxygen. After filtration of the initiator, the polymerization was brought to completion, and vinyl chloride was introduced in large excess under nitrogen pressure and polymerized.

Some researchers synthesized telechelic polymers with terminal carboxyl groups by the polymerization of monomers with azobiscyano pentanoic acid [58–61]. After it was converted to its acid chloride by means of $SOCl_2$ and treated with t-butyl hydroperoxide in the presence of pyridine, polymers with two peroxide terminal groups was obtained. Beylen and Smets [60] reported that they could transform the carboxylic acid end groups of polystyrene into perester groups in only very low yields (13% and 8.5%). The reaction pathway can be shown as follows:

$$\text{HOO—R—N=N—R—COOH} \xrightarrow{+\text{styrene}} \text{HOOC—R} \sim \text{polystyrene} \sim \text{R—COOH}$$

$$\xrightarrow{+SOCl_2} \text{Cl—}\overset{O}{\underset{\|}{C}}\text{—R} \sim \text{polystyrene} \sim \text{R—}\overset{O}{\underset{\|}{C}}\text{—Cl}$$

$$\xrightarrow{+t\text{-BuOOH}} t\text{-BuOO—}\overset{O}{\underset{\|}{C}}\text{—R} \sim \text{polystyrene} \sim \text{R—}\overset{O}{\underset{\|}{C}}\text{—OOBu-}t$$

Reed has prepared telechelic polydienes by using azobiscyanopentanoic acid as initiator in the polymerization of butadiene, isoprene, and chloroprene [61].

The incorporation of soft segments (e.g., polyether and polyester) into the backbone of a rigid polymer (e.g., polystyrene) can have a very positive effect on the flexibility of the polymer. Many attempts have been made to combine the polyether structure with that of a vinyl monomer. In one of these approaches, the terminal hydroxyl groups of polyethers are converted to perester groups by a condensation reaction of mono-t-butylperoxysuccinyl chloride [62] or by using N,N'-carbonyl diimidazole with t-butyl hydroperoxide producing imidazoyl percarboxylate with high yield [63].

$$\text{Im—}\overset{O}{\underset{\|}{C}}\text{—Im} + (CH_3)_3COOH \xrightarrow{THF}$$

$$\text{Im—NH} + \text{Im—N-}\overset{O}{\underset{\|}{C}}\text{-OOC}(CH_3)_3 + (CH_3)_3\text{COO}\overset{O}{\underset{\|}{C}}\text{OOC}(CH_3)_3$$

Imidazoyl percarboxylate,   dibutyl percarbonate (less than 10%)

$$-2 \underset{N}{\overset{N\nearrow}{\bigsqcup}} \Bigg\downarrow +HO-Z-OH$$

$$(CH_3)_3COO\overset{O}{\underset{\|}{C}}-O-Z-O-\overset{O}{\underset{\|}{C}}-OOC(CH_3)_3$$

Macrobisperoxide, Z=poly(oxyethylene)
with $M_n = 400, 6000$ g mol^{-1}

ABA-type block copolymers (polystyrene–polyoxyethylene–polystyrene) are obtained by the polymerization of styrene in the presence of macrobisperoxide as initiator at 130°C. At this temperature, it completely decomposes during polymerization in 3 hr and block copolymers having up to 30% polyether soft segment can be obtained.

Another method originated by Tobolsky and Rembaum in 1964 involved coupling reactions of diisocyanates with polyether glycols or polyesters [64]. The coupling reaction pathways can be shown as follows:

$$HO\text{\textasciitilde}OH + 2OCN-R-NCO \rightarrow OCN-R-NH-\overset{O}{\underset{\|}{C}}-O\text{\textasciitilde}O-\overset{O}{\underset{\|}{C}}-NH-R-NCO$$

Polyetherglycol   diisocyanate

$$\xrightarrow{+2R'OOH} R'OO\overset{O}{\underset{\|}{C}}-NH-R-NH-\overset{O}{\underset{\|}{C}}-O\text{\textasciitilde}O-\overset{O}{\underset{\|}{C}}-NH-R-NH-\overset{O}{\underset{\|}{C}}OOR'$$

$$\downarrow \Delta$$

$$R'O\cdot + \cdot O\overset{O}{\underset{\|}{C}}-NH-R-NH-\overset{O}{\underset{\|}{C}}-O\text{\textasciitilde}O-\overset{O}{\underset{\|}{C}}-NH-R-NH-R-\overset{O}{\underset{\|}{C}}-O\cdot + \cdot OR'$$

$$\downarrow + \text{vinyl monomer}$$

ABA-type block copolymer

This macroperoxycarbamate was then used to initiate polymerization of some vinyl monomers. ABA block copolymers obtained this way are thermoplastic elastomers when A block is a glassy segment such as polystyrene and B is a soft segment such as polyoxyethylene glycol.

To investigate and expand this basic system, numerous studies concerned with new initiating systems, effects of reaction conditions on physical properties, and attempts to increase block copolymer formation have been performed [65–72]. Raw materials for preparing different types of macrobisperoxycarbamates are listed in Table 3.

Macroazo Initiators

The common characteristic of the reaction of macroazo initiators is that a thermally labile azo group along a polymer backbone is cleaved to yield a polymer radical, which then initiates a block copolymerization in the presence of another monomer:

$$\text{\textasciitilde}N=N\text{\textasciitilde} \xrightarrow[-N_2]{\text{heat}} \text{\textasciitilde}\cdot + \cdot\text{\textasciitilde}$$

Table 3 Raw Materials for Preparing Macrobisperoxycarbamates

Diisocyanate	Polyether or polyester	Hydroperoxide	Reference
2,4-tolylene-	Poly(oxyethylene) glycol Poly(oxypropylene) glycol Poly(ethyleneadipate)	*t*-butylhydroperoxide	64
2,4-tolylene-	Poly(ethyleneadipate)	Cumenehydroperoxide	65
2,4-tolylene-	Poly(tetramethyleneoxide) glycol	*t*-butyl-hydroxymethyl-hydroperoxide	66
Bis(*p*-isocyanatocyclo-hexyl) methane	Poly(oxyethylene) glycol Poly(oxypropylene) glycol	2,5-dimethyl-2,5-dihydroperoxy hexane Cumenehydroperoxide	67, 69–71 72
Bis(*p*-isocyanatocyclo-hexyl) methane	Poly(oxypropylene) glycol	Bis(β-hydroxyethyl) disulfide*	68

*Gives macrobisdisulfides.

$$\xrightarrow{+\text{monomer}} \begin{array}{l} \nearrow \text{\textasciitilde\textasciitilde}(M)_x + (M)_y\text{\textasciitilde\textasciitilde} \\ \quad \text{termination by disproportionation} \\ \searrow \text{\textasciitilde\textasciitilde}(M)_{x+y}\text{\textasciitilde\textasciitilde} \\ \quad \text{termination by combination} \end{array}$$

Macroazo initiators can be divided into several groups: polymeric azocarbamates, polymeric azoamides, polymeric azoesters, and various other azo catalysts.

Polyazocarbamates. Polymeric azocarbamates were first prepared by Furukawa and co-workers [73]. The method is based on a chain-extended coupling reaction between poly(oxypropylene) glycol (M_n = 2000 and 4000 g mol^{-1}) or poly(oxytetramethylene) glycol (M_n = 2200 g mol^{-1}) with diisocyanate terminal groups and azobiscyano-*n*-pentanol as the following reaction schemes:

$$\text{HO}\text{\textasciitilde\textasciitilde}\text{OH} \xrightarrow{+\text{diisocyanate}} \text{OCN}(\!\!-\!\!\text{R}\!-\!\text{NHCO}\text{\textasciitilde\textasciitilde}\text{O}\!-\!\overset{\overset{\text{O}}{\|}}{\text{C}}\!-\!\text{NH}\!)_m\text{R}\!-\!\text{NCO}$$

$$\xrightarrow{+\text{azobisalcohol}} (\!\!-\!\!\text{R}'\!-\!\text{N}\!\!=\!\!\text{N}\!-\!\text{O}\!-\!\overset{\overset{\text{O}}{\|}}{\text{C}}\!-\!\text{NH}(\!\!-\!\!\text{R}\!-\!\text{NH}\!-\!\overset{\overset{\text{O}}{\|}}{\text{C}}\!-\!\text{O}\text{\textasciitilde\textasciitilde}\text{O}\!-\!\overset{\overset{\text{O}}{\|}}{\text{C}}\!-\!\text{NH}\!)\text{R}\!-\!\text{NH}\!-\!\overset{\overset{\text{O}}{\|}}{\text{C}}\!)_n$$

where R and R' are hydrocarbon residues of diisocyanate and azobisalcohol, respectively.

Vinyl acetate, styrene, methyl methacrylate, and vinyl chloride were polymerized with macroazo initiators as initiator in benzene solution at 85°C. ABA-type block copolymers of polyether and the vinyl monomers mentioned previously, except vinyl chloride, were obtained with high conversion (86–98%). But the conversion in the case of vinyl chloride did not exceed 44%. Block copolymer synthesis by a similar method has been described and the kinetics of styrene polymerization by the macroazocarbamate was studied at 80°, in bulk [74].

Polyazoamides. Polymeric azoamides containing —N=N— units have been synthesized by the interfacial polycondensation between sebacoyl chloride (or adipoyl chloride) and hexamethylenediamine in the presence of small amounts of azobiscyanopentanoyl chloride at room temperature. CCl_4 or nitrobenzene was used as organic phase and NaOH aqueous solution as an aqueous phase [75–77]. A typical polyamide-polystyrene block copolymers then can be prepared by decomposition of the azo units of polyazoamid, initiating radical polymerization of styrene in cresol. A general polyazoamide formula can be designated as indicated below:

$$+\!\!\!\begin{array}{c}O\\\|\\C\end{array}\!\!-R_1-N\!=\!N-R_1-\begin{array}{c}O\\\|\\C\end{array}\!\!+\!NH-R_2-NH-\begin{array}{c}O\\\|\\C\end{array}\!\!-R_3-\begin{array}{c}O\\\|\\C\end{array}\!\!\!\!+_n\!\!+_m$$

polyazoamide

$$R_1 = -CH_2CH_2-\underset{\underset{CN}{\|}}{\overset{\overset{CH_3}{|}}{C}}-,\ R_2 = +CH_2+_6,\ R_3 = +CH_2+_8$$

Polyazoesters. Laverty and Gardlund [78] prepared block copolymers of poly(oxyethylene) (M_n = 300, 600, 1000, 4000, 6000, 12,000 g mol^{-1}) and polyvinyl chloride by first condensing 4,4'-azobis(4-cyanopentanoylchloride with dihydroxy-terminated poly(oxyethylene) to form a polyazoester with several azo groups in the chain:

$$\underset{CN}{\overset{O}{\|}}{ClCCH_2CH_2}\underset{|}{\overset{CH_3}{|}}{C}-N\!=\!N-\underset{|}{\overset{CH_3}{|}}{C}CH_2CH_2\overset{O}{\overset{\|}{CCl}} + HO-\text{polyoxyethylene}-OH$$

$$\rightarrow +CCH_2CH_2\underset{CN}{\overset{CH_3}{|}}{C}-N\!=\!N-\underset{CN}{\overset{CH_3}{|}}{C}CH_2CH_2\overset{O}{\overset{\|}{CO}}-\text{polyoxyethylene}+_m$$

Polyazoester, m = 5–24

Subsequent thermal initiation of the polyazoester in the presence of a vinyl monomer led to the formation of a block copolymer [79, 80]. Multiblock copolymers of the $(BA)_n$–B-type, where A is poly(dimethylitaconate) and B is polybutadiene, were synthesized using a polyazoester similarly prepared from hydroxyl-terminated polybutadienes instead of poly(oxyethyleneglycol) [81].

Nitriles react with alcohols in the presence of hydrochloric acid to form iminoester hydrochlorides which are hydrolyzed to the esters (Pinner synthesis). Heitz and coworkers [82–90] published several papers on the polyazoester synthesis from the reaction of a series of poly(oxyethylene) glycol or poly(oxypropylene)glycol and azobisisobutironitrile (AIBN) in the presence of dry hydrochloric acid at 0–5°C according to Pinner synthesis [89]. The reaction pathways of the polyazoester synthesis can be shown as follows:

$$HO-R-OH + AIBN \xrightarrow{HCl} +O-R-O-\underset{\underset{CH_3}{|}}{\overset{\overset{HCl \cdot HN}{\|} \overset{CH_3}{|}}{C}}-\underset{\underset{CH_3}{|}}{C}-N=N-\underset{\underset{CH_3}{|}}{\overset{\overset{CH_3}{|} \overset{NH \cdot HCl}{\|}}{C}}-C+_n$$

$$\xrightarrow[-NH_4Cl]{+H_2O} +O-R-O-\underset{\underset{CH_3}{|}}{\overset{\overset{O}{\|} \overset{CH_3}{|}}{C}}-C-N=N-\underset{\underset{CH_3}{|}}{\overset{\overset{CH_3}{|} \overset{O}{\|}}{C}}-C+_n$$

polyazoester
R = di-, tri-, tetra-, poly(ethyleneoxide) glycol, and 1,6 hexanediol; $n < 12$

Partial decomposition of this polyazo initiator in the presence of different vinyl monomers results in an active polymer containing azo groups. This active polymer can initiate the polymerization of a second monomer to give AB-, ABA-, or (AB)$_n$-type block copolymer depending on the kind of termination reaction of monomer used [85]. This type of synthesis allows combinations of hard and soft segments (thermoplastic elastomers) [83] and of hydrophylic and hydrophobic segments (surface-active polymers) [84, 86, 87].

Other azo catalysts. Sheppard and McLeay [91, 92] synthesized an azo catalyst having two different thermal stabilities:

$$H_3C-\underset{\underset{CH_3}{|}}{\overset{\overset{CH_3}{|}}{C}}-N=N-\underset{\underset{CH_3}{|}}{\overset{\overset{CH_3}{|}}{C}}-S-CH_2-CH_2-O-\overset{\overset{O}{\|}}{C}-CH_2-CH_2-\underset{\underset{CN}{|}}{\overset{\overset{CH_3}{|}}{C}}-N=N-\underset{\underset{CH_3}{|}}{\overset{\overset{CH_3}{|}}{C}}-CH_3$$

The less stable azo group adjacent to the nitrile moiety first decomposes in the presence of styrene, and subsequent thermolysis of the more stable azo group occurs in the presence of monomeric methyl methacrylate.

The amide groups can be converted into diazo compounds by diazotation with HNO_2:

$$R_1-NH-CO-R_2 + HNO_2 \xrightarrow{-H_2O} R_1-\underset{\underset{NO}{|}}{N}-CO-R_1 \rightarrow R_1-N=N-OCO-R_2$$
$$\text{N-nitroso-N-acylamine} \qquad \text{Diazoester}$$

Berlin et al. [93, 94] carried out the preparation of polyazophenylene and copolymers containing conjugated fragments. Aromatic N,N'-bis(nitrosoacetyl) diamine gives aromatic biradicals, by splitting CO_2 and N_2:

$$CH_3-\overset{\overset{O}{\|}}{\underset{\underset{NO}{|}}{C}}-N-\underset{}{\bigcirc}-R-\underset{}{\bigcirc}-\underset{\underset{NO}{|}}{N}-\overset{\overset{O}{\|}}{C}-CH_3 \rightarrow CH_3-\overset{\overset{O}{\|}}{C}-O-N=N-\underset{}{\bigcirc}-R-\underset{}{\bigcirc}-\approx$$

$$-N=N-O-\overset{\overset{O}{\|}}{C}-CH_3 \xrightarrow{\Delta} 2CH_3\cdot +\cdot\underset{}{\bigcirc}-R-\underset{}{\bigcirc}\cdot +CO_2 +N_2$$
$$\text{biradical}$$

R = $-O-$, $-CH_2-$, $-SO_2-$ or direct bond.

Emulsion polymerization of isoprene initiated by this aromatic biradical gives a block copolymer consisting of polyphenylene and polyisoprene blocks. The polyazophenylene units in the block copolymer are formed from the polyrecombination of the biradicals during the polymerization of isoprene.

Some new monomeric and polymeric azo initiators were prepared by Nuyken and colleagues [95, 96]:

$$\text{Monomeric azoinitiator} \xrightarrow{\text{vinyl monomer}} \text{Polymeric azoinitiator}$$

Nuyken and Weidner recently reviewed polymeric azo initiators [96].

Azoperoxidic Initiators

Azoperoxide compounds are particularly important in block copolymer synthesis since the thermal stabilities of the azo groups and the peroxide groups are different. This difference is only possible when the azo and peroxy groups are attached to different carbon atoms. If the azo and peroxy groups are attached to the same carbon atom, azoperoxy initiators play the role of one-step decomposable initiators [13]. In block copolymer synthesis, the azo groups are first thermally activated in the presence of a vinyl monomer, followed by a second-stage polymerization with the other monomer. Since active polymers obtained in the first step have peroxy group in the polymer backbone, they can initiate the other vinyl polymerization, or the reverse procedure can be applied to obtain block copolymers by using azoperoxidic initiators. In these compounds, as the azo groups are thermally activated in order to produce free radicals peroxygen groups can be activated either thermally or by reducing agent (e.g., tetraethylene pentamine) at room temperature. A common synthesis of azoperoxidic initiators is carried out by the condensation reaction of 4,4'-azobis(4-cyanovaleric acid dichloride) and *t*-butyl hydroperoxide, peracetic acid, or perbenzoic acid [97–101].

$$\underset{\text{Azodiacyl peroxide,}}{\text{Cl}-\overset{\text{O}}{\overset{\|}{\text{C}}}(\text{CH}_2)_3\overset{\text{CN}}{\underset{\text{CH}_3}{\overset{|}{\text{C}}}}-\text{N}=\text{N}-\overset{\text{CN}}{\underset{\text{CH}_3}{\overset{|}{\text{C}}}}(\text{CH}_2)_2\overset{\text{O}}{\overset{\|}{\text{C}}}-\text{Cl} + 2\text{ROOH}} \xrightarrow{+\text{KOH}}$$

$$\text{ROOC}(\text{CH}_2)_2\overset{\text{CN}}{\underset{\text{CH}_3}{\overset{|}{\text{C}}}}-\text{N}=\text{N}-\overset{\text{CN}}{\underset{\text{CH}_3}{\overset{|}{\text{C}}}}(\text{CH}_2)_2\overset{\text{O}}{\overset{\|}{\text{C}}}-\text{OOR}$$

Azodiacyl peroxide, R = $\text{CH}_3-\overset{\text{O}}{\overset{\|}{\text{C}}}-$

$\text{C}_6\text{H}_5-\overset{\text{O}}{\overset{\|}{\text{C}}}-$

$(\text{CH}_3)_3\text{C}-$

In the first stage of the polymerization, azoperoxidic initiators can produce free radicals by decomposition of either azo or peroxy groups in the presence of a vinyl monomer, and active polymers are formed:

$$\text{ROOC}-\text{R}-\text{N}=\text{N}-\text{R}-\text{COOR}$$

$\Delta, 60°C \swarrow \qquad \searrow +\text{tetraethylenepentamine, } 25°C$

$2\text{ROOC}-\text{R}\cdot \qquad\qquad \cdot\text{OC}-\text{R}-\text{N}=\text{N}-\text{R}-\text{CO}\cdot + 2\text{RO}\cdot$

$\downarrow +\text{monomer} \qquad\qquad\qquad \downarrow +\text{monomer}$

$\text{ROOCR}\sim\sim\text{RCOOR} \qquad\qquad \sim\sim\text{OCRNNRCO}\sim\sim$

Active polymer having   Active polymer having
terminal peroxy groups   azogroup in the middle
                                        of polymer backbone

In the second stage, active polymers initiate the polymerization of a second vinyl monomer. In this way, poly(styrene-*b*-methyl methacrylate) and poly(butylacrylate-*b*-methyl methacrylate) block copolymers were synthesized with high homopolymer formation (50%) [100, 101].

The synthesis and decomposition studies of some azoacyl peroxides were described elsewhere [102–104]. A new type of polyazo–peroxy initiator used in multiblock copolymer synthesis was described by Hazer [105]. It can be prepared by the coupling reaction of polymeric peroxycarbamate [106] having isocyanate groups and a polyazoester [85] having hydroxyl terminal groups. To synthesize multiblock copolymers using this initiator, a procedure with three stages is necessary. In the first stage, peroxyazo initiators partly decomposed in styrene give an active polymer. The amount of peroxygen in the active polymer decreases as the polymerization time increases (Fig. 3). In the second stage, active polymers are used in the polymerization of a second monomer to obtain block copolymers which still have undecomposed peroxygen groups in the polymer backbone. Peroxide and azo groups remaining in the active polymers and block copolymers are arranged in the middle positions in the main chain as understood from the data in the Table 4. Since the active block copolymers have undecomposed peroxygen and azo groups, they were used in the third vinyl monomer polymerization. Poly(styrene-*b*-methyl methacrylate-*b*-butyl methacrylate) and poly(styrene-*b*-methyl methacrylate-*b*-acrylonitrile) block copolymers were obtained this way [105].

Block Copolymerization via the Step-Growth Procedure

Any condensation reaction such as the reaction between acid chloride and alcohol (or amine) or isocyanate and alcohol can be applied for block copolymer synthesis. Step-growth polymerizations give block copolymers having sequences with predictable molecular weights, and a pure reaction medium is not necessary for block copolymerization. But polycondensation reactions are performed only when block lengths are short. Block

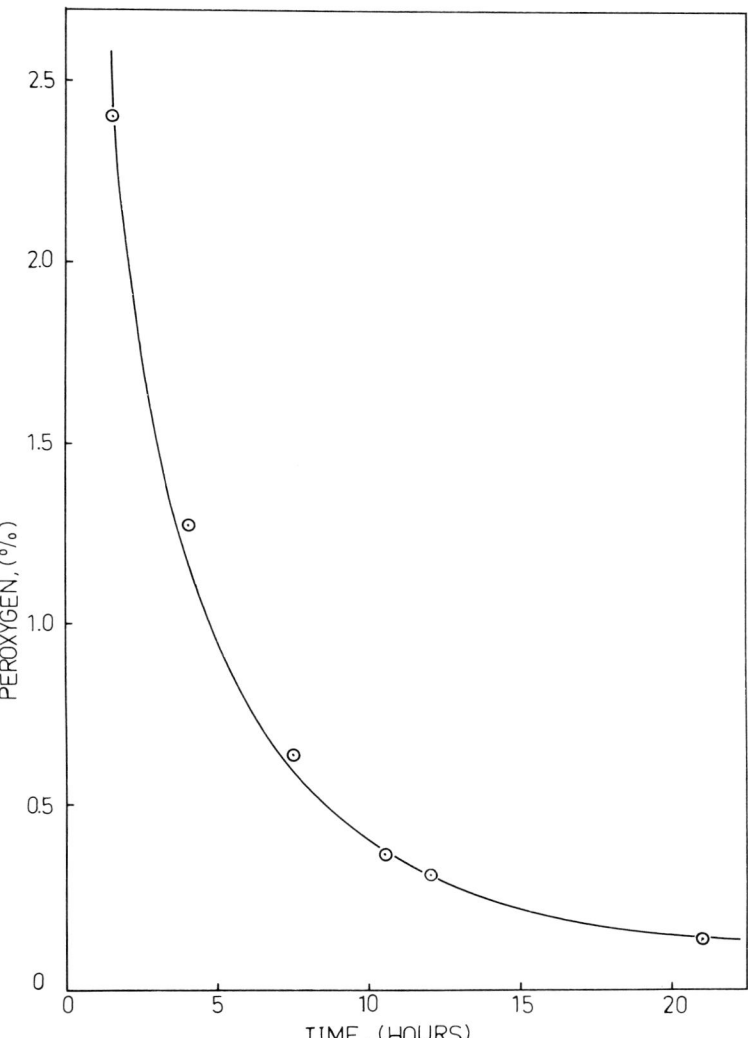

Figure 3 Variation of the peroxygen amount in active PS samples with the time of styrene copolymerization at 80°C. Copolymerization with 10 wt% peroxyazoinitiator, II-*e*. (From Ref. 105.)

copolymers with short block lengths usually arranged in $-(AB)-n$ · AB, ABA, and BAB block copolymers also can be obtained by the step-growth procedure. The step-growth procedure can be divided into three types: esterification, amidation, and urethane formation.

Esterification

Esterification is the reaction between acid chlorides (or acids) and alcohols. For this, poly(oxyethylene), polystyrene, or polybutadiene with dicarboxylic acid chloride terminal

Table 4 Intrinsic Viscosities before and after Decomposition of the Active Polymers (PS-II-e-1 and PS-II-e-2) and Active Block Copolymers (cop-26–29)

Active polymer	η of active polymer (dl g^{-1})	
	Before decomposition	After decomposition
PS-II-e-1	0.72	0.68
PS-II-e-2	0.36	0.28
Cop-26	2.20	1.80
Cop-27	2.80	1.30
Cop-28	3.30	2.10
Cop-29	2.40	1.10

Source: Data reprinted with permission from Ref. 105. Copyright 1985 Hüthig and Wepf Verlag.

groups are mainly reacted with polyethers, aromatic or aliphatic polyesters with hydroxyl terminal groups, and hydroxy-terminated polystyrene [107–110].

$$Cl-\underset{\underset{O}{\|}}{C}\sim\underset{\underset{O}{\|}}{C}-Cl + HO\sim\sim\sim OH \rightarrow Cl + (\underset{\underset{O}{\|}}{C}\sim\underset{\underset{O}{\|}}{C}-O\sim\sim\sim O)_n H$$

$+(AB)_n$-type block copolymer

When polysulfones with acid chloride terminal groups are reacted with α,ω-di(sodium phenolate) oligosulfone, block copolymers based on polysulfones are obtained [111–114]:

$$NaO\sim\sim\bigcirc-SO_2-\bigcirc\sim\sim ONa + Cl-\underset{\underset{O}{\|}}{C}-\text{polybutadiene}-\underset{\underset{O}{\|}}{C}-Cl$$

$$\downarrow -NaCl$$

$$+(O\sim\sim\bigcirc-SO_2-\bigcirc\sim\sim O\underset{\underset{O}{\|}}{C}-\text{polybutadiene}-\underset{\underset{O}{\|}}{C})_n$$

polysulfon-polybutadien block copolymer

Siloxane block copolymers displaying thermoplastic elastomer properties can be obtained by polycondensation reactions of hydroxyl-terminated polymers and functionalized (COOH, –Cl-terminated) dimethyl siloxanes [115–119]. Polycarbonate–polysiloxane block copolymers show good film properties and hydrolytic stability:

$$\sim\sim\left[\underset{\underset{O}{\|}}{C}-O-\bigcirc-\underset{\underset{CH_3}{|}}{\overset{\overset{CH_3}{|}}{C}}-\bigcirc-O\right]_m\left[\underset{\underset{CH_3}{|}}{\overset{\overset{CH_3}{|}}{Si}}-O\right]_n\sim\sim$$

Polycarbonate–polydimethylsiloxane block copolymer

Amidation

Amidation reactions lead to block copolyamides. Diacid chlorides also react with diamine compounds at low temperatures to give block copolyamides [120–125]:

$$HO-\text{polyoxyethylene}-OH \xrightarrow{ClC(O)(CH_2)_4C(O)Cl} Cl-\overset{O}{\underset{\|}{C}}-\text{polyoxyethylene}-\overset{O}{\underset{\|}{C}}-Cl \quad (I)$$

$$x(I) + y Cl-\overset{O}{\underset{\|}{C}}(CH_2)_4\overset{O}{\underset{\|}{C}}-Cl + (y+x)H_2N(CH_2)_6NH_2 \rightarrow$$

$$-(-\overset{O}{\underset{\|}{C}}-\text{polyoxyethylene}-\overset{O}{\underset{\|}{C}}+NH(CH_2)_6NH\overset{O}{\underset{\|}{C}}(CH_2)_4\overset{O}{\underset{\|}{C}}+_n NH(CH_2)_6NH-)_m$$

Block copolyamides have unusual properties such as high strengths, high melting points (350°C and higher), and low solubilities (soluble only in very polar solvents such as sulfuric acid) and therefore constitute a remarkable class of materials.

Polyoxyethylene–polymetaphenylene isophthalamide–polyoxy–ethylene triblock copolymers are similarly synthesized in three steps [122]:

Step 1: $nH_2N-C_6H_4-NH_2 + nCl-\overset{O}{\underset{\|}{C}}-C_6H_4-\overset{O}{\underset{\|}{C}}-Cl \rightarrow Cl-\overset{O}{\underset{\|}{C}}\sim NH_2$

Step 2: $Cl-\overset{O}{\underset{\|}{C}}\sim NH_2 + Cl-\overset{O}{\underset{\|}{C}}-C_6H_4-\overset{O}{\underset{\|}{C}}-Cl \rightarrow Cl-\overset{O}{\underset{\|}{C}}\sim\overset{O}{\underset{\|}{C}}-Cl$

Step 3: $Cl-\overset{O}{\underset{\|}{C}}\sim\overset{O}{\underset{\|}{C}}-Cl + HO\sim\sim OH \rightarrow H(-O\sim\sim O-\overset{O}{\underset{\|}{C}}\sim\overset{O}{\underset{\|}{C}}-O\sim\sim O-)H$
          polyoxyethylene   (predominantly
          (large excess)   triblock structure)

Polyetheramide block copolymers having high structural regularity were synthesized by the direct polycondensation reactions of amines and carboxylic acids in the presence of pyridine and triphenylphosphine [126–128]:

$$\underset{\text{Polyoxyethylene}}{HOOC\sim\sim COOH} + \underset{\text{Polyaramide}}{H_2N\sim\sim\sim NH_2} \xrightarrow[\text{pyridine}]{P(OC_6H_5)_3} H[\underset{\text{Polyoxy-ethylene}}{OOC\sim\sim C-}\underset{\text{Polyaramide}}{NH\sim\sim\sim NH+}]_n H$$

These reactions have been studied by Yamazaki et al. in detail [129].

Transesterification is also used to obtain block copolymers of polyether ester thermoplastic elastomer and block copolymers of bisphenol A polycarbonate and polybutylene terephthalate [130, 131]. ABA-type block copolymers having different properties are synthesized in similar ways [132, 133]. Wondraczeck and Kennedy [134] synthesized di-, tri-, and starblock copolymers based on the reaction hydroxy-terminated polyisobutylene, PIB(OH)$_n$, and nylon-6.

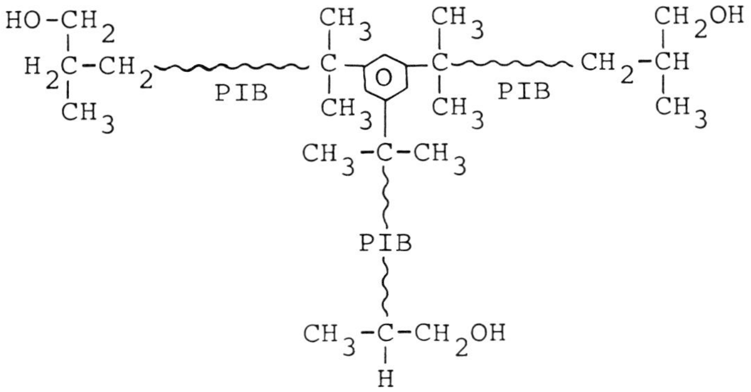

Trihydroxy-terminated polyisobutylene

After these hydroxyl groups were converted into carboxylic acid chlorides on isocyanates, block copolymers are formed with polyamide blocks having terminal amine groups.

Urethane Formation

Hydroxyl- or carboxyl-terminated polymers (telechelics) [135–138] can be obtained by the polymerization of styrene, butadiene, or isoprene in the presence of initiators having functional terminal groups such as polyazoester [82] or azobiscyanovaleric acid [59].

Urethane formation generated from the reaction of isocyanate and alcohol is successfully applicable in the preparation of block copolymers. Grezlak and Wilkes [139] synthesized ABA-type block copolymers from the reaction of hydroxyl-terminated poly(methyl methacrylate) (PMMA) and diisocyanate-terminated polyesters (PEst). Here, hydroxyl-terminated PMMAs were obtained by the free radical polymerization of methyl methacrylate in the presence of azobisisobutyronitrile and mercaptoethanol (as chain transfer agent):

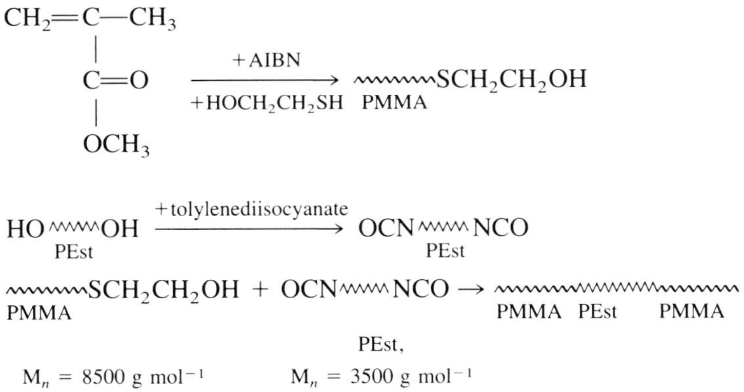

Block copolymers based on polyoxyethylene were also synthesized this way [140–146]. Rahman and Avny [143–145] synthesized AB-, ABA-, and BAB-type block copolymers based on poly(oxyethylene) glycol and hydroxy- or carboxy-terminated polybutadiene via isocyanate coupling reaction. Mechanical, viscoelastic, and physical properties of these block copolymers were studied [143].

Block Copolymerization via the Anionic Mechanism

Since Szwarc et al. [12, 147] discovered that the well-defined structure of living polymers could be produced in an anionic mechanism, major advances in anionic polymerization have been carried out. By this discovery, in particular $(AB)_n$ and ABA block copolymers (thermoplastic elastomers) arose a new type material. These block copolymers consist of glassy blocks (A) and soft blocks (B). Polybutadiene or polyisoprene (B) is usually combined with polystyrene (A), whose softening point is around 90°C and thus limits the range of service temperature. These block copolymers, however, consisting of completely linear chains with no chemical crosslinks between them, exhibit mechanical behavior similar to that of rubber vulcanizates.

Anionic block copolymerization was carried out by two different types of catalyst systems: monofunctional initiators (e.g., butyl lithium) and difunctional initiators (e.g., sodium naphthalene and dilithio compounds). Cyclic oxides, cyclic esters, cyclic amides, and some vinyl monomers can undergo anionic polymerization in very pure reaction media with negligible homopolymer formation.

Monofunctional Initiators

sec-Butyl, *iso*-butyl, *tert*-butyl, and some organolithium initiators are monofunctional anionic initiators and they are used to obtain ABA-type block copolymers in a three-stage process [148–161]. They have different stabilities for different vinyl monomers. For instance, relative reactivities of butyllithium initiators for styrene are

$$\text{sec-BuLi} > \text{iso-BuLi} > \text{tert-BuLi}$$

The three-stage process leading to ABA block copolymers by these initiators can be written as follows:

Step 1: *sec*-BuLi + styrene → *sec*-Bu∼∼⁻Li⁺
   (polystyryl anion)

Step 2: polystyryl anion + butadiene → *sec*-Bu∼∼—⁻Li⁺

Step 3: *sec*-Bu∼∼—⁻LiPL $\xrightarrow[\text{2) termination}]{\text{1) +styrene}}$ PS∼∼PB∼∼PS (ABA block copolymer)

Difunctional Initiators

Sodium naphthalene difunctional initiators. The initiation step in the polymerization of styrene can be considered as follows [12, 147]:

At low temperatures, the radical ends dimerize and difunctional initiator is formed:

$$\text{Na}^+ \ \bar{C}H\text{-}CH_2\text{:} \ (Ph) \quad + \quad \text{:}CH_2\text{-}\bar{C}H \ \text{Na}^+ \ (Ph) \quad \longrightarrow \quad \text{Na}^+ \ \bar{C}H\text{-}CH_2\text{-}CH_2\text{-}\bar{C}H\text{Na}^+ \ (Ph)(Ph)$$

Difunctional initiator

$$\xrightarrow{+\text{styrene}} \quad {}^-\!\!\sim\!\!\sim\!\!\sim\!\!\underset{PS}{}\!\!\sim\!\!\sim\!\!\sim\text{CH-}CH_2\text{-}CH_2\text{-}CH\!\sim\!\!\sim\!\!\sim\!\!\underset{PS}{}\!\!\sim\!\!\sim\!\!\sim{}^-$$
(with Ph groups on the CH carbons)

$$\xrightarrow[\text{termination}]{+\text{second monomer}} \quad \sim\!\!\sim\!\!\sim\text{CH-}CH_2\text{-}CH_2\text{-}CH\!\sim\!\!\sim\!\!\sim$$
(with Ph groups on the CH carbons)

Then the living block copolymer is terminated with water or any polar solvent. Poly(α-methylstyrene-*b*-styrene-*b*-α-methylstyrene) [162], poly(α-methyl styrene-*b*-butadiene-*b*-α-methylstyrene) [163], ABA triblock copolymers, and poly(oxyethylene-*b*-isoprene-*b*-styrene-*b*-isoprene-*b*-oxyethylene) ABCBA block copolymers [164] were synthesized by this initiator system.

Dilithio initiators. Dilithio initiators can be prepared from some vinyl compounds and lithium [165–170]. One of these is 1,1,4,4-tetraphenyl butane prepared from 1,1-diphenylethylene and lithium [165, 166]:

$$2\underset{\underset{Ph}{|}}{\overset{\overset{Ph}{|}}{C}}\!\!=\!\!CH_2 + 2Li \rightarrow Li^{+\,-}\underset{\underset{Ph}{|}}{\overset{\overset{Ph}{|}}{C}}\!\!-\!CH_2\!-\!CH_2\!-\!\underset{\underset{Ph}{|}}{\overset{\overset{Ph}{|}}{C}}{}^-Li^+$$

1,4,-Dilithio-1,1,4,4-tetraphenyl butane

Block copolymers are prepared from dilithium initiators by at least two different routes. These can be shown in the following generalized schematic equations for polymerizations involving styrene and butadiene [167]:

$$Li^{+\,-}R^-Li^+ \xrightarrow{\text{butadiene (B)}} Li^{+\,-}\!\!-\!\!\text{polybutadiene}\!\!-\!\!{}^-Li^+ \xrightarrow[\text{2) termination}]{\text{1) styrene (S)}} \text{PS---PB---PS}$$
"pure" blocks

and

$$Li^+RLi^+ \xrightarrow[\text{2) termination}]{\text{1) styrene (S) + butadiene (B)}} \text{SS . . . SBSBSBB . . . BSBSBSS . . . SS}$$
"tapered" blocks

Polyoxyethylene (POE) with amide terminal groups also behaves as a difunctional anionic initiator in the presence of sodium [171]:

$$\text{PhCH}_2\text{NH}\text{\textasciitilde}\text{NHCH}_2\text{Ph} + 2\text{Na} \rightarrow \underset{\underset{\text{Ph—CH}_2}{|}}{\text{N}^-}\text{\textasciitilde}\underset{\underset{\text{CH}_2\text{—Ph}}{|}}{\text{N}^-} \xrightarrow{+\text{methyl methacrylate (MMA)}}$$
POE, Mn = 980, Na⁺ Na⁺

$$\underset{\underset{\text{Ph—CH}_2}{(PMMA)|}}{\text{———N}}\text{\textasciitilde}\underset{\underset{\text{CH}_2\text{—Ph}}{|(PMMA)}}{\text{N———}}$$
(POE)

Anionic Block Copolymerization of Cyclic Monomers

Organolithium-initiated anionic polymerizations of cyclic sulfides lead to "living" systems, which are therefore suitable for the synthesis of block copolymers of predetermined structure:

$$\text{RLi} + \overset{S}{\overset{/\backslash/\backslash}{-\text{C}-\text{C}-}} \rightarrow \text{R—C—C—S—Li}$$

Morton and Mikesell reported that ethylene sulfide–butadiene triblock copolymers were much weaker and less extensible than the corresponding styrene–diene–styrene block copolymers [172].

Hshieh successfully prepared polydiene–polyester block copolymers by using butyllithium as initiator and by polymerizing vinyl monomer first followed by addition and polymerizing of lactone [173]:

$$\text{\textasciitilde}^-\text{Li}^+ + n\underset{\underset{\underset{\text{Lactone}}{(\text{CH}_2)_n}}{|}}{\overset{\overset{O}{\|}}{\text{C}}\text{———O}} \rightarrow \text{\textasciitilde}[\overset{\overset{O}{\|}}{\text{C}}(\text{CH}_2)_x\text{O}]_n \overset{\overset{O}{\|}}{\text{C}}(\text{CH}_2)_x\text{—O}^-\text{Li}^+$$
Polyester

Block copolymerization of lactams usually is carried out via polymeric activators [174]. Polystyrene-bis-acyllactams activators [175, 176], α,ω-dicarbonylcaprolactam–polyoxyethylene activators [177], and polysiloxane-based activators [178] are all used in the anionic block copolymerization of ε-caprolactam [175]:

$$\underset{(\text{CH}_2)_m}{\overset{\overset{O}{\|}}{\text{C}}\text{—N}}\overset{\overset{O}{\|}}{\text{—C}}\text{—CH}_2\text{CH}_2\underset{\underset{\text{CH}_3}{|}}{\overset{\overset{\text{CN}}{|}}{\text{—C}}}\text{—N}=\text{N}\underset{\underset{\text{CH}_3}{|}}{\overset{\overset{\text{CN}}{|}}{\text{—C}}}\text{—CH}_2\text{CH}_2\overset{\overset{O}{\|}}{\text{—C}}\text{—N}\underset{(\text{CH}_2)_m}{\overset{\overset{O}{\|}}{\text{—C}}}$$

$$\downarrow +\text{styrene}$$

$$\underset{(\text{CH}_2)_m}{\overset{\overset{O}{\|}}{\text{C}}\text{—N}}\overset{\overset{O}{\|}}{\text{—C}}\text{—CH}_2\text{CH}_2\underset{\underset{\text{CH}_3}{|}}{\overset{\overset{\text{CN}}{|}}{\text{—C}}}\text{—polystyrene}\underset{\underset{\text{CH}_3}{|}}{\overset{\overset{\text{CN}}{|}}{\text{—C}}}\text{—CH}_2\text{CH}_2\overset{\overset{O}{\|}}{\text{—C}}\text{—N}\underset{(\text{CH}_2)_m}{\overset{\overset{O}{\|}}{\text{—C}}}$$

Polystyrene-bis-acyllactam

$$\underset{(CH_2)_m}{\overset{O}{\underset{\|}{C}}-N}\left(\begin{array}{c}\\\end{array}\right)\underset{\text{Polyamide block}}{[C(CH_2)_5NH]_x}\overset{O}{\underset{\|}{C}}-CH_2CH_2-\underset{CH_3}{\overset{CN}{\underset{|}{C}}}-\text{polystyrene}-$$

$$\approx-\underset{CH_3}{\overset{CN}{\underset{|}{C}}}-CH_2CH_2-\overset{O}{\underset{\|}{C}}\underset{\text{Polyamide block}}{[NH(CH_2)_5 C]_x}\underset{(CH_2)_m}{\overset{O}{\underset{\|}{N}}}\overset{O}{\underset{\|}{C}}$$

The activators are recommended to accelerate the initiation of the ε-caprolactam polymerization with a strong base. As the activator residue is anchored to the growing polyamide chain, the use of a macromolecular activator provides an attractive pathway to block copolymerization [178].

Block Copolymers by Cationic Polymerization

Since the discovery by Szwarc of living anionic polymerization, a great deal of research has been conducted in this area. But the living cationic mechanism leading to block copolymers was only recently considered in block copolymer syntheses. The most important disadvantage is chain transfer reactions, especially to the heterocyclic monomers. Some vinyl monomers such as styrene, isobutylene, ethyl vinyl ethers, and cyclic monomers such as tetrahydrofurane, 1-*tert*-butyl aziridine, 1,3-dioxalone, and 1,3-dioxepane can be copolymerized cationically in the presence of several catalysts.

One of the first attempts to get block copolymer from tetrahydrofurane and oxetane via a living cationic mechanism was made by Dreyfuss and Dreyfuss [179]. They used $PhN_2^+ PF_6^-$ as catalyst in the polymerization of tetrahydrofurane and subsequently added oxetane. Kennedy and Melby [180] then prepared block copolymers of isobutylene and styrene AB block copolymers via living cationic mechanism in the presence of 2,6-dimethyl 2-bromo 6-chloroheptane/Et_3Al catalyst system. An HI/I_2 initiator system also leads to well-defined living polymers of a very narrow molecular weight distribution. The first example of this was in the work of Sawamoto et al. [181]. They prepared block copolymers of ethyl propenyl ether and isobutyl vinyl ether in this way. The catalysts for living cationic polymerization are listed in the Table 5. A promising development in block copolymer synthesis via living cationic polymerization is the inifer technique of Kennedy [190–192]. This is a new synthesis technique for the preparation of telechelic polymers that involves controlled initiation by and chain transfer to special bifunctional initiating and transfer agents (inifer). Since the transfer to catalyst always predominates the transfer to monomer during the polymerization, it will be possible to overcome the disadvantages of chain transfer to monomer, or monomer transfer for short. Polymerization steps in this technique can be shown as follows (*p*-dicumylchloride + BCl_3 inifer system) [191]:

Ion generation:

$$Cl-\underset{CH_3}{\overset{CH_3}{\underset{|}{\overset{|}{C}}}}-\underset{\bigcirc}{}-\underset{CH_3}{\overset{CH_3}{\underset{|}{\overset{|}{C}}}}-Cl \quad +BCl_3 \longrightarrow Cl-\underset{CH_3}{\overset{CH_3}{\underset{|}{\overset{|}{C}}}}-\underset{\bigcirc}{}-\underset{CH_3}{\overset{CH_3}{\underset{|}{\overset{|}{C}}}}{}^+ \quad BCl_4^-$$

Table 5 Some Catalysts Leading Living Cationic Polymerization

Catalyst	Reference
$PhN_2^+ PF_6^-$	179
$CH_3-\underset{CH_3}{\underset{|}{\overset{Br}{\overset{|}{C}}}}-CH_2CH_2-\underset{CH_3}{\underset{|}{\overset{Cl}{\overset{|}{C}}}}-CH_3 + Et_3Al$	180
$HI + I_2$	181
$Et_3O^+ BF_4^-$	182
$Ph_2C^+\!-\!\bigcirc\!-\!(CH_2)_2\!-\!\bigcirc\!-\!C^+Ph_2 \cdot 2SbCl_6^-$	183, 184
$Ph_3C^+ SbCl_6^-$	185
CF_3SO_3H	186, 187
$Cl-\underset{CH_3}{\underset{|}{\overset{CH_3}{\overset{|}{C}}}}-\bigcirc-\underset{CH_3}{\underset{|}{\overset{CH_3}{\overset{|}{C}}}}-Cl + AgSbF_6$	188
I_2	189

Cationation:

$$Cl-\underset{CH_3}{\underset{|}{\overset{CH_3}{\overset{|}{C}}}}-\bigcirc-\underset{CH_3}{\underset{|}{\overset{CH_3}{\overset{|}{C}}}}^+ + M\,(Monomer) \longrightarrow Cl-\underset{CH_3}{\underset{|}{\overset{CH_3}{\overset{|}{C}}}}-\bigcirc-\underset{CH_3}{\underset{|}{\overset{CH_3}{\overset{|}{C}}}}-M^+$$

Propagation:

$$Cl-\underset{CH_3}{\underset{|}{\overset{CH_3}{\overset{|}{C}}}}-\bigcirc-\underset{CH_3}{\underset{|}{\overset{CH_3}{\overset{|}{C}}}}-M^+ + nM \longrightarrow Cl-\underset{CH_3}{\underset{|}{\overset{CH_3}{\overset{|}{C}}}}-\bigcirc-\underset{CH_3}{\underset{|}{\overset{CH_3}{\overset{|}{C}}}}-M\!\sim\!\sim\!M^+$$

Chain transfer to inifer:

$$Cl-\underset{CH_3}{\underset{|}{\overset{CH_3}{\overset{|}{C}}}}-\bigcirc-\underset{CH_3}{\underset{|}{\overset{CH_3}{\overset{|}{C}}}}-M\!\sim\!\sim\!M^+ + Cl-\underset{CH_3}{\underset{|}{\overset{CH_3}{\overset{|}{C}}}}-\bigcirc-\underset{CH_3}{\underset{|}{\overset{CH_3}{\overset{|}{C}}}}-Cl \longrightarrow$$

$$Cl-\underset{CH_3}{\underset{|}{\overset{CH_3}{\overset{|}{C}}}}-\bigcirc-\underset{CH_3}{\underset{|}{\overset{CH_3}{\overset{|}{C}}}}-M\!\sim\!\sim\!M-Cl + Cl-\underset{CH_3}{\underset{|}{\overset{CH_3}{\overset{|}{C}}}}-\bigcirc-\underset{CH_3}{\underset{|}{\overset{CH_3}{\overset{|}{C}}}}^+$$

Chain transfer to organic halides yields a polymer with a halide end group and, provided carbenium ion continues the kinetic chain, a polymer with chlorine terminal groups [191]. The chlorine-terminated polyisobutylene obtained by this inifer technique has been used to initiate the subsequent polymerization of styrene or α-methylstyrene in the presence of diethylaluminum chloride [192]:

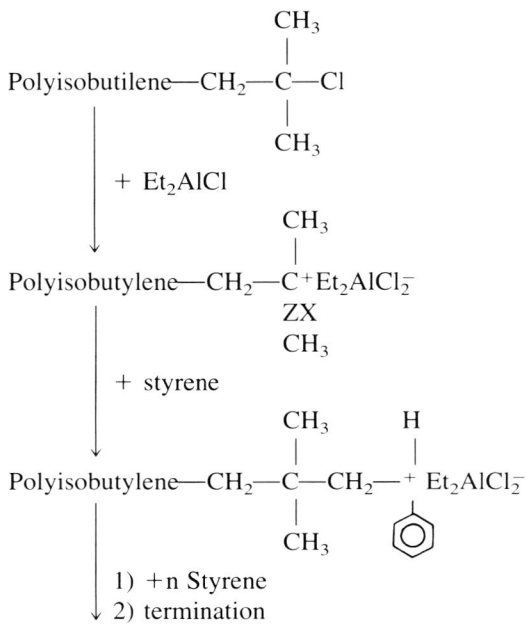

Poly(isobutylene-b-styrene)
AB type block copolymer

Star-shaped (radial, palmtree) polymers having different rheological behavior and mechanical properties from respective linear polymers are also synthesized by this living cationic mechanism. A series of multifunctional cationic initiators was synthesized from 1,5,9-cyclododecatriene [193]:

By using the inifer technique, Kennedy et al. also prepared three-arm–star poly-isobutylene carrying exactly two or three anisol terminals [194]. This telechelic polymer leads to three-armed–star block copolymer.

Block Copolymers by Coordination Polymerization

On the production of polypropylene–polyethylene block copolymer, many patents are found in the literature [7, 195]. Polypropylene gains high rigidity, impact strength, and fluidity when polyethylene blocks are bonded to polypropylene backbone [196–204]. But the synthesis of block copolymers has never been obtained easily, as in the case of block copolymers of the styrene/methyl methacrylate type, prepared by the anionic process. This is most probably due to the very short lifetime of growing polyolefin chains.

In this system, coordination (or Ziegler) catalysts of an organoaluminum compound such as $Al(C_2H_5)_3$ or $Al(C_2H_5)_2Cl$ and $TiCl_3$ (or $TiCl_4$) are used for living coordination polymerization catalyst. Gilbert and colleagues [200] used the catalyst combination of $Al(C_2H_5)Cl$ and $TiCl_3$ to obtain ethylene–propylene block copolymer. They proved that the catalyst remained active during the copolymerization and that chain transfer was limited. In a later publication, they reported the preparation of ABA block copolymer of ethylene–propylene–ethylene by the same catalyst system [201]. Agouri and co-workers observed the high efficiency of adding $Zn(C_2H_5)_2$ into the catalyst system combination of $(C_2H_5)_2AlCl$ and $TiCl_3$ in obtaining polyethylene–polymethyl methacrylate block copolymers [205].

Doi and co-workers [206, 207] found that the soluble catalyst system of V(acetylacetonate) with $Al(C_2H_5)_2Cl$ initiates a living polymerization of propylene at temperatures below $-65°C$ to afford syndiotactic polypropylenes with narrow molecular weight distributions. Apart from this, they found that anisole has an activation effect on the living coordination polymerization of propylene.

Block Copolymers by Coupling Reactions

In the synthesis of block copolymers, a combination of polymerization types has been used successfully. This route involves the coupling reactions of polymer anion, the coupling reactions of polymer cation, and coupling reactions of polymer anion to polymer cation.

Coupling Reactions of Polymer Anion

Living polymer anion can be coupled to some molecules leading to block copolymer formation via free radical or step-growth reactions. Polystyryl anion obtained anionic polymerization of styrene in the presence of *sec*-butyllithium coupled to a free radical initiator [208–211].

When polystyryl anion is deactivated with azobisisobutyronitrile (AIBN), polystyrene-containing azo groups can be prepared; this is used as initiator in vinyl polymerization [208]:

$$\text{wwww}^- \underset{\text{polystyryl anion (PS}^-)}{} + \text{CH}_3-\underset{\underset{N}{\overset{C}{\|}}}{\overset{\overset{CH_3}{|}}{C}}-N=N-\underset{\underset{N}{\overset{C}{\|}}}{\overset{\overset{CH_3}{|}}{C}}-\text{CH}_3 \rightarrow$$

AIBN

$$\text{CH}_3-\underset{\underset{\overset{\|}{N^-}}{\overset{C}{\text{wwwC}}}}{\overset{\overset{CH_3}{|}}{C}}-N=N-\underset{\underset{\overset{\|}{N^-}}{\overset{C}{\text{Cwww}}}}{\overset{\overset{CH_3}{|}}{C}}-\text{CH}_3 \xrightarrow[-N_2]{+MMA} \underset{\text{PMMA}}{\text{CH}_3-\overset{\overset{CH_3}{|}}{C}\text{wwww}} \quad 2\text{wwwC}\underset{N^-}{\overset{\|}{}}$$

Block copolymer

Similarly, the deactivation of polystyryl anion was carried out with 2,2'-azobis(2-chloropropane) [209] or benzoyl peroxide [210, 211] in order to obtain block copolymers via a free radical mechanism.

When lead-trialkyl chloride is used for deactivator of polymer anion, block copolymers are formed via a free radical mechanism [212, 213]:

$$\underset{\text{Polyisopropenyl lithium}}{\text{wwwwwww}^- \text{Li}^+} + \text{ClPb(CH}_3)_3 \xrightarrow{-\text{LiCl}} \underset{\text{Polyisopren}}{\text{wwwwPb(CH}_3)_3}$$

$$\xrightarrow{\Delta, 60°C} \underset{\underset{\text{radical}}{\text{polyisoprene}}}{\text{wwww}} + \cdot\text{Pb(CH}_3)_3 \xrightarrow{+\text{styrene}} \underset{\text{polyisoprene polystyrene}}{\text{wwwwwwwwww}}$$

Because homopolymer formation in the case of block copolymer synthesis via free radical mechanism is unavoidable, homopolymer formation is also unavoidable in this situation.

Eastmond and co-workers [214,215] terminated the living anionic styrene (or methyl methacrylate) polymerization by bromine via Grignard reaction or a group transfer polymerization process based on silyl ketyl acetal chemistry. Then they used redox reactions between metal carbonyl derivatives and polystyrene organic halides extensively to produce organic radicals (including macroradicals from polymers containing reactive halogen) capable of initiating free radical polymerization. The reaction scheme is

$$M^0 + R - X \rightarrow M^1X + R\cdot$$

where M^0, M^1 = metal carbonyl in its zeroth or first oxidation states and X = Cl, Br.

Polystyryl anion is converted to polystyrene having carboxyl terminal groups by coupling with carbon dioxide. Carboxylate-terminated polystyrene can be used in the block copolymer synthesis via step-growth polymerization [216, 217]:

Step 1: $\sim\sim\sim^-Na^+ + CO_2 \rightarrow \sim\sim COONa \xrightarrow{+\text{pivalolactone}} \sim\sim \underset{PS}{C}-O-CH_2-\underset{CH_3}{\overset{CH_3}{C}}-COONa$
         PS$^-$ PS

(A polymer activator for step-growth polymerization of pivalolactone)

Step 2: $\underset{PS}{\sim\sim COOH} \xrightarrow{+H_2N(CH_2)_6NH_2} \underset{PS}{\sim\sim\sim\sim NH_2} \xrightarrow{\text{N-carboxyanhydride of carbobenzoxy-L-Lysine}}$ Poly(styrene-b-L-Lysin) block copolymer

The hydroxyl-terminated polystyrene obtained by deactivating polystyril anion with ethylene oxide was reacted phosgene, methylcarbazate respectively. Then oxidation of this polymer, azoterminated polymer is obtained [218].

Similarly, functional polymers are obtained by combination of polystyryl anion with isocyanates [219, 220] and ethyl chloroformate [221] in order to prepare block copolymers via step-growth polymerization.

A disadvantage of the anionic polymerization technique is that none of the electrolyte monomers (e.g., methacrylic acid) can be employed directly. To prepare block copolymers of such monomers, 9-vinyl phenantrene and trimethylsilyl methacrylate were sequentially copolymerized via a living anionic mechanism and then trimethyl–silyl ester groups were completely hydrolyzed under mild conditions as follows [222]:

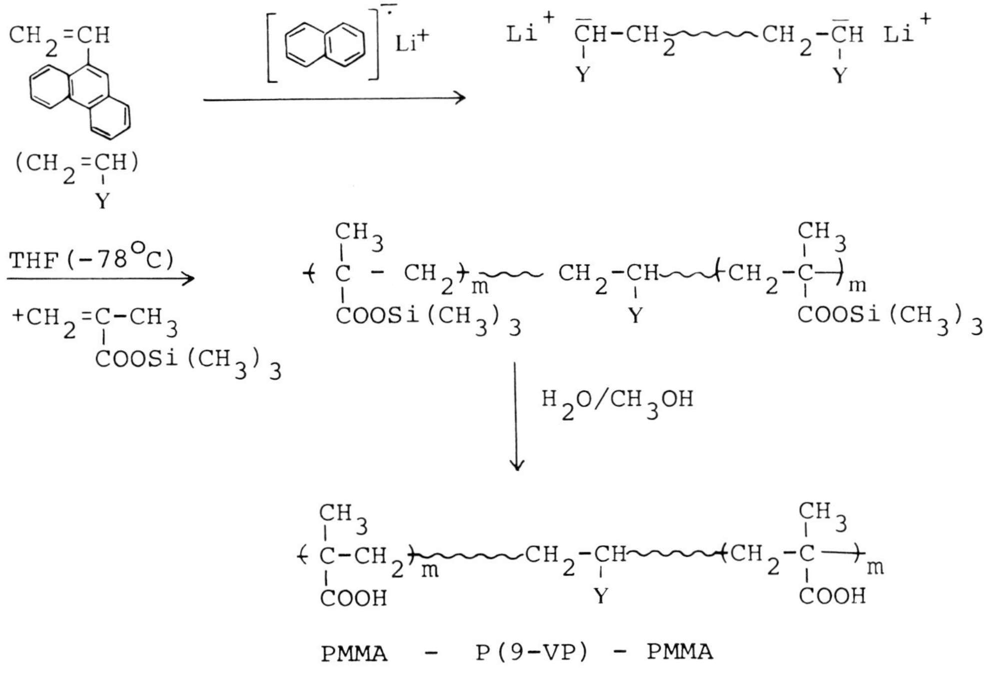

$CH_2=CH$ (9-vinyl phenanthrene structure) $\xrightarrow{[\text{naphthalene}]^- Li^+}$ $Li^+ \ \bar{C}H-CH_2\sim\sim\sim CH_2-\bar{C}H \ Li^+$
                                                                            $\quad\quad\quad Y \quad\quad\quad\quad\quad\quad Y$

$(CH_2=CH)$
$\quad\ |$
$\quad\ Y$

$\xrightarrow{\text{THF }(-78°C)}_{+CH_2=C-CH_3,\ COOSi(CH_3)_3}$ $\left(\underset{COOSi(CH_3)_3}{\overset{CH_3}{\underset{|}{C}}-CH_2}\right)_m \sim\sim CH_2-\underset{Y}{CH}\sim\sim\left(CH_2-\underset{COOSi(CH_3)_3}{\overset{CH_3}{\underset{|}{C}}}\right)_m$

$\downarrow H_2O/CH_3OH$

$\left(\underset{COOH}{\overset{CH_3}{\underset{|}{C}}-CH_2}\right)_m \sim\sim CH_2-\underset{Y}{CH}\sim\sim\left(CH_2-\underset{COOH}{\overset{CH_3}{\underset{|}{C}}}\right)_m$

PMMA   -   P(9-VP)   -   PMMA

α-Methylstyrene-butadiene-α-methylstyrene linear block copolymer with a pure block structure was prepared by *sec*-butyllithium and dichlorodiphenyl silane coupling agent in a four-step process [223].

Similarly, AB block copolymer of hexamethyl cyclotrisiloxane and 2-isopropenyl naphthalene were coupled to ABBA block copolymers by addition of dichlorodimethyl silane as follows [224]:

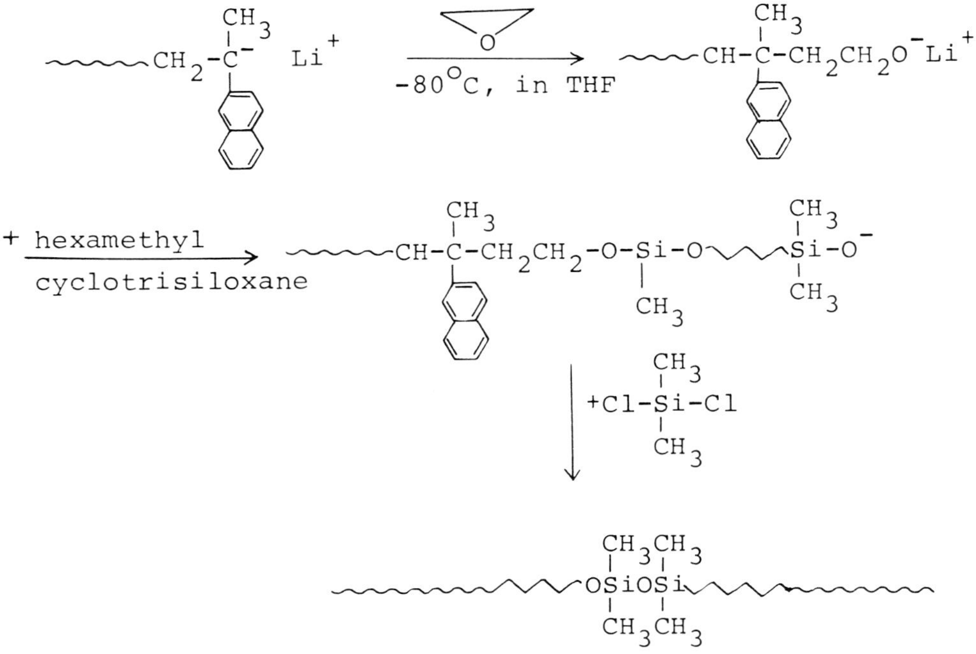

ABBA block copolymer

Polysiloxanes can be obtained by coupling reactions between living anionic silicone polymers with silicone oligomers having chloride terminal groups [225]:

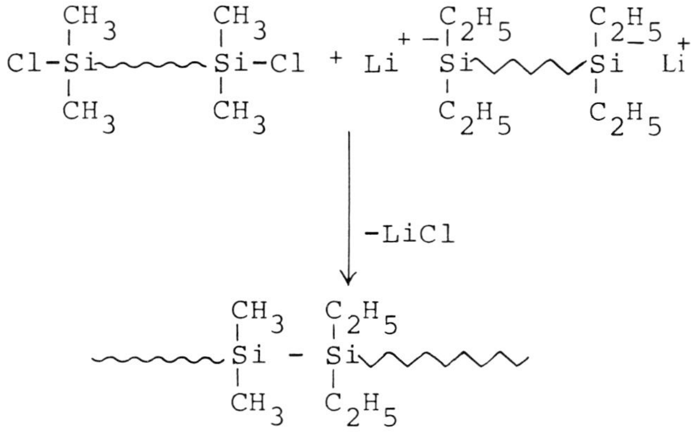

Star block copolymers can be also prepared by coupling reaction. Polybutadienyllithium chain end reacts in a virtually complete fashion with silicon tetrachloride [226]:

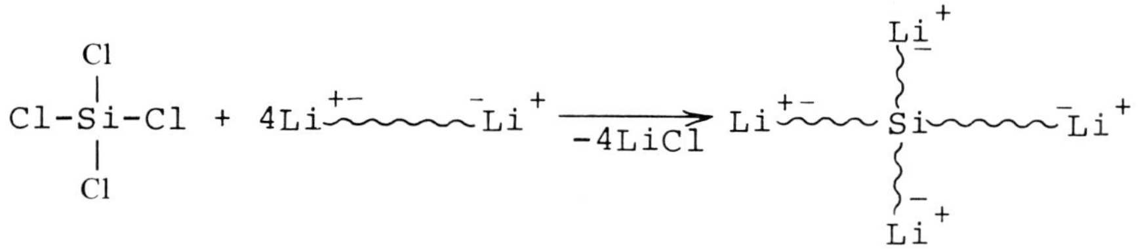

$$\text{Cl-Si-Cl} + 4\text{Li}\sim\sim\sim\text{Li}^+ \xrightarrow{-4\text{LiCl}} \text{Li}^+\sim\sim\sim\underset{\underset{\text{Li}^+}{\overset{\overset{\text{Li}^+}{|}}{\text{Si}}}}{}\sim\sim\sim\text{Li}^+$$

Kennedy [227] described a similar way to get star block and H block copolymer by the reaction of the dichlorosilane and living anionic polystyrene:

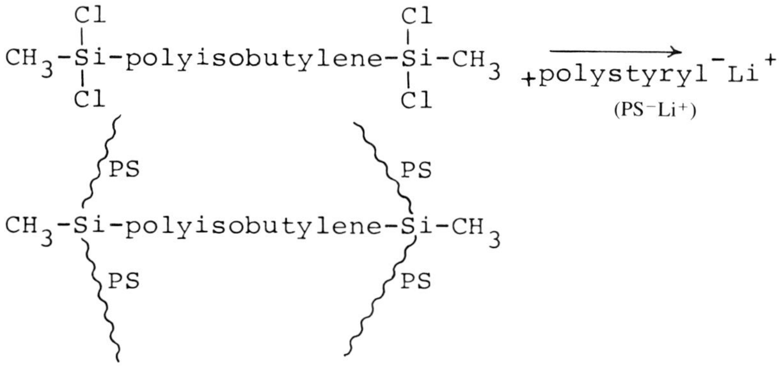

$$\underset{\text{Cl}}{\overset{\text{Cl}}{\text{CH}_3\text{-Si-polyisobutylene-Si-CH}_3}} \xrightarrow{+\text{polystyryl}^-\text{Li}^+}$$
$$(\text{PS}^-\text{Li}^+)$$

$$\text{CH}_3\text{-Si-polyisobutylene-Si-CH}_3$$
with PS branches

H-block copolymer

A star-branched polymer was prepared in high yield by first treating polystyryllithium with 1,3-bis(1-phenyl vinyl) benzene. Then the living coupling product was treated with isoprene to give a star-branched polymer [228]. Several acid groups are collected at one or both polydimethylsiloxane end chains by grafting β-mercaptopropionic acid on very short polymer end blocks. After acid end groups are neutralized with tetramethylammonium hydroxide, this can be used in the polymerization of pivalolactone leading to palmtree (star block) copolymer [229].

Another coupling reaction of polymer anion is the coupling of living polymer anion to chlorinated polyisobutylene [230]. A cationic polymerization initiator also can be obtained by a coupling reaction of polymer anion [231]:

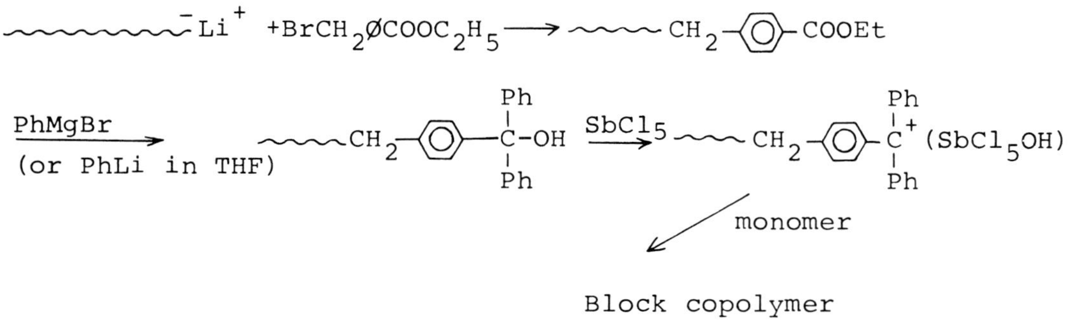

$$\sim\sim\sim\text{Li}^+ + \text{BrCH}_2\phi\text{COOC}_2\text{H}_5 \longrightarrow \sim\sim\sim\text{CH}_2\text{-}\phi\text{-COOEt}$$

$$\xrightarrow[\text{(or PhLi in THF)}]{\text{PhMgBr}} \sim\sim\sim\text{CH}_2\text{-}\phi\text{-}\underset{\text{Ph}}{\overset{\text{Ph}}{\text{C-OH}}} \xrightarrow{\text{SbCl}_5} \sim\sim\sim\text{CH}_2\text{-}\phi\text{-}\underset{\text{Ph}}{\overset{\text{Ph}}{\text{C}^+}}(\text{SbCl}_5\text{OH})^-$$

$$\downarrow \text{monomer}$$

Block copolymer

α,ω-p-Toluenesulfonic acid–ester-terminated polystyrene, which initiates 2-methyl-2-oxazoline polymerization cationically, is obtained by the coupling reaction of polymer anion [232]:

$$\text{PS} \xrightarrow{\text{CH}_2\overset{O}{-}\text{CH}_2} \text{PS}\text{---CH}_2\text{CH}_2\text{OH} \xrightarrow{\text{Tosilium-Cl}} \text{PS}\text{---CH}_2\text{CH}_2\text{OTS}$$

In a similar manner, in the transformation from polymer anion to Ziegler catalyst, a polymeric coordination catalyst can be obtained [233, 234]:

$$\text{---}^-\text{Li}^+ + \text{AlCl}_3 \xrightarrow[-\text{LiCl}]{} \text{---AlCl}_2$$

Coupling Reactions of Polymer Cations

Transfer of polyoxyethylene glycol to the growing polytetrahydrofurane cationic chain would give a block copolymer [235]:

$$\text{HO}\text{---O}^+ \cdot \text{+HO}\text{---OH} \rightarrow \underset{\text{block copolymer}}{\text{HO}\text{---O}\text{---OH}} \text{+HBF}_3\text{OROH}$$

The synthesis of a propylene-tetrahydrofurane block copolymer can be performed via living coordination polymerization. Its synthesis is based on the transformation of living cationic polypropylene ends which initiate the living polymerization of THF as schematically represented by reaction below [236]:

$$\text{---CH---}\underset{|}{\overset{\text{CH}_3}{\text{CH}}}\text{---V} \xrightarrow[-\text{VI}]{I_2} \text{---CH---}\underset{|}{\overset{\text{CH}_3}{\text{CH}}}\text{---I} \xrightarrow[-\text{AgI}]{+\text{AgClO}_4}$$

$$\text{---CH}_2\text{---}\underset{|}{\overset{\text{CH}_3}{\text{CH}}}^+\text{ClO}_4^- \xrightarrow{\text{THF}} \underset{\text{Polypropylene}}{\text{---CH}_2\text{---}\underset{|}{\overset{\text{CH}_3}{\text{CH}}}\text{---O---(CH}_2)_4}\underset{\text{Polytetrahydrofuran}}{\text{---O(CH}_2)_4\text{O}^+} \quad \text{ClO}_4^-$$

In contrast to living tetrahydrofurane, thiolanium or azetidinium polymers can be isolated in pure form and can be stored under normal conditions for an indefinite time. Therefore, polytetrahydrofurane-terminated thiolane (or 1,3,3-trimethylazetidine) was condensed with disodium salt of 4,4'-azobis(4-cyanopentanoic acid) in order to prepare polytetrahydrofurane having an azo group in the main chain [237]:

$$2\underset{\text{poly-THF}}{\text{---S}^+}\;\text{CF}_3\text{SO}_3^- + \text{NaOOCCH}_2\text{CH}_2\underset{|}{\overset{\text{CH}_3}{\underset{\text{CN}}{\text{C}}}}\text{---N}=\text{N---}\underset{|}{\overset{\text{CH}_3}{\underset{\text{CN}}{\text{C}}}}\text{---CH}_2\text{CH}_2\text{COONa}$$

$$\downarrow -\text{CF}_3\text{SO}_3\text{Na}$$

$$\underset{\text{poly-THF}}{\text{---}}\text{N}=\text{N}\underset{\text{poly-THF}}{\text{---}}$$

An azooxocarbenium initiator was described by Yağcı [238]. In order to synthesize block copolymers in two steps by this initiator, first tetrahydrofurane is polymerized by an oxocarbenium ion with low nucleophilic counterions prepared by reacting 4,4'-azobis(4-cyanopentanoyl chloride) with $AgClO_4$, $AgBF_4$, or $AgPF_6$ [239, 240]:

$$\underset{\substack{\text{Cl}-\overset{\text{O}}{\underset{\|}{\text{C}}}\text{CH}_2\text{CH}_2-\underset{\underset{\text{CN}}{|}}{\overset{\overset{\text{CH}_3}{|}}{\text{C}}}-\text{N}=\text{N}-\underset{\underset{\text{CN}}{|}}{\overset{\overset{\text{CH}_3}{|}}{\text{C}}}-\text{CH}_2\text{CH}_2\overset{\text{O}}{\underset{\|}{\text{C}}}-\text{Cl}}}{\underset{+2\text{AgBF}_4 \quad -2\text{AgCl}}{+ \text{THF} \quad \downarrow}}$$

$$\text{BF}_4^-\overset{+}{\text{O}}(\text{CH}_2)_4\text{O}\sim\sim\text{N}=\text{N}\sim\sim\text{O}(\text{CH}_2)_4\overset{+}{\text{O}}\text{BF}_4^-$$

The resultant polytetrahydrofurane having an azo group in the main chain can initiate the polymerization of another vinyl monomer in order to obtain block copolymers. Chloromethylester of azobiscyanopentanoic acid and $AgBF_4$ initiator system is also useful for preparing block copolymers by this technique [241].

Coupling Reactions of Polymer Anion to Polymer Cation

Block copolymer formation is carried out by the coupling reaction of living anionic polymer and living cationic polymer. This principle was first employed by Berger et al. [242] as well as by Asami and Chikazawa, who combined living polystyrene dianion with living cationic poly-THF to obtain block copolymer [243]. If dianion and dication are treated, multiblock copolymers are obtained [244]:

$$\sim\!\!\sim\!\!\sim^{-}\;\;^{-} + \;\;^{+}\!\!\sim\!\!\sim\!\!\sim^{+} \;\rightarrow\; (\sim\!\!\sim\!\!\sim\!\!\sim\!\!\sim\!\!\sim)_n$$

$(AB)_n$-type
block copolymer

Many studies of the reaction between living polystyrene and living poly-THF have been performed [245–249]. By treating polystyrene having a basic terminal group (e.g., tertiary amine) with living THF polymer, a block copolymer can be obtained [250].

CHARACTERIZATION OF BLOCK COPOLYMERS

After every synthesis of block copolymers, the product must be identified to provide answers to questions such as whether the product is a homopolymer blend or block copolymer and whether the product is a mixture of homopolymer and block copolymer. After this identification, other investigations may be needed to determine block copolymer structure, molecular weight, and mechanical and viscoelastic properties. Because of this, references cited in the "Synthesis of Block Copolymers" section offer one or more block copolymer characterization methods.

The presence of chain transfer and/or termination processes in free radical, cationic coordination or some coupling systems will inevitably result in products which are mixtures. The existence of homopolymers changes the properties of block copolymers. Grezlak and Wilkes showed that the modulus increases with added poly(methyl methacrylate) and decreases with added polyester for polymethyl methacrylate polyester ABA-type block copolymer [139] (Table 6).

A rapid distinction of block copolymers from their homopolymer blends is based on optical characteristics. Because of the gross incompatibility of most homopolymer blends,

Table 6 Homopolymer Effect on Stress–Strain Behavior of Polymethylmethacrylate/Polyester/Polymethylmethacrylate Block Copolymer

Sample	% Homopolymer	E (kg cm^{-2})	σ_B (kg cm^{-2})	ϵ_B
A	15.3 PMMA	4585	103.5	0.03
B	10.5 PMMA	3901	117.4	0.97
C	5.9 PMMA	4892	179.9	2.49
3A	0.0	4322	237.6	5.17
D	6.4 polyester	3194	187.3	5.14
E	14.3 polyester	1635	97.1	4.42
F	28.5 polyester	936	66.4	5.01

Key: E = modulus
σ_B = ultimate tensile strength
ϵ_B = elongation at break
PMMA = poly(methyl methacrylate)
3A = polymethyl methacrylate/polyester/polymethyl methacrylate block copolymer
Source: Data reprinted with permission from Ref. 139. Copyright 1975 John Wiley & Sons.

opaque films result due to a high degree of light scattering at the interface between the two phases. Block copolymers produce transparent films, since they usually exist in microphase-separated morphological states in which the domains are too small to scatter visible light. However, to obtain pure block copolymers, it is necessary to separate homopolymers from the block copolymer mixture. For this purpose, selective solvent extraction and fractional precipitation methods are useful [47, 54, 97, 100, 105, 251].

In selective solvent extraction, each homopolymer is sequentially extracted from the block copolymer mixture by appropriate solvents that are nonsolvents for block copolymers. As a typical example, for a mixture of PMMA, PS, and poly(S-b-MMA) block copolymer, the purification can be made in two steps. In a soxhlet apparatus, the mixture is first extracted by cyclohexane to dissolve polystyrene (PS); the remaining polymer mixture is then extracted by acetonitrile to dissolve polymethyl methacrylate (PMMA); and the remaining insoluble part is pure block copolymer.

The fractional precipitation method readily distinguishes block copolymers from their homopolymer blends. In this technique, sample is dissolved in a solvent and then polymer is precipitated by gradually adding a nonsolvent. Precipitation curves are established in a plot of precipitated polymer versus γ (nonsolvent, ml/solvent, ml). As shown in Figure 4, precipitation curves of block copolymers are situated between those of related homopolymers. The turbidimetric titration based on the solvent–precipitant system is also a useful technique for determination of the *amount* of homopolymer in the block copolymer mixture [21]. Turbidimetric titration can be made on block copolymers using a turbidity titrator. In this case, sample is dissolved in a solvent. Precipitant is added at a given rate. The reflectance at 90° is measured [77].

After homopolymer separation from block copolymer, structural analysis of pure block copolymer can be made by chemical, optical, spectrometric, mechanical, and thermal methods. Block copolymer micrographs obtained by optical methods give some informa-

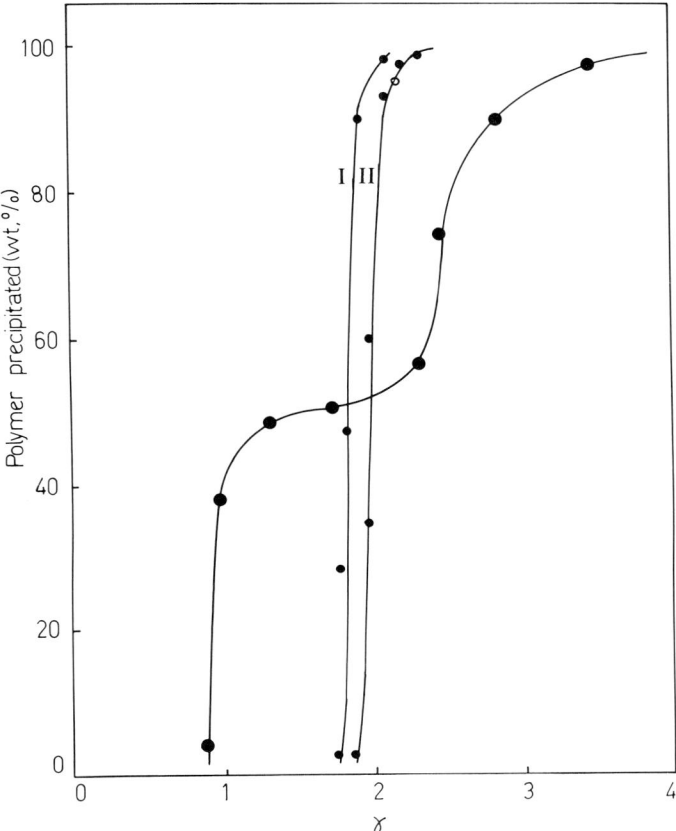

Figure 4 Fractional precipitations of a PS-PnBMA blend and poly(S-b-nBMA) block copolymers I (49% PS) and II (35% PS). γ indicates the volume ratio of precipitant to solvent. Precipitant: methanol; solvent: tetrahydrofurane. (From Ref. 54.)

tion about block copolymer morphology such as domain shapes (e.g., rodlike, spherulite, and lamellar) [9, 134, 252–258].

Determination of the block length and the sequence of blocks in the block copolymer is difficult without the synthesis method. An experimental technique for determining block length can be applicable only to block copolymers having polyester or polyether blocks, which can be easily hydrolized [80, 87]. After complete hydrolization of polyester or polyether blocks, the remaining vinyl blocks can be obtained and their molecular weight can be measured. Structural analysis in the block copolymers can be made by spectrometric techniques (NMR, IR, UV) [28, 37, 54, 68, 78, 259]. For molecular weight determinations of block copolymers, the gel permeation chromatography (GPC) technique [260] has been used successfully more frequently than viscosity and osmometry.

When two polymer blocks are incompatible, microphase separation has occurred in the block copolymer. Because of this microphase separation, block copolymers show two different glass transition temperatures which are similar to those of their corresponding homopolymers. Thermal analysis such as modulus–temperature relationships, differential scanning calorimetry, and rheological measurements can show multiple glass transition and/or melting behavior [261–266]. A typical modulus–temperature relationship for poly(S-*b*-*n*BMA) block copolymer is plotted in Figure 5. Stress–strain diagrams indicate thermoplastic behavior [267], and radial block and triblock copolymers behave rather differently from diblock copolymers [134]. Figure 6 shows stress–strain diagrams of styrene-*n*-butylmethacrylate block copolymers.

CONCLUSION

The discovery of block copolymers is the most important step in polymer chemistry. The necessity of block copolymer synthesis derives from the process of preparing the materials having several different properties. By the Szwarc's living anionic polymerization tech-

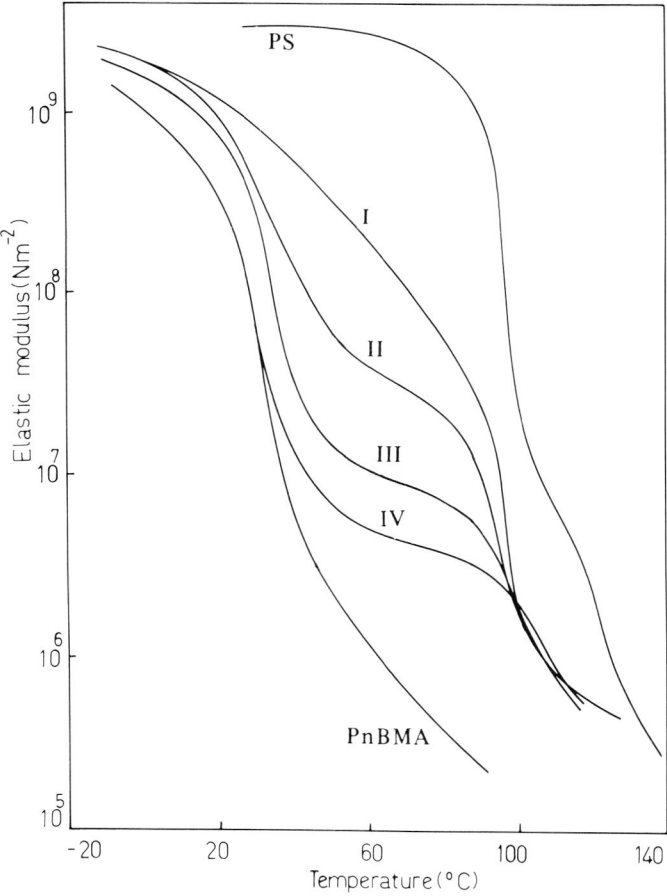

Figure 5 Elastic modulus–temperature relationship for poly(S-*b*-*n*BMA) block copolymers I (76% PS), II (53% PS), III (35% PS), and IV (23% PS). Curves for PS and P*n*BMA are included. (From Ref. 54.)

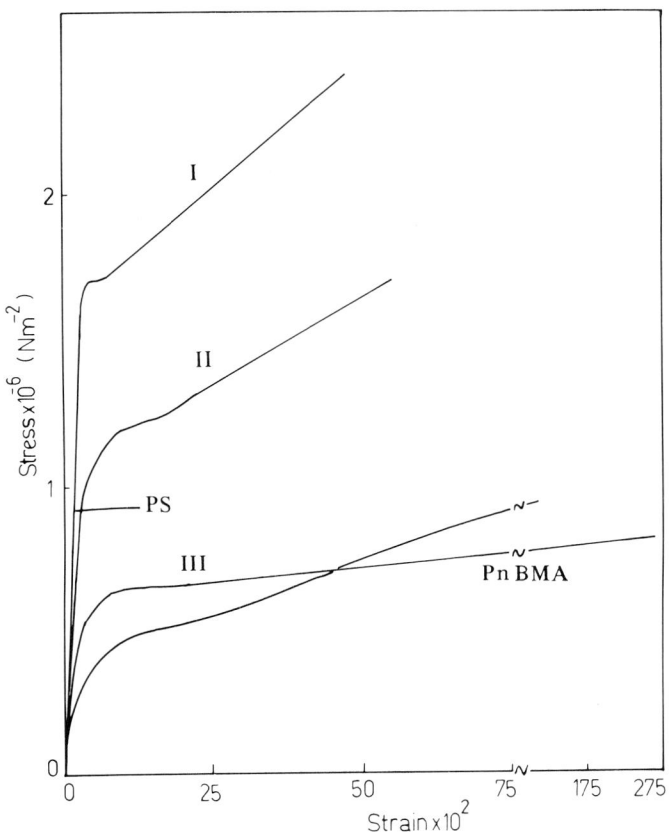

Figure 6 Stress–strain relationship for various copolymers of poly(S-b-nBMA) I (53% PS), II (35% PS), and III (23% PS). (From Ref. 54.)

nique, ABA, $-(AB)_n$-, and radial block copolymers arose as thermoplastic elastomer, which was a new type of material when A and B blocks are hard and soft segment, respectively. Therefore, the goal of most synthesis technique has been to obtain thermoplastic elastomers. At ambient temperatures, these block copolymers exhibit mechanical behavior similar to that of rubber vulcanizates, although they consist of completely linear chains with no chemical crosslinks between them, they are thermoplastic, can be easily molded, and are soluble in common solvents. However, they have no commercial applications as pure materials [5]. Polypropylene, polyethylene, or ethylene–vinyl acetate copolymer can be blended with these block copolymers. Large amounts of inert fillers such as whiting, talc, and clays can be used in these compounds. They are used as adhesives, sealents, coatings, footwear, mechanical goods, nonionic surfactants, and automotive applications.

The study of the synthesis and characterization of block copolymer synthesis will of course continue. New block copolymers with higher service temperatures or different properties may be obtained by new initiator systems in the future.

ACKNOWLEDGMENT

Most papers considered in the preparation of this chapter were supplied by the library of the Middle East Technical University and the Higher Education Documentation Center in Ankara. I would like to thank our polymer research group for helping me in the literature search from CA abstracts. I would like to thank my wife, Kadriye, and our children, Ebru and Burcu, who supported me during this work.

The author thanks Butterworth and Co. (Publishers) Ltd. (Figs. 2, 4, 5, 6), Hüthig and Wepf Verlag, Basel (Figure 3 and Table 4), and John Wiley & Sons, Inc. (Table 6) for permission to use the figures and tables indicated.

REFERENCES

1. D. C. Allport and W. H. Janes (eds.), *Block Copolymers*, Applied Science Publishers, Barking, England (1973).
2. A. Noshay and J. E. McGrath, *Block Copolymers: Overview and Critical Survey*, Academic Press, New York (1977).
3. R. J. Ceresa (ed.), *Block and Graft Copolymerization*, Vols. 1 and 2, Wiley, London (1976).
4. S. L. Aggarwal (ed.), *Block Polymers*, Plenum, New York (1970).
5. N. R. Legge, S. Davison, H. E. De La Mare, G. Holden, and M. K. Martin, *ACS Symp. Ser.*, 285: 175 (1985).
6. J. E. McGrath, *J. Chem. Educ.*, 58: 914 (1981).
7. R. Jerome, R. Fayt, and T. Ouhadi, *Prog. Polym. Sci.*, 10: 87 (1984).
8. P. Rempp and E. Franta, *Vysokomol. Soyedin.*, A21(11): 2529 (1979).
9. S. L. Aggarwal, *Polymer*, 17: 938 (1976).
10. J. V. Dawkins, in *Block Copolymers* (D. C. Allport and W. H. Janes, eds.), Applied Science Publishers, Barking, England, p. 370 (1973).
11. J. H. Bolland and H. W. Melville, "Some Remarks on the Photochemical Polymerization of Chloroprene and Some Related Molecules," *Proceedings of the First Rubber Tech. Conf. London*, W. Heffer, London, p. 239 (1938); *Chem. Abst.*, 32: 8396g (1938).
12. M. Szwarc, M. Levy, and R. Milkovich, *J. Am. Chem. Soc.*, 78: 2656 (1956).
13. C. R. Simionescu, E. Comanita, M. Pastravanu, and S. Dumitriu, *Prog. Polym. Sci.*, 12: 1 (1986).
14. A. Ueda and S. Nagai, *Kogaku to Kogyo* (Osaka), 60(2): 57 (1986); *Chem. Abst.*, 105: 173071p (1986).
15. G. Smets and T. Doi, in *New Trends in Photochemical Polymers: Proceedings, International Symposium in Honour of Professor Bengt Raanby* (N. S. Allen and J. F. Rabek, eds.), Elsevier Applied Science Publishers, Barking, England, p. 113 (1985).
16. H. W. Melville, *J. Chem. Soc.*, 414 (1941).
17. B. L. Funt and E. Collins, *J. Polym. Sci.*, 28: 359 (1958).
18. C. H. Bamford and A. D. Jenkins, *J. Chem. Phys.*, 56: 798 (1959).
19. J. Hiemeleers and G. Smets, *Makromol. Chem.*, 47: 7 (1961).
20. G. Smets, *J. Polym. Sci.*, 52: 1 (1961).
21. P. E. M. Allen, J. M. Downer, G. W. Hastings, H. W. Melville, P. Molyneux, and J. R. Urwin, *Nature*, 177: 910 (1956).
22. S. Küçükyavuz and B. M. Baysal, *J. Polym. Sci. Polym. Chem. Ed.*, 16: 2901 (1978).
23. T. Otsu, *J. Polym. Sci.*, 26(13): 236 (1957).
24. T. Otsu and M. Yoshida, *Polym. Bull.*, 7: 192 (1982).
25. T. Otsu and A. Kyriyama, *Polym. Bull.*, 11: 135 (1984).
26. G. Smets, W. Winter, and G. Delzenne, *J. Polym. Sci.*, 55: 767 (1961).
27. T. Sato, Y. Yutani, and T. Otsu, *Polymer*, 24: 1018 (1983).

28. T. Sato, T. Iwaki, S. Mori, and T. Otsu, *J. Polym. Sci. Polym. Chem. Ed.*, *21*: 819 (1983).
29. T. Sato, T. Yamamoto, T. Ogawa, and T. Otsu, *J. Polym. Sci. A, Polym. Chem.*, *24*: 1519 (1986).
30. R. B. Seymour and G. A. Stahl, *J. Macromol. Sci. Chem.*, *A11:* 53 (1977).
31. A. V. Olenin, M. B. Lachinov, V. A. Kasaikin, V. P. Zubov, and V. A. Kabanov, *Vysokomol. Soyedin.*, *A, 20*: 407 (1978).
32. C. H. Bamford and S. U. Mullik, *Polymer,* *17*: 94 (1976).
33. C. H. Bamford and X. Han, *Polymer, 22*: 1299 (1981).
34. V. A. Kargin, B. M. Kovarskaya, L. I. Glubenkova, M. S. Akutin, and G. L. Slonimsky, *Khim. Prom.*, 77(1957); *Chem. Abst., 51*: 18686i (1957).
35. D. J. Angier, R. J. Ceresa, and W. F. Watson, *J. Polym. Sci., 34*: 699 (1959).
36. Y. Minoura, T. Kasuya, S. Kawamura, and A. Nakano, *Kogyo Kogaku Zasshi, 69*(3): 549 (1966); *Chem. Abst., 65*: 10684 (1966).
37. Y. Minoura, T. Kasuya, S. Kawamura, and A. Nakano, *J. Polym. Sci. A-1, 5*: 43 (1967).
38. M. S. Akutin, N. Y. Parlashkevich, I. N. Kogan, S. P. Kalinina, and L. I. Menes, *Primen. Ul'traakust. Issled. Vesh., 10*: 47 (1960); *Chem. Abst., 56*: 13086 (1962).
39. C. Keqiang, S. Ye, L. Huilin, and X. Xi, *J. Macromol. Sci. Chem., A22*: 455 (1985).
40. D. C. Allport, in *Block Copolymers* (D. C. Allport and W. H. Janes, eds.), Applied Science Publishers, Barking, England, p. 35 (1973).
41. H. V. Pechmann and L. Vanino, *Bericht, 27*: 1510 (1894).
42. A. E. Woodward and G. Smets, *J. Polym. Sci., 17*: 51 (1955).
43. N. S. Tsvetkov and E. S. Beletskaya, *Ukr. Khim. Zh., 29*: 1072 (1963).
44. N. S. Tsvetkov and R. F. Markovskaya, *Vysokomol. Soyedin., 7*(1): 169 (1965).
45. W. E. Parker, L. P. Witnauer, and D. Swern, *J. Am. Chem. Soc., 79*: 1929 (1975).
46. Yu. L. Zherebin, S. S. Ivanchev, and V. I. Galibei, *Zh. Organ. Khim., 7*: 1660 (1971).
47. Yu. L. Zherebin, S. S. Ivanchev, and N. M. Domareva, *Vysokomol. Soyedin., A16*(4): 893 (1974).
48. S. S. Ivanchev, L. R. Uvarova, M. S. Matveyentseva, and I. P. Zyat'kov, *Vysokomol. Soyedin., A25*(9): 1838 (1983).
49. B. Hazer, *J. Polym. Sci. A, Polym. Chem., 25,* (1987).
50. S. S. Ivanchev, T. A. Kuznetsova, V. Konovalenko, A. P. Yuvchenko, K. L. Moiseichuk, and Y. A. Ol'dekop, *Vestsi Akad. Nauk. BSSR, Ser. Khim. Nauk., 2*: 58 (1979); *Chem. Abst., 91*: 21214s (1979).
51. A. I. Prisyazhnyuk and S. S. Ivanchev, *Vysokomol. Soyedin., A12*(2): 450 (1970).
52. T. A. Tolpygina, V. I. Galibei, and S. S. Ivanchev, *Vysokomol. Soyedin., A14*(5): 1027 (1972).
53. N. S. Tsvetkov, R. F. Markovskaya, Yu. A. Saprykin, and V. Ya. Zhukovskii, *Vysokomol. Soyedin., A14*(9): 2072 (1972).
54. B. Hazer and B. M. Baysal, *Polymer, 27*: 961 (1986).
55. R. J. Ceresa, *Polymer, 1*: 397 (1960).
56. G. Smets, G. Weinand, and S. Deguchi, *J. Polym. Sci. Polym. Chem. Ed., 16*: 3077 (1978).
57. G. Weinand and G. Smets, *J. Polym. Sci. Polym. Chem. Ed.,* 16: 3091 (1978).
58. C. H. Bamford and A. D. Jenkins, *Nature, 176*: 78 (1955).
59. R. M. Haines and W. A. Waters, *J. Chem. Soc.* (*London*), 4256 (1955).
60. M. Van Beylen and G. Smets, *Makromol. Chem., 69*: 140 (1963).
61. S. F. Reed, *J. Polym. Sci. A-1, 9*: 2147 (1971).
62. S. S. Ivanchev, T. A. Kuznetsova, A. A. Syrov, and A. A. Berlin, *Vysokomol. Soyedin., B17*: 104 (1975); *Chem. Abst., 83*: 10937p (1975).
63. A. Ladousse, C. Filliatre, B. Maillard, C. Manigand, and J. J. Villenave, *Eur. Polym. J., 15*: 987 (1979).
64. A. V. Tobolsky and A. Rembaum, *J. Appl. Polym. Sci., 8*: 307 (1964).
65. E. Zaganiaris and A. Tobolsky, *J. Appl. Polym. Sci., 14*: 1997 (1970).

66. N. Z. Erdy and C. F. Ferraro, *J. Polym. Sci. A-1*, 8: 763 (1970).
67. B. M. Baysal, W. T. Short, and A. V. Tobolsky, *J. Polym. Sci. A-1*, 10: 909 (1972).
68. F. J. T. Fildes and A. V. Tobolsky, *J. Polym. Sci. A-1*, 10: 151 (1972).
69. B. M. Baysal, *METU J. Pure Appl. Sci.*, 5: 193 (1972).
70. B. M. Baysal, E. H. Orhan, and I. Yılgör, *J. Polym. Sci. Symp.*, 46: 237 (1974).
71. E. H. Orhan, I. Yılgör and B. M. Baysal, *Polymer*, 18: 286 (1977).
72. I. Yılgör and B. M. Baysal, *Makromol. Chem.*, 186(3): 463 (1985).
73. J. Furukawa, S. Takamori, and S. Yamashita, *Angew. Macromol. Chem.*, 1, 92 (1967).
74. H. Yürük, A. B. Özdemir, and B. M. Baysal, *J. Appl. Polym. Sci.*, 31: 2171 (1986).
75. A. Ueda, Y. Shiozu, Y. Hidaka, and S. Nagai, *Kobunshi Ronbunshu*, 33: 131 (1976); *Chem. Abst.*, 85: 22051z (1976).
76. A. Ueda and S. Nagai, *J. Polym. Sci. Polym. Chem. Ed.*, 22: 1783 (1984).
77. A. Ueda and S. Nagai, *J. Polym. Sci.*, 22: 1611 (1984).
78. J. J. Laverty and Z. G. Gardlund, *J. Polym. Sci. Polym. Chem. Ed.*, 15: 2001 (1977).
79. S. H. Davies and M. W. Thomson, U.K. Pat. 2,115,810 (1983).
80. A. Ueda and S. Nagai, *J. Polym. Sci. A, Polym. Chem.*, 24: 405 (1986).
81. J. M. G. Cowie and M. Yazdani-Pedram, *Brit. Polym. J.*, 16: 127 (1984).
82. R. Walz, B. Bömer, and W. Heitz, *Makromol. Chem.*, 178: 2527 (1977).
83. R. Walz and W. Heitz, *J. Polym. Sci. Polym. Chem. Ed.*, 16: 1807 (1978).
84. H. R. Dicke and W. Heitz, *Makromol. Chem., Rapid Commun.*, 2: 83 (1981).
85. C. Oppenheimer and W. Heitz, *Angew. Makromol. Chem.*, 98: 167 (1981).
86. P. S. Anand, H. G. Stahl, W. Heitz, G. Weber, and L. Bottenbruch, *Makromol. Chem.*, 183: 1685 (1982).
87. H.-R. Dicke and W. Heitz, *Colloid Polym. Sci.*, 260: 3 (1982).
88. W. Heitz, M. Lattekamp, CH. Oppenheimer, and P. S. Anand, *ACS Symp. Ser.*, 212: 337 (1983).
89. W. Heitz, C. Oppenheimer, P. S. Anand, and X.-Yü Qiu, *Makromol. Chem., Suppl. 6*: 47 (1984).
90. X.-Yü Qiu, W. Ruland, and W. Heitz, *Angew. Makromol. Chem.*, 125: 69 (1984).
91. C. S. Sheppard and R. F. McLeay, B.R.D. Pat. 1,940,473, 10.02 (1970); *Chem. Abst.*, 72: 90983m (1970).
92. R. E. McLeay and C. S. Sheppard, U.S. Pat. 4,075,286 (1978); *Chem. Abst.*, 88: 170790s (1978).
93. A. A. Berlin, B. G. Gerasimov, A. A. Ivanov, and L. P. Beregovykh, *J. Macromol. Sci. Chem.*, A11: 811 (1977).
94. A. A. Berlin, B. G. Gerasimov, A. A. Ivanov, A. P. Muslukov, and N. I. Sheludchenko, *J. Macromol. Sci. Chem.*, A11: 821 (1977).
95. O. Nuyken, H. Schuster, and R. Kerber, *Makromol. Chem.*, 184: 2285 (1983).
96. O. Nuyken and R. Weidner, *Adv. Polym. Sci.*, 73–74: 145 (1986).
97. A. S. Shaikh, S. Dumitriu, E. Comanita, and C. I. Simionescu, *Polym. Bull.*, 3: 363 (1980).
98. S. Dumitriu, A. S. Shaikh, E. Comanita, and C. R. Simionescu, *Eur. Polym. J.*, 19: 263 (1983).
99. C. R. Simionescu, K. G. Sik, E. Comanita, and S. Dumitriu, *Angew. Makromol. Chem.*, 126: 73 (1984).
100. I. Piirma and L.-P. H. Chou, *J. Appl. Polym. Sci.*, 24: 2051 (1979).
101. B. Z. Gunesin and I. Piirma, *J. Appl. Polym. Sci.*, 26: 3103 (1981).
102. M. Schulz, G. West, and S. Ourk, *J. Prakt. Chem.*, 317: 463 (1975).
103. H. C. Haas and E. M. Idelson, U.S. Pat. 3,271,384 (1966).
104. R. McLeay and C. S. Sheppard, U.S. Pat. 3,917,744 (1975); *Chem. Abst.*, 84: 45386f (1976).
105. B. Hazer, *Angew. Makromol. Chem.*, 129: 31 (1985).
106. B. Hazer, Ph.D. dissertation, Karadeniz Technical University, Trabzon (1978).

107. M. B. Polk, K. B. Bota, E. C. Akubuiro, and M. Phingbodhipakkiya, *Macromolecules, 4*: 1626 (1981).
108. N. A. Memon and H. L. Williams, in *Macromolecular Synthesis*, Vol. 8 (E. M. Pearce, ed.), Wiley, New York, p. 31 (1982).
109. Z. G. Gardlund and M. A. K. Bator, *J. Polym. Sci. Polym. Chem. Ed., 21*: 1251 (1983).
110. A. Ueda, Y. Hidaka, and S. Nagai, *Kobunshi Ronbunshu, 36*: 123 (1979). (From its English abstract.)
111. T. C. Chiang and S. L. Ng, *Polymer, 22*: 3 (1981).
112. A. Pourdjavadi, P. J. Madec, and E. Marechal, *Eur. Polym. J., 20*: 305 (1984).
113. A. Pourdjavadi, P. J. Madec, and E. Marechal, *Eur. Polym. J., 20*: 311 (1984).
114. D. C. Webster and J. E. McGrath, *Contemp. Top. Polym. Sci., 4*: 959 (1984).
115. A. Noshay, M. Matzner, G. Karoly, and G. B. Stampa, *J. Appl. Polym. Sci., 17*: 619 (1973).
116. J. S. Riffle, R. G. Freelin, A. K. Banthia, and J. E. McGrath, *J. Macromol. Sci. Chem., A15*: 967 (1981).
117. J. Patricia, D. C. Webster, and J. E. McGrath, *Polym. Prepr., 25*: 91 (1984).
118. J. J. O'Malley, T. I. Pacansky, and W. J. Stauffer, *Macromolecules, 10*: 1197 (1977).
119. D. D. Stewart, E. N. Peters, C. D. Beard, R. B. Moffit, G. T. Kwiatkowski, J. J. Bohan, and E. Hedaya, *J. Appl. Polym. Sci., 24*: 115 (1979).
120. A. Ya. Yakubovich, A. N. Flerova, and V. S. Yakubovich, *Vysokomol. Soyedin., A13*: 994 (1971).
121. A. Ya. Yakubovich, A. N. Flerova, V. S. Yakubovich, G. F. Shalygin, V. S. Naumov, and L. B. Sokolov, *Vysokomol. Soyedin., A14*: 1838 (1972).
122. R. J. Zdrahala, E. M. Firer, and J. F. Fellers, *J. Polym. Sci. Polym. Chem. Ed., 15*: 689 (1977).
123. R. Castaldo, G. Maglio, and R. Palumbo, *J. Polym. Sci. Polym. Lett. Ed., 16*: 643 (1978).
124. J. M. Huet and E. Marechal, *Eur. Polym. J., 10*: 771 (1974).
125. F. de Candia, V. Petrocelli, R. Russo, G. Maglio, and R. Palumbo, *Polymer, 27*: 797 (1986).
126. G. N. Babu and S. Samant, *Eur. Polym. J., 17*: 421 (1981).
127. J. Preston, W. R. Krigbaum, and R. Kotek, *J. Polym. Sci. Polym. Chem. Ed., 20*: 3241 (1982).
128. Y. Imai, M. Kajiyama, S. Ogata, and M. Kakimoto, *Polym. J., 16*: 267 (1984).
129. N. Yamazaki, M. Matsumoto, and F. Higashi, *J. Polym. Sci. Polym. Chem. Ed., 13*: 1373 (1975).
130. G. K. Hoeschele and W. K. Witsiepe, *Angew. Makromol. Chem., 29/30*: 267 (1973).
131. J. Devaux, P. Godard, and J. P. Mercier, *Polym. Eng. Sci., 22*: 229 (1982).
132. Y. Yamashita, Y. Iwaya, and K. Ito, *Makromol. Chem., 176*: 1207 (1975).
133. S. C. Shit and S. Maiti, *J. Polym. Sci. Polym. Lett. Ed., 24*: 383 (1986).
134. R. H. Wondraczek and J. P. Kennedy, *J. Polym. Sci. Polym. Chem. Ed., 20*: 173 (1982).
135. S. F. Reed, *J. Polym. Sci. A-1, 9*: 2029 (1971).
136. S. F. Reed, *J. Polym. Sci. A-1, 10*: 2025 (1972).
137. S. F. Reed, *J. Polym. Sci. A-1, 10*: 2493 (1972).
138. B. B. Idega, S. P. Vernekar, and N. D. Ghatge, *J. Polym. Sci. Polym. Chem. Ed., 21*: 385 (1983).
139. J. H. Grezlak and G. L. Wilkes, *J. Appl. Polym. Sci., 19*: 769 (1975).
140. Y. Shimura and N. Ikeda, *J. Polym. Sci. A-1, 11*: 1271 (1973).
141. E. Bayer, I. Gatfield, H. Mutter, and M. Mutter, *Tetrahedron, 34*: 1829 (1978).
142. B. Masar, P. Cefelin, and J. Sebenda, *J. Polym. Sci. Polym. Chem. Ed., 17*: 2317 (1979).
143. R. Rahman and Y. Avny, *J. Macromol. Sci. Chem., A13*: 953 (1979).
144. R. Rahman and Y. Avny, *J. Macromol. Sci. Chem., A13*: 971 (1979).
145. R. Rahman and Y. Avny, *J. Macromol. Sci. Chem., A14*: 581 (1980).

146. J. J. Laverty and Z. G. Gardlund, *J. Appl. Polym. Sci.*, *26*: 3657 (1981).
147. M. Szwarc, *Nature*, *24*: 1168 (1956).
148. W. E. Smith, F. R Galliano, D. Rankin, and G. J. Manteli, *J. Appl. Polym. Sci.*, *10*: 1659 (1966).
149. R. E. Cunningham and M. R. Treiber, *J. Appl. Polym. Sci.*, *12*: 23 (1968).
150. P. E. Black and D. J. Worsfold, *J. Appl. Polym. Sci.*, *14*: 1671 (1970).
151. C. W. Brown and I. F. White, *J. Appl. Polym. Sci.*, *16*: 2671 (1972).
152. R. E. Cunningham and M. L. Wise, *J. Appl. Polym. Sci.*, *16*: 107 (1972).
153. J. G. Zilliox, J. E. Roovers, and S. Bywater, *Macromolecules*, *8*: 573 (1975).
154. L. J. Fetters, E. M. Firer, and M. Dafauti, *Macromolecules*, *10*: 1200 (1977).
155. T. N. Hay and J. F. McCabe, *J. Polym. Sci. Polym. Chem. Ed.*, *16*: 2893 (1978).
156. C. Goett, M. Lambla, and C. Wippler, *Makromol. Chem.*, *180*: 1865 (1979).
157. K. Arai, T. Kotaka, Y. Kitano, and K. Yoshimura, *Macromolecules*, *13*: 1670 (1980).
158. K. Arai, T. Kotaka, Y. Kitano, and K. Yoshimura, *Macromolecules*, *13*: 455 (1980).
159. Y. Lu, R. Xu, and G. Jin, *Coaodeng Xuexiao Huaxue Xuebao*, *4*: 248 (1983); *Chem. Abst.*, *99*: 38871s (1983).
160. S. Küçükyavuz, Dozent thesis, Middle East Technical University, Ankara (1981).
161. Z. Küçükyavuz and S. Küçükyavuz, *Makromol. Chem.*, *187*: 2469 (1986).
162. M. Baer, *J. Polym. Sci. A*, *2*: 417 (1964).
163. G.-H. Hsiue and W.-R. Yang, *J. Polym. Sci. Polym. Chem. Ed.*, *22*: 1525 (1984).
164. D. W. Koetsier, A. Bantjes, J. Feijen, and D. J. Lyman, *J. Polym. Sci. Polym. Chem. Ed.*, *16*: 511 (1978).
165. L. J. Fetters and M. Morton, *Macromolecules*, *2*: 453 (1969).
166. M. Mouzali, J. Lacoste, and M. J. M. Abadie, *Eur. Polym. J.*, *17*: 685 (1981).
167. R. E. Cunningham, M. Auerbach, and W. J. Floyd, *J. Appl. Polym. Sci.*, *16*: 163 (1972).
168. L. H. Tung, G. Y.-S. Lo, and D. E. Beyer, *Macromolecules*, *11*: 616 (1978).
169. P. Guyot, J. C. Favier, M. Fontanille, and P. Sigwalt, *Polymer*, *23*: 73 (1982).
170. A. D. Broske and J. E. McGrath, *Polym. Prepr.*, *26*: 241 (1985); *Chem. Abst.*, *103*: 6752p (1985).
171. T. Suziki, Y. Murakami, and Y. Takegami, *J. Polym. Sci. Polym. Let. Ed.*, *17*: 241 (1979).
172. M. Morton and S. L. Mikesell, *J. Macromol. Sci.*, *A7*: 1391 (1973).
173. H. L. Hsieh, *J. Appl. Polym. Sci.*, *22*: 1119 (1978).
174. Y. Yamashita, H. Matsui, and K. Ito, *J. Polym. Sci. Polym. Chem. Ed.*, *10*: 3577 (1972).
175. J. Stehlicek and J. Sebenda, *Eur. Polym. J.*, *13*: 949 (1977).
176. J. Stehlicek and J. Sebenda, *Eur. Polym. J.*, *13*: 955 (1977).
177. J. Stehlicek and J. Sebenda, *Eur. Polym. J.*, *18*: 535 (1982).
178. P. M. Lefebvre, R. Jerome, and P. Teyssié, *Makromol. Chem.*, *183*: 2453 (1982).
179. M. P. Dreyfuss and P. Dreyfuss, *Polymer*, *6*: 93 (1965).
180. J. P. Kennedy and E. G. Melby, *J. Polym. Sci. Polym. Chem. Ed.*, *13*: 29 (1975).
181. M. Sawamoto, K. Ebra, A. Tanizaki, and T. Higashimura, *J. Polym. Sci. A, Polym. Chem.*, *24*: 2919 (1986).
182. Y. Yamashita, M. Okada, and M. Hirota, *Angew. Makromol. Chem.*, *9*: 136 (1969).
183. Y. Yamashita, M. Hirota, H. Matsui, A. Hirao, and K. Nobutoki, *Polym. J.*, *2*: 43 (1971).
184. M. Villesange, G. Sauvet, J. P. Vairon, and P. Sigwalt, *Polym. Bull.*, *2*: 131 (1980).
185. J. M. Rooney, D. R. Squire, and V. T. Stannet, *J. Polym. Sci. Polym. Chem. Ed.*, *14*: 1877 (1976).
186. W. Chwialkowska, P. Kubisa, and S. Penczek, *Makromol. Chem.*, *183*: 753 (1982).
187. E. J. Goethals, M. Van de Velde, and F. D'Haese, *Polym. Prepr.*, *25*: 140 (1984).
188. M. Sawamoto and J. P. Kennedy, *J. Macromol. Sci. Chem.*, *A18*: 1293 (1982–1983).
189. T. Higashimura, M. Mitsuhashi, and M. Sawamoto, *Macromolecules*, *12*: 178 (1979).
190. J. P. Kennedy, *Makromol. Chem. Suppl.*, *3*: 1 (1979).
191. J. P. Kennedy and R. A. Smith, *J. Polym. Sci. Polym. Chem. Ed.*, *18*: 1523 (1980).
192. J. P. Kennedy and R. A. Smith, *J. Polym. Sci. Polym. Chem. Ed.*, *18*: 1539 (1980).

193. G. Cai and D. Yan, *Makromol. Chem., 187*: 553 (1986).
194. M. K. Mishra, B. Sar-Mishra, and J. P. Kennedy, *Polym. Bull., 16*: 47 (1986); *Chem. Abst., 105*: 153613p (1986).
195. T. G. Hegg, in *Block Copolymers* (D. C. Allport and W. H. Janes, eds.), Applied Science Publishers, Barking, England, p. 493 (1973).
196. A. Toyoda and N. Kashiwa, Jpn. Pat. 6,169,823 (1986); *Chem. Abst., 105*: 98138k (1986).
197. T. Nishikawa, Ger. Pat. 3,346,326 (1984); *Chem. Abst., 101*: 131328q (1984).
198. W. Gruber, K. D. Ruempler, and G. Schweier, Ger. Pat. 3,324,793 (1985); *Chem. Abst., 102*: 132672e (1985).
199. Y. Tajima, Y. Ganno, V. Uchida, S. Sugita, K. Kawabe, and Y. Iwasaki, U.K. Pat. 2,167,423 (1986); *Chem. Abst., 105*: 115580t (1986).
200. P. Prabhu, A. Schindler, M. H. Theil, and R. D. Gilbert, *J. Polym. Sci. Polym. Chem. Ed., 19*: 523 (1981).
201. P. Prabhu, A. Schindler, M. H. Theil, and R. D. Gilbert, *J. Polym. Sci. Polym. Lett. Ed., 18*: 389 (1980).
202. V. Busico, P. Carradini, P. Fontana, and V. Savino, *Makromol. Chem. Rapid. Commun., 6*: 743 (1985).
203. K. Endo and T. Otsu, *J. Polym. Sci. A, Polym. Chem., 24*: 1615 (1986).
204. S. Ueki, C. Imai, and T. Makishima, Eur. Pat. 166,536 (1986); *Chem. Abst., 104*: 187068 (1986).
205. E. Agouri, C. Parlant, P. Mornet, J. Reau, and J. F. Teitgen, *Makromol. Chem., 137*: 229 (1970).
206. Y. Doi and S. Ueki, *Makromol. Chem. Rapid Commun., 3*: 225 (1982).
207. Y. Doi, T. Koyama, and K. Soğa, *Makromol. Chem., 186*: 11 (1985).
208. Y. Vinchon, R. Reeb, and G. Riess, *Eur. Polym. J., 12*: 317 (1976).
209. S. Hoering, A. Mueller, A. Blazejewski, and J. Ulbricht, *Plaste Kautsch., 32*: 326 (1985).
210. L. V. Zamoiskaya and E. B. Milovskaya, *Vysokomol. Soyedin., A21*: 1974 (1979).
211. E. B. Milovskaya and L. V. Zamoiskaya, *Polymer, 23*: 884 (1982).
212. T. Souel, F. Schué, M. Abadie, and D. H. Richards, *Polymer, 18*: 1292 (1977).
213. M. J. Abadi, F. Schue, T. Souel, and D. H. Richards, *Polymer, 22*: 1076 (1981).
214. C. H. Bamford, G. C. Eastmond, J. Woo, and D. H. Richards, *Polymer, 23*: 643 (1982).
215. G. C. Eastmond and J. Grigor, *Makromol. Chem. Rapid Commun., 7*: 375 (1986).
216. Y. Yamashita and T. Hane, *J. Polym. Sci. A-1, 11*: 425 (1973).
217. J. P. Billod, A. Douy, and B. Gallod, *Makromol. Chem., 177*: 1889 (1976).
218. D. S. Campbell, D. E. Loeber, and A. J. Tinker, *Polymer, 25*: 1141 (1984).
219. R. J. Ambrose and W. L. Hergenrother, *J. Polym. Sci. Polym. Lett. Ed., 14*: 603 (1976).
220. Y. Nitadori, E. Franta, and P. Rempp, *Makromol. Chem., 179*: 927 (1978).
221. D. Petit, R. Jêrôme, and P. Teyssié, *J. Polym. Sci. Polym. Chem. Ed., 17*: 2903 (1979).
222. Y. Morishima, T. Hashimoto, Y. Itoh, M. Kamachi, and S. Nozakura, *Makromol. Chem. Rapid Commun., 2*: 507 (1981).
223. A. Gandini, G. Perego, A. Raggere, R. Vitali, and A. Zazzetta, *Polym. Bull., 12*: 71 (1984).
224. T. Rhein and R. C. Schulz, *Makromol. Chem., 186*: 2301 (1985).
225. J. P. Wesson and T. C. Williams, *J. Polym. Sci. Polym. Chem. Ed., 19*: 65 (1981).
226. Le-Khac-Bi and L. J. Fetters, *Macromolecules, 9*: 732 (1976).
227. J. P. Kennedy, *J. Polym. Sci. Polym. Symp., 72*: 73 (1985).
228. P. R. Quirck, F. H. Lignatz, and C. W. Chih, *Polym. Prepr., 21*: 188 (1986).
229. P. M. Lefebvre, R. Jêrôme, and P. Teyssié, *J. Polym. Sci. Polym. Chem. Ed., 21*: 789 (1983).
230. J. P. Kennedy, D. Y. Chung, and A. Guyot, *J. Polym. Sci. Polym. Chem. Ed., 19*: 2737 (1981).
231. M. Villasange, G. Sauvet, J. P. Vairon, and P. Sigwalt, *Polym. Bull., 2*: 131 (1980).
232. K. Ishizu, S. Ishikawa, and T. Fukutomi, *J. Polym. Sci. Polym. Chem. Ed., 23*: 445 (1985).

233. P. Cohen, M. J. M. Abadie, F. Schué, and D. H. Richards, *Polymer*, 22: 1316 (1981).
234. M. A. Darzewinski and R. E. Cohen, *J. Polym. Sci. A Polym. Chem.*, 24: 2457 (1986).
235. J. M. Hammond, J. F. Hooper, and W. G. P. Robertson, *J. Polym. Sci. A-1*, 9: 295 (1971).
236. Y. Doi, Y. Watanabe, S. Ueki, and K. Soga, *Makromol. Chem. Rapid Commun.*, 4: 533 (1983).
237. F. D'Haese and E. J. Goethals, *Makromol. Chem. Rapid Commun.*, 7: 165 (1986).
238. Y. Yağcı, *Polymer Commun.*, 26: 7 (1985).
239. Y. Yağcı, *Polymer Commun.*, 27: 21 (1986).
240. A. Akar, A. C. Aydoğan, N. Talınlı, and Y. Yağcı, *Polymer Bull.*, 15: 293 (1986).
241. K. Nakagawa, M. Murakami, T. Yamamoto, K. Tabuchi, K. Inoue, and T. Tanigaki, *Memoirs of the Niihama National College of Technology* (*Science and Engineering*), 22: 54 (1986).
242. G. Berger, M. Levy, and D. Vofsi, *J. Polym. Sci. Polym. Lett. Ed.*, 4: 183 (1966).
243. R. Asami and M. Chikazawa, Papers presented at the 19th Annual Meeting of the Chemical Society of Japan, No. 4, p. 1022 (1966). (From Yamashita et al. Ref. 244.)
244. Y. Yamashita, K. Nobutoki, Y. Nakamura, and M. Hirota, *Macromolecules*, 4: 548 (1971).
245. D. H. Richards, S. B. Kingston, and T. Souel, *Polymer*, 19: 68 (1978).
246. D. H. Richards, S. B. Kingston, and T. Souel, *Polymer*, 19: 806 (1978).
247. M. Kucera, F. Bozek, and K. Majerova, *Polymer*, 20: 1013 (1979).
248. M. Kucera, F. Bozek, and K. Majerova, *Polymer*, 23: 207 (1982).
249. S. S. Tseng, H. Z. Zhang, and X. D. Feng, *Polym. Bull.*, 8: 219 (1982).
250. D. H. Richards, *Polym. Prepr.*, 25: 216 (1984).
251. G. Smets, L. Convent, and X. Van Der Borght, *Makromol. Chem.*, 23: 162 (1957).
252. S. Bywater, *Polym. Eng. Sci.*, 24: 104 (1984).
253. H. Hasegawa, T. Hashimoto, H. Kawai, T. P. Lodge, E. J. Amis, C. J. Glinka, and C. C. Han, *Macromolecules*, 18: 67 (1985).
254. J. T. Koberstein and R. S. Stein, *Polymer*, 25: 171 (1984).
255. E. Pedemonte and G. C. Alfonso, *Macromolecules*, 8: 85 (1975).
256. T. Soen, T. Inoue, K. Miyoshi, and H. Kawai, *J. Polym. Sci. A-2*, 10: 1757 (1972).
257. P. R. Lewis and C. Price, *Nature*, 223: 494 (1969).
258. G.-H. Hsiue and S. W. F. Shih, *J. Appl. Polym. Sci.*, 30: 1659 (1985).
259. A. V. Tobolsky, A. Eisenberg, and K. F. O'Driscoll, *Anal. Chem.*, 31: 203 (1959).
260. L. H. Tung, *J. Appl. Polym. Sci.*, 24: 953 (1979).
261. M. Girolamo and J. R. Urwing, *Eur. Polym. J.*, 7: 225 (1971).
262. L. D. Guidice, R. E. Cohen, G. Attalla, and F. Bertinotti, *J. Appl. Polym. Sci.*, 30: 4305 (1985).
263. G. Kraus, C. W. Childers, and J. T. Gruver, *J. Appl. Polym. Sci.*, 11: 1581 (1967).
264. N. Overbergh, G. Weinand, and G. Smets, *J. Polym. Sci. Polym. Chem. Ed.*, 16: 3107 (1978).
265. S. V. Conjeevaram, R. S. Benson, and D. J. Lyman, *J. Polym. Sci. Polym. Chem. Ed.*, 23: 429 (1985).
266. K. E. Hardenstine, C. J. Murphy, R. B. Jones, L. H. Sperling, and G. E. Manser, *J. Appl. Polym. Sci.*, 30: 2051 (1985).
267. G. Holden, E. T. Bishop, and N. R. Legge, *J. Polym. Sci. C*, 26: 37 (1969).

5

Synthesis and Characterization of Aromatic Polyesters

Yoshio Imai and Masa-aki Kakimoto
Tokyo Institute of Technology
Meguro-ku, Tokyo, Japan

INTRODUCTION	177
SYNTHESIS OF AROMATIC POLYESTERS	178
Acid Chloride Route	178
Phenyl Ester Route	181
Phenol Acetate Route	182
Phenol Silyl Ether Route	183
Direct Polyesterification Method	183
STRUCTURE–PROPERTY RELATIONSHIPS IN POLYARYLATES	185
THERMOTROPIC LIQUID CRYSTALLINE POLYARYLATES	191
Introduction of Flexible Spacer Units	195
Modification with Substitution on Aromatic Rings	197
Modification with Unsubstituted Rodlike Components	199
Modification with Rigid Kinks	201
COMMERCIAL DEVELOPMENT OF POLYARYLATES	204
REFERENCES	207

INTRODUCTION

Linear polyesters derived from dicarboxylic acids and diols can be classified into four categories, as seen in Table 1.

Aromatic polyesters are designated in a broad sense as polyesters that contain one or more aromatic components in the main chains and therefore include aliphatic polyarylates and aromatic ones, as well as aliphatic–aromatic polyesters such as commercially important poly(ethylene terephthalate) and poly(butylene terephthalate).

A number of aromatic polyarylates have been synthesized to date in various combinations of bisphenols with different aromatic dicarboxylic acids and AB-type aromatic hydroxy acids. Polyesters of this category are associated with many valuable properties such as a high glass transition temperature, a good dielectric constant, and capability of forming strong films and moldings. These surpass polycarbonates, one of the big-five engineering plastics, with respect to some characteristics.

Table 1 Classification of Polyesters

Diol	Dicarboxylic acid	Polymer
Aliphatic	Aliphatic	Aliphatic polyester
Aliphatic	Aromatic	Aliphatic–aromatic polyester
Aromatic (bisphenol)	Aliphatic	Aliphatic polyarylate
Aromatic (bisphenol)	Aromatic	Aromatic polyarylate

This chapter deals with mainly aromatic polyarylates with emphasis on high-temperature engineering plastics and thermotropic liquid crystalline polyarylates.

SYNTHESIS OF AROMATIC POLYESTERS

Many reactions can be used for the synthesis of aromatic polyarylates but the most useful are (1) acid chloride route, (2) phenyl ester route, (3) phenol acetate route, (4) phenol silyl ether route, and (5) direct esterification method [1, 2].

Acid Chloride Route

Generally applicable and mostly used for the preparation of polyarylates is the polycondensation of dicarboxylic acid chlorides with bisphenols (Eq. 1) or that of AB-type monomers (Eq. 2), because diacid chlorides are the most reactive of the derivatives of dicarboxylic acids. Therefore, polyarylates are readily obtainable in high molecular weights by interfacial and low-temperature solution methods at ordinary temperature, although two high-temperature procedures, solution and melt/solid phase, are employed as well.

$$\text{HO-Ar-OH} + \text{ClC(=O)-Ar'-C(=O)Cl} \longrightarrow \left[\text{-O-Ar-OC(=O)-Ar'-C(=O)-} \right]_n + 2\,\text{HCl} \quad (1)$$

$$\text{HO-Ar-C(=O)Cl} \longrightarrow \left[\text{-O-Ar-C(=O)-} \right]_n + \text{HCl} \quad (2)$$

Interfacial Polycondensation

The reaction of acid chlorides with phenols in the form of their active phenolate anions is a rapid, essentially irreversible reaction giving quantitative yields of esters at normal temperature. The classical Schotten–Baumann reaction has been applied to polyarylate synthesis since the early 1950s [1].

The general procedure for carrying out the interfacial polycondensation is as follows. A bisphenol is dissolved in aqueous sodium hydroxide to form a solution of the bisphenolate anion, which is treated by stirring a solution of a dicarboxylic acid chloride in a water-immiscible solvent such as dichloromethane. Although the water-soluble sodium bisphenolate has low solubility in the water-immiscible solvent, the polycondensation, which is diffusion-controlled, takes place at the boundary of the two liquid phases and the liberated chloride ion is neutralized by the alkali in the aqueous phase.

The two-phase reaction can be greatly accelerated and improved by the use of a small amount of catalyst such as quaternary ammonium and phosphonium salts or crown ethers, affording higher molecular weight polyarylates [3]. Such catalysts are called phase-transfer catalysts. The effect results from the formation of organic solvent–soluble bis-phenolates and subsequent transport of the chloride ion to the aqueous phase as the quaternary onium salts, which in turn promote further reaction.

This two-phase polycondensation, frequently termed a phase-transfer–catalyzed polycondensation [4], is currently used for the preparation of a variety of condensation polymers in addition to polyarylate synthesis [5]. Table 2 summarizes the effect of phase-transfer catalysts on the inherent viscosity of polyarylate **1** [6], as well as the structurally related aromatic polyesters, polysulfonate **2** and polyphosphonate **3** [4].

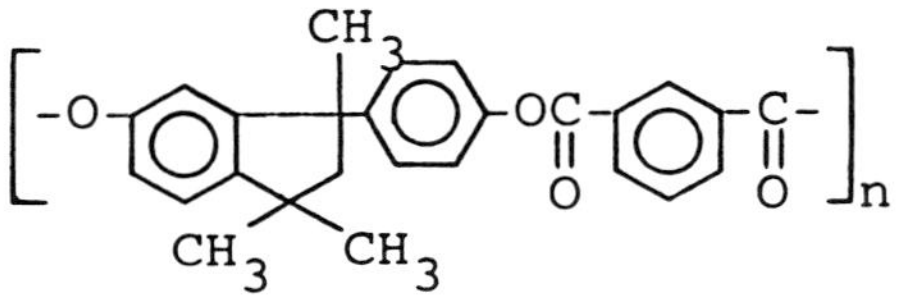

1

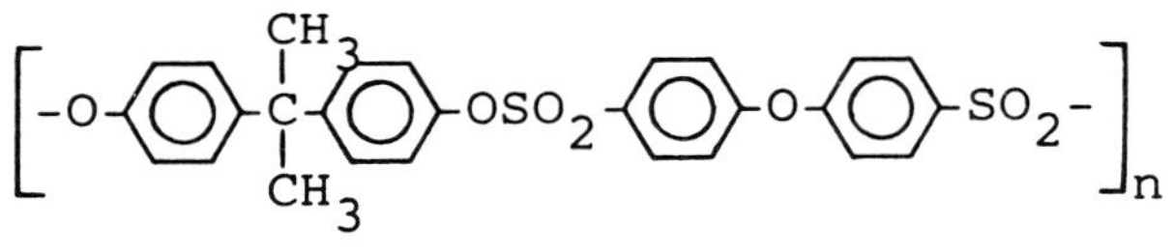

2

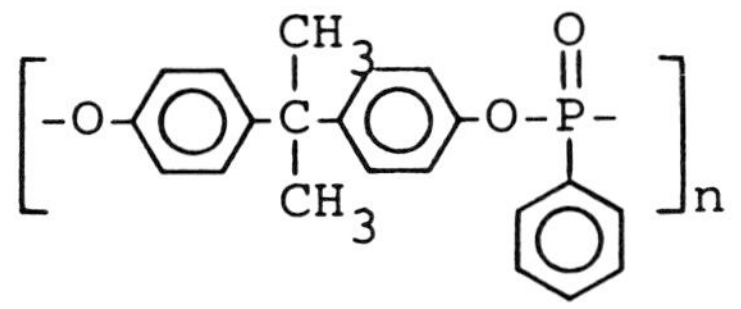

3

The selection of the reaction medium is also important. Polyarylates are successfully prepared in water-immiscible organic liquids which dissolve or at least highly swell the polyarylates. Chlorinated hydrocarbons such as dichloromethane, chloroform, and 1,2-dichloroethane, and aromatic solvents like nitrobenzene are conveniently used organic solvents, since they dissolve many polyarylates [1, 4–7]. Other reaction variables such as reactant concentration, stirring rate, and reaction temperature and time are discussed in detail in the literature [1].

In contrast to the high-temperature polycondensation methods described later, the interfacial polycondensation is irreversible and no exchange reactions occur between the macromolecules, so that the molecular weight distributions of the products may differ markedly from those from the high-temperature processes. The polymer yields are generally 80% to quantitative, and in favorable cases molecular weights as high as 80,000 can be attained.

Table 2 Effect of Phase Transfer Catalysts in the Preparation of Polyarylate **1** and Related Polymers **2** and **3** by Interfacial Polycondensation [4,6]

Catalyst	Inherent viscosity of polymer		
	1	**2**	**3**
None	0.32	0.20	0.05
Tetrabutylammonium chloride	0.73	1.44	0.36
Benzyltriethylammonium chloride	1.11	1.02	0.21
Cetyltrimethylammonium chloride	0.85	1.19	0.70
Benzyltriphenylphosphonium chloride	0.81		0.88
Cetyltributylphosphonium bromide	0.77		0.28
18-Crown-6	0.79	0.99	0.30
Dibenzo-18-crown-6	0.45	0.95	0.59
Dicyclohexyl-18-crown-6	0.82	1.28	0.64

The interfacial polycondensation can be widely used on a laboratory scale for the preparation of a variety of polyarylates [1]. Moreover, this method has been realized in the commercial production of polycarbonate **4**, derived from phosgene, which is the most simple diacid chloride [8], and polyarylate **5** having isophthalate–terephthalate structures.

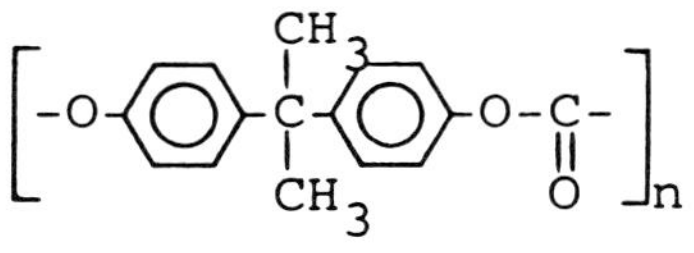

4

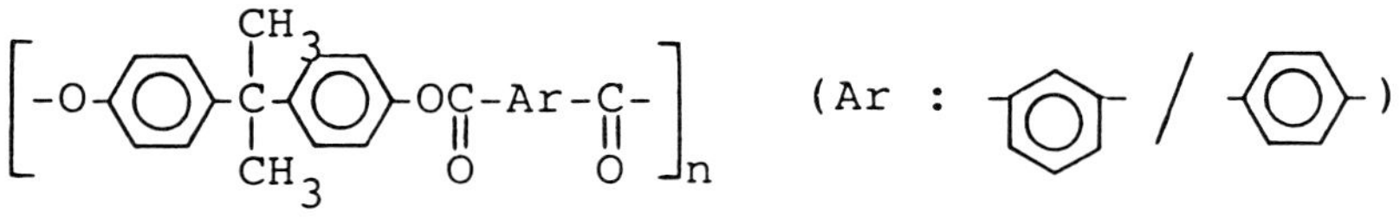

5

Low-Temperature Solution Polycondensation

The low-temperature solution polycondensation is carried out at ordinary temperature in a single liquid phase inert to both reactants [1]. The liquid phase can contain one or more organic solvents. With this process, it is necessary to use a stoichiometric or excess amount of an organic base, such as pyridine or triethylamine, which combines with the liberated hydrogen chloride. Polyarylates of high molecular weights are successfully prepared in pyridine alone or in combinations of tertiary amines and inert organic solvents, such as triethylamine-1,2-dichloroethane [9] and pyridine-*sym*-tetrachloroethane [10].

The procedure most often starts with all the reactants fully dissolved, but this is not always an essential requirement. The polyarylates may remain in solution or precipitate at any time. The advantage of this process is that low-boiling solvents and normal pressure can be applied. The analogous process for polycarbonate **4** is a large-scale production.

High-Temperature Solution Polycondensation

The diacid chloride route is further applicable to high-temperature techniques in addition to the two low-temperature procedures. Although the polycondensation of diacid chlorides with bisphenols without any hydrogen chloride acceptors is slow, the reaction rate increases with increasing temperature. The high-temperature solution polycondensation is therefore carried out at an elevated temperature around 200°C in inert high-boiling solvents such as nitrobenzene and *o*-dichlorobenzene [11]. The evolved hydrogen chloride is continuously removed from the refluxing solvent with the aid of an inert gas sweep, and no acid acceptor is needed.

To obtain high molecular weight products, it is essential that the reactants, solvent, and carrier gas are well dried, since traces of moisture hydrolyze the diacid chlorides, thus blocking the reaction. The polyarylates which are soluble or highly swollen in the hot solvents are successfully obtained in high molecular weights [9]. Catalysts such as amines and metallic magnesium play a significant role in polyarylate synthesis [2, 12].

Melt/Solid Phase Polycondensation

Polyarylates of high molecular weights are produced by the polycondensation of dicarboxylic acid chlorides with bisphenols in anhydrous melt or solid phase system at a temperature ranging from 100 to 300°C [13]. One has the problem of high viscosity of the reaction mass and the demand to remove the by-product, hydrogen chloride. Therefore, a vacuum must be applied to remove the last traces of hydrogen chloride at the final stage of the reaction. In this process, a considerable discoloration problem often arises.

Phenyl Ester Route

High-melting and poorly soluble aromatic dicarboxylic acids, which are neither readily purified nor easily polymerized directly, are usually converted to polyarylates by way of their phenyl esters, whose lower melting points permit the production of homogeneous melts of the reactants. The polycondensation through ester exchange reaction of the diphenyl esters of aromatic dicarboxylic acids with bisphenols (Eq. 3) or that of AB-type monomers (Eq. 4) can be performed easily at an elevated temperature. Since phenyl esters are much less reactive than the corresponding acid chlorides, a higher reaction temperature and a longer reaction time are needed to prepare high molecular weight polyarylates.

$$HO-Ar-OH + C_6H_5O\underset{O}{\overset{\|}{C}}-Ar'-\underset{O}{\overset{\|}{C}}OC_6H_5 \longrightarrow \left[-O-Ar-O\underset{O}{\overset{\|}{C}}-Ar'-\underset{O}{\overset{\|}{C}}-\right]_n + 2\ C_6H_5OH \quad (3)$$

$$HO-Ar-\underset{O}{\overset{\|}{C}}OC_6H_5 \longrightarrow \left[-O-Ar-\underset{O}{\overset{\|}{C}}-\right]_n + 2\ C_6H_5OH \quad (4)$$

Many of the polycondensations are carried out favorably in the melt state or in the slurry with inert high-boiling media at a temperature ranging from 200 to 300°C [2, 14]. The diffusion rate of the reaction by-product, phenol, determines the rate of polycondensation, since ester-exchange processes are inherently reversible. A remarkable aspect of

Phenol Acetate Route

The preparation of polyarylates by high-temperature polycondensation through an acidolysis ester-exchange reaction between diacetates of bisphenols with free dicarboxylic acids (Eq. 5) or by that of AB-type monomers (Eq. 6) has been known since the 1950s [3, 15], and the phenol acetate route has become increasingly important since the 1970s [14, 16].

$$CH_3\underset{O}{\overset{O}{C}}-Ar-O\underset{O}{\overset{O}{C}}CH_3 + HO\underset{O}{\overset{O}{C}}-Ar'-\underset{O}{\overset{O}{C}}OH \longrightarrow \left[-O-Ar-O\underset{O}{\overset{O}{C}}-Ar'-\underset{O}{\overset{O}{C}}-\right]_n + 2\ CH_3\underset{O}{\overset{O}{C}}OH \quad (5)$$

$$CH_3\underset{O}{\overset{O}{C}}-Ar-\underset{O}{\overset{O}{C}}OH \longrightarrow \left[-O-Ar-\underset{O}{\overset{O}{C}}-\right]_n + CH_3\underset{O}{\overset{O}{C}}OH \quad (6)$$

Polycondensation is usually performed by heating in the melt or solid state at a temperature ranging from 200 to 350°C in a nitrogen atmosphere with or without catalysts [17]. The reaction with no catalyst is preferable, because catalysts may affect the color and the thermal stability of the final products. In the later stage, polycondensation is a diffusion-controlled process, and a reactor is desirable to improve the mixing and to accelerate the evolution of the reaction by-product, acetic acid. A vacuum is then applied to the reaction in order to remove the acetic acid from the melt or solid mass and to increase the molecular weight of the polyarylates.

In the melt or solid-phase process, the problems of undesirable side reactions sometimes arise. To avoid these problems, a high-temperature solution process using appropriate high-boiling inert liquid media is preferable; this provides the necessary control of the reaction, yielding polyarylates of higher quality [4, 18–21]. Thus the monomer purification and the reaction conditions applied play a very important role in the overall properties and appearance of the final products. Polyarylate **6** and the related thermotropic liquid crystalline polyarylates **7** and **8**, as well as polyarylate-modified polyester **9** [22], are successfully produced commercially by the high-temperature acetate method.

Phenol Silyl Ether Route

The phenol silyl ether method for the synthesis of polyarylates was developed in 1979. Bisphenol disilyl ethers can be used for polycondensation with dicarboxylic acid chlorides at an elevated temperature around 250°C or above, yielding polyarylates of high molecular weights (Eq. 7) [23]. Other polyarylates are also obtainable for AB-type monomers (Eq. 8) [24, 25].

$$(CH_3)_3SiO-Ar-OSi(CH_3)_3 + ClC(O)-Ar'-C(O)Cl$$
$$\longrightarrow \left[-O-Ar-OC(O)-Ar'-C(O)-\right]_n + 2\,(CH_3)_3SiCl \quad (7)$$

$$(CH_3)_3SiO-Ar-C(O)Cl \longrightarrow \left[-O-Ar-C(O)-\right]_n + (CH_3)_3SiCl \quad (8)$$

High-temperature polycondensation is carried out in high-boiling inert liquid media and proceeds in solution or suspension with the evolution of the by-product, trimethylsilyl chloride. Polyarylate **6**, having molecular weights in the range of 20,000–50,000, is readily produced by this procedure [25]. In addition, the melt/solid-state process is also applicable for polyarylate preparation.

Direct Polyesterification Method

The direct polyesterification procedure is not applicable to the simplest monomer pairs, bisphenols and aromatic dicarboxylic acids (Eq. 9), because the phenol components are normally too unreactive to be esterified directly. This is also true for the AB-type hydroxy acids (Eq. 10). Therefore, to promote direct polyarylate formation, it is essential to use suitable condensing reagents, XY. These reagents act as activating agents for dicarboxylic acids, giving the activated diacid intermediates (Eq. 11), which in turn are condensed with the bisphenols producing the polyarylates (Eq. 12). Equation 13 can thus be obtained as the sum of Eqs. 11 and 12. Direct polyesterification is also applicable to AB-type monomers (Eq. 14).

$$HO-Ar-OH + HOC(O)-Ar'-C(O)OH \longrightarrow \left[-O-Ar-OC(O)-Ar'-C(O)-\right]_n + 2\,H_2O \quad (9)$$

$$HO-Ar-C(O)OH \longrightarrow \left[-O-Ar-C(O)-\right]_n + H_2O \quad (10)$$

$$HOC(O)-Ar'-C(O)OH + 2\,XY \longrightarrow XC(O)-Ar'-C(O)X + 2\,YOH \quad (11)$$

$$HO-Ar-OH + XC(O)-Ar'-C(O)X \longrightarrow \left[-O-Ar-OC(O)-Ar'-C(O)-\right]_n + 2\,HX \quad (12)$$

$$HO-Ar-OH \; + \; HOC-Ar'-COH \; + \; 2\,XY$$
$$\underset{O}{\overset{\|}{}}\;\;\underset{O}{\overset{\|}{}}$$

$$\longrightarrow \left[-O-Ar-O\underset{O}{\overset{\|}{C}}-Ar'-\underset{O}{\overset{\|}{C}}-\right]_n + \; 2\,YOH \; + \; 2\,HX \quad (13)$$

$$HO-Ar-\underset{O}{\overset{\|}{C}}OH \; + \; XY \; \longrightarrow \; \left[-O-Ar-\underset{O}{\overset{\|}{C}}-\right]_n + \; YOH \; + \; HX \quad (14)$$

This type of direct polyesterification is a new method that has been extensively studied since 1980 [26].

The general one-pot two-step procedure for carrying out direct polycondensation is as follows. A dicarboxylic acid is reacted with a condensing reagent in a solvent such as pyridine, and the resulting activated diacid intermediate is condensed with a bisphenol in the solution under relatively mild conditions.

Table 3 lists effective condensing reagents among those investigated so far for the preparation of polyarylate **5** of high inherent viscosities. A variety of polyarylates with high molecular weights, including **5**, can be readily obtained by the use of these reagents [27–33]. The condensing reagents presumably afford highly reactive diacid derivatives such as acyloxyphosphonium salts or mixed anhydrides by the in situ reaction with dicarboxylic acids.

Direct polycondensation is also sensitive to reaction medium, aging conditions (temperature and time) for the first-stage reaction, and reaction temperature for the second-stage polycondensation. The most preferable reaction medium is pyridine, and the temperature suitable for the latter polycondensation is in the range of 80–120°C. In some cases, polyarylates of high molecular weights can be obtained by a simple one-step procedure that involves a simple mixing of a mixture of monomer reactants with a condensing reagent in solution. The direct polyesterification method is useful for the laboratory-scale preparation of polyarylates.

Table 3 Reaction Conditions for Direct Polycondensation of Bisphenol A with Iso- and Terephthalic Acids (50:50) by Means of Various Condensing Reagents

Condensing reagent	Reaction medium	Inherent viscosity of polyarylate **5**	Reference
$(C_6H_5)_3PCl_2$ + triethylamine	Chlorobenzene	1.42[a]	27
$(C_6H_5)_2POCl$	Pyridine	1.08	28
$(C_6H_5)_2POCl$ + LiBr	Pyridine	1.48	29
$POCl_3$ + LiCl	Pyridine	1.26	30
$CH_3C_6H_4SO_2Cl$	Pyridine	1.20	31
$CH_3C_6H_4SO_2Cl$ + dimethylformamide	Pyridine	1.29	32
$SOCl_2$	Pyridine	2.28	33

[a]Reduced specific viscosity.

STRUCTURE–PROPERTY RELATIONSHIPS IN POLYARYLATES

The basic mechanical property profile of amorphous polyarylates is generally similar to their polycarbonate counterparts, except the higher glass transition extends the higher temperature utility. The excellent ultraviolet stability is another characteristic of polyarylates; they undergo a photochemical Fries rearrangement to yield a thin surface of polymer backbones containing significant amounts of o-hydroxybenzophenone unit that acts as an ultraviolet stabilizer [34]. This results in better property and transparency retention of polyarylates upon ultraviolet exposure. Other subtle advantages of polyarylates include improved flammability resistance, improved environmental stress rupture resistance, and significantly better notched toughness.

The glass transition temperatures T_g of various types of polyarylates are given in Tables 4–6 [35–44], and the melting points T_m as well as T_g of the polyarylates containing ether linkages are listed in Table 7 [45], all of which can be used to illustrate the following discussion of the effects of some structural variations on thermal properties.

In general, the effects of the nature of the chain repeating units on T_g as well as T_m are closely related to chain stiffness and symmetry. The bisphenol A-based polyisophthalate and polyterephthalate, which are the most representative polyarylates, have T_g of 180 and 210°C, respectively [7]. The fact that these T_g values are much higher than that of the corresponding polycarbonate ($T_g = 149°C$) can be explained on the basis of the introduction of a rigid phthaloyl group into the polymer backbones. A decrease in chain stiffness by the incorporation of ether linkages causes a marked reduction of T_g of the polyarylates (Table 7).

The substitution of unsymmetrical *meta*-oriented phenylene groups for symmetric *para*-oriented groups has a significant and cumulative effect in reducing the T_g of the polyarylates (Tables 4–6). It appears to make little difference whether substitution takes place with the diacid component or with the bisphenol component (Table 6).

Another important factor affecting the T_g is hindrance to free rotation along the polymer chain resulting from the presence of unsymmetrically fused ring structures like 1,1,3-trimethyl-3-phenylindane unit, or bulky cyclic side groups such as 9,9-diphenylfluorene moiety (Tables 4 and 5). The introduction of these fused ring and bulky groups into the polymer main chain causes a marked rise in the T_g of the polyarylates. It is interesting to note that such polyarylates have high solubility characteristics in addition to high T_g.

Amorphous polyarylates having T_g of 250°C or below as well as crystalline polyarylates that melt in the temperature range of about 250–340°C can be melt-processed without thermal degradation to produce fibers, films, and moldings. High T_g polyarylates that are soluble in organic solvents are fabricated into films by solution casting.

Aromatic polysulfonates, sulfur-containing analogs of polyarylates, are usually amorphous and have considerably lower T_g than those of the corresponding polyarylates with CO–O in place of SO_2–O (Table 8) [46, 47]. For example, the T_g of the polysulfonate derived from bisphenol A and *m*-benzenedisulfonyl chloride is 79°C, whereas the analogous polyisophthalate has a T_g of 180°C. The T_g of the aromatic polysulfonates respond to structural changes in the arylene units in the same way as do those of polyarylates, and they are mostly in the range of 80–160°C. None of these polysulfonates has yet acquired technical importance.

Aromatic polyphosphonates, phosphorus-containing analogs of polyarylates, resemble aromatic polysulfonates in structure–property relationships [48–50]. The T_g of aromatic polyphosphonates are mostly in the range of 80–150°C and much lower than those of the

Table 4 Glass Transition Temperatures of Aromatic Polyisophthalates and Polyterephthalates

$$\left[-O-Ar-O\underset{\underset{O}{\|}}{C}-Ar'-\underset{\underset{O}{\|}}{C}- \right]_n$$

Ar	Ar' (meta-C₆H₄)	Ar' (para-C₆H₄)	Reference
–C₆H₄–C(CH₃)₂–C₆H₄–	180	210	7
–C₆H₄–C(CF₃)₂–C₆H₄–	200		35
–C₆H₄–C(C₆H₅)₂–C₆H₄–		280	36
indane-CH₃/CH₃/CH₃ structure	235	253	6
spirobiindane (CH₃)₄ structure	260		37
binaphthyl		285	38
fluorene-diphenyl structure	270	370*	39, 36*
phenolphthalein structure	318		40

Table 4 (*Continued*)

Ar	Ar'		Reference
[structure: diphenyl isoindolinone with NH, C=O]	325	327	40

Table 5 Glass Transition Temperatures of Aromatic Polyarylates

$$\left[-O-Ar-O\underset{\underset{O}{\|}}{C}-Ar'-\underset{\underset{O}{\|}}{C}-\right]_n$$

Ar	Ar'		
	[tetrafluorobenzene] a	[trifluorobenzene] b	[tetraphenylthiophene] c
[bisphenol A: -C(CH$_3$)$_2$- bridging two phenyls]	150		235
[hexafluoroisopropylidene: -C(CF$_3$)$_2$-]	145	215	
[indane-type with CH$_3$, CH$_3$, CH$_3$]	185	185	335
[9,9-fluorene]		255	295

[a]From Ref. 41.
[b]From Ref. 42.
[c]From Ref. 43.

Table 6 Glass Transition Temperatures of Methylene-Bridged Aromatic Polyarylates

Polymer	T_g (°C)
-O-⟨C₆H₄⟩-OC(O)-⟨C₆H₄⟩-CH₂-⟨C₆H₄⟩-C(O)-	108
-O-⟨C₆H₄⟩-CH₂-⟨C₆H₄⟩-OC(O)-⟨C₆H₄⟩-CH₂-⟨C₆H₄⟩-C(O)-	101
-O-⟨C₆H₄⟩-CH₂-⟨C₆H₄⟩-OC(O)-⟨C₆H₄⟩-CH₂-⟨C₆H₄⟩-C(O)-	123
-O-⟨C₆H₄⟩-C(CH₃)₂-⟨C₆H₄⟩-OC(O)-⟨C₆H₄⟩-CH₂-⟨C₆H₄⟩-C(O)-	149
-O-⟨C₆H₄⟩-CH₂-⟨C₆H₄⟩-OC(O)-⟨C₆H₄⟩-CH₂-⟨C₆H₄⟩-C(O)-	123
-O-⟨C₆H₄⟩-CH₂-⟨C₆H₄⟩-OC(O)-⟨C₆H₄⟩-CH₂-⟨C₆H₄⟩-C(O)-	185
-O-⟨C₆H₄⟩-C(CH₃)₂-⟨C₆H₄⟩-OC(O)-⟨C₆H₄⟩-CH₂-⟨C₆H₄⟩-C(O)-	212
-O-⟨C₆H₄⟩-C(CH₃)₂-⟨C₆H₄⟩-OC(O)-⟨C₆H₄⟩-C(O)-⟨C₆H₄⟩-C(O)-	165
-O-⟨C₆H₄⟩-C(CH₃)₂-⟨C₆H₄⟩-OC(O)-⟨C₆H₄⟩-C(O)-⟨C₆H₄⟩-C(O)-	218

Source: Ref. 44.

Table 7 Melting Points and Glass Transition Temperatures of Ether-Linked Aromatic Polyarylates

Polymer	T_m (°C)	T_g (°C)
	275	160
	291	147
	280	134
	188	103
	405	
	378	
	264	
		119
	296	
	438	
	232	159
	340	
	339	
	312	
	259	119

Table 7 (*Continued*)

Polymer	T_m (°C)	T_g (°C)
structure 1	285	
structure 2		125
structure 3		104
structure 4	188	131
structure 5	334	
structure 6	450	
structure 7	325	
structure 8	327	
structure 9	282	
structure 10	312	
structure 11		101
structure 12	251	119
structure 13	345	

Source: Ref. 45.

Table 8 Glass Transition Temperatures of Aromatic Polysulfonates

$$\left[-O-\underset{CH_3}{\overset{CH_3}{\underset{|}{\overset{|}{C}}}}--OSO_2-Ar-SO_2- \right]_n$$

Ar	T_g (°C)
—⟨phenyl⟩—	49,[a] 79[b]
—⟨phenyl⟩—⟨phenyl⟩—	160–165,[a] 111[b]
—⟨phenyl⟩—CH$_2$—⟨phenyl⟩—	114–116[a]
—⟨phenyl⟩—O—⟨phenyl⟩—	118–120,[a] 125[b]
—⟨phenyl⟩—SO$_2$—⟨phenyl⟩—	100–104[a]

[a]From Ref. 46.
[b]From Ref. 47.

corresponding polycarbonates (Table 9). These aromatic polyphosphonates are of commercial interest because of their flame–retardant characteristics. The properties of the representative polyphosphonate moldings are summarized in Table 10.

THERMOTROPIC LIQUID CRYSTALLINE POLYARYLATES

Thermotropic liquid crystalline polyarylates have received great attention since 1976 when the first indication of the thermotropic liquid crystalline character of poly-*p*-oxybenzoyl-modified poly(ethylene terephthalate) **9** was found by Jackson and co-workers [22].

A wide variety of thermotropic liquid crystalline polyarylates, especially those based on rigid rodlike aromatic ester "mesogenic" units with flexible "spacer" units, have been prepared and characterized by various academic laboratories for their ability to form liquid crystalline melt, the type of phase formed (smectic or nematic), their transition temperatures, and the morphology of the mesophase [51–57]. However, more technical information from the viewpoint of engineering materials is scanty. In this section, the

Table 9 Glass Transition Temperatures of Aromatic Polyphosphonates

$$\left[-O-Ar-O-\underset{R}{\overset{\overset{O}{\|}}{P}}- \right]_n$$

Ar	T_g (°C)	
	R = CH$_3$	R = C$_6$H$_5$
–C$_6$H$_4$–C(CH$_3$)$_2$–C$_6$H$_4$–	90[a]	104,[a] 99[b]
–C$_6$H$_4$–O–C$_6$H$_4$–	80[a]	
–C$_6$H$_4$–S–C$_6$H$_4$–	78[a]	86[a]
–C$_6$H$_4$–SO$_2$–C$_6$H$_4$–		146[c]
–C$_6$H$_4$–C$_6$H$_4$–	137[a]	146,[a] 120[b]
indane-type (1,1,3-trimethyl)		124[b]
spirobifluorene-type		188[b]

[a]From Ref. 48.
[b]From Ref. 49.
[c]From Ref. 50.

Table 10 Properties of Aromatic Polyphosphonate Molding

$$\left[-O-\underset{}{\bigcirc}-\underset{}{\bigcirc}-O-\underset{\underset{CH_3}{|}}{\overset{\overset{O}{\|}}{P}}- \right]_n$$

Density	1.32 g cm^{-3}
Tensile strength	44 MPa
Elongation at break	21%
Flexural modulus	2.6 GPa
Glass transition temperature	137°C
Vicat B temperature	132°C
Limiting oxygen index	75

Source: Ref. 48.

effects of composition on the melting point and properties of plastics and fibers of thermotropic liquid crystalline polyarylates are described.

Thermotropic liquid crystalline aromatic polyarylates generally have rigid, rodlike structures. The basic structures are benzene rings attached to para positions through ester groups, i.e., poly-*p*-oxybenzoyl **6** and poly(*p*-phenylene terephthalate) **10**.

$$\left[-O-\underset{}{\bigcirc}-O\underset{\underset{O}{\|}}{C}-\underset{}{\bigcirc}-\underset{\underset{O}{\|}}{C}- \right]_n$$

10

Polyarylates **6** and **10** have high melting points of 610 and 600°C, respectively, which are well above their decomposition temperatures [58]. Therefore, these polyarylates cannot be melt-processed by conventional techniques. The high melting points of the basic structures can be decreased to a temperature below 400°C, at which the polyarylates are sufficiently stable to be extruded, melt-spun, or injection-molded.

Thermotropic liquid crystalline polyarylates can be obtained by (a) introduction of flexible "spacer" units into the basic rodlike aromatic "mesogenic" structures, (b) substitution on the aromatic rings, (c) modification of the polymers with certain rodlike comonomers, or (d) introduction of rigid kinks into the straight polymer chains. It is often necessary to use a combination of two of these approaches to reduce the melting points of the polyarylates to a desirable range [58, 59]. The most frequently used monomers for the synthesis of thermotropic liquid crystalline polyarylates are listed in Table 11 according to their class [10, 56].

The remarkable characteristic of thermotropic liquid crystalline polyarylates is that the melt-spun and injection-molded products have very high strength and modulus and a very low coefficient of linear thermal expansion. The exceptional mechanical and thermal properties are due to the presence of rigid, rodlike polymeric chains, which become highly oriented during melt-spinning or injection-molding, and long relaxation times, which

Table 11 Suitable Monomers for the Preparation of Thermotropic Liquid Crystalline Polyarylates

Bisphenol	Aromatic dicarboxylic acid	Hydroxy acid

permit the polymer chains to retain their orientation while the polymers are cooling. Another important feature of thermotropic liquid crystalline polyarylates is that they have low melt viscosities compared with non–liquid crystalline polyarylates, and this is advantageous for melt-processing.

Introduction of Flexible Spacer Units

The high melting point of poly-*p*-oxybenzoyl (POB) **6** can be reduced below 400°C by modification with poly(ethylene terephthalate) (PET) or poly(ethylene 2,6-naphthalenedicarboxylate) (PEN) molecule [22, 60]. As shown in Table 12, the melting point of POB-modified PET **9** decreases with increasing PET content. The copolyesters containing 40–90 mol % POB are highly anisotropic and produce opaque melts. Minimum melt viscosities are obtained with the polymers containing 60–70 mol % POB. At this composition range, the mechanical properties of the injection-molded copolyesters such as tensile strength, flexural modulus, and impact strength are at a maximum and the molding shrinkage is zero, when the moldings are measured along the flow direction of the polymer melts, indicating the occurrence of molecular orientation during molding. The tensile properties are superior to those of commercial glass-fiber–reinforced PET. In this regard, copolyesters of this class are termed self-reinforced polyesters. The properties obtained in the perpendicular direction, however, are similar to or somewhat lower than the properties of polyesters that are not liquid crystalline.

In addition to the POB-modified PET **9**, the POB-modified PEN **11** containing 60–80 mol % POB is also a thermotropic liquid crystalline polyester [60] and gives injection-molded plastics having high tensile and flexural properties and a zero molding shrinkage (Table 12). Because of the high glass transition temperature of PEN (111°C), the copolyester with 60 mol % POB has a high heat distortion temperature of 120°C, compared to 64°C for the POB-modified PET with similar composition.

Another POB copolyester with thermotropic liquid crystallinity is copolyester **12** based on hydroquinone/4-carboxyphenylpropionic acid polyester (T_m = 425°C) and POB (T_m = 610°C). The copolyester with 80 mol % POB has a melting point of 300°C and is melt-

Table 12 Properties of Poly-*p*-oxybenzoyl–Modified Poly(ethylene terephthalate) **9**

		Polymer **9**			
Property	PET	*x*/*y* = 60:40	40:60	20:80	GF-PET[a]
Melting point (°C)	245	226		293	
Injection-molding temperature (°C)	275	250	260	340	
Tensile strength (MPa)	55	197	232	240	134
Elongation at break (%)	240	10	20	24	3
Flexural modulus (GPa)	2.3	7.6	12.5	9.6	9.3
Notched Izod impact strength (J m^{-1})	16	64	416	117	80
Glass transition temperature (°C)	69	81			
Heat deflection temperature (°C)	66	71	64	154	
Mold shrinkage (%)	0.6	0	0	0	

[a]Glass-fiber–reinforced (30 wt%) PET.
Source: Ref. 22.

$$\left[\left(\text{OCH}_2\text{CH}_2\text{O}\underset{\text{O}}{\overset{\text{C}}{\text{C}}}\text{-}\bigcirc\!\bigcirc\text{-}\underset{\text{O}}{\overset{\text{C}}{\text{C}}}\right)_x\!\!\left(\text{O}\text{-}\bigcirc\text{-}\underset{\text{O}}{\overset{\text{C}}{\text{C}}}\right)_y\right]_n$$

11

Table 13 Properties of Poly-*p*-oxybenzoyl–Modified Poly(ethylene 2,6-naphthalenedicarboxylate) **11**

Property	PEN	Polymer **11** $x/y = 40:60$	20:80
Injection-molding temperature (°C)	280	245	360
Tensile strength (MPa)	82	170	153
Elongation at break (%)	16	14	14
Flexural modulus (GPa)	2.6	13.2	11.3
Notched Izod impact strength (J m^{-1})	37	37	32
Heat deflection temperature (°C)	91	120	226
Mold shrinkage (%)	1.0	0	0

Source: Ref. 60.

spun at 335°C into fibers. The tensile strength to elongation at break to tensile modulus values of the as-spun fibers are 6.9 g d^{-1} : 1.7% : 470 g d^{-1}, whereas high-strength and high-modulus fibers of 20 g d^{-1} : 2.9% : 960 g d^{-1} are obtained by heat treatment of the fibers at 320°C. It is known that heat treatment of polyester fibers with rigid structures increases their molecular weights and tensile properties, and the same effect is observed with this copolyester fiber.

$$\left[\left(\text{O}\text{-}\bigcirc\text{-}\text{O}\underset{\text{O}}{\overset{\text{C}}{\text{C}}}\text{CH}_2\text{CH}_2\text{-}\bigcirc\text{-}\underset{\text{O}}{\overset{\text{C}}{\text{C}}}\right)_x\!\!\left(\text{O}\text{-}\bigcirc\text{-}\underset{\text{O}}{\overset{\text{C}}{\text{C}}}\right)_y\right]_n$$

12

Injection-molded plastics with moderately high tensile properties are also obtained from some of the thermotropic liquid crystalline homopolyesters with flexible methylene spacer units; polyesters **13** and **14** are typical (Table 14) [58, 61]. In general, the properties tend to be downward as the length of the flexible spacer increases.

$$\left[-\text{O}\text{-}\bigcirc(\text{CH}_3)\text{-}\text{O}\underset{\text{O}}{\overset{\text{C}}{\text{C}}}\text{-}\bigcirc\text{-}\text{ORO}\text{-}\bigcirc\text{-}\underset{\text{O}}{\overset{\text{C}}{\text{C}}}-\right]_n$$

13

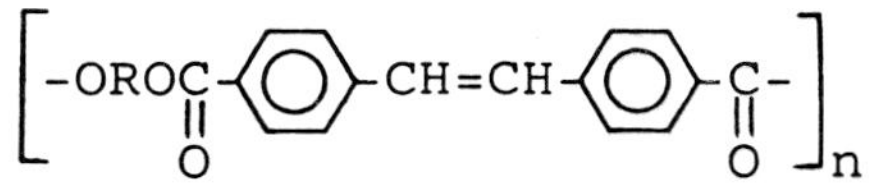

$$\left[-\text{OROC}-\bigcirc-\text{CH}=\text{CH}-\bigcirc-\underset{\underset{O}{\|}}{C}- \right]_n$$
$$\overset{\|}{O}$$

14

Table 14 Properties of Thermotropic Liquid Crystalline Polyesters

Property	Polymer 13[a]		Polymer 14[b]	
	R = —(CH$_2$)$_2$—	—(CH$_2$)$_4$—	R = —CH$_2$CH— $\quad$ CH$_3$	—(CH$_2$)$_4$—[c]
Melting point (°C)	316	212	263	272
Injection-molding temperature (°C)			280	280
Tensile strength (MPa)	259	74	199	263
Elongation at break (%)			5	13
Flexural modulus (GPa)	11.4	10.3	11.5	4.8
Notched Izod impact strength (J m^{-1})			139	278
Heat deflection temperature (°C)	177	148	230	107

[a]From Ref. 58.
[b]From Ref. 61.
[c]The polyester was modified with 20 mol % of 2,6-naphthalenedicarboxylic acid.

Modification with Substitution on Aromatic Rings

The second approach for reducing the high melting point of basic poly(p-phenylene terephthalate) **10** (T_m = 600°C) is substitution on the aromatic rings. When a substituted hydroquinone is used as a monomer and the substituent is phenyl, the corresponding polyterephthalate **15** melts at 340°C and is low enough to be injection-molded or melt-spun without thermal degradation [58, 60]. The melting point of the homopolyarylate **16** derived from phenylhydroquinone and 2,6-naphthalenedicarboxylic acid is 390°C [60].

Table 15 lists the properties of the injection-molded polyisophthalate of phenylhydroquinone **17**, which is not liquid crystalline because of the existence of kinks in the polymer backbone, the polyterephthalate homopolymer **15**, and its copolyarylates **18** and

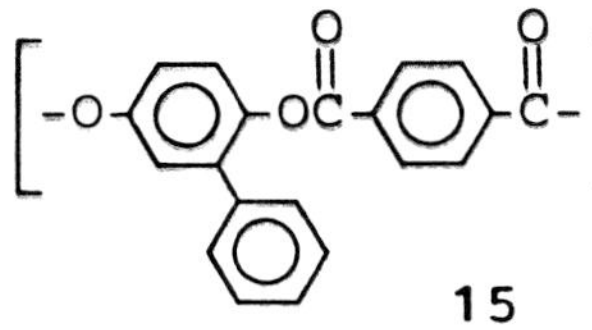

15 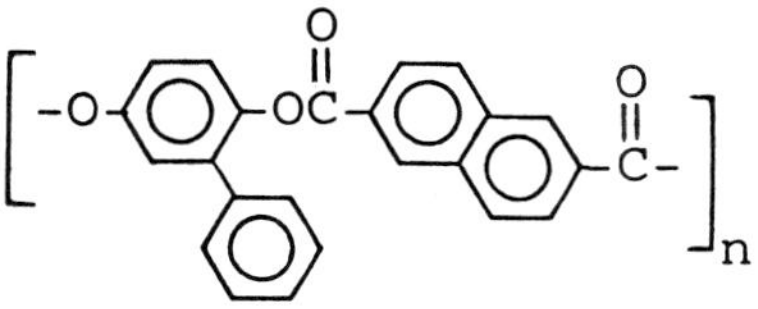 $\qquad\qquad$ **16**

19 modified with hydroquinone or 2,6-naphthalenedicarboxylic acid [59]. The tensile properties of the copolyterephthalate **18** are much higher than those of the homopolyterephthalate **15**. The results also indicate a slight increase in the tensile properties of the copolyarylates **19** modified with 2,6-naphthalenedicarboxylic acid compared with the homopolyterephthalate **15**. The heat-treated fibers of the copolyarylates **19**, on the other hand, do not have higher tensile strength and modulus than those of the homopolyterephthalate **15** (Table 16) [59, 60].

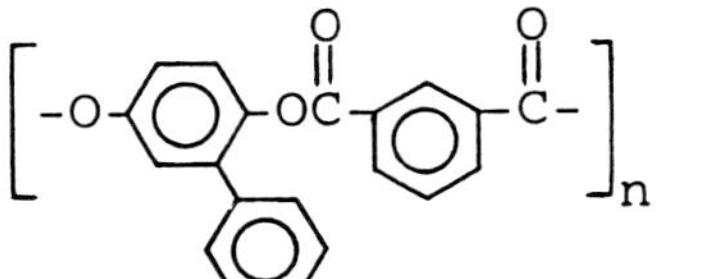

17

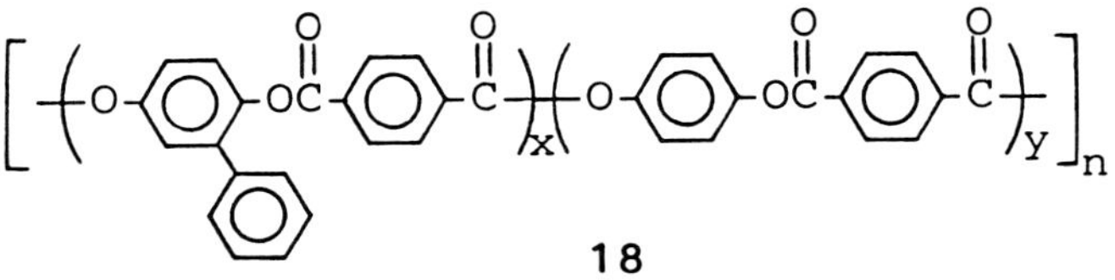

18

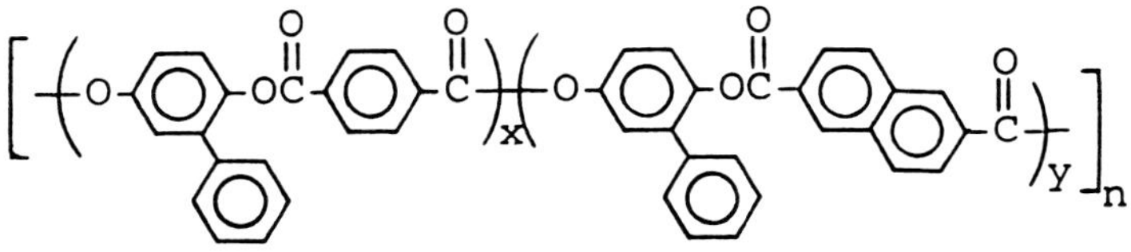

19

Table 15 Properties of Thermotropic Liquid Crystalline Polyarylates

	Polymer			Polymer **19**	
Property	**15**	**17**	**18**	$x/y = 70:30$	20:80
Melting point (°C)	340		370	278	350
Injection-molding temperature (°C)	360	340	400	350	380
Tensile strength (MPa)	179	87	220	193	181
Elongation at break (%)	6	35	8	9	7
Flexural modulus (GPa)	13.3	3.3	18.4	11.9	11.0
Notched Izod impact strength (J m^{-1})	85	27	107	267	160
Heat deflection temperature (°C)	258	114	269	158	189

Source: Ref. 59.

Table 16 Properties of Heat-Treated Fibers of Thermotropic Liquid Crystalline Polyarylates

		Polymer **19**	
Property	Polymer **15**	$x/y = 90{:}10$	30:70
Melting point (°C)	340	310	330
Tensile strength (g d^{-1})	32	20	23
Elongation at break (%)	4.3	5.2	6.3
Tensile modulus (g d^{-1})	910	350	370

Source: Refs. 59 and 60.

Modification with Unsubstituted Rodlike Components

The third approach for reducing the high melting points of the rigid, rodlike polyarylates is modification with certain unsubstituted rodlike comonomers which, because of their rodlike structure, do not reduce the degree of crystallinity (e.g., 4,4'-biphenyl and 2,6-naphthalene derivatives).

This approach was effectively used by Economy when he produced the copolyarylates **7** based on poly-*p*-oxybenzoyl (POB) **6** and poly(4,4'-biphenylene terephthalate) **20** [16]. The POB homopolymer **6** and the copolyarylate **7** with 67 mol % POB have been marketed since 1972 by Carborundum under the tradenames of Ekonol and Ekkcel I-2000, respectively. The properties of injection-molded copolyarylates **7** are listed in Table 17. Although the tensile properties are not high, these properties can be improved by molding at higher temperature; tensile strength and flexural modulus values of 156 MPa and 10.5 GPa, respectively, are obtainable [58].

20

Table 17 Properties of Thermotropic Liquid Crystalline Polyarylates

	Polymer **7**[a]		Polymer **22**[b]	
Property	$x/y = 33{:}67$	67:33	$x/y = 50{:}50$	20:80
Tensile strength (MPa)	97	69	210	226
Elongation at break (%)	8	7–9		
Flexural modulus (GPa)	4.8	3.2	15.4	12.4
Notched Izod impact strength (J m^{-1})	53		128	336
Heat deflection temperature (°C)	293	300	>230	>230
Mold shrinkage (%)	1.2		0	0

[a]From Ref. 16.
[b]From Ref. 58.

The other most effective monomers are 2,6-naphthalene derivatives such as diol, dicarboxylic acid, and hydroxy acid. Even though poly(p-phenylene 2,6-naphthalenedicarboxylate) **21** melts at 580°C, compared with 600°C for poly(p-phenylene terephthalate) **10**, the melting point of copolyarylate **22** based on **21** and POB **6** is appreciably lower than that of copolyarylate **23** with **6** and **10** structures, and hence these naphthalene-containing copolyarylates **22** having melting points as low as 325°C are readily melt-processable [58]. The injection-molded copolyarylates **22** have high strength, high modulus, and zero shrinkage, which are characteristic of liquid crystalline copolyarylates (Table 17). The 2,6-naphthalene ring structure does introduce a crankshaft conformation in the polymer chain; however, because the chain continues in a direction parallel to the original direction, there appears to be no loss in liquid crystallinity.

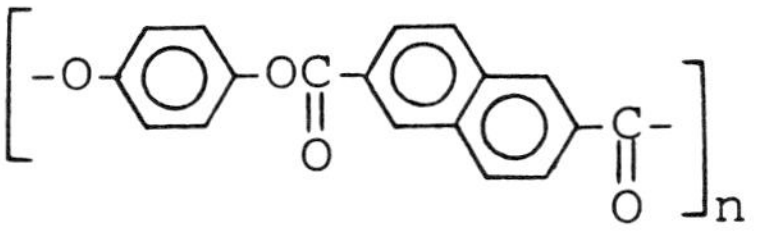

21

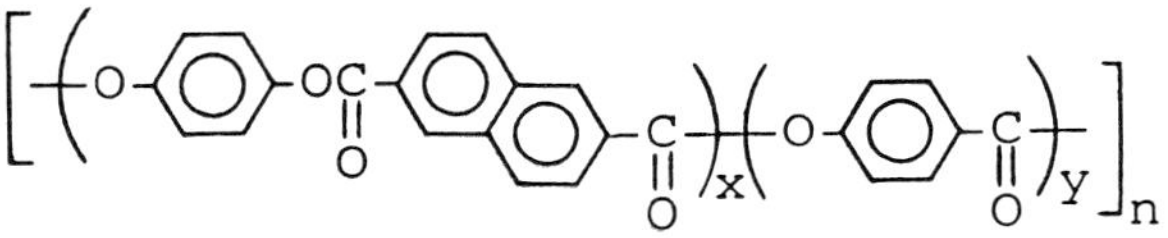

22

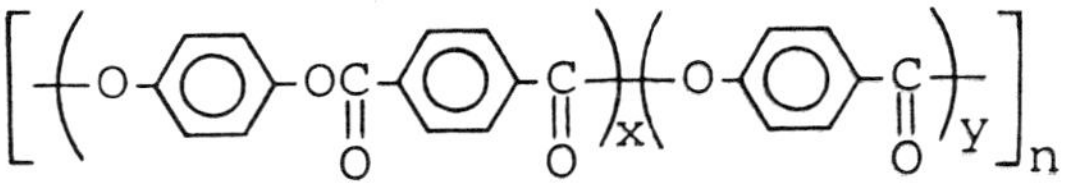

23

Thermotropic liquid crystalline copolyarylates **24** with melting points below 400°C are also obtained from hydroquinone, terephthalic acid, 2,6-naphthalenedicarboxylic acid, and p-hydroxybenzoic acid [60]. When the proportions of $x/y/z$ in the copolyarylates are 25:25:50, the melting point is 325°C. High-strength and high-modulus fibers are readily produced by melt-spinning at 350°C and subsequent heat treatment at 320°C; the tensile strength, elongation at break, and tensile modulus are 22 g d^{-1}, 3.1%, and 990 g d^{-1}, respectively.

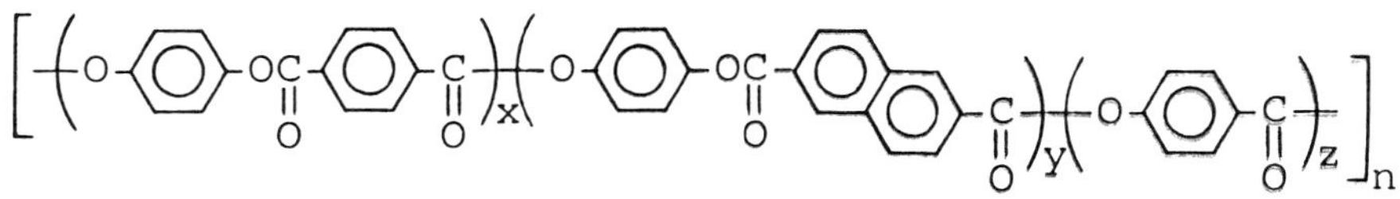

24

The melting point of the POB homopolyarylate **6** is reduced drastically by the modification with 6-hydroxy-2-naphthoic acid, and the copolyarylates **8** having melting points as low as 240°C are obtained. This type of thermotropic liquid crystalline copolyarylates are marketed by Celanese under the trade name Vectra.

Liquid crystalline polyarylates also can be prepared with 1,4- and 1,5-oriented naphthalene monomers, but the polymers have higher melting points than those based on 2,6-oriented naphthalene monomers [60]. If the bonds attached to the naphthalene ring are not coaxial (1,4-orientation) or parallel (1,5- and 2,6-orientation), the naphthalene monomer will not contribute to the liquid crystalline nature of the polymer. In fact, the homopolyarylate **25** derived from 2,7-naphthalenediol and terephthalic acid, which has rigid kink structure, melts at 418°C and apparently is not liquid crystalline.

25

Modification with Rigid Kinks

The fourth approach for reducing the high melting point of the basic poly(*p*-phenylene terephthalate) structure is introduction of rigid kinks into the polymer backbone, which causes the polymer chain to continue at an angle to the original extended-chain conformation. Consequently, as the amount of kinking component is increased, the thermotropic liquid crystallinity and the orientability of the polyarylates from the melt are decreased; therefore, the levels of tensile properties that can be attained are decreased. Examples of kinking components are listed in Table 18, which includes bisphenols and dibenzoic acids with one atom (such as carbon or oxygen) linking the two aromatic rings, and *meta*-oriented modifiers such as resorcinol, isophthalic acid, and *m*-hydroxybenzoic acid.

Table 19 shows the maximum amount of each bisphenol monomer which can be copolymerized with retention of thermotropic liquid crystallinity of the copolyterephthalates **26** [10]. The results clearly demonstrate that the greater the bulkiness of the central substituents, X, in the bisphenols, the lower the threshold comonomer amount which can be introduced in the copolyterephthalate without completely losing the thermotropic liquid crystalline characteristics.

Therefore, very high plastic and fiber tensile properties can be attained if the level of modification by these comonomer components is low, but these properties decrease as the kinking component increases. In addition to reducing the crystalline melting point, kinked groups also reduce the degree of crystallinity of the copolyarylates; consequently, heat deflection temperatures of the injection-molded plastics are reduced.

26

Table 18 Kinking Comonomers Used in the Preparation of Thermotropic Liquid Crystalline Polyarylates[a]

Bisphenol	Aromatic dicarboxylic acid	Hydroxy acid
(structures)	(structures)	(structures)

[a] X = CH_2, $C(CH_3)_2$, O, S, C=O, SO_2, etc.

Table 19 Effect of Amount of Bisphenols on the Properties of Thermotropic Liquid Crystalline Copolyarylate **26**

Substituent X	Maximum amount y (mol %)
$-C(CH_3)_2-$	40
$-SO_2-$	50
$-CH_2-$	60
$-S-$	60
$-O-$	70
Resorcinol	60

Source: Ref. 10.

An example is given for the copolyarylates **27** derived from hydroquinone, terephthalic acid, 2,6-naphthalenedicarboxylic acid, and *m*-hydroxybenzoic acid [60]. Although the homopolyarylate **28** of *m*-hydroxybenzoic acid is not liquid crystalline because of the existence of a large number of rigid kinks in the polymer chain (T_m = 181–185°C and T_g = 145°C) [24], the copolyarylate **27** with the composition of 25:40:35 is liquid crystalline because only 35 mol % of the kinking component is present. As shown in Table 20, this copolyarylate has higher tensile strength and higher flexural modulus, because of the thermotropic liquid crystallinity, which permits extended-chain orientation on injection molding [60].

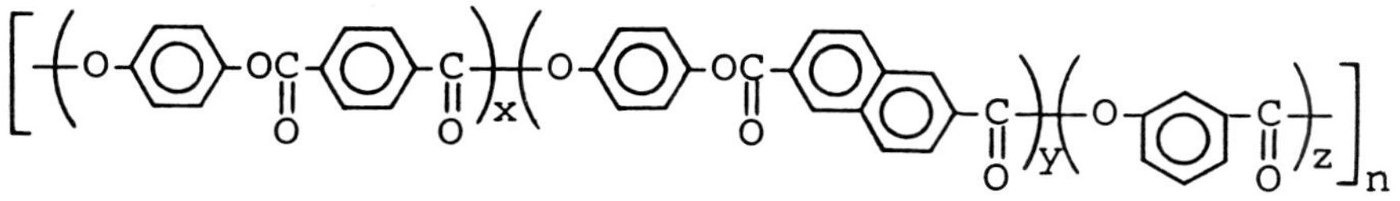

27

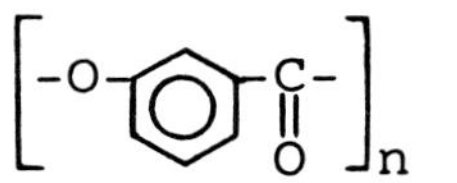

28

Table 20 Properties of Thermotropic Liquid Crystalline Polyarylates

Property	Polymer **28**	Polymer **27**[a]
Tensile strength (MPa)	87	249
Elongation at break (%)	93	11
Flexural modulus (GPa)	2.7	8.7
Notched Izod impact strength (J m^{-1})	59	107
Heat deflection temperature (°C)	122	120

[a]$x/y/z$ = 25:40:35.
Source: Ref. 60.

High-strength and high-modulus fibers of the thermotropic liquid crystalline homopolyarylate **29** and copolyarylates **30–32** having rigid kinks in the polymer chains are listed in Table 21. These are obtained by melt-spinning and subsequent heat treatment [62, 63].

29

[Structures 30, 31, 32 shown]

Table 21 Properties of Heat-Treated Fibers of Thermotropic Liquid Crystalline Polyarylates

Property	Polymer			
	29	**30**[a]	**31**[b]	**32**[c]
Melting point (°C)	356	334	311	312
Glass transition temperature (°C)	121			
Tensile strength (g d^{-1})	10.6	15.4	24.1	24.8
Elongation at break (%)	1.6	2.5	2.9	2.9
Tensile modulus (g d^{-1})	671	608	683	818

[a]$x/y = 20:80$.
[b]$x/y/z = 10:10:80$.
[c]$x/y/z = 20:2.5:77.5$.
Source: Refs. 62 and 63.

COMMERCIAL DEVELOPMENT OF POLYARYLATES

The most important member of the polyarylate family is based on bisphenol A and various ratios of iso- and terephthalic acids. The properties of a series of random copolyarylates **5** are summarized in Table 22 [7].

The first commercial production of **5**-type copolyarylates was introduced in 1973 by Unitika, Japan. Since then, copolyarylates of this class have been offered commercially by several suppliers in the United States and Europe (Table 23).

The polyarylates are clear and amorphous thermoplastics that combine clarity, high

Table 22 Properties of Copolyarylates **5**

Iso-/terephthalic acid ratio	T_g (°C)	Crystallinity	Solubility	Film toughness
100/0		50	+	Brittle
90/10	181	45	++	Tough
80/20	188	25	+++	Very tough
70/30	188	5	++++	Very tough
60/40	188	5	++++	Very tough
50/50	194	5	++++	Very tough
40/60	191	22	+++	Tough
30/70	192	30	+++	Tough
20/80	203	33	++	Brittle
10/90		35	++	Very brittle
0/100		40	+	Very brittle

Source: Ref. 7.

Table 23 World Major Suppliers of Polyarylates

Polymer type	Country	Supplier	Trade name	Entry
5	Japan	Unitika	U-Polymer	1973
	United States	Union Carbide	Ardel	1978
		Amoco		
		Hooker Chem.	Durel	1979
		Celanese		
		Du Pont	Arylon	1986
	Belgium	Solva	Arylef	
	West Germany	Bayer	APE	1979
6	United States	Carborundum	Ekonol	1970
	Japan	Sumitomo Chem.	Ekonol	1979
7	United States	Carborundum	Ekkcel	1972
		Dartco	Xydar	1984
	Japan	Sumitomo Chem.	Ekonol	1979
8	United States	Celanese	Vectra	1985
9	United States	Eastman	X7G	1976
	Japan	Unitika	LC	1985
		Mitsubishi Chem.	EPE	1985

heat deflection temperatures, high impact strength, and good electrical properties with inherent ultraviolet stability and flame retardance. The polymers are injection-molded and extruded, and the moldings are used for a variety of applications. Table 24 summarizes their physical, mechanical, and electrical properties [64].

The other important polyarylates are polyarylate **6** based on *p*-hydroxybenzoic acid and its copolyarylates **7**, which were commercially produced in 1970 by Carborundum under the trade names of Ekonol and Ekkcel, respectively. The latter copolyarylates exhibit

Table 24 Properties of Polyarylates **5**

Property	ASTM test method	U-Polymer (U-100)	Durel (400)	APE (KLI-9300)
Density (g cm^{-3})	D792	1.21	1.21	1.20
Tensile strength (MPa)	D638	70	69	70
Elongation at break (%)	D638	55	50	56
Flexural modulus (GPa)	D790	1.9	2.3	2.3
Notched Izod impact strength (J m^{-1})	D256	196	294	274
Heat deflection temperature (°C)	D648	175	171	165
Coefficient of linear thermal expansion (cm cm^{-1} °C^{-1})	D696	6.1×10^{-5}	6.3×10^{-5}	
Dielectric constant (1 MHz)	D150	3.0	3.0	3.2
Dissipation factor (1 MHz)	D150	0.015	0.022	0.017
Volume resistivity (ohm · cm)	D257	2.0×10^{16}	2.0×10^{16}	$>10^{16}$
Dielectric strength (kV mm^{-1})	D149	39	18	>30

Source: Ref. 64.

thermotropic liquid crystalline behavior and thus a unique property profile. Thermotropic copolyarylates of another series **8** and nonarylate type thermotropic polyesters **9** have also been marketed (Table 23).

Poly-*p*-oxybenzoyl **6** exhibits a high degree of crystallinity and does not melt below its decomposition temperature of 450°C. The polyarylate can be fabricated by high-temperature sintering techniques used routinely in the ceramics and metal industries. In fact, the polymers show some flow under high pressures of 35–70 MPa at temperatures of about 400°C. The compression-sintered specimens of Ekonol have a density of 1.44 g cm^{-3} and exhibit excellent mechanical properties; the flexural strength and modulus are 74 MPa and 7 GPa, respectively. The low coefficient of friction of 0.10–0.16 is most unusual considering the high bulk modulus of the polymer.

The polymer material is available as a finely divided powder and can be blended with various powdered metals such as aluminum and with polytetrafluoroethylene, and is used in flame spray compounds. The plasma-sprayed coatings are self-lubricating, wear-resistant, and thermally stable.

Thermotropic liquid crystalline polyarylates **7** and **8** are high-temperature engineering plastics that provide high levels of performance properties such as heat distortion temperatures and can be melt-processed on conventional equipments.

Thermotropic liquid crystalline melts provide two immediate benefits. The low melt viscosity of these polyarylates results in good injection-molding characteristics, comparable to commodity plastics like polypropylene, although they have much higher melt temperatures, around 400°C. In addition, the densely packed, fibrous structure of molded parts affords exceptional mechanical properties. Table 25 lists the properties of representative polyarylates [65].

High-strength high-modulus fibers of modified Ekonol *D3* can be produced by melt-spinning; their properties are tensile strength of 3.4–4.1 GPa, elongation at break of 2.4–3.1%, and tensile modulus of 130–140 GPa. These tensile values are well above those of reputable Kevlar aramid fibers.

Table 25 Properties of Thermotropic Liquid Crystalline Polyarylates

Property	Ekkcel (I2000)	Ekonol (E2000)	Xydar (SRT300)	Vectra (A950)	LC (2000)
Density (g cm^{-3})	1.40	1.40	1.35	1.40	1.29
Tensile strength (MPa)	100	74	137	206	108
Elongation at break (%)	8	6	4.9	3	4.4
Flexural modulus (GPa)	4.8	4.6	14	8.8	8.3
Notched Izod impact strength (J m^{-1})		39	127	431	127
Heat deflection temperature (°C)	293	293	355	180	64
Coefficient of linear thermal expansion (cm cm^{-1} °C^{-1})		2.9×10^{-5}	10^{-6}	10^{-6}	10^{-6}
Dielectric constant (1 MHz)		2.9	2.9	3.6	3.3
Dissipation factor (1 MHz)		0.025	0.038	0.014	0.038
Volume resistivity (ohm · cm)		10^{15}	1.1×10^{15}	6×10^{16}	1.2×10^{17}
Dielectric strength (kV mm^{-1})		14	31	39	20

Source: Ref. 65.

Most of the molded polyarylates display very high tensile and flexural properties at ambient and extreme temperatures. For example, tensile strength and flexural modulus for Xydar at 300°C are 26 MPa and 5.5 GPa, respectively. They also provide excellent dimensional stability and electrical properties. The most obvious applications of this class of polyarylates are those that take advantage of the material's resistance to high temperatures, fire, and chemicals. The polyarylates are capable of replacing materials such as ceramics, metals, composites, and other engineering plastics.

REFERENCES

1. P. W. Morgan, *Condensation Polymers by Interfacial and Solution Methods,* Interscience, New York (1965).
2. G. Bier, *Polymer, 15*: 527 (1974).
3. A. Conix, *Ind. Eng. Chem., 51*: 147 (1959).
4. Y. Imai, *J. Macromol. Sci.-Chem., A15*: 833 (1981).
5. Y. Imai and M. Ueda, in *Crown Ethers and Phase Transfer Catalysis in Polymer Science* (L. J. Mathias and C. E. Carraher, Jr., eds.), Plenum, New York, p. 121 (1984).
6. Y. Imai and S. Tassavori, *J. Polym. Sci. Polym. Chem. Ed., 22*: 1319 (1984).
7. W. M. Eareckson, *J. Polym. Sci., 40*: 399 (1959).
8. H. Schnell, *Ind. Eng. Chem., 51*: 157 (1959).
9. P. W. Morgan, *J. Polym. Sci. A, 2*: 437 (1964).
10. J.-I. Jin, S. Antoun, C. Ober, and R. W. Lenz, *Brit. Polym. J., 12*: 132 (1980).
11. E. Turska, L. Pietrzak, and R. Jantas, *J. Appl. Polym. Sci., 23*: 2409 (1979).
12. M. Matzner, R. Barclay, and C. N. Merriam, *J. Appl. Polym. Sci., 9*: 3337 (1965).
13. H.-R. Dicke and R. W. Lenz, *J. Polym. Sci. Polym. Chem. Ed., 21*: 2581 (1983).
14. J. Economy, R. S. Strom, V. I. Matkovich, S. G. Cottis, and B. E. Nowak, *J. Polym. Sci. Polym. Chem. Ed., 14*: 2207 (1976).
15. M. Levine and S. C. Temin, *J. Polym. Sci., 28*: 179 (1958).
16. J. Economy, *J. Macromol. Sci.-Chem., A21*: 1705 (1984).

17. T.-S. Chung, *Polym. Eng. Sci., 26*: 901 (1986).
18. R. W. Stackman, *Ind. Eng. Chem. Prod. Res. Dev., 20*: 336 (1981).
19. W. Volksen, J. R. Lyerla, Jr., J. Economy, and B. Dawson, *J. Polym. Sci. Polym. Chem. Ed., 21*: 2249 (1983).
20. H. R. Kricheldorf and G. Schwarz, *Polymer, 25*: 520 (1984).
21. Y. Kato, S. Endo, K. Kimura, Y. Yamashita, H. Tsugita, and K. Monobe, *Kobunshi Ronbunshu, 44,* 35 (1987).
22. W. J. Jackson, Jr., and H. F. Kuhfuss, *J. Polym. Sci. Polym. Chem. Ed., 14*: 2043 (1976).
23. H. R. Kricheldorf and G. Schwarz, *Polym. Bull., 1*: 383 (1979).
24. H. R. Kricheldorf, Q.-Z. Zang, and G. Schwarz, *Polymer, 23*: 1821 (1982).
25. H. R. Kricheldorf and G. Schwarz, *Makromol. Chem., 184*: 475 (1983).
26. F. Higashi, *Polym. Appl. (Japan), 33*: 171 (1984).
27. S. Kitayama, K. Sanui, and N. Ogata, *J. Polym. Sci. Polym. Chem. Ed., 22*: 2705 (1984).
28. F. Higashi, A. Hoshio, and J. Kiyoshige, *J. Polym. Sci. Polym. Chem. Ed., 21*: 3241 (1983).
29. F. Higashi, A. Hoshio, Y. Yamada, and M. Ozawa, *J. Polym. Sci. Polym. Chem. Ed., 23*: 69 (1985).
30. F. Higashi, Y. Fujiwara, and Y. Yamada, *J. Polym. Sci. Polym. Chem. Ed., 24*: 589 (1986).
31. F. Higashi, N. Akiyama, and T. Koyama, *J. Polym. Sci. Polym. Chem. Ed., 21*: 3233 (1983).
32. F. Higashi, N. Akiyama, I. Takahashi, and T. Koyama, *J. Polym. Sci. Polym. Chem. Ed., 22*: 1653 (1984).
33. F. Higashi, T. Mashimo, and I. Takahashi, *J. Polym. Sci. Polym. Chem. Ed., 24*: 97 (1986).
34. S. M. Cohen, R. H. Young, and A. H. Markhart, *J. Polym. Sci. A-1, 9*: 3263 (1971).
35. Y. Maruyama, M. Kakimoto, and Y. Imai, *J. Polym. Sci. Polym. Chem. Ed., 24*: 3555 (1986).
36. V. V. Korshak, S. V. Vinogradova, and Y. S. Vygodskii, *J. Macromol. Sci.-Rev. Macromol. Chem., C11*: 45 (1974).
37. K. C. Stueben, *J. Polym. Sci. A, 3*: 3209 (1965).
38. D. Sek and Z. Jedlinski, *Eur. Polym. J., 15*: 103 (1979).
39. A. M. Usmani, *J. Macromol. Sci.-Chem., A18*: 251 (1982).
40. P. W. Morgan, *J. Polym. Sci. A, 2*: 437 (1964).
41. M. Kakimoto, S. Harada, Y. Oishi, and Y. Imai, *J. Polym. Sci. Polym. Chem. Ed.* (accepted).
42. M. Kakimoto, S. Harada, Y. Oishi, and Y. Imai, unpublished.
43. M. Kakimoto, Y. S. Negi, and Y. Imai, *J. Polym. Sci. Polym. Chem. Ed., 24*: 1511 (1986).
44. S. J. Havens and V. L. Bell, *J. Polym. Sci. Polym. Chem. Ed., 24*: 901 (1986).
45. T. Jinda, M. Noyori, and T. Matsuda, *J. Soc. Fiber Sci. Tech. Jpn., 39*: T-160 (1983).
46. A. Conix and U. L. Laridon, *Angew. Chem., 72*: 116 (1960).
47. Y. Imai, M. Ueda, and M. Ii, *Kobunshi Ronbunshu, 35*: 807 (1978).
48. M. Schmidt, D. Freitag, L. Bottenbruch, and K. Reinking, *Angew. Makromol. Chem., 132*: 1 (1985).
49. Y. Imai, H. Kamata, and M. Kakimoto, *J. Polym. Sci. Polym. Chem. Ed., 22*: 1259 (1984).
50. K.-S. Kim, *J. Appl. Polym. Sci., 28*: 1119 (1983).
51. A. Blumstein, ed., *Liquid Crystalline Order in Polymers,* Academic Press, New York (1978).
52. A. Ciferri, W. R. Krigbaum, and R. Meyer, eds., *Polymer Liquid Crystals,* Academic Press, New York (1982).
53. L. L. Chapoy, ed., *Recent Advances in Liquid Crystalline Polymers,* Elsevier, New York (1985).
54. A. Blumstein, ed., *Polymeric Liquid Crystals,* Plenum, New York (1985).
55. C. K. Ober, J.-I. Jin, and R. W. Lenz, *Adv. Polym. Sci., 59*: 103 (1984).
56. J. L. White, *J. Appl. Polym. Sci. Appl. Polym. Symp., 41*: 3 (1985).
57. N. Koide, *Mol. Cryst. Liq. Cryst., 139*: 47 (1986).

58. W. J. Jackson, Jr., *Brit. Polym. J.*, *12*: 154 (1980).
59. W. J. Jackson, Jr., *J. Appl. Polym. Sci. Appl. Polym. Symp.*, *41*: 25 (1985).
60. W. J. Jackson, Jr., *Macromolecules*, *16*: 1027 (1983).
61. W. J. Jackson, Jr., and J. C. Morris, *J. Appl. Polym. Sci. Appl. Polym. Symp.*, *41*: 307 (1985).
62. T. Jinda and T. Matsuda, *J. Soc. Fiber Sci. Tech. Jpn.*, *41*: T-87 (1985).
63. T. Jinda, M. Noyori, and T. Matsuda, *J. Soc. Fiber Sci. Tech. Jpn.*, *40*: T-1 (1984).
64. H. Hasegawa, *Jpn. Plast.*, *38*(1): 131 (1987).
65. T. Okada and J. Suenaga, *Jpn. Plast.*, *37*(3): 111 (1986).

6
Nylon Polymerization

Santosh K. Gupta*
University of Notre Dame
Notre Dame, Indiana

INTRODUCTION	211
POLYMERIZATION OF ε-CAPROLACTAM	212
HYDROLYTIC POLYMERIZATION OF ε-CAPROLACTAM IN BATCH REACTORS	214
NYLON-6 POLYMERIZATION IN HOMOGENEOUS CONTINUOUS FLOW STIRRED TANK REACTORS AND IN CASCADES OF REACTORS	221
EFFECT OF NONIDEALITIES	223
SIMULATION OF INDUSTRIAL REACTORS	227
OPTIMIZATION OF NYLON-6 REACTORS	236
CYCLIZATION REACTIONS IN NYLON-6 POLYMERIZATION	240
NYLON-6,6 POLYMERIZATION	242
OTHER NYLONS	245
CONCLUSIONS	245
REFERENCES	245

INTRODUCTION

Nylons are polymers having amide

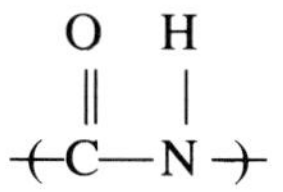

$$\underset{\substack{\| \quad | \\ }}{\text{+C—N+}}^{\text{O} \quad \text{H}}$$

linkages [1], and so they are called polyamides. Since proteins are polyamides of various amino acids, they too fall into this category, but the discussion here is limited to industrially important *synthetic* polyamides. There are two classes of synthetic nylon. One of these is formed from cyclic monomers, for example, nylon-6 {H[—HN—$(CH_2)_5$—CO—$]_n$—OH}, which has six carbon atoms per repeat unit and is normally manufactured from the monomer ε-caprolactam. Another example in this class is nylon-12 {H—[NH—$(CH_2)_{11}$—CO—$]_n$—OH}, having 12 carbon atoms in the repeat unit, and made from the lactam of 12-amino dodecanoic acid. The single index used in describing these nylons

---
*On leave from Indian Institute of Technology, Kanpur, India

indicates the number of carbon atoms in the repeat unit. The second class of synthetic nylon is formed from various diamines and diacids, and their nomenclature involves two indices indicating the number of carbon atoms in the diamine and diacid units respectively, e.g., nylon-6,6 {H—[—NH—(CH$_2$)$_6$—NH—CO—(CH$_2$)$_4$—CO—]$_n$—OH}, made from hexamethylene diamine and adipic acid.

Of the various nylons available, nylon-6 and nylon-6,6 are the most important and the literature offers more information on their polymerization [2–6] than on that of other nylons. This chapter reviews the analysis and design of reactors used for these two systems and presents some qualitative information on the other nylons.

POLYMERIZATION OF ε-CAPROLACTAM

Two routes are used for the polymerization of ε-caprolactam [C_1, or CL : HN—(CH$_2$)$_5$—CO]. The most commonly used industrial process is the hydrolytic polymerization of CL,

Table 1 Kinetic Scheme for Nylon-6 Polymerization[a]

1. Ring opening

$$C_1 + W \underset{k_1' = k_1/K_1}{\overset{k_1}{\rightleftharpoons}} P_1$$

2. Polycondensation

$$P_n + P_m \underset{k_2' = k_2/K_2}{\overset{k_2}{\rightleftharpoons}} P_{n+m} + W, \quad n, m = 1, 2, \ldots$$

3. Polyaddition

$$P_n + C_1 \underset{k_3' = k_3/K_3}{\overset{k_3}{\rightleftharpoons}} P_{n+1}, \quad n = 1, 2, \ldots$$

4. Ring opening of cyclic dimer

$$C_2 + W \underset{k_4' = k_4/K_4}{\overset{k_4}{\rightleftharpoons}} P_2$$

5. Polyaddition of cyclic dimer

$$P_n + C_2 \underset{k_5' = k_5/K_5}{\overset{k_5}{\rightleftharpoons}} P_{n+2}, \quad n = 1, 2, \ldots$$

6. Reaction with monofunctional acid

$$P_n + P_{m,x} \underset{k_2'}{\overset{k_2}{\rightleftharpoons}} P_{m+n,x} + W, \quad n, m = 1, 2, \ldots$$

[a] C_1 = ε-caprolactam;
C_2 = H—N(CH$_2$)$_5$—[—C(=O)—N(H)—(CH$_2$)$_5$—]$_n$—C(=O)—; W = water;
P_n = H[—N(H)—(CH$_2$)$_5$—C(=O)]$_n$—OH; $P_{n,x}$ = X—[—C(=O)—N(H)—(CH$_2$)$_5$—]$_{n-1}$—C(=O)—OH; X = unreactive group.

in which water (W) is used [7, 8] to open the ε-caprolactam ring (in the temperature range of 220–270°C) to give a linear molecule, amino caproic acid {P_1 or ACA: H_2N—(—CH_2)$_5$—COOH}. Polymerization then proceeds by the step-growth mechanism of the bifunctional ACA, giving linear polymers, P_n {H—[HN—(—CH_2)$_5$—CO—]$_n$—OH}, with water as the condensation by-product. In addition to these reactions, the caprolactam ring can be directly opened by the amino end group [9] of any linear polymer molecule, P_n. This reaction leads to the growth of the molecule by one monomeric unit at a time, and so has characteristics of chain growth polymerization. Thus the major reactions in the hydrolytic process can be represented schematically [2–4, 6–9] by the first three equations (ring opening, polycondensation, and polyaddition) in Table 1. The reversible nature of these reactions was recognized by Carothers [7] very early in his work on polyamides. The equilibrium product formed at conditions usually encountered in industrial reactors has a number-average chain length, μ_n, of about 140–180, and contains about 8–10% of unreacted monomer. The latter is removed by various techniques: evaporation in falling film evaporators under vacuum, drying of solid chips in fluidized beds or in tumble driers, leaching of the solid polymer chips in hot water, etc. In these *finishing* operations, further polymerization also occurs as the CL diffuses through to the interface, and concepts of mass transfer with chemical reaction are required for modeling them. The product from this hydrolytic route is suitable for use as tire cords, fibers for apparel, and as resins for molded articles. A typical flow chart for the polymerization of nylon-6 using this scheme is seen in Figure 1 [2].

The second route of polymerization of nylon-6 is by the ionic chain growth mechanism, using, for example, initiators like sodium hydroxide, lactamates of alkali metals, and pentaalkyl guanidine [5]. The advantage of this route is that it can be carried out below the melting point (220°C) of nylon-6 and leads to the formation of very high molecular weight polymers. The polymer thus formed is usually employed in making large cast articles. Most nylon-6 is manufactured industrially using the hydrolytic route, and in view of this attention is focused on the design of reactors for nylon-6 through the first mechanism.

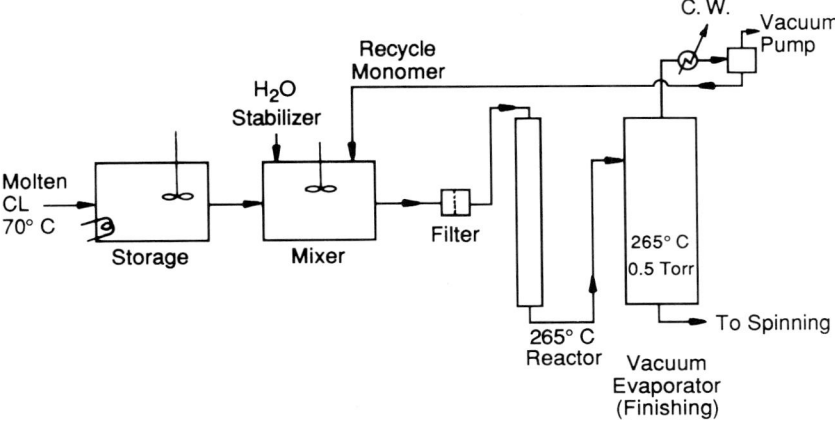

Figure 1 Process flowsheet [2] for nylon-6 manufacture. Vacuum evaporation of unreacted caprolactam and higher cyclic oligomers is being used.

HYDROLYTIC POLYMERIZATION OF ε-CAPROLACTAM IN BATCH REACTORS

In this section the isothermal polymerization of nylon-6 in sealed tubes (with negligible vapor space) is discussed. The results can be used to estimate rate constants from experimental data. Since mass balance equations for nonvaporizing batch reactors or plug-flow (idealized tubular) reactors (PFRs) are identical, the results in this section are also valid for these cases. In later sections, the polymerization of nylon-6 in continuous flow stirred tank reactors (CSTRs), and in cascades of different types of reactors used industrially, is discussed.

The kinetic scheme for nylon-6 polymerization includes the three major reactions—ring opening, polycondensation, and polyaddition—given in Table 1. In addition, the formation of higher cyclic oligomers [8]

$$C_2, C_3, \ldots \left\{ \begin{array}{c} H \quad\quad\quad O \;\; H \quad\quad\quad\quad O \\ | \quad\quad\quad\; \| \quad | \quad\quad\quad\quad \| \\ N-(CH_2)_5 + C-N-(-CH_2)_5 \!\!\!+_{n-1} C \end{array} \right\}$$

is an important side reaction. Even though the total amount of these compounds formed is small (below about 2–3% by weight), it is known that they cause problems in the spinning and molding of the final polymer. As a result, hot water extraction is used before these processing operations to remove the unreacted monomer as well as the higher cyclic compounds. Since this is an expensive, energy-intensive process, a design engineer endeavors to keep the formation of these compounds below a critical value to minimize cost. The incorporation of these side reactions is thus necessary in any realistic kinetic model. The cyclic oligomers, C_n, can also undergo ring opening and polyaddition reactions as follows [10]:

$$C_n + W \rightleftarrows P_n, \quad n = 2, 3, \ldots \quad (1a)$$

$$C_n + P_m \rightleftarrows P_{n+m}, \quad n = 2, 3, \ldots; \quad m = 1, 2, \ldots \quad (1b)$$

Usually, the reactions with C_2 are incorporated in the kinetic scheme only, because it has been found experimentally that the formation of the cyclic dimer predominates and because incorporating the reactions for the higher cyclic oligomers makes the analysis more difficult.

Sometimes stabilizers like monofunctional acids (e.g., acetic acid) are added to the reaction mass to control the molecular weight of the polymer formed (and thus produce different grades of the polymer from the same reactor) or to increase the rate of polymerization. The acid end groups of the monofunctional compounds,

$$P_{m,x}\{X + \overset{O}{\underset{\|}{C}} - \overset{H}{\underset{|}{N}} - (CH_2)_5 +_{m-1} COOH, \text{ with } X \text{ unreactive}\}$$

react with the amino groups of P_n but do not participate in ring opening or polyaddition reactions. The complete kinetic scheme of nylon-6 polymerization including reactions with C_2 and $P_{m,x}$ is given in Table 1. At times, monofunctional amines like n-butyl amine are used for the same purpose, and appropriate modifications to the kinetic scheme in this table must be made since these compounds can also react with $C_1, C_2, \ldots$. Other side reactions like decarboxylation [2, 11], desamination [2, 11], and peroxidation of caprolactam [2, 12] also occur but have not been incorporated in Table 1, since their effects appear to be less important.

The various reactions of Table 1 have been found to be catalyzed by the carboxyl groups present in the reaction mass and the forward rate constants can be represented empirically (accounting for all the mechanistic steps) by Arrhenius-type terms

$$k_i = A_i^o \exp\left(\frac{-E_i^o}{RT}\right) + A_i^c \exp\left(\frac{-E_i^c}{RT}\right)[-COOH]$$

$$\equiv k_i^o + k_i^c \sum_{n=1}^{\infty} ([P_n] + [P_{n,x}]), \qquad i = 1, 2, \ldots, 5 \qquad (2)$$

where square brackets represent concentration (in g mol kg^{-1}). The equilibrium constants, K_i, are given by the standard thermodynamic equation

$$K_i = \exp\left(\frac{\Delta S_i - \Delta H_i/T}{R}\right), \qquad i = 1, 2, \ldots, 5 \qquad (3)$$

Table 2 gives the constants required for obtaining the various rate and equilibrium constants, as determined by Tai and colleagues [4, 13]. These workers performed an extensive and detailed series of experiments [14, 15] and their constants in Table 2 are more precise than earlier values [8, 9, 16–18].

Mass balance equations can now be written for an isothermal batch reactor and are available in a recent review [6]. These equations, when integrated numerically using the Runge-Kutta technique, give the concentrations of the individual molecular species C_1, C_2, W, P_1, P_2, ..., $P_{1,x}$, $P_{2,x}$, ..., as functions of time. The molecular weight distributions (called exact MWDs) so obtained for some common feed and reactor conditions are seen in Figure 2 [19, 20]. Such a numerical integration of the equations for the individual molecular species is extremely cumbersome and time-consuming. To simplify the analysis, it is possible to study the MWD in terms of its moments:

$$\lambda_i \equiv \sum_{n=1}^{\infty} n^i [P_n]; \qquad \lambda_{i,x} \equiv \sum_{n=1}^{\infty} n^i [P_{n,x}], \qquad i = 0, 1, 2, 3, \ldots \qquad (4)$$

The equations for the molecular species can be summed to give equations for the moments. The final set of equations is available in Ref. 21. Closure conditions for the moments are required to solve this set of equations. These arise because of the reversible nature of the reactions in Table 1. Gupta and colleagues [21–23] have demonstrated that the equations

$$[P_2] = [P_3] = [P_1] \qquad (5a)$$

$$\lambda_3 = \frac{\lambda_2(2\lambda_2\lambda_0 - \lambda_1^2)}{\lambda_1\lambda_0} \qquad (5b)$$

$$\lambda_{3,x} = \frac{\lambda_{2,x}(2\lambda_{2,x}\lambda_{0,x} - \lambda_{1,x}^2)}{\lambda_{1,x}\lambda_{0,x}} \qquad (5c)$$

used along with the moment equations give numerical results that are close to exact values for an extremely wide range of reactor types and conditions.

Figures 3 to 8 show some typical results for nonvaporizing isothermal polymerizations of nylon-6 in batch (or plug-flow) reactors under conditions described in Table 3. The

Table 2 Data for Rate and Equilibrium Constants for Nylon-6 Polymerization

Reaction[a]	A_i^o [kg (gmol·hr^{-1})]	E_i^o (cal gmol^{-1})	A_i^c [kg^2 (gmol^2·hr)$^{-1}$]	E_i^c (cal gmol^{-1})	ΔH_i^b (cal gmol^{-1})	ΔS_i^b (eu)
1	5.9874×10^5	1.9880×10^4	4.3075×10^7	1.8806×10^4	1.9180×10^3	-7.8846×10^0
2	1.8942×10^{10}	2.3271×10^4	1.2114×10^{10}	2.0670×10^4	-5.9458×10^3	9.4374×10^{-1}
3	2.8558×10^9	2.2845×10^4	1.6377×10^{10}	2.0107×10^4	-4.0438×10^3	-6.9457×10^0
4	8.5778×10^{11}	4.2000×10^4	2.3307×10^{12}	3.7400×10^4	-9.6000×10^3	-1.4520×10^1
5	2.5701×10^8	2.1300×10^4	3.0110×10^9	2.0400×10^4	-3.1691×10^3	5.8265×10^{-1}

[a]Reaction numbers from Table 1.
[b]ΔH_i and ΔS_i are assumed independent of temperature.
Source: Ref. 4. Copyright 1983, American Chemical Society, Washington, D.C.

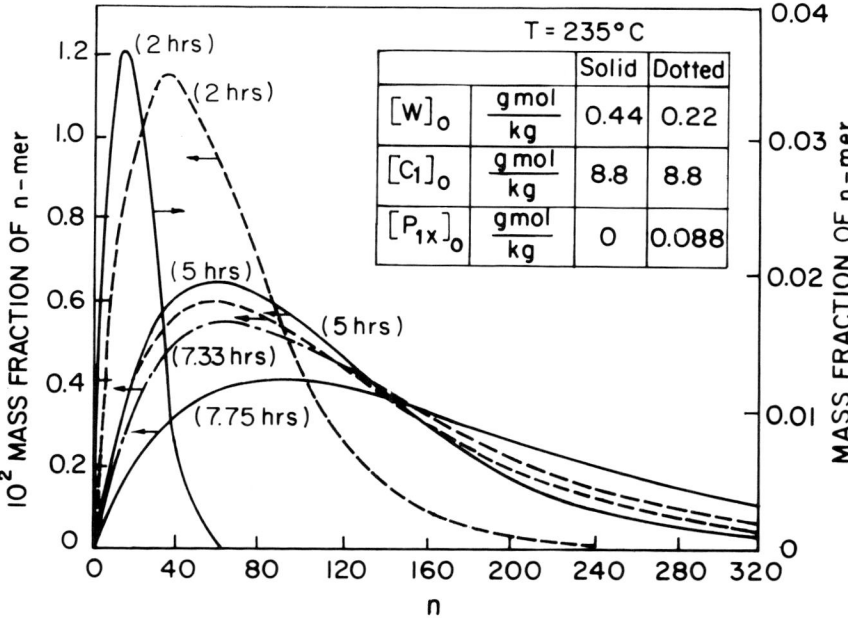

Figure 2 Molecular weight distributions for nylon-6 formed in isothermal batch reactors [19, 20].

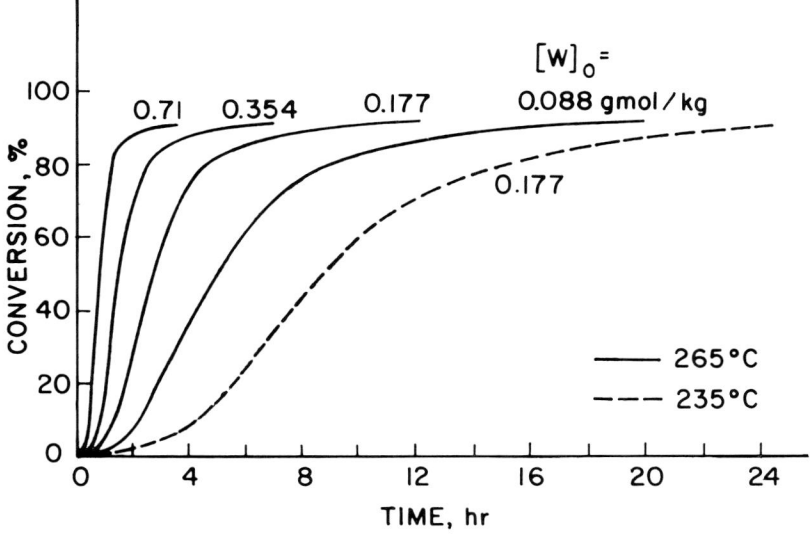

Figure 3 Monomer conversion [2, 18, 24] as a function of time for isothermal polymerization of C_1 with water in batch reactors. No monofunctional species, no cyclization.

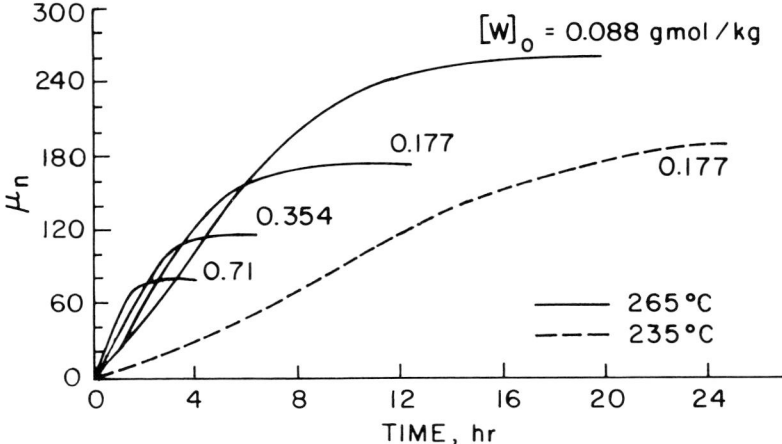

Figure 4 Number-average chain length, μ_n, as a function of time [2, 18, 24]. Conditions same as for Figure 3.

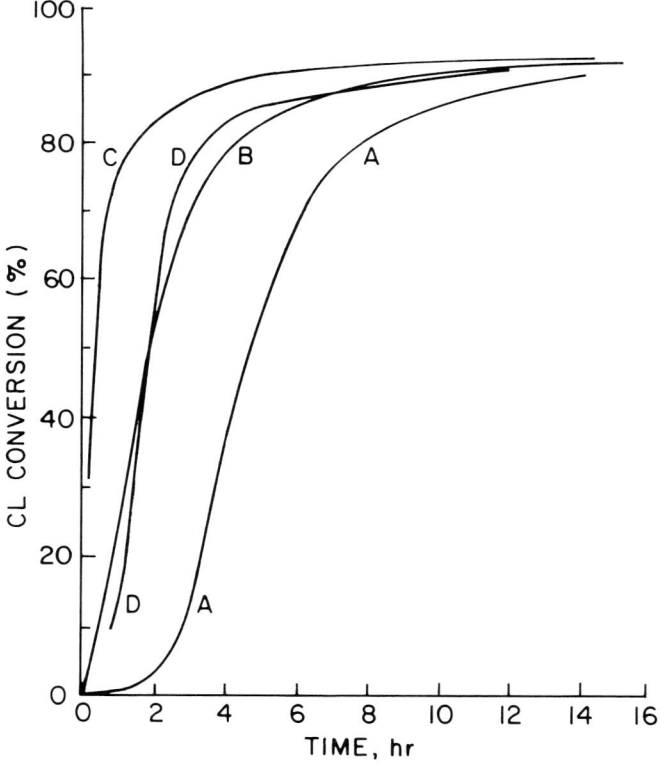

Figure 5 Caprolactam conversion [22] for various conditions given in Table 3.

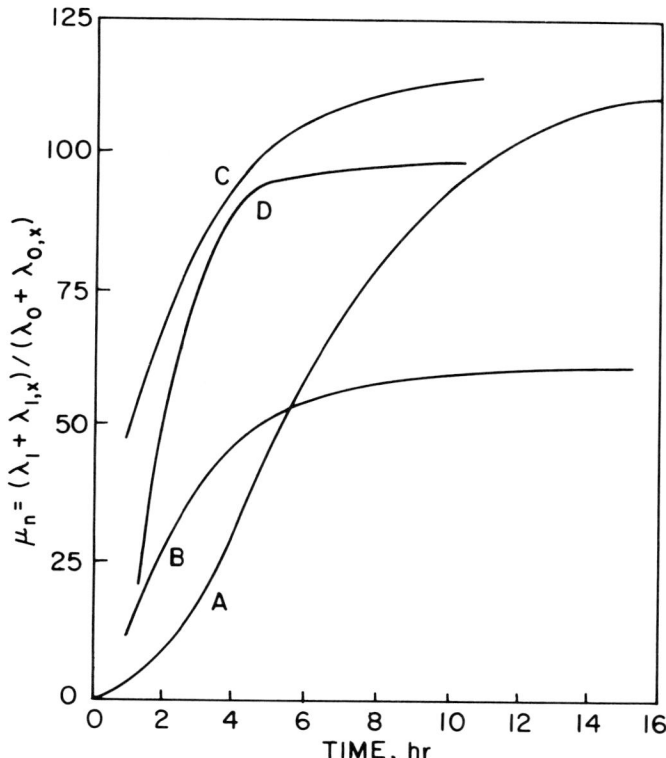

Figure 6 Degree of polymerization for various conditions shown in Table 3 [22].

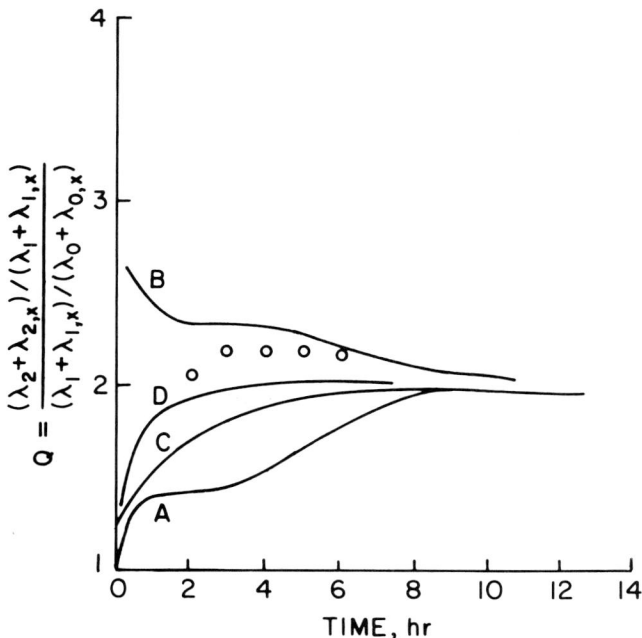

Figure 7 Polydispersity index for various conditions shown in Table 3 [22]. Circles represent experimental values of Tai et al. [25] for $[W]_0 = 0.8$ gmol kg^{-1}, $T = 240°C$.

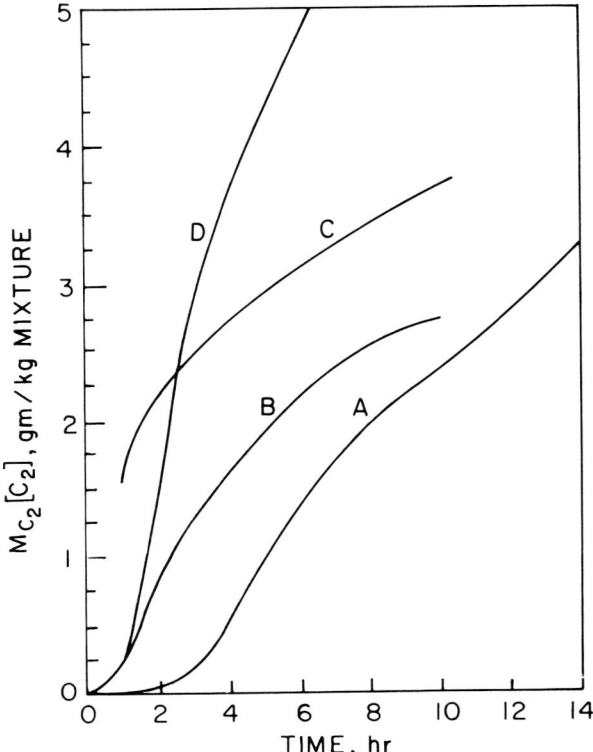

Figure 8 Grams of cyclic dimer per kilogram of mixture for various conditions shown in Table 3 [22]. M_{C_2} is the molecular weight of C_2.

Table 3 Conditions Used for Generating Figures 4 to 7[a]

Curve	$[W]_0$ (gmol kg^{-1} mixture)	$[P_{1,x}]_0$ (gmol kg^{-1})	$[P_1]_0$[b] (gmol kg^{-1})	T (°C)
A	0.44	0	0	235
B	0.44	0.088	0	235
C	0	0	0.44	235
D	0.44	0	0	265

[a]Data from Ref. 22.
[b]Aminocaproic acid (P_1) can be used, if available, with CL in the feed.

number and weight average chain lengths, μ_n and μ_w, shown in these figures, are defined by

$$\mu_n = \frac{\lambda_1 + \lambda_{1,x}}{\lambda_0 + \lambda_{0,x}} \qquad (6a)$$

$$\mu_w = \frac{\lambda_2 + \lambda_{2,x}}{\lambda_1 + \lambda_{1,x}} \qquad (6b)$$

and the polydispersity index, Q, by

$$Q = \frac{\mu_w}{\mu_n} \qquad (7)$$

Careful study of these results [2, 18, 22, 24, 25] indicates the following:

1. The polymerization (conversion and μ_n) is speeded up by an increase in the temperature and initial water concentration and by the addition of monofunctional acids or ACA, but the equilibrium monomer conversion is almost unaffected by these changes.
2. The equilibrium value of μ_n is lowered by an increase in temperature or the initial water concentration and by the addition of monofunctional acids, but it is almost unaffected by the addition of ACA.
3. The equilibrium polydispersity index is approximately 2.0, irrespective of initial conditions, although values higher than 2 may be observed briefly in the presence of monofunctional acids.
4. The formation of cyclic dimers is speeded up substantially by an increase in temperature, and equilibrium concentrations of this side-product are not attained in the time scales leading to near-equilibrium conditions for monomer conversion, μ_n and Q.

Some of these isothermal results lead to interesting ramifications in the design of optimal and suboptimal reactor cascades, as discussed later.

NYLON-6 POLYMERIZATION IN HOMOGENEOUS CONTINUOUS FLOW STIRRED TANK REACTORS AND IN CASCADES OF REACTORS

It was observed in the preceding section that the use of ACA in the feed to the batch reactor speeds up polymerization while keeping the equilibrium value of μ_n unchanged. One technique that could be exploited to have increased amounts of ACA present in the reaction mass is to use a continuous flow stirred tank reactor (CSTR). It is possible that a sequence of a CSTR and a plug-flow reactor (PFR), or sequences of several CSTRs, may be better for nylon-6 manufacture than a PFR alone. Indeed, several industrial processes with cascades of stirred tank reactors are currently being used [27, 28]. Such reactors have the added advantage of good heat and mass transfer rates.

An idealized mathematical description of a CSTR is a perfectly mixed or micromixed reactor, called the homogeneous CSTR (HCSTR). Mass balance equations can easily be written and solved for such a reactor operating at steady state. Figure 9 gives some typical results [23] as a function of the average residence time, $\bar{\theta}$ (= volume/flow rate). Values of

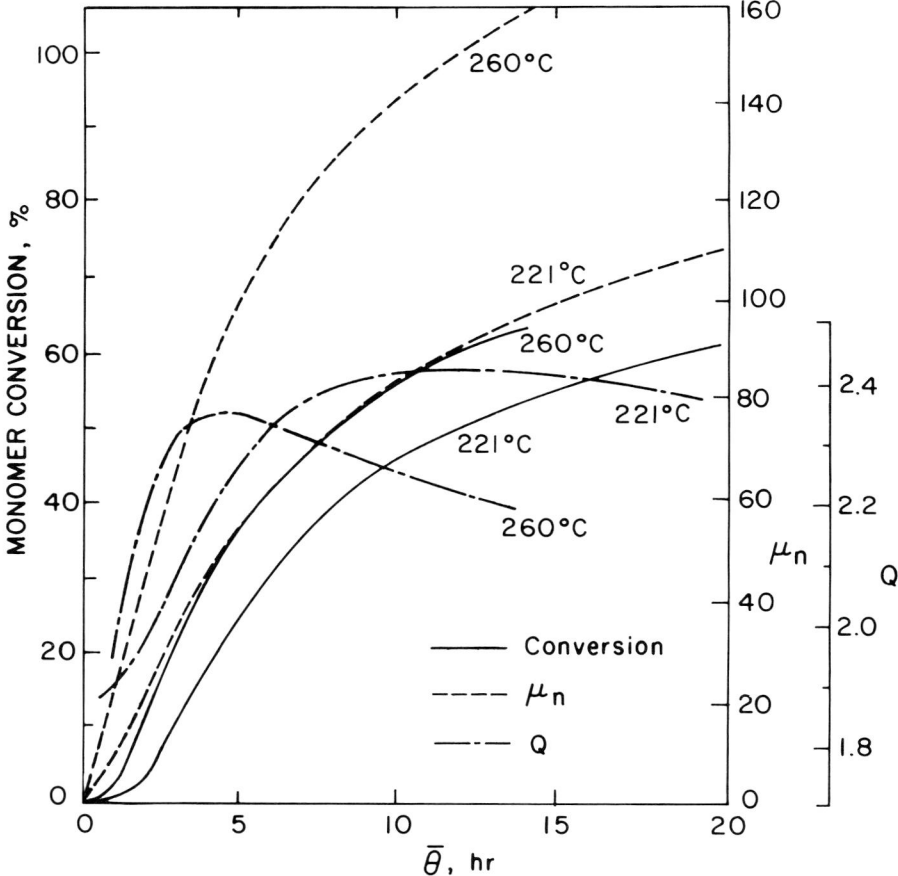

Figure 9 Nylon-6 polymerization [23] in a single (nonvaporizing) isothermal HCSTR, with feed-containing caprolactam ($[C_1]_0 = 8.8$ gmol kg^{-1}) and water only. $[W]_0 = 0.2411$ gmol kg^{-1} at 221°C and 0.111 gmol kg^{-1} at 260°C. $[P_{1,x}]_0 = 0$.

the polydispersity index, Q, above 2.0 are encountered. Figure 9 indicates that monomer conversions are lower than those encountered in batch or plug-flow reactors. Therefore, cascades of CSTRs and PFRs are preferred over single CSTRs. Figure 10 shows how one can obtain lower *total* residence times in cascades of reactors producing the same final polymer [29]. In this case, some of the water is removed from the reaction mass at a later stage. The reason for this is apparent from Figures 3 and 4—high initial water concentrations lead to faster polymerizations in the beginning, whereas low water concentrations in the later reactors lead to higher equilibrium values of μ_n. The use of HCSTRs in the beginning helps reduce the total residence time of the cascade because of the higher concentrations of ACA.

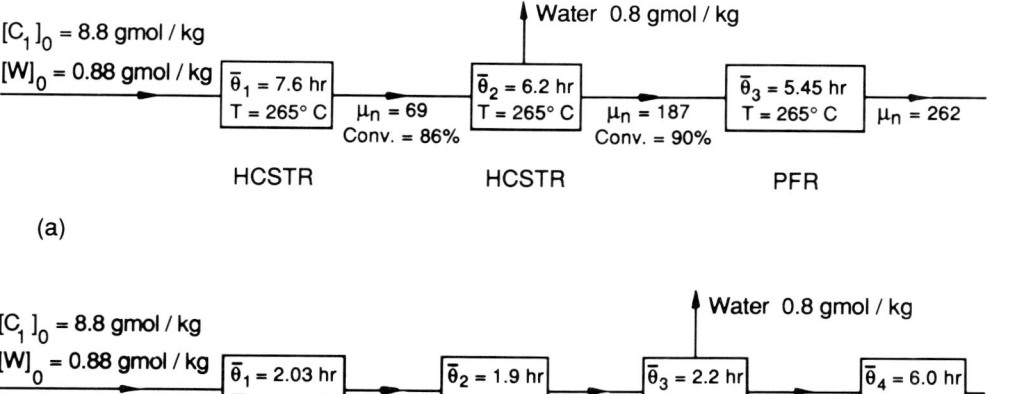

Figure 10 Polymerization of nylon-6 in two typical trains of reactors, leading to the same final value of μ_n. Total residence times are (a) 19.25 hr and (b) 12.13 hr. (Reprinted from Ref. 29 with permission of John Wiley & Sons, New York.)

EFFECT OF NONIDEALITIES

Even though most of the theoretical studies on nylon-6 polymerization have been carried out for a single ideal reactor or a cascade of ideal reactors, it is now well established that several nonidealities exist in real tubular reactors or CSTRs. The effects of some of these nonidealities are studied in this section.

The effect of imperfect micromixing in CSTRs is studied first. This is done using the mathematical idealization called the segregated continuous flow stirred tank reactor (SCSTR) [30, 31]. This is similar to an HCSTR, but the fluid elements are not well mixed on a microscopic level. Each fluid "packet" is assumed to be segregated and "insulated" from its neighbors, and so each behaves like an infinitesimal batch reactor. Actual CSTRs show behavior intermediate between HSCTRs and SCSTRs. Tai et al. [32] have simulated the performance of SCSTRs and have presented their results in the form of contour maps. Figure 11 shows some of these results in which the hatched region represents the range of industrially encountered values of the variables. Trains of SCSTRs have been studied by Nagasubramanian and Reimschuessel [29], and Figure 12 shows results for the same system as shown in Figure 10b with HCSTRs replaced by SCSTRs. It is found that a lower total residence time is required in a cascade of SCSTRs than in a cascade of HCSTRs, to produce nylon-6 of the same number-average chain length.

Another nonideality that can be easily accounted for is the presence of a nonuniform (i.e., non–plug-flow) velocity profile in real tubular reactors [33]. This is expected to improve the agreement between the reactor model and the actual reactor performance. To

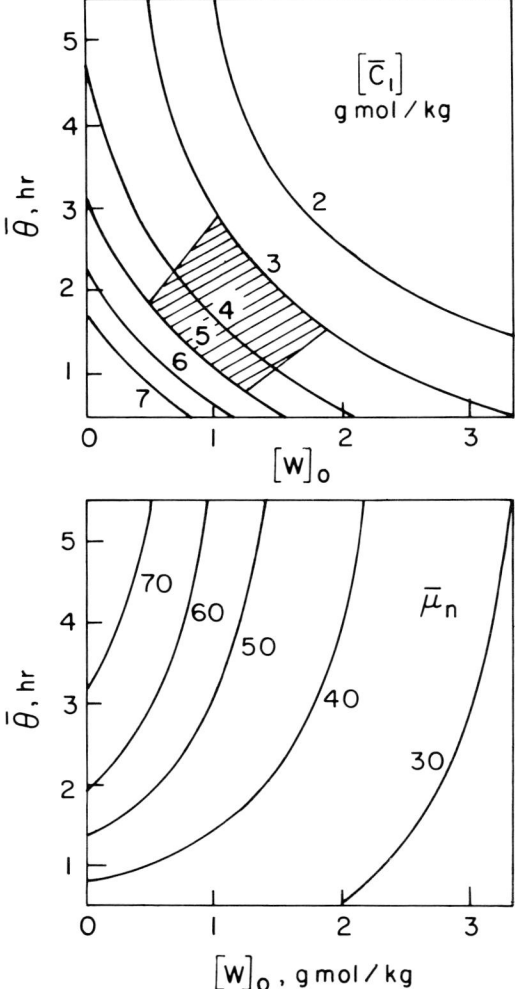

Figure 11 Results for an SCSTR at 260°C. Hatched region represents industrially important values. $[P_{1,x}]_0 = 0$. ($\bar{\mu}_n$ and $[\tilde{C}_1]$ represent product characteristics.) (Reprinted from Ref. 32 with permission of John Wiley & Sons, New York.)

account for the velocity profile, it is assumed that each fluid element is completely segregated. Batch reactor equations can then be solved for each fluid element (as was also done for an SCSTR) and the mean values of the concentrations of the various molecular species at the exit may be computed. Tai et al. [32] obtained average product conditions ($[\tilde{C}_1]$ and $\bar{\mu}_n$) for various values of $\bar{\theta}$ and T and once again presented their results in the form of contour plots for tubular reactors having a parabolic velocity profile. Figure 13 shows some typical results. This figure also shows results for a PFR. The results on the laminar flow reactor have been found to match [32] with some data available in patents. It is found that lower μ_n products are obtained when laminar flow is present than when plug

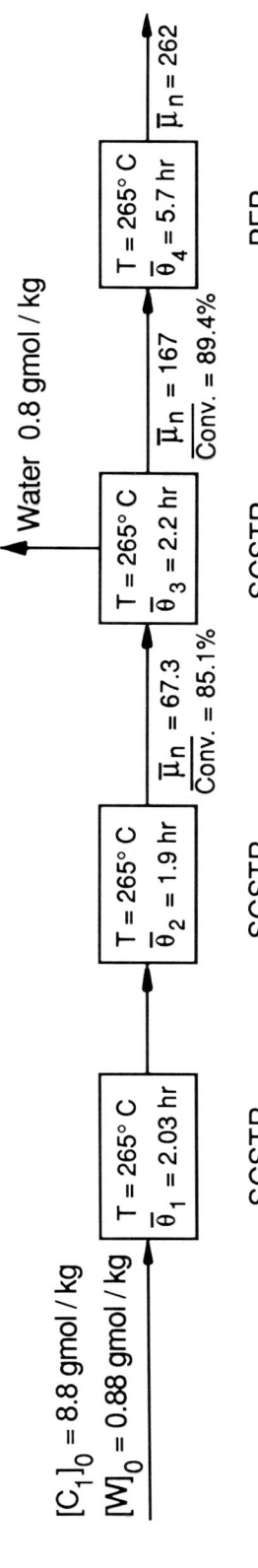

Figure 12 Nylon-6 polymerization in a train of SCSTRs and a PFR. Product same as in Fig. 10. (Reprinted from Ref. 29 with permission of John Wiley & Sons, New York.)

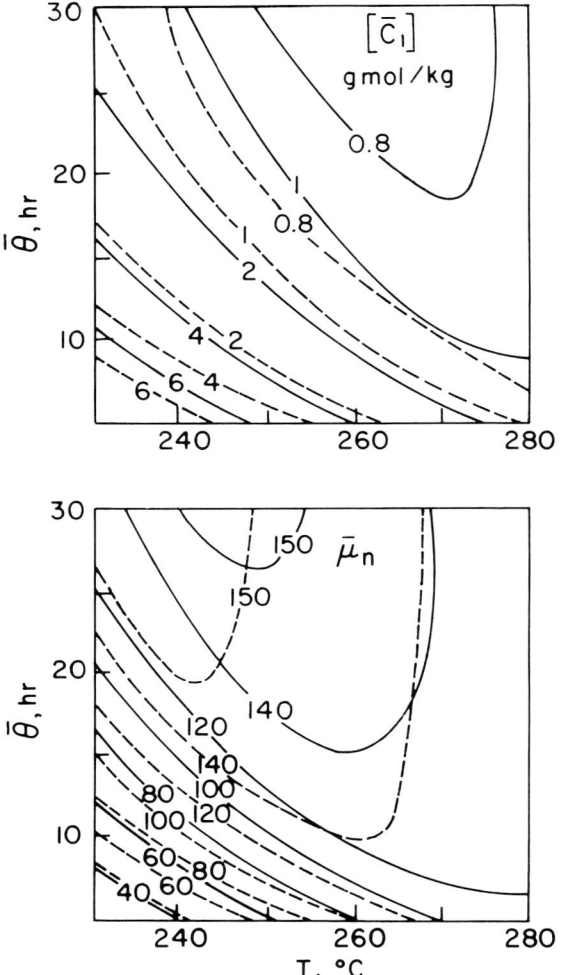

Figure 13 Conditions at the outlet of an isothermal tubular reactor under (*a*) laminar flow conditions [32] (solid lines) and (*b*) plug flow conditions (dashed lines). $[W]_0 = 0.22$ gmol kg^{-1}. $[P_{1,x}]_0 = 0$.

flow exists in a reactor having the same average residence time. Because of mixing of products from different radial locations, the average polydispersity index at the exit will be higher than that in a PFR.

In this section, the effect of two important nonidealities, micromixing and residence time distribution, have been studied. The models used represent real reactors far more closely than do ideal models like the PFR and HCSTR. In fact, tracer studies can be carried out on industrial reactors or pilot plants to obtain the actual residence time distribution function, and this may be used to obtain the mean values of the various concentrations and moments in the product stream. Alternatively, if it is possible to carry out experimental studies on nonreacting systems in industrial reactors, one can approximate the latter as sequences of PFRs and HCSTRs or as a PFR with a recycle [34, 35]. Appropriate mass balance equations for these models can then be solved to predict the performance of real reactors much more closely than possible with a single ideal reactor. In the next section, one such model for an industrial reactor is presented.

SIMULATION OF INDUSTRIAL REACTORS

The reactor models (including nonidealities) discussed in the previous section explain the behavior of several industrial reactors and reactor cascades fairly well, provided there is no evaporation of water. As mentioned earlier, high concentrations of W are normally used in the initial stages to speed up the slow ring-opening step, and thereafter W is removed to yield a product having high μ_n. As a result, industrial reactors for nylon-6 polymerization incorporate at least one stage (or have some zone) wherein W is removed from the reaction mass [36]. Figure 14 shows how the use of 0.577 g mol kg^{-1} of water in

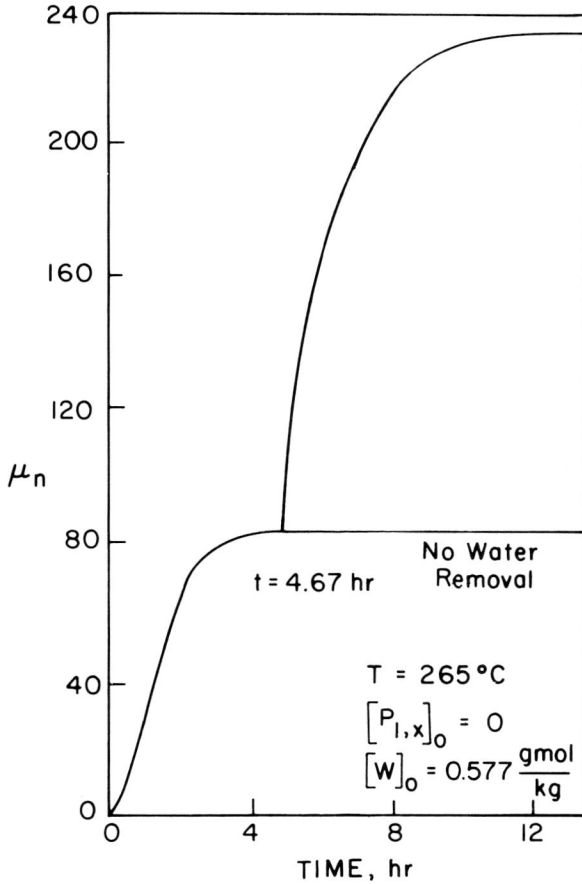

Figure 14 μ_n vs. time [22] for a two-stage PFR, both at 265°C with all water present at $t = 4.67$ hr removed instantaneously. No water is evaporated inside either of the two reactors. $[P_{1,x}]_0 = 0$. $[W]_0 = 0.577$ gmol kg^{-1}.

the feed leads to an equilibrium value μ_n of about 80 if no W is removed. But if *all* W present in the reaction mass is removed instantaneously (so that [W] becomes zero before the reaction starts again) at $t = 4.67$ hr, much higher values of μ_n are obtained. The instantaneous removal of all W is an oversimplification of the physical process and it is necessary to model the mass transfer in nylon-6 polymerization appropriately, accounting for the resistances of the diffusion and chemical reaction steps.

In addition to accounting for effects of mass transfer, heat transfer aspects of industrial reactors must also be studied, particularly at high conversions where the viscosity of the reaction mass becomes high and the thermal conductivity becomes low. In fact, adiabatic operation of tubular reactors is a better description of the reactor under such conditions than is the isothermal model. In this section, both mass and energy transfer effects are considered.

A commonly used industrial reactor for nylon-6 polymerization is the VK column (Vereinfacht Kontinuierliches Rohr) [2–4]. This consists of a vertical tube operating at atmospheric pressure. The feed enters the top of the column and is heated to about 220–270°C using heat exchangers in the form of internal gratings. In this top region, water and ε-caprolactam evaporate continuously. The bubbles of vapor formed in the reaction mass cause intense agitation as they rise up to a reflux condenser, which condenses and returns the caprolactam to the column. The hydrostatic head some distance downstream and the increase in the boiling point due to the polymerization prevent further vaporization, and so the remainder of the column is a nonvaporizing tubular reactor. Gratings are also used in this section both to facilitate heat removal and to ensure almost plug-flow velocity profiles.

Jacobs and Schweigman [34] carried out extensive pilot plant tests and also collected

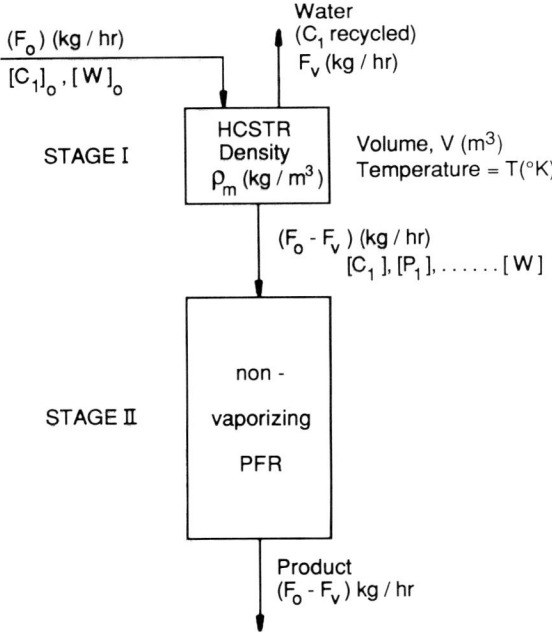

Figure 15 Simple model of a VK column [34].

data on industrial VK columns. Based on temperature measurements and tracer studies, they showed that the top zone of the column is well mixed owing to the agitation caused by the bubbles. They find that the concentration of W in this zone is a function of temperature alone and is given by the following empirical relation:

$$W = \frac{1.76 - 0.0060T}{1.8} \tag{8}$$

where W is in gmol kg^{-1} and T is in °C. [W] is found to be independent both of the residence time in this zone and of the feed conditions. This indicates that in the top zone of the VK column, the water concentration is determined solely by vapor–liquid equilibrium considerations, and that there is negligible mass transfer resistance.

Based on residence time distribution measurements, Jacobs and Schweigman [34] suggest that the top zone of the VK column be modeled either as a single HCSTR (with appropriate vaporization terms) or as a series of two HCSTRs with backmixing and vaporization. They find a small difference between the performances predicted by these two models. Figure 15 shows the first model used by these workers. Mass balance equations can easily be written for the two stages of this model, incorporating Eq. 8. A set of algebraic equations is obtained for Stage I, while a set of ordinary differential equations is obtained for Stage II. These can be solved easily. Figure 16 shows some typical results

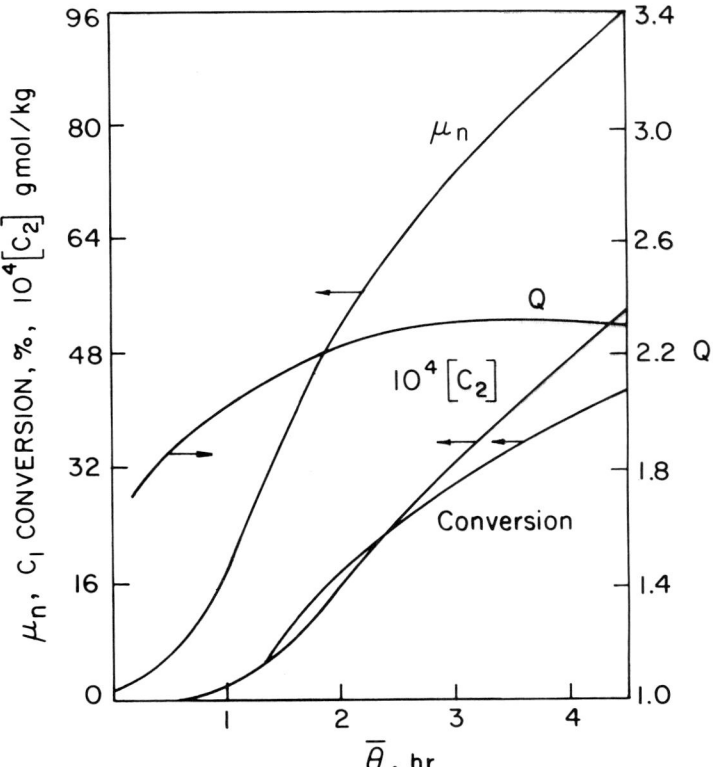

Figure 16 Results [23] for the top zone of the VK column as modeled in Figure 16. Temperature = 260°C, $[C_1]_0$ = 8.407 gmol kg^{-1}, [W] = 0.11 gmol kg^{-1} (equilibrium). $[P_{1,x}]_0 = 0$.

[23] for Stage I. Similar results obtained earlier by Jacobs and Schweigman [34] have been found to agree with some industrial data, details of which, however, are not presented. The variation of the polydispersity index with $\bar{\theta}$ for the HCSTR with vaporization of W differs from that in the absence of vaporization, as shown in Figure 9.

The foregoing model for the top of the VK column can be improved by using either better or more fundamental equations for vapor–liquid equilibrium instead of the empirical Eq. 8. Fukumoto [37] suggests the use of the following equation based on experimental thermodynamic equilibrium data on nylon-6–C_1–W systems:

$$[W] = P_T \exp\left[\frac{8220}{T} - 24.0734\right] \tag{9}$$

where P_T is the total pressure in mm Hg, [W] is in gmol kg^{-1}, and T is in K. It is observed that Eqs. 8 and 9 yield similar values of [W], particularly at high temperatures. However, since the number-average chain length, μ_n, is extremely sensitive to the water content in the reaction mass, there is a need to develop more precise correlations.

Recently, some research [32, 38, 39] has been directed toward developing a better theoretical analysis of vapor–liquid equilibrium in nylon-6 reactors. Tai et al. [32] used the following relationships:

$$\log_{10} P_W^\circ = -\frac{2080}{T} + 8.55 \tag{10a}$$

$$\log_{10} P_{C_1}^\circ = -\frac{3280}{T} + 9.03 \tag{10b}$$

$$\log_{10}\left(\frac{x_W}{P_W}\right) = \frac{3570}{T} - 11.41 \tag{10c}$$

$$\log_{10} P_{C_1} = -\frac{4100}{T} + 9.6 \tag{10d}$$

$$x_W = \frac{[W]}{[W] + [C_1] + [C_2] + \lambda_0 + \lambda_{0,x}} \tag{10e}$$

$$P_T = P_W + P_{C_1} \tag{10f}$$

where P_W° and $P_{C_1}^\circ$ (in mm Hg) are the vapor pressures of pure water and ϵ-caprolactam, P_W and P_{C_1} (mm Hg) are the partial pressures of caprolactam and water over a nylon-6–C_1–W mixture (Eq. 10c represents some average over several practical concentration ranges of the individual components), and x_W is the mole fraction of water in the liquid phase. Equations 10c, d, and f can be solved to obtain x_W or [W] at a desired total pressure, P_T, and temperature, T. For $P_T = 1$ atm and $T = 260°C$, this gives $x_W = 0.01318$ or $[W] \simeq 0.105$ gmol kg^{-1} (for a caprolactam conversion of about 8–10% usually present at the top of the VK column). This may be compared with the equilibrium value of 0.111 gmol kg^{-1} using Eq. 8. The need for more detailed studies on vapor–liquid equilibrium is once again apparent in view of the extreme sensitivity of μ_n to [W].

Gupta and Gandhi [38, 39] suggest the use of

$$\ln(P_{C_1}^\circ) = 13.006 - \frac{7024.023}{T} \tag{11a}$$

$$\ln(P_W^\circ) = 11.6703 - \frac{3814.44}{T - 46.13} \tag{11b}$$

where P is in atm and T in K to represent experimental vapor pressure data [40, 41]. These are much better than Eqs. 10a and b. These may be used along with Eq. 10f and the following, more detailed equations:

$$P_W = x_W P_W^\circ(T)\gamma_W(T, x_W, x_{C_1}, \lambda_0, \lambda_{0,x}) \tag{12a}$$

$$P_{C_1} = x_{C_1} P_{C_1}^\circ(T)\gamma_{C_1}(T, x_W, x_{C_1}, \lambda_0, \lambda_{0,x}) \tag{12b}$$

where x_{C_1} is the mole fraction of caprolactam and γ_W and γ_{C_1} represent the activity coefficients of W and C_1 in the liquid phase. The activity coefficients are functions of the temperature, mole fraction of W and C_1, and the concentration of the polymer, $\lambda_0 + \lambda_{0,x}$, in the liquid phase, and are much lower [42, 43] than the value of unity given by Raoult's law. Wilson's equation for the γ's has been used, with the constants determined by curve fitting some experimental vapor–liquid equilibrium data of Giori and Hayes [42]. Unfortunately, there is a large amount of scatter in the data so plotted, and much more work needs to be done to get good expressions for γ_W and γ_{C_1}.

Another possible improvement in the reactor model of the top of the VK tube is the use of a sequence of three HCSTRs with backmixing (recycle) as well as vapor flow [39]. Figure 17 shows the notation used. The compositions of the liquid in the three HCSTRs differ, and consequently there would be a difference in the compositions of the equilibrium vapors. An amount $\alpha_1 F_1$ kg hr^{-1} of liquid having the composition of the liquid in HCSTR 1 (in addition to F_1) flows into the second HCSTR, while an equal amount of recycle liquid having the composition of the liquid in HCSTR 2 flows back to reactor 1. These two streams represent the backmixing present in the top of the VK column. Similar flows exist among the second and third HCSTRs in the model. Mass balance equations can be written for this system [39] and solved using equilibrium relationships predicted by a modified Raoult's law (with volume fractions used instead of mole fractions for the liquid phase) as suggested by Jacobs and Schweigman [34]:

$$P_W = \frac{[W]}{\lambda_1 + \lambda_{1,x}} P_W^\circ \tag{13a}$$

$$P_{C_1} = \frac{[C_1]}{\lambda_1 + \lambda_{1,x}} P_{C_1}^\circ \tag{13b}$$

Figure 18 shows some typical results. It is observed that backmixing leads to an increase in both the monomer conversion and μ_n but does not affect the (equilibrium) value of 2.0 for the polydispersity index of the product from the VK column. It is also observed that significantly higher conversions and μ_n can be obtained from the VK column than from a single nonvaporizing isothermal PFR, starting with identical feed streams. It may be added that these results are extremely useful for making qualitative inferences, but they need to be tested against actual plant data before being used for design.

Several other types of industrial nylon-6 reactors in use have been described by Sittig [44], but simulation studies on only a few of these are available in the literature. Figure 19 is a schematic of one such reactor. Feed of caprolactam and water flows through a coiled tube within the reactor and gets preheated. After the material exits from the coil, at A, it flows down its outside. Most of the polymerization occurs outside the coils. The pressure is maintained sufficiently high to avoid any vaporization. The advantage of such a reactor configuration is that it is heat-integrated, since a considerable amount of the heat of reaction is used to preheat the feed. Mass and energy balance equations for such a reactor configuration can be written for the inside and outside of the coils, assuming no radial

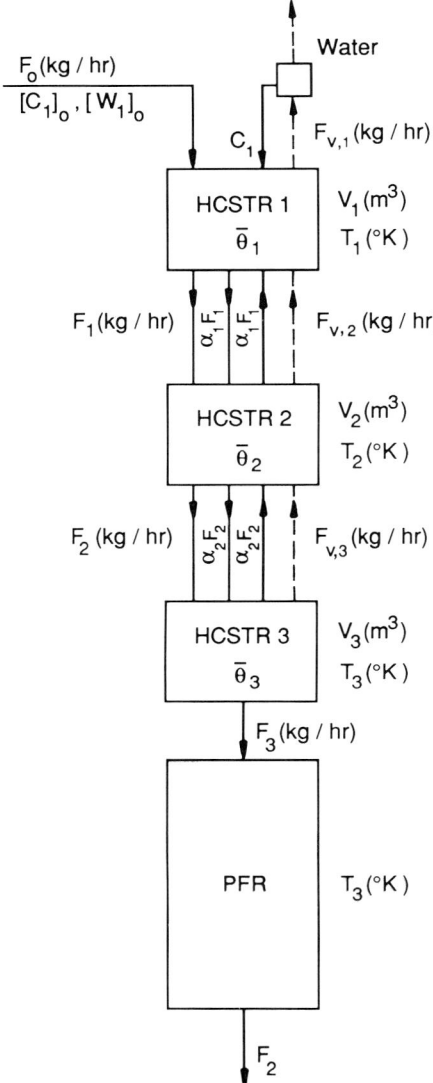

Figure 17 Improved model of the VK column [38, 39]. Solid lines represent liquid flow; dashed lines represent vapor flow. $[W]_0 = 0.05$ mol/mol C_1; $\theta_1 = \theta_2 = \theta_3 = 1$ hr; $F_0 = 1000$ kg hr^{-1}.

gradients in temperature or concentrations, and solved iteratively [45]. Results are shown in Figures 20 and 21 for some typically encountered conditions. A temperature maximum is observed outside of the coils and this could be extremely sensitive [45] to some of the parameters under certain conditions. Such a reactor thus needs to be operated under well-controlled conditions. Also, the sensitivity of the temperature profile to the heat transfer coefficients necessitates the estimation of the latter with high precision. The following correlations [34] for the physical properties have been used for this and most earlier studies:

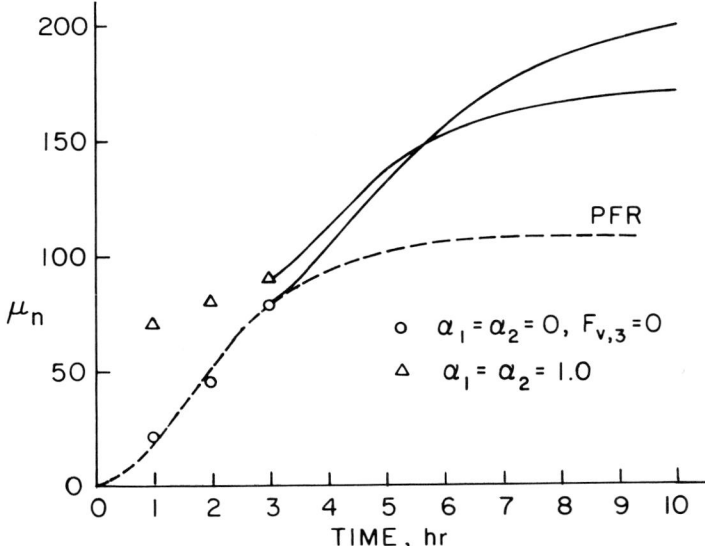

Figure 18 μ_n vs. time (or residence time) [38, 39] for the model shown in Figure 17. T_1, T_2, and T_3 are 240, 253.5, and 265°C respectively. Dashed curve shows results on a single nonvaporizing PFR at 253.5°C, using same feed as to HCSTR 1.

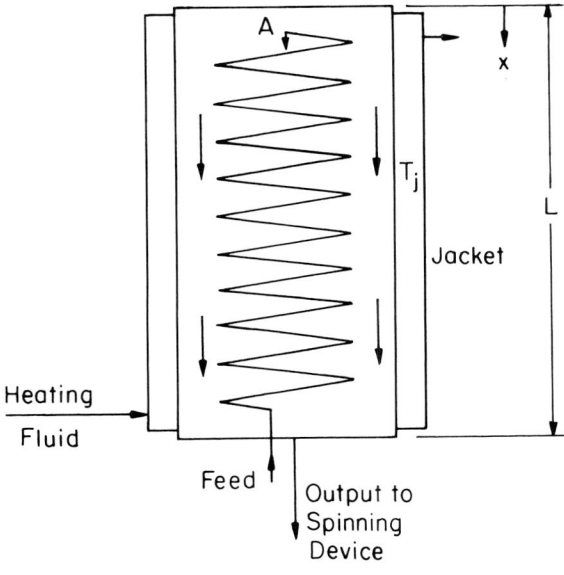

Figure 19 Schematic diagram of another nylon-6 reactor [68–70].

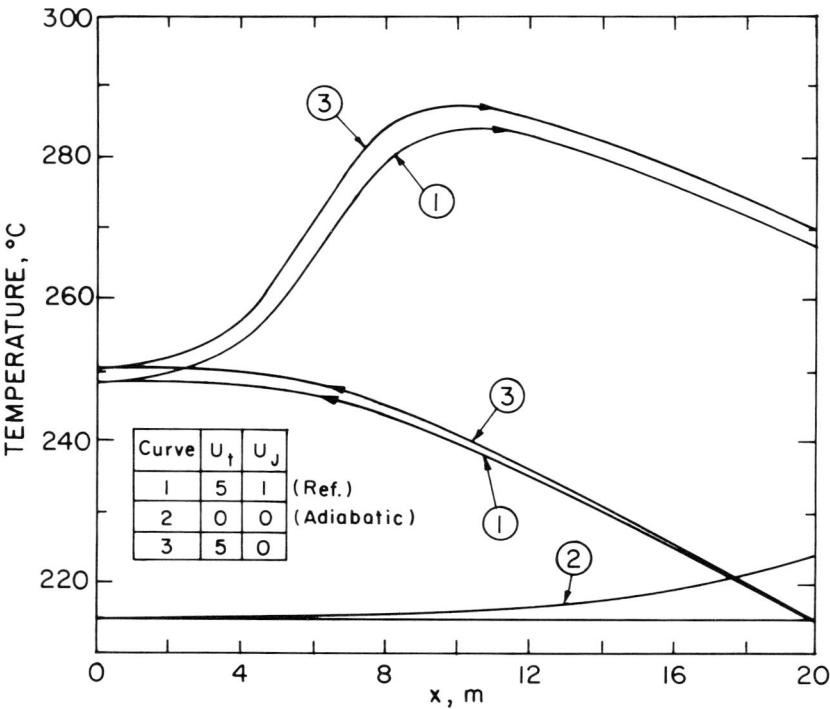

Figure 20 Temperature profiles under typical conditions for the reactor shown in Figure 19. U_t and U_J are the tube and jacket overall heat transfer coefficients. Feed is caprolactam with 0.15 gmol of water per kilogram, at 488K. $T_J = 530$K. Other conditions available in Ref. 45.

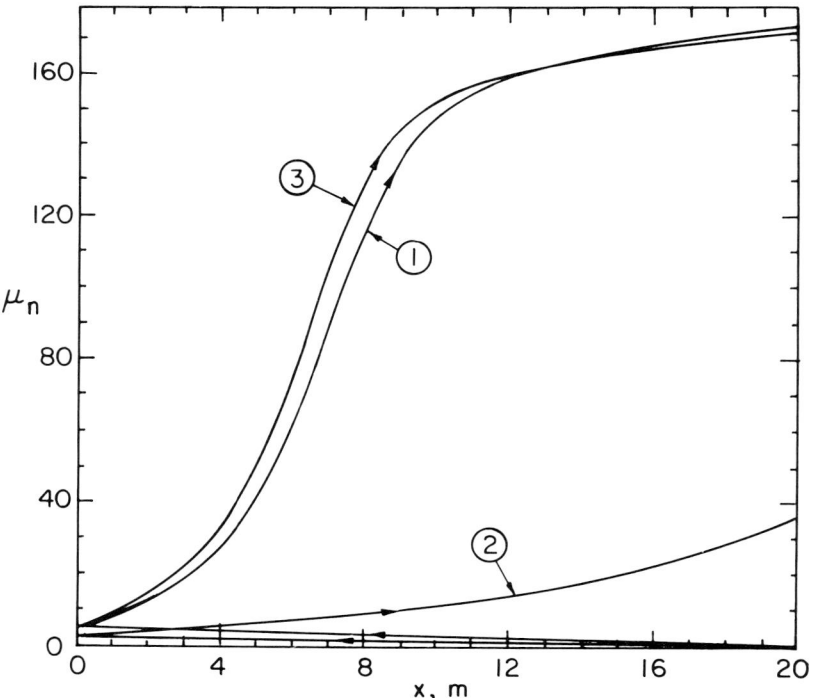

Figure 21 μ_n profiles for the reactor in Figure 19. Conditions same as in Figure 20.

$$k_m = 0.21 \tag{14a}$$

$$\rho_m = 1000\{1.0065 + 0.0123[C_1] + (T - 495)(0.00035 + 0.00007[C_1])\}^{-1} \tag{14b}$$

$$c_{p,m} = 659.3\frac{[C_1]}{[C_1]_0} + \left(1 - \frac{[C_1]}{[C_1]_0}\right)(486.1 + 0.337T) \tag{14c}$$

$$\alpha_m = \frac{k_m}{\rho_m c_{p,m}} \tag{14d}$$

where k_m is thermal conductivity in W m^{-1} K^{-1}, ρ is in kg m^{-3}, and $c_{p,m}$ is in cal (kg K)$^{-1}$. These equations are used along with standard equations for heat transfer coefficients [46]. These correlations, which differ from those recently suggested by Tai et al. [32, 47], emphasize the need for obtaining better correlations, particularly for the viscosity of the reaction mixture.

Yet another industrial reactor that has been studied recently [48] is a semibatch reactor in which feed of caprolactam and water is charged at 100°C under a nitrogen atmosphere. The reaction mass is first heated to about 250–270°C, during which time vaporization of water and some caprolactam takes place, gradually building up the pressure. When the pressure in the reactor exceeds a value, P_{set}, the relief valve opens and releases the vapors into a condensor line. The relief valve keeps the pressure in the reactor at P_{set} for some time, after which the pressure is lowered in a programmed manner to slightly above atmospheric. The reaction mass is equilibrated at this pressure for some time, after which it is extruded out in the form of strands for postreactor processing. Gupta et al. [48] neglect mass transfer resistances in modeling this sequence of operations in the semibatch reactor and use the modified Raoult's law (Eq. 13) to describe vapor–liquid equilibrium. A typical set of results is shown in Figure 22, along with some industrial data under

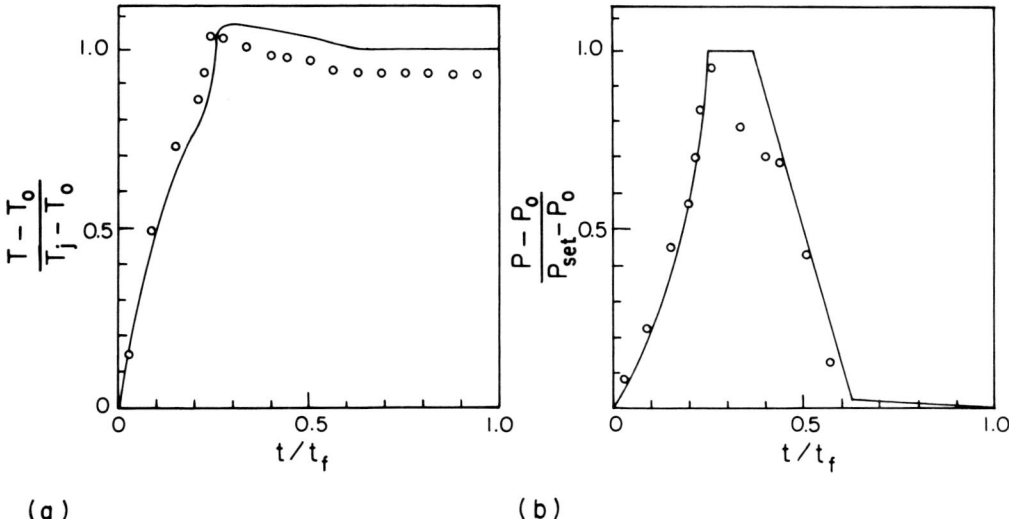

Figure 22 Dimensionless (a) temperature and (b) pressure profiles for the semibatch reactor described in detail in Ref. 48. Water concentration in feed = 2.52%. Solid lines represent results of simulation; circles represent data. (Reprinted from Ref. 48 with permission from copyright holders.)

similar conditions. It is observed that the temperature and pressure histories are described quite well by the simulation results. However, the number-average molecular weight of the final product as predicted by the model is higher than obtained from the actual plant, indicating the presence of mass transfer resistance. Further work along these lines is in progress.

OPTIMIZATION OF NYLON-6 REACTORS

In the preceding sections it was shown that high water concentrations must be used at the beginning of nylon-6 polymerization to speed the conversion, followed by low concentrations of water to obtain a high value ($\sim$150–200) of μ_n. Some typical results on cascades of several reactors accomplishing this have already been presented. In these studies [18, 29, 49], several reactor sequences were considered and the corresponding total reaction time required to give polymer of a desired μ_n was computed, thus giving an idea of the optimal setup.

In this section, the optimization of tubular reactors is considered, since these reactors are more common now. Most studies [50–52] which have been carried out in the past on such reactors have determined optimal profiles for the temperature and the water concentration to minimize a chosen objective function. These studies are preliminary in nature and mostly report semiquantitative results because of proprietary reasons. In addition, the kinetic schemes used in these early optimization studies incorporate several assumptions, particularly for the rate of formation of cyclic oligomers. With more complete and precise kinetic information now available [4, 13], a more systematic optimization study can be carried out [26, 53–55].

Table 4 presents a summary of the objective functions, constraints, and *qualitative* optimal temperature (and water concentration) profiles from the various studies. A direct comparison of results of the first three studies is not possible since different workers assumed different kinetic schemes and rate constants. The remaining studies used the recent data of Tai and Tagawa [4] and so are consistent. Qualitatively different temperature profiles are obtained for different choices of the objective function and constraints. The results of Gupta et al. [54] assume a fixed value for the reaction time, t_f, and so are appropriate for optimal *operation* of existing reactors. On the other hand, the more recent study by Ray and Gupta [26] makes no such assumption for t_f and so is useful for the optimization of reactors at the design stage. In most of these studies, the temperature is constrained to lie between two limits, typically 220 and 270°C. The former value represents the normal melting point for nylon-6; the latter represents the normal boiling point for caprolactam. These constraints thus ensure single-phase polymerization. Detailed effects of varying the several parameters—e.g., feed water concentration, $[W]_0$, desired chain length of the product, $\mu_{n,d}$, desired concentration of cyclic dimer in the product, $[C_2]_d$—are available in the original references [26, 53–55].

In all the optimization studies just discussed, the optimal temperature (and [W]) profiles are obtained without consideration of heat (and mass) transfer aspects in the reactor. In other words, the reaction mass is maintained at these computed temperatures somehow by supplying or removing heat (or water) at rates computed by appropriate energy balance (or mole balance) equations. Relatively little work exists on the optimization of nylon-6 tubular reactors with both mass and energy transfer accounted for. The computed optimal temperature profiles may also not be practically feasible. Suboptimal profiles approximating the calculated ones may then be used.

Table 4 Nylon-6 Tubular Reactor Optimization Problems Studied

Control variable	Objective function	End-point constraints[a]	Qualitative profiles	Comments	Reference
$T, [W]$	t_f	μ_{n,t_f}		No C_2 equation (t_f minimized)	52
T	$EV \equiv 11.3\left\{[C_1]\right.$ $\left. + \sum_{2}^{\infty} n[C_n]\right\}_{t_f}$ $+ 12.2[P_{1,x}]_{t_f}$	$\mu_{n,t_f} \geq \mu_{n,d}$ $(\eta_{t_f} \geq \eta_0)$		Evaporation in earlier part at T_0, for time $t = 0$–t_0. PFR from t_0 to t_f. t_f fixed.	50
T	t_f	$EV_{t_f} \geq EV_d$ $\eta_{t_f} \geq \eta_d$		t_f not fixed	

Table 4 (Continued)

Control variable	Objective function	End-point constraints[a]	Qualitative profiles	Comments	Reference
T	$[C_1]_{t_f}$	$\mu_{n,t_f} = \mu_{n,d}$ $[C_2]_{t_f} = [C_2]_d$		t_f fixed C_2 reactions empirical	51
T	$\alpha_3 [C_1(t_f)] + \int_0^{t_f} \left\{ \dfrac{\alpha_2}{\mu_{n,d}^2} \right.$ $\times (\mu_{n,t_f} - \mu_{n,d})^2$ $\left. + \alpha_1 [C_2]^2 \right\} dt$			t_f fixed $220°C < T < 270°C$	53
T	$[C_1]_{t_f}^2$	$[C_2]_{t_f} = [C_2]_d$ $\mu_{n,t_f} = \mu_{n,d}$		t_f fixed C_2 reactions of Tai and Tagawa [4, 13] $220°C \leq T \leq 270°C$	54

T	$[C_1]^2_{t_f}$	$\mu_{n,t_f} = \mu_{n,d}$ (stopping)	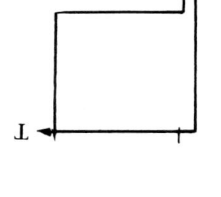 t_f floating $220°C \leq T \leq 270°C$
T	$[C_2]_{t_f}$	$\mu_{n,t_f} = \mu_{n,d}$ (stopping)	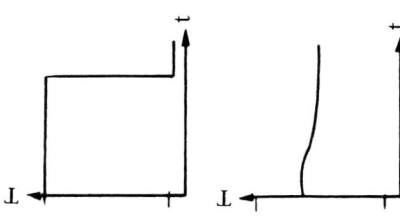 t_f floating $220°C \leq T \leq 270°C$
T	$[C_2]_{t_f}$	$[C_1]_{t_f} = [C_1]_d$ $\mu_{n,t_f} = \mu_{n,d}$ (stopping)	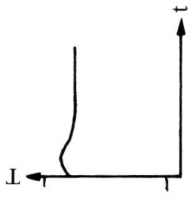 t_f floating $220°C \leq T \leq 270°C$
T	$[C_1]^2_{t_f}$	$[C_2]_{t_f} = [C_2]_d$ $\mu_{n,t_f} = \mu_{n,d}$ (stopping)	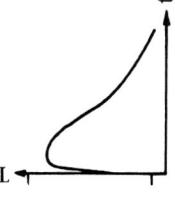 t_f floating $220°C \leq T \leq 270°C$

[a]Subscript t_f indicates conditions at the end of the reactor, subscript d indicates desired values.

CYCLIZATION REACTIONS IN NYLON-6 POLYMERIZATION

In the discussion thus far, the reactions associated with only the cyclic dimer have been considered. It is known that higher cyclic oligomers C_n are also present in the reaction mass and these need to be removed by an energy-intensive, hot water extraction process. Any simulation or optimization study therefore must incorporate reactions associated with these compounds. Relatively little kinetic information exists on these reactions and most optimization studies have used very crude models to account for them [50, 51, 56]. In this section, the available information on the kinetic and equilibrium constants for cyclization of nylon-6 oligomers is presented.

Semlyen and colleagues [57, 58] and other workers [59–61] have obtained the equilibrium constants for the following cyclization reaction:

$$P_{n+m} \stackrel{K'_{C_n}}{\rightleftharpoons} P_m + C_n \tag{15}$$

(which can replace Eq. 1a), for n up to 6, using gel permeation chromatography measurements. The values of K'_{C_n} are shown in Figure 23. The data can be curve-fitted by equations of the following types [2, 20]:

$$\log_{10} K'_{C_n} = \frac{a(n)}{T} + b(n), \quad T > 483 \text{ K} \tag{16}$$

$$K'_{C_n} = \frac{a^*(T)}{n^2} \tag{17}$$

with $a(n)$, $b(n)$, and $a^*(T)$ obtained empirically and K'_{C_n} in gmol kg^{-1}.

Recently, some theoretical modeling has been attempted to explain the variation of K'_{C_n} with n [63–65]. Jacobson and Stockmayer [63] mentioned that to obtain C_n by reaction (15), P_{n+m} must first break into P_n and P_m (reaction 2 of Table 1) and then P_n must cyclize. The entropy of the first step, ΔS_1, can be written as

$$\Delta S_1 = k \ln\left(\frac{V}{V_s}\right) \tag{18}$$

where k is the Boltzmann constant. The term in parentheses in Eq. 18 represents the fact that after reaction, the cleaved chain ends can be anywhere in the volume V of the reaction mass, whereas just before reaction, they must lie within a small volume V_s. To estimate the entropy change for the cyclization of P_n, it is observed that the two chain ends of P_n must turn around and lie within a small volume V_s of each other. By using concepts of chain statistics [33] the fraction of molecular configurations that satisfy this criterion can be found to be

$$\text{Prob} = \left(\frac{3}{2\pi \nu n l^2}\right)^{3/2} V_s \tag{19}$$

where ν is the number of backbone *bonds* in a repeat unit, each of length l. The entropy change of the cyclization of P_n is then

$$\Delta S_2 = k \ln\left(\frac{\text{Prob}}{n}\right) \tag{20}$$

where n must be used because of the n symmetrical —CONH— groups in C_n. The total entropy for reaction (15) thus is the sum of ΔS_1 and ΔS_2 and is $k \ln\{(3/2\pi\nu)^{3/2} V/(l^3 n^{5/2})\}$.

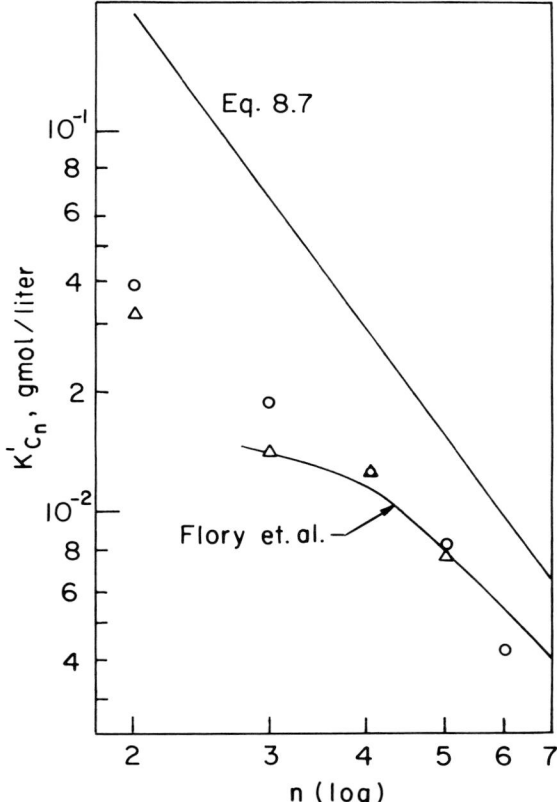

Figure 23 Experimental values of K'_{C_n} at 252°C. ○, data of Semlyen et al. [57, 58]; △, data of Zahn et al. [59–61]. To obtain K'_{C_n} in gmol kg^{-1}, divide by 0.93. Plots of Eq. 21 and computed results of Flory and colleagues [62] also shown. (Reprinted with permission from Ref. 62. Copyright 1976, American Chemical Society, Washington, D.C.)

If the energy change associated with Eq. 15 is neglected (since the same number of bonds is involved on both sides), the entropy change is related to the change in free energy, and thus to the equilibrium constant, as

$$K'_{C_n} = \exp\left(\frac{\Delta S}{k}\right) = \left(\frac{3}{2\pi \nu}\right)^{3/2} \frac{1}{l^3 n^{5/2} N_{Av}} \tag{21}$$

where the Avogadro number N_{Av} has been substituted for $1/V$ to obtain K'_{C_n} in conventional units of gmol L^{-1}. Equation 21 is plotted in Figure 23 and is found to overestimate K'_{C_n}.

Equation 21 assumes that the backbone bonds are completely free to rotate about each other. Flory and colleagues [62, 65, 66] used the rotational isomeric state model to replace Eq. 19. In their model, they accounted for the presence of bond angle restrictions and steric hindrances and also incorporated the constraint that for cyclization to occur, the first bond and the last bond must lie at an angle, θ, the bond angle (by postulating an extra hypothetical bond which must lie at angle θ to the last one and must be "almost" parallel to the first bond). Their computed results are also shown in Figure 23 and are found to explain experimental results much better.

It should be emphasized that K'_{C_n} as used in this section is written in terms of molecular species such that

$$K'_{C_n} = \frac{[P_m][C_n]}{[P_{n+m}]} \qquad (22)$$

If Eq. 15 is written as

$$P_{n+m} \underset{nk'_{C_n}}{\overset{k_{C_n}}{\rightleftharpoons}} P_m + C_n \qquad (23)$$

such that k_{C_n} and k'_{C_n} represent the rates associated with functional groups, the corresponding equilibrium constant is obtained as [19, 20]

$$K_{C_n} \equiv \frac{k_{C_n}}{k'_{C_n}} = nK'_{C_n} \qquad (24)$$

Some preliminary simulations using the empirical Eq. 17 have been carried out [20]. It is observed that the conversion of the monomer is essentially unchanged by the incorporation of the cyclization reactions but the μ_n reduces by almost 10%. Figure 24 shows the variation of the cyclic oligomer concentrations for a typical isothermal, nonvaporizing batch reactor [20]. Since k'_{C_n} (Eq. 23) has been assumed equal to k_3 of Table 1 in this study, the time variation of $[C_2]$, $[C_3]$, . . . , may be slightly in error. The total cyclic compounds are observed to be around 6–7% by weight, which is slightly higher than industrially encountered values. Other experimental HPLC data on the equilibrium concentrations of C_3 to C_6 at different values of T and $[W]_0$ have been reported [10] but these have not yet been used to obtain rate and equilibrium constants for the various cyclic oligomerization reactions (except for the cyclic dimer). This is because of mathematical complexities associated with the use of functionalities of the type given in Eq. 21, which lead to fractional moments in the moment equations.

NYLON-6,6 POLYMERIZATION

In addition to nylon-6, several other nylons (e.g., nylon-11, 12, 6,6; 6,10; 6,12) are of commercial importance. Kinetic data on these are generally scarce [1, 67] and very little simulation work has been reported on their polymerization, except on nylon-6,6. This polymer is prepared in a multistage process; the first step consists of the preparation of nylon-6,6 salt:

$$H_2N\text{—}(\text{—CH}_2)_6\text{—}NH_2 + HOOC\text{—}(CH_2)_4\text{—}COOH \rightarrow$$

$$H_2N\text{—}(CH_2)_6NH_3^+ \ {}^-O\text{—}\overset{\overset{O}{\|}}{C}\text{—}(CH_2\text{—})_4\text{—}COOH \qquad (25)$$

Dry adipic acid (AA) is mixed with an aqueous solution of hexamethylene diamine (HMDA) and the salt (MP 195°C) is precipitated by the addition of methanol. The salt is then centrifuged and an aqueous solution of about 60% salt concentration is sent to the polymerization stage. The use of the salt enables the manufacture of high molecular weight polymer, since equimolar amounts of —NH_2 and —COOH groups are thereby ensured.

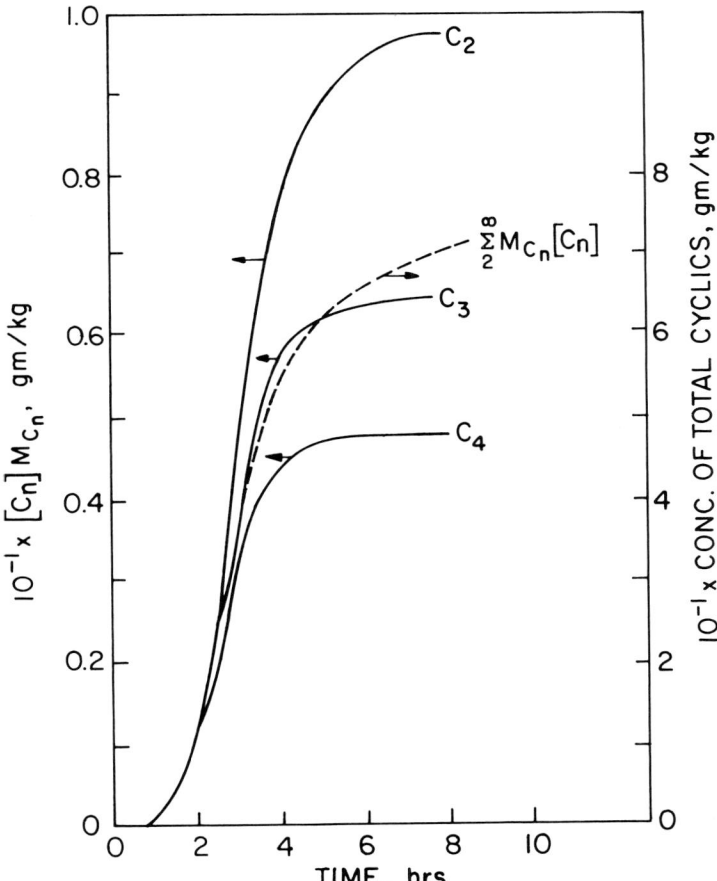

Figure 24 Variation of cyclic oligomer concentrations [20] in nylon-6 polymerization in isothermal sealed tubes. $[W]_0 = 0.44$ gmol kg^{-1}, $[P_{1,x}]_0 = 0$, $T = 235°C$, $a^*(T) = 0.1692$.

The exact metering of the diamine and the adipic acid, and the subsequent makeup of vaporized diamine, is extremely cumbersome, even though several recent processes have been commercialized in which molten HMDA and AA are used directly and salt manufacture is avoided [1, 67].

The aqueous salt solution is heated to about 260–270°C at pressures of about 250 psi in an oxygen-free atmosphere. This promotes rapid polymerization. The number-average molecular weight increases to about 15,000 under these conditions. Thereafter, the pressure is reduced to atmospheric values in a flash tank, and the material is then held at about 270°C for about 0.5 hr to obtain nylon-6,6 of molecular weights about 16,000–18,000, which have commercial usage. Figure 25 is a schematic flow chart of the process used for nylon-6,6 manufacture [68–70].

The rate of polymerization of the nylon salt has been studied by Ogata [71, 72] and is a typical example of reversible ARB polycondensation. The polymerization is represented by

$$—COOH + —NH_2 \underset{k/K}{\overset{k}{\rightleftharpoons}} —CONH— + H_2O \tag{26}$$

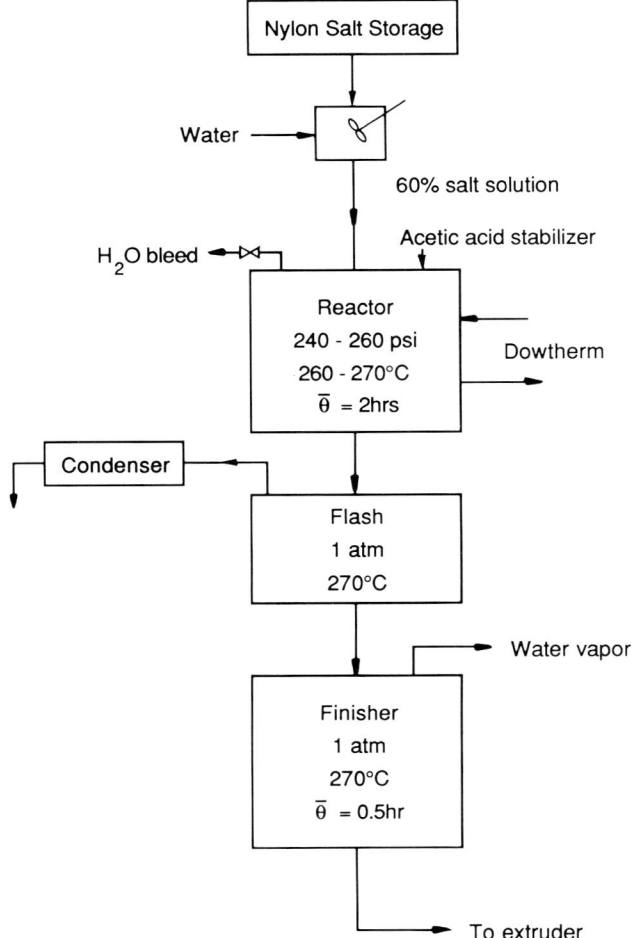

Figure 25 Flow chart for nylon-6,6 manufacture.

Ogata found from sealed tube experiments that both k and K are functions of temperature as well as the *initial* water concentration. He suggests the use of

$$\log_{10} K = \frac{5800}{T} - 8.32 \tag{27}$$

(T in K) even though the experimental values of K vary by almost ±20% around the value given by this equation. Recently, Ogata's data have been empirically curve-fitted [73, 74] and it has been found that K can be represented in terms of the initial water concentration and temperature while the rate constant, k, is represented in terms of the water concentration in the reaction mass *at the time of interest* and the temperature. The exact correlations for k and K are available in Refs. 73 and 74. Some isothermal simulations have been carried out for the polymerization of nylon-6,6 salt, and it is found that equilibrium is attained at extremely low values of μ_n (below 10.0). Simulations of industrial reactors for this nylon have not yet been reported.

Considerable attention has been focused on continuous polymerizations on nylon-6,6 in the recent past, using molten AA and HMDA. The reactions are carried out at very high pressures (1000 psi) and at 275°C for about 15–30 min, to prevent any boiling. After this, the pressure is decreased (using valves or tubes of increasing diameter) to about 250–350 psi, to prevent freezing during the vaporization of steam (several kinds of equipment have been patented to achieve this) [1]. The melt is then polymerized to equilibrium at atmospheric pressure, the usual residence times being about 1 hr for the last finishing stage. Again, there is very little available in the literature on the simulation of industrial reactors under actual conditions of operation.

OTHER NYLONS

Very little information is available in the literature on the polymerization of other nylons. Nylon-7 (from 7-aminoheptanoic acid), nylon-8 (from caprylactam), and nylon-9 are expensive and offer little to justify their industrial manufacture. Nylon-11, from ω-amino undecanoic acid (made from castor oil), is manufactured by heating the monomer under nitrogen at 215°C for about 3 hr, using catalysts [68–70]. Almost complete conversions are obtained, and so finishing operations are not required. This polymer has excellent electrical insulation properties, besides offering the attractive possibility of manufacture from an agricultural (renewable) raw material. Nylon-12, from lauryl lactam, can also be polymerized by heating the monomer with water, along with catalysts [68–70]. Almost 100% conversions can be obtained, as for nylon-11. Copolyamides of nylon-6,6, nylon-6,10, and nylon-6 can also be prepared [69]. These copolymers are soluble in mixtures of water and alcohol, and they are useful as abrasion and oil-resistant coatings for textiles, paper, leather, etc.

CONCLUSIONS

In this chapter, the polymerization of two important nylons has been discussed. It was found that the reaction mechanism is more complex for nylon-6 than for nylon-6,6. Detailed modeling of industrial reactors for nylon-6 was presented; important side reactions were included and experimentally determined rate and equilibrium constants were used. An attempt was made to optimize these reactors and it was observed that different temperature profiles are obtained depending on the objective function used. The modeling of postreactor or finishing operations for nylon-6, however, was not discussed since detailed modeling of the solid-state polymerization involved is still in its infancy [75]. Very little modeling has been reported for the polymerization of nylon-6,6, though its kinetic scheme is simple and involves the polymerization of a simple A-R-B-type monomer molecule (where A and B are reactive functional groups). This is probably because of the lack of experimental rate data and the lack of trustworthy data for vapor–liquid equilibrium.

REFERENCES

1. D. B. Jacobs and J. Zimmerman, in *Polymerization Processes* (C. E. Schildknecht and I. Skeist, eds.), Wiley, New York, pp. 424–467 (1977).
2. H. K. Reimschuessel, *J. Polym. Sci. Macromol. Rev., 12*: 65 (1977).
3. H. K. Reimschuessel, in *Ring Opening Polymerization* (K. S. Frisch and S. L. Reegan, eds.), Marcel Dekker, New York, pp. 303–326 (1969).

4. K. Tai and T. Tagawa, *Ind. Eng. Chem. Prod. Res. Dev.*, 22: 192 (1983).
5. J. Sebenda, *Prog. Polym. Sci.*, 6: 123 (1978).
6. S. K. Gupta and A. Kumar, *J. Macromol. Sci. Revs. Macromol. Chem. Phys.*, C26: 183 (1986).
7. W. H. Carothers and G. T. Berchet, *J. Am. Chem. Soc.*, 52: 5289 (1930).
8. P. H. Hermans, D. Heikens, and P. F. van Velden, *J. Polym. Sci.*, 30: 81 (1958).
9. C. A. Kruissink, G. M. van der Want, and A. J. Staverman, *J. Polym. Sci.*, 30: 67 (1958).
10. K. Tai and T. Tagawa, *J. Appl. Polym. Sci.*, 27: 2791 (1982).
11. H. K. Reimschuessel and G. J. Dege, *J. Polym. Sci. A-1*, 8: 3265 (1970).
12. G. J. Dege and H. K. Reimschuessel, *J. Polym. Sci. Polym. Chem. Ed.*, 11: 873 (1973).
13. Y. Arai, K. Tai, H. Teranishi, and T. Tagawa, *Polymer*, 22: 273 (1981).
14. K. Tai, H. Teranishi, Y. Arai, and T. Tagawa, *J. Appl. Polym. Sci.*, 24: 211 (1979).
15. K. Tai, H. Teranishi, Y. Arai, and T. Tagawa, *J. Appl. Polym. Sci.*, 25: 77 (1980).
16. F. Wiloth, *Z. Phys. Chem., N.F.*, 11: 78 (1957).
17. S. M. Skuratov, A. A. Strepichejev, and E. N. Kanarskaja, *Faserforsch. Textiltech.*, 4: 390 (1953).
18. H. K. Reimschuessel and K. Nagasubramanian, *Chem. Eng. Sci.*, 27: 1119 (1972).
19. S. K. Gupta, A. Kumar, P. Tandon, and C. D. Naik, *Polymer*, 22: 481 (1981).
20. S. K. Gupta, C. D. Naik, P. Tandon, and A. Kumar, *J. Appl. Polym. Sci.*, 26: 2153 (1981).
21. S. K. Gupta and A. Kumar, *Reaction Engineering of Step Growth Polymerization*, Plenum, New York (1987).
22. S. K. Gupta, A. Kumar, and K. K. Agrawal, *J. Appl. Polym. Sci.*, 27: 3089 (1982).
23. A. Ramagopal, A. Kumar, and S. K. Gupta, *Polym. Eng. Sci.*, 22: 849 (1982).
24. S. K. Gupta and A. Kumar, *Chem. Eng. Commun.*, 20: 1 (1983).
25. K. Tai, Y. Arai, H. Teranishi, and T. Tagawa, J. Appl. Polym. Sci., 25: 1789 (1980).
26. A. K. Ray and S. K. Gupta, *Polym. Eng. Sci.*, 26: 1033 (1986).
27. Allied Chem. Corp., Brit. Pat. 938,652 (1963).
28. S. C. Chu and I. C. Twilley, 6th Annual Synthetic Fibers Symposium, AIChE, Virginia (1969).
29. K. Nagasubramanian and H. K. Reimschuessel, *J. Appl. Polym. Sci.*, 16: 929 (1972).
30. O. Levenspiel, *Chemical Reaction Engineering*, 2nd ed., Wiley, New York, Chapter 13 (1972).
31. J. J. Carberry, *Chemical and Catalytic Reaction Engineering*, McGraw-Hill, New York (1976).
32. K. Tai, Y. Arai, and T. Tagawa, *J. Appl. Polym. Sci.*, 27: 731 (1982).
33. A. Kumar and S. K. Gupta, *Fundamentals of Polymer Science and Engineering*, Tata McGraw Hill, New Delhi, India (1978).
34. H. Jacobs and C. Schweigman, "Mathematical Model for the Polymerization of Caprolactam to Nylon-6," Proc. 5th Eur./2nd Intl. Symp. Chem. Rxn. Eng., Amsterdam, pp. B7.1–26 (1972).
35. S. K. Gupta, D. Kunzru, A. Kumar, and K. K. Agrawal, *J. Appl. Polym. Sci.*, 28: 1625 (1983).
36. Vereinigte Glanzstoff Fabriken, Germ. Pat. 1,167,021 (1962).
37. O. Fukumoto, *J. Polym. Sci.*, 22: 263 (1956).
38. A. Gupta and K. S. Gandhi, in *Frontiers of Chem. Rxn. Eng.* (L. K. Doraiswamy and R. A. Mashelkar, eds.), Wiley Eastern, New Delhi, India, pp. 667–681 (1984).
39. A. Gupta and K. S. Gandhi, *Ind. Eng. Chem. Prod. Res. Dev.*, 24: 327 (1985).
40. *Encyclopedia of Industrial Chemical Analysis*, Vol. 8, Wiley, New York, p. 115 (1971).
41. *International Critical Tables*, Vol. 3, McGraw-Hill, New York, p. 233 (1928).
42. C. Giori and B. T. Hayes, *J. Polym. Sci. A-1*, 8: 335 (1970).
43. J. P. Roos, *Adv. Chem. Ser. 133*: 303 (1974).
44. M. Sittig, *Polyamide Fiber Manufacture*, Noyes Data Corp., Park Ridge, NJ (1972).

45. S. K. Gupta and M. Tjahjadi, *J. Appl. Polym. Sci.*, 33:933 (1987).
46. O. Levenspiel, *Engineering Flow and Heat Exchange*, Plenum, New York (1984).
47. K. Tai, Y. Arai, and T. Tagawa, *J. Appl. Polym. Sci.*, 28: 2527 (1983).
48. A. Gupta, S. K. Gupta, K. S. Gandhi, B. V. Ankleswaria, M. H. Mehta, M. R. Padh, and A. V. Soni, "Simulation of a Semibatch Nylon 6 Reactor," Proc. 2nd Intl. Chem. Rxn. Eng. Conf., Pune, India (1987).
49. M. V. Tirrell, G. H. Pearson, R. A. Weiss, and R. L. Laurence, *Polym. Eng. Sci.*, 15: 386 (1975).
50. W. F. H. Naudin ten Cate, "Application of the Maximum Principle of Pontryagin to Optimize a Nylon 6 Continuous Polymerization Process," Proceedings of the International Congress on the Use of Electronic Computers in Chemical Engineering, Paris (1973).
51. S. Mochizuki and N. Ito, *Chem. Eng. Sci.*, 33: 1401 (1978).
52. P. J. Hoftyzer, J. Hoogschagen, and D. W. van Krevelen, "Optimization of Caprolactam Polymerization," Proceedings of the 3rd European Symposium on Chemical Reaction Engineering, Amsterdam, pp. 247–253 (1964).
53. A. Ramagopal, A. Kumar, and S. K. Gupta, *J. Appl. Polym. Sci.*, 28: 2261 (1983).
54. S. K. Gupta, B. S. Damania, and A. Kumar, *J. Appl. Polym. Sci.*, 29: 2177 (1984).
55. A. K. Ray and S. K. Gupta, *J. Appl. Polym. Sci.*, 30: 4529 (1985).
56. A. Mochizuki and N. Ito, *Chem. Eng. Sci.*, 28: 1139 (1973).
57. J. A. Semlyen and G. R. Walker, *Polymer*, 10: 597 (1969).
58. J. M. Andrews, F. R. Jones, and J. A. Semlyen, *Polymer*, 15: 420 (1974).
59. H. Spoor and H. Zahn, *Z. Anal. Chem.*, 168: 190 (1959).
60. H. Zahn and G. B. Gleitsman, *Agnew. Chem.*, 75: 772 (1963).
61. M. Rothe, *J. Polym. Sci.*, 30: 227 (1958).
62. M. Mutter, U. W. Suter, and P. J. Flory, *J. Am. Chem. Soc.*, 98: 5745 (1976).
63. J. Jacobson and W. H. Stockmayer, *J. Chem. Phys.*, 18: 1600 (1950).
64. J. A. Semlyen, *Adv. Polym. Sci.*, 22: 41 (1976).
65. P. J. Flory, U. W. Suter, and M. Mutter, *J. Am. Chem. Soc.*, 98: 5733 (1976).
66. P. J. Flory, *Statistical Mechanics of Chain Molecules*, Wiley, New York (1969).
67. D. C. Jones and T. R. White, in *Step Growth Polymerization* (D. H. Solomon, ed.), Marcel Dekker, New York, pp. 41–94 (1972).
68. O. E. Snider and R. J. Richardson, in *Encyclopedia of Polymer Science and Technology*, Vol. 10 (H. F. Mark and N. G. Gaylord, eds.), Wiley, New York, pp. 347–460 (1969).
69. E. C. Schule, in *Encyclopedia of Polymer Science and Technology*, Vol. 10 (H. F. Mark and N. G. Gaylord, eds.), Wiley, New York, pp. 460–482 (1969).
70. W. Sweeny and J. Zimmerman, in *Encyclopedia of Polymer Science and Technology*, Vol. 10 (H. F. Mark and N. G. Gaylord, eds.), Wiley, New York, pp. 483–597 (1969).
71. N. Ogata, *Makromol. Chem.*, 42: 52 (1960).
72. N. Ogata, *Makromol. Chem.*, 43: 117 (1961).
73. A. Kumar, S. Kuruville, A. R. Raman, and S. K. Gupta, *Polymer*, 22: 387 (1981).
74. A. Kumar, R. K. Agarwal, and S. K. Gupta, *J. Appl. Polym. Sci.*, 27: 1759 (1982).
75. F. C. Chen, R. G. Griskey, and G. H. Beyer, *AIChEJ*, 15: 680 (1969).

7

The Radical Polymerization of Vinyl Ethers

Mikiharu Kamachi
*Osaka University,
Toyonaka, Osaka, Japan*

INTRODUCTION	249
RADICAL REACTIVITY OF VINYL ETHERS	250
Radical Reactivity of Vinyl Ethers Toward Carbon-Centered Radicals	250
Radical Reactivity of Vinyl Ethers Toward Oxy and Phosphonyl Radicals	251
Ab Initio Calculation	252
RADICAL POLYMERIZATION OF VINYL ETHERS	253
Homopolymerization	253
Structure of Propagating Radical	255
Copolymerization	257
RADICAL POLYMERIZATION OF DIVINYL ETHERS	260
Cyclic Polymerization of Divinyl Ethers	260
Cyclic Polymerization of Divinyl Acetals	262
Other Divinyloxy Compounds	265
Copolymerization	265
REFERENCES	267

INTRODUCTION

Vinyl monomers such as styrene and methyl acrylate undergo radical polymerization to yield high polymers. However, vinyl ethers are allowed to polymerize very slowly to low molecular weight product by radical initiators such as 2,2'-azobisisobutyronitrile (AIBN) and benzoylperoxide. Therefore, radical polymerizations of vinyl ethers are of very little commercial importance, and cationic initiators are used for the preparation of high molecular weight polymers [1, p. 181]. Although vinyl ethers are not homopolymerized readily by radical initiators, copolymerizations proceed well, especially with vinyl monomers that have electron-accepting substituents to give high polymers [2].

Radical cyclopolymerizations of divinyl ethers have been known to give high polymers with the unsaturated monocyclic unit and the bicyclic unit. Similarly, divinyl formal and divinyl acetal can be polymerized by radical initiators to high polymers which contain 1,3-dioxolane rings as the predominant structural unit in the main chain.

In this chapter, the radical polymerization of vinyl ethers and their homologs is reviewed along with basic data on the radical addition reactions of vinyl ethers.

RADICAL REACTIVITY OF VINYL ETHERS

Radical Reactivity of Vinyl Ethers Toward Carbon-Centered Radicals

No absolute rate constants for radical additions to vinyl ethers have been determined thus far. However, the radical reactivity of vinyl ethers can be qualitatively discussed by the relative rate constants for the radical addition to their double bonds. Radical additions of methyl radical to double bonds of vinyl compounds [3, 4] and the monomer reactivity ratios r_1 and r_2 in their copolymerizations [1, p. 453] are usually used for the estimation of the radical reactivity of vinyl monomers.

The addition of methyl radical, which was produced by thermal decomposition of acetyl peroxide, to unsaturated compounds, A, was studied in isooctane by Szwarc et al. [3, 4].

$$CH_3COOOOCCH_3 \rightarrow 2CH_3COO \cdot \rightarrow 2CH_3 \cdot + 2CO_2 \tag{1}$$

$$CH_3 \cdot + A \xrightarrow{k_{add}} CH_3A \cdot \tag{2}$$

Since hydrogen abstraction of methyl radical from isooctane takes place competitively with the addition reaction,

$$CH_3 \cdot + C_8H_{18} \xrightarrow{k_{abs}} CH_4 + C_8H_{17} \cdot \tag{3}$$

the ratio k_{add}/k_{abs}, which is called methyl affinity [3, 4], is estimated by measurements of CH_4 and CO_2 produced in the presence and absence of vinyl compounds.

The methyl affinity for ethyl vinyl ether is shown in Table 1 along with those of other vinyl monomers. This result shows that vinyl ethers are less reactive toward methyl radical than ethylene. This finding is supported by molecular reactivity ratios (r_1 and r_2) in the copolymerization of ethylene (M_1) and ethyl vinyl ether (M_2); the molecular

Table 1 Relative Reactivities of Various Monomers Toward Methyl Radical and *ter*-Butoxyl Radical

Monomer	$CH_3 \cdot$	$(CH_3)_3O \cdot$
Vinyl ether	6[a]	9.6[b]
Ethylene	26	
Vinyl acetate	37	0.28
Methyl acrylate	800	1.54
Methyl methacrylate	1420	1.73
Acrylonitrile	1540	0.52
Styrene	796	29.8

[a]Ethyl vinyl ether.
[b]Isobutyl vinyl ether.

reactivity ratios for the copolymerization of ethyl vinyl ether with ethylene toward polyethylene radical have been reported to be $r_1 = 2.7\text{--}5.4^6$ and $r_2 = 0^6$:

$$\text{---}CH_2CH_2\cdot + CH_2{=}CH_2 \xrightarrow{k_{11}} \text{---}CH_2CH_2\cdot \quad (4)$$

$$\text{---}CH_2CH_2\cdot + CH_2{=}CH(OC_2H_5) \xrightarrow{k_{12}} \text{---}CH_2\dot{C}H(OC_2H_5) \quad (5)$$

$$\text{---}CH_2\dot{C}H(OC_2H_5) + CH_2{=}CH_2 \xrightarrow{k_{21}} \text{---}CH_2CH_2\cdot \quad (6)$$

$$\text{---}CH_2\dot{C}H(OC_2H_5) + CH_2{=}CH(OC_2H_5) \xrightarrow{k_{22}} \text{---}CH_2\dot{C}H(OC_2H_5) \quad (7)$$

where $r_1 = k_{11}/k_{12}$ and $r_2 = k_{22}/k_{21}$.

Since $1/r_1 < 1$, we can conclude that the reactivity of ethyl vinyl ether with polyethylene radicals is less than that of ethylene.

Radical Reactivity of Vinyl Ethers Toward Oxy and Phosphonyl Radicals

An ESR study of the reactivity of vinyl monomers toward *tert*-butoxyl radical was performed by Satoh and Otsu [6] using the spin trapping method. When di-*tert*-butyl peroxalate-initiated polymerization of vinyl monomers was carried out in the presence of xylene (X) and 2-methyl-2-nitrosopropane (BNO) (the former is an internal standard and the latter is a spin-trapping reagent), the following reaction took place competitively:

$$((H_3C)_3COOCO)_2 \rightarrow 2CO_3 + 2(H_3C)_3CO\cdot \quad (8)$$
(DBPOX)

$$(H_3C)_3CO\cdot + CH_2{=}CH(X) \xrightarrow{k_1} (H_3C)_3CO\text{---}CH_2\text{---}\dot{C}H(X) \quad (9)$$
I

$$(H_3C)_3CO\cdot + H_3C\text{---}C_6H_4\text{---}CH_3 \xrightarrow{k_2} (H_3C)_3COH + CH_3\text{---}C_6H_4\text{---}\dot{C}H_2 \quad (10)$$
II

$$\text{I} + (CH_3)_3CNO \xrightarrow{\text{fast}} (H_3C)_3CO\text{---}CH_2\text{---}CH(X)\text{---}N(C(CH_3)_3)\text{---}O\cdot \quad (11)$$
(BNO)
III

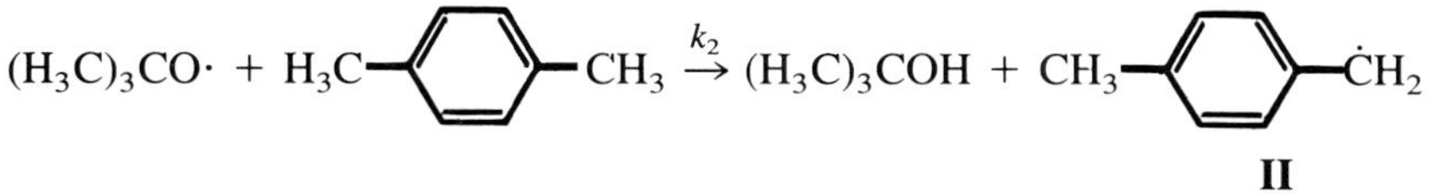

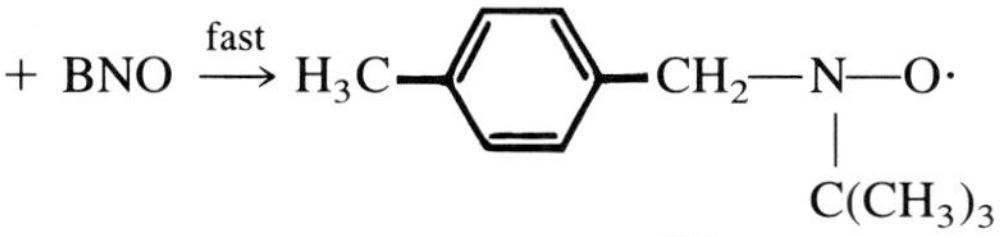

$$\text{II} + BNO \xrightarrow{\text{fast}} H_3C\text{---}C_6H_4\text{---}CH_2\text{---}N(C(CH_3)_3)\text{---}O\cdot \quad (12)$$
IV

Ratios of rate constants for the addition reaction of *tert*-butoxyl radical with vinyl monomers to that for hydrogen abstraction from xylene, k_1/k_2, were estimated from the relation

$$\frac{[\text{III}]}{[\text{IV}]} = \frac{k_1[M]}{k_2[X]} \quad (13)$$

Table 2 Bimolecular Rate Constants of Phosphonyl Radical with Various Monomers in Cyclohexane at Room Temperature

Monomer	$k_{R\cdot+M}$ (L mol^{-1} sec^{-1})	
	$O=\dot{P}(C_6H_5)_2$	$O=\dot{P}(OC_2H_5)_2$
Butyl vinyl ether	4.0×10^6	1.4×10^6
Vinyl acetate	1.6×10^6	1.8×10^6
Methyl acrylate	3.5×10^7	1.6×10^7
Acrylonitrile	2.0×10^7	2.6×10^7
Methyl methacrylate	8.0×10^7	5.3×10^7
Styrene	6.0×10^7	2.5×10^8

Results are shown in Table 1. The order of reactivity of vinyl ethers with the *tert*-butoxyl radical differs from that with methyl radical: vinyl ethers is more reactive toward *tert*-butoxyl radical than MMA and acrylonitrile. This result shows that the polar effect plays a more important role in the addition reaction of *tert*-butoxyl radical.

Phosphonyl radicals of the structure $O=\dot{P}R_1(R_2)$ [R_1, R_2 = C_6H_5, C_6H_5; C_6H_5, $OCH(CH_3)_2$; CH_3, OCH_3; OCH_3, OCH_3; and OC_2H_5, OC_2H_5] were generated by UV photolysis of appropriate acylphosphine oxide and acylphosphonates [7]:

$$O=\dot{P}R_1(R_2) + CH_2=CHX \xrightarrow{k_{R\cdot+M}} O=P\underset{R_2}{\overset{R_1}{|}}-CH_2-CHX\cdot \qquad (14)$$

The rate constants, as shown in Table 2, for the addition of these radicals to vinyl compounds were determined by laser photolysis techniques [8]. Comparison of the rate constants shows that the reactivity of *tert*-butyl vinyl ether toward phosphonyl radicals is 10–100 times less than those of styrene and methyl methacrylate, and similar to that of vinyl acetate. Sumiyoshi and Schnabel [8] explained the difference in reactivity between vinyl monomers on the basis of the Q and e values of the monomers.

Ab Initio Calculation

It was recently reported that the reaction paths for chemical reactions of small organic molecules can be energetically outlined [9] and, in some cases, predicted by the ab initio molecular orbital method combined with the energy gradient technique [10]. For example, the activation energy and the heat of reaction for the addition reaction of methyl radical to ethylene were estimated to be 6.7 and 22.7 kcal mol^{-1}, respectively [11], being in satisfactory agreement with the corresponding experimental data of 7.7 and 25 kcal mol^{-1} [12]. This agreement indicates that the ab initio molecular orbital calculation is useful for an understanding of the free radical addition reactions. Fueno and Kamachi [13] applied this method to the addition reaction of methyl radical to vinyl compounds and estimated the minimum-energy path for the reactions of the resulting radicals. The activation energy and heat of reaction were estimated for several vinyl compounds; the results are shown in Table 3 along with those for ethylene. Greater heat of reaction and smaller activation energy were estimated in the addition reaction of the methyl radical to acrylonitrile and

Table 3 Activation Energy and Heat of Reaction in Addition Reactions of Methyl Radical to Vinyl Compounds[a]

Monomers	$E\ddagger$ (kcal mol^{-1})	ΔH (kcal mol^{-1})
Methyl vinyl ether	10.1	−21.9
Propylene	7.3	−23.9
Ethylene	6.7[b]	−24.4[c]
Acrylic acid	3.6	−29.9
Acrylonitrile	1.6	−34.6

[a]UHF SCF (3–21 G basis).
[b]Experimental $E\ddagger = 7.9$ kcal mol^{-1}.
[c]Experimental $\Delta H = -25.5$ kcal mol^{-1}.

methacrylic acid as compared with ethylene, which indicates that cyano and carbonyl groups stabilize the radical center through resonance stabilization. On the other hand, lower heat of reaction and larger activation energy were estimated in methyl vinyl ether as compared to ethylene, which predicts that the methoxy group leads to a destabilization of the radical end of the transition state. Comparison of the activation energy shows that the reactivity of the vinyl compounds toward methyl radical decrease in the following order: acrylonitrile > methacrylic acid > ethylene > propylene > methyl vinyl ether. This order agrees in trend with that in methyl affinity (Table 1). Since no methyl affinity for methyl vinyl ether has been reported, that for ethyl vinyl ether [3] was used instead of that for methyl vinyl ether. Geometries of the transition state and the resulting radicals were anticipated by the SCF procedure (3–21 G basis) combined with the energy gradient technique. The optimized geometries of radicals obtained by the addition reaction of the methyl radical to acrylonitrile, methacrylic acid, ethylene, and methyl vinyl ether are shown in Figure 1. These results show that the radicals obtained from acrylonitrile and ethylene are composed of the sp^2 hybrid structure while methyl vinyl ether gives radicals deviating from this hybrid structure. The theoretical information on the geometry of these radicals is consistent with the structure anticipated by hyperfine splitting constants due to α-proton in the ESR spectrum of the propagating radicals [14] (see section on Structure of Propagating Radical).

RADICAL POLYMERIZATION OF VINYL ETHERS

Homopolymerization

It has long been thought that monovinyl ethers are not allowed to polymerize to high molecular weight polymers by free radical initiators. However, no systematic study on the radical polymerization of monovinyl ethers has been performed. Matsumoto et al. [15] performed a systematic study on the radical polymerization of butyl vinyl ether (BVE) and found that BVE undergoes radical polymerization to yield an oligomer using a considerable amount of radical initiator as required for the polymerization of allylic compounds and dialkyl fumarates. The polymerization of BVE was effectively inhibited by radical scavengers such as hydroquinone and 4-hydroxy-2,2,6,6-tetramethylpiperidine-1-oxyl.

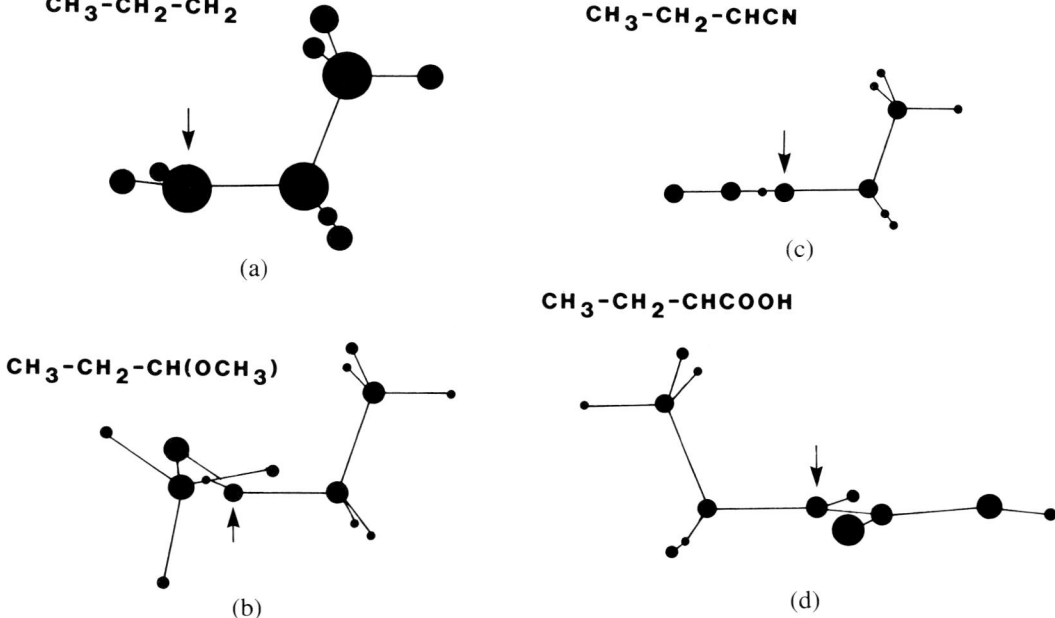

Figure 1 The optimized geometries of model radicals of propagating ends. (*a*) Ethylene, (*b*) methyl vinyl ether, (*c*) acrylonitrile, and (*d*) methacrylic acid.

Gel permeation chromatography (GPC) measurements showed that the molecular weight of the radically obtained samples lies in the oligomer range. The ^{1}H NMR spectrum of the oligomer was identical to that of high molecular weight poly(BVE) prepared by $BF_3 \cdot O(C_2H_5)_2$. These results clearly demonstrate that BVE can be polymerized radically to give an oligomer. Table 4 shows typical polymerization results of BVE compared with those of allyl acetate (AAc) [16] and diethyl fumarate (DEF) [17]; the polymerizability of BVE is between those of AAc and DEF. The polymerization behavior of BVE is similar to that of allylic compounds which have 10^2–10^3 times larger monomer transfer constants

Table 4 Results of the Bulk Polymerizations of BVE, AAc, and DEF Initiated with 0.1 mol L^{-1} of Radical Initiators (polymerization time = 5 hr)

Monomer	Initiator[a]	Temperature (°C)	Conversion (%)	$\bar{P}_n$	$\bar{P}_w/\bar{P}_n$
BVE	AIBN	70	7.1		
AAc	AIBN	70	3.6		
DEF	AIBN	70	28.1		
BVE	DTBPO	110	18.8		
AAc	DTBPO	110	6.0		
DEF	DTBPO	110	74.5		

[a] AIBN, 2,2'-azobisisobutyronitrile; DTBPO, di-*tert*-butylperoxide.

than vinyl acetate, styrene, and alkyl acrylate. This suggests that a chain transfer reaction is more likely to occur in vinyl ethers than in other polymerizable vinyl monomers such as vinyl acetate and methyl acrylate.

Polymer formation has been reported by the radical polymerization of vinyloxytrimethylgermane, although the molecular weight of polymers obtained was not mentioned [18]. The Q and e values of vinyloxytrimethylsilane have been determined by the copolymerization of this monomer with vinyl monomers such as acrylonitrile, vinyl acetate, and MMA, being similar to those of isobutyl vinyl ether [19]. Accordingly, polymer is possibly formed by the radical polymerization of vinyloxytrimethylsilane.

Structure of Propagating Radical

Kamachi et al. [14] obtained ESR spectra of the propagating radicals in the radical polymerization of several vinyl monomers by ESR spectrometry with a TM_{110} mode cavity improved for photoreaction [20, 21]. Figure 2a is the spectrum taken during UV

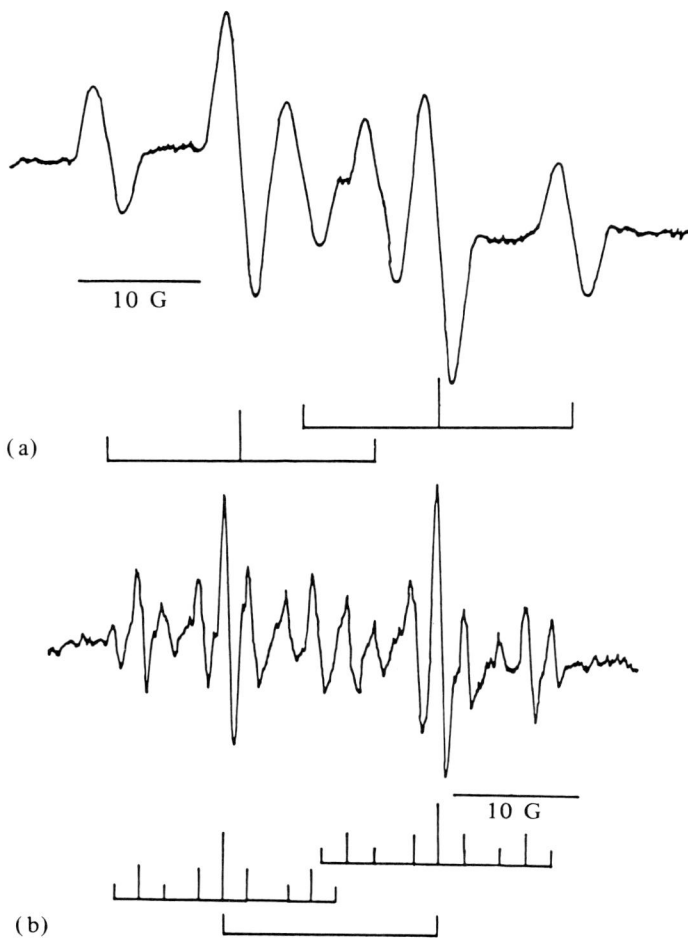

Figure 2 ESR spectra for propagating radicals of *tert*-butyl vinyl ether and isobutyl vinyl ether in their bulk polymerizations at room temperature. [DTBPO] = 1.3 M.

irradiation for *tert*-butyl vinyl ether. The well-resolved six-line spectrum can be assigned to the propagating radical —CH$_2$ĊH[OC(CH$_3$)$_3$], in which the hyperfine splitting constant is 16.1 G for one proton and 10.4 G for two methylene protons. The well-resolved spectrum seen in Figure 2b was obtained in the radical polymerization of isobutyl vinyl ether. This is more complicated than that for *tert*-butyl vinyl ether, reflecting a long-range interaction of the unpaired electron with the protons of the substituent of the compound. It is assignable to the propagating radical —CH$_2$ĊH[OCH$_2$CH(CH$_3$)$_2$], in which the hyperfine splitting constants are 16.8, 7.4, and 2.2 G for α-proton, two equivalent β-methylene protons, and two equivalent protons of the side group, respectively. The equal hyperfine

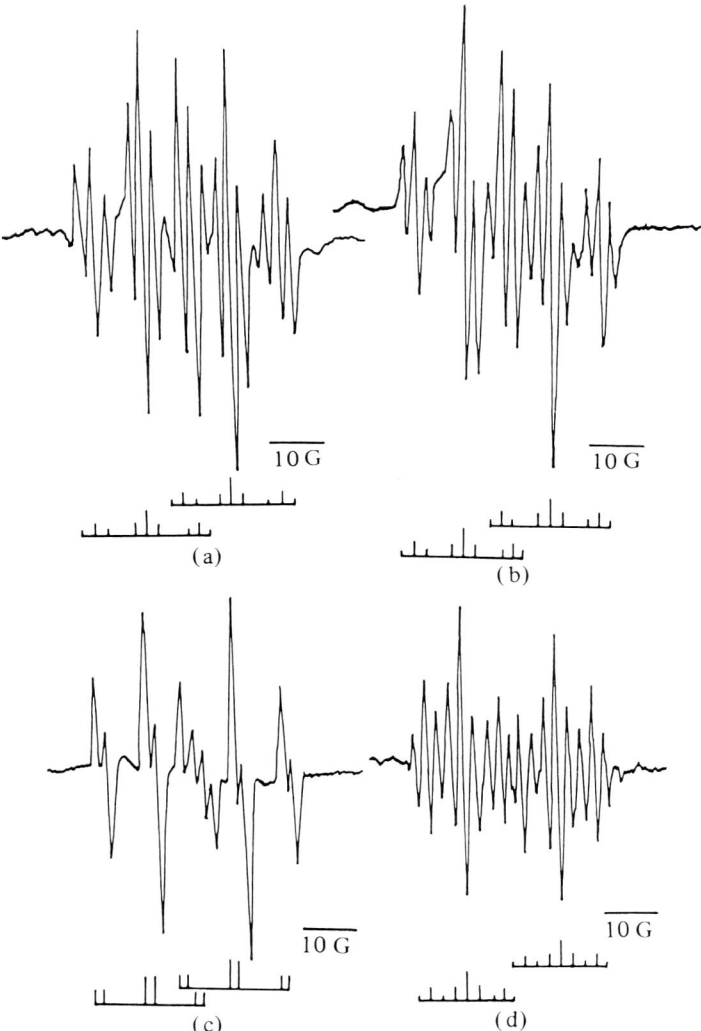

Figure 3 ESR spectra for propagating radicals of various vinyl ethers in their bulk polymerizations at −50°C. (*a*) Ethyl vinyl ether; (*b*) butyl vinyl ether; (*c*) isopropyl vinyl ether; and (*d*) isobutyl vinyl ether. [Di-*tert*-butylperoxide] = 0.87 M.

Table 5 g-Values and Hyperfine Splitting Constants of Propagating Radicals of Various Vinyl Ethers (—$CH_2CH\cdot$)
|
OR

R (substituents)	g	a (α-H) (G)	a (β-H) (G)	a (γ-H) (G)
—CH_2CH_3	2.003	16.1	9.5	2.2
—$CH_2CH_2CH_2CH_3$	2.003	16.1	9.5	2.2
—$CH_2CH(CH_3)_2$	2.003	16.8	7.4	2.1
—$CH(CH_3)_2$	2.003	15.7	10.7	1.4
—$C(CH_3)_3$	2.003	16.1	10.6	

splitting constants of the two β-methylene protons indicate that the propagating radicals of vinyl ethers rotate freely about the C_α-C_β bond, as in the case for vinyl acetate and methyl methacrylate. Kamachi et al. [14] found that similar ESR spectra were obtained by the reaction of *tert*-butoxyl radical with vinyl ethers at −50°C. These spectra are seen in Figure 3. They are more complicated than that from *tert*-butyl vinyl ether because of a long-range hyperfine interaction with protons of the substituent of these compounds. The splitting constants for the propagating radicals of vinyl ethers are presented in Table 5. The hyperfine splitting constants of α-protons of vinyl ethers were 4–6 G smaller than that of ethyl acrylate and vinyl acetate, which are 22.5 and 20.5 G, respectively [22].

The following two possibilities may account for the difference: (a) a decrease in the spin density of the unpaired electron [23]; and (b) an increase in the S character of the propagating radical [24, 25]. It has been shown by an ab initio molecular orbital calculation that the spin densities in vinyl ether radicals are higher than in the radical of methyl acrylate [26]. This rules out possibility a. Possibility b is consistent with the ab initio calculation which indicated the conformation of the model radical of the propagating end of methyl vinyl ether [$CH_3CH_2\dot{C}H(OCH_3)$] radical deviated from the planar sp^2 conformation (Fig. 1). The deviation of the propagating radical from the planar sp^2 conformation leads to σ-radical, which has higher reactivity than π-radical. The higher reactivity of the propagating radical increases the possibility of side reactions such as the chain transfer reaction as well as propagation. Since methyl affinity and ab initio calculation show that the reactivity of vinyl ethers toward carbon-centered radicals is lower than those of other vinyl compounds (see Table 1), the side reactions are considered more likely to take place in the radical polymerization of vinyl ethers than in conventional vinyl monomers such as methyl methacrylate, styrene, and vinyl acetate. This concept agrees with the experimental results that polymerization behavior of vinyl ether is similar to that of allylic compounds which have reactivities 10^2–10^3 times larger than those of styrene, vinyl acetate, and methyl methacrylate [16]. Accordingly, the reason why high polymers are not obtained from vinyl ethers is possibly ascribable to such side reactions.

Copolymerization

The copolymerization of vinyl ethers with other vinyl monomers is of practical interest. The Q and e values of vinyl ethers [27] and vinyloxytrimethyl silane [19] are listed in Table 6. The e value indicates that these compounds have electron-donating olefinic bonds. Accordingly, cationic polymerization is less desirable because of the high ten-

Table 6 The Q and e Values for Vinyl Ethers (CH_2=CHOR)

Monomers	Q	e
Ethyl	0.032	−1.17
Propyl	0.014	−1.52
Butyl	0.087	−1.20
Isobutyl	0.015–0.018	−1.59 to −2.11
tert-Butyl	0.087–0.15	−1.55 to −1.58
Octyl	0.061	−0.79
Trimethylsilyl	0.018	−2.39

dency of the vinyl ethers to homopolymerize. Fortunately, radical copolymerization can be used for the preferred method of preparing the industrially important copolymers. It is especially suitable for copolymerization with monomers having electron-withdrawing groups, such as the acrylates and methacrylates, maleic anhydride, and vinyl chloride [28]. Solution, emulsion, suspension, or bulk polymerization technique can be used, depending on the nature of the reactants. Reactivity ratios are known for some comonomer systems; those of commercial interest are summarized in Table 7 where M_1 and M_2 are comonomer and alkyl vinyl ether, respectively [28]. The r_2 values for the alkyl vinyl ethers approach zero, suggesting that the reactivity of alkyl vinyl ethers toward their propagating radicals is less than those of other monomers. To explain copolymer forma-

Table 7 Reactivity Ratios for Radical Polymerization of Vinyl Ethers (ROCH=CH_2) with Other Vinyl Monomers

Comonomer (M_1)	R	r_1	r_2
Vinyl acetate	Ethyl	3.0	0
Acrylonitrile		5.0	0
Methyl acrylate		3.3	0
Ethylene		2.7–5.4	0
Maleic anhydride	Butyl	0.045	0
Methyl acrylate		3.6	0
Methyl methacrylate		1.6	0.2
Styrene		15	0
Acrylonitrile	Isobutyl	0.85	0.02
Methacrylonitrile		3.88	0
Vinyl acetate		2.54	0
Methyl methacrylate		12.9	0.25
Acrylonitrile	Chloromethyl	1.0	0
Methyl acrylate		5.0	0.15
Acrylonitrile	Trimethylsilyl	0.53	0
Methacrylonitrile		4.35	0
Vinyl acetate		2.72	0.20
Methyl methacrylate		11.6	0

tion, charge transfer complexes were pointed out between alkyl vinyl ether and vinyl monomers having electron-withdrawing groups [29–35]. Vinyl ethers are alternately copolymerized with N-substituted maleimides [36]. By comparison of the ^{13}C NMR chemical shifts of the model compounds with those of the copolymers, Olson and Butler [36–38] showed that the predominant structure at the succinimide unit is erythro in the copolymer. Copolymer conditions favoring the formation of the complex invariably gave higher erythro contents in the resulting copolymers [36–38]. The results were interpreted as indicating that the stereoregularity of the succinimide units is dependent on the fraction of maleimide monomer in complex form. The copolymerization of vinyl ether with maleic anhydride provides additional evidence that charge-transfer complexes form between vinyl ethers and maleic anhydride and that these complexes participate in the copolymerization [39–42].

In the copolymerization of butyl vinyl ether (M_1) and methyl methacrylate (M_2), the values of r_2 in the absence of aromatic hydrocarbon and in the presence of naphthalene, anthracene, and chrysene were 11.7 ± 0.5, 3.8 ± 0.2, 1.1 ± 0.1, and 2.2 ± 0.1, respectively, while r_1's were zero in all cases [43, 44]. No aromatic moieties were found in these copolymers. Further investigation on the influence of aromatic additives indicated that the presence of aromatic molecules lowered the termination rate constants more than the propagation rate constants, thereby increasing the overall rates of polymerization [45]. The effect of the aromatic additives was ascribed to the formation of charge transfer complexes between the growing macroradical and the aromatic hydrocarbons. Since only a small aromatic solvent effect on the radical polymerization of MMA has been found [46–48], these results suggest that the presence of aromatic molecules lowered the reactivity of radicals terminated with vinyl ether units.

Organometallic compounds were used as additives for improving the composition ratio of the copolymer. In the case of copolymerization of butyl vinyl ether with 4-vinyl pyridine or 2-methyl-5-vinyl pyridine [49], the reaction rate and the content of vinyl ether units increased with addition of $Zn(OCOCH_3)_2$. Similarly, the addition of triisobutylaluminum to the copolymerization system of methyl methacrylate and butyl vinyl ether increased the content of the latter [50]. The influence of the additives on the copolymer composition can be explained by the change in polarization and stabilization of double bond as the result of complexation of monomers.

Schmitt and Schuerch [51] carried out the copolymerization of optically active 1-α-methylbenzyl vinyl ether with maleic anhydride with benzoyl peroxide in ethyl acetate to obtain a levorotatory copolymer. After removal of the optically active group by hydrolysis, a dextrorotatory copolymer was produced. This result shows that asymmetric induction took place in the polymer chain. Similar asymmetric induction has been observed in the radical copolymerization of optically active 1-methyl vinyl ethers with N-phenylmaleimide [52], N-(p-tolyl)-maleimide [53], MAn [54], dimethyl maleate [54], dimethyl fumarate [54], acenaphthylene [55], vinylene carbonate [56], and indene [56]. Two concepts have been proposed on the origin of asymmetric induction in the copolymers of 1-methyl vinyl ether and α,β-disubstituted monomers: (a) stereoregular propagation of charge transfer complexes and (b) preferential attack of α,β-disubstituted monomer on the propagating end of chiral 1-methyl vinyl ether from the one direction of the smallest steric hindrance. Stereoselective radical copolymerization was reported in the copolymerization of racemic α-methylbenzyl vinyl ether with (S)-(−)- or (R)-(+)-N-(α-methylbenzyl)-maleimide using AIBN [57]. In this case, the optical activity of the nonpolymerized α-methylbenzyl vinyl ether was of opposite sign to that of the optically active comonomer, clearly indicating that the copolymerization is stereoselective. Optically active copolymers were also obtained in the radical copolymerization of isobutyl vinyl ether with maleic anhydride in 1-menthol [58]. The specific rotation of the copolymers decreased with an increase in the reaction temperature. This asymmetric induction might be caused by the hydrogen-bonding solvation of the polar group of a vinyl monomer in the ground state and the transition state of a propagation step.

Recently, monomethoxy polyethylene glycol vinyl ether macromonomers [Me(EG)$_n$ VE, n = 1–39] were synthesized and copolymerized with MAn to give alternating copolymers [59]. The copolymerization rates significantly depended on the ethylene glycol chain length of the Me(EG)$_n$VE, resulting in a relatively lower polymerization rate of higher molecular weight macromonomer. The use of nonelectron donor solvent such as carbon tetrachloride led to higher yield and higher molecular weight. The origin of the solvent effect still remains unsolved.

Terpolymers and quaterpolymers were prepared by radical polymerizations from points of their industrial applications such as latexes [28].

RADICAL POLYMERIZATION OF DIVINYL ETHERS

Cyclic Polymerization of Divinyl Ethers

Divinyl ether undergoes radical polymerization to produce soluble polymers with highly cyclized structure [60]. The polymerization results are summarized in Table 8 [61]. It has been reported that gelation occurred frequently at higher polymerization temperature or at greater conversion.

Many kinds of structural units are conceivable for the polymer, as seen in Scheme 1. Aso et al. [62] proposed that the polymer contained bicyclic structures S_{55}, S_{56}, and unsaturated structure S_0 on the basis of the telomerization study, and Guanta et al. [63] suggested the presence of the monocyclic structures S_{50} and S_{60} as well as the bicyclic structures S_{55} and S_{56} from kinetic and stereochemical considerations. Based on ^{13}C NMR spectra of this polymer and its model compounds, Tsukino and Kunitake [64] showed that among many conceivable structural units seen in Scheme 1, the polymer contained an unsaturated unit S_{50} and a bicyclic unit S_{55} in the 1:1 ratio. This structure was confirmed by comparison of the ^{13}C NMR spectrum of this polymer with that of the polymer whose pendant vinyloxy groups were completely removed by hydrolysis.

Table 8 Selected Examples of Radical Polymerization of Divinyl Ethers[a]

AIBN[b] (mol %)	Temperature (°C)	Time (hr)	Conversion (%)	PDB[c] (%)	M_n
0.45	70	6	28	28	61,500
0.50	50	44	24	26	32,000
1.00	60	7	23	24	19,000

[a]20 wt% in benzene.
[b]Mol % of monomer.
[c]Content of the pendant double bond. PDB is 100% when each monomer unit contains one vinyl group.

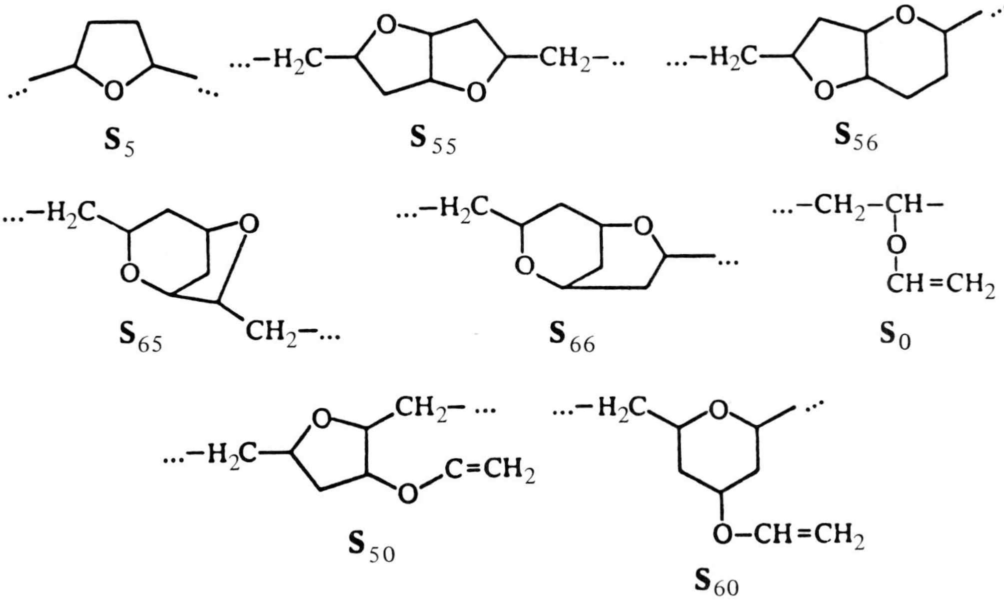

[Scheme 1 structures: S_5, S_{55}, S_{56}, S_{65}, S_{66}, S_0, S_{50}, S_{60}]

Scheme 1

The preceding analysis of the ^{13}C NMR spectrum indicates that the polymer chains involve approximately equimolar amounts of the monocyclic unit (S_{50}-1 and/or S_{50}-2) and the bicyclic unit (S_{55}-1, S_{55}-7, and/or S_{55}-8).

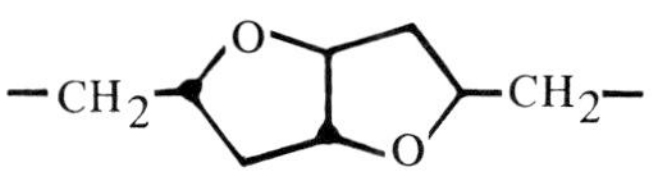

Monocyclic unit:

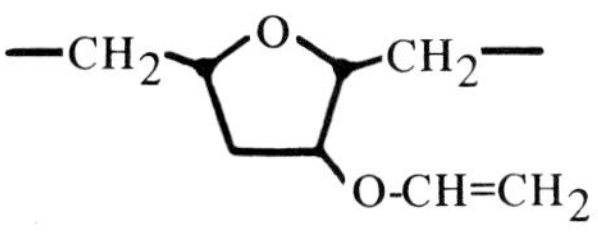

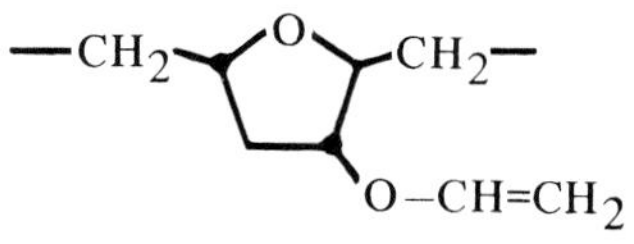

[Structures S_{50}-1 and S_{50}-2]

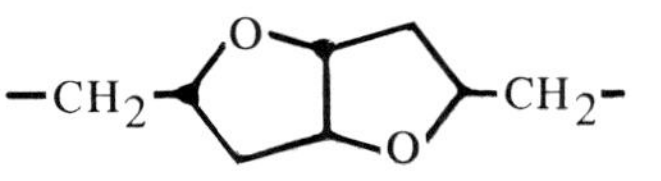

Bicyclic unit:

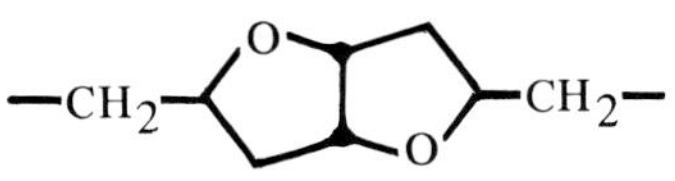

[Structures S_{55}-1, S_{55}-7, and S_{55}-8]

The ^{13}C NMR carbon chemical shifts expected for all the possible stereoisomers of these structural units were estimated using a number of model compounds. Consequently, the final spectral assignment shows that the monocyclic unit and the bicyclic unit are S_{50}-1 and S_{55}-7, respectively. The polymer structure is

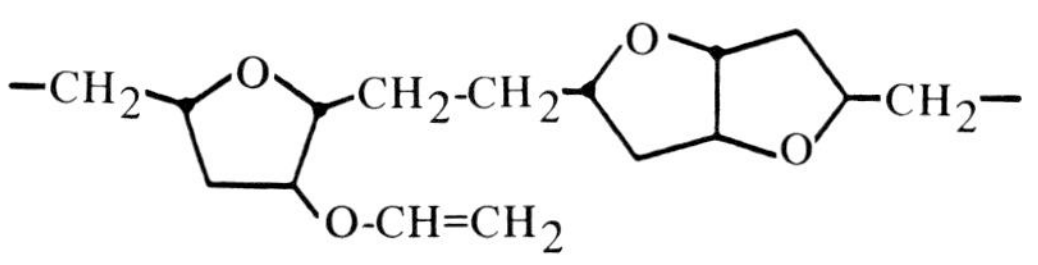

Similar stereoregular polymers were obtained by the radical polymerization of *cis*-propenyl vinyl ether and 2-methylpropenyl vinyl ether.

Scheme 2 summarizes the cyclopolymerization process of divinyl ether and its homologs, including stereochemistry. In this process, monomers always react with growing radicals at the unsubstituted side to form radicals **1**. The head-to-tail addition of monomer to the uncyclized radical **1** gives radical **2**, which cyclizes stereoselectively to monocyclic radical **3**. This radical either adds to monomer or cyclizes stereoregularly to form a bicyclic radical.

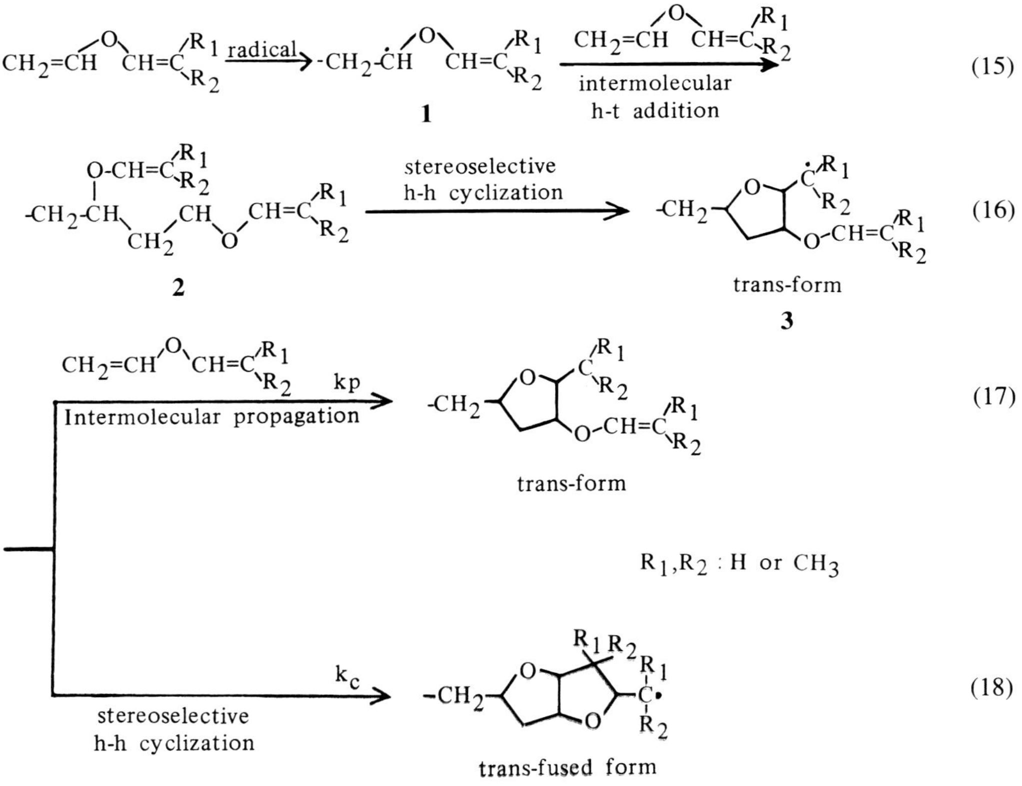

R_1, R_2 : H or CH_3

Scheme 2

Cyclic Polymerization of Divinyl Acetals

Divinyl formal (DVF), acetoaldehyde divinyl acetals (CH_3-DVA), and acetone divinyl acetal (2CH_3-DVA) also readily undergo radical polymerizations to produce soluble polymers [65–71].

```
CH₂=CH  CH=CH₂        CH₂=C H  CH=CH₂      CH₂=C H  CH=CH₂
    \   /                 \   /                 \   /
     O O                   O O                   O O
     \ /                   \ /                   \ /
     CH₂                    CH                    C
                            |                    / \
                           CH₃                 CH₃  CH₃

     DVF                 CH₃-DVA              2CH₃-DVA
```

Results of the polymerizations are listed in Table 9. The determination of the polymer structures was performed by chemical means (hydrolysis and the subsequent determination of the 1,2-glycol content), and by IR and ^{1}H NMR spectroscopies [67–71]. These data showed that the polymers involved the cyclized structure. However, the conclusion was not necessarily definitive.

The polymer structures were determined by the detailed analysis of ^{13}C NMR spectra of these polymers [72, 73]. Comparison of the ^{13}C NMR spectra of the polymers with those of model compounds of the cyclic units indicated that these polymers were composed of the cis-4,5 disubstituted 1,3-dioxolane unit as the predominant structural unit in the main chain. Trans-4,5 disubstituted 1,3-dioxolane and pendant dioxolane units were involved as minor structural units in the polymers. The pendant dioxolane unit is considered to be formed by hydrogen abstraction of propagating methylene dioxolane radicals from the neighboring methylene group [see Scheme 3, (21) and (22)]. The polymerization processes of divinyl formal are summarized in Scheme 3.

Table 9 Radical Polymerizations of Divinyl Formal Divinyl Acetal and Acetone Divinyl Acetal

Monomer	Polymerization temperature (°C)	AIBN (mol L^{-1})	Polymerization time (hr)	Conversion (%)	M_n
DVF[a]	−10	5 × 10^{-2}	15.0	7.5	
DVF[b]	10	1 × 10^{-1}	8.0	8.6	5,500
DVF[b]	30	5 × 10^{-2}	2.0	2.0	4,500
DVF[b]	50	1 × 10^{-1}	20	7.0	14,000
DVF[b]	70	2 × 10^{-2}	1.5	8.1	8,100
CH₃DVA[a]	−10	5 × 10^{-2}	19.0	8.8	11,000
CH₃DVA[a]	10	2 × 10^{-1}	0.5	6.4	8,600
CH₃DVA[a]	30	5 × 10^{-2}	0.6	8.7	18,000
CH₃DVA[a]	50	2 × 10^{-2}	7.0	6.0	30,000
CH₃DVA[a]	70	2 × 10^{-2}	0.5	6.0	26,000
2CH₃DVA[a]	−10	5 × 10^{-2}	7.0	5.0	9,000
2CH₃DVA[c]	10	2 × 10^{-1}	0.6	6.3	8,100
2CH₃DVA[c]	30	5 × 10^{-1}	0.4	9.3	13,000
2CH₃DVA[c]	50	2 × 10^{-2}	16.0	8.7	48,000
2CH₃DVA[c]	70	2 × 10^{-2}	1.5	9.9	54,000

[a][M] = 5.0M in toluene.
[b][M] = 2.5M in toluene.
[c][M] = 5.0M in benzene.

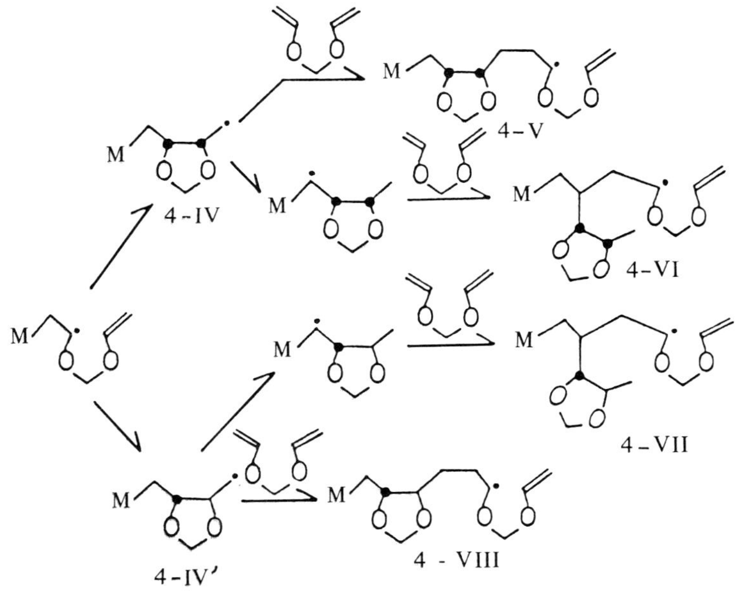

Scheme 3

(Scheme showing reactions 19–22 with structures 4-IV, 4-V, 4-VI, 4-VII, 4-IV', 4-VIII)

The content of *cis*-4,5 disubstituted 1,3-dioxolane units increased with lowered polymerization temperature and with increased monomer concentration. Further study of the ^{13}C NMR spectrum of polyDVF showed that the methylene dioxolane unit had stereoisomerism in the connection of two *cis*-dioxolane rings: meso and racemic form.

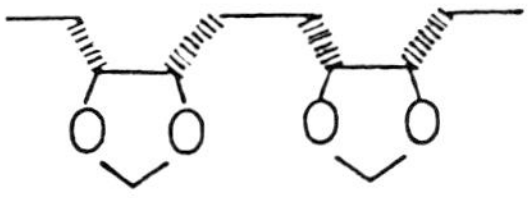

Meso (m) 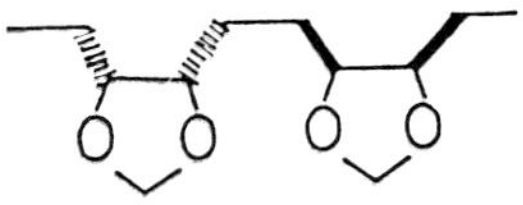   Racemic (r)

The mode of connection is constant [meso (m) to racemic (r) = 1:1], in spite of the change in the polymerization conditions. The equal amounts of the meso and racemic connections might be randomly distributed [74]. The following four types (a–d) are conceivable:

...rrrrrrrr...   (a)

...mmmmmmmm...   (b)

...mmrrmmrr...   (c)

...mrmrmrmr...   (d)

Type c appears least sterically demanding among the four types of the connections on the basis of the CPK molecular model. It is expected that highly stereoregular polymers are obtainable by selecting polymer conditions.

Furue [75] showed that trivinyloxymethane gave soluble polymers with AIBN even at high conversion.

Other Divinyloxy Compounds

Nozakura and colleagues [18, 19, 76] showed that methyl vinyloxysilane and methylvinyloxygermane [$(CH_3)_nM(OCH=CH_2)_{4-n}$, M = Si or Ge] were polymerized by radical initiators, and that the rate of polymerization increased with increasing number of vinyloxy groups. IR spectra of polymers showed clearly the presence of pendant vinyl group due to an intermolecular propagation without cyclization. The structure of these polymers is different in the presence of the pendant vinyl group from that of polydivinyl acetal shown in the preceding section. The content of the pendant vinyl group was estimated to be about 20% in both polydivinyloxysilane [76] and polydivinyloxygermane [18] which were isolated at less than 10% conversion. The content of the pendant vinyl group was much lower than expected if cyclopolymerization did not occur, indicating that high extent of cyclopolymerization took place in the polymerization process. The polymers were converted into poly(vinyl alcohol) by methanolysis, which has a considerable amount of the 1,2-glycol structure. The presence of the 1,2- structure was ascribed to the formation of the five-membered ring in the polymerization of the vinyloxy compounds. The sum of the content of the pendant vinyl group and five-membered ring unit was less than 50% of monomer units estimated from the degree of polymerization of poly(vinyl alcohol) derived. Accordingly, the residual units were ascribed to six-membered ring formation, which gave vinyl alcohol units by hydrolysis. The polymer structure is given by

$$\text{(structure 23)}$$

where M = Si or Ge. This difference in the polymer structure between polydivinyloxysilane and polydivinylacetal was found in the stereoregularity of the polymers. It was explained on the basis of the difference in bond length between Si–O and C–O bonds and in the Q and e values between vinyloxysilane and vinylacetal. The extent of the cyclization was found to increase with lowered monomer concentration and with increased polymerization temperature. Variation of relative abundance of the five-membered ring to the six-membered ring with polymerization temperature was explained by the difference in the preexponential factors for ring closures.

Yokota and Takada [77] prepared cyclic polymers with six-membered rings by the radical polymerization of o-isopropenylphenyl vinyl ether.

Radical polymerizations of perfluorooxoalkylenedivinyl ethers {$CF_2=CF[OCF_2CF(CF_3)]_nO(CF_2)_6O[CF(CF_3)CF_2O]_mCF=CF_2$ ($n + m$ = 2,4)} were performed by 0.5–4 mol % peroxide catalyst to form crosslinked, transparent glassy polymers (gel fraction 50–60%) [78].

Copolymerization

Butler [79, 80] performed radical copolymerizations of divinyl ether (DVE) and maleic anhydride (MAn) to obtain soluble polymers. Results of the typical copolymerizations are

Table 10 Radical Copolymerization of DVE (M_1) and MAn (M_2)[a]

M_2 mole fraction	Total monomer (M)	AIBN (M)	Polymerization time (min)	Conversion (%)	m_2[c] mole fraction
0.15	2.00	5.0×10^{-3}	13	9.5	0.65
0.30	2.00	5.0×10^{-3}	13	9.4	0.65
0.50	2.00	5.0×10^{-3}	13	11.2	0.66
0.67	2.00	5.0×10^{-3}	12	9.6	0.66
0.75	2.00	5.0×10^{-3}	12	7.3	0.66
0.90	2.00	5.0×10^{-3}	20	6.0	0.67
0.33	2.00	1.0×10^{-2}	25	40.0	0.67
0.50	2.00	1.0×10^{-2}	25	43.7	0.66
0.67	2.00	1.0×10^{-2}	25	36.3	0.67
0.67[b]	2.00	1.0×10^{-2}	25	86.1	0.65

[a]Polymerization conditions: 60°C, benzene medium unless stated otherwise.
[b]Polymerization solvent: acetone.
[c]m_2: mole fraction of M_2 in the copolymer.

given in Table 10. The rate of polymerization seems to increase with decreasing solvent polarity. The copolymer composition (DVE : MAn) was always determined to be 1 : 2 by elementary analysis. In this pioneering study, Butler proposed the presence of the [4,3,0]-bicyclic structure **9**, which is derived by the addition (head-to-tail–type addition) of the cyclizing radical. This structure was later reconfirmed by ^{1}H NMR study [81]. However, Aso et al. [82] suggested the possibility of the formation of **10** due to the head-to-head–type addition on the basis of an oligomerization study, and more recently Samuels [83] suspected the formation of the [3,3,0]bicyclic unit **10** from measurements of polymer solution properties and molecular model studies.

Kunitake and colleagues [84, 85] investigated the structure of copolymers of DVE with MAn, which were prepared in different solvents, by ^{13}C NMR spectroscopy, and found that it was dependent on polymerization solvents. The main structure of the cyclopolymer prepared in chloroform was consistent with the presence of the symmetrical bicyclic unit with cis junction and the trans monocyclic anhydride unit (see **11** in Scheme 4), while the structure of the cyclocopolymer prepared in a polar medium (acetone + CS_2) was different from that of the polymer obtained in chloroform. In fact, it was estimated that the former contained 90% of **9** and 10% of **10** (see **12** in Scheme 4). Therefore, it is concluded that the ring closure in this case mainly took place by the head-to-tail addition of the cyclizing radical. The origin of the solvent effect on the polymer structure remains unsolved. The major copolymerization processes are shown in Scheme 4.

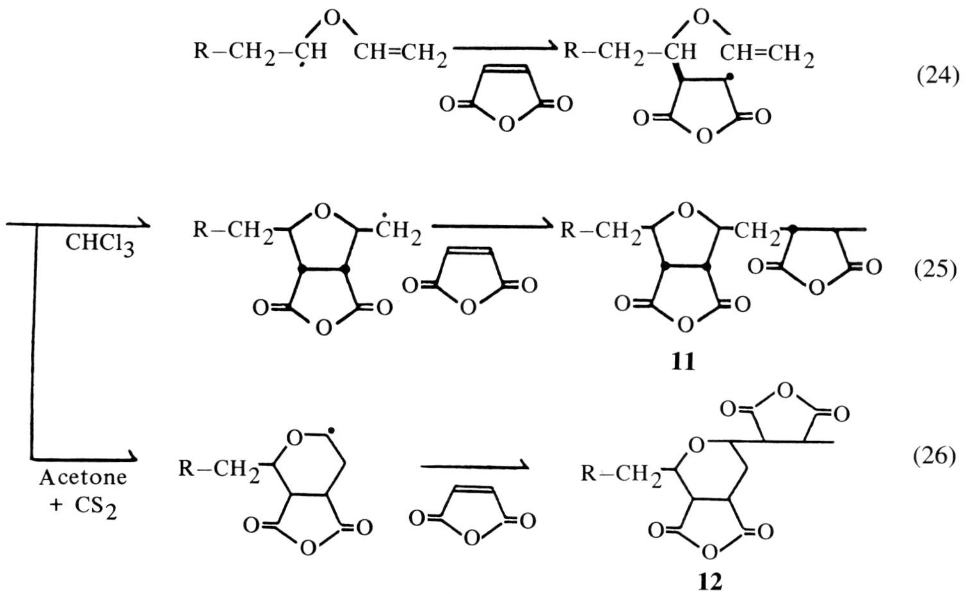

[Scheme 4 with equations (24), (25), (26); structures with CHCl₃ and Acetone + CS₂ conditions, compounds 11 and 12]

Scheme 4

Similarly, radical copolymerizations of *cis*-propenyl vinyl ether (PVE), 2-methylpropenyl vinyl ether (CH$_3$-PVE), and *cis*-dipropenyl ether (DPE) with MAn were performed in benzene. The polymer structures examined by ^{13}C NMR spectroscopy were similar to that of polyDVF [85].

Tsarik and colleagues [86, 87] performed radical copolymerizations of divinyl ether of hydroquinone (MQ) and MAn to obtain crosslinked copolymers with MQ:MAn units approaching 1:2. Porous crosslinked copolymers were prepared by the copolymerization of this vinyl ether with methyl acrylate, butyl acrylate, MMA, and acrylonitrile.

REFERENCES

1. G. Odian, *Principles of Polymerization*, 2nd ed., Wiley-Interscience, New York (1981).
2. G. E. Ham, *Copolymerization*, Interscience, New York (1964).
3. M. Szwarc and J. H. Binks, in *Theoretical Organic Chemistry*, Butterworth, London, pp. 271–272 (1958).
4. C. Walling, in *Free Radical in Solution*, Wiley, New York, p. 498 (1957).
5. L. J. Yong, in *Polymer Handbook*, 2nd ed. (J. Brandrup and E. H. Immergut, eds.), Wiley, New York (1975).
6. T. Sato and T. Otsu, *Polymer, 11*: 389 (1975).
7. T. Sumiyoshi, A. Henne, P. Lechtken, and W. Schnabel, *Z. Naturforsch, A39*: 434 (1984).
8. T. Sumiyoshi and W. Schnabel, *Makromol. Chem., 186*: 1811 (1985).
9. W. J. Hehre, L. Random, P. R. Schleyer, and J. A. Pople, *Ab Initio Molecular Orbital Theory*, Wiley, New York (1986).
10. S. Kato and K. Morokuma, *J. Chem. Phys., 72*: 206 (1980).
11. S. Nagase, T. Fueno, and K. Morokuma, *J. Am. Chem. Soc., 101*: 5849 (1979).
12. R. J. Cvetanovic and R. S. Irwin, *J. Chem. Phys., 46*: 1694 (1967).
13. T. Fueno and M. Kamachi, *Macromolecules*, in press. No. 3 (1988).
14. M. Kamachi, K. Tanaka, and Y. Kuwae, *J. Polym. Sci. Polym. Chem. Ed., 24*: 925 (1986).

15. A. Matsumoto, K. Iwanami, and M. Oiwa, *Makromol. Chem. Rapid Commun.*, *4*: 277 (1983).
16. A. Matsumoto, K. Iwanami, and M. Oiwa, *J. Polym. Sci. Polym. Lett. Ed.*, *19*: 497 (1981).
17. T. Otsu, O. Ito, N. Toyoda, and S. Mori, *Makromol. Chem. Rapid Commun.*, *2*: 725 (1981).
18. S. Kida, S. Nozakura, and S. Murahashi, *Polym. J.*, *3*: 234 (1972).
19. R. Ohno, S. Nozakura, and S. Murahashi, *Kobunshikagaku*, *25*: 188 (1968).
20. M. Kamachi, *Adv. Polym. Sci.*, *82*: 209 (1987).
21. M. Kamachi, M. Kohno, Y. Kuwae, and S. Nozakura, *Polym. J.*, *14*: 749 (1982).
22. B. Ranbey and J. F. Rabek, *ESR Spectroscopy in Polymer Research*, Springer Verlag, Berlin-Heidelberg-New York (1977).
23. J. E. Wertz and J. R. Bolton, *Electron Spin Resonance*, McGraw-Hill, New York (1972).
24. K. L. Lee and K. Ching, *J. Chem. Soc.*, Perkin Tras. 2, 161 (1985).
25. R. W. Fessenden and R. H. Schuler, *J. Chem. Phys.*, *43*: 2704 (1965).
26. M. Imoto, S. Sakai, and T. Ouchi, *J. Chem. Soc., Japan*, 91 (1985).
27. L. J. Young, in *Polymer Handbook*, 2nd ed. (J. Brandrup and E. H. Immergut, eds.), Wiley, New York (1975).
28. N. D. Field and D. H. Lorenz, in *Vinyl and Diene Monomers*, Part 1 (E. D. Leonard, ed.), Wiley-Interscience, New York, pp. 365–411 (1970).
29. M. G. Baldwin, *J. Polym. Sci., A*, *3*: 703 (1965).
30. M. L. Hallensleben, *Makromol. Chem.*, *144*: 267 (1970).
31. B. Zeegers and G. G. Butler, *J. Macromol. Sci. Chem.*, *A6*: 1569 (1972).
32. J. W. Schwertert and G. B. Butler, *J. Polym. Sci. Polym. Chem. Ed.*, *16*: 1375 (1978).
33. S. Iwatsuki and T. Itoh, *Macromolecules*, *12*: 208 (1979).
34. S. Iwatsuki and Y. Yamashita, *J. Chem. Soc. Japan Ind. Chem. Sect.*, *68*: 1138 (1965).
35. H. A. A. Rasoul and H. K. Hall, *J. Org. Chem.*, *47*: 2080 (1982).
36. K. G. Olson and G. B. Butler, *Macromolecules*, *16*: 707 (1983).
37. K. G. Olson and G. B. Butler, *Macromolecules*, *17*: 2480 (1984).
38. K. G. Olson and G. B. Butler, *Macromolecules*, *17*: 2486 (1984).
39. K. Fujimori, *Polym. Bull.*, *13*: 459 (1985).
40. K. Fujimori, N. A. Wickrammasinghe, and A. A. Nanda, *Aust. J. Chem.*, *33*: 189 (1980).
41. G. B. Butler and K. Fujimori, *J. Macromol. Sci. Chem.*, *A6*: 1533 (1972).
42. T. Nishiuchi, Y. Kijima, H. Yamamoto, and T. Takahashi, *Nippon Kagaku Kaishi*, 1972 (1975).
43. E. M. Shaikhutdinov, B. A. Zhubanov, and S. Kh. Khusainova, *Vysokomol. Soedin.*, *B15*: 869 (1973).
44. E. M. Shaikhutdinov, B. A. Zhubanov, S. R. Rafikov, and S. Kh. Kusanova, *Vysokomol. Soedin.*, *A19* (1977).
45. Zh. E. Eginbaev, A. Zh. Mashrapova, I. I. Mai, K. A. Ayapbergenov, and Z. M. Muldakhmetov, *Izv. Akad. Nauk. Kaz. SSR Ser. Khim.*, *1983*: 37 (1983).
46. C. H. Bamford and S. Brumby, *Makromol. Chem.*, *105*: 222 (1967).
47. M. Kamachi, *Adv. Polym. Sci.*, *38*: 56 (1981).
48. M. Kamachi, D. J. Liaw, and S. Nozakura, *Polym. J.*, *11*: 921 (1979).
49. E. M. Shaikhutdinov, Z. S. Nurkeeva, and S. V. Chebeiko, *Izv. Akad. Nauk. Kaz. SSR Ser. Khim.*, *28*: 53 (1978).
50. E. M. Shaikhutdinov, Z. S. Nurkeeva, and S. V. Chebeiko, *Izv. Akad. Nauk. SSR Ser. Khim.*, *26*: 25 (1976).
51. G. J. Schmitt and C. Schuerch, *J. Polym. Sci.*, *45*: 313 (1960).
52. M. Kurosawa, T. Doiuchi, and Y. Minoura, *J. Polym. Sci. Polym. Chem. Ed.*, *16*: 129 (1978).
53. K. Matsuzaki and T. Sugimoto, *Makromol. Chem.*, *164*: 127 (1973).
54. M. Kurosawa and Y. Minoura, *J. Polym. Sci. Polym. Chem. Ed.*, *17*: 470 (1979).

55. M. Kurosawa, H. Yamaguchi, and Y. Minoura, *J. Polym. Sci. Polym. Chem. Ed.*, *17*: 485 (1979).
56. M. Kurosawa and Y. Minoura, *J. Polym. Sci. Polym. Chem. Ed.*, *17*: 3297 (1979).
57. T. Doiuchi and H. Yamaguchi, *Eur. Polym. J.*, *20*: 831 (1984).
58. J. Asakura, M. Yoshihara, and T. Maeshima, *J. Macromol. Sci. Chem.*, *A18*: 285 (1982).
59. T. Suzuki and T. Tomono, *J. Polym. Sci. Polym. Chem. Ed.*, *22*: 2829 (1984).
60. C. Aso and S. Ushio, *Kogyo Kagaku Zasshi*, *65*: 2085 (1962).
61. M. Tsukino and T. Kunitake, *Macromolecules*, *12*: 387 (1979).
62. C. Aso, S. Ushio, and M. Sogabe, *Makromol. Chem.*, *100*: 100 (1967).
63. M. Guanta, G. Camino, and L. Trossarelli, *Makromol. Chem.*, *131*: 237 (1976).
64. M. Tsukino and T. Kunitake, *Polym. J.*, *11*: 437 (1979).
65. T. Miyake, *J. Chem. Soc. Japan Ind. Chem. Sect.*, *64*: 1272 (1961).
66. S. G. Matsoyan, *J. Polym. Sci.*, *52*: 189 (1961).
67. Y. Minoura and M. Mitoh, *J. Polym. Sci. A*, *3*: 2149 (1965).
68. A. A. Arbuzova, T. T. Borsova, O. B. Iv., G. P. Milkailov, A. S. Nigmankhajaev, and K. S. Sultanov, *Vysokomol. Soedin.*, *8*: 926 (1966).
69. C. Aso, T. Kunitake, and S. Ando, *J. Macromol. Sci. Chem.*, *A5*: 167 (1971).
70. H. J. Dietrich and M. A. Raymond, *J. Macromol. Sci. Chem.*, *A6*: 191 (1972).
71. M. A. Raymond and H. J. Dietrich, *J. Macromol. Sci. Chem.*, *A6*: 206 (1972).
72. M. Tsukino and T. Kunitake, *Polym. J.*, *17*: 657 (1985).
73. M. Tsukino and T. Kunitake, *Polym. J.*, *17*: 943 (1985).
74. M. Tsukino and T. Kunitake, *ACS Symp. Ser.*, *195*: 73 (1982).
75. M. Furue, Doctoral dissertation (1969).
76. M. Furue, S. Nozakura, and S. Murahashi, *Kobunshikagaku*, *24*: 522 (1967).
77. K. Yokota and Y. Takada, *Kobunshikagaku*, *30*: 217 (1973).
78. V. A. Ponomarenko, A. A. Glazkov, A. V. Ignatenko, V. N. Shelgaev, S. P. Krukovskii, E. R. Androsyuk, and A. I. Rakhimov, *Vysokomol. Soedin.*, *B21*: 794 (1979).
79. G. B. Butler, *J. Polym. Sci.*, *48*: 279 (1960).
80. G. B. Butler, *J. Macromol. Sci. Chem.*, *5*: 219 (1971).
81. G. B. Butler and Y. C. Chu, *J. Polym. Sci. Polym. Chem. Ed.*, *17*: 877 (1979).
82. C. Aso, S. Ushio, and M. Sogabe, *Makromol. Chem.*, *100*: 100 (1967).
83. R. J. Samuels, *Polymer*, *18*: 452 (1977).
84. M. Tsukino and T. Kunitake, *Polym. J.*, *13*: 671 (1981).
85. T. Kunitake and M. Tsukino, *J. Polym. Sci. Polym. Chem. Ed.*, *17*: 877 (1979).
86. Ya. L. Tsarik and L. I. Antyiferova, *Vysokomol. Soedin.*, *A27*: 486 (1985).
87. Ya. L. Tsarik, D. G. Vokena, N. I. Skobeava, O. A. Edel'shleim, and A. V. Kababina, *Plast. Masz.*, 10 (1984).

8

Electroinitiated Cationic Polymerization

Levent Toppare
Middle East Technical University
Ankara, Turkey

INTRODUCTION	271
General Aspects	271
Comparison with Other Techniques of Initiation	272
Developments in Electrochemical Cationic Polymerization	272
The Relationship Between the Highest Occupied Molecular Orbital Energies and the Anodic Peak Potentials of Monomers	274
Initiation Mechanism	275
EXPERIMENTAL METHODS	276
General Aspects	276
Electrolysis	281
Electroinitiated Polymerization	282
Electroinitiated Copolymerization	291
CONCLUSIONS	301
REFERENCES	302

INTRODUCTION

General Aspects

Electroinitiated polymerization may be defined as polymerization initiated by species formed in an electrode reaction. The initiator may be the oxidized (or reduced) form of the monomer or another solution component of electrolysis. Electroinitiated polymerization reactions provide a new method of polymerization with a fine control of the initiation step. For each electron transferred through the solution, a corresponding chemical reaction must occur at the electrode. Since several competing processes may take place simultaneously, the determination of the particular reaction is dependent upon the exact potential of the electrode in question; thus selectivity can be exercised over the possible chemical reactions.

Factors such as overpotential at an electrode, migration of ions, adsorption–desorption as well as film coating on the electrode surface, interactions on the electrode material, and diffusion can play important roles in the determination of the products of an electrochemical reaction. Polymerization by electrochemical techniques may be studied from different standpoints:

1. The rate of the reaction may be controlled by programming the current.
2. The molecular weight distributions can be controlled.
3. The molecular structure—i.e., stereoregularity—due to the adsorption on the electrode surface of the electrode may be influenced by electrolysis conditions.
4. Electrochemically generated coupling reactions may induce new polymers.
5. The properties of polymers may be modified by crosslinking, grafting, or degradation reactions, which can be induced by electrochemical means.
6. Selective voltage control can play an important role in the production of several polymers and especially of copolymers with desired properties such as copolymer composition.
7. Selectivity of an electrochemical method may be advantageous when compared to other experimental polymerization techniques.

Comparison with Other Techniques of Initiation

An analogy to photochemical polymerization can be made as both techniques make use of an external source for the activation energy. The number of photons in photolytic reactions is related directly to the number of molecules activated; similarly, calculation of the number of electrons transferred at the electrode is possible through the application of Faraday's laws. Control of the potential may lead to certain reactions that can be carried out at a particular energy and exclude the reactions that require a higher activation potential. In a similar manner, control of the activating wavelength or energy can result in specific chemical reactions with the exclusion of others in photolytic initiation. Once initiation through a cation and/or radical cation is achieved, whether electrochemically or photolytically, further propagation of the reaction is essentially unaffected by the passage of current or light. On the other hand, one important advantage of electrochemical activation over the latter is that it is possible to reverse the current, thus changing the chemical environment of the electrode, which will yield a change in the course of the reaction. It is far easier to produce a large number of electrons by the passage of current than it is to generate the same number of photons.

The problem is more difficult if flux is taken into account. Even high-intensity lamps do not provide photon fluxes greater than the electron fluxes that can be easily generated electrochemically. Further improvements can be achieved in electrochemistry by increasing the electrode areas and current densities. It is also a common fact that in general only few electrons are responsible for yielding a polymer molecule, which makes electrochemical polymerization more favorable than electrochemical organic synthesis where at least one electron is required for the chemical transformation of each molecule. On the other hand, this technique has its own experimental difficulties. Maintenance of high conductivity and solubility; choice of appropriate solvents, supporting electrolytes, and electrode material; and rather low molecular weight products are the main limitations of this method.

Developments in Electrochemical Cationic Polymerization

The electrochemical polymerization method was used by Szarvasy [1] for the first time in 1900. In spite of that early observation, no report on the subject appeared until a study on electroinitiated polymerizations of some vinylic monomers [2, 3] in the late 1940s. It took several more years for the first definite evidence of an electrochemically initiated polymerization proceeding by a cationic mechanism to be reported [4]. In a copolymerization study of a system containing acrylonitrile and styrene, Breitenbach reported the formation of polystyrene in the anode compartment of a divided cell. Polymerizations of styrene, isobutylvinyl ether, and vinyl carbazole were the next to be carried out readily by this technique [5]. The difficulty of obtaining reproducible kinetic results in cationic polymerizations seems to be the main factor in limiting the number of reports in this field. Several workers published their results in late 1960s [6–8]. The first detailed investigation of the kinetics of cationic mechanism is considered to be the work of Strobel and Schulz [9].

Funt and Blain investigated the cationic polymerization of styrene [10] and isobutylvinyl ether [11]. Their study shed light on the electrochemical cationic initiation method. It was proved, first, that no polymer formation occurred in the absence of current, and second, there was a precise control of current over the course polymerization. This showed that the reaction did not occur through the initiation of some species in the solution since any impurities present had been removed by the passage of current. In addition, the possibility of the electric field effect over the rate of propagation was ruled out. Probably the most important contribution to future work was made when they tried to explain the mechanism of the initiation step. They reported that it involves the oxidation of the anion of the supporting electrolyte or the direct oxidation of the monomer to a propagating radical cation. Although it is now known that to carry out a direct electron transfer from the monomer to the anode an electroinert solvent–electrolyte system is required, their idea led future workers to employ different electrochemical systems to achieve direct initiation. An excellent survey by Breitenbach [12] and a general review by Baizer [13] were yet to come before some workers who utilized the constant potential instead of the constant current as the controlled parameter in electrochemical cationic polymerization studies polymerized styrene by a direct electron transfer initiation, which was carried out at the anodic peak potential of the monomer [14]. In doing so, Akbulut et al. utilized the cyclic voltammetry (CV) method. They claimed the following mechanism involving adsorbed styrene monomer on the anode surface:

$$M_{ads} \rightarrow M_{ads}^+ + e$$
$$M_{ads}^+ + M \rightarrow M - M_{ads}^+$$

This group [15] also confirmed the validity of Giusti's [16] conclusions on electric field effects on cationic polymerization. That is to say, the enhancement of the polymerization rate when the field is applied was simply due to the electrolytic phenomena which generate active species at the anode in addition to those already formed by the system.

Electroinitiated cationic polymerization of several vinyl and etheric monomers in nonaqueous media has been of interest for some time. Acenaphthylene [17], indene [18], and trioxane [19] were polymerized with electrolytes containing perchlorate as the anion. A general survey describing electroinitiation as a tool in polymer chemistry was also given by Parravano [20]. During the last decade, it seems that most of the polymer chemists continue to employ current-controlled polymerization reactions. Nevertheless, there seems to be an increase in the number of studies and groups interested in electrochemical initiation. The use of quaternary ammonium salts was employed widely for a variety of monomers [21–23]. One striking aspect is the use of electrochemistry as a tool for polymer coatings [24]. There are many publications on modified electrodes as conductors and semiconductors [25, 26]. The discussion of electrochemically induced conducting films is outside the scope of this chapter.

In recent years the use of potential control over the electroinitiated cationic polymerization process gained interest. Akbulut et al. [14] showed that controlled potential electrolysis (CPE) had some advantages over constant-current electrolysis (CCE) in terms of initiation mechanisms. To achieve a direct initiation, that is, a direct electron transfer from the monomer to the anode, electrochemical initiation must be carried out at or below the measured anodic peak potentials of the monomers. Determination of the exact peak potentials with respect to a given reference electrode can be accomplished by cyclic voltammetry. This method can be regarded as out of use only when the activation of monomer is possible only beyond the discharge potential of a solvent–electrolyte couple.

The Relationship Between the Highest Occupied Molecular Orbital Energies and the Anodic Peak Potentials of Monomers

Anodic peak potentials ($E_{p.a}$) of monomers can be correlated with the highest occupied molecular orbital energies (HOMO) of monomers in the case of electroinitiated cationic polymerization. Cationic homopolymerization of the related monomer is accomplished by an electron transfer from HOMO of the monomer to the working electrode (anode) as soon as the energy band of the anode becomes comparable to the HOMO of the monomer in a short time of electrolysis (Fig. 1).

$E_{1/2}(\text{ox})$ potentials of some organic compounds were reported to be directly proportional to ionization energies [27, 28]. Theoretical relations between $E_{1/2}(\text{ox})$ and the HOMO energies, ϵ_m, had been proposed earlier by Hoijtink [29]. It has been reported that modified Hückel molecular orbital (HMO) calculations predict the observed difference of the $E_{1/2}(\text{ox})$ of related olefins [30]. The oxidation potentials of various systems were correlated with perturbational Hückel molecular orbital considerations by Akbulut et al. [31]. The model chosen was the styrene molecule. The rest of the substituted styrenes were treated as perturbed structures. To visualize the effects which caused variations in peak potentials, Hückel molecular orbital considerations were applied [32].

In the case of the cationic copolymerization, initiation primarily involves the activation of the monomer with lower $E_{p.a}$ value. When the polymerization was started, after a certain time, the energy band of the electrode and HOMO energy of the monomer became comparable, i.e., the energy gap between ϵ_m of the monomer and energy band of the

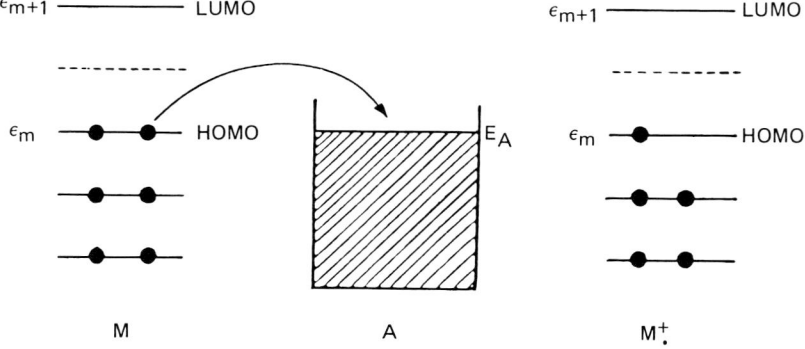

Figure 1 Schematic representation of molecular energies of the monomer (M), the radical cation of monomer ($M^+_\cdot$) and the energy band of the anode (A).

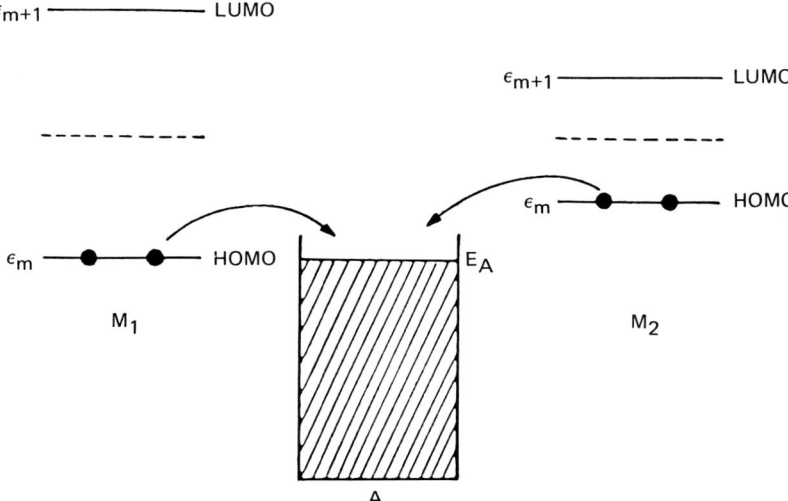

Figure 2 Schematic representation of molecular energies of the monomers (M_1), (M_2) and the energy band of the anode (A).

electrode became smaller when compared with that of monomer with higher $E_{p.a}$ value. An electron transfer from the related monomer to the anode yields a radical cation–dication which propagates the polymerization (Fig. 2).

Initiation Mechanism

Electroinitiated cationic initiation may take place in two ways: indirect and direct initiation.

Indirect Cationic Initiation

Indirect cationic initiation may occur either through the application of voltages higher than the oxidation potentials of all species present in the system, leading to uncontrolled electrolyses, or through the use of electroactive salts as the supporting electrolytes and/or solvents in current-controlled electrolyses:

$$ClO_4^- \rightarrow ClO_4\cdot + e$$
$$ClO_4\cdot + RHC=CH_2 \rightarrow RHC^+-CH_2\cdot + ClO_4^-$$
$$2RHC^+-CH_2\cdot \rightarrow RHC^+-CH_2-CH_2-C^+HR$$

and

$$BF_4^- \rightarrow BF_4\cdot + e$$
$$BF_4\cdot + RHC=CH_2 \rightarrow RHC^+-CH_2\cdot + BF_4^-$$
$$2RHC^+-CH_2\cdot \rightarrow RHC^+-CH_2-CH_2-^+CHR$$

The radical cation or dication can propagate cationic polymerization.

However, cyclic voltammetry measurements of the conjugated compounds indicated that they could easily be oxidized compared to tetrabutylammonium fluoroborate. This supporting electrolyte is inert toward electrooxidation up to $+3.5$ V vs. Ag°/Ag^+ reference electrode in acetonitrile, whereas most of the monomers in question have lower $E_{p.a}$ with respect to the same reference electrode.

The mechanism of indirect initiation was discussed by Funt and Blain [10, 11].

Direct Cationic Initiation

Monomers are directly involved in electron transfer phenomena when cations or radical cations of their own are generated. In direct cationic initiation an electron is transferred from the HOMO level to the anode:

$$RCH=CH_2 \rightarrow RHC^+—CH_2 \cdot + e$$
$$2RCH^+—CH_2 \cdot \rightarrow RHC^+—CH_2—CH_2—{}^+CHR$$

If the polymerization potential is higher than the oxidation peak potential of the monomer, the second electron transfer may take place to produce a dication. However, the less stable primary carbonium ion may be terminated and propagation may proceed through a secondary carbonium ion:

$$RHC^+—CH_2 \cdot \rightarrow RHC^+—CH_2^+ + e$$

EXPERIMENTAL METHODS

General Aspects

Voltammetry

The most frequently used record of electrochemical behavior is the voltammogram. This represents a plot of current versus potential. The potential of the working electrode is varied over a certain range and the current flow is observed as the electrode potential reaches a value at which electrolysis of the substrate takes place. The general convention for representation of voltammetric data is a graph on which potential is plotted on the abscissa and current on the ordinate. Anodic voltammograms are scanned from right to left and the resulting anodic currents are by convention negative and downward (Fig. 3).

Polarography, one of the most widely used voltammetric methods, can provide a considerable amount of information. The working electrode is the dropping mercury electrode (DME), which is a long, thin capillary attached to a mercury reservoir of certain height. The potential of DME may be scanned gradually to more negative values and the current flowing between the drop and the auxiliary electrode is recorded as a function of the working electrode potential. The voltammogram in this case is known as a polarogram. The main drawback of this method is that it cannot be used for the determination of oxidative behavior of substrates. On the other hand, there are several advantages: the simplicity of the experimentation, reproducibility of drop time and drop size, and constant renewal of mercury surface.

Cyclic Voltammetry

Cyclic voltammetry (CV) has become a popular tool in the last two decades in studying the electrochemical behavior of electroactive species. It is extensively used in the fields of electrochemistry, inorganic chemistry, physical organic chemistry, and biochemistry. The effectiveness of cyclic voltammetry results from its capability for rapidly observing the redox behavior of substrates over a wide potential range.

In this technique, the voltage applied to the working electrode is varied as a cyclic

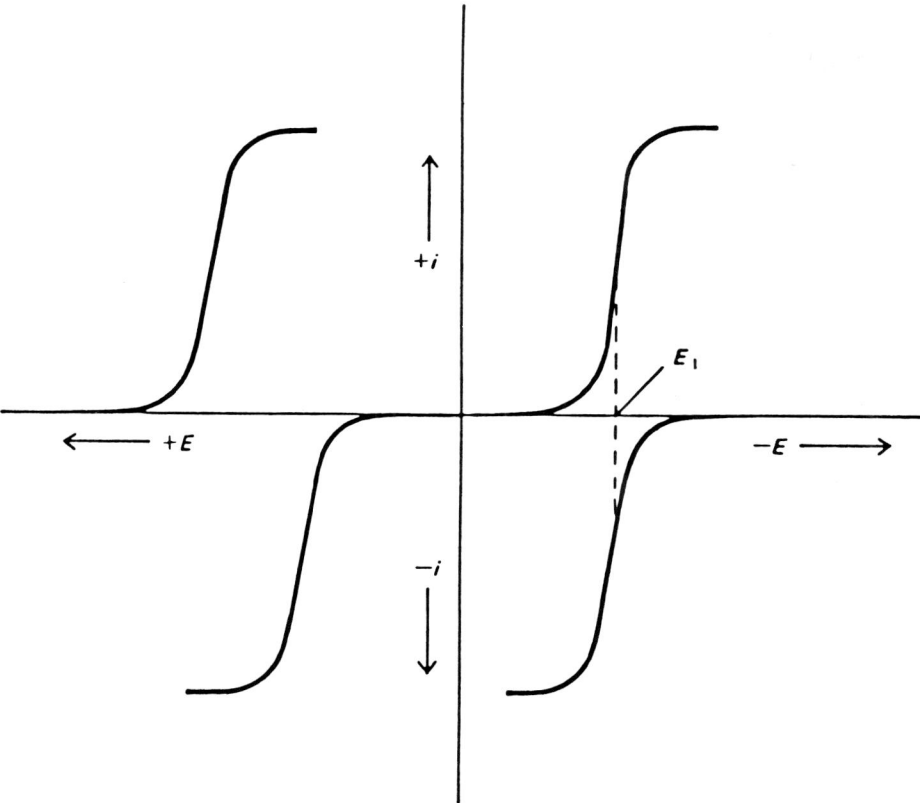

Figure 3 Representative voltammograms with conventional current and potential indications.

triangular wave (Fig. 4a). In cyclic voltammograms current response is plotted as a function of applied potential by a recorder. The current depends on the diffusion of electroactive material to the electrode surface, the number of electron transfers, the electrode area, and the voltage scan rate. In addition to the redox properties of the substrate, cyclic voltammetry gives information about the reversibility of the electron transfer process.

A typical voltammogram of a reversible process is given in Figure 4b. Reversibility of the electron transfer means that the oxidized and the reduced forms are in equilibrium at the electrode surface. During the forward scan, the oxidized form of substrate is reduced to R at cathode potential $E_{p.c}$, and at reverse scan the reduced form on the electrode surface is reoxidized to the original form at anodic potential $E_{p.a}$,

$$O + ne \rightleftharpoons R$$

where the diffusion coefficients of oxidized and reduced forms are the same. For a reversible reaction, the following relation is valid:

$$E_{p.a} - E_{p.c} = \frac{0.056}{n} \text{ V}$$

where n is the number of electrons transferred.

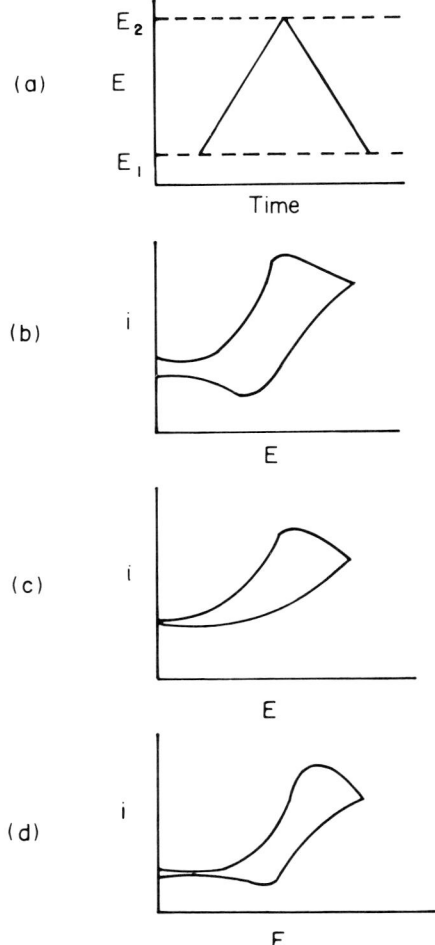

Figure 4 Cyclic voltammograms. (*a*) Triangular waveform, (*b*) reversible process, (*c*) irreversible process, and (*d*) reversible process followed by a chemical reaction.

A cyclic voltammogram of an irreversible electron transfer is given in Figure 4*c*. The product cannot be recycled electrochemically to give back the original reactant,

$$O + ne \rightarrow R$$

In some cases a reversible electron transfer can be followed by a chemical reaction. The cyclic voltammogram of such an electrochemical behavior is illustrated in Figure 4*d*:

$$O + ne \rightleftharpoons R$$

$$R \xrightarrow{k} Z$$

In this type of voltammogram, the substrate shows an intense peak on the forward scan, whereas the reverse peak is diminished effectively. If the scan rate is much slower than the rate of the reverse reaction, the reverse peak disappears totally and it resembles irrevers-

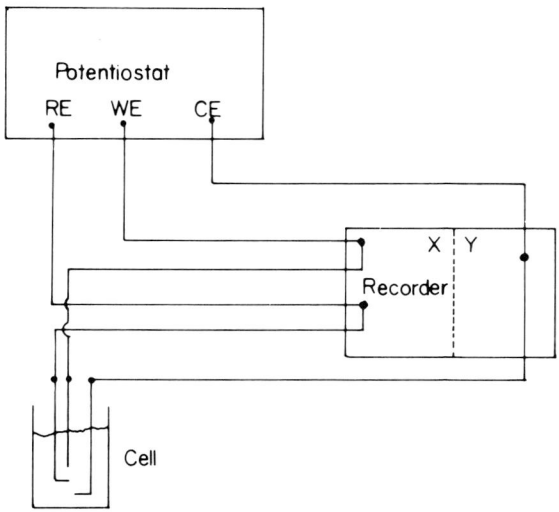

Figure 5 Circuit arrangement in constant potential electrolysis system.

ible voltammograms. However, at high scan rates a certain amount of O is reduced back, which is indicated by the reverse peak.

A typical cyclic voltammetry system is composed of a potentiostat, an X-Y recorder, and a function generator (Fig. 5). A cyclic voltammetry cell consists of three electrodes, one being a reference electrode. For practical purposes saturated calomel electrode (SCE)

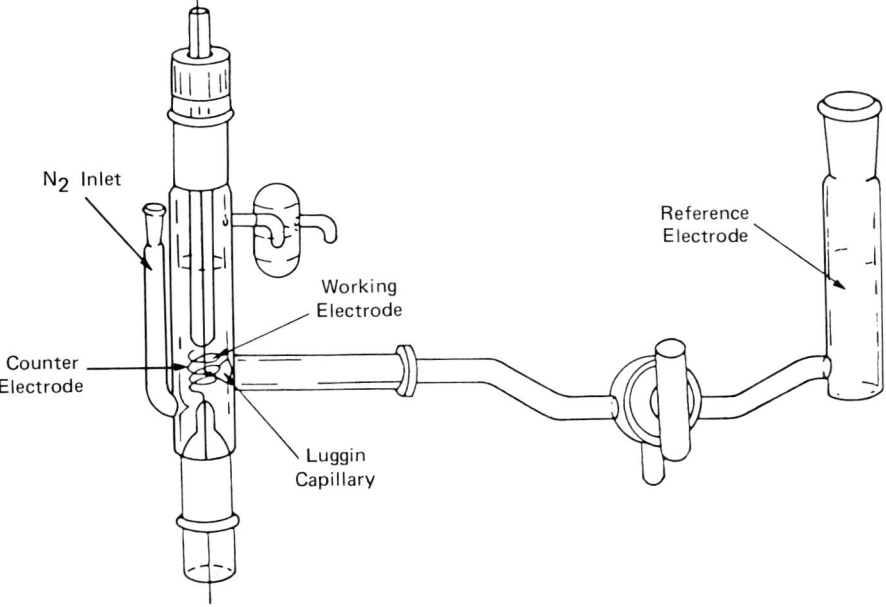

Figure 6 Cyclic voltammetry cell.

Table 1 Redox Potentials of Some Monomers with Respect to $Ag°/AgBF_4$ Reference Electrode[a]

Monomer	$E_{p,a}$ (V)	$E_{p,c}$ (V)	Reference
Acrolein	+2.95	−2.30	14
Acrylamide	+2.60	−2.80	14
Acrylonitrile	+2.50	−2.90	14
Allyl bromide		−2.90	14
Carbon disulfide	+3.00	−2.40	14
t-Cinnamaldehyde	+2.00	−2.00	14
Indene	+1.40	−2.90	14
α-Methylstyrene	+1.85		14
Propylene oxide	+3.20		14
Propylene sulfide	+2.90		14
Styrene	+1.80	−3.20	14
2-Vinylpyridine	+2.55	−2.95	14
4-Vinylpyridine	+3.30	−2.60	14
N-Vinyl-α-pyrrolidone	+1.30		14
2-Chlorostyrene	+2.00	−2.80	31
3-Chlorostyrene	+2.05	−2.70	31
4-Chlorostyrene	+1.90	−2.80	31
Acrylic acid		−1.50	33
N-Benzilidene aniline	+1.50	−2.30	33
	+2.45		
3-Bromostyrene	+2.00	−2.70	33
4-Bromostyrene	+2.05	−3.20	33
t-Cinnamic alcohol	+1.50		33
	+1.90		
t-Cinnamic acid	+1.90	−1.50	33
		−2.30	
Ethyl acrylate		−2.90	33
Ethyl methacrylate		−3.20	33
Methyl methacrylate		−3.10	33
Ethyl, β-methylcinnamate	+2.05	−2.50	33
		−3.10	
Ethyl propenyl ether	+1.30		33
4-Methoxystyrene	+1.10		
	+1.40		33
	+2.00		
β-Nitrostyrene	+2.45	−1.35	33
Phenylisocyanate	+1.80		
	+2.25		33
4-Fluorostyrene	+2.10		34
4-Methylstyrene	+1.70		34
β-Methyl styrene	+1.65		34
1,2-Epoxy-4-epoxyethylcyclohexane	+2.70		
	+3.00		81
Epoxycyclohexane	+3.00		81
Epoxycyclopentane	+2.90		81
Epoxystyrene	+2.40		81
Isoprene	+2.00		99
β-Bromostyrene	+2.00		113

[a] In acetonitrile–tetrabutylammonium fluoroborate couple, working electrode is a platinum bead and auxiliary electrode is a platinum wire.

and silver salts mainly constitute the reference electrodes. Several materials such as platinum, graphite, lead, mercury, mercurized platinum, and steel can be employed as working and/or auxiliary electrodes. Figure 6 represents an example of one of the several designs of CV cell currently in use.

Electrochemical redox behavior of some well-known monomers is tabulated in Table 1. Voltammograms were taken in acetonitrile (AN) tetrabutylammonium fluoroborate (TBAFB) solvent electrolyte couple. Working and auxiliary electrodes were platinum bead and platinum wire, respectively. The reference electrode was $Ag°/Ag^+$ (0.01 M).

Electrolysis

Electroinitiated polymerization can be attained by constant-current and constant-potential electrolysis.

Constant Current Electrolysis

Constant current electrolysis (CCE) is carried out in a cell containing two electrodes as shown in Figure 7. Current is kept constant throughout the electrolysis and the potential is allowed to vary. Although simple in application, it has some disadvantages. Since the potential is a variable parameter, the nature of the generated species may be unknown. The involvement of species present in the system in addition to the monomer is generally inevitable; thus complications may arise in the initiation and propagation steps.

Constant Potential Electrolysis

Constant Potential Electrolysis (CPE) is achieved with a three-electrode system. The potential of the working electrode with respect to a reference electrode is adjusted to a desired value and kept constant by a potentiostat while current is allowed to vary (Fig. 8). The voltage between the working and the reference electrode may be called the polymer-

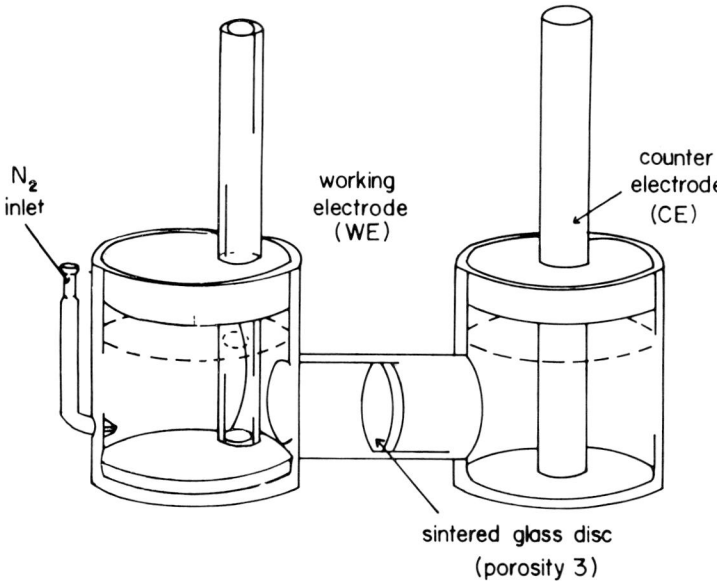

Figure 7 Constant-current electrolysis cell.

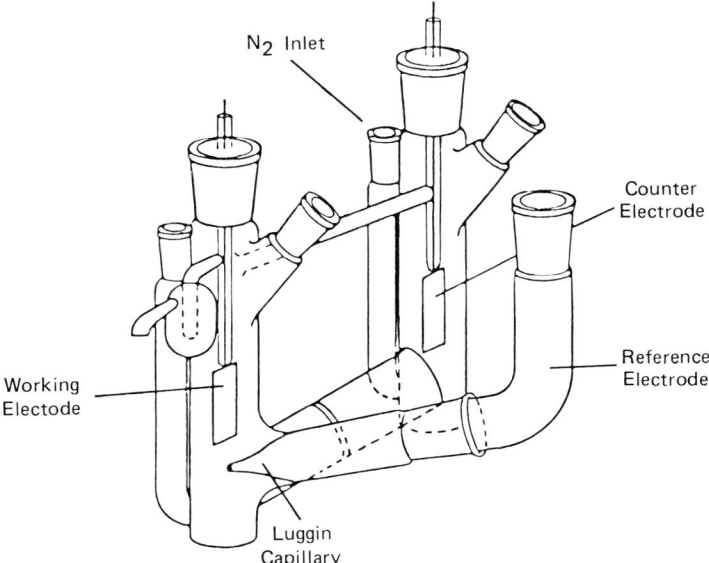

Figure 8 Constant-potential electrolysis cell with platinum electrodes. (From Ref. 96.)

ization potential (E_{pol}). By keeping the potential constant, creation of undesired species is prevented; hence the initiation becomes selective, i.e., through the monomer itself.

Electroinitiated Polymerization

Styrene and Derivatives

In 1962 Breitenbach and Srna [5] were able to polymerize styrene cationically, although the molecular weights were very low. In similar experiments with α-methylstyrene only very low molecular weight polymers were formed. Funt and Blain [10] performed detailed kinetic studies on the electroinitiation of styrene in methylene chloride (DM) as the solvent and tetrabutylammonium perchlorate as the supporting electrolyte. The use of DM as a cationic polymerization solvent was introduced and in terms of the chain transfer constants, this solvent proved ideal. Tidswell and Doughty [35] electrolyzed a solution of styrene and sodium borofluoride in sulfolane. A cationic polymerization of styrene was initiated by BF_3 cocatalyzed by HF and/or H_2O. The first controlled potential electrolysis at the anodic potential of styrene was carried out by Akbulut et al. [14]. The electrolyte used was tetrabutylammonium fluoroborate, which was not electroactive at the electrolysis potential, and the reference electrode was an $Ag°/Ag^+$ electrode. The initiation step was given as the direct electron transfer from styrene. They also suggested that the growth of the polymer takes place partially on the electrode surface. The current behavior and the effect of stirring supported this point. Their approach was to measure the anodic peak potentials of the monomers in question by CV prior to electrolysis.

The electroinitiated polymerization of α-methylstyrene was also studied by the same group [15] using the same technique, the controlled potential method. Electric field effects were also studied. The initial rate of disappearance of the monomer was found to be the same both in the absence and in the presence of the external electric field. The rate of formation of the polymer at high conversions decreased when the electric field was imposed. However, detailed analysis of polymers suggested that such observed electric field effects were due to electrochemical reactions rather than the second Wien effect. The electroinitiated cationic polymerization of α-methylstyrene in a tetrabutylammonium fluoroborate–dichloromethane system was investigated at various temperatures by controlled potential electrolysis [36]. It was established that polymerization occurs at low temperatures with higher yields. Apparent activation energy of the polymerization was also given.

The cationic electroinitiation of styrene derivatives such as 4-fluoro-, 2-chloro-, 3-chloro-, and 4-chloro- was studied at platinum electrodes in a TBAFB–DM system at low temperatures. No polymers of those monomers could be obtained at room temperature. The molecular weights of the polymers were low [34].

Electrochemically initiated polymerization of methoxystyrenes was studied by Cerrai and his group [37, 38]. They pointed out that polymerization was carried out without the control of the anodic potential and some of their kinetic results indicated that two parallel initiation reactions (direct and indirect) could occur. A further study was made of the cationic electropolymerization of a methoxystyrene (*trans*-anethole) initiated by a direct electron transfer at a platinum electrode [39]. This time the electrochemical behavior of the system had been preliminarily investigated by voltammetry. An ECE oxidation process of the substrate through a dication was proposed.

Acrylamide and Derivatives

Koval'chuk and colleagues [40–42] reported the polymerization of acrylamide and methacrylamide anodically in the presence of ethylene glycol and sodium acetate. Fernandez and Larrocha [43] showed that acrylamide oxidation occurred on the semiactivated iridium electrode at potentials >850 mV. This oxidation promoted a surface polymerization with a partial reduction of the surface oxides. Electrode coating began with monomer oxidation. A sufficient anodic potential resulted in the electrode passivation caused by the formation of the polyacrylamide coating.

Electroinitiated polymerization of acrylamide was also attained in solutions of some zinc salts [44]. Electrolysis of acrylamide in dimethylformamide (DMF) or acetone in the presence of $Zn(NO_3)_2$ or $ZnCl_2$ led to high molecular weight polymer formation both in the anode and in the cathode. The molecular weight of the polymer increased with increasing monomer concentration, and the monomer conversion increased with increasing current. A remarkable postpolymerization was observed even after the termination of the electric current. A possible mechanism for the anodic initiation reaction based on current–potential measurements was given.

Carboxylic Acids

Although the process cannot be considered cationic, Kolbe-type oxidations at platinum anodes were of interest in the study of various carboxylic acids. It was a general hope to prepare linear polymers from dicarboxylic acids via Kolbe electrolysis. Garrison [45] was able to report success in this respect. Toy [46] reported the preparation of a polymethylene from the electrolysis of solutions of a glutaric acid with sodium in methanol. The electrochemical polymerization of a diacid via Kolbe reaction was investigated by Gozlan and Zilkha [47] with regard to the oligomeric and side products of the reaction. The gaseous and volatile compounds were identified and the less volatile compounds were separated into fractions.

Amino Aromatic Compounds

Electrical properties of polymers obtained by anodic oxidation of aniline were discussed by Langer [48]. The DC electrical conductivity, current–voltage character, and thermoelectric effects were measured for aniline black, which was obtained by the anodic oxidation of $PhNH_2 \cdot HCl$.

Benzene and Derivatives

Wisdom [49–51] prepared benzene polymers by electrolyzing benzene in a ternary complex solution which consisted of ~20-carbon atom compound (RH), a hydrogen halide (HX), and aluminum trihalide (AlX_3). Polymerizations were achieved at platinum electrodes. Shepard and Dannels [52, 53] polymerized benzene in aqueous hydrogen fluoride–benzene solutions. Benzene and its derivatives were also used in the preparation of conductive or semiconductor polymers, which involves electrochemical oxidation in the presence of an electron acceptor and a dopant [54–56].

Heteroatomic Molecules

Electrically conducting heteroatomic polymers of bicyclic monomers were prepared by anodic electrochemical polymerization in acetonitrile–tetraethylammonium fluoroborate couple in a cell having three electrodes at a constant anode potential of +0.9 V vs. SCE [57]. There are several examples of polypyrolle polymers synthesized by anodic oxidation in the literature. Polymethylene pyrroles [58] were obtained by electrodeposition of its monomer in a solvent containing an alkylammonium salt. β-Halopyrroles [59] gave polymers with conductivities ranging from 10^{-5} to 10^{-2} $\Omega^{-1}cm^{-1}$. Polymers derived from 3,4-diiodo-, 3,4-dibromo-, 3,4-dichloro-, and 3-bromopyrrole were reported to form stable electrodes. A process was described for producing polyheteroatomic compounds by the electrochemical oxidation of monomeric heteroatomic compounds in an aqueous electrolyte in which the monomers were dispersed in the presence of conductive salts and dispersing agents [60]. The filmlike products obtained had good mechanical properties and high electrical conductance. The electrodes consisted of platinum or nickel. At an anodic current density of 200 μA cm^{-2} and a cell voltage of 2.2 V, following an electrolysis period of 12 hr an 80-μm-thick polypyrrole film was removed from the anode.

Electropolymerizations of some oxyheterocyclic monomers were carried out under amperostatic conditions at platinum electrodes with various electrolytes by Andruzzi et al. [61]. Both CCE and CPE electrolyses revealed that the acidity responsible in the polymerization was due to only one of the various products formed by the reaction between chlorine dioxide and the solvent. Cyclic voltammetry studies carried out on various lactones showed the impossibility of performing polymerization via direct initiation of the monomer.

Phenols and Derivatives

In 1969 a patent was published concerning the electrolytic polymerization of phenol with electrode deposition [62]. An aqueous solution of phenol was placed in the anode compartment of an electrolytic cell. During the process iron anode had a uniform coating of polyphenylene oxide. Borman [63] reported the polymerization of substituted phenols using copper salts as catalysts in aqueous solution to give polyarylene oxides. Electrochemical oxidation of phenols in nonaqueous solvents was also reported to give high molecular weight products [64].

Different azophenols were oxidized by electrochemical methods as well as chemical means on iron and copper anodes in water–methanol mixture [65]. Poly(oxy-4-phenylazo-1,2-phenylene) and its 3-allyl derivatives were obtained from 4-hydroxyazobenzene and 3-allyl-4-hydroxybenzene, respectively. Adherent films having very good protective features were obtained in short electrolysis time. Polymers produced via electrochemical oxidation of phenols were utilized in the determination of permselectivity of films [66]. It was also demonstrated that platinum electrodes coated with such films could be used as pH sensors since they responded selectively to H^+ ion. The anodic oxidation of 1,3-disubstituted phenols was reported using transition metal anodes [67]. The passivating effect was attributed to the film formation on the electrode. The physical properties of the polymeric film indicated its usefulness as a protective coating for metals. Poly(2,6-disubstituted phenols) were prepared by electrolytic oxidative polymerization [68]. A solution of dichloromethane–methanol mixture containing tetraalkylammonium salt and the monomer were electrolyzed at a current density of 10 mA cm^{-2} using platinum electrodes. Synthesis of oligo(phenylene oxide) by electrooxidative polymerization was reported by Tsuchide et al. [69]. The mechanism of the electrooxidative polymerization was elucidated by electrochemical and ESR measurements.

Polycyclic Hydrocarbons

Pyrene, fluorene, fluoranthrene, and triphenylene were electrooxidized into polymeric films which are polymer–anion composites containing ~12% anion. The polymers were reported to be rich in hydrogen and good conductors [70].

Ethers

Breitenbach and Srna [5] reported the electroinitiated polymerization of isobutylvinyl ether in nonaqueous solvents. More details on the polymerization of the monomer with various supporting electrolytes were given by Funt and Blain [11].

The polymerization of tetrahydrofuran (THF) has been studied extensively by various authors [6, 8, 71]. Mengoli and Vidotto [72–74] reported further studies of cationic polymerization of THF. Dey and Rudd [75] polymerized THF by anodic electrolysis on platinum electrodes in a three-compartment cell. The rate of polymerization was followed by measuring the electrical resistance of the electrolyte solution. The polymer was found to be living in nature. The initiation was proposed to be a result of the oxidation of THF rather than the electrolyte. Yamazaki's group [76] previously reported the initiation of THF through the interaction of THF and the proton formed via H-abstraction of ClO_4. Both studies revealed that the polymer yield was increased with the reaction time and the anodic charge passed.

The cationic polymerization of trioxane [77–79] showed characteristic behaviors in different solvents. In particular in chlorinated hydrocarbons the current effects were related to the reaction kinetics and the polymer molecular weight.

In the electrochemically initiated polymerization of 1,3-dioxolane, the anode process was attributed to the monomer [80] and accordingly its intermediate oxidation products were believed to be the initiating species. An equilibrium was proposed which was independent of the amount of the initial charge. The conversion at equilibrium depended on temperature and initial monomer concentration. The process exhibited living features and autocatalysis at temperatures lower than 50°C.

The electrochemical polymerization of cyclohexylvinyl ether was investigated in chlorinated hydrocarbons using various tetraalkylammonium salts by Cerrai et al. [21]. Initiation was attributed to electrochemically formed protons. Anodic polarography measurements had excluded the oxidation of solvent and supporting salts and the authors believed that the initiating protons arose from the oxidation of residual water. On the other hand,

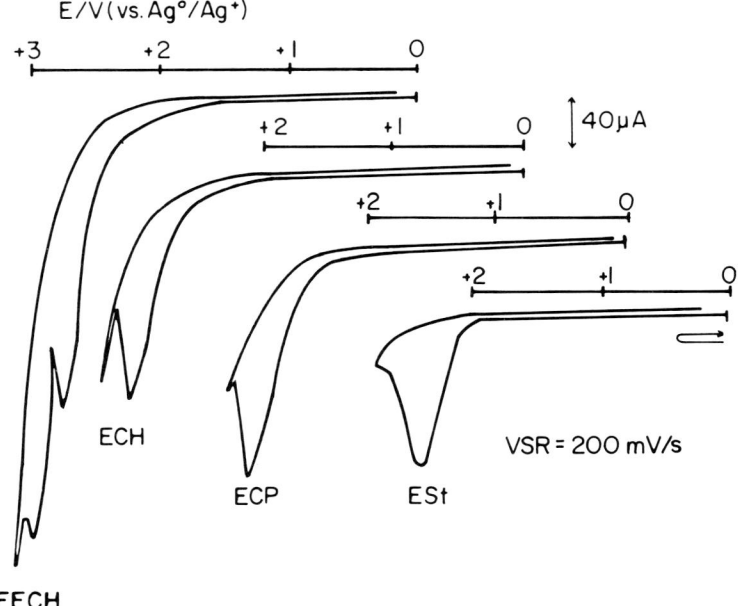

Figure 9 Cyclic voltammograms of some epoxides obtained in acetonitrile–tetrabutylammonium fluoroborate, at 25°C on a platinum wire vs. Ag°/AgBF$_4$. (From Ref. 81.)

Table 2 Electroinitiated Cationic Polymerization of Some Epoxides Carried Out in Dichloromethane on a Platinum Electrode[a]

Monomer[b]	$E_{p,a}$ (V)[c]	E_{pol}[c]	Maximum current (mA)	T (°C)	Monomer[d] reacted (%)	MP of polymer	[η] (dl g^{-1})
EECH	+2.70 +3.00	+2.30	2.0	−37	10	>365	
ECH	+3.00	+2.30	4.5	−37	68.8	93	0.071[e]
ECP	+2.90	+2.20	5.0	−35	69.0	−10	0.13[e]
Est	+2.40	+2.20	5.0	−20	>95	15	0.092[f]

[a]Working and auxiliary electrodes were 5 cm² platinum foils.
[b]EECH = 1,2-epoxy-4-epoxyethylcyclohexane; ECH = epoxycyclohexane; ECP = epoxycyclopentane; Est = epoxystyrene.
[c]Versus Ag°/AgBF$_4$ electrode.
[d]Measured by gas chromatography.
[e]Measured in benzene at 35°C.
[f]Measured in benzene at 30°C.
Source: Ref. 81.

cyclic voltammetric measurements proved that polymerization can be initiated through the monomer itself [81]. In acetonitrile–tetrabutylammonium fluoroborate couple the oxidation peak potentials of some epoxides could be determined (Fig. 9). The voltammograms show that $E_{p.a}$ of monomers are below the solvent–electrolyte discharge potential and also ruled out the presence of water. Potentiostatic electrolyses (CPE) carried out in dichloromethane yielded polymers of those epoxides via cationic mechanism by direct electron transfer (Table 2). Electroinitiated polymerization of 1,2-epoxy-4-epoxyethylcyclohexane (EECH) yielded crosslinked polymers on the surface of the anode [82]. The polymerization propagated via opening of the epoxy rings. The two oxidation peaks of EECH, +3.00 and +2.70 V versus Ag°/Ag^+, were assigned to the epoxy ring on the cyclohexane and epoxyethyl groups, respectively. The polymerization of epoxycyclohexane (ECH) was also achieved by constant-potential electrolyses [83]. The effect of the temperature on the polymerization rate is given in Table 3.

Electroinitiated polymerization of epoxycyclopentane (ECP) showed almost the same features as epoxides activated at constant potential [84]. A possible mechanism is given in Figure 10.

Table 3 Polymer Yields and Intrinsic Viscosities of Electroinitiated Polymerization of ECH Carried Out in Dichloromethane[a]

T (°C)	E_{pol}[b] (V)	I_{max}[c] (mA)	$[\eta] \times 10^2$ (dl g^{-1})	$\bar{M}_n$	Polymer[d] yield (%)
−35	2.30	4.5	7.30	9.8	68.7
−20	2.30	4.5	7.60	10.4	51.7
+5	2.30	4.5	5.65	7.3	39.6

[a]Data from Ref. 83.
[b]E_{pol} is polymerization potential, which was kept lower than $E_{p.a}$ of monomer (+3.00 V).
[c]I_{max} is maximum current observed.
[d]At the end of 90 min.

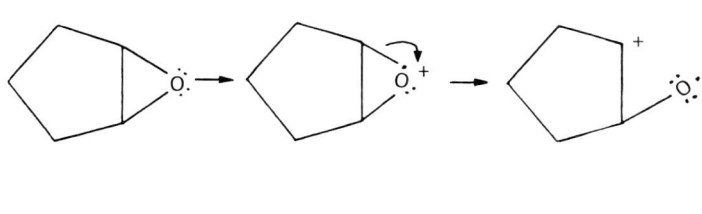

Figure 10 Direct initiation in the polymerization of epoxycyclopentane via controlled potential electrolysis. (Based on Ref. 84.)

Aldehydes

Strobel and Schulz [9] in 1970 polymerized acrolein in nitromethane solutions of lithium perchlorate. Spectral studies revealed that only vinyl-type units were encountered in the polymerization process.

$$n\text{CH}_2=\text{CH}-\text{C}(=\text{O})\text{H} \rightarrow \sim(\text{CH}_2-\text{CH}_2)_{\widetilde{n}}-\text{C}(=\text{O})\text{H}$$

The electrooxidation of furfural at constant current was achieved by Goncalves et al. [85] in acidic media at 30°C. Polymers had thermosetting characteristics. The spectroscopic analyses suggested a tridimensional structure.

N-Vinylcarbazole

In Breitenbach and Srna's 1962 paper [5] the cationic polymerization of N-vinylcarbazole (NVC) in nonaqueous solution was mentioned. Although the yield was good, molecular weights were reported to be low. Poly(N-vinylcarbazole) with an average molecular weight of 2000–5000 was synthesized electrochemically in the presence of zinc bromide by Philips et al. [86]. The reaction apparently proceeded with the formation of a monomer $ZnBr_2$ complex. At a mole fraction of 5:1 (monomer to $ZnBr_2$), ~70% conversion was reported. Polymer-modified electrodes were prepared by electropolymerization of NVC in acetonitrile [87]. The poly(N-vinylcarbazole) film coating on a semiconductor was followed in situ by parallel UV-visible absorption spectroscopy. The film was dark green resulting from the association of carbazolylium cation radicals with the anions of the supporting electrolyte.

Acrylic Acid and Derivatives

The anode-initiated polymerization of acrylic acid and methyl methacrylate derivatives was reported to be due to a free radical mechanism in a Kolbe reaction by several authors. The idea of a possible cation radical being responsible for the initiation of polymerization was proposed by Arnold and Swift [88]. A bipyridylium salt (BP^{2+}) was used as the source of the radicals:

$$BP^{2+} \xrightarrow{O_2} BP^+ \rightarrow BP^+ - O_2$$

Acenaphthylene

Alekseeva et al. [89] investigated the electroinitiated polymerization of acenaphthylene in acetonitrile solutions of various electrolytes. Monomer conversion increased with increasing initial monomer concentration and current density and it decreased by addition of traces of water into the acetonitrile. Conversion was also influenced by the nature of the supporting electrolyte. This direct effect of the electrolyte on the conversion is proof of an indirect method of initiation. A cyclic voltammogram of acenaphthylene yields the possibility of direct initiation through the monomer [90] (Fig. 11). Electrolyses carried out at +1.50V vs. $Ag°/Ag^+$ and −35°C in dichloromethane reached 55% conversion at the end of 2 hr [90].

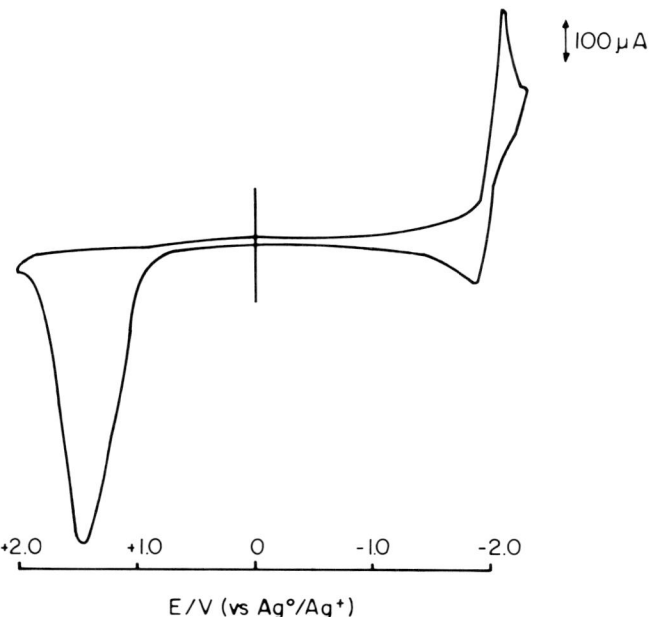

Figure 11 Cyclic voltammogram of acenaphthylene measured in acetonitrile–tetrabutylammonium fluoroborate on a platinum wire vs. Ag°/AgBF$_4$ reference electrode. Voltage scan rate was 200 mV sec^{-1}.

Indene

Several solutions of indene in the presence of lithium perchlorate were electrolyzed to prepare polyindene [91, 92] under controlled current conditions. With no reasonable explanation, solvents were described as useful or not in the electroinitiated polymerization of indene. Since there was no cyclic voltammetry prior to electrolyses, most probably the discharge potentials of the solvents played an important role in the initiation mechanism. Further anodic polymerization attempts at different temperatures were discussed by Mano and Calafate [22], who utilized dichloromethane–quaternary ammonium salts. The electroinitiated cationic polymerization was extensively studied by Bhadani and colleagues [93–95]. Effects of preelectrolysis, monomer concentration, electrolyte concentration, and electrode materials were discussed. A polymerization at constant electrode potential was also carried out. A detailed study of the electroinitiated cationic polymerization of indene appeared by Akbulut and co-workers [96].

Temperature independence of the anodic peak potential of indene was proved by cyclic voltammetry (Fig. 12). The apparent shift in the place of the $E_{p.a}$ was attributed to the film formation on the anode surface. CV measurements obtained by flaming the electrode surface before each run revealed that there was no change at the $E_{p.a}$ of the monomer (+1.40 vs. Ag°/Ag$^+$). The polymerization rate dropped very quickly during constant potential electrolysis at +1.40 V due to film formation on the anode. At +2.00 V the large IR drop due to coating was overcome and the polymerization rate was enhanced. The kinetics of polymerization showed the temperature dependence of the reaction with a positive apparent activation energy, which is rare in cationic polymerization.

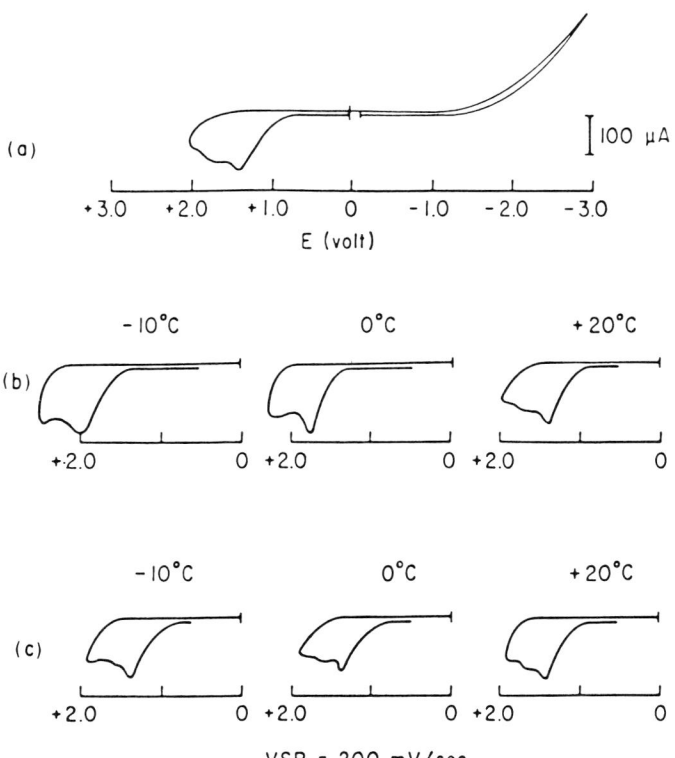

Figure 12 Cyclic voltammograms of indene measured on platinum wire vs. Ag°/AgBF$_4$. (*a*) Cyclic voltammogram obtained in acetonitrile–tetrabutylammonium fluoroborate, (*b*) cyclic voltammogram measurements in dichloromethane obtained successively without clearing the anode surface, and (*c*) cyclic voltammogram measurements obtained by flaming the anode before each run. (From Ref. 96.)

Isocyanates

Although polymerization of phenylisocyanate propagates through an anionic mechanism, one aspect in the electroinitiation technique of this monomer is worth mentioning here. Electroinitiated polymerization of isocyanate was reported by Shapoval et al. [97], who claimed that crystalline polymers of this monomer could be prepared via electrochemical methods. However, a later study by Akbulut [98] revealed that the product deposited on the cathode surface was a cyclic trimer. As the cathodic peak potential of the monomer could not be measured in CV measurements, Akbulut carried out an anodic process by constant-potential electrolysis at the oxidation peak potential of the monomer (+2.25 V vs. Ag°/Ag$^+$). He obtained no polymer in the anolyte but a crystalline product (trimer) on the cathode (Pt) surface as well as the amorphous polymer of phenylisocyanate in the catholyte. This sort of electroinitiation can be called indirect, although the anodic reaction is a direct electron transfer process.

Isoprene

Although there are many publications on the polymerization of isoprene via chemical methods, few studies have appeared on electroinitiation. Most of these studies discuss electroinitiated polymerization achieved by anionic mechanism via constant-current electrolysis. The application of constant-potential electrolysis to the polymerization of isoprene with a cationic mechanism was introduced by Akbulut et al. [99]. Prior to electrolysis the $E_{p.a}$ of isoprene was measured in acetonitrile by cyclic voltammetry (Fig. 13). To check the mechanism, i.e., to be sure that it was purely cationic, polymerization utilizing the same conditions in the presence of a radical inhibitor was accomplished. No decrease in the rate of polymerization was observed. Nevertheless, it was found that the polymerization rate decreased as the polymerization temperature was decreased.

Isoprene is yet another monomer with a positive apparent activation energy in cationic polymerization. The IR spectra of the polymer displayed a cyclized structure. This property of the resultant polymer was also attributed to the cationic initiation. It was suggested that when the monomer was absorbed on the anode surface, an electron would be transferred from the monomer to the electrode, possibly yielding a radical cation. The cation end then initiated the polymerization on the anode surface, yielding cyclic structures (Fig. 14).

Electroinitiated Copolymerization

Electroinitiated copolymerization reactions succeeded mainly with the aid of donor–acceptor complexes. Numerous examples of such copolymerizations are cited in literature. Funt and co-workers [100, 101], by using zinc bromide, succeeded in preparing

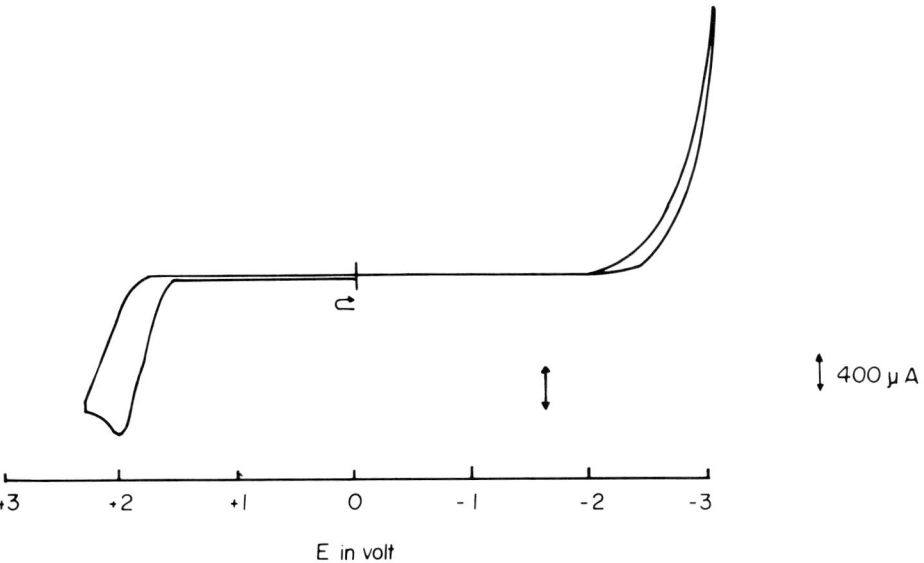

Figure 13 Cyclic voltammogram of isoprene, measured in acetonitrile–tetrabutylammonium fluoroborate on a platinum wire vs. Ag°/AgBF$_4$ reference electrode. (Based on Ref. 99.)

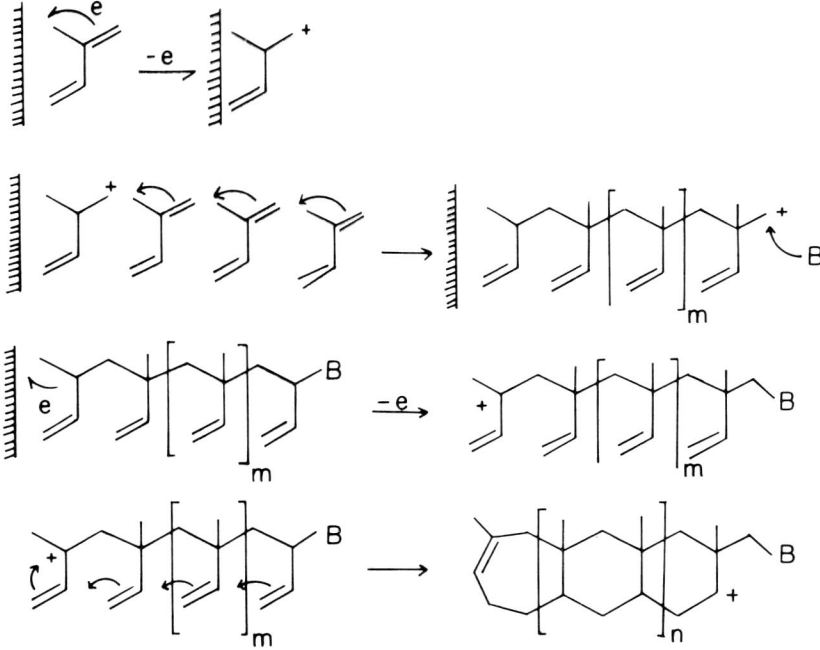

Figure 14 The proposed mechanism of direct initiation on a platinum anode in the constant-potential electrolysis of isoprene. (Based on Ref. 99.)

alternating copolymers of styrene and acrylonitrile. The yield was reported to be dependent on the current passed through the cell. The molecular weights were high. Phillips and colleagues [102–106] showed extensive kinetic and product correlations between photo- and electroinitiation in zinc salt–catalyzed vinyl aromatic systems. The first electroinitiated copolymerizations via direct electron transfers based on the oxidation peak potentials of the monomers were studied by Akbulut et al [107]. In principle three aspects were considered as a whole:

1. The applied polymerization potential (E_{pol}) effect on the resultant copolymer composition.
2. The E_{pol} effect on the reactivity ratios.
3. The possibility of keeping the electrode surface clean by means of methods other than stirring the solution or rotating the working electrode, which had previously been seen to have negative effects on the polymer molecular weight.

To determine the effects, if any, of E_{pol} on copolymer composition, copolymerization reactions were studied at various applied potentials [107]. In the copolymerization of α-methylstyrene and β-bromostyrene at applied potentials +2.20, +2.50, and +2.80 (vs. Ag°/Ag$^+$), the energy band of the anode reaches values lower than those corresponding $E_{p.a}$ of the monomers. As E_{pol} increases, the gap between the zero level and the energy band of anode increases (Fig. 2). The energy band of the anode during polymerization may fall below HOMO energies of the monomers at potentials above +2.00 V. As a

Table 4 Electroinitiated Copolymerization of α-Methylstyrene and β-Bromostyrene[a]

E_{pol}[b] (V)	T (°C)	% Composition[c]		Conversion[d] (%)	Reactivity ratio		$[\eta] \times 10^2$[e] (dl g^{-1})
		α-MetSt	β-BrSt		α-MetSt	β-BrSt	
+2.00	−30	91.0	9.0	7.6	1.40	1.12	
+2.20	−30	84.0	16.0	8.9	1.02	1.21	0.32
+2.50	−30	68.5	31.5	11.7	1.02	1.16	2.44
+2.80	−30	61.5	38.5	36.9	0.99	0.76	

[a]Data from Ref. 107.
[b]E_{pol} is the copolymerization potential.
[c]Calculated by IR spectroscopy.
[d]Measured by gas chromatography at the end of 90 min of electrolysis and calculated by $([M_1]_t + [M_2]_t/[M_1]_0 + [M_2]_0) \times 100$.
[e]$[\eta]$ measured in benzene at 30°C.

result, as evident from Table 4, more α-methylstyrene ($E_{p.a} = +1.85$ V) should be incorporated into the polymer. The contribution of β-bromostyrene ($E_{p.a} = +2.00$ V) can be increased only if E_{pol} is increased.

The reactivity ratios of the monomers were calculated by the integrated Lewis–Mayo equation [108]. This method requires gas chromatographic analysis of polymer solutions taken from the electrolysis cell at definite time intervals. One important aspect is that reactivity ratios were not found to be the same at different polymerization potentials. This clearly shows that in copolymerizations via constant-current electrolysis (CCE), r_1 and r_2 values represent only average values.

Electroinitiated cationic copolymerizations of indene-4-methoxystyrene and indene-β-bromostyrene monomer couples were studied by Toppare et al. [109]. As can be seen in Table 5, at the same temperature and applied copolymerization potential, the compositions of the two copolymers differ greatly. In the case of indene (M_1)–β-bromostyrene

Table 5 Copolymerization of Indene, β-Bromostyrene, and 4-Methoxystyrene by Constant Potential Electrolysis at 20°C

E_{pol}[a] (V)	%Composition[b]		Conversion[c] (%)	Reactivity ratio		$[\eta]$[d] (dl g^{-1})
	Indene	Monomer		Indene	Monomer	
+2.50	89.5	11.5[e]	8.8	0.88 ± 0.01	0.90 ± 0.01[e]	0.0712
+2.50	58.0	42.0[f]	13.3	0.84 ± 0.01	0.84 ± 0.01[f]	0.1745

[a]Copolymerization potential versus Ag°/Ag$^+$ reference electrode.
[b]Calculated by IR spectroscopy.
[c]Calculated by gas chromatography at the end of 150 min.
[d]Measured in benzene at 30°C.
[e]β-Bromostyrene.
[f]4-Methoxystyrene.
Source: Ref. 109.

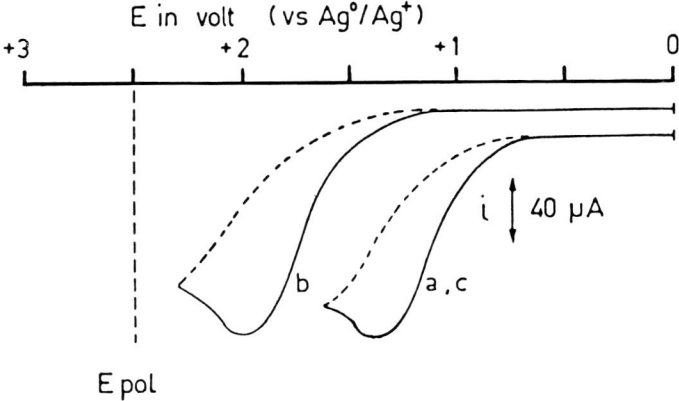

Figure 15 Cyclic voltammograms of indene (A), β-bromostyrene (B), and 4-methoxystyrene (C) measured in acetonitrile–tetrabutylammonium fluoroborate on a platinum electrode. (From Ref. 109.)

(M_2) copolymerization, the amount of M_1 in the copolymer is much higher than M_2. Since there is a large difference between the oxidation peak potentials of the monomers (Fig. 15), the observed composition is not surprising. On the other hand, when the $E_{p.a}$ values for both monomers are identical, as in the case of M_1–M_3(4-methoxystyrene) copolymerization, the incorporation of both monomers is comparable.

Electroinitiated copolymerization of indene and α-methylstyrene was discussed in another study by Toppare et al. [110]. Table 6 shows the expected increase of the higher $E_{p.a}$ monomer (α-methylstyrene) content in the copolymer with increasing E_{pol} for a variety of polymerization potentials. The temperature dependence of the copolymer com-

Table 6 Copolymerization of Indene and α-Methylstyrene[a]

E_{pol}[b] (V)	T (°C)	% mole composition[c]		Conversion[d] (%)	Reactivity ratios		$[\eta]$[e] (dl g^{-1})
		I	MS		I	MS	
+2.20	−30	73.0	27.0	15.3	0.84 ± 0.01	0.91 ± 0.01	
+2.50	−30	65.0	35.0	23.9	0.92 ± 0.01	0.95 ± 0.01	
+2.80	−30	50.0	50.0	24.5	0.94 ± 0.01	0.96 ± 0.01	0.077
+2.20	−40	53.0	47.0	11.9	1.03 ± 0.01	1.00 ± 0.01	
+2.50	−40	51.5	49.5	17.7	0.87 ± 0.01	0.92 ± 0.01	0.131
+2.80	−40	49.0	51.0	15.5	0.90 ± 0.01	0.94 ± 0.01	0.147

[a]Data from Ref. 110.
[b]Copolymerization potential vs. Ag°/Ag$^+$ reference electrode.
[c]Measured by IR spectoscopy.
[d]Measured by gas chromatography at the end of 120 min of electrolysis calculated by ([M_2]$_t$ + [M_2]$_t$/[M_1]$_0$ + [M_2]$_0$) × 100.
[e]Measured in benzene at 30°C.

position was also emphasized for such a couple. As discussed earlier, the rate of polymerization of indene [96] was increased with increasing polymerization temperature, whereas the reverse is true for α-methylstyrene polymerization [36]. In Table 6, it is shown that the temperature effect may counterbalance the applied potential effect. At any low E_{pol} (+2.00 V) the expectation is that mole composition of indene (lower $E_{p.a}$ monomer) will be greater than that of α-methylstyrene in the copolymer at a given temperature. With decreasing temperature it is observed that at the same E_{pol} (+2.00 V), the percentage compositions are comparable. Also with increasing E_{pol}, α-methylstyrene incorporation increases at both temperatures ($-30°$ and $-40°C$).

Electroinitiated cationic copolymerization of indene and styrene was also investigated [111]. Effects of copolymerization potential and temperature on the copolymer composition and the reactivity ratios are listed in Table 7.

Electroinitiated copolymerizations of styrene-4-methoxystyrene [112] and styrene-β-bromostyrene [113] revealed the same trends (Tables 8 and 9). This behavior of copolymer composition could be emphasized if the applied potentials (E_{pol}) were to be selected between the two oxidation peak potentials of the monomers M_1 and M_2. In Figure 12 it was shown that the working electrodes were coated with a polymer film which caused the $E_{p.a}$ to shift to more anodic potentials. To eliminate the passivation effect, either the electrodes should be cleaned continuously, which prevents kinetic study of copolymerization, or—better—the minimum copolymerization potentials should be found at which the resistance of the film can be overcome for each monomer couple. The latter was chosen in all studies mentioned. Lately another solution has been offered by Akbulut et al. [114]: the ultrasonic cleaning of the electrode with the help of ultrasonic waves, which do not

Table 7 Copolymerization of Indene and Styrene[a]

E_{pol}[b] (V)	T (°C)	% Composition[c] Indene	Styrene	Reactivity ratio Indene	Styrene	$[\eta]$[d] (dl g^{-1})	Conversion[e] (%)
+2.00	20	85.0	15.0	1.14 ± 0.01	1.21 ± 0.01		21.7
+2.20	20	74.5	25.5	0.90 ± 0.01	0.85 ± 0.01		24.7
+2.50	20	60.0	40.0	0.90 ± 0.01	0.88 ± 0.01	0.0425	23.8
+2.70	20	51.0	49.0	1.00 ± 0.01	1.05 ± 0.01	0.0413	31.1
+3.00	20	42.0	58.0	1.23 ± 0.01	1.30 ± 0.01	0.0468	34.0
+3.20	20	42.0	58.0	0.75 ± 0.01	0.63 ± 0.01	0.0594	26.9
+2.20	0	64.5	35.5	0.88 ± 0.01	0.83 ± 0.01		22.4
+2.50	0	61.0	39.0	1.03 ± 0.01	1.12 ± 0.01		27.6
+2.70	0	56.0	44.0	0.90 ± 0.01	0.85 ± 0.01		31.3
+2.20	−20	89.0	11.0	0.88 ± 0.01	0.84 ± 0.01		12.1
+2.50	−20	87.0	13.0	0.67 ± 0.01	0.64 ± 0.01		15.1
+2.70	−20	80.0	20.0	1.12 ± 0.02	1.16 ± 0.02		25.8

[a]Data from Ref. 111.
[b]Copolymerization potential Ag°/Ag$^+$ reference electrode.
[c]Calculated by IR spectroscopy.
[d]Measured in benzene at 30°.
[e]Calculated by gas chromatography at the end of 90 min of electrolysis calculated by $\{[M_1]_t + [M_2]_t/[M_1]_0 + [M_2]_0\} \times 100$.

Table 8 Electroinitiated Copolymerization of Styrene 4-Methoxystyrene at Different Applied Voltages[a]

E_{pol}[b] (V)	% Composition		Total[e] conversion (%)	r'_1[f] 4-MOS	r_2 ST	$[\eta]$[g] (dl g^{-1})
	4-MOS[c]	ST[d]				
+2.00	71.5	28.5	17.44	0.88 ± 0.01	0.67 ± 0.01	0.017
+2.20	67.5	32.5	10.88	0.80 ± 0.01	0.42 ± 0.01	0.168
+2.50	57.5	42.5	7.25	1.59 ± 0.02	2.50 ± 0.02	0.190

[a]$E_{p.a}$ of styrene is +1.80 V and that of 4-methoxystyrene is +1.40 V.
[b]Electrolysis potential vs. Ag°/Ag$^+$ reference electrode.
[c]4-Methoxystyrene.
[d]Styrene.
[e]Measured by gas chromatographic analyses of electrolysis solutions at the end of 150 min of electrolysis, calculated from $([M_1]_t + [M_2]_t)/([M_1]_0 + [M_2]_0) \times 100$.
[f]Calculated by integrated Lewis–Mayo equation.
[g]Measured in benzene at 30°C.
Source: Ref. 112.

Table 9 Copolymerization of Styrene and β-Bromostyrene by Constant-Potential Electrolysis in 1,2-Dichloroethane at 0°C[a]

E_{pol}[b] (V)	% Mole composition[c]		Conversion[f] (%)	$[\eta]$[g] (dl g^{-1})
	St[d]	β-BrSt[e]		
+2.50	82.5	17.5	7.9	0.0437
+2.80	72.0	28.0	21.1	
+3.00	68.0	32.0	25.8	

[a]Data from Ref. 113.
[b]Polymerization potential vs. Ag°/Ag$^+$ reference electrode.
[c]Calculated by IR spectroscopy.
[d]Styrene.
[e]β-Bromostyrene.
[f]Calculated by gas chromatographic analyses at the end of 150 min of electrolysis.
[g]Measured in benzene at 30°C.

cause appreciable convection in the electrolysis cell. Ultrasound introduced to the polymerization system prevents the polymer film coating on the electrode surface, thus enabling polymerization potentials comparable to $E_{p.a}$ of monomers to be applied [114]. In the copolymerization of isoprene and α-methylstyrene in the absence of ultrasound the concentration of reacted isoprene decreased with increasing E_{pol}, whereas that of the reacted α-methylstyrene remained almost constant. This behavior persists throughout polymerization and is related to film formation on the anode surface. At lower $E_{p.a}$ values the formation of the thin polymeric film requires a longer time than at high E_{pol} values

(Fig. 16). However, in the presence of ultrasound, which brings about rapid adsorption–desorption, the electrode surface was kept clean almost until the end of the electrolysis. As a result, a slight increase in the reacted monomer concentrations versus E_{pol} was observed.

The effect of polymerization potential on the total percentage composition is shown in

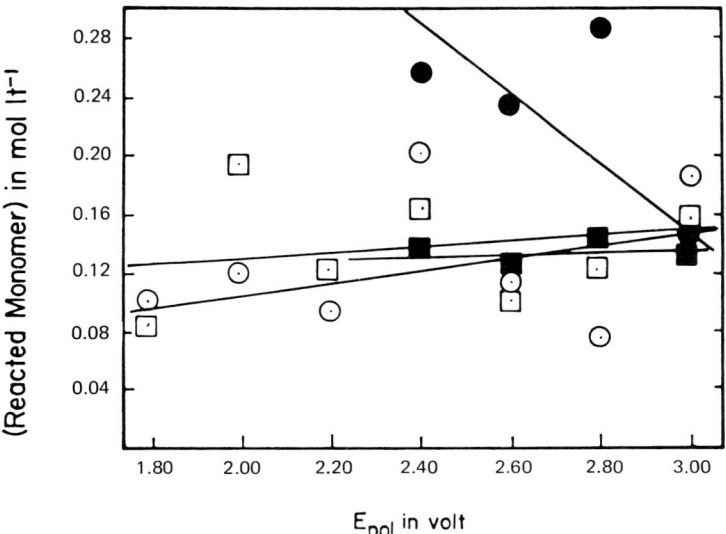

Figure 16 Effect of polymerization potential on the reacted monomer concentration in the copolymerization of isoprene with α-methylstyrene at −30°C. ● Reacted isoprene concentration in the absence of ultrasonic waves; ⊙ in the presence of ultrasonic waves; ■ reacted α-methylstyrene concentration without ultrasound; ⊡ with ultrasound. (From Ref. 114.)

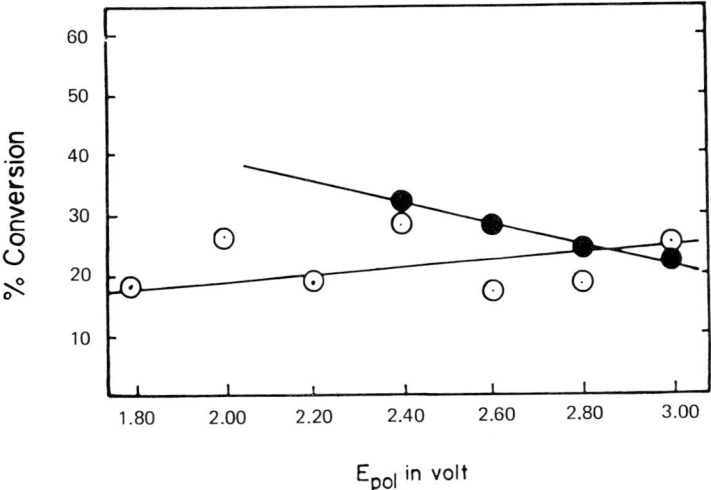

Figure 17 Effect of polymerization potential on conversion in the cationic copolymerization of isoprene with α-methylstyrene. ● in the absence and ⊙ in the presence of ultrasonic waves. (From Ref. 114.)

Figure 17. Since the total conversion was calculated from reacted isoprene and α-methylstyrene concentrations and reacted monomer concentration decreases with increasing E_{pol}, the total percentage conversion decreases as expected. On the other hand, with the presence of waves the conversion shows a slight increase with increasing E_{pol} owing to the clear electrode surface.

It was also observed that E_{pol} affected the copolymer composition (Fig. 18). When $E_{p.a}$ values of the monomers are considered, the increase of E_{pol} enhances the incorporation of isoprene in copolymers. Nevertheless, this trend in the presence of ultrasonic waves does not continue at higher E_{pol} values (Tables 10 and 11).

The effects of E_{pol} on reacted monomer concentration, reactivity ratios, and composition of copolymers in the absence of ultrasound are illustrated in Figure 19. Reacted monomer concentrations have a convergence point at $E_{pol} = +3.00$ V, which may be

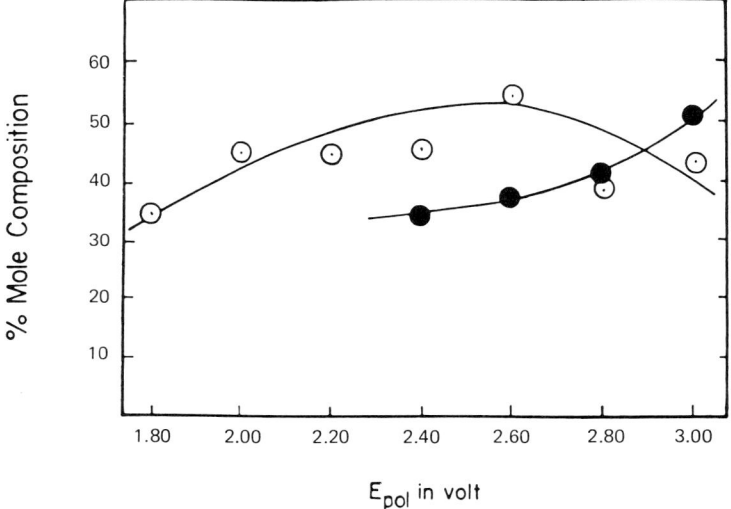

Figure 18 Effect of polymerization potential on the isoprene content (mol %) of isoprene-α-methylstyrene copolymers. ● without ultrasound and ⊙ with ultrasound. (From Ref. 114.)

Table 10 Effect of Polymerization Potential on Electroinitiated Cationic Copolymerization of Isoprene with α-Methylstyrene in the Absence of Ultrasonic Vibration at $-30°C$

E_{pol} (V)	% Mole composition[a]		Reactivity ratio		Conversion[b] (%)	Yield (%)	$[\eta]^a$ (dl g^{-1})
	Isoprene	α-Methylstyrene	Isoprene	α-Methylstyrene			
+2.40	33.0	67.0	0.26 ± 0.02	0.19 ± 0.02	32.8	0.4	0.112
+2.60	36.0	64.0	0.36 ± 0.01	0.23 ± 0.01	28.8	3.3	0.110
+2.80	39.5	60.5	0.48 ± 0.01	0.50 ± 0.01	24.5	5.5	0.098
+3.00	50.0	50.0	0.80 ± 0.01	0.75 ± 0.01	22.9	14.4	0.044

[a]Calculated for 5 hr of electrolysis.
[b]Calculated for 2 hr of electrolysis.
Source: Ref. 114.

Table 11 Effect of Polymerization Potential on Electroinitiated Cationic Copolymerization of Isoprene with α-Methylstyrene in the Presence of Ultrasonic Vibration at 30°C

E_{pol} (V)	% Mole composition[a]		Reactivity ratio		Conversion[b] (%)	Yield[a] (%)	$[\eta]$[a] (dl g^{-1})
	Isoprene	α-Methylstyrene	Isoprene	α-Methylstyrene			
1.80	34.0	66.0	0.53 ± 0.01	0.62 ± 0.01	18.4	3.9	0.051
2.00	44.5	55.5	0.97 ± 0.01	0.72 ± 0.01	26.2	36.7	0.046
2.20	43.5	57.0	0.94 ± 0.03	0.89 ± 0.03	19.3	23.6	0.045
2.40	44.0	56.0	0.81 ± 0.01	0.78 ± 0.01	28.4	21.7	0.049
2.60	53.5	46.5	0.77 ± 0.02	0.77 ± 0.02	17.1	34.9	0.057
2.80	38.0	62.0	0.70 ± 0.02	0.98 ± 0.02	18.3	41.5	0.045
3.00	42.0	58.0	0.95 ± 0.01	0.97 ± 0.01	20.5	36.9	0.050

[a]Calculated for 5 hr of electrolysis.
[b]Calculated for 2 hr of electrolysis.
Source: Ref. 114.

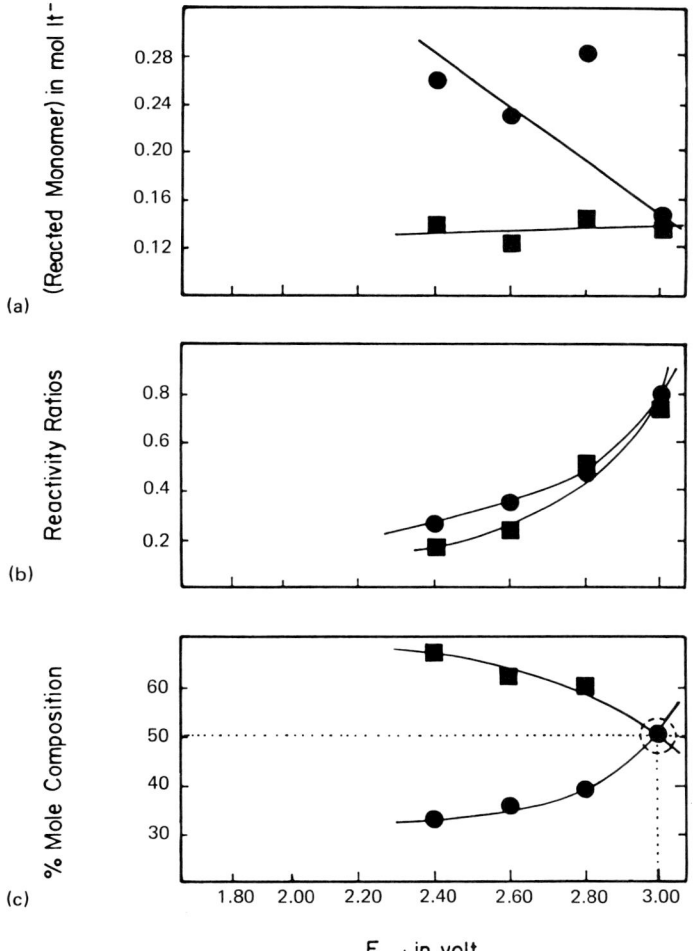

Figure 19 Effect of polymerization potential on (*a*) reacted monomer concentration, (*b*) reactivity ratios, (*c*) mol % composition in the copolymerization of isoprene (●) with α-methylstyrene (■) in the absence of ultrasound at −30°C. (From Ref. 114.)

called the critical polymerization potential (E_{crit}). This behavior can also be seen for reactivity ratios and percentage composition of the copolymer.

An argument similar to that of Ref. 114 can be made concerning the copolymerization of isoprene and styrene [115] (Tables 12 and 13). Percentage compositions of copolymers at various E_{pol} show that at $E_{pol} = +3.30$ V isoprene content in the copolymer becomes equal to that of styrene (Fig. 20). Hence this potential where r_1 approaches r_2 and reacted monomer concentrations become equal to each other can be called the critical polymerization potential of styrene–isoprene couple. As a result, ultrasonic vibration can be counted as a means of keeping the electrode surface clean, provided that the correlation between E_{pol} and copolymer composition is lost. Yet in these studies the existence of critical polymerization potential suggests that the reactivity ratios and copolymer composition can be controlled by the choice of E_{pol}.

Table 12 Effect of Polymerization Potential on Electroinitiated Cationic Copolymerization of Isoprene with Styrene in the Absence of Ultrasonic Vibration at 0°C[a]

E_{pol} (V)	% Mole composition[b]		Reactivity ratio		Conversion[c] (%)	Yield[b] (%)	$[\eta]^b$ (dl g^{-1})
	Isoprene	Styrene	Isoprene	Styrene			
+2.30	26.0	74.0	0.50 ± 0.03	0.36 ± 0.03	38.9	5.8	0.092
+2.70	44.5	55.5	0.57 ± 0.01	0.39 ± 0.01	37.2	12.9	0.055
+3.00	48.5	51.5	0.65 ± 0.02	0.55 ± 0.02	24.4	38.3	0.049
+3.30	50.5	49.5	1.17 ± 0.01	1.29 ± 0.01	14.9	67.3	0.844

[a]Data from Ref. 115.
[b]Calculated for 5 hr of electrolysis by using the precipitated copolymers.
[c]Calculated for 2 hr of electrolysis by gas chromatographic analysis of polymer solutions removed from the cell.

Table 13 Effect of Polymerization Potential on Electroinitiated Cationic Copolymerization of Isoprene with Styrene in the Presence of Ultrasonic Vibration at 0°C[a]

E_{pol} (V)	% Mole composition[b]		Reactivity ratio		Conversion[c] (%)	Yield (%)	$[\eta]^b$ (dl g^{-1})
	Isoprene	Styrene	Isoprene	Styrene			
+1.80	46.0	54.0	0.46 ± 0.01	0.58 ± 0.01	21.2	3.9	0.082
+2.00	22.0	78.0	0.45 ± 0.02	0.96 ± 0.02	13.4	2.4	0.086
+2.30	56.5	43.5	0.62 ± 0.01	0.63 ± 0.01	14.3	4.0	0.072
+2.60	66.5	33.5	0.89 ± 0.03	0.62 ± 0.03	25.7	3.1	0.055
+2.80	75.0	25.0	0.67 ± 0.01	0.63 ± 0.01	23.4	29.1	0.058
+3.00	61.0	39.0	1.14 ± 0.03	0.90 ± 0.03	35.1	34.2	0.084
+3.30	49.0	51.0	1.32 ± 0.02	1.05 ± 0.02	38.1	34.3	0.068

[a]Data from Ref. 115.
[b]Calculated for 5 hr of electrolysis by using the precipitated copolymers.
[c]Calculated for 2 hr of electrolysis by gas chromatographic analysis of polymer solutions removed from the cell.

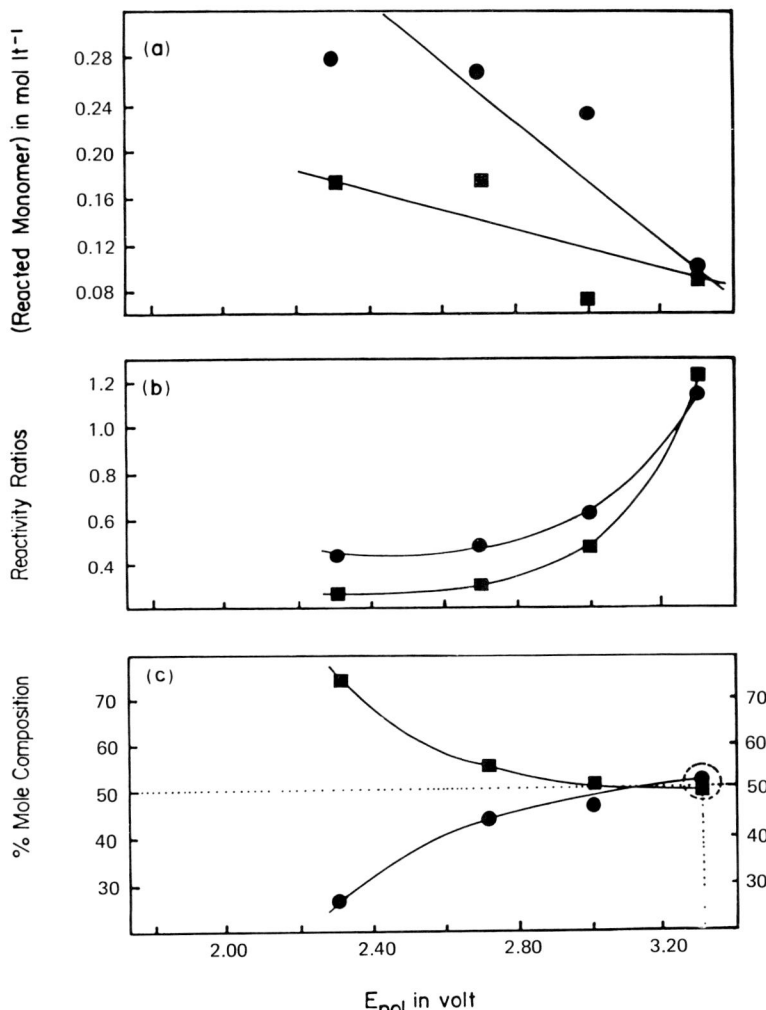

Figure 20 Effect of polymerization potential on (*a*) reacted monomer concentration, (*b*) reactivity ratios, and (*c*) % composition, in the copolymerization of isoprene (●) with styrene (■) in the absence of ultrasound at 0°C. (Based on Ref. 115.)

CONCLUSIONS

A comparison of the number of papers published in the five years following 1982 with the five-year period preceding 1982 indicates the growing interest in the field of electroinitiated cationic polymerization. It is appropriate to suggest the wide employment of the controlled potential method rather than CCE for the sake of mechanistic determination in electroinitiated polymerization.

Electroinitiated polymerization has not yet led to the synthesis of new polymers that are unattainable by other techniques. Although the electroinitiated polymerization process for some monomers has been proven to be more favorable experimentally, the monomers can be polymerized by conventional methods as well. The main advantage of electroinitiation most probably lies in the selectivity of electrochemistry. Similar chemical environments of very similar monomers may mean very different electron transfer potentials; thus they may be regarded as having electrochemically different environments.

There is a possibility of using electrochemistry as a tool of polymerization to build up a polymer by a series of successive steps in which the subtraction or addition of each electron would result in the linking of further monomer units to the chain. Another highlight of electroinitiation can be the production of particular polymeric configurations such as block polymers. It is also possible to couple two methods of initiation, photopolymerization and electropolymerization. One method can be used in the initiation process while the other can be the terminating method. Molecular weight control along with the desired molecular weight distribution can be achieved by electrochemical means. Within the electrochemical limits it also seems possible that the copolymers with desired compositions and reactivity ratios can be synthesized. Another favorable aspect of this technique is probably the economics of the process. In general, in an electroorganic synthesis the maximum efficiency can represent one mole of substrate per faraday of electricity, whereas in electropolymerization few electrons account for a polymer chain.

New areas of fundamental interest could be surface coatings and conducting polymers. In the last decade considerable developments were made in these areas. It is also possible to produce polymers with flame-resistant properties. For some halophenols several oxidation peak potentials can be measured by cyclic voltammetry in a single run where every electron transfer process may bring different characteristics of polymers such as linear, branched, or crosslinked. These will surely affect the physical properties of the resultant polymers coated on the electrode as heat or flame resistant. Experimentally, by being aware of such oxidation potentials, it is feasible to carry out such polymerization reactions, whereas with chemical initiators or other techniques control of the structure may entail problems.

REFERENCES

1. E. C. Szarvasy, *J. Chem. Soc.*, 77: 207 (1900).
2. E. A. Rembold, Ph.D. dissertation, Ohio State University, Columbus (1947).
3. C. L. Wilson, *Rec. Chem. Prog.*, 10: 25 (1949).
4. J. W. Breitenbach, *Makromol. Chem.*, 2: 171 (1960).
5. J. W. Breitenbach and C. Srna, *Pure Appl. Chem.*, 4: 245 (1962).
6. C. F. Heins, *J. Polym. Sci. B,* 7: 625 (1969).
7. S. N. Bhadani, Ph.D. dissertation, University of Manitoba, Winnipeg (1966).
8. N. Yamazaki, *Adv. Polym. Sci.*, 6: 377 (1969).
9. W. Strobel and R. C. Schulz, *Makromol. Chem.*, 133: 303 (1970).
10. B. L. Funt and T. J. Blain, *J. Polym. Sci. A-1,* 8: 3339 (1970).
11. B. L. Funt and T. J. Blain, *J. Polym. Sci.*, 9: 115 (1971).
12. J. W. Breitenbach, *Adv. Polym. Sci.*, 9: 47 (1972).
13. M. M. Baizer, *Nature,* 56: 405 (1969).
14. U. Akbulut, J. E. Fernandez, and R. L. Birke, *J. Polym. Sci. Polym. Chem.*, 13: 133 (1975).
15. U. Akbulut, R. L. Birke, and J. E. Fernandez, *Makromol. Chem.*, 179: 2507 (1978).
16. P. Giusti, *Makromol. Chem.*, 175, 1157 (1974).

17. K. Yasuo and M. Hiroso, *J. Macromol. Sci. Chem.*, *8*(3): 573 (1984).
18. K. Yasuo and M. Nakoishi, *Nippon Kagaku Kaishi*, *11*: 2221 (1972).
19. G. Mengoli and G. Vidotto, *Makromol. Chem.*, *165*: 137 (1973).
20. G. Parravano, in *Organic Electrochemistry* (M. M. Baizer, ed.), Marcel Dekker, New York, p. 947 (1973).
21. P. Cerrai, G. Guerra, and M. Tricoli, *Eur. Polym. J.*, *15*(2): 153 (1979).
22. E. B. Mano and A. L. B. Calafate, *Rev. Quim. Ind.*, *50*(586): 10 (1981).
23. E. B. Mano and A. L. B. Calafate, *J. Polym. Sci. Polym. Chem.*, *19*(12): 3325 (1981).
24. E. P. Kovalchuk, L. A. Murkind, and L. P. Laurishchev, *Lakokras. Mater. Ikhtiol. Primen.*, *2*: 15 (1980).
25. J. E. Dubois and P. C. Lacaze, *J. Electrochem. Soc.*, *130*(2): 346 (1983).
26. A. F. Diaz, J. I. Castillo, J. A. Logan, and W. Y. Lee, *J. Electroanal. Chem. Int. Electrochem.*, *129*(1–2): 115 (1981).
27. W. C. Neikam, G. R. Diemeler, and M. Desmond, *J. Electrochem. Soc.*, *111*: 1191 (1964).
28. M. Fleishmann and D. Pletcher, *Tetrahedron Lett.*, *60*: 6255 (1968).
29. G. J. Hoijtink, *Rev. Trav. Chim.*, *77*: 555 (1958).
30. M. Katz and H. Wendt, *Electrochim. Acta*, *21*: 215 (1976).
31. U. Akbulut, L. Toppare, and L. Türker, *J. Polym. Sci. Polym. Chem.*, *23*: 1631 (1985).
32. A. Streitwieser, *Molecular Orbital Theory for Organic Chemists*, Wiley, New York (1966).
33. U. Akbulut, Assoc. Prof. dissertation, Middle East Technical University, Ankara (1978).
34. L. Toppare, Ph.D. dissertation, Middle East Technical University, Ankara (1982).
35. B. M. Tidswell and A. G. Doughty, *Polymer*, *12*: 241 (1971).
36. U. Akbulut, S. Eren, and L. Toppare, *Polymer*, *25*: 1028 (1984).
37. P. Cerrai, P. Giusti, G. Guerra, and M. Tricoli, *Eur. Polym. J.*, *11*: 101 (1974).
38. P. Cerrai, G. Guerra, and M. Tricoli, *Eur. Polym. J.*, *12*: 247 (1975).
39. P. Cerrai, G. Guerra, M. Tricoli, and L. Nucci, *Eur. Polym. J.*, *16*: 867 (1980).
40. M. S. Tsvetkov and E. P. Koval'chuk, *Ukr. Khim. Zh.*, *35*: 1217 (1969).
41. M. Y. Fioshin, E. P. Koval'chuk, and M. S. Tsvetkov, *Electrokhimiya*, *5*: 1188 (1969).
42. N. S. Tsvetkov and E. P. Koval'chuk, *Vysokomol Soedin.*, *B11*: 42 (1969).
43. O. T. Fernandez and R. Larrocha, *Pitture Vernici*, *60*(7): 93 (1984).
44. S. N. Bhadani and S. Kundu, *J. Polym. Mater.*, *1*(1): 31 (1984).
45. W. E. Garrison, *Diss. Abstr.*, *20*: 77 (1959).
46. S. M. Toy, *J. Electrochem. Soc.*, *114*: 1042 (1967).
47. A. Gozlan and A. Zilkha, *Eur. Polym. J.*, *20*(12): 1199 (1984).
48. J. J. Langer, *Mater. Sci.*, *10*(1–2): 173 (1984).
49. N. E. Wisdom, U.S. Pat. 3,437,570 (1967).
50. Esso, Br. Pat. 1,136,699 (1968).
51. Esso, Fr. Pat. 1,534,181 (1968).
52. A. F. Shepard and B. F. Dannels, *J. Polym. Sci. A-1*, *4*: 551 (1966).
53. A. F. Shepard and B. F. Dannels, U.S. Pat. 3,386,899 (1968).
54. Matsushita Elec. Ind. Co. Ltd. Jpn. Pat. 58,196,233 (1983).
55. Matsushita Elec. Ind. Co. Ltd. Jpn. Pat. 58,187,432 (1983).
56. Matsushita Elec. Ind. Co. Ltd. Jpn. Pat. 59,207,932 (1984).
57. Facultes Universitaires Notre-Dame de la Paix. Belg. Pat. 899,925 (1984).
58. B. F. Goodrich Co. U.S. Pat. 4,487,667 (1984).
59. P. Audebert and G. Bidan, *Mol. Cryst. Liq. Cryst.*, *118*(1–4): 187 (1985).
60. G. Wegner and W. Werner (BASF A.G.). Ger. Pat. DE 3,318,856 (1984).
61. F. Andruzzi, P. Cerrai, G. Guerra, L. Nucci, A. Pescia, and M. Tricoli, *Eur. Polym. J.*, *18*: 685 (1982).
62. W. R. Grace and Co. Br. Pat. 1,156,309 (1968).
63. W. F. H. Borman, Can. Pat. 800,592 (1968).
64. R. Kijkstra and J. De Jonge, Extended Abstract, IUPAC, Boston (1971).

65. M. M. Musiani, G. Mengoli, B. Pelli, and C. Folnari, *Makromol. Chem.*, *183*(8): 1869 (1982).
66. Y. Ohnuki, H. Matsuda, and T. Ohsaka, *J. Electroanal. Chem. Int. Electrochem.*, *158*(1): 55 (1983).
67. F. Bruno, M. C. Phan, and J. E. Dubois, *Electrochim. Acta*, *22*(4): 451 (1977).
68. Mitsubishi Chem. Ind. Co. Ltd. Jpn. Pat. 59,56,426 (1984).
69. E. Tsuchide, H. Nishida, and T. Maekawa, *J. Macromol. Sci. Chem.*, *A21*(8–9): 1081 (1984).
70. R. J. Waltman and A. F. Diaz, *J. Electrochem. Soc.*, *132*(3): 631 (1985).
71. R. W. Dyck, M.Sc. Thesis, Simon Fraser University, Burnaby, B.C. (1969).
72. G. Mengoli and G. Vidotto, *Makromol. Chem.*, *50*: 277 (1971).
73. G. Mengoli and G. Vidotto, *Makromol. Chem.*, *53*: 57 (1972).
74. G. Mengoli and G. Vidotto, *Eur. Polym. J.*, *8*: 671 (1972).
75. A. N. Dey and E. J. Rudd, *J. Electrochem. Soc. Electrochem. Sci. Tech.*, 1294 (1974).
76. S. Nakahama, S. Hino and N. Yamazaki, *Polym. J.*, *2*(1): 56 (1971).
77. G. Mengoli and G. Vidotto, *Makromol. Chem.*, *165*: 137 (1973).
78. G. Mengoli and G. Vidotto, *Makromol. Chem.*, *165*: 145 (1973).
79. G. Mengoli and S. Valcher, *Eur. Polym. J.*, *10*: 959 (1974).
80. G. Mengoli and S. Valcher, *Eur. Polym. J.*, *11*: 169 (1975).
81. U. Akbulut, L. Toppare, A. Usanmaz, and A. Önal, *Makromol. Chem. Rapid Commun.*, *4*: 259 (1983).
82. A. Önal, A. Usanmaz, U. Akbulut, and L. Toppare, *Br. Polym. J.*, *15*: 187 (1983).
83. U. Akbulut, A. Önal, A. Usanmaz, and L. Toppare, *Br. Polym. J.*, *15*: 179 (1983).
84. A. Önal, A. Usanmaz, U. Akbulut, and L. Toppare, *Br. Polym. J.*, *16*: 102 (1984).
85. S. R. Goncalves, M. A. Tronco, A. Colla, P. Janissek, and T. Rabockai, *An. Simp. Bras. Electroquim. Electroanal.* 4th, (1984).
86. D. C. Phillips, D. H. Davies, and J. D. B. Smith, *Macromolecules*, *5*(6): 674 (1972).
87. P. J. Lacaze, J. E. Dubois, A. Desbene-Monvernay, P. L. Desbene, J. J. Basselier, and D. Richard, *J. Electroanal. Chem. Int. Electrochem.*, *147*(1–2): 107 (1983).
88. R. Arnold and D. A. Swift, *Aust. J. Chem.*, *22*: 859 (1969).
89. T. A. Alekseeva, E. G. Olkhovskaya, and V. D. Bezuglyi, *Zh. Prikl. Khim.*, *54*(9): 2107 (1981).
90. L. Toppare, Middle East Technical University, unpublished results (1984).
91. Y. Kikuchi and H. Fukuda, *J. Polym. Sci. Polym. Chem.*, *11*(10): 2709 (1973).
92. Y. Kikuchi and N. Masahiko, *Nippon Kagaku Kaishi*, *11*: 2221 (1972).
93. S. N. Bhadani and P. P. Baranwal, *Makromol. Chem.*, *178*: 1049 (1977).
94. S. N. Bhadani and P. P. Baranwal, *Makromol. Chem.*, *178*: 2637 (1977).
95. S. N. Bhadani, P. P. Baranwal, and Y. K. Prasad, *Makromol. Chem.*, *179*: 1623 (1978).
96. U. Akbulut, S. Eren, and L. Toppare, *J. Macromol. Sci. Chem.*, *A21*(3): 335 (1984).
97. G. S. Shapoval, E. M. Skobetz, and N. P. Markova, *Vysokomol. Soedin.*, *8*: 1313 (1966).
98. U. Akbulut, *Makromol. Chem.*, *180*: 1073 (1979).
99. U. Akbulut, L. Toppare, and B. Yurttas, *J. Polym. Sci. Polym. Lett.*, *24*: 185 (1986).
100. B. L. Funt, I. McGregor, and J. Tanner, *J. Polym. Sci. B*, *8*(10): 699 (1970).
101. B. L. Funt, I. McGregor, and J. Tanner, *J. Polym. Sci. B*, *8*(10): 695 (1970).
102. D. C. Phillips, D. H. Davies, and J. D. B. Smith, *Makromol. Chem.*, *154*: 32 (1972).
103. D. C. Phillips, J. D. B. Smith, and D. H. Davies, *J. Polym. Sci. A-1*, *10*: 3267 (1972).
104. D. H. Davies, D. C. Phillips, and J. D. B. Smith, *J. Polym. Sci. A-1*, *10*: 3253 (1972).
105. D. C. Phillips, J. D. B. Smith, and D. H. Davies, *J. Polym. Sci. Polym. Chem.*, *11*: 1867 (1973).
106. D. H. Davies, D. C. Phillips, and J. D. B. Smith, *J. Polym. Sci. Polym. Chem.*, *17*: 1153 (1979).
107. L. Toppare, S. Eren, and U. Akbulut, *Br. Polym. J.*, *16*: 71 (1984).

108. R. V. Der Meer, H. N. Linssen, and A. L. German, *J. Polym. Sci. Polym. Chem.*, *16*: 2915 (1978).
109. L. Toppare, S. Eren, L. Türker, and U. Akbulut, *Polymer, 25*: 1655 (1984).
110. L. Toppare, S. Eren, and U. Akbulut, *J. Polym. Sci. Polym. Chem.*, *22*: 2941 (1984).
111. L. Toppare, S. Eren, and U. Akbulut, *Br. Polym. J.*, *17*: 257 (1985).
112. L. Toppare, S. Eren, Ö. Özel, and U. Akbulut, *J. Macromol. Sci. Chem. A21*(10): 1281 (1984).
113. L. Toppare, S. Eren, and U. Akbulut, *J. Polym. Sci. Polym. Chem.*, *23*: 303 (1985).
114. U. Akbulut, L. Toppare, and B. Yurttas, *Polymer, 27*: 803 (1986).
115. U. Akbulut, L. Toppare, and B. Yurttas, *Br. Polym. J.*, *18*: 273 (1986).

9
Radiation-Induced Reactions of Polystyrene Derivatives

Katsumi Tanigaki and Kazuo Tateishi
*NEC Corporation,
Kawasaki City, Japan*

INTRODUCTION	308
FUNDAMENTAL ASPECTS OF PHOTOLYSIS FOR POLYSTYRENE DERIVATIVES	308
CONSIDERATION OF ENERGETIC GROUNDS	310
Primary Photochemical Reaction	310
Dynamics of the Reaction	311
Secondary Photochemical Reaction	312
Difference Between Photolysis and Radiolysis	313
Quantum Yields	314
SPECTROSCOPIC APPROACHES TO THE ANALYSIS OF RADIATION-INDUCED CHANGES IN POLYMER FILMS	317
Photooxidation	317
Photocoloration of Polystyrene Derivatives	319
Radiation-Induced Chemical Reactions	319
LOW MOLECULAR WEIGHT MODEL COMPOUND STUDY	320
Unresolved Problems	320
General Considerations for Low Molecular Weight Model Compounds	321
FINAL PRODUCT ANALYSES OF GAS CHROMATOGRAPHY–MASS SPECTROMETRY	322
Halogen-Free Substituent Effect on Photolysis	324
Halogen Substituent Effect on Photolysis	324
ESR ANALYSES AT 77 K	328
RADIATION-INDUCED REACTIONS OF LOW MOLECULAR WEIGHT MODEL COMPOUNDS	330
NH_2, OH, OCH_3, CH_3, H, and $COCH_3$ Derivatives	330
F, Cl, Br, and I Derivatives	330
CH_2Cl Derivative	331
Hydrogen Abstraction by Chlorine Radicals from Each Derivative	332
Direct Evidence of Photocoloration of Polystyrene Derivatives	334
Intramolecular Rearrangements of α-Radicals	336
CONSIDERATION OF RADIATION-INDUCED REACTIONS OF POLYSTYRENE DERIVATIVE POLYMER FILMS	336
CONSIDERATION OF THE DIFFUSION OF ACTIVE GASES IN A POLYMER FILM	336

APPLICATION OF POLYSTYRENE DERIVATIVES TO LITHOGRAPHIC FIELDS	338
CONCLUSIONS	338
APPENDIX: MASS DATA AND FINAL PRODUCT	340
REFERENCES	343

INTRODUCTION

Polystyrene derivatives are part of the famous family of vinyl polymers [1–7], which have been used for a long time in many scientific and engineering fields. Many researchers have been attracted to these polymers because of their variety of syntheses, simple molecular structures, and interesting characteristics. Several researchers are still engaged in the study of the radiation-induced reactions of polystyrene derivatives, which show complicated reaction features even when their structures are simple.

In the early stages of research, gas product analyses were conducted to monitor the photolysis of polystyrene derivatives. Grassie and Weir made intensive efforts to clarify the photolysis mechanism using their own apparatus specially designed for this [8–12]. Burlant and Serment measured gas products of the γ-radiolysis of p-substituted polystyrenes [13]. Mass analyses of gaseous products as well as ultraviolet and infrared absorption analyses have given important information on the radiation-induced reactions of polystyrene derivatives [14–37]. Electron spectroscopy (especially ESCA) has also been applied to analyze the changes in polystyrene derivative film surfaces exposed to radiation [38–40]. Recently, time-resolved spectroscopies, laser flash photolysis, and pulse radiolysis have been used to investigate the reaction kinetics of polystyrene derivatives [41–51]. Even now, these methods are yielding important information. In general, however, the radiation-induced reactions of polymer systems are difficult to understand. In particular, radiation-induced reactions of each monomer unit have eluded direct and/or indirect observations.

We try in this chapter to elucidate the radiation-induced reactions of monomer units, using low molecular weight model compounds whose structures are similar to the repeating units of polystyrene derivatives [52, 53]. To relate the analysis results for the model compounds to those in polymer systems, we consider the diffusion problem of gaseous products in a polystyrene film.

The purpose of this chapter is to review some of the progress in radiation-induced reactions of polystyrene derivatives. The problems still unresolved are presented. The details of our approach to the problems remaining in the grey area are stated and we present the evidence used to clarify unresolved phenomena. Further, the use of polystyrene derivatives in the promising areas of photolithography and high-energy radiation lithography is covered. Design methods are presented from the viewpoint of sensitivity, on the basis of the clarified radiation-induced reaction mechanisms.

FUNDAMENTAL ASPECTS OF PHOTOLYSIS FOR POLYSTYRENE DERIVATIVES

Polystyrene derivatives absorb light at wavelengths below 300 nm. Polymers irradiated by light around 250 nm are excited to their higher energy levels and undergo several reactions: oxidation, coloration, crosslinking, degradation, etc.

When polystyrene derivatives are irradiated by 254-nm light (energy = 471 kJ mol^{-1}), they are excited into a level S_n^* (usually $\pi \to \pi^*$ transition), only about 30 kJ mol^{-1} above the S_1. The excited molecules drop to the S_1 level, then very rapidly lose their energy through an $S_1 \to S_0$ radiationless transition. Deexcitation of the excited phenyl singlet and the $S_1 \to S_0$ radiationless transition occur with quantum yields in excess of 0.5 [54]. Further deactivation occurs through monomer and excimer fluorescence processes, the rates of which are 1×10^9 sec^{-1} and 5.3×10^7 sec^{-1}, respectively [55, 56]. In neat polystyrene derivative films the excimer fluorescence is more frequently seen [44], as the concentration of monomer is notably higher than in solution. The S_1 molecules undergo chemical reactions or otherwise move to the T_1 level through intersystem crossing in competition with S_1 to S_0 deactivation. Molecules in the T_1 state also react photochemically or lose their energy through radiationless or phosphorescence processes. These processes are seen in Figure 1.

The photochemical reactions of polystyrene derivatives take place from S_1 and/or T_1 states. The following equation is applied:

$$\Sigma \Phi_i = \Phi_F + \Phi_P + \Phi_{IC} + \Phi_R = 1$$

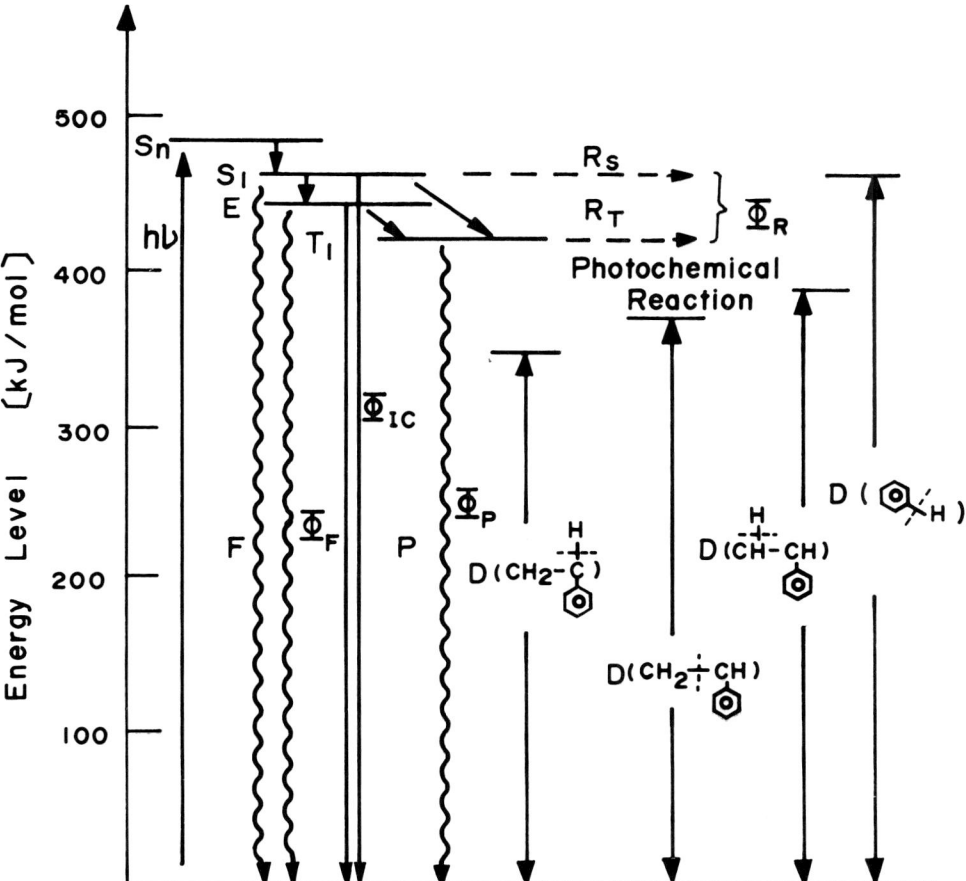

Figure 1 Photochemical reaction processes of polystyrene derivatives.

where Φ_F is the quantum yield of fluorescence, Φ_P is the quantum yield of phosphorescence, Φ_{IC} is the quantum yield of radiationless transition (internal conversion), and Φ_R is the quantum yield of photochemical reaction (from S_1 and/or T_1). Some part of the absorbed energy $(1 - \Phi_F - \Phi_P - \Phi_{IC})$ is used for the photochemical reaction, and the quantum yields Φ_R are reported to typically range from 10^{-5} to 10^{-3} [14–37].

CONSIDERATION ON ENERGETIC GROUNDS

Primary Photochemical Reaction

Initial photochemical reactions are associated with bond fissions adjacent to phenyl rings. The dissociation energy of each bond of polystyrene derivatives is seen in Figure 2. The energy equivalent to a 254-nm wavelength is 471 kJ mol^{-1}, which is greater than almost all of the bond dissociation energies. Not all of the energy absorbed, however, is transferred to a photochemical reaction, some of it being lost with emission or nonradiatively. The actual amount of energy transferred to the photochemical reaction is not so large, which explains why bond fission occurs selectively.

A great number of researchers have been studying the photochemical reactions of polystyrene for many years. The most frequently observed bond fission is reported to be β-fission (type 3) [58–68]. This result seems to reflect that the bond dissociation energy C–H of type 3 is smaller than the other bond dissociation energies. When an α-hydrogen is replaced by a methyl group [poly(α-methylstyrene)], bond fission of type 2 and/or 1 is generally reported, radicals generated from which degrade high molecular weight polymers to low molecular weight polymers [69–73]. This result also seems to reflect that the bond dissociation energies C–C and C–H of types 1 and 3 are considerably less than the other chemical bonds.

When the reactivity of substituents X is high enough compared with that of C–H (type 3), the features of photochemical reactions change markedly. In addition to α-hydrogen cleavage from polystyrene backbones, other bond fissions from substituents take place. In the caption of Figure 2, bond dissociation energies are listed for typical substituents. As can be seen among the energies listed, halogens (especially Cl, Br, I), CH$_2$–Cl, and CO–

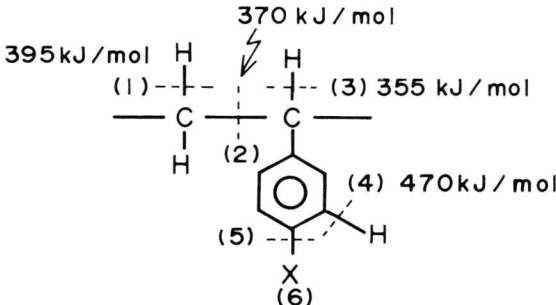

Figure 2 Bond dissociation energies E_b of polystyrene derivatives. $E_b[(5)]$: 426 kJ mol^{-1} for X = CH$_3$, 426 kJ mol^{-1} for X = OCH$_3$, 468 kJ mol^{-1} for X = OH, 418 kJ mol^{-1} for X = NH$_2$, 522 kJ mol^{-1} for X = F, 334 kJ mol^{-1} for X = Br, 271 kJ mol^{-1} for X = I, 405 kJ mol^{-1} for X = COCH$_3$; $E_b[(6)]$: 288 kJ mol^{-1} for CH$_2$–Cl, 350 kJ mol^{-1} for CO–CH$_3$, 355 kJ mol^{-1} for CH$_2$–H. (Data from Ref. 57.)

CH_3 have small bond dissociation energies. Complicated reactions different from those of polystyrene are expected in the case of poly(halogenostyrene), poly(chloromethylstyrene), and poly(acetylstyrene).

Dynamics of the Reaction

In addition to energetic considerations, reaction kinetics is another important factor to be taken into account; i.e., consideration of the reaction rates in each step is very important for an understanding of the reaction features observed. Recently computer technology has been much advanced and the dynamics of the reactions have been pursued directly by many researchers [41–51].

The dynamics of polystyrene irradiated with 266-nm wavelength light is seen in Figure 3. Polystyrene is excited by light into its first excited singlet state, forming an excimer with ground-state polystyrene. Photochemical reactions seem to occur from this excimer state; the contributions of the excited triplet and ionization are small [44]. The most frequently observed α-radicals are produced with a rate of $k_R = 3 \times 10^8$ sec^{-1} through homolytic splittings. These macro α-radicals are relatively stable, live for as long as 20–30 μs, and give rise to a crosslinking reaction with a rate of about 4×10^4 sec^{-1}.

Compared to the reaction rate of the α-radicals with oxygen, the crosslinking reaction rate of macro α-radicals is quite slow. During the crosslinking reaction, oxygen attacks the macro α-radicals to form peroxyradicals. Oxidation usually occurs in competition with the crosslinking reaction via this process [74]. Excited singlet oxygen is reported to be responsible for the oxidation [75–77]. Oxidized products are formed via these macroperoxyradicals. The accepted oxidation processes of polystyrene are summarized in Figure 4. From the macroperoxyradicals formed, aldehyde, carboxylic acid, and phenol-type products are formed during continuous irradiation.

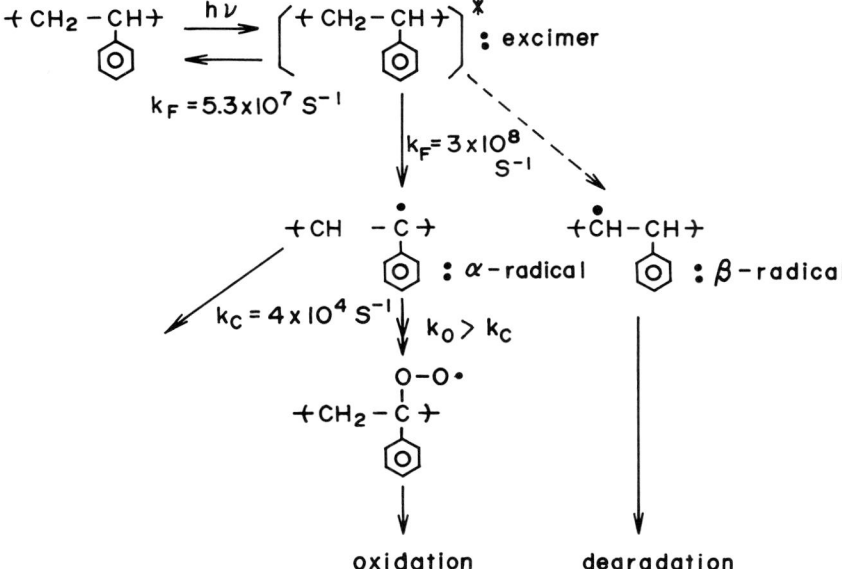

Figure 3 Kinetics of the photolysis of polystyrene.

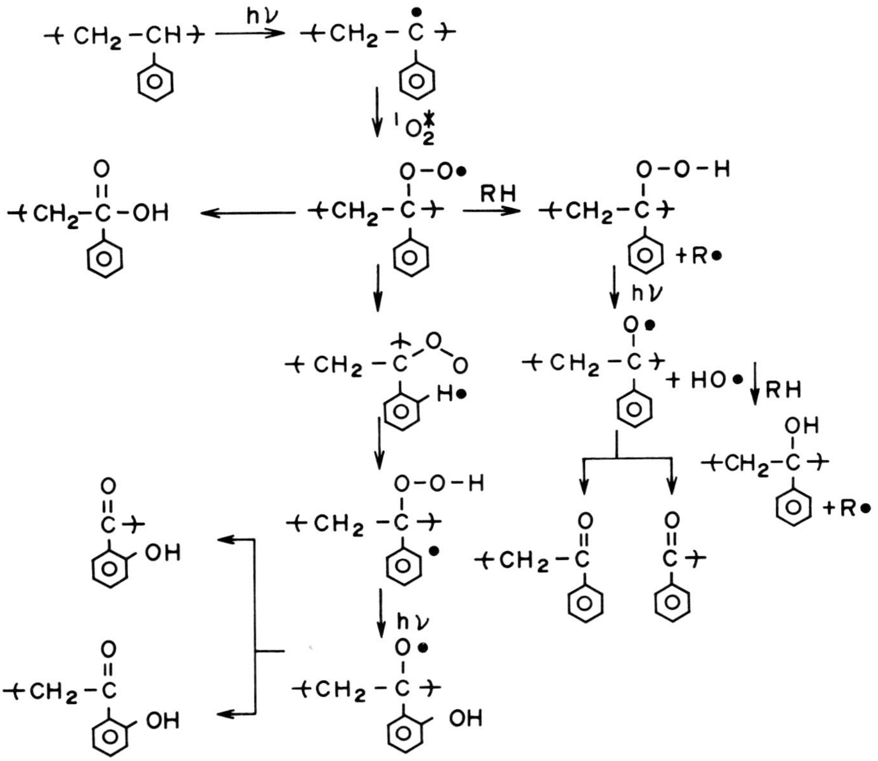

Figure 4 Typical oxidation processes of polystyrene.

Secondary Photochemical Reaction

In addition to macroradicals produced, the primary photochemical reactions give rise to gaseous free radicals at the same time. The gaseous free radicals abstract hydrogen from the monomer units of polystyrene derivatives, subsequently giving rise to other macroradicals besides the primary macroradicals [26, 28]. This is a well-known secondary photochemical reaction.

The secondary macroradicals can also be responsible for a crosslinking reaction. If the macroradicals are produced through a secondary reaction (a subsequent hydrogen abstraction reaction) before primary macroradicals undergo a crosslinking reaction, the secondary macroradicals produced will be involved in the crosslinking reaction in the same manner as the primary macroradicals. The macroradicals produced through a subsequent hydrogen abstraction often play an important role in photochemical reactions.

Coloration phenomena usually observed in the photolysis of polystyrene derivatives (absorption of polystyrene derivatives at long wavelengths increases in proportion to the time for which they are irradiated [11, 21]) are considered to be partly due to the secondary hydrogen abstraction reaction. The double bonds conjugated with benzene rings, which are formed from primary macroradicals or secondary macroradicals, are believed to be one of the origins of the photocoloration.

Difference Between Photolysis and Radiolysis

The photon energy of 254-nm wavelength light cannot ionize organic molecules, but high-energy radiations, electron beams, x-rays, γ-rays, etc., ionize polystyrene derivatives. The cations $P^{+\cdot}$ are produced through the ionization process, accompanying the ejection of electrons. The electrons ejected move very fast as quasi-free electrons and induce the primary dissociation reaction through dissociative electron attachment. In another way, they collide with the produced cations to give rise to excited molecules through a geminate recombination process. This is a major difference between photolysis and radiolysis. The differences are summarized in each reaction step classified in Figure 5: first stage, primary reaction, secondary reaction, and final stage.

At the first stage of a reaction photolysis proceeds through excitation of polystyrene derivatives, but radiolysis begins with their ionization. In the case of polystyrene, mainly the excimer fluorescence of polystyrene is observed on photolysis [43]. On the other hand, the absorptions of cations, radicals, and an excited triplet are also observed on radiolysis [44] in the same time range monitored, as seen in Figure 6. That is, on radiolysis cations are produced through direct ionization, and excited singlet/triplet molecules are produced through direct excitation and/or geminate recombination.

The first macroradicals are produced through dissociation in a primary reaction. Free radicals are produced very fast in radiolysis through a dissociative electron attachment as well as through a homolytic splitting of the excited state singlet and/or triplet molecules. The excited molecules also interact with ground-state molecules and produce free radicals.

Gaseous anions and radicals are simultaneously produced through dissociative electron attachment or homolytic splitting. These anions participate in a secondary reaction in such a way that they interact with the macrocation produced in the first ionization step and form

	Photolysis	Radiolysis
First stage	$P \xrightarrow{h\nu} P^x$	$P \rightsquigarrow P_\bullet^+ + e^-$ (ionization) $P \rightsquigarrow P^x$ (excitation) $P_\bullet^+ + e^- \rightarrow P^*$ (geminate recombination)
Primary reaction	$P^x \rightarrow R_1\bullet + X\bullet$ $P^x + P \rightarrow R_1\bullet + \lvert P\text{--}X \rvert$	$P^x \rightarrow R_1\bullet + X\bullet$ (homolitic splitting) $P + e^- \rightarrow R_1\bullet + X^-$ (dissociative electron attachment) $P^x + P \rightarrow R_1\bullet + (P\text{---}X)$
Secondary reaction	$P\bullet + X\bullet \rightarrow (P\text{---}X)$ $(P\text{---}X) \rightarrow R_2\bullet + HX$	$P + X\bullet \rightarrow (P\text{---}X)$ $P_\bullet^+ + X^- \rightarrow (P\text{---}X)$ $(P\text{---}X) \rightarrow R_2\bullet + HX$
Final stage	$R_1\bullet + R_1\bullet \rightarrow R_1\text{-}R_1$ $R_1\bullet + R_2\bullet \rightarrow R_1\text{-}R_2$ $R_2\bullet + R_2\bullet \rightarrow R_2\text{-}R_2$ $X\bullet + X\bullet \rightarrow X_2$	$R_1\bullet + R_1\bullet \rightarrow R_1\text{-}R_1$ $R_1\bullet + R_2\bullet \rightarrow R_1\text{-}R_2$ $R_2\bullet + R_2\bullet \rightarrow R_2\text{-}R_2$ $X\bullet + X\bullet \rightarrow X_2$

Figure 5 Photolysis and radiolysis of polystyrene derivatives, P = polymer, R_1 = primary produced macroradical, R_2 = secondary produced macroradical, X = active species ($X\cdot$ = radical, X^- = anion), [P---X] = charge transfer complex produced through the interaction between P and X.

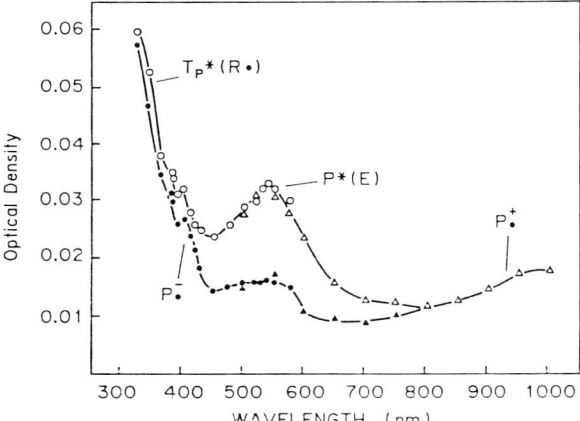

Figure 6 Transient absorption spectra observed in the pulse radiolysis of polystyrene solid films at 10 ns (○,△) and 30 ns (●,▲) after 2 ns electron pulses, T_{P*} : triplet, $R\cdot$: macroradical, $P^{\cdot-}$: anion radical, $P^*(E)$: excimer. (Data from Ref. 44.)

some transients [P----X]. The gaseous free radicals produced through homolytic splitting also interact with parent molecules, forming the same transients. Free radicals involved in a final stage reaction are subsequently produced from these transients in a secondary reaction.

The free radicals produced in the primary and secondary reactions collide to react and give rise to final products in a final stage reaction. Photolysis and radiolysis show no differences in the final stage of a reaction.

Quantum Yields

In general, the quantum yields of deexcitation and radiationless deactivation of polystyrene derivatives are very large. Thus the observed quantum yield of the photochemical reactions is quite low at $\Phi_R = 10^{-5}$ to 10^{-3}, which reflects this. The summarized quantum yields of photolysis for typical polystyrene derivatives are listed in Table 1. The G values of radiolysis by ^{60}Co-γ rays for polystyrene derivatives are also listed in this table. The order of the quantum yield of each derivative seems to show a correlation between photolysis and radiolysis.

Comparing the quantum yields of $\Phi(H_2)$ and $\Phi(HX)$ among these derivatives, it is found that electron-donating groups (CH_3, OCH_3, OH, NH_2) increase $\Phi(H_2)$, and conversely that electron-abstracting groups ($COCH_3$, NO_2) decrease $\Phi(H_2)$. The quantum yield $\Phi(H_2)$ is increased by the presence of halogen substituents (especially Cl, Br, I) and is accompanied by liberated hydrogen halogenide gases observed as $\Phi(HX)$.

The quantum yields of photochemical crosslinking Φ_C and G values of γ-ray radiation-induced crosslinking G_C are listed in Table 2. The quantum yields and G values of crosslinking can be related to those of released gases as follows:

$$\Phi_C = 2\Phi(H_2) + 2\Phi(HX) + \cdots$$

and

$$G_C = 2G(H_2) + 2G(HX) + \cdots$$

Table 1 Quantum Yields and G Value of Polystyrene Derivatives on Photolysis and Radiolysis

$-\!\!+\!\mathrm{CH_2-CH}\!\!+_n$ phenyl-X, X =	Φ at 254 nm	G $^{60}Co-\gamma$ ray
H	$\Phi_{H_2} = 0.0006$	$G_{H_2} = 0.039$
CH_3	$\Phi_{H_2} = 0.0010$ $\Phi_{C_2H_6} = 1.4 \times 10^{-6}$ $\Phi_{CH_4} = 6 \times 10^{-6}$	$G_{H_2} = 0.043$
OCH_3	$\Phi_{H_2} = 0.0016$ $\Phi_{CH_3OH} = 0.0008$ $\Phi_{CH_4} = 0.0004$ $\Phi_{C_2H_6} = 0.0003$	$G_{H_2} = 0.041$ $G_{CH_4} = 0.064$
OH		
NH_2	$\Phi_{H_2} = 0.0018$	
$COCH_3$		
F	$\Phi_{H_2} = 0.0006$ $\Phi_{HF} = 0.00006$	
Cl	$\Phi_{H_2} = 0.0024$ $\Phi_{HCl} = 0.0030$	$G_{H_2} = 0.017$
Br	$\Phi_{H_2} = 0.0060$ $\Phi_{HBr} = 0.0021$	$G_{H_2} = 0.017$
I		
NO_2	$\Phi_{H_2} = 0.0003$	$G_{H_2} = 0.026$
Et	$\Phi_{H_2} = 0.0007$ $\Phi_{C_2H_6} = 0.00004$ $\Phi_{CH_4} = 0.00007$	
iPr	$\Phi_{H_2} = 0.0008$ $\Phi_{C_2H_6} = 0.00007$ $\Phi_{CH_4} = 0.00015$	
tBu	$\Phi_{H_2} = 0.0010$ $\Phi_{C_2H_6} = 0.0001$ $\Phi_{CH_4} = 0.0002$ $\Phi_{C_4H_{10}} = 0.00005$	
CN		$G_{H_2} = 0.014$

Table 2 Quantum Yields and G Value of Crosslinking Units for Polystyrene Derivatives

$-(CH_2-CH)-$ phenyl-X, X =	Φ_c	G_c	ρ density
H	—	0.05	1.06
F	—	—	1.18
Cl	0.004	0.61	1.23
Br	0.008	0.41	1.58
I	—	—	1.73
OH	—	0.04	1.20

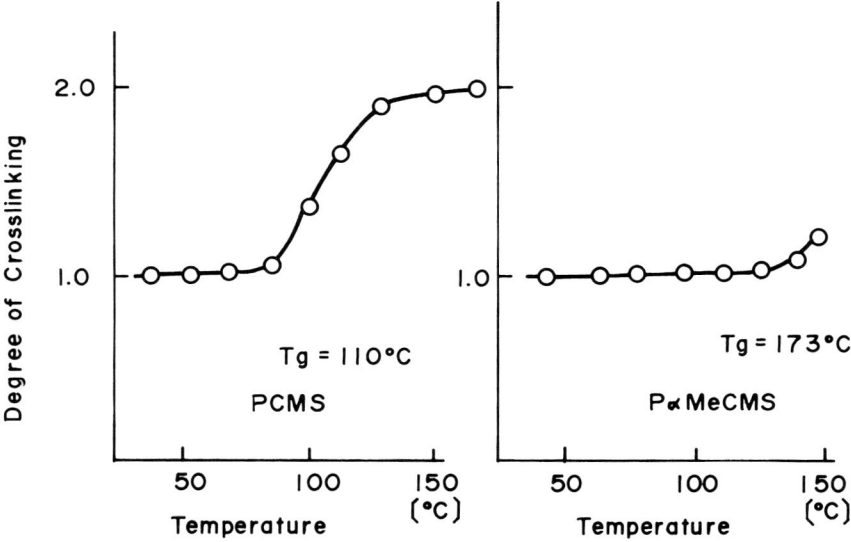

Figure 7 Dependence of degree of crosslinking on temperature for poly(chloromethylstyrene) and poly(α-methylchloromethylstyrene).

The calculated value [e.g., in the case $X = Cl$, $2\Phi(H_2) + 2\Phi(HCl) = 0.0108$], however, is much higher than the observed one ($\Phi_C = 0.004$), probably because the macroradicals produced remain in solid films without reacting. There are experiments indicating that not all of the free radicals produced participate in a crosslinking reaction. In Figure 7, the dependence of the degree of crosslinking reactions on temperature is shown for poly-(chloromethylstyrene) and poly(α-chloromethylstyrene), which have different glass transition temperatures: $T_g = 110$ and $173°C$. Notably large changes are seen around the glass transition temperatures. This result supports the view that the free radicals remain without reacting below the T_g of the film matrices.

SPECTROSCOPIC APPROACHES TO THE ANALYSIS OF RADIATION-INDUCED CHANGES IN POLYMER FILMS

To date, several spectroscopic methods have been applied to measure the changes occurring in polymer films on photolysis or radiolysis. These have provided important information on some aspects of photolysis and radiolysis of polystyrene derivatives: photooxidation, photocoloration, and radiation-induced reactions.

Photooxidation

Photooxidation of polystyrene is an important phenomenon from the viewpoint of polymer stability and degradation. Ultraviolet (UV) absorption and infrared (IR) spectroscopy have given important information on this [8–12, 74–78].

The UV absorption spectra of polystyrene before and after exposure to light in air are shown in Figure 8. The spectra show that the absorption increases around 240 nm and at wavelengths longer than 290 nm. These results indicate that double bonds conjugating with phenyl rings are formed.

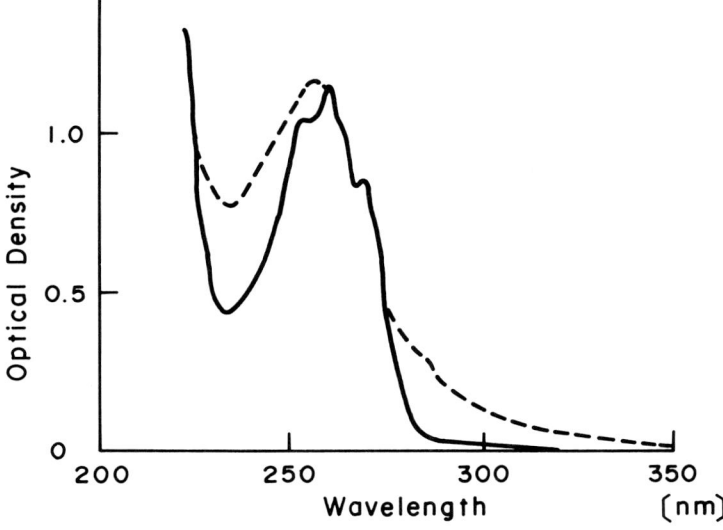

Figure 8 Ultraviolet spectra of a polystyrene film before (———) and after (------) exposure to light.

The IR spectra of polystyrene before and after exposure in air are shown in Figure 9. In this figure, the OH absorption band (about 3500 cm^{-1}) and the C=O absorption band (about 1700 cm^{-1}) clearly appear after exposure to UV light for 5 hr. These results suggest that the photoinduced radicals react with oxygen to form aldehyde and carboxylic acid.

To clarify the oxidation products, final product analyses were conducted using *sec*-butylbenzene [75]. Some aldehyde, carboxylic acid, and phenol-type products were detected.

Electron spectroscopy for chemical analysis (ESCA) has also been conducted to observe oxidation of polystyrene derivatives [40]. The ESCA spectra of polystyrene before and after exposure to light in air are shown in Figure 10. The signals corresponding to

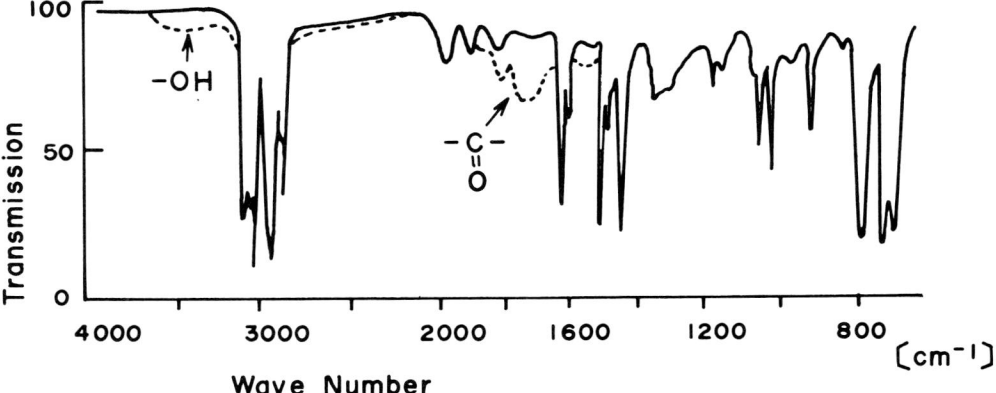

Figure 9 Infrared spectra of a polystyrene film before (———) and after (------) exposure to light.

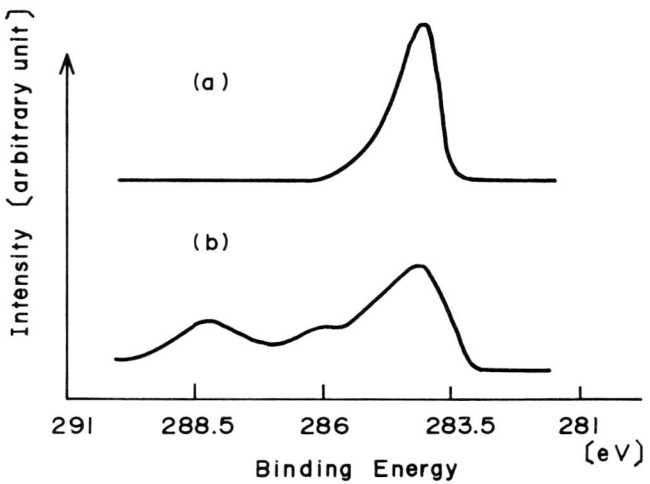

Figure 10 ESCA spectra of C_{1s} core level for a polystyrene film (*a*) before and (*b*) after exposure to light in air. (Data from Ref. 40.)

oxidized carbon appear in a higher bonding energy region than that of the original carbon C_{1S} core signal.

From these experiments, the photooxidation of polystyrene is considered to occur as shown in Figure 4. The oxidation proceeds through the interaction of the photoinduced α-radical with the photoexcited singlet oxygen.

Photocoloration of Polystyrene Derivatives

It is frequently observed that irradiated polystyrene derivatives change from colorless to yellow. This is generally called a photocoloration phenomenon. The measured UV absorption spectra before and after exposure to light clearly indicate that absorption at longer wavelengths increases as the degree of the coloration increases. For instance, such spectra of poly(p-bromostyrene) are shown in Figure 11. The origin of photocoloration has been ascribed to double bonds which are formed through subsequent hydrogen abstraction.

Radiation-Induced Chemical Reactions

Polymers irradiated by light or high-radiation beams change their chemical and physical properties. Some gases are released from the polymer films. The liberated gases can be monitored by a mass spectrometer [13–34], and their amounts can also be measured by a specially designed vacuum apparatus (the first intensive study was conducted by Grassie and Weir around 1970 using their own apparatus [8–12]).

Figure 12 shows volatile gaseous product formation from poly(halogenostyrene)s exposed to light at 254 nm under high vacuum. These results indicate that hydrogen gas and hydrogen halogenide gases are liberated from the films. Thus both hydrogen and halogens are cleaved on photolysis.

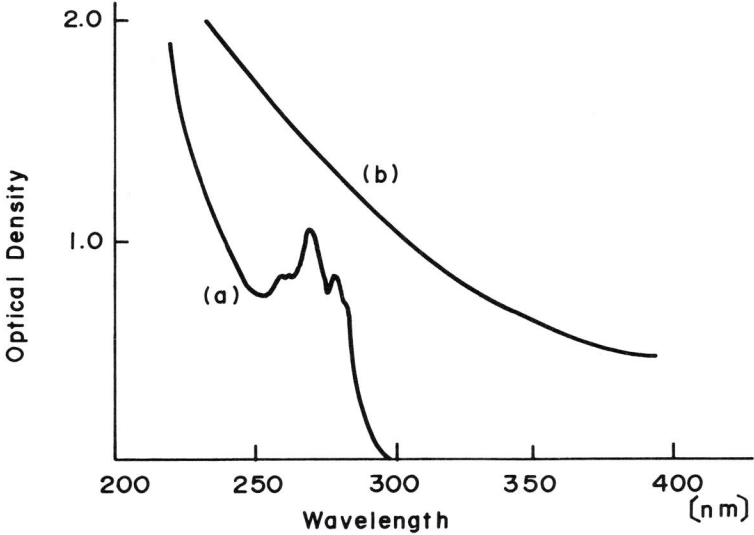

Figure 11 Ultraviolet spectra of a poly(4-bromostyrene) film (a) before and (b) after exposure to light. (Data from Ref. 21.)

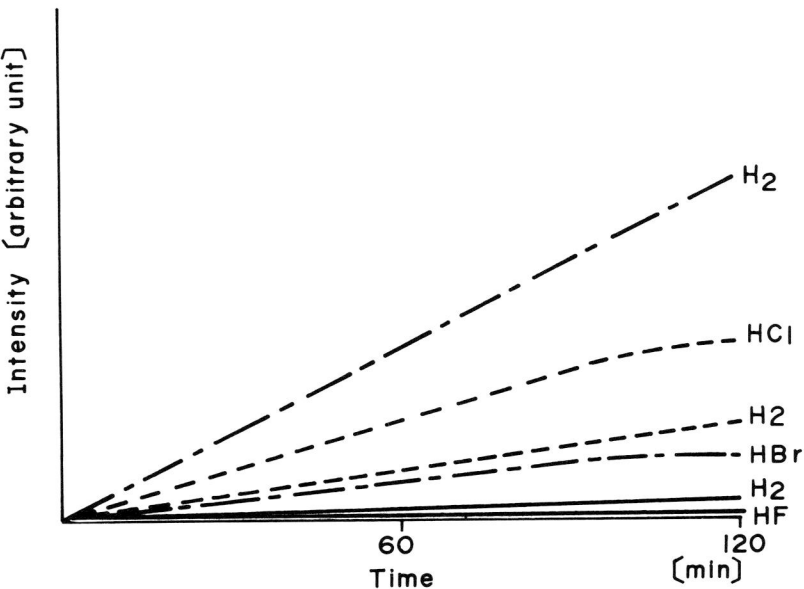

Figure 12 The production yields of gases from exposed poly(halogenated styrene)s, poly(4-fluorostyrene) (———), poly(4-chlorostyrene) (------), poly(4-bromostyrene) (– – –). (Data from Ref. 21.)

The composition of polystyrene derivatives is changed by exposure to light of high-energy beams. These changes in polymer films can be examined by ESCA. ESCA spectra before and after irradiation by electron beams are shown for poly(p-chlorostyrene) and poly(p-bromostyrene) in Figure 13. After irradiation, the relative intensities of Cl_{1S}, $Cl_{2P1/2}$, and $Cl_{2P2/3}$ signals decrease for poly(p-chlorostyrene), and those of Br_{1S}, $Br_{2P1/2}$, and $Br_{2P2/3}$ signals decrease for poly(p-bromostyrene). These results reveal that halogens are liberated from poly(halogenostyrene)s by electron beam irradiation.

LOW MOLECULAR WEIGHT MODEL COMPOUND STUDY

Unresolved Problems

Photolysis and radiolysis mechanisms of polystyrene derivatives have been studied by several spectroscopic analysis methods as stated in the previous section. There are, however, many ambiguities remaining in the reaction mechanisms:

1. What kind of free radical produced is the most important one involved in a crosslinking formation in each polystyrene derivative?
2. What is the probability that free β-radicals produced are involved in a crosslinking reaction?
3. What is the ratio of halogen cleavage to α-hydrogen cleavage in poly(halogenostyrene)s?

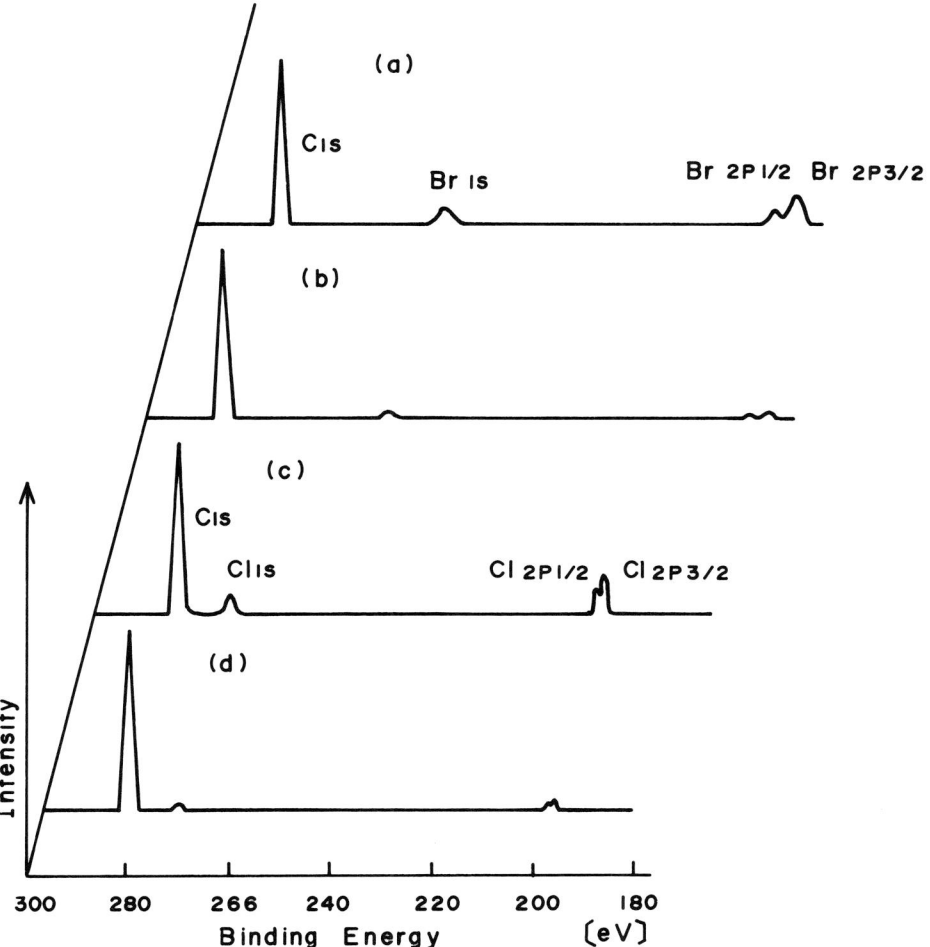

Figure 13 ESCA spectra of poly(4-bromostyrene) (*a*) before and (*b*) after irradiation by electron beams, and of poly(4-chlorostyrene) (*c*) before and (*d*) after irradiation by electron beams. (Data from Ref. 40.)

4. Do photoinduced phenyl radicals formed through halogen release participate in a crosslinking reaction?
5. Is there any direct evidence of photocoloration of polystyrene derivatives which clarifies the actual mechanism of the coloration?

These ambiguities are very difficult to resolve by conventional spectroscopic methods. We have tried to examine this grey area using low molecular weight model compounds.

General Considerations for Low Molecular Weight Model Compounds

The backbone structure of polystyrene derivatives is shown in Figure 14. From the viewpoint of the bond dissociation energies listed in the legend of Figure 2, the α-hydro-

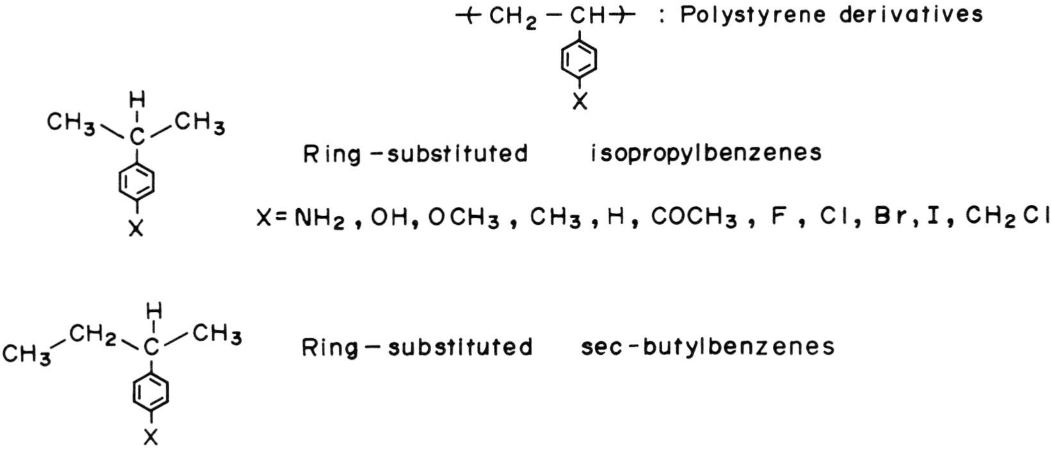

Figure 14 The molecular structures of polystyrene derivatives, ring-substituted *sec*-butylbenzenes, and ring-substituted isopropylbenzenes.

gen is considered to be the most sensitive when the substituent X is not very sensitive. Experiments conducted to date support this belief [13–33, 58–73]. The bond dissociation energy of β-hydrogen is not particularly small, but it is much larger than that of α-hydrogen. The reactivity of the β-hydrogen is thus considered to be lower. Therefore, isopropylbenzene analogues (shown in Fig. 14) can be used as low molecular weight model compounds, though *sec*-butylbenzene derivatives are the most suitable model compounds.

Methyl, methoxy, hydroxy, and amino groups are used as electron-donating substituents, and acetyl is used as an electron-abstracting one. To study the reactions of halogen-containing derivatives, fluorine, chlorine, bromine, iodine, and chloromethyl are used as substituents.

FINAL PRODUCT ANALYSES BY GAS CHROMATOGRAPHY–MASS SPECTROMETRY

Low molecular weight model compounds were irradiated with UV light at 254 nm and then introduced into an SE-30 column (Shimadzu), the temperature of which was raised from 70 to 230°C at a rate of 10°C min^{-1} to separate the products formed. The products formed were identified by their mass spectra after separation. Figures 15–17 are gas chromatograms that were thus obtained and mass spectra of the products used for analyses are listed in the appendix.

Gas chromatograms of neat *sec*-butylbenzene before and after exposure to UV light are seen in Figure 15 (*a, b*). Mainly the product assigned to the coupling product of two α-radicals was observed. The same experiment was conducted for *tert*-butylbenzene, in which α-hydrogen is missing. No strong intensity signal was seen under the same experimental conditions. This indicates that α-hydrogen in the polystyrene backbone structure is very sensitive to light, which supports the view that isopropylbenzene derivatives can be used as suitable model compounds of polystyrene derivatives.

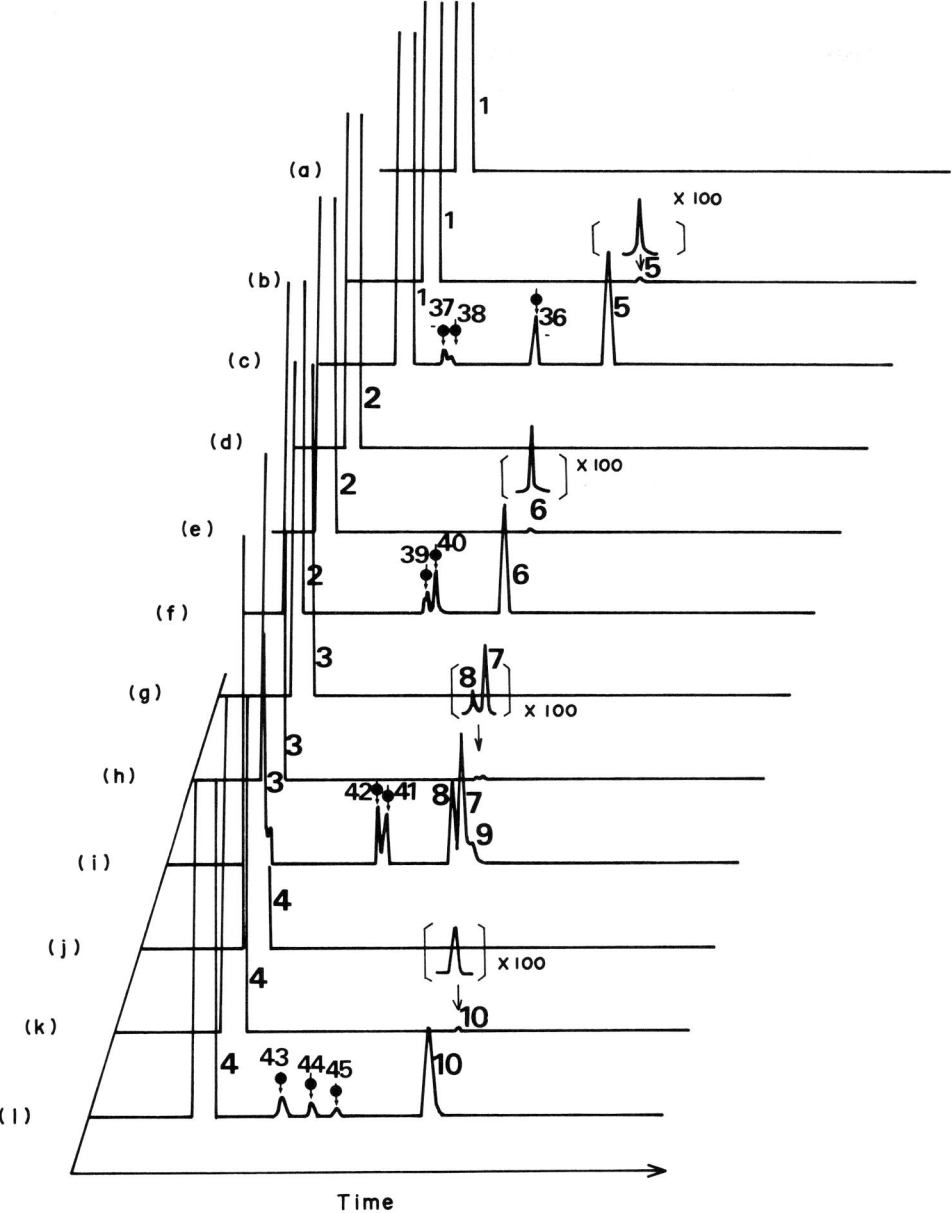

Figure 15 Gas chromatograms of (*a*) *sec*-butylbenzene; (*b*) *sec*-butylbenzene exposed to light; (*c*) a mixture of *sec*-butylbenzene with CCl_4 exposed to light; (*d*) IPBZ; (*e*) IPBZ exposed to light; (*f*) a mixture of IPBZ with CCl_4 exposed to light; (*g*) 4MeIPBZ; (*h*) 4MeIPBZ exposed to light; (*i*) a mixture of 4MeIPBZ with CCl_4 exposed to light; (*j*) 4MeOIPBZ; (*k*) 4MeOIPBZ exposed to light; and (*l*) a mixture of 4MeOIPBZ with CCl_4.

Halogen-Free Substituent Effect on Photolysis

Gas chromatograms of neat isopropylbenzene (IPBZ) and 4-methoxyisopropylbenzene (4MeOIPBZ) before and after exposure to light are shown in Figure 15 (*d, e, j, k*). As before, the radical coupling products of two α-radicals were mainly obtained. These results indicate that substituents themselves are less sensitive to light than α-hydrogen. Thus the sensitivity increase observed is attributable to the sensitivity change in α-hydrogen by substituent effects. However, in the case of 4-methylisopropylbenzene (4MIPBZ), a product related to the benzylradical, formed through hydrogen liberation from a methyl group, was formed besides the α-radical–α-radical coupling product, as shown in Figure 15 (*g, h*). This reveals that the methyl group is also sensitive to light, like α-hydrogen. This is because a benzyl-type radical is quite stable.

Gas chromatograms of 4-hydroxyisopropylbenzene (4OHIPBZ), 4-aminoisopropylbenzene (4NH$_2$IPBZ), and 4-acetylisopropylbenzene (4-MeCOIPBZ) before and after exposure to light are shown in Figure 16 (*a, b, d, e, g, h*). The features of the chromatograms are different from those of IPBZ.

In the case of 4OHIPBZ, another product concerning the phenoxy radical is observed besides the main α-radical–α-radical coupling product. This indicates that the α-radical is the main free radical involved in a free radical coupling reaction, but that the phenoxy radical is also involved to some extent.

In the case of exposed 4NH$_2$IPBZ mainly one product was given, formed through the coupling of the radical from amino groups. This result shows that the amino group is much more sensitive than the α-hydrogen. This is in good agreement with the result reported on the basis of gaseous product analyses [19].

In the case of 4MeCOIPBZ, a product related to the acetyl radical formed through the release of a methyl radical from an acetyl group was produced besides the coupling product of two α-radicals. The amount of the former product is greater than that of the latter. This reveals that the acetyl group is more highly sensitive than the α-hydrogen, and methyl liberation from the acetyl group predominates.

Halogen Substituent Effect on Photolysis

The gas chromatograms of 4-halogenoisopropylbenzenes, 4-fluoroisopropylbenzene (4FIPBZ), 4-chloroisopropylbenzene (4-ClIPBZ), 4-bromoisopropylbenzene (4BrIPBZ), 4-iodoisopropylbenzene (4IIPBZ), and 4-chloromethylisopropylbenzene (4CMIPBZ) before and after exposure to UV light are shown in Figure 17. Mass spectra of the final products used for analyses are listed in the appendix. Each derivative is discussed.

Figure 17 (*a, b*) are gas chromatograms of 4FIPBZ before and after exposure to light. The final product observed was the coupling product of two α-radicals. This feature of the reaction is very similar to that of IPBZ. The fission of fluorine could not be judged in this study, though a mass spectrometry study reported that a poly(fluorostyrene) film showed a small amount of fluorine liberation as a product of HF [21]. The sensitivity was nearly the same as that of IPBZ.

The chromatograms of 4ClIPBZ before and after exposure to light are also shown in Figure 17 (*c, d*). Isopropylbenzene and an α-radical–α-radical coupling product were mainly formed. This result indicates that both chlorine liberation and α-hydrogen liberation occur during exposure. The produced α-radical was involved in a coupling reaction. On the other hand, the phenyl radical produced through chlorine liberation was very short-lived and immediately converted into isopropylbenzene.

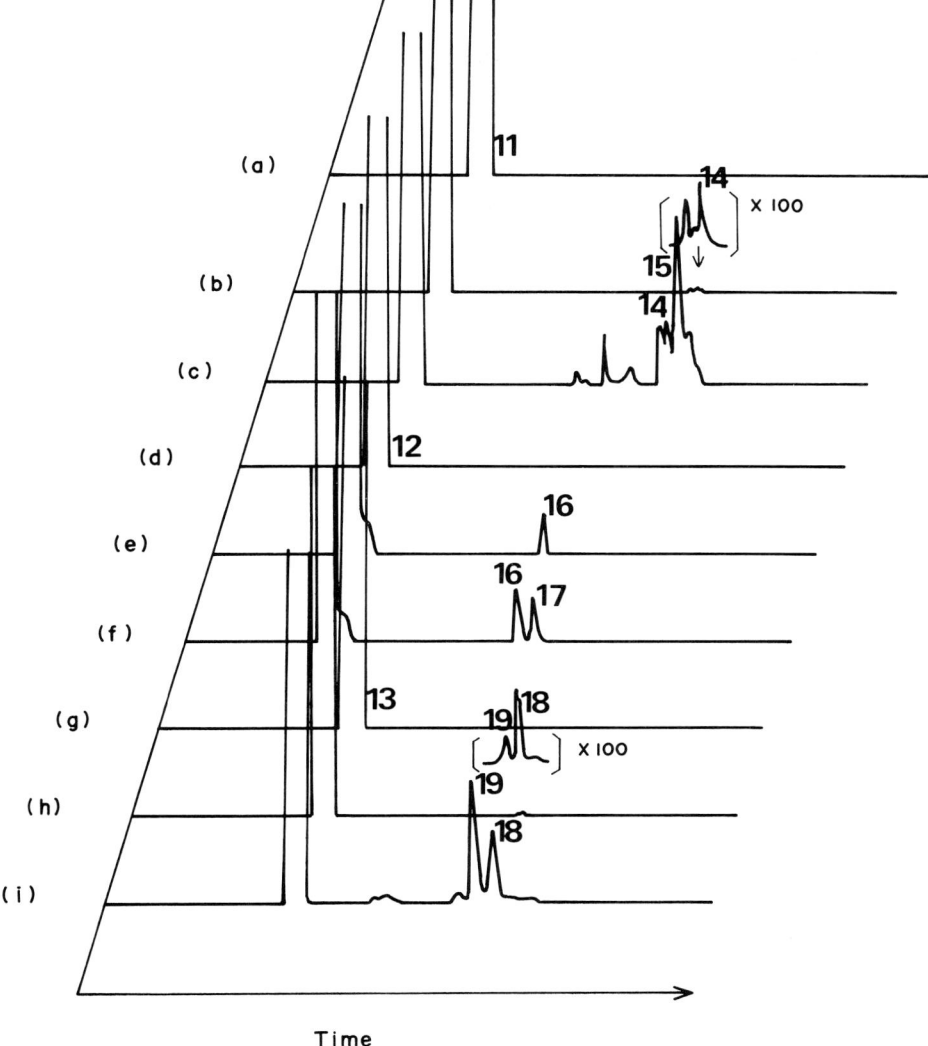

Figure 16 Gas chromatograms of (a) 4OHIPBZ; (b) 4OHIPBZ exposed to light; (c) a mixture of 4OHIPBZ with CCl_4; (d) 4NH$_2$IPBZ; (e) 4NH$_2$IPBZ exposed to light; (f) a mixture of 4NH$_2$IPBZ with CCl_4; (g) 4MeCOIPBZ; (h) 4MeCOIPBZ exposed to light; and (i) a mixture of 4MeCOIPBZ with CCl_4 exposed to light.

The gas chromatograms of 4BrIPBZ before and after exposure to light are also shown in Figure 17 (e, f). The exposed 4BrIPBZ also gave both isopropylbenzene and an α-radical–α-radical coupling product. The result shows that this reaction feature of 4BrIPBZ is almost the same as that of 4ClIPBZ. The main difference is in the production of halogeno α-methylstyrene. This is formed through subsequent hydrogen abstraction by the bromine radical and is direct evidence of a double bond formation, which has been considered to result in photocoloration. The photocoloration problem will be discussed later.

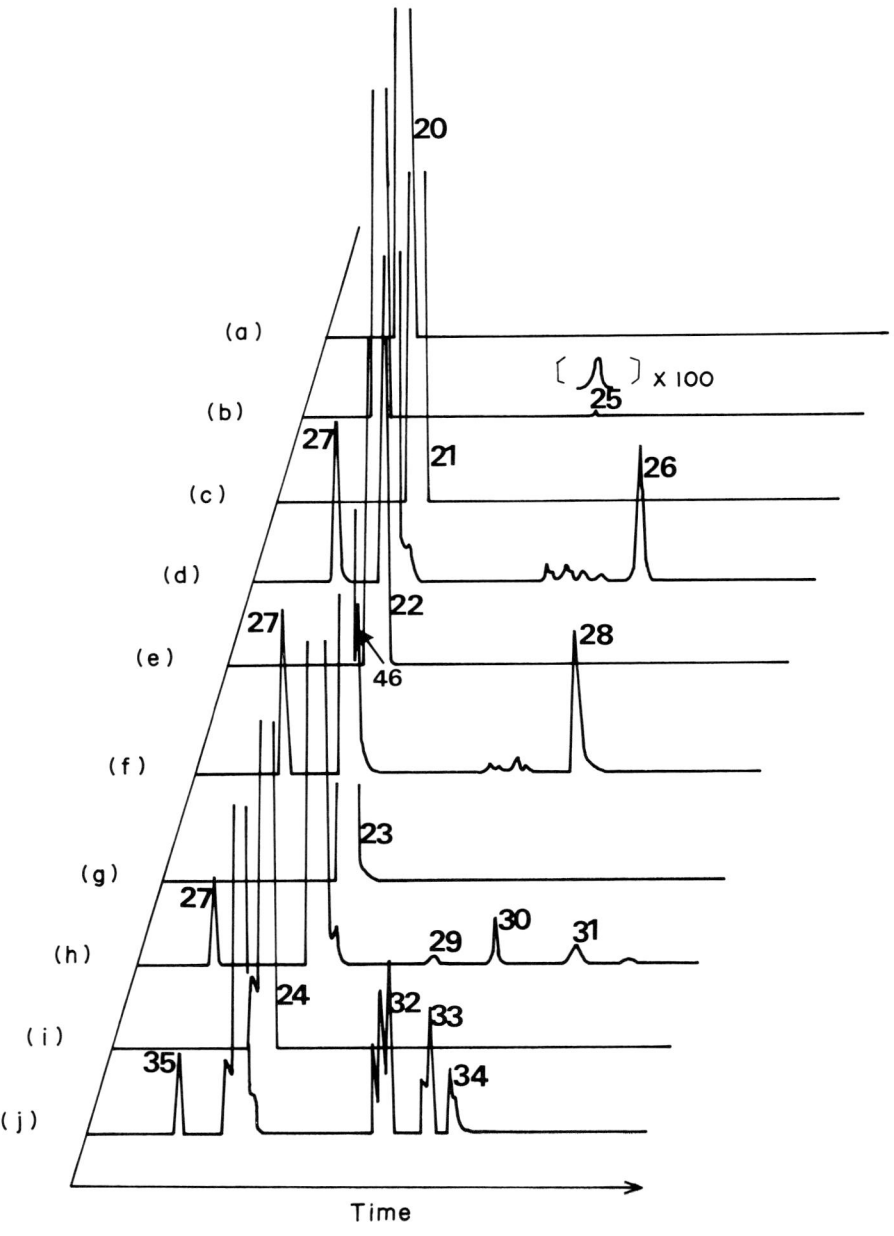

Figure 17 Gas chromatograms of (*a*) 4FIPBZ; (*b*) 4FIPBZ exposed to light; (*c*) 4ClIPBZ; (*d*) 4ClIPBZ exposed to light; (*e*) 4BrIPBZ; (*f*) 4BrIPBZ exposed to light; (*g*) 4IIPBZ; (*h*) 4IIPBZ exposed to light; (*i*) 4ClCH$_2$IPBZ; and (*j*) 4ClCH$_2$IPBZ exposed to light.

Figure 17 also presents the gas chromatograms of 4IIPBZ before and after exposure to light (*g, h*). The final products observed are isopropylbenzene, the coupling product of two isopropylphenyl radicals, the coupling product of an isopropylphenyl radical and an α-radical, and the coupling product of two α-radicals. The final products in the case of 4IPBZ are quite different from those in the case of 4ClIPBZ and 4BrIPBZ. The phenyl radical produced through iodine release is included in the coupling reaction. This difference might arise from the very slow liberation of iodine.

The gas chromatograms of 4CMIPBZ before and after exposure to light are shown in Figure 17 (*i, j*). 4-Methylisopropylbenzene, a coupling product of two benzyl radicals, a coupling product of a benzyl radical and an α-radical, and a coupling product of two α-radicals were obtained as the final products. 4-Methylisopropylbenzene was produced through hydrogen abstraction by the produced benzyl radical. The amount of hydrogen abstraction by the benzyl-type radical is not very large, estimating from the amount of 4-methylisopropylbenzene. It is well known that benzyl-type radicals abstract hydrogen [79, 80]. The reaction begins with the fission of a C–Cl bond to produce benzyl-type radicals. The chlorine radicals produced simultaneously bring about hydrogen abstraction from α-hydrogen, through which α-radicals are obtained. Both the primary and secondary free radicals produced are involved in coupling product formation in the final stage of a reaction.

The relative production yields of final products for the halogen-containing isopropylbenzene derivatives are listed in Table 3. Among the halogen-containing isopropylbenzene derivatives studied here, the chloromethyl-substituted derivative shows high sensitivity to UV light. The order of the sensitivity to 254-nm light is $CH_2Cl > Br > Cl > I > F = H$. The low sensitivity of the I-substituted derivative is due to iodine formation, which reduces the transparency of the sample. Thus light can reach the sample only near the surface.

The relative amount of free radicals produced is estimated from the relative production

Table 3 Relative Production Yields for Halogen-Containing Isopropyl Benzenes

Product / X (X=)	⌬	⌬-⌬	⌬-⌬-X	X-⌬-⌬-X	Σ •⌬	Σ •⌬-X
F	—	0	0	≪0.1	0	≪0.2
Cl	28.5	0	0	21.8	40	60
Br	33.3	0	0	13.6	55	45
I	11.3	0.2	3.7	1.0	73	27
Product / CH₂X	CH₃	⌬-⌬	⌬-CH₂Cl	CH₂Cl-⌬-CH₂Cl	•CH₂	•CH₂Cl
X = Cl	10	35	30	15	65	35

yields of the final products. It can be seen that the ratio of halogen elimination to total elimination is in the order I > Br > Cl > F. This result seems to reflect the bond dissociation energies: C–I < C–Br < C–Cl < C–F. On the other hand, the ratio of α-radical elimination to the total elimination is on the order of Cl > Br > I. This result is interpreted in terms of the strength of hydrogen abstraction: Cl > Br > I. It is noted that the chloromethyl derivative shows higher sensitivity than the chlorine derivative. This is because of the stability of the benzyl radical.

ESR ANALYSES AT 77 K

ESR studies at 77 K have been conducted in order to clarify the primary stage of a reaction. Neat samples of isopropylbenzene derivatives as low molecular weight model compounds are introduced into quartz sample tubes, exposed to 254-nm light at 77 K, and ESR spectra are measured. The ESR spectra measured are shown in Figure 18.

Methyl, methoxy, and hydroxy derivatives show the same simple ESR spectrum. A typical ESR spectrum is shown in Figure 18 (a). This simple singlet signal is assigned to an α-radical. This implies that α-hydrogen cleavage is a predominant reaction in primary radical formation.

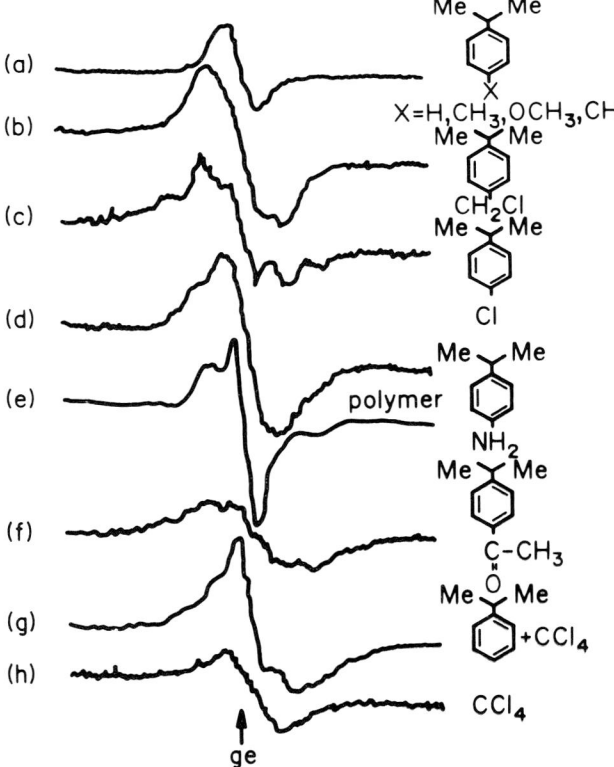

Figure 18 ESR spectra at 77 K after exposure to light for (a) IPBZ, 4MeIPBZ, 4MeOIPBZ, and 4OHIPBZ; (b) 4CMIPBZ; (c) 4ClIPBZ; (d) NH$_2$IPBZ; (e) poly(4-aminostyrene); (f) 4MeCOIPBZ; (g) a mixture of IPBZ with CCl$_4$; and (h) CCl$_4$.

Spectra with structures are observed from amino, acetyl, chloromethyl and chloro-derivatives. The deconvoluted and analyzed spectra are shown in Figure 19.

The observed spectrum of 4MeCOIPBZ [Fig. 19 (f)] is interpreted to result from a methyl radical signal (1:3:3:1 quartet -×--×-) and a singlet signal of a phenyl ketone-type radical (-●--●-). The deconvoluted spectra were a 1:3:3:1 quartet (hyperfine splitting, 2.1 mT) and a singlet of 1.7 mT half-width. Thus the main free radicals are formed through methyl liberation from an acetyl group.

For amino derivatives, a more highly resolved spectrum was obtained from poly(aminostyrene) [Fig. 18 (e)] than from 4NH$_2$IPBZ [Fig. 18 (d)]. Both spectra are assigned to an amino-type radical. In such a radical, hyperfine splittings are given by the interaction of an electron with a hydrogen nucleus and a nitrogen nucleus. The interaction with a hydrogen nucleus gives a doublet, and that with a nitrogen nucleus gives a 1:1:1 triplet. The hyperfine coupling constants, a_H = 3.0 mT for hydrogen and a_N = 1.5 mT for nitrogen, give rise to a 1:1:2:1:1 quintet spectrum and explain the spectrum observed in Figure 19 (d) reasonably well.

The spectrum of 4CMlIPBZ [Fig. 18 (b)] is not assigned to only one radical, as it has a structure that cannot be interpreted as one component. To analyze the observed spectrum, experiments were carried out on a mixture of IPBZ with carbon tetrachloride. The mixture showed a quintet feature signal with a splitting of 1.4 mT as shown in Figure 18 (g), while

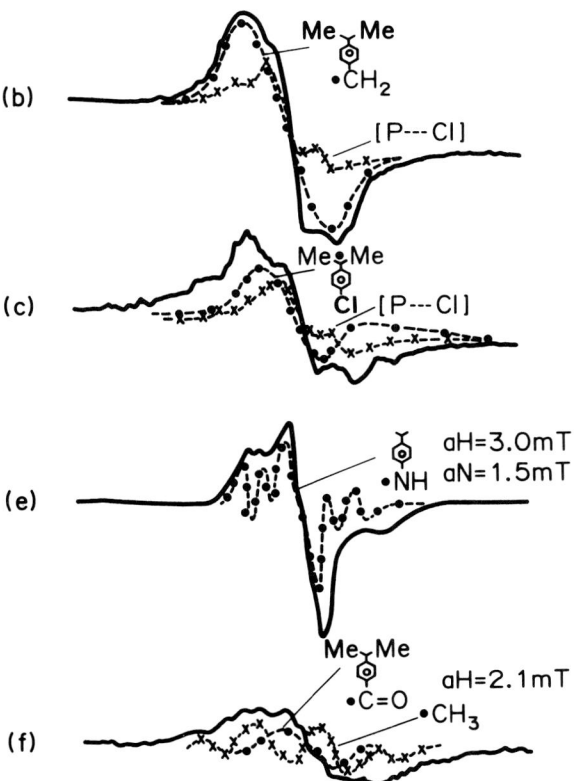

Figure 19 ESR spectra observed (———) and simulated (-- · -- · -- and -- X -- X --).

carbon tetrachloride itself showed a singlet signal of 3.0 mT half-width as shown in Figure 18 (*h*). This quintet signal is considered to arise from an interaction between a photo-induced chlorine radical and IPBZ. Here it is expressed as [P---Cl], which would be related to a charge transfer complex. Tabata's group reported the formation of a charge transfer complex on the basis of time-resolved spectroscopy [47].

The spectrum observed is deconvoluted to a broad singlet signal of a 4.4 mT half-width and the quintet feature signal observed in a mixture of IPBZ with CCl_4. The deconvoluted broad singlet signal (-●--●-) is assigned to the benzyl-type radical formed through chlorine cleavage from a chloromethyl group. Bridge observed only a broad singlet signal of 4.6 mT half-width from γ-irradiated toluene at 77 K and assigned it to a benzyl radical [81]. The deconvoluted broad singlet signal is in good agreement with his result. Therefore, the primary reaction of 4CMIPBZ is chlorine release from a chloromethyl group. This is believed to result from the large stabilization of the benzyl-type radical.

The observed spectrum of 4ClIPBZ is considered to result from an α-radical signal and a [P---C]-type signal. The deconvoluted spectra are shown in Figure 19 (*c*). The observed ESR spectrum indicates that both chlorine cleavage and α-hydrogen cleavage take place in the primary reaction of 4ClIPBZ.

RADIATION-INDUCED REACTIONS OF LOW MOLECULAR WEIGHT MODEL COMPOUNDS

The radiation-induced reactions of low molecular weight model compounds for polystyrene derivatives are discussed, based on gas chromatography–mass spectrometry and ESR at 77 K.

NH_2, OH, OCH_3, CH_3, H, and $COCH_3$ Derivatives

Amino, hydroxy, and methoxy groups are typical electron-donating substituents. An acetyl group is a typical electron-abstracting group. Their sensitivities to 254-nm light were in the order $NH_2 \gg OH > OCH_3 > CH_3 > H > COCH_3$, which has a good correlation with the order of electron donation.

The reaction schemes are shown in Figure 20. In general, an α-radical is produced in a primary reaction, as an α-hydrogen is the most sensitive to light. The sensitivity is highly dependent on substituents. Electron-donating substituents increase the sensitivity of α-hydrogen, while electron-abstracting substituents decrease the sensitivity. However, when a substituent itself is more sensitive to light than the α-hydrogen, other reactions are observed. From the amino derivative, an amino-type radical is predominantly produced. The acetyl derivative gives more methyl cleavage from an acetyl group than α-hydrogen cleavage. A phenoxy radical is accompanied by the release of α-hydrogen in the case of the hydroxy derivative. A benzyl-type radical, resulting from hydrogen release from a methyl group of the methyl derivative, is also accompanied by the liberation of α-hydrogen, because of the high stability of the benzyl radical.

F, Cl, Br, and I Derivatives

The halogen–carbon bonds of halogenoisopropylbenzenes are very weak and halogen release, as well as α-hydrogen release, occurs, except for the fluorine derivative. Almost equal amounts of cleavage of chlorine and α-hydrogen were observed in the chlorine derivative. The iodine derivative showed a markedly higher iodine cleavage than α-hydro-

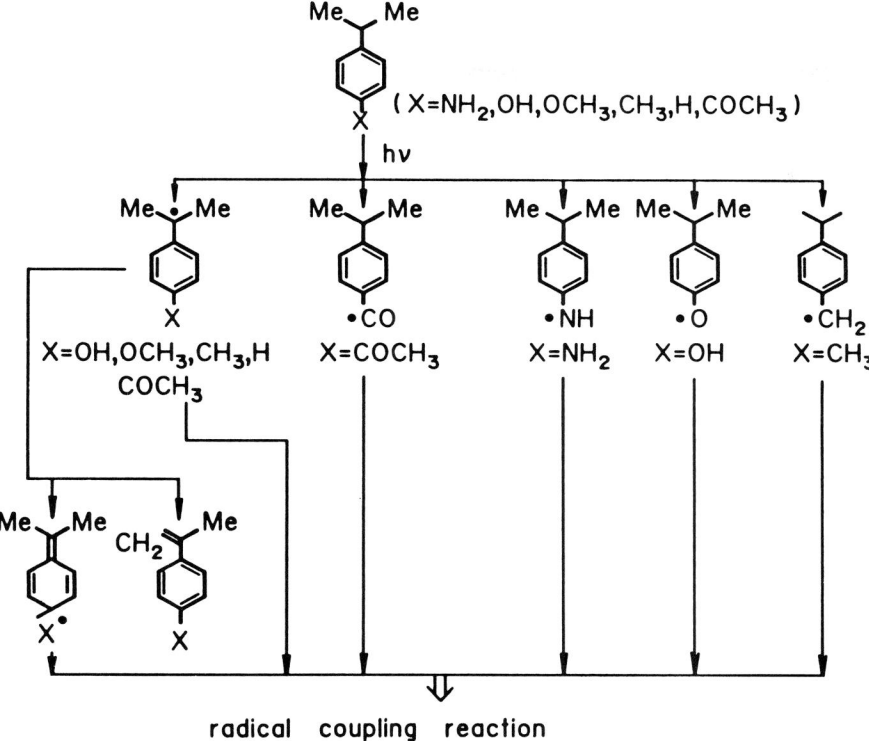

Figure 20 Radiation-induced reactions of ring-substituted isopropylbenzenes.

gen cleavage. The order of the amount of halogen cleavage on photolysis of halogeno derivatives is I > Br > Cl > F, which corresponds to the order of bond dissociation energies.

The reaction schemes of halogenoisopropylbenzenes are shown in Figure 21. The produced phenyl radical is generally very short-lived and immediately converts into the benzene ring, not being involved in a radical–radical coupling reaction. However, the iodine derivative gives final products from the phenyl radicals.

The produced halogen radicals abstract hydrogen to produce free radicals in a subsequent secondary reaction, through a charge transfer complex between a benzene ring and a halogen radical [45]. The power of the chlorine radical to abstract hydrogen is very great; the order is considered to be Cl > Br > I. The final product yields of α-radical–α-radical coupling products support this consideration.

CH_2Cl Derivative

The chloromethyl derivative gives a large number of benzyl-type radicals in a primary reaction because of the remarkably high sensitivity of a chloromethyl group. The ESR study gave evidence for this. The benzyl-type radical has a long lifetime of about 25 μs [45] because of its high stability. A simultaneously produced chlorine radical abstracts hydrogen from the α-position to give an α-radical within 200 ns [45]. Thus these two

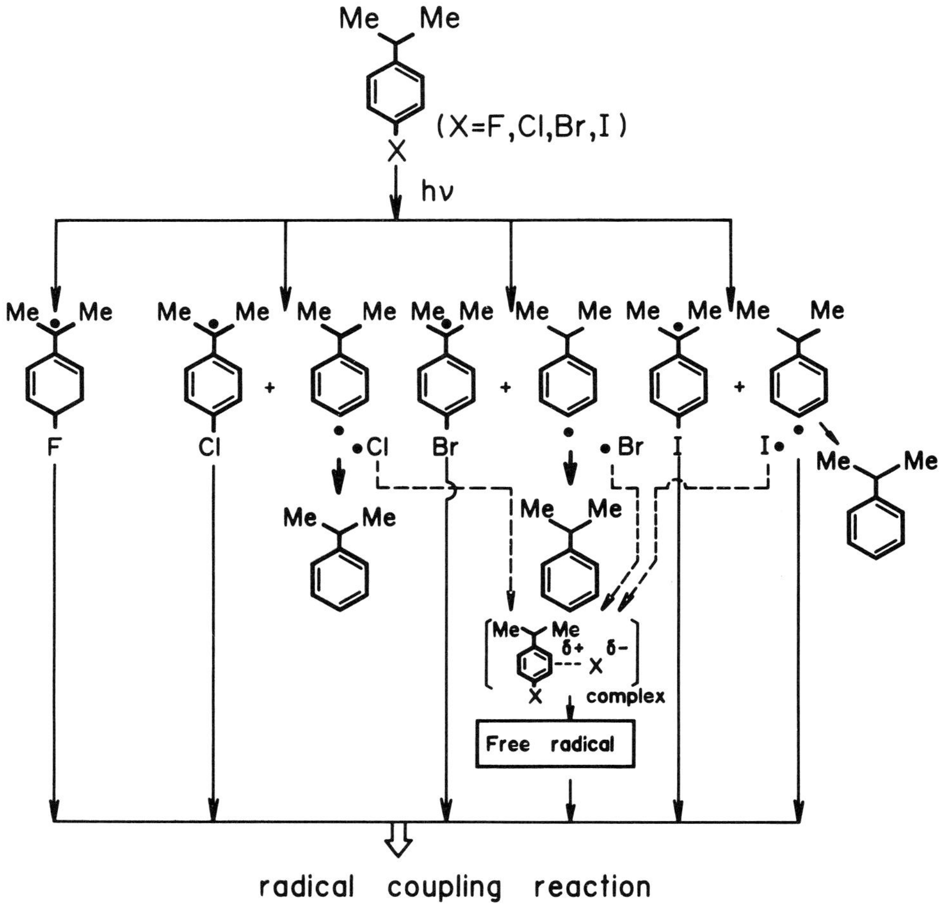

Figure 21 Radiation-induced reactions of halogenated isopropylbenzenes.

radical types, an α-radical and a benzyl-type radical, are involved in a homo- or cross-coupling reaction, giving rise to the three types of final product observed. Benzyl-type radicals also abstract hydrogen, transforming to methylisopropylbenzene, but the amount is small. The reaction schemes are seen in Figure 22.

Hydrogen Abstraction by Chlorine Radicals from Each Derivative

The chlorine radicals produced by light are very active in the hydrogen abstraction reaction. The hydrogen abstraction reaction produces free radicals in a secondary reaction, which are also involved in the final stage free radical coupling reaction in a similar way to the primary free radicals produced. Thus hydrogen abstraction by chlorine radicals is important. The susceptibility of each derivative to hydrogen abstraction has been checked. CCl_4 is added to a mixture of IPBZ and its derivative (1 mol/1 mol mixture), and then exposed to 254 nm light. The susceptibilities were measured by the relative homo-coupling final product yields of CCl_4-added 1 : 1 mixtures.

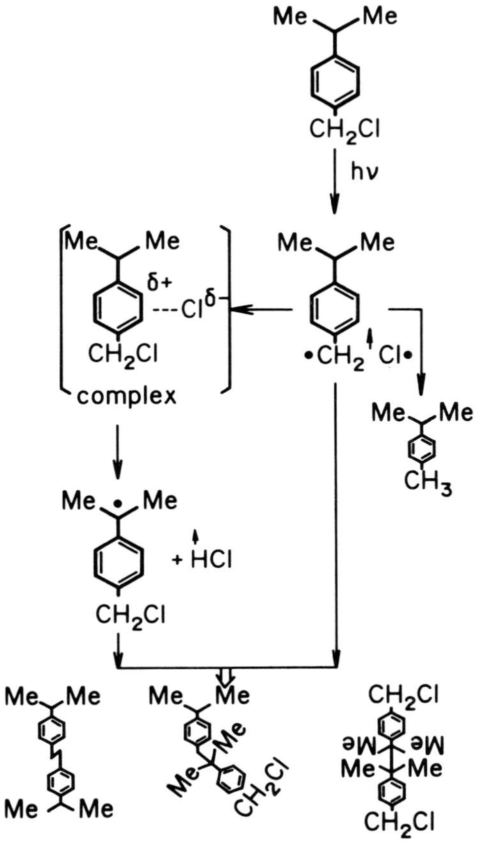

Figure 22 Radiation-induced reactions of chloromethylisopropylbenzene.

Figure 23 shows the susceptibility of each derivative to hydrogen abstraction. The abscissa displays the Hammett constants of the substituents. A good correlation was obtained between the susceptibilities and the Hammett constants. There is a tendency for the susceptibility to increase as the electron donation of substituent increases. This result is interpreted in terms of the stability of the charge transfer complex of each derivative with a chlorine radical. The stabilization increases as the electron donation of a substituent increases, because electrons flow from a derivative to chlorine in the charge transfer complex.

Hydrogen abstraction occurs on substituents as well as on α-hydrogen. The gas chromatograms for a mixture of each derivative with CCl_4 are shown in Figure 15 (c, f, i, l) and in Figure 16 (c, f, i). The amino derivative and the hydroxy derivative show a large amount of hydrogen abstraction from amino and hydroxy groups. The markedly high susceptibility to hydrogen abstraction observed in 4NH$_2$IPBZ and 4OHIPBZ is ascribed to abstraction from a substituent itself. Hydrogen is also abstracted from a methyl group of 4MeIPBZ, where the α-radical–benzyl-type radical crosscoupling product and the benzyl-type radical–benzyl-type radical homocoupling product were given, besides the α-radical–α-radical homocoupling product, as shown in Figure 15 (i). The ratio of hydrogen

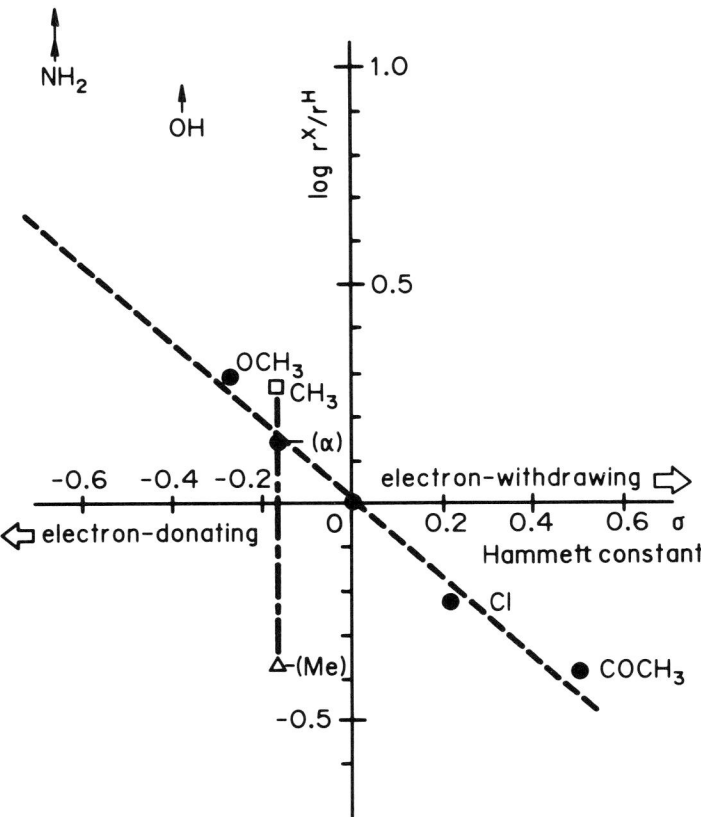

Figure 23 Susceptibilities of isopropylbenzene derivatives to hydrogen abstraction by photo-induced chlorine radicals.

abstraction from the α position to that from the methyl group was found to be about 3 : 1 on the basis of relative final product yields. The susceptibility to hydrogen abstraction from the α position is in good agreement with the expectation from the Hammett constant. From the 4MeCOIPBZ–CCl$_4$ mixture, a great number of α-radical–α-radical homocoupling products were observed, even though the photoinduced reaction of the neat acetyl derivative is phCO–CH$_3$ cleavage as shown in Figure 16 (*i*). This indicates that hydrogen abstraction by chlorine radicals takes place mainly on the α position in the acetyl derivative.

Direct Evidence of Photocoloration of Polystyrene Derivatives

The mass spectrum of an α-radical–α-radical homocoupling product usually shows spectra corresponding to that of the α-radical, which is exactly half of the product. The base peak observed shows the mass of the α-radical. However, if the intramolecular rearrangement reaction to double bond formation is predominant, the α-radical transforms into α-methylstyrene. The probability depends on the kind of substituents.

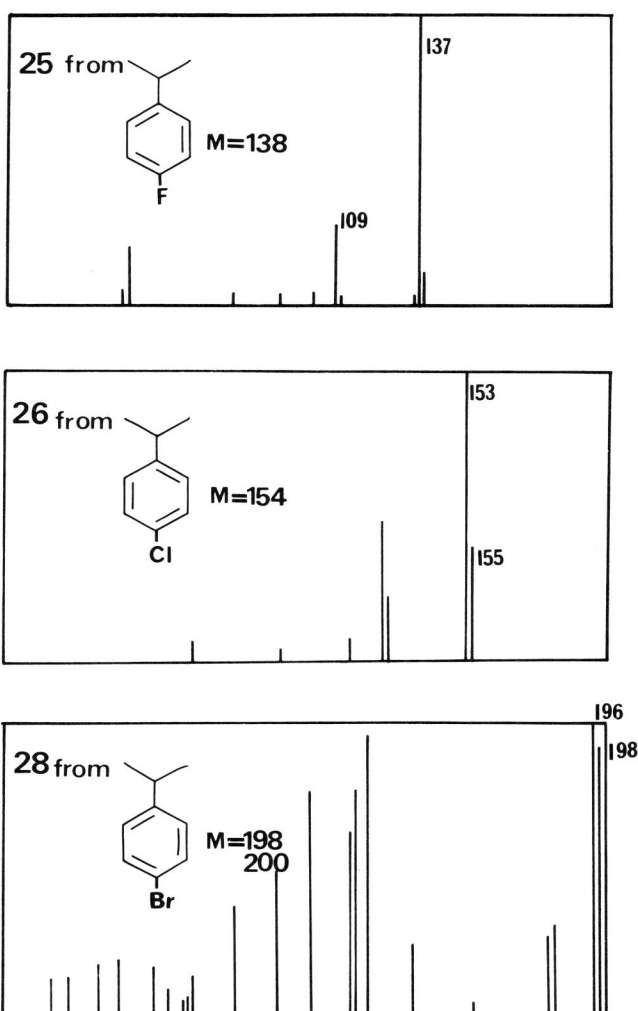

Figure 24 Mass spectra of the homocoupling products of two α-radicals for IPBZ and its halogenated analogues.

Figure 24 shows the mass spectra of α-radical homocoupling products observed for halogeno derivatives. Among these spectra, the bromo derivative shows evidence of considerable double bond formation, such that the base peak occurs at a mass fragment corresponding to a two-hydrogen–eliminated molecule derived from the parent molecule. Further important evidence is that exposed 4BrIPBZ shows α-methylstyrene formation [**4b**] to a great extent, as shown in Figure 17 (*f*). These are direct indications of the double bond formation conjugating with a benzene ring. These results indicate that the photocoloration of polystyrene derivatives results from double bond formation through intramolecular rearrangement of the produced α-radicals, which competes with crosslinking formation.

Intramolecular Rearrangements of α-Radicals

Two important rearrangements of an α-radical must be considered. One is an intramolecular rearrangement that leads to a cyclohexadienyl radical, and the other is α-methylstyrene formation through hydrogen abstraction, both of which are shown in Figure 20. The former intramolecular rearrangement has been a problem for a long time [58–65]. The observed final products as shown in Figure 15 by ↓ are considered to be associated with this rearrangement. The results of the gas chromatograms indicate that this rearrangement is greatly dependent on the substituents. The latter intramolecular rearrangement is associated with photocoloration phenomena described in the preceding section.

CONSIDERATION OF RADIATION-INDUCED REACTIONS OF POLYSTYRENE DERIVATIVE POLYMER FILMS

Irradiated polymers are excited to their high-energy states by direct excitation or ionization. Primary macro-free radicals are produced through different types of dissociation: homolytic splitting, dissociative electron attachment, etc. The most important thing in this early stage of the two reactions—i.e., initiation and primary radical formation—is the lifetime of the macro-free radicals produced. The very short-lived macroradicals, such as the phenyl radical of poly(chlorostyrene) and poly(bromostyrene), cannot participate in crosslinking formation.

The simultaneously produced active gaseous species in a primary macroradical formation play an important role in producing secondary macroradicals through the subsequent hydrogen abstraction reaction. Chlorine radicals are known to show a strong hydrogen abstraction ability. A large number of secondary free radicals are produced through a charge transfer complex. This is shown for the photolysis of poly(chloromethylstyrene) in Figure 25. The photoproduced chlorine radical forms a charge transfer complex and produces an α-radical at a rate of 5×10^6 sec^{-1}, which is quite fast compared with the crosslinking reaction rate of the primary produced benzyl radical (4×10^4 sec^{-1}). The relationship between the lifetime of primary radicals and the rate of secondary free radical formation is very important. This relationship determines how the secondary macroradicals are involved in a crosslinking reaction. As a result, the primary macroradicals and the secondary macroradicals participate in a crosslinking reaction in the case of poly(chloromethylstyrene).

The photoinduced chlorine radical also abstracts hydrogen from the β position at a rate less than one-tenth that from the α position [44]. The β-macroradical produced is considered to result in the degradation of polystyrene derivatives [44].

When the lifetime of a macroradical is long enough for it to be attacked by oxygen, oxidation occurs through a macroperoxyradical.

When the macroradical produced is very susceptible to hydrogen abstraction to form a double bond, photocoloration is observed. Poly(bromostyrene) shows a large coloration phenomenon due to the bromine effect.

CONSIDERATION OF THE DIFFUSION OF ACTIVE GASES IN A POLYMER FILM

To discuss radiation-induced reactions, the diffusion of the produced active gases must be taken into account. This is because the diffusion lengths of radiation-induced active gases determine how closely macroradicals are produced in a polymer film. Diffusivity coefficients of several gases in polystyrene, dependent on the size of gases, are listed in Table 4.

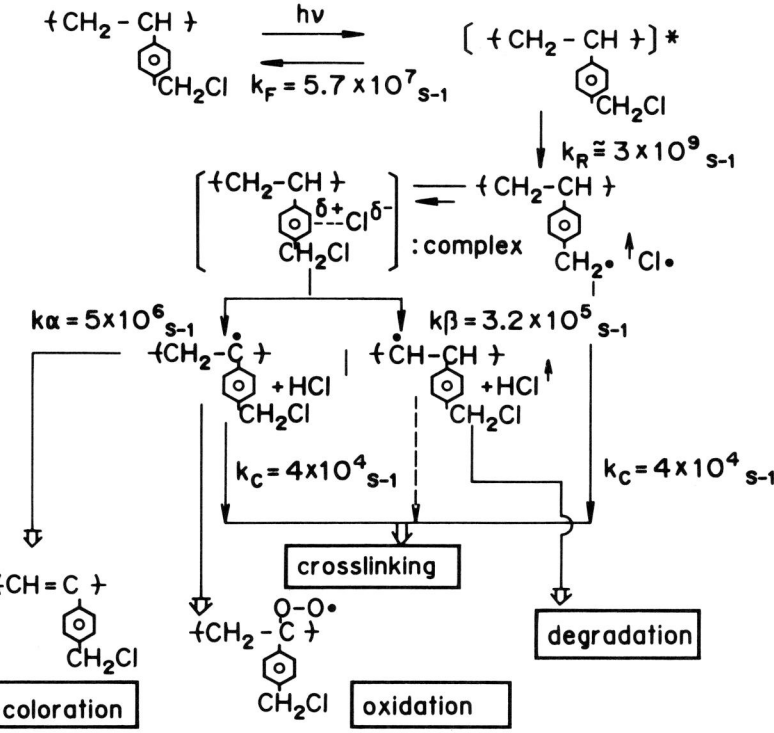

Figure 25 Radiation-induced reactions of poly(chloromethylstyrene).

Table 4 Diffusivity Data for Gases in Polystyrene

Gas	D [cm²/s] at 25°C	Do [cm²/s]	Ed [kcal/mol]
He	1.04×10^{-5}	0.0019	3.1
H_2	4.36×10^{-6}	0.0036	4.0
O_2	1.1×10^{-7}	0.125	8.3
CO_2	5.8×10^{-8}	0.128	8.7

D : diffusivity at 25°C
Ed : activation energy
D = Do exp (−Ed/RT)

The diffusion problem of active gases is discussed for poly(chloromethylstyrene). The diffusivity of a chlorine radical in polystyrene ($D = 9 \times 10^{-8}$ cm^2 sec^{-1}) was used estimate the diffusion length [$= (D_\tau)^{1/2}$] during hydrogen abstraction. Here τ denotes the lifetime of the charge transfer complex, which is reported to be 200 ns. The diffusion length is estimated as 1.3×10^{-7} cm. Thus hydrogen abstraction by a chlorine radical is estimated to extend to more than five monomer units. On the other hand, one benzyl-type radical is estimated to be formed per 10 to 100 monomer units by radiation from Charlesby's gel formation theory [83]. Thus these two macroradicals are both involved in crosslinking formation in the case of poly(chloromethylstyrene). Consequently, it is concluded that both primary macroradicals and secondary macroradicals are involved in a crosslinking reaction in a similar way, if the diffusion length of radiation-induced active gases is large enough to produce secondary macroradicals over a large region.

APPLICATION OF POLYSTYRENE DERIVATIVES TO LITHOGRAPHIC FIELDS

One of the most useful applications of polystyrene derivatives is for photolithography and high-energy radiation lithography [84–100]. Polystyrene derivatives are the best materials for resists, as they have a high resistance to plasmas and to physical sputtering [100], and their sensitivity to light and electrons is easily increased by introducing highly sensitive functional groups, such as a chloromethyl, chlorine, bromine, and iodine [84–100].

The relationship between sensitivity increase and the ratios of highly sensitive functional groups is simply derived on the basis of Charlesby's gel formation theory [101]:

$$\frac{1}{D_g(\text{M–N})M_w(\text{M–N})} = \frac{x_\text{M}}{D_g(\text{M})M_w(\text{M})} + \frac{x_\text{N}}{D_g(\text{N})M_w(\text{N})}$$

$D_g(\text{M–N})$, $D_g(\text{M})$, and $D_g(\text{N})$ denote the doses where first gelation occurs for a copolymer M–N and homopolymers M and N; $M_w(\text{M–N})$, $M_w(\text{M})$, and $M_w(\text{N})$ denote the weight average molecular weight for a copolymer M–N and homopolymers M and N; x_M and x_N are the molar ratio of monomer units M and N. According to this derivation, the sensitivity of copolymers may be evaluated at any component ratio using the sensitivity of homopolymers.

The comparison of experimental data and calculated results is displayed in Figure 26. Calculated sensitivities are shown as a function of mole fraction for a highly sensitive monomer N. A good correlation can be seen. Sensitivity saturation essentially exists. Some discrepancies are seen in the high mole fraction of N region. The discrepancy is more prominent in poly(styrene-co-iodostyrene) than in poly(styrene-co-chlorostyrene). This is due to the difference in hydrogen abstraction strength between a chlorine radical and an iodine radical. This is not taken into account in the derivation. The derived sensitivity dependence is useful in designing copolymer crosslinking negative resists.

The susceptibility of monomer units to hydrogen abstraction by a halogen radical or an anion is also one of the important factors in achieving high sensitivity to radiation. Electron-donating substituents give a large susceptibility as stated. Therefore, a monomer with an electron-donating substituent is preferable as a component of a halogen-containing copolymer resist [102].

CONCLUSIONS

Recent progress in radiation-induced reactions of polystyrene derivatives has been reviewed and the problems remaining unresolved stated. The uses of low-molecular weight

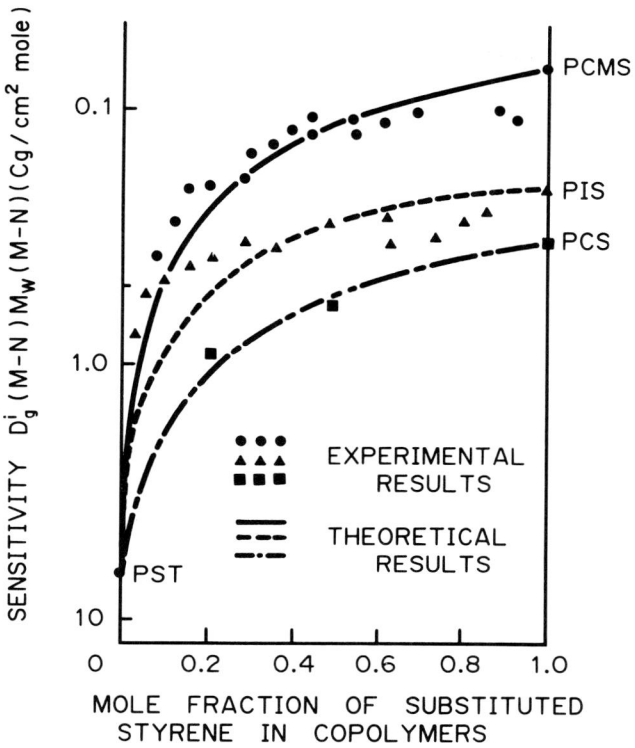

Figure 26 Dependence of copolymer sensitivity to electron beams on mole fraction of highly sensitive monomer units. (From Ref. 101.)

model compounds for elucidation of the photolysis and radiolysis of ring-substituted polystyrenes in the grey area were mentioned. The results obtained from final product analyses by gas chromatography–mass spectrometry and from free radical detection by ESR at 77 K were discussed in detail for 4-amino, 4-hydroxy, 4-methoxy, 4-methyl, and 4-acetyl derivatives and for halogen-containing derivatives, such as 4-fluoro, 4-chloro, 4-bromo, and 4-iodo derivatives. In addition to the radiation-induced reactions, the susceptibilities of each monomer unit to hydrogen abstraction by a chlorine radical were determined. The processes causing photocoloration of polystyrene derivatives were deduced from studies of low molecular weight model compounds. One of the most useful applications of polystyrene derivatives, to lithography, was discussed from the viewpoint of sensitivity to radiation.

ACKNOWLEDGMENT

The first author (K. T.) is grateful to Mr. M. Suzuki, who made many suggestions and shared his time in the experiments in the first stage of the low molecular weight compound studies. K. T. also offers his grateful acknowledgment to Drs. K. Nakaoka, T. Haruta, and H. Nishida of Toray Industry Inc. for discussion of the syntheses of the fluoro, chloro, bromo, and iodo derivatives, and to Mr. Y. Saotome of M. D. Research Lab. for the synthesis during his time at NEC of 1,2-bis(4-isopropylphenyl)ethane from 4-chloromethylisopropylbenzene, which was used for confirmation of the assignment for one of the final products. K. T. would like to express sincere appreciation to Prof. S. Tagawa and Dr. M. Washio of Tokyo University for the continual helpful discussions of kinetics of the radiation-induced reaction of poly(chloromethylstyrene) and their courtesy in allowing the use of flash photolysis and pulse radiolysis spectra. Both authors are indebted to Mr. H. Adachi for help with the gas chromatography–mass spectrometry measurements, to Dr. H. Makino for affording them the chance to make the ESR measurements, and to Mrs. H. Honda for her technical assistance in ESR measurements. The authors also gratefully acknowledge the editing of the manuscript by Dr. M. T. Fowler. In order to cover every relevant paper, a search for papers related to radiation-induced reactions of polystyrene derivatives was conducted by Miss A. Iwasaki of the NEC library center, to whom the authors express their appreciation.

This manuscript is dedicated to Mrs. Tomoko Tanigaki, who continuously encouraged her husband during the preparation of the manuscript.

APPENDIX: MASS DATA AND FINAL PRODUCT

sec-**Butylbenzene 1** (purchased from Tokyo Kasei, distilled bp = 167°C): 134 (M^+, 25), 119 (base peak, 100), 91 (5), 77 (12); 3,4-dimethyl-3,4-diphenylhexane; **5:** MS (70 eV), m/e (rel. intensity), 133 (base peak, 100; M^+ not observed), 117 (9), 105 (12), 91 (92), 79 (3), 77 (4); **36:** (70 eV), m/e (rel. intensity), 133 (base peak 100, M^+ not observed), 129 (5), 115 (18), 105 (19), 91 (86), 78 (5), 77(7); **37:** MS (70 eV), m/e (rel. intensity), 168 (10), 105 (base peak, 100), 103 (5), 91 (7), 78 (3); **38:** MS (70 eV), m/e (rel. intensity), 168 (12), 105 (base peak, 100), 103 (7), 91 (8), 79 (7), 77 (9).

IPBZ 2 isopropylbenzene (purchased from Tokyo Kasei, distilled bp = 152–153°C): MS (70 eV), m/e (rel. intensity), 120 (M^+, 30), 105 (base peak, 100), 91 (5), 79 (11), 77 (13); 2,3-dimethyl-2,3-diphenylbutane **6:** MS (70 eV), m/e (rel. intensity), 119 (base peak, 100; M^+ not observed), 103 (14), 91 (60), 77 (14); **39:** MS (70 eV), m/e (rel. intensity), 105 (base peak, 100; M^+ not observed), 77 (12); **40:** MS (70 eV), m/e (rel. intensity), 119 (base peak, 100; M^+ not observed), 115 (8), 91 (27), 78 (4), 79 (4).

4MeIPBZ 3 4-methylisopropylbenzene (purchased from Tokyo Kasei, distilled bp = 176°C): MS (70 eV), m/e (rel. intensity), 134 (M^+, 29), 119 (base peak, 100), 91 (18), 77 (5); 2,3-dimethyl-2,3-di(4-methylphenyl)butane **7:** MS (70 eV), m/e (rel. intensity), 133 (base peak, 100), 105 (17), 91 (6), 79 (1), 77 (2); 1,1-dimethyl-1-(4-methylphenyl)-2-(4-isopropylphenyl)ethane **8:** MS (70 eV), m/e (rel. intensity), 133 (base peak, 100) 117 (6), 105 (6), 91 (5), 77 (3); 1,2-bis(isopropylphenyl)-ethane **9:** MS (70 eV), m/e (rel. intensity), 266 (M^+, 7), 133 (base peak, 100), 118 (11), 105 (11), 91 (6), 79 (2), 77 (2); **41:** MS (70 eV), m/e (rel. intensity), 252 (2), 250 (2), 133 (base peak, 100), 115 (5), 105 (20), 93 (6), 91 (6); **42:** MS (70 eV), m/e (rel. intensity), 252 (11), 250 (11), 237 (7), 235 (7), 133 (base peak, 100), 105 (7), 91 (5).

4MeOIPBZ 4 4-methoxyisopropylbenzene (purchased from Tokyo Kasei, distilled bp = 103–104°C/27 mm Hg): MS (70 eV), m/e (rel. intensity), 150 (M^+, 30), 135 (base peak, 100), 120 (4), 105 (15), 91 (13), 79 (5), 77 (7); 1,2-dimethyl-1,2-(4-methoxyphenyl)butane **10:** MS (70 eV), m/e (rel. intensity), 149 (base peak, 100; M^+ not observed), 121 (5), 91 (10); **43:** MS (70 eV), m/e (rel. intensity), 256 (10),

Table 5 Detected Final Products for Irradiated Isopropylbenzene Derivatives

Me—Me ⬡ X X=	Final product	
	Main product	other product
NH$_2$ 12	⊱⬡-NH-NH-⬡⊰ 16	NH$_2$-⬡-()-⬡-NH$_2$ 17
OH 11	HO-⬡-()-⬡-OH 14	⊱⬡-O• → 15
OCH$_3$ 4	CH$_3$O-⬡-()-⬡-OCH$_3$ 10	
CH$_3$ 3	CH$_3$-⬡-()-⬡-CH$_3$ 7	⊱⬡-CH$_2$-⬡-CH$_3$ 8 ⊱⬡-CH$_2$-CH$_2$-⬡⊰ 9
H 2	⬡-()-⬡ 6	
COCH$_3$ 13	⊱⬡-C(=O)• → 18	CH$_3$CO-⬡-()-⬡-COCH$_3$ 19
F 20	F-⬡-()-⬡-F 25	
Cl 21	⊱⬡ 27	Cl-⬡-()-⬡-Cl 26
Br 22	⊱⬡ 27	Br-⬡-()-⬡-Br 28
I 23	⊱⬡ 27	⊱⬡-⬡⊰ 29 ⊱⬡+⬡-I 30 I-⬡-()-⬡-I 31
CH$_2$Cl 24	⊱⬡-CH$_3$ 35	⊱⬡-⬡⊰ 32 ⊱⬡(CH$_2$Cl)$_2$ 33 ClCH$_2$-⬡-()-⬡-CH$_2$Cl 34
Me—Me ⬡ 1	⬡-()-⬡ 5	

254 (45), 252 (45), 241 (10), 239 (40), 237 (45), 221 (3), 219 (25), 217 (40), 135 (base peak, 100), 119 (35), 117 (25), 107 (45), 91 (65), 77 (18); **44:** MS (70 eV), m/e (rel. intensity), 270 (10), 268 (30), 266 (30), 253 (90), 251 (90), 149 (15), 135 (10), 121 (base peak, 100), 103 (15), 91 (35), 77 (15); **45:** MS (70 eV), m/e (rel. intensity), 268 (4), 196 (4), 161 (4), 149 (base peak, 100), 121 (12), 115 (5), 109 (7), 91 (7), 77 (10).

4OHIPBZ 4-hydroxyisopropylbenzene **11** (purchased from Tokyo Kasei, mp = 62°C): MS (70 eV), m/e (rel. intensity), 136 (M^+, 26), 121 (base peak, 100), 107 (4), 103 (12); **14:** MS (70 eV), m/e (rel. intensity), 270 (52), 255 (base peak, 100), 239 (33), 213 (20), 169 (25), 120 (30), 91 (12), 77 (7); **15:** MS (70 eV), m/e (rel. intensity), 298 (28), 239 (base peak, 100), 134 (8), 121 (10), 119 (10), 91 (20), 77 (5).

4NH$_2$IPBZ 12 4-aminoisopropylbenzene (purchased from Aldrich, distilled bp-86°C/ 667 Pa: MS (70 eV), m/e (rel. intensity), 135 (M^+, 30), 120 (base peak, 100), 103 (5), 93 (59), 91 (7), 77 (5); N,N'-bis(4-isopropylphenyl)hydrazine **16:** MS (70 eV), m/e (rel. intensity), 253 (55), 238 (base peak, 100), 222 (12); **17:** MS (70 eV), m/e (rel. intensity), 266 (50), 253 (10), 238 (24), 147 (20), 135 (14), 119 (base peak, 100), 91 (29).

4MeCOIPBZ 13 4-acetylisopropylbenzene (pure para isomer was synthesized by the acetylation of IPBZ using acetylchloride in CCl$_4$ according to the Perrier method [103], bp = 137–139°C/2533 Pa: MS (70 eV), m/e (rel. intensity), 162 (M^+, 30), 147 (base peak, 100), 119 (11), 91 (15), 77 (10); **18:** MS (70 eV), m/e (rel. intensity), 306 (58), 162 (base peak, 100), 145 (90), 263 (15), 291 (10); **19:** MS (70 eV), m/e (rel. intensity), 265 (base peak, 100), 162 (80), 119 (70), 91 (30), 79 (4), 77 (4).

4FIPBZ 20 4-fluoroisopropylbenzene (synthesized by the Sandmeyer reaction from 4-aminoisopropylbenzene using fluoroboric acid bp-95–97°C/1.7 × 10^4 Pa): MS (70 eV), m/e (rel. intensity), 138 (M^+, 30), 123 (base peak, 100), 103 (45), 101 (7), 77 (10); 2,3-dimethyl-2,3-bis(4-fluorophenyl)butane **25:** 137 (base peak, 100; M^+ not observed), 109 (35), 101 (5), 83 (2), 75 (2).

4ClIPBZ 21 4-chloroisopropylbenzene (synthesized by the Sandmeyer reaction from 4-aminoisopropylbenzene using cuprous chloride): MS (70 eV), m/e (rel. intensity), 154 (M^+, 59), 139 (base peak, 100), 125 (8), 119 (19), 103 (86), 77 (32); 2,3-dimethyl-2,3-bis(4-chloroisopropyl)butane **26:** MS (70 eV), m/e (rel. intensity), 153 (base peak, 100; M^+ not observed), 137 (6), 125 (47), 117 (11), 115 (11), 103 (15), 91 (10), 77 (9); isopropylbenzene **27:** MS (70 eV), m/e (rel. intensity), 120 (M^+), 105 (base peak, 100), 91 (5), 79 (11), 77 (13).

4BrIPBZ 22 4-bromoisopropylbenzene (synthesized by the Sandmeyer reaction from 4-aminoisopropylbenzene using cuprous bromide): MS (70 eV), m/e (rel. intensity), 200 & 198 (M^+, 35), 95 & 93 (95), 119 (20), 104 (base peak, 100), 91 (15), 77 (28); 2,3-dimethyl-2,3-bis(4-bromoisopropylphenyl)butane **28:** 198 & 196 (95, M^+ not observed), 193 & 191 (35), 156 (8), 136 (25), 121 (base peak, 100), 117 (70), 115 (50), 102 (65), 91 (45), 77 (40); isopropylbenzene **27:** MS (70 eV), m/e (rel. intensity), 120 (M^+), 105 (base peak, 100), 91 (5), 79 (11), 77 (13); 4-bromo-α-methylstyrene **46:** MS (70 eV), m/e (rel. intensity), 198 & 196 (M^+ base peak, 100), 183 & 181 (30), 117 & 115 (75), 102 (63), 91 (28), 77 (20), 75 (25).

4IIPBZ 23 4-iodoisopropylbenzene (synthesized by the Sandmeyer reaction from

4-aminoisopropylbenzene using potassium iodide): MS (70 eV), m/e (rel. intensity), 246 (M$^+$, 78), 231 (base peak, 100), 119 (7), 104 (70), 91 (15), 77 (20); 2-(4-iodophenyl)-2-(4-isopropylphenyl)propane **30**: 364 (base peak, 100), 349 (83), 222 (15), 207 (15), 179 (48), 165 (25), 152 (10), 89 (10); 2,3-dimethyl-2,3-bis(4-iodophenyl)butane **31**: 362 (55), 347 (30), 245 (95), 207 (base peak, 100), 77 (40).

CMIPBZ 24 chloromethylisopropylbenzene: MS (70 eV), m/e (rel. intensity), 168 & 170 (M$^+$, 50, 15), 153 & 155 (base peak, 100, 30), 134 (60), 119 (38), 117 (70), 115 (12), 91 (20), 77 (5); 1,2-bis(isopropylphenyl)ethane **32**: MS (70 eV), m/e (rel. intensity), 266 (M$^+$, 7), 133 (base peak, 100), 118 (11), 105 (11), 91 (6), 79 (2), 77 (2); 2-methyl-1-(isopropylphenyl)-2-(chloromethylphenyl)propane **33**: 300 (3), 285 (3), 167 & 169 (base peak, 100, 30), 133 (20), 117 (10), 105 (5), 91 (10); 2,3-dimethyl-2,3-(chloromethylphenyl)butane **34**: MS (70 eV), m/e (rel. intensity), 167 & 169 (base peak, 100, 30), 133 (7), 132 (7), 131 (7), 117 (8), 91 (5); 4-methylisopropylbenzene **35**: MS (70 eV), m/e (rel. intensity), 134 (M$^+$, 29), 119 (base peak, 100), 91 (18), 77 (5).

REFERENCES

1. Articles appearing in T. Davidson, ed., *Polymers in Electronics,* ACS Symposium Series No. 242, American Chemical Society, Washington (1984).
2. W. W. Kaeding, L. B. Young, and A. G. Prapas, *Am. Chem. Soc. Div. Fuel Chem. Prepr.,* 28: 158 (1983).
3. J. D. Lytle, G. W. Wilkerson, and J. G. Jaramillo, *Appl. Opt., 18*: 1842 (1979).
4. O. A. Gunder and A. F. Korunova, *Stisintill. Org. Lyuminofory, 3*: 74 (1974).
5. W. P. Bishop, E. T. Snow, E. P. Royer, J. L. Benson, and J. Plimpton, *Nucl. Sci. Abstr., 30*: 18721 (1974). (Report of SAND-745089.)
6. L. A. Harrash, *IEEE Trans. Nucl. Sci., 17*: 278 (1970).
7. L. H. Lee, *Appl. Polym. Symp., 23*: 167 (1974).
8. N. Grassie and N. A. Weir, *J. Appl. Polym. Sci., 9*: 963 (1965).
9. N. Grassie and N. A. Weir, *J. Appl. Polym. Sci., 9*: 975 (1965).
10. N. Grassie and N. A. Weir, *J. Appl. Polym. Sci., 9*: 987 (1965).
11. N. Grassie and N. A. Weir, *J. Appl. Polym. Sci., 9*: 999 (1965).
12. R. Greenwood and N. A. Weir, *J. Appl. Polym. Sci., 19*: 1409 (1975).
13. W. Burlant and V. Serment, *J. Polym. Sci., 58*: 491 (1962).
14. N. A. Weir, *J. Appl. Polym. Sci., 17*: 401 (1973).
15. N. A. Weir, T. H. Milkie, and D. Nicholas, *J. Appl. Polym. Sci., 23*: 609 (1979).
16. N. A. Weir and T. H. Milkie, *Eur. Polym. J., 16*: 141 (1979).
17. N. A. Weir and T. H. Milkie, *J. Polym. Sci. Polym. Chem. Ed., 17*: 3723 (1979).
18. N. A. Weir, M. Rujimethabhas, and T. H. Milkie, *Polym. Photochem., 1*: 205 (1981).
19. N. A. Weir and T. H. Milkie, *Makromol. Chem., 180*: 1729 (1979).
20. N. A. Weir and T. H. Milkie, *Polym. Degrad. Stab., 1*: 105 (1979).
21. N. A. Weir and T. H. Milkie, *J. Polym. Sci. Polym. Chem. Ed., 17*: 3735 (1979).
22. N. A. Weir and T. H. Milkie, *Eur. Polym. J., 16*: 141 (1980).
23. N. A. Weir and T. Milkie, *Polym. Degrad. Stab., 2*: 225 (1980).
24. N. A. Weir, M. Rujimethabhas, and P. Q. Clothier, *Eur. Polym. J., 17*: 431 (1981).
25. N. A. Weir and M. Rujimethabhas, *Eur. Polym. J., 18*: 813 (1982).
26. M. Rujimethabhas and N. A. Weir, *Eur. Polym. J., 19*: 779 (1983).
27. N. A. Weir, J. Arct, and M. Farahani, *Polym. Degrad. Stab., 13*: 361 (1985).
28. N. A. Weir and M. Rujimethabhas, *Eur. Polym. J., 21*: 493 (1985).
29. M. Rappon, *Eur. Polym. J., 22*: 319 (1986).

30. U. Ramelow and B. M. Baysal, *J. Appl. Polym. Sci.*, *32*: 5865 (1986).
31. Z. A. Smirnova, A. S. Khachaturov, and A. F. Dokukina, *Vysokomol. Soedin.*, *B14*: 96 (1972).
32. L. Monnerie and F. Laupretre, in *Structural Dynamic Molecular Systems*, vol. 2 (R. R. Daudel and N. Dordrecht, ed.), pp. 129–154 (1986).
33. A. R. Monahan and A. Vanlaeken, *Am. Chem. Soc. Div. Polym. Chem. Polym. Prepr.*, *10*: 356 (1969).
34. B. Y. Byl'skill, O. F. Pozdnyakov, V. R. Regel, and B. P. Redkov, *Mekh. Polim.*, *5*: 835 (1973).
35. A. Sournia, *J. Chem. Ohys.*, *64*: 1805 (1967).
36. M. M. Coleman, E. J. Moskala, S. E. Howe, and P. C. Painter, *Polym. Mater. Sci. Eng.*, *51*: 286 (1984).
37. S. Mohanraji and W. T. Ford, *Macromolecules*, *18*: 351 (1985).
38. M. C. Burell, Y. S. Liu, and H. S. Cole, *J. Vac. Sci. Technol.*, *A4*: 2459 (1986).
39. D. T. Clark and H. R. Thomas, *J. Polym. Sci. Polym. Chem. Ed.*, *16*: 791 (1978).
40. H. Hiraoka and L. W. Welsh, Jr., "Radiation Chemistry and Resist Applications of Poly(p-Substituted Styrenes)," Proceedings of Symposium on Electron Ion Beam Scientific Technology, 10th International Conference, Montreal, Canada (1982).
41. S. W. Beaven, G. Beck, and W. Schnabel, *Eur. Polym. J.*, *14*: 385 (1978).
42. W. Schnabel, *Makromol. Chem.*, *180*: 1487 (1979).
43. S. Tagawa and W. Schnabel, *Polym. Photochem.*, *3*: 203 (1983).
44. S. Tagawa, W. Schnabel, M. Wshio, and Y. Tabata, *Radiat. Phys. Chem.*, *18*: 1087 (1981).
45. Y. Tabata, S. Tagawa, and M. Washio, *Mater. Microlithogr.*, *ACS Symp. Ser.*, *266*: 151 (1984).
46. S. Tagawa, "Radiation Chemistry and Photochemistry of Polystyrene and Chloromethylated Polystyrene as Deep UV, Electron and Ion Beam Resists," Proceedings of the International Ion Engineering Congress, Vol. 3, Kyoto, Japan, p. 1681 (1983).
47. Y. Tabata, S. Tagawa, and M. Washio, *Am. Chem. Soc. Div. Polym. Chem. Prepr.*, *25*: 287 (1984).
48. T. Ikeda, S. Okamura, and H. Yamaoka, *J. Polym. Sci. Polym. Chem. Ed.*, *15*: 2971 (1977).
49. T. Ikeda, K. Kawaguchi, H. Yamaoka, and S. Okamura, *Macromoles*, *11*: 735 (1978).
50. T. Ikeda, H. Yamaoka, T. Matsuyama, and S. Okamura, *J. Phys. Chem.*, *82*: 2329 (1978).
51. W. Schnabel, *Makromol. Chem.*, *180*: 1487 (1979).
52. K. Tanigaki, M. Suzuki, Y. Saotome, Y. Ohnishi, and K. Tateishi, *J. Electrochem. Soc.*, *132*: 1678 (1985).
53. K. Tanigaki, K. Tateishi, and Y. Ohnishi, *Polym. Eng. Sci.*, *26*: 1116 (1986).
54. R. B. Cundall, L. Pereira, and D. A. Robinson, *J. Chem. Soc., Faraday*, *11*: 69, 701 (1973).
55. K. P. Ghiggino, R. D. Wright, and D. Phillips, *J. Polym. Sci. Polym. Phys. Ed.*, *16*: 1499 (1978).
56. N. S. Karasash and V. A. Krongauz, *Dokl. Akad. Nauk SSSR*, *197*: 836 (1971).
57. S. L. Murov, in *Handbook of Photochemistry*, Marcel Dekker, New York, pp. 79–81 (1973).
58. R. B. Ingalls and L. A. Wall, *J. Chem. Phys.*, *35*: 370 (1961).
59. L. A. Wall and R. B. Ingalls, *J. Chem. Phys.*, *41*: 1112 (1964).
60. R. E. Florin, L. A. Wall, and D. W. Brown, *J. Polym. Sci. A*, *1*: 1521 (1969).
61. R. F. Cozzens, W. B. Moniz, and R. B. Fox, *J. Chem. Phys.*, *48*: 581 (1968).
62. R. E. Florin, L. A. Wall, and D. W. Brown, *Trans. Faraday Soc.*, *56*: 1304 (1960).
63. H. Fisher, *Kolloid-Z.*, *180*: 64 (1962).
64. J. Tino, M. Capla, and F. Szocc, *Eur. Polym. J.*, *6*: 397 (1970).
65. R. J. Abraham and D. H. Whiffen, *Trans. Faraday Soc.*, *54*: 1291 (1958).

66. K. E. Wilzbach and L. Kaplan, *J. Am. Chem. Soc.*, *86*: 2307 (1964).
67. G. Porter and E. Strachan, *Trans. Faraday Soc.*, *54*: 1595 (1958).
68. J. Lucki and B. Ranby, *Polym. Degrad. Stab.*, *1*: 1 (1979).
69. I. K. Chernova, V. P. Golikov, S. S. Leshchenko, V. I. Muromtsev, and V. L. Karpov, *Khim. Vys. Energ.*, *8*: 342 (1974).
70. I. K. Chernova, V. P. Golikov, S. S. Leshchenko, V. L. Muromtsev, and V. L. Karpov, *Khim. Vys. Energ.*, *8*: 265 (1974).
71. O. A. Gunder and V. A. Koba, *Khim. Vys. Eng.*, *8*: 83 (1974).
72. Z. Hlouskova, J. Placek, and F. Szocs, *Eur. Polym. J.*, *22*: 387 (1986).
73. T. Ikeda, H. S. Euler-Chelpin, S. Okamura, and H. Yamaoka, *Macromolecules*, *17*: 1655 (1984).
74. J. F. Rabek and B. Ranby, *J. Polym. Sci. Polym. Chem. Ed.*, *12*: 273 (1974).
75. B. Ranby and J. Lucki, *Pure Appl. Chem.*, *52*: 295 (1980).
76. J. F. Rabek, J. Sanetra, and B. Ranby, *Macromolecules*, *19*: 1674 (1986).
77. T. G. Tessier, J. M. J. Frechet, C. G. Wilson, and H. Ito, *Mater. Microlithgr.*, *ACS Symp. Ser. 266*: 269 (1984).
78. I. K. Chernova, S. S. Leshchenko, V. P. Golikov, and V. L. Karpov, *Vysokomol. Soedin.*, *A22*: 2175 (1980).
79. A. Bromberg, K. H. Schmidt, and D. Meisel, *J. Am. Chem. Soc.*, *106*: 3056 (1984).
80. B. Bockrath, E. Bittner, and J. McGrew, *J. Am. Chem. Soc.*, *106*: 135 (1984).
81. N. K. Bridge, *Nature*, *185*: 31 (1960).
82. V. Stannett, in *Diffusion in Polymers* (J. Crank and G. S. Park, eds.), Academic Press, London, pp. 41–73 (1968).
83. A. Charlesby, *Proc. R. Soc. London, A*, *222*: 542 (1954).
84. R. G. Brault, R. L. Kubena, and J. E. Jensen, *Polym. Eng. Sci.*, *23*: 941 (1983).
85. E. D. Feit, L. F. Thompson, C. W. Wilkins, Jr., M E. Wurtz, E. M. Doerries, and L. E. Stillwagon, *J. Vac. Sci. Technol.*, *16*: 1997 (1979).
86. M. A. Hartney, A. E. Novembre, and F. S. Bates, *J. Vac. Sci. Technol.*, *B3*: 1346 (1985).
87. M. A. Hartney, R. G. Tarascon, and A. E. Novembre, *J. Vac. Sci. Technol.*, *B3*: 360 (1985).
88. R. Tarascon, M. Hartney, and M. J. Bowden, *ACS Symp. Ser. 266*: 361 (1984).
89. F. Sche, B. Holil, R. Sagnes, C. Montginoul, L. Giral, F. Buiguez, B. Serre, and C. Rosilio, *Microelectron. Eng.*, *5*: 315 (1986).
90. M. Hashimoto, T. Iwayanagi, H. Shiraishi, and S. Nonogaki, *Polym. Eng. Sci.*, *26*: 1090 (1986).
91. H. Shiraishi, N. Hayashi, T. Ueno, O. Suga, F. Murai, and S. Nonogaki, *Polym. Mater. Sci. Eng.*, *55*: 279 (1986).
92. J. M. J. Frechet, T. G. Tessier, C. G. Wilson, and H. Ito, *Macromolecules*, *18*: 317 (1985).
93. H. Ito and C. G. Wilson, *ACS Symp. Ser. 242*: 11 (1984).
94. J. Liutkus, M. Harzakis, J. Shaw, and J. Paraszczak, *Polym. Eng. Sci.*, *23*: 1047 (1983).
95. J. C. Jagt and P. W. Whipps, *Philips Tech. Rev.*, *39*: 346 (1980).
96. R. G. Braut and L. J. Miller, *Polym. Eng. Sci.*, *20*: 1064 (1980).
97. G. N. Babu, P. H. Lu, S. L. Hsu, and J. C. W. Chien, *J. Polym. Sci. Polym. Chem. Ed.*, *22*: 213 (1984).
98. Y. Kamoshida, M. Koshiba, H. Yoshimoto, Y. Harita, and K. Harada, *J. Vac. Sci. Technol.*, *B1*: 1156 (1983).
99. K. Sukegawa and S. Sugawara, *Jpn. J. Appl. Phys.*, *20*: L583 (1981).
100. H. Gokan, K. Tanigaki, and Y. Ohnishi, *Solid State Technol.*, *28*: 163 (1985).
101. K. Tanigaki, Y. Ohnishi, and S. Fujiwara, *ACS Symp. Ser. 242*: 177 (1984).
102. K. Tanigaki, M. Suzuki, and Y. Ohnishi, *J. Electrochem. Soc.*, *133*: 977 (1986).
103. D. T. Mowry, M. Ronell, and W. F. Huber, *J. Am. Chem. Soc.*, *68*: 1105 (1946).

10
Advances in PVC Polymerization

Bertil Törnell
University of Lund
Lund, Sweden

INTRODUCTION	347
DEVELOPMENT OF THE PVC INDUSTRY	348
MANUFACTURE, PROPERTIES, AND USES OF PVC RESINS	349
Polymerization Processes	349
Main Outlets of PVC	349
Characteristic Properties of PVC	349
Quality Aspects	350
POLYMERIZATION—GENERAL CONSIDERATIONS	351
Physics	351
Chemistry and Kinetics	352
MASS POLYMERIZATION	356
General	356
Particle Formation at Low Conversion	356
SUSPENSION POLYMERIZATION	358
General	358
Grain Structure and Suspension Stabilizers	359
EMULSION POLYMERIZATION	360
Normal Emulsion Polymerization	360
Microsuspension or Hybrid Processes	361
KINETICS OF SUSPENSION AND EMULSION POLYMERIZATION AT CONSTANT RATE OF INITIATION	362
Suspension Polymerization	362
Emulsion Polymerization	365
ECONOMIZATION OF PVC PRODUCTION	368
REFERENCES	372

INTRODUCTION

The scientific and patent literature on polymerization of vinyl chloride (VCM) is extensive. A comprehensive review covering the literature up to about 1984 was recently published [1]. Some less comprehensive but very useful monographs have also appeared

[2, 3]. This chapter deals primarily with fundamental work on VCM polymerization. No attempts have been made to include in this report contributions which are solely reported in the patent literature. The report concentrates on those aspects of VCM polymerization with which recent work has provided us a better understanding or new insight. The progress in the field made during the last few years is mainly the result of new or strongly improved experimental techniques (e.g., more powerful spectroscopic methods) that have become available.

Besides a discussion of recent work in the field, this chapter briefly introduces PVC and describes the main polymerization processes used for PVC production. This serves mainly as background for those who have not been active in the field.

DEVELOPMENT OF THE PVC INDUSTRY

Poly(vinyl chloride), PVC, is one of our oldest thermoplastic polymers. It has a long history [4]. The preparation of PVC in the form of a white, seemingly useless powder was described as early as 1835 by Regnault. It was not until 1912, however, that Fritz Klatte determined, by dissolution and what we today refer to as plasticization, that PVC could be used for the preparation of industrially valuable materials [5], although his method was never used for production. Commercial production of PVC was begun in Germany (1931), soon followed by the United States (1933). The early U.S. production was based on work carried out by W. L. Semon, of the B.F. Goodrich Co. Semon had accidentally found that PVC, when heated in certain high boiling solvents, gave a flexible, elastic, and chemically rather inert material [6]. Plasticized PVC was used initially as a sealing material, for cable insulation, and for the manufacture of raincoats and shower curtains. According to Semon, the PVC used by Goodrich at that time was a high molecular weight polymer prepared by mass polymerization to low conversion.

The early commercial progress of PVC stimulated work that resulted in the acetylene process for vinyl chloride, the preparation of PVC by emulsion and suspension polymerization, new efficient plasticizers, and better techniques for stabilizing PVC toward thermal degradation. With the advent of effective stabilizers it became possible to process PVC in unplasticized form. Further development resulted in the mass route to PVC and new PVC resins with improved processing properties. A significant step in promoting the growth of the PVC industry was the development of the oxychlorination process, which made it possible to use ethylene as feedstock for VCM production.

In the year 1935 the world consumption of PVC amounted to 11,000 metric tons. In 1950 the production had increased to 220,000 tons. At this time the PVC consumption exceeded the total consumption of all other thermoplastic materials together. During the period from 1950 to 1980 the world production of PVC increased to 11 million tons—a factor of 50 [1]. Currently, next to polyethylene, PVC is the most frequently used thermoplastic material.

The enormous expansion of the PVC industry to a large extent is due to the high versatility of PVC as a plastic raw material. In unplasticized form PVC is hard and tough. It has a glass transition temperature close to that of polystyrene. Because of its polar nature, it is resistant to most organic solvents and can be made flexible by plasticization using high boiling polar solvents. The rapid expansion of the PVC industry, however, is also due to the fact that PVC to about 57% consists of chlorine. This makes PVC an important outlet for chlorine, which to some extent can be regarded as a by-product of sodium hydroxide production.

MANUFACTURE, PROPERTIES, AND USES OF PVC RESINS

Polymerization Processes

The commercial production of PVC is based on three main processes: suspension, emulsion, and mass polymerization. The emulsion process is the oldest. It is now used mainly for preparing specialty resins (E-PVC) for paste applications. The suspension process is now the dominating route to PVC (S-PVC). The mass process gives resins (M-PVC) with properties similar to those of the suspension process. Small amounts of PVC, mainly copolymers, are prepared by solution polymerization. The solution process is not dealt with here.

Of the 1980 total world PVC capacity (16 million tons/a), the S-, E-, and M- processes accounted for about 82, 10, and 8%, respectively. Large differences exist between regions. Thus in the United States, S-, E-, and M-PVC accounted for 80, 10, and 10%, respectively, whereas in West Germany, the corresponding figures were 60, 28, and 12% [1].

Main Outlets of PVC

PVC is processed either in unplasticized (hard PVC) or plasticized form (flexible PVC). Hard PVC is a hard and tough material. It is often pigmented but can also be obtained in transparent form. The relative importance of hard PVC is increasing. In western Europe, hard PVC in 1980 accounted for about 62% of the total PVC market [1]. The main outlets for hard PVC are pipes and fittings, extruded profiles, films, bottles, and phonograph records.

Flexible PVC is used mainly as film, cable insulation, flooring, leather substitutes, flexible tubing, and extruded profiles. Most flexible PVC is prepared from dry blends. This requires porous PVC resins, having the capacity to suck into its pore structure the necessary amount of plasticizer. A minor fraction of flexible PVC (about 6%) is prepared by paste technology. In this case, fairly compact resins must be used. A paste is first prepared by dispersing the resin, pigments, and other additives in liquid plasticizer. The paste is then applied to a substrate (e.g., a release paper or a fabric) and converted to a homogeneous material by heating to about 160°C.

Characteristic Properties of PVC

In pure form, PVC is a hard thermoplastic material with a T_g of about 80°C. It is basically an amorphous polymer. Because of the presence in the polymer of syndiotactic sequences of sufficient length, PVC shows some crystallinity. The crystalline melting point is high (about 225°C) and considerably exceeds normal processing temperatures of 160–180°C.

Intrinsically, PVC is a thermally unstable polymer and gives off hydrochloric acid when heated. The dehydrochlorination process reaches a measurable rate already at temperatures just above 100°C. The reaction is autocatalytic and results in discoloration, ascribed to the occurrence of polyene sequences in the molecules. As dehydrochlorination starts at temperatures much lower than normal processing temperatures, PVC processing requires the use of efficient stabilizers.

The presence of crystallinity and the high melting point of PVC offer some explanation for the peculiar melt rheology of PVC. It is generally agreed that unplasticized PVC does not form a true melt as other thermoplastic materials do. The flow units in a PVC melt are not individual molecules but small, sticky particles (domains or microdomains, which by some authors are believed to derive their origin from molecular aggregates formed in the polymerization process) formed by successive crushing and grinding of the PVC resin grains. Although not completely understood, the size and structure of the flow units most likely are related to the structure of the resin particles and hence to the processes of primary particle nucleation and agglomeration that occur during the polymerization process. This, and the fact that heat build-up during processing is mainly due to the friction between the small particles obtained by successive disintegration of the resin grains, explains the great interest devoted to grain structure control in suspension and mass polymerization processes.

The successful processing of hard PVC requires proper and balanced use of additives. Important constituents of hard PVC formulations are the lubricants. These are substances added in order to control heat build-up by adjusting the internal friction in the melt (internal lubricants) and to prevent the PVC melt from sticking to the steel surfaces of the processing equipment (external lubricants). Each resin quality requires its own formulation as the resins normally contain polymerization processing aids, which may serve as or interact with the lubricants.

Quality Aspects

As with other thermoplastic materials, the molecular weight and molecular weight distribution (MWD) are important quality properties. With PVC, the molecular weight and MWD are determined almost exclusively by chain transfer to monomer, i.e., by the polymerization temperature (see below). Molecular weight control therefore does not present a difficult problem.

Some quality properties—e.g., electrical properties, color, and thermal stability of the polymer—may be affected by the particular choice of processing aids and by the exact way in which the process is conducted. The knowledge of effects of this type generally remains a secret within the PVC industry.

Except for the properties discussed above, all other PVC properties of practical importance are determined by the physical form and structure of the PVC resin. This goes with about equal weight for S-, E-, and M-PVC. With paste resins (mainly E-PVC resins), the grain size distribution and the structure of the grains determine the rheology of the paste and affect the process of gelation. The resin properties in this case are controlled mainly by the particle size distribution of the parent PVC latex and the techniques used for drying, grinding, and sieving. A paste resin should be fairly compact. This means that the latex particles must be able to pack closely during drying. For this reason, paste resins are often prepared from PVC lattices with bimodal particle size distributions. The idea is that the small particles should fit into the interstices of the close-packed large particles. Drying, grinding, and sieving reportedly strongly affect the rheological properties of the final pastes. It is possible to prepare resins which either give pseudoplastic or dilatant pastes.

The preparation of hard and flexible PVC from S- and M-PVC via dry blending requires porous resins of uniform porosity. The grain size distribution should be fairly narrow. Small particles cause dust problems and oversize particles melt less readily. Moreover, in dry blending, a narrow particle size distribution should favor a homogeneous distribution of plasticizers and other additives over the polymer grains.

Porosity requirements vary with application. For flexible PVC, the total pore volume of course should be large enough to absorb the required amount of plasticizer. The nature of the pore structure is also important as it determines the rate by which the plasticizer is absorbed into the resin particles. Resins for hard PVC should also possess a uniform pore structure in order to facilitate an even distribution of lubricants and other liquid additives during dry blending. Under conditions where the extruder capacity is limited by the volumetric capacity of the feed zone, resins with high bulk densities and hence moderate porosities are preferred. The resin must not be too compact, however, as this would impede melting by successive disintegration and interparticle friction.

POLYMERIZATION—GENERAL CONSIDERATIONS
Physics

At normal temperatures, VCM is a gas. Its boiling temperature at normal pressure is $-13.8°C$. The solubility of VCM in water at temperatures in the range 15–60°C amounts to $8.8 \times 10^{-3} P/P_0$ (kg/kg water), where P is the partial pressure and P_0 the saturation pressure of VCM [7].

The liquid density at 25°C is 901.3 kg m^{-3}. The density of technical PVC falls into the range 1390–1400 kg mg^{-3} [8]. The large difference between the densities of the monomer and its polymer implies that the volume decrease during polymerization is rather large. Data on the heat of polymerization of VCM are rather scattered [9]. A recent determination of ΔH_p using direct calorimetric measurements during polymerization gave a value of -97.6 kJ mol^{-1}, with a standard deviation (six determinations) of 0.8 kJ mol^{-1} [10].

PVC is insoluble in its monomer. From mass polymerization studies, Boissel and Fischer [11] found that the solubility of the polymer formed at the start of the polymerization was lower than 10^{-3}%. Cotman et al. from studies of the solubility of low molecular weight telomers estimated that PVC polymer radicals lose their solubilities in monomer when reaching a size of between 25 and 32 monomer units [12].

Studies of the swelling of PVC by its monomer have shown that the relation between the monomer concentration in the polymer phase and the partial pressure of the monomer is rather complicated. The monomer uptake in PVC at low partial pressures has been thoroughly studied by Berens [13], who found that the last traces of VCM were strongly held by the polymer. Berens described the swelling of PVC at low pressures as the sum of a normal dissolution process and a process which leads to condensation of monomer in the holes present in the glassy PVC polymer phase. In the low partial pressure range, the extent of swelling varied with temperature and the prehistory of the polymer, suggesting that the hole-filling process was affected by volume relaxation effects. According to Berens, the swelling at high partial pressures is athermal and can be described by the Flory–Huggins equation, using a χ value of 0.98. Other swelling studies on PVC latex particles in water confirmed the observation by Berens that PVC has a high affinity toward VCM at low partial pressures [7]. At higher partial pressures, the latex swelling experiments showed that χ varied considerably with composition and took on different values for PVC samples prepared at different polymerization temperatures. A low polymerization temperature resulted in a high χ value. This latter effect reflects the influence of syndiotacticity on the swelling process. At saturation pressure the swelling of a flat sample (no surface tension effects) of PVC prepared at 50°C was estimated as 0.304 kg VCM per kg PVC. The experiments also showed that small amounts of VCM were solubilized in the emulsifier adsorbed at the surface of the latex particles.

Chemistry and Kinetics

General Considerations

Commercially, all PVC is prepared by free radical polymerization. The polymerization of VCM follows the same general reaction scheme as that obeyed by other monosubstituted ethylenes. The propagation is mainly head-to-tail with an occasional head-to-head arrangement, the consequence of which will be dealt with later. The tendency toward syndiotactic placements increases with a decrease in the polymerization temperature. The difference in activation energy for isotactic and syndiotactic propagation is almost 3 kJ mol^{-1} [14]. The increase in syndiotacticity with decreasing polymerization temperature is accompanied by an almost linear increase in the glass transition temperature [15] and in crystallinity [16]. The crystallinity of PVC prepared at normal polymerization temperatures (50–70°C) is low. Crystallization leads to the formation of molecular aggregates, which persist when the polymer is dissolved in normal PVC solvents, unless the solutions are heated to about 200°C (see, e.g., Ref. 17). This has to be observed in molecular weight determinations of PVC by GPC or light scattering, for example.

The kinetics of vinyl chloride polymerization is to a large extent independent of the particular reaction conditions used, i.e., whether mass, suspension, or emulsion processes are concerned. This goes for the molecular weight and its distribution, the effect of initiator concentration and conversion on polymerization rate, and the overall reaction rate profile. These peculiarities can be rationalized in the following way. In all types of processes, the majority of the polymer is formed by reaction in a monomer-swollen polymer phase, the composition of which is determined by the swelling equilibrium of the VCM–PVC system and hence does not depend on process type as long as minor differences due to particle size effects are neglected. Another important factor is the unusually high frequency of chain transfer to the monomer. The importance of this reaction explains why the molecular weight is determined by reaction temperature and is almost independent of the rate of initiation and conversion. It also explains the fact that the MWD of PVC prepared under isothermal conditions closely approaches the most probable distribution ($M_w/M_n = 2$).

Moreover, irrespective of the process conditions used, the rate of polymerization is proportional to the square root of initiator concentration. This is the normal situation for mass and suspension polymerization but is atypical for emulsion processes. Why, then, is the rate of emulsion polymerization with VCM, at given particle concentration, proportional to the square root of initiator concentration? This problem was first explained by Ugelstad et al. [18]. The observed square root dependence on initiator concentration is also a consequence of the chain transfer to monomer reaction. With VCM this reaction is rapid and occurs via the formation of a chlorine radical (see below). This very small radical undergoes rapid desorption from and readsorption into the latex particles. The interparticle exchange of radicals makes interparticle termination an important process. The effect of subdividing the polymerizing phase, so important in polymerizations following the Smith–Ewart case II kinetics, thus disappears (at low particle numbers) or becomes considerably reduced [18]. Because of the rapid exchange of radicals between particles, the number of radicals per particle is very low and the radical population in the particles adjusts itself rapidly to changes in the rate of generation of radicals in the aqueous phase.

During polymerization, a separate phase of liquid monomer exists up to a conversion of about 75%. This point is observed as a pressure drop in the reactor. From this instant on, the monomer concentration in the polymer phase decreases with conversion. As polymerization proceeds, the rate continues to increase, however, and passes through a more or less pronounced peak (Fig. 1). This is caused by the well-known Trommsdorff or gel effect, which is ascribed to a decrease in the termination rate constant with increasing viscosity of the polymerizing medium. With PVC, the termination is viscosity-controlled also in the fully monomer-swollen polymer phase. The termination rate constant is about 2500 times higher in the swollen monomer phase than in the liquid monomer [19]. This explains why PVC polymerizations behave autocatalytically almost from the very beginning. Mass and suspension polymerizations of VCM are two phase polymerization processes. As the specific rate of polymerization is about 50 times more rapid in the gel phase than in the monomer phase (when compared at equal rates of initiation), the total rate of polymerization increases in proportion to the amount of gel phase. The latter applies of course also to emulsion polymerization where no polymerization at all goes on in the monomer phase. The decrease in polymerization rate beyond the rate maximum is due to the decrease in monomer concentration and to the fact that also the rate of propagation becomes viscosity-controlled at sufficiently high viscosity. The limiting conversion reached corresponds to a monomer–polymer composition, the glass transition temperature of which is equal to the temperature of polymerization. In commercial operation, the polymerization is interrupted at a conversion of below 90%, i.e., a long time before the limiting conversion is reached.

As the molecular weight of PVC is determined by the polymerization temperature, there is little incentive to use chain transfer agents. The only obvious reason to use a chain

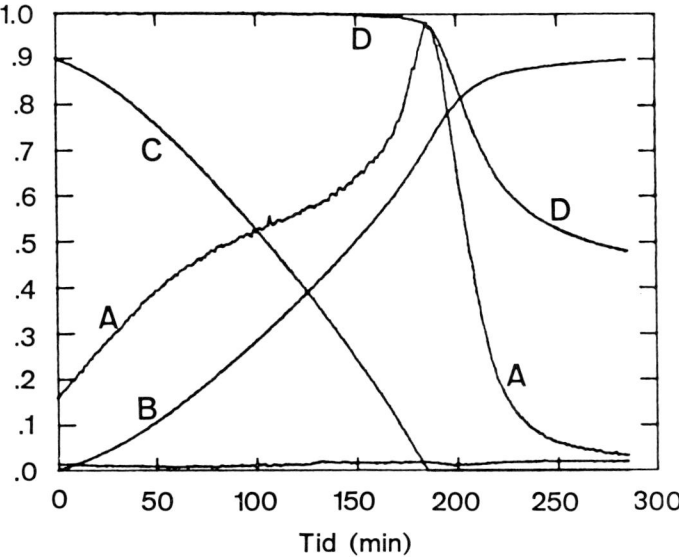

Figure 1 The kinetics of a typical suspension polymerization of VCM. A represents the rate of polymerization (arbitrary units), B is the conversion, C the amount of VCM present in the form of liquid monomer (relative to the total initial amount of VCM), and D is the pressure in the reactor (1.0 in the pressure scale is the sum of the saturation pressures of VCM and water at the temperature of polymerization, 50°C). The unindexed curve following the abscissa is the stirring power (in the same power scale as A).

transfer agent would be in the preparation of low molecular weight PVC in reactor equipment which cannot sustain the high pressure associated with a high reaction temperature.

Chain Transfer to Monomer and the Development of Anomalous Structures

The nature and mechanism of formation of so-called anomalous structures in PVC have been the subjects of extensive studies over the years. The results are of interest in the present context as they have led to an understanding of the mechanism behind the process of chain transfer to monomer. A review of work published up to about 1980 was given by Caraculacu [20]. Two more recent reviews by Hjertberg and Sörvik [21] and by Maddams [22] were recently published. The reader is referred to these review articles for references to the original literature.

The progress made in understanding the nature of chain transfer to monomer and of the formation of anomalous structures in PVC is to a large extent due to the rapid development of new and sensitive NMR techniques. The now accepted mechanism of chain transfer to the monomer was first confirmed by Starnes et al. (e.g., Ref. 23). It is illustrated by the reaction scheme in Figure 2, which appeared in a recent paper by Hjertberg and Sörvik [24].

According to this mechanism, the chain transfer to monomer starts with a head-to-head addition to the growing radical giving **2**. This new radical can take part in a propagation reaction which results in the formation of internal head-to-head arrangements in the

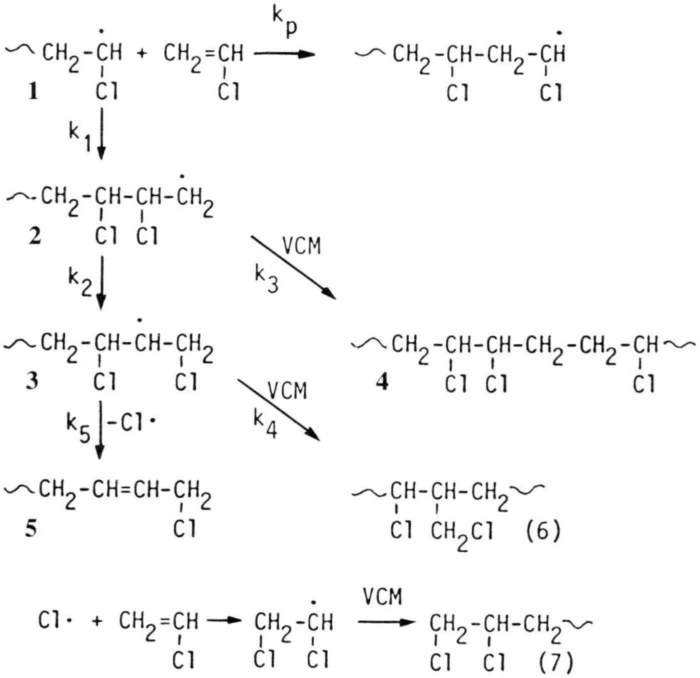

Figure 2 Reaction scheme for chain transfer to monomer and for the formation of important "anomalous" structures in PVC.

polymer. The radical **2** is very reactive and can stabilize itself by rearranging to **3**. Propagation of this radical leads to the formation of chloromethyl side chain branching. The concentration of chloromethyl branches in technical PVC is about 4–5 per 1000 monomer units. By expulsion of chlorine, radical **3** gives a polymer with a 1-chloro-2-alkene end group. The chain transfer to monomer is then completed when the chlorine atom adds monomer and starts a new growing chain, which will have a 1,2-dichloro alkane end group (**7**). Structures **5** and **7** should be the most frequent end groups in PVC. In ordinary PVC, the content of these structures is somewhat less than one per molecule. The content of internal head-to-head arrangements in PVC is very low and not higher than 0.2 per 1000 monomer units.

As the monomer concentration in the polymer phase decreases toward the end of the polymerization, the balance between the two reactions from radical **3** should change toward chlorine expulsion. This would lead to a decrease in the molecular weight. Based on the reaction scheme in Figure 2, Hjertberg and Sörvik [24] showed that the chain transfer coefficient C_M ($= K_{tr}/K_p$) should be expressed as

$$C_M = \frac{k_5 k_1}{k_p(k_5 + k_4[M])} \tag{1}$$

where the rate constants are defined by Figure 2 and [M] is the concentration of monomer at the locus of polymerization.

The decrease in molecular weight with a decrease in monomer concentration was proven in seed emulsion polymerizations carried out under subsaturation conditions [24]. In the actual experiments the monomer was fed to the reactor from a monomer reservoir kept at a lower temperature than that of the reactor.

Chlorine radicals can also take part in chain transfer to polymer. This is illustrated by the reaction scheme of Figure 3. The first reaction in this scheme gives rise to the formation of one molecule of hydrochloric acid. This reaction may possibly explain the observation that the PVC primary particles formed in mass and suspension polymerization are negatively charged (see below). Attack by the chlorine atom on the methylene group with subsequent expulsion of chlorine offers an explanation of the formation of internal

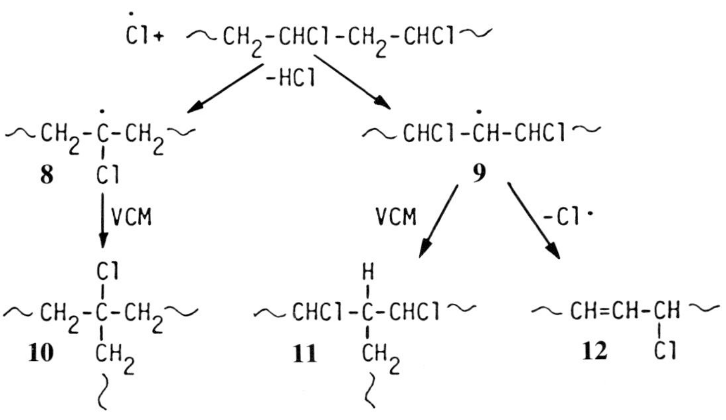

Figure 3 Reaction scheme for chain transfer to polymer by chlorine atoms and the formation of long chain branches and internal unsaturation in PVC.

unsaturation in PVC (**12**). Propagation of the intermediate radicals **8** and **9** leads to the formation of long chain branches (< 0.5 per 1000 monomer units). These propagations will give long chain branches associated with a tertiary chlorine (**10**) or a tertiary hydrogen (**11**). The attack by chlorine on the methylene group seems to be the preferred route.

Long chain branches in PVC may also arise by attack of a macroradical on the polymer. In this case the attack is primarily on the chloromethylene group, giving **8**. Other anomalous structures in PVC can be formed by backbiting. This gives 2,4-dichlorobutyl branches (about 1 per 1000 monomer units). According to Starnes et al., backbiting can also give rise to 2-chloroethyl branches [25]. Most tertiary chlorine in PVC is associated with short chain branches. The amount of both long and short chain branches increases as the monomer concentration in the polymer phase decreases. According to Hjertberg and Sörvik, tertiary chlorine, because of its relatively high abundance, is the most important thermally labile structure in PVC [24].

MASS POLYMERIZATION

General

The mass polymerization process was introduced by Pechiney St. Gobain in the mid-1960s. This is the only mass process in use. The process is discontinuous and is carried out in two steps. The first step uses a stirred tank reactor. This step can be described as a precipitation polymerization in liquid monomer and is interrupted at a conversion below 15%. The reaction mixture now consists of a VCM suspension containing open-structured polymer grains formed by agglomeration of primary PVC particles. The grain size distribution of the suspension is determined by the stirring conditions (higher stirring speeds give smaller grains) and corresponds to the grain size distribution of the final resin. The second step is carried out in an autoclave which has a stirrer arrangement suitable for powder mixing. During this step the suspension is transformed to an apparent dry powder which contains imbibed and dissolved monomer. Reflux condensors are said to be used for cooling. The polymerization is allowed to proceed to a conversion of between 80 and 90%.

The mass process gives pure resins (except for the initiator, no or very small amounts of additives are used) with a narrow size distribution. The grains are skin-free. Some problems with dusting during dry mixing, ascribed to the presence of loosely bound, fine particles at the surface of the grains, have been reported.

Particle Formation at Low Conversion

The only aspects of mass polymerization to be dealt with here are the formation, growth, stability, and agglomeration of the primary PVC particles formed at low conversion. These processes play an important—perhaps decisive—role in determining grain structure buildup during mass and suspension polymerization.

Results from a recent study of mass polymerization by photon correlation spectroscopy covering conversions up to a few tenths of a percent showed that primary particles with a diameter in the range 0.08–0.1 μm were formed immediately after initiation [26]. The number of these particles remained constant during a certain fairly well defined growth period. When the primary particles reached a size of about 0.2 μm, they started to coagulate.

It is obvious that the primary particles observed during the very first part of the polymerization (at conversions on the order of 0.01%) must have a history. These particles are probably formed by coagulation of a large number of basic particles having a size of about 10–20 nm [11]. A basic particle is believed to result from association of the polymer molecules formed from a single primary radical. Basic particles have never been observed experimentally. Unstabilized particles of this size are too unstable to be observed by photon correlation spectroscopy. Rance and Zichy [27] have shown that the particle size and number density of primary particles formed by diffusion-controlled coagulation of basic particles should approach equilibrium within 10–20 sec after initiation. This time is shorter than the time (about 30 sec) required to generate an autocorrelation function.

The fact that the primary particles are monodisperse [28] and stable toward coagulation for a considerable growth period suggests that they are stabilized. Electrophoretic measurements have shown that the primary PVC particles are negatively charged [29] and hence should be electrostatically stabilized. This conclusion is supported by the fact that stable primary PVC particles in vinyl chloride coagulate on addition of VCM-soluble electrolytes [30]. In this context it is interesting to note that Davidson and Witenhafer have shown that agglomerates of primary particles formed in monomer droplets during suspension polymerization possess electrophoretic mobility [31]. These authors also showed that such agglomerates formed regular arrays in the droplets, an observation that could be explained only by the presence of long-range stabilizing forces.

The nature of the stabilizing charges on the PVC particles is a matter of discussion. It has been suggested that the charges arise from impurities present in the monomer or by adsorption of chlorine ions formed by dehydrochlorination during polymerization [25].

Results from studies of primary particle formation and growth in mass polymerization of VCM at conversions up to 7% have recently been published. In these studies, polymerization was carried out in a stirred reactor. The particle size was determined by scanning electron microscopy using a sample preparation technique which prevented particle coalescence. The number of primary particles was calculated from the particle size distribution and the conversion as determined by calorimetric measurement. In agreement with the results discussed in the preceding paragraphs, there was a first stage during which a constant number of primary particles grew at equal rates. The particle number density during this stage was in good agreement with previously published data. When the particles reached a certain limiting size, they started to agglomerate. The limiting size varied with the stirring speed and was 0.18 μm at a stirrer tip speed of 1.5 m sec^{-1} and 0.35 μm at the lower tip speed 0.4 μm sec^{-1}. These sizes were reached at conversions of 0.2 and 1%, respectively [32]. These results showed that the stability of the primary particles decreased with size and that the particles were shear sensitive. In agglomerates formed at high stirring speed, i.e., from relatively stable particles, multiple particle contacts were much less frequent than in agglomerates formed at low stirring speed. This is in agreement with the assumption that the particles were electrostatically stable and that their stability decreased with size [33].

As the primary particles started to agglomerate, a second stage of particle nucleation commenced. The presence of a second nucleation step was previously suggested by Boissel and Fischer [11]. The particles formed in the second nucleation step grew to the same limiting size as those formed initially. The agglomerated particles did not show a measurable growth. The agglomerates obtained using a stirrer tip speed of 1.5 m sec^{-1} consisted of loosely packed agglomerates of about 100 μm in diameter. When the stirrer was stopped these agglomerates sedimented rapidly and a sharp boundary was formed between the sediment and an upper turbid monomer phase containing unagglomerated primary particles. In an experiment where the agglomerated particles had a diameter of 0.18 μm, the average size of the unagglomerated particles was 0.16 μm. Their number concentration was 2.3×10^{13} mol^{-1}, in good agreement with the number of primary particles per mole of monomer formed in the first nucleation step. The sediment volume increased with conversion in a manner indicating that new agglomerates were formed and that the agglomerates became somewhat more compact with conversion. At a conversion of 6%, the sediment volume corresponded to about 90% of the total volume. From these and other results it could be concluded that nucleation, growth, and stability of the primary particles formed in the first and second nucleation steps were equal. This implies that the stabilizing negative charges on the primary particles cannot derive from impurities. The charged species responsible for particle stabilization must have been formed as a consequence of the polymerization process itself.

Suspensions of large primary particles grown under unstirred conditions agglomerated on shaking. In the absence of additives, the agglomeration of the primary particles was trigged by the stirring conditions. It was found, however, that addition of VCM-soluble quaternary ammonium salts caused a stable dispersion of primary particles in VCM to coagulate. Such salts when present during a normal polymerization experiment (in a stirred reactor) destabilized the primary particles and gave agglomerates composed of smaller particles [30]. Similar results were obtained by adding small amounts of the VCM-soluble fraction of Span 20 (described as sorbitol monolaurate). The effect in this case was probably due to the presence of impurity electrolytes in the product. Certain high molecular weight additives (EVA, PVAc, and PMMA) were found to stabilize the particles, probably by conferring steric stability to the particles [28]. For further discussion of the electrostatic stabilization of PVC primary particles in VCM, the reader is referred to Refs. 26 and 34.

SUSPENSION POLYMERIZATION

General

Suspension polymerization of VCM is carried out batchwise, mostly in stainless steel reactors having highly polished internal surfaces and sizes up to 200 m³. Continuous operation has not proven feasible. An important reason for this is that the residence time distribution has to be very narrow in order to avoid the formation of compact, hardmelting grains. In film production such grains would appear as fisheyes.

In suspension polymerization, liquid monomer (100 parts) is emulsified in water (150–200 parts). The emulsion is stabilized by small amounts of surface-active polymers (primary suspension stabilizers), often in combination with a low molecular weight emulsifier (secondary suspension stabilizer). The total amount of stabilizers is low (about 1 part per 100 monomer). Polymerization is carried out using one or several monomer-soluble initiators. The aqueous phase may be buffered. The exact procedures followed in technical polymerization processes are not known. It should be understood, however, that the processes may be complicated.

The stirring requirements are stringent. The reactors are jacketed and cooled through the reactor wall. This may be combined with reflux condenser cooling. An important technical and hygienic problem is fouling of the reactor surface. This must be avoided in order to reduce fisheyes and minimize reactor cleaning. For this reason the reactor surface must be very smooth and stagnant zones must be avoided. Baffles and internal coolers cannot be used.

The postpolymerization operations include steam stripping to remove residual monomer, recovery of resin by centrifugation and washing, drying (mostly in rotating ovens), mild milling, sieving, and packaging. Until about 1980, all commercial suspension resins consisted of grains surrounded by a skin composed of suspension stabilizer and partly coalesced primary PVC particles. The first process for preparing skin-free suspension resins was developed by the Swedish company KemaNord, presently Norsk Hydro [35].

Grain Structure and Suspension Stabilizers

The product quality of suspension PVC is to a large extent controlled by the particular suspension stabilizer system used and the precise manner in which the polymerization process is conducted. The suspension stabilizer system affects the shape of the grains, the grain size distribution, resin porosity, and bulk density of the resin.

Traditionally the primary suspension stabilizer is a water-soluble polymer; cellulose derivatives (e.g., methyl hydroxypropyl cellulose) and poly(vinyl alcohols) are important examples. Recently cold water–soluble poly(vinyl alcohols) with relatively high acetate contents (degree of hydrolysis about 70–75%) as well as poly(vinyl alcohols) of still higher acetate content, which are insoluble in both water and VCM, have become popular.

Secondary emulsifiers may be anionic or nonionic surface-active agents. The interaction between primary and secondary suspension stabilizers as well as the effect of VCM on these interactions [36, 37] are probably important factors in S-PVC production. These types of interactions are not particularly well studied and the choice of a suitable suspension stabilizer system is empirical.

The mechanism by which the suspension stabilizers exert their influence is still mainly unknown. It is often assumed that the stabilizers exert their action only at the interface between water and monomer, but this cannot be taken for granted. A study of the commercial product Span 20 from ICI/Atlas, a commonly used secondary suspension stabilizer, showed that this product contained 95% VCM-soluble matter. In a 1 : 263 : 193 weight mixture of Span 20/VCM/water, 67% of the emulsifier was found in the monomer phase [33]. It cannot be excluded, therefore, that components of the suspension stabilizers are present in the monomer droplets, where they can affect primary particle stability. The suspension stabilizers, when adsorbed at the VCM–water interface, manipulates the rheology of that interface. This should affect the "stirring conditions" inside the drops. In view of the fact that the primary particles are shear sensitive, such an effect might influence primary particle agglomeration in the droplets and thereby the resin properties.

Measurements of the effect of suspension stabilizers on the interfacial tension at the VCM–water interface have been reported [38]. The interfacial tension was found to be strongly time dependent, particularly with the poly(vinyl alcohols). With the cellulose ether tested (Methocel F 50), mechanically strong, solid films were formed at the interface. This film formation could be delayed by addition of secondary suspension stabilizers. Although Methocel formed solid skins at the interface, Methocel-stabilized emulsions were considerably less stable toward coalescence than poly(vinyl alcohol)-stabilized emulsions [39]. With both types of system, grain formation during polymerization seemed to involve either droplet agglomeration or droplet break-up and coalescence [39, 40]. In any case the droplet size increased during the first stage of the process. In experiments with Methocel, the final grain size distribution was found to correspond to the drop size distribution observed at 5% conversion [41].

A study by Davidson and Whitenhafer [31] on suspension polymerization using poly(vinyl alcohol) as stabilizer showed that a very thin skin composed of PVC-grafted stabilizer was formed around the monomer droplets. When studied at low conversion, this skin was found to be reinforced by an adhering layer of closely packed, small, primary PVC particles, extractable by THF. It is possible that in many S-PVC systems, the grains are formed by agglomeration of a number of VCM–PVC droplets surrounded by a skin of this type. Grafting of the stabilizer by PVC should affect the solution properties of the interfacial layer and make it less hydrophilic. This might promote flocculation of droplets. In cases where this would be true, the S-PVC grain structure should be controlled by the following processes:

- Formation and agglomeration of primary particles in the individual VCM droplets, followed by gradual cementing together of the agglomerates (these processes should proceed as in a mass process).
- Formation of a PVC-reinforced skin around the monomer droplets.
- Agglomeration of a number of skin-covered droplets.
- Shrinking of the composite drop/grain because of volume changes during polymerization and invasion of the pore system by water. (The extent to which the package of agglomerate-filled droplets can resist shrinking may be important for the final porosity of the resin.)

No reports on grain formation during preparation of skin-free S-PVC resins seem to have been published.

The previous discussion presupposes that the initiator is evenly distributed over the monomer droplets. If the initiator is introduced in the reactor after emulsification of the monomer, this need not to be the case. Under such conditions a mechanism in which initiator-stung droplets repeatedly swell by monomer (because of the lower monomer vapor pressure in droplets containing dissolved initiator) and break up by stirring may be operative [39].

EMULSION POLYMERIZATION

Normal Emulsion Polymerization

Emulsion polymerization of VCM is practiced mainly for the preparation of specialty paste resins. Polymerization is generally by batch processes, but continuous processes are also used. Batch polymerization is carried out in glass-lined (small reactors) or stainless steel autoclaves. Stirring requirements are not as critical as in suspension processes.

In normal emulsion processes, water-soluble initiators, both thermal and redox systems, are used. As compared to suspension polymerization, fairly large quantities of emulsifiers are needed, up to 3 parts per 100 parts of monomer. Emulsifiers are normally low molecular weight anionic or nonionic surface-active agents. Nonionics are seldom used alone. Soaps are often used as their ammonium salts because these decompose during drying and give resins with very low electrolyte contents. In many technical processes, the emulsifier is used at submicellar concentrations. In these cases, particle initiation should be homogeneous. Because of the high solubility of vinyl chloride in water, homogeneous nucleation may be important also in other cases. The influence of emulsifier type has been studied by Hopff and Fackla [42].

In traditional emulsion processes, seed polymerization is used to obtain a particle size distribution suitable for paste resins. The maximum particle size which can be reached by seed polymerization is fairly small, about 1 μm. This and the fact that multistep seed polymerizations are expensive have led to the introduction of new processes referred to as dispersion, microsuspension, or hybrid processes (see below).

The product of an emulsion polymerization is a latex, the particle size of which falls in the range from below 0.1 μm to about 1 μm. The latex is steam stripped to remove residual monomer. The polymer is then recovered by flocculation or spray drying. Flocculation can give resins in the size range typical for S-PVC [41]. Most E-PVC resins are prepared by spray drying. The technique used for spray drying, grinding, and sieving is reported to be crucial for the end use properties of the final resin.

Microsuspension or Hybrid Processes

To simplify the preparation of paste resins with a suitable particle size distribution, new alternatives to the traditional emulsion process have been developed. These processes are based on polymerization in submicron- or micron-sized monomer droplets, using monomer-soluble initiators. The crucial point in these processes is the preparation of sufficiently stable emulsions of fine monomer droplets. This requires very high shear rates but above all efficient stabilization of the droplets. High shear rates can be obtained by pumping through narrow slits or openings, using high-pressure pumps. Colloidal mills or supersonic waves can also be used.

Monomer droplets of the actual size range have to be stabilized toward flocculation and coalescence, but also toward competitive growth. The reason for competitive growth is the increase in vapor pressure over small drops with decreasing drop size (consider the Kelvin equation). Unless the discontinuous phase in an emulsion of fine drops is not completely insoluble in the continuous phase, the larger droplets will grow at the expense of smaller. Because of the high water solubility of VCM, this mechanism would lead to a rapid increase in the average drop size. Stabilization toward competitive growth has been discussed by Ugelstad et al. [43]. The solution to this problem is to control the vapor pressure of the drops by dissolving in the monomer a substance X, which is completely insoluble in water and therefore cannot be transported between the individual drops. In this case, if a drop becomes smaller because of competitive growth, the concentration of the substance X in the drop increases. This results in an increase in the mole fraction of X and hence in a decrease in the vapor pressure of monomer over that particular drop. Eventually the vapor pressure will decrease to a level which makes the drop stable toward further shrinking.

The water-insoluble compound X must of course be present during the emulsification process. It is not necessary, however, to use all or any of the monomer at this stage. If one succeeds in preparing a very fine emulsion of X in water, the X drops will swell spontaneously on addition of vinyl chloride. The substance X should be a low molecular weight compound in order to be an effective vapor pressure suppressant. According to Ugelstad et al. [44], hexadecane is a suitable candidate for X. Submicron droplets of hexadecane will be able to take up 100–500 parts of VCM. From the patent literature it seems that completely water-insoluble initiators can play the role of X.

An alternative to the foregoing microsuspension processes which does not require intensive homogenization has also been described by Ugelstad (see Ref. 45). In this case a long-chain fatty alcohol is first mixed with an anionic emulsifier and water in amounts giving rise to the formation of liquid crystals and mixed micelles. When vinyl chloride is added to this mixture, it will almost simultaneously form an emulsion of very fine monomer droplets. These can be polymerized to particles with monomer-soluble initiators present in the vinyl chloride or by water-soluble initiators. In this case the fatty alcohol functions as compound X. The "spontaneous" emulsification is probably due to the transient decrease of the interfacial tension between vinyl chloride and water as the fatty alcohol is transferred from the aqueous phase to the monomer phase.

KINETICS OF SUSPENSION AND EMULSION POLYMERIZATION AT CONSTANT RATE OF INITIATION

The theory of emulsion polymerization of vinyl chloride has been treated by Ugelstad et al. [44, 46], Nomura and Harada [47], and by Friis and Hamielec [45]. Theoretical treatments of polymerization in mass or suspension have been published by Abdel-Alim and Hamielec [48], Kuchanov and Bort [49], and Kelsall and Maitland [50], among others. The two latter treatments consider the distribution of initiators and the transport of radicals between the liquid monomer phase and the monomer-swollen polymer phase. These factors may affect the reaction rate development during the process. The impression is that the mathematical modeling of emulsion and suspension polymerization is much ahead of its experimental testing.

This section is restricted to a discussion of reaction rate profiles observed in typical suspension and emulsion polymerizations carried out at constant rate of initiation. A comparison of theoretical models for emulsion polymerization with experimental results is included. The present discussion is based on experiments carried out in a calorimetric reactor [19]. This technique gives rate data directly, without resorting to derivation of conversion versus time curves. The sensitivity and time resolution of the particular instrument used allowed a fairly detailed resolution of the reaction rate profiles and of the variation of the pressure in the reactor.

Suspension Polymerization

Figure 4 gives results from two suspension polymerizations, one (a) with di-lauryl-peroxide (DLPO), the other (b) with azo-bis-isobutyronitrile (AIBN) as initiator [19]. Methocel F 50 and ammonium laurate were used as suspension stabilizers. To emphasize the difference between suspension and emulsion polymerization kinetics, the results from two emulsion polymerization experiments are included in Figure 4. These experiments were carried out using persulfate as initiator and (c) sodium dodecyl sulfate (SDS) or (d) ammonium laurate (AL) as emulsifier. All of the experiments were carried out at 50°C. The diagrams give the reaction rate (arbitrary units), the pressure in the reactor (in the pressure scale used, 1 is equal to the sum of the saturation pressures of VCM and water at 50°C), and the relative amount of VCM present in the form of a liquid phase (from mass balances and solubility data) as a function of conversion.

The results in Figure 4 show that with VCM the reaction rate profile of a suspension and an emulsion polymerization may be quite similar. In both cases distinct pressure drops occurred at conversions around 70%, i.e., just before the pronounced rate maxima.

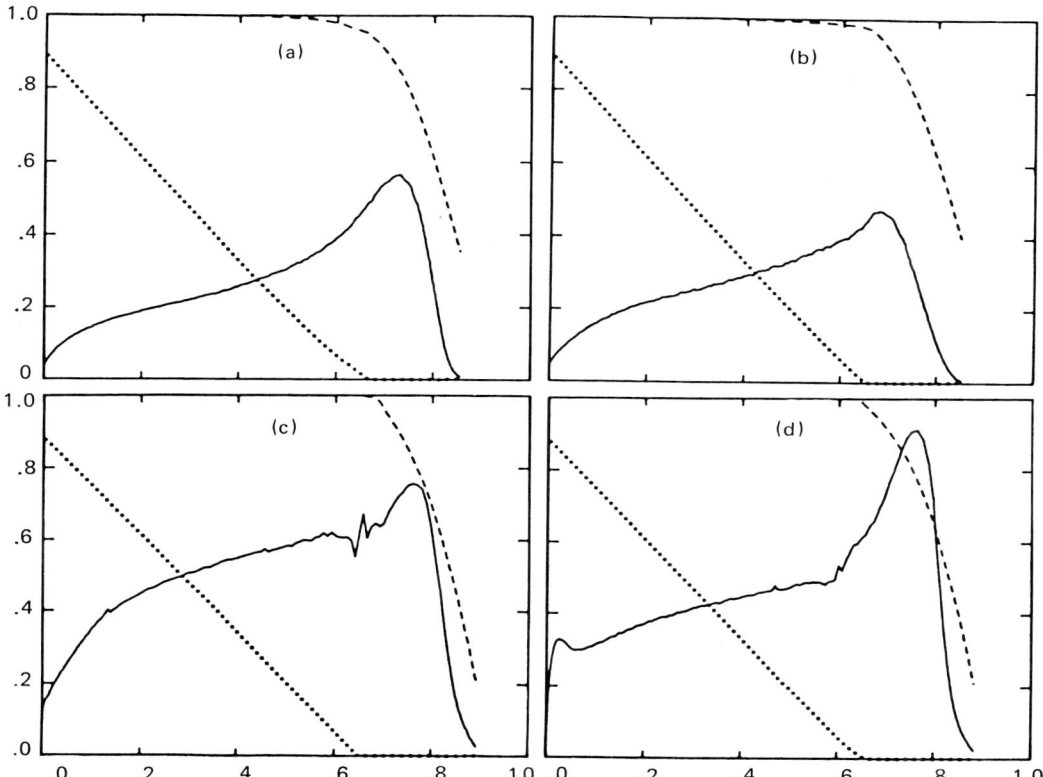

Figure 4 Comparison of suspension and emulsion polymerization of VCM. Solid curves represent the rate of polymerization in arbitrary units; dotted curves, the amount of liquid VCM relative to the total initial amount of monomer; and dashed curves, the gas phase pressure relative to the initial pressure in the reactor. (*a*) and (*b*) are from suspension polymerizations using different initiators, DLPO and AIBN, respectively, whereas (*c*) and (*d*) are from emulsion polymerizations using different emulsifiers, SDS and AL, respectively. All experiments at 50°C.

In the suspension polymerization experiments, however, the pressure started to decrease slowly already at conversions around 30–50%. In the suspension polymerizations the first sign of a pressure drop occurred at a time when a considerable amount of liquid monomer was still present. A rough estimate shows that at this time the particles consisted of one-third liquid VCM, the rest being monomer-swollen PVC. This pressure drop obviously was caused by capillary condensation in the fine pores of the particles. In this conversion range, the pressure curve should be related to the morphology of the PVC grains. The change from a constant to a slowly decreasing pressure at conversions below 70% is responsible for the occurrence of the inflection points observed at about 50% conversion in the suspension polymerization rate curves. This inflection point is caused by a slow continuous decrease in the termination rate constant as the activity of VCM decreases due to capillary condensation. Similar effects were not present in the emulsion polymerization experiments, where the activity of VCM remained constant as long as VCM droplets are present.

A comparison of the results in Figures 4a and b shows that the maximum peak was much more pronounced in the DLPO run than in the AIBN run. The rate development at low conversion was also different in the two runs. Since the decomposition rate constants for AIBN and DLPO at 50°C are rather low and of the same magnitude, the observed differences suggest the presence of specific initiator effects. The specific initiator effects observed at high conversions may be due to a difference in the variation of the initiator

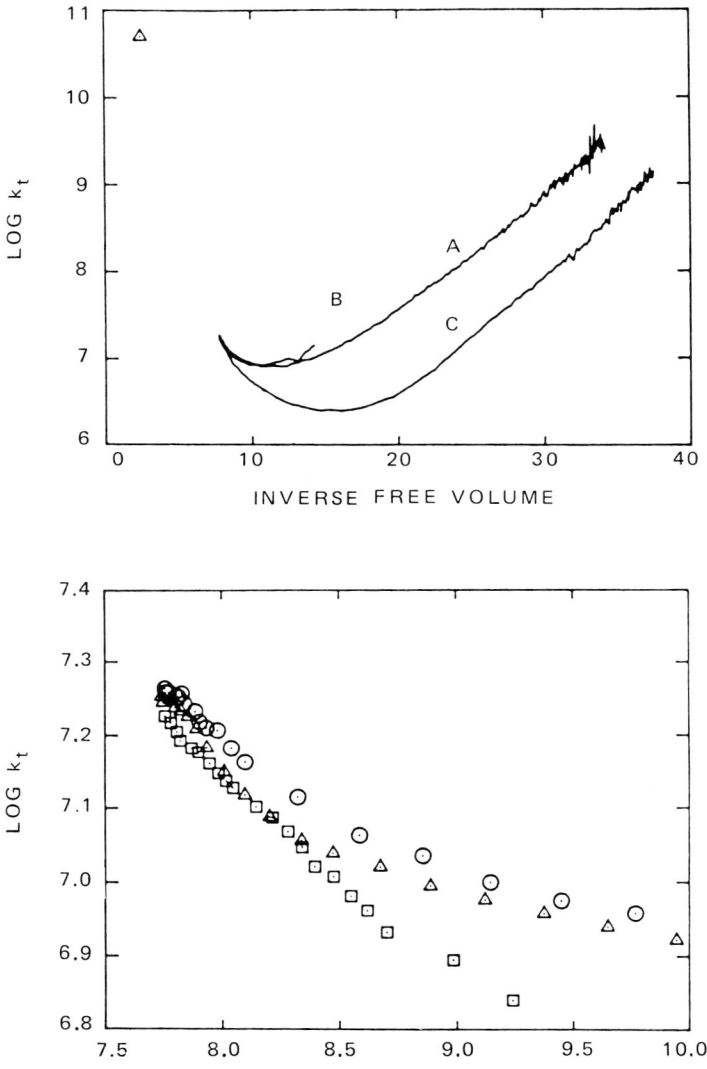

Figure 5 The termination rate constant at 50°C, as evaluated from suspension polymerizations, versus the inverse free volume. The curves refer to experiments with AIBN (A and B) and DLPO (C) as initiators. The lower diagram has expanded scales. The triangle included in the upper diagram represents k_t for termination in liquid monomer.

efficiency with viscosity. A general decrease in initiator efficiency with increasing viscosity may explain the fact that a lower limiting conversion was reached in the suspension polymerizations than in the emulsion polymerizations (note the difference in the final VCM pressure reached in these cases).

Figure 5 shows a plot of the termination rate constant as a function of the inverse free volume, as calculated from the results of Figures 4a and b and from a similar suspension polymerization experiment with AIBN as initiator but with magnesium dodecyl sulfate instead of AL as secondary suspension stabilizer. In calculating k_t, it was assumed that k_p and the initiator efficiency ϵ were independent of conversion. Hence the plotted data reflect the combined effect of conversion on k_t, k_p, and ϵ. Clearly, an unambiguous calculation of the true variation of k_t with conversion cannot be made. For the purpose of simulation of emulsion polymerization (see below) it was assumed that k_t varied with the free volume f according to Eq. 2:

$$k_t = 10^{[7.3 - 0.3(1/f - 7.5)]} + \text{const} \quad (\text{dm}^3 \text{ mol}^{-1} \text{ sec}^{-1}) \tag{2}$$

This equation was obtained from the slope of the log k_t versus inverse free volume curve representing the DLPO experiment in Figure 5b. The constant included in Eq. 2 represents a limiting value of k_t corresponding to termination by small radicals.

Emulsion Polymerization

Generally, the rate of emulsion polymerization can be written

$$R_p = \frac{\bar{n} N k_p [M]_p}{N_A} \tag{3}$$

where $\bar{n}$ is the average number of radicals per particle, N the total number of particles, $[M]_p$ the monomer concentration in the particles, and N_A is Avogadro's number. This equation can be used to calculate $\bar{n}$ from experimental data. In the treatment of emulsion polymerization by Ugelstad and Hansen [51] the calculation of $\bar{n}$ is based on a radical balance over the continuous phase, assuming a quasi–steady-state concentration of radicals. The calculation takes into account radical formation by decomposition of the initiator and radical adsorption by and desorption from the particles as well as the disappearance of radicals by termination in the particles and in the aqueous phase. The final expression for the radical balance is given as

$$\alpha = \alpha' + m\bar{n} - Y\alpha^2 \tag{4}$$

where α, α', m, and Y are dimensionless numbers, α represents the rate of radical adsorption into the particles, α' the rate of radical formation, and m the rate of radical desorption from the particles relative to the rate of disappearance of radicals by termination in the particles. Y is a parameter which expresses the relative importance of termination in the continuous phase. These dimensionless numbers are defined as follows:

$$\alpha = \frac{\rho_a V_p}{N^2 k_t^*} \tag{5}$$

$$\alpha' = \frac{\rho_i V_p}{N^2 k_t^*} \tag{6}$$

$$m = \frac{k_d V_p}{N k_t^*} \tag{7}$$

$$Y = \frac{2N^2 k_t^* k_{tw}^*}{(k_a)^2 V_p} \qquad (8)$$

In these equations ρ_a and ρ_i are the rate of radical adsorption in the particles and the rate of radical production by initiator decomposition, V_p is the total volume of particles per cubic decimeter of water. The k's are rate constants for radical adsorption in the particles (subscript a), radical desorption from the particles (subscript d), termination in the particles (subscript t) or in the aqueous phase (subscript tw). An asterisk means that the rate constant is in molecular units, i.e., the normal second-order rate constant divided by Avogadro's number, N_A. The "rate constants" for radical adsorption and desorption (dimension, sec^{-1}) are defined as

$$\rho_a = k_a [R]_w^* \qquad (9)$$
$$\rho_d = k_d \Sigma n N_n \qquad (10)$$

where $[R]_w^*$ is the concentration of radicals in the water phase (in units of mol dm^{-3}), N_n the number of particles per cubic decimeter of water containing n radicals. According to Ugelstad et al. [52], $\bar{n}$ can be calculated from the expression

$$\bar{n} = 1/2(2\alpha/m + 2\alpha/(1 + m + 2\alpha)/(2 + m + 2\alpha)/(3 + m + \cdots) \qquad (11)$$

which converges very fast. Thus if α', m, and Y are known, the corresponding values of α and $\bar{n}$ can be obtained from Eqs. 4 and 11. An approximate closed expression for the rate of polymerization has also been given, which should be applicable to emulsion polymerization of VCM during interval II, i.e., after particle nucleation is ended and before the separate VCM phase has disappeared. It reads

$$R_p = \left(\frac{k_p [M]_p}{N_A}\right) (\rho_i)^{1/2} \left(\frac{V_p}{k_t^*} + \frac{N}{2k_d}\right)^{1/2} \qquad (12)$$

where $k_d = k_d'(N/V_p)^{2/3}$ and k_d' should be independent of size.

Equation 12 was derived assuming termination in the aqueous phase to be unimportant and the number of particles containing more than two radicals to be negligibly small. If termination in the water phase is dominating while normal conditions are still prevailing, the appropriate expression would be

$$R_p = k_p [M]_p \left(\frac{k_a}{k_d}\right) \left(\frac{\rho_i}{2k_{tw}^*}\right)^{1/2} \qquad (13)$$

Figure 6 presents results from four emulsion polymerization experiments. These were carried out using various concentrations of SDS, A-C, and with one concentration of AL, D. The polymerization temperature was 50°C, and initiation was by equal amounts of potassium persulfate. As recommended by Mörk and Ugelstad, citrate was used to complex impurity catalysts [53]. The total particle number per cubic decimeter of water ($\times 10^{-18}$) in the different experiments was 1.9 (a), 0.88 (b), 0.20 (c), and 0.094 (d). The experimental data are represented by the dots. As can be seen from Figure 6d, with AL as emulsifier the rate of polymerization increased very rapidly on addition of initiator and went through a slight maximum after about 10 min. This maximum was always observed in experiments with AL. Also, in the experiments with SDS, the results indicated a sudden change in the variation of reaction rate with time after about 10 min. These phenomena are most likely connected with the end of the particle nucleation interval. In

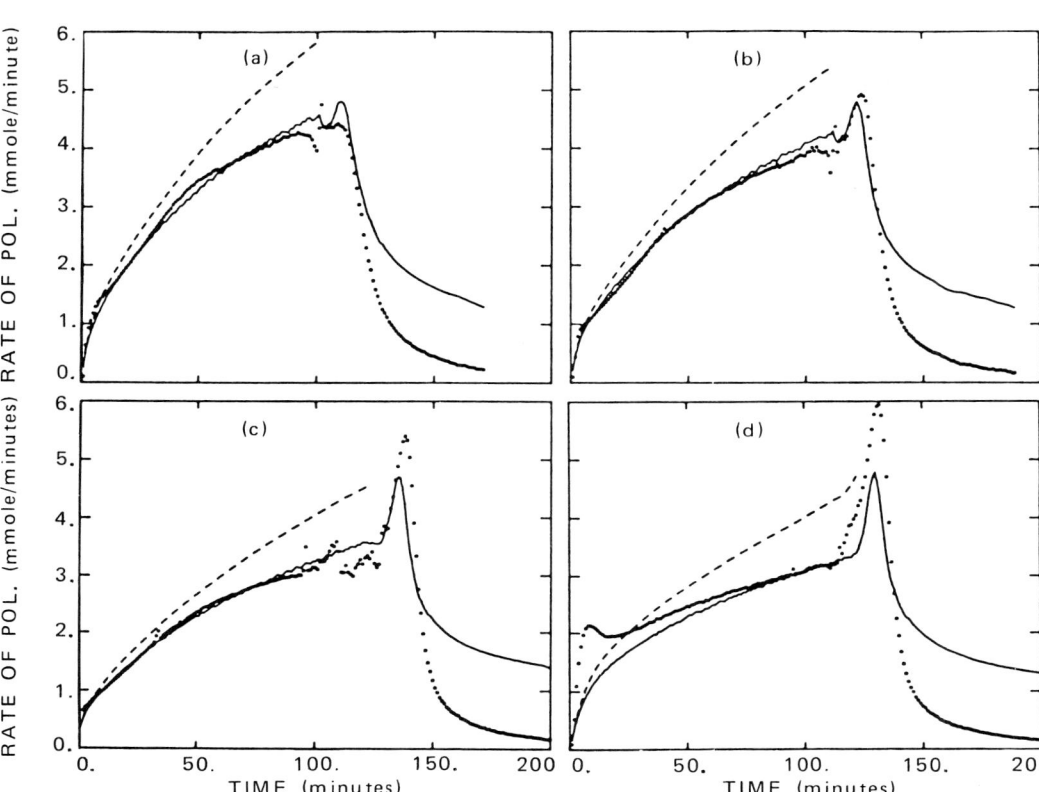

Figure 6 The rate of polymerization versus time for four emulsion polymerizations giving different numbers of particles. The emulsifier was SDS in experiments (a), (b), and (c), and AL in experiment (d). The dots represent the experimental data, the solid lines were calculated according to the theory of Ugelstad et al. (Eqs. 4 and 11), the dashed lines by Eq. 12. The temperature (50°C) and the amount of initiator (potassium persulfate) were the same in all of the experiments.

the experiment with AL, the drop in rate after the maximum suggests that the number of particles decreased by agglomeration.

The kinetics during intervals II and III were analyzed according to the theoretical model developed by Ugelstad et al. It was found that the experimental data obtained during interval II could not be explained by Eqs. 4 and 11, assuming that termination in the aqueous phase was negligible (i.e., that $Y\alpha^2 = 0$). Such an approach required different values of k'_d for each polymerization.

As the value for k_t in the PVC phase decreases rapidly with decreasing concentration of VCM (Fig. 5), the relative importance of water phase termination would increase rapidly after the disappearance of the liquid VCM phase. If termination in the aqueous phase is assumed to dominate already at the conversion corresponding to the top of the maximum peak, Eq. 13 and the fact that R_p at the maximum peak seemed to be rather insensitive to variations in particle number would indicate that the ratio k_a/k_d should be almost constant. Although a physical explanation for such a condition cannot be provided, a constant value of k_a/k_d gave a very good fit between experimental data and the data calculated using Eqs.

4 and 11. This is seen by comparing the calculated data (solid curves) in Figure 6 with the experimental data (dots). The solid curves were calculated using k_t from Eq. 2 and the following values for the other rate constants:

$k_i = 1 \; 10^{-6} \; \text{sec}^{-1}$

$k'_d = 1.1 \; 10^{11} \; \text{dm}^2 \; \text{sec}^{-1}$

$k_p = 1 \; 10^4 \; \text{dm}^3 \; \text{mol}^{-1} \; \text{sec}^{-1}$

$\dfrac{k_a}{k_d} = 50.1$

$k_{tw} = 1 \; 10^{10} \; \text{dm}^3 \; \text{mol}^{-1} \; \text{sec}^{-1}$

As can be seen from Figure 6 the theoretical model was applicable over the whole of interval II and most of interval III. The deviations that occurred toward the end of interval III can be ascribed to the influence of the increasing viscosity in the PVC phase on k'_d and k_p. It is also possible that k_a varies as the amount of VCM solubilized in the emulsifier layer surrounding the particles decreases with decreasing VCM activity. An error in the variation of k_t with conversion should also give rise to deviations between model and experiment in interval III.

Data for the rate of polymerization calculated according to Eq. 12 have also been included in Figure 6 (dashed lines). In these calculations the rate constants were given the same values as above. As can be seen, the deviation between the experimental and calculated rates increased with increasing conversion. This is in agreement with the results of a test of the approximation made in the derivation of Eq. 12 using the present set of rate constants. This test showed that the approximation was not entirely justified at higher conversions. This, of course, is consistent with the observation that the term $Y\alpha^2$ of Eq. 4 cannot be neglected.

In the derivation of Eq. 4, a quasi–steady-state concentration of radicals was assumed. This condition was fulfilled in the experiments of Figure 6. It was only in the experiment with AL and then only for a few minutes just before the top of the maximum peak that the total number of radicals ($\bar{n}N$) increased at a rate exceeding 0.001 of the rate of radical formation.

It would of course be interesting to try to obtain a still better agreement between theory and experiment than that demonstrated in Figure 6. However, more detailed modeling is presently not meaningful. Such modeling requires more accurate data on the variation of k_p and k_t with conversion and independent studies of radical transport from and to polymer particles.

ECONOMIZATION OF PVC PRODUCTION

As was demonstrated in the previous section, when VCM is polymerized at constant rate of initiation, a very pronounced rate peak occurs toward the end of the polymerization process. At industrial scale and for safety reasons, the power of heat evolved by the reaction (proportional to the rate of polymerization) must not at any instant exceed the cooling capacity of the reactor. The productivity of the reactor equipment would thus be very badly utilized by polymerizing at constant rate of initiation.

To take full advantage of the cooling capacity of the reactor, the rate of initiation should vary with time. The ideal situation would be to keep the polymerization rate constant, at a level that gives a heat effect which, by a certain safety margin, approaches the maximum cooling capacity at hand. The dominating approach to this problem is to use a combination of rapid and slow initiators. It is also possible to adjust the temperature during the process, or to use nonisothermal conditions in combination with initiator packages. Theoretical analysis of various alternatives have been published by Hamielec et al. (e.g., Ref. 54).

To obtain a rectangularlike reaction rate profile at normal polymerization temperatures, very rapid initiators must be combined with slower ones. An important group of rapid initiators is the peroxydicarbonates. Many of these are not stable at ambient temperature and must be stored and shipped at low temperatures. However, peroxydicarbonates of long-chain fatty alcohols like cetyl alcohol have been found to be stable as long as they remain in the crystalline state. Such initiators, which do not require low-temperature handling, have found extensive use in certain geographical areas. It should be understood that a charged polymerization reactor containing the amount of initiator required to reach the final conversion is a potential bomb. As a consequence, the safety margin with respect to the cooling system must not be too low, and efficient inhibitor systems must be available in case of electricity dropout.

Laboratory-scale experiments have recently shown that it is possible to design initiator systems for both emulsion and suspension polymerization of VCM, which permit the rate of initiation to be varied in a controlled fashion during the polymerization process. Systems of this type should make it possible to combine a safe production with full utilization of the cooling capacity of the reactors. Examples illustrating this are seen in Figures 7 and 8 [55]. In both cases rapid redox systems were used. In the polymerization experiments, the free radical precursors, *tert*-butyl hydroperoxide (Fig. 7) and cumyl hydroperoxide (Fig. 8), were fed to the reactor at a rate controlled by a regulator, which tried to keep the reactor temperature at 50°C. The reactor was cooled by a Peltier cooler fed with a constant current. In the emulsion polymerization experiment, the water phase contained a small amount of vanadium(IV)-oxysulfate and an excess of ascorbic acid (with respect to the expected consumption of hydroperoxide). In this system radicals were formed by reaction between *tert*-butylhydroperoxide and vanadium(IV). The five-valent vanadium formed in this reaction was immediately reduced back to vanadium(IV) by the ascorbic acid. As can be seen from Figure 7, the set temperature value was reached after about 20 min and was then kept constant until the vinyl chloride pressure had decreased to about 0.6 MPa. Figure 7c shows the flow rate of hydroperoxide solution required to give the temperature profile of Figure 7a.

In the suspension polymerization experiment, the vinyl chloride phase contained the reducing component of the redox system. It was a 1/1 complex between triethyl boron and ammonia. The cumyl hydroperoxide was added as a 0.1 M weakly alkaline aqueous solution. The hydroperoxide, which was stable toward thermal decomposition at the temperature used, was extracted into the monomer droplets. Radicals were produced in the monomer phase by reaction between the hydroperoxide and the triethyl boron complex. As seen from Figure 8a, the reaction rate could be kept constant almost from the start up to and somewhat beyond the pressure drop. Figure 8c gives the consumption of hydroperoxide as a function of time. The product in this case was a porous suspension resin of narrow size distribution having an average particle size of about 150 μm.

The initiator systems used in preparing Figures 7 and 8 may not be quite ideal for full-scale production of PVC. The results demonstrate, however, that it is possible to control the rate of polymerization by using rapid redox initiation systems, where one of the redox components is continuously added to the reactor at a controlled rate. This can obviously be done without conflicting the colloid stability requirements of emulsion and suspension polymerization.

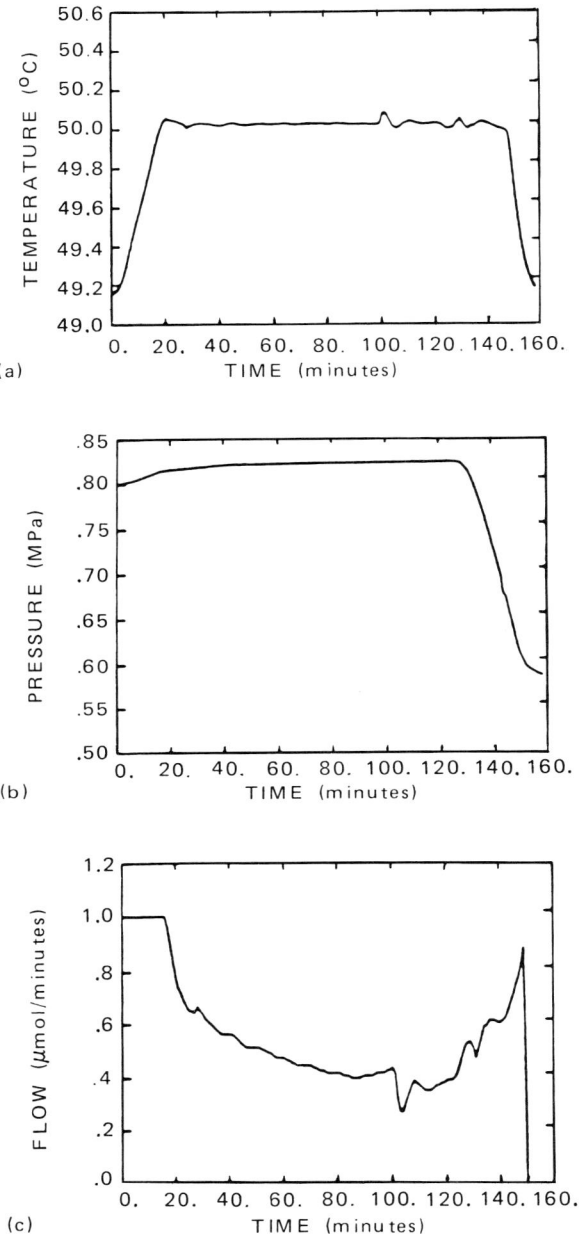

Figure 7 Emulsion polymerization of VCM using a vanadium-catalyzed redox initiator system. One of the components of the initiator system, *tert*-butylhydroperoxide, was continuously added at a controlled rate to keep the polymerization rate constant. (*a*) plots the reactor temperature, (*b*) the gas phase pressure in the reactor, and (*c*) the feed rate of hydroperoxide during the run.

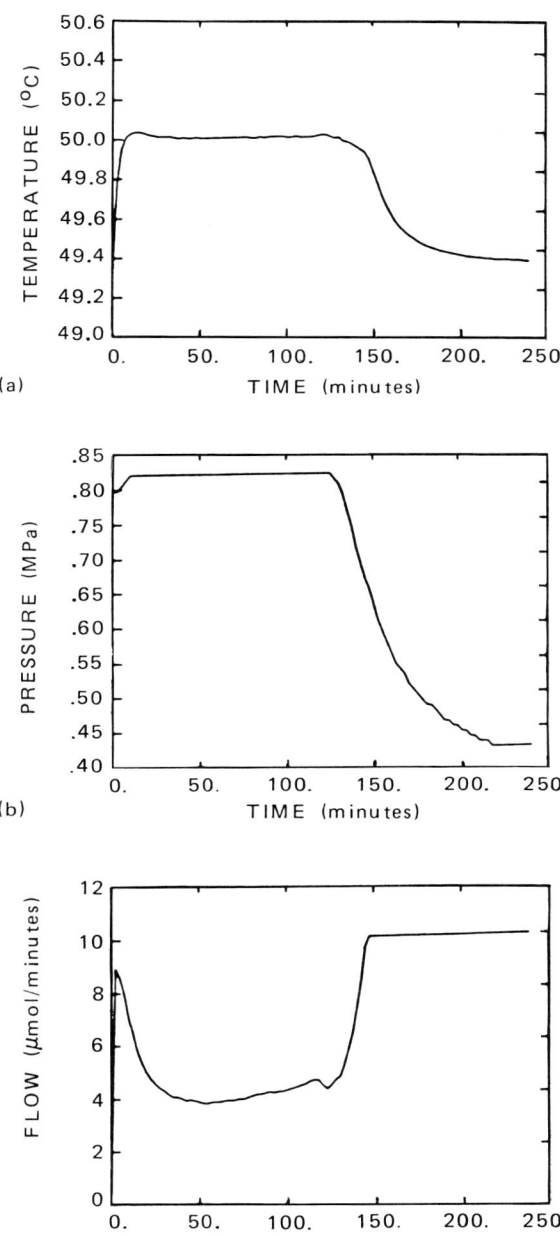

Figure 8 Suspension polymerization of VCM using redox initiation with triethyl boron (present in the monomer) and continuous addition of cumyl hydroperoxide through the aqueous phase. The hydroperoxide addition was controlled to give a constant rate of polymerization. (*a*) plots the reactor temperature, (*b*) the gas phase pressure in the reactor, and (*c*) the feed rate of hydroperoxide.

ACKNOWLEDGMENT

Thanks are due to my former graduate students Drs. Holger Nilsson, Christer Silvegren, and Jaan Uustalu for very stimulating discussions on various aspects of VCM polymerization and to Sten Porrvik, Odd Bjerke, and Sven Pettersson, of the former Swedish company KemaNord, for many years of cooperation on problems related to PVC production. Thanks are also due to The Swedish Board for Technical Development, the founder of most of our work on PVC.

REFERENCES

1. H. Felger, ed., *Polyvinylchlorid*, Carl Hanser Verlag, Munich (1986).
2. R. H. Burgess, ed., *Manufacture and Processing of PVC*, Applied Science Publishers, London (1982).
3. G. Butters, ed., *Particulate Nature of PVC*, Applied Science Publishers, London (1982).
4. M. Kaufmann, *The History of Polyvinyl Chloride*, Maclaren, London (1969).
5. F. Klatte, Ger. Pat. 281,877 (1912); Griesheim-Electron.
6. W. L. Semon and A. G. Stahl, in *History of Polymer Science and Technology*, (R. B. Seymour, ed.), Marcel Dekker, New York (1982).
7. H. Nilsson, C. Silvegren, and B. Törnell, *Eur. Polym. J.*, *14*: 737 (1978).
8. W. Wenig, *J. Polym. Sci. Polym. Phys. Ed.*, *16*: 1635 (1978).
9. J. Brandrup and E. H. Immergut, eds., *Polymer Handbook*, Interscience, New York (1975).
10. H. Nilsson, C. Silvegren, and B. Törnell, *Chem. Scripta*, *19*: 164 (1982).
11. J. Boissel and N. Fischer, *J. Macromol. Sci.*, *A11*(7): 1249 (1977).
12. J. D. Cotman, Jr., M. F. Gonzales, and C. Claver, *J. Polym. Sci. A-1*, *5*: 1137 (1967).
13. A. R. Berens, *Angew. Makromol. Chem.*, *47*: 97 (1975).
14. A. H. Abdel-Alim and A. E. Hamielec, *J. Polym. Sci. Polym. Chem. Ed.*, *12*: 483 (1974).
15. G. Ceccorulli, M. Pizzoli, and G. Pezzin, *J. Macromol. Sci. Phys.*, *14*: 499 (1977).
16. J. A. Manson, S. A. Iobst, and R. Acosta, *J. Macromol. Sci. Phys.*, *B9*: 301 (1974).
17. J. Lyngaae-Jörgensen, *Polym. Prepr.*, *14*(2): 738 (1974).
18. J. Ugelstad, P. C. Mörk, P. Dahl, and P. Rangnes, *J. Polym. Sci. C*, *27*: 49 (1969).
19. H. Nilsson, C. Silvegren, and B. Törnell, *Angew. Makromol. Chem.*, *112*: 125 (1983).
20. A. A. Caraculacu, *Pure Appl. Chem.*, *53*: 398 (1981).
21. T. Hjertberg and E. M. Sörvik, in *Degradation and Stabilization of PVC* (E. D. Owen, ed.), Elsevier, Applied Science Publishers, London (1984).
22. W. F. Maddams, in *Degradation and Stabilization of PVC* (E. D. Owen, ed.), Elsevier Applied Science Publishers, London (1984).
23. W. H. Starnes, F. C. Schilling, I. M. Plitz, R. E. Cais, D. J. Freed, R. L. Hartless, and F. A. Bovey, *Macromolecules*, *16*: 790 (1983).
24. T. Hjertberg and E. M. Sörvik, *J. Polym. Sci., A: Polym. Chem.*, *24*: 1313 (1986).
25. W. H. Starnes, Jr., F. C. Schilling, K. B. Abbås, R. E. Cais, and F. A. Bovey, *Macromolecules*, *12*(4): 556 (1979).
26. F. M. Willmouth, D. G. Rance, and K. M. Henman, *Polymer*, *25*: 1185 (1984).
27. D. G. Rance and E. L. Zichy, *Pure Appl. Chem.*, *53*: 377 (1981).
28. E. B. Törnell, and J. M. Uustalu, *J. Vinyl Techn.*, *4*: 53 (1982).
29. D. G. Rance and E. L. Zichy, *Polymer*, *20*: 266 (1979).
30. E. B. Törnell and J. M. Uustalu, *Polymer*, *27*: 250 (1986).
31. J. A. Davidson and D. E. Witenhafer, *J. Polym. Sci. Polym. Phys. Ed.*, *18*: 51 (1980).
32. E. B. Törnell and J. M. Uustalu, *J. Appl. Polym. Sci.* (in press).
33. J. M. Uustalu, Dissertation, Lund Institute of Science and Technology, Lund (1985).
34. B. Törnell, J. M. Uustalu, and B. Jönsson, *Colloid Polym. Sci.*, *264*: 439 (1986).
35. J. Bystedt and I. Lundquist, *J. Vinyl Techn.*, *2*: 209 (1980).

36. I. Aladjoff, H. Nilsson, C. Silvegren, and B. Törnell, *Acta Chem. Scand., 36*: 259 (1982).
37. I. Aladjoff, H. Nilsson, C. Silvegren, and B. Törnell, Acta Chem. Scand., 36: 267 (1982).
38. H. Nilsson, C. Silvegren, and B. Törnell, *J. Vinyl Techn., 7*: 112 (1985).
39. H. Nilsson, T. Norviit, C. Silvegren, and B. Törnell, *J. Vinyl Techn., 7*: 119 (1985).
40. H. Nilsson, C. Silvegren, and J. M. Uustalu, *Polym. Commun., 24*: 268 (1983).
41. H. Hassander, H. Nilsson, C. Silvegren, and B. Törnell, in *Polymer Colloids*, vol. 2 (R. M. Fitch, ed.), Plenum, New York (1980).
42. H. Hopff and I. Fakla, *Br. Polym. J., 2*: 40 (1970).
43. J. Ugelstad, P. C. Mörk, K. H. Kaggerud, T. Ellingsen, and A. Berge, *Adv. Colloid. Interf. Sci., 13*: 101 (1980).
44. J. Ugelstad, P. C. Mörk, F. K. Hansen, K. H. Kaggerud, and T. Ellingsen, *Pure Appl. Chem., 53*: 323 (1981).
45. N. Friis and A. E. Hamielec, *J. Appl. Polym. Sci., 19*: 97 (1975).
46. J. Ugelstad and P. C. Mörk, *Br. Polym. J., 2*: 31 (1970).
47. M. Nomura and M. Harada, *J. Chem. Eng. Jpn., 4*: 54, 160 (1971).
48. A. H. Abdel-Alim and A. E. Hamielec, *J. Appl. Polym. Sci., 16*: 783 (1972).
49. S. I. Kuchanov and D. N. Bort, *Polym. Sci. USSR, 15*: 2712 (1973).
50. D. G. Kelsall and G. C. Maitland, in *Polymer Reaction Engineering* (K. H. Reichert, ed.), Carl Hanser Verlag, Munich (1983).
51. J. Ugelstad and F. K. Hansen, *Rubber Chem. Techn., 49*: 536 (1976).
52. J. Ugelstad, P. C. Mörk, and J. O. Aasen, *J. Polym. Sci. A-1, 5*: 2281 (1967).
53. P. C. Mörk and J. Ugelstad, *Makromol. Chem., 128*: 83 (1969).
54. A. E. Hamielec, J. L. Brash, and P. Coroyannakis, Non-Isothermal Polymerization of Vinyl-Chloride–Multiple Initiator Systems, Proceedings 2nd International Symposium on PVC, Lyon (1976).
55. L. Jansson, H. Nilsson, C. Silvegren, and B. Törnell (forthcoming).

11
Kinetic Modeling of Polymerization Reactions

Anil Kumar and Pankaj K. Khandelwal
Indian Institute of Technology–Kanpur
Kanpur, India

INTRODUCTION	375
STEP GROWTH POLYMERIZATION	377
Kinetic Model of Reversible Linear Step Growth Polymerization	377
Kinetic Model for Reversible Nonlinear Step Growth Polymerization	384
RADICAL POLYMERIZATION	403
Initiation	403
Propagation Reaction	404
Termination of Polymer Radicals	405
Average Molecular Weight in Radical Polymerization	408
Gel Effect in Radical Polymerization	410
Temperature Effects in Radical Polymerization	418
REVERSIBLE RADICAL POLYMERIZATION	419
CONCLUSIONS	423
APPENDIX: ANALYTICAL SOLUTION OF ALGEBRAIC EQUATIONS INVOLVED IN ADDITION POLYMERIZATION	424
REFERENCES	426

INTRODUCTION

A polymer is a high molecular weight material; polymerization is the series of reactions leading to the formation of polymers from low molecular weight material (called monomers). These can be broadly classified as step growth and chain growth polymerizations [1, 2]. In the former, the growth of molecules occurs through the reaction of reactive groups, e.g., —COOH, —NH$_2$, —OH, located on the molecules. Polymer formation can occur through this mechanism only when the starting monomer has at least two reactive groups, when there are more than two of them in the monomer, the resultant polymer is either branched or crosslinked in structure. As opposed to this, linear polymers are formed by the polymerization of bifunctional monomers. In chain growth polymerization there are growth centers in the reaction mass, to which monomers add on successively until either all monomers are consumed or some external agent terminates the polymerization.

Depending on the nature of these growth centers, chain growth polymerization can be further classified into radical, cationic, anionic, and stereoregular polymerizations [1–8]. It has been found that different mechanisms of polymerization have different characteristics. As an example, in step growth polymerization, the weight average molecular weight of polymer formed increases with the reaction time, whereas for chain growth polymerization it reaches a certain maximum value only.

Like all other reactions in nature, polymerization is reversible. However, in almost all the studies, it is assumed to be irreversible in order to simplify the analysis. In reversible polymerization, the overall conversion is limited by equilibrium; in practice, the reverse step is suppressed by applying high vacuum and higher temperatures. Unlike reactions of low molecular weight compounds, in polymerization there is a formation of several homologs which are chemically the same but differ in their molecular weights. The reaction mass can thus be characterized by a molecular weight distribution (MWD). This is nothing but a plot of the concentration of oligomer of chain length n (denoted by $[P_n]$) versus n. The different physical properties of the polymers are to a large extent dependent upon MWD and a given MWD can be equivalently represented by its moments. For polymeric systems, only the first three moments can be experimentally measured [1, 3], and they have been shown to characterize the MWD completely. The general definition of moments is given by

$$\lambda_K = \sum_{n=1}^{\infty} n^K [P_n]; \quad K = 0,1,2,\ldots \quad (1)$$

where $[P_n]$ is the concentration of P_n species of chain length n in the reaction mass. The number and the weight average chain lengths μ_n and μ_w are the polydispersity index ρ of the polymer are defined in terms of these moments as

$$\mu_n = \frac{\lambda_1}{\lambda_0} \quad (2a)$$

$$\mu_w = \frac{\lambda_2}{\lambda_1} \quad (2b)$$

$$\rho = \frac{\mu_w}{\mu_n} \quad (2c)$$

In Eq. 2c, ρ measures the breadth of the MWD and is equal to 1 for monodisperse material. On multiplying μ_n and μ_w with the molecular weight of the repeat unit, one obtains the number average molecular weight $\bar{M}_n$ and the weight average molecular weight $\bar{M}_w$. These are known to determine the physical properties of the polymers. For example, the viscosity of molten polymers is proportional to the weight average molecular weight and above a certain value of M_w, the viscosity usually varies [3] as $\bar{M}_w$ [3, 4]. This is important in estimating the power requirements for pumping the reaction mass, say, through a tubular reactor, a spinnerette to obtain fibers, or extruders and molds. In another example, the aliphatic polyester formed from ω-hydroxydecanoic acid has little strength or spinnability when μ_n is about 25 but it gives long, extremely weak fibers that can be cold-drawn when μ_n is about 55 [2]. However, for μ_n above 100, it can be spun easily and cold-drawn to strong fibers. In the following section we present the analysis of batch reactors forming polymers through the reversible step growth and radical mechanisms.

STEP GROWTH POLYMERIZATION [1–3]

Step growth polymerization occurs through reaction of two or more reactive groups. If the starting monomer is bifunctional, the resulting polymer is linear in structure; otherwise the polymer is branched. The following analysis of batch reactors is presented for both cases.

Kinetic Model of Reversible Linear Step Growth Polymerization

It is assumed that the starting monomer is an ARB type where A and B are the reacting functional groups. On polymerization, larger chains are formed but no matter what the chain length, it has one unreacted A and one unreacted B at its ends. The growth of the polymer can be schematically written as

$$P_m + P_n \underset{k_p'}{\overset{k_p}{\rightleftharpoons}} P_{m+n} + W, \qquad m, n = 1, 2, \ldots \tag{3}$$

where W is the condensation product and P_m is a polymer chain having m repeat units. It is observed that polymerization is represented by a set of infinite elementary reactions given in Eq. 3 and it is desired to model these kinetically to determine the MWD of the polymer as a function of time. Based on the experiments of Bhide and Sudborough, Flory proposed the equal reactivity hypothesis in which the reactivity of the reactive groups is independent of the chain length of the polymer. This serves as the most basic assumption in the modeling of step growth polymerization and is described below.

A chemical reaction can occur only when the reacting molecules collide. The rate of reaction, R, can thus be written in terms of the product of the collision frequency, $\omega_{m,n}$, between P_m and P_n and the probability of their reaction, $Z_{m,n}$. Therefore,

$$R = \alpha \omega_{m,n} Z_{m,n} \tag{4}$$

where α is a constant of proportionality. From the equal reactivity hypothesis [1], the probability of reaction, $Z_{m,n}$, is independent of m and n. Thus, if Z is the probability of reaction between two reacting groups and P_m and P_n can react in s distinct ways, $Z_{m,n}$ would be equal to sZ. Additionally, the collision frequency $\omega_{m,n}$ between two dissimilar molecules is proportional to $[P_m][P_n]$, whereas that between P_m and P_n is proportional to $[P_m]^2/2$. Consequently, if k_p is the rate constant associated with the reaction of functional groups, the kinetic model under the equal reactivity hypothesis can be written as

$$k_{p,mn} = \frac{R}{[P_m][P_n]} = sk_p, \qquad m \neq n; m, n = 1, 2, \ldots \tag{5a}$$

$$= \frac{R}{[P_m]^2} = \frac{sk_p}{2}, \qquad m = n; m = 1, 2, \ldots \tag{5b}$$

For linear chains, the reacting functional groups are always located at the end of the chain. This means that in Eq. 5, s is always 2, as seen in Figure 1. Any polymer species P_n ($n \geq 2$) is formed in the forward step by reacting a $P_r(r < n, r = 1, 2 \ldots)$ with a P_{n-r}. In the reverse reaction, it is formed by the attack of the condensation product, W, on the bonds of the polymer chain. W can react on any position of the chain but there are two locations on it which would give a small chain of the same length. For a linear polymer, P_n, the total number of bonds is $(n - 1)$. Consequently, the mole balance relation for P_n in a batch reactor is given by

$$\frac{d[P_1]}{dt} = 2k_p[P_1] \sum_{i=1}^{\infty} [P_i] + 2k_p'[W] \sum_{n=2}^{\infty} [P_n] \tag{6a}$$

$$\frac{d[P_n]}{dt} = -2k_p[P_n] \sum_{m=1}^{\infty} [P_m] + k_p \sum_{r=1}^{n-1} [P_r][P_{n-r}]$$

$$+ 2k_n'[W] \sum_{m=n+1}^{\infty} [P_m] - k_p'(n-1)[P_n][W], \quad n \geq 2 \tag{6b}$$

where k_p and k_p' are the forward and the reverse reaction rate constants, respectively. The relation for the kth moment can be derived by multiplying the preceding mole balance relations for P_n by n^k ($k = 0,1,2$) and summing the various terms appropriately. The ultimate moment generation relates are given by [10]

$$\frac{d\lambda_0}{dt} = -k_p\lambda_0^2 + k_p'[W](\lambda_1 - \lambda_0) \tag{7a}$$

$$\frac{d\lambda_1}{dt} = 0 \tag{7b}$$

$$\frac{d\lambda_2}{dt} = 2k_p\lambda_1^2 + \frac{k_p'}{3}[W](\lambda_1 - \lambda_3) \tag{7c}$$

To be able to solve Eq. 6, the MWD of the feed to the batch reactor must be known. Let us assume that at $t = 0$ it is given by

$$[P_n]\bigg|_{t=0} = [P_n]_0, \quad n = 1, 2, 3, \ldots \tag{8}$$

Correspondingly, its moments can be found and are assumed to be given by

$$\lambda_k\bigg|_{t=0} = \sum_{n=1}^{\infty} n^k[P_n]_0 = \lambda_k^0, \quad k = 0,1,2 \tag{9}$$

Equation 7b is the easiest to integrate and it shows that the first moment λ_1 is time invariant, equal to λ_1^0. It may be observed that the first moment is the same as the total count of repeat units, which is time invariant during polymerization. Equation 7c gives λ_2 but it involves λ_3 and can be integrated only when some information about it is known. However, if the moment generation relation for λ_3 is written, it is found that it involves λ_4

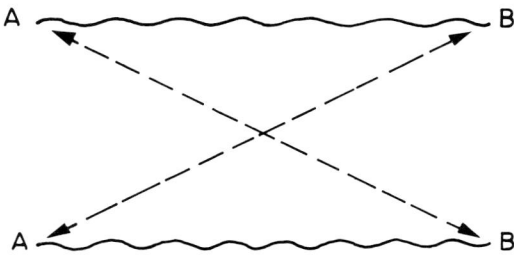

Figure 1 Schematic representation of reaction between linear chains.

and the moment generation relation for λ_4 involves λ_5, and so on. This hierarchy of relation has been broken by using a suitable closure relation. This approximate closure relation for the bifunctional step growth polymerization was proposed in the literature by assuming the MWD to be given by the Schulz–Zimm relation for which λ_3 is related to λ_0, λ_1, and λ_2 by the following relation [11, 12]:

$$\lambda_3 \cong \frac{\lambda_2(2\lambda_2\lambda_0 - \lambda_1^2)}{\lambda_1\lambda_0} \tag{10}$$

In the literature, computations of moments have been made and results have been compared with those found from the exact MWD [11, 13–17]. The match between the two has been good and the approximation given in Eq. 10 yields satisfactory results in the entire range of computation.

The major difficulty of computing the MWD of reversible ARB polymerization lies in the fact that the chain growth step consists of an infinite number of elementary reactions as written in Eq. 6. The various mole balance relations for species P_i for any i in batch reactors are highly nonlinear and coupled differential equations. Computationally one solves an arbitrarily large number (say N) of these equations such that the concentration of P_N is close to zero. However, some truncation error is introduced in the repeat count variable because higher chain length species are neglected. In actual practice, one would like to determine the first three moments of the distribution—λ_0, λ_1, and λ_2—and the concentrations of the first few oligomers, which cannot be determined without solving the entire MWD. It is possible to decouple these equations using the following transformation [18]:

$$\sum_{i=n+1}^{\infty} [P_i] = \lambda_0 - [P_n] - \sum_{i=1}^{n-1} [P_i] \tag{11}$$

By using Eq. 11, the MWD relations given in Eq. 6 can be modified:

$$\frac{d[P_1]}{dt} = -2k_p[P_1]\left(\frac{\lambda_0 + k_p'[W]}{k_p}\right) + 2k_p'[W]\lambda_0 \tag{12a}$$

$$\frac{d[P_n]}{dt} = -2k_p[P_n]\lambda_0 + k_p \sum_{r=1}^{n-1} [P_r][P_{n-r}] - 2k_p'[P_n][W](n+1)$$

$$+ 2k_p'[W]\left\{\lambda_0 - \sum_{i=1}^{n-1} [P_i]\right\}, \quad n \geq 2 \tag{12b}$$

and at $t = 0$;

$$[P_n] = [P_n]_0; \quad n = 1,2,3,\ldots \tag{12c}$$

Equations 12 are now in the decoupled form and the concentration of any species can be determined sequentially by first determining the zeroth moment of distribution, λ_0, through Eq. 7a and the problem of truncation error is completely eliminated.

To illustrate the method outlined above, formation of polyethylene terephthalate (PET) in batch reactors is analyzed with ethylene glycol (EG) evaporating [19–21]. If it is assumed that the polymer formed cannot be evaporated and the vapor–liquid equilibrium is governed by Raoult's law, then

$$p_r = p_r^*(T) \frac{[W]}{[W] + \lambda_0} \tag{13}$$

where p_r and $p_r^*(T)$ are the pressure applied on the reactor and the vapor pressure of the condensation product and [W] is the concentration of the condensation product. Experiments have shown that PET has an equilibrium constant $K(=k_p/k_p')$ of 0.5, which is independent of temperature, and conversion of 99% is needed to obtain fiber-grade polymer. In order to push the polymerization in the forward direction, high vacuum and temperature are applied. The effect of the polymerization pressure and the temperature can be evaluated through the use of Eqs. 7, 12, and 13 as follows. To obtain the solution of Eqs. 12 and 13, they are written in the nondimensional form using the following variables [18]:

$$P_n = \frac{[P_n]}{\lambda_1^0} \qquad (14a)$$

$$\beta = \frac{k_p'}{k_p} \equiv \frac{1}{K} \qquad (14b)$$

$$t' = k_p \lambda_1^0 t \qquad (14c)$$

$$w = \frac{[W]}{\lambda_1^0} \qquad (14d)$$

where

$$\lambda_1^0 = \sum_1^\infty n[P_n]_0 \qquad (14e)$$

It is assumed that the feed to the batch reactor has its MWD given by Flory's distribution with conversion p_A of functional groups as 0.3; i.e., at $t = 0$,

$$p_n = (1 - 0.3)(0.3)^n$$

The moment equations can be derived; they are

$$\frac{d\lambda_0}{dt} = -\lambda_0^2 + \beta W(\lambda_1 - \lambda_0) \qquad (15a)$$

$$\frac{d\lambda_1}{dt} = 0 \qquad (15b)$$

$$\frac{d\lambda_2}{dt} = 2\lambda_1^2 + \frac{\beta}{3} W(\lambda_1 - \lambda_3) \qquad (15c)$$

These can be solved using the closure relation given in Eq. 10. Once the pressure and the temperature are specified, λ_0, λ_2. P_1, and P_2 can be found from Eqs. 15 and 16 using the Runge–Kutta numerical technique. The concentration of the condensation product W is obtained from Eq. 13 assuming the escaping vapors and the liquid are in thermodynamic equilibrium.

By using the computation method outlined here, it is possible to determine λ_0, λ_2, P_1, and P_2 as a function of the dimensionless reaction time and the μ_n and ρ are computed using Eq. 2. At low pressures, a large amount of EG is evaporated and μ_n and ρ increase continuously with time (as shown in Fig. 2). This implies that the polymers formed at low reaction pressures are highly polydispersed long-chain species. As opposed to this, at high pressures, the situation is found to be just the opposite. The dimensionless concentrations of the first two species P_1 and P_2 are presented in Figure 3 at different system pressures.

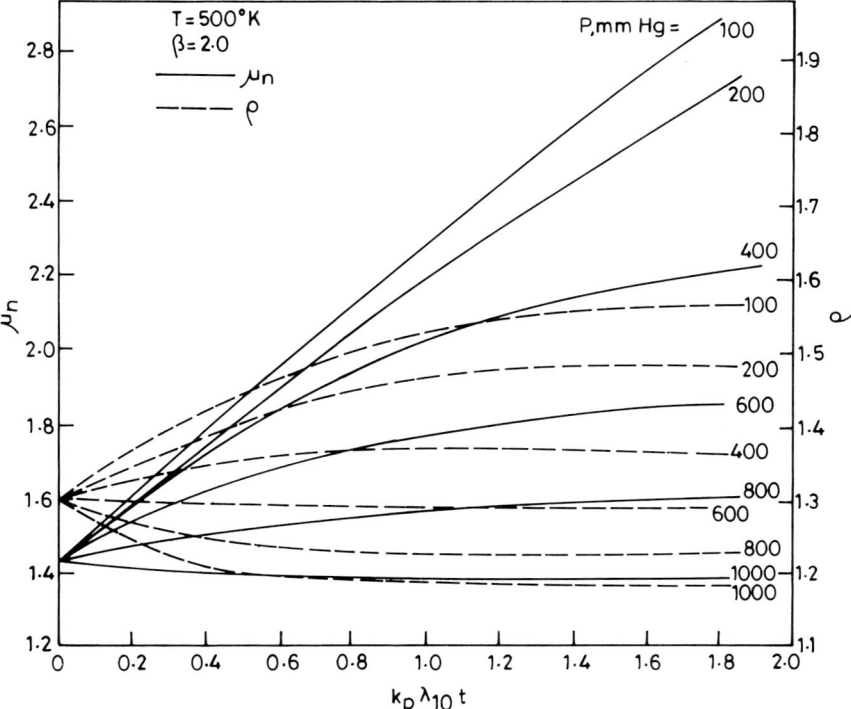

Figure 2 Number average chain length, μ_n, and polydispersity index, ρ, versus time, for various values of reaction pressures ($T = 500$ K, $\beta = 2.0$).

As the reaction proceeds the monomer, P_1, converts into the higher oligomers. The consumption of the monomer depends on the extent of the forward reaction. It is thus seen that the reactor pressure affects the consumption of P_1 in the reaction mass. Similarly, the concentration of the dimer, P_2, increases with the residence time of the reactor. The equilibrium point is specified when the concentration of any species does not change with time. We thus see from Figure 3 that the equilibrium is delayed as the system's pressure is reduced.

When the temperature of the reactor is increased, the vapor pressure p^* in Eq. 13 goes up, thus reducing the reaction mass. This in turn reduces the effect of the reverse step in step growth polymerization. In PET reactors, it is seen that the equilibrium constant, K, is unaffected by the temperature. Figures 4 and 5 present data for P_1, P_2, μ_n, and ρ with the reaction times for different temperatures. It is found that at high temperatures, higher chain length species are formed and the polymer mass is also highly polydispersed. The effect of higher temperature is similar to that of the reduction of the total pressure and the equilibrium is delayed.

In the analysis presented above, it was assumed that the functional groups of various oligomers have the same reactivity, independent of the chain length, n, of the molecules they are located on. The comparison of computed results with the assumption of equal reactivity with experimental data has sufficiently indicated that the overall polymerization

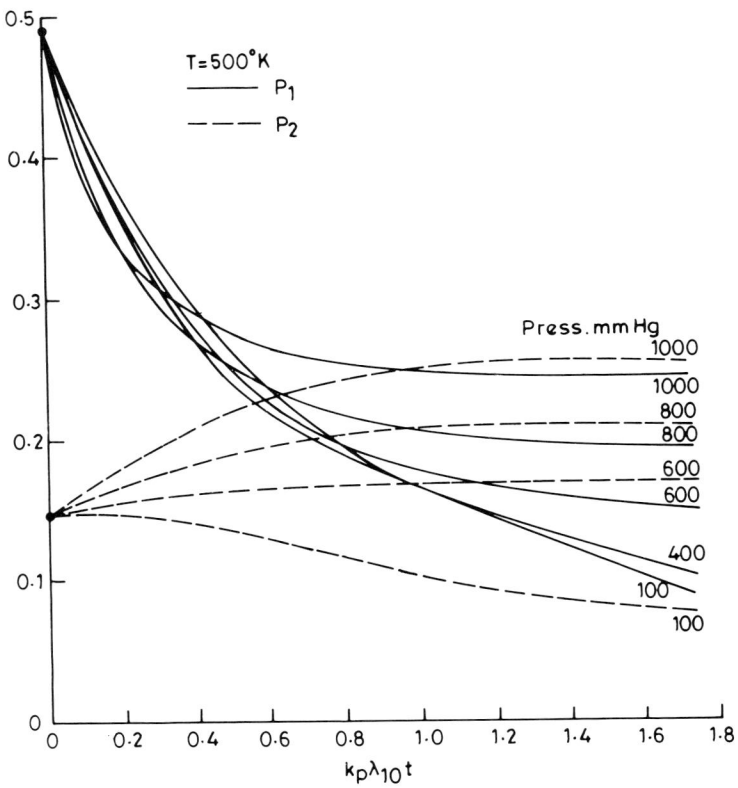

Figure 3 Effect of reactor pressure on concentrations of P_1 and P_2. $P_{10} = 0.49$, $P_{20} = 0.143$.

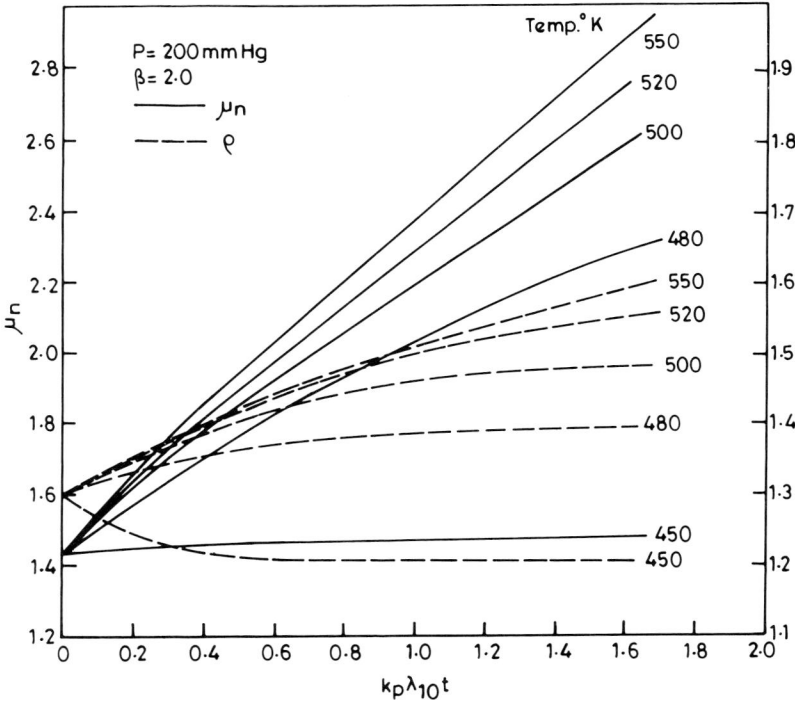

Figure 4 Number average chain length, μ_n, and polydispersity index, ρ, versus time, t, for various values of reaction temperatures ($P = 200$ mm Hg, β 2.0).

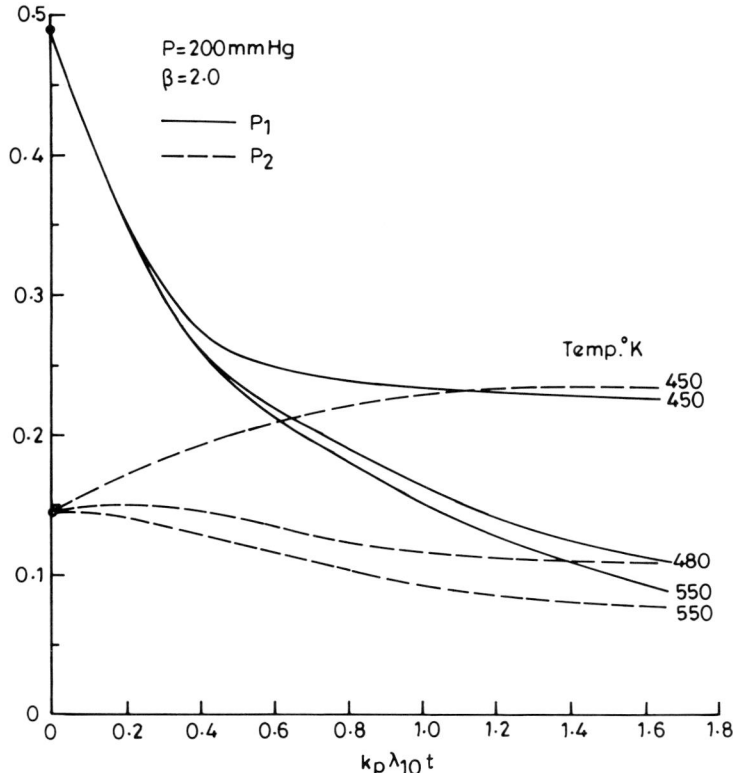

Figure 5 Effect of reactor temperature on concentration of P_1 and P_2.

is far more complex and the assumption of equal reactivity is a considerable simplification [22]. In addition, as the polymerization progresses, the viscosity of the reaction mass increases several fold and the overall reaction, at some stage depending on the reactor used, becomes mass transfer controlled [13, 23–25].

In the reaction-controlled region, unequal reactivity of functional groups is commonly observed. This can arise when the reacting groups are not kinetically equivalent, as in the polymerization of a diacid with a glycol having primary and secondary OH groups which have different reactivities. These asymmetric monomers are usually represented symbolically as the polymerization of (AA + BC) monomers where functional groups B and C react with A with different rate constants. The ring-opening reaction, as found in anhydrides, generates two functional groups which particpate in the chain-building process and these two steps generally have different reactivities. The formation of epoxy resins can be represented by the following equations:

$$\sim\!\!ROH + CH_2\!-\!\!\overset{O}{\overset{\diagup\ \diagdown}{CH}}\!-\!CH_2Cl \overset{k_1}{\rightarrow} \sim\!\!R\!-\!O\!-\!CH_2CH\!-\!CH_2 \quad \text{(16a)}$$

$$\sim\!\!ROCH_2\overset{O}{\overset{\diagup\ \diagdown}{CH}}\!-\!CH_2 + HOR\!\!\sim \overset{k_2}{\rightarrow} \sim\!\!ROCH_2\!-\!\overset{\overset{OH}{|}}{CH}\!-\!CH_2\!-\!OR\!\!\sim \quad \text{(16b)}$$

where k_1 and k_2 are the different rate constant values. In urethane polymerization, the unequal reactivity is explained by examining the effect of charge densities of the isocyanate groups on the phenyl diisocyanate monomer, OCN—⟨O⟩—NCO. As one of the isocyanate group reacts, the positive charge on the phenyl group reduces from its initial value and the reactivity of the second —NCO group falls. This phenomenon is known as induced asymmetry and can be modeled kinetically as

$$BB + AA \xrightarrow{k_a} A + B \tag{17a}$$

$$\sim\!\!B + AA \xrightarrow{k_a} A\!\!\sim \tag{17b}$$

$$\sim\!\!B + A\!\!\sim \xrightarrow{k_b} \sim\!\!BA\!\!\sim \tag{17c}$$

where AA represents the diisocyanate. Case analyzed various situations of asymmetry and induced asymmetry and derived the molecular weight distribution in terms of the probabilities of reaction of the various functional groups [26].

There is another class of unequal reactivity polymerization in which the rate constants of the various oligomers are dependent on their chain lengths. This is chain length dependent reactivity [16, 27–36], which is found, for example, in the polymerization of sodium-p-fluorothiophenoxide [37]. In this, the reaction mass has more unreacted monomer than that predicted by the equal reactivity hypothesis. This observation suggests that the monomer has a lower reactivity. In the formation of polyamides, on the other hand, polymer molecules are found to have lower reactivity [38].

Kinetic Model for Reversible Nonlinear Step Growth Polymerization

Nonlinear step growth polymerization takes place with monomers RY_f having functionality, $f > 2$. In this Y is the reactive group which on reaction gives rise to chemical bonds. There are several industrially important nonlinear step growth polymerization systems and some of these are polyesters from adipic acid or phthalic anhydride and glycerol or pentacrythritol (alkyl resins), curing of epoxy propolymers with diamines, curing of phenol formaldehyde polymers with hexamethylene tetramine, etc. In this class of polymerization, branched molecules are formed at low conversions of reactive groups. At a well-defined conversion [$= 1/(f - 1)$], some of these branched molecules are found to combine into an infinite network structure of macroscopic dimensions, called a gel [1]. This phenomenon occurs long before the reactive groups are completely consumed, and the point at which this occurs is referred to as the critical or gel point. Experimentally, the gel point is recognized as the state when the viscosity of the reaction mass becomes infinite and "gas bubbles" fail to rise through the reaction mass.

The study of nonlinear step growth polymerization is more complex than that of the linear case, and several approaches have been taken by different workers in this area to model the nonlinear. Flory [1, 39, 40] and Stockmayer [41–44] have approached this problem by determining the probabilities of finding various branched molecular structures in the reaction mass. Thereafter, they used these probability distributions to compute the number and weight average molecular weights of the polymer before gelation. Their approach, however, becomes exceedingly complex for systems of industrial importance. Other workers [45–47] have attempted to derive the average molecular weights directly, without first obtaining the detailed distributions. However, the kinetic approach presented in this section is more advantageous than the other approaches because it can be used easily to predict the behavior of nonlinear polymerization in homogeneous continuous stirred tank reactors (HCSTRs). In addition, it can be more easily adapted to account for intramolecular reactions.

The analysis using the kinetic approach of the self-polymerization of RY_f monomers, with Y functional group reacting with Y, is presented in this section. It is assumed that P_n represents an n-mer in the reaction mass. If there are no intramolecular reactions, there are $nf - 2(n - 1)$ unreacted Y groups on P_n since every bond in the molecule is formed by reaction of two Y groups. Similar to the linear step growth polymerization, kinetic analysis has been made for the reversible nonlinear step growth polymerization with the monomer of functionality, f. In the forward reaction of step growth polymerization, two functional groups of different molecules react to give higher molecular weight species. When the reactive groups of the same molecule react, cyclic molecules are formed. The modeling of the intramolecular reaction is extremely complex and is still an active area of research [58–70]. Jacobson and Stockmayer [64] analyzed the formation of intramolecular bonds and observed that two reacting groups in a given chain must turn around and lie within a small volume, V_s, of each other. However, it has been shown that this represents only a crude description of the cyclization process. Mutter, Sater, and Flory [71] used the theory of chain statistics to describe this process. In this model they accounted for the presence of bond angle restrictions as well as steric hindrance. In addition, they included the cyclization constraints that the bonds of reacting sides must approach each other so that they are parallel to each other.

The reverse step to this involves the reaction of the bonds of a macromolecule with the molecules of the condensation product, W. In multifunctional polymerization, the polymer chains are in general branched and as a result of this the modeling of the reverse reaction becomes considerably more complex [72–74] compared to those for linear chains because when W reacts with any on the bonds of the polymer chain, it forms oligomers of small chains. Let us now consider the reverse reaction of a P_{10} molecule with W. Regardless of its structure, it has nine bonds where W can react. If P_{10} is linear, there are two equivalent sites on it which would form P_1, P_2, up to P_9 when W reacts with it (see Fig. 6a). It is thus seen through this example that the number of sites forming species of lower molecular weight out of the higher ones is independent of the chain structure in linear polymers. However, for branched molecules this is not found to be so and the reverse step is found to be chain structure–dependent as demonstrated below. In Figure 6b, P_{10} is assumed to be of the star type with four branches of equal length. From the figure it is seen that a star P_{17} cannot break to form P_5 to P_{12}. However, it has four equivalent sites where, by the reverse reaction, smaller molecules P_1, P_2, P_3, P_4, P_{13}, P_{14}, P_{15}, and P_{16} can be formed. In Figure 6c, a general branched molecule is shown with one branch at a given branch point of the chain backbone. It is assumed that there exists an equal number of repeat units between the two branch points in this structure. Suppose this molecule has a total of r repeat units and we wish to determine the number of sites on it where P_n will be formed on reaction with W. A term $\delta_{r \to n}$ is defined which gives the number of sites on P_r that would give P_n (where $n < r$) through the reverse reaction. This term accounts for the structural effects; for linear polymers it is equal to 2 for all feasible n. If in Figure 6c, P_r has b branch points with each segment having x repeat units, it is found that

$$\delta_{r \to i} = \delta_{r \to (r-i)} = b + 2 \quad \text{if } k \leq x \tag{18a}$$

$$\left.\begin{array}{l}\delta_{r \to i} = 0 \\ \phantom{\delta_{r \to i}} = 2\end{array}\right\} \quad \begin{array}{l}\text{if } (mx + 1) \leq i \leq (m + 1)x \\ \text{if } 1 + (m + 1)x \leq i \leq (m + 2)x\end{array} \tag{18b}$$
$$\tag{18c}$$

where $m = 1, 2, \ldots$

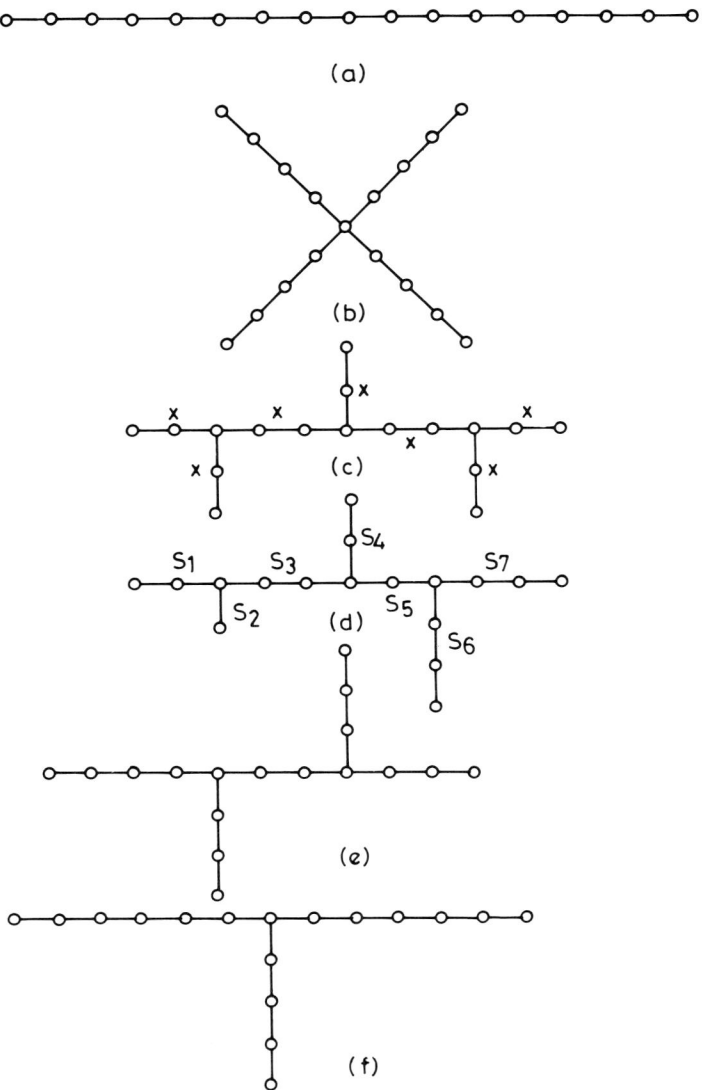

Figure 6 Structures of branched polymer chains.

The structure shown in Figure 6c can be further generalized by assuming unequal lengths of the segments. For any general chain structure, to find $\delta_{r \to n}$, the (a) location, number, and lengths of various branches and (b) chain length of the molecule must be known. It is then possible to work out the length of the backbone and determine $\delta_{r \to n}$ using a computer program as follows.

We first characterize any branched molecule by dividing it into different segments in Figure 6d where the molecule is divided into seven segments s_1 to s_7. If each branch is treated as one of the segments, it is then observed that every alternate segment is a branch. The total number of segments, n_s, and the number of branches, b, are related by

$$n_s = 2b + 1 \tag{19}$$

Since the segments lengths s_i ($i = 1, 2, \ldots, n_s$) are known, the backbone chain length, n_B, and the location of branch points K_m ($m = 1, 2, \ldots, b$) in P_r are related by

$$n_B = r - \sum_{i=1}^{b} s_{2i} \tag{20a}$$

$$K_m = K_{m-1} + s_{2m-1} + 1, \qquad K_0 = 0 \tag{20b}$$
$$m = 1, 2, \ldots, b$$

For example, in Figure 6d, $s_1 = 2$, $s_2 = 1$, $s_2 = 2$, $s_4 = 2$, $s_5 = 1$, $s_6 = 3$, and $s_7 = 3$. For this case backbone chain length n_B is 11 and branch point locations K_m are positions 3, 6, and 8 on the backbone.

To find $\delta_{r \to n}$, backbone indices I_p and I'_q for the forward and the reverse directions are defined. These specify the number of repeat units if the backbone is broken at the pth and qth positions from the forward and reverse directions respectively. As an example, in Figure 6d, if the chain scission occurs at the seventh bond of the backbone, molecule P_{10} is formed from this end, and which gives $I_7 = 10$ in the forward direction. As opposed to this, in the reverse direction, the same bond breakage gives P_7 and the corresponding index $I'_4 = 7$. For symmetric chains, the values of I_p and I'_q for same p and q are identical; otherwise they are not. The flow chart for generating these indices is given in Table 1. The information so generated is used to find $\delta_{r \to n}$; the algorithm for this is given in Table 2. To avoid the counting of the available sites twice, scanning of the molecule is done from both ends up to its midpoint, $N_{1/2}$ given by

$$N_{1/2} = \frac{r}{2}, \qquad r \text{ even}$$

$$= \frac{r-1}{2}, \qquad r \text{ odd} \tag{21}$$

It is evident that the number of possible sites for forming P_n ($n = 1, 2, \ldots, r-1$) and P_{r-n} out of P_r is equal and

$$\delta_{r \to n} = \delta_{r \to (r-n)} \tag{22}$$

The algorithm presented in Table 2 is divided into two major sections. In the first, branches are scanned to find $\delta_{r \to n}$. In the second, scanning of the backbone is done from the forward and reverse directions. Results for $\delta_{17 \to n}$ ($n = 1, 2, \ldots, 16$) have been generated from various chain structures seen in Figure 6 and summarized in Table 3.

For all branched molecules, the following two relations are found to hold always:

$$\sum_{n=1}^{r-1} \delta_{r \to n} = 2(r - 1) \tag{23a}$$

$$\sum_{n=1}^{r-1} n \delta_{r \to n} = r(r - 1) \tag{23b}$$

Table 1 Generation of the Forward and Reverse Backbone Indices

READ

   r : TOTAL CHAIN LENGTH OF POLYMER
   b : TOTAL NUMBER OF BRANCHES
   s_i : SEGMENT LENGTHS, $i = 1, 2, \ldots n_s$

COMPUTE TOTAL NUMBER OF SEGMENTS, n_s, AND BACKBONE CHAIN LENGTH, n_B

$$n_s = 2b + 1$$

$$n_B = r - \sum_{i=1}^{b} s_{2i}$$

COMPUTE BRANCH POINT POSITION, $K_n (n = 1, 2, \ldots b)$ BY

$$K_n = K_{n-1} + S_{2n-1} + 1 \;,\; K_0 = 0$$

DEFINE BACKBONE INDEX I_l IN THE FORWARD DIRECTION
   $l = 1, 2, \ldots n_B$
   $I_0 = 0,\; i = 1,\; l = 1$

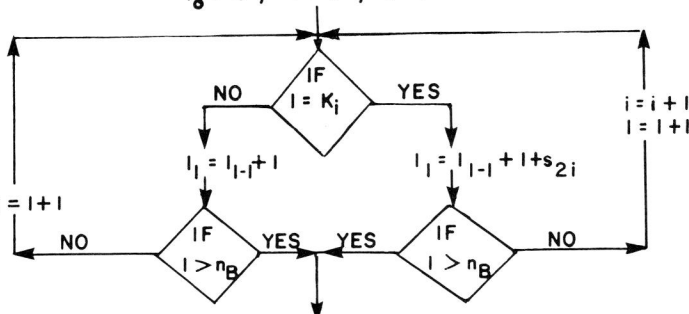

COMPUTE BACKBONE CHAIN IN REVERSE DIRECTION
   $I'_m,\; m = 1, 2, \ldots n_B$

   $I'_0 = 0,\; i_s = 2b,\; i_t = b$

   $m = 1$

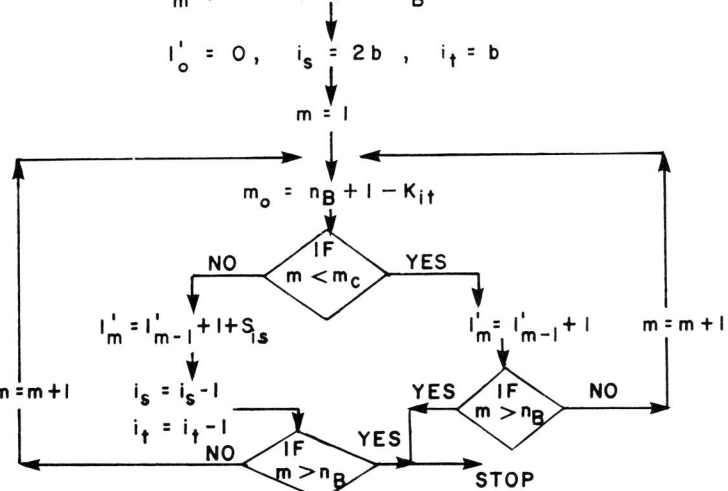

Table 2 Calculation of $\delta_{r \rightarrow n}$ from the Knowledge of Backbone Indices Generated from Table 1

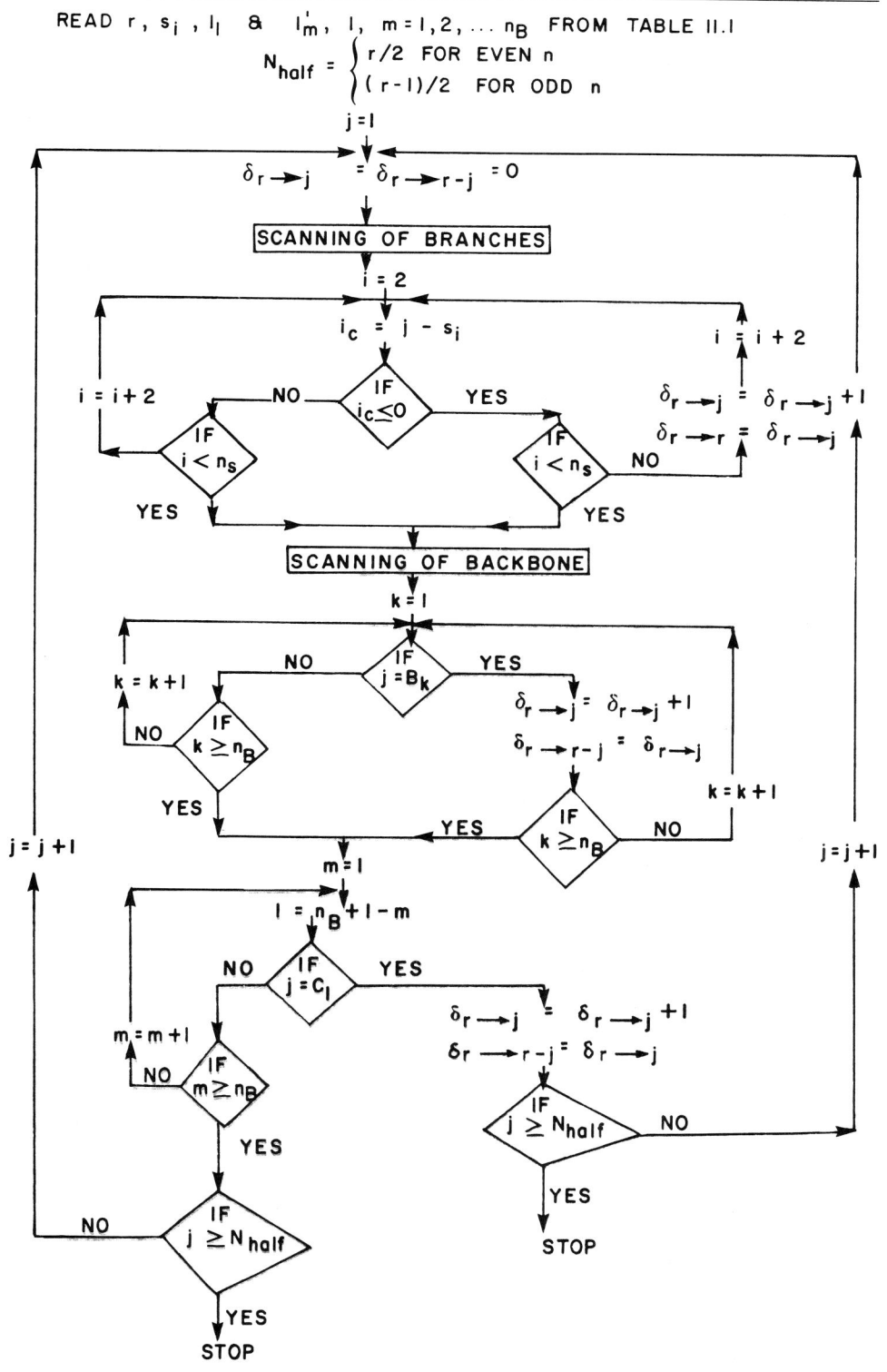

Table 3 Value of $\delta_{r \rightarrow n}$ for Different Branched Structures Seen in Figure 6

n	$\delta_{17 \rightarrow n}$ for Fig 6a	$\delta_{17 \rightarrow n}$ for Fig 6b	$\delta_{17 \rightarrow n}$ for Fig 6c	$\delta_{17 \rightarrow n}$ for Fig 6d	$\delta_{17 \rightarrow n}$ for Fig 6e	$\delta_{17 \rightarrow n}$ for Fig 6f
1	2	4	5	5	4	3
2	2	4	5	4	4	3
3	2	4	0	2	4	3
4	2	4	0	1	1	3
5	2	0	2	1	0	2
6	2	0	2	1	0	2
7	2	0	2	1	1	0
8	2	0	0	1	2	0
9	2	0	0	1	2	0
10	2	0	2	1	1	0
11	2	0	2	1	0	2
12	2	0	2	1	0	2
13	2	4	0	1	1	3
14	2	4	0	2	4	3
15	2	4	5	4	4	3
16	2	4	5	5	4	3

However, the summation $\sum_{n=1}^{r-1} n^2 \delta_{r \rightarrow n}$ depends on the structure. If the chain structure is that given in Figure 6c, then

$$\sum_{n=1}^{r-1} n^2 \delta_{r \rightarrow n} = A_1 r^3 + A_2 r^2 + A_3 r + A_4 \qquad (24)$$

where

$$A_1 = \frac{(20b^3 + 24b^2 + 14b + 2)}{3(2b + 1)^3} \qquad (25a)$$

$$A_2 = -\frac{(8b^4 + 64b^3 + 88b^2 + 32b + 6)}{6(2b + 1)^3} \qquad (25b)$$

$$A_3 = \frac{(16b^4 + 30b^3 + 26b^2 + 16b + 2)}{6(2b + 1)^3} \qquad (25c)$$

$$A_4 = -\frac{(2b^4 + 9b^3 + b^2)}{6(2b + 1)^3} \qquad (25d)$$

It is observed that the relations given in Eqs. 25 are independent of the functionality f of the starting monomer. If there are no intramolecular reactions, there would be $nf - 2(n - 1)$ unreacted Y groups on P_n because for every bond formed, two Y groups are consumed. Two reacting molecular species P_n and P_m have $f_1 = [nf - 2(n - 1)]$ and $f_2 = [mf - 2(m - 1)]$ unreacted functional groups on them. When they undergo reaction, there are $f_1 f_2 / 2$ distinct ways by which they can react. The factor of $1/2$ is included because the reactive groups Y are identical. For any general species P_n produced according to Eq. 3

the mole balance relation for the step growth polymerization in the batch reactor can now be derived as

$$\frac{d[P_1]}{dt} = 2k_p(a + 1)[P_1]\left\{\sum_{m=1}^{\infty}(am + 1)[P_m]\right\} + k'_p[W]\sum_{r=2}^{\infty}\delta_{r\to 1}[P_r] \tag{26a}$$

$$\frac{d[P_n]}{dt} = k_p\sum_{r=1}^{n-1}(ar + 1)\{a(n - r) + 1\}[P_r][P_{n-r}]$$

$$-2k_p(an + 1)[P_n]\sum_{m=1}^{\infty}(am + 1)[P_m]$$

$$-k'_p[P_n][W](n - 1) + k'_p[W]\sum_{r=n+1}^{\infty}\delta_{r\to n}[P_r] \quad \text{for } n \leq 2 \tag{26b}$$

$$a = \frac{f - 2}{2} \tag{26c}$$

Equations 26 involve the term $\delta_{r\to n}$, which is dependent on the number of branches in the reacting polymer chains. This implies that Eqs. 26 cannot be solved without knowledge of the chain structure.

A functional group approach is now discussed in order to find the average branching of polymer in the reaction mass. Consider a hexafunctional monomer ($f = 6$). At every collision of two oligomers, two reactive groups are eliminated to form a bond between them. Whenever reactive groups in the middle of the chain undergo chemical reaction, a branch is produced at that point. A general chain structure produced by the polymerization of a hexafunctional monomer is shown in Figure 7. It is possible to follow the course of polymerization through the functional group approach, which is a generalization of Flory's analysis of bifunctional step growth polymerization.

In the functional group approach, one defines ($f + 1$) functional group species in order to follow the polymerization of RY_f monomers. As an example, one would require seven species A to G as shown in Figure 8. These differ from each other in terms of the number of unreacted reactive groups Y on them. For example, A has all reactive groups unreacted and is the same to the monomer while B has one reacted reactive group, C has two, and so on. As species B has only one of its reactive group reacted, it can be situated only at chain ends. Species C forms the linear part of the chain, while the other species lead to branching. It is possible to represent the chain structure of Figure 7 in terms of species B to G very conveniently.

Various forward and reverse reactions leading to polymer formation involving these

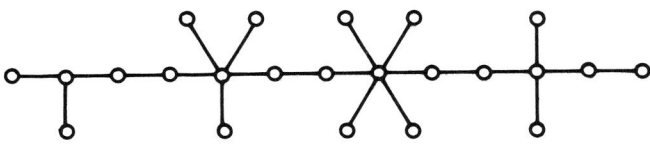

Figure 7 General structure of polymer chain in the polymerization of hexafunctional monomer.

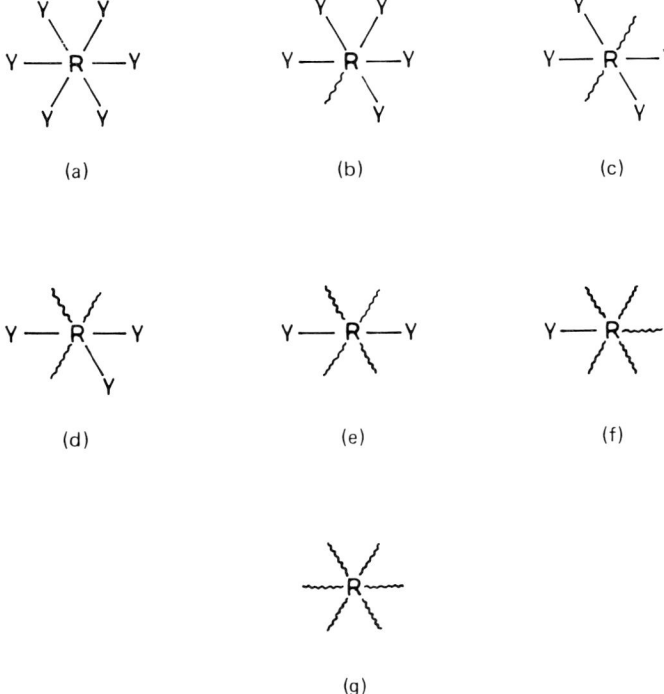

Figure 8 Structures of functional group species.

functional group species can be easily written and are summarized in Table 4. In writing these, it has been assumed that all species are available for chemical reaction with equal likelihood. It is assumed that species A reacts with C. There is a chemical reaction of reactive groups Y on species A with those of species C; when this occurs, species A would become B because one of its reactive group is reacted while C becomes D for the same reason. In view of this, the forward step of the reaction between species A and C can be represented as

$$A + C \xrightarrow{k_3} D + (B) + W \tag{27}$$

on the right-hand side B appears in parentheses to indicate that B is not free but is in the combined state as B–D.

To determine the reactivity k_3 (see Table 4), it is observed that reaction between any two species occurs only when they collide. If the collision frequency between species A and C is represented Z_{AC} and Q_{AC} as the probability of reaction between A and C, then rate $\mathcal{R}_{AC}$ is given as

$$\mathcal{R}_{AC} = \alpha Z_{AC} Q_{AC} \tag{28}$$

where α is the constant of proportionality. For equal reactivity of reactive groups Y, Z_{AC} would be given by $[(6 \times 4)/2Z]$ where Z is the probability of reaction between Y's of the two species A and C. The factor $(6 \times 4)/2$ arises because species A has six and species C

Table 4 Various Reactions of Species A to G

Forward Reactions

$$A + A \xrightarrow{k_1} B + (B) + W, \qquad k_1 = \frac{6 \times 6}{2} k$$

$$A + B \xrightarrow{k_2} C + (B) + W, \qquad k_2 = \frac{6 \times 5}{2} k$$

$$A + C \xrightarrow{k_3} D + (B) + W, \qquad k_3 = \frac{6 \times 4}{2} k$$

$$A + D \xrightarrow{k_4} E + (B) + W, \qquad k_4 = \frac{6 \times 3}{2} k$$

$$A + E \xrightarrow{k_5} F + (B) + W, \qquad k_5 = \frac{6 \times 2}{2} k$$

$$A + F \xrightarrow{k_6} G + (B) + W, \qquad k_6 = \frac{6 \times 1}{2} k$$

$$B + B \xrightarrow{k_7} G + (C) + W, \qquad k_7 = \frac{5 \times 5}{2} k$$

$$B + C \xrightarrow{k_8} D + (C) + W, \qquad k_8 = \frac{5 \times 4}{2} k$$

$$B + D \xrightarrow{k_9} E + (C) + W, \qquad k_9 = \frac{5 \times 3}{2} k$$

$$B + E \xrightarrow{k_{10}} F + (C) + W, \qquad k_{10} = \frac{5 \times 2}{2} k$$

$$B + F \xrightarrow{k_{11}} G + (C) + W, \qquad k_{11} = \frac{5 \times 1}{2} k$$

$$C + C \xrightarrow{k_{12}} D + (D) + W, \qquad k_{12} = \frac{4 \times 4}{2} k$$

$$C + D \xrightarrow{k_{13}} E + (D) + W, \qquad k_{13} = \frac{4 \times 3}{2} k$$

$$C + E \xrightarrow{k_{14}} F + (D) + W, \qquad k_{14} = \frac{4 \times 2}{2} k$$

$$C + F \xrightarrow{k_{15}} G + (D) + W, \qquad k_{15} = \frac{4 \times 1}{2} k$$

$$D + D \xrightarrow{k_{16}} E + (E) + W, \qquad k_{16} = \frac{3 \times 3}{2} k$$

$$D + E \xrightarrow{k_{17}} F + (E) + W, \qquad k_{17} = \frac{3 \times 2}{2} k$$

$$D + F \xrightarrow{k_{18}} G + (E) + W, \qquad k_{18} = \frac{3 \times 1}{2} k$$

$$E + E \xrightarrow{k_{19}} F + (F) + W, \qquad k_{19} = \frac{2 \times 2}{2} k$$

$$E + F \xrightarrow{k_{20}} G + (F) + W, \qquad k_{20} = \frac{2 \times 1}{2} k$$

Table 4 (*Continued*)

$$F + F \xrightarrow{k_{21}} G + (G) + W, \qquad k_{21} = \frac{1 \times 1}{2} k$$

Reverse Reactions

$$B + (B) + W \xrightarrow{k'_1} A + A, \qquad k'_1 = \frac{[B]}{XX} k'$$

$$B + (C) + W \xrightarrow{k'_2} A + B, \qquad k'_2 = \frac{[C]}{XX} k'$$

$$B + (D) + W \xrightarrow{k'_3} A + C, \qquad k'_3 = \frac{[D]}{XX} k'$$

$$B + (E) + W \xrightarrow{k'_4} A + D, \qquad k'_4 = \frac{[E]}{XX} k'$$

$$B + (F) + W \xrightarrow{k'_5} A + E, \qquad k'_5 = \frac{[F]}{XX} k'$$

$$B + (G) + W \xrightarrow{k'_6} A + F, \qquad k'_6 = \frac{[G]}{XX} k'$$

$$C + (C) + W \xrightarrow{k'_7} B + B, \qquad k'_7 = 2 \frac{[C]}{XX} k'$$

$$C + (D) + W \xrightarrow{k'_8} B + C, \qquad k'_8 = 2 \frac{[D]}{XX} k'$$

$$C + (E) + W \xrightarrow{k'_9} B + D, \qquad k'_9 = 2 \frac{[E]}{XX} k'$$

$$C + (F) + W \xrightarrow{k'_{10}} B + E, \qquad k'_{10} = 2 \frac{[F]}{XX} k'$$

$$C + (G) + W \xrightarrow{k'_{11}} B + F, \qquad k'_{11} = 2 \frac{[G]}{XX} k'$$

$$D + (D) + W \xrightarrow{k'_{12}} C + C, \qquad k'_{12} = 3 \frac{[D]}{XX} k'$$

$$D + (E) + W \xrightarrow{k'_{13}} C + D, \qquad k'_{13} = 3 \frac{[E]}{XX} k'$$

$$D + (F) + W \xrightarrow{k'_{14}} C + E, \qquad k'_{14} = 3 \frac{[F]}{XX} k'$$

$$D + (G) + W \xrightarrow{k'_{15}} C + F, \qquad k'_{15} = 3 \frac{[G]}{XX} k'$$

$$E + (E) + W \xrightarrow{k'_{16}} D + D, \qquad k'_{16} = 4 \frac{[E]}{XX} k'$$

$$E + (F) + W \xrightarrow{k'_{17}} D + E, \qquad k'_{17} = 4 \frac{[F]}{XX} k'$$

$$E + (G) + W \xrightarrow{k'_{18}} D + F, \qquad k'_{18} = 4 \frac{[G]}{XX} k'$$

Table 4 (*Continued*)

$$F + (F) + W \xrightarrow{k'_{19}} E + E, \quad k'_{19} = 5 \frac{[F]}{XX} k'$$

$$F + (G) + W \xrightarrow{k'_{20}} E + F, \quad k'_{20} = 5 \frac{[G]}{XX} k'$$

$$G + (G) + W \xrightarrow{k'_{21}} F + F, \quad k'_{21} = 6 \frac{[G]}{XX} k'$$

Note: XX = [B] + [C] + [D] + [E] + [F] + [G].

has four reactive groups and the distinct combinations of two Y groups between them is $(6 \times 4)/2$. Since the collision frequency between species A and C is proportional to [A][C], it implies that the reactivity k_3 for this step is given by

$$k_3 = \frac{6 \times 4}{2} k \tag{29}$$

where k is the reactivity of Y groups. Using similar logic the rate constants of all the forward steps have been derived and included in Table 4.

The reverse reaction in multifunctional polymerization involves the reaction of the bonds in functional group species with the condensation product, W. However, complexity arises because we must have knowledge of the neighborhood species. For example, one considers bond B–D reacting with W. In the reverse reaction step, B would become A, but simultaneously D would become C. Thus it appears to be a trimolecular reaction even though in reality it is only bimolecular in nature. If attention is focused on the reverse reaction of any given functional group species, knowledge of the connecting species must be available. The reverse reaction step can be written as

$$B + (D) + W \xrightarrow{k'_3} A + C \tag{30}$$

The rate constant for this step in Table 4 has been shown to be k'_3 and it is desired to write this in terms of the reverse rate constant k' between given reaction Y group and the condensation product W. To do so, the reaction is taken as that between B and W. Since species B has only one reacted bond in it, its reactivity would therefore be k' and the reverse rate $\mathcal{R}'_{B(D)}$ is given by

$$\mathcal{R}'_{B(D)} = k'[B][W]P_{B(D)} \tag{31}$$

where $P_{B(D)}$ is the probability of finding D adjacent to species B. If it is assumed that any species can be found adjacent to B, this probability term can be approximated as

$$P_{B(D)} = \frac{[D]}{[B] + [C] + [D] + [E] + [F] + [G]} \tag{32}$$

With this background and the kinetic scheme given in Table 4, it is possible to write the mole balance relations for these functional group species in batch reactors. These are given by

$$\frac{d[A]}{dt} = -3k[A][M] + k'[B][M] \tag{33a}$$

$$\frac{d[B]}{dt} = 3k[A][M] - \frac{5}{2}k[B][M] + 2k'[C][W] - k'[B][W] \tag{33b}$$

$$\frac{d[C]}{dt} = \frac{5}{2}k[B][M] - 2k[C][M] + 3k'[D][W] - 2k'[C][W] \tag{33c}$$

$$\frac{d[D]}{dt} = 2k[C][M] - \frac{3}{2}k[D][M] + 4k'[E][W] - 3k'[D][W] \tag{33d}$$

$$\frac{d[E]}{dt} = \frac{3}{2}k[D][M] - k[E][M] + 5[F][W] - 4k'[E][W] \tag{33e}$$

$$\frac{d[F]}{dt} = k[E][M] - \frac{k}{2}[F][M] + 6k'[G][W] - 5k'[F][W] \tag{33f}$$

$$\frac{d[G]}{dt} = \frac{k}{2}[F][M] - 6k'[G][W] \tag{33g}$$

$$\frac{d[W]}{dt} = k[M]\left\{\frac{3}{2}[A] + \frac{5}{4}[B] + [C] + \frac{3}{4}[D] + \frac{[E]}{2} + \frac{[F]}{4}\right\}$$
$$-k'[W]\left\{\frac{[B]}{2} + [C] + \frac{3}{2}[D] + 2[E] + \frac{5}{2}[F] + 3[G]\right\} \tag{33h}$$

where

$$[M] = 6[A] + 5[B] + 4[C] + 3[D] + 2[E] + [F] \tag{33i}$$

Once the concentrations of the functional group species are known the average branches present in the reaction mass can be determined in the following way. Species D, E, F, and G have 1, 2, 3, and 4 branches, respecitvely. The average number of branches present in the reaction mass can be given by

$$\bar{b} = \frac{[D] + 2[F] + 3[F] + 4[G]}{\Sigma_{n=1}^{\infty}[P_n]} \tag{34}$$

Computations have shown that in multifunctional polymerization, the concentration of the higher functional group species E to G are present in negligible concentrations compared to B to D all the way up to the gel point. Thus an average structure (as shown in Fig. 6c) can be assumed without introducing any error in computations.

In multifunctional polymerization one is mainly interested in the moments of the distribution instead of the entire MWD. Equation 21 can be added according to Eq. 1 for all n to obtain the moment generation results. With the help of Eq. 47, the following relations are derived:

$$\frac{d\lambda_0}{dt} = k(a\lambda_1 + \lambda_0)^2 + k'[W](\lambda_1 - \lambda_0) \tag{35a}$$

$$\frac{d\lambda_1}{dt} = 0 \tag{35b}$$

$$\frac{d\lambda_2}{dt} = 2k(a\lambda_2 + \lambda_1)^2 + k'[W]\{(A_2 + 1)\lambda_2 + A_3\lambda_1 + A_4\lambda_0\}$$
$$+ k'[W](A_1 - 1)\lambda_3 \tag{35c}$$

Equation 35c is dependent on the higher moment λ_3 and can be solved only after assuming some moment closure relation as is done for linear polymerization.

Using the computer algorithm for $\delta_{r \to n}$, r was fixed at 17. This was calculated for various n ($n = 1, 2, \ldots, 16$), and results are presented in Table 3 for different chain structures. The table shows an extreme sensitivity to the chain structure. Since the mole balance relations involve $\delta_{r \to n}$, the MWD of the polymer is expected to be greatly affected by the branching. The computation technique outlined here has been demonstrated for the batch polymerization of hexafunctional monomers. For this we need seven functional group species (Fig. 8); their numerical solution shows that the total branching is small for the conversion of reactive groups below the gel point. With the information on average branching so generated, it is now possible to determine the MWD of the polymer. In Figure 9, concentrations of P_n species versus chain length n were plotted with the time of polymerization t' as a parameter.

Examination of Figure 9 reveals that as the reaction time t' increases, the MWD becomes broader. This is expected with increasing time, higher oligomers are formed in larger concentrations and both the average chain length and the polydispersity index ρ are expected to increase. In Figure 10, the MWD of a given time of polymerization is plotted for different equilibrium constants, K. As K is increased, the reaction is closer to irreversible polymerization and for a given time of polymerization, a higher conversion of reactive groups is achieved. This would in turn imply that the MWD would become

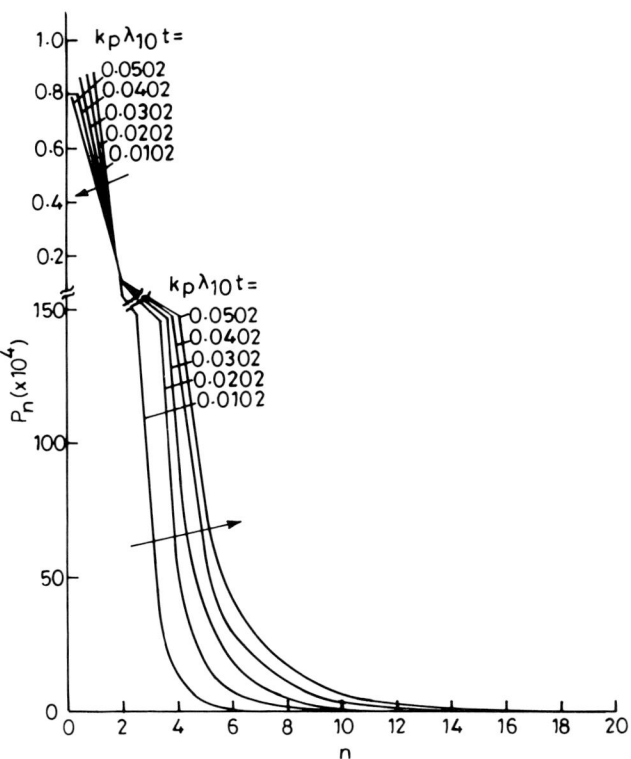

Figure 9 Dimensionless concentration of P_n species versus n (t varies from 0.0102 to 0.0502).

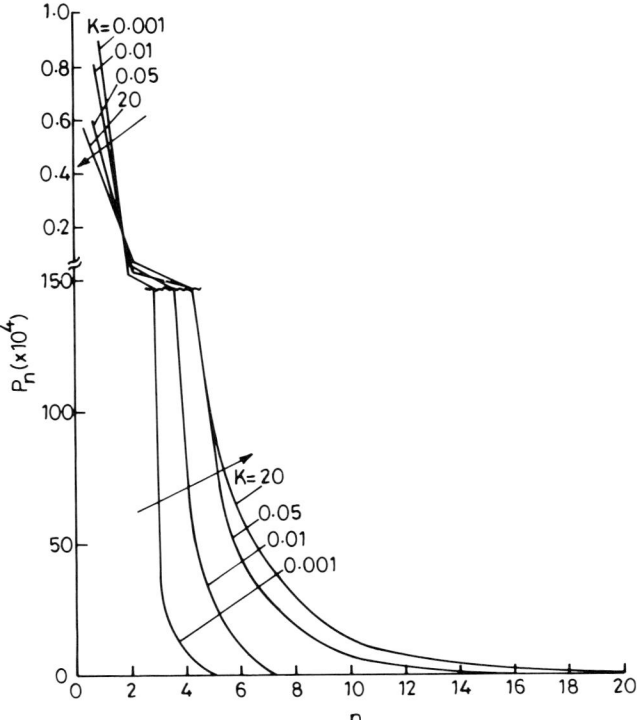

Figure 10 Dimensionless concentration of P_n species versus n (K varies from 0.0 to 20).

broader for increasing K, which is indeed seen in this figure. From the MWD results, it is possible to calculate the moments λ_0 and λ_2 (defined in Eq. 1). Theoretically (also found computationally), the first moment λ_1 is time invariant; thus the number and weight average chain lengths μ_n and μ_w and the polydispersity index ρ can be determined. It is found that the overall polymerization is limited by equilibrium conversion, which increases with increasing K. It is further observed from Eqs. 18 that for a given r, $\delta_{r \to n}$ is $(b + 2)$ for small n, which implies that as the degree of branching b increases, smaller chains (and therefore large chains due to symmetry) are formed preferentially. This would give rise to a higher polydispersity index compared to those formed from irreversible polymerization at the same conversion of reactive groups. An an example, for $K = 0.1$, at conversion of 13.87% the polydispersity index is 2.10; for the same conversion for irreversible polymerization it is 1.89.

In deriving an expression for $\Sigma_{n=1}^{r} n^2 \delta_{r \to n}$ in Eqs. 25, a specific chain structure as in Figure 6c was assumed. Computations for noninteger segment length x and noninteger number of branches were tested against wide variations of parameters. Noninteger x in actuality amounts to stating that all segment lengths are not of equal length. If the chain length r and the number of branches b (whole numbers here) are given, x can be calculated as

$$x = \frac{r - b}{2b + 1} \qquad (36)$$

If x calculated this way is a noninteger (say 6.7), the lower integer value x^* of x is chosen (in this case 6) and r^* is calculated as

$$r^* = (2b + 1)x^* \tag{37}$$

The remainder $(r - r^*)$ repeat units are distributed on various chain segments. For this chain structure, $\delta_{r \to n}$ can be computed using the algorithm of Table 2. On comparing the exact $\sum_{r=1}^{n} n^2 \delta_{r \to n}$ with that determined from Eq. 24, results are found to be within 1%. On similar testing for fractional branching b, the discrepancy between the exact results and those from Eq. 24 is negligible.

The generation relation for the second moment λ_2 in Eqs. 26 involves third moment λ_3. To solve it, a moment closure relation for λ_3 is needed. If the polymerization is assumed irreversible, the MWD in multifunctional self-polymerization can be solved exactly and is given by

$$\frac{[P_n]}{[P]_0} = \frac{\{n(f-1)\}!}{n!\,\{(f-2)n + 2\}!} f p^{n-1}(1-p)^{n(f-2)+2} \tag{38}$$

where p is the conversion of reactive groups. From this, the third moment for the polymerization of hexafunctional monomers can be derived as

$$\lambda_3 = \frac{(1 + p - 7p^2 + 5p^3) + (1 - 5p)^2 2p\lambda_2}{(1 - 5p)^3} \tag{39}$$

which can serve as a moment closure approximation. From the numerical solution of the MWD relations given in Eqs. 26, one can also find λ_0, λ_1, λ_2, and λ_3 and curve-fit these results using the empirical relation

$$\ln \lambda_3 = a_1 + a_2 \ln \lambda_0 + a_3 \ln \lambda_2 \tag{40}$$

where a_1 and a_3 are the curve-fit constants to be determined. On doing this we find that

$$\lambda_3 = \frac{1.036\, \lambda_2^{1.45}}{\lambda_0^{2.67}} \tag{41}$$

With the moment closure approximations in Eqs. 39 and 41, Eq. 35c has been solved for λ_2 and its values compared with those found from the MWD results. It is found that for a given K, Eq. 41 serves as a good approximation provided the conversion is well below the gel point conversions. As the gel point conversion is approached, Eq. 39 serves as a good representation, as can be seen in Table 5.

With the developments outlined in this work, it is now possible to analyze the equilibrium in multifunctional step growth polymerization. Since at equilibrium the MWD is stationary, their moments would also be time invariant, or $d\lambda_i/dt = 0$ $(i = 1, 2 \ldots)$. From Eq. 35a, for the zeroth moment, we obtain

$$K = \frac{k}{k'} = \frac{[W](\lambda_1 - \lambda_0)}{(a\lambda_1 + \lambda_0)^2} \tag{42}$$

Since the condensation product does not leave the reaction mass, the stoichiometry of the polymerization for λ_0^0 moles of monomer initially present is given by

$$[W] = \lambda_0^0 - \sum_{n=1}^{\infty} [P_n] = \lambda_0^0 - \lambda_0 \tag{43}$$

Table 5 Comparison of λ_2 Values Obtained by Using Three Different λ_3 Expressions Given by Eqs. 39–41

			λ_2 Values		
Time	Conversion (%)	By MWD	Using Flory-Schulz relation for λ_3	Using Stockmayer relation for λ_3	Using Eq. 41 for λ_3
For $K = 0.1$					
0.0012	0.3586	1.02191	1.02191	1.02192	1.02191
0.0102	2.9600	1.20845	1.20860	1.20852	1.20841
0.0202	5.6483	1.47215	1.47379	1.47255	1.47217
0.0302	8.1048	1.81644	1.82477	1.81760	1.81718
0.0402	10.3214	2.27229	2.30387	2.27445	2.27703
0.0502	12.2970	2.87917	2.98653	2.88106	2.90465
0.0552	13.1950	3.25250	3.44561	3.24940	3.30720
0.0592	13.8720	3.58946	3.89488	3.57228	3.68610
For $K = 0.02$					
0.0012	0.358623	1.02191	1.02191	1.02192	1.02191
0.0102	2.92437	1.20550	1.20619	1.20583	1.20530
0.0202	5.40353	1.44374	1.45101	1.44552	1.44374
0.0302	7.40084	1.70129	1.73204	1.70549	1.70354
0.0402	8.92173	1.95266	2.03790	1.95538	1.96110
0.0502	10.03150	2.17232	2.35367	2.15412	2.19195
0.0602	10.81690	2.34623	2.66384	2.25357	2.37890
0.0702	11.3609	2.47397	2.95522	2.19925	2.51586
0.0802	11.7348	2.56466	3.21877	1.92928	2.60580
0.0902	11.9599	2.61093	3.44994	1.38786	2.65666
0.1052	12.1603	2.65038	3.73493	0.065054	2.67934
For $K = 0.01$					
0.0012	0.3585	1.02190	1.02191	1.02192	1.02190
0.0102	2.88118	1.20194	1.20329	1.20258	1.20155
0.0202	5.13143	1.41337	1.42601	1.41643	1.41322
0.0302	6.70479	1.60084	1.64567	1.60678	1.60344
0.0402	7.70824	1.74085	1.84022	1.74353	1.7483
0.0502	8.31215	1.83239	1.99886	1.86508	1.84459
0.0602	8.66409	1.88767	2.12117	1.89569	1.90176
0.0702	8.85234	1.91398	2.21235	1.94470	1.93246
0.0802	8.95715	1.92876	2.27917	1.90999	1.94689
0.1024	9.05435	1.94282	2.36902	1.86951	1.95117
0.1194	9.07846	1.94649	2.40623	1.64685	1.94443
For $K = 0.0025$					
0.0012	0.35802	1.02187	1.02188	1.02192	1.02183
0.0102	2.65117	1.18330	1.18779	1.18540	1.18185
0.0202	4.00993	1.29997	1.32615	1.30607	1.29796
0.0302	4.51773	1.34784	1.40186	1.35533	1.34683
0.0402	4.68372	1.36373	1.43904	1.36922	1.36377
0.0502	4.73626	1.36872	1.45737	1.36957	1.36941
0.0602	4.75310	1.37031	1.46673	1.36426	1.37150
0.0702	4.7580	1.37084	1.47167	1.35543	1.37223
0.0762	4.76011	1.37098	1.47346	1.34855	1.37243

For $\lambda_0^0 = 1$, the equilibrium conversion p_{eq} can be obtained as

$$p_{eq} = \frac{\sqrt{K}}{1 + \sqrt{K}} \tag{44}$$

The second moment λ_2 at equilibrium can similarly be obtained by setting $d\lambda_2/dt = 0$, or

$$2(a\lambda_2 + \lambda_1)^2 + \frac{\lambda_0^0 - \lambda_0}{K}\{(A_2 + 1)\lambda_2 + A_3\lambda_1 + A_4\lambda_0\}$$
$$+ \frac{(\lambda_0^0 - \lambda_0)}{K}(A_1 - 1)\lambda_3 = 0 \tag{45}$$

In Figure 11, equilibrium conversion p_{eq} has been plotted as a function of K. Equation 44 reveals that it is independent of branching and is a monotonic function of K. As K is increased p_{eq} is increased, and for $K = 0.0625$, $p_{eq} = 0.20$, which would mean that gelation would occur for hexafunctional monomers for any k greater than this value.

In Table 6, equilibrium weight average chain length μ_{Weq} has been computed for the two-moment closure relation in Eqs. 34 and 41. This table reveals that results are ex-

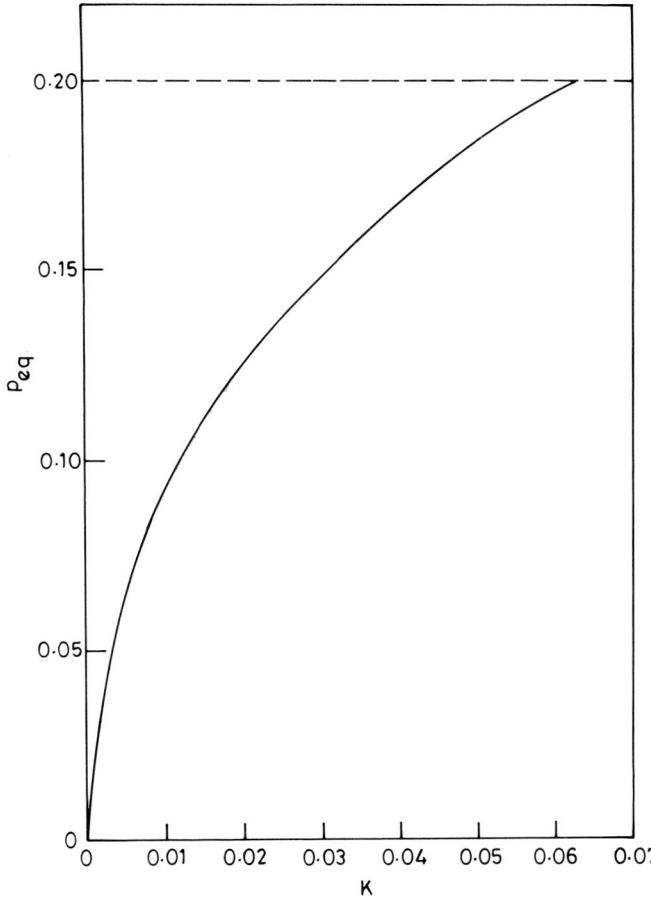

Figure 11 Equilibrium conversion, p_{eq}, versus K.

Table 6 Comparison of $\mu_{w,eq}$ Using Three Different Correlations for λ_3 Given by Eqs. 39–41

K	$\mu_{w,eq}$ using Eq. 39	$\mu_{w,eq}$ using Eq. 40	$\mu_{w,eq}$ using Eq. 41
$b = 0.02$			
0.010	2.586	−4.293	−0.661
0.012	2.933	−10.293	1.579
0.020	4.910	111.668	3.168
0.030	11.585	111.879	4.626
0.040	44.029	112.202	6.609
0.050	213.104	113.728	9.291
0.060	216.323	308.177	12.998
0.062	216.611	4470.830	13.901
$b = 0.03$			
0.010	2.652	−4.230	5.774
0.012	3.010	−10.390	5.861
0.020	5.117	75.489	6.244
0.030	12.753	75.721	6.854
0.040	20.243	76.118	7.729
0.050	275.248	78.217	9.119
0.060	277.955	302.607	12.564
0.062	278.211	4560.400	13.444
$b = 0.05$			
0.010	2.759	−4.0890	6.273
0.012	3.147	−10.4786	6.358
0.020	5.511	706.251	6.723
0.030	15.401	706.423	7.292
0.040	106.837	706.601	8.083
0.050	125.055	707.048	9.295
0.060	127.121	739.031	11.848
0.062	127.333	3491.830	12.688
$b = 0.1$			
0.010	2.947	−3.686	5.653
0.012	3.393	−10.048	5.751
0.020	6.317	114.000	6.125
0.030	24.130	114.204	6.662
0.040	92.917	114.502	7.390
0.050	99.406	115.878	8.480
0.060	100.803	289.511	10.679
0.062	100.961	4244.030	11.454

tremely sensitive to the closure approximation. It was observed earlier that as K approaches a value of 0.0625, for hexafunctional polymerization gelation would occur and $\mu_W \to \infty$. According to this table the moment closure in Eq. 41 gives finite $\mu_{W_{eq}}$ while that in Eq. 39 gives a high value. This implies that the empirical relation in Eq. 39 does not yield satisfactory results near the gel point.

RADICAL POLYMERIZATION

Unlike step growth polymerization in which the reaction occurs between the functional groups, in chain reaction polymerization the monomer polymerizes in the presence of compounds called initiators. The initiator continually generates growth centers in the reaction mass, and monomer molecules are rapidly added. It is this sequential addition of monomer molecules to growing centers which differentiates chain reactions from step growth polymerization.

The growth centers can be ionic (cationic and anionic), free radical, or coordinational in nature depending on the kind of initiator system used. Based on the nature of the growth centers, chain reaction polymerization is further classified as radical, cationic, anionic, or coordination (or stereoregular) polymerization [75]. Radical polymerization is utilized considerably more in industry and the discussion in this section is confined to radical polymerization.

Initiators for radical polymerization generate free radicals in the reaction mass. There are two types of radicals present in the reaction mass during polymerization:

1. Primary radicals, which are generated by initiator molecules directly.
2. Growing chains radicals, which are generated by the reaction between the primary radicals and the monomer molecules.

The growing chain radicals keep adding monomer molecules sequentially; this type of reaction is known as propagation. Reaction between a primary radical and a polymer radical or between two polymeric radicals makes polymer radicals unreactive by destroying their radical nature; such reactions are called termination. There are thus five kinds of species in the reaction mass at any time: initiator molecules, monomer molecules, primary radicals, growing chain radicals, and terminated polymer molecules. In order to model radical polymerization kinetically, the various reactions—initiation, propagation, and termination—must be understood.

Initiation

The molecules of initiator (denoted by I_2) can generate radicals by a homolytic decomposition of covalent bonds on absorption of energy, which can be in the form of heat, light, or high-energy radiation, depending on the initiator employed. Commercially, heat-sensitive initiators such as azo or peroxide compounds are employed. Radicals could also be generated between a pair of compounds, called redox initiators, one of which contains an unpaired electron. During the initiation, the unpaired electron is transferred to the other compound (called the acceptor), which undergoes bond dissociation. An example of redox initiators is a ferrous salt with hydrogen peroxide. In this section, however, only heat-sensitive initiators are discussed primarily because of their extensive use in industry.

The homolytic decomposition of initiator molecules can be represented schematically as

$$I_2 \xrightarrow{k_I} 2I \tag{46}$$

where I_2 is the initiator molecule and I, the primary radical. The rate of production of primary radicals, r'_i according to Eq. 46, is

$$r'_i = 2k_1[I_2] \tag{47}$$

where $[I_2]$ is the concentration of the initiator in the system at any time.

The primary radicals, I, combine with a monomer molecules, M, according to the schematic reaction

$$I + M \xrightarrow{k_1} P_1 \tag{48}$$

where P_1 is the polymer chain radical having one monomeric unit on it and k_1 is the rate constant of reaction. The rate of production, r_i, of the polymer radicals, P_1, can be written as

$$r_i = k_1[I][M] \tag{49}$$

where [I] and [M] are the concentrations of the primary radical and the monomer in the reaction mass, respectively.

Equations 47 and 49 imply that all the radicals generated by the homolytic decomposition of initiator molecules are used in generating the polymer chain radicals and no primary radicals are wasted by any other reaction. This is not true in reality and an initiator efficiency is defined to take care of the waste of primary radicals. The initiator efficiency, f, is the fraction of the total primary radicals produced by reaction 46 which are used in generating polymer radicals by reaction 48. Thus the rate of generation of polymer radicals is given by

$$r_i = 2fk_1[I_2] \tag{50}$$

Sometimes a solvent is added to the monomer for better temperature control. This has been shown to affect the initiator efficiency and is explained in terms of the cage theory [76, 77]. On supplying energy to initiator molecules, cleavage of a covalent bond occurs as shown in Eq. 46. According to this theory, the two dissociated fragments are surrounded by the reaction mass, which forms a sort of cage around them. The two fragments stay inside the cage for a finite amount of time during which they can recombine to give back the initiator molecule. The fragments that do not recombine diffuse, and the separated fragments are called primary radicals. If the monomer molecule is highly reactive, it can also react with a fragment inside a cage. The characteristics of the reaction medium determine how long the dissociated fragments stay inside the cage and they also affect the initiator efficiency. It is therefore expected that, if all other conditions are equal, a more viscous reaction mass would lead to a lower initiator efficiency.

Propagation Reaction

The propagation reaction is defined as the addition of monomer molecules to the growing polymer radicals. In the reaction mass there are polymer radicals of all possible sizes. In general, a polymer radical, P_n, indicates that there are n monomeric units joined together by covalent bonds in the chain radical. The propagation reaction can be written schematically as

$$P_n + M \xrightarrow{k_{pn}} P_{n+1}, \quad n = 1,2,\ldots \tag{51}$$

where k_{pn} is the rate constant for the reaction between P_n and a monomer molecule, M. In general this would depend on the size of the chain radical. It is not difficult to foresee the increasing mathematical complexity resulting from the multiplicity of the rate constants. As a good first approximation, the principle of equal reactivity is assumed to be valid even in the case of polymer radicals, which means that

$$k_{p1} = k_{p2} = k_{p3} = \cdots = k_{pn} = k_p \tag{52}$$

Termination of Polymer Radicals

The termination reaction is the one in which polymer chain radicals are destroyed. This can occur only when a polymer radical reacts with another polymer radical or with a primary radical. The former is called the mutual termintion and the latter primary termination. These reactions can be written as

$$P_m + P_n \xrightarrow{k_{t(m,n)}} M_{n+m} \tag{53a}$$

$$P_m + I \xrightarrow{k_{t,I(m)}} M_m \tag{53b}$$

where $m, n = 1,2,3,\ldots$. M_{n+m} signifies a dead polymer chain, i.e., it cannot undergo any further propagation reaction. In the case of mutual termination the inactive polymer chains can be formed either by combination or by disproportionation. In combination termination, two chain radicals just combine to give an inactive chain, whereas in disproportionation, one chain radical gives up an electron to the other and both chains become inactive. These two types of termination can be represented by

$$P_m + P_n \xrightarrow{k_{tc(m,n)}} M_{m+n} \quad \text{(combination)} \tag{54a}$$

$$P_m + P_n \xrightarrow{k_{td(m,n)}} M_m + M_n \quad \text{(disproportionation)} \tag{54b}$$

where $k_{tc(m,n)}$ and $k_{td(m,n)}$ are the rate constants for termination by combination and disproportionation, respectively. Once again, the principle of equal reactivity is assumed to hold: all the rate constants are independent of the chain lengths of the polymer radicals, i.e.,

$$k_{tc(m,n)} = k'_{tc} \tag{55a}$$

$$k_{td(m,n)} = k'_{td} \quad \text{for } m \neq n \tag{55b}$$

$$k_{tc(m,n)} = \frac{k'_{tc}}{2} \tag{55c}$$

$$k_{td(m,n)} = \frac{k'_{td}}{2} \quad \text{for all } m = n \tag{55d}$$

Various reactions occurring in radical polymerization are summarized in the Table 7 [1, 3]. The intermediate species present at the time in the reaction mass are I, P_1, P_2, P_3, etc., and the mole balance equations for each of these and the initiator molecule I_2 in a batch reactor can be written as

$$\frac{d[I_2]}{dt} = k_I[I_2] \tag{56a}$$

$$\frac{d[\mathrm{I}]}{dt} = 2fk_1[\mathrm{I}_2] - k_1[\mathrm{I}][\mathrm{M}] - k_{tp}[\mathrm{I}]\sum_{m=1}^{\infty}[\mathrm{P}_m] \tag{56b}$$

$$\frac{d[\mathrm{P}_1]}{dt} = k_1[\mathrm{I}][\mathrm{M}] - k_p[\mathrm{P}_1][\mathrm{M}] - k_{tp}[\mathrm{P}_1][\mathrm{I}]$$
$$-k_t'\{[\mathrm{P}_1]^2 + [\mathrm{P}_1][\mathrm{P}_2] + [\mathrm{P}_1][\mathrm{P}_3] + \cdots\} \tag{56c}$$

$$\frac{d[\mathrm{P}_n]}{dt} = k_p[\mathrm{M}][\mathrm{P}_{n-1}] - k_p[\mathrm{M}][\mathrm{P}_n] - k_{tp}[\mathrm{P}_n][\mathrm{I}]$$
$$-k_t'\{[\mathrm{P}_n][\mathrm{P}_1] + [\mathrm{P}_n][\mathrm{P}_2] + \cdots + [\mathrm{P}_n]^2 + \cdots\} \tag{56d}$$

where

$$k_t' = k_{tc}' + k_{td}' \tag{56e}$$

If [P] is defined as the total concentration of polymer radicals in the reaction mass, then

$$[\mathrm{P}] = [\mathrm{P}_1] + [\mathrm{P}_2] + \cdots = \sum_{n=1}^{\infty}[\mathrm{P}_n] \tag{57}$$

and Eqs. 56c, d, and e can be added to give the rate of production of [P] as

$$\frac{d[\mathrm{P}]}{dt} = \frac{d\sum_{n=1}^{\infty}[\mathrm{P}_n]}{dt} = k_1[\mathrm{I}][\mathrm{M}] - k_{tp}[\mathrm{I}][\mathrm{P}] - k_t'[\mathrm{P}]^2 \tag{58}$$

Table 7 Reactions Occurring in Radical Polymerization

Initiation:
$$\mathrm{I}_2 \xrightarrow{k_I} 2\mathrm{I}$$
$$\mathrm{I} + \mathrm{M} \xrightarrow{k_1} \mathrm{P}_1$$

Propagation:
$$\mathrm{P}_m + \mathrm{M} \xrightarrow{k_p} \mathrm{P}_{m+1} \qquad m = 1, 2, \ldots$$

Termination:
  Primary:
$$\mathrm{P}_m + \mathrm{I} \xrightarrow{k_{tp}} \mathrm{M}_m \qquad m = 1, 2, \ldots$$
  Combination:
$$\mathrm{P}_m + \mathrm{P}_n \xrightarrow{k_{tc}'} \mathrm{M}_{m+n} \qquad m \neq n, \; m,n = 1, 2, \ldots$$
$$\mathrm{P}_m + \mathrm{P}_m \xrightarrow{k_{tc}'/2} \mathrm{M}_{2m} \qquad m = 1, 2, \ldots$$

Disproportionation:
$$\mathrm{P}_m + \mathrm{P}_n \xrightarrow{k_{td}'} \mathrm{M}_m + \mathrm{M}_n \; m \neq n, \; m,n = 1, 2, \ldots$$
$$\mathrm{P}_m + \mathrm{P}_m \xrightarrow{k_{td}'/2} 2\mathrm{M}_m \qquad m = 1, 2, \ldots$$

The corresponding relation for the initiation steps can be written from Eqs. 56a and b as

$$\frac{d[I_2]}{dt} = -k_I[I_2] \tag{59}$$

$$\frac{d[I]}{dt} = 2fk_I[I_2] - k_1[I][M] - k_{tp}[I][P] \tag{59b}$$

The study of Eqs. 58 and 59 reveals that these involve only [P]. An identical set of equations can be derived from the following simplified mechanism of the radical polymerization.

Initiation:

$$I_2 \xrightarrow{k_I} 2I \tag{60a}$$

$$I_2 + M \xrightarrow{k_1} P \tag{60b}$$

Propagation:

$$P + M \xrightarrow{k_p} P \tag{60c}$$

Termination:

$$P + I \xrightarrow{k_{tp}} \text{dead chains} \tag{60d}$$

$$P + P \xrightarrow{k'_{tc}/2} \text{dead chains} \tag{60e}$$

$$P + P \xrightarrow{k'_{td}/2} \text{dead chains} \tag{60f}$$

The kinetic equivalence of the mechanisms given in Table 7 and Eqs. 60 is thus established. The replacement of a complex series of equations given in Table 7, where each molecular species in the reaction mass is distinguished by the far simpler Eqs. 60, is analogus to simplification of the complex reactions involving individual species in step growth polymerization by the reaction between reactive groups. Indeed, both these simplifications are a direct consequence of the equal reactivity hypothesis.

The mole balance equations can be simplified even beyond the simplification achieved in Eqs. 58 and 59. Since the number of monomer molecules in the reaction mass is generally much higher than the number of polymer radicals,

$$k_1[M] \gg k_{tp}[P] \tag{61}$$

Eq. 59b can be written as

$$\frac{d[I]}{dt} \cong 2fk_I[I_2] - k_1[I][M] \tag{62}$$

Also, in radical polymerization the slowest reaction is the dissociation of initiator molecules, and as soon as a primary radical is produced, it is consumed by reactions b and c of Eqs. 60. Thus the concentration of I is expected to be much less than that of P, i.e.,

$$[M] \gg [P] \gg [I] \tag{63}$$

and Eq. 58 can be rewritten as

$$\frac{d[P]}{dt} = k_1[I][M] - k_t'[P]^2 \tag{64}$$

Equations 62 and 64 imply that the primary termination step, Eq. 60d, can be neglected in the kinetic mechanism.

The rate of monomer consumption r_p for radical polymerization can now be derived from Eqs. 60b and c and approximated as

$$r_p \simeq k_p[P][M] \tag{65}$$

To find the total concentration of chain radicals, P, in the reaction mass, the steady-state approximation is used [78]. From Eq. 62,

$$\frac{d[I]}{dt} = 2fk_1[I_2] - k_1[M][I] = 0 \tag{66}$$

$$[I] = \frac{2fk_1[I_2]}{k_1[M]} \tag{67}$$

Similarly, from Eq. 64,

$$\frac{d[P]}{dt} = k_1[M][I] - k_t'[P]^2 = 0 \tag{68}$$

and from Eqs. 67 and 67,

$$[P] = \frac{2fk_1[I_2]^{1/2}}{k_t'} \tag{69}$$

The rate of propagation after some induction time is thus given by

$$r_p = k_p \left\{ \frac{2fk_1[I_2]^{1/2}}{k_t'} \right\} [M] \tag{70}$$

Average Molecular Weight in Radical Polymerization

Average molecular weight in radical polymerization can be found from the kinetic model, Eqs. 60, as follows. The kinetic chain length, ν, is defined as the average number of monomer molecules reacting with a polymer chain radical during the latter's entire lifetime. This is the ratio of the rate of consumption of the monomer, to the rate of generation of polymer radicals, r_i:

$$\nu = \frac{r_p}{r_i} \tag{71}$$

From the steady-state approximation, the rate of initiation, r_i, should be equal to the rate of termination, r_t. Therefore,

$$\nu = \frac{r_p}{r_t} = \frac{k_p[M][P]}{k_t'[P]} = \frac{k_p[M]}{k_t'[P]} \tag{72}$$

On eliminating [P] with the help of Eq. 69, ν is given by

$$\nu = \frac{k_p}{(2fk_1k_t')^{1/2}} \frac{[M]}{[I_2]^{1/2}} \tag{73}$$

Equation 73 shows that the kinetic chain length reduces with increasing initiator concentration. This is expected since an increase in $[I_2]$ would lead to more chains being produced.

The quantity that is of interest is the average chain length, μ_n. This is directly related to ν because μ_n gives the average number of monomer molecules per dead polymer chain whereas ν gives the average number of monomer molecules per growing polymer radical. To be able to find the exact relationships between the two, the mechanism of termination must be carefully analyzed. If the termination of polymer radicals occurs only by combination, then each of the dead chains would consist of 2ν monomer molecules. If termination occurs only by disproportionation, each of the inactive polymer molecules would consist of ν monomer molecules. If termination occurs by both these mechanisms, then

$$\mu_n = \alpha \nu \tag{74}$$

where α would be between 1 and 2.

As the initiator concentration is increased, the rate of polymerization goes up (Eq. 70), but ν, and therefore μ_n, goes down (Eq. 73). Therefore, control of the initiator concentration is one way of monitoring the molecular weight of the polymer.

Another method for controlling the molecular weight of the polymer is the use of a transfer agent. The polymer chain radicals react with the transfer agent and lose the capacity to add any further monomer molecules, the molecule of the transfer agent acquiring radical character in this process. The latter can grow by adding monomer molecules like any other growing chain in the reaction mass. The transfer reaction is represented schematically as

$$P_n + S \xrightarrow{k_{tr,S}} M_n + P_1, \quad n > 1 \tag{75}$$

where S is a molecule of the transfer agent.

Equation 75 should be added to the kinetic model if radical polymerization is being carried out with a transfer agent. Since the transfer reaction occurs with equal likelihood in the reaction mass, it is not necessary to distinguish chain radicals kinetically from each other once again, and Eq. 75 can be included in the following form in Eqs. 60:

$$P + S \xrightarrow{k_{tr,S}} M_d + P \tag{76}$$

Since this reaction does not reduce or increase the total number of chain radicals in the reaction mass, from Eq. 65 it follows that the presence of a transfer agent does not affect r_p. However, the kinetic chain length, ν, changes drastically depending on the value of $k_{tr,S}$ and [S]. Equation 72 can easily be modified to account for the presence of transfer agents:

$$\nu = \frac{r_p}{k_t'[P]^2 + k_{tr,S}[P][S]} \tag{77}$$

or on taking the reciprocal

$$\frac{1}{\nu} = \frac{k_{tr,S}[S]}{k_p[M]} + \frac{(2fk_1k_t')^{1/2}}{k_p} \frac{[I_2]^{1/2}}{[M]} \tag{78}$$

Equation 78 predicts a decrease in μ_n with increasing concentration of the transfer agent.

Chain transfer reactions occur quite commonly in radical polymerization with initiator as well as monomer as follows:

With initiator:

$$P_n + I_2 \xrightarrow{k_{tr,I_2}} M_n\text{---}I + I \tag{79}$$

With monomer:

$$P_n + M \xrightarrow{k_{tr,M}} M_n + P_1 \tag{80}$$

These reactions can be similarly incorporated into the kinetic model and the average chain lengths can be found to be

$$\frac{1}{\alpha\mu_n} = C_M + C_S \frac{[S]}{[M]} + \frac{k_t}{k_p^2} \frac{r_p}{[M]^2} + C_I \left(\frac{k_t'}{2fk_p^2 k_1}\right) \frac{r_p^2}{[M]^3} \tag{81}$$

where

$$C_M = \frac{k_{tr,M}}{k_p} \tag{82a}$$

$$C_I = \frac{k_{tr,I_2}}{k_p} \tag{82b}$$

$$C_S = \frac{k_{tr,S}}{k_p} \tag{82c}$$

Gel Effect in Radical Polymerization

The considerable increase in the rate of polymerization (as shown in Fig. 12 [75]) and the average chain length, μ_n (Figure 13 [80]) is a phenomenon common to all monomers

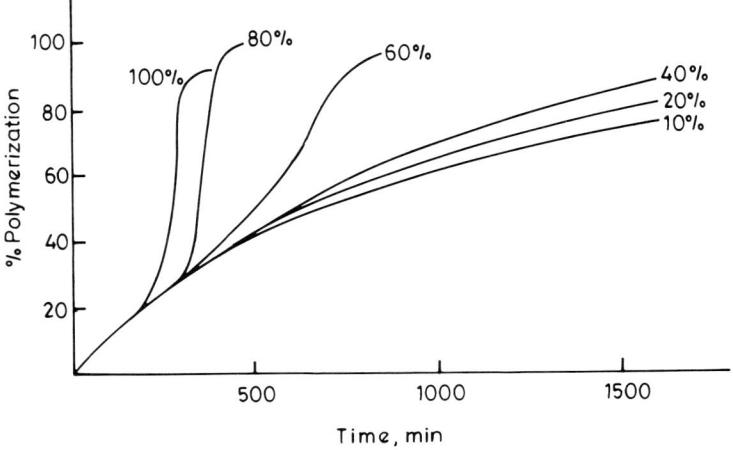

Figure 12 Polymerization of methyl methacrylate at 50°C with benzoyl peroxide initiator at various monomer concentrations (benzene as diluent).

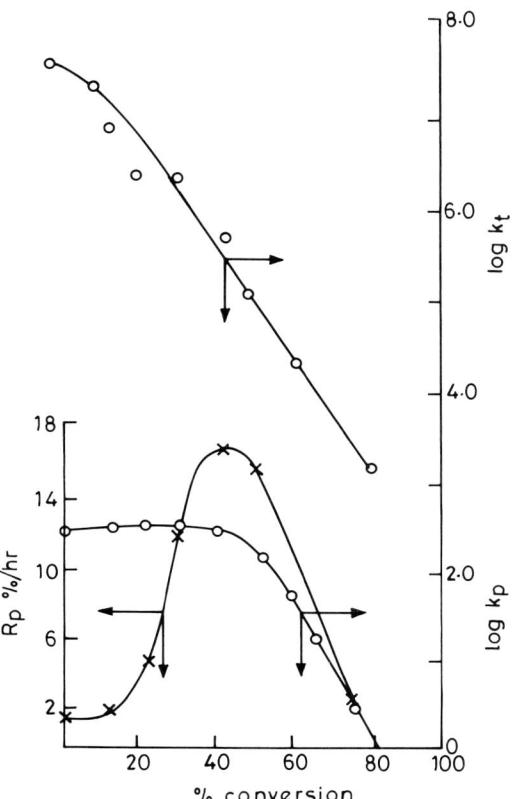

Figure 13 Average molecular weight as a function of conversion in the polymerization of methyl methacrylate at 60°C and 0.0045 mol % initiator. (Reprinted from Ref. 83, with permission of ACS, Washington.)

undergoing radical polymerization at low temperatures and is called the autoacceleration, or gel effect. This has been a subject of several studies [81–95] and has been attributed to the fall in values of the rate constants, k_p and k_t, as shown for methyl methacrylate in Figure 14. It is also seen that the rate constant k_t is affected first and falls in magnitude considerably faster than k_p. The large fall in k_t can be explained by observing that the termination step involves the diffusion of two polymer molecules from the bulk to each other's proximity. Once the proximate pair has been formed, the radical segments must undergo segmental diffusion before the termination reaction can occur. The fall in k_t occurs because the segmental diffusion becomes sluggish with the increase in the viscosity of the reaction mass. On the other hand, the propagation reaction involves the diffusion of a small molecule (i.e., monomer) from the bulk to the radical segment. There is little segmental diffusion involved in this step, thus leaving k_p unaffected up to considerably higher conversions.

During the course of free radical polymerization from the bulk monomer to complete or limiting conversion, the movement of polymer radicals toward each other goes through several regimes of changes. To demonstrate this, attention is focused on a solution consisting of dissolved, nonreacting polymer molecules. When the solution is very dilute,

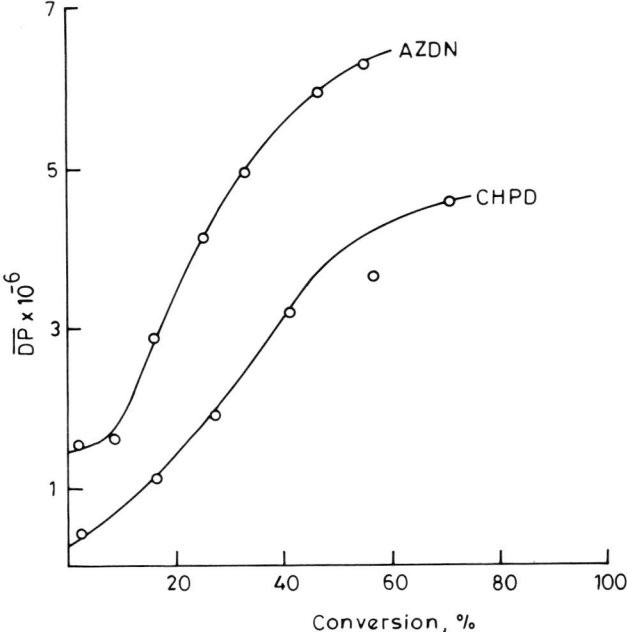

Figure 14 Bulk polymerization 25°C with AzDN initiator. The rate of initiation is 8.36×10^{-9} moles L sec^{-1}.

the polymer molecules exist in a highly coiled state and they behave like hydrodynamic spheres. In this regime, polymer molecules can undergo translational motion easily and the overall diffusion is completely governed by polymer–solvent interactions. As the polymer concentration is increased (say beyond a critical conversion C^*), the translational motion of molecule begins to be affected by the presence of other molecules. This effect, which was absent earlier, constitutes the second regime. On increasing the concentration of the polymer still further (say beyond C^{**}), in addition to the intermolecular interactions in translational motion, polymer chains begin to impose topological constraints upon the motion of surrounding molecules due to their long-chain nature. In other words, polymer molecules begin to be entangled. DeGennes modeled the motion of polymer chains in this regime through a "tube" defined by the points of entanglement. A polymer molecule can move through this only by a snakelike wriggling motion along its length; this mode of motion is sometimes called reptation. Finally, at very high concentrations (say beyond C^{***}), polymer chains begin to exert direct friction upon each other. The values of C^*, C^{**}, and C^{***} have been experimentally shown to depend on the molecular weight of the polymer.

To demonstrate the correspondence between the polymer solvent system described above and free radical polymerization, Turner [90] and Driscoll's group [82, 91–93] demonstrated that gelation starts at the polymer concentration of C^{**}. In fact, it has been shown that k_t changes continuously as the polymerization progresses [89, 95], first increasing slightly but subsequently reducing drastically at higher conversions. Tulig and Tirrell [89, 95] have argued, based on their experimental and theoretical analysis, that similar regimes for k_t must exist in the entire range of conversion.

Based upon the foregoing physical picture of the gel effect, a model was proposed by Cardenas and O'Driscoll [82, 91, 92] in which two populations of radicals are assumed to exist in the reaction mass. The first are those which are physically entangled (denoted by P_{ne}) and therefore have lower termination rate constant (k_{te}), compared to that (k_t) of the second population (denoted by P_n), which are untangled. Whenever a polymer radical grows in chain length beyond a critical value n_c, it is assumed that it becomes entangled and its termination rate constant falls from k_t to k_{te}. If it is assumed that the propagation rate constant, k_p, is not affected, the overall mechanism of radical polymerization can then by represented as follows:

Initiation:

$$I_2 \xrightarrow{fk_1} 2I \qquad (83a)$$

$$I + M \xrightarrow{k_1} P_1 \qquad (83b)$$

Propagation:

$$P_n + M \xrightarrow{k_p} P_{n+1} \qquad (83c)$$

$$P_{ne} + M \xrightarrow{k_p} P_{(n+1)e} \qquad (83d)$$

Termination:

$$P_m + P_n \xrightarrow{k_t} M_m + M_n \qquad (83e)$$

$$P_m + P_{ne} \xrightarrow{k_{tc}} M_m + M_n \qquad (83f)$$

$$P_{me} + P_{ne} \xrightarrow{k_{te}} M_m + M_n \qquad (83g)$$

In the above scheme, the reaction between the entangled and the unentangled radicals has been assumed to occur with rate constant k_{tc} lying between k_t and k_{te}. The mole balance equations for the various radicals can be easily written from the kinetic mechanism given above, and for batch reactors

$$\frac{d[P_1]}{dt} = 2fk_1[I_2] - \{k_p[M] + 2k_t[P] + 2k_{tc}[P_e]\}[P_1] \qquad (84)$$

$$\frac{d[P_n]}{dt} = [M][P_{n-1}] - \{k_p[M] + 2k_t[P] + 2k_{tc}[P_e]\}[P_n], \qquad 2 \leq n \leq n_c \qquad (85)$$

$$\frac{d[P_{ne}]}{dt} = k_p[M][P_{(n-1)e}] - \{k_p[M] + 2k_{tc}[P] + 2k_{te}[P_e]\}[P_{ne}], \qquad n_e > n_c \qquad (86)$$

where

$$[P] = \sum_{n=1}^{n_c} [P_n] \qquad (87a)$$

$$[P_e] = \sum_{n=n_c+1}^{\infty} [P_{ne}] \qquad (87b)$$

Since the total reaction time is usually much larger than the individual radical lifetime, the steady-state approximation can be assumed to valid here. On summing Eqs. 84 and 85 for values of n up to n_c, one gets

$$2fk_1[I_2] + k_p[M] \sum_{n=2}^{\infty} [P_{n-1}]$$

$$- \{k_p[M] + 2k_t[P] + 2k_{tc}[P_e]\} \sum_{n=1}^{\infty} [P_n] = \frac{d[P]}{dt} = 0 \tag{88}$$

or

$$2fk_1[I_2] - \{2k_t[P] + 2k_{tc}[P_e]\}[P] = k_p[M][P_{nc}] \tag{89}$$

It is assumed that k_{tc} is the geometric mean of k_t and k_{te}, i.e.,

$$k_{tc} = (k_t k_{te})^{1/2} = \delta k_t \tag{90}$$

where

$$\delta = \left(\frac{k_{te}}{k_t}\right)^{1/2} \tag{91}$$

With the help of Eq. 90, Eq. 89 is simplified to

$$2fk_1[I_2] - 2k_t\{[P] + \delta[P_e]\}[P] = k_p[M][P_{nc}] \tag{92}$$

Similarly, Eq. 85 is added for all values of n above n_c assuming the steady-state approximation is valid. On doing this, the following relation is obtained:

$$k_p[M] \sum_{n=n_c+1}^{\infty} [P_{(n-1)e}] - \{k_p[M] + 2k_{tc}[P] + 2k_{te}[P_e]\} \sum_{n=n_c+1}^{\infty} [P_{ne}]$$

$$= \frac{d[P_e]}{dt} = 0 \tag{93}$$

which simplifies to

$$2\delta k_t\{[P] + \delta[P_e]\} = k_p[M][P_{nc}] \tag{94}$$

Equating the left-hand sides of Eqs. 92 and 94,

$$[P] + \delta[P_e] = \left\{\frac{fk_1[I_2]}{k_t}\right\}^{1/2} \tag{95}$$

It is now necessary to define the probability of propagation to find the distribution of P_n. After a polymer chain radical is formed by the reaction between a primary radical and a monomer molecule, it can either propagate by reaction 83c and d or be terminated by transfer or mutual termination reactions 83e, f, and g. Therefore, the probability of propagation, β, for chain length less than n_c is found as

$$\beta = \frac{k_p[M]}{k_p[M] + 2k_t[P] + 2k_{tc}[P_e]} = \frac{1}{1 + \{k_t fk_1[I_2]/k_p^2[M]^2\}^{1/2}} \tag{96}$$

where the transfer reactions have been neglected. Under the steady-state approximation, $d[P_n]/dt$ is set equal to zero in Eq. 85. It is recognized that P_n is formed only when there is a propagation reaction with P_{n-1}, or

$$[P_n] = \beta[P_{n-1}] \quad \text{for } n \le n_c \tag{97}$$

Similarly, from the steady-state approximation, $d[P_1]/dt$ is equal to zero and Eq. 84 gives

$$[P_1] = \frac{2fk_1[I_2]}{k_p[M]} \tag{98}$$

With the help of Eqs. 97 and 98, one finds

$$[P_n] = \beta^n \frac{2fk_1[I_2]}{k_p[M]} \tag{99}$$

It is observed that β is a quantity close to unity, which implies that $\ln \beta$ can be approximated by $(\beta - 1)/\beta$ and the term β^n can be rewritten as

$$\beta^n = \exp(n \ln \beta) = \exp\left\{-n \frac{(-\beta + 1)}{\beta}\right\} = \exp\left(-\frac{n}{\nu}\right) \tag{100}$$

where

$$\frac{1}{\nu} = 2\left\{\frac{k_t fk_1[I_2]}{k_p^2[M]^2}\right\}^{1/2} \tag{101}$$

and is the same as the kinetic chain length defined in Eq. 73. $[P_{n_c}]$ can now be calculated from Eq. 99 and substituted in Eqs. 94 and 95 to obtain

$$[P] = \left\{\frac{fk_1[I_2]}{k_t}\right\}^{1/2}\left[1 - \exp\left(-\frac{n_c}{\nu}\right)\right] \tag{102a}$$

and

$$[P_e] = \left\{\frac{fk_1[I_2]}{k_t}\right\}^{1/2} \exp\left(-\frac{n_c}{\nu}\right) \tag{102b}$$

The rate of polymerization, r_p, can now be determined as

$$r_p = k_p[M]\{[P] + [P_e]\} = k_p[M]\left\{\frac{fk_1[I_2]}{k_t}\right\}^{1/2}\left[1 + \frac{1-\delta}{\delta}\exp\left(-\frac{n_c}{\nu}\right)\right] \tag{103}$$

The monomer conversion, x, as radical polymerization is defined in terms of monomer concentration initially present in the reaction mass, $[M]_0$, and that at any reaction residence time, $[M]$. This is given by

$$x = \frac{[M]_0 - [M]}{[M]_0} \tag{104}$$

Equation 103 can now be written in terms of the monomer conversion, x, as

$$r_p = \frac{dx}{dt} = k_p(1-x)\left\{\frac{fk_1[I_2]}{k_t}\right\}^{1/2}\left[1 + \frac{1-\delta}{\delta}\exp\left(-\frac{n_c}{\nu}\right)\right] \tag{105}$$

To solve for the rate of polymerization at a given conversion, x, it is necessary to determine δ and n_c. From rheological studies, the relation between critical number average chain length μ_{nc} and the volume fraction ϕ_P of polymer present in solution is given by

$$K_c = \mu_{nc}^\gamma \phi_P \tag{106}$$

K_c is a constant which is (almost) independent of temperature but has different value for different polymers. In this equation γ is a constant that lies between 0.5 and 1.0, and it is found that the computed rate curves are relatively insensitive to the value of γ chosen. It is postulated that k_{te} is inversely proportional to the entanglement density d_e or

$$k_{te} = \frac{b}{d_e} \tag{107}$$

where b is a constant of proportionality and d_e is given by

$$d_e = a\left(\frac{\phi_P \mu_n}{K_c}\right) = d_e \theta \left(\frac{K_c}{\phi_P \mu_n^\gamma}\right)^{1/2} \tag{108}$$

Equations 106–108 can be combined to give

$$de_0 = \left(\frac{b}{ak_t}\right)^{1/2} \tag{109}$$

To be able to solve average molecular weights, the mode of termination must be known. If it is assumed that termination occurs largely by the disproportionation mechanism,

$$\frac{d[M_n]}{dt} = \{2k_t[P] + 2k_{tc}[P_e]\}[P_n], \qquad n \leq n_c \tag{110a}$$

$$\frac{d[M_n]}{dt} = \{2k_{tc}[P] + 2k_{te}[P_e]\}[P_{ne}], \qquad n > n_c \tag{110b}$$

From Eqs. 110, it is possible to derive the zeroth (λ_0), first (λ_1), and second (λ_2) moments of the inactive polymer chains as follows. For example, λ_2 is given by

$$\frac{d\lambda_2}{dt} = \sum_{n=1}^{n_c} n^2[P_n]\{2k_t[P] + 2k_{tc}[P_e]\} + \sum_{n=n_c+1}^{\infty} n^2[P_{ne}]\{2k_{tc}[P] + 2k_{te}[P_e]\} \tag{111}$$

The moments of inactive polymers are thus related to the moments of the radicals, which can now be determined by Eqs. 84–86. Assuming the steady-state approximation, it is possible to derive from these relations the following:

$$\sum_{n=1}^{n_c} n^2[P_n] = \frac{2fk_1[I_2] + k_p[M]\{2\sum_{n=1}^{n_c-1} n[P_n] + \sum_{n=1}^{n_c-1}[P_n] - n_c^2[P_{nc}]\}}{2k_t[P] + 2k_{tc}[P_e]} \tag{112a}$$

and

$$\sum_{n=n_c+1}^{\infty} n^2[P_{ne}] = \frac{k_p[M]\{2\sum_{n=n_c}^{\infty} n[P_{ne}] + \sum_{n=n_c}^{\infty}[P_{ne}] + n_c^2[P_{nc}]\}}{2k_{tc}[P] + 2k_{te}[P_e]} \tag{112b}$$

Substituting these in Eq. 107, one obtains

$$\frac{d\lambda_2}{dt} = k_p[M]\{[P] + [P_e]\} + 2k_p[M]\sum_{n=1}^{\infty} n[P_n] \tag{113}$$

Similarly, it can be found that

$$\frac{d\lambda_0}{dt} = 2fk_1[I_2] \tag{114}$$

and

$$\frac{d\lambda_1}{dt} = k_p[M]\{[P] + [P_e]\} \qquad (115)$$

Driscoll and coworkers [82] have treated K_c and $d_{e\theta}$ in Eqs. 106 and 108 as parameters and have solved these equations simultaneously. From the moments calculated by Eqs. 113–115, the number and weight average molecular weights can be found. Using Eqs. 107 and 108, k_{te} and γ are computed. On substituting these in Eq. 103, the next incremental conversion is obtained by integration. For the set of parameters for methyl methacrylate polymerization given in Table 8, these equations have been integrated and compared with experimental values of conversion and molecular weights in Figures 15 and 16. The theory is found to describe the experimental data very well.

Table 8 Parameters for the Radical Polymerization of Methyl Methacrylate

Parameter	Temperature (°C)	
	70	90
$k_p (fk_1/k_t)^{1/2}$	7.5×10^{-4}	3.8×10^{-3}
k_t/k_p^2 (mol-sec L^{-1})	28	16
k_1 (sec^{-1})	3.5×10^{-5}	5.0×10^{-4}
K_c	8.97	5.69
α_p	1.11×10^{-1}	1.67×10^{-1}
$1/f$	0.1596	0.1731

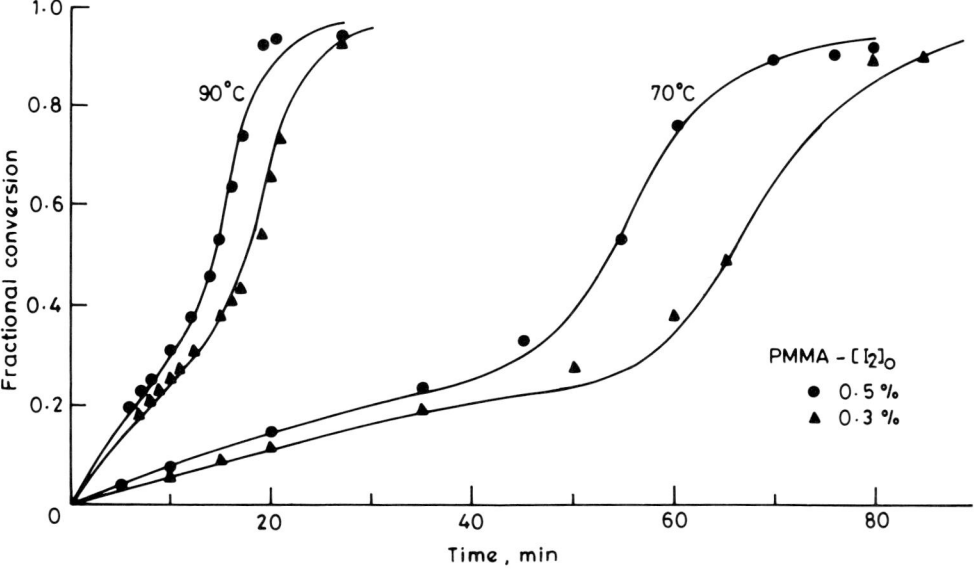

Figure 15 Experimental and predicted values of conversion versus time of polymerization for methyl methacrylate $[I_2]_0$ for (●) 0.5% and (▲) 0.3%. (Reprinted from Ref. 82, with permission of John Wiley and Sons, New York.)

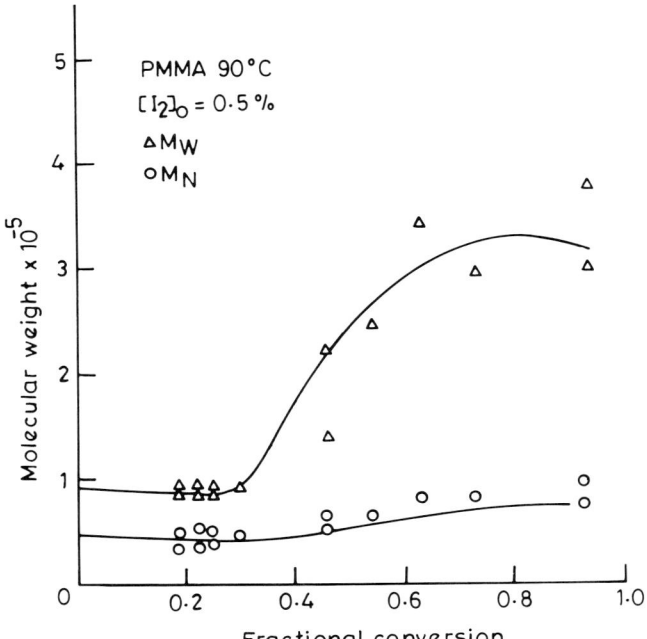

Figure 16 Experimental and predicted values of molecular weight versus fractional conversion for methyl methacrylate with initiator concentration 0.5%. (Reprinted from Ref. 82, with permission of John Wiley and Sons, New York.)

Temperature Effects in Radical Polymerization

In the initial stages (i.e., before the gel effect sets in) the temperature dependence of various rate constants can be expressed through the Arrhenius law:

$$k_I = k_{I0} e^{-E_I/RT}$$
$$k_p = k_{p0} e^{-E_p/RT}$$
$$k_t = k_{t0} e^{-E_t/RT}$$

In this representation, E_I, E_p, and E_t are activation energies of the initiation, propagation, and termination steps respectively and are tabulated extensively in polymer handbooks [76]. The temperature dependence of r_p and μ_n can be derived as

$$r_p = \frac{k_{p0} k_{I0}^{1/2}}{k_{t0}^{1/2}} \exp\left(-\frac{E_p - E_t/2 + E_I/2}{RT}\right) \{f[I_2]^{1/2}[M]\}$$

$$\mu_n = \frac{k_{p0}}{2k_{I0}^{1/2} f^{1/2} k_{t0}^{1/2}} \frac{[M]}{[I_2]^{1/2}} \exp\left(-\frac{E_p - E_I/2 - E_t/2}{RT}\right)$$

The activation energies are such that the overall polymerization for thermally dissociating initiator is exothermic, i.e., $(E_p - E_t/2 + E_p/2)$ is positive and rate increases with temperature. As opposed to this, $\{E_p - E_I/2 - E_t/2\}$ is normally negative for such cases and μ_n decreases with increasing temperature.

After the gel point sets in, the temperature dependence of k_p and k_t can be represented by

$$k_p = k_{p0} \exp\left\{-E\left(\frac{1}{V_F} - \frac{1}{V_{Fc1}}\right)\right\}$$

$$k_t = k_{t0} \left(\frac{\mu_{wc}}{\mu_w}\right)^{1.75} \exp\left\{-A\left(\frac{1}{V_F} - \frac{1}{V_{Fc2}}\right)\right\}$$

where A and B are constants, whereas V_F is free volume fraction defined as

$$V_F = \{0.025 + \alpha_P(T - T_{gP})\}\frac{V_P}{V_T} + \{0.025 + \alpha_M(T - T_{gM})\}\frac{V_M}{V_T}$$

$$+ \{0.025 + \alpha_S(T - T_{gS})\}\frac{V_S}{V_T}$$

In these equations, subscripts P, M, and S denote polymer, monomer, and solvent, respectively. T is the polymerization temperature, T_g is the glass transition temperature, and α_g is the thermal expansion coefficient for the glassy state. V_{Fc2} is determined by the point where the gel point starts and is given by

$$K_c = \mu_\omega^{0.5} \exp\left(\frac{A}{V_{Fc2}}\right)$$

V_{Fc1} is determined by the point where the propagation rate constant becomes diffusion controlled. V_T is the specific volume of the reaction mass.

REVERSIBLE RADICAL POLYMERIZATION

All polymerization reactions are reversible and have an equilibrium for a given temperature at which point the Gibbs free energy is zero. In the preceding section, various reaction steps were assumed to be irreversible in radical polymerization, which would serve as good approximation only for low temperatures. Let us first consider the thermodynamic equilibrium for addition polymerization before presenting the kinetic model made for reversible addition polymerization.

It is assumed that the formation of the polymer can be represented by

$$nM \rightleftharpoons M_n \tag{116}$$

which states that n monomer molecules combine to form a polymer molecule of chain length n. The change in the Gibbs free energy ΔG for this process is defined by

$$\Delta G = \frac{1}{n} G_{\text{polymer}} - G_{\text{monomer}}$$

$$= \left(\frac{1}{n} H_{\text{polymer}} - H_{\text{monomer}}\right) - T\left(\frac{1}{n} S_{\text{polymer}} - S_{\text{monomer}}\right)$$

$$\triangleq \Delta H_p - T\Delta S_p \tag{117}$$

where H and S represent enthalpy and entropy and ΔH_p and ΔS_p are the changes in enthalpy and entropy of polymerization per monomer unit, respectively.

Almost all addition polymerizations are exothermic. Moreover, polymerization is the process of joining monomer molecules by covalent bonds, something equivalent to the stringing of beads into a necklace, the final state being more ordered and consequently having a lower entropy. Thus ΔS_p is always negative and is normally a large negative number, which cannot be neglected. Equation 117 therefore suggests that there is a ceiling temperature T_c at which the addition polymerization is at equilibrium (or $\Delta G = 0$), i.e.,

$$T_c = \frac{\Delta H_p}{\Delta S_p} \tag{118}$$

For a given monomer concentration [M], ΔS in this equation is equal to

$$\Delta S_p = \Delta S_p^0 + R \ln [M] \tag{119}$$

where ΔS_p^0 is the entropy change when the polymerization is carried out at standard state and is independent of [M]. The standard state of a liquid monomer is defined to be that at which monomer concentration is one molar at the temperature and pressure of the polymerization. If $[M]_e$ is the monomer concentration at equilibrium, then

$$\ln [M]_e = \frac{\Delta H_p^0}{RT} - \frac{\Delta S_p^0}{R} \tag{120}$$

where ΔH_p is the same as ΔH_p^0 by definition. As an example, for styrene being polymerized at 60°C, ΔH_p^0 is 16,700 cal mol^{-1} and ΔS^0 25 cal mol-K^{-1}. $[M]_e$ for styrene at 60°C is given by

$$[M]_e = \exp\left\{-\frac{16,700}{1.987(273 + 60)} + \frac{25}{1.987}\right\} = 3.7 \times 10^{-6} \text{ mol L}^{-1} \tag{121}$$

It is thus seen that at 60°C, polymerization would go to complete conversion. Polystyrene has a ceiling temperature of 670 K and as the polymerization temperature approaches this value, various reaction steps would become reversible and the analysis of the preceding section must be modified as follows.

In the irreversible mechanism of radical polymerization given in Table 7, the propagation step has considerably lower activation energies. Consequently, this step would become reversible first and the polymerization then can be represented as follows.

Initiation:

$$I_2 \xrightarrow{k_I} 2I \tag{122a}$$

$$M + I \underset{k_1'}{\overset{k_1}{\rightleftarrows}} P_1 \tag{122b}$$

Propagation:

$$P_n + M \underset{k_p'}{\overset{k_p}{\rightleftarrows}} P_{n+1} \tag{123}$$

Termination:

$$P_n + P_m \xrightarrow{k_{td}} M_n + M_m, \quad n \geq 1 \tag{124a}$$

$$P_n + P_m \xrightarrow{k_{tc}} M_{n+m} \quad n \geq 1 \tag{124b}$$

$$P_n + S \xrightarrow{k_{tr,S}} P_1 + M_n \qquad n \geq 2 \qquad (124c)$$

$$P_n + M \xrightarrow{k_{tr,M}} P_1 + M_n \qquad n \geq 2 \qquad (124d)$$

Reactions in Eq. 124c and d are transfer reactions, which can occur to a monomer molecule or any transfer agent that may be present in the reaction mass. If the steady-state approximation is assumed to hold, the mole balance for P_n is given by

$$\frac{d[P_1]}{dt} = k_1[M][I] - k_1[P_1] - k_p[M][P_1] + k_p'[P_2] + k_{tr,M}[M] \sum_{n=2}^{\infty} [P_n]$$

$$+ k_{tr,S}[S] \sum_{n=2}^{\infty} [P_n] - (k_{tc} + k_{td})[P_1][P] = 0$$

$$\frac{d[P_2]}{dt} = k_p[M]\{[P_1] - [P_2]\} + k_p'\{[P_3] - [P_2]\}$$

$$- \{k_t[P] + k_{tr,M}[M] + k_{tr,S}[S]\}[P_2] = 0$$

$$\vdots \qquad \vdots \qquad \vdots$$

$$\frac{d[P_n]}{dt} = k_p[M]\{[P_{n-1}] - [P_n]\} + k_p'\{[P_{n+1}] - [P_n]\}$$

$$- \{k_t[P] + k_{tr,M}[M] + k_{tr,S}[S]\}[P_n] = 0$$

$$\vdots \qquad \vdots \qquad \vdots \qquad (125)$$

where

$$k_t = k_{tc} + k_{td} \qquad (126a)$$

and

$$[P] = \sum_{i=1}^{\infty} [P_i] \qquad (126b)$$

On adding all relations from P_2 on, one gets

$$\frac{d\sum_{n=2}^{\infty}[P_n]}{dt} = 0 = k_p[M][P_1] - k_p'[P_2]$$

$$- \{k_t[P] + k_{tr,M}[M] + k_{tr,S}[S]\} \sum_{n=2}^{\infty} [P_n] \qquad (127)$$

or

$$\sum_{n=2}^{\infty} [P_n] = \frac{k_p[M]}{k_t[P] + k_{tr,M}[M] + k_{tr,S}[S]} [P_1]$$

$$- \frac{k_p'}{k_t[P] + k_{tr,M}[M] + k_{tr,S}[S]} [P_2] \qquad (128)$$

Substituting this in the balance for P_1, one has

$$\frac{d[P_1]}{dt} = k_1[M][I] - k_1'[P_1] + k_{tr,S}[S][P] + k_{tr,M}[M][P] + k_p'[P_2] \qquad (129)$$

Balance for [I] combined with the steady-state hypothesis yields

$$\frac{d[I]}{dt} = 0 - 2fk_I[I_2] + k_1'[P_1] - k_1[M][I] = 0 \qquad (130)$$

On adding all balance relations, one obtains [P] as

$$\frac{d[P]}{dt} = 0 = k_1[M][I] - k_1[P_1] - k_t[P]^2 \qquad (131)$$

If we substitute

$$\alpha = k_p u[M] \qquad (132a)$$
$$\beta = k_p' \qquad (132b)$$
$$\gamma = k_p[M] + k_p' + k_{tr,M}[M] + k_{tr,S}[S] + k_t[P] \qquad (132c)$$
$$\theta = {}_{tr,M}[M] + k_{tr,S}[S] + k_t[P] \qquad (132d)$$

Eq. 125 reduces to the set of algebraic equations given in Eq. A1 in the appendix. Then in terms of P_1, P_n can be determined as

$$[P_n] = [P_1](\alpha x_\infty)^{n-1} \qquad (133)$$

But

$$[P] = \sum_{n=1}^{\infty} [P_n] = [P_1]\{1 + \alpha x_\infty + (\alpha x_\infty) + \cdots\} = \frac{[P_1]}{(1 - \alpha x_\infty)}$$

or

$$[P_1] = (1 - \alpha x_\infty)[P] \qquad (134)$$

Equation 133 therefore reduces to

$$[P_n] = [P](1 - \alpha x_\infty)(\alpha x_\infty)^{n-1} \qquad (135)$$

The mole balance of inactive polymer M_n is given by

$$\frac{d[M_n]}{dt} = \{k_{tr,M}[M] + k_{tr,S}[S] + k_{td}[P]\}[P_n] + \frac{k_{tc}}{2}\sum_{r=1}^{n-1}[P_r][P_{n-r}] \qquad (136)$$

One defines the following kth moment of M_n distributions as

$$\lambda_k = \sum_{1}^{\infty} n^k[M_n] \qquad (137)$$

and the moment generation relations for λ_0, λ_1, and λ_2 can be easily derived as

$$\frac{d\lambda_0}{dt} = \{k_{tr,S}[S] + k_{tr,M}[M] + k_{td}[P]\}(1 - \alpha x_\infty)[P]\frac{\alpha x_\infty}{1 - \alpha x_\infty} + \frac{1}{2}k_{tc}[P]^2 \qquad (138a)$$

$$\frac{d\lambda_1}{dt} = \{k_{tr,S}[S] + k_{tr,M}[M] + k_{td}[P]\}[P_1] \frac{(\alpha x_\infty)(2 - \alpha x_\infty)}{(1 - \alpha x_\infty)^2} + \frac{k_{tc}[P_1]^2}{(1 - \alpha x_\infty)^3} \quad (138b)$$

$$\frac{d\lambda_2}{dt} = \{k_{tr,S}[S] + k_{tr,M}[M] + k_{td}[P]\}[P_1] \frac{4(\alpha x_\infty) - 3(\alpha x_\infty)^2 + (\alpha x_\infty)^3}{(1 - \alpha x_\infty)^3}$$

$$+ k_{tc}[P_1]^2 \frac{2 + \alpha x_\infty}{(1 - \alpha x_\infty)^4} \quad (138c)$$

Equations 138 reveal that the moments of the polymer formed by reversible radical polymerization are a function of a term (αx_∞) in which x_∞ involves the reverse rate constant. Irreversible polymerization has been completely analyzed in the literature. On comparison with Eqs. 138, it is found that irreversible polymerization has a form identical to that seen in these equations if (αx_∞) is replaced by probability of propagation, α^*, defined as

$$\alpha^* = \frac{k_p[M]}{k_p[M] + k_{tr,S}[S] + k_{tr,M}[M] + k_t[P]} \quad (139)$$

It can be shown that as $k_p' \to 0$, (αx_∞) reduces to α^* and in this regard Eqs. 138 give the most general form of the result.

CONCLUSIONS

Polymerization can in general be classified as step growth and chain growth type. In this chapter it is observed that polymer formation, like any other reaction in nature, is reversible. In any reasonable modeling, this must be taken into account.

Step growth polymerization of bifunctional monomers leads to the formation of linear chains, whereas for that of monomers having functionality more than 2, the resultant polymer is either branched or network. The reversible step growth polymerization of bifunctional monomers has been modeled and it is shown that under a certain transformation, the MWD equations can be decoupled and are considerably easier to solve numerically. The reversible polymerization of multifunctional monomers can be modeled only after the reaction of branched molecules is written. In order to do that, one needs to know the number of sites on a P_r molecule whcih would give P_n ($n < r$) on cleavage. This has been denoted by $\delta_{r \to n}$ and it is shown that the MWD equations for reversible multifunctional polymerization involves this factor. $\delta_{r \to n}$ is dependent on the chain structure, and there is a computer program that gives the value for specified chain length and structure. Numerical solution of the MWD shows that the polydispersity index of the polymer formed is higher than that of the polymer formed through irreversible mechanism.

Chain growth polymerization can occur through an ionic as well as a radical mechanism, and the discussion is focused on the latter in view of its industrial importance. It is shown that radical polymerization involves three steps: initiation, propagation, and termination. If the polymerization is carried out at low temperatures, the polymerization is essentially irreversible and can be roughly divided into initial and advanced stages which are separated by a gel point. Beyond the gel point the termination rate constants become diffusion controlled and there is a considerable drop in magnitude, which gives a very high rate of polymerization with a corresponding increase in the average chain length of the polymer. The overall polymerization at low temperatures is not limited by the equilibrium but by the occurrence of a glass state of the reaction mass. To control molecular weight in radical polymerization, one is forced to use high temperatures to suppress the formation of the undesirable glassy state, but some of the reaction steps then become reversible. A kinetic scheme for radical polymerization has been discussed and analyzed for the average chain length and the polydispersity index of polymerization.

APPENDIX: ANALYTICAL SOLUTION OF ALGEBRAIC EQUATIONS INVOLVED IN ADDITION POLYMERIZATION

Let us consider the following infinite set of algebraic equations which must be solved for p_n for all n:

$$\frac{p_2}{\theta} = \frac{p_{20}}{\theta} + \alpha(p_1 - p_2) + \beta(p_3 - p_2)$$

$$\frac{p_3}{\theta} = \frac{p_{30}}{\theta} + \alpha(p_2 - p_3) + \beta(p_4 - p_3)$$

$$\vdots$$

$$\frac{p_n}{\theta} = \frac{p_{n0}}{\theta} + (P_{n-1} - P_n) + (p_{n+1} - p_n) \tag{A1}$$

where p_{i0} (for $i = 1,2,3, \ldots$), α, θ, and β are constants.

In order to solve for p_n for all n, it is assumed that p_{N_1+1} is known precisely, where N_1 is any integer greater than 2. Equation A1 yields p_{N_1} as

$$p_{N_1} = x_{N_1} P_{N_1-1} + y_{N_1} P_{N_1} + \frac{1}{\theta} Z_{N_1 N_1} P_{N_{10}} \tag{A2}$$

where

$$x_{N_1} = \frac{1}{\gamma} \tag{A3a}$$

$$y_{N_1} = \frac{1}{\gamma} \tag{A3b}$$

$$Z_{N_1 N_1} = \frac{1}{\gamma} \tag{A3c}$$

$$\gamma = \left(\alpha + \beta + \frac{1}{\theta}\right) \tag{A3d}$$

The expression for p_{N_1-1} involves p_{N_1}, which can be eliminated using Eq. A2. This way p_{N_1-1} can be written in terms of p_{N_1-2} and p_{N_1+1}. Proceeding in this way, it is possible to derive for any n ($< N_1$)

$$p_n = \alpha x_n p_{n-1} + \beta y_n p_{N_1+1} + \frac{1}{\theta} \sum_{i=n}^{N_1} Z_{in} p_{i0} \tag{A4}$$

where

$$x_n = \frac{1}{\gamma - \alpha \beta x_{n+1}} \tag{A5a}$$

$$y_n = \frac{\beta y_{n+1}}{\gamma - \alpha\beta x_{n+1}} = \frac{1}{\beta} \prod_{i=n}^{N_1} (\beta x_i), \qquad 2 \leq n \leq N \tag{A5b}$$

$$Z_{in} = \begin{cases} x_n & \text{for } i = n \\ \dfrac{\gamma}{\beta} y_{N_1+n-i} & \text{for } n + 1 \leq i \leq N_1 - 1 \\ y_n & \text{for } i = N_1 \end{cases} \tag{A5c}$$

Proceeding this way, it is possible to determine p_1 in terms of p_{N_1+1}.

We next examine the properties of the solution as N_1 approaches ∞. The value of x_n in Eq. A5a can be written in the form of an infinite successive fraction as

$$\lim_{\substack{N_1 \to \infty \\ n \ll N_1}} x_n \triangleq x_\infty = \cfrac{1}{\gamma - \cfrac{\alpha\beta}{\gamma - \cfrac{\alpha\beta}{\gamma - \cfrac{\alpha\beta}{\gamma - \cdots}}}} \tag{A6}$$

As $N_1 \to \infty$, x_n for $n \ll N_1$ approaches an asymptotic value independent of n, and Eq. A6 gives

$$x_\infty = \cfrac{1}{\gamma - \cfrac{1}{\gamma_1 - x_\infty}} \tag{A7}$$

where

$$\gamma_1 = \frac{\gamma}{\alpha\beta} \tag{A8}$$

Equation A7 can be solved for x_∞ as

$$x_\infty = \frac{\gamma - \sqrt{\gamma^2 - 4\alpha\beta}}{2\alpha\beta} \tag{A9}$$

where the lower root serves as a physically relevant quantity.

To study the properties of the infinite continued fraction formed by y_n in Eq. A5, it is first recognized that no matter what the values of α, β, and θ, βx_∞ is always less than unity:

$$\begin{aligned}
\beta x_\infty &= \beta \frac{\gamma}{2\alpha\beta}\left\{1 - \sqrt{1 - \frac{2\alpha}{\gamma}\frac{2\beta}{\gamma}}\right\} \\
&= \left(\frac{\gamma}{2\alpha}\right)\left\{\frac{1}{2}\frac{2\alpha}{\gamma}\frac{2\beta}{\gamma} + 0\left(\frac{\alpha^2}{\gamma_2}\frac{\beta^2}{\gamma_2}\right)\right\} = \frac{\beta}{\gamma} + 0\left(\frac{\alpha^2}{\gamma^2}\frac{\beta^2}{\gamma^2}\right)
\end{aligned} \tag{A10}$$

i.e., since α/γ and β/γ both are less than unity, (βx_∞) is always less than unity. If it is recalled from Eq. A5 that $y_n = y_{n+1}(\beta x_\infty)$ and if we denote by y_n^* the value of n at which $x_n = x_\infty$ is a good approximation, then y_{n-1}, y_{n-2}, etc., can be written as

$$y_{n-i} = y_n^*(\beta x_\infty)^i, \qquad i = 1, 2, \ldots, n - 1 \tag{A11}$$

Since $(\beta x_\infty) < 1$, y_n can be made sufficiently close to zero by choosing N_1 sufficiently large. Furthermore, as $N_1 \to \infty$ and $n \ll N_1$, the following are also true:

$$Z_{i'} \rightarrow x_\infty \tag{A12a}$$

$$Z_{N_1 n} \rightarrow 0 \tag{A12b}$$

or more generally

$$Z_{in} \rightarrow x_\infty (\beta x_\infty)^{i-n}, \quad \text{for } n \leq i \tag{A13}$$

As $N_1 \rightarrow \infty$, p_n in Eq. A1 is given by

$$p_n = \frac{x}{\theta} \sum_{i=n}^{\infty} (\beta x_\infty)^{i-n} p_{i0} + (\alpha x_\infty) p_{n-1} \tag{A14}$$

and this gives p_n for all n in Eq. A1 as follows. Using this, p_2 is written in terms of p_1, which is substituted in the relation of p_1, and p_n is determined by successively evaluating p_2, p_3, etc.

REFERENCES

1. P. J. Flory, *Principles of Polymer Chemistry*, Cornell University Press, Ithaca, N.Y. (1953).
2. R. W. Lenz, *Organic Chemistry of Synthetic High Polymers*, Wiley, New York (1967).
3. A. Kumar and S. K. Gupta, *Fundamentals of Polymer Science and Engineering*, Tata McGraw-Hill, New Delhi (1978).
4. P. E. M. Allen and C. R. Patrick, *Kinetics and Mechanism of Polymerization Reactions*, Ellis Horwood, Chichester (1974).
5. J. Furukawa and O. Vogl, *Ionic Polymerization, Unsolved Problems*, Marcel Dekker, New York (1976).
6. T. Keii, *Kinetics of Ziegler–Natta Polymerization*, Kodansha, Tokyo (1972).
7. J. Boor, *Ziegler–Natta Catalysts and Polymerizations*, Academic Press, New York (1979).
8. J. C. W. Chien, *Coordination Polymerization*, Academic Press, New York (1975).
9. G. Odian, *Principles of Polymerization*, 2nd ed., Wiley, New York (1981).
10. A. Kumar, *J. Appl. Polym. Sci.*, 34: 571 (1987).
11. S. K. Gupta, A. Kumar, and K. K. Agarwal, *J. Appl. Polym. Sci.*, 27: 3089 (1982).
12. K. Tai, Y. Arai, H. Teranishi, and T. Tagawa, *J. Appl. Polym. Sci.*, 25: 1789 (1980).
13. S. K. Gupta and A. Kumar, *Chem. Eng. Comm.*, 20: 1 (1983).
14. H. Kilkson, *Ind. Eng. Chem. Fundam.*, 7: 354 (1968).
15. J. A. Biesenberger, *AIChEJ*, 11: 369 (1965).
16. A. Kumar, P. Rajora, N. L. Agarwalla, and S. K. Gupta, *Polymer*, 23: 222 (1982).
17. H. M. Hulbert and S. Katz, *Chem. Eng. Sci.*, 19: 555 (1964).
18. A. Kumar, *Macromolecules*, 20: 220 (1987).
19. A. Kumar, S. N. Sharma, and S. K. Gupta, *J. Appl. Polym. Sci.*, 29: 1045 (1984).
20. A. Kumar, S. N. Sharma, and S. K. Gupta, *Polym. Eng. Sci.*, 24: 1205 (1984).
21. A. Kumar, S. K. Gupta, and D. Kunzru, *J. Appl. Polym. Sci.*, 27: 4421 (1982).
22. D. H. Solomon, ed., *Step Growth Polymerization*, Marcel Dekker, New York (1972).
23. M. Amon and C. D. Denson, *Ind. Eng. Chem. Fundam.*, 19: 415 (1980).
24. S. K. Gupta, N. L. Agarwalla, and A. Kumar, *J. Appl. Polym. Sci.*, 27: 1217 (1982).
25. S. K. Gupta, A. Kumar, and K. K. Agarwal, *Polymer*, 23: 1367 (1982).
26. L. C. Case, *J. Polym. Sci.*, 29: 455 (1958).
27. V. S. Nanda and S. C. Jain, *J. Chem. Phys.*, 49: 1318 (1968).
28. G. B. Taylor, *J. Am. Chem. Soc.*, 69: 638 (1947).
29. S. I. Kuchanov, M. L. Keshtov, P. G. Halatur, V. A. Vasnev, S. V. Vinogradova, and V. V. Korshak, *Macromol. Chem.*, 184: 105 (1983).
30. S. K. Gupta, N. L. Agarwalla, P. Rajora, and A. Kumar, *J. Polym. Sci. Polym. Phys. Ed.*, 20: 933 (1982).

31. R. Goel, S. K. Gupta, and A. Kumar, *Polymer, 18*: 851 (1977).
32. S. K. Gupta, A. Kumar, and A. Bhargava, *Eur. Polym. J., 15*: 557 (1979).
33. S. K. Gupta, A. Kumar, and A. Bhargava, *Polymer, 20*: 305 (1979).
34. A. Kumar, S. K. Gupta, and R. Saraf, *Polymer, 21*: 1323 (1980).
35. S. K. Gupta, A. Kumar, and R. Saraf, *J. Appl. Polym. Sci., 25*: 1049 (1980).
36. A. S. Gupta, A. Kumar, and S. K. Gupta, *Br. Polym. J., 13*: 76 (1981).
37. R. W. Lenz, C. E. Hanlovitz, and H. A. Smith, *J. Polym. Sci., 58*: 351 (1962).
38. J. H. Hodkin, *J. Polym. Sci. Polym. Chem. Ed., 14*: 409 (1976).
39. P. J. Flory, *J. Am. Chem. Soc., 63*: 3083 (1941).
40. P. J. Flory, *Chem. Rev., 39*: 137 (1949).
41. W. H. Stockmayer, *J. Chem. Phys., 11*: 45 (1943).
42. W. H. Stockmayer, *J. Chem. Phys., 12*: 125 (1944).
43. W. H. Stockmayer, *J. Polym. Sci., 9*: 69 (1952).
44. W. H. Stockmayer, *J. Polym. Sci., 11*: 424 (1953).
45. M. Gordon, *Proc. R. Soc. London A, 268*: 240 (1962).
46. D. S. Hutler, G. N. Malcolm, and M. Gordon, *Proc. R. Soc. London A, 295*: 29 (1966).
47. M. Gordon and T. G. Parker, *Proc. R. Soc. Edinburgh, A69*: 181 (1970).
48. M. Gordon, T. C. Ward, and R. S. Whitney, in *Polymer Networks* (A. J. Chompt and S. Newman, eds.), Plenum, New York, pp. 1–21 (1971).
49. M. Gordon and G. R. Scantlebury, *J. Chem. Soc. London B*, 1 (1967).
50. M. Gordon and M. Judd, *Nature, 234*: 96 (1971).
51. K. Dusek, M. Gordon, and S. B. Ross-Murphy, *Macromolecules, 11*: 236 (1978).
52. T. E. Harris, *Theory of Branching Processes,* Springer-Verlag, Berlin, Chap. 1 (1963).
53. C. W. Macosko and D. R. Miller, *Macromolecules, 9*: 199 (1976).
54. D. R. Miller and C. W. Macosko, *Macromolecules, 9*: 206 (1976).
55. D. R. Miller and C. W. Macosko, *Macromolecules, 11*: 656 (1978).
56. D. R. Miller, E. M. Valles, and C. W. Macosko, *Polym. Eng. Sci., 19*: 272 (1979).
57. D. R. Miller and C. W. Macosko, *Macromolecules, 13*: 1063 (1980).
58. R. F. T. Stepto, in *Developments in Polymerization,* Vol. 3 (R. N. Haward, ed.), Applied Science Publishers, Barking, U.K., p. 81 (1982).
59. J. L. Stanford and R. F. T. Stepto, *Br. Polym. J., 9*: 124 (1977).
60. R. F. T. Stepto, *Polymer, 20*: 1324 (1979).
61. A. B. Fasina and R. F. T. Stepto, *Makromol. Chem., 182*: 2479 (1981).
62. K. Dusek and W. Prins, *Adv. Polym. Sci., 6*: 1 (1969).
63. W. B. Temple, *Makromol. Chem., 160*: 277 (1972).
64. H. Jacobson and W. H. Stockmayer, *J. Chem. Phys., 18*: 1600 (1950).
65. M. Gordon and W. B. Temple, *Makromol. Chem.,* 263 (1972).
66. N. A. Plate and O. V. Noah, *Adv. Polym. Sci., 31*: 133 (1979).
67. I. I. Romanstova, Yu. A. Taran, O. V. Noa, A. M. Yelyashevich, Yu. Ya. Gotlib, and N. A. Plate, *Vysokomol. Soedin. A, 19*: 2800 (1977).
68. S. K. Gupta, S. Nath, and A. Kumar, *J. Appl. Polym. Sci., 30*: 557 (1985).
69. M. Abramowitz and J. A. Stegun, *Handbook of Mathematical Functions,* Dover, New York (1965).
70. A. Kumar and S. K. Gupta, *J. Macromol. Sci., Rev. Chem. Phys., C26*: 183 (1986).
71. M. Mutter, U. W. Suter, and P. J. Flory, *J. Am. Chem. Soc., 98*: 5745 (1976).
72. A. Kumar and P. K. Khandelwal, *J. Appl. Polym. Sci., 33*: 1835 (1987).
73. A. Kumar and P. K. Khandelwal, *Polym. Commun., 28*: 48 (1987).
74. A. Kumar and P. K. Khandelwal, Modelling of Reversible Multifunctional Step Growth Polymerization, M. Tech. Thesis, 11 T Kanpur, India, 1986.
75. R. W. Lenz, *Organic Chemistry of Synthetic High Polymers,* Interscience, New York (1967).
76. J. Brandrup and E. H. Immergent, *Polymer Handbook,* 2nd ed., Wiley-Interscience, New York (1975).

77. G. Odian, *Principles of Polymerization*, 2nd ed., McGraw-Hill, New York (1982).
78. O. Levenspiel, *Chemical Reaction Engineering*, 2nd ed., Wiley, New York (1972).
79. W. H. Ray, On the Mathematical Modelling of Polymerization Reactors, *J. Macromol. Sci. Rev. Macromol. Chem.*, *C8*: 1 (1973).
80. S. L. Liu and N. R. Amundson, *Rubber Chem. Tech.*, *34*: 995 (1961).
81. P. G. Gladyshev and S. R. Ratikov, *Russ. Chem. Rev.*, *35*: 405 (1966).
82. J. Cardenas and K. F. O'Driscoll, *J. Polym. Sci. Polym. Chem. Ed.*, *14*: 883 (1976).
83. F. L. Marten and A. E. Hamielec, *ACS Symp. Ser.*, *104* (1979).
84. S. K. Soh and D. C. Sundberg, *J. Polym. Sci. Polym. Chem. Ed.*, *20*: 1299 (1982).
85. S. K. Soh and D. C. Sundberg, *J. Polym. Sci. Polym. Chem. Ed.*, *20*: 1315 (1982).
86. S. K. Soh and D. C. Sundberg, *J. Polym. Sci. Polym. Chem. Ed.*, *20*: 1331 (1982).
87. S. K. Soh and D. C. Sundberg, *J. Polym. Sci. Polym. Chem. Ed.*, *20*: 1345 (1982).
88. W. Y. Chiu, G. M. Carratt, and D. S. Soong, 16: 348 (1983).
89. T. J. Tulig and M. Tirrell, *Macromolecules*, *14*: 1501 (1981).
90. D. T. Turner, *Macromolecules*, *10*: 221 (1977).
91. J. Cardenas and K. F. O'Driscoll, *J. Polym. Sci. Polym. Chem. Ed.*, *15*: 1883 (1977).
92. J. Cardenas and K. F. O'Driscoll, *J. Polym. Sci. Polym. Chem. Ed.*, *15*: 2097 (1977).
93. K. F. O'Driscoll, J. M. Dionisio, and H. K. Mahabadi, in *Polymerization Reactors and Processes* (J. N. Henderson and T. C. Bouton, eds.), American Chemical Society, Washington (1979).
94. F. L. Marten and A. E. Hamielec, *J. Appl. Polym. Sci.*, *27*: 489 (1982).
95. T. J. Tulig and M. Tirrell, *Macromolecules*, *15*: 459 (1981).

12

Optimization of Polymerization Reactors

Jorge N. Farber*
Planta Piloto de Ingenieria Quimica
UNS–CONICET
Bahía Blanca, Argentina

INTRODUCTION	429
SOME ASPECTS OF OPTIMIZATION OF POLYMERIZATION REACTORS	430
Polymerization Kinetics and Reactor Design	430
Reactor Dynamics and Stability	433
Mathematical Tools of Optimal Control Theory	435
FORMULATION OF THE OBJECTIVES	437
The Multiobjective Nature of the Problems	438
Multiobjective Decision Analysis and the Concept of Noninferiority	439
OVERVIEW OF PREVIOUS CONTRIBUTIONS	442
BATCH CHAIN POLYMERIZATION	444
The Minimum Time Problem	444
Modification of Molecular Weight Distribution Properties	451
STEP REACTIONS	452
COPOLYMERIZATION REACTORS	454
OPTIMIZATION OF TRANSIENT OPERATION	459
CONCLUSION	463
NOTATION	464
REFERENCES	465

INTRODUCTION

The relevance of optimization of polymerization processes and the interest of specialists are easily justified by the economic importance of polymerization reaction engineering. The manufacture of polymeric materials makes up a considerable fraction of the overall volume of chemical production, both in cost and mass terms. Other important reasons in support of optimization are found in the particular nature of polymeric products. A polymeric material cannot be considered a simple chemical substance. Chemical composition by itself does not determine unequivocally a number of other properties directly related to the applicability, quality, and economic value of the product. Some of these properties, like processability and mechanical performance, are strong functions of the molecular weight distribution (MWD) and are largely determined by reaction conditions.

Current affiliation: Tremco Ltd., Toronto, Ontario, Canada

Consequently, conversion by itself cannot be considered an important objective if it is separated from the specification of product properties, mainly derived from the existence of a MWD. This establishes an essential difference with current optimization problems in typical chemical engineering processes involving "short molecules."

Several important product specifications are characterized by their sensitivity to the prevailing reactor conditions. In consequence, any perturbation or departure from prescribed reactor states may affect critical molecular properties like MWD, copolymer composition and sequence distributions (CCD, CSD), degree of branching, crosslinking, and stereoregularity. Some of these factors or their combination determine the rheological behavior, morphology, mechanical performance, size distribution in particulate systems, etc.

In general, there is little flexibility after synthesis. The monomer which has been irreversibly converted to off-specification products cannot be easily recovered through any simple operation following the reaction. The economic impact of introducing optimal criteria in the operation of polymerization processes can be easily recognized.

SOME ASPECTS OF OPTIMIZATION OF POLYMERIZATION REACTORS

Polymerization Kinetics and Reactor Design

Polymers are produced through a wide spectrum of kinetic mechanisms, which normally combine many parallel and consecutive reaction steps [1–4]. A very general classification

Table 1 Classification of Simple Polymerization Reactions

Monomer coupling with termination (eg. radical polymerization)	$I \xrightarrow{k_d} 2R^*$ $R^* + M \xrightarrow{k_p} M_1^*$ $M_n^* + M \xrightarrow{k_p} M_{n+1}^*$ $M_n^* + M_m^* \begin{matrix} \xrightarrow{k_t^d} M_n + M_m \\ \xrightarrow{k_{t,c}} M_{n+m} \end{matrix}$	initiation propagation disproportionation termination by combination
Monomer coupling without termination (e.g. living polymerization)	$I + M \xrightarrow{k_i} M_1^*$ $M_n^* + M \xrightarrow{k_p} M_{n+1}^*$	initiation propagation
Polymer coupling (e.g. polycondensation)	$M_n + M_m \xrightarrow{k} M_{n+m}$	propagation

Source: Adapted from Ref. 1.

of polymerization mechanisms can be found in Table 1. Three main groups are emphasized: monomer coupling with termination (chain reactions), monomer coupling without termination (ionic or living polymerization), and polymer coupling (step reactions). Many industrial processes are run without a complete understanding of the kinetic mechanisms. Additional complexity normally arises from physical interactions of a diffusional nature like the Trommsdorff or gel effect [5].

The design of adequate reaction environments for the different polymerization systems leads to the wide spectrum of equipment currently used in industry. Even though industrial reactors may adopt many different configurations, they can be classified into the idealized types described in Table 2.

The particular choice of reaction environment determines the MWD. Continuous chemical reaction in a stirred tank differs in general from batch reaction owing to the presence of a residence time which may be caused by hydrodynamic or diffusional effects, or both. As pointed out in Denbigh's classical work [6], the MWD is determined by the relative magnitude of the lifetime of the propagating species compared to the mean residence time in the reactor. This effect is visualized in Figure 1, comparing the characteristics of distributions obtained in ideal batch and continuous reactors.

Concentration history is an additional effect that may influence the outcome of a reaction and the MWD. The degree of mixing or segregation influences the way reactive molecules (of statistically distributed chain length) encounter other molecules for eventual

Table 2 Kinetic Mechanisms and Reaction Media in Polymerization Processes

Polymerization Reactions and Processes / Reactors	Monomer Coupling With Termination				Monomer Coupling Without Termination		Polymer Coupling		
	Solution	Precipitation	Suspension	Emulsion	Solution	Precipitation	Solution or Melt Polycondensation	Interfacial Polycondensation	Solid Phase Polycondensation
BR — Batch Reactor	•	•	•	•	•	•	•	•	•
BR — Semibatch Reactor	•		•	•					
CPFR — Continuous Plug Flow Reactor	•					•	•		•
CSTR — Stirred Tank Cascade	•		•	•	•	•			
CSTR — Continuous Stirred Tank Reactor	•	•		•	•				

Source: Adapted from Ref. 2.

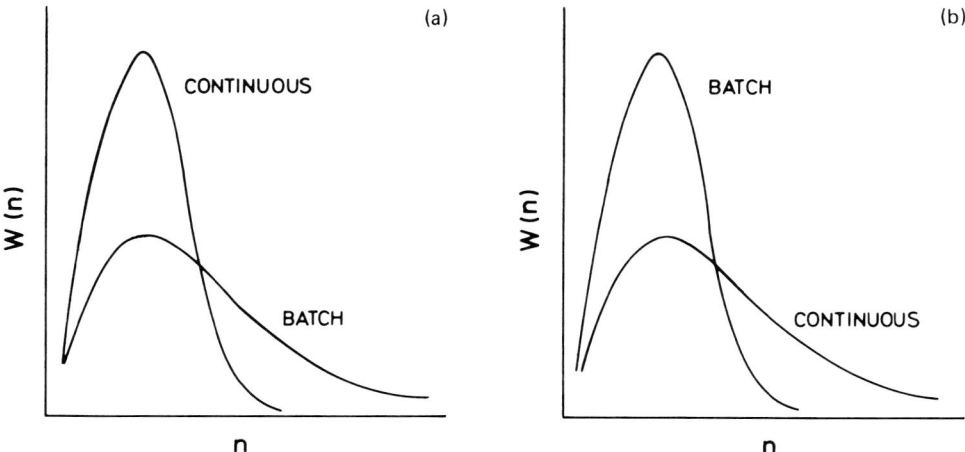

Figure 1 Effect of residence time distribution on chain length distribution in ideal batch and continuous reactors: (a) life of active polymer is short compared to mean residence time (radical polymerization); (b) life of active polymer is long compared to mean residence time (living polymerization). (Adapted from Ref. 6.)

reaction during their sojourn inside the vessel. Figure 2 compares the MWD and polydispersity (PD) obtained in a batch reactor (BR), segregated continuous stirred tank reactor (SCSTR), and homogeneous continuous tank reactor (HCSTR). Bearing in mind the dynamic equivalence of the ideal batch reactor and plug flow reactor (PFR) in steady state, and considering simple kinetic schemes in linear homopolymerization, Gerrens [1, 2] illustrated nine possibilities for the MWD resulting from the combination of different reactions and reactors. A similar classification of these idealized situations, including bounds for PD, is presented in Table 3.

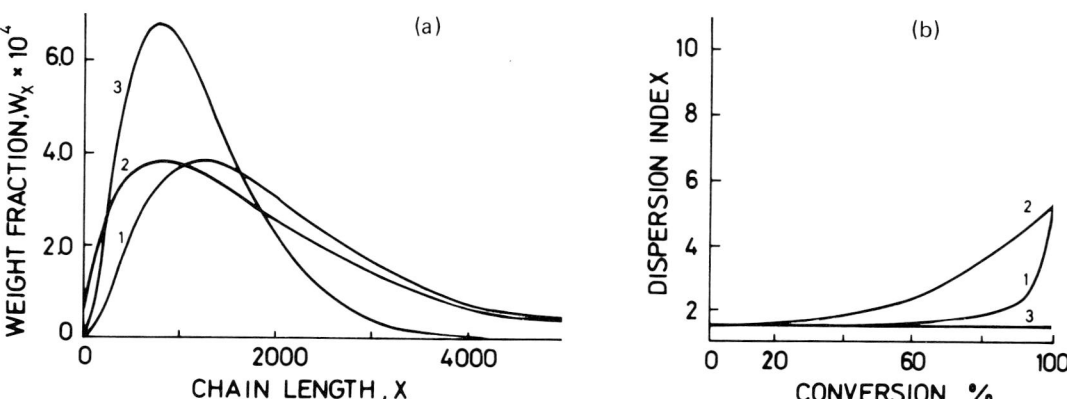

Figure 2 Effect of segregation on chain polymerization: (a) chain length distribution (60% conversion); (b) polydispersity vs. conversion; 1, BR; 2, SCSTR; 3, HCSTR. (From Ref. 7.)

Table 3 Molecular Weight Distributions Resulting from Different Kinetic Mechanisms and Reactors

REACTION \ REACTOR	CPFR OR BR	HCSTR	SCSTR
MONOMER COUPLING / MONOMER COUPLING WITH TERMINATION	1.1 BROADER THAN SCHULZ-FLORY	1.2 SCHULZ-FLORY DISTRIBUTION	1.3 BROADER THAN 1.1
MONOMER COUPLING WITHOUT TERMINATION	2.1 POISSON OR GOLD DISTRIBUTION	2.2 SCHULZ-FLORY DISTRIBUTION	2.3 BETWEEN 2.1 AND 2.2
POLYMER COUPLING	3.1 SCHULZ-FLORY DISTRIBUTION	3.2 MUCH BROADER THAN SCHULZ-FLORY	3.3 BETWEEN 3.1 AND 3.2

Source: Ref. 1.

The formulation of an accurate mathematical model for the process, with capacity to predict changes in the state of the system over a wide range of parameters, is essential for any useful optimization study.

In principle, all problems of polymerization, including those related to optimization, are described by an infinite set of nonlinear simultaneous equations. This complex description of the state of the system is essential for the determination of the MWD and its evolution during the process. However, generating function techniques can be used to collapse the infinite set into a single characteristic equation from which all moments of the MWD can be derived [4, 8, 9]. Fortunately, a few moments often suffice to characterize these distributions.

Modeling of transport phenomena offers a particular challenge. Most polymerization processes are heterogeneous multiphase systems with surface and volume ratios changing during the evolution of the reaction. High nonidealities have to be taken into account in the description of compatibility among phases, diffusion-controlled reactions, mixing patterns, and heat removal with the associated control problems (runaway) [10].

Reactor Dynamics and Stability

The mathematical analysis of system dynamics and stability is a highly complex task, given the higher dimensions of the models and the presence of several sources of unstable response. Multiplicity of steady states is known to occur even in the simplest polymerization systems in solution.

Figure 3 illustrates a typical phase plane for a copolymerization reactor in limit-cycle operation [11]. For a constant heat transfer capability (constant cooling water temperature and flow rate), an oscillatory output is obtained with the inherent quality control problem.

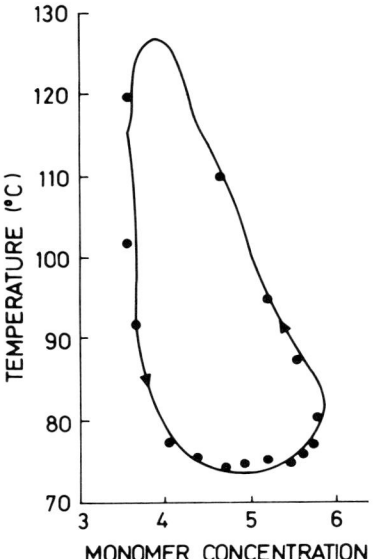

Figure 3 Phase plane plot for continuous solution copolymerizer. Limit-cycle operation under constant coolant temperature and flow rate. (From. Ref. 11.)

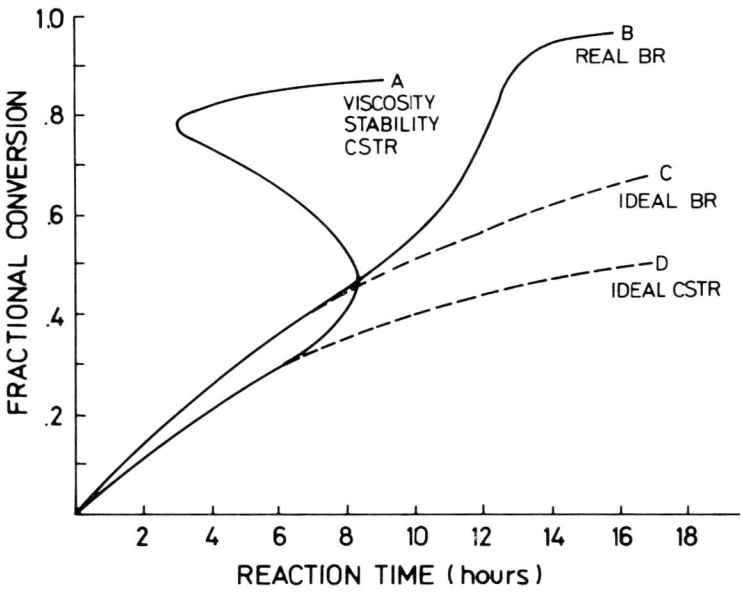

Figure 4 Comparison of reaction paths (conversion vs. time) for ideal and nonideal reactors. (From Ref. 12.)

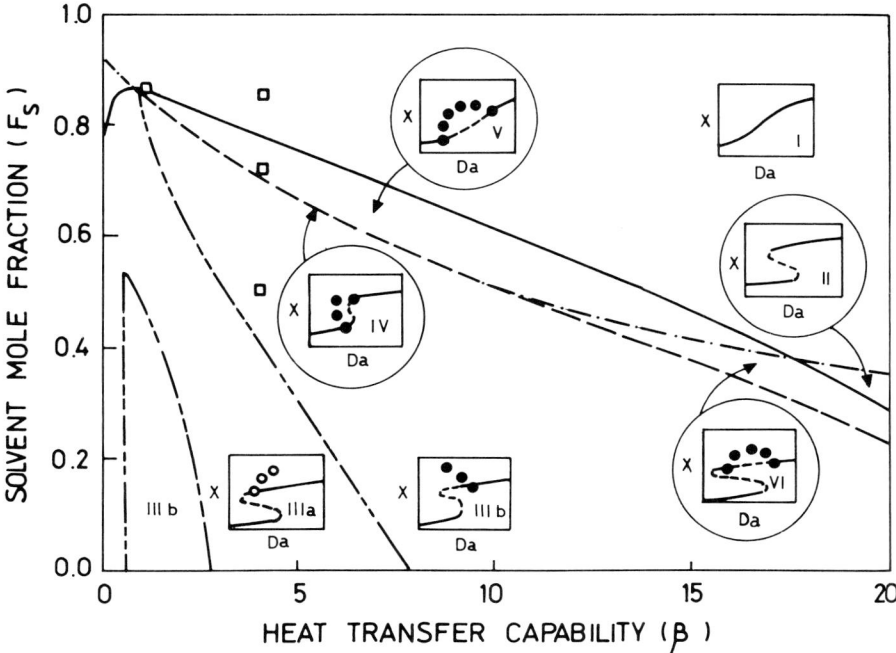

Figure 5 Dynamic behavior regions for continuous vinyl acetate polymerization. (From Ref. 16.)

Instabilities of thermal origin are not surprising, given the large heat evolution and difficult heat transfer characteristics of polymerization reactors. However, multiplicity and associated stability problems have been demonstrated for perfectly isothermal operation in CSTRs. The problem is analogous to concentration instability, which is characteristic of autocatalytic and heterogeneous catalytic reactions.

Knorr and O'Driscoll [12] demonstrated the multiplicity of solutions in bulk radical polymerization due to the Trommsdorff effect (Fig. 4). Further evidence has been presented for anionic polymerizations wherever there is an autocatalytic term in the initiation and propagation balances [13].

A detailed mathematical treatment of dynamics and stability analysis in several polymerization systems has been presented by Ray and co-workers (14–17). These authors determined and classified the dynamic regions according to the formalism proposed by Uppal et al.[18–19]. Figure 5 illustrates the dynamic regions in a vinyl acetate polymerization in solution, as obtained through changes in solvent concentration and heat transfer capability. As discussed later, a complete understanding of the reactor dynamics must precede any optimization study.

Mathematical Tools of Optimal Control Theory

In every branch of science, engineering, and economics, there exist systems that are controllable, that is, systems that can be made to behave in different ways depending on the will of the operator. Each time the operator of a system exerts an option, a choice in the distribution of the quantities controlling the system, he produces a change in the distribution of the states occupied by the system and hence a change in the final state. The important task is to individualize, among all admissible options, the particular option that will render the system optimum. In mathematical terms, this option will minimize the difference between the final value and the initial value of an arbitrarily chosen function of the state of the system. The body of knowledge covering such problems is known as optimal control theory.

A general approach of these optimization problems should encompass three aspects: modeling, optimal design, and optimal regulation.

1. *Modeling:* The first step in the analysis of the process is to derive and verify the mathematical equations which describe the dynamic change in the state and control parameters.

2. *Optimal Design:* The optimal design problem to be considered for both batch and continuous systems is to determine the path followed by the control variables as a function of time (or principal coordinate), in order to translate the state of the system from its initial to its final condition achieving the maximum (or minimum) values of an objective function.

3. *Optimal regulation:* This problem deals with how to regulate the process along the established optimal trajectories in the face of disturbances, such that the deviation from a prescribed final state can be minimized.

This chapter deals exclusively with the second aspect, optimal design. Dynamic optimization problems originate in relation to batch and plug flow reactors and transient operation of continuous stirred tanks. Details of mathematical aspects of optimal control theory can be found in Refs. 20–24.

The basic ideas of optimization of dynamic systems can be stated as follows. For a given system described by the state equations

$$\dot{\bar{X}}(t) = \bar{f}[\bar{X}(t), \bar{U}(t), t] \tag{1}$$

a control U^* is sought that will guide an initial state $\bar{X}(t_0)$ to a final $\bar{X}(t_f)$ minimizing some generalized performance measure or objective functional of the form

$$J(\bar{U}) = h[\bar{X}(t_f), t_f] + \int_{t_0}^{t_f} g[\bar{X}(t), \bar{U}(t), t]\, dt \tag{2}$$

where J is the final objective, h is a linear or nonlinear function of the final state, t_f is a specified or unspecified final time, and g is the integrand in the integral objective function.

The minimization of the particular form $J(\bar{U})$ is termed the Bolza problem [25]. The first term h in the objective J constitutes by itself a Mayer problem, defined by a function of the final state of the system.

The integral term is a Lagrange problem and accounts for the variation of the integrand along the whole system trajectory. The Bolza problem includes the Mayer and Lagrange problems as special cases, and the three can be shown to be completely equivalent through some basic theorems of the calculus of variations [26]. Further generalization accounting for constraint functions on the state and/or control variables can be found in the work of Bryson and Ho [24] and Kirk [22].

In terms of the Hamiltonian, defined as

$$H[\bar{X}(t), \bar{U}(t), \bar{P}(t), t] = g[\bar{X}(t), \bar{U}(t), t] + \bar{P}^T(t)\{\bar{f}[\bar{X}(t), \bar{U}(t), t]\} \tag{3}$$

Pontryagin's minimum principle establishes that the necessary conditions for U^* to be optimal are

$$\dot{\bar{X}}^*(t) = \frac{\partial H}{\partial \bar{P}} [\bar{X}^*(t), \bar{U}^*(t), \bar{P}^*(t), t] \qquad (4)$$

$$\dot{\bar{P}}^*(t) = -\frac{\partial H}{\partial \bar{X}} [\bar{X}^*(t), \bar{U}^*(t), \bar{P}^*(t), t] \qquad \text{for all } t \in t_0, t_f \qquad (5)$$

$$0 = \frac{\partial H}{\partial \bar{U}} [X^*(t), U^*(t), P^*(t)]$$

and

$$\left\{ \frac{\partial H}{\partial \bar{X}} [\bar{X}^*(t_f), t_f] - \bar{P}^*(t_f) \right\}^T \delta x_f$$

$$+ \left\{ H[\bar{X}^*(t_f), \bar{U}(t_f), \bar{P}(t_f), t_f] + \frac{\partial h}{\partial t} [\bar{X}^*(t_f), t_f] \right\} \delta t_f = 0 \qquad (6)$$

Equation 6, stating necessary conditions for an extremum of J, is presented in very general form and must be adapted to the boundary conditions of each particular problem. A large combination of boundary conditions arise due to the existence of free or fixed final times, the presence of constraint functions, etc. [21, 22, 24].

These equations provide necessary conditions for optimality. The complexity of typical problems in chemical engineering virtually precludes establishment of sufficiency conditions. A review of sufficiency theorems is presented by Peterson and Zalkind [27], but it is limited to simplified cases, generally not applicable to current problems.

The application of Pontryagin's minimum principle typically results in the generation of two-point-boundary value problems (TPBVP), whose methods of solution can be classified into two main categories, indirect and direct [21–24].

In indirect, or shooting, methods, optimality conditions are stated at every point, and state and adjoint equations are solved simultaneously. Trial and error procedures on the boundary conditions are the bases of the method. Shooting methods are particularly inadequate for high-sensitivity models, i.e., those related to chain polymerization systems.

In direct methods, on the contrary, optimality conditions are not imposed. The optimum is sought through successive approximations, and state and adjoint equations are solved separately. This is the basis of gradient (steepest descent) methods.

FORMULATION OF THE OBJECTIVES

The formulation of an objective is challenging in many ways. One challenge is the need to count with a realistic kinetic model, capable of describing the complex interactions among the reaction environment, the chemical structure, the resulting morphology, and the final properties [28].

A second challenge is posed by the need to correlate a variety of product specifications derived from empirical measurements (e.g., performance in mechanical tests, chemical resistance and stability, plasticizer absorption) with more formally established parameters related to the chemical structure and the MWD.

Translation of these aspects into the mathematical expression of an objective function is a largely unresolved problem. Consequently, earlier researchers have traditionally dealt with simple policies related to (a) the speed of the process (minimum time problem), (b) modification of properties (breadth of the MWD), and (c) achievement of specified properties (end-point constraints on conversion, molecular weight, PD, etc.).

The mathematical expressions of all these objectives can be considered as particular cases of the more general Bolza problem (Eq. 2).

The Multiobjective Nature of the Problems

Until very recently, virtually all problems in chemical engineering have been considered as single-objective optimization problems. In addition to the economic efficiency, the performance of many process systems requires an evaluation through a variety of criteria. Since most of these criteria are noncommensurable, they cannot be used to formulate a scalar-valued overall performance criterion.

Therefore, the most important aspect, economic efficiency in most cases, has been traditionally considered as a single objective, and the other criteria have been incorporated as inequality constraints indicating permissible levels. However, the complexity of modern chemical plants led to a reevaluation of the importance of performance criteria. Reliability, safety, control performance, environmental quality, etc., have become as important as is their economic efficiency [29]. In order to obtain such a goal of process engineering, process systems analysis had to deal with multiobjective optimization in which two or more noncommensurable objectives are considered simultaneously.

This situation obviously extends to polymerization processes. In this case, however, the multiobjective nature of optimization problems is more intrinsic and has to be related to the distinctive features exhibited by polymeric materials.

As stated earlier, a polymer cannot be considered a simple substance, given that the chemical composition by itself does not permit a meaningful characterization. Many other properties normally correlated with the molecular weight distribution will determine the applicability and market value of a product. In copolymerization the situation is particularly complex because the product properties depend not only on the molecular size, but also on the way the two comonomers are distributed along the chain (CCD and CSD). Typical conflicting trends occur when some of the previous criteria are optimized simultaneously. For example, economic performance generally can be improved by operation at high conversions. However, this may have a detrimental effect on the molecular weight and composition, and consequently, on the product specifications.

An overview of contributions indicates that until recently, optimization studies related to polymerization processes were based on scalar approaches. A variety of static and dynamic optimization problems have been formulated by adding several individual criteria in a single final expression f:

$$f = \sum_{i=1}^{n} w_i f_i(\bar{X}) \tag{7}$$

where f_i are the individual objective terms and $\bar{X}$ is the decision vector. A set of arbitrary normalized weights w_i is used to stress the relative importance of the different component terms [55]. However, since the result of the optimization problem can vary significantly as the weighting coefficients change, and since very little is usually known about how to choose these weights, a necessary approach is to solve the same problem for many different sets of values of the coefficients. Still confronted with all these solutions, the designer must choose among them, presumably on the basis of some additional criteria. Further complications arise when the individual criteria cannot be expressed in equivalent terms, that is, they are noncommensurable.

In some particular cases, the minimization of an objective of the form of Eq. 7 will yield a noninferior point, or tradeoff, among objectives, as obtained through a formal vector approach. However, this is true only if the solution to Eq. 7 is coincident with its convex hull [30, 31]. Unfortunately, this is not always the case in current chemical engineering problems, and convexity requirements generally cannot be easily verified in large-dimensional systems.

With few exceptions, the classical method of treating multiple objective problems has been to reduce them to equivalent scalar-valued objective problems which permit the use of standard optimization techniques. Nevertheless, a perennial difficulty lies in the notion of "scalarization." In many complex systems it is neither easy nor desirable to summarize the polyvalent consequences of a problem by a single numerical value. The price one pays for scalarization is that the meaning of optimization is no longer clear.

Multiobjective Decision Analysis and the Concept of Noninferiority

The first formulation of a problem of optimizing vector-valued criteria was given in 1896 in a publication by the economist Pareto [32]. This field has evolved in the last few decades into a recognized specialty of operations research, and has rapidly spread to other areas in engineering. A historical survey of major contributions in optimal control theory was presented by Salukvadze [33].

In the problem of specifying simultaneously several optimal characteristics there is an implicit notion that exact optimization of a scalar functional is virtually unattainable. More precisely, if a control is chosen to optimize a scalar functional, then it is almost impossible to optimize a second scalar functional using the same control parameters (Fig. 6).

A general vector optimization problem is formulated as follows:

Find

$$\min_{\bar{X}} [f_1(\bar{X}), f_2(\bar{X}), \ldots, f_n(\bar{X})] = \min_{\bar{X}} \bar{f}(\bar{X}) \tag{8}$$

subject to

$$g_k(\bar{X}) \leq 0, \quad k = 1, 2, \ldots, m$$

where $\bar{X} \in R^N$ is the decision vector, $\bar{f}: R^N \rightarrow R^n$ is the objective function vector, $\bar{g}: R^N \rightarrow R^m$ is the constraint vector, and $\bar{o} \in R^m$ is a vector whose elements are all zero.

The constraints $g(\bar{X}) \leq 0$ determines a feasible set T of values for the decision vector $\bar{X}$, $T = \bar{X}[\bar{g}|(\bar{X}) \leq 0]$. Also, each vector $\bar{X} \in T$ determines a unique value $f(\bar{X})$; so there exists a set S of feasible values for $f(\bar{X})$, $S = [\bar{f}(\bar{X})|X \in T]$. This duality permits a view of the multiobjective problem as

Min $\bar{f}(\bar{X})$ s.t. $\bar{f}(\bar{X}) \in S$

or as

Min $\bar{f}(\bar{X})$ s.t. $\bar{X} \in T$

The first step in the solution of a multiobjective problem consists of determining where a compromise or tradeoff between competing objectives must be made, and what alterna-

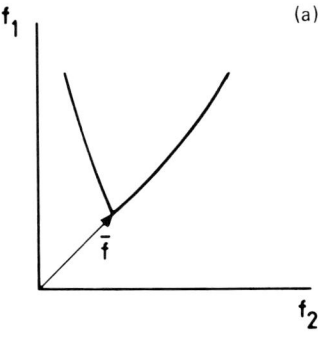

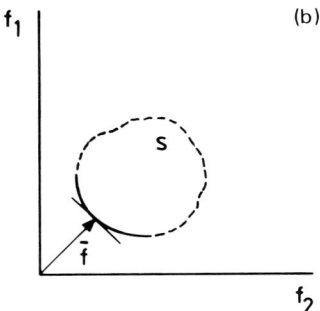

Figure 6 The noninferior set. (*a*) Exact minimization of a multiobjective problem. (*b*) The vectors which terminate on the marked segment of the boundary of S comprise the set of noninferior solutions. (From Ref. 66.)

tives are available. The ambiguity in the meaning of an optimal solution arises because there is no unique way to reach a compromise. Therefore, a logical prerequisite to any decision-making activity is to fully understand the interdependence of the objectives. In a vector objective formulation, the condition characterizing a compromise solution can be expressed in analytical form through the concept of noninferiority, which can be summarized as follows [34]: No improvement in any component index can be achieved except at the expense of the degradation in at least one of the other component indices. An index vector for which this condition holds is called a noninferior index, and it is a point where a compromise or tradeoff should be made. Furthermore, the loci of all such vectors describe the complete range of compromise solutions available. As indicated in Figure 6, the noninferior set of tradeoffs must lie on the boundary of the feasible region.

In the decision-making process, the determination of the noninferior set is the only step that can be treated on an entirely objective basis. It is understood that the decision maker will not be relieved from the decision itself; however, the method will facilitate a solution by eliminating from the analysis all those options that can be improved (inferior ones).

Many techniques for multiobjective decision analysis are available, with applicability to different disciplines. Cohon proposed a categorization of techniques according to the information flow in the decision-making context [35]. A complete review of methods was presented by Hwang et al. [36].

As seen in Table 4, two main types of flow are identified: from decision maker to

Table 4 Multiobjective Decision Analysis. Classification of Numerical Methods According to the Information Flow

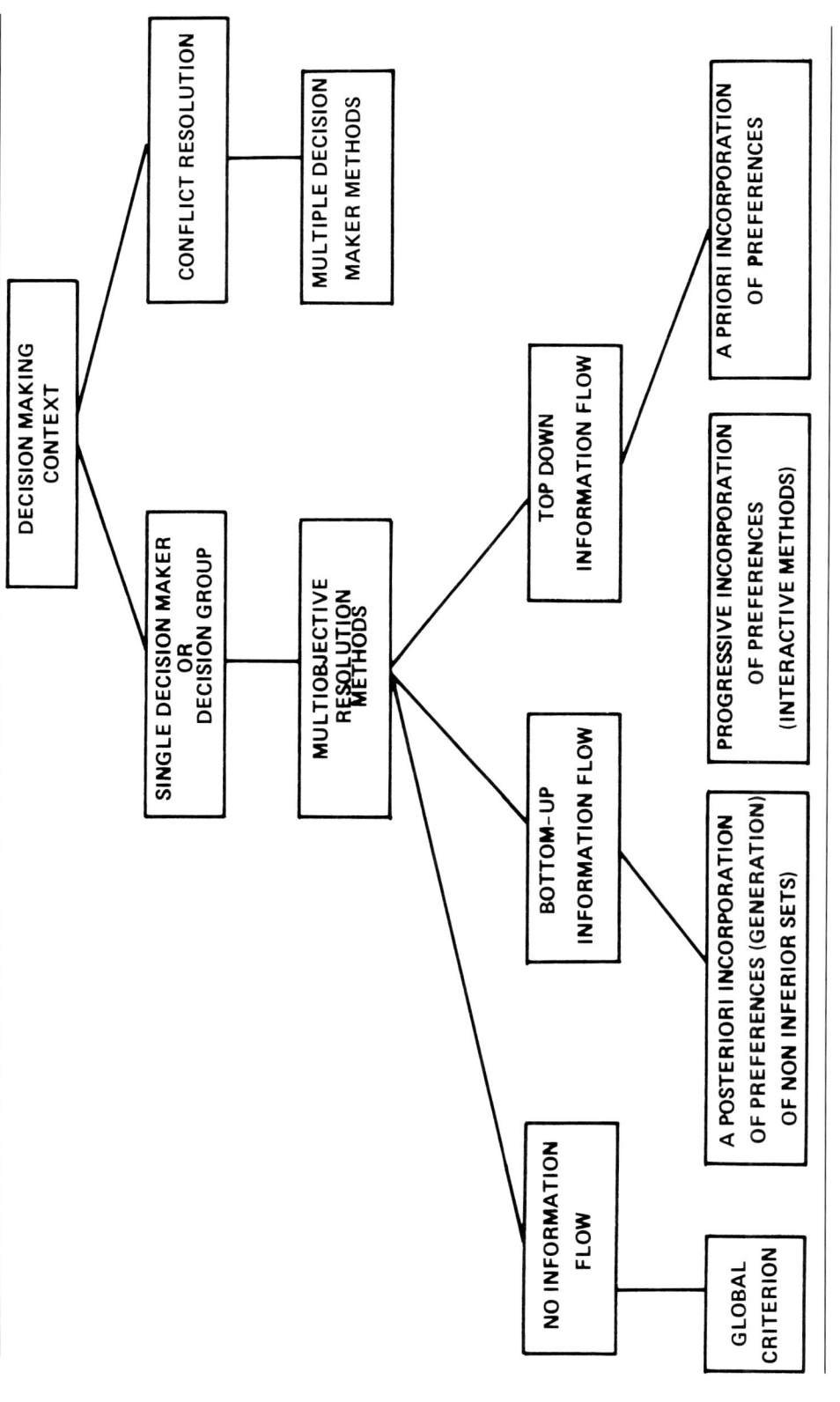

Source: Ref. 66.

analyst (top-down), and from analyst to decision maker (bottom-up). In the first case (top-down), decision makers explicitly articulate preferences so that a best compromise solution can be found. In the second case (bottom-up), analysts apply generating techniques to determine the noninferior sets and tradeoffs among objectives. This information, which can be obtained objectively through mathematical tools, is then given to decision makers. While some techniques allow for only one of these types of information flow, interactive methods combine elements of both at different stages of the solution process [35, 36].

For low-dimensional problems in optimal design, as discussed here, the logical procedure is to understand the interdependence among objectives as a first step through calculation of the noninferior sets. The decision process will then be completed by reducing the analysis to the noninferior set.

Several methods available for the determination of the noninferior sets include the parametric method and the ϵ-constraint approach. Details can be found in Refs. 34–36.

OVERVIEW OF PREVIOUS CONTRIBUTIONS

For a number of years, the technology of polymeric materials evolved mainly as an empirical body of knowledge. Related aspects, described in previous sections, apparently became reasons for the slow progress in modeling and optimization.

The improper understanding of complex interactions led to simplified models lacking the required accuracy for predictive purposes. These basic limitations were obviously translated to optimization studies. A poor model performance may cause drastic changes so as to invalidate important conclusions drawn in relation to an optimal policy.

A review indicates that practically all previous contributions have dealt with rather simple policies, basically related to the speed of the process (minimum time) or modification of MWD properties. It is only recently that more relevant contributions became available, thanks to better understanding of kinetic mechanisms and reactor dynamics, as well as development of suitable numerical methods. Table 5 presents a summary of major contributions in optimization of polymerization processes. The different types of controls or manipulated variables are noted.

Temperature is obviously the most important factor, as it governs both the rate of the process and the product properties. Each optimal policy will affect in a particular way the relative rates among the intervening reaction steps.

Fewer contributions deal with initiator feed rate (or concentration in reactor), either as a single control parameter or in combination with temperature. Serious limitations arise for the practical implementation of optimal concentration histories due to complex mixing patterns in high-viscosity media and multiphase systems. Some practical verifications have been limited to solution processes.

Consideration of optimal initial initiator concentration as a constant input parameter is instead fundamental in the formulation of minimum time policies.

Manipulation of monomer feed, either as a single control or simultaneously with temperature, has been considered in start-up policies minimizing off-specification products in continuous operation, and in composition control in semibatch copolymerization [65].

Polydispersity as a measure of the breadth of the MWD is a particularly insensitive function. In continuous operation, forced oscillations in reactor temperature, flow rate or concentration of any reagent, etc., have been studied in light of their advantages over steady-state operation. In some cases, considerable flexibility has been reported in the

Table 5 Summary of Contributions to Study of Optimization of Polymerization Reactors

REACTOR–REACTION / TYPE OF CONTROL	BATCH – SEMIBATCH BR (PFR in SS) –SBR			CONTINUOUS CSTR		
	MONOM. COUPLING (CHAIN)		POLYM.COUPLING (STEP)	MONOM. COUPLING (CHAIN)		POLYM.COUPLING (STEP)
	WITH TERM. (RADICAL)	WITHOUT TERM. (IONIC)		WITH TERM. (RADICAL)	WITHOUT TERM. (IONIC)	
TEMPERATURE	37–58, 63–67			55,81 84–86		55
TEMPERATURE + INITIATOR (I_0, I)	39,41 44–47 49,55		55 74–78	55		
INITIATOR	41,43,49 53,57,58	59–61		85		
TEMPERATURE + MONOMER (or reactant)	65 72,73					
MONOMER (or reactant)	57,58 62,65					55
OPTIMAL PERIODIC OPERATION					68–71	
OPTIMAL REACTOR CONFIGURAT.	72,73 79–82			79–82		

control of MWD properties in CSTRs. A recent review on this area was presented by Meira [68]. For a particular mechanism, when the reactor type has been (optimally) chosen in advance to produce the narrowest possible MWD in the steady state, the effect of periodic operation seems to invariably be a broadening of the MWD. If, however, the reactor type is not optimal in the sense of a narrow distribution, then narrowings in the MWD are feasible.

It is important to remark that most contributions in periodic operation of polymerization reactors are limited to studies on the system performance in relation to possible modification of the MWD. The true optimal periodic control problem, however, has been approached only in Refs. 69–71.

The type of polymerization reactor and the reactor system configuration or arrangement can significantly influence the conversion level and the MWD properties. This leads to an optimal reactor configuration selection problem, without even introducing considerations on optimal system performance by control of manipulated variables.

Most optimization studies in polymerization have been limited to single reactors. With few exceptions [72, 73, 79–82], the choice of conventional types of reactors and the alternatives for their interconnection have been arbitrarily or empirically determined.

Nonlinear systems synthesis or process synthesis techniques provide systematic procedures for the selection of the reactor types and the system configuration for optimal performance. Further details on these tools can be found in Refs. 87–91.

BATCH CHAIN POLYMERIZATION

In the area of chain polymerization, by far the largest number of contributions have related to batch radical systems.

Temperature is the most important factor affecting the rate of the process. In general, the reaction rate increases with temperature, and at the same time the molecular weight tends to decrease with the concomitant deterioration in a number of properties. The optimal temperature conditions must therefore be chosen in light of two conditions: the process should take the shortest possible time; and it should produce a polymer with useful properties.

To meet these requirements previous contributors have dealt mainly with two categories of problems or policies, minimizing the batch time and narrowing the MWD. In the minimum time problem, a final conversion or conversion and a cumulative average molecular weight are specified as end-point constraints. In narrowing the MWD, polydispersity alone has been normally specified, with free molecular weight averages.

Only a few relevant aspects are included to illustrate this topic. A discussion and comparison of previous contributions can be found in Refs. 44, 47, and 48. The minimum time problem is discussed first.

The Minimum Time Problem

Major progress in this topic is described in Refs. 37–50 and 67 (also see Table 5).

One of the simplest possible policies is the one that minimizes the time involved in translating the state of an ideal polymerization system from an initial to a final condition. Some emphasis is given to this particular topic because elementary conclusions drawn in relation to ideal systems can be applied, at least qualitatively, to more complex chain polymerization systems.

The ideal system is described by only two balance equations related to the monomer and initiator. Because of its simplicity, it admits a single analytical solution.

Krichevskaya [92] found an exact solution to the problem of optimization of two parallel reactions by following the formalism of the calculus of variations and the solution of the Euler–Lagrange equations. Shatkan [37] applied the same procedure to an ideal polymerization. The two balances for monomer and initiator can be written as

$$\frac{dM}{dt} = A_1 e^{-E_1/RT} M I^\beta \tag{9}$$

$$\frac{dI}{dt} = A_2 e^{-E_2/RT} I \tag{10}$$

where M and I are monomer and initiator concentrations, β is the reaction order with respect to initiator, A_1 and A_2 are preexponential factors, and E_1 and E_2 are the activation energies. Minimization of the time to reach a desired conversion, or

$$\min J = \int_0^{t_f} dt \tag{11}$$

implies the solution of the Euler–Lagrange equations, which can be achieved through simple analytical manipulation. A phase-plane plot for monomer and initiator concentration is illustrated in Figure 7a. The optimal temperature history vs. time is presented in Figure 7b. These results indicate that temperature should increase at a definite rate, so as to maintain a constant reaction rate (path $o-m$). Then temperature should be set up at the highest allowable value (point m), and react isothermally at that upper bound (path $m-f$).

These results describing a first-order decay in initiator concentration are equivalent with the work by Szepe and Levenspiel [93] for a heterogeneously catalyzed batch reactor with first-order catalyst deactivation. The common conclusion is that the "effective rate coefficient" (or global polymerization rate coefficient, here defined as $A_1 e^{-E_1/RT} I^\beta$), should remain constant along the optimal path. This finding does not hold any more if the right-hand side of Eq. 10 depends on M [94]. Yoshimoto et al. [38] restated this problem for an ideal system in the presence of initiators by following the formalism of Pontryagin's minimum principle. Similar results were obtained.

The characteristics of an ideal radically initiated polymerization can be summarized as follows:

1. The temperature profile is determined by the relative magnitude of the activation energies for propagation and initiation (E_1 and E_2 respectively).
2. If $E_1 > E_2$, the system will be controlled by the propagation reaction. Consequently, the temperature should be set up at its upper bound from the very onset of the process. If $E_1 < E_2$ (usually the case in radical polymerization), then the temperature should rise at a definite rate and reach the upper limit, after which the reaction will proceed at constant temperature. This indicates that at the start of the process (nonisothermal portion), the propagation centers should be made available by initiator decomposition at a particular rate so that the propagation reaction can proceed in optimal fashion.

Sacks et al. [43] applied the minimum principle to determine temperature and initiator addition policies that minimize the time to reach a desired conversion and number average molecular weight. The Trommsdorff effect was included in the model, but no transfer

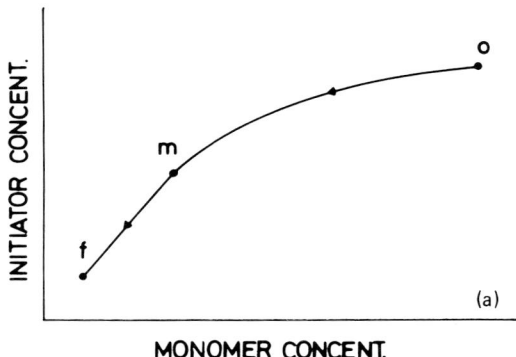

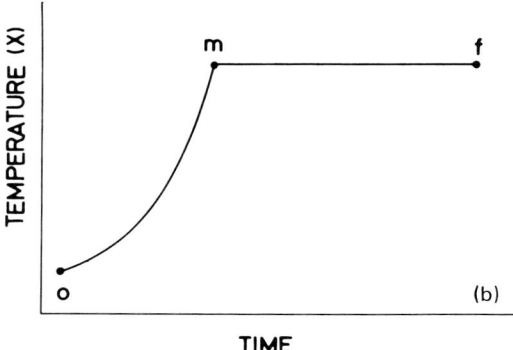

Figure 7 Minimum time problem in ideal batch radical polymerization. (*a*) Phase plane for monomer and initiator concent. (*b*) Optimal temperature vs. time. (Adapted from Ref. 37.)

mechanisms or volume change was accounted for. The calculated rising temperatures resembled the results obtained by Szepe and Levenspiel [93] for heterogeneously catalyzed reactions with first-order catalyst deactivation. The common conclusion is that for a policy to be optimal, the global rate coefficient (polymerization rate coefficient) should remain constant. This result appeared in all models that included a first-order decay in the initiator concentration.

The optimal initiator addition policy for a batch isothermal chain addition polymerization is one that maintains a constant initiator concentration in the reactor. Yoshimoto et al. [42] studied optimal policies for the thermally initiated polymerization of styrene. As no initiator was included in the model, it was possible, with only two state equations, to calculate minimum times to reach a specified conversion and molecular weight. The authors demonstrated that for this policy, there is a unique initial temperature and thermal history that permits one to reach an a priori selected final state (a conversion and a molecular weight).

The minimum time problem for thermal polymerization of styrene was restated by Wu et al. [48]. Significant improvements were introduced by a realistic kinetic model including volume contraction and gel effect. Both kinetic model and optimal trajectories were experimentally verified, with excellent agreement. Application of the minimum principle leads to a convenient solution as a disjoint problem. Figure 8 illustrates temperature versus time policies for different target conversions and molecular weights.

In a series of papers, Chen and co-workers [44–46] restated the minimum time problem in batch polymerization. The development followed closely the work by Sacks et al. [43], with consideration of gel effect through empirical functions and with constant volume. Isothermal and nonisothermal policies, including consideration of the best initial initiator concentration and optimal initiator feed rate, were calculated and compared. The authors showed that for single initiator loadings, there is a unique optimal loading for each desired molecular weight [44]. The optimal policy was not necessarily better than the isothermal policy if the optimal initiator concentration was not employed simultaneously.

Figure 9 illustrates temperature versus time trajectories and the influence of initial initiator concentration on minimum time.

Practically all contributions dealing with minimum time included specification of a number average molecular weight, but did not consider a weight average molecular weight (or PD) as an additional requirement. The reasons are the difficulties in establishing efficient algorithms for the TPBVPs characterized by severe sensitivity problems [50]. Also, for the highly nonlinear stiff systems of equations, it is not easy to prove that the systems are controllable in a technical sense, i.e., if the state can be transferred from an initial to a final condition in a finite time.

Thomas and Kiparissides [49] solved the problem of including polydispersity as an additional end-point constraint by implementation of a discrete gradient method and the concept of target set. Accordingly, the final desired output was allowed to lie anywhere within a ±5% band defining acceptable ranges for conversion, molecular weight, and PD. Small tolerances will eventually preclude convergence to a solution.

Farber and Laurence [50] restated the minimum time problem to reach a specified

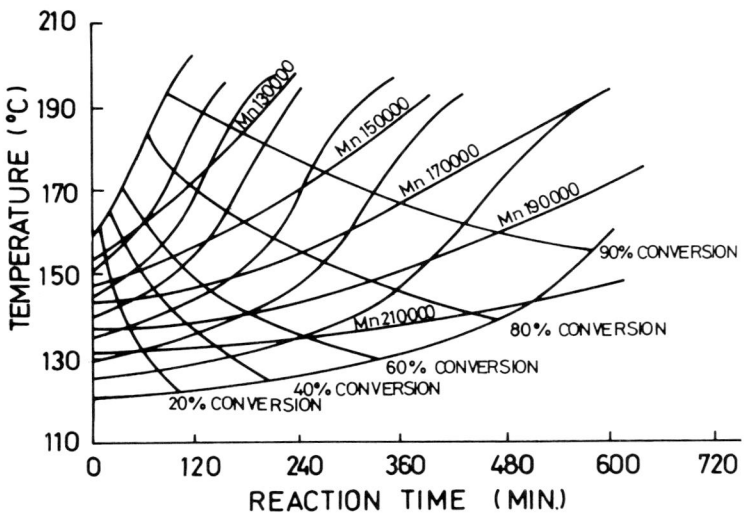

Figure 8 Batch bulk thermal polymerization of styrene. Optimal temperature vs. time for a target molecular weight. (From Ref. 48.)

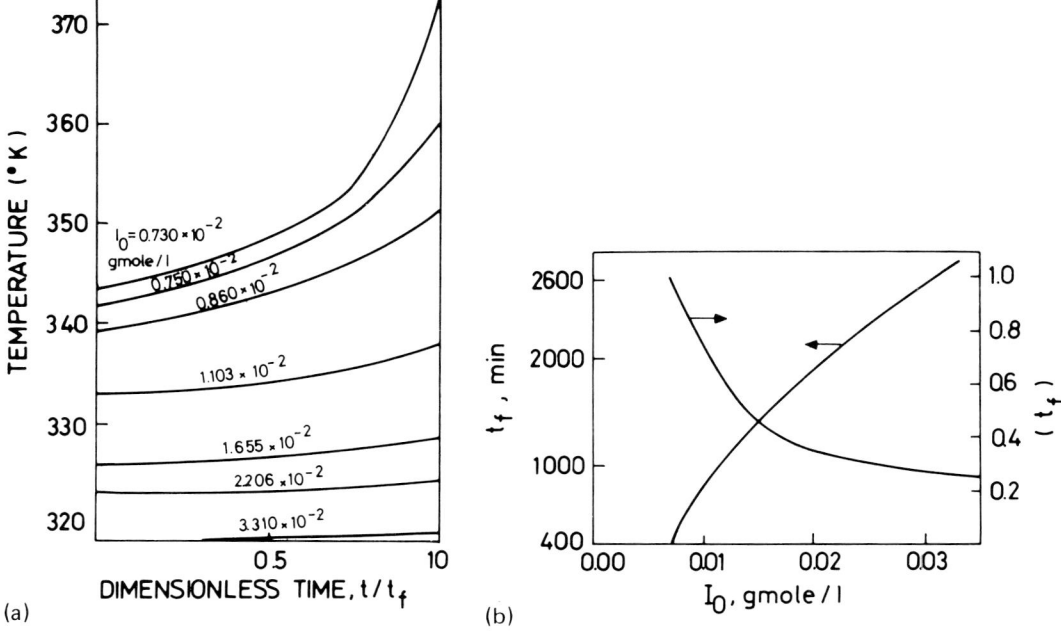

Figure 9 Minimum time problem. (*a*) Optimal temperatures vs. time. (*b*) Dependence of reaction end time and final initiator concentration in optimal policies (DP = 1000, 50% conversion). (From Ref. 44.)

conversion and molecular weight. The kinetic model for polymerization of styrene accounted for chemical as well as thermal initiation, volume contraction, and gel effect. Experimental verification was previously performed by one of the authors [48]. The following balance equations result:

Monomer:

$$\frac{dM}{dt} = (K_p + K_f)M\lambda_0 + \frac{M_0 + \epsilon M}{M_0} = f_1(\bar{X}, \bar{U}) \quad (12)$$

Initiator:

$$\frac{dI}{dt} = -K_d I - (K_p + K_f)\epsilon\lambda_0 \frac{MI}{M_0} = f_2(\bar{X}, \bar{U}) \quad (13)$$

where $X_1 = M$, $X_2 = I$, $dM/dt = f_1$, $dI/dt = f_2$, $\bar{U}$ is the control vector (temperature), ϵ is the volume contraction coefficient, and λ_0 is the zeroth-order moment of the radical chain length distribution.

The necessary conditions for optimality are stated in several theorems of the Pontryagin's theory. Some of those theorems are specific for the particular case of time optimality [20, 95]. Accordingly, a pseudo-Hamiltonian is defined as

$$H^*[\psi(t), \bar{X}(t), \bar{U}(t)] = \sum_{i=1}^{2} \psi_i f_i \quad (14)$$

where the adjoint functions ψ satisfy the relations

$$\frac{d\psi_i}{dt} = -\frac{\partial H^*}{\partial X_i}, \quad i = 1,2 \tag{15}$$

The boundary conditions for the system can be deduced by following the routine of first preparing the variations of the objective, and then seeking convenient conditions that make those variations equal to zero. For this particular problem, the existence of an extremum requires that

$$\begin{aligned}
X_1(0) &= M_0 & X_1(t_f) &= \text{fixed} \\
X_2(0) &= I_0 & X_2(t_f) &= \text{unspecified} \\
\psi_1(0) &= \text{unspecified} & \psi_1(t_f) &= \text{unspecified} \\
\psi_2(0) &= \text{unspecified} & \psi_2(t_f) &= 0
\end{aligned} \tag{16}$$

The necessary conditions are

$$H^*[\bar{\psi}(t), \bar{X}(t), \bar{U}(t)] = \max = M$$
$$H^*[\bar{\psi}(t_f), \bar{X}(t_f), \bar{U}(t_f)] \geq 0 \tag{17}$$

for all $t_0 < t < t_f$. Along the trajectories where the control U is unconstrained, the maximum of H^* implies

$$\frac{\partial H^*}{\partial U} = 0, \quad \frac{\partial^2 H^*}{\partial U^2} < 0 \tag{18}$$

Solution of the TPBVP defined by Eqs. 12, 13, and 15, subject to the conditions 16 and 17, will provide the temperature path that minimizes the time for a given conversion.

For the minimum time problem under study, it is possible to devise a convenient numerical solution based on the disjoint character of the system equations. Since the number of state equations does not exceed by more than one the number of controlled variables, it is possible to deduce from the stationary condition on the Hamiltonian and its time invariance

$$\frac{\partial H^*}{\partial U} = 0, \quad \frac{d}{dt}\frac{\partial H^*}{\partial U} = 0 \tag{19}$$

a form of Euler's equation [21] given by

$$\dot{U} = \frac{dU}{dt} \tag{20}$$

Consequently, an explicit form for the time change of temperature is made available through Eq. 20.

Equations 12, 13, and 20 constitute an initial value problem. The simultaneous solution will provide the trajectory of the controlled variable (temperature) satisfying the stationary conditions. Determination of the optimal path implies the simultaneous solution of the previous set and the adjoint equations. The task is then to detect initial conditions for ψ_1 and ψ_2 (if they exist) that will satisfy the boundary conditions. In some cases the search for the proper boundary conditions can be facilitated by defining a new function

$$\xi = \frac{\psi_2}{\psi_1} \tag{21}$$

$$\frac{d\xi}{dt} = \frac{1}{\psi_1}\frac{d\psi_2}{dt} - \xi\frac{d\psi_1}{dt}$$

or

$$\frac{d\xi}{dt} = -\frac{\partial f_1}{\partial X_2} - \xi \frac{\partial f_2}{\partial X_2} + \xi \frac{\partial f_1}{\partial X_1} + \xi^2 \frac{\partial f_2}{\partial X_1} \quad (22)$$

From the simplified equations for an ideal polymerization, it is easy to deduce that ξ (or ψ_1 and ψ_2) must belong to the third quadrant of the ψ_1–ψ_2 plane in order to satisfy the stationary condition and the maximality of the Hamiltonian. So, by application of the minimum principle, the original variational problem was transformed in a two-point boundary value problem, which was then solved as an initial value problem.

Details on the simulation are given in Ref. 50. Figure 10 contains optimal trajectories as a function of time. Each curve constitutes the locus of all those trajectories that, starting at the same initial conditions, will minimize the time for a desired conversion (to be achieved at some point along the trajectory). In other words, all those policies that are referred to the same initial conditions will have coincident portions.

The dots along the curve lines indicate the depletion of the initiator. At that point, the temperature must take its maximum value, and minimum time or maximum reaction rate in the absence of initiator is achieved by maximum temperature.

The consideration of volume contraction is very important to the model, if any predictive capacity is expected. Although molecular weights are not altered, the conversion versus time relation is affected by errors in time prediction on the order of 50% at high conversions. The magnitude of these errors may well mask all improvements achieved through optimal criteria.

The previous policy that minimizes the time to reach a desired conversion was compared with another policy which contains, besides conversion, a number average molecular weight as an additional requirement [50]. In order to make a comparison feasible, the initial and final conditions for the second problem were set equal to those of the first problem. The mathematical equivalence of both policies was established, although sen-

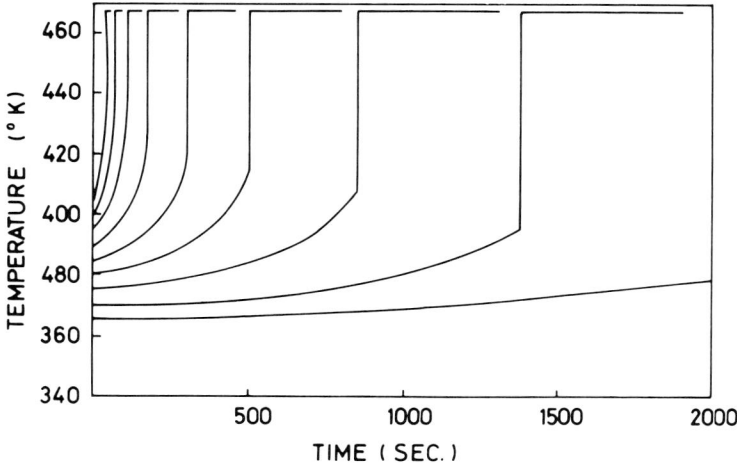

Figure 10 Minimum time problem. Optimal temperature vs. time. (From Ref. 50.)

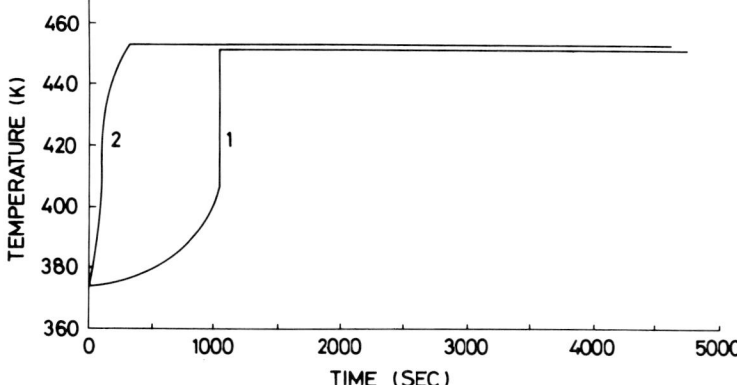

Figure 11 A comparison of two policies. Optimal temperature vs. time. 1, Minimum time to reach a desired conversion; 2, minimum time to reach a desired conversion and number average molecular weight. (From Ref. 50.)

sitivity problems did not allow a complete numerical verification. Figure 11 compares the first policy solved as an initial value problem with the second policy solved by a gradient method after 58 iterations. It was concluded that given the extreme insensitivity of molecular weight due to the presence of two initiation mechanisms, its inclusion in a policy is not justified. For the particular styrene system and conditions for simultaneous thermal initiation, the equivalent simpler policy is a much better choice [50].

Modification of Molecular Weight Distribution Properties

The work of Hoffman et al. [53] is one of the earliest references on narrowing the MWD. In order to eliminate the drift in chain length due to changes in monomer and initiator concentrations, these authors calculated monomer and initiator feed rates that maintain the DP constant. Analytical equations were obtained for simple polymerization models.

Sacks et al. [56] studied the effect of temperature on MWDs for batch chain addition polymerizations. By use of the minimum principle, these authors found that the variations which minimize the breadth of the MWD keep the instantaneous number average chain length constant. Such variations are decreasing temperatures when the isothermal number average degree of polymerization $\bar{X}_n$ decreases with time and increasing temperatures when $\bar{X}_n$ increases with time (i.e., gel effect).

The temperature variations that maximize the breadth of the MWD are step changes in temperature between a maximum and a minimum allowable temperature for the system. Such variations produce a bimodal distribution with the maximum separation between modes. Figure 12a illustrates optimal policies for the case of an isothermal $\bar{X}_n$ increasing with time. The minimum PD policy is one of increasing temperature, which keeps $\bar{X}_n$ constant. The maximum PD is a step decrease in temperature resulting in a bimodal distribution (Fig. 12b).

Further details and references in this area can be found in work by Louie and Soong [57, 58].

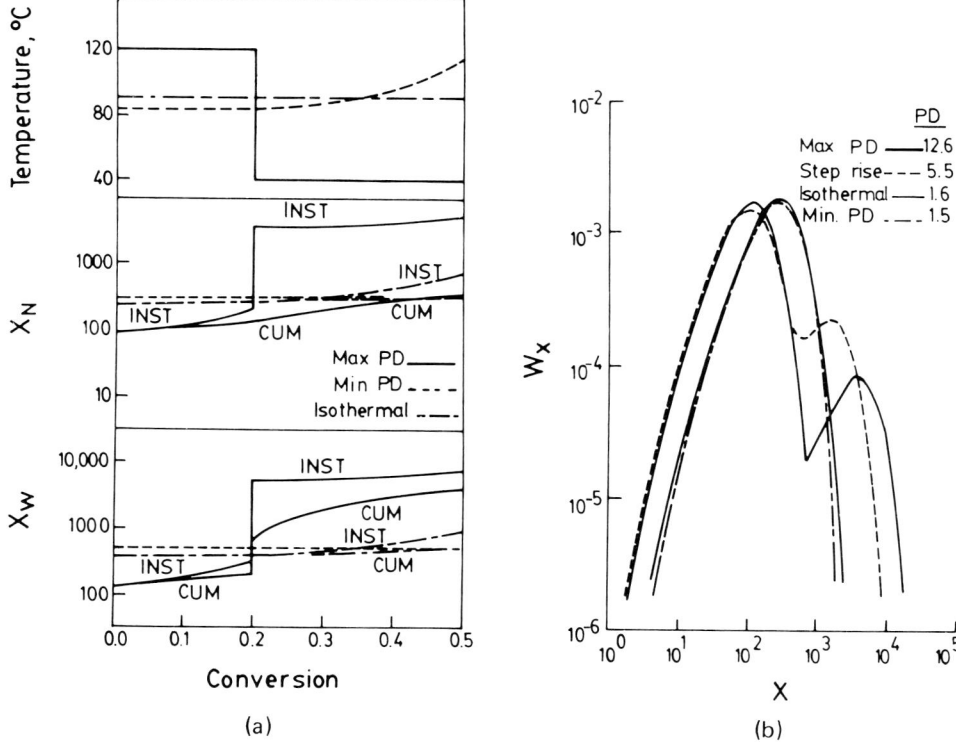

Figure 12 (*a*) Effect of temperature on MWD. Temperature policies for maximum and minimum breadth for the case of isothermal $\bar{X}_n$ increasing with time. (*b*) Comparison of MWD. (From Ref. 56.)

STEP REACTIONS

As indicated in Table 5, the number of contributions in optimization of step polymerization systems is relatively small. In simple polycondensation systems of ARB-type monomers, optimal criteria lead to almost trivial results.

Most of the available contributions deal with production of polyamides (nylon-type products) and polyesters (e.g., polyethylene terephthalate) [72–78].

Practically all generalities on modeling and optimization of polymerization reactors also apply to step reaction systems. Related references can be found in previous reviews.

Once the time is fixed (batch time in BR, or residence time in CSTR) there are two degrees of freedom, the temperature history and the feed composition, which can be manipulated independently as control variables.

A possible useful objective for these problems may take the form

$$J = w_1|M_1(t_f)|^2 + w_2|\bar{M}_w(t_f) - \bar{M}_w^*|^2 \tag{23}$$

The first term maximizes conversion or minimizes the concentration of unreacted monomer (M_1 at the end of the process), as well as the separation costs that originate after the reaction. The second term minimizes the deviation from a prescribed weight average molecular weight.

It can be shown that for the particular case of simple polycondensation of ARB monomers there is a single degree of freedom in the system, which does not permit arbitrary choice of design variables. The reaction is represented by a single kinetic step of the form (Table 1)

$$M_n + M_m \xrightarrow{k} M_{n+m} + W \tag{24}$$

The balance equations for species of any length are given by

$$\frac{dM_1}{dt} = -2kM_1 \sum_{i=1}^{\infty} M_i \tag{25}$$

$$\frac{dM_n}{dt} = -2k_p M_n \sum_{i=1}^{\infty} M_i + k_p \sum_{r=1}^{n-1} M_r M_{n-r} \tag{26}$$

where $\lambda_0 = \sum_{i=1}^{\infty} M_i$ is the zeroth-order moment. The result of balance equations for the first three moments [5] are

$$\frac{d\lambda_0}{dt} = -k_p \lambda_0^2 \tag{27}$$

$$\frac{d\lambda_1}{dt} = 0 \quad \text{or} \quad \lambda_1 = \lambda_{10} \quad \text{(constant feed value)} \tag{28}$$

$$\frac{d\lambda_2}{dt} = 2k\lambda_1^2 = 2k\lambda_{10}^2 \tag{29}$$

Useful design variables are the concentration M_1 (related to separation costs), the degree of polymerization $\bar{X}_n = \lambda_1/\lambda_0$, and the polydispersity $PD = \lambda_2\lambda_0/(\lambda_{10})^2$. Another possible choice would be M_1, λ_0, and λ_2. As shown below, however, these three variables are not independent.

On dividing Eq. 27 by Eq. 29 and integrating with the condition $\lambda_0 = \lambda_{00}$ when $\lambda_2 = \lambda_{20}$, it is found that

$$\lambda_0 = \frac{2\lambda_{00}}{2 + (\lambda_{00}/\lambda_{10}^2)(\lambda_2 - \lambda_{20})} \tag{30}$$

Analogously, dividing Eq. 25 by Eq. 29 and integrating with the condition $M_1 = M_{10}$ when $\lambda_2 = \lambda_{20}$,

$$\frac{M_1}{M_{10}} = \frac{4}{2(\lambda_{00}/\lambda_{10}^2)(\lambda_2 - \lambda_{20})} \tag{31}$$

From the last two expressions it is deduced that for a general batch irreversible polycondensation of ARB monomers there is only one degree of freedom in the system. Once λ_2 is specified, M_1 and λ_0 are uniquely determined. Analogously, a choice of either M_1 or λ_0 determines the other two. This is valid for any arbitrary feed, which can be pure monomer or a mixture of oligomers, and any arbitrary temperature history.

Several optimization problems can be posed. A current one is the achievement of a desired number average chain length or $\bar{X}_n$ in the minimum time. A choice of λ_2 will determine M_1 and λ_0. From Eq. 29 it is seen that λ_2 is a monotonically increasing function of time, with slope $2K_p(\lambda_{10})^2$,

$$\lambda_2 = \lambda_{20} + 2(\lambda_{10})^2 t_f K(T) \tag{32}$$

The minimum of t_f or the maximum of $K(T)$ is achieved by allowing the highest possible temperature in the reactor. It can be concluded that the minimum time problem in batch polycondensation is a maximum temperature policy with final time used to control the number average chain length [55].

The foregoing restrictions are eliminated in more complex reaction mechanisms with competition of several kinetic steps.

COPOLYMERIZATION REACTORS

Simultaneous polymerization of two or more monomers is characterized by the flexibility of different combinations of reactants and reactor conditions. Consequently, a wide variety of product properties can be tailored through proper understanding of the process and adequate control. As in homopolymerization, copolymerization reactions produce polymers with a distribution of molecular weights. The same mathematical techniques can be extended to the description of the MWD and its evolution during the process. Additional complexity arises from kinetic mechanisms augmented with cross-reaction terms [97, 98].

However, MWD is not the only distributed property of copolymers that can be influenced by the design conditions. The relative amounts of the comonomers result in a copolymer composition distribution (CCD), and the ordering of the different monomeric units along the chain produces a sequence distribution (CSD). These two distributions are important measures of product quality to be considered in the optimal design and operation of a copolymerization reactor.

The associated control problems are very important. Some copolymers must be produced within narrow tolerances in the reaction conditions, which may otherwise affect the phase compatibility, morphology, and final properties.

From the kinetic viewpoint, copolymerization reactions can also be classified in step and chain growth. The scope of step growth is limited because the comonomers containing the same type of functional group exhibit nearly equal reactivity toward propagation. Thus the extent and nature of incorporation of different comonomers into the copolymer is a function of the concentration and functional group characteristics [99]. As in step homopolymerization, high molecular weight is not obtained until monomer conversion is quite high. The average composition of a step growth copolymer is basically that of the initial monomer mixture. Thus CCD is not a crucial or important problem in step growth copolymerization. CSD control, however, is a significant problem in relation to the final properties.

In chain growth, monomer reactivities are dependent on the monomer structure and the reaction conditions. By use of appropriate monomers and polymerization techniques, a wide variety of copolymer structures (and properties) can be designed. The most common types are random, alternating, block, graft, and network copolymers.

For a chain reaction in a well-mixed BR or PFR, the instantaneous composition can be approximated by

$$F_1 = \frac{(r_1 - 1)f_1^2 + f_1}{(r_1 + r_2 - 2)f_1^2 + 2(1 - r_2)f_1 + r_2} \tag{33}$$

where F_1 is the mole fraction of monomer M_1 in the polymer, $f_1 = M_1/(M_1 + M_2)$ is the mole fraction of monomer M_1 in solution, $r_1 (= K_{11}/K_{12})$ is the reactivity ratio of monomer M_1, $r_2 (= K_{22}/K_{21})$ is the reactivity rates of monomer M_2, and where the K_{ij} are the propagation rate constants for the propagation steps

$$P_{n,m} + M_1 \xrightarrow{K_{11}} P_{n+1,m}$$

$$P_{n,m} + M_2 \xrightarrow{K_{12}} Q_{n,m+1}$$

$$Q_{n,m} + M_2 \xrightarrow{K_{22}} Q_{n,m+1}$$

$$Q_{n,m} + M_1 \xrightarrow{K_{21}} P_{n+1,m} \quad (34)$$

where $P_{n,m}$ is the concentration of growing polymer with terminal M_1, and $Q_{n,m}$ is the concentration of polymers with terminal M_2.

The dynamic behavior of a BR may affect the uniformity of properties of the copolymer product. Probably the most important problem is the composition drift phenomenon. As the intervening comonomers do not react with the growing chain at the same rate, one monomer in particular may become preferentially depleted from the reaction medium. Consequently, the copolymer formed at that instant will also become progressively depleted in that monomer.

Except for the particular case in which M_1 and M_2 disappear in such a way that f_1 remains constant, the CCD will not be monodisperse due to variations in f_2 with total monomer conversion.

In order to produce a polymer with monodisperse CCD, there are two main alternatives: (a) to convert to semibatch operation with addition of the proper amounts of the most reactive monomer as the reaction proceeds, in order to maintain f_1 constant; or (b) to adjust the reactor temperature (affecting the values of r_1 and r_2) so as to keep F_1 constant in the face of changes in f_1.

Control scheme (a) is quite simple, as it would only require addition of one monomer species. However, its practical implementation is seriously limited because the monomer must be perfectly mixed with the reacting mass in order to produce a monodisperse CCD. Otherwise, any imperfection in the mixing will produce a widened CCD.

Control scheme (b) is not as simple conceptually, as it requires the selection of a temperature at each f_1 to give the correct combination of r_1 and r_2 for constant F_1. In fact, this cannot always be achieved for all systems. Nevertheless, it has the enormous practical advantage that no material is added to the polymer, so that only temperature uniformity is required at all times. In fact, the design of reactors for high-viscosity systems with adequate temperature control is probably an easier task than the achievement of perfect micromixing [63].

Ray and Gall [63] studied in detail the alternative of temperature control. The authors indicated that in cases where the activation energies of the four propagation reactions have the proper relative values, there exist temperature versus time policies which in principle eliminate drift without converting to semibatch operation. They delineated the necessary conditions on the activation energies. Their results can be summarized as follows:

1. A single temperature for each f_1 is always guaranteed for

$$F_{10} < 0.5, n > 1, \quad F_{10} > 0.5, 0 \leq n < 1 \quad (35)$$

and for all F_{10}, $n < 0$.

2. Either two temperatures or no temperature exists for each f_1 when

$$F_{10} > 0.5, n > 1, \quad \text{or} \quad F_{10} < 0.5, 0 \le n < 1 \tag{36}$$

where $n = (E_{11} - E_{12})/(E_{22} - E_{21})$ and E_{ij} is the activation energy for addition of monomer M_j to growing polymer terminating in M_i.

It is obvious that the proposed temperature variations for CCD control tend to widen the MWD. Therefore, the above temperature policies offer best practical advantages whenever MWD modifications do not become critical.

The effectiveness of the temperature control schemes proposed by Ray and Gall [63] have been experimentally investigated by Tirrell and Gromley [64] for batch radical systems. For styrene–acrylonitrile (ST–AN) systems the authors determined temperature drift in CCD. Calculations were performed without resort to variational calculus, given the disjoint character of the problem. Figure 13 illustrates one of these calculations for a particular value of initial temperature and initiator concentration (T_0 and I_0), and several target values for F_1 (copolymer composition). It is found that increasing temperature policies result for f_1^0 below the azeotropic composition and decreasing temperature policies occur above that critical composition.

All these contributions on CCD control through temperature manipulation have not considered the energy balances in the models. The practical implementation of optimal temperature histories in real scale reactors, where thermal lags may become important, is an aspect requiring further investigation.

All previous problems, including those related to homopolymerization, have been approached as scalar dynamic optimization problems by determination of a proper time-varying manipulation of the control as the best solution. The failure of scalarization, already discussed in a previous section, becomes particularly notorious in copolymerization. In this case, each of the three distributions, MWD, CCD, and CSD, has its unique and profound influence on the final properties of the product. An optimal trajectory

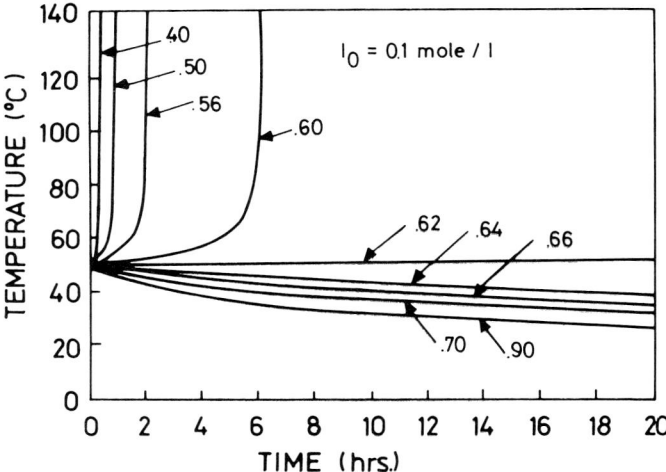

Figure 13 Optimal temperature vs. time policies for maintaining constant composition (F_1). (From Ref. 64.)

established in relation to one objective in particular may have an adverse effect on one or more of the other component objectives.

A scalar approach will not make obvious the formulation of an objective containing several performance measures in a single expression. No criteria are generally available a priori on the choice of weighting factors and the solutions being generated. The final result, summarized in a single numerical value, gives no insight about the overall competition or conflict among objectives (tradeoffs). Recognition of the multiobjective nature of these problems implies a more correct formulation as vector optimization problems.

Tsoukas et al. [65] applied the tools of multiobjective optimization to the semibatch copolymerization of styrene–acrylonitrile (ST–AN) and demonstrated the advantages of a correct formulation through the concept of Pareto optimality. Noninferior sets of optimal solutions (tradeoff points) were calculated for dual objectives of the form.

$$f_1 = \int_0^{tf} [F_1^* - F_1(t)]^2 \, dt$$

and

$$f_2 = \text{PD}(t_f) \tag{37}$$

the first minimizing deviations from a target composition F_1^* and the second minimizing PD or the breadth of the MWD.

The authors showed that for several sets of control variables (temperature, monomer, temperature and monomer) it is not convenient to control with respect to one objective functional exclusively. Almost always, compromise solutions can be found which are closer to ideal solutions for both objectives.

Figure 14 illustrates noninferior sets obtained with different controls. The absolute minima for the above objectives are $f_1 = 0$ and $f_2 = 1.5$. The curved tradeoff locus corresponds to temperature. It is also found that monomer addition alone permits a much closer approach to ideal solutions than can be obtained with other controls. Temperature may be of advantage only when PD is not an important objective.

Farber [66] extended the analysis to a continuous free radical polymerization reactor in solution. Conflicting trends among conversion, molecular weight, and composition have been analyzed by calculation of the noninferior sets.

Dual objectives of the form $\bar{f} = (f_1, f_2)$ were considered where the component terms f_1, f_2 are related to pairs among a total number average molecular weight, the monomer conversion of component 1, and a composition objective component defined as the mole fraction of monomer 2 in the binary copolymer. Two copolymerization systems, methyl methacrylate–vinyl acetate (MMA–VA) and styrene–acrylonitrile (ST–AN) were simulated. In all cases it was found that this continuous system allows little control of polydispersity, which remains practically insensitive to all parameter changes.

For the MMA–VA system, the copolymer composition presents more flexibility to the reactor conditions, but with the concomitant negative effect on total average molecular weight or conversion.

For the ST–AN system and the particular choice of reactor parameters, composition cannot be modified in any considerable degree through temperature changes. Figure 15 illustrates the effect of temperature on conversion, composition, molecular weight, and polydispersity. Figure 16 contains a noninferior set for simultaneous maximization of molecular weight and conversion. Two independent parameters, temperature and residence time, were considered.

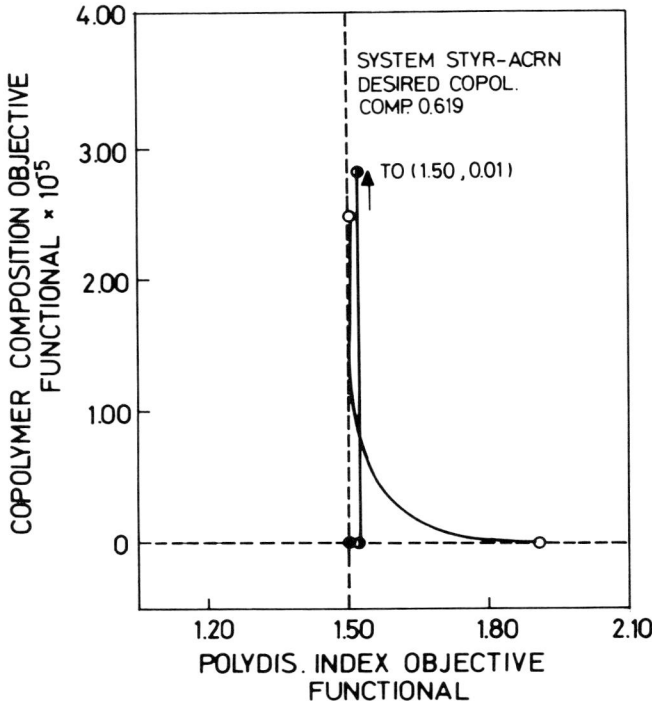

Figure 14 Noninferior sets for composition and polydispersity under different controls. ○, temperature; ◑, monomer addition; ●, temperature and monomer addition. (From Ref. 65.)

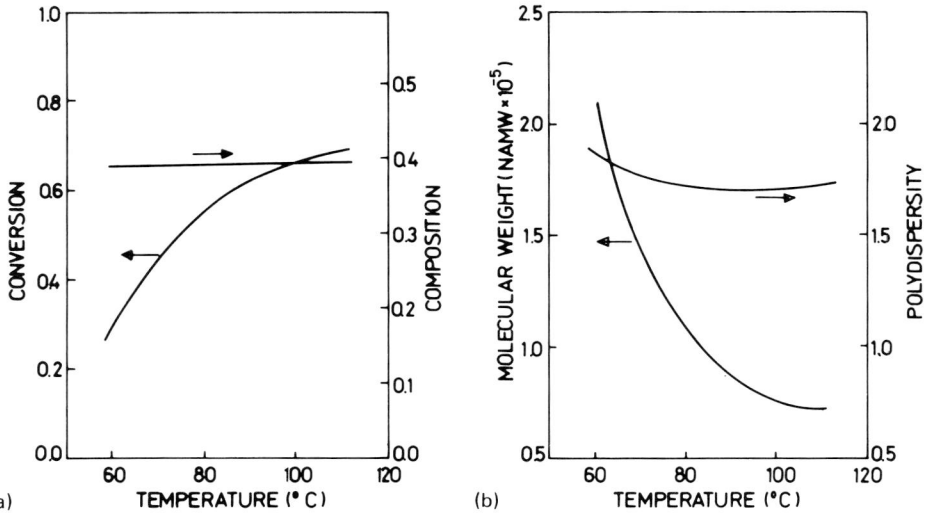

Figure 15 ST–AN system. Continuous copolymerization in solution. (*a*) Conversion and composition vs. temperature. (*b*) Total number average molecular weight and polydispersity vs. temperature. (From Ref. 66.)

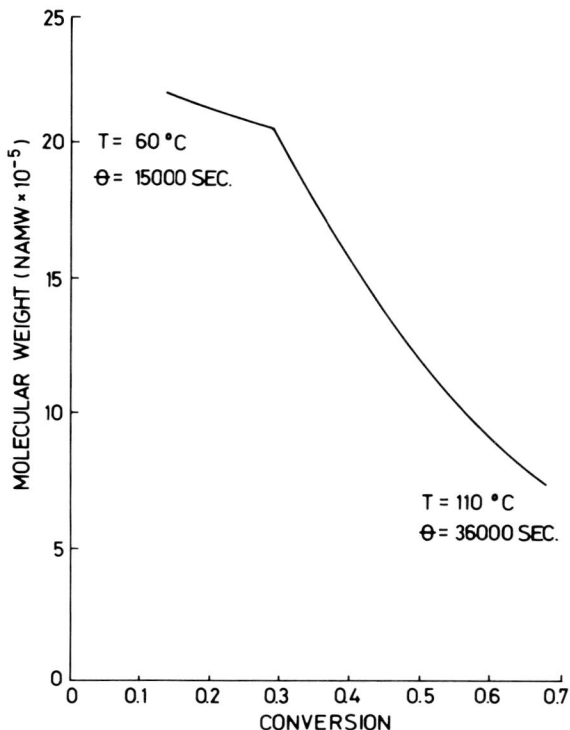

Figure 16 ST–AN system. Noninferior set for simultaneous maximization of total number average molecular weight and conversion. Two independent parameters, temperature and residence time. (From Ref. 66.)

OPTIMIZATION OF TRANSIENT OPERATION

Although optimization of transient operation is an important problem in chemical engineering, it becomes particularly relevant in continuous polymerization processes. A selected operation condition will determine, besides a conversion level, a variety of important properties defining the applicability of the polymeric material. Therefore, it becomes crucial to minimize disturbances from a prescribed reactor state.

A common feature to many polymerization processes is the long residence times (as large as 10^6 sec) required to reach high conversion and adequate molecular weight. Consequently, reactor disturbances are accompanied by very long transients, with the concomitant production of a significant amount of polymer far from the desired specifications.

Despite the inherent economic importance, many of these problems have not been resolved. In general, operation policies for these situations have been based on criteria established on a particular reacting system, which is an expensive proposition.

The start-up problem is the attainment of a desired state at the beginning of continuous operation. The change of specification problem occurs when the state of the system evolves from an initial condition to a final condition defining new product specifications.

Both problems are mathematically equivalent and cannot be considered as linearizations around a steady state; they must be seen as large disturbances to be described by the complete equations.

In most problems related to transient operation in continuous reactors, the final conditions represent steady states which also define a number of product specifications through a set of simultaneous uncoupled equations (i.e., related to the MWD properties). The selection of these target states is obviously conditioned by their accessibility and stability. Therefore, knowledge of the reactor dynamics and stability must precede any optimization study.

In addition, it is necessary to verify that a final condition is indeed an accessible target in relation to a given initial state. This statement is valid for any regular trajectory as well as for the case of optimal trajectories. In this second situation, however, the accessibility of certain regions may be still more difficult. In addition to the restrictions dictated by the reactor dynamics, the evolution of the system is constrained by the particular optimality criteria adopted in the problem.

Other effects may also prevent the accessibility of a desired operation condition: constraints in the state and control parameters, depletion of reactants, etc. [85].

Outside the field of polymerization, the problem of monitoring large disturbances (i.e., start-up) has been treated by Siebenthal and Aris [100] for the case of time optimality. Jackson [101] studied the role of the reactor dynamics and stability in the optimal start-up of autothermal reactors. Han [102] used frequency domain criteria to solve this problem of accessibility of a selected steady state in simple irreversible kinetic schemes. However, the importance of these studies has not been clearly recognized. In these simple systems, the typical ignition phenomenon is characterized by only three steady states normally related to very different reactor conditions; therefore, their accessibility may be achieved without resort to complicated analysis.

In polymerization reactors the situation is different. The large dimension of the state system of equations and the existence of several sources for unstable response (besides the traditional thermal sources) result in a complex dynamic behavior with many possible steady states. Some aspects of polymerization reactor dynamics in large disturbances (start-up) are illustrated in Refs. 103–106 without considering the optimization problem.

One of the earliest references to the problem of large disturbances in continuous polymerization reactors is due to Taylor [83]. This author studied the evolution of the state of the reactor after an isothermal step change in initiator concentration with the purpose of reducing the specified final molecular weight (Fig. 17).

Lee et al. [84] simulated the start-up problem for an ideal polymerization in a CSTR. The optimality criterion consisted of maintaining the average kinetic chain length constant. No considerations were given to the possible steady states for the system. Hicks et al. [55] studied some start-up policies in continuous stirred tanks for chain and step reactions. All these policies were developed under the condition of no monomer in the feed, reducing the case to a batch reactor problem.

Farber and Laurence [85] simulated the time optimality problem in continuous operation for the thermal polymerization of styrene in solution. This constitutes one of the simplest possible polymerization models. The state of the system is described by two balance equations related to monomer conversion and energy. The task is to find the values of the global heat transfer coefficient or manipulated variable, optimal in guiding the state of the system to a desired target steady state in the minimum time. The solution describes both large disturbances (start-up problem) and small disturbances (regulator

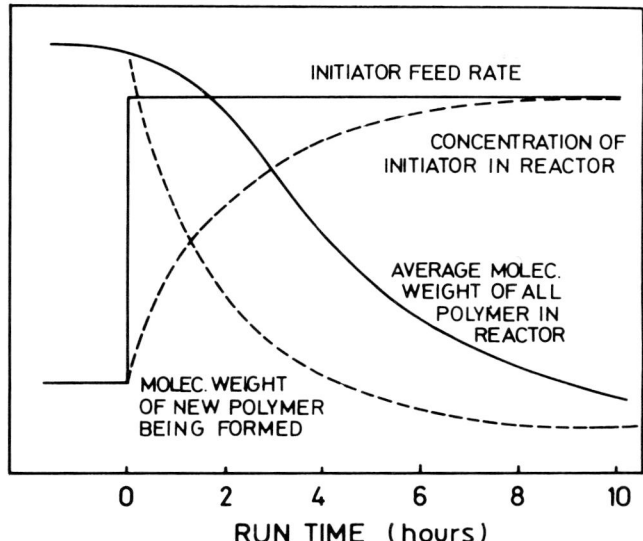

Figure 17 Illustration of the way initiator concentration and molecular weight changes lag behind step change in initiator feed rate. (From Ref. 83).

problem). Application of the minimum principle to a problem that is linear in the control variable (coolant flow rate) yields solutions with bang-bang control, singular control, or a combination of the two.

The analytical procedure to determine the switching times is given in full detail in Ref. 100. The analysis is completely equivalent to superimposing the phase planes for adiabatic and full cooling operation (Fig. 18). The result is presented in Figure 19. The switching curves are those two single curves that lead to the selected steady state (paths CBA and FED).

The determined optimal control policy is an open loop control. The extension to a closed loop is immediate and is based on the fact that for autonomous systems, theorems guarantee the uniqueness of the solutions.

For a given value of the control parameter h, trajectories in the concentration vs. temperature plane do not intersect at any point, except the steady state corresponding to that particular value of h. Consequently, any point representing the natural state of the system will undergo the same policy in order to reach the origin in the minimum time, whether it is an open or closed loop analysis.

Any extra complexity in the model resulting in a more complex objective or an increased number of state equations will preclude use of the phase plane analysis. In general, for dimensions larger than 2, the method becomes ineffective and the influence of reactor dynamics has to be considered through different procedures.

Farber and Laurence [85] also studied the start-up and change of specification problems in a CSTR for the solution polymerization of styrene in the presence of an initiator. The objective to be minimized is the unweighted sum of squared deviations of the state of the system $\bar{X}(t)$ from their target values $\bar{X}^*$ defining a steady state:

$$J = \sum_{i=1}^{3} [X_i(t_f) - X_i^*] + w \int_0^{t_f} U_Q \, dt \tag{38}$$

The last term allows for minimization of monomer consumption during the transient.

Application of Pontryagin's minimum principle to this class of problems typically results in a Hamiltonian that is linear in the control variable but highly nonlinear in the state variables:

$$H = \bar{f} + \phi(\bar{X}, t)U \tag{39}$$

In all these situations the solution is characterized by the presence of singular arcs or singular controls.

If $\phi(\bar{X}, t)$ becomes identically zero over a finite interval, the Hamiltonian ceases to be an explicit function of the control variable, and the problem is referred to as singular. When this identity is verified, the minimum principle and the calculus of variations fail to provide effective optimality conditions in singular controls, and they do not have a general procedure to derive optimal singular extremals.

In case of singular controls, optimality conditions have to be sought through a variety of approaches based on a generalized Legendre–Clebsch condition [85].

Optimal bang-bang solutions, which can occur in conjunction with optimal singular solutions, are known to appear in a large number of design and control problems in chemical engineering. Under certain parameter conditions, stirred tanks and tubular reactors systematically present singular controls, even in the case of simple reactions.

The knowledge of the singular surface (defined by $H_U = 0$, $H_{UU} = 0$) may facilitate the analysis and synthesis of a given control problem. Though a formula defining a singular surface may be obtained, its solution is not immediate, because analytical solu-

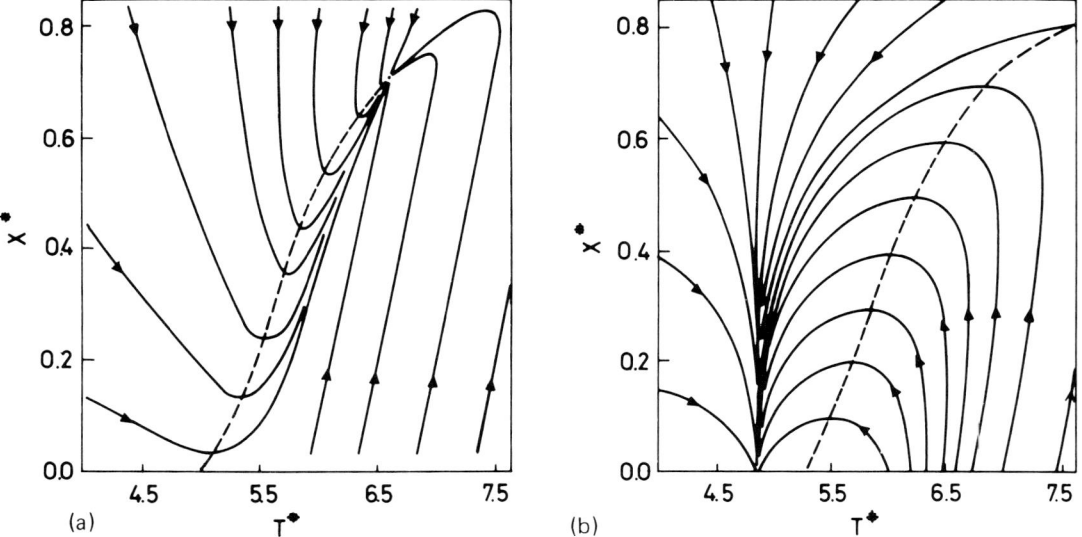

Figure 18 Phase plane analysis. Scaled conversion vs. dimensionless temperature. (*a*) Adiabatic operation. (*b*) Full cooling operation. (From Ref. 85.)

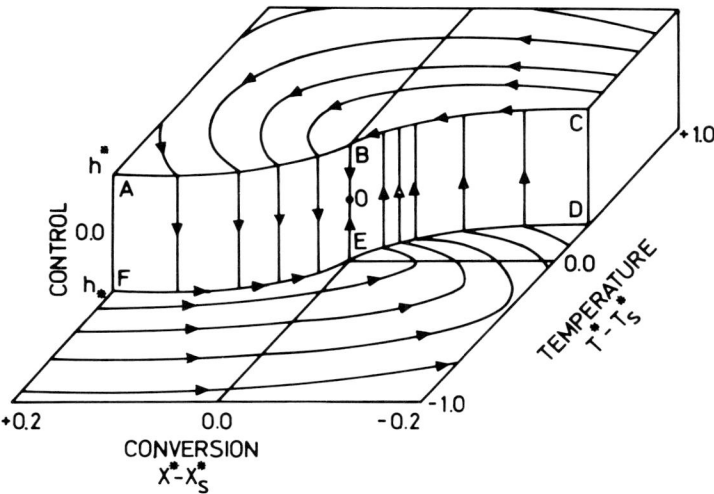

Figure 19 Phase plane analysis. Time optimal control trajectories.

tions for state and adjoint functions are not generally known a priori [85]. As a consequence, and as a general rule, it is usually impossible to establish the existence of singular controls without obtaining the actual computational solution to the problem.

For the synthesis of the optimal singular solution, gradient methods were used. Details on numerical aspects can be found in Refs. 85 and 107.

The start-up problem was simulated for a target steady state representing 70% conversion. Minimization of the objective given by Eq. 38 was sought by simultaneous manipulation of temperature and monomer feed rate. Numerical evaluation of policies involving bang-bang and/or singular controls was performed with a version of the piecewise minimum-H strategy, combined with a limit process to remove the singular condition. Besides the good convergence properties, the method requires a minimum of computer time and storage in comparison with related algorithms [107].

The evolution of control and state variables is presented in Fig. 20. The sensitivity of the initiator becomes evident with a concentration decay of almost four orders of magnitude.

All those fixed time policies that include minimization of the feed flow rate during the transient indicate that the process should be initiated as a batch reaction, or without monomer in the feed. When bounds are present, the flow rate should remain at its lower bound.

CONCLUSION

An overview of optimization of polymerization reactors has been presented. The applicability of tools of optimal control theory, previous contributions, and some important problems and trends have been discussed.

Progress in this area has been conditioned by the comprehension of complex phenomena characterizing polymerization reaction engineering and the availability of realistic mathematical models. For a number of years, most of the work has been characterized by

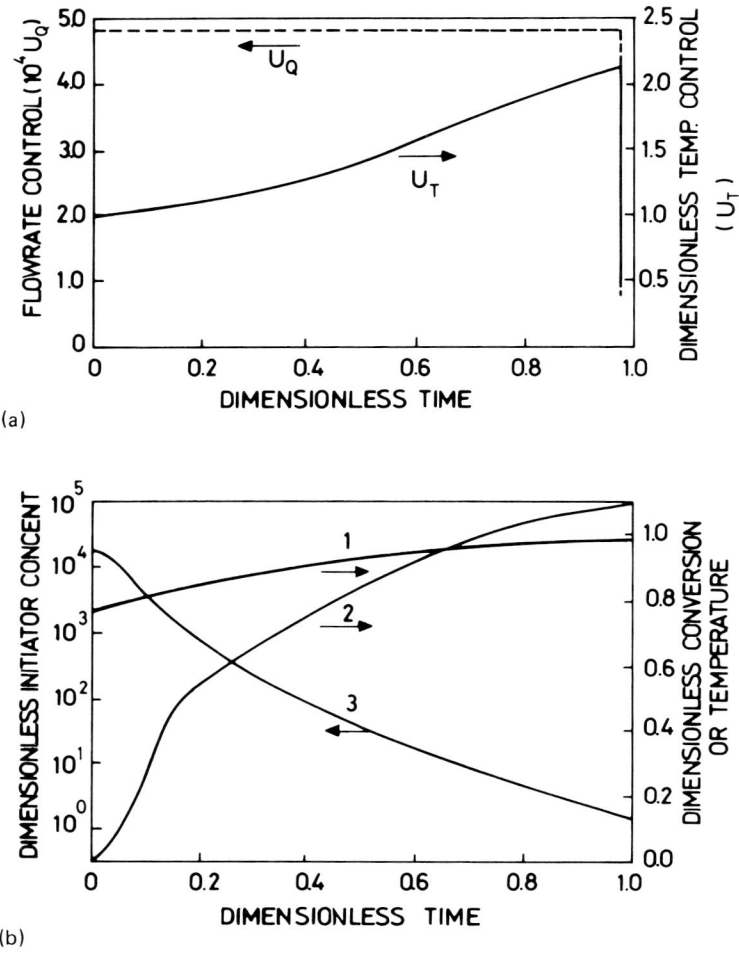

Figure 20 (a) Start-up problem. Simultaneous temperature control (U_T) and flow rate control (U_Q) vs. time. (b) Initiator concentration, conversion, and temperature vs. time. (From Ref. 85.)

simplified models and objectives, with optimization studies confined to the reactor unit itself. A variety of relevant problems remain unresolved. It is only recently that the focus in this area has been changed through better understanding of kinetic mechanisms, structure–property relationships, dynamic behavior, and development of adequate algorithms.

NOTATION

BR	batch reactor
CCD	copolymer composition distribution
CPFR	continuous plug flow reactor
CSD	copolymer sequence distribution
E_i	activation energies

E_{ij}	activation energy for addition of monomer M_j to growing polymer terminating in M_i
$\bar{f}$	state vector, or vector objective
f_i	vector components, mole fraction of monomer M_i in copolymer
F_i	copolymer composition, mole fraction of M_i
F_{i0}	initial copolymer composition
H	Hamiltonian function
HCSTR	homogeneous continuous stirred task reactor
I	initiator concentration
J	objective functional
K	step polymerization rate constant
K_d	initiation rate constant
K_f	transfer to monomer rate constant
K_p	propagation rate constant
M_i	monomer concentration of component i
$\bar{M}_n$	number average molecular weight
$\bar{M}_w$	weight average molecular weight
MWD	molecular weight distribution
PD	polydispersity index $\bar{X}_w/\bar{X}_n$
PFR	plug flow reactor
$P_{n,m}$	concentration of growing polymer with terminal M_1
$Q_{n,m}$	concentration of growing polymer with terminal M_2
r_i	reactivity ratio
SBR	semibatch reactor
SCSTR	segregated continous stirred tank reactor
t	time
T	temperature
$\bar{U}$	control vector
U_Q	monomer flowrate control parameter, $U_Q = \theta^{-1}$
U_r	temperature control parameter (heat transfer capability)
W	condensation by-product
$\bar{X}$	state vector
$\bar{X}_n$	number average degree of polymerization
$\bar{X}_w$	weight average degree of polymerization
β	reaction order for initiator in global rate
ϵ	volume contraction coefficient
ξ	ratio of adjoint functions
λ_i	ith-order moment of active polymer chain length distribution
λ_{i0}	initial ith-order moment of active polymer chain length distribution
ψ	adjoint function vector
θ	residence time

REFERENCES

1. H. Gerrens, "Polymerization Reactors and Polyreactions. A Review," Proceedings of the 4th International/6th European Symposium on Chemical Reaction Engineering, Heidelberg, FRG, Vol. 2: Survey Papers, DECHEMA, p. 585 (1976).
2. H. Gerrens, *Ger. Chem. Eng.*, 4: 1 (1981).

3. C. E. Schildknecht and I. Skeist, *Polymerization Processes,* Interscience, New York (1977).
4. D. H. Sebastian and J. A. Biesenberger, *Polymerization Engineering,* Wiley, New York (1983).
5. K. F. O'Driscoll, *Pure Appl. Chem., 53*: 617 (1981).
6. K. G. Denbigh, *Trans. Faraday Soc., 43*: 648 (1947).
7. Z. Tadmor and J. A. Biesenberger, *Ind. Eng. Chem. Fund., 5*: 336 (1966).
8. W. H. Ray, *J. Macromol. Sci. Rev. Macromol. Sci., C8*: 1 (1972).
9. W. H. Ray and R. L. Laurence, in *Chemical Reactor Theory: A Review* (L. Lapidus and N. Amundson, eds), Prentice-Hall, Englewood Cliffs, N.J., p. 532 (1976).
10. W. H. Ray, *ACS Symp. Ser., 226*: 101 (1983).
11. T. Keane, "Single Phase Polymerization Reactors," Proceedings 2nd Chemical Reaction Engineering Symposium, paper A 7-1 (1972).
12. R. S. Knorr and K. F. O'Driscoll, *J. Appl. Polym. Sci., 14*: 2683 (1970).
13. A. Ahmad, M. N. Treybig, and R. G. Anthony, *J. Appl. Polym. Sci., 21*: 2021 (1977).
14. W. H. Ray, in *Modelling of Chemical Reactors* (Ebert, Deuflhard, and Jager, eds.), Springer, New York, p. 337 (1981).
15. R. Jaisinghani and W. H. Ray, *Chem. Eng. Sci., 32*: 811 (1977).
16. J. W. Hamer, T. A. Akramov, and W. H. Ray, *Chem. Eng. Sci., 36*: 1897 (1981).
17. A. D. Schmidt, A. B. Clinch, and W. H. Ray, *Chem. Eng. Sci., 39*: 419 (1984).
18. A. Uppal, W. H. Ray, and A. B. Poore, *Chem. Eng. Sci., 29*: 967 (1974).
19. A. Uppal, W. H. Ray, and A. B. Poore, *Chem. Eng. Sci., 31*: 205 (1976).
20. L. S. Pontryagin, V. C. Boltianskii, R. V. Gamkrelidze, and E. V. Mischenko, *The Mathematical Theory of Optimal Processes,* Interscience, New York (1962).
21. M. Denn, *Optimization by Variational Methods,* McGraw-Hill, New York (1969).
22. D. E. Kirk, *Optimal Control Theory: An Introduction,* Prentice-Hall, Englewood Cliffs, N.J. (1970).
23. W. H. Ray and J. Szekely, *Process Optimization,* Wiley, New York (1973).
24. A. E. Bryson and Y. C. Ho, *Applied Optimal Control,* 2d. ed., Blaisdell, Waltham, Mass. (1976).
25. A. Miele, in *Optimization Techniques* (G. Leitman, ed.), Academic Press, New York, Ch. 4 (1962).
26. L. Lapidus and R. Luus, *Optimal Control of Engineering Processes,* Blaisdell, Waltham, Mass. (1967).
27. D. W. Peterson and J. H. Zalkind, *Int. J. Control, 28*: 589 (1978).
28. R. W. Nunes, J. R. Martin, and J. F. Johnson, *Polym. Eng. Sci., 22*: 205 (1982).
29. N. Takama and T. Umeda, *Chem. Eng. Sci., 36*: 129 (1980).
30. D. G. Luenberger, *Introduction to Linear and Non-Linear Programming,* Addison-Wesley, Reading, Mass. (1973).
31. T. L. Vincent and W. J. Grantham, *Optimality in Parametric Systems,* Wiley, New York (1981).
32. V. Pareto, *Cours d'Economie Politique,* Lausanne (1896).
33. M. Salukvadze, *Vector Valued Optimization Problems in Control Theory,* Academic Press, New York (1979).
34. Y. Y. Haimes, W. A. Hall, and H. T. Freedman, *Multiobjective Optimization of Water Resources Systems,* Elsevier, Amsterdam (1975).
35. J. L. Cohon, *Multiobjective Programming and Planning,* Academic Press, New York (1978).
36. C. L. Hwang, S. R. Paidy, K. Yoon, and A. S. Masud, *Comp. Oper. Res., 7*: 5 (1980).
37. F. A. Shatkan, *Vysokomol. Soedin., 7*: 449 (1965). [*Polymer Sci. USSR, 7*: 498 (1965).]
38. Y. Yoshimoto, H. Yanagawa, T. Susuki, Y. Inaba, and T. Araki, *Kagaku Kogaku, 32*: 595 (1968).
39. H. Nishimura and F. Yokoyama, *Kagaku Kogaku, 32*: 601 (1968).

40. P. E. King and J. M. Skaates, *Ind. Eng. Chem. Proc. Des. Dev.*, *8*: 114 (1969).
41. K. Osakada and L. T. Fan, *J. Appl. Polym. Sci.*, *14*: 3063 (1970).
42. Y. Yoshimoto, H. Yamagawa, T. Suzuki, T. Araki, and Y. Inaba, *Int. Chem. Eng.*, *11*: 147 (1971).
43. M. E. Sacks, S. Lee, and J. A. Biesenberger, *Chem. Eng. Sci.*, *27*: 2281 (1972).
44. S. A. Chen and W. F. Jeng, *Chem. Eng. Sci.*, *33*: 735 (1978).
45. S. A. Chen and K. F. Lin, *Chem. Eng. Sci.*, *35*: 2325 (1980).
46. S. A. Chen and N. Wang, *Chem. Eng. Sci.*, *36*: 1295 (1981).
47. B. Boorga Rao and R. D. Mhaskar, *Polymer*, *22*: 1593 (1981).
48. G. Z. A. Wu, L. A. Denton, and R. L. Laurence, *Polym. Eng. Sci.*, *22*: 1 (1982).
49. I. Thomas and C. Kiparissides, *Can. J. Chem. Eng.*, *62*: 184 (1984).
50. J. N. Farber and R. L. Laurence, *Chem. Eng. Commun.*, *46*: 347 (1986).
51. B. J. Yoon and H. K. Rhee, *Chem. Eng. Commun.*, *34*: 253 (1985).
52. H. Mavridis and C. Kiparissides, *Polym. Proc. Eng.*, *3*: 263 (1985).
53. R. E. Hoffman, S. Schreiber, and G. Rosen, *Ind. Eng. Chem.*, *56*: 51 (1964).
54. L. F. Beste and H. K. Hall, *J. Macromol. Sci. (Chem.)*, *1*: 121 (1966).
55. J. Hicks, A. Mohan, and W. H. Ray, *Can. J. Chem. Eng.*, *47*: 590 (1969).
56. M. E. Sacks, S. Lee, and J. A. Biesenberger, *Chem. Eng. Sci.*, *28*: 241 (1973).
57. B. M. Louie and D. S. Soong, *J. Appl. Polym. Sci.*, *30*: 3707 (1985).
58. B. M. Louie and D. S. Soong, *J. Appl. Polym. Sci.*, *30*: 3825 (1985).
59. F. Banderman, *Angew. Macromol. Chem.*, *18*: 137 (1971).
60. F. Banderman and R. Dunsing, *Chem. Ing. Tech.*, *46*: 617 (1974).
61. F. Banderman, R. Dunsing, and K. Elgert, *Chem. Ing. Tech.*, *46*: 615 (1974).
62. R. Hanna, *Ind. Eng. Chem.*, *49*: 208 (1957).
63. W. H. Ray and C. E. Gall, *Macromolecules*, *2*: 425 (1969).
64. M. Tirrell and K. Gromley, *Chem. Eng. Sci.*, *36*: 367 (1981).
65. A. Tsoukas, M. Tirrell, and G. Stephanopoulos, *Chem. Eng. Sci.*, *37*: 1785 (1982).
66. J. N. Farber, *Polym. Eng. Sci.*, *26*: 499 (1986).
67. S. A. Chen and S. T. Lee, *Polym. Eng. Sci.*, *25*: 987 (1985).
68. G. R. Meira, *J. Macromol. Sci. Rev. Macromol. Chem.*, *20*: 207 (1981).
69. B. Langner and F. Bandermann, *Angew. Macromol. Chem.*, *71*: 101 (1978).
70. G. L. Frontini, G. E. Elicabe, D. A. Couso, and G. R. Meira, *J. Appl. Polym. Sci.*, *31*: 1019 (1986).
71. G. L. Frontini, G. E. Elicabe, and G. R. Meira, *J. Appl. Polym. Sci.* (1987).
72. P. J. Hoftyzer, J. Hoogschagen, and D. W. Van Krevelen, "Optimization of Caprolactam Polymerization," *Proceedings 3rd European Symposium on Chemical Reaction Engineering*, p. 247 (1964).
73. H. K. Reimschuessel and K. Nagasubramanian, *Chem. Eng. Sci.*, *27*: 1119 (1972).
74. A. Ramagopal, A. Kumar, and S. K. Gupta, *J. Appl. Polym. Sci.*, *28*: 2261 (1983).
75. A. Kumar, S. N. Sharma, and S. K. Gupta, *Polym. Eng. Sci.*, *24*: 1205 (1984).
76. S. K. Gupta, B. S. Damania, and A. Kumar, *J. Appl. Polym. Sci.*, *29*: 2177 (1984).
77. A. K. Ray and S. K. Gupta, *J. Appl. Polym. Sci.*, *30*: 4529 (1986).
78. A. K. Ray and S. K. Gupta, *Polym. Eng. Sci.*, *26*: 1033 (1986).
79. J. S. Shastry, L. T. Fan, and L. E. Erikson, *J. Appl. Polym. Sci.*, *17*: 3101 (1973).
80. J. S. Shastry, L. T. Fan, and L. E. Erikson, *J. Appl. Polym. Sci.*, *17*: 3127 (1973).
81. J. S. Shastry, Ph.D. dissertation, Kansas State University, 1973.
82. J. J. Davis and R. I. Kermode, *Ind. Eng. Chem. Proc. Des. Dev.*, *14*: 459 (1975).
83. J. M. Taylor, *Chem. Eng. Prog.*, *58*: 42 (1962).
84. S. I. Lee, J. Nozue, and T. Imoto, *Kagaku Kogaku*, *32*: 601 (1968).
85. J. N. Farber and R. L. Laurence, *Macromol. Chem. Symp.*, *2*: 193 (1986).
86. R. Thiele, *Chem. Eng. Sci.*, *41*: 1123 (1986).

87. D. Rudd and C. C. Watson, *Strategy of Process Engineering,* Wiley, New York (1968).
88. D. F. Rudd, G. J. Powers, and J. J. Siirola, *Process Synthesis,* Prentice-Hall, Englewood Cliffs, N.J. (1973).
89. N. Nishida, G. Stephanopoulos, and A. W. Westerberg, *AIChE J., 27*: 321 (1981).
90. S. P. Chitra and R. Govind, *AIChE J., 31*: 177 (1985).
91. S. P. Chitra and R. Govind, *AIChE J., 31*: 185 (1985).
92. E. L. Krichevskaya, *Zh. Fiz. Khim., 26*: 3 (1952).
93. S. Szepe and O. Levenspiel, *Chem. Eng. Sci., 23*: 881 (1968).
94. S. I. Lee and C. M. Crowe, *Chem. Eng. Sci., 25*: 743 (1970).
95. V. G. Boltyanskii, *Mathematical Methods of Optimal Control,* Holt, Rinehart and Winston, New York (1971).
96. Y. D. Kwon and L. B. Evans, *AIChE J., 21*: 1158 (1975).
97. W. H. Ray, *Macromolecules, 4*: 162 (1971).
98. W. H. Ray, T. L. Douglas, and E. W. Godsalve, *Macromolecules, 4*: 166 (1971).
99. C. G. Overberger, *J. Polym. Sci. Polym. Symp., 72*: 67 (1985).
100. C. D. Siebenthal and R. Aris, *Chem. Eng. Sci., 19*: 729 (1964).
101. R. Jackson, *Chem. Eng. Sci., 21*: 241 (1966).
102. C. D. Han, *Ind. Eng. Chem. Fundam., 9*: 634 (1970).
103. B. W. Brooks, *Chem. Eng. Sci., 34*: 1417 (1979).
104. N. F. Umoh, P. Harriott, and R. Hughes, *Chem. Eng. J., 21*: 85 (1981).
105. B. W. Brooks, *Chem. Eng. Sci., 36*: 589 (1981).
106. E. E. Badder and B. W. Brooks, *Chem. Eng. Sci., 39*: 1499 (1984).
107. N. Nishida, Y. A. Liu, L. Lapidus, and S. Hiratzuka, *AIChE J., 22*: 505 (1976).

II
POLYMER CHARACTERIZATION AND MOLECULAR STRUCTURE

13

Techniques for Polymer Property Characterization

Nicholas P. Cheremisinoff
Exxon Chemical Company
Linden, New Jersey

INTRODUCTION	471
ATOMIC ABSORPTION SPECTROSCOPY	471
CHROMATOGRAPHY	473
INDUCTIVELY COUPLED PLASMA ATOMIC EMISSION SPECTROSCOPY	476
ION CHROMATOGRAPHY	478
ION SELECTIVE ELECTRODES	480
MASS SPECTROMETRY	481
NEUTRON ACTIVATION ANALYSIS	485
NUCLEAR MAGNETIC RESONANCE SPECTROMETRY	486
THERMAL ANALYSIS	487
ULTRAVIOLET, VISIBLE, AND INFRARED SPECTROMETRY	496
FOURIER TRANSFORM INFRARED SPECTROMETRY	497
X-RAY FLUORESCENCE SPECTROMETRY	498
SUMMARY	501

INTRODUCTION

This chapter highlights analytical tools used widely in the study and characterization of polymer molecular/chemical structures and their important physical characteristics. The descriptions of various techniques are not intended to be comprehensive, but rather to highlight the measurement principles and to provide a working knowledge of the type of information obtainable as a summary of the principal characterization tools used in polymer property studies and product development activities. It will help to orient the encyclopedia user in terms of the measurement tool terminology used in later chapters of this and subsequent volumes.

ATOMIC ABSORPTION SPECTROSCOPY

Atomic absorption is used for the determination of ppm levels of metals. It is not normally used for the analysis of the light elements such as H, C, N, O, P, and S, halogens, and noble gases. Higher concentrations can be determined by prior dilution of the sample. It is not recommended if a large number of elements are to be measured in a single sample.

The technique of atomic absorption (AA) spectrometry involves vaporizing a sample so that the element of interest is atomized at high temperatures. The element concentration is determined based on the attenuation or absorption by the analyte atoms of a characteristic wavelength emitted from a light source. The light source can be a hollow cathode lamp containing the element to be measured. Separate lamps are used for each element. The detector can consist of a photomultiplier tube. A monochromator is used to separate the element line and the light source is modulated to reduce the amount of unwanted radiation reaching the detector. Figure 1 illustrates a conventional setup which employs a flame atomization system for liquid sample vaporization. An air–acetylene flame is used for most elements and reaches temperatures of approximately 2300°C. A higher temperature nitrous oxide–acetylene flame (~2900°C) is used for more refractory oxide-forming elements. Electrothermal atomization techniques such as a graphite furnace can be used for the direct analysis of solid samples.

Although AA is a valuable technique, its use in recent years has declined in favor of ICP and XRF methods of analysis. The most common application of AA is for the determination of boron and magnesium in oils.

Conventional instruments will analyze liquid samples only with dilute acid and xylene solutions. The volume of solution needed depends on the number of elements to be determined. The technique provides excellent sensitivity for most elements with limited spectral interferences. For some elements sensitivity can be extended into the sub-ppb range with flameless methods.

The disadvantages with conventional AA arrangements are that they require a liquid sample and the determination of several elements per sample is slow and requires larger

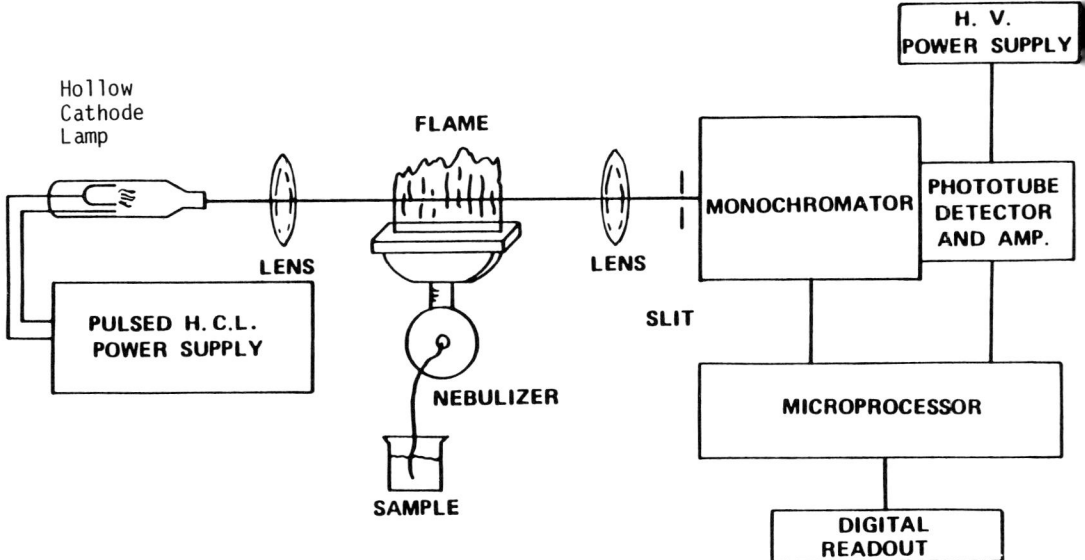

Figure 1 Schematic of atomic absorption spectrometer.

volumes of solution due to the sequential nature of the method. Chemical and ionization interferences must be corrected by modification of the sample solution. Also, calibration curves have a limited linear range.

It is important to note that chemical interference can arise from the formation of thermally stable compounds such as oxides in the flame. The use of electrothermal atomization, a hotter nitrous oxide–acetylene flame, or the addition of a releasing agent such as lanthanum can help reduce the interference. Flame atomization produces ions as well as atoms. Since only atoms are detected, it is important that the ratio of atoms to ions remain constant for the element being analyzed. This ratio is affected by the presence of other elements in the sample matrix. The addition of large amounts of an easily ionized element such as potassium to both the sample and standards helps mask ionization interference.

CHROMATOGRAPHY

Chromatography involves the separation of molecular mixtures by distribution between two or more phases, one phase being essentially two-dimensional (a surface) and the remaining phase, or phases, being a bulk phase brought into contact in a countercurrent fashion with the two-dimensional phase. The two main branches of chromatography are gas chromatography and liquid chromatography. Liquid chromatography can be further subdivided as shown in Figure 2. A chromatographic separation involves the following. A sample is placed at the top of a column where its components are sorbed and desorbed by a carrier. This partitioning process occurs repeatedly as the sample moves toward the outlet of the column. Each solute travels at its own rate through the column; consequently, a band representing each solute will form on the column. A detector attached to the column's outlet responds to each band. The output of the detector response versus time is referred to as a chromatogram. The time of emergence identifies the component, and the peak areas define its concentration, based on calibration with known compounds.

When the moving phase is a gas, the technique is called gas chromatography (GC). Here the sample is usually injected at high temperature to ensure vaporization, and as such, only materials volatile at this temperature can be analyzed.

There are two main types of GC. First, if the stationary phase is a solid, the technique is called gas–solid chromatography. The separation mechanism is principally one of

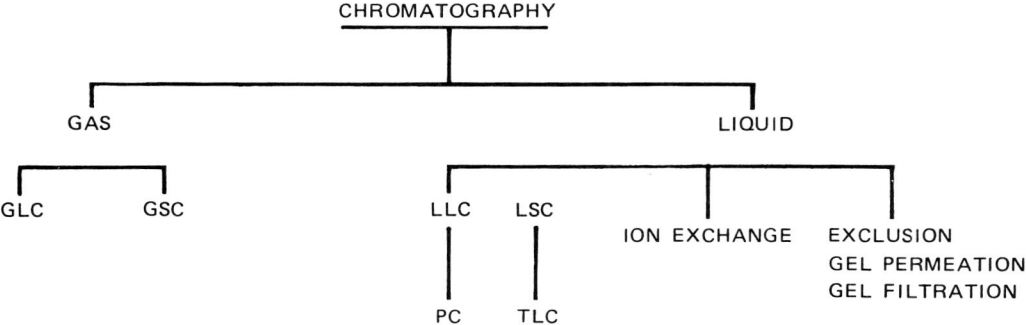

Figure 2 Classification of chromatographic techniques.

adsorption. Those components more strongly adsorbed are held up longer than those which are not.

In the second case, if the stationary phase is a liquid, the technique is referred to as gas–liquid chromatography and the separation mechanism is principally one of partition (solubilization of the liquid phase).

Gas chromatography has developed into one of the most powerful analytical tools available to the organic chemist. The technique allows separation of extremely small quantities of materials (10^{-6} g).

The characterization and quantitation of complex mixtures can be accomplished with this process. The introduction of long, narrow columns, both megabore and capillary, produces a greater number of theoretical plates, increasing the efficiency of separation beyond that of any other technique available. The technique is applicable over a wide range of temperatures (~40–350°C), making it possible to chromatograph materials covering a wide range of volatilities. The laboratory uses packed columns along with megabore and capillary. In this way the broadest range of gas chromatographic problems can be addressed. The detector used to sense and quantitate the effluent provides the specificity and sensitivity for the analytical procedure. Table 1 lists significant detector characteristics.

When the moving phase is a liquid, the technique is called liquid chromatography (LC). Here, the sample is first dissolved in the moving phase and injected at ambient temperature; hence there is no volatility requirement for samples. However, the sample must dissolve in the moving phase. Note that LC has an important advantage over GC: the solubility requirement can usually be met by changing the moving phase. The volatility requirement is not so easily overcome.

There are basically four types of LC, depending on the nature of the stationary phase and the separation mechanism:

- **Liquid–Liquid Chromatography (LLC)** This is partition chromatography or solution chromatography. The sample is retained by partitioning between mobile liquid and stationary liquid. The mobile liquid cannot be a solvent for the stationary liquid. A subgroup of liquid/liquid chromatography is paper chromatography.
- **Liquid–Solid Chromatography (LSC)** This is adsorption chromatography. Adsorbents such as alumina and silica gel are packed in a column and the sample components are displaced by a mobile phase. Thin layer chromatography and most open column chromatography are considered liquid–solid chromatography.
- **Ion Exchange Chromatography** This technique employs zeolites and synthetic organic and inorganic resins to perform chromatographic separation by an exchange of ions between the sample and the resins. Compounds which have ions with different affinities for the resin can be separated.
- **Exclusion Chromatography** This is another form of liquid chromatography where a uniform nonionic gel is used to separate materials according to their molecular size. The small molecules get into the polymer network and are retarded, whereas larger molecules cannot enter the polymer network and will be swept out of the column. The elution order is the largest molecules first, medium next, and the smallest molecules last. The term gel permeation chromatography (GPC) has evolved to describe separations performed on polymers which swell in organic solvent.

The trend in liquid chromatography has tended to move away from open column toward what is called high-pressure liquid chromatography (HPLC) for analytical as well as

Table 1 Detector Characteristics

Detector	Principle of operation	Selectivity	Sensitivity	Linear range	MDQ[a]	Stability
Thermal conductivity	Measures thermal conductivity of gas	Universal	6×10^{-10}	10^4	10^{-5} g of CH_4 per volume of detector effluent	Good
Flame ionization	H_2–O_2 flame	Responds to organic compounds, not to H_2O or fixed gases	9×10^{-3} for alkane	10^7	2×10^{-11} g for alkane	Excellent
Electron capture	$N_2 + B \rightarrow e^-$ $e^- +$ sample $\rightarrow$	Responds to electron-adsorbing compounds, e.g., halogen	2×10^{-14} for CCl_4	10^5		Good
Hall electrolytic conductivity detector		In halogen mode responds to halogens		10^0	1×10^{13} g of Cl sec^{-1}	Poor

preparative work. The change in technique is due to the development of high-sensitivity, low–dead volume detectors. This has resulted in high-resolution, high-speed, and better sensitivity liquid chromatography. Both GPC and HPLC techniques used in polymer characterization are discussed in detail in subsequent chapters.

The output of a chromatographic instrument can be of two types, either a plot of retention time versus detector response, where the peak areas represent the amount of each component present in the mixture, or a computer printout giving the names of components and the concentration of each in the sample. The units of concentration can be reported as weight percent or ppm by weight, volume percent or ppm by volume, and/or mole percent.

INDUCTIVELY COUPLED PLASMA ATOMIC EMISSION SPECTROSCOPY

In inductively coupled plasma (ICP) atomic emission spectroscopy, the sample is vaporized and the element of interest atomized in an extremely high-temperature ($\sim 7000°C$) argon plasma, generated and maintained by radiofrequency coupling. The atoms collide with energetically excited argon species and emit characteristic atomic and ionic spectra that are detected by a photomultiplier tube. Separation of spectral lines can be accomplished in two ways: a sequential or scanning ICP, as illustrated in Figure 3a, where a scanning monochromator with a movable grating is used to bring the light from the wavelength of interest to a single detector; or the method shown in Figure 3b, which uses simultaneous or direct reader ICP. Here a polychromator with a diffraction grating is used to disperse the light into its component wavelengths and a separate detector is used for each element wavelength. Detectors for the elements of interest are set by the vendor during manufacture. Occasionally a scanning channel is added to a direct reader to allow measurement of an element not included in the main polychromator.

ICP is used for the determination of ppm levels of metals in liquid samples. It is not suitable for the noble gases, halogens, or light elements such as H, C, N, and O. Sulfur requires a vacuum monochromator. A direct-reader ICP excels at the rapid analysis of multielement samples. Common sample types analyzed by ICP include trace elements in polymers, wear metals in oils, and numerous one-of-a-kind catalyst species. Usually instruments are limited to the analysis of liquids only. Solid samples require some sort of dissolution procedure prior to analysis. The final volume of solution should be at least 25 ml. The solvent can be either water, usually containing 10% acid, or a suitable organic solvent such as xylene.

The technique provides good detection limits and a broad linear range for most elements. With a direct-reading instrument multielement analysis is very fast. Chemical and ionization interferences frequently found in atomic absorption spectroscopy are suppressed in ICP analysis. Since all samples are converted to simple aqueous or organic matrices before analysis, the need for a standard matched to the matrix of the original sample is eliminated. A disadvantage with ICP is that relatively extensive sample preparation is needed to meet solution requirements. More than one sample preparation method may be needed per sample depending on the range of elements requested. Spectral interferences can complicate the determination of trace elements in the presence of other major metals. Also, ICP instruments are not rugged. Constant attention by a trained operator, especially to the sample introduction and torch systems, is required. Spectral interferences, such as line overlaps, are prevalent and must be corrected for accurate quantitative analysis. With

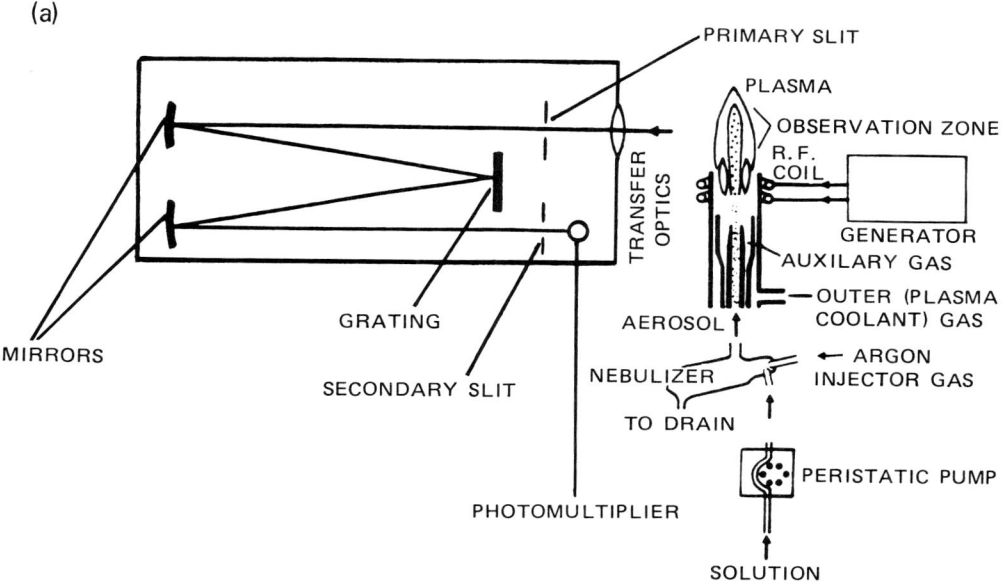

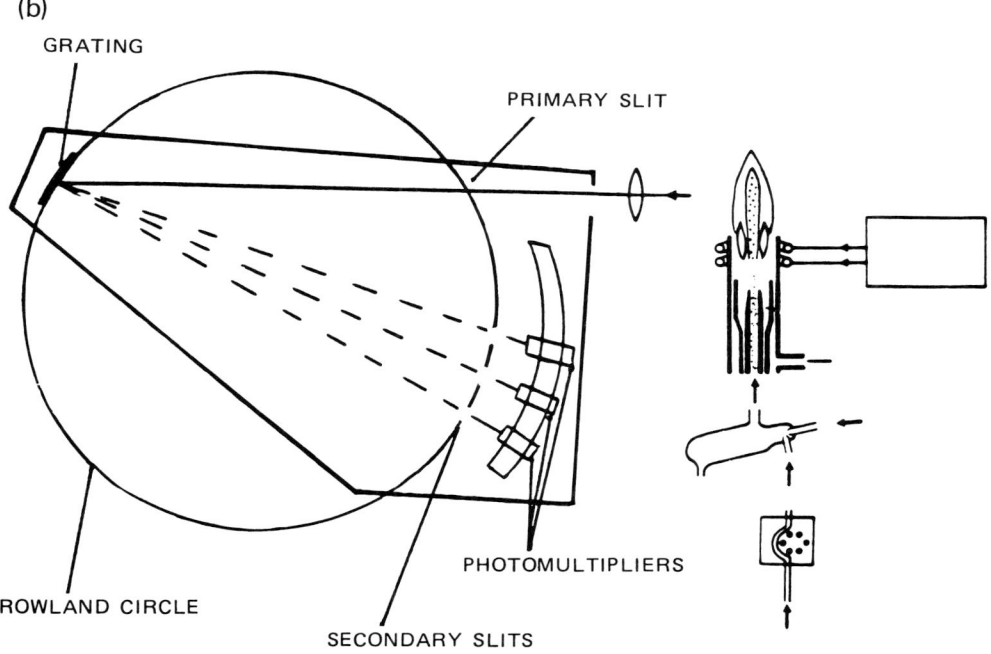

Figure 3 (*a*) Schematic of scanning ICP; (*b*) direct reader ICP.

ION CHROMATOGRAPHY

Commercial ion chromatograph instruments have evolved through the 1980s. Ion chromatography (IC) is a combination of ion exchange chromatography, eluent suppression, and conductimetric detection. For anion analysis, a low-capacity anion exchange resin is used in the separator column and a strong cation exchange resin in the H^+ form is used in the suppressor column. A dilute mixture of $Na_2CO_3 NaHCO_3$ is used as the eluent, because carbonate and bicarbonate are conveniently neutralized to low-conductivity species and the different combinations of carbonate–bicarbonate give variable buffered pH values. This allows the ions of interest in a large range of affinity to be separated. The anions are

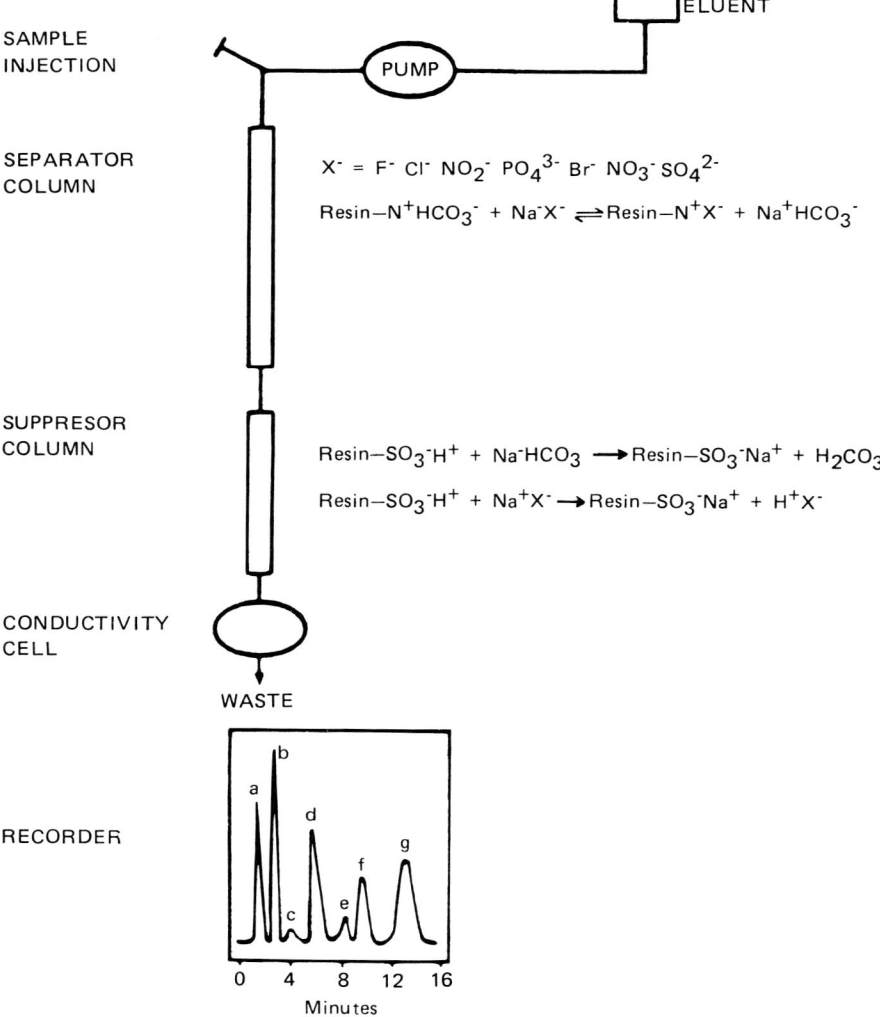

Figure 4 Ion chromatography flow scheme.

eluted through the separating column in the background of carbonate–bicarbonate and conveniently detected based on electrical conductivity. The following reactions take place on these two columns for an anion X.

Separator column:

$$\text{Resin–N}^+\text{HCO}_3^- + \text{Na}^+\text{X}^- \rightleftharpoons \text{Resin–N}^+\text{X} + \text{Na}^+\text{HCO}_3^- \tag{1}$$

Suppressor column:

$$\text{Resin–SO}_3^- + \text{Na}^+\text{HCO}_3^- \rightleftharpoons \text{Resin–SO}_3^-\text{Na}^+ + \text{H}_2\text{CO}_3 \tag{2a}$$
$$\text{Resin–SO}_3^-\text{H}^+ + \text{Na}^+\text{X}^- \rightleftharpoons \text{Resin–SO}_3^-\text{Na}^+ + \text{H}^+\text{X}^- \tag{2b}$$

As a result of these reactions in the suppressor column, the sample ions are presented to the conductivity detector as H^+X^-, not in the highly conducting background of carbonate–bicarbonate but in the low-conducting background of H_2CO_3.

Figure 4 is a schematic of the IC system. Dilute aqueous sample is injected at the head of the separator column. The anion exchange resin selectively causes the various sample anions of different types to migrate through the bed at different respective rates, thus effecting the separation. The effluent from the separator column then passes to the suppressor column where the H^+ form cation exchange resin absorbs the cations in the eluent stream. Finally, cation exchange resin absorbs the cations in the eluent stream. The suppressor column effluent passes through a conductivity cell. The highly conductive anions in a low-background conductance of H_2CO_3 are detected at high sensitivity by the conductivity detector. The nonspecific nature of the conductimetric detection allows several ions to be sequentially determined in the same sample. The conductimetric detection is highly specific and relatively free from interferences. Different stable valence states of the same element can be determined. It is important to note that due to the nonspecific nature of the conductivity detector, the chromatograph peaks are identified only by their retention times. Thus two ions having the same or close retention times will be detected as one broad peak, giving erroneous results.

Figure 5 illustrates a typical chromatogram for the standard common anions F^-, Cl^-,

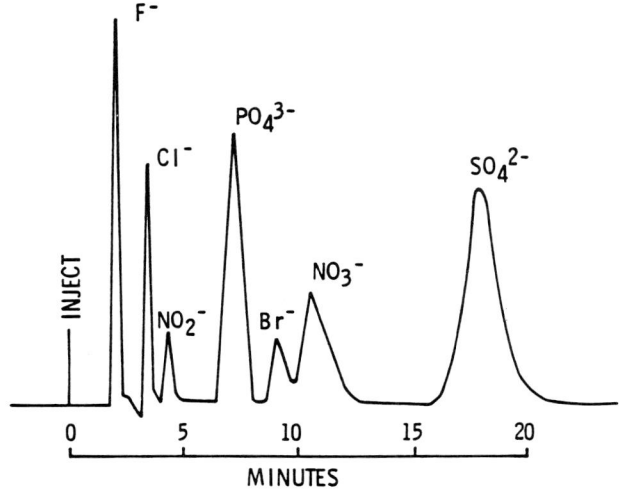

Figure 5 Analysis of standard inorganic anions by ion chromatography.

NO_2^-, PO_4^{3-}, NO_3^-, SO_4^{2-} and Br^-. Numerous applications of ion chromatography have been illustrated in the literature for a variety of complex matrices.

ION SELECTIVE ELECTRODES

The ion selective electrode (ISE) method measures the ion activities or the thermodynamically effective free ion concentrations. ISE has a membrane construction that serves to block the interfering ions and permits the passage of only those ions for which it was designed. However, this rejection is not perfect; hence some interferences from other ions occur. The electrode calibration curves are good over four to six decades of concentration. The typical time per analysis is about a minute. Although designed as a single-element technique, many elements can be determined sequentially by changing electrodes, provided calibration curves are prepared for all ions. Also, the instrument is portable and is thus useful for field studies. Sample volumes needed are typically about 5 ml, although 300 µl or less can be measured with special modifications. An accuracy of 2–5% is achieved. The instrument measures the activity of the ions in solution. This activity is related to concentration and thus, in effect, the instrument measures the concentration. However, if an ion such as fluoride, which complexes with metals like iron or aluminum, is to be measured, it must be decomplexed from these cations by the addition of a reagent such as citric acid or EDTA. ISEs for at least 22 ionic species are commercially available.

An example is described here for the measurement of fluoride ions in solution. The fluoride electrode uses a LaF_3 single crystal membrane and an internal reference, bonded into an epoxy body. The crystal is an ionic conductor in which only fluoride ions are mobile. When the membrane is in contact with a fluoride solution, an electrode potential develops across the membrane. This potential, which depends on the level of free fluoride ions in solution, is measured against an external constant reference potential with a digital pH/mV meter or specific ion meter. The measured potential corresponding to the level of fluoride ions in solution is described by the Nernst equation:

$$E = E_0 - S \log A \qquad (3)$$

where

E = measured electrode potential
E_0 = reference potential (a constant)
S = electrode slope
A = fluoride level in solution

The level of fluoride, A, is the activity, or effective concentration, of free fluoride ions in solution. The total fluoride concentration, C_t, may include some bound or complexed ions as well as free ions. The electrode responds only to the free ions, whose concentration is

$$C_f = C_t - C_b \qquad (4)$$

where C_b = concentration of fluoride ions in all bound or complexed forms. The fluoride activity is related to free fluoride concentration by the activity coefficient,

$$A = \gamma C_f \qquad (5)$$

Ionic activity coefficients are variable and largely depend on total ionic strength. Ionic strength is defined as

$$\text{Ionic strength} = \frac{1}{2}\Sigma C_i Z_i^2 \tag{6}$$

where

C_i = concentration of ion i

Z_i = charge of ion i

If the background ionic strength is high and constant relative to the sensed ion concentration, the activity coefficient is constant and activity is directly proportional to concentration. Since the electrode potentials are affected by temperature changes, the sample and standard solutions should be close to the same temperature. At the 20-ppm level a 1°C change in temperature gives a 2% error. The slope of the fluoride electrode response also changes with the temperature. The electrode can be used at temperatures from 0 to 100°C, provided that the temperature has equilibrated, which may take as long as an hour. In general, it is best to operate near room temperature.

ISEs are subject to two types of interference: method interference and electrode interference. In the former, some property of the sample prevents the electrodes from sensing the ion of interest; e.g., in acid solution fluoride forms complexes with H^+ and the fluoride ISE cannot detect the masked fluoride ions. In the electrode interference, the electrode responds to ions in solution other than the one being measured; e.g., bromide ion presents severe interference in using chloride ISE. The extent of interference depends on the relative concentration of analyte to interfering ions. The interfering ions can be complexed by changing pH or adding a reagent to precipitate them. In the example of fluoride determination, the fluoride forms complexes with aluminum, silicon, iron, and other polyvalent cations as well as hydrogen. These complexes must be destroyed in order to measure total fluoride, since the electrode will not detect complexed fluoride. This is achieved by adding a total ionic strength adjustment buffer which contains the reagent CDTA (cyclohexylene dinitrilotetraacetic acid), which preferentially complexes the cations and releases the fluoride ions. The carbonate and bicarbonate anions interfere by making the electrode response slow; hence these ions are eliminated by heating the solution with acid until all CO_2 is removed. At pH > 7, hydroxyl ions interfere, while at pH < 5, the H^+ ions form complexes such as HF_2^-, thus producing low fluoride results. Addition of TISAB to both samples and standards and further adjustment of pH to between 5.0 and 5.5 are necessary to eliminate the hydroxide interference and the formation of hydrogen–fluoride complexes. Other common anions such as other halides, sulfate, nitrate, phosphate, or acetate do not interfere in the fluoride measurement.

MASS SPECTROMETRY

Mass spectrometry is a measurement technique that is concerned with both the chemistry and the physics of molecules, in particular with gaseous ions. In the conventional setup the ions of interest are positively charged ions. The system has three functions:

1. To produce ions from the molecules under investigation.
2. To separate these ions according to their mass-to-charge ratio.
3. To measure the relative abundances of each ion.

The methodology of mass separation is governed by both the kinetic energy of the ion and the ion's trajectory in an electromagnetic field. There exists a balance between the centripetal and centrifugal forces which the ion experiences. Centripetal forces are caused by kinetic energy and centrifugal forces, by the electromagnetic field. From a force balance

$$\frac{mU^2}{r} = qUB \tag{7}$$

where

m = ion mass
U = ion velocity
r = radius of ion trajectory in the magnetic field
q = ion charge
B = magnetic field strength

The right-hand side of this expression represents the centripetal force, and the left-hand side is the centrifugal force.

Solving for the mass-to-charge ratio gives the following expression:

$$\frac{m}{q} = \frac{Br}{U} \tag{8}$$

The kinetic energy of the ion is

$$qV = \frac{1}{2}mU^2 \tag{9}$$

where

m = ion mass
q = ion charge
U = ion velocity
V = accelerating potential

Solving for U,

$$U = \sqrt{\frac{2qV^{1/2}}{m}} \tag{10}$$

Substituting Eq. 10 into Eq. 8 results in

$$\frac{m}{q} = \frac{Br}{2qV^{1/2}/m} \tag{11}$$

and squaring each side of the expression yields

$$\frac{m^2}{q^2} = \frac{B^2r^2m}{2qV} \tag{12a}$$

or

$$\frac{m}{q} = \frac{B^2r^2}{2V} \tag{12b}$$

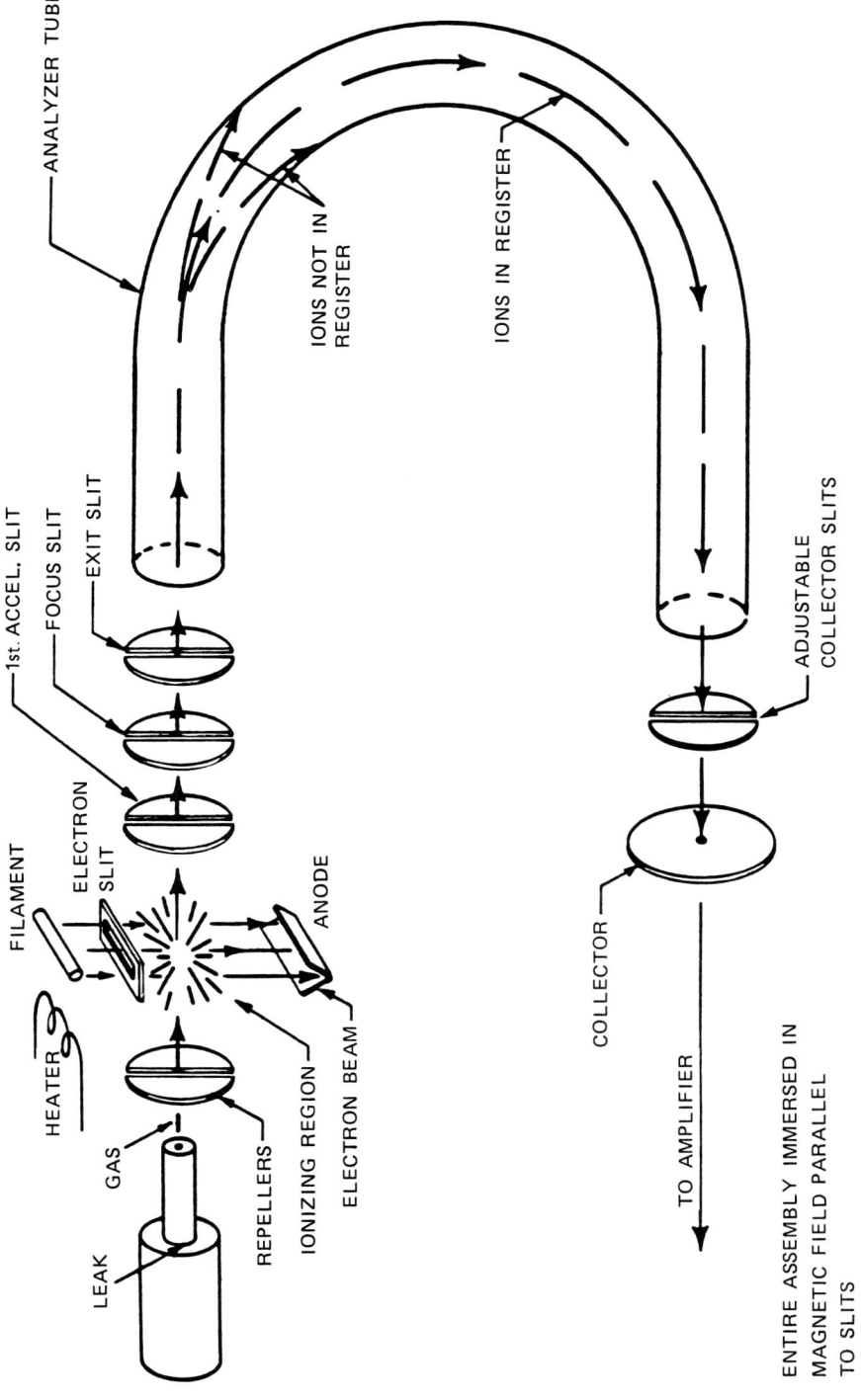

Figure 6 Schematic of mass spectrometer analyzer.

These expressions show that the mass-to-charge ratio can be determined from information on B, r, and V. Since r is constant for a mass passing through two stationary slits, scanning a spectrum is achieved by varying either B or V, keeping the other constant. The instrument arrangement is illustrated in Figures 6 and 7. Figure 6 is a schematic of the operation of a mass spectrometer analyzer. When combined with a GC as shown in Figure 7, the system can be automated to obtain a wide body of information concerning the structure of organic compounds, their elemental composition, and compound types in complex mixtures.

A detailed interpretation of the mass spectrum frequently allows the positions of the functional groups to be determined. Moreover, mass spectrometry is used to investigate reaction mechanisms and kinetics and is also used in tracer work. The mass spectrum may be either in analog form (chart paper) or digital form (printed paper). Analyses are calculated to give mole percent, weight percent or volume percent. Individual components, compound types by carbon number, or total compound types are reported. This is determined by the nature of the sample and the requirements of the submitter.

A wide variety of materials from gases to solids and from simple to complex mixtures can be analyzed. The molecular weight and atomic composition are generally determined. Only a very small amount of sample is required. Most calibration coefficients can be used for long periods of time.

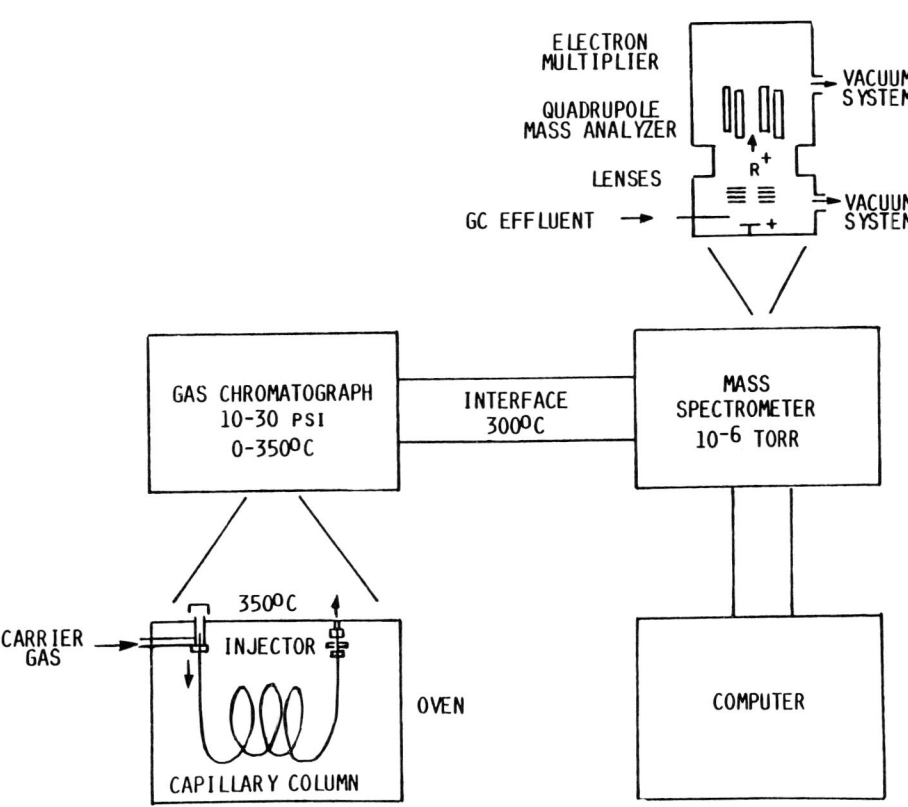

Figure 7 Schematic of GC/MS instrument.

The primary disadvantages with this measurement technique are as follows. Some compounds such as long-chain esters and polyethers decompose in the inlet system and the spectrum obtained is not that of the initial substance. Calibration coefficients are required for quantitative analyses. Also, the sample introduced to the instrument usually cannot be recovered. There are also some classes of compounds, such as olefins and naphthenes, which give very similar spectra and cannot be distinguished except by analysis before and after hydrogenation or dehydrogenation.

NEUTRON ACTIVATION ANALYSIS

Neutron activation analysis is a method of elemental analysis in which nonradioactive elements are converted to radioactive elements by neutron bombardment, and the elements of interest are determined from their resulting radioactivity. This system is illustrated in Figure 8. High-energy (14 MeV) neutrons are generated by the reaction of medium-energy deuterium ions with tritium. For oxygen analysis, the carefully weighed sample is irradiated for 15 sec to convert a small amount of the oxygen-16 to nitrogen-16, which emits gamma rays with a half life of 7.4 sec. The irradiated sample is transferred to a scintillation detector where the gamma rays are counted for 30 sec to ensure that all usable radioactivity has been counted and that no significant radioactivity remains in the sample. The system is calibrated with standards of known oxygen content. The specific information obtained is the counts per 30 sec, from a digital counter, and this is converted to weight percent of the element of interest.

This technique offers several advantages:

1. It may be the best method available for determining total oxygen directly.
2. It is fast (about 10 min per analysis; repeat analysis on weighed sample requires only 1 min).
3. It is nondestructive.
4. It offers moderate sensitivity.

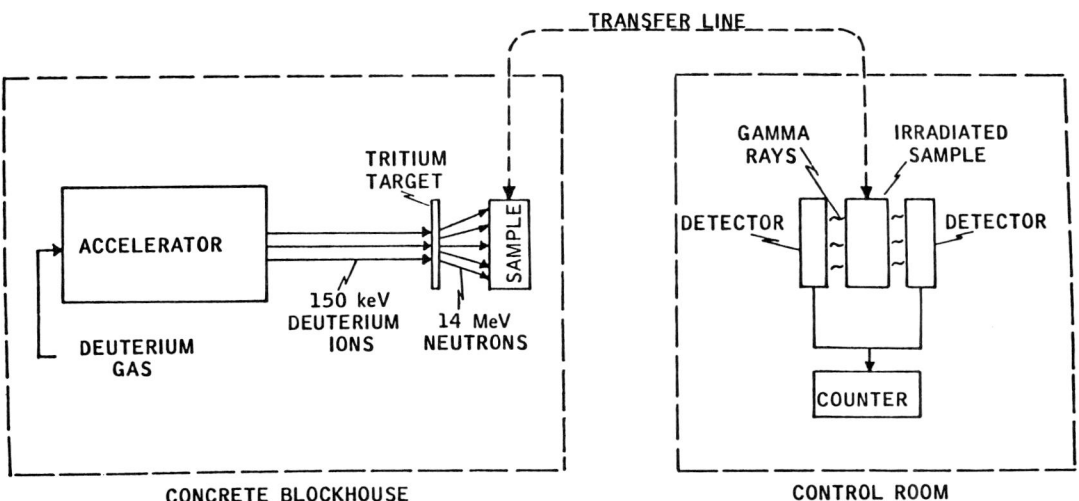

Figure 8 Neutron activation analysis system.

NUCLEAR MAGNETIC RESONANCE SPECTROMETRY

Nuclear magnetic resonance (NMR) is a spectrometric technique for determining chemical structures. When an atomic nucleus with a magnetic moment is positioned in a magnetic field, it will align with the applied field. The energy required to reverse this alignment depends on the strength of the magnetic field and to a lesser extent on the environment of the nucleus, i.e., the nature of the chemical bonds between the atom of interest and its immediate vicinity in the molecule. This reversal is a resonant process and occurs only under certain conditions. By determining the energy levels of transition for all of the atoms in a molecule, one can determine various properties of the structure. The energy levels can be expressed in terms of frequency of electromagnetic radiation; typically they fall in the range of 5–600 mHz electromagnetic radiation for high magnetic fields. The minor spectral shifts due to chemical environment are the primary features for interpreting structure. These are expressed in terms of ppm shifts from the reference frequency of a standard (e.g., tetramethyl silane).

The most common nuclei examined by NMR are ^{1}H and ^{13}C. These are NMR-sensitive nuclei of the most abundant elements in organic constituents. ^{1}H represents more than 99% of all hydrogen atoms, while ^{13}C is slightly more than 1% of all carbon atoms. In addition, ^{1}H is more sensitive than ^{13}C on an equal nuclei basis. Until recently, instruments did not have sufficient sensitivity for routine ^{13}C NMR, and ^{1}H was the only practical technique. Most of the time it is solutions that are characterized by NMR, although ^{13}C NMR is possible for some solids, but at substantially lower resolution than for solutions. Resonant frequencies can be used to determine molecular structures. ^{1}H resonances are specific for the types of carbon they are attached to, and to a lesser extent to the adjacent carbons. These resonances may be split into multiplets, as hydrogen nuclei can couple to other nearby hydrogen nuclei. The magnitude of the splittings, and the multiplicity, can be used to determine the chemical structure in the vicinity of a given hydrogen. When all of the resonances observed are similarly analyzed, it is possible to determine the molecular structure. It is important to note that since only hydrogen is observed, any skeletal feature without an attached hydrogen can only be inferred. Complications arise if the molecule is very complex, because then the resonances can overlap significantly and it becomes difficult or impossible to resolve.

^{13}C resonances can be used to directly determine the skeleton of an organic molecule. The resonance lines are narrow and the chemical shift range (in ppm) is much larger than for ^{1}H resonances. Furthermore, the shift depends on the structure of the molecule for up to three bonds in all directions from the site of interest. This means each shift becomes specific, and the structure can be easily assigned, frequently without any ambiguity. Usually the sample of interest is not a pure compound, but is a complex mixture. As a result, a specific structure determination for each molecular type is not practical, although it is possible to determine an average chemical structure. Features that may be determined include the hydrogen distribution between saturate, benzylic, olefinic, and aromatic sites. The carbon distribution is usually split into saturate, heterosubstituted saturate, aromatic + olefinic, carboxyl, and carbonyl types. More details are possible, but they depend greatly on the nature of the sample and the information desired.

Any gas, liquid, or solid sample that can be dissolved in solvents, such as CCl_4, CH_3Cl, acetone, or DMSO to the 1% level or greater can be analyzed by NMR. Samples of ~0.1 g or larger of pure material are usually sufficient. Solids can also be analyzed in their solid state; however, special arrangements are required. In either case, the analysis is nondestructive so that samples can be recovered for further analysis if necessary.

NMR analysis can be conducted in a temperature range from $-209°C$ (liquid nitrogen) to $+150°C$. This gives the experimenter the ability to slow down rapid molecular motions to observable rates or to speed up very slow or viscous motions to measurable rates. The technique is a very powerful analytical tool. It provides the best characterization of compound structure and may provide absolute identification of specific isomers in simple mixtures. It may also provide a general characterization by functional groups which cannot be obtained by any other technique. As is typical with many spectroscopic methods, adding data from other techniques (such as mass or infrared spectrometry) can often provide greatly improved characterizations.

THERMAL ANALYSIS

Thermal analysis includes several techniques in which a property of a material is continuously measured as the sample is programmed through a predetermined temperature profile. Among the most common techniques are thermal gravimetric analysis (TGA) and differential scanning calorimetry (DSC).

In TGA the mass loss versus increasing temperature of the sample is recorded. The basic instrument requirements are a precision balance, a programmable furnace, and a recorder. Figure 9 illustrates the arrangement. Modern instruments, however, tend to be automated and include software for data reduction. In addition, provisions are made for surrounding the sample with an air, nitrogen, or oxygen atmosphere. A typical TGA spectrum for calcium oxalate is seen in Figure 10.

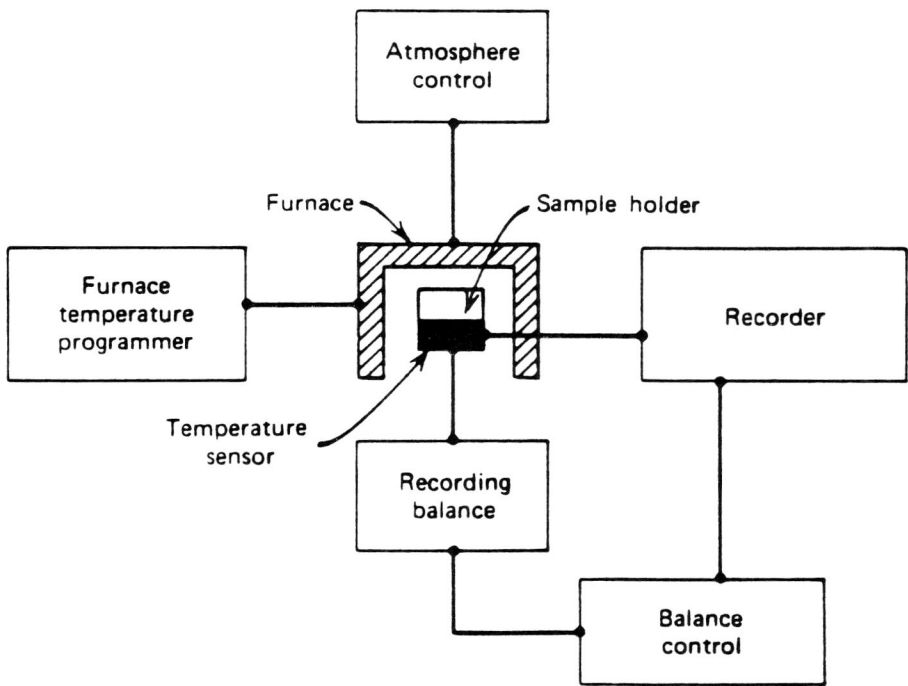

Figure 9 Block diagram of a typical TGA instrument.

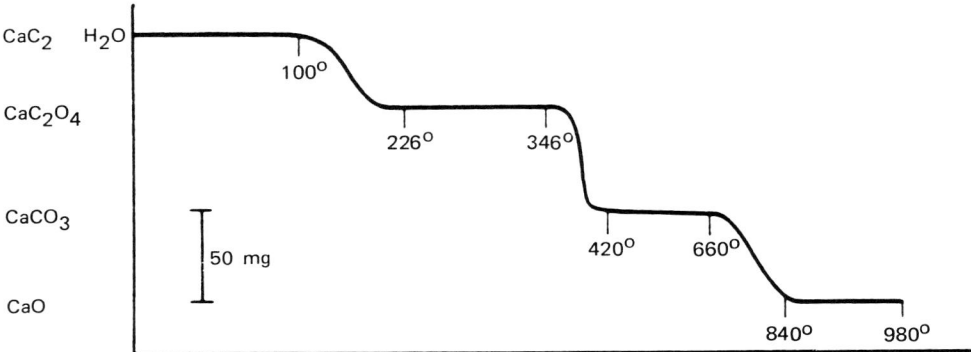

Figure 10 TGA spectrum of calcium oxalate.

A thermal analysis curve is interpreted by relating the measured property versus temperature data to chemical and physical events occurring in the sample. It is frequently a qualitative or comparative technique. In TGA the mass loss can be due to such events as the volatilization of liquids and the decomposition and evolution of gases from solids. The onset of volatilization is proportional to the boiling point of the liquid. The residue remaining at high temperature represents the percentage of ash content of the sample.

In DSC the measured energy differential corresponds to the heat content (enthalpy) or the specific heat of the sample. DSC is often used in conjunction with TGA to determine if a reaction is endothermic—such as melting, vaporization, and sublimation—or exothermic—such as oxidative degradation. It is also used to determine the glass transition temperature of polymers. Figure 11 is a block diagram of a typical DSC.

The measurement principle involves bringing the average temperature of the sample holder to a desired initial temperature and then, by means of a microprocessor, a desired change in average temperature of the sample holder to a final temperature is achieved. When a transition such as melting, boiling, dehydration, or crystallization occurs in the sample, an endothermic or exothermic reaction takes place. The change in power needed to maintain the sample holder at the same temperature as a reference holder (i.e., its programmed temperature) during the transition is recorded as a peak. Usually the chart abscissa denotes the transition temperature and the peak area indicates the total energy transfer to or from the sample.

The sample holders are comprised of two platinum alloy cups which are mounted in the cavities of a solid aluminum block, and which contain a heater and sensor. For measurements above ambient conditions the instrument is usually capable of circulating a cooling fluid through the aluminum block to maintain it at a constant temperature, independent of the temperature range of the measurement. Since this is a direct calorimetric measurement, each sample holder must have its own built-in heater and a temperature sensor. A high-gain closed-loop electronic arrangement provides differential electrical power to the heaters to compensate for temperature fluctuations. This means that for practical purposes, the holder temperature is unaffected by the sample behavior. The differential power required to maintain the balanced condition can be recorded directly in millicalories per second on an appropriate recorder. This measurement is equivalent to the rate of energy absorption or evolution of the sample.

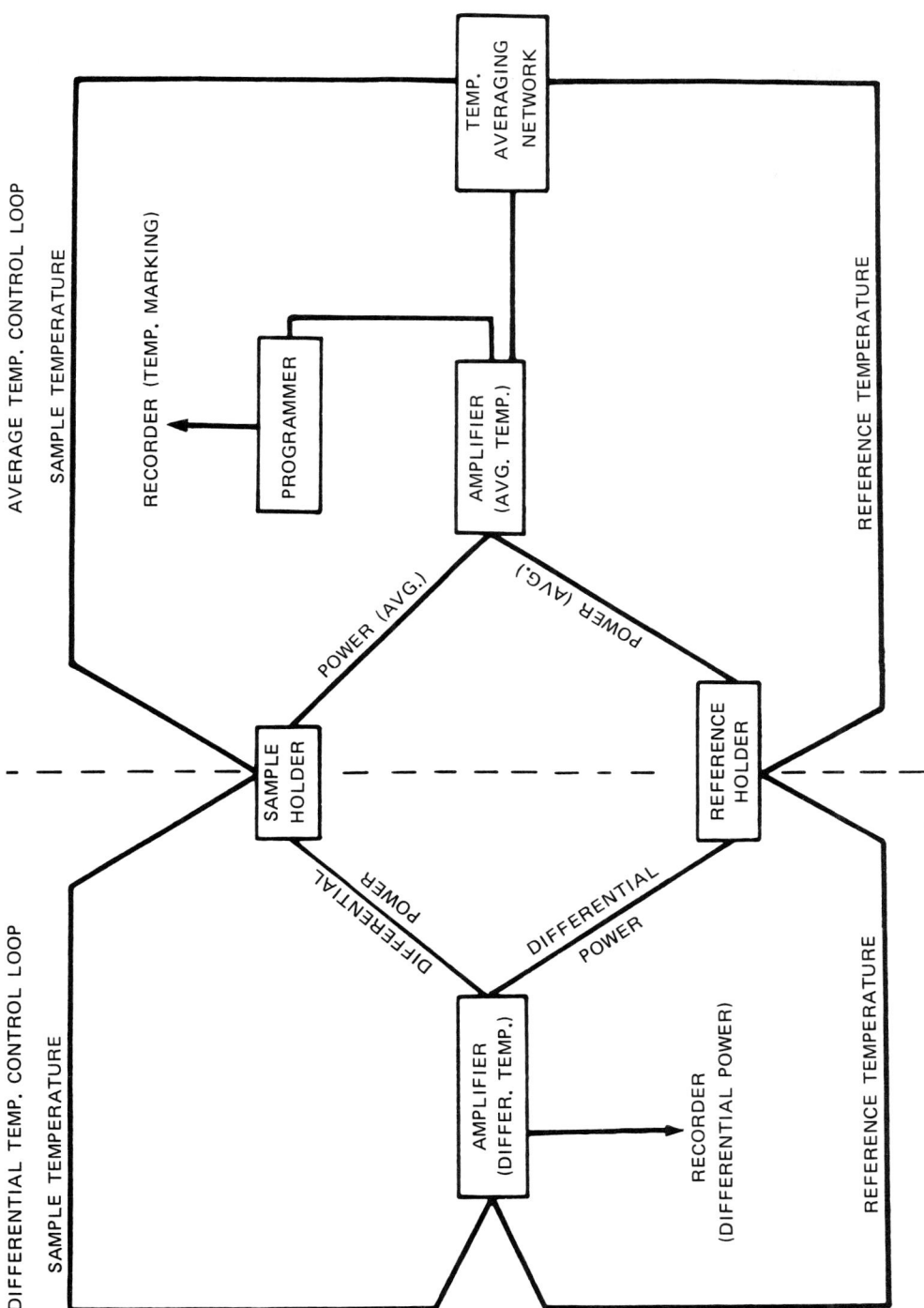

Figure 11 Block diagram of a typical DSC instrument.

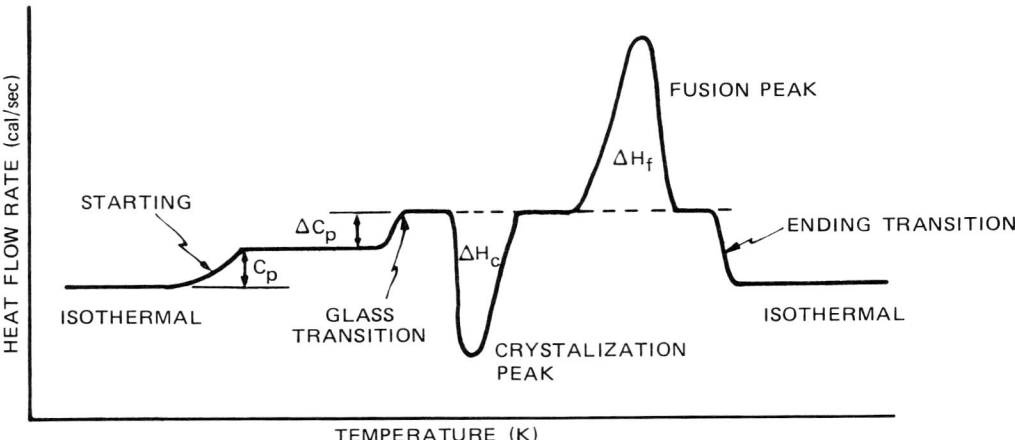

Figure 12 Typical DSC spectrum.

Figure 12 illustrates typical calorimetric data derived from DSC measurements. Endothermic transitions and increases in heat capacity are represented by upscale departures from the ordinate baseline, whereas exothermic reactions are represented by downside departures from the baseline. The abscissa is usually calibrated in Kelvins.

A DSC instrument will measure the rate of energy absorption or evolution by the sample in units of millicalories per second. In some experiments, e.g., in kinetic reaction studies, this rate can be measured directly. In measurements of heat capacity (or specific heat) the ordinate displacement is measured directly since $C_p = dH/dT$, which is related to the heat of absorption dH/dt:

$$\frac{dH}{dt} = \frac{dH}{dT} \times \frac{dT}{dt} \tag{13}$$

where dT/dt = temperature scanning rate.

The total energy associated with a transition or reaction is obtained by integrating over the peak with respect to time:

$$\Delta H = \int \frac{dH}{dt}\, dt \tag{14}$$

Both solid and liquid samples can be analyzed. Solid samples can be in film, powder, crystal, or granular form. Quantitative accuracy is usually unaffected by sample shape; however, the qualitative appearance of the thermogram may be affected by the sample configuration. For maximum peak sharpness and resolution, a configuration which provides good contact between the sample surface and pan is desirable. Thin films or disks of sample or fine granules spread in a thin layer on the pan bottom give good instrument performance. Polymer films can be sampled by cutting out sections of the film with a cork borer.

A sharp first-order transition, such as the melting point of a pure material, displays an isothermal behavior. Since the temperature of the sample does not increase during the transition, the temperature of the transition is that temperature at which the transition is first observed. Note that this temperature is not influenced by variations in sample size, whereas the peak temperature can be.

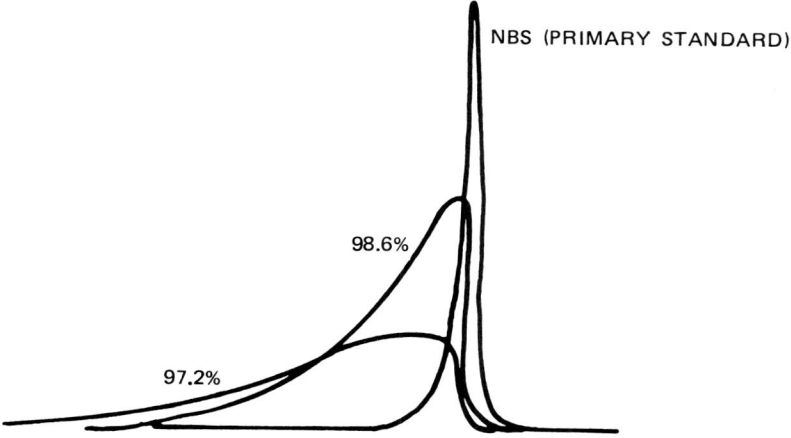

Figure 13 Effect of purity on melting peak shapes of benzoic acid.

Crystalline or semicrystalline materials usually undergo melting. Exceptions may occur if the material is sublimed or decomposed prior to reaching the melting point. Melting is the most commonly encountered thermal behavior and is observed as a sharp endothermic peak on a DSC thermogram. The area under the melting peak is a direct measure of the heat of fusion for the sample. The sharpness of the melting peak is affected by such factors as purity and crystalline perfection.

For nonpolymeric materials, an impurity will depress its melting point and significantly broaden its melting range. Figure 13 illustrates typical melting peaks for different purity samples of benzoic acid. These types of curves enable one to distinguish samples of different purities. In contrast, polymers inherently melt over broad temperature ranges. Figure 14 shows the thermogram for PTFE, illustrating that the material has a relatively

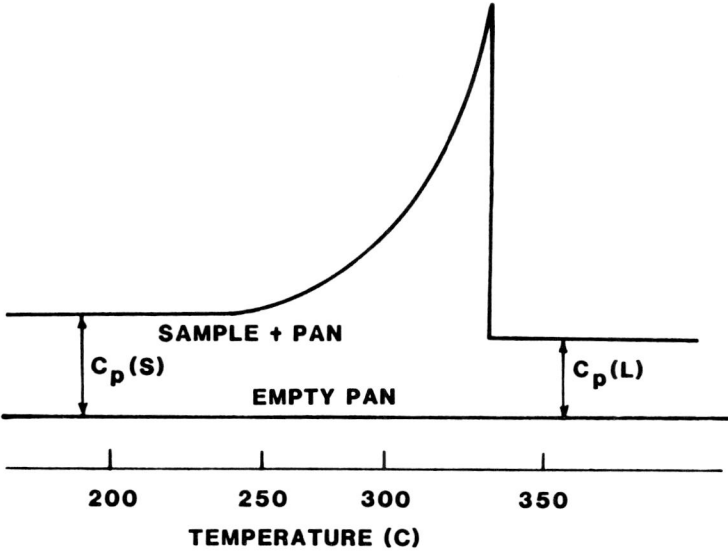

Figure 14 Melting peak of PTFE.

sharp melting peak but melts over a 150°C temperature range. The broad melting characteristic is attributed to crystallite size or imperfections through the polymer, and is not evidence of impurity. The smaller, less perfectly formed crystals tend to melt at lower temperatures because of surface free energy effects. The melting characteristics of a polymer are influenced by a number of factors that will affect the degree of perfection of crystallinity, such as thermal history. From the measured heat of fusion, the percentage of crystallinity of the polymer can be determined.

An important aspect to consider is the presence of multiple melting peaks. This can be attributed to several factors. First, crystallization is very sensitive to the thermal history that the polymer sample is subjected to during processing. If the material is maintained at different temperatures, crystal growth will be promoted in those regions. As a result, on heating, a minimum can be detected at the annealing temperature since those crystals formed on annealing will melt above the annealing temperature and those crystals formed on cooling from the annealing temperature will melt below that temperature. An example of processing pretreatment is illustrated in Figure 15. The three EPDMs are chemically and structurally identical, but they were processed and annealed under different conditions. Thermograms of this type offer the possibility of controlling processing conditions. Also, the influence of thermal history can be distinguished from structural differences by introducing a constant thermal history to the samples under study, such as cooling at 20°C min^{-1} from above the melt. If the rescans between samples are then identical, the initial differences may be attributed to thermal history. On the other hand, if the rescans continue to show differences, then the samples are likely to be structurally different.

Multiple melting peaks can be observed in blends and mixtures. When two polymers are mixed together, they may retain their individual thermal characteristics. Figure 16 illustrates the thermogram of a blend of a high- and low-density polyethylene. Branching tends to impede crystallite perfection; hence the low-density constituent melts at the lower temperature, being almost completely resolved from melting of the crystallites formed from the linear molecules. Scans such as waxes are comprised of mixtures of chemically similar but structurally different molecules. Melting thermograms of these materials can show multiple peaks as well. Wax thermograms in general are highly characteristic and can be used as a means of identification.

Some comments concerning copolymers are warranted. There are basically two types, random and block. Random copolymers are different from their homopolymers, whereas a block copolymer usually retains the characteristics of the homopolymer. A block copolymer thermogram resembles that of a physical blend.

In terms of characterizing polymers and evaluating their processability, one of the

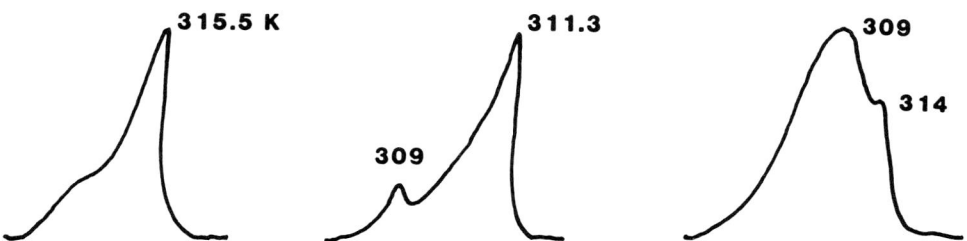

Figure 15 Effect of pretreatment on an EPDM's melting behavior.

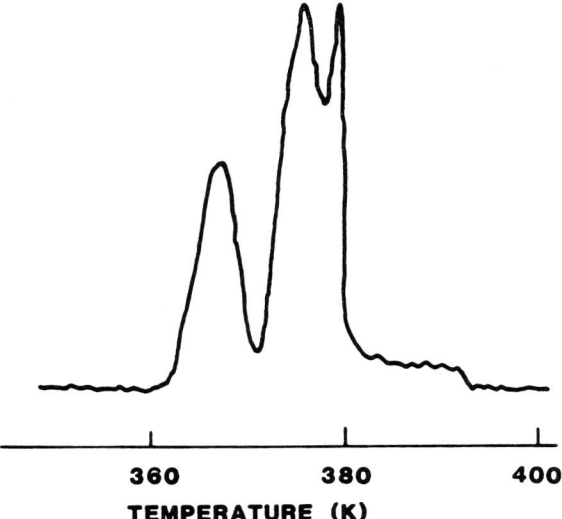

Figure 16 Multiple melting peaks from a physical blend of branched and linear polymers.

more important properties to define is crystallinity. Crystallization is of four general types with respect to DSC measurements: program-cooled crystallization, cold crystallization, recrystallization, and isothermal crystallization.

Program-cooled crystallization refers to materials that melt on heating and then crystallize on cooling as witnessed by a sharp exotherm on the DSC trace. All materials will supercool and hence the rate of crystallization may be faster than the rate of melting. In this case, the crystallization exotherm may be taller than the melting endotherm. If complete crystallization has occurred, the peak areas (i.e., energies) will, however, be the same. A problem in crystallization studies is that a number of materials do not crystallize completely, or even partially, on cooling. For example, organic constituents that have been crystallized from solution often tend not to crystallize from the melt. With polymers, the process of crystallization is not instantaneous but instead involves the processes of nucleation and propagation. As a result, the extent of crystallization depends also on the rate of cooling.

In cold crystallization, polymers crystallize very slowly so that they can be rapidly cooled to a completely amorphous glassy state. When they are heated through the glassy region, molecular mobility increases and the substance crystallizes. This behavior is referred to as cold crystallization since it occurs at temperatures far below the melting point. An example of this is shown in Figure 17 for a copolymer sample that has been shock cooled (solid line) from the melt. The cold-crystallization exotherm is at 296 K.

As noted earlier, polymers tend to melt over a broad temperature range because of the surface free energy of the smaller, less perfectly formed crystals. Once these crystals are melted, however, the free energy of the liquid can be higher than that of the more perfectly formed crystals, which are stable in the temperature range. Consequently, it is within thermodynamic principles for the liquid to crystallize as shown in Figure 17. Controlled cooling of the copolymer resulted in partial crystallization, as witnessed by the absence of the cold-crystallization exotherm observed for the amorphous sample. On the

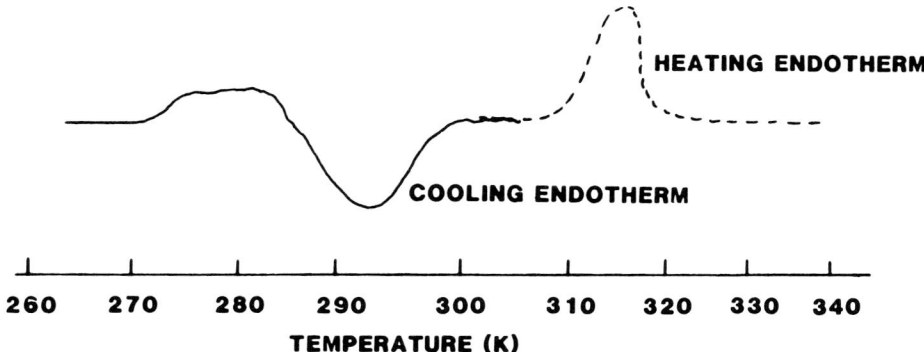

Figure 17 Typical thermogram of a copolymer.

other hand, the morphologies of the crystals formed give rise to recrystallization, as evidenced by the exothermic peak at about 315 K, during the melting cycle of the experiment. These crystallization effects on heating influence determination of the true crystallinity of the sample. Simply considering the area under the fusion peak as a measure of crystallinity would likely result in large errors.

A final note on crystallization concerns tests that are conducted under isothermal conditions. It is important to realize that isothermal operation can affect the degree of crystallization as well as the size distribution of the melting curve. It is important to standardize measurements to account for the degree of supercooling on the rate of crystallization.

Another property of polymeric materials of importance is the glass transition. At the glass transition temperature (T_g) the material changes from a brittle glassy state to a plastic or liquidlike state. The glass transition is thought of as a second-order transition since it involves a discontinuous change in a secondary thermodynamic quantity such as the specific heat. Note that a DSC thermogram is essentially a continuous plot of specific heat as a function of temperature, and hence the glass transition will be observed as a discontinuity or step change in the ordinate of the trace. The magnitude of the step change will be proportional to the amorphous content of the polymer. The T_g is influenced by several factors such as molecular weight, the degree of cure, plasticizer, and polymerization. Table 2 gives typical values of T_g and melting temperatures for various polymers.

The DSC is used routinely in polymer characterization studies. Specific thermodynamic properties measured include specific heats, heats of fusion, and quantitative and qualitative analysis of polymer blends and copolymers for distinguishing folded-chain from extended-chain morphologies, for evaluating the effects of comonomers or chain substitutes on morphology and general thermal behavior, for crystallinity determinations, thermal stability, and crystallization rates. In addition, the DSC technique can be used for evaluating the effects and effectiveness of additives such as plasticizers, antioxidants, and accelerators, as well as catalyst and nucleating effects on the polymer's thermodynamic properties. Although DSC methods are not generally thought of in terms of processability testing, the technique enables one to establish useful correlations between DSC-derived data and finished product quality. There are many polymer processing operations that are critically dependent on the control of thermal treatments; examples are curing, heat-

Table 2 Glass Transition and Crystalline Melting Points of Different Polymers

Polymer	T_G (K)	T_m (K)
Silicone rubber	150	215
Polybutadiene	173	253
Butyl rubber (polyisobutylene)	205	
Natural rubber (polyisoprene)	203	245
Polyvinylidene fluoride	234	483
Polypropylene	253	449
Polyvinyl acetate	303	
Nylon-6	320	498
PVC	355	453
Polystyrene	373	503

setting, annealing, shock-cooling, and hot-drawing. Each one of these operations can be studied and controlled using DSC techniques. An example of this occurs when heat of fusion data generated by the DSC are correlated with compound green strength and extrusion rates of polymers. Another example is seen in Figure 18. Figure 18 shows the thermograms of two polymers made with the same catalyst system (solution polymerization) but at greatly different reactor mixing conditions. Note that both polymers show the same heats of fusion, and therefore have the same level of crystallinity. In addition, both show near-identical melting ranges and peaks, but upon cold crystallization, polymer B (well mixed in reactor) crystallizes more than 20° below polymer A. We could conclude that polymer B is a more homogeneous polymer in terms of its ethylene distribution. In a molding operation we may expect polymer A to display the problem of mold shrinkage. In simple mixing/milling tests both polymers were compounded identically, rolled into sheets, and allowed to set for 8 hr. It was observed that polymer A showed nearly twice as much shrinkage as polymer B. Hence we may conclude that DSC can be employed routinely for processability testing in a manner that involves empirically relating thermodynamic properties to product processing performance.

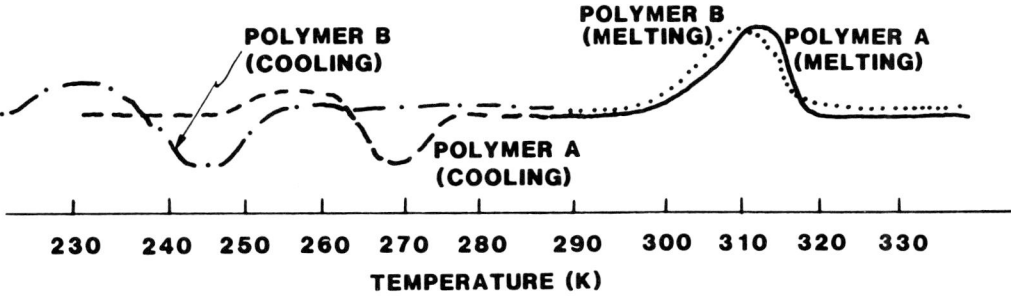

Figure 18 Cold crystallization detected by DSC. Differences between polymers can be related to mold shrinkage.

It is important to try to relate thermodynamic observations to the molecular properties of the polymer in order to establish processability correlations. For elastomers, the key molecular requirements are:

Weakly interacting noncrystalline unstrained chains above the glass transition temperature.
Presence of both chemical and physical crosslinks.
Presence of a small number of chain ends—both high molecular weight ends and reactive chain ends—after crosslinking.
Presence of a two-phase system: display crystallization on extension; and display polymeric domains.

ULTRAVIOLET, VISIBLE, AND INFRARED SPECTROMETRY

When electromagnetic radiation is passed through a sample, some wavelengths are absorbed by the molecules of the material. Energy is transferred from the radiation to the sample. In this case the molecules of the sample are elevated to an excited energy state. The total energy state of the ensemble of molecules may be regarded as the sum of the four kinds of energy: electronic, vibrational, rotational, and translational. Translational energy is associated with an elevation of the temperature of the sample. Rotational energy is derived from the absorption of very high wavelengths of infrared radiation (25–500 μm) and is manifested by an increase in the rotational energy of the sample molecules. Vibrational energy arises when radiation in the midinfrared region is absorbed (2–25 μm) and is manifested by an increase in the vibrational energies of functional groups within the sample molecule. Electronic energy is gained by an ensemble of molecules when an electron is promoted to a higher molecular orbital by absorption in the ultraviolet and visible regions of the spectrum (0.2–0.8 μm).

Pure transitions between rotational states represent small energy changes (i.e., high wavelength). Absorption spectra observed in the far infrared are generally "pure" in the sense that the energy absorbed by the molecule is entirely converted into pure rotational motion. This is not the case in the other regions of the spectrum. Thus when higher amounts of energy are absorbed by molecules, the vibrational motions generated are not restricted to those for which the rotational properties of the molecule remain constant. The absorption band therefore represents a composite of vibrational motions, each occurring in molecules of different rotational levels. The same is true for electronic absorptions, where both rotational and vibrational properties of the molecules are impressed on the electronic transitions.

Interpretation of absorption spectra is somewhat complex. If a molecule vibrates with pure harmonic motion and the dipole moment is a linear function of the displacement, then the absorption spectrum will consist of fundamental transitions only. If either of these conditions is not met, the spectrum will contain overtones (multiples of the fundamental) and combination bands (sums and differences). Most of these overtones and combination bands occur in the near-infrared (0.8–2.0 μm).

Not all vibrations and rotations are infrared-active. If there is no change in dipole movement, then there is no oscillating electric field in the motion, and there is no mechanism by which absorption of electromagnetic radiation can take place. An oscillation, or vibration, about a center of symmetry therefore will not be observed in the infrared spectrum (absorption) but can be observed in the Raman spectrum (scattering).

To summarize, the five regions of the electromagnetic spectrum of interest are:

Ultraviolet (electronic):   0.2–0.4 μm
Visible (electronic):   0.4–0.8 μm
Near-IR (overtones):   0.8–2.0 μm
Mid-IR (vibrational):   2.0–25.0 μm
Far-IR (rotational):   25.0–500.0 μm

Electronic transitions (UV, visible spectra) generally give information about unsaturated groups in the sample molecules. Olefins absorb near 0.22 μm, aromatics near 0.26–0.28 μm, carbonyls near 0.20–0.27 μm, polynuclear aromatics near 0.26–0.50 μm, and conjugated C=S groups near 0.62 μm. Any material which is colored will generally show absorption in the visible region. The intensity of the absorption is proportional to the number of chromophores giving rise to the absorption band.

Overtones (near-IR) are useful for studying the presence of groups containing hydrogen. Fundamentals involving hydrogen vibration tend to congregate near the same frequencies in the mid-IR, but they are easier to distinguish and study in the overtone region. Vibrational transitions (mid-IR) are the most useful to study. These give information about the presence or absence of specific functional groups in a sample. Practically all functional groups (that have an infrared-active fundamental) display the fundamental over a very narrow range of wavelength in the mid-infrared region. Moreover, the whole spectrum, containing fundamentals, overtones, and combination bands, constitutes a fingerprint of the sample. This means that although we might not know what a sample is, we will always know it later if it occurs again. Finally, the absorption intensity of any band, whether fundamental or overtone, is proportional to the number of functional groups giving rise to the signal.

FOURIER TRANSFORM INFRARED SPECTROMETRY

In dispersive spectrometry, the wavelength components of light are physically separated in space (dispersed) by a prism of grating as in the Michelson interferometer seen in Figure 19. Modern dispersive spectrometers divide the incident beam into two beams: one beam goes through the sample and the other goes through a suitable reference material. The intensity of both beams is monitored by a suitable detector, and final data output can be displayed in either transmittance or absorbance:

$$\text{Transmittance} = \frac{I_s}{I_R} \tag{15}$$

$$\text{Absorbance} = -\log\left(\frac{I_s}{I_R}\right) \tag{16}$$

where I_s and I_R refer to the intensities in the sample beam and reference beam, respectively. This rationing occurs at each wavelength element, and the final plot is a graphical display with transmittance or absorbance on the Y axis and wavelength or frequency on the X axis.

In Fourier transform spectrometry, the wavelength components of light are not physically separated. Instead, the light is analyzed in the time frame of reference (the time domain) by passing it through a Michelson interferometer. This interferometer operates so that light is separated into two beams by a beamsplitter. One beam strikes a stationary

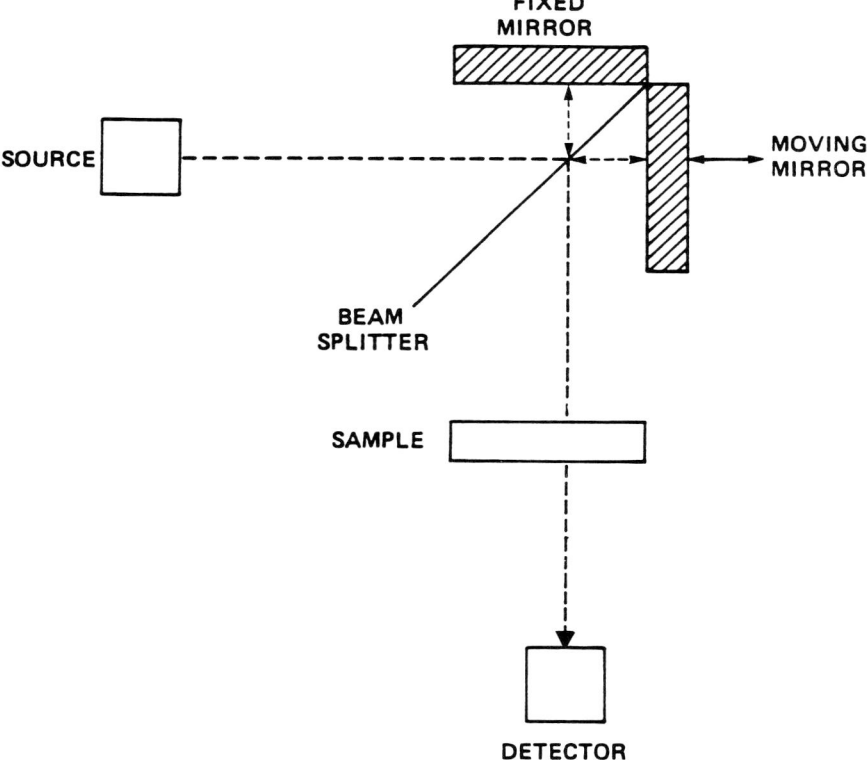

Figure 19 Schematic of Michelson interferometer.

mirror and is reflected back to the beamsplitter. The other beam strikes a moving mirror and is reflected back to the beamsplitter. The two beams are recombined at the beamsplitter and proceed on to the sample and the detector. Note that upon combination the two beams will interfere constructively or destructively, depending on whether the difference in path length of the two beams is an integral multiple of the wavelength. The difference in path length is called the retardation, and when the retardation is an integral multiple of the wavelength, the interference is maximally constructive. A plot of retardation vs. intensity measured at the detector is essentially the time domain function of the intensity. This can be transformed by Fourier transform mathematical techniques into a frequency function of the intensity.

X-RAY FLUORESCENCE SPECTROMETRY

X-ray fluorescence spectrometry (XRF) is a nondestructive technique of elemental analysis. XRF is based on the principle that each element emits its own characteristic x-ray line spectrum. When an x-ray beam impinges on a target element, orbital electrons are ejected. The resulting vacancies or holes in the inner shells are filled by outer shell electrons. During this process, energy is released in the form of secondary x-rays known as fluoroscence. The energy of the emitted x-ray photon depends on the distribution of electrons in the excited atom. Since every element has a unique electron distribution, every element produces a unique secondary x-ray spectrum, whose intensity is proportional to the concentration of the element in the sample. The excitation process and resulting x-ray spectrum are illustrated for calcium in Figure 20.

X-ray fluorescence instrumentation is divided into two types, wavelength dispersive (WDXRF) and energy dispersive (EDXRF) spectrometry. Both techniques can be automated with extensive computer systems for unattended operations that include data collection, reduction, and presentation. In a wavelength dispersive spectrometer (Fig. 21), radiation emitted from the sample impinges on an analyzing crystal. The crystal diffracts the radiation according to Bragg's law and passes it on to a detector which is positioned to collect a particular x-ray wavelength. Most spectrometers have two detectors and up to six crystals to allow optimization of instrument conditions for each element.

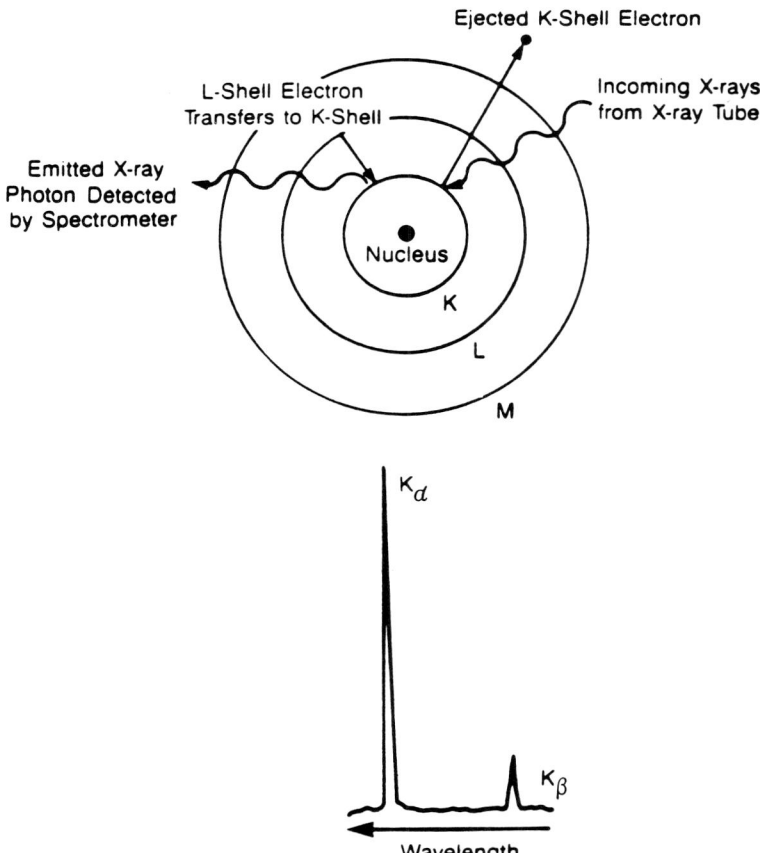

Figure 20 XRF excitation process and resulting spectrum for calcium.

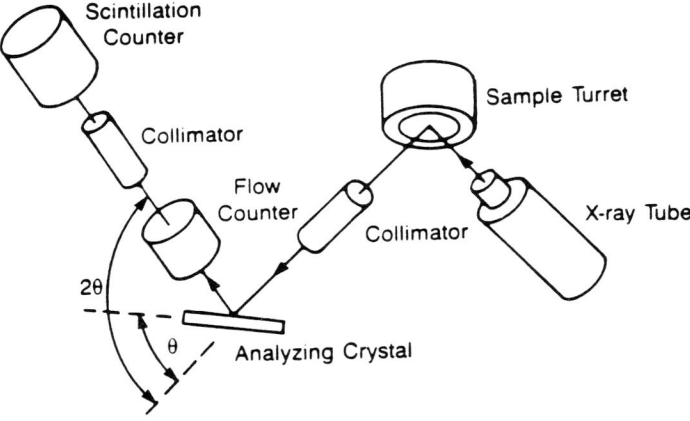

Figure 21 Wavelength-dispersive XFR spectrometer.

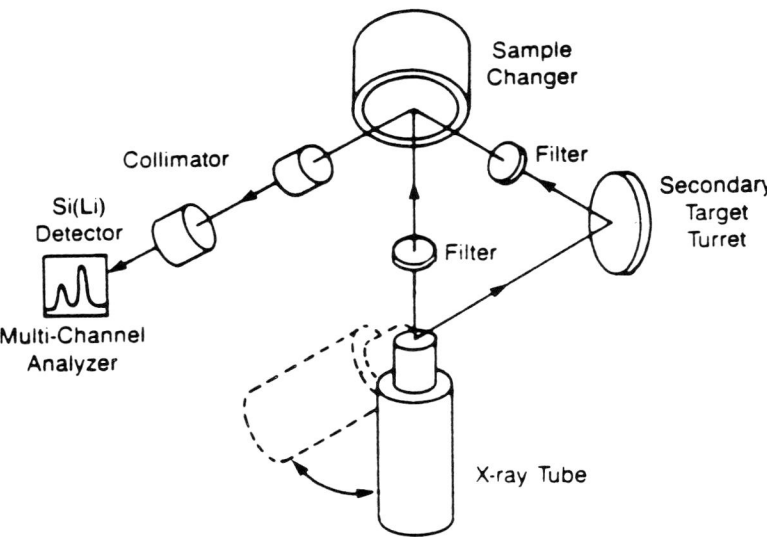

Figure 22 Energy-dispersive spectrometer.

In an energy dispersive spectrometer (see Fig. 22) the emitted x-ray radiation from the sample impinges directly on a solid-state lithium drifted silicon detector. This detector is capable of collecting and resolving a range of x-ray energies at one time. Therefore, the elements in the entire periodic table or a selected portion can be analyzed simultaneously. Optimization for specific elements is accomplished through the use of secondary targets and/or filters. In some cases, radioisotope sources are used in place of x-ray tubes in instruments designed for limited elemental applications.

XRF offers rapid nondestructive elemental analysis of liquids, powders, and solids. Although the first-row transition elements are the most sensitive, elements from atomic number 12 (magnesium) and greater can be measured over a dynamic range from trace (ppm) to major (%) element concentrations. EDXRF is well suited for qualitative elemental identification of unique samples, while WDXRF excels at high-precision quantitative analysis.

Whenever quantitative analysis is desired, care must be taken to use proper standards and account for interelement matrix effects since the inherent sensitivity of the method varies greatly between elements. Methods to account for matrix effects include standard addition and internal standard and matrix dilution techniques as well as numerous mathematical correlation models.

XRF is used by both research and plant laboratories to solve a wide variety of elemental analysis problems. Among the most common are the quantitative analysis of additive elements (Ba, Ca, Zn, P, and S) in additives and lubricating oils, and halogens in polymers. The high analytical precision of WDXRF has enabled the development of methods for the precious metal assay of fresh reforming catalyst that rival the precision of classical wet chemical methods. The percentage of active metals such as Pt, Ir, Re, or Ru as well as trace contaminants such as S and Cl has also been determined on numerous catalyst types.

Aqueous and organic liquids, powders, polymers, papers, and fabricated solids can all be analyzed directly by XRF. The method is nondestructive, so unless dilution is required, the original sample is returned to the submitter. Although the method can be applied to the analysis of materials ranging in size from milligram quantities to bulk parts such as engine pistons, a minimum of 5 g of sample is usually required for accurate quantitative analysis.

Important advantages of XRF over other methods of elemental analysis are that it is nondestructive and requires minimal sample preparation. WDXRF, when properly calibrated, offers precision and accuracy comparable to wet chemical methods of analysis. EDXRF offers rapid qualitative analysis of total unknowns.

SUMMARY

This chapter has provided an overview of analytical tests used in the characterization and analysis of polymeric materials. There are of course other analytical tools used to obtain various types of information, but it is not practical to list every technique that the various analytical disciplines are capable of applying. Table 3 has been compiled as a qualitative guide for selecting various measurement techniques and tools for sample characterization. It is important to keep in mind that the purpose of analytical techniques is to provide accurate compilation of information which characterizes the properties and chemical/physical structure of samples. These data are invaluable to the design and development of new products as well as in establishing product quality control specifications. Table 3 organizes the most widely used analytical tools into four main areas:

Elemental Analysis Useful for the determination of some or all of the elements in the sample, without regard to how they are structurally bound together. There are both single-element and multielement techniques. Single-element methods measure only one element at a time. In contrast, multielement techniques can measure many elements at one time.

Molecular Structure Analysis Useful for the determination of the chemical and/or physical arrangement of the atoms in a molecule. Usually a combination of two or

Table 3 Guide to Analytical Test Methods for Polymer Characterization

Type of analysis	Measurement technique	Type of information obtained
Single-element techniques	Atomic absorption	Quantitative, for a limited number of elements
	Neutron activation	Quantitative, used for oxygen
	Wet chemical methods	Quantitative and qualitative
	Combustion	Converts C, H, S, N to forms readily measured by other methods. Burns organic materials away from metals
Multielement techniques	Emission spectroscopy	Qualitative and quantitative
	X-ray fluorescence	Qualitative and quantitative
	High-resolution mass spectrometry	Quantitative for elements in a single molecular or a single molecular fragment
Molecular structure analysis	Infrared spectrometry	Determines functional groups present and fingerprints individual constituents
	Nuclear magnetic resonance	Determines functional groups and their order in the molecule
	Mass spectrometry	Determines composition of fragments and of total molecule; relative intensities related to structure
	Electron-spin resonance	Determines arrangement of atoms which influence unpaired electrons
	X-ray diffraction	Determines spatial arrangements of atoms
Multicomponent analysis	Gas chromatography	Analysis of complex mixtures of moderately volatile materials
	Liquid chromatography	Analysis of materials not suitable for 6C
	Infrared spectrometry	Analysis of solids, liquids, or gases of moderate complexity

General characterization	Mass spectrometry	Analysis of complex mixtures of moderately volatile materials
	Nuclear magnetic resonance	Analysis of simple mixtures only
Chemical characterization	Nuclear magnetic resonance	General tool for samples in liquid and solid states
	Mass spectrometry	Good for hydrocarbons and their O, N, and S derivatives
	Infrared spectrometry	Good for detecting carbonyls, OH, NH, and carbon chains
	Ultraviolet spectrometry	Good for detecting aromatic compounds
	Visible spectrometry	Shows transmission or absorption of visible light
	Electrometric titration	Determines titratable groups
	Voltametry (polarography)	Determines certain groups which are susceptible to electrically induced reactions in solution
Physical characterization	Light microscopy	Determines shape, size, texture, color, and transparency of samples magnified up to 2000×
	Scanning electron microscopy	Determines shape, size, and texture of samples magnified up to 100,000×
		Can show distribution of most elements at surface of sample
	Transmission electron microscopy	Provides internal as well as surface characteristics of very thin sections
	Gel permeation chromatography	Determines distribution of sample molecules according to their size and shape
	Thermogravimetry	Measures changes in sample weight with temperature; various thermal methods measure changes in other properties with temperature
	Gas chromatographic distillation	Determines standardized distillation characteristics

more of these techniques are required for complete characterization.

Multicomponent Analysis   Useful for determining some or all of the individual compounds present in a sample.

General Characterization   Used in the determination of the character or properties of the sample, rather than providing detailed information on chemical composition or structure. Techniques can be divided into chemical and physical characterization. In chemical characterization, the chemical function groups or compound composition or structure is determined. Physical characterization provides useful information on physical characteristics.

14
Optical Microscopy for Studying Molecular Ordering in Polymers

Christopher Viney[*]
*IBM Almaden Research Center
San Jose, California*

INTRODUCTION	506
OPTICAL ANISOTROPY IN POLYMERS	506
Optical Anisotropy	506
The Optical Indicatrix	507
Sources of Birefringence	509
Uniaxial and Biaxial Indicatrices: The Effect of Symmetry	509
Optic Sign	511
Uniaxial and Biaxial Polymeric Systems	511
Sources of Extinction Between Crossed Polars	512
Measurement of Birefringence	513
The Ray Velocity Surface	514
Linear, Circular, and Elliptical Retarders	515
PRACTICAL CONSIDERATIONS	516
Microscope Hardware	516
Specimen Thickness	517
INTERPRETATION OF CONTRAST	520
Interpretation of Orientation Birefringence Changes	520
Measurement and Interpretation of Extinction Directions	521
Use of Circularly Polarized Light to Identify Biaxial Ordering	523
Use of Optical Pleochroism to Identify the Local Chain Axis Orientation	523
EXAMPLES OF OPTICAL MICROSCOPY APPLIED TO THE STUDY OF ORDERING IN POLYMERS	524
Spherulites	524
Uniaxially Deformed Conventional Polymers	528
Fibers and Uniaxially Sheared Films of Liquid Crystalline Polymers	529
Biaxially Deformed Polymers	531
Form Birefringence	531
Liquid Crystalline Polymer Textures	532
REFERENCES	533

[*]*Current affiliation:* University of Washington, Seattle, Washington

INTRODUCTION

Local molecular order in polymers can be characterized by optical microscopy if that order changes over distances comparable to or greater than the wavelength of light. Used for this purpose, the microscope does more than fulfill the role of a powerful magnifying glass: the nature of intermolecular correlations is deduced by observing the way in which the polymer specimen changes the polarization state, intensity, and direction of light passing through it. The approaches adopted for interpreting contrast are mostly based on established techniques used by optical crystallographers [1–5].

OPTICAL ANISOTROPY IN POLYMERS

Optical Anisotropy

The refractive index of a medium can be expressed in terms of molecular polarizability [6] or in terms of dielectric permittivity [3, 7]. Either formalism quantifies the extent to which the electric field associated with a linearly polarized light wave can *polarize* the distribution of electrons in the molecules constituting the medium. Individual polymer molecules are highly anisotropic. Referring to the simple extended hydrocarbon molecule shown in Figure 1, we see that a light wave vibrating parallel to direction Z' will induce a greater polarization of bond electrons than a disturbance vibrating parallel to Y', which in turn will induce a greater polarization than one vibrating parallel to X'. Consider now a large number of such molecules, all having a similar orientation relative to the external laboratory axes, collected into an optically resolvable volume. We expect linearly polarized light vibrating parallel to Z' to experience a higher refractive index than light polarized parallel to Y', which should experience a higher refractive index than light polarized parallel to X'.

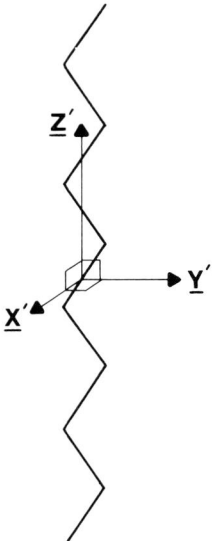

Figure 1 Schematic representation of extended hydrocarbon chain. The zigzag backbone is in the plane of the paper ($Y'Z'$).

If the polymer molecules were instead collected with their Z' axes still similarly oriented but with individual molecules rotated randomly about this axis, there would be no correlation between the axes X' or Y'. There would still be anisotropy of the refractive index, though of a character different from that observed in the previous case. We continue to expect that light which is linearly polarized parallel to Z' will experience the greatest refractive index. However, light that is linearly polarized normal to Z' would now experience a constant refractive index that is independent of its azimuth in the plane normal to Z'. If the molecules were to be arranged with no orientational correlation between any of the axes X', Y', Z', the refractive index would be independent of direction.

The variation of refractive index with direction is therefore related to the anisotropy of the polarizability of individual molecules within an optically resolvable region and to the ordering of molecules within this region. It is described graphically by a device known as the optical indicatrix [1–7].

The Optical Indicatrix

The indicatrix can be derived empirically by following the sequence of steps listed below and illustrated in Figure 2:

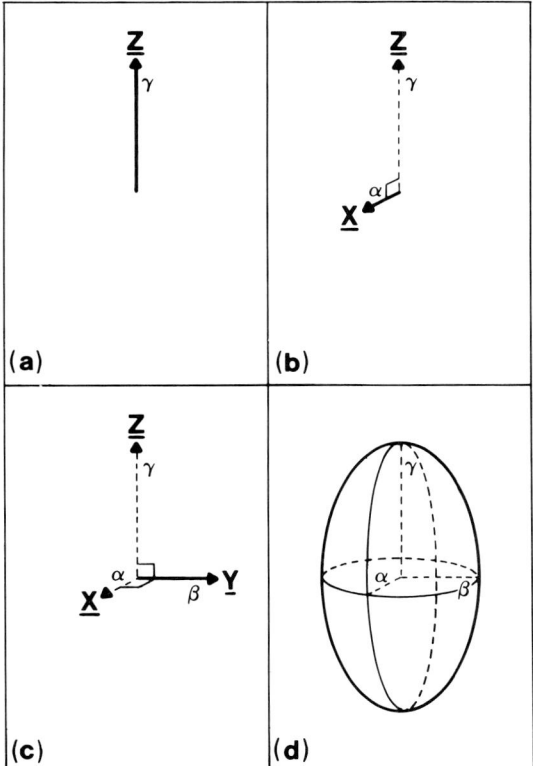

Figure 2 Sequence illustrating the empirical derivation of the optical indicatrix for an anisotropic medium. α, β, and γ are principal refractive indices, associated with reference directions X, Y, and Z respectively. The steps are described in detail in the text.

a. Identify the direction (Z) along which the refractive index is a maximum (γ).
b. Identify the direction (X) along which the refractive index is a minimum (α). One always finds X to be perpendicular to Z.
c. Identify the direction (Y) that is normal to both X and Z. The refractive index in that direction (β) is always found to be larger than any other measured in the XY plane.
d. Draw the ellipsoid defined by the equation

$$\frac{x^2}{\alpha^2} + \frac{y^2}{\beta^2} + \frac{z^2}{\gamma^2} = 1, \qquad \gamma \geq \beta \geq \alpha$$

This ellipsoid is the optical indicatrix. The quantities α, β, and γ are known as principal refractive indices.

More formal derivations lead to the indicatrix by considering it as the representation surface for relative dielectric impermeability [7].

The indicatrix has a particularly useful property, derived rigorously from Maxwell's equations [7] but presented here in the form of a graphical construction. When light is propagated in an arbitrary direction through an optically anisotropic specimen, it consists of two separate waves which are linearly polarized in mutually orthogonal directions. The polarization azimuth and associated refractive index of each wave is found as follows (Fig. 3):

a. Identify the plane which is perpendicular to the direction (ll') in which the light is propagated through the material and which passes through the center of the indicatrix. In general, this plane will intersect the indicatrix in an elliptical section.
b. Identify the semiaxes of this section. The direction of each defines the polarization azimuth of the associated wave, in the plane containing the semiaxis and ll'. The lengths of the semiaxes (n_{max}, n_{min}) define the corresponding refractive indices. They are referred to respectively as the slow and fast (vibration) directions of the

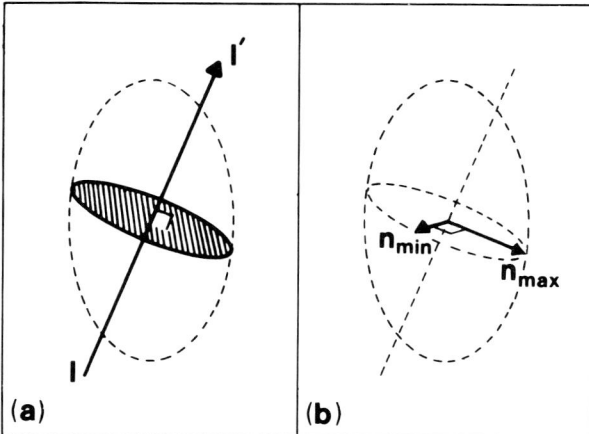

Figure 3 Optical indicatrix used to identify the polarization azimuth and refractive index associated with each component wave when light is propagated (along direction ll') through an anisotropic medium. The steps are described in detail in the text.

material, since the propagation velocity of each ray depends on the associated refractive index.

Since the light encounters two refractive indices, the material is said to be birefringent. There are three principal birefringences (Δn):

$\Delta n_{YX} = \beta - \alpha$
$\Delta n_{ZX} = \gamma - \alpha$
$\Delta n_{ZY} = \gamma - \beta$

Sources of Birefringence

The foregoing discussion of the optical indicatrix was accomplished with implicit reference to orientation birefringence. One can additionally identify two other sources of birefringence.

Stress birefringence [8] arises when inter- and intramolecular bonds become distorted in response to a stress state. The stress may be applied externally or it may be a residual internal stress. It does not depend on the individual molecules having anisotropic polarizability and, most important, does not rely on ordering of the molecules—either before application of the stress or as a result of it. Stress birefringence is alternatively known as deformation birefringence or as strain birefringence. It has little significance in the context of ordering in polymers.

Form birefringence [8–9] may arise when two or more phases coexist in a system. The following conditions apply:

1. The separate phases are optically isotropic.
2. The individual isotropic phases have different refractive indices.
3. The molecules in at least one of the phases are organized into similar particles which are large compared to the size of the molecules but small compared to the wavelength of light (so that they are not themselves optically resolvable). However, these particles must be orientationally ordered over an optically resolvable distance.

Thus we may (for example) be concerned with oriented rodlike, platelike, or elliptical particles embedded in a matrix consisting of the other phase.

Uniaxial and Biaxial Indicatrices: The Effect of Symmetry

A central section through the optical indicatrix drawn in Figure 2d will in general be an ellipse, with semiaxes both $\geq \alpha$ and both $\leq \gamma$. Since $\alpha \leq \beta \leq \gamma$, the α–γ plane contains two noncollinear radius vectors which have length β (Fig. 4), so that the indicatrix has two circular central sections (we recall that the radius vector normal to the α–γ plane also has length β). The normals to these circular sections are known as optic axes; because there are two of them, this indicatrix is described as being biaxial.

There is a useful axiom in crystallography, known as Neumann's principle [3, 7], which states that the symmetry of a physical property must include at least the nontranslational symmetry elements of the volume of material exhibiting that property. This result places constraints on the orientation and shape of the indicatrix. If the structural symmetry is lower than that of the biaxial indicatrix, only the orientation of the indicatrix is affected by that symmetry. The orthorhombic symmetry of the indicatrix is then a property of the

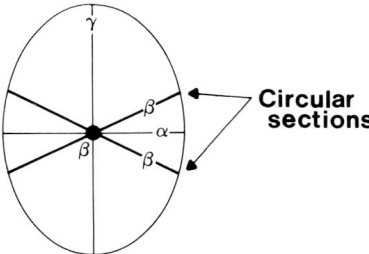

Figure 4 View of the α–γ plane in a (positive) biaxial indicatrix. The radii of length β and the two circular sections are identified.

interaction between light and the material, and not a property of the material itself. If the material has structural symmetry that is higher than the symmetry of the biaxial indicatrix, both the orientation and shape of the indicatrix are constrained. In particular, the presence of a 3-, 4-, 6-, or ∞-fold rotation axis dictates that the elliptical central section normal to this axis is a circle. The indicatrix is therefore an ellipsoid of revolution about the axis of high symmetry. The existence of only one central section gives rise to a single optic axis, and the indicatrix is now properly described as being uniaxial. The refractive index normal to the optic axis is denoted o, and is known as the ordinary refractive index, because rays with vibration directions normal to the optic axes obey Snell's law. The refractive index parallel to the optic axis is denoted e; it is known as the extraordinary refractive index, because a ray with its vibration direction parallel to the optic axis does not obey Snell's law. Indeed, a ray associated with any intermediate refractive index (denoted e') is also an extraordinary ray. With reference to the construction in Figure 3, we see that light propagated in an arbitrary direction through optically uniaxial material is always resolved into an ordinary ray and an extraordinary ray (Fig. 5).

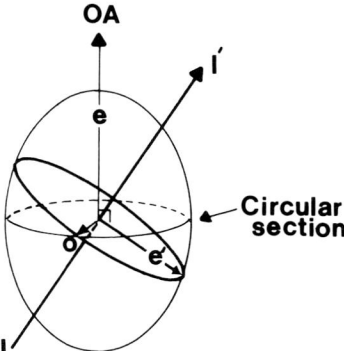

Figure 5 Light propagated in an arbitrary direction (II', in plane of diagram) through optically uniaxial material is resolved into an ordinary ray (refractive index o, normal to plane of diagram) and an extraordinary ray (refractive index e, in plane of diagram).

Optic Sign

For optically uniaxial material, we define the optic sign to be positive when $e > o$ and negative when $e < o$. If the indicatrix is biaxial, the optic sign [3] is positive when $\gamma^2(\beta^2 - \alpha^2) < \alpha^2(\gamma^2 - \beta^2)$ and negative when $\gamma^2(\beta^2 - \alpha^2) > \alpha^2(\gamma^2 - \beta^2)$. This approximates to $(\beta - \alpha) < (\gamma - \beta)$ (positive) and $(\beta - \alpha) > (\gamma - \beta)$ (negative).

Uniaxial and Biaxial Polymeric Systems

Typical polymer single crystals are optically biaxial. The greatest perfection of ordering in such crystals is achieved by allowing them to grow slowly from solution. The crystals usually have a thickness that is considerably smaller than their length or breadth. The molecules run from one large surface to the other [9–10], though not necessarily normal to them [11–12]. Also, the crystal thickness is typically much smaller than the molecular length, which requires chain folding. Crystals grown from the melt are less perfect; molecules either fold back on themselves at the surface and reenter the same crystal, or they continue into an adjacent crystal. Both the configuration and preferred conformation will influence whether the shape of the molecules is locally rodlike or lathlike. If lathlike, the molecules in these crystals will tend to pack into unit cells that have orthorhombic or lower symmetry. Monoclinic or even triclinic symmetry is to be expected for crystals consisting of all but the most regular types of polymer molecule; it is especially important to note that in these cases the principal axes of the indicatrix cannot be assumed to lie either parallel or perpendicular to the polymer chain axes.

Most commercially produced polymer fibers are drawn uniaxially from isotropic melts or solutions. Individual fibers are therefore also optically uniaxial. Uniaxial or biaxial crystalline regions may form locally on cooling. However, they tend to be small and may not be optically resolvable, since the drawing process gives molecules little chance to develop long-range correlation of position or of rotations about their chain axes. While a uniaxial assembly of simple aligned molecules (e.g., polyethylene) may be expected to be optically positive, this is not necessarily the case for all uniaxially aligned polymers. Nitrocellulose fibers, for example, are optically positive if there are on average less than 2.39 nitrate group substituents on each glucose monomer residue, and they are optically negative if the nitration level is greater than this [13–15]. It is the orientation of the anisotropically polarizable nitrate groups that directly dominates the optical characteristics of the fibers in this case, and not the orientation of the polymer backbone.

The biaxial properties of polymer films arise from the fabrication process. At its simplest, this involves extruding polymer between closely spaced rollers—aligning the polymer along the so-called machine direction, normal to the rollers—and then drawing the sheet along a direction parallel to the rollers. An alternative route starts with an extruded tube of the polymer, which is then blown into a wider diameter and thinner walled tube while the material is still sufficiently plastic. The tube is then cut and opened flat to produce a sheet.

In liquid crystalline polymers, long-range orientational ordering of the molecules is present as an equilibrium property. Long-range positional ordering may also occur, but never in all three spatial dimensions. The least ordered case is that which is found in a simple nematic: the rodlike chain segments point in approximately the same direction locally, but their centers of gravity are not positioned on any lattice. The characteristic symmetry element is an infinite-fold rotation axis parallel to the direction of preferred molecular alignment; the corresponding optical indicatrix is therefore uniaxial. In terms of molecular ordering, the simple nematic state resembles a drawn amorphous polymer fiber, the only significant difference being that extended conformation and local alignment of molecules are equilibrium properties of the liquid crystalline state. If the molecules in a nematic are lathlike, long-range lateral correlation of rotations about their chain axes may develop without crystallization occurring: the infinite-fold symmetry axis disappears, and the optical properties are then biaxial. This ordering scheme is described as biaxial nematic. Only in cases where the infinite-fold symmetry axis is replaced by a diad can one necessarily consider one principal axis of the indicatrix to lie parallel to the preferred direction of chain axis orientation [16]. Until recently, biaxial nematic ordering had been detected only in a few lyotropic small molecule liquid crystals [17, 18], since crystallization is favored in most substances. However, there is evidence that the random copolymerization of two or more monomer types can depress the crystallization temperature of the resulting polymer sufficiently to enable biaxially ordered nematic mesophases to form before the melt solidifies on cooling [19–23]. Biaxial nematic ordering has also been identified in polymers comprised of molecules that have planar mesogenic sidegroups attached to the backbone [24].

The ordering scheme in cholesteric systems is derived from that in a simple nematic by superimposing a twist along a direction normal to the optic axis. Planes of molecules rotated by 2π are separated by a distance p called the pitch. A simple nematic can therefore be regarded as a cholesteric of infinite pitch. The optical properties of cholesterics are uniaxial.

In smectic liquid crystals there is a more complex ordering pattern than in nematics. In addition to there being long-range chain axis orientation, the extended segments of molecules are constrained to lie in layers. They may also be positionally ordered within the layers, and there may be some relationship between the directions of molecular tilt in successive layers. The optical indicatrix will be uniaxial if the preferred direction of molecular orientation is normal to the plane of the layers—and if the extended segments of molecules are either rodlike or lathlike with no long-range lateral correlation of rotations about their long axes.

Sources of Extinction Between Crossed Polars

When a birefringent specimen is observed between crossed polars, linearly polarized light from the polarizer is resolved into two orthogonal components as described previously. Interference of these components occurs when they emerge from the top surface of the specimen. The resultant is resolved into the transmission azimuth of the analyzer, at 90° to that of the polarizer. The transmitted intensity I is given by [3]

$$I = A^2 \sin^2 2\theta \, \sin^2\left[\frac{\pi d}{\lambda}(n_{\max} - n_{\min})\right] \tag{1}$$

where A = amplitude transmitted by polarizer, θ = angle between the transmission azimuth of the polarizer and the slow vibration direction in the specimen, λ = wavelength of light used, and $n_{\max}$, $n_{\min}$ = refractive indices associated with the slow and fast vibration directions, respectively.

One can identify four possible conditions under which a specimen might show extinction when viewed in monochromatic light between crossed polars:

1. The specimen is optically isotropic.
2. The specimen is optically anisotropic and is being viewed along a direction parallel to an optic axis ($n_{\max} = n_{\min}$).

3. The specimen stage or crossed polars have been rotated to an orientation where the vibration directions lie parallel to the transmission azimuths of the crossed polars ($\theta = m\pi/2$, m an integer).
4. The combination of specimen thickness and birefringence causes the two refracted rays to emerge from the specimen with a phase difference equal to a whole number of wavelengths [$d(n_{max} - n_{min}) = m\lambda$]. The resultant of interference then has the same polarization azimuth as the incident light.

Condition 4, where only specific wavelengths are extinguished, does not arise in white light; one instead observes the color complementary to the one extinguished. The interference colors are most vivid when $\theta = \pi/4 \pm \pi/2$, when the specimen is described as being in a 45° orientation between the crossed polars. Condition 3 applies if the observation of extinction is found to depend on the angle θ. If extinction occurs in white light, for all rotations of the crossed polars, one can test for condition 2 by tilting the specimen so that light is propagated through it in a different direction. If extinction still persists, one knows that it must be due to condition 1. An alternative test for conditions 1 and 2 involves the use of a circularly polarizing polarizer and analyzer [25]. In this case, extinction can no longer arise from the incident light being polarized parallel to a vibration direction in the specimen.

Measurement of Birefringence

An approximate measure of $|n_{max} - n_{min}|$ can be determined from the interference color exhibited by a specimen [1–5]. Quantitative measurements require the use of a compensator. In most cases this involves placing one or more pieces of birefringent material (the compensator), of known birefringence and thickness, above the specimen. If the slow vibration direction of the compensator coincides with the fast vibration direction of the specimen, one can obtain extinction (compensation) if the product of birefringence and thickness is the same for both. In this situation, the compensator has precisely removed the phase difference that the specimen introduced between the components of light vibrating along its slow and fast directions. The thickness of compensator above a particular region of the specimen can be changed if the compensator contains one or more wedge-shaped elements. Both thickness and birefringence may be changed if one or more elements of the compensator can be tilted; the tilt axis is usually designed to lie parallel to a vibration direction of the element. Examples of the former type are the quartz wedge [3] and the Babinet [26] and Soleil [26] compensators, while the most common among the latter type are the Berek [1, 16] and Ehringhaus [27] compensators. A quartz wedge is the simplest to use but is also the least accurate. The most versatile is the Ehringhaus compensator: it introduces a uniform phase difference across the field of view, it permits accurate compensation for phase differences as high as 260π (i.e., 130 wavelengths), and it can easily be tailored to have dispersion characteristics that are similar to those of the specimen.

An entirely different route to compensation is afforded by the de Senarmont method [4, 28], which applies to monochromatic light. A birefringent plate that introduces a $\lambda/4$ retardation (a quarter-wave plate) is placed above the specimen; the specimen is in a 45° orientation between the crossed polars, while the vibration directions of the quarter-wave plate are parallel to the transmission azimuths of the polars. The resultant polarization state of light transmitted by the polarizer–specimen–quarter-wave plate combination is linear, and the retardation due to the specimen can be expressed in terms of the analyzer azimuth required to extinguish the light. This compensator gives unambiguous measurements for retardations up to $\lambda/2$.

The Ray Velocity Surface

When light is incident *normally* on a parallel-sided optically anisotropic specimen (orthoscopic illumination), the two orthogonally polarized transmitted waves share a common wave normal, parallel to the propagation direction and normal to the vibration direction of each. For an extraordinary wave, the ray direction differs from the wave normal (Fig. 6). Snell's law holds for extraordinary rays in anisotropic media if the angle of refraction is measured to the wave normal instead of to the ray direction. In situations where it is necessary to consider the two separate ray directions as light travels through a specimen, it is useful to work with the ray velocity surface. This is defined to have the property that the perpendicular distance from the origin to the tangent plane at any point is proportional to the velocity of propagation of a wave front that is parallel to that tangent plane [3]. Thus, referring again to Figure 6, we see that the radius of the ray velocity surface at any point defines the ray direction of the wave, while the normal to the tangent plane at that point is the associated wave normal. Along principal vibration directions, the radius to the surface coincides with the tangent-normal. Since wave propagation velocity is proportional to n^{-1} (the reciprocal of refractive index), the radius of the ray velocity surface is also proportional to n^{-1} along the principal refractive index directions.

The indicatrix construction "The Optical Indicatrix," above, shows that a given propagation direction can be associated with two possible refractive indices or velocities, so that the ray velocity surface is in fact a pair of surfaces. In uniaxial materials, the ordinary ray always travels with a velocity proportional to o^{-1}, so that the corresponding ray velocity surface is a sphere of radius o^{-1}. The extraordinary ray travels with a velocity that lies between limits proportional to e^{-1} and o^{-1}, and the corresponding ray velocity surface is an ellipsoid of revolution about the radius having length o^{-1}. The equation of the ellipsoid is

$$\frac{x^2}{o^2} + \frac{y^2}{o^2} + \frac{z^2}{e^2} = \frac{1}{e^2 o^2}$$

Figure 7 shows central cross sections through the ray velocity surfaces and the indicatrix, cut parallel to the optic axis, for optically positive and optically negative uniaxial materials.

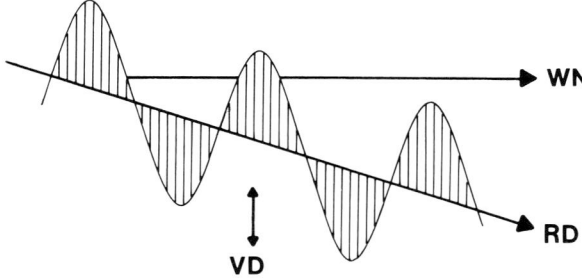

Figure 6 Ray direction **RD**, vibration direction **VD**, and wave normal **WN** for an extraordinary wave in an anisotropic medium. All three vectors lie in the plane of the diagram.

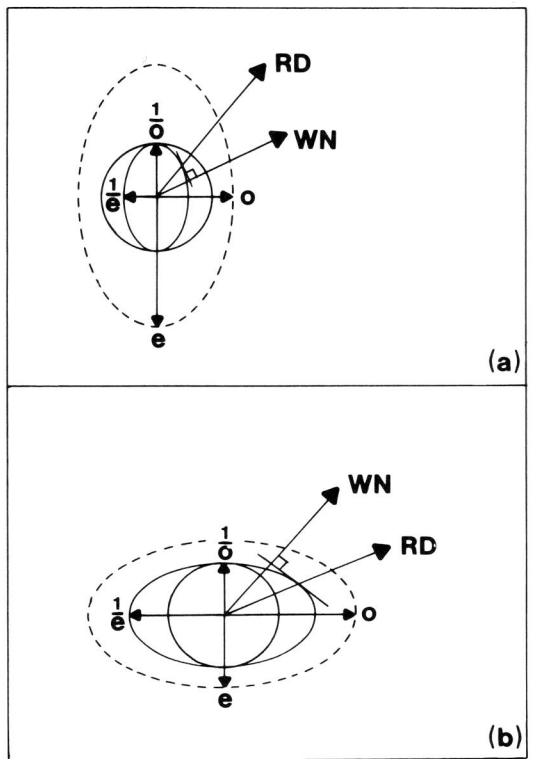

Figure 7 Ray velocity surface pairs for (*a*) a positive uniaxial material and (*b*) a negative uniaxial material. In each case, the broken elliptical outline indicates the shape (but not the relative scale) of the indicatrix; it is geometrically similar to the extraordinary ray velocity surface. Each diagram also shows how the extraordinary ray velocity surface relates a given ray direction **RD** to the corresponding wave normal **WN**.

The form of the ray velocity surfaces for optically biaxial materials is less straightforward, making it impractical to draw simple diagrams [5]. The principle of their derivation, however, is exactly the same as for uniaxial materials.

Linear, Circular, and Elliptical Retarders

For most of this chapter the type of specimen under consideration can be described as a linear retarder. The property distinguishing this class of retarder from any other is that one can identify two mutually perpendicular linear polarization states which are propagated through it without change: these are linear polarizations parallel to either of the two vibration directions, as discussed above.

A circular retarder, sometimes instead called a rotator, will alter all but the two possible circular polarization states. An example of a specimen with this property would be a film of optically active material (such as a cholesteric liquid crystalline polymer) with surfaces normal to the axis of high symmetry. It is important to avoid confusion between the circular retarder described here and a $\lambda/4$ or $\pi/2$ linear retarder (quarter-wave plate).

The elliptical retarder is the most general type; a thin parallel-sided slab of such a material will change the polarization state of any light propagated through it, with the exception of two mutually perpendicular elliptical states [29]. Their azimuths can be identified as "fast" or "slow" according to which is propagated more quickly. A film of optically active material could again serve as an example, if its surfaces were normal to some arbitrary direction.

PRACTICAL CONSIDERATIONS

Microscope Hardware

There are essential practical requirements that microscopists in any field must consider when selecting the most appropriate equipment [30, 31]. For work on molecular ordering in polymers, the following are especially relevant. The microscope should be equipped with both linear and circular [25] polars. It is essential to be able to rotate *both* polarizer and analyzer relative to the specimen. The preferred way is by rotating the polars (and also the compensator slot if present), rather than the specimen stage, since this allows a constant field of view to be maintained. Both polars should have graduated scales, with divisions every 1° if possible. One should be able to withdraw the polars from the optical path in the microscope.

Most polarizing microscopes are fitted with a Bertrand lens, to allow the back focal plane of the objective to be observed. If the field of view is entirely filled by a single crystal or domain, and convergent illumination conditions apply, one then sees an interference figure, sometimes also called the conoscopic image or directions image [1–5]. The term "conoscopic image" is also used loosely to describe the intensity distribution in the objective back focal plane—in other words, the optical diffraction pattern [32]—when polycrystalline or polydomain specimens are being examined. However, the light incident on the specimen should be parallel (instead of convergent) in this case. The microscope should be designed to allow the conoscopic image/optical diffraction pattern to be photographed, as well as to be observed at the eyepiece(s).

The versatility of the microscope is increased if phase contrast and/or interference contrast is also available, allowing refractive index gradients in the specimen to be imaged. Since it may be desirable to observe a given field of view under more than one set of imaging conditions, it should be possible to change from one contrast mode to another without disturbing the specimen.

If specimens are to be viewed above or below ambient, to make in situ observations of the effect of temperature on molecular ordering, the microscope should accommodate a heating/cooling stage. This in turn necessitates the use of long working distance objectives. Regarding the stage itself, one should be able to change the field of view easily; the windows should not affect the polarization state of light passing through them, and they should be easily replaceable in case they become contaminated by degradation products of the polymer specimen. Heating/cooling should be possible in a controlled manner at a variety of rates.

A universal stage allows the specimen to be tilted easily about a number of independent axes [33], permitting the user to observe the optical properties of the specimen for any propagation direction of the light. As for use with a heating/cooling stage, the microscope should be equipped with a set of long working distance objectives.

Objectives and eyepieces used with polarized light should be fabricated so as to be strain-free; otherwise the lenses themselves can contribute (via strain birefringence) to the contrast observed in the image.

Microscope manufacturers can usually supply a range of adaptors to allow most makes of camera to be fitted. A Polaroid camera will suffice for "once-off" photography where the aim is to record the general features in a given field of view. The finest detail resolved by a high numerical aperture objective is lost, however, since there is no photographic negative from which to produce enlargements at a scale where the eye can perceive this detail. Also, if an experiment necessitates observation of a constant field of view under different illumination conditions or different polarization states of the incident light, use of such a camera may be unsuitable. The mechanical disturbance occasioned by changing the film may be sufficient to cause the specimen to move; the resulting change in the field of view may be significant, especially if the specimen is being observed at high magnifications. In such instances, it is preferable to use a 35 mm camera back. Advancing the film then causes minimal vibration and may be achieved with a motor drive attachment. If the camera is equipped with a mirror release and self-timer, these can be used to withdraw the camera mirror from the light path several seconds before the shutter is opened, and thus cut down on vibrations during exposure. If persistent vibrations do result in blurred negatives at high magnifications, a convenient solution is afforded by attaching aluminum bracing struts to the microscope or camera casing and to the benchtop or wall with Plasticene [34].

A video camera and digitizer may be used instead of a film-holding camera. At the expense of some loss in resolution, such a system can be useful in a detailed point-to-point analysis of the image intensity, in comparing the appearance of a given field of view under a variety of illumination conditions, and for long-term recording of images. If the specimen undergoes rapid microstructural changes, it is important to ensure that the video camera has an adequately high scan rate.

A heat filter may be needed between the light source and the specimen if the latter contains highly volatile solvent and is held between an unsealed slide–coverslip combination. Heat from the light source, focused by the condenser lenses, may otherwise be sufficient to accelerate solvent loss.

The user should become familiar with the polarizer and analyzer scale readings that correspond to crossed orientations of the polars. Also, an accessory should be inserted into the accessory slot above the specimen stage, while the slot is oriented in either the NE–SW or NW–SE direction, and the direction of movement as it enters the field of view should be noted. This observation should be made at the normal viewing eyepiece(s), at the camera focusing eyepiece, and at the camera viewfinder. In each case, the user should check whether the accessory enters the field of view along the same diagonal as the one intuitively associated with the direction of the accessory slot. Different microscopes have different configurations of mirrors and/or prisms to direct light to an eyepiece or to the camera, and some configurations produce inverted or reflected perceptions of diagonal directions.

Specimen Thickness

The first motivation for working with thin specimens is to ensure that they are transparent. Furthermore, unambiguous interpretation of contrast in the polarizing microscope requires one to work with specimens so thin that there is no through-thickness change in optical orientation. Practically, this can mean that the specimen thickness should be less than the scale of significant in-plane variations in optical orientation.

Various methods exist for predicting how a series of superimposed birefringent layers affects the polarization state of transmitted light. One may successively resolve the light along the vibration directions of each layer [35, 36] and thus eventually calculate the amplitude and vibration directions of the light emerging from the top surface. Some more elegant approaches involve the use of matrix notation to represent the optical orientation and properties of each layer. The polarization state of the incident light is also represented by a matrix, and the emergent light is then described by the result of matrix multiplication. Various matrix representations exist, the most commonly used being those due to Stokes [29, 37, 38], Muller [29, 37, 38], Jones [39–47], and Yeh [48–50]. Another elegant (and still much neglected) device is the Poincaré sphere construction [4, 28, 29, 38, 51, 52]. Historically, it provided the first simple solution to qualitative problems involving the propagation of polarized light through a series of birefringent layers; much laborious spherical trigonometry may be needed, however, if a quantitative answer is required.

Using either the method of resolving components [20, 53] or the Poincaré sphere [53, 54], one can show that two or more superimposed linearly birefringent layers will in general not show extinction between crossed polars in white light, unless the optical orientation of successive layers differs by a multiple of $\pi/2$. Indeed, if the layers are discrete, extinction will in general not be observed for *any* polarizer or analyzer azimuth, if the specimen is viewed in *white* light.

Additional results relevant to polarized light microscopy of polymers follow. All can be derived in terms of the properties of Jones matrices; most are also demonstrable on the Poincaré sphere.

1. If a specimen consists of a sequence of linear retarders, the intensity of monochromatic light transmitted between crossed polars varies smoothly on rotation of either the specimen or the crossed polars. Extinction is not observed in general, but there are four intensity *minima* per 360° rotation of the specimen or crossed polars, occurring at 90° intervals.

2. Under the conditions described for 1 above, the transmitted intensity observed for *any* crossed polar azimuth is reproduced on rotating the crossed polars by a multiple of 90°.

3. If a specimen consists of a sequence of linear retarders, the intensity of monochromatic light transmitted between crossed polars remains unchanged if the sequence is reversed. Consequently, the intensity is also independent of which way up the specimen is viewed between crossed polars, if the specimen is inverted by rotation about an axis parallel to the transmission azimuth of one of the polars.

4. If a specimen consists of a single linear retarder, and if a compensator of given thickness and optical orientation is also present in the light path, the intensity transmitted between crossed polars is not changed if the positions of the specimen and the compensator are interchanged.

5. If the specimen in 4 consists of a sequence of linear retarders, the intensity transmitted between crossed polars does depend on the relative positions of the specimen and compensator.

6. If a specimen consists of a sequence of linear retarders, its effect on linearly polarized light can be duplicated by a system consisting of a single linear retarder and a single pure rotator. In the limit of the individual layers of the specimen being only weakly birefringent (e.g., due to the layers being very thin), the rotator component of the equiva-

lent system becomes a second-order consideration, even if the optical *orientations* of adjacent layers are widely dissimilar [51, 54]. In other words, the specimen is then optically equivalent to a single linear retarder, having extinction orientations between *crossed* polars in monochromatic linearly polarized light. The extinction orientations may vary significantly enough with wavelength to allow such a specimen to be distinguished from one in which the through-thickness optical orientation is constant. At a given wavelength, the transmitted intensity is practically independent of the *order* in which light encounters the individual layers. The optical properties described here are likely to be of concern only to a microscopist working with weakly birefringent (polymer) liquid crystals.

7. Another type of limiting behavior arises when there is an infinitesimal difference in optical orientation between adjacent layers in a sequence of infinitesimally thin linear retarders (arbitrary birefringence). In this case, the rotator component of the equivalent system dominates. If the polarizer azimuth is parallel to a vibration direction in the bottom layer of such a specimen, the polarization azimuth of light is rotated continuously as it is propagated upward through the specimens [55]. It can therefore be extinguished by an analyzer. Such guiding of light can be effected by liquid crystals in which the preferred direction of molecular orientation varies continuously with changing depth in the specimen.

Situations intermediate between those discussed have also been modeled [56, 57]. Again it is found that intensity minima, rather than extinction, are predicted when the specimen is rotated between crossed polars in white light.

From the foregoing, we note that areas of specimen which never extinguish between crossed polars in white light indicate nonconstant optical orientation through the specimen thickness.

The need to work with thin specimens also arises from the practicality of not wanting to resolve the separate transmitted rays as distinct images. The separation of the two refracted rays in a given domain will depend on the shape of the indicatrix, the orientation of the indicatrix relative to the incident light beam, and the thickness of the material. If the shape of the indicatrix is known—requiring knowledge of the principal refractive indices—one may calculate the maximum divergence between radius and tangent-normal for points on the ray velocity surface. For an optically uniaxial material, this will be equal to the maximum divergence between the *o*-ray and the *e*-ray, for light incident normally on the specimen surface as would occur during standard orthoscopic microscopy. The maximum separation of the rays as they exit the specimen is then given by a simple relationship involving the maximum divergence and the specimen thickness.

For a specimen consisting of several adjacent birefringent domains, unambiguous interpretation of contrast in the orthoscopic image requires that some significant area of the top surface of each domain should be so remote from all the neighboring domains that it is not entered by light that has been deviated out of one of these neighboring domains (Fig. 8). This condition is less likely to be met as the specimen thickness increases. If the microscope is to be used to observe single-domain detail at its theoretical resolution limit, typically taken [58] as λ/NA for a given objective having numerical aperture NA, it can be shown [59] that the specimen thickness should not exceed

$$\frac{\lambda eo}{\times \mathrm{NA}|e^2 - o^2|} \quad (2)$$

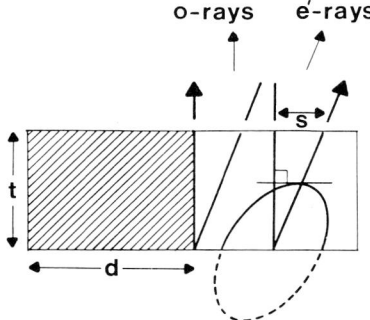

Figure 8 The unshaded domain has an optical orientation giving rise to maximum divergence between ordinary and extraordinary rays. Some of the extraordinary radiation may enter from the shaded neighboring domain (optical orientation arbitrary). The distance s increases with increasing specimen thickness t. Equation 2 in the text sets an upper limit on the acceptable magnitude of t, if s is not to exceed half the domain width d, when a given objective is being used at the limit of its resolution.

This criterion becomes practically significant when one is working with a highly birefringent specimen and an objective of high numerical aperture: care must then be taken to ensure that the specimen is adequately thin. The combination of high birefringence and fine microstructure may arise in the context of polymer liquid crystals.

It is appropriate to define here what is precisely meant by the term "domain." In this chapter it refers to a region in which the optical orientation is effectively constant over at least a microscopically resolvable distance—in other words, it is a region for which the concept of the indicatrix has practical meaning. The regions between domains are narrow compared to the lateral extent of a domain, and they can be the site of significant point-to-point changes in the orientation of the local symmetry axes.

INTERPRETATION OF CONTRAST

Interpretation of Orientation Birefringence Changes

It is apparent that birefringence can give an indication of molecular ordering, and that changes in birefringence reflect changes in molecular ordering. When measurements of orientation birefringence are referred to in the polymer literature, it usually is understood that an optically uniaxial system is under consideration. Justification lies in the implied assumption that the birefringence is due to local alignment of molecular chain axes, with no correlated rotations about the chain axes on an optically resolvable scale; it is assumed that multiaxial molecular correlations exist on such a scale only when specifically forced to do so by biaxial mechanical drawing [9, 60] or by an applied magnetic field [61, 62].

For this reason, it is common to relate birefringence changes to a change in the properties (shape, orientation) of a uniaxial optical indicatrix, without further qualification. Also, an observed birefringence change is often taken to be the consequence of a change in the relative orientation of the chain axes (either in three dimensions or at least within the specimen plane), again without further justification. In practice this need not be the explanation (Fig. 9). A change of birefringence consequent to thermal or radiation treatment, for example, might simply reflect a change in the orientation (but not the

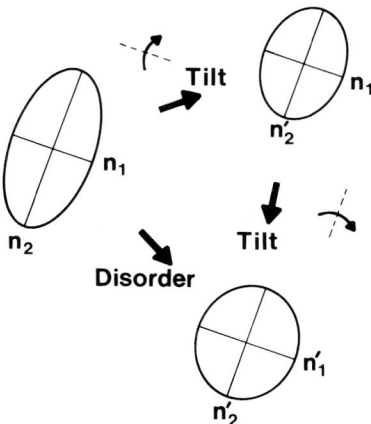

Figure 9 A given change in birefringence can arise either from a change of indicatrix shape (change in local molecular order) or from a change of indicatrix orientation. In both cases, the final in-plane birefringence $(n'_2-n'_1)$ may be identical, as shown.

shape) of the optical indicatrix [63, 64]. To check for this possibility, one might examine at least two nonparallel sections through the specimen. Alternatively, one could use a universal stage to tilt the specimen so that different sections through the indicatrix can be sampled [33], or one could supplement the optical observations with selected area diffraction [65, 66].

Should it be apparent that the indicatrix is after all biaxial, a change in birefringence may be due to one or more of the following: (a) a change in the relative orientation of the chain axes; (b) a change in the perfection of the correlation of rotations about the chain axes; (c) a change in orientation of the indicatrix.

Measurement and Interpretation of Extinction Directions
Uniaxial Indicatrix

We shall assume that specimens are sufficiently thin for the constituent domains to exhibit unambiguous extinction at appropriate orientations between crossed polars. The term "extinction direction" refers synonymously to either vibration direction. Authors often also use the term "optic axes" or "optical axes" as a synonym for extinction directions [67–71]. This conflicts with the previous (and better established) definition of an optic axis as the normal to a circular section of the optical indicatrix, and it should therefore be avoided.

If the domains are optically uniaxial, one vibration direction is parallel to the projection of the local preferred chain axis orientation (often called the director) onto the specimen plane. Whether this is the slow or fast vibration direction depends on the distribution of polarizable bonds in the domain. A compensator can be used to distinguish between the fast and slow directions. Point-to-point changes of the in-plane director orientation can be mapped out by following the corresponding extinction direction as a function of crossed polar orientation [72, 73]. Use of a compensator to distinguish between fast and slow vibration directions may also be necessary if one is to distinguish local molecular orientation from global director orientation [74]. It should be noted that, strictly, a director is defined [75] as the axis of symmetry for the orientation distribution of any consistently chosen molecular axis (as distinct from the preferred direction of chain axis orientation). Thus there need be no immediate association of the director and the chain axis.

Biaxial Indicatrix

Optical biaxiality in a polymer can be induced by biaxial mechanical processing, or it can occur spontaneously as in a liquid crystalline nematic or smectic phase of appropriate symmetry. The optical orientation information that can be accessed without use of a tilting stage is more limited than that obtainable from optically uniaxial material. Also, under equilibrium conditions, the multiaxial correlations typically extend over a maximum distance of only a few micrometers. It is therefore impractical to use the conoscopic image to help with local optical orientation determination; this technique [2, 3] is restricted to domains that occupy a significant area of the field of view. For convenience of discussion, we distinguish between two cases.

Biaxial indicatrix; one principal axis (Q) confined to specimen plane. One extinction direction (one cannot, without further information, say which one) defines the direction of Q. The other defines the intersection of the specimen plane and the plane containing the two remaining principal axes of the indicatrix. If it is known that Q is either the shortest or longest principal axis, it can be identified by using a compensator to establish which vibration direction is fast and which is slow. However, if Q is the principal axis of intermediate length, distinguishing between fast and slow vibration directions is not sufficient to allow identification of Q (Fig. 10). In the case of an optically uniaxial indicatrix, one principal axis (o) always lies in the specimen plane (Fig. 5). If the principal refractive indices are known, one can combine a measurement of birefringence with a determination of slow and fast directions to define the angle between the specimen normal and the unique axis of the indicatrix [3]. However, for an optically biaxial indicatrix, such measurements do not suffice to define uniquely the angle between the specimen normal and any of the principal axes. This point can be appreciated readily by noting the forms of the isochromes [3]—each of which corresponds to a contour of constant optical path difference—in uniaxial and biaxial interference figures.

Biaxial indicatrix; no principal axes confined to the specimen plane. This situation is realized if, for example, a longitudinal slice is cut from an extruded pellet of polymer in which the molecules have the local correlations appropriate to a biaxial indicatrix. In the

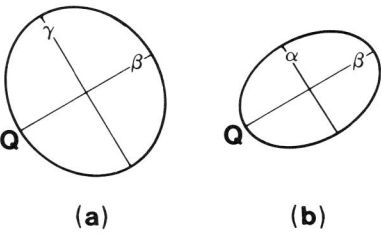

Figure 10 Two views of a given optical indicatrix, with the principal axis (Q) of intermediate length lying in the specimen plane. In case (a), Q is the fast vibration direction; in case (b), it is the slow vibration direction. Without additional information, it is not possible to use a compensator to identify Q unambiguously.

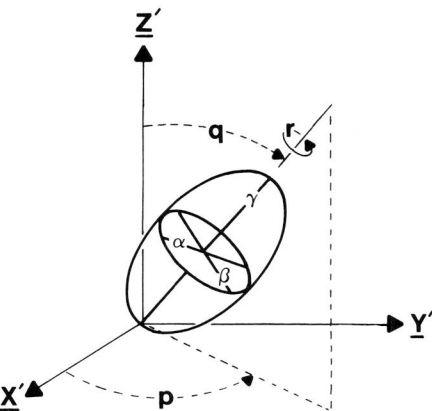

Figure 11 Geometry relating the optical indicatrix to the specimen plane ($X'Z'$), when the local chain axis tilt is held constant relative to this plane (angles p and q fixed) and the molecular segments rotate about their chain axes while preserving any local rotational correlation (angle r variable).

absence of perfect molecular alignment along the extrusion direction, the chain axes in any domain are likely to be inclined at some angle to the specimen plane. This and/or molecular asymmetry ensures that no principal refractive index is necessarily confined to the specimen plane. If the local chain axis tilt is held constant relative to this plane, and the molecular segments are rotated about their chain axes while preserving the rotational correlation (Fig. 11), the extinction directions in the specimen plane may be made to adopt any orientation relative to the macroscopic extrusion axis, provided that the axial ratios of the indicatrix lie within certain limits [20]. This can result in the specimen exhibiting a high degree of chain axis alignment as determined from x-ray diffraction, while the point-to-point optical orientation can appear to vary randomly.

Use of Circularly Polarized Light to Identify Biaxial Ordering

A specimen viewed in white light between crossed circular polarizers shows extinction only in localities where the indicatrix is intersected parallel to a circular section. If a uniaxially drawn polymer specimen consists of domains in which the molecules are uniaxially correlated, it is to be expected that the majority of optic axes will be aligned parallel (or nearly parallel) to the macroscopic draw axis. Should the molecular correlation within domains be biaxial, most optic axes will be inclined away from the draw axis (Fig. 12). The two cases can therefore be distinguished by sectioning the specimen at various angles to the draw axis and using crossed circular polarizers to identify the section that contains the greatest fraction of areas exhibiting extinction [21]. A similar argument would apply to a specimen consisting of homeotropically aligned molecules.

Use of Optical Pleochroism to Identify the Local Chain Axis Orientation

The term "optical pleochroism" refers to anisotropic light absorption in an optically resolvable region of a specimen: if an optically pleochroic specimen is observed in

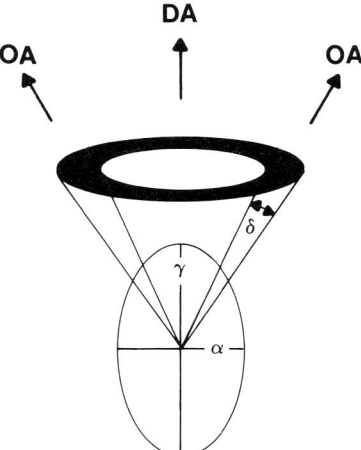

Figure 12 Sketch of α–γ section through a (positive) biaxial optical indicatrix, showing the optic axis directions (OA). If the principal refractive index direction γ lies within an angular range δ centered on the macroscopic draw axis (DA), the range of optic axis directions will be bound by the surfaces of two cones as shown.

linearly polarized light, the extent of absorption will change as the polarizer is rotated [76, 77]. If the molecular correlations within a domain have symmetry lower than orthorhombic, the principal axes of the representation surface for absorption need not coincide with the principal axes of the indicatrix. However, in specimens which are only a few micrometers thick, the presence of a sufficient density of light-absorbing states to have a significant effect on contrast is a property of the chain axis direction alone—or at least its projection onto the specimen plane. One may therefore be able to identify this direction, if the absorption is sufficiently marked [22]. It may be found that this direction does not coincide with either extinction direction in the same area of specimen, in which case one can infer the existence of biaxial molecular correlation without having to compare results from optical microscopy and x-ray diffraction.

EXAMPLES OF OPTICAL MICROSCOPY APPLIED TO THE STUDY OF ORDERING IN POLYMERS

Spherulites

The basic structural unit of a spherulite is a fibril (Fig. 13). The organization of polymer chains in an idealized fibril is a 1-D analog of that in a polymer crystal, so that one pair of opposite faces can be thought of as consisting of chain folds. This structure is described as being idealized because, in practice, a given chain may either fold back on itself at the surface and remain confined to the same fibril or may enter the amorphous zone between fibrils and then continue into a neighboring fibril or reenter the original fibril. Depending on whether the chains are normal to the chain fold surfaces, and on whether the "straight" sections of the molecules are rodlike, fibrils have the potential to be optically uniaxial or biaxial, though in practice they are too small to be resolved individually. A spherulite

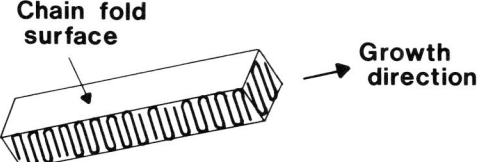

Figure 13 Molecular organization and chain fold surfaces in a polymer fibril.

consists of a very large number of fibrils which have all grown outward from a central nucleus. The spherulite size depends on the number of nuclei which formed in the melt or solution during crystallization. Diameters ranging from submicroscopic to several centimeters are possible.

The optimum contrast between crossed polars is obtained if the specimen is crystallized between glass surfaces separated by only a few micrometers, or if a thin section is microtomed from bulk crystallized material. In these specimens, the view of the spherulites is essentially a two-dimensional one. The orthoscopic image consists of a black Maltese cross, usually (but not necessarily) with arms parallel to the polarizer and analyzer transmission azimuths (Fig. 14). If the specimen stage is rotated, the orientation of the Maltese cross remains fixed relative to the viewer. One therefore deduces that the molecular ordering *in the spherulite as a whole* has spherical symmetry. The Maltese cross contrast can be interpreted [10, 78–81] by considering a narrow sector of the spherulite, consisting of a bundle of fibrils: the fibrils extend radially, while the chain direction is predominantly tangential.

On the optically resolvable scale of the bundle, we are usually dealing with a uniaxial indicatrix. One principal axis (denoted n_l) is then parallel to the average chain direction; the other (denoted n_t) is transverse to the average chain direction (Fig. 15). The birefringence of the bundle therefore can lie between zero and $|n_l - n_t|$, and the bundle will have extinction directions which are parallel and perpendicular to its length. Thus extinction between crossed polars will occur where the fibrils are approximately parallel to the polarizer and analyzer transmission azimuths.

A biaxial indicatrix could lead to the situation shown in Figure 16; locally, the semi-axes of the indicatrix as sectioned by the specimen plane do not lie radially and tangen-

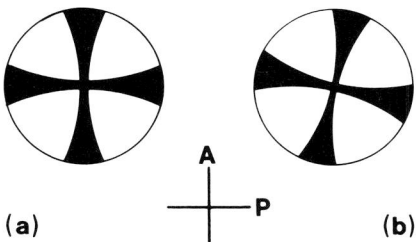

Figure 14 Schematic representation of Maltese cross contrast from spherulites. The most commonly observed cross orientation, relative to the transmission azimuths of the crossed polars, is shown in (*a*). A few polymers consist of spherulites which exhibit the extinction shown in (*b*). P = polarizer transmission azimuth; A = analyzer transmission azimuth.

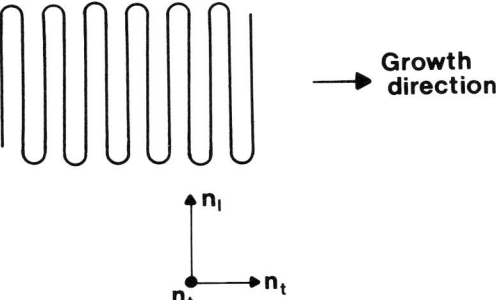

Figure 15 Average chain direction and principal refractive index directions in an optically uniaxial bundle of fibrils.

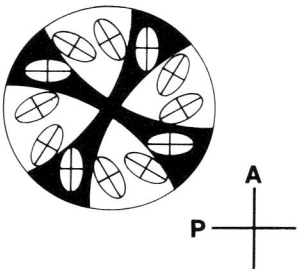

Figure 16 Schematic of extinction directions in a spherulite when the constituent bundles of fibrils are optically biaxial. P = polarizer transmission azimuth; A = analyzer transmission azimuth.

tially, though globally their orientation is still consistent with the overall symmetry of the spherulite. The arms of the Maltese cross do not lie parallel to the transmission azimuths of the polars in this case.

The fibrils in a real spherulite do not simply grow continuously outward from the nucleus. If they did, the density of a spherulite would decrease continuously from the center to the perimeter. In practice, the fibrils branch [82] in order to maintain an approximately constant polymer density. As a result of this branching, the arms of the extinction cross become wider and more diffuse toward the perimeter of a spherulite. In many polymers, bundles of fibrils twist or tilt about the radial direction [80, 83], the pitch (but not necessarily the sense) of the twist/tilt being independent of direction as measured from the center of the spherulite. The bundles therefore exhibit extinction at periodic intervals along their length, at positions where the light is propagated along a direction associated with zero birefringence (Fig. 17). Because of the circular symmetry of the structure as seen in the microscope, one observes a series of dark rings centered on the middle of the spherulite. (Their spacing tends to decrease with decreasing temperature of spherulite formation. This has led to the suggestion that the twist/tilt results from stresses set up during crystallization. Detailed causes and mechanisms are still the subject of

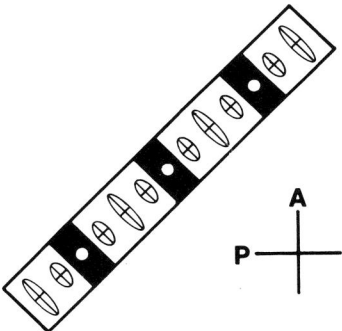

Figure 17 Schematic of extinction bands crossing a twisted bundle of fibrils, the extinction being due to the periodic occurrence of an optical orientation associated with zero birefringence. P = polarizer transmission azimuth; A = analyzer transmission azimuth.

research.) The banding in spherulites is most clearly observed by viewing a specimen between crossed polars and crossed quarter-wave plates, so that the Maltese cross is not present in the image [10].

It is of course possible for an optically resolvable region of uniaxially correlated polymer molecules to have the slow vibration direction lying normal to the average chain direction. One can usually check whether this is the case for a particular material by drawing some into a fiber: the molecules are then necessarily extended and aligned along the length of the fiber, and one can measure whether the axial or radial refractive index is greater. We deduce that an optic sign can be defined for spherulites (Fig. 18) according to whether it is the tangential or radial vibration direction that is slow—which can easily be determined with a compensator. Although most crystalline polymers yield negative spherulites, the optic sign is a useful initial clue when attempting to identify an "unknown" polymer by observing its properties.

Polarized light microscopy can be used as a complement to electron microscopy and x-ray diffraction in detailed studies of spherulite crystallography [84]. Used together with a heating stage, the optical microscope enables data on crystal nucleation and growth rates and mechanisms to be obtained [78, 85, 86]. Examination of thin sections microtomed from fabricated products [74] can indicate the extent to which an amorphous (and possibly oriented) surface "skin" has replaced the desired fine spherulitic microstructure.

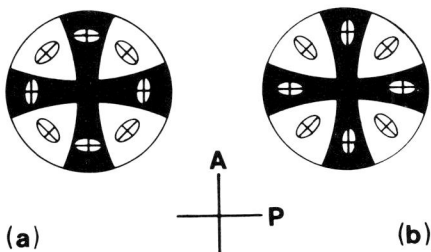

Figure 18 Schematic of fast and slow vibration directions in uniaxial spherulites: (*a*) positive; (*b*) negative. P = polarizer transmission azimuth; A = analyzer transmission azimuth.

A size distribution for spherulites in a given field of view can be estimated from the position and size of intensity maxima in the objective back focal plane, as observed with a Bertrand lens [9].

Uniaxially Deformed Conventional Polymers

A drawn polymer fiber is optically uniaxial. If the fiber axis (parallel to the optic axis) is the slow vibration direction, we speak of an optically *positive* fiber; if it is the fast vibration direction, the fiber is optically *negative*.

If we can assume that fiber drawing merely changes the degree of orientation of polymer chain segments, *without affecting the molecular conformation*, one can use optical birefringence as a measure of the degree of orientation [8]:

$$|n_a - n_r| = \Delta n = \Delta n_{max} \cdot \tfrac{1}{2}[3\langle\cos^2\theta\rangle - 1] = \Delta n_{max} P$$

where n_a = axial refractive index, n_r = radial refractive index, Δn = fiber birefringence, θ is the angle between an individual chain segment axis and the macroscopic fiber axis, P is known as an orientation parameter, and angular brackets "$\langle\rangle$" denote average. The quantity Δn_{max} is known as the intrinsic birefringence. It is the birefringence that would be observed for perfect orientational ordering of the molecules and has to be calculated from first principles (e.g., using individual bond polarizabilities) or estimated experimentally. The latter approach is usually more convenient and accurate. It may be effected by matching measurements of birefringence change to changes in orientation as determined by an alternative technique such as wide-angle x-ray diffraction [87] or infrared dichroism [88]; data for perfect orientational ordering are then estimated by extrapolation.

Measurement of fiber birefringence is used to characterize the molecular ordering introduced by fiber spinning and drawing processes [89–91]; it constitutes a simple technique for comparing the effectiveness of different processing routes in producing molecular alignment. The development of birefringence can additionally be used to monitor stress-induced crystallization of polymer melts [92], and studies of flow birefringence [93] have been used to identify optimal jet-flow geometries for use in producing high chain extension in polymer melts. These and related applications of the technique [94, 95] may not always involve use of the microscope, depending on the scale over which "global" molecular alignment is anticipated.

When an application demands a drawn polymer of zero birefringence, compensating blends of optically positive and negative polymers [96] may be useful. Combined measurements of stress, strain, and orientation birefringence are used in testing models [97–102] that describe polymer deformation at the molecular level.

Values of P do not describe the molecular orientation distribution fully. Any orientation distribution function can be expressed as a summed series of spherical harmonics (analogous to the manner in which a wave can be expressed as a sum of Fourier components); only the first term of the series is described by the orientation parameter P [97].

Polymers for which drawing changes the molecular chain conformation, or which have polarizable sidegroups whose ordering depends on the degree of chain alignment, do not obey the simple relationship given above. It is then necessary to measure chain segment orientation by a diffraction technique and to prepare a reference plot of orientation versus birefringence.

There are sets of more complex formulas which define order parameters for units in which the molecular correlations are biaxial [76].

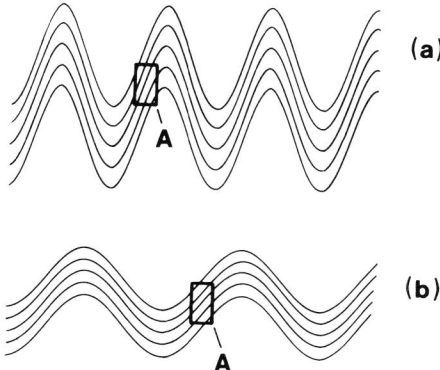

Figure 19 Schematic of how the global order parameter may change while the local ordering (e.g., in the region marked A) remains constant. The curves indicate the point-to-point variation of optical orientation in the specimen plane. The global order parameter is higher in (b) than in (a).

When supramolecular ordering is present in a specimen, it is necessary to be cautious about the scale at which an observed change in birefringence is related to a change in the order parameter. The specimen may, for example, have the supramolecular order shown in Figure 19a. Locally, in a region such as that marked A, the order parameter would be much higher than the global order parameter as determined from the birefringence of the specimen as a whole. Thermal or radiation treatment might result in the structure changing to that shown in Figure 19b. While the global order parameter has changed, the local ordering has not. If the "pleats" in the microstructure were below the resolution limit of the objective, one would observe that the structural change is associated with a change in birefringence. If, on the other hand, an individual pleat were to fill the field of view, the birefringence of regions such as A would remain constant. The structural change, however, could be detected via the associated change in the local orientation of the vibration directions. If the local birefringence were to change too, this could be interpreted in terms of a change in local order parameter (indicatrix shape), or it might simply arise from a change in the tilt of the indicatrix relative to the specimen plane. In cases of doubt, the ambiguities may have to be resolved by point-to-point electron microdiffraction.

If absolute values of refractive index (rather than birefringence) are required, immersion techniques such as the Becke line method [3] or dispersion staining [103–105] can be employed.

Fibers and Uniaxially Sheared Films of Liquid Crystalline Polymers

A common property of these specimens is the one-dimensional periodic banded texture which they exhibit when viewed between crossed polars [23, 106–113]. The changing position of the bands as a function of crossed polar orientation is due to the direction of preferred molecular orientation following a serpentine path relative to the shear axis—a deduction confirmed by electron microdiffraction studies. This is seen as limiting the axial stiffness of such materials. The banded texture itself, and also the form of the spatially periodic variation in optical orientation, can be characterized by means of the corresponding optical diffraction pattern as observed with a Bertrand lens [114, 115]. The diffraction pattern is sensitive to small changes in periodicity, such as might occur on partial annealing [115], and to any residual periodicity that might survive almost complete annealing [62].

When viewing these specimens between crossed polars, it is usual to start with a standard orientation where the polarizer transmission azimuth is parallel to N–S or E–W and the shear axis is parallel to the transmission direction of either polar (Fig. 20). If the microstructural periodicity is fine, i.e., approaching the resolution limit of the microscope, one may observe that the image differs in all four cases [116], even though Eq. 1 predicts that it should be the same. A number of factors can contribute to this type of effect:

1. Predominant are the anisotropic transmission properties at the boundaries between media of different optical densities [6, 26]. In effect, the transmitted beam is depleted in light polarized normal to the plane of incidence. A specimen with a one-dimensional periodic microstructure can be regarded as a diffraction grating; a significant fraction of light incident normally on the bottom surface will emerge from the top surface at a well-defined angle to the straight-through direction. Depending on the refractive indices of the specimen and the surrounding medium, a greater or lesser degree of Brewster polarization can occur at the top specimen surface. Thus, according to whether the analyzer orientation is chosen to transmit light parallel to or normal to the plane of incidence at this surface, a greater or lesser amount of light will contribute to the image.

2. If the microscope is equipped with a camera, it is necessary for there to be some means for directing light toward either the focusing eyepiece or the camera turret. On some microscopes, light is allowed to reach both camera and eyepiece at the same time by means of a beam splitter, consisting of two glass prisms separated by a thin layer of low refractive index material. Here the anisotropic transmission properties of the interface between the prisms leads to the image intensity being dependent on the analyzer orientation relative to the beam splitter. Also, the contrast as recorded by the camera may be different from that observed at the focusing eyepiece, so the camera viewfinder should be used to check the image if it is to be photographed.

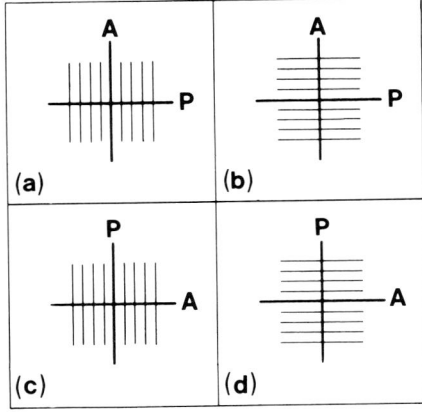

Figure 20 Schematic of the four possible standard orientations for a specimen with a 1-D periodic microstructure when viewed between crossed polars.

3. Polaroid sheet has anisotropic transmission properties (as a function of both angle of incidence and obliquity), even for light with an initial linear polarization parallel to the transmission azimuth of the sheet [117]. This factor is especially relevant if the periodicity of the microstructure is sufficiently fine to ensure that the specimen diffracts light at a large angle to the straight-through direction. Light then passes through the analyzer at a high obliquity, for which the anisotropy is most marked.

Microscopic studies of birefringence changes as a function of time have been used to monitor the relaxation behavior of sheared films of liquid crystalline polymer melts [118] and solutions [112].

Biaxially Deformed Polymers

As noted earlier, biaxial properties of polymer films arise from the nature of the fabrication process. The three principal refractive indices typically lie in the directions shown in Figure 21. The two in the plane of the film are parallel to and transverse to the machine direction; the third is normal to the plane of the film. Two quantities are of particular interest to manufacturers:

1. The in-plane birefringence of the film, given by n_3-n_2. This parameter provides a measure of the balance of the film, i.e., the extent to which the in-plane optical properties are isotropic [9]. This in turn provides an indication of the extent to which the in-plane mechanical properties are isotropic—an optimum condition which the physical processing is usually designed to achieve. Polymer films with high in-plane birefringence are sometimes tailored [119] for use in polarizers and filters.

2. The degree of planar orientation, a measure of the extent to which molecules are constrained to lie in the plane of the film, and hence an indirect measure of the in-plane strength and modulus. It is quantified by determining the average of the two principal birefringences which do not lie in the specimen plane:

$$\frac{1}{2}[(n_2 - n_1) + (n_3 - n_1)] \equiv \frac{n_2 + n_3}{2} - n_1$$

Form Birefringence

Conditions necessary for form birefringence to arise were listed in "Sources of Birefringence," above. They are met by polymeric systems such as the following:

1. Block co- or terpolymers, where long runs of a particular monomer in different molecules can become phase-separated into rodlike or platelike domains. The

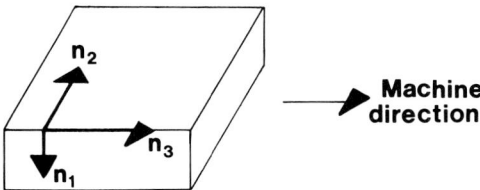

Figure 21 Principal refractive index directions in a polymer film.

individual domains may be too small to resolve, but their presence may be inferred from form birefringence if they are aligned over optically resolvable distances [9].
2. Crazes in glassy polymers [120]. Craze formation in glassy polymers accompanies deformation to failure. A craze is essentially a crack, crossed by a large number of polymer fibrils that typically have diameters of a few hundredths of a micrometer. To a first approximation, one can therefore model a craze as a collection of rodlike domains (the polymer) in a matrix of air. The situation can be complicated by the fibrils additionally exhibiting orientation birefringence, due to molecular alignment introduced while they are "drawn" across the craze as it opens.
3. Polymerized gels [121]. Here form birefringence is due to local association of monomers in aspherical groups.

For simple systems, the form birefringence can be related quantitatively to structure. In the formulas that follow the volume fractions of domain and matrix are denoted f_1 and f_2 respectively (note: $f_1 + f_2 = 1$). Their respective refractive indices are n_1 and n_2. For platelike domains, it can be shown that

$$e^2 - o^2 = \frac{f_1 f_2 (n_1^2 - n_2^2)^2}{(1 + f_1)n_2^2 + f_2 n_1^2}$$

A collection of oriented submicroscopic isotropic platelike domains in an isotropic matrix behaves optically as though it were a negative uniaxial crystal.

For rodlike domains, it can be shown that, if $f_1 \ll f_2$,

$$e^2 - o^2 = -\frac{f_1 f_2 (n_1^2 - n_2^2)^2}{(1 + f_1)n_2^2 + f_2 n_1^2}$$

The collection of oriented submicroscopic isotropic rodlike domains in an isotropic matrix therefore behaves optically as though it were a positive uniaxial crystal.

More complex formulas exist to describe form birefringence when the domains are ellipsoidal [122]. The sign of the birefringence is always consistent with the results quoted above: it is negative for oblate domains and positive for prolate domains. The form birefringence is zero if the domains are spherical.

Measurement of the sign of the birefringence can therefore provide an indication of the aspect ratio of the domains. A measurement of the magnitude of the birefringence may allow an estimate of the volume fraction occupied by the domains if the values of n_1 and n_2 are known.

One may be able to vary n_2 in order to test whether a system is exhibiting form birefringence. If only form birefringence is present, the specimen will be optically isotropic when $n_2 = n_1$. If other sources contribute to the birefringence, a plot of $|e^2 - o^2|$ versus n_2 will have a minimum at $n_2 = n_1$. The latter result is obtained if the birefringence of a polymer craze is measured with the specimen viewed in each of a variety of immersion oils [120].

Liquid Crystalline Polymer Textures

Microscopists working with liquid crystals use the term optical texture, or simply texture, to describe the image obtained when a thin liquid crystalline specimen is observed microscopically, ordinarily but not necessarily between crossed polars. Here "thin" implies that the separation of the upper and lower surfaces is smaller than the lateral extent of a domain. Textures are observed mostly with the specimen confined between glass surfaces (e.g., microscope slide and cover slip), though sometimes the upper glass surface is absent.

The qualitative features of liquid crystal textures result from a combination of defects in the scheme of molecular ordering and changes in the local orientation of the optical indicatrix around these defects. If a material is observed under conditions where it is truly a liquid crystal (i.e., at an appropriate temperature, or in sufficient concentration in a suitable solvent), features in the texture may be mobile, reflecting the fluid nature of the liquid crystalline state. Molecular ordering characteristic of high temperature liquid crystalline phases in polymers may often be preserved to room temperature by rapid quenching—a consequence of the relatively slow kinetics of phase transformations in polymers. It is convenient to the microscopist if specimens are quenchable in this way, because it allows one to work at magnifications and resolutions which are inaccessible when the specimen is held on a heating stage. The textures displayed by different liquid crystal modifications are discussed in detail in various specialty texts [70, 123–126] which collectively also offer a large enough number of micrographs for a reader to gain some appreciation of what textures are "typical."

Observation of textures, and the way in which they change as the crossed polars are rotated, may allow one to deduce the symmetry of the local molecular order [72, 73, 127–129], and hence to identify the liquid crystal phase. When relying on texture for the purposes of phase identification, one needs to be aware of the problem of paramorphoses. At a molecular level, the transition between two liquid crystalline phases will often occur via a route that leads to the smallest possible immediate change in the structure. Since the appearance of textures is dependent on the structure, the immediate post-transition texture may not easily be distinguishable from its pre-transition counterpart. Stable textures that are characteristic of the new phase may require a long time, sometimes months, to form. Smectics are especially likely to favor paramorphoses, particularly if the transition involves highly ordered smectic phases. More specific textures can be encouraged to form by evaporating an optically isotropic (dilute) solution of the material at a temperature where the liquid crystalline phase is to be studied [124]. Also, one can usually resolve persistently ambiguous cases by using the law of selective miscibility [130, 131]: If two liquid crystals are fully miscible for all binary compositions, they can be assigned to the same phase. One can build up phase diagrams for a liquid crystal by observing how the texture changes as a function of temperature, solvent concentration, or concentration of other liquid crystalline components in a system [131, 132].

The pitch of a cholesteric polymer can be measured directly from "fingerprint" textures [133].

ACKNOWLEDGMENT

The author is grateful for the use of text processing facilities at the IBM Almaden Research Center.

REFERENCES

1. P. Gay, *An Introduction to Crystal Optics,* Longmans, London (1982).
2. F. D. Bloss, *An Introduction to the Methods of Optical Crystallography,* Holt, Rinehart and Winston, New York (1961).

3. D. McKie and C. H. McKie, *Crystalline Solids*, Nelson, London, Chap. 12 (1974).
4. E. E. Wahlstrom, *Optical Crystallography*, Wiley, New York (1969).
5. N. H. Hartshorne and A. Stuart, *Crystals and the Polarising Microscope*, Arnold, London (1960).
6. M. Born and E. Wolf, *Principles of Optics*, Pergamon, Oxford (1983).
7. J. F. Nye, *Physical Properties of Crystals*, Oxford University Press, Oxford (1957).
8. R. S. Stein and G. L. Wilkes, in *Structure and Properties of Oriented Polymers* (I. M. Ward, ed.), Applied Science Publishers, London (1975).
9. D. A. Hemsley, *The Light Microscopy of Synthetic Polymers*, Oxford University Press, Oxford (1984).
10. D. C. Bassett, *Principles of Polymer Morphology*, Cambridge University Press, Cambridge (1981).
11. A. Keller, *Philos. Mag.*, 6: 329 (1961).
12. D. C. Bassett and A. Keller, *Philos. Mag.*, 7: 1553 (1962).
13. T. J. Lewis, *J. Appl. Polym. Sci.*, 23: 2661 (1979).
14. T. J. Lewis, *Polymer*, 23: 710 (1982).
15. C. Ash and T. J. Lewis, *Polymer*, 26: 643 (1985).
16. C. Viney, *Transmitted Polarised Light Microscopy*, Microscope Publications, Chicago. (in press)
17. L. J. Yu and A. Saupe, *Phys. Rev. Lett.*, 45: 1000 (1980).
18. A. Saupe, P. Boonbrahm, and L. J. Yu, *J. Chim. Phys. Phys. Chim. Biol.*, 80: 7 (1983).
19. C. Viney, G. R. Mitchell, and A. H. Windle, *Polym. Commun.*, 24: 145 (1983).
20. C. Viney, G. R. Mitchell, and A. H. Windle, *Mol. Cryst. Liq. Cryst.*, 129: 75 (1985).
21. A. H. Windle, C. Viney, R. Golombok, A. M. Donald, and G. R. Mitchell, *Faraday Disc. Chem. Soc.*, 79: 55 (1985).
22. A. M. Donald, C. Viney, and A. H. Windle, *Philos. Mag. B*, 52: 925 (1985).
23. C. Viney, A. M. Donald, and A. H. Windle, *Polymer*, 26: 870 (1985).
24. F. Hessel and H. Finkelmann, *Polym. Bull.*, 15: 349 (1986).
25. A. W. Hendry, *Photoelastic Analysis*, Pergamon, Oxford (1966).
26. R. S. Longhurst, *Geometrical and Physical Optics*, Longmans, London (1967).
27. A. Ehringhaus, *Z. Kristallogr.*, 76: 315 (1931).
28. G. N. Ramachandran and S. Ramaseshan, in *Encyclopedia of Physics*, Vol. 25/1 (S. Flügge, ed.), Springer, Berlin (1961).
29. W. A. Shurcliff, *Polarised Light*, Harvard University Press, Cambridge, Mass. (1962).
30. R. Haynes, in *Optical Microscopy of Materials*, Blackie and Son, Glasgow, Chaps. 1–7 (1984).
31. J. Chandler, *Lab. Pract.*, 24: 25 (1985).
32. S. G. Lipson and H. Lipson, *Optical Physics*, Cambridge University Press, Cambridge (1981).
33. I. D. Muir, *The 4-Axis Universal Stage*, Microscope Publications, Chicago (1981).
34. G. Newlands, *Lab. Pract.*, 28: 508 (1979).
35. G. Szivessy and W. Herzog, *Z. Instrumentenkd.*, 57: 49 (1937).
36. R. D. Mindlin, *J. Opt. Soc. Am.*, 27: 288 (1937).
37. E. Hecht and A. Zajac, in *Optics*, Addison-Wesley, Reading, Mass., Chap. 8 (1974).
38. W. A. Shurcliff and S. S. Ballard, *Polarized Light*, Van Nostrand, Princeton, N.J. (1964).
39. R. C. Jones, *J. Opt. Soc. Am.*, 31: 488 (1941).
40. H. Hurwitz and R. C. Jones, *J. Opt. Soc. Am.*, 31: 493 (1941).
41. R. C. Jones, *J. Opt. Soc. Am.*, 31: 500 (1941).
42. R. C. Jones, *J. Opt. Soc. Am.*, 32: 486 (1942).
43. R. C. Jones, *J. Opt. Soc. Am.*, 37: 107 (1947).
44. R. C. Jones, *J. Opt. Soc. Am.*, 37: 110 (1947).
45. R. C. Jones, *J. Opt. Soc. Am.*, 38: 671 (1948).
46. R. C. Jones, *J. Opt. Soc. Am.*, 46: 126 (1956).

47. H.-Y. Hsü, M. Richartz, and Y.-K. Liang, *J. Opt. Soc. Am., 37*: 99 (1947).
48. P. Yeh, *J. Opt. Soc. Am., 69*: 742 (1979).
49. P. Yeh, *Surf. Sci., 96*: 41 (1980).
50. P. Yeh, *J. Opt. Soc. Am., 72*: 507 (1982).
51. F. Pockels, *Lehrbuch der Kristalloptik,* Teubner, Leipzig (1906). [rpt. Johnson Reprint Corporation, New York (1969)]
52. F. Ratajczyk and I. Scierski, *Optik, 71*: 7 (1985).
53. C. Viney and A. H. Windle, *Faraday Disc. Chem. Soc., 79*: 105 (1985).
54. F. C. Frank, *Faraday Disc. Chem. Soc., 79*: 104 (1985).
55. S. Chandrasekhar, *Liquid Crystals,* Cambridge University Press, Cambridge, Chap. 4 (1980).
56. E. P. Raynes and R. J. A. Tough, *Mol. Cryst. Liq. Cryst. Lett., 2*: 139 (1985).
57. E. P. Raynes, *Mol. Cryst. Liq. Cryst. Lett., 4*: 69 (1987).
58. H. W. Zieler, *Microscope, 17*: 249 (1969).
59. C. Viney, *Microscope* vol. 36 (1), in press (1988).
60. R. D. Deanin, in *Polymer Structure, Properties, and Applications,* Cahners, Boston, Chap. 5 (1972).
61. P. G. de Gennes, in *The Physics of Liquid Crystals,* Oxford University Press, Oxford, Chap. 3 (1979).
62. C. Viney, *Faraday Disc. Chem. Soc., 79*: 228 (1985).
63. D. T. Grubb and A. Keller, *J. Mater. Sci., 7*: 822 (1972).
64. D. T. Grubb, *J. Mater. Sci., 9*: 1715 (1974).
65. A. M. Donald, *J. Mater. Sci. Lett., 3*: 44 (1984).
66. A. M. Donald and A. H. Windle, *J. Mater. Sci., 19*: 2085 (1984).
67. J. H. Wendorff, in *Liquid Crystalline Order in Polymers* (A. Blumstein, ed.), Academic Press, New York (1978).
68. G. L. Wilkes, *J. Polym. Sci. Polym. Lett. Ed., 10*: 935 (1972).
69. R. B. Meyer, *Molec. Cryst. Liq. Cryst., 16*: 355 (1972).
70. H. Sackmann and D. Demus, *Molec. Cryst. Liq. Cryst., 21*: 239 (1973).
71. G. H. Brown, *Anal. Chem., 41*: 26 (1969).
72. M. R. Mackley, F. Pinaud, and G. Siekmann, *Polymer, 22*: 437 (1981).
73. C. Viney and A. H. Windle, *J. Mater. Sci., 17*: 2661 (1982).
74. C. Viney, *Polym. Eng. Sci., 26*: 1021 (1986).
75. F. C. Frank, *Philos. Trans. R. Soc. London A, 309*: 71 (1983).
76. H. Kawai and S. Nomura, in *Developments in Polymer Characterisation,* Vol. 4 (J. V. Dawkins, ed.), Applied Science Publishers, London (1983).
77. B. E. Read, in *Structure and Properties of Oriented Polymers* (I. M. Ward, ed.), Applied Science Publishers, London (1975).
78. A. Sharples, *Introduction to Polymer Crystallisation,* Arnold, London (1966).
79. J. Schultz, *Polymer Materials Science,* Prentice-Hall, Englewood Cliffs, N.J. (1974).
80. A. Keller, *J. Polym. Sci., 17*: 291 (1955).
81. A. Keller, *J. Polym. Sci., 17*: 351 (1955).
82. A. Keller and J. R. S. Waring, *J. Polym. Sci., 17*: 447 (1955).
83. F. P. Price, *J. Polym. Sci., 37*: 71 (1959).
84. D. R. Norton and A. Keller, *Polymer, 26*: 704 (1985).
85. D. C. Bassett and R. H. Olley, *Polymer, 25*: 935 (1984).
86. D. C. Bassett and A. S. Vaughan, *Polymer, 26*: 717 (1985).
87. M. Pick, R. Lovell, and A. H. Windle, *Polymer, 21*: 1017 (1980).
88. M.-S. S. Wu, *J. Appl. Polym. Sci., 32*: 3263 (1986).
89. G. L. Wilkes, *Adv. Polym. Sci., 8*: 91 (1971).
90. A. E. Zachariades, W. T. Mead, and R. S. Porter, in *Ultra-High Modulus Polymers* (A. Ciferri and I. M. Ward, eds.), Applied Science Publishers, London (1979).
91. W. T. Mead and R. S. Porter, in *Flow-Induced Crystallization in Polymer Systems* (R. L. Miller, ed.), Gordon and Breach, New York (1979).

92. R. S. Stein, M. Hashiyama, and M. K. Parpart, in *Flow-Induced Crystalization in Polymer Systems* (R. L. Miller, ed.), Gordon and Breach, New York (1979).
93. M. R. Mackley and G. S. Sapsford, in *Developments in Oriented Polymers*, Vol. 1 (I. M. Ward, ed.), Applied Science Publishers, London (1982).
94. H. Block, E. M. Gregson, W. D. Ions, G. Powell, R. P. Singh, and S. M. Walker, *J. Phys. E (Sci. Instrum.)*, *11*: 251 (1978).
95. H. Janeschitz-Kriegl, *Polymer Melt Rheology and Flow Birefringence*, Springer, Berlin (1983).
96. B. R. Hahn and J. H. Wendorff, *Polymer*, *26*: 1619 (1985).
97. D. J. Brown and A. H. Windle, *J. Mater. Sci.*, *19*: 1997 (1984).
98. D. J. Brown and A. H. Windle, *J. Mater. Sci.*, *19*: 2013 (1984).
99. D. J. Brown and A. H. Windle, *J. Mater. Sci.*, *19*: 2039 (1984).
100. D. J. Brown, *Polym. Commun.*, *26*: 42 (1985).
101. B. Erman and P. J. Flory, *Macromolecules*, *16*: 1601 (1983).
102. B. Erman and P. J. Flory, *Macromolecules*, *16*: 1607 (1983).
103. L. Forlini and W. C. McCrone, *Microscope*, *19*: 243 (1971).
104. W. C. McCrone, *Microscope*, *23*: 213 (1975).
105. W. C. McCrone, *Microscope*, *23*: 221 (1975).
106. S. C. Simmens and J. W. S. Hearle, *J. Polym. Sci. Polym. Phys. Ed.*, *18*: 871 (1980).
107. G. Kiss and R. S. Porter, *Molec. Cryst. Liq. Cryst.*, *60*: 267 (1980).
108. D. G. Graziano and M. R. Mackley, *Molec. Cryst. Liq. Cryst.*, *106*: 73 (1984).
109. M. Horio, S. Ishikawa, and K. Oda, *J. Appl. Polym. Sci. Appl. Polym. Symp.*, *41*: 269 (1985).
110. A. M. Donald, C. Viney, and A. H. Windle, *Polymer*, *24*: 155 (1983).
111. C. Viney, A. M. Donald, and A. H. Windle, *J. Mater. Sci.*, *18*: 1136 (1983).
112. A. M. Donald, C. Viney, and A. P. Ritter, *Liq. Cryst.*, *1*: 287 (1986).
113. P. Navard, *J. Polym. Sci. Polym. Phys. Ed.*, *24*: 435 (1986).
114. C. Viney, *Microscope*, *32*: 93 (1984).
115. C. Viney and A. H. Windle, *Polymer*, *27*: 1325 (1986).
116. C. Viney and A. H. Windle, *Philos. Mag. A*, *55*: 463 (1987).
117. L. Baxter, *J. Opt. Soc. Am.*, *46*: 435 (1956).
118. N. J. Alderman and M. R. Mackley, *Faraday Disc. Chem. Soc.*, *79*: 149 (1985).
119. H. G. Rogers, R. A. Gaudiana, W. C. Hollinsed, P. S. Kalyanaraman, J. S. Manello, C. McGowan, R. A. Minns, and R. Sahatjian, *Macromolecules*, *18*: 1058 (1985).
120. H. R. Brown, *J. Polym. Sci. Polym. Phys. Ed.*, *17*: 1417 (1979).
121. E. Geissler, A.-M. Hecht, and J. Torbet, *Polymer*, *27*: 1489 (1986).
122. W. L. Bragg and A. B. Pippard, *Acta Crystallogr.* *6*: 865 (1953).
123. N. H. Hartshorne, *The Microscopy of Liquid Crystals*, Microscope Publications, London (1974).
124. D. Demus and L. Richter, *Textures of Liquid Crystals*, Verlag Chemie, Weinheim (1978).
125. H. Kelker and R. Hatz, *Handbook of Liquid Crystals*, Verlag Chemie, Weinheim (1978).
126. G. W. Gray and J. W. G. Goodby, *Smectic Liquid Crystals*, Leonard Hill, Glasgow (1984).
127. S. B. Warner and M. Jaffe, *J. Cryst. Growth*, *48*: 184 (1980).
128. Y. Bouligand, P. E. Cladis, L. Liebert, and L. Strzelecki, *Molec. Cryst. Liq. Cryst.*, *25*: 233 (1974).
129. M. Kléman, L. Liebert, and L. Strzelecki, *Polymer*, *24*: 295 (1983).
130. B. Fayolle, C. Noel, and J. Billard, *J. Phys. Colloq. C3*, *40*: 485 (1979).
131. G. Sigaud, M. F. Achard, F. Hardouin, M. Mauzac, H. Richard, and H. Gasparoux, *Macromolecules*, *20*: 578 (1987).
132. C. Viney and A. H. Windle, *Liq. Cryst.*, *1*: 379 (1986).
133. R. S. Werbowyj and D. G. Gray, *Macromolecules*, *17*: 1512 (1984).

15

Microscopy Techniques for Polymer Characterization

L. Bartosiewicz and C. J. Kelly
Ford Motor Company
Dearborn, Michigan

INTRODUCTION	538
BASIC PROPERTIES AND APPEARANCE OF PLASTICS	538
POLYMEROGRAPHY	540
Terminology	540
Scope	541
OVERVIEW OF MICROSCOPY	541
Stereomicroscopy	542
Polarized Light Microscopy	542
Fluorescence Microscopy	543
Nomarski Differential Interference Contrast	543
Phase Contrast Microscopy	543
Sample Preparation	543
Metal Shadowing	546
Measurement of Surface Relief	546
Microtomy	547
EXPERIMENTAL RESULTS	547
Unfilled Nylon-6	547
Nylon-6	549
Nylon-6 with Additives	549
Nylon-66	552
Heat-Treated Nylon-66	552
Polyethylene	556
Polychlorotrifluoroethylene	556
Liquid-Phase Etching of Polypropylene	558
Laser Etching of Polypropylene	563
Electron Beam Etching of Polypropylene	563
APPLICATIONS OF FRACTOGRAPHY	564
OPTICAL MICROSCOPE HEATING STAGE	574
Polycarbonate–Polybutylene Terephthalate	578
Polycarbonate–Polybutylene Terephthalate Photoelasticity	581
Polyurethane	583
Polyurethane Examination with the Heating Stage	583
Polycarbonate–Polyarylate Blend	586
Wood Flour–Filled Copolymer	586

SCANNING ELECTRON MICROSCOPY	591
TRANSMISSION ELECTRON MICROSCOPY	591
Extruded Wood Flour–Filled Copolymer with TEM	592
Rubber Latex with Polystyrene	593
IMAGE ENHANCEMENT	593
Paint Contaminant Study	593
Polycarbonate and Polymethyl Methacrylate	598
MECHANICAL STRAINING STAGE FITTED TO THE SCANNING ELECTRON MICROSCOPE	600
REFERENCES	603

INTRODUCTION

Microscopic examination of plastics is widely used in detailed examination for broad understanding to aid polymer science, including polymer chemistry, polymer physics, basic research, and failure analysis. To permit these investigations the sample preparation must be specific and suit the intended purpose. In this chapter the primary concern is microscopic examination.

The subject of polymerography, or resinography, is not new, but the investigating techniques for sample preparation are by no means routine or uniform. Hence it is prudent and appropriate to begin by sorting, classifying, and clarifying the methods. The first and perhaps the simplest division is between the uncomplicated and the complex procedures. In uncomplicated procedures there are no interferences, or only foreseeable ones. The preparation process must be stable and reproducible. This presupposes that preparations and results remain unchanged with a specific procedure, although some small modifications in the parameters may be necessary. In complex procedures the ad hoc and empirical approach dominates. The main focus is on specialized thin sample preparation techniques that include abrasive thinning, room temperature or cryogenic microtomy, and thin film casting. Success often is elusive.

In the following survey we discuss the various methods to maintain and preserve the material's integrity and overall characteristics during sample preparation. But before embarking on the specifics, some of the available simple tests to classify, if not fully identify, the material must be examined so that processing unknown materials with unsuitable or outright detrimental techniques can be avoided. It is often possible to determine the type of plastic with simple methods. These include the *Plastics Identification Table* by Hansjurgen Saechtling [1], for years the benchmark of plastics identification, or the *Modern Plastics Encyclopedia,* which lists and describes the properties of commercially available materials [2]. But numerous industrial copolymers defy simple methods of identification and only more advanced and experimentally demanding methods will succeed [3–7].

BASIC PROPERTIES AND APPEARANCE OF PLASTICS

Plastics are organic materials obtained by chemically reacting or polymerizing monomer substances of low molecular weight to form polymers and compounds of high molecular weight. The process mechanisms include condensation, crosslinking, polymerization, and copolymerization. Chemically similar plastics can be produced from different monomers.

Another noteworthy distinction must be made between a plastic and a polymer. The term plastic is commonly understood to relate to properties, whereas the term polymer relates to composition. Actually, a plastic can be best described as an organic material that solidifies from a melt; polymer is defined as a chemically bonded repeating molecular structure. Both plastics and polymers are manufactured and sold under various trade names. The material's investigator must rely on published sources to translate trade names into meaningful chemistry or polymer science [2]. The converse, identifying a specific trade name from the microstructure, may be difficult, if not impossible.

The interacting macromolecules of these plastics are responsible for the material's properties including hardness, strength, modulus, impact properties, and softening and melting temperatures. Plastics with linear chainlike molecules with lengths of several hundred nanometers (nm) or with few crosslinked macromolecules are prone to soften when heated. The group of plastics that soften and flow on heating are thermoplastics. Some of the more widely used are acrylonitrile butadiene styrene, nylon, acrylic polyethylene, polypropylene, polystyrene, polychlorotrifluoroethylene, polycarbonate, polyethylene terephthalate, and polybutylene terephthalate. Depending on the type of monomer used, materials are produced which may be rigid or rubbery at ambient temperatures. Polymers have some universal characteristics as shown schematically in Figure 1. These materials will solidify if cooled and the heating and cooling cycle is repeatable. A notable

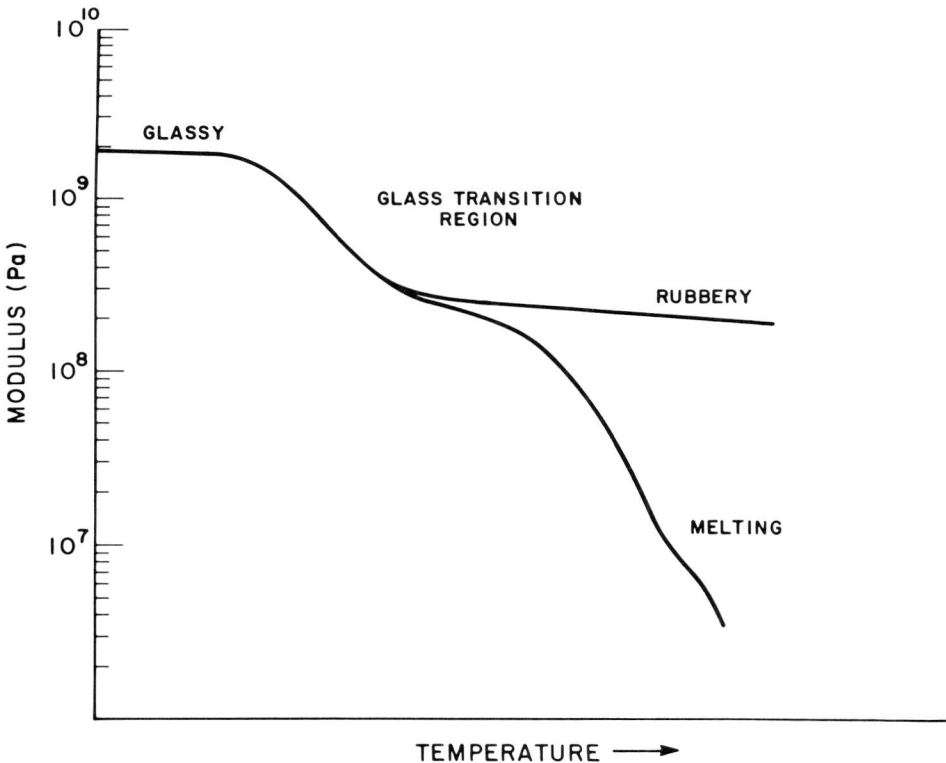

Figure 1 Generalized polymer characteristics.

exception to this behavior for thermoplastics is temperature-related chemical stability. Such materials will chemically decompose below the softening temperature, because their chemical stability with temperature is lower than the cohesion between the molecular chains. Brief mention of the organic glasses (e.g., polystyrene) is also important, because these glasses transform within a relatively narrow range of temperature. This glass transition temperature T_g is near the midpoint of the temperature range where the transition to the vitreous state occurs.

These linear or branched thermoplastics are soluble in organic solvents. Amorphous polymers usually dissolve in solvents specific to a given class. Semicrystalline or crystalline polymers usually dissolve with substantial difficulty [8, 9]. Crosslinked polymers do not dissolve but only swell in solvents. Specific behavior with respect to solvents greatly aids in identification [1] and is widely used to elucidate the morphology of polymer blends.

Thermosetting plastics, in contrast to thermoplastics, cannot be reprocessed. Typical examples are phenolic, epoxy, polyester, polyurethane, allyl, and amino. The macromolecules are crosslinked by heating or chemical reaction with additives and the plastic will neither dissolve nor melt. Only by chemically destroying the crosslinks can the plastic be decomposed. In contrast to the highly crosslinked thermoset materials, rubberlike elastomers have few crosslinked macromolecules. The elastomers will soften before degrading. This behavior can be used to distinguish elastomers from other elastic thermoplastics.

Within the main groups of plastics, a large number of different properties can result by using chemical modification, copolymerization, and filling or reinforcing agents. The physical appearance may be sufficient to distinguish between the main groups.

POLYMEROGRAPHY

Terminology

To avoid ambiguous or unique terminologies to describe the results of microscopical observations, standard ASTM terminology [10] should be used. Another informative and quick reference for plastic or polymer definitions covers the terminologies and terms used by the industry [2]. Occasionally even the most basic of these terms, structure and morphology, are incorrectly used. Suffice to say that *structure* relates to construction or arrangement of the atoms or molecules, and *morphology* describes the size, shape, and development of the various features formed by the molecules, e.g., spherulite size.

In polymerography, the term materials includes a large number of composites. Composites consist of various but distinct phases which may contribute to or detract from the composite's properties. The major distinction between different composites is the various materials that are incorporated. Some composites contain only organic materials, i.e., resin and dispersed polymer. Other composites contain minerals, glass fibers, metals, or wood flours in addition to a polymer base. The first group does not present the same challenge to sample preparation, because the properties of the components are similar. The preparation techniques for the second group of composites must consider the diversity of the fillers. The grinding and polishing as well as the examination process must suit all the constituents of the composite. The processes used to examine these materials are amply documented [11]. In general, the ASTM methods and procedures enjoy the cooperative support of the federal government [12] and the textile industry [13].

Scope

The scope of polymerography includes failure analysis, forensic analyses, materials research, development, quality control, and teaching [14]. The appropriate relevant methods include physical, optical, and chemical determinations. The main subject of this report, however, is optical light polymerography, but on occasion examples of electron microscopy are also presented. In polymerography, the microscope is used to interpret structure and requires coordination between the eyes and the brain [15]. The microscope may also be used in conjunction with physical test accessories such as the hot stage, tensile fixture, or three-point bending apparatus. The morphological assessment, however, is limited by the resolving power of the optical light microscope.

Finally, the scope of polymerography involves the microscopist. The choice of sample preparation and the applied microscopy techniques are critical to the interpretation of morphology. The polymerographer must be certain that the plastic specimen is truly representative. The polymerographer must also anticipate the introduction of some artifacts, because changes are unavoidable during sample preparation. Without experience, knowledge, and care, the subsequent interpretation of morphology will suffer, especially if artifacts mimic morphological details. In extreme cases the results will be questionable or invalid.

OVERVIEW OF MICROSCOPY

There are two major aspects to microscopic examination of plastics. The first is the overall visual inspection with the intent to gain a general impression. This may be accomplished either with the stereomicroscope, if low magnification will suit the purpose, or with the compound microscope. Indeed many of the basic morphologies of interest in the polymer's matrix (e.g., fillers, fibers, fracture surfaces, cracks, or even crazes) may be resolved by the stereomicroscope. Occasionally the observations can relate impressions or provide estimates to describe specific characteristics of the samples. In fact, all the necessary observations needed to relate morphology, composition, behavior, properties, and in some instances history of the material are also possible. This type of visual inspection with the stereomicroscope is qualitative.

The second major use of microscopy is to obtain quantitative information. Quantitative microscopy has been aided by developments in automation. In basic research, the application of quantitative microscopy to analytical techniques is well known. Engineering and development activities are increasing the use of these quantitative microscopy measurements.

The different quantitative techniques utilize various analytical procedures or methods. Methods involving counting the number of features or measuring the length, size, area, volume, distance, or mean free path between particles are applications with which the physical dimensions of microstructures are quantitatively derived. Other quantitative methods include optical determination of refractive index, birefringence, or reflectance. In addition, there are methods to derive parameters such as surface roughness or relief, crystallization rates, strain measurement with gradients, transparent coating thickness, diffusion zones between crystalline and amorphous materials, and distribution or uniformity of multiphase systems [16–18]. These quantitative techniques may differ from typical microscopy methods. An example of a unique technique is the optical measurement of stress in glass or polymers. The measuring techniques are transferable from one material to another and the evaluation methods are similar [19].

These quantitative microscopy measurements are rarely absolute but rely on the use of standards or reference materials (e.g., stage micrometers, melting point standards, or standard reference materials). These standards help to eliminate all but some systematic errors. Systematic errors such as nonuniform thin sections, focusing incompatibility of mixed feature sizes with magnification, or hot stage specimen chamber temperature gradients, however, can present other difficulties. Some of these systematic errors, or biases, may be minimized with numerical corrections or with alternative techniques if the cause of the error is recognized. Some, however, are difficult to detect.

Both reflected and transmitted light are used in polymerography. The illumination mode is generally dictated by the sample. Reflected illumination is the only choice if the specimen is opaque. If metals, minerals, wood, or other fillers are mixed with the polymer, the choice between illumination modes is less clear-cut because only some of the filler particles may be truly opaque. However, with some polymer matrices, the filler particles produce light scattering that restricts or prohibits the use of transmitted light. These heterogeneous polymer mixtures are best viewed with reflected light. Differences in the refractive indices and the associated critical angles for each component may produce good image contrast. Another advantage of reflected light microscopy is the reduced sample preparation time. A large variety of plastics cannot be easily prepared for transmitted light microscopy, but the surface polishing for reflected illumination is not too difficult.

The detail resolution possible with the optical microscope depends on the type and intensity of illumination, specimen contrast, optics, and the human eye. The ability to distinguish between two image points is called resolving power and depends on the numerical aperature of the lens and the wavelength of the illuminating light. The practical limit for the optical microscope is 0.2 μm. The most widely used optical microscopy techniques include reflected light with or without oblique illumination, transmitted light, polarized light, differential interference contrast, and phase contrast. Only a brief description of these techniques is included; more detailed information can be found in Refs. 20–25. With combinations of these techniques, the morphologies of most plastics can be studied.

Stereomicroscopy

The original stereoscopical microscope was proposed by Greenough, an American biologist. The idea was later transformed into reality [26]. Surface examinations with the stereomicroscope are quick and easy. The relatively high magnification, comfortable working distance, convenient illumination, realistic image appearance, large field of view, and accommodating depth of field all make the instrument practical. Resolution between 3.5 and 85 μm is obtainable.

Polarized Light Microscopy

Polarized light microscopy is useful for studying anisotropic crystalline materials. Many plastics are anisotropic; therefore, polarized light observations yield information not otherwise available. The basic principle of polarized light microscopy which produces the birefringence of crystalline materials is the phase shift of polarized light when transmitted through anisotropic crystalline materials. Different crystals and different crystal axes within a crystal will produce phase shifts, which produce different colors. Strain in the material also produces phase shifts, which are easily viewed in polarized light.

Fluorescence Microscopy

For fluorescence microscopy, the mercury vapor lamp is a suitable light source, because it emits energy in the blue UV part of the spectrum. When some molecules are excited with this energy, the molecules retain some of the energy for a period of time before returning to a ground state via the release of fluorescent radiation. The wavelength of the emitted radiation, which is always longer than the exciting mercury light, is analyzed with various filters. The choice and details depend upon the material and the microscopy technique (i.e., brightfield or darkfield).

Nomarski Differential Interference Contrast

Differential interference contrast (DIC) is a qualitative technique ideal to enhance details that may otherwise lack contrast. DIC is used with both incident and transmitted illumination and is especially helpful to examine very small surface elevation differences. Images obtained with DIC are characterized by a three-dimensional appearance. This three-dimensional appearance, however, can be misleading and details that appear elevated may not be. Each detail has an arbitrary bright or dark side, depending on the DIC setting. Image interpretation requires great care.

Phase Contrast Microscopy

The previously discussed phase shift phenomenon is recalled here. Light travels as a sine wave with specific amplitude and wavelength. Amplitude and wavelength changes are detected by the human eye as intensity and color changes, respectively. Light passing through a polymer or glass will emerge retarded compared to the direct light. The retarded light will be out of phase with the nonretarded wave. Neither the human eye nor the photographic emulsions can detect these phase changes, and specimen features with very small refractive index differences remain practically invisible. Phase contrast microscopy is used to examine these specimens. The phase contrast technique converts the specimen-caused light phase differences to amplitude differences with a quarter-wave plate. When the beams are recombined, the interference produces amplitude differences which are visible. Optical alignment is critical, because loss of some of the beam will introduce errors.

Phase contrast microscopy is used for both reflected and transmitted light. For transmitted light, only very thin sections are suitable. With reflected light, fine surface structures, fillers, boundaries, phase separations, surface imperfections, and etching results are within the scope of the technique. Phase contrast can provide interesting information on polished surfaces when the components have the same reflecting power and thus are indistinguishable in brightfield. If the optical constant differences are sufficient to produce phase changes during the reflection, the result is visible as brightness contrast.

Sample Preparation

The major concern with all sample preparation techniques for polymerography is to provide artifact-free representation of the microstructure. The basic techniques—sectioning, embedding, grinding, polishing, and etching—are modified metallographic procedures [27–29]. Sectioning of plastics can be accomplished using various equipment. Careful location of the desired cross section, representative of the plastic being examined, is the critical first step. A wet cutoff machine, equipped with a silicon–carbide abrasive wheel, is preferred. During sectioning, the sample must not be allowed to reach temperatures that will soften or in extreme cases melt the surface. For cooling, pure water is preferred, but if an aqueous emulsion of commercial cutting oil is used, the sample's surface must be removed with subsequent grinding. If dry sectioning is required, a band saw, jeweler's saw, or hacksaw may be used.

As for metals in metallography, plastic materials can also be embedded prior to grinding, polishing, and etching. An excellent choice, but dependent on the specimen–medium compatibility, is to mount the samples in transparent epoxy. Clear epoxy offers obvious advantages for light optical microscopy. Most epoxy embedding resins are low molecular weight long-chain polymers, and basically only the end groups react; i.e., the end of the hardener chain reacts with the end of the resin chain. There are some minor side reactions on the chain, but neither the process nor the result of these reactions is significant to the embedding process. The epoxy curing reaction is exothermic; hence room-temperature curing epoxy, which cures essentially overnight, is the most widely used. Also, epoxy resins are resistant to most acids, bases, or solvents that may be used for subsequent etching.

For porous specimens, vacuum embedding is the recommended procedure. Plastic embedding materials with vacuum processing techniques require care to avoid ''boiling'' the absorbed moisture and the dissolved nitrogen or oxygen in the epoxy. The reason for boiling, even at moderately low vacuum such as 100 mm Hg, is that a volume of 1 at atmospheric pressure will increase approximately eight times at 100 mm Hg. If this vapor pressure is not controlled and the embedding medium is allowed to boil, the specimen will be filled with gas bubbles instead of embedding medium.

The viscosity of epoxy resins increases with elevated temperature during curing, which reduces the capillary rise. Cooling the epoxy resin to $+5°C$ slows the polymerization but increases the viscosity. Therefore, cooling is not useful for increasing the capillary rise. These factors are observed for samples with rough surfaces. When air is removed from rough surfaces, epoxy fills the void. When the solvent evaporates, the surface tension of the polymer increases, resulting in greater capillary action. As shown in Figure 2, the rise of the epoxy, through the capillary action, is about the same at 100 or 110 mm Hg pressure. But at atmospheric pressure the rise is considerably less.

To be useful for microscopy, the specimen surface must be further refined after the initial sectioning. This is done by mechanical wet grinding and polishing. Routine grinding consists of three steps using 320, 400, and 600 grit silicon–carbide abrasives. Occasionally specialized grinding stones or surface grinding may be appropriate. Polishing is more critical and must be adapted to suit the plastic. A scratch-free surface is required. Particular attention to the final polishing with a 1 μm diamond polishing compound, followed by a submicrometer-sized polishing suspension, is necessary. For best results, 50% liquid soap and water lubrication of the short-nap polishing cloth is recommended. One to two volume percent solvent such as trichloroethylene or acetone may be mixed in a slurry with submicrometer-sized alumina or chrome oxide. With aromatic or aliphatic hydrocarbons the polishing must be intermittent, using very light pressure, to avoid localized heating. Localized heating accelerates smearing, swelling, or surface dissolution, which will obscure fine surface features. Even nominal pressures will increase the surface temperature and cause undesirable solvent effects. When compared to metals, polymers are soft and polish easily with very light pressure.

The next step in the morphology delineation process is etching. After the polishing is

Figure 2 Epoxy capillary action: 760, 100, and 110 mm Hg.

complete, the surface has a high luster without visible details of the microstructure. In semicrystalline polymers, the etching makes the crystalline texture visible. For amorphous polymers, etching is usually used to dissolve the polymer to make visible the underlying structure of the material. Some very faint difference in surface reflectance may be visible when viewed under the microscope. The morphology is revealed only by exposing the surface to liquid or vaporized aromatic or chlorinated hydrocarbons. Radiating energy that converts to heat may also be used. The choice of hydrocarbon or the radiating energy level is dependent on the surface chemistry of the sample. Etching with oxidizing agents is also useful. Chromic acid is widely used but extreme care is required during etching and structure interpretation. Stress cracking patterns from frozen-in molding stresses may be revealed and misinterpreted as structural details [30].

During the etching process, the polish-produced surface luster is dulled; this is the first-stage indicator. Presumably the various stages relate to the polymer undergoing partial and selective swelling of the less ordered regions. Consequently, the chains or their segments are reordered by partial recrystallization. These changes are microscopic and sufficient detail is retained to permit optical observation of the morphology. This is accomplished with proper control of the etching parameters, which include etchant selection, temperature, exposure time, or energy exposure. Unless otherwise noted, the etched surfaces were dried with filtered low-pressure compressed cold or warm air. High-pressure air is avoided to prevent embedding undesirable airborn particulates in the sample surface.

Fine control of the interaction between a polymer and an etchant is difficult, because of the imperceptible rate with which the changes occur. Moderately overetched surface morphology is often difficult to recognize, because overall it is similar in detail to properly delineated morphology. The degree of surface swelling is perhaps the best indicator of the integrity of an etch-delineated morphology. This is especially true for polymers with large spherulites (e.g., polypropylene or nylon). As a general rule, with surface swelling less than 0.5 μm the structural visibility of fine details does not appear to suffer when observed at magnifications typical of optical microscopy.

To minimize etching-produced artifacts, the process must operate at compatible temperatures. If the operating temperature of the etchant or the radiation is above the recrystallization temperature, large-scale microstructural distortions will result. This distorted surface morphology may show little resemblance to the true morphology. To minimize swelling at higher temperatures, the sample must be exposed to etchants only for very short times with intermittent cooling. A suitable etchant must be a relatively poor solvent to suppress large-scale swelling, loosening, or eventual dissolution of the crystallized phase. But it still must retain selectivity with respect to the polymer phases. Actual removal of polymer material by dissolution with an etchant is not necessary to successful morphology delineation. Carefully designed experiments indicate that morphology delineation is still possible by vapor-induced swelling of the surface. Various polymeric materials were examined with different etching techniques to illustrate the possible results. The plastics selected are of commercial interest.

Microtomed thin sections were also prepared for transmitted light microscopy for comparison with the etch-delineated surface morphology. Incident light fluorescence microscopy was used for the same purpose when microtomy proved impractical. To avoid the introduction of uncertainty, only those specimen regions considered essentially identical in composition and morphology were compared.

Metal Shadowing

To distinguish between proper shadowing and "decorating" surface effects with the microscope may present some problems. Shadowing is relatively easy to define but difficult to practice [31]. The mechanical aspects seem simple enough because the atoms of the evaporated material become instantly immobilized when they strike the sample's surface. This seemingly simple process of thin film deposition is very complex, however, because of thermal conditions, surface diffusion, crystallization, coalescence, and possible chemical interaction with the host surface. Selective interaction between sites of an etched sample may result from a possible affinity with the shadowing material. This may cause noticeable or unnoticeable artifacts because the heavily deposited material will tend to highlight or distort the fine image details. Too much deposited material will obscure all fine details.

In addition to the specimen-caused artifacts or decorative shadowing, other experimental factors can cause misleading enhancement of features. Vacuum condition, deposition rate, distance, specimen temperature, and the presence of ions may contribute, or the final result may suffer from the interplay between all these factors. Experimental control is very important but may not always assure successful results. Overall cleanliness, sufficiently high vacuum, and low specimen temperature all help to reduce unwanted effects.

Measurement of Surface Relief

To measure height or depth of microscopic surface structures, coatings, or surface reliefs—i.e., roughness—with the optical microscope, interferometers are used. For practical purposes, these devices are distinguished according to their application. Most instruments are equipped with polychromatic and monochromatic light sources. The polychromatic light source is used to determine the location of the zero-order fringe. This zero-order fringe is used as a reference location for estimating surface elevation or depression. Monochromatic light is then used to quantify the surface elevations. Only a very thin surface coating on an otherwise uniformly flat surface can be measured with monochromatic light. To measure thicker coatings or surface roughness the polychromatic light source must be used.

There are other practical difficulties with surface-related interference measurements. Some polymers, especially if dark-colored pigments are added, have a very weak tendency to reflect light, which is one of the primary requirements of the technique. The surface reflectivity can be changed with an artificial but simple method to measure the surface elevations of an orange peel or embossed surface. The surface is replicated using an acetate replicating film. The light absorption of the replicated surfaces is not uniform and direct interference measurements will not be accurate; however, thin evaporated aluminum coating produces sufficient reflection to image the interference.

The surface of a polymer-painted plaque was replicated and partly covered with an evaporated aluminum film, as seen in Figure 3. On the upper part of Figure 3A, the uncoated surface interference image is barely visible, whereas the image of the lower portion is suitable for making accurate surface elevation measurements. The highest point of elevation above the lowest surface point is 7 μm. The replicating and aluminizing process was repeated using an adjacent part of the plaque to test the reproducibility of the technique. The second replica is seen in Figure 3B, and the measurements for the two samples are almost identical.

Microtomy

Preparing thin sections at room or reduced temperatures is tedious and time-consuming. The tedium is associated with obtaining artifact-free thin section for end use and includes the necessary preliminaries (i.e., sizing, encapsulating, or trimming). The section thickness must suit the contemplated microscopy [32, 33]. For light microscopy up to 500×, sections approximately 3 μm thick are sufficient, but high magnifications require much thinner sections. And transmission electron microscopy requires sections 0.1 μm or less. The techniques are borrowed and modified from biological applications.

Only a few of the industrially used plastics have suitable consistency to be microtomed at room temperature. With cooling, however, these elastic materials can be microtomed because their hardness increases (see Fig. 1). In most cases a temperature between $-20°$ and $-70°C$ is sufficient. For even hard polymers, the low temperatures have advantages because they help to inhibit plastic deformation and the consequent artifacts. There is no simple rule of thumb, but the main parameters, besides temperature of the specimen, include knife temperature, cutting speed, and clearance angle [34]. A common difficulty, knife frosting, can be reduced by using ethylene glycol or a 50% glycol–water mixture, provided it is compatible with the sample. Special silicon or fluoride compounds require temperatures of $-150°C$ or less. All low-temperature microtomy, however, requires care to eliminate the introduction of moisture-related artifacts.

The danger of introducing artifacts is always present, and producing truly representative thin sections is rarely routine. The ability to recognize and distinguish subtle artifacts from real morphology details comes only with experience. This microtomy experience and familiarity with the material's morphology is important to recognize large-scale but inconspicuous changes introduced during sectioning.

EXPERIMENTAL RESULTS

Unfilled Nylon-6

Injection-molded unfilled nylon-6 test bars were etched with high-purity xylene at 65°C for 2–3 min [35]. Rigid temperature control was not necessary, and a temperature be-

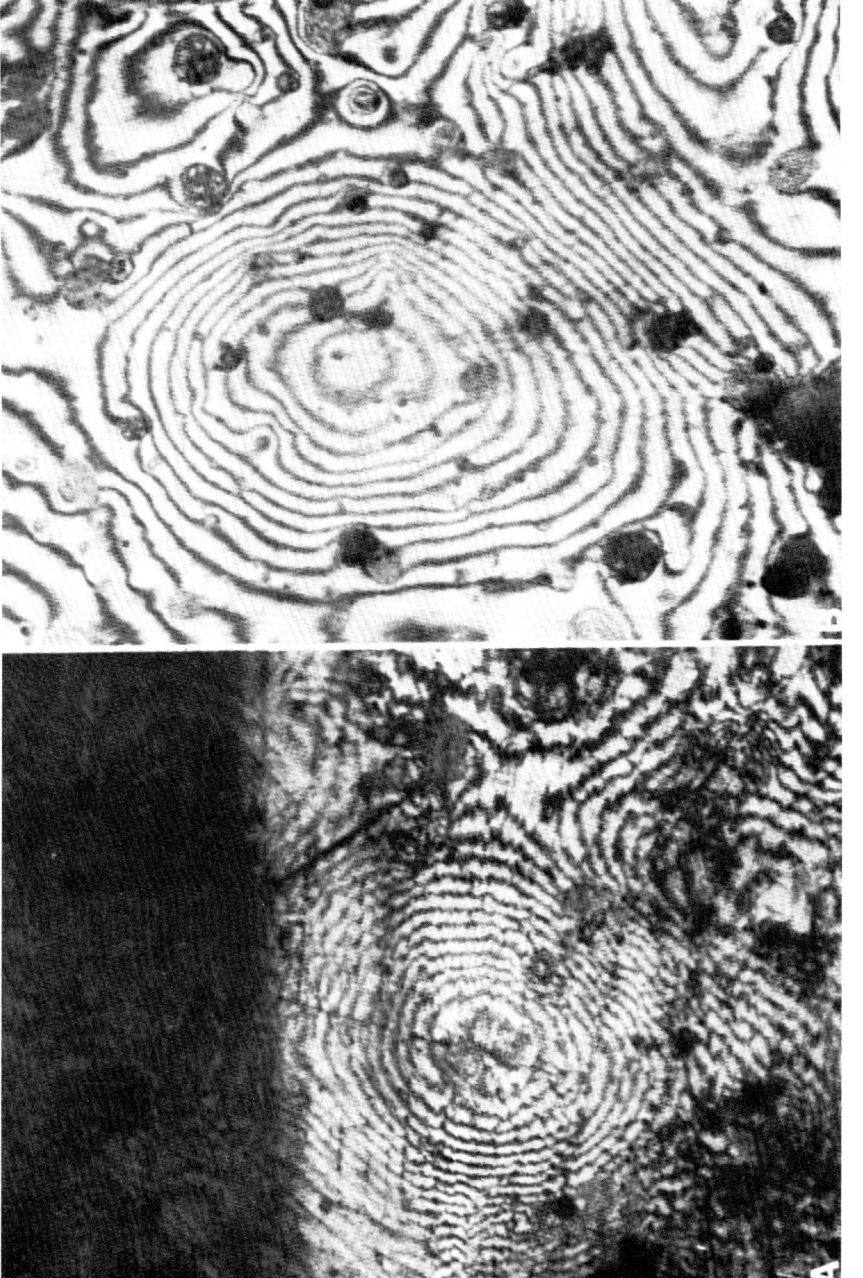

Figure 3 Measurement of surface relief on polymer painted plaque replica coated with aluminum film: A, lower half aluminum coated, upper half not coated (250×); B, aluminum coating showing the interference pattern which indicates the surface elevations (400×).

tween 65 and 70°C was found suitable. The etched and microtomed nylon-6 structures are shown in Figures 4A and B. The microtomed cross section shows the spherulite structure using transmitted polarized light. The spherulite size, boundaries, and overall arrangement are similar for both the etched and the microtomed samples. This agreement between etched and microtomed samples using optical microscopy indicates that detrimental artifacts have not been introduced during the etching process.

To document the effect of overetching, the 70°C etching temperature was exceeded. Figure 4C shows the same specimen with a much deeper etch. This slight to moderately overetched condition occurred with the xylene etch at 80°C for 1 min. The major structural details are still clearly resolved. However, even to the casual observer, the strong shadow patterns indicate significant elevation difference between borders and centers of the spherulites. These elevation differences are between 2.6 and 3.2 μm and present only minor focusing problems. Also, some structural details are lost by swelling or localized dissolution. The coarser surface texture and slight wrinkling result from the onset of a slight but arrested dissolution. The diminished spherulite borders and the unnatural elongation caused by the onset of localized dissolution are especially noticeable (upper right corner of Fig. 4C).

The sample was exposed to xylene for an additional 30 sec and the micrograph is shown in Figure 4D. Most surface details are lost. The shadows are much stronger and the height difference between the borders and the centers of the spherulites has increased to between 7 and 9 μm. Focusing on fine surface details is impossible even with intermediate magnifications. The first xylene exposure has sensitized the surface, because freshly prepared surfaces exposed to xylene for 1.5 min at 80°C show less drastic overetching.

Nylon-6

The etched topography of nylon-6 was studied with DIC [35]. The sample was purposely overetched and overheated to eliminate surface wrinkling and to expose the entire surface to moderate dissolution. This washes out much surface detail. With the wide contrast range available, DIC highlights minute surface details even at minimum brightness. With further adjustment, the field gradually brightens and passes through the complete range of interference colors up to and beyond the first order. The most appropriate position to show the fine surface detail is prior to the appearance of interference colors.

The spherulite appearance is clear with a three-dimensional effect. One spherulite side is lighter and the other side is darker than the background. The spherulite shown in Figure 5A has a dark outline on the top side. This image can be easily reversed with contrast adjustment and Figure 5B shows the reverse view. The spherulite has actually swelled and is in fact raised slightly higher than the flatter surface details. The reverse view shows the false image of a recessed spherulite. The false image occurs because the observer tends to assume that the illumination is always from the same direction. Careful observation shows other differences. Notice that the numerous etch pits in Figure 5B are missing in Figure 5A. Other minor discrepancies between the images are for the most part the result of the illumination direction. It is up to the polymerographer to choose the optimum image condition with the most details.

Nylon-6 with Additives

Polymer chemists know that adding fillers to a polymer will change the appearance of the morphology. A 2 vol % talc and 15 vol % glass fiber–filled nylon-6 is an excellent

Figure 4 Nylon-6: A, etched in xylene, polarized reflected light (180×); B, microtomed thin section, polarized transmitted light (180×); C, overetched with xylene at 80°C, 1 min (180×); and D, etched an additional 30 sec (180×). C and D use polarized reflected light.

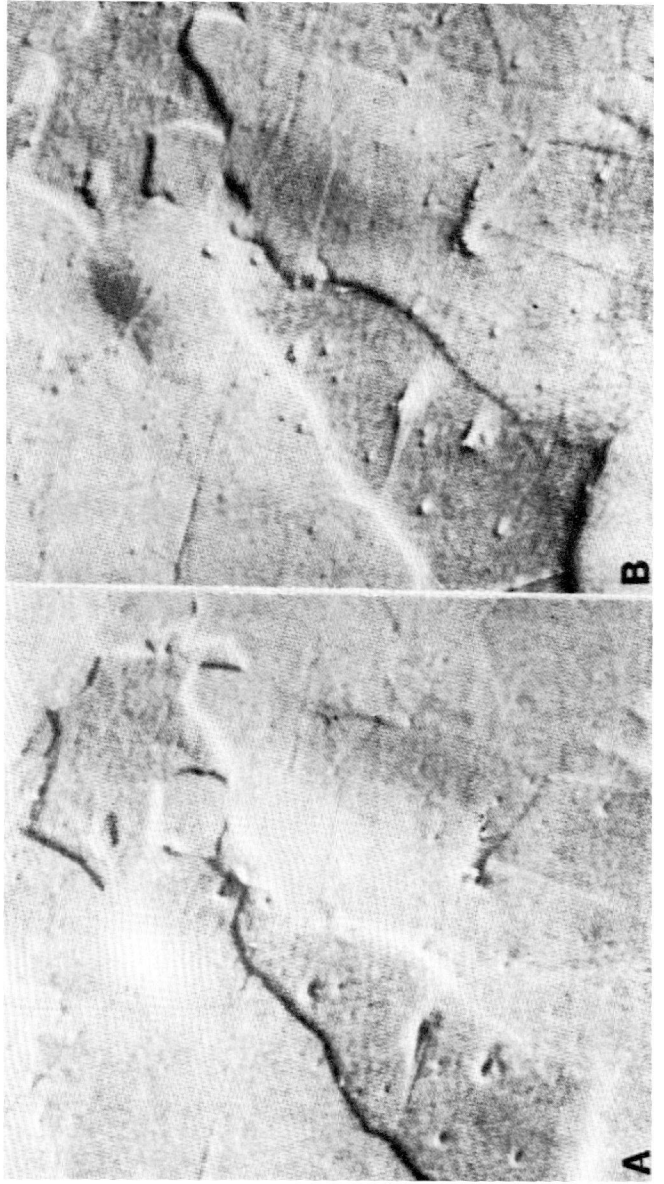

Figure 5 Nylon-6, DIC incident light: A and B show contrast adjustment (200×).

example [35]. The surface was sectioned at an angle of 25° to the predominant glass fiber orientation to accentuate their presence. The freshly prepared surface was etched with xylene at 50°C for 12 min. The fine morphology of the polymer matrix is clearly resolved as shown in Figure 6A. For comparison, a cryomicrotomed cross section was prepared. Microtoming parallel to the glass fibers was abandoned because unpredictable glass fiber dislocations of the matrix obscured the morphology. The glass-filled nylon-6 sample, microtomed normal to the glass fibers, is seen in Figure 6B. Because of the finite thickness of the microtomed section, occasionally some of the dislodged glass debris has perturbed the matrix. However, the matrix is clearly resolved over relatively large regions and the microstructure compares favorably to the etch-delineated morphology. Artifacts from polishing or etching were not detected.

Nylon-66

The next example, nylon-66, etched 4–5 min in xylene at 75°C and dried at room temperature, is shown in Figure 7A [35]. Overall the DIC image overall appears totally flat. And the black and white alternating image details, although with sharp contrast and resolution, are void of any indication of the three-dimensional character of the surface. The etched-surface appearance with DIC could almost be mistaken for the transmitted light image of a microtomed thin section. For comparison, this nylon-66 was also microtomed. The transmitted light optical morphology, shown in Figure 7B, is in good agreement with the etch-delineated surface structure. The general agreement of the features with the microtomed thin section image of Figure 7B signifies the absence of etch-induced artifacts and at the same time reveals the technique's limitation; i.e., the three-dimensional image appearance is absent. Observe the optical heterogeneity of the nylon thin section as well as the DIC imaged surface structure.

To further document the absence of artifacts, the internal morphology and the etch-delineated surface morphology were studied with image analysis. To use uniform discrimination, it was assumed that the spherulite boundaries were consistent. Using polarized light, the spherulites are easy to recognize by the Maltese cross. When the polarizer and the analyzer are rotated, the cross also rotates. Each spherulite shows only one Maltese cross, so the rotation helps to delineate individual spherulites. However, only linear measurements of the spherulite size are obtainable, and the number of recognizable spherulites cannot be related to percentage of crystallinity. More than 100 recognizable spherulites were measured in several microtomed sections. Objective area selection is very important, but the selection process, operator discretion, or sample quality may still introduce biases. Nevertheless, a good approximation can be obtained to confirm or reject the similarities between the transmission and incident light morphologies. The average spherulite size for the microtomed cross sections is between 6 and 9 μm, which is in acceptable agreement with the 4–7 μm size obtained for the etched-surface structure.

Heat-Treated Nylon-66

Heat treatment is known to affect the structure of polymers to an appreciable degree. The sample was annealed for 18 hr at 125°C and subsequently repolished and etched with xylene for 3 min at 65°C [35]. As the direct result of the heat treatment, the spherulites have greatly increased in size so that they clearly delineated, as shown in Figure 7C. At this size, the features imaged with DIC show again the three-dimensional appearance. The minute details of the surface topography are clearly visible and the individual spherulites are well

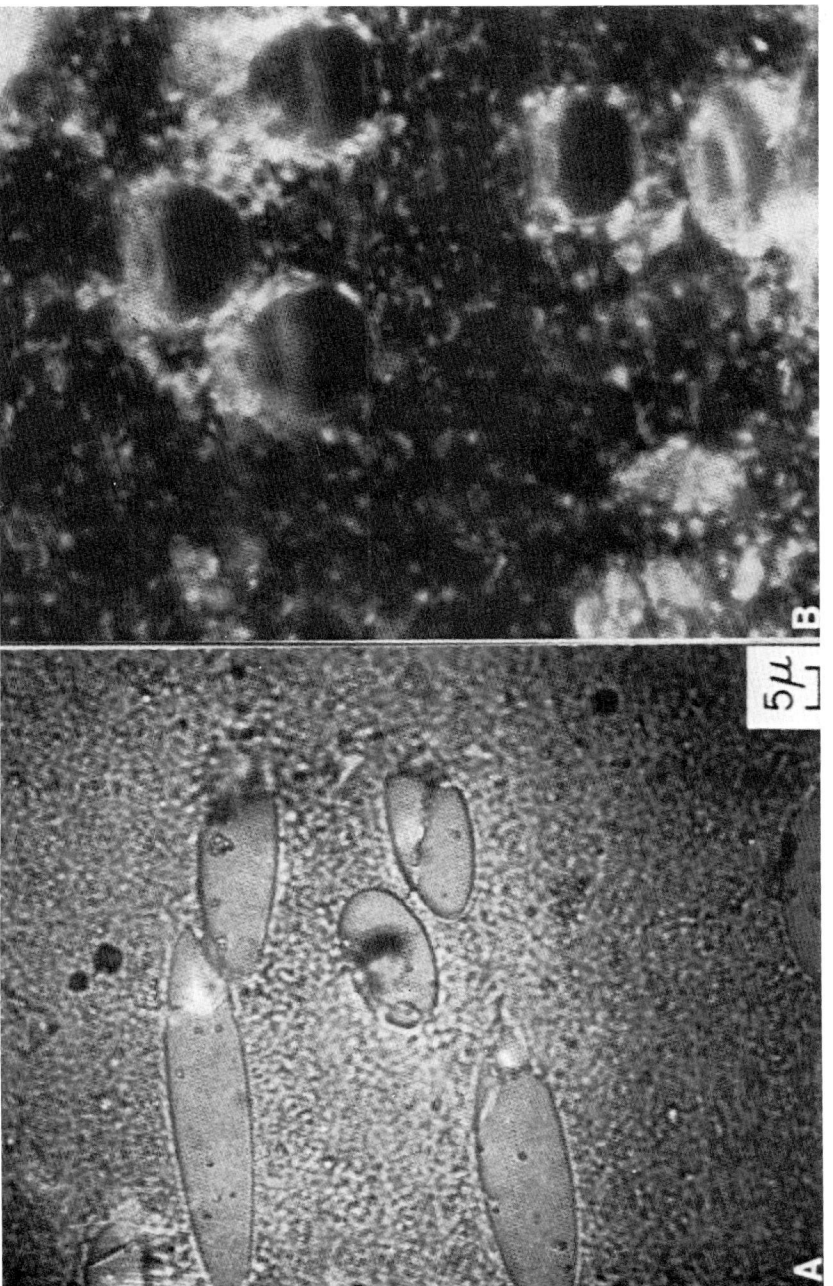

Figure 6 Glass-filled nylon-6: A, polished cross section etched with xylene at 50°C, 12 min, polarized reflected light (1000×); B, microtomed thin section, polarized transmitted light (1600×).

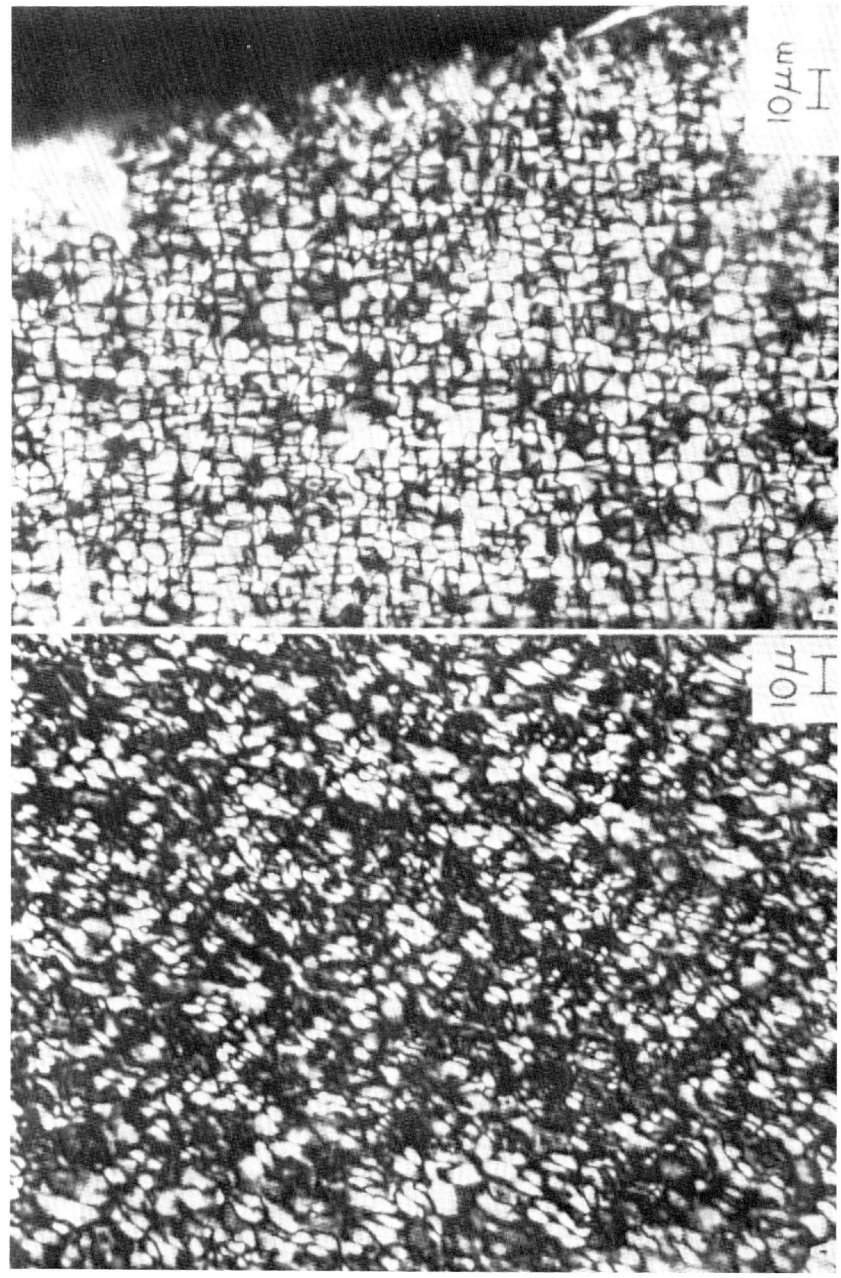

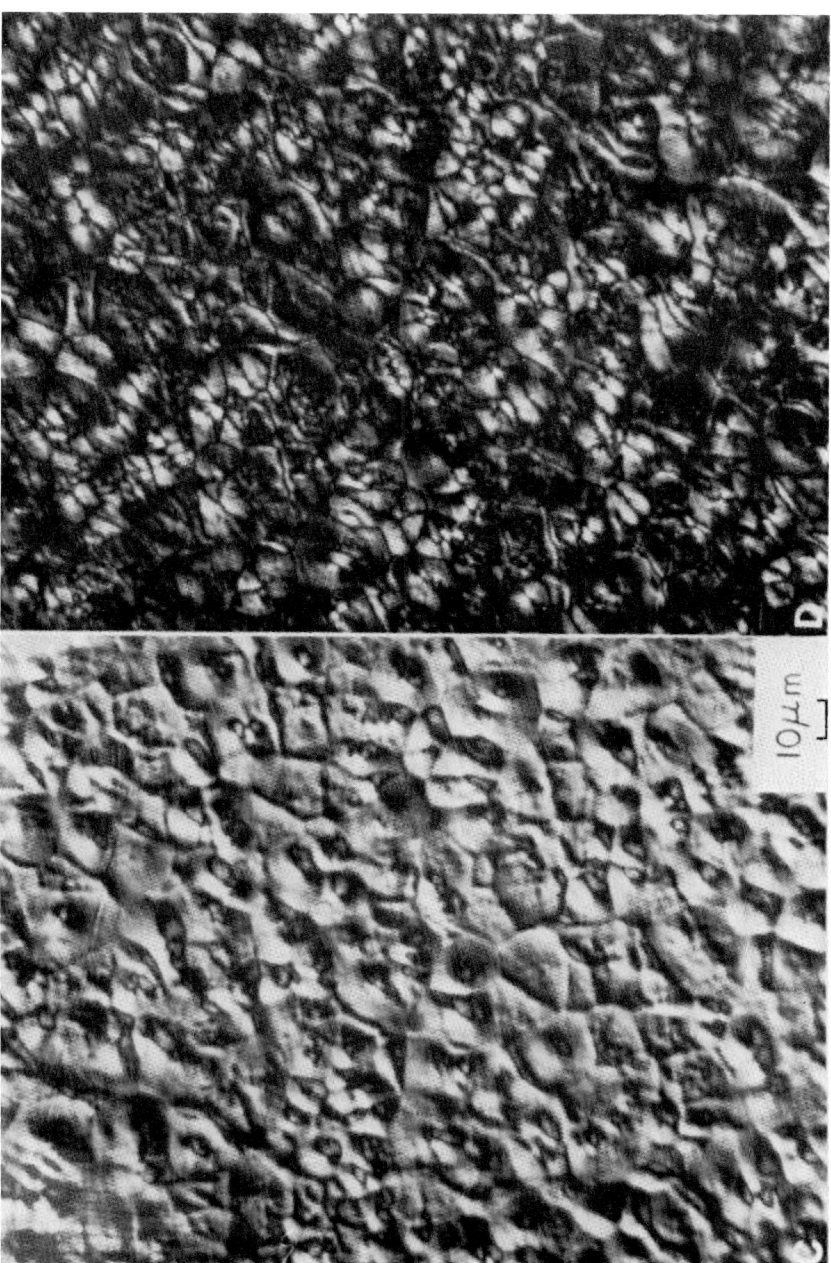

Figure 7 Nylon-66: A, polished cross section, etched in xylene at 75°C, 4.5 min, DIC-reflected light (500×); B, microtomed thin section, polarized transmitted light (500×); C, etched cross section annealed at 125°C, 18 hr, polarized reflected light (500×); and D, microtomed thin section of C, polarized transmitted light (500×).

delineated. Note the presence of a few much larger spherulites that may have been the result of unintentional localized heat retention.

A microtomed thin section of the annealed sample is shown in Figure 7D, for comparison. The individual spherulites are clearly delineated and agree with the DIC image in Figure 7C. Using this image comparison, the measured average spherulite size of the microtomed and surface-delineated morphology agree within 2–3 µm. Of course the larger spherulite size helps the measurement.

Polyethylene

With the following experiment, the appearance of polishing-induced artifacts on a highly polished polyethylene surface was studied. The surface was prepared using 0.05-µm alumina polishing slurry. The polished surface quality was critical to subsequent etching studies of very fine surface details. Because preparing the required surface quality was tedious, the surface, prior to etching, was thoroughly examined. Reflected light showed only dubious details about the surface quality in some areas. In addition, many important surface details were invisible. Attempts to show the minute surface details with oblique illumination were also fruitless. The surface in small areas appeared as if it were smeared or torn instead of cleanly cut by the abrasive slurry. The reflected light image seen in Figure 8A "clearly" and "convincingly" shows, without the benefit of phase contrast, that the surface has only minimal damage caused by localized smearing of some small torn debris. This misleading, improper observation arises from the limited depth of field. Since phase contrast highlights minute variations in height as brightness differences, this technique proved to be useful.

With phase contrast the brightness increases with increasing difference in level and the elevated details become brighter while the low-lying areas appear darker. If the difference exceeds one quarter of the wavelength, or approximately 1300 Å, the contrast is decreased but surface topography is still well delineated. Beyond this point, the direction of the differences in level cannot be ascertained. With phase contrast, Figure 8B shows the true surface quality of the same area (Fig. 8A). The pulpy appearance is replaced with definite outlines of the damaged surface. The polishing damage appears as a fine lamellar structure. Indeed, the surface damage by the polishing is extensive and the cause was traced to the polishing cloth.

Polychlorotrifluoroethylene

Successful etching experiments with polychlorotrifluoroethylene were conducted. This material was selected for its known poor solubility, and therefore presumably difficult etching characteristics [35]. The material's fine morphology was successfully delineated on the polished surface using carbon tetrachloride at 25–35°C for 2–5 min. Temperature is a less important etching factor for this material. Slight to moderate structure deterioration was observed at 50°C with a 1-min etching time. At 50°C it was necessary to divide the 1-min etching time into four equal periods. After each 15-sec period, the sample was allowed to cool to room temperature. The etched surface morphology for polychlorotrifluoroethylene is seen in Figure 9. All surface etching starts with a smooth flat plane. If no swelling or minimal swelling occurs, as with this material, good focusing and excellent detail visibility are retained. Microtomed sections of this polymer were also examined in polarized light. The images, however, failed to show clear identifiable features, presumably because large-scale crystallinity is not present. This example illustrates a case where the

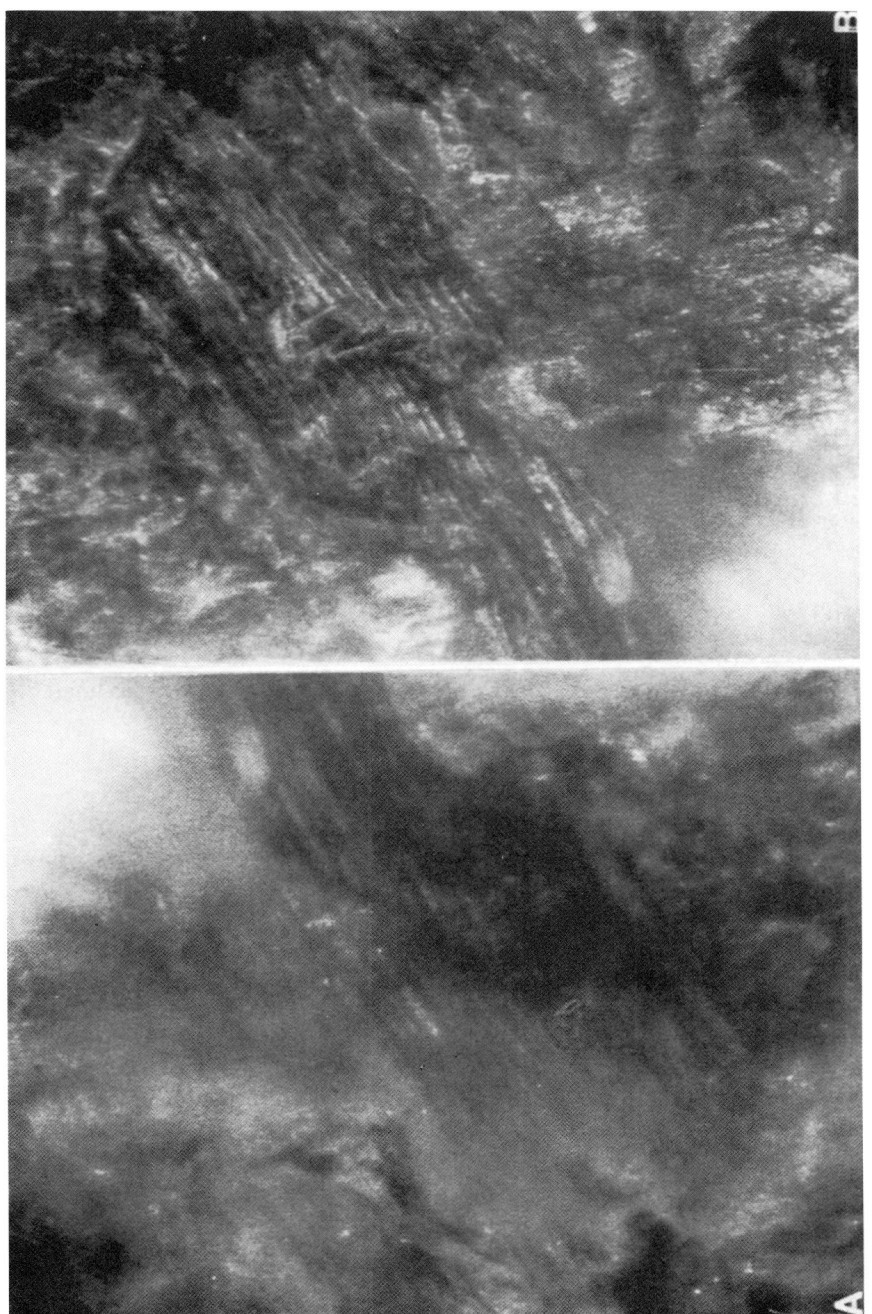

Figure 8 Polyethylene with 0.05-μm polish: A, reflected polarized light (1200×); and B, reflected light phase contrast (1200×).

Figure 9 Polychlorotrifluoroethylene polished cross section etched with carbon tetrachloride, incident oblique polarized light (500×).

surface delineation technique reveals the morphology unobtainable with microtomed thin sections.

Liquid-Phase Etching of Polypropylene

The polypropylene sample used for etching experiments was cut from an injection-molded plaque and included the molded outer surface [35]. Finding a proper etchant for polypropylene was complex. The best result was obtained with a mixture of equal parts of benzene, chloroform, and xylene. Temperatures between 80 and 85°C with a total etching time of 2–3 min produced optimum results. The sample was submerged in the etchant for periods of 20–30 sec. To avoid overheating, the sample was withdrawn and allowed to cool to room temperature before continuing. At the higher temperatures, the etching is very rapid, difficult to control, and in some instances instantaneous; a sample volume of only 3–5 g will rapidly attain the temperature of the etchant.

With incident oblique light, some minute details of the polypropylene morphology are shown in Figure 10A. This micrograph represents a depth of more than 0.5 mm from the outer mold surface. This cross section clearly shows the heterogeneity of the material. The spherulite size at the outer mold surface is considerably smaller.

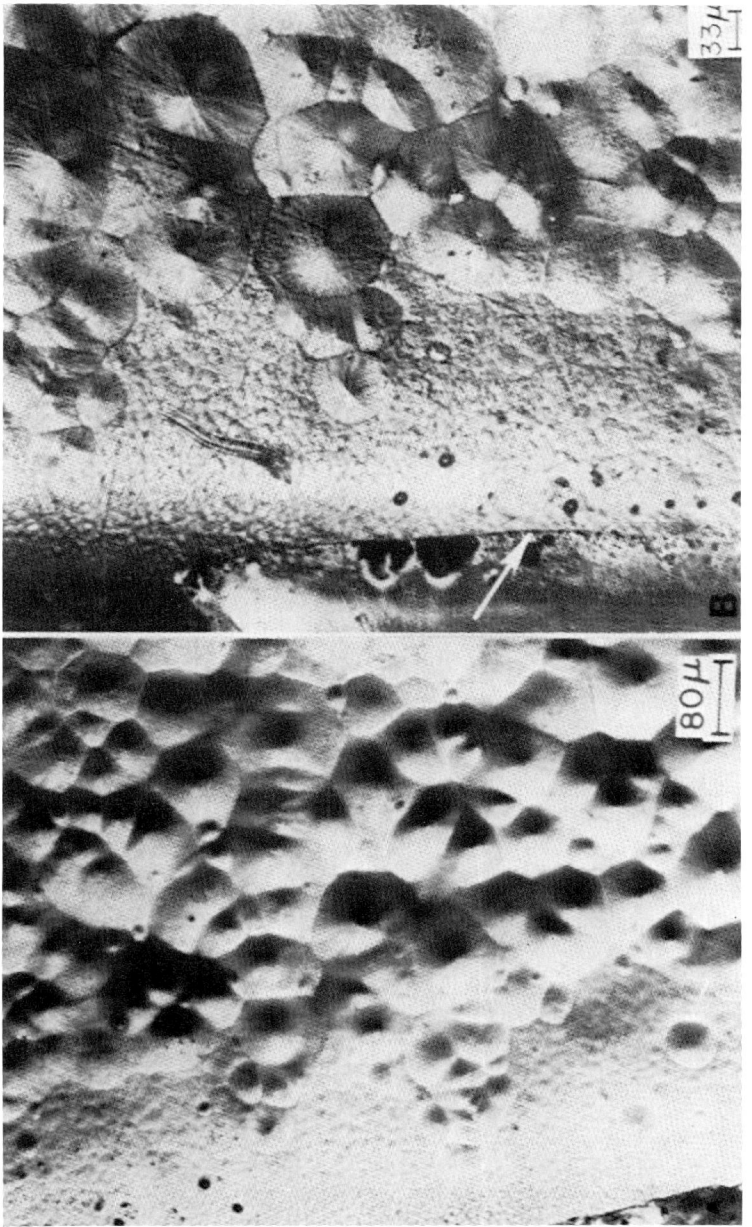

Figure 10 Polypropylene polished cross section hydrocarbon vapor phase etched: A, incident oblique light (125×); and B, DIC incident light (150×). The arrow indicates the outer mold surface.

The incident light image with DIC is similar to images obtainable with incident oblique illumination. To illustrate this resemblance, the surface topography of etched polypropylene was examined with both methods. Surface details and the spherulite boundaries are shown with excellent contrast in Figure 10B; the arrow indicates the outer mold surface. The sharp boundary outlines and pointed tips of the spherulites are clear indications that the sample was neither overetched nor overheated. While details of the spherulites are clearly resolved with DIC, the surface topography is perhaps more evident using the oblique incident illumination as shown in Figure 10A. Both micrographs show identical features but the DIC image clearly illustrates the high quality of the etch-delineated spherulites. Conversely, the blunted spherulite tips and the featureless surface of the spherulites seen in Figure 10A give the wrong impression of a poorly etched surface. However, these two techniques complement each other; the DIC provides superior information about surface details and by using the obliquely illuminated image spherulite elevation may be estimated.

In etched polypropylene, the presence of a few hexagonal spherulites among many monoclinic spherulites can be highlighted with DIC. A lightly etched sample, coated with a vapor-deposited layer of aluminum approximately 1000 Å thick, was examined under the microscope. The hexagonal spherulite, easily identified by its shape and its surface pattern, is shown in Figure 11B. Repeated experiments disclosed that the hexagonal spherulites were more easily attacked by the reagent. To follow this preferential attack, the same area of the sample was repolished and exposed to the etchant for 2-sec periods. The total etching time was 8 sec. Indeed, the hexagonal spherulite, seen in Figure 11A, is preferentially attacked and shows the beginnings of within-spherulite markings. At this stage of the etching the spherulite type can not be ascertained.

The hexagonal polypropylene spherulites appear flatter when compared to the surrounding monoclinic spherulites. To confirm this impression, the surface was examined with incident light oblique illumination. Figure 11C clearly illustrates the slightly raised tip of the hexagonal spherulite surrounded by the higher conical monoclinic spherulites. Thus the hexagonal spherulite is not as flat as it appears in the DIC image seen in Figure 11B. The monoclinic spherulites are approximately 0.4–0.6 μm above the background.

As mentioned earlier, removing polymer material by direct dissolution is probable during the solvent etching. But dissolution may not be necessary to successfully delineate the morphology. To perfect the polypropylene spherulitic morphology delineation, a new approach was designed. Any contact between the polymer surface and the etchant's liquid phase was eliminated, which automatically excludes dissolving any surface layer. A new sample, from the same injection-molded polypropylene plaque, was exposed to an air flow saturated with the vapor of the hydrocarbon mixture at 80°C for 15 min. Before exposure, the sample was preheated to prevent surface condensation. After the 15-min exposure only minimal relief was visible on the sample surface reminiscent of a weak unsatisfactory etch. The spherulite morphology was nearly impossible to discern. Subsequently, when the swelling-sensitized surface is exposed to heat or laser energy, an excellent structure delineation is obtained. The polypropylene spherulite, seen in Figure 12A, is the result of the vapor phase–etching technique and subsequent heating at 85°C in air for 10 min. Details are clearly resolved and no deformation or artifacts are in evidence in this relatively high-magnification optical image. For comparison, a microtomed thin section, also vapor etched, is shown in Figure 12B. This high-magnification transmission electron micrograph compares favorably with the etch-delineated cross section. Some etch-produced localized artifacts (bubbles) are in evidence at this high magnification (right side of Fig. 12B).

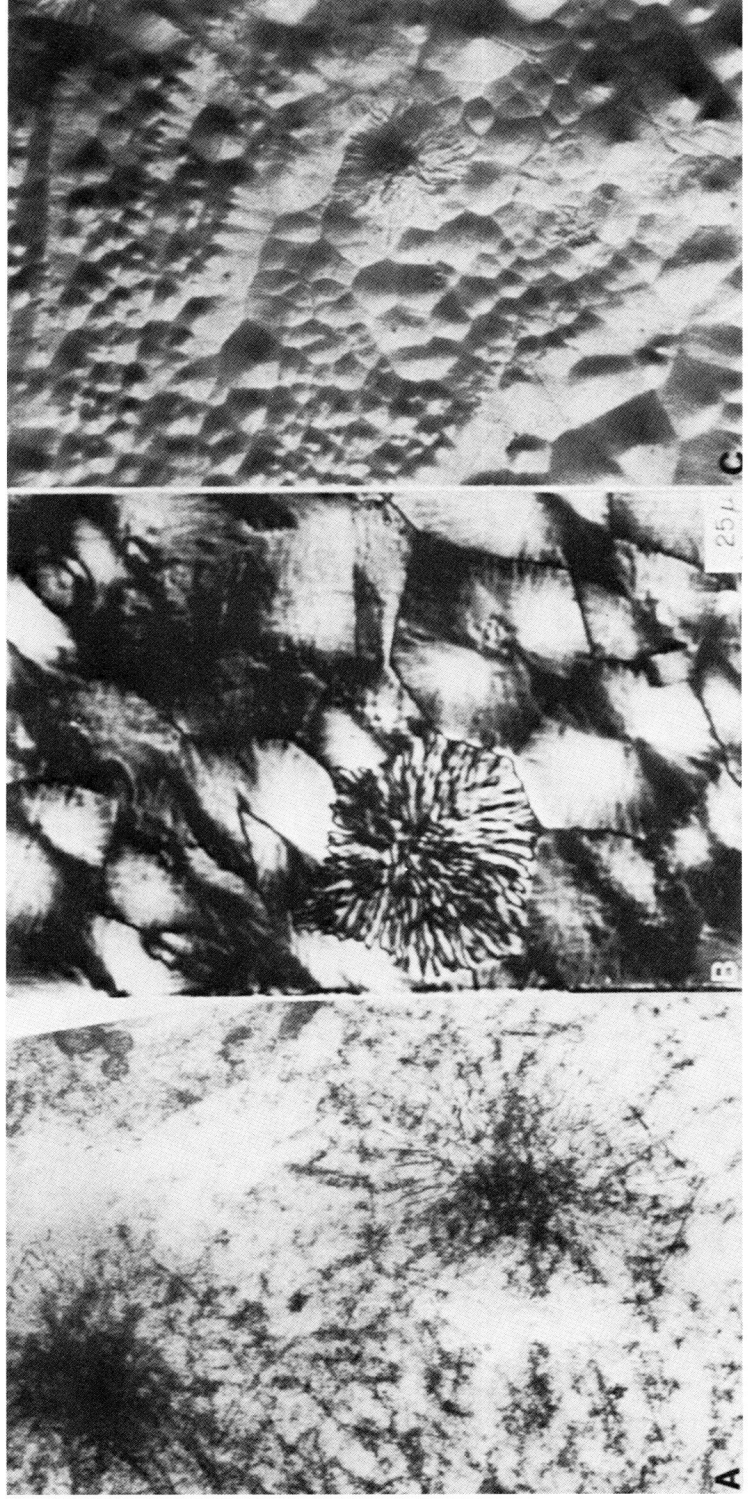

Figure 11 Polypropylene: A, polished cross section, etched 8 sec, DIC (550×); B, lightly etched cross section, aluminum coated, DIC incident light (400×); and C, incident oblique polarized light (125×).

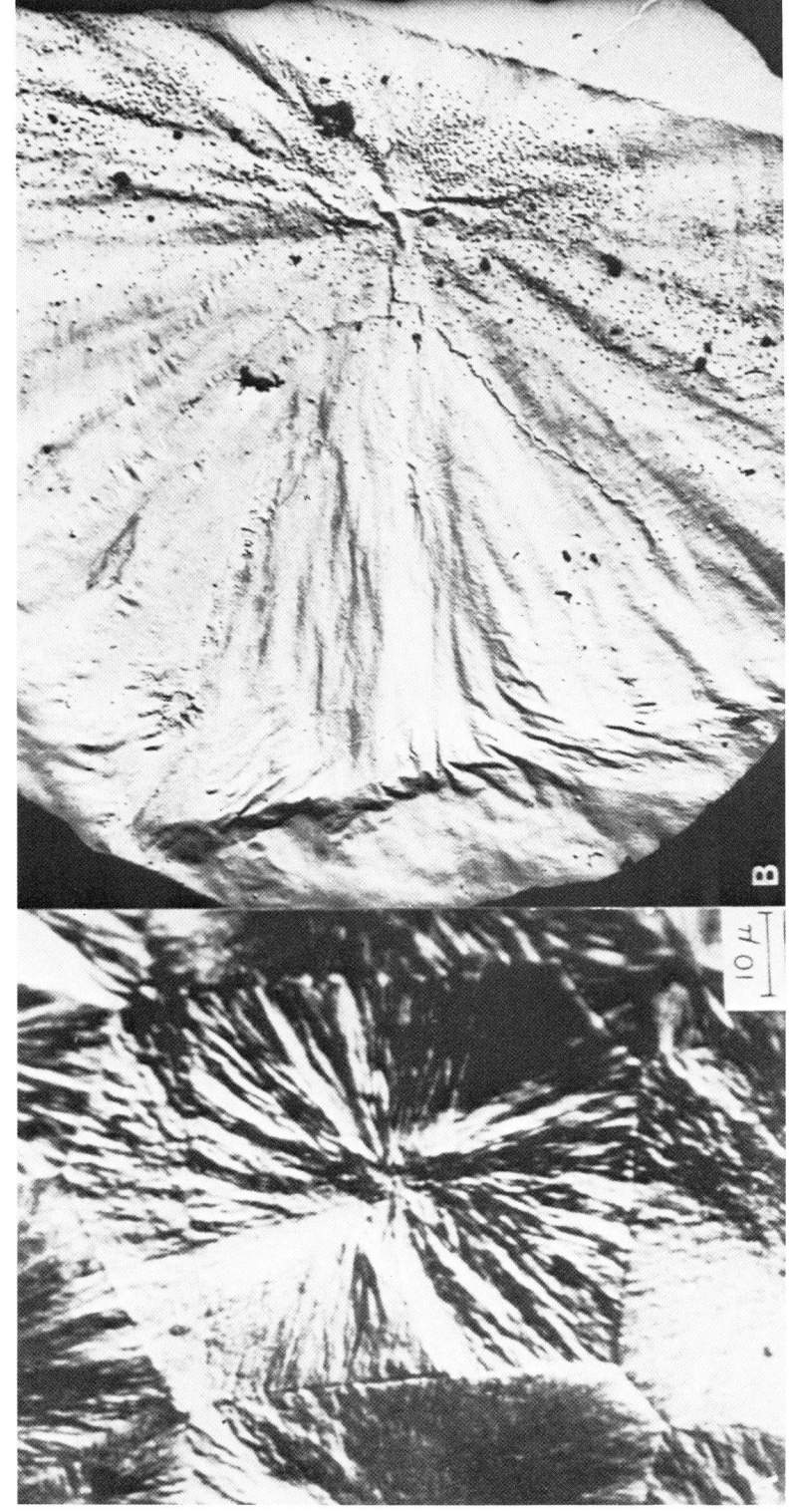

Figure 12 Polypropylene hydrocarbon vapor phase etched cross section, heated at 85°C showing spherulite: A, DIC incident light (1200×); and B, microtomed thin section, transmission electron micrograph (2000×).

There are two interesting areas for further examination of this high-resolution incident light image: (a) comparison of etch qualities at high-magnification liquid phase–versus vapor phase–delineated morphology and (b) microscopy limitations. Microscopy limitations are related to the polymer's ability to undergo partial and selective swelling; this may be detrimental to delail resolution because of the limited field depth of high-resolution objectives. The image in Figure 12A shows sharply delineated features and well-preserved details which are clear even at this higher magnification. This may quantitatively indicate the amount of swelling, because the depth of field is on the order of 0.6 μm.

Laser Etching of Polyprophlene

Using an argon laser beam (λ = 514.5 nm, 100 mW for 17 sec) as an alternative energy source, another polypropylene sample previously exposed to the vapor mixture was irradiated. The familiar structural pattern associated with polypropylene is shown with incident light in Figure 13A. The same area is shown in Figure 13B using the DIC method; the well-delineated spherulites are clearly visible here. The morphology observed with this laser-etching technique is comparable, if not superior, to that obtained with the liquid-phase etching. Controlling the laser energy, and the resulting surface temperature, is the only critical parameter. This parameter is relatively easy to control with attenuating filters. Duplication of the laser-delineated morphology was not possible without the surface preexposure to the hydrocarbon vapor. Repeated experiments with different lasers, energy levels, or exposure times yielded no structural delineation without the preexposure to the hydrocarbon vapor. Instead unfamiliar patterns or damaged surfaces were observed.

The laser beam–etched polypropylene morphology was compared with DIC and incident light. The fine structural delineation of the individual spherulites is very good when viewed with reflected light only, as seen in Figure 13A. The DIC image (Fig. 13B), however, accentuates fine surface details with contrast. This DIC image is very important to evaluating etch quality. The sharp spherulite tips and boundaries indicate that neither overheating nor underetching has occurred. The incident light image, Figure 13A, perhaps shows more within-spherulite details than the DIC image. A microtomed thin section is shown in Figure 13C. The general agreement of this thin section morphology with the previous figures confirms that no artifacts or distortion has been introduced during etching. This agreement, however, does not assure an artifact-free surface at magnifications higher than obtainable with optical microscopy.

Electron Beam Etching of Polypropylene

Electron beam irradiation in the scanning electron microscope (using 15–20 kV and 10^{-9}–10^{-11} mA) also produced good to excellent delineation of the morphology for the samples presensitized with hydrocarbon. With vapor, the presensitization (surface swelling) is necessary for electron beam etching even though the sample is examined in a vacuum, which then may extract some of the vaporized hydrocarbons. The presensitized surface, in spite of hydrocarbon reduction, is still sufficiently sensitive to the electron beam. Figure 14 presents incident light plane-polarized images of the surfaces that were irradiated by the electron beam. Figure 14A shows slight underetching, 14B slight overetching, and 14C a perfect etch. The advantage of this in-situ etching procedure is the ability to monitor and observe morphology changes on a microscopic scale. The vapor surface presensitization is more critical for the electron beam delineation procedure than

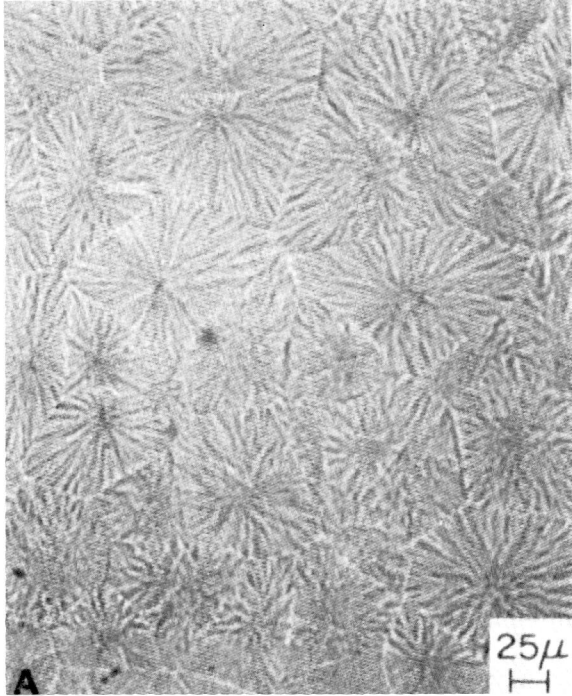

Figure 13 Polypropylene cross section, exposed to 80°C hydrocarbon vapor for 15 min, then laser irradiated for 17 sec: A, incident polarized light (200×); B, DIC reflected light (200×); and C, microtomed thin section, transmitted polarized light (250×).

for the laser or heat etching procedures. The quality of the structure delineation is similar for all three etching methods, if carefully performed.

The results of these vapor exposures show that the removal of surface material is not necessary to elucidate the morphology. This fact raises questions about using the term "etching" to describe the surface delineation process.

APPLICATIONS OF FRACTOGRAPHY

The conditions that cause fractures are numerous, and the actual reason for a fracture can be complex, at times defying solution. However, if the various modes of fracture are understood, some possible causes of failure can be eliminated, thus narrowing the choices. The remaining information helps to identify the fracture mode. Although the causes of failure may be complex, the fracture path characteristics and the ductile or brittle markings are relatively easy to recognize. These fracture mode characteristics are well documented in the literature [36, 37]. Conversely, information about impact-caused stress waves, interactions with the fracture tip, and the resultant markings or the incidence of secondary fractures and their characteristics require detailed study.

Fractography is used to distinguish between fracture modes, and the diversity of fracture analyses preclude a rigid empirical approach and requires many different tech-

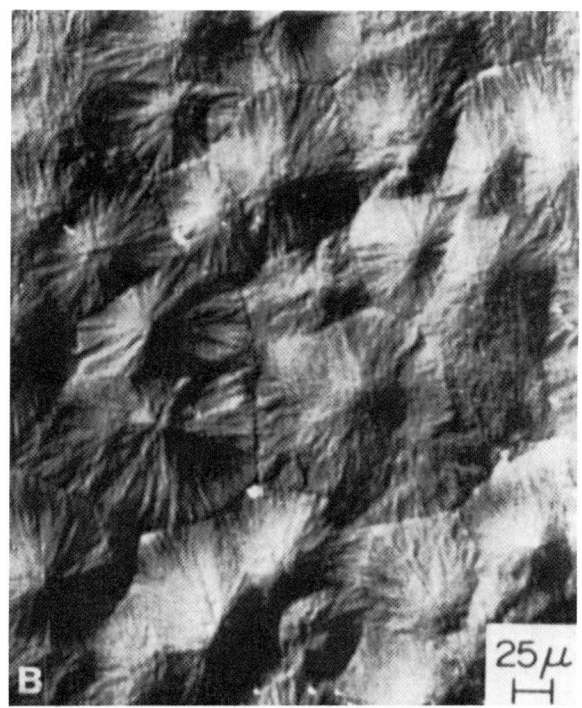

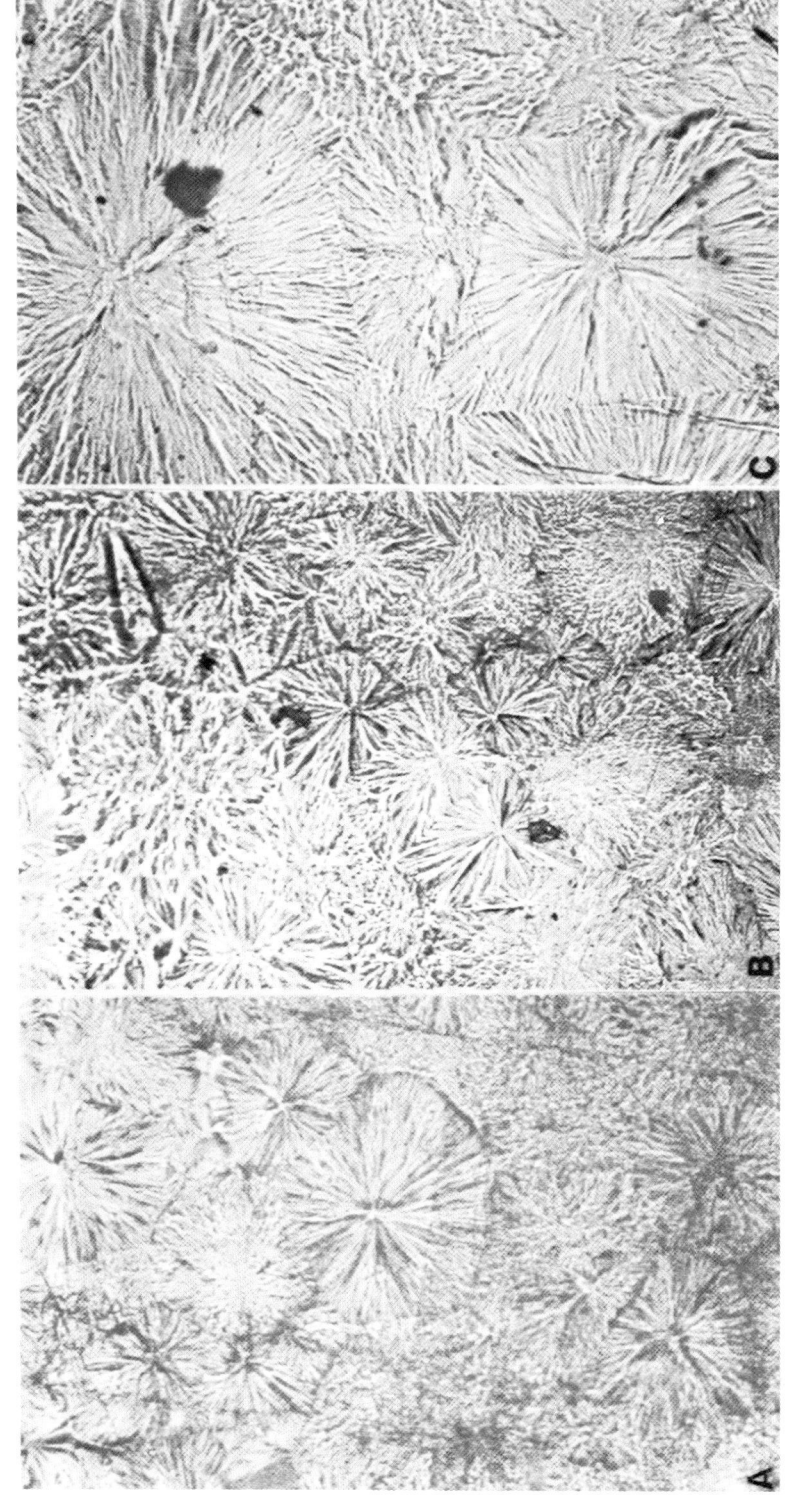

Figure 14 Polypropylene cross section exposed to 80°C hydrocarbon vapor for 15 min, then irradiated in the SEM electron beam, reflected polarized light: A, slightly underetched (200×); B, slightly overetched (200×); and C, perfect etch (500×).

niques. The applied methods and techniques must vary to suit the sample and the information desired. Some fracture analyses require the observation of the finest details using high-magnification microscopy, but others may be adequately examined at low magnification. Most tedious sample preparations can be omitted and the fracture surface can be examined directly with the scanning electron microscope (SEM). For examination of fine structural details, care must be exercised to prevent possible artifacts induced by the electron beam.

Fracture markings reveal the fracture's deviation from the planar form. With no deviation, the fracture would take the form of a perfect plane, and no marking would be visible. The primary causes for deviations are inhomogeneities in the polymer, e.g., fillers. Hard filler particles in plastics will not be cleaved by the fracture front because the front will pass around them taking the path of least resistance. This leaves markings which appear as a bump in one fracture face and a depression in the other. The resultant fracture surface and the polymer's structure are clearly related. Thus the fracture surface reveals some of the microstructure. These observations have limitations and require careful interpretation, but filler structures, some crystal morphology, and even phase separations can be identified with proper care.

Fracture markings are characteristic of the fracture mode. Smooth surfaces are usually produced by fast, brittle crack propagation. Roughness is caused by ductile tearing or crazing, or is the result of the polymer's composition. Fatigue fractures usually show striations, which can be used to explain the deformation history. Detailed interpretation of these fracture markings can be found in the work of Hertzberg and Manson [38]. In most instances, the fracture front behavior must be considered together with the structure. Of course, for homogeneous polymers, the fracture front behavior involves the material's deformation characteristics, i.e., plasticity or viscoelasticity effects.

An injection-molded 50–50 wt% blend of polycarbonate (PC) and polybutylene terephthalate (PBT) with plasticizer and carbon black was selected for this example [39, 40]. Polymer blends like PC–PBT can have a very complex morphology and it is useful to classify the structure in terms of the size of the domains. Domain sizes larger than approximately 5 μm are the result of flow history in injection molding. Typical flow-induced structures are seen in Figure 15; notice the structure of the flow-induced vortex shown at higher magnification in Figure 15B. The more regular morphology shown in Figures 16A and B reveals a different topography that also is related to the flow but follows the streamlines. Details smaller than 5 μm are usually related to the blend's "ultimate" structure. In Figure 16C the ultimate structure is shown where the PC has been removed in the white areas and also shows the morphology of the PBT phase. At this level, evidence may also be found of toughening additives such as rubber particles (black dots). These composition factors complicate the interpretation of fracture surfaces.

Fracture surface roughness does not always arise from crack propagation. When the primary fracture front is diverted through secondary fractures, a rough surface will result. Furthermore, the reduced-velocity fracture follows the morphology and causes dimpled surfaces. Conversely, high-velocity fractures produce smooth surfaces. The PC–PBT blend fracture surfaces at room temperature and −30°C show evidence of both fracture types, i.e., the low- and high-velocity fracture markings [41]. Figures 17A–D are typical of an ambient temperature ductile fracture. Figure 17A shows typical Wallner line–caused markings that result from the disturbances of secondary crack initiations [37]. The factors influencing the size and shape of the dimples seen in Figure 17B are numerous. The simplest and most obvious explanation is ductility and temperature. However, the mechanism is more complex. The dimples result from the joining and subsequent stretching of

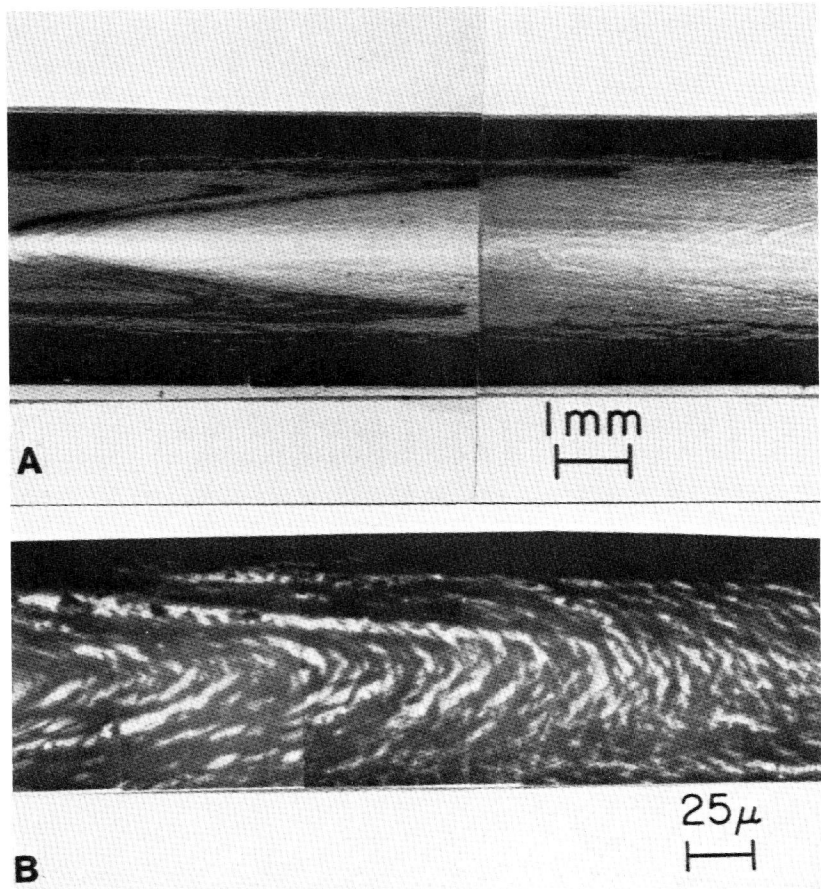

Figure 15 Polycarbonate–polybutylene terephthalate molding flow vortex induced structure, polished etched cross sections, reflected light: A, 10×; and B, 400×.

small voids which intersect the free surface of the growing microvoids [41]. Temperature plays a key role, and the dimple size changes with the fracture temperature. Figure 17C shows the enlarged SEM view near the fracture initiation site. For contrast, the low-temperature fracture face is shown in Figures 17E–H and the smoothness indicates that the fracture occurred at high velocity. Figure 17E shows a convoluted microstructural appearance that may be caused by structure, temperature, or the combined effect of the two. The dimpled surface of the fracture near the initiation area is shown in Figure 17F; this morphology indicates localized thermal conditions that may have been induced with the high-velocity fracture rate.

The fracture markings of the room- and reduced-temperature fractured PC–PBT blend are numerous. They result when the fracture front deviates from the planar path when it encounters structural features. The secondary fractures produce the most common markings and propagate in a semicircular pattern with the planes parallel to the main fracture plane. In Figure 18A, the secondary fracture markings show a circular pattern on the

Figure 16 Polycarbonate–polybutylene terephthalate: A, streamline-induced coarse surface structure, polished etched surface, reflected light (500×); B, streamline-induced fine surface structure, polished etched surface, reflected light (1000×); and C, ultimate fine structure, transmission electron micrograph, microtomed thin section (13,500×).

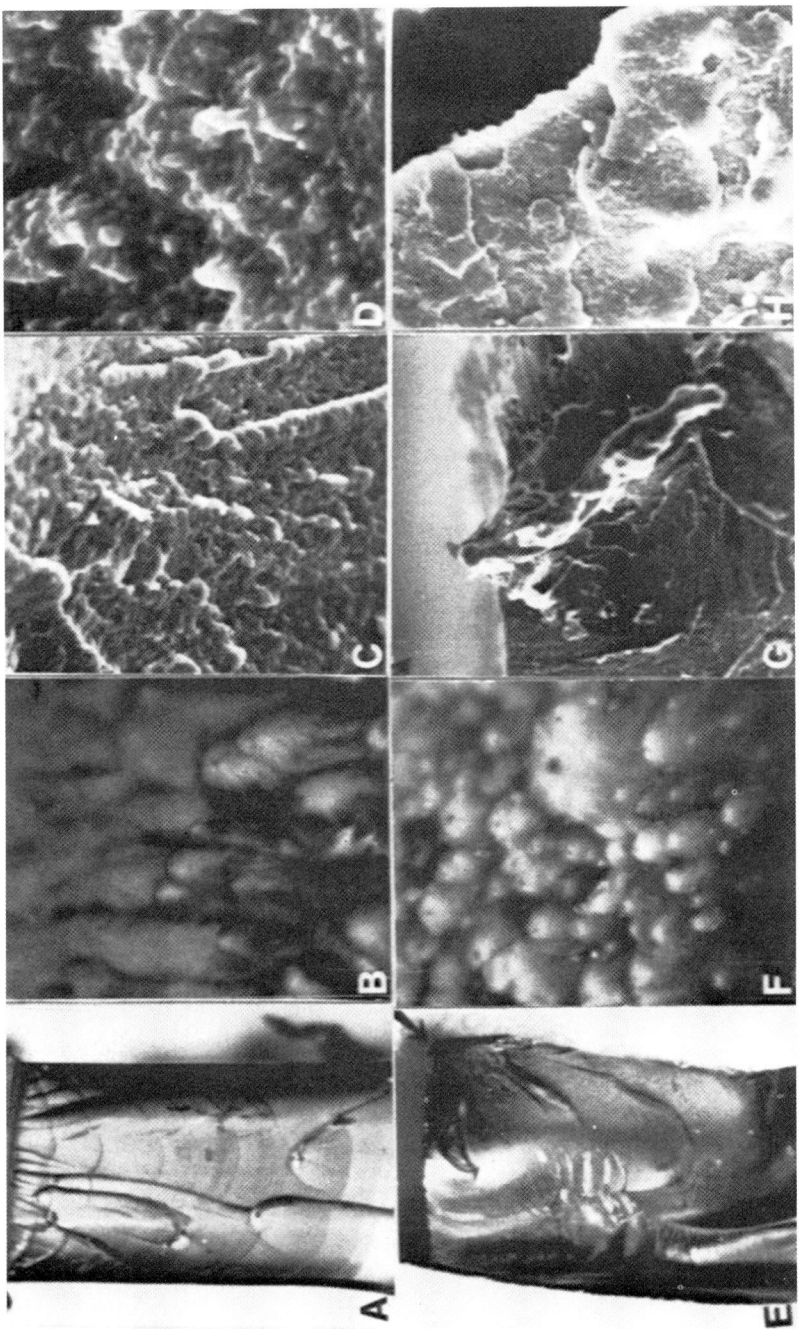

Figure 17 Polycarbonate–polybutylene terephthalate room temperature fracture markings at fracture origin: A, 7.5×; B, 100×; C, 250×; and D, 1500×; and low-temperature (−30°C) brittle fracture markings at fracture origin: E, 7.5×; F, 50×; G, 250×; and H, 1500×. A, E, incident light; B, F, polarized incident light; C, D, G, H, scanning electron micrographs.

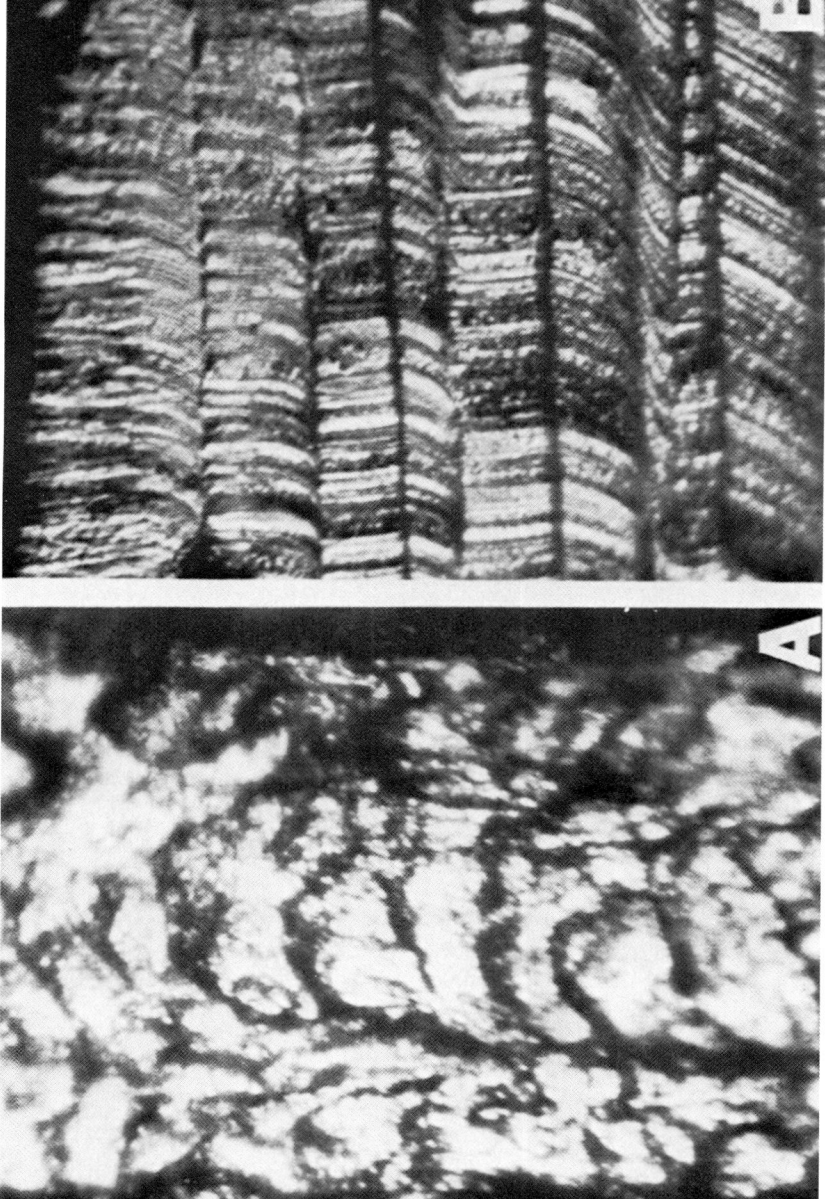

Figure 18 Polycarbonate–polybutylene terephthalate: A, secondary fracture markings in oblique reflected light (350×); and B, thin cross section microtomed normal to the fracture surface with transmitted polarized light (350×).

Figure 19 Polycarbonate–polybutylene terephthalate fracture surface, SEM: A, 2000×; B, 10,000×; and C, 20,000×.

fracture surface. If the main and the secondary fracture planes are close, they may be joined by a small step. Figure 18B shows a microtomed cross section of this secondary fracture using transmitted polarized light. The secondary fracture planes are joined and the surface is relatively rough. However, the roughness increases with the distance from the fracture initiation because new secondary fractures occur. These microcracks have a high stress concentration at the tip, and the stress at a given distance from the tip in any direction is directly related to the crack's length.

More than just occasionally the examination of polymer surface fractures requires high magnification to reveal all the necessary details. In the case of the PC–PBT blend, the dispersed elastomer particles are of particular interest. In Figure 19A, the individual particles are very difficult to see and could be easily confused with the fine details of this brittle fracture. Some of these fine fracture details may indeed resemble the shape of the elastomer particles. Only with increased magnification can the particles be distinguished from the islands of PC, as shown in Figure 19B. With still higher magnification, the image reveals the presence of smaller particles. Figure 19C shows clearly the dispersion of these particles. At this high magnification all the dispersed elastomer particles are clearly resolved and easy to distinguish from the rounded tips of the fracture planes.

Engineering thermoplastics are usually a blend of different materials, and to delineate the individual phases by etching is a challenge. To clearly see the fine structural details, a

Microscopy Techniques for Polymer Characterization / 573

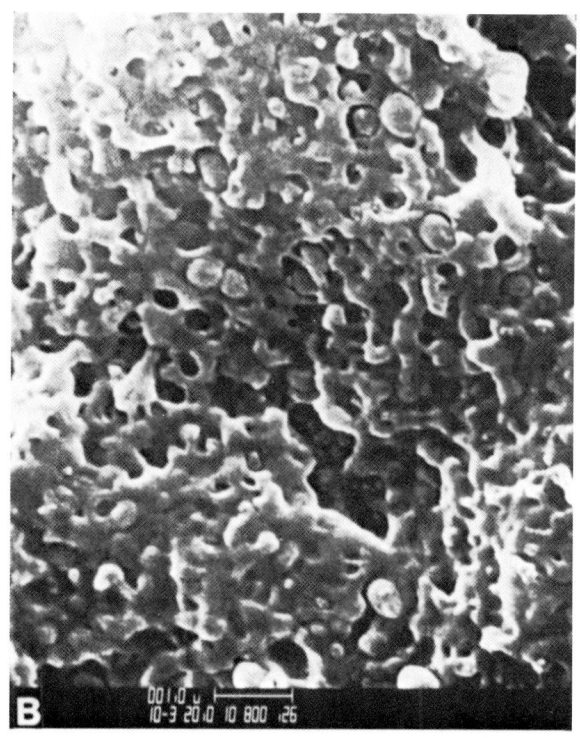

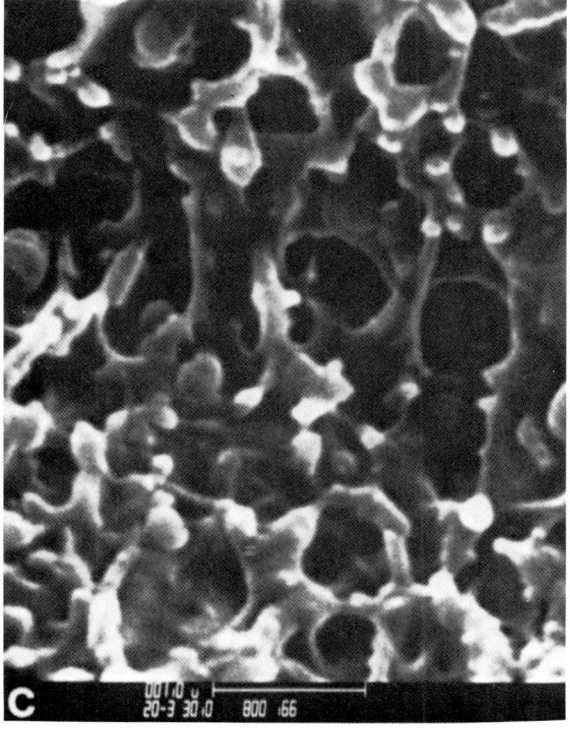

smooth surface prior to etching is critical. Therefore, the abrasive used for the final surface polishing should not exceed 0.5 μm. The polished surface of this blend, in the unetched condition, is void of any recognizable morphology, as seen in Figure 20A. An etch containing methylene chloride and toluene was used. A ratio of 10:1 or 5:1 was the most suitable. The exposure time was adapted to the dissolution rate of the two phases and is between 5 and 10 sec. The delineation of the PC and the PBT phases results from the solubility of the PC. This etch clearly shows the morphology, including the dark PBT rich weld joint as shown in Figure 20B.

If high-resolution microscopy is contemplated to resolve some particularly fine details, the addition of 1–3% glycerine is recommended to moderate the dissolution rate and preserve the amorphous polycarbonate (white phase) details. Figures 20C and D illustrate etching with the glycerine additive, which reveals fine PC details.

Phase delineation with etching is a powerful technique to identify manufacturing or processing conditions that may affect the material's properties. Etching this blend reveals the arrangements and detail of both the coarse and fine features of the PC phase. The relationship between the fine and coarse structures of the brittle PC phase is important, because structure refinement is a function of flow during molding. The higher the shear stress, the finer the PC morphology becomes. Figure 21 shows examples of the etch-delineated morphologies that result from these shear stresses.

An attempt to check for solvent-induced artifacts by comparison with the microstructural appearance of the microtomed thin sections was abandoned because of the opaqueness of the thin sections. Instead, fluorescence microscopy was used to ascertain the integrity of the solvent-delineated microstructures. The fluorescing properties of polycarbonate makes this approach feasible. Figure 22A shows a freshly polished unetched sample. For comparison, the same area etched without glycerine additive is shown in Figure 22B. This fluorescence image reveals the true ribbonlike PC morphology. Ideally, the etch-delineated morphology must preserve the frozen-in pattern that originated during solidification (e.g., welding or molding). Of course, a compromise between image fidelity and visibility is necessary because the PC phase reacts with the solvent in stages. The first stage is swelling and the intentional slight overetching is detectable if the morphologies seen in Figure 22 are carefully compared. This comparison provides sufficient proof that at room temperature, using the recommended etch conditions, no significant artifacts are observed with optical microscopy.

The examples of etching regimes for different polymeric materials show that preventing the introduction of artifacts is very difficult, because the methods, solvents, phases, temperature, and recrystallization process all interact. It is clear, however, that the morphology revealed by the various etching techniques produces a reasonable representation of the process or manufacturing-induced characteristics. Thus etching enhances the morphology's visibility. Obviously, a compromise between morphology fidelity and visibility is necessary.

OPTICAL MICROSCOPE HEATING STAGE

A heating stage consists of three basic parts and is mounted directly on the microscope stage. The components include a specimen carrier, heating chamber, and stage movement mechanism. Modern heating stages are equipped with automated controls which are necessary to reproduce microstructures that are influenced by the heating rate. Heating stages are compatible with most transmitted light optical techniques and adapt to research microscopes. Reflected light can also be used.

Microscopy Techniques for Polymer Characterization / 575

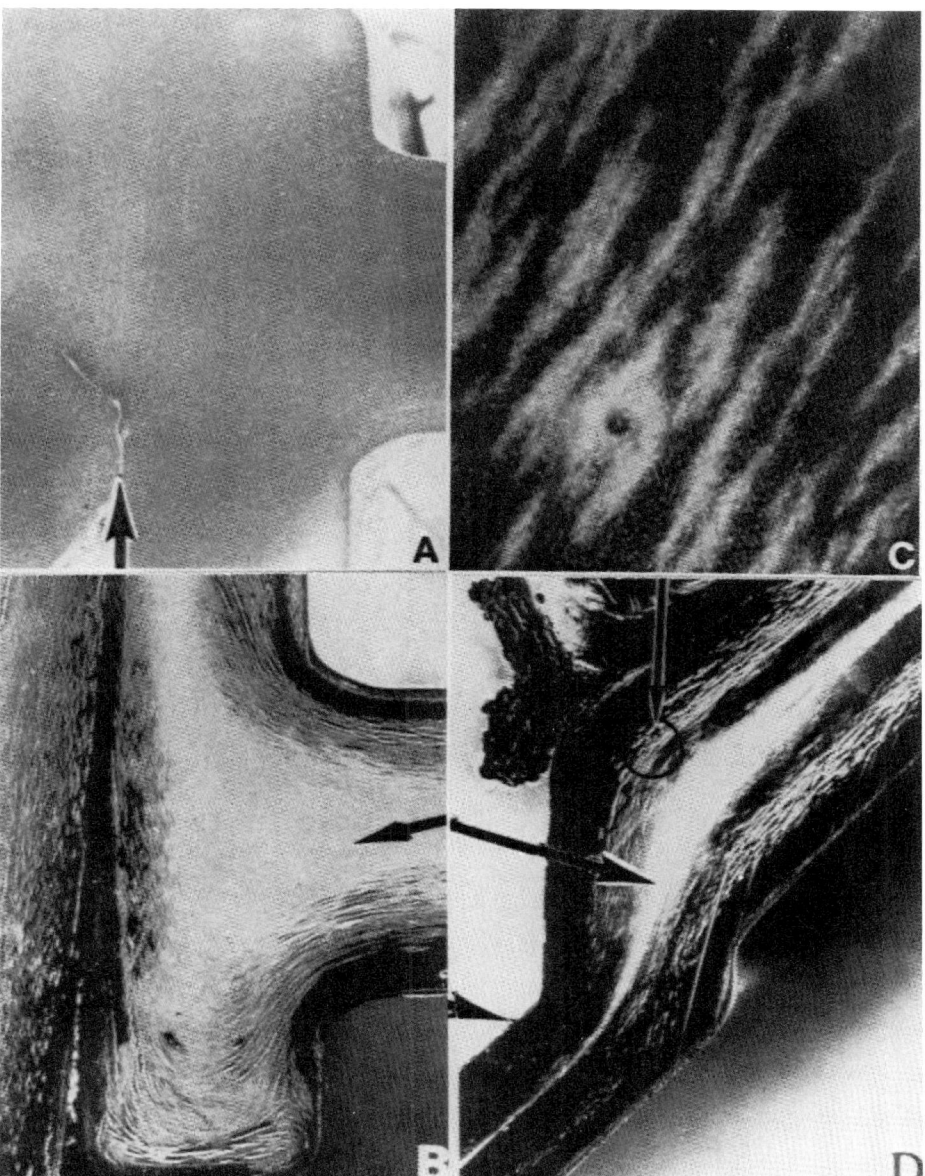

Figure 20 Polycarbonate–polybutylene terephthalate friction-welded interface: A, polished unetched cross section, reflected light (10×); B, etched cross section, reflected light (10×); C, etched cross section, reflected light (500×); and D, Same as C (10×). Arrows point to white PC-rich areas.

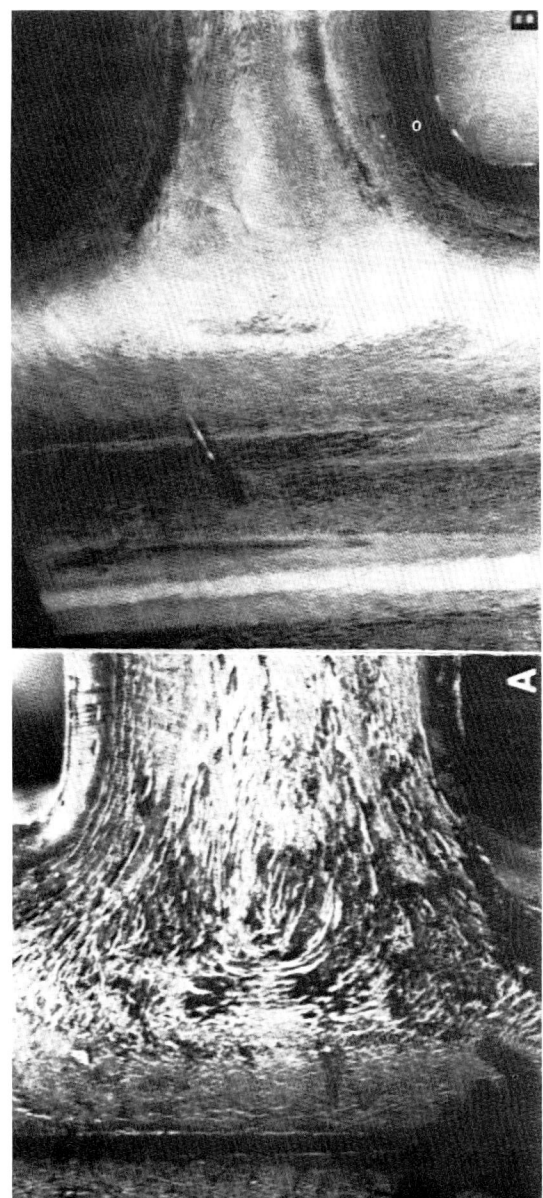

Figure 21 Polycarbonate–polybutylene terephthalate–etched polished cross sections, reflected light (10×): A, coarse PC morphology; and B, finer PC morphology from higher molding shear stress.

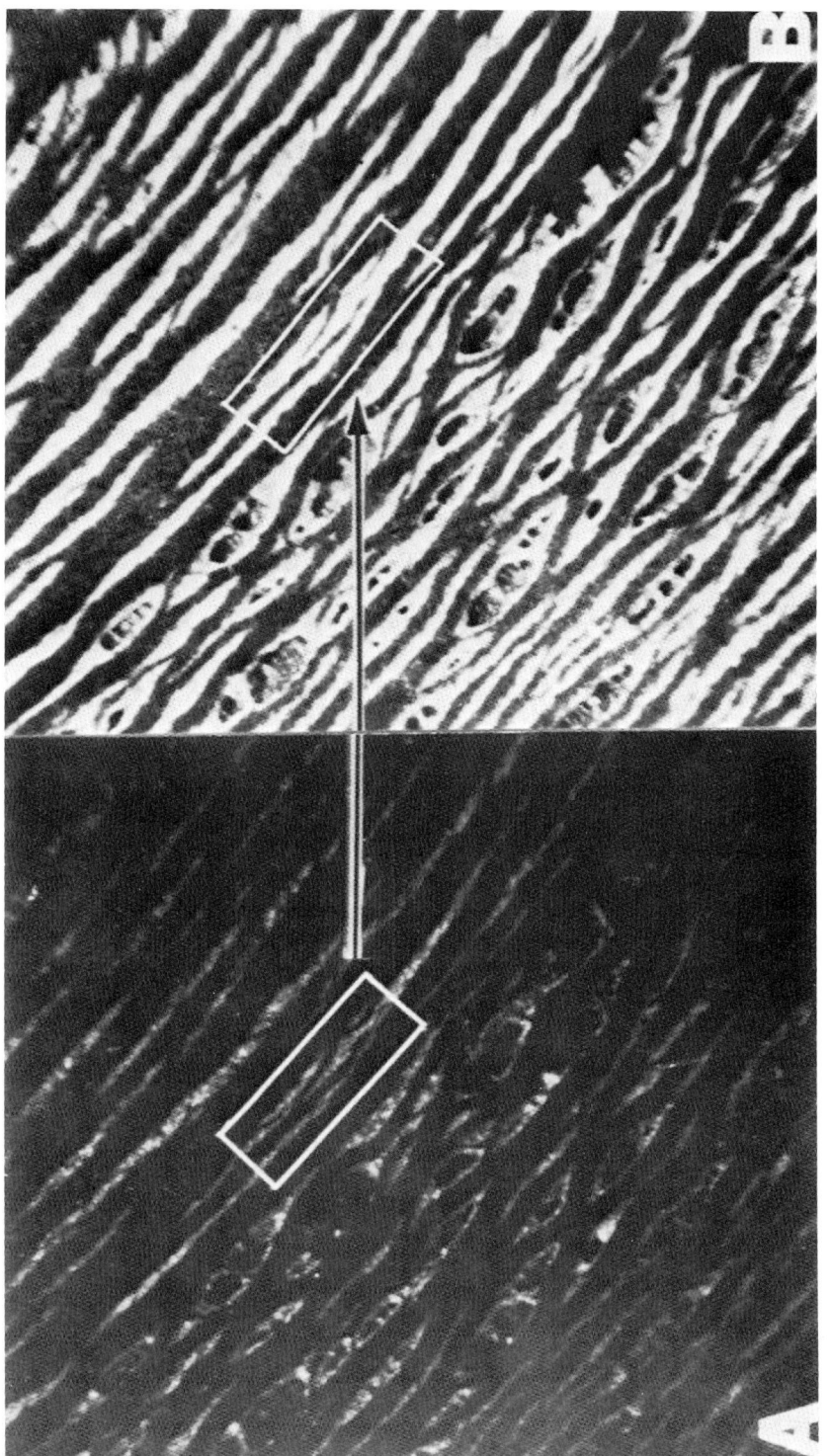

Figure 22 Polycarbonate–polybutylene terephthalate: A, polished cross section, fluorescence image at 250×; and B, etched cross section, reflected light (250×).

Microscopic examination of plastics with a heating stage can provide visual information about temperature-related properties. These include stress relaxation, swelling, softening, flowing, coalescing, melting, sintering, foaming, phase separation, and cooling rate effects on the microstructure. If the sample can not be observed during these changes, related questions can be answered only through long, systematic experimentation. If temperature phenomena can be surmised only after the conclusion of systematic experiments, the direct observations possible with a hot stage are invaluable. This direct observation is even more important to material development and failure analysis. Furthermore, the hot stage use may quickly solve problems related to inadequate blending, particle size, shape, or other morphology development. Another advantage is that the heating chamber prevents contamination of the specimen.

Polycarbonate–Polybutylene Terephthalate

Examination of thin sections of the friction-welded interface with transmitted polarized light disclosed the presence of large-scale birefringence in the weld zone [39, 40]. During welding, the material is distorted by forces that act on the material. Upon cooling, the distortion is preserved and the deformation-induced strain produces optical anisotropy or refractive index differences. It was assumed that the welding, not molding, caused the birefringence and thus with annealing it would be eliminated. The sample's birefringent appearance was typical of an elongated tensile pattern and similar to that obtained in tensile tests. The microtomed thin section of the weld interface was examined in the hot stage with transmitted polarized light. Before annealing, the weld interface birefringence was recorded and is shown in Figure 23A. The birefringence occupies the entire weld seam and indicates that the tension extends along the entire length of the weld interface. This thin section was then annealed. An annealing temperature of 130°C for 30 min did not appreciably affect the birefringence. Only after 2.5 hr at 145°C were noticeable changes observed to indicate stress relaxation. This temperature is slightly above the T_g of the PC phase. The thin section was examined at 30-min intervals and gradual but continuous changes were observed. After 4.5 hr at 145°C the birefringence was hardly noticeable and the experiment was terminated at 5 hr. Figure 23B shows the final result of the annealing process. The tensile components seen in Figure 23A are more obvious and clearly demonstrate the anisotropy. The stress birefringence is apparently the result of shrinkage and is caused partly by the weld-produced structural deformations that were frozen-in during solidification. These frozen-in stresses could be relaxed only at temperatures above the glass transition temperature of polycarbonate. Indeed the result of shrinkage has played an important role as well as the friction welding process. However, the birefringence was mainly caused by the PC–PBT arrangement and the phase's widely different thermal expansion that created high stress in the weld interface. The complete disappearance of birefringence indicates the total relaxation of the weld-induced stresses.

The usefulness of the optical heating stage may be extended beyond conventional applications. Vapor etching and metal shadowing in conjunction with the heating stage can be used to measure thermal expansions. Although the PC and PBT phases expand at highly different rates, the differences can be documented. A microtomed section of the weld interface was vapor etched to delineate the phases. A 3-mm measuring grid was placed on this etched surface and gold was evaporated onto the surface. Upon removing the grid, the gold grid outline remained. The weld interface section, positioned on the heating stage, is shown before heating in Figure 24E. Notice the undistorted gold grid

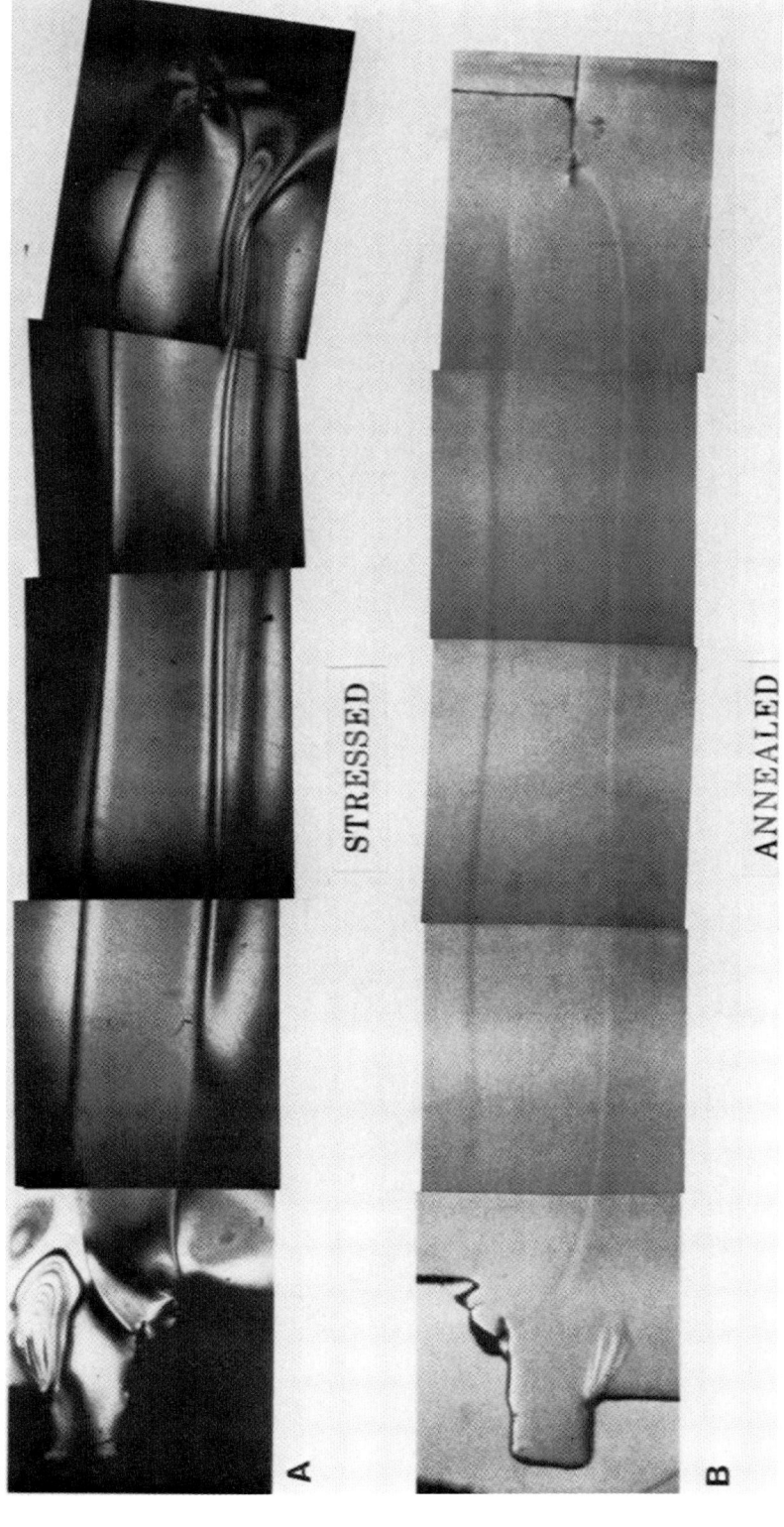

Figure 23 Polycarbonate–polybutylene terephthalate weld interface: A, before annealing, transmitted polarized light (25×); and B, after annealing 5 hr at 145°C, transmitted polarized light (25×).

579

Figure 24 Polycarbonate–polybutylene terephthalate weld interface: A, 180°C (50×); B, 220°C (50×); C, 225°C (50×); D, 225°C (325×); and E, room temperature (50×).

outline. Incidentally, to avoid burning the thin section when evaporating the gold, the specimen must be kept at least 10–12 cm from the gold source. The heat load radiated from the gold source may be hazardous to all plastics but especially to polymers with low melting points.

The sample is then heated in the hot stage and the distances between the grid bars are measured. The distortion of the weld interface upon heating is clearly visible in Figure 24A–C. The arrangement of the phases with heating becomes increasingly noticeable because the apparent crystallinity of the PBT-rich side increases as seen in Figures 24A–C. The crystalline appearance is especially noticeable in Figure 24D. The amorphous appearance of the PC-rich side remains essentially unchanged. On the PBT-rich side, birefringence due to crystallinity of the PBT increases. This could be caused by further crystallization of the initially quenched PBT-rich surface. With increased temperature the distance between the grid lines changes, which allows the calculation of the two thermal expansions. The two phases interact at the weld interface significantly; therefore, precise measurement of the individual phase expansions is not possible. However, the two phases expand with highly different rates and in different directions. The difference in direction is due to unpredictable phase arrangements, the difference in composition, and the further crystallization of the PBT phase. In the temperature range between 180 and 225°C the PBT-rich side expanded 3–4%, while the PC-rich side expanded only 0.2–0.5%. This investigation elucidated and clarified the interaction of the phases during welding and the presence of stress shown in the previous hot-stage experiment.

Polycarbonate–Polybutylene Terephthalate Photoelasticity

Birefringence measurement is a convenient and reliable technique to determine the stress state of polymers. In noncrystalline transparent solid polymers, strain produces optical anisotropy, which is the result of variation in refractive index. The birefringence caused by orientation of the polymer molecules during flow or by quenching is not directly related to the internal stress. The flow-induced birefringence can be separated from the quenching-induced stress birefringence by carefully preparing thin sections from the center of the molded specimen. The internal stresses induced by quenching are relieved to a large degree by repeated cutting.

The PC–PBT was joined with linear friction welding and thus was exposed to deforming forces. The strain associated with the welding operation distorts the material at and near the weld interface. The distortion is permanently frozen-in and prevents the elastically deformed portions from returning to the predistortion state. The area from which the thin section was microtomed parallel to the weld interface was randomly selected.

The strain patterns arising from the deformation were visible with transmitted light illumination along the entire weld interface. Figure 25A shows the weld cross section with the interface oriented at 45° to the vibration planes of the crossed polars. The pattern varies with position along the interface and has such a complexity that interpretation, in spite of the excellent resolution, is very difficult. It was postulated that not only the intense but also the less noticeable strain patterns resulted from the welding-caused deformation.

For interpretation purposes, the important elements of these patterns are the dark lines, called isoclines, along which the strain is oriented in the vibration direction of one of the polars. The lines or areas of equal color are the isochromes, which represent lines of equal strain. The sample was rotated and the shift of the isoclines noted to trace approximately

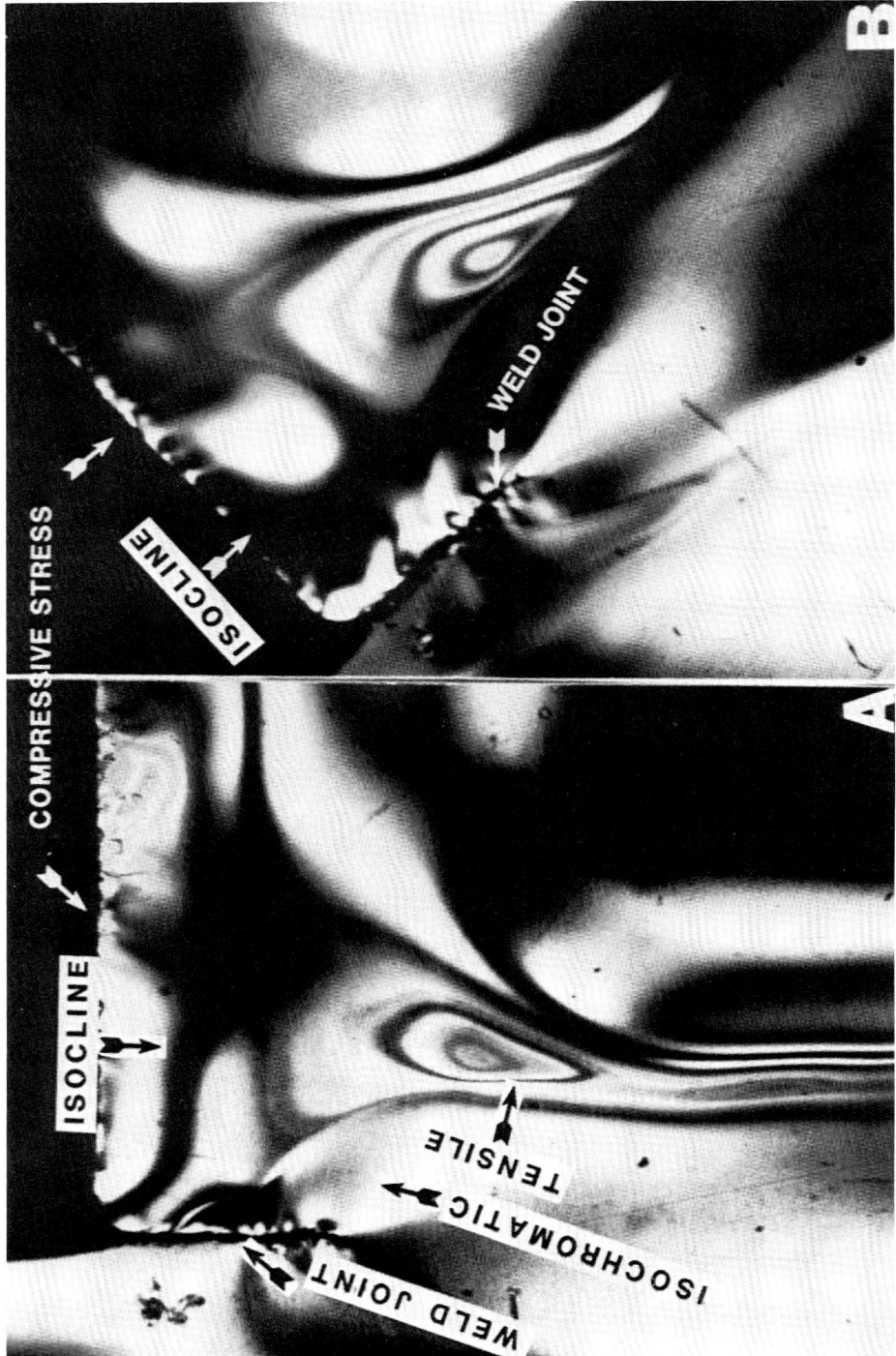

Figure 25 Polycarbonate–polybutylene terephthalate weld interface thin section, polarized transmitted light: A, at 45°; and B, specimen rotated 45° (100×).

these complex strain orientations. Figure 25B shows the sample in the 45° rotated position. In this position, the isocline separation shows the approximate areas that are under predominant compressive stress and those which are principally under tensile load. The intensity of the tensile and compressive components in the weld interface was assumed to be the direct result of welding. The most intense strain patterns were at both ends of the weld interface (only one end is shown) and the localized stress levels in both regions are between 21 and 34 MPa (3000 and 5000 psi) depending on the position. The stress near the weld midpoint (not shown) was in the 21 MPa range. Because stress intensities elsewhere in the molded part have not been seen to equal or even approach these high stress levels, the assumption was logical. However, based on the previous experiment, which showed the PC–PBT arrangement and the thermal expansion rates of the phases in the weld, reevaluation was necessary. The effect of compressive clamping pressure during welding is clearly visible in Figure 25B where the compressed region is partially relaxed near and at the fractured weld seam.

Polyurethane

Thin sections from reaction injection-molded (RIM) polyurethane disks were prepared using abrasive thinning and cryomicrotomy. In the RIM process two streams of the reactive components are mixed rapidly before injection into the mold. The components are typically reactive isocyanate prepolymer (with birefringent properties) and polyol. Hence the degree of mixing can be revealed using birefringence. The cryomicrotomed thin sections are valuable for evaluating the grinding- and polishing-induced strains on abrasively thinned samples. Sample disks were selected at random from each of two different lots. One lot contained used and failed parts. The other lot contained used but not failed parts. The failed part was microtomed. The comparison part from the lot containing no failures was mechanically thinned. Both samples were examined in transmitted polarized light to detect and compare the birefringent strain pattern and the flow lines that originate during molding. The dense, coarse-textured mold lines cover the entire cross section in Figure 26A, which indicates poor molding conditions. Some minor strain patterns are visible near the fastener hole and next to the fracture. The strain pattern, shown in Figure 26B, may have been introduced during the mechanical thinning. With the exception of some isolated and fine flow lines the section has no other distinct features. Comparing the section's morphologies, the difference is substantial. Identifying the origin of these differences with respect to the flow lines is obvious, but the origin of the birefringence is not certain. The mechanically thinned sample was annealed at 126°C for 15 hr to separate the mechanically induced strain from the molded-in strain. The result appeared inconclusive, and due to the difficult and time-consuming sample preparation, further analysis was pursued with optical hotstage microscopy.

Polyurethane Examination With the Heating Stage

During failure examination, hotstage microscopy was necessary to study the morphology of the RIM polyurethane disks seen in Figure 26. Cryomicrotomed uniform thin sections were used to examine the birefringence around the fastener holes, which may have been caused by stress. The polyurethane was examined in the hotstage using polarized light. Because the refractive index changes with thickness, the birefringence could not be conclusively related to stress. It was postulated that stress from microtomy was responsible for the observed birefringence. To confirm this postulation, the sample was annealed

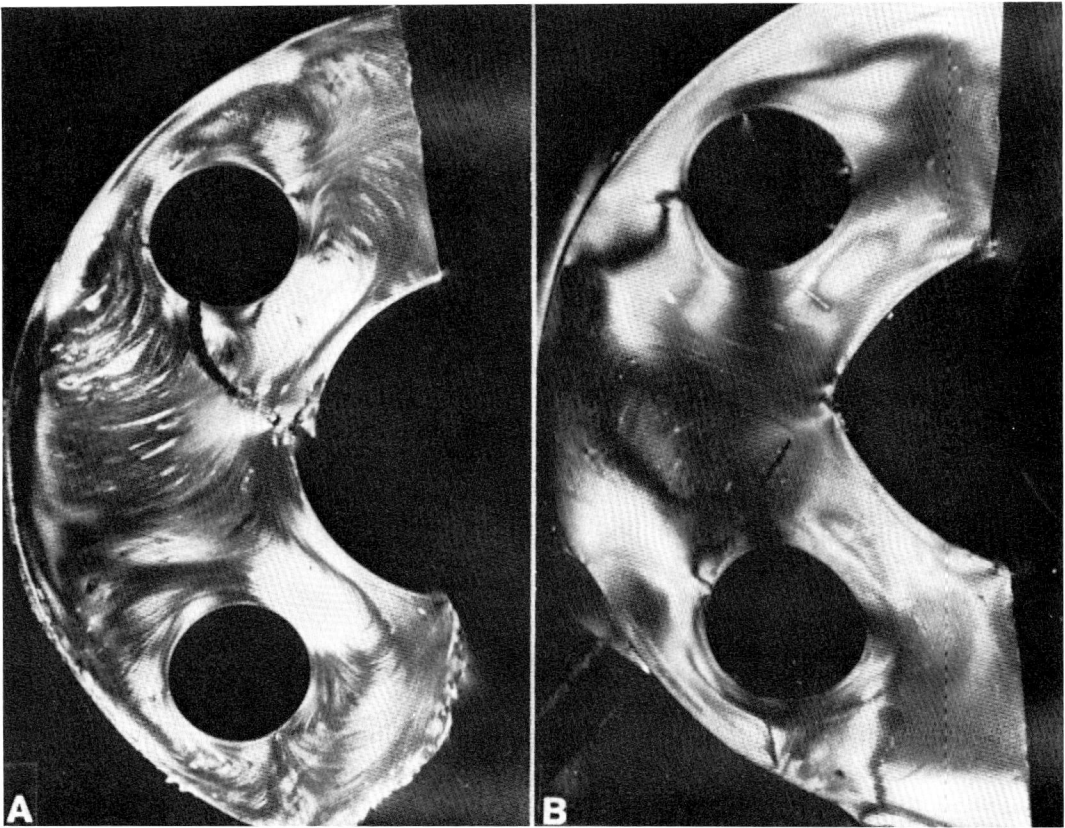

Figure 26 Polyurethane, thin sections showing birefringence with transmitted polarized light at 1.5×: A, poor molding; and B, mechanical thinning.

at 126°C for 16 consecutive hours to relax the stresses. The polyurethane sample position was fixed during the entire examination to avoid any thickness related variations which could cause changes in birefringence. The birefringence image was photographed before annealing; Figure 27A shows the pattern. The sample was photographed again after annealing; the change is seen in Figure 27B. This minor birefringence change indicates that no excess stress was relieved and the large-scale birefringence is the result of poor initial mixing.

The birefringence is the result of molecular orientation. If physical separation of the components occurs in the mold, because of insufficient premixing, the molecular arrangement of the individual noncrosslinked components may create stress in the material. If stress had been induced during microtomy, the annealing would relieve the stress and a substantial change in birefringence would have been observed. Any heterogeneity from improper mixing, however, is permanently preserved by chemical crosslinking and will not respond to annealing, as was shown with the annealing experiment.

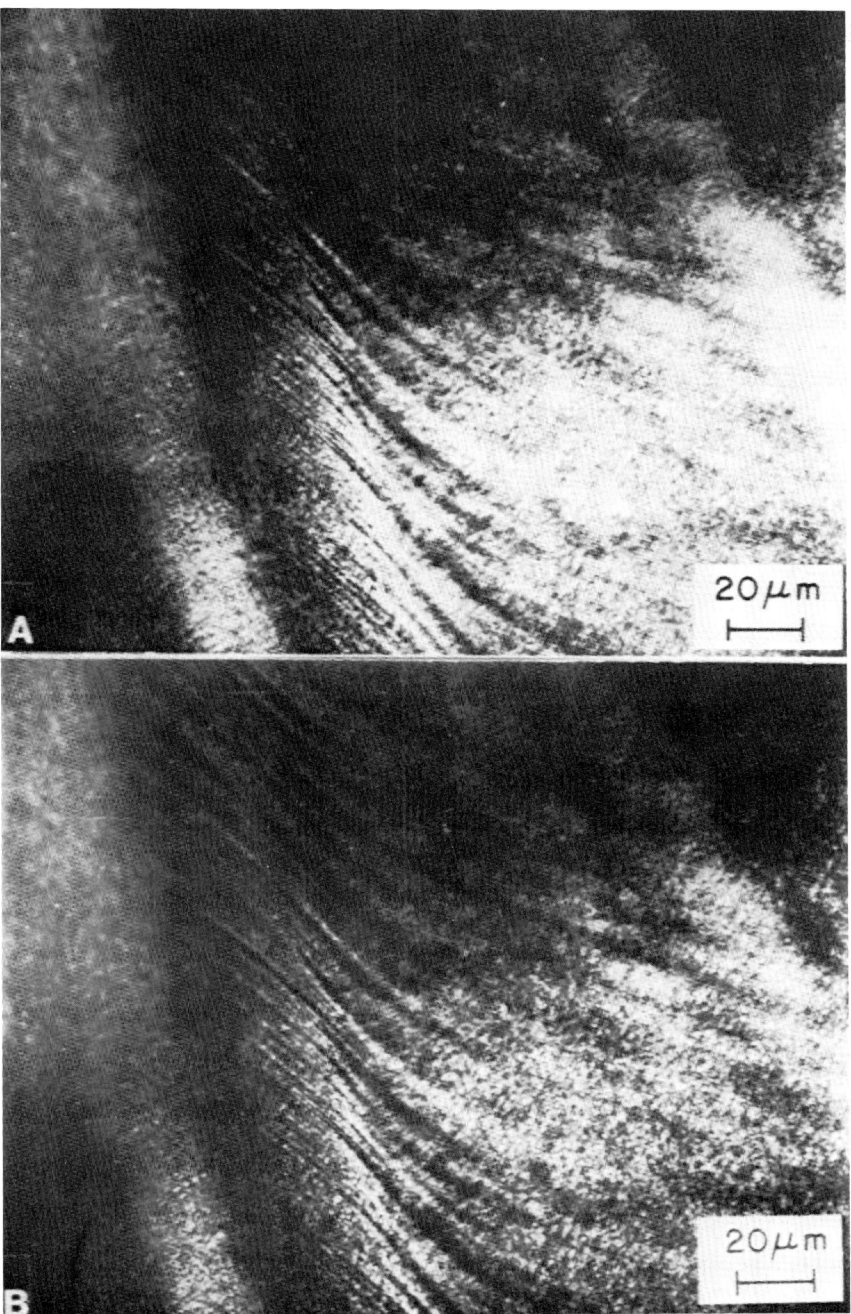

Figure 27 Polyurethane polished thin sections, polarized transmitted light: A, before annealing (500×); and B, after annealing at 260°F for 16 hr (500×).

Polycarbonate–Polyarylate Blend

Fractures of transparent materials can be examined with transmitted light without replication or preparation of thin sections. This type of observation complements the incident light investigations. It is especially useful for brittle fractures that also exhibit some ductile appearance. Perhaps the best way to characterize this type of fracture is to borrow the term "quasi-cleavage" from metallurgy. Figure 28 shows the fracture surfaces with two types of illumination. The reflected light image, Figure 28A, does not show all the details of the propagating crack front in focus. However, the transmitted light image, Figure 28B, clearly shows the plane of the fracture. This observation clarifies and reinforces the evaluation that the distinct fracture pattern is the result of the interaction between the simultaneously propagating fracture front and an elastic shock wave in the material.

Another fracture area using the same approach is seen in Figure 29. Again, the incident light image is not fully satisfactory to analyze the fracture. With Figure 29A, the details of the rolling tear ridge marks (the plastic deformation component) are not clear and the river markings are out of focus. The image is poor to marginal for properly classifying the fracture. Only the rolling tear markings show indications of an impact-caused cleavage fracture. The fracture characteristics needed to identify the fracture mode, however, become indisputable with the transmitted light image seen in Figure 29B. All the fine details of the fracture are in sharp focus at this relatively high magnification. The river markings, the rolling tear, and the brittle microcracks provide unmistakable evidence of a brittle fracture.

Wood Flour–Filled Copolymer

Examination of this extruded 10% wood flour–filled polypropylene–polyethylene copolymer began with the stereomicroscope [42]. The great depth of field of the stereomicroscope benefits the examination of rough fracture face topographies. Fibril wetting, orientation, shape, and wood species are important considerations [43]. These variables can all significantly affect the physical properties of the composite. Of course, these parameters cannot be fully examined with the limited resolution of the stereomicroscope, but information on fibril wetting, alignment, distribution, and geometry are extractable. A large agglomerate and adjacent loosely attached fibers are shown in Figure 30A. In this low-magnification view, the smaller agglomerates are difficult to see. These agglomerates, although out of focus, still convey valuable information about the fibril distribution. The fibril distribution is important because it is the result of process-related variables, especially mixing. A loosely attached wood fiber is seen in Figure 30B. In spite of the stereomicroscope's limited resolution, observations about wetting were related to the composite's physical properties. This partially wet fiber can be easily pulled out of the polymer matrix and offers no beneficial reinforcement to the composite. This observation (Fig. 30B) suggested a wetting problem, which was later confirmed. The complex wetting phenomenon is partially explained by rheology, the extrusion pressure, and the wood filler size reduction process [44].

Cryomicrotomed thin sections of the extruded and 10% wood flour–filled polypropylene and 12% polyethylene copolymer were examined with transmitted light microscopy. Polarized light was essential to distinguish even the larger details of the morphology. The transmitted light image of the copolymer thin section, seen in Figure 31A, does not reveal the details of the morphology. To a large extent, the wood filler blends with the background and is virtually undistinguishable from other features. The large dark features

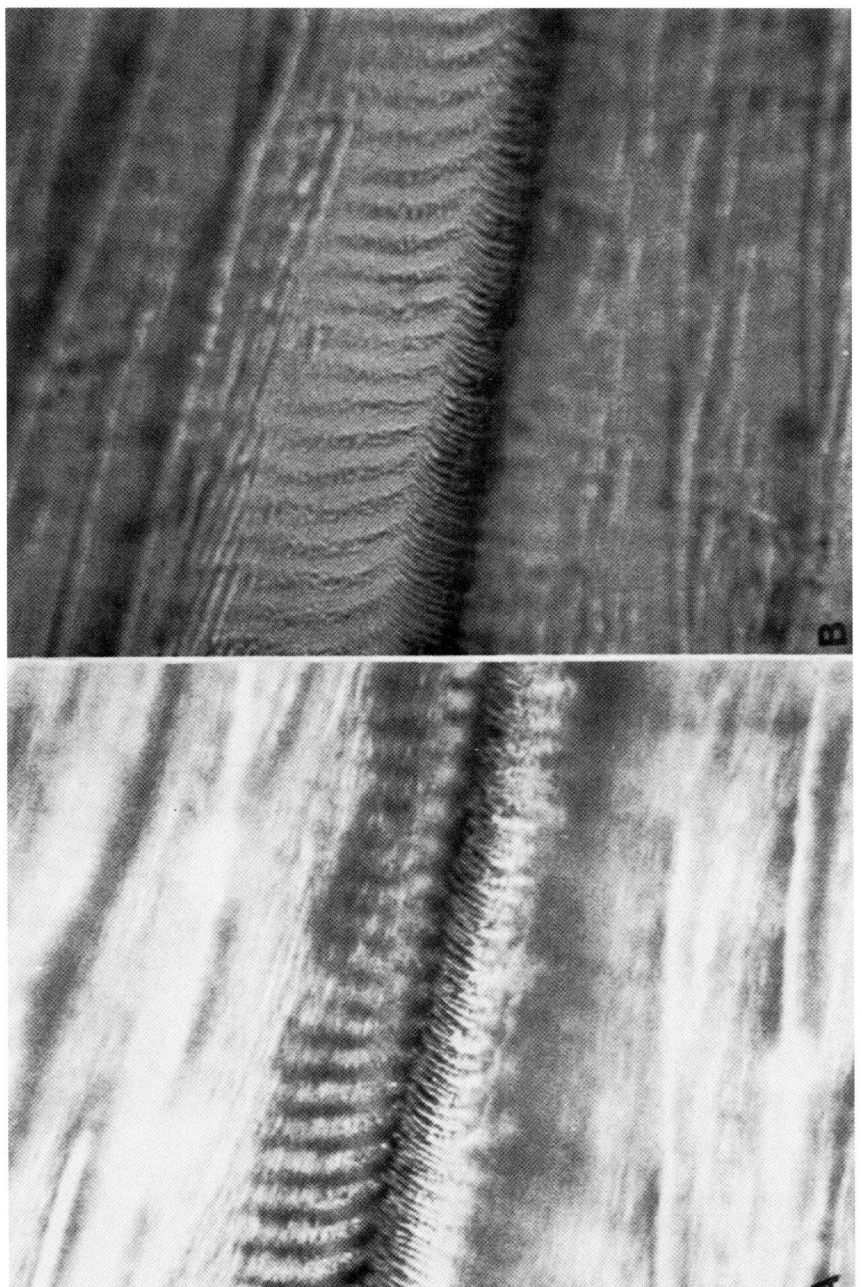

Figure 28 Polycarbonate–polyarylate blend: A, reflected polarized light (350×); and B, transmitted polarized light (350×).

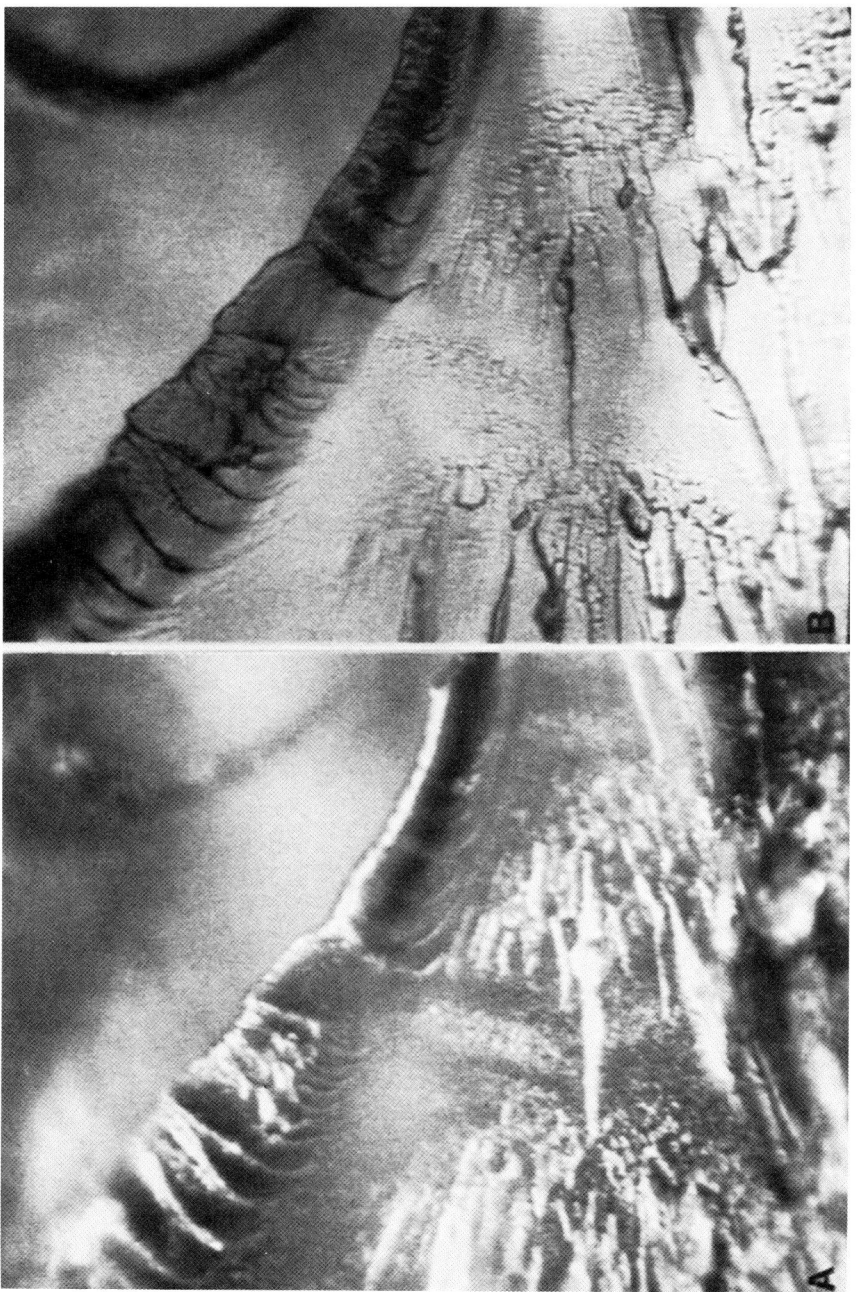

Figure 29 Polycarbonate–polyarylate blend: A, reflected polarized light (350×); and B, transmitted polarized light (350×).

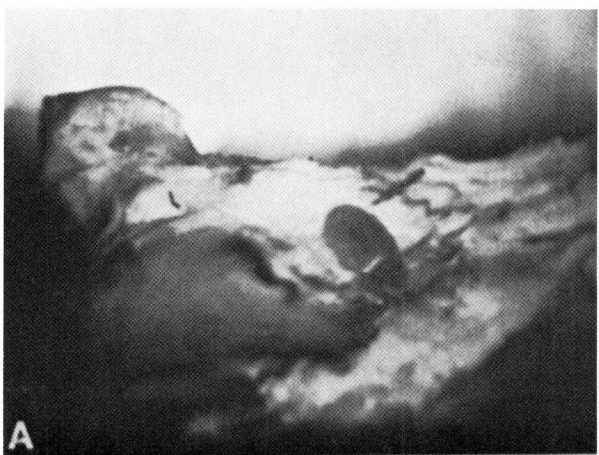

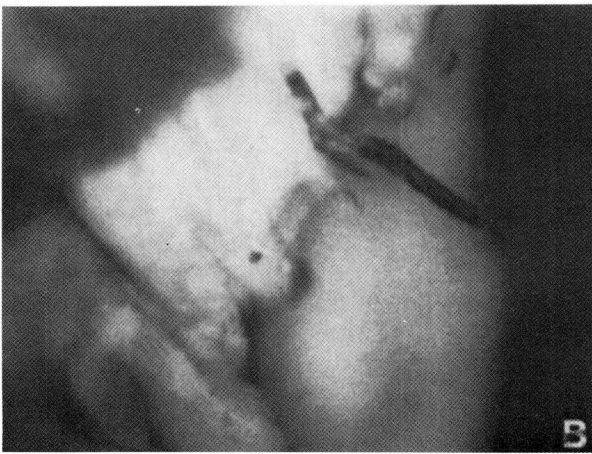

Figure 30 Wood flour–filled copolymer in reflected polarized light showing wetting problem: A, 25×; and B, 120×.

could be mistaken for the wood flour but they are intentionally introduced artifacts to identify specific sample areas. Figure 31B shows the wood flour fibrils with slightly uncrossed polars as bright features against the dark background. The three large wood flour agglomerates and a large rectangular fiber are the most conspicuous. The various colors in the microscope enhance many details that are lost with the black and white image. Further enhancement of the wood flour visibility is possible with the use of a quarter-wave plate as shown in Figure 31C. The fine fiber details are more visible, but the image still lacks numerous details that can be seen with the aid of color contrast.

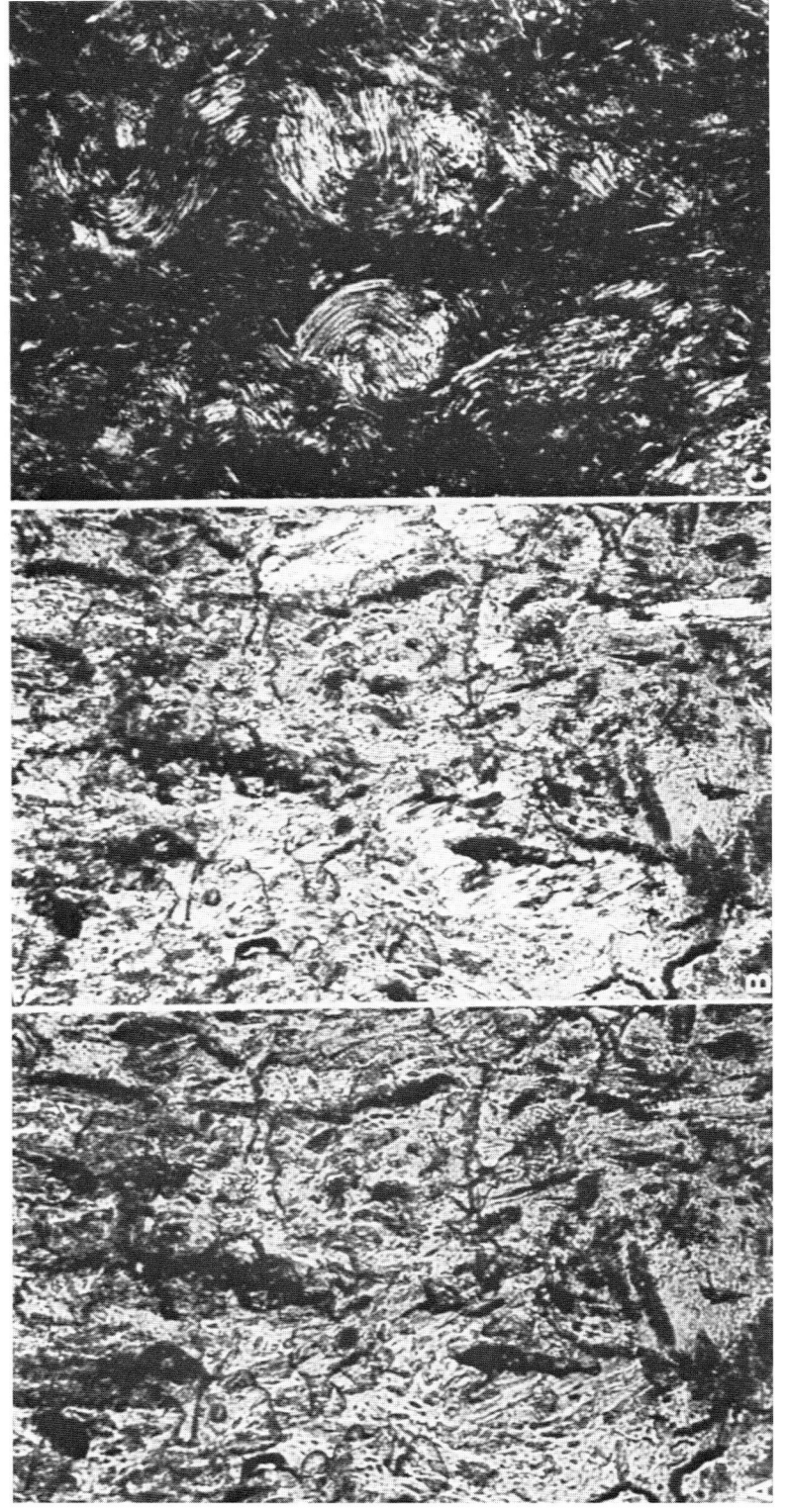

Figure 31 Wood flour–filled copolymer: A, microtomed thin section in polarized transmitted light (200×); B, same field with partial uncrossed polarized light (200×); and C, microtomed thin section, polarized light, quarter-wave plate (200×).

SCANNING ELECTRON MICROSCOPY

In polymerography, as in metallography, the scanning electron microscope (SEM) extends the limited range of resolution of the light optical microscope. The high resolution, the enormous depth of field, and the relatively simple specimen preparation make the instrument's routine use desirable. Since the late 1970s the SEM has become a routine tool in the laboratory.

The SEM image is formed by scanning a finely focused electron beam in a raster mode across the sample surface. The electron beam and specimen interaction produce secondary electron emission and electron-induced x-ray fluorescence. The instrument is highly sensitive to surface morphology because low-energy electrons escape only from the upper atomic layers. These escaping electrons are detected and used in the image-forming process. Because of the high surface sensitivity of the electrons, the spatial resolution is excellent. The average SEMs routinely attain a lateral spatial resolution of less than 100 Å. Recent instrument advances have improved the resolution capability of the instrument [45].

For SEMs without low-voltage scanning capability, the most important requirement is that the sample be conductive, since nonconductive samples will collect a surface charge that will distort the image. Coating the nonconductive samples with an evaporated or sputtered layer of conductive material will conduct the surface charges to the ground. Since gold is a good conductor and an excellent emitter of escaping electrons, it is routinely used to coat samples. Imaging with low-voltage electrons (1.0–1.5 keV) is not without problems; perhaps the biggest disadvantage is the reduced beam brightness and the limited resolution.

In studying wood flour–filled copolymer, the wood fibrils in the copolymer were earlier examined with the stereomicroscope. The limited resolution of the stereomicroscope did not permit close-up observation. Since the lack of fibril wetting by the matrix can significantly affect the ultimate properties of this material, it was necessary to pursue the investigation with the SEM.

After evaporating gold on the previously examined fracture surface, the partially pulled-out fibril was examined. Figure 32 shows the fibril's intimate details. Except for a small, loosely attached fragment of the matrix the entire fibril appears to be uncoated. The lack of wetting extends to the smallest surface details of the fibril as indicated by the well-preserved fine-line surface markings. Even at the branching or in the fracture vicinity, the lack of matrix infiltration—and its conspicuous absence—is evident. The improved resolution provided by the SEM was invaluable for the final analysis.

TRANSMISSION ELECTRON MICROSCOPY

The development of transmission electron microscopy (TEM) has accelerated with astonishing rapidity since the late 1970s. The conventional TEM technique has been greatly expanded and transformed into an all-around analytical tool. This gain was made possible with the development of numerous analytical techniques that can detect and quantify the various signals that a high-voltage electron beam yields after striking a thin, transparent specimen [46]. The extent of the analytical capability includes small-particle characterization, monolayer-type segregation effects in metals, microscopic identification of phases, and microchemical analysis. These features greatly enhance the general understanding of

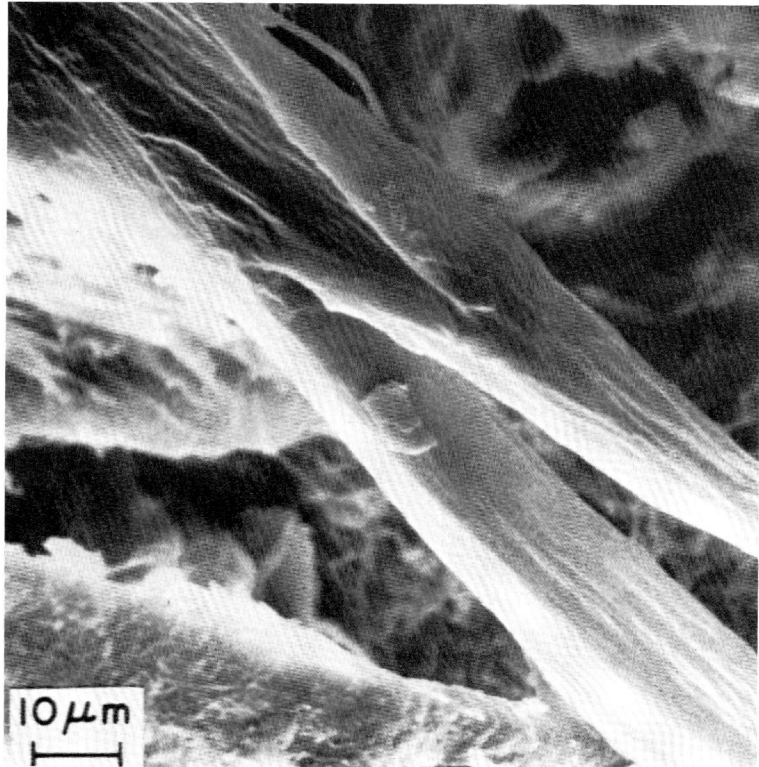

Figure 32 Wood flour fibril, SEM (1200×).

the relationships among structure, property, and chemistry. In addition, the examination of beam-sensitive polymeric materials is vastly improved with modern TEMs. The added ability of the instrument has already had a profound effect on material science.

The improved analytical capability of TEMs with electron-transparent samples comes from scanning the electron beam as opposed to the stationary beam of the conventional TEM. The specimen is illuminated with a finely focused scanning electron beam, similar to the approach used in SEM. This is an advantage for polymeric materials that may degrade under the stationary beam. They can be examined for sufficiently long periods with the scanning mode, thus permitting full use of TEM-attached microanalytical tools. In addition, all the conventional SEM imaging techniques are available for use.

Extruded Wood Flour–Filled Copolymer With TEM

To the original formulation of 10% wood flour and 12% polyethylene, in a polypropylene base, 8% EPDM (terpolymer of ethylene, propylene, and a diene monomer)-type rubber was added to improve the flexibility of the material. The broad class of EPDM is noncrystalline; however, slightly crystalline EPDM rubbers are also available. Thermoviscoelasticity of these materials is very complex over a wide temperature range, indicating that these materials are not completely amorphous [47].

The dispersion of the rubber particles was of interest. Since the resolving power of the light optical microscope is insufficient to resolve the rubber particles, TEM was chosen. With cryomicrotomy, normal to the extruded surfaces, thin sections were prepared for the examination. The thin sections were stained with osmium tetroxide vapor to alter the density of the rubber particles and provide the necessary contrast for microscopy. This selective deposition of heavy atoms on the material is an obvious method to reveal structure that cannot be observed without staining. Figure 33A shows the unstained section. Except for the wood flour fibers and particles, the image contains no other recognizable features. The dispersion of stained rubber particles is shown in Figure 33B. No pattern of segregation is observable. The image reveals that the material contains a moderate concentration of spherical rubber particles with a fairly uniform dispersion. The particles' shape indicates that the extruded sheet is isotropic in the plane of the microtomed surface.

Rubber Latex With Polystyrene

A molded rubber–polystyrene mixture was cryomicrotomed for transmitted light phase contrast examination of the dispersed phase. Previously, both the polarized transmitted light and the TEM failed to produce useful information. It was postulated that polystyrene was dispersed in the rubber latex matrix. These materials, with nearly the same refractive indices, can be visually separated using phase contrast. Figure 34A shows the transmitted polarized light image, which is totally lacking any discernible details. The slight contrast variations observed are due to the uneven sample thickness and possible contamination. For comparison, Figure 34B, the phase contrast image, clearly shows the dispersed polystyrene phase. This is a surprising observation since latex particles are below the resolution of the optical microscope; thus these are islands of coagulated latex. A wide variety of particle sizes was observed in several areas. The particle size is important because it can be related to the material's strength.

IMAGE ENHANCEMENT

The light optical microscope is mainly limited by lateral resolution, and in spite of the numerous and desirable attributes over other imaging devices, its use is restricted in many instances. Around 1970, a video device was first coupled to the microscope, promising new directions were opened. The capability of the microscope was enormously enhanced and many of the previously limited attributes were expanded. Today, enhancing the poor contrast by electronically increasing the gain and suppressing stray background signals or intensifying weak images into clearly visible pictures is routine. The improvements in electronic imaging methods have advanced the applications of the optical microscope [48].

Paint Contaminant Study

The paint examination was aided by image enhancement because the thickness of microtomed sections was not suitable for the high optical magnifications needed. Numerous attempts to produce sections 0.5 μm thick were frustrated by the porous and loose interconnection of the paint film and by the interaction of the microtome knife with the hard and well embedded paint contaminants. The contaminant particles were of interest and the object was to determine if these contaminants were only on the surface or throughout the bulk; Figure 35B shows the overall appearance of the defect in reflected light. The collapsed craters are approximately 10–15 μm deep and partially covered by the deformed paint film.

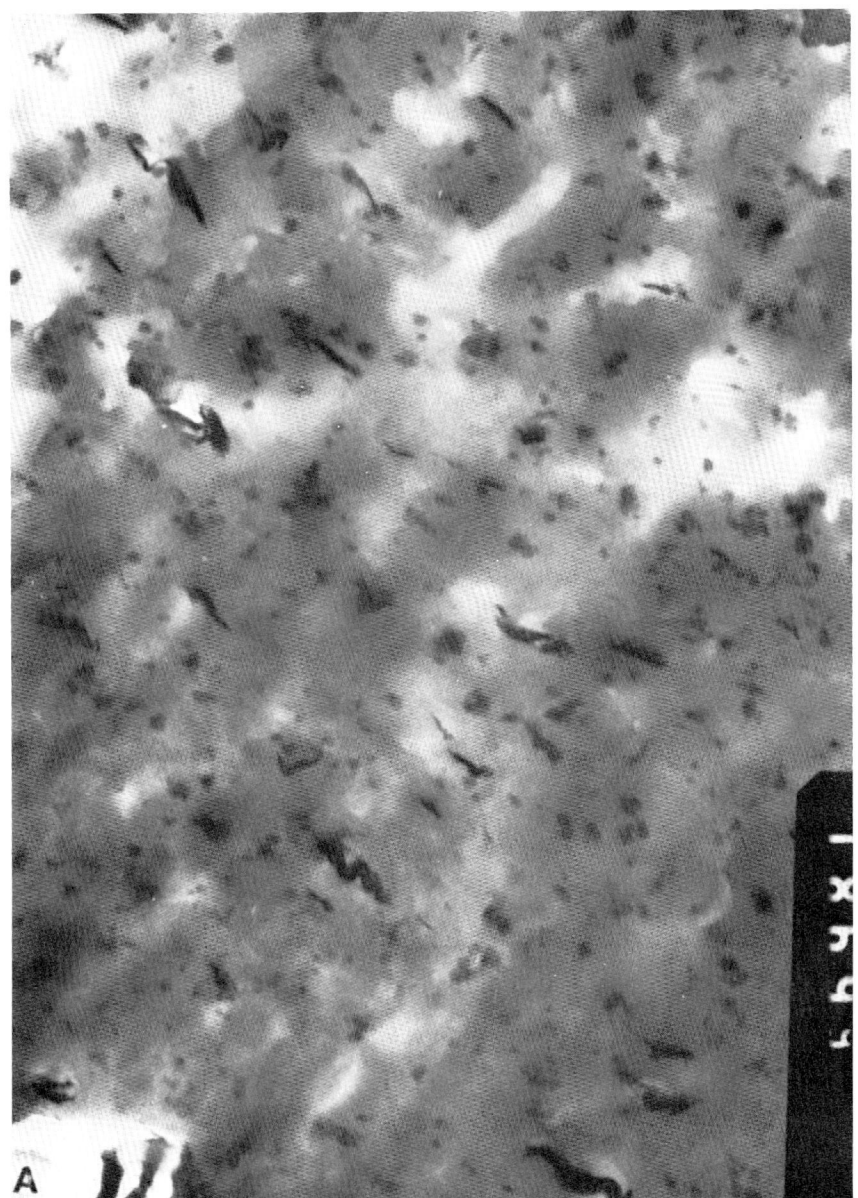

Figure 33 Wood flour–filled copolymer with EPDM additives, TEM: A, before staining (5000×); and B, after osmium tetroxide stain (25,000×).

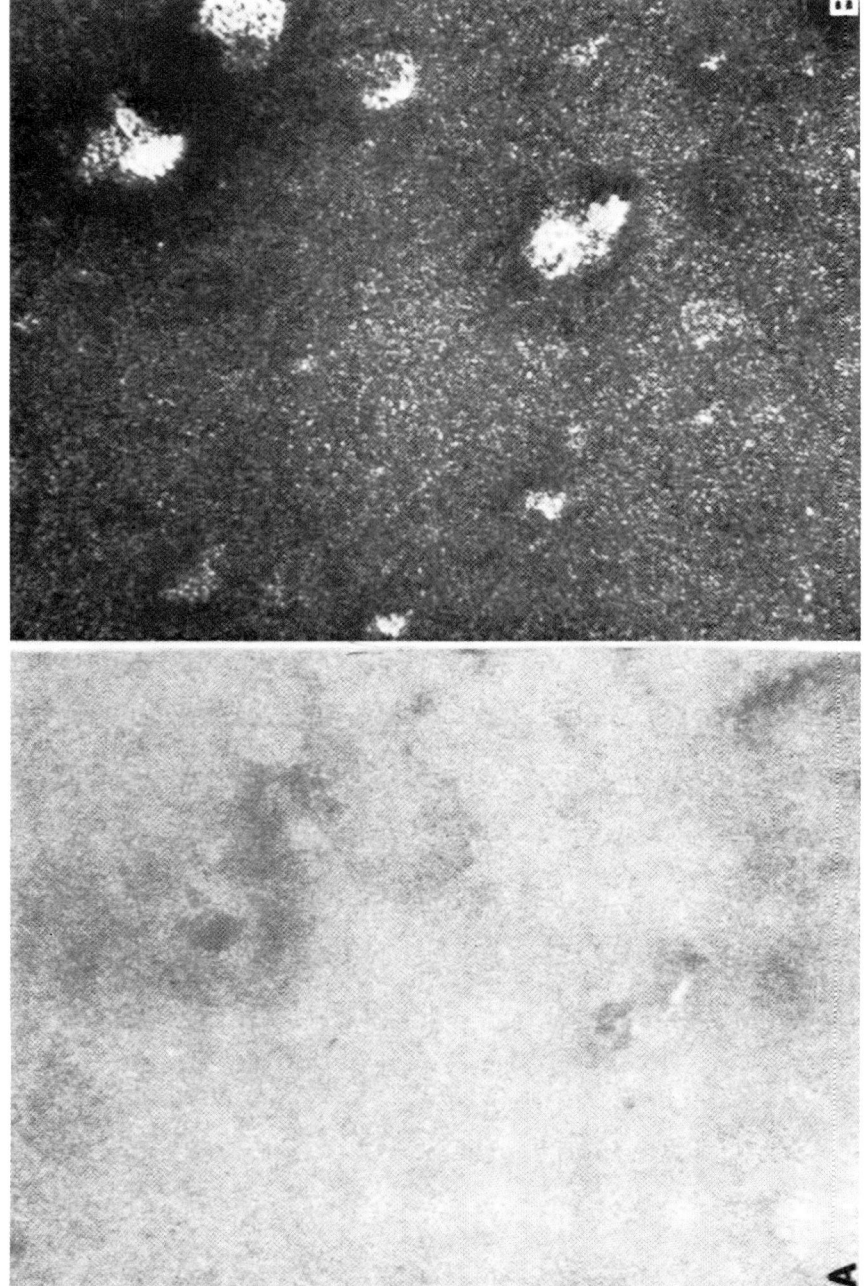

Figure 34 Polystyrene latex rubber: A, transmitted polarized light (1500×); and B, transmitted light phase contrast (1500×).

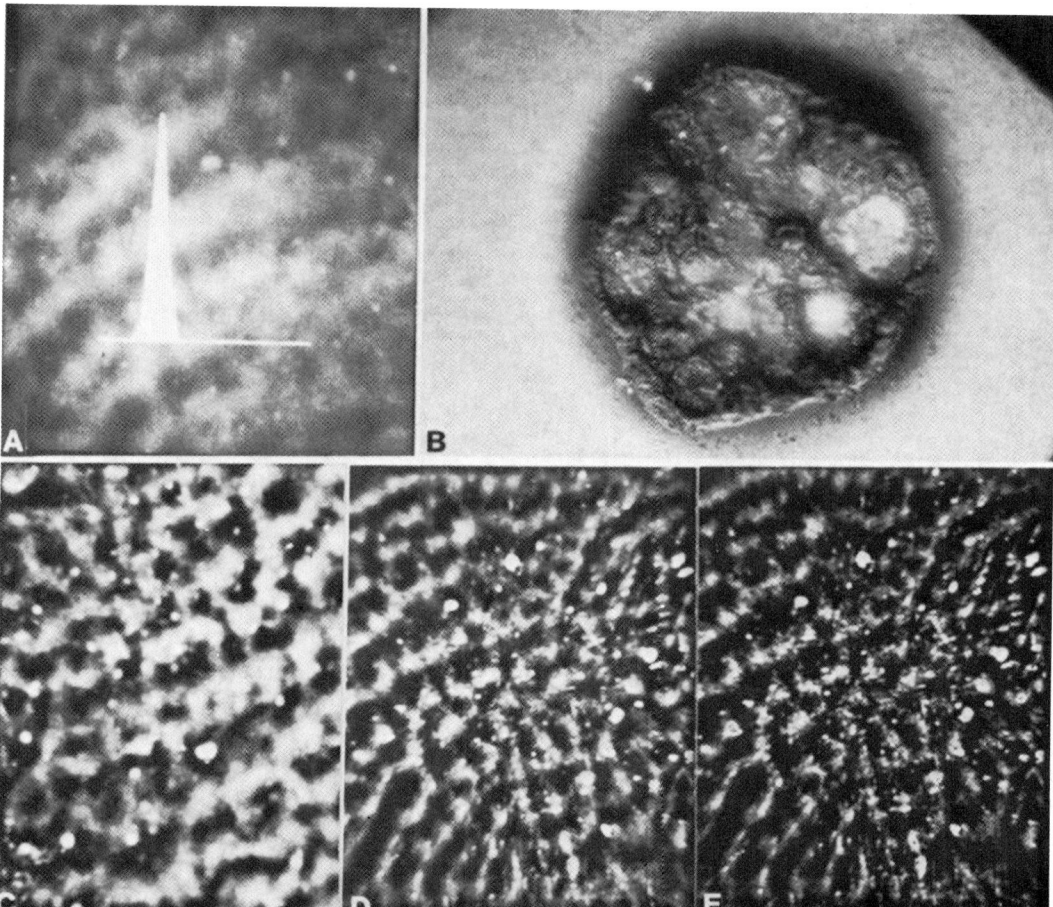

Figure 35 Paint defect: A, transmitted light thin film with histogram of gray levels (1200×); B, reflected light (100×); C, first enhancement; D, second enhancement; and E, third enhancement (1200×).

The first transmitted light optical image of the thin section, seen in Figure 35A, gives a vague and uncertain indication of the paint film matrix and the contaminants. The superimposed histogram indicates the gray values measured with the Zeiss Contron image processor. The image details of the contaminants are insufficient to study the particles' shape or size. This unclear image was digitized and electronically enhanced with the image processor. With Figures 35C–E, the step-by-step results of the three-stage enhancements are shown. Notice the continuously increasing clarity and number of the visible contaminants. In the final stage, the enhanced image details are clear and sufficient for analysis. The final enhanced image must be carefully interpreted because only the most highly reflecting facets of the particles have been preserved by the enhancement process. Thus the size and shape may not truly represent the actual particles.

Polycarbonate and Polymethyl Methacrylate

The numerous industrial applications of these materials are well known for their unique properties and characteristics. Polymethyl methacrylate (PMMA) is also known as Plexiglas. The stereomicroscope was fitted with an optical attachment that consisted of mirrors and a beam combiner to study the effect of fatigue on the material's appearance. Photographs with a quasi-three-dimensional effect were obtained.

Longitudinal cross sections were cut and polished from round test bars that were fatigued. The striations and crazes are important for assessing failures of these glassy materials. The interpretation is relatively simple, because the damage is directly related to the amount of fatigue. Fatigue-produced crazes and other damage in the form of fine matrix lines and surface roughness were examined with the stereomicroscope. These crazes and fine matrix lines relate directly to fatigue properties. Furthermore, the fine band size is inversely related to the square of the yield strength [49–51].

If the fatigue properties of PC and PMMA are compared, the result is surprising, because the modulus of elasticity of polycarbonate and PMMA are the same. With identical test conditions (i.e., load, temperature, frequency of cycles) the polycarbonate will absorb approximately ten times more fatigue cycles than the PMMA before similar visible signs of fatigue appear. Figure 36A shows crazes and fine matrix lines for polycarbonate that has absorbed a much higher amount of fatigue than the rougher appearing

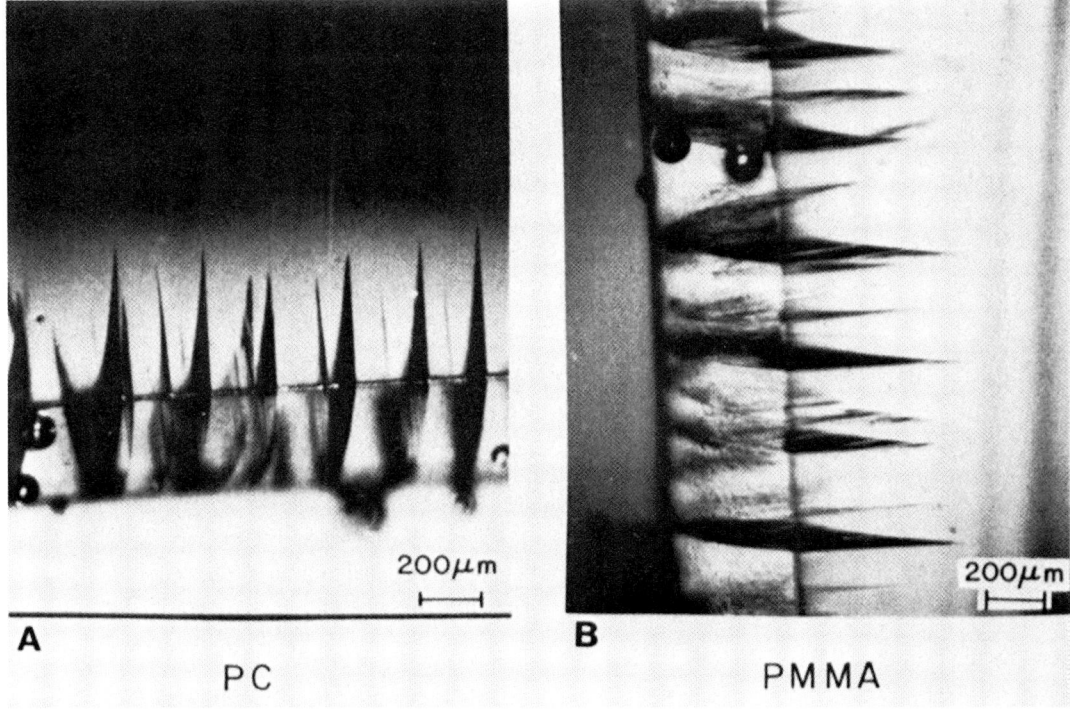

Figure 36 A, Polycarbonate and B, polymethyl methacrylate showing fatigue crazes and fine matrix lines with transmitted polarized light (50×).

Figure 37 Mechanical straining stage for SEM.

PMMA sample seen in Figure 36B. At this low magnification, only the large crazes are distinct in both figures. However, the difference in surface roughness is substantial. The ultimate fatigue life of these materials favors the polycarbonate.

MECHANICAL STRAINING STAGE FITTED TO THE SCANNING ELECTRON MICROSCOPE

The test methods for polymers also include mechanical testing, and the most direct approach is to observe the physical processes of deformation with the SEM. To follow the changes either in a dynamic or static mode is very desirable. For this purpose a mechanically operated three-point straining stage was designed and fitted to the SEM stage. The stage design is relatively complex due to the various movement-related requirements. The sample and electron beam interaction must be accommodated during the tilt or lateral motions, and the incrementally applied load must be drift-free at all load levels. The space restriction of the sample chamber is the major limiting factor. The load capacity of the mechanical straining stage (MSS) is a function of the space restriction and is 360 kg. Figure 37 shows the MSS assembled on the microscope stage with a sample in strained position. The sample size is approximately 12 mm by 50 mm with a thickness of 3 mm. The maximum tilt angle from the horizontal is 13° and the axial sample translation in the X and Y directions accommodates the freedom of movement of the microscope stage. All materials of construction are nonmagnetic stainless steel to avoid possible field effects.

From a polyethylene sheet containing 15 vol % glass fibers, an appropriate-sized sample was cut to fit the MSS stage. The sample surface exposed to the electron beam was covered with a sputtered gold layer. Occasionally, and especially if adjacent edges are to be examined, these surfaces must also receive conductive coating to eliminate charging at the edges. The direct observation of the process leading to the nucleation and propagation of cracks was of interest. In addition, the gradual effect of the incrementally applied but continuously increasing load was also of interest. Figure 38A is a composite photograph of the surface fracture at relatively low magnification. Since only the surface is covered with conductive gold layer, the fracture is bright due to the charging effect of the freshly opened uncoated surfaces.

Relaxing the imposed strain will reduce the charging and the visibility of the crack is decreased; in some instances, and for very small cracks, it will entirely disappear, although occasionally and at high magnification some vague or slight indications may remain even for the smallest crack. The reason for this is that after relaxation, the microdetails of the two returning surfaces are mismatched. Figure 38B shows the relaxed surface. Very small fracture sites after relaxation must be located with care, because charging effects highlight the fracture site by the absence of the conductive coating, and the same effect may be produced with peeling, flaking, or insufficient coating. The surface produced by mechanical straining must be interpreted carefully. Figure 39 shows detailed portions of the fracture surface. The entire fraction process can be videotaped while the matrix and the fiberglass separation occur. Thus examining both dynamic and static modes is possible.

Tilting the strained or fractured sample permits stereoscopic imaging. These three-dimensional images show important damage or deformation details. The depth of fiber debonding, wetting, fiber fracture displacements, and matrix tearing become spatially visible and dimensionally measurable. In general, the perceptual deficits of two-dimensional representation can be overcome with the stereoscopic image.

Figure 38 Glass-reinforced copolymer laminate SEM: A, mechanically strained (150×); and B, strain relaxed (150×).

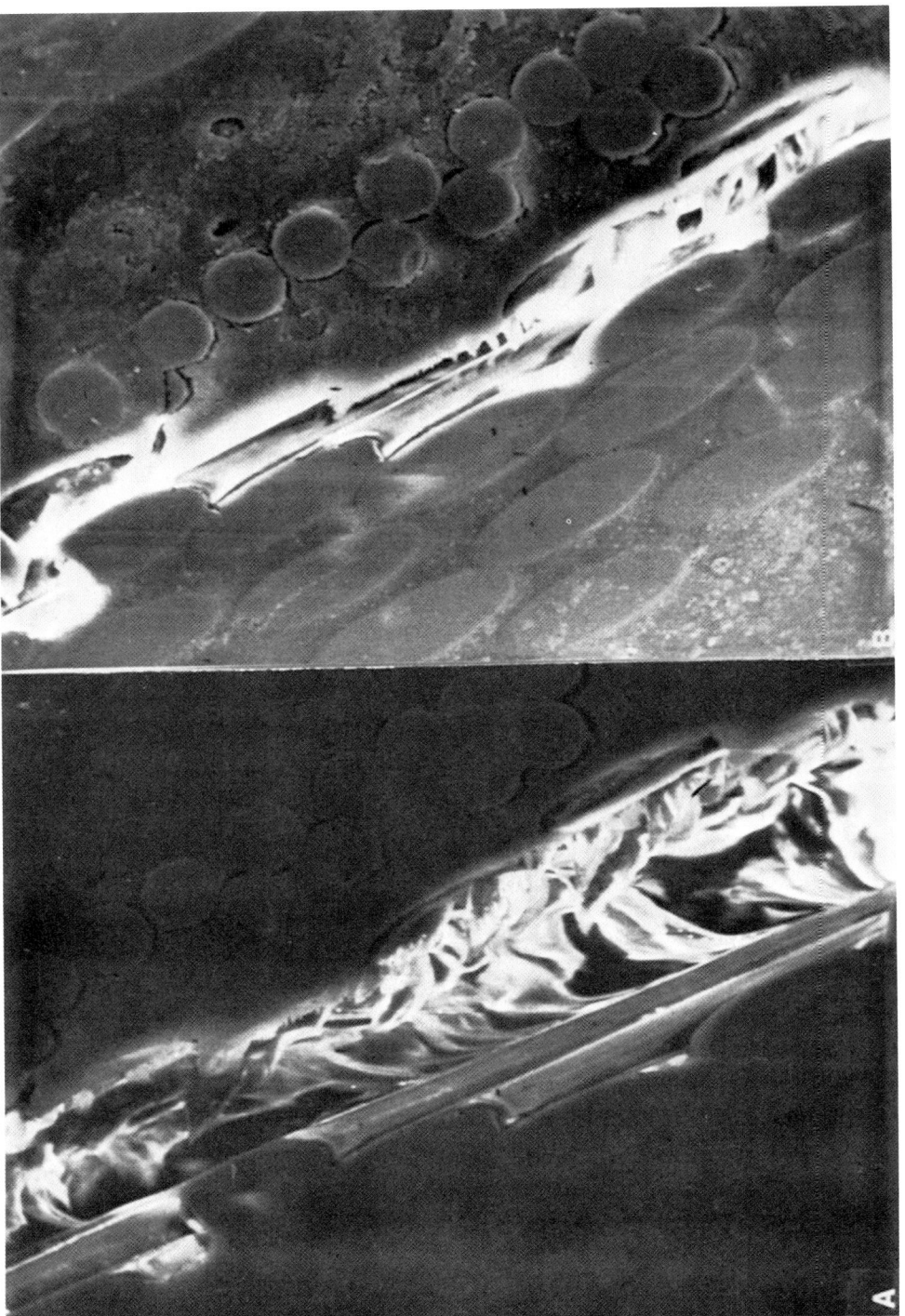

Figure 39 Glass-reinforced copolymer laminate SEM: A, mechanically strained (750×); and B, strain relaxed (750×).

ACKNOWLEDGMENTS

Several colleagues are owed gratitude for helping, compiling, and criticizing this paper into a cohesive and comprehensive document for publication. Of special note are the technical contributions of Dr. P. Beardmore, Dr. D. Mencik, K. Plumer, V. Mindroiu, F. A. Alberts, and W. Allie. The guidance and counsel of Dr. H. Van Oene were invaluable. In addition, we thank Dr. S. S. Labana, Dr. D. Schutzle, Dr. R. A. Pett, and Dr. M. A. Roberts for the critical review of this paper.

REFERENCES

1. Hansjurgen Saechtling, *Plastics Identification Table*, 8th ed., Hanser, Munich (1979).
2. S. S. Siegel, A Plastics Glossary, in *Modern Plastics Encyclopedia*, Vol. 46, No. 10a, pp. 37–52, McGraw-Hill, New York (1969–70).
3. A. Krause, A. Lange, and M. Ezrin, *Chemical Analysis of Plastics*, Hanser, Munich (1982).
4. H. A. Willis and D. C. M. Squirrel, *Identification and Analysis of Plastics*, Butterworth, London (1972).
5. W. C. Wake, *Analysis of Rubbers and Rubber-like Polymers*, 2nd ed., Maclaren, London (1969).
6. T. R. Crompton, *Chemical Analysis of Additives in Plastics*, 2nd. ed., Pergamon, Oxford (1977).
7. D. O. Hummel and F. Scholl, *Atlas of Polymer and Plastics Analysis*, 3 vols., Hanser, Munich (1978, 1981, 1982).
8. D. A. S. Ravens, *Polymer*, Vol. 1, 375 (1960).
9. A. Myagi and B. Wunderlich, *J. Polym. Sci. Polym. Phys. Ed.*, 10: 2073–2083 (1972).
10. *Compilation of ASTM Standard Definitions*, 4th ed., American Society for Testing and Materials, Philadelphia (1979).
11. *ASTM Yearbook*, American Society for Testing and Materials, Philadelphia. (annual)
12. H. E. Witt, *Interaction with Commercial Standards Setting Bodies*, Federal Register Publication (December 1976).
13. *Technical Manual of the American Association of Textile Chemistry and Colorists*, Research Triangle Park, N.C. (annual)
14. T. G. Rochow, *Symposium on Resinographic Methods*, ASTM STP No. 348, Philadelphia (1964).
15. H. W. Zieler, *The Optical Performance of the Light Microscope*, Vols. 14 and 15, Microscope Publications, Chicago (1972, 1973).
16. J. Serra, *Image Analysis and Mathematical Morphometry*, Academic Press, New York (1981).
17. R. E. Miles, "On Estimating Aggregate and Overall Characteristics from Thin Sections by Transmission Microscopy," Proceedings of the 4th International Congress for Stereology, NBS Special Publication 431, Government Printing Office, Washington, p. 1 (1976).
18. R. T. DeHoff, "Quantitative Image Processing," Proceedings of the 21st Annual Conference of the Microbeam Analysis Society, Albuquerque, N.M. (A. D. Romig, Jr. and W. F. Chamber, eds.), pp. 479–481 (1986).
19. R. T. Dehoff, *Metallography*, pp. 71–91, Vol. 4, American Elsevier, New York (1971).
20. W. Pepperhoff, *Sci. Techn. Inf.* 17: 199 (1979).
21. W. C. McCrone, L. B. McCrone, and J. G. Delly, *Polarized Light Microscopy*, McCrone Research Institute Edition, Chicago (1984).
22. S. K. Kobos, *Am. Lab.*, Vol. 18, 64 (1986).
23. George F. Vander Voort, in *Metallography Principles and Practice*, McGraw-Hill, New York, pp. 60–164, 267–333 (1984).
24. George F. Vander Voort, in *Metals Handbook*, 9th ed. (K. Mills, ed.), American Society for Metals, Metals Park, Ohio, pp. 70–88 (1985).

25. L. Bartosiewicz, "Materialography, The Non-metal Metallography," ASM Technical Report System No. 76-26, American Society for Metals, Metals Park, Ohio, pp. 1–6 (1976).
26. S. Czapski and W. Gebhardt, *Z. wiss. Mikrosk.*, *14*: 289 (1897).
27. L. E. Samuels, in *Metals Handbook*, 9th ed. (K. Mills, ed.), American Society for Metals, Metals Park, Ohio, pp. 33–47 (1985).
28. G. Petzow and G. Elssner, in *Metals Handbook*, 9th ed. (K. Mills, ed.), American Society for Metals, Metals Park, Ohio, p. 70 (1985).
29. L. Bartosiewicz and R. E. LaRose, in *Materials Engineering*, Penton/IPC Reinhold Publication, pp. 41–45 (December 1977).
30. A. Mencik and D. Fitchmun, "Texture of Injection-Molded Polypropylene," Technical Report No. SR 71-143 (October 20, 1971).
31. W. J. Henderson and K. Griffiths, in *Principles and Techniques of Electron Microscopy*, Vol. 2 (M. A. Havat, ed.), Van Nostrand, New York, pp. 151–196 (1972).
32. L. Seveus, "LKB Application Note," MLU/an-4, Available from LKB Produkter AB, Stockholm, Sweden (1971).
33. H. Schafer, O. D. Hennemann, and J. Rickel, *Sonderh. Prakt. Metallogr.*, *17*: 144–152 (1986).
34. J.-J. Chang et al., "Freezing Sectioning and Observation of Artifacts of Frozen Hydrated Sections for Microscopy," Proceedings of the 21st Annual Conference of the Microbeam Analysis Society, Albuquerque, N.M., pp. 231–233 (1986).
35. L. Bartosiewicz and Z. Mencik, *J. Polym. Sci.*, *12*: 1163 (1974).
36. P. H. Geil, in *Polymer Single Crystals*, Wiley, New York, pp. 232–240 (1963).
37. H. Schordin, in *Fractures* (B. L. Averback, ed.), Wiley, New York, p. 298 (1959).
38. R. W. Hertzberg and J. A. Manson, *Fatigue of Engineering Plastics*, Harcourt Brace Jovanovich, New York (1980).
39. L. Bartosiewicz and C. J. Kelly, in *Advances in Polymer Technology*, Vol. 6, Wiley, New York, pp. 185–192 (1986).
40. L. Bartosiewicz and C. J. Kelly, *Res. Dev.*, Vol. 27, 80 (1985).
41. H. Schordin, "Velocity Effects in Fracture," Proceedings of the Swampscott Conference, 1954.
42. L. Bartosiewicz and C. J. Kelly, in *Advances in Polymer Technology*, Vol. 7, Wiley, New York, pp. 21–33 (1987).
43. E. Jayne, in *Theory and Design of Wood Fiber Composite Materials*, Syracuse University Press, Syracuse, N.Y., p. 21 (1972).
44. G. R. Lightshev, P. H. Short, K. S. Kalasinsky, and L. Mann, *J. Mississippi Acad. Sci.*, *24*: 77 (1979).
45. M. Sato et al., "Development of the Ultrahigh Resolution SEM," Proceedings of the 21st Annual Conference of the Microbeam Analysis Society, Albuquerque, N.M. (A. D. Roming, Jr. and W. F. Chambers, eds.) (1986).
46. J. J. Hren, J. I. Goldstein, and D. C. Joy, eds., *Introduction to Analytical Electron Microscopy*, Plenum, New York (1979).
47. R. J. R. Scholtens et al., *J. Polym. Sci. Polym. Phys. Ed.*, *22*: 1223 (1984).
48. S. Inoue, *Video Microscopy*, Plenum, New York (1986).
49. P. Beardmore and S. Rabinowitz, "Cyclic Deformation and Craze Growth in Polycarbonate," *Fracture 1977*, Vol. 3, ICF4, Waterloo, Canada (June 19–24, 1977).
50. P. Beardmore and J. Fellers, *Mater. Sci. Eng.*, *5*: 120 (1969/70).
51. S. Rabinowitz, A. R. Krause, and P. Beardmore, *J. Mater. Sci.*, *8*: 11 (1972).

16
Viscoelastic and Molecular Characterization of Commercial Polymers

I. P. Briedis
Institute of Polymer Mechanics
Latvian SSR Academy of Sciences
Riga, USSR

MOLECULAR PARAMETERS OF POLYMERS	605
Molecular Mass Distribution	606
Average Molecular Mass	607
Molecular Mass Distribution Functions	609
Accuracy of MMD Parameter Determination	610
VISCOELASTIC PROPERTIES OF A POLYMER MELT	611
Linear Viscoelasticity	611
Narrow Molecular Mass Distributions	612
Broad Molecular Mass Distributions	617
Relaxation Time Spectrum	624
INTERRELATION BETWEEN MOLECULAR AND VISCOELASTIC PARAMETERS	628
Averaging of Viscosities	628
Extension of the Addition Law to the Relaxation Spectrum	629
Formation of Viscoelastic Functions from Molecular Structure Parameters	629
Dependence of Viscoelastic Parameters of Melt on Molecular Structure	631
Application of Analogies Between Viscosity and Viscoelasticity Parameters in the Study of Rheological and Molecular Parameter Dependences	634
APPROXIMATED DEPENDENCES FOR TAKING INTO ACCOUNT RELATIVE CHANGES IN MOLECULAR STRUCTURE IN TECHNOLOGICAL PROCESSES	635
Linear Polymers	635
Branched Polymers	636
Determination of Viscoelasticity Parameters η_0 and τ_r	638
REFERENCES	639

MOLECULAR PARAMETERS OF POLYMERS

Not a single synthetic high-molecular compound is a chemically pure substance in the strict sense of the word. Linear and branched polymers are a mixture of macromolecules of different lengths owing to the random nature of their formation. This circumstance leads to the basic difference between high-molecular and low-molecular substances. Linear polymers, the molecules of which consist of equal consecutively connected links, are mixtures of homologues with various chain lengths. Branched polymers are inhomogeneous in both chain length and structure. Polydispersity and branching affect a number of physical properties of polymers, in the first-line properties of their melts and solutions.

Polymeric materials are characterized by values of high molecular mass, i.e., by large sizes of the molecules. However, along with very large molecules there exist relatively small ones in the polymer. A considerable proportion of the physical properties of a polymer is determined by its molecular mass. The properties of polydisperse polymers depend on certain average values of molecular mass. Average values accepted for molecular mass distribution (MMD) may be considered comprehensive if they make reproducibility of MMD possible—in other words, in a slightly narrower sense, if they make it possible to determine certain physical characteristics of the material.

It ought to be stressed that certain properties of polymers do not depend on molecular mass but are determined by the width of the MMD. It is therefore necessary to indicate the factors that characterize the latter.

The manner in which compositional inhomogeneity affects the general characteristics of a test sample has not yet been clarified. Thus, for instance, properties of a polymer depend on MMD, on mutual position of side groups, on the nature of their distribution along the chain, and on other peculiarities of molecular structure.

Frequently cleavage of a hydrogen atom from the molecular chain takes place during polymerization: the chain starts growing in a different direction, thus forming a branched macromolecule. As a result we get a change in the mechanical properties of the polymer; hence branching indices form an important characteristic of the molecular structure of these materials.

The polymer frequently contains microgels, crosslinked particles of a very high molecular mass. It also contains a considerable amount of low-molecular fractions and oily admixtures. In the processing and estimating of experimental data sometimes the presence of low-molecular fractions and admixtures and, still more frequently, that of microgels is not taken into account. This lends considerable uncertainty to the value of the average MMD. A necessity arises to restore these components of MMD which have been lost in measurements [1].

Molecular Mass Distribution

Commercial polymers usually have a rather broad MMD. Figure 1 shows MMD data for commercial LDPE, as obtained from gel permeation chromatographic measurements. The test samples of LDPE consist of molecules with minimum molecular mass of $\sim 10^3$. At corresponding stages of polymerization of the molecule no continuous net is formed, hence the material is not a polymer in the usual sense of the word. At $M < M_c$ the polymer does not undergo elastic stress and forms low-molecular waxes.

Usually the width of MMD is denoted by the ratio between two average molecular mass values. One of the most widespread characteristics is inhomogeneity, introduced by Schulz [2]: $u = (\bar{M}_w/\bar{M}_n) - 1$, where $\bar{M}_n$ is the average numerical, and M_w is the weight average molecular mass, respectively (cf. the following section).

It is generally assumed that MMD must affect the shape of the relaxation spectrum of the polymer. Figure 2 shows the frequency relaxation spectra of test samples, the MMD values for which are presented in Figure 1. The high-frequency regions of the spectra, representing the motion of molecular segments, practically merge into a common dependence. At low frequencies the motion of the melt takes place mainly with molecules, and

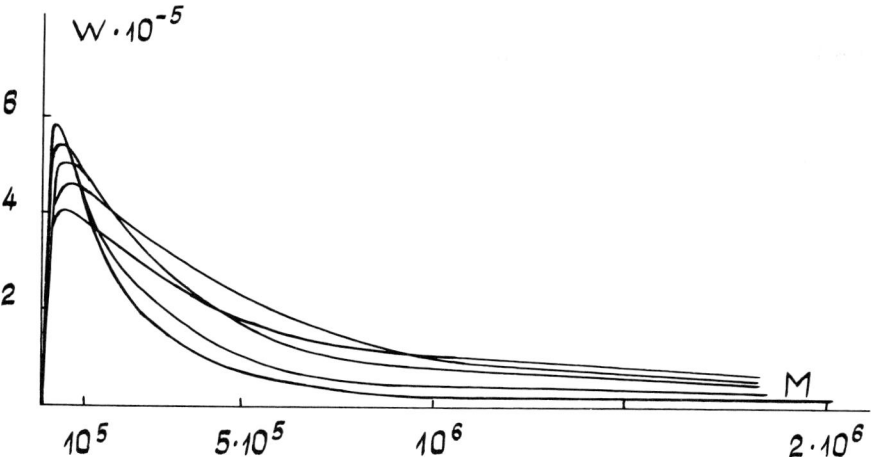

Figure 1 MMD of a number of commercial polymers (LDPE) synthesized in the same assembly. The samples differ in parameters $\bar{M}_w$, $\langle g \rangle_w$, and u.

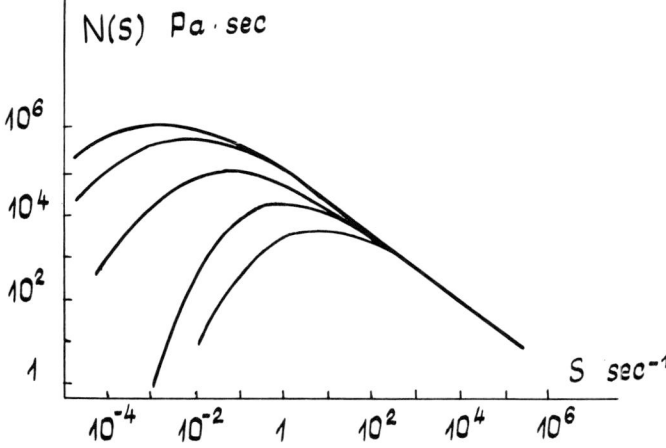

Figure 2 Frequency spectra of relaxation of MMD test samples, the MMD of which are presented in Figure 1.

in this case considerable differences are observed in the low-frequency part of the spectrum.

Average Molecular Mass

As stated previously, the polymer consists of molecules of different molecular mass. Hence there are various ways of averaging the molecular mass of the polymer. The method to be used is determined by the index of averaging, p, and by the form of the MMD function. The generalized equation of mean molecular weights has the form

$$\bar{M}_{pq} = \left[\int_0^\infty M^p f_q(M) dM\right]^{1/p} \tag{1}$$

where M_{pq} is the average molecular mass; $f_q(M)$ is the MMD function of order q, normalized in such a way that

$$\int_0^\infty f_q(M) dM = 1 \tag{2}$$

The expression in brackets in Eq. 1 is the moment of MMD of order p, denoted by S_q^p (lower index indicates form of MMD function). At $p = 1$ we obtain q, average molecular masses:

$$\bar{M}_q = \int_0^\infty M f_q(M) dM \tag{3}$$

The function $f_q(M)$ can be found by different means: if it is proportional to the number of molecules lying between M and $M + dM$, we write $f_n(M)$; if it is proportional to the mass of molecules NM within the same range, then we have $f_w(M)$; if $f_q(M) NM^2$, we write $f_z(M)$; and so forth. According to the type of averaging, q, the mean molecular masses are the arithmetical mean of the molecular masses of order q, i.e., we have average numerical, average weight, and z-average, and other averages:

$$\bar{M}_n = \int_0^\infty M f_n(M) dM$$

$$\bar{M}_w = \int_0^\infty M f_w(M) dM$$

$$\bar{M}_z = \int_0^\infty M f_z(M)$$

$$\bar{M}_{z+1} = \cdots \tag{4}$$

Therefore MMD functions are mostly expressed through a mass function obtained directly from fractionating experiments $f_w(M) = W_{(M)}$.

Average molecular weights are expressed by means of a recurrence formula

$$\bar{M}_q = \frac{S_q}{S_{q-1}} \tag{5}$$

where

$$S_q = \int_0^\infty M_q W(M) dM \tag{6}$$

is the moment of the MMD.

In light of Eq. 5, Eq. 4 may be expressed through the weight function:

$$\bar{M}_n = \left(\int_0^\infty M^{-1} W(M) dM\right)^{-1} \tag{7}$$

$$\bar{M}_w = \int_0^\infty M W(M) dM \tag{8}$$

$$\bar{M}_z = \frac{\int_0^\infty M_2 W(M) dM}{\int_0^\infty M W(M) dM} \tag{9}$$

From the practical viewpoint the moments of MMD of various orders are not equivalent, even if the accuracy of determination within the whole range of molecular masses is the same. This is a result of accumulation of errors in the calculation of moments, the error being larger the more the order of the moment differs from unity. In the determination of mean molecular weight values the accuracy of the result largely depends on the value of mass value intervals into which the whole range of molecular masses is being split. The characteristics of the MMD under investigation change to a diminishing degree: the mean molecular masses M_n and M_z increase, M_w decreases tending toward their true values. It was found that when taking an unlimited interval, from 0 to ∞, masses of molecular in a real polymer instead of a limited one is not essential.

One of the most important rheological characteristics of a melt—the maximum Newtonian viscosity—is in a rather simple dependence on molecular mass average value. It follows a potential equation. Hence one mostly employs the mass average value for characterizing molecular mass.

Molecular Mass Distribution Functions

There have been many attempts to describe MMD by means of analytical functions. There are several distribution functions of molecular mass; some are obtained from the kinetics of the polymerization process, and others are found empirically.

The Schulz distribution function [3] is one that is frequently applied:

$$W(M) = \frac{(-\ln a)^{b+2}}{\Gamma(b+2)} M^{b+1} \alpha^M \tag{10}$$

where a and b are parameters determining the molecular mass average value (the parameter value b is inversely proportional to distributional width) and Γ is a gamma function.

From the Schulz distribution we obtain the ratio between molecular mass average values:

$$\frac{\bar{M}_w}{\bar{M}_n} = \frac{b+2}{b+1} \tag{11}$$

and

$$\frac{\bar{M}_z}{\bar{M}_w} = \frac{b+3}{b+2}$$

It follows that at narrow distributions M_w/M_n is a characteristic which is more sensitive to MMD width than the ratio M_z/M_w. In practice a distribution dependence proposed by Tung [4] is also used:

$$W(M) = yz^{-yM^2} M^{z-1} \tag{12}$$

Here z, just like b in Eq. 10, is inversely proportional to distribution width. Distributions 1 and 12 are alike, differing only in experimental error in determining MMD [5].

A function of normal logarithmic distribution, or the Kraemer–Lansing function [6], is also frequently used:

$$W(M) = \frac{1}{sM\sqrt{2\pi}} \exp\left(-\frac{1}{2s^2} \ln^2 \frac{M}{M_0}\right) \tag{13}$$

Normal logarithmic distribution has been successfully used in the description of MMD in polyethylene obtained on Ziegler catalysts [7]. Considering the amount of long-chain branching of the molecules, this distribution can also be employed for representing MMD in low-density polyethylene [8]. However, cases have been reported in which the description applies only for narrow MMD [9]. Several other distributions are known, such as those of Flory, Poisson, Gauss, Bisley, and Wesslaw.

Accuracy of MMD Parameter Determination

For elucidating the problem of accuracy in the various methods of averaging let us consider the distribution in Eq. 13. Maximum density of this distribution is obtained at $M = M_0 e^{s^2}$ and equals

$$W_{max} = \frac{2e^{s^2/2}}{sM_0} \qquad (14)$$

Let us construct two distributions $W(M)$ and $W'(M)$ connected by the condition

$$W'_{max} = aW_{max} \qquad (15)$$

The constants of both distributions are interrelated:

$$se^{s^2/2} = as' e^{s'^2/2} \qquad (16)$$

$$M'_0 = M_0 e^{s'^2 - s} \qquad (17)$$

Calculations performed for MMD of medium width show that lowering by 5% of the height of distribution at the peak leads to a change in $\bar{M}_n$ of 7%, in $\bar{M}_w$ of 12%, and in M_z by ~90%.

With increase in molecular mass the accuracy of its determination decreases. To a certain extent this phenomenon may be simulated by a shift of the horizontal distribution axis. A shift of only 1% of the horizontal axis from W_{max} produces a 30% change in $\bar{M}_w$ and a change by a factor of ~12 in $\bar{M}_z$.

The preceding transformations of MMD, the magnitude of which does not exceed the accuracy of its determination in experiment, indicates that averaging of high powers is not applicable for identifying MMD. These averagings are dramatically affected by the high-molecular part of the distribution, which is determined with little accuracy in practice. Hence one must refrain from using in the description of MMD molecular mass average values higher than $\bar{M}_w$. This is particularly important in the description of branched polymers, since the value of molecular mass obtained depends on the way of accounting for branching. From this point of view only two values of molecular mass, $\bar{M}_n$ and $\bar{M}_w$, may be used.

The average weight molecular mass is the mathematical expectancy M^* of distribution:

$$\bar{M}_w = M^* \qquad (18)$$

Dispersion of MMD is given by the equation

$$D^2 = \int_0^\infty (M - M^*)W(M)dM \qquad (19)$$

from which, after some transformations and taking into account $\int_0^\infty W(M)dM = 1$, we get

$$D^2 = \int_0^\infty M^2 W(M)dM - \bar{M}_w^2 \qquad (20)$$

It follows from Eqs. 10 and 11 that

$$\bar{M}_w \cdot \bar{M}_z = \int_0^\infty M^2 W(M) dM \tag{21}$$

Expression 21 permits transformation of Eq. 20 into the following relation:

$$D^2 = \bar{M}_w(\bar{M}_z - \bar{M}_w) \tag{22}$$

Expression 22 determines the dispersion of MMD as a function of two molecular mass average values. The ratio between dispersion and mathematical expectancy,

$$\left(\frac{D}{M^*}\right)^2 = \frac{\bar{M}_z}{\bar{M}_w} - 1 \tag{23}$$

is an analogue to inhomogeneity of dispersion $(\bar{M}_w/\bar{M}_n) - 1$. If $\bar{M}_w/\bar{M}_n = \bar{M}_z/\bar{M}_w$, we have

$$\left(\frac{D}{M^*}\right)^2 = u \tag{24}$$

It follows from Eqs. 18 and 23 or 24 that the description of MMD through one molecular mass average value and the polydispersity index is analogous to an estimate of distribution through mathematical expectancy and dispersion, which makes sense only for identifying a similar MMD.

In this connection one of the most important problems is setting conditions of similarity between MMD, which is necessary for determining the effects of molecular structure on properties of a polymer.

VISCOELASTIC PROPERTIES OF A POLYMER MELT

Viscoelastic properties manifest themselves in the process of nonstationary flow in which strain, stress, or both are time functions, and the deformation rate is comparable to characteristic relaxation time.

In the case of shearing flow, deformation of macromolecules takes place causing partial orientation of the macromolecules. This change leads to partial accumulation inside the material of the work done by the external force. This accumulated work does not get dispersed owing to resistance to viscous flow, which corresponds to a transition of the macromolecule into conformations that correspond to a decrease in probability of state, i.e., entropy [10].

In macroscopic observations this effect manifests itself first in the appearance of large elastic deformations that can be detected in the course of transient processes preceding the onset of stable flow regime and after its termination, which manifests itself in time dependences of strain, i.e., rate of deformation. Second, the presence of normal strain components indicates elasticity of the melt during stationary shearing flow.

Linear Viscoelasticity

Under conditions of stationary flow linear viscoelasticity may be observed within the range of stresses (rate of deformation) corresponding to Newtonian flow. In this case stresses and deformation rates are small and difficult to measure. Thus to obtain a relaxation curve it is necessary to change the sensitive range of the measuring device within a sufficient time interval in the course of the experiment, which, at present, is a difficult technical problem.

The most reliable method of determining linear viscoelastic characteristics is that of linear periodic deformations. In the course of measuring the resistance of the melt to linear periodic deformation, changes in frequency range and in sensitivity of the measuring device are much simpler.

Results of experiments obtained under conditions of periodic stress at circular frequency ω are qualitatively equivalent to results of experiments performed in transitional regimes at times $t = 1/\omega$. There exist analytical dependences connecting complex dynamic characteristics with the relaxation modulus [11]. The linear viscoelastic indices are measured under conditions of settled oscillations. According to Faitelson and Tsiprin [12], linearity of deformation amplitude for LDPE melts within the circular frequency range $157 < \omega < 377$ is $\sim 100\%$ ($\gamma = 1$). It may be expected that at lower frequencies the range of amplitude linearity will increase. Under conditions of linear periodic deformation the components of the complex shear modulus $G^* = G' + iG''$, the complex compliance $J^* = (G' - iG'')/(G'^2 + G''^2)$, or the complex viscosity G^*/ω are determined.

The material may be subjected to periodic deformation in the flowing, viscoelastic, or glassy state, which cannot be achieved at steady-state flow. It is this circumstance that constitutes the basic advantage of periodic deformation as a method of identifying molecular structure of a polymer, as compared to stationary flow.

Narrow Molecular Mass Distributions

The basic characteristic of a viscoelastic melt of a polymer is its maximum Newtonian (initial) viscosity η_0 and the characteristic time of relaxation τ_r.

Newtonian (Initial) Viscosity

The first measurements on test samples with narrow MMD were performed by Fox and Flory [13], yielding two regions of molecular mass M. In the first region we have for each homological series of polymers $M \leq M_c$

$$\eta_0 = k_0 M^{\beta_F} \tag{25}$$

in the second, for $M > M_c$, we have

$$\eta_0 = kM^\theta \tag{26}$$

The time-dependent constants k_0 and k are determined by their affiliation to the given homological series, β is close to unity, and $\theta = 3.5$. The boundary between the two regions is not clearly defined; hence M_c is determined only approximately. According to the Fox theory [13, 14], viscosity of polymers is determined basically by the structure of the chain molecule. Results obtained by Fox do not represent molecular theory as such but are only a generalization of experimental data. The dependence of viscosity on molecular mass was established by Debye, but the expression he proposed contains the friction coefficient of the molecule, whose dependence on molecular mass is not known. The problem of finding the dependence of viscosity on molecular mass was solved by Bueche [15, 16]. The result obtained shows that for monodisperse polymers we have for $M > M_c$

$$\eta_0 \sim M^{3.5} \tag{27}$$

Characteristic Relaxation Time

Studies of the effects of molecular structure of polymers on their mechanical properties began with investigations of their diluted solutions, so that observations were independent of effects of molecular interactions. Owing to permanent chaotic motion a polymer molecule constantly changes its conformation in solution. Resistance to this motion is due to intramolecular effects, which may be regarded as internal viscosity and hydrodynamic resistance of the solvent. For complete characterization of conformation it is necessary to have full information on size and shape of the monomeric links and on effects of stacking and interaction between macromolecules with molecules of the stack. This knowledge must be more complete than is available at present. For this reason the basic assumption for quantitative description of viscoelastic properties of a polymer consists of the possibility of regarding a polymer molecule as a total consisting of identical elements joined consecutively, so that each element is deformed independently of the others. Such a splitting up of the polymeric chain into elements is of a conditional nature and requires an acceptance of assumptions whose viability can be judged only from agreement between the adopted model and predicted experimental results [17]. Various ways of accounting for external and intermolecular forms of resistance of the "bead" model have been discussed by Kargin and Slonimsky [18, 19], by Gottlieb, Volkenstein, and Ptitsin [20, 21], and by Rouse [22]. Rouse's theory assumes the absence of hydrodynamic interaction between the segments, which corresponds to the case of the free-draining coil of macromolecules.

For a chain consisting of $n + 1$ segments we have n relaxation times, all of which can be expressed through the maximum relaxation time. The latter is described, according to Kargin–Slonimsky–Rouse theory, in the following way:

$$\tau_m = \frac{6(\eta_0 - \eta_c)}{\pi^2 NKT} \tag{28}$$

where η is the viscosity of the solution, π is the viscosity of the solvent, K is the Boltzmann constant, T is the absolute temperature, and N is the number of molecules per cubic centimeter of solution. It can be shown that the maximum relaxation time increases with chain length, while minimum relaxation time is independent of the latter. The spectrum of relaxation time is described by the relation

$$H(\tau) = NKT \sum_{p=1}^{n} \tau_p \delta(\tau - \tau_p) \tag{29}$$

where $\tau_p = \tau_m p^{-2}$. From Eq. 29, the components of the complex module G^* are

$$G' = NKT \sum_{p=1}^{n} \frac{(\omega \tau_p)^2}{1 + (\omega \tau_p)^2} \tag{30}$$

$$G'' = NKT \sum_{p=1}^{n} \frac{\omega \tau_p}{1 + (\omega \tau_p)^2} + \omega \eta_c \tag{31}$$

The first attempts to apply this result for concentrated solutions and melts of polymers were based on the assumption that the molecule performs the same motions here as in dilute solution. The resistance to motion is expressed through the mean monomer coefficient of friction, whereby the following expression is obtained:

$$\tau_m = \frac{6\eta_0 M}{\pi^2 \rho RT} \tag{32}$$

In Eq. 32 the number of molecules per cubic centimeter is replaced by the magnitude $\rho N_0/M$, where ρ is the density of the polymer mass, N_0 is the Avogadro number, and R is the universal gas constant.

In the Kargin–Slonimsky–Rouse model the behavior of the stretched chain is studied under the action of longitudinal forces. Bueche [23] considers a macromolecule rolled into a coil and placed into a field of shearing force. The motion of segments with respect to the center is studied, each segment being considered as a Maxwell element. Displacement of the coil consists of translational shift of the mass center and rotation of the coil around it. Bueche theory yields the expression

$$\tau_p = \frac{12\eta_0 M}{\pi^2 p^2 \rho RT}$$

In Zimm's theory [24, 25], which was developed from ideas proposed by Kirkwood and Riseman [26], a chain molecule is analyzed (as in Refs. 18–21), but hydrodynamic interaction between segments is also introduced.

Superimposing in dimensionless coordinates viscoelastic frequency functions for monodisperse polymers of different molecular mass [27] showed that at sufficiently large molecular mass the relative shift along the frequency axis is proportional to η_0. It follows that the relaxation time is proportional to Newtonian viscosity, i.e., $\tau \sim M^\theta$. This discrepancy with experimental results may be due to the effect of macromolecular entanglement on the motion of their segments.

The macromolecule is always in the surrounding of others, and therefore frequently the motions of short and long segments of the chain are considered separately [29, 30]. Short chains are not affected by entanglement, which determines a major part of the relaxation spectrum. The boundary between the regions of segments is characterized by the magnitude of the critical molecular mass M_c, which divides the dependence of viscosity on molecular mass into two regions.

Several models were proposed, based on various assumptions about the manner whereby the elastic and inelastic resistances are summed in the process of building up a mechanical model [31–34]. These theories also apply the idea of dividing the relaxation spectrum into two regions. It is assumed in these theories that the resistance to displacement of the macromolecules depends on entanglement, the friction coefficient of the segment being determined only by the value of molecular mass.

Pokrovsky assumes that the resistance to displacement of segments depends on their location in the chain [35].

Network models constitute a special class of models connecting molecular and mechanical properties of a polymer [36–39]. In these models interaction is not taken into account through the effective coefficient of friction but is localized at molecular junctions where interaction between macromolecules takes place. The junctions are of a fluctuational nature. Dependences obtained in Refs. 36 and 37 are the same as those in the bead model.

These theories assume the presence of special points, i.e., entanglements in which molecular interaction takes place [40]. It may be mentioned that models exist [41–43] which consider molecular interaction without resorting to the concept of entanglement.

Elasticity Parameters of the Melt

The equation of state for the case of a linear viscoelastic body in a convective coordinate system is of the following form:

$$\sigma(t) = \int_{-\infty}^{t} \dot{\gamma}[\varphi(t - t') + G_\infty] dt' \tag{33}$$

After changing to a stationary coordinate system X_1, X_2, X_3 we obtain [44]

$$\sigma_{ij}(x_1 t) = \int_{-\infty}^{t} 2\varphi(t - t') \frac{\partial X_m}{\partial x_i} \frac{\partial X_n}{\partial x_j} \dot{\gamma}_{mn} dt' \tag{34}$$

If we set the relaxation function in the form

$$\varphi(t) = \int_{0}^{\infty} N(s) e^{-ts} ds \tag{35}$$

$N(s)$ being the spectrum of relaxation frequencies, which is a generalized form of bead theory concepts, then the shearing stress is

$$\sigma_{12} = \dot{\gamma} \int_{0}^{\infty} \frac{N(s)}{s} (1 - e^{-t/s}) ds \tag{36}$$

and the first normal stress difference is

$$\sigma_{11} - \sigma_{22} = 2\dot{\gamma}^2 \int_{0}^{\infty} \frac{N(s)}{s^2} [1 - e^{-t/s}(1 + s)] ds \tag{37}$$

For stationary flow regime ($t \to \infty$) we obtain for Newtonian viscosity and normal stress coefficient

$$\eta_0 = \frac{\sigma_{12}}{\dot{\gamma}} = \int_{0}^{\infty} \frac{N(s)}{s} ds$$

$$\zeta_0 = \frac{\sigma_{11} - \sigma_{22}}{2\dot{\gamma}^2} = \int_{0}^{\infty} \frac{N(s)}{s^2} ds \tag{38}$$

Expressions for periodic deformation may be obtained if we consider the rate of deformation in Eq. 34 as a harmonic time function $-\dot{\gamma} = \dot{\gamma}_0 e^{-i\omega t}$. Defining similarly the ratio between stress and strain as $\sigma_{12}/\dot{\gamma} = G^*$ we obtain well-known expressions

$$G' = \int_{0}^{\infty} N(s) \frac{\omega^2 ds}{\omega^2 + s^2} \quad \text{and} \quad G'' = \int_{0}^{\infty} N(s) \frac{\omega s\, ds}{\omega^2 + s^2} \tag{39}$$

Compliance under periodic deformation can be defined as

$$I^* = \frac{1}{G^*} = I' - iI'' \tag{40}$$

where

$$I' = \frac{G'}{G^*} \quad \text{and} \quad I'' = \frac{G''}{G^*}$$

The value of instantaneous compliance is obtained at $\omega \to 0$ and $I'' \to 0$ while

$$I' \to I_0 = \frac{\int_0^\infty [N(s)/s^2]ds}{[\int_0^\infty [N(s)/s]ds]^2} \tag{41}$$

Considering Eqs. 37 and 38 we obtain

$$\zeta_0 = I_0 \eta_0^2 \tag{42}$$

From theoretical considerations obtained for polymers with narrow MMD [45] the dependence expresses the inverse proportionality of the initial value of the module (at $\dot{\gamma} \to 0$) with molecular mass

$$G_0 = \frac{K_9 \rho RT}{M} \tag{43}$$

Usually Eq. 43 is written for conditions of instantaneous compliance

$$I = \frac{2}{5} \frac{M}{\rho RT} \tag{44}$$

Equation 44 is analogous to the fundamental equation of entropic rubberlike elasticity [46, 47]. The formula expresses the situation in which the larger the molecular mass the easier it is to get the macromolecule out of its most probable conformation.

The tendency toward direct proportionality between compliance and molecular mass in experiments with polydisperse polymers can be observed only in the region of low molecular mass. With increase in molecular mass G_0 becomes independent of the molecular mass [48–50]. It thus appears that elastic properties of a polymer are determined not by the length of the molecular chain but by the length of the interval between entanglements.

If compliance J_0 does not depend on molecular mass, then it follows from Eq. 42 that

$$\zeta \sim M^{2\theta} \tag{45}$$

which means that the coefficient of normal stress depends on molecular mass at a power twice as large as viscosity.

Frequency Dependences of Viscoelasticity Parameters

Frequency dependences of the storage modulus G' and of the loss modulus G'' in Figure 3 are typical of narrow MMD [51–53]. An important feature of frequency dependences of the storage modulus is a well-defined plateau region of viscoelasticity, the height of which does not depend on molecular mass. The length of the plateau region along the frequency axis increases with molecular mass of the polymer. The $G''(\omega)$ dependence shows a peak whose height does not depend on molecular mass. The position of the peak shifts toward the low-frequency region with increases in the molecular mass of the polymer. The frequency dependences $G'(\omega)$ and $G''(\omega)$ are positioned in two regions of physical state of the polymers: in the region of flow state at low frequencies, in which the storage modulus increases rapidly with frequency; and in the region of high-elasticity state where G' remains constant, while the loss modulus passes through a minimum trough. With increases of molecular weight of the test sample the transition into the high-elasticity state takes place at lower frequencies.

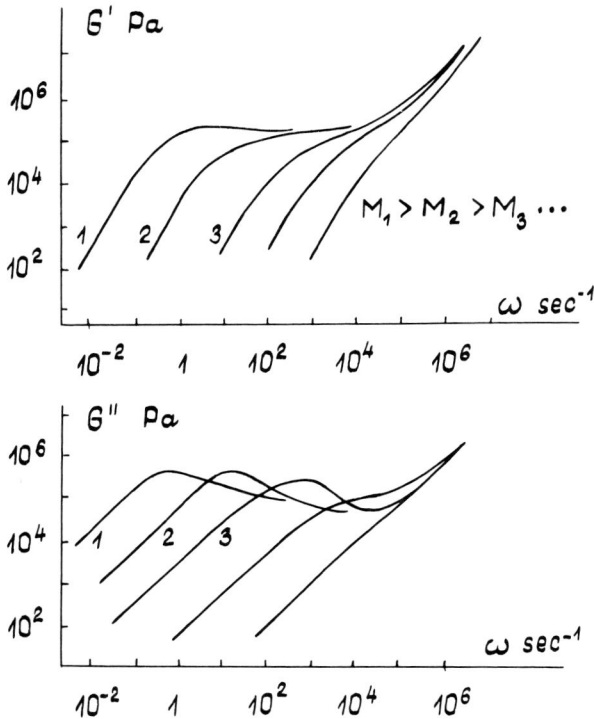

Figure 3 Frequency dependences of the components of the complex modulus of polymers with narrow MMD.

Broad Molecular Mass Distributions

Newtonian (Initial) Viscosity

The effect of MMD width on Newtonian viscosity is of second-order smallness, compared to that of $\bar{M}_w$. It was therefore neglected for some time, and it was thought that that polydispersity could be accounted for by introducing molecular mass average into the equation [55, 56]. There also exists an opinion that divergence of viscosity vs. molecular mass dependence from the power law occurs because η_0 is determined not by $\bar{M}_w$, but by some other molecular mass value which is larger than $\bar{M}_w$ but smaller than $\bar{M}_z$. Some workers consider that the MMD dependence of η_0 cannot be described by average molecular weight only, but is a function of several $\bar{M}_q$ values, i.e.,

$$\eta_0 = \kappa\varphi(\bar{M}_{q1}, \bar{M}_{q2}, \ldots) \tag{46}$$

under the condition that if $\bar{M}_{q1} \to \bar{M}_{q2} \to \cdots \to M$, then $\varphi(M_{q1}, M_{q2}, \ldots) \to M^{3.4}$.

According to Palma et al. [56] the MMD dependence of η_0 may be described in the first approximation if the effect of MMD is accounted for by mass average averaging of molecular mass. However, it is necessary, as a next approximation, to introduce into the expression for η_0 ratios between two different MMD moments [57, 58]. Existence of polydispersity lowers the Newtonian viscosity of the polymer. The $\bar{M}_w$ dependence of η_0

is presented in Figure 4 for practically monodisperse fractions of linear polyethylene using data from Refs. 59–61. The same dependence for linear polyethylene using data from Refs. 61–65 is presented in Figure 5.

For the fractions the coefficient of linear correlation between log η_0 and log $\bar{M}_w$ equals $r = 0.991$, and the index of $\bar{M}_w$ dependence of $\bar{M}_w$ is 3.30. For polydisperse polyethylene (Fig. 5) the index = 3.374, which is also rather close to the "classical" value, but the linear correlation coefficient for log η_0 and log $\bar{M}_w$ is, in this case, considerably lower than that for narrow fractions ($r = 0.8232$); maximum deviation reaches 350% for η_0.

To account for the polydispersity effect [57, 66] the following equation was used:

$$\eta_0 = \kappa \bar{M}_w^\theta (u + 1)^m \tag{47}$$

where $\theta = 3.36$ and $m = 0.51$. Applying Eq. 47 to the results presented in Figure 5 we obtain the values $\theta = 4.12$ and $m = 0.463$.

To account for polymolecularity the Newtonian viscosity has been represented [67] in the following form:

$$\eta_0 = cF_1F_2 \tag{48}$$

where c is determined by the mobility of the links and is independent of mobility at sufficiently large molecular mass [68]. The factor F_{1i} accounts for correlation of links of the selected ith molecule.

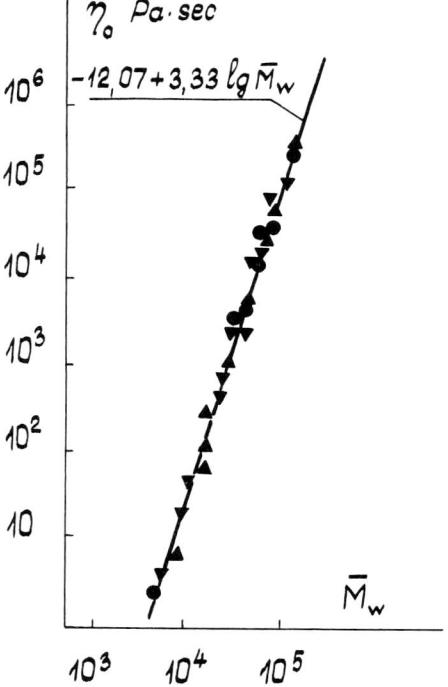

Figure 4 Dependence of Newtonian viscosity of a melt of an HDPE fraction on mass average molecular mass, as traced from data [59–61].

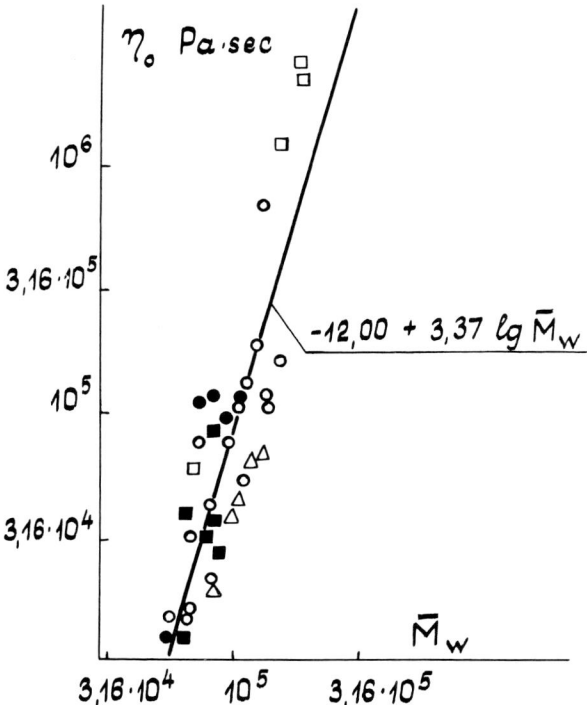

Figure 5 Dependence of Newtonian viscosity of linear polydisperse polymers on mass average molecular mass, according to Refs. 61–65.

F_2 characterizes the independence of motion of long-chain molecules and is determined by hydrodynamic intermolecular interaction, being dependent on molecular mass of the chain and on average molecular mass of the test sample M_n, i.e., $F_{2i} \sim (M_i M_n)^{(\beta-\alpha)/2}$. Considering the Ninomia–Ferry law of addition of fraction viscosity the following expression was obtained for $\alpha = 1$ and $\beta = 3.5$:

$$\eta_0 = k\bar{M}_w^{0.25} \int_0^\infty f_n(M) M^{3.25} dM \tag{49}$$

The expression takes into account the effect of polymolecularity on η_0 and passes to a value of $\eta_0 = kM^{3.5}$ at sufficiently narrow distributions.

Allowance for Long-Chain Branching of Molecules

The degree of branching of molecules of the polymer depends on the length of the side groups (branches). If the length of the side groups is comparable to the length of the basic chain, then one speaks of long side chains. The molecular mass of such branches is rather large; hence they have a substantial effect on the relaxation time of the molecule as a whole. The existence of long branches is essential when large relaxation times play the main part in the process under study. This may take place in processes which are determined by flow at low shearing velocity and at low-frequency periodic deformation. Hence branching has a notable effect on Newtonian velocity and initial elasticity.

The first attempt at evolving a quantitative method of evaluating the degree of branching was undertaken by Zimm and Stockmayer [70]. Branching is estimated by means of the conformational parameter g, which equals the ratio between the mean square radii of gyration of the branched and the linear macromolecules of the same molecular mass. The mean square radius of gyration of the branched molecule is always smaller than that of the linear one. Another method of estimation consists of the determination of the number of branches per molecule.

For polymolecular polymers the mass average value $\langle g \rangle_w$ of coefficient g is taken into consideration.

The existence of short side branches which do not cause entanglement also enhances diminution of the volume of the molecular coil, which leads to a lowering of viscosity [71]. However, the effect of short-chain branches is considerably smaller than that of long-chain branches. It may be noted that some authors [59] did not observe any effect of short-chain branches on physical properties of polymers. The contribution of short-chain branches can be demonstrated by an example. For polyethylene with $\bar{M}_w = 10^6$, possessing 5 long-chain and 20 short-chain branches per molecule, the conformational parameter equals 0.559 without considering short branches and 0.523 when they are taken into account [72]. The value of the conformational parameter of a short-chain branching g_k, as calculated from theoretical models according to Stockmayer, Berry, Orofine, and Altenburg was found to be in linear dependence on length and number of branchings [73]. Branching has practically no effect on rheological parameters if molecular mass lies below 10^4 [74].

Attempts have been made to characterize branching, particularly of the short-chain kind, by the methyl group content per molecule. A comparison between the total number of CH_3 groups and the number of those positioned at the terminals of the macromolecules appears to suggest that their total content, e.g., in commercial LDPE, cannot be considered characteristic of branching of the polymer, since 60–80% of the CH_3 groups are situated at the terminals of the macromolecules [75].

Newtonian viscosity depends on the conformational parameters to the same degree as on molecular mass [60, 62, 63], hence we have

$$\eta_0 = k[\langle g \rangle_w \bar{M}_w f(u)]^\theta \tag{50}$$

If Eq. 50 holds for a set of test samples of different molecular structure, then the results can be represented in the form of a straight line in $\log \eta_0 - \log x$ coordinates if we put $x = \langle g \rangle_w \bar{M}_w f(u)$ (Fig. 6). For the polydispersity function the condition $f(u) < 1$ is valid if $u > 0$, and $f(u) = 1$ for $u = 0$. Therefore, the $\log \eta_0 - \log x$ data assume a series of points lying to the right of the straight line K determined by Eq. 51 if we put $x = \langle g \rangle_w \bar{M}_w$

$$\log \eta_0 = \log k + \theta \log [\langle g \rangle_w \bar{M}_w f(u)] \tag{51}$$

Let us draw a straight line L through the points $\log \eta_0 - \log \langle g \rangle_w \bar{M}_w$ applying the least squares method. This line corresponds to Eq. 52:

$$\log \eta_0 = \log k_0 + \theta_0 \log (\langle g \rangle_w \bar{M}_w) \tag{52}$$

The polydispersity index u does not depend on the product $\langle g \rangle_w \bar{M}_w$, hence deviation Δ of the points from the straight line (Eq. 52) are of random nature. If deviations Δ within the $[0, \log f(u_{\max})]$ range are equally probable, then at the sufficiently large number of test samples the regression equation (52) determines a straight line parallel to Eq. 51, i.e., $\theta = \theta_0$.

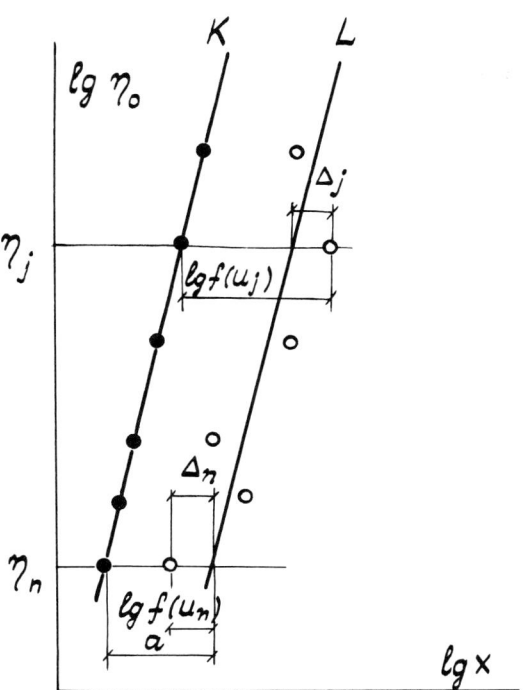

Figure 6 K dependence of Newtonian viscosity on molecular parameters, accounting for polydispersity log $\eta_0 - \log [\langle g \rangle_w \bar{M}_w f(u)]$.

It may be seen from Figure 6 that

$$\log f(u) = a + \Delta \tag{53}$$

Experimental results presented in Figure 7 show that in the first approximation Δ depends linearly on the polydispersity index:

$$\Delta = a_0 + k_{\eta 0} u \tag{54}$$

It follows from Eqs. 53 and 54 that

$$\log f(u) = a + a_0 + k_{\eta 0} u \tag{55}$$

If $u = 0, f(u) = 1$, then $a + a_0 = 0$ and

$$f(u) = \exp(k_\eta u) \tag{56}$$

where $k = 2.303 k_{\eta 0}$; the coefficient $k_{\eta 0}$ can be found from the linear correlation between Δ and the polydispersity index.

The dependence of η_0 on the product $\langle g \rangle_w \bar{M}_w$, as found for 54 test samples of commercial branched polyethylene [60, 62, 63, 77–79], is presented in Figure 8. These results do not yield any functional dependence, since they occupy a large region in the log η_0–log $\langle g \rangle_w \bar{M}_w$ coordinate system. In these coordinates the equation of linear regression is expressed as follows:

$$\log \eta_0 = -13.00 + 3.72 \log [\langle g \rangle_w \bar{M}_w] \tag{57}$$

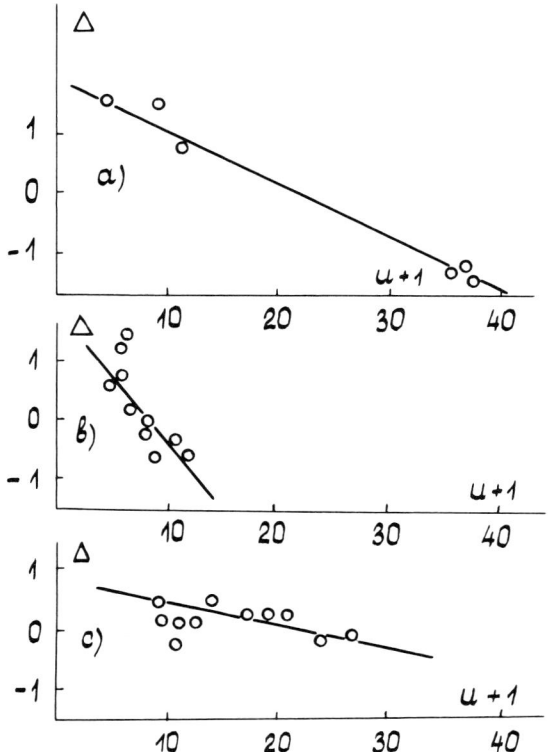

Figure 7 Dependence of polydispersity function on polydispersity index u.

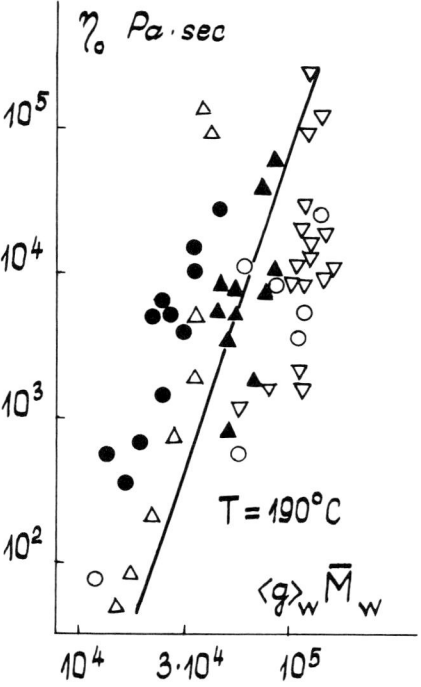

Figure 8 Dependence of Newtonian viscosity on parameter $\langle g \rangle_w \bar{M}_w$ for six sets of LDPE obtained in various reactors according to Refs. 60, 62, 63, 77–79.

The correlation coefficient 0.6185 indicates weak interdependence between η_0 and $\langle g \rangle_w \bar{M}_w$.

The experimentally determined values of η_0 differ from the values calculated according to Eq. 57 by a factor of 30; the inapplicability of this equation even for approximate estimates of η_0 dependence on $\langle g \rangle_w \bar{M}_w$ follows from the low value of the correlation coefficient.

Attempts to describe the experimental results presented in Figure 8 by means of dependence (Eq. 50), using the polydispersity function (Eq. 56), were not successful; however, application of these formulas for each set of test samples of this polymer, of which each set was synthesized in its own experimental outfit, yielded satisfactory results (Fig. 9).

A comparison of the curves presented in Figure 9 shows that for each of the sets of polyethylene test samples studied by means of Eq. 50, the numerical values of the coefficients must be different. A sufficiently close correlation for molecular structure indices in each set of samples produced by means of the same technology indicates that these sets possess different properties accounted for by constant coefficients pertaining to samples with similar technologies of preparation.

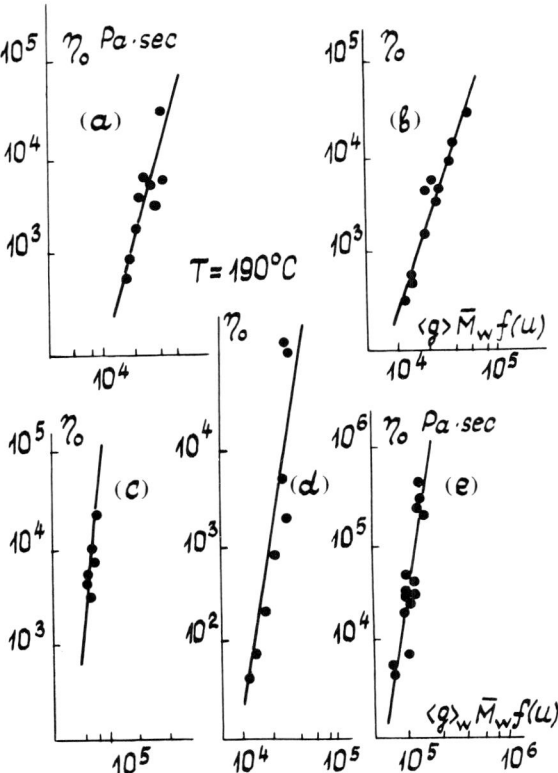

Figure 9 Dependences of η_0 on $\langle g \rangle \bar{M}_w f(u)$ for sets of LDPE obtained in one reactor according to (a) Ref. 60, (b) Ref. 62, (c) Ref. 63, (d) Ref. 77, and (e) Ref. 78. For each group the constants in equation (50) are different.

Relaxation Time Spectrum

A large amount of experimental work confirms the view that the distribution density of relaxation times reflects the nature of the MMD. It may be assumed that each particle of the system is characterized by its own relaxation time or, in the case of a macromolecule, by a certain set of relaxation time values. There exists, accordingly, a definite relationship between the relaxation time spectrum and MMD.

In molecular theories that describe all types of motion by means of the same friction coefficient, it is easy to find the friction coefficient for a polymer mixture (Eq. 54). This effect is assumed equal for all components of the mixture. Ninomiya and Ferry used these assumptions for evolving the laws of addition of viscosities and viscoelasticity parameters for two monodisperse polymers [80–82]. The relaxation spectrum $H(\tau)$ can, in this case, be described in the following way:

$$H(\tau) = W_1 H_1 \frac{\tau \eta_{01} \bar{M}_w}{\eta_0 M_1} + W_2 H_2 \frac{\tau \eta_{02} \bar{M}_w}{\eta_0 M_2} \tag{58}$$

where W_1 and W_2 are the weight contributions of the components.

Bueche attempted to extend Eq. 58 for the case of continuous distribution of molecular mass $f_n(M)$:

$$H(\tau) = \frac{\rho RT}{\bar{M}_n} \int_0^\infty \Sigma \tau_p \delta(\tau - \tau_p) f_n(M) dM \tag{59}$$

where

$$\tau_p = \frac{6 \eta M^2}{\pi^2 p^2 \rho \bar{M}_w RT} \tag{60}$$

$f_n(M)$ being the numerical function of molecular mass distribution, and δ the Dirac delta function.

On the basis of the work of Rouse and Zimm, Peticolas and Meneffi evolved a generalized molecular theory for polydisperse polymers and obtained dependences for building up a relaxation spectrum from the relaxation curve of stress. Experimental tests have shown that calculated MMD curves are narrower than experimental ones.

Transformations based on the theory of numbers allow us to obtain $W(M)$ from time (frequency) dependence of a certain parameter of viscoelasticity. Other attempts at determining the MMD curve from the shape of rheologic functions are also known. However, existing theoretical work does not yield the necessary accuracy of results which could be used in practice. The calculated MMD curve is always narrower than the experimental one.

The Greassley network theory [83] was extended for the case of polydisperse polymers, but the results have not been subjected to detailed experimental testing. Some of the theoretical conclusions do not agree with experimental data [17]. The theory exaggerates the degree of effect of polydisperity on elastic properties of polymers.

In some cases satisfactory agreement has been achieved between the MMD curve and that of the flow [84].

Several phenomenologic theories [85–88] make use of the idea of cutoff of the relaxation spectrum at shear flow. Bersted [89, 90] applies this theory in MMD calculations from the flow curve. He proposed the hypothesis that in molecular flow all the masses that lie above a certain critical value (which, in its turn, depends on the shear rate) should be considered as having relaxation times equal to that of molecules of critical mass. The effective viscosity and compliance in shear flow can be described by the expression

$$\eta(\dot{\gamma}) = \int_0^{\tau_c(\gamma)} H(\tau)d\tau \tag{61}$$

$$I(\dot{\gamma}) = \frac{\int_0^{\tau_c(\gamma)} \tau H(\tau) d\tau}{[\eta(\dot{\gamma})]^2} \tag{62}$$

Effective viscosity may be expressed through molecular mass

$$\log \eta(\dot{\gamma}) = A + B \log \bar{M}_w^* \tag{63}$$

where

$$\bar{M}_w^* = \sum_{i=1}^{i=c-1} W_i M_i + M_c(\dot{\gamma}) \sum_{i=c}^{i=\infty} W_i$$

The Bersted theory has been developed by Pedersen and Ramm for the case of branched molecules [91]. In terms of this theory the average molecular mass value of the parameter $(gM)_w$ was found; it is determined by the root mean square radius of the macromolecular tangle. The agreement between calculations according to Eqs. 61 and 62 and experimental results has been observed only within a limited range of shear rates, particularly in the case of Eq. 62.

Attempts to establish a connection between the flow curve and MMD are also based on the theoretically calculated dependence for polydisperse polymers, in which the relaxation time is expressed through molecular mass value. The dependence $\eta(\dot{\gamma})$, found through molecular mass, is determined by summation of dependences for all fractions, each of them acting in proportion to its part in the distribution. Middleman [92] thus obtained

$$\eta(\dot{\gamma}) = \frac{1}{\bar{M}_w \bar{M}_n} \int_0^\infty M^2 F[\dot{\gamma}, \tau(M)] f_n(M) dM \tag{64}$$

where $F[\dot{\gamma}, \tau(M)]$ is the dependence of viscosity on the shear rate for monodisperse polymers.

Frequency Dependences of Viscoelasticity Parameters

Frequency dependences of the components of the complex modulus of a melt of polydisperse polymers differ from corresponding dependences of monomolecular polymers. In the low-frequency region numerically corresponding to shear rate, at which Newtonian flow of the melt takes place, we have G'' proportional to frequency and G' proportional to its square value, as in the monomolecular case. For a polydisperse polymer, the $\bar{M}_w$ value of which is equal to M of a monodisperse polymer, the divergence of G' from quadratic and of G'' from linear frequency dependence starts at considerably lower frequencies than in the monodisperse case, and the $G'(\omega)$ and $G''(\omega)$ dependences have a smaller slope. The broader the MMD curve, the slower the rate of increase of G' and of G'' with frequency.

The $G''(\omega)$ dependence curve does not show any peak within the high elasticity range, which always appears in the case of narrow-distribution polymers. In the plateau region the modulus components are not constant, the value increasing slowly with deformation frequency. In this region sometimes changes are observed in the mutual position of the curves representing $G'(\omega)$ or $G''(\omega)$ in the case of some polymer homologs, which is due to the shape of the MMD dependence. The high elasticity plateau of polymers with broad MMD as well as that of polymers with narrow MMD lengthens with increase in molecular mass. $G'(\omega)$ and $G''(\omega)$ show a faster rate of increase with frequency in the case of long-chain branching than in the case of linear polymers. In the plateau region these functions reveal weak dependence not only on molecular mass, but also on long-chain branching of the molecules.

Figure 10 presents $G'(\omega)$ and $G''(\omega)$ dependences for LDPE [78]. The shape of these dependences is typical of industrial polymer melts; similar results have been obtained by Schroff and Shida [62, 79] and others. Figure 11, constructed according to data of Schroff and Shida, shows that dynamic viscosity $\eta'(\omega)$ decreases faster with frequency in the case of a branched polymer melt than in that of linear one.

Effect of Synthesis Reactor on Monotypeness of MMD

Calculation of the MMD curve from kinetic data is of extreme interest from the point of view of molecular structure research. The problem reduces to calculation of the MMD function on the basis of the known reaction mechanism and of kinetic constants. MMD has been calculated for various mechanisms of chain growth by several workers [93, 94].

For each polymerization type the MMD function can be calculated. Changes in the parameters of the polymerization process (monomer content, initiator and transfer agents, reaction rate constants) leave the unimodal nature of MMD unaltered but produce a change in its width. This holds only for the case of smooth rate of temperature change.

Frenkel suggests that a large number of theoretical distributions may be reduced to one preliminary distribution whose parameters have a clear meaning [95]. An analysis of these parameters can be performed by estimating the ratio $\bar{M}_w/\bar{M}_n$, which leads to a conclusion on the monotypeness of distribution, i.e., to a possibility of describing them by means of one function, in which a change of parameters produces a change in the $\bar{M}_w/\bar{M}_n$ ratio. Under these conditions polymers of one homologous series may be described by means of one common MMD function.

This theory applies to the case of ideal polymerization. For a tubular reactor an ideal polymerization scheme can be achieved if there is no distribution of flow rates along the cross section of the tube and we have piston flow; all molecules spend equal stretches of time in the polymerization zone, and all are under the same conditions. In a real reactor we always have a distribution in the time spent by the reagent in the reactor owing to distribution in flow rate over the cross section of the tube. The gradient of flow rate also determines the conditions of mixing of the reagents. In addition, there exists a critical value of temperature difference between the axis of the reactor and the wall. If this is exceeded, the reaction proceeds in a nonstationary regime. These peculiarities of a real reactor affect the polymerization process and thus also the nature of the MMD.

Imposed distortions do not correspond to normal kinetics of the reaction and are characteristic of the given reactor only. Accordingly a qualitative relationship between synthesis parameters and MMD is determined by the type of reaction and does not depend on individual peculiarities of the reactor. On the other hand, these peculiarities may have considerable effect on quantitative relationships. Hence the MMD of products obtained from one and the same reactor must be similar, as may be seen from a comparison of molecular and rheological parameters. Studies performed lead to the conclusion that polymers obtained from different assemblies of synthesis possess unaccountable parameters $\bar{M}_w$, $\langle g \rangle_w$, and u. For one set these characteristics may be accounted for with

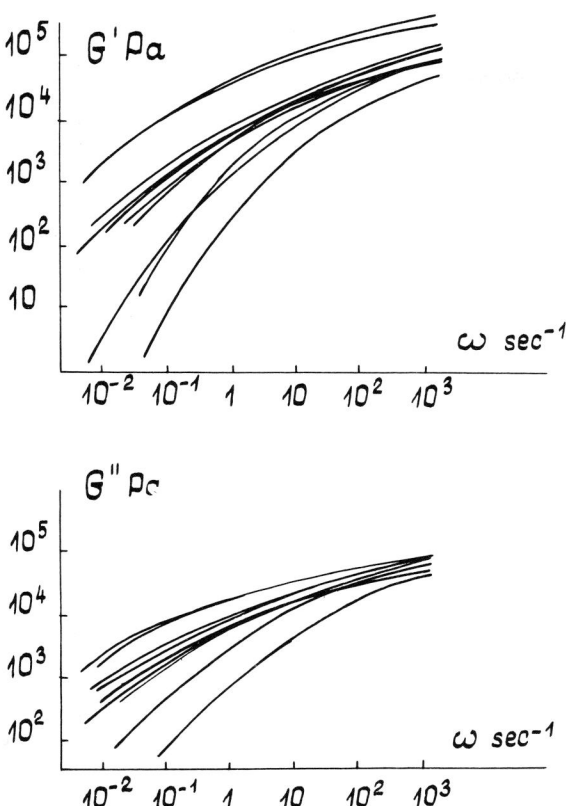

Figure 10 Frequency dependence of components of the complex modulus for polydisperse polymers. (Data from Ref. 78.)

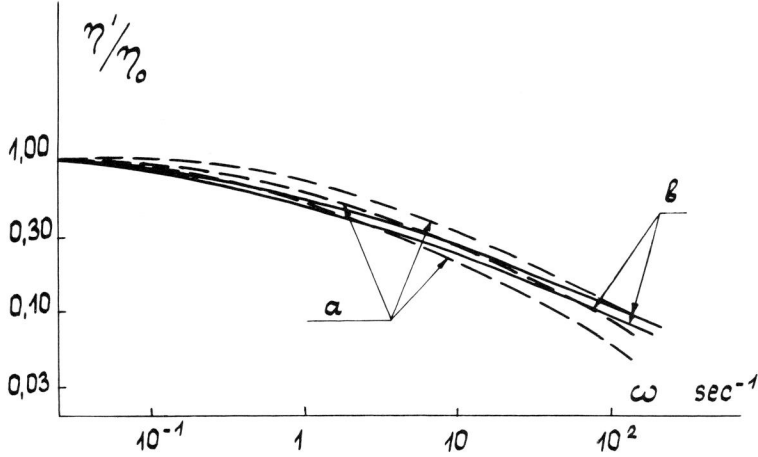

Figure 11 Frequency dependences of the dynamic viscosity for (a) linear and (b) branched polymers.

accuracy sufficient for practical purposes by using the coefficients of the equations relating rheological and molecular parameters [96]. The effects of methods of synthesis on molecular structure and on the nature of MMD have been studied in many investigations on commercial polymers (e.g., Ref. 97).

INTERRELATION BETWEEN MOLECULAR AND VISCOELASTIC PARAMETERS

Averaging of Viscosities

To ascertain the relation between viscoelastic parameters and molecular structure indices of commercial polymers it is necessary to know, in addition to the MMD, the distribution of long-chain branching over molecular masses. As a characteristic of long-chain branching of macromolecules the conformation parameter g for monodisperse fractions and its mass average value $\langle g \rangle_w$ for polydisperse test samples may be used. The $\langle g \rangle_w$ and g values depend on the branching index: the number of knots of the long-chain branching $\langle n \rangle_w$ and n per molecule.

Initial Newtonian viscosity of a polymer melt can be expressed through its molecular structure indices by means of Eq. 50. A polymer possesses a continuous unimodal molecular mass distribution $W(M)$, the mass average molecular mass of which is described according to Eq. 8.

Let us consider a polymer as a mixture of an infinite number of elementary fractions of molecular mass M in the M and $M + dM$ interval. Expressing its mass average molecular mass in Eq. 50 according to Eq. 8, we have

$$\eta_0 = k \left[\langle g \rangle_w f(u) \int_0^\infty MW(M)dM \right]^\theta \tag{65}$$

Maximum Newtonian viscosity may be expressed by a formula following from Eq. 50 for $u = 0$ (then $\bar{M}_w = M$, $\langle g \rangle_w = g$), and

$$\eta_{0f} = k(gM)^\theta \tag{66}$$

This makes it possible to replace the molecular mass of the elementary reaction in Eq. 65 by the corresponding viscosity:

$$\eta_0 = \left[\langle g \rangle_w f(u) \int_0^\infty \eta_{0f}^{1/\theta} g^{-1} W(M)dM \right]^\theta \tag{67}$$

The parameter of long-chain branching is found by means of the Zimm–Stockmayer formula [70] from the branching index. According to the Mendelson–Drott hypothesis [98] the long-chain branching index is proportional to the molecular mass

$$n = \ell M \tag{68}$$

The mass average value of the conformational parameter of a branched molecule can be calculated from

$$\langle g \rangle_w = \frac{\int_0^\infty gMW(M)dM}{\int_0^\infty MW(M)dM} \tag{69}$$

Extension of the Addition Law to the Relaxation Spectrum

The law obtained for averaging of viscosity (Eq. 67) is applied to dynamic viscosity. Through averaging frequency dependences of dynamic viscosities η'_f of elementary fractions the frequency dependence of the dynamic frequency of a polymolecular polymer was obtained as

$$\eta'(\omega) = \left[\langle g \rangle_w f(u) \int_0^\infty \eta'^{1/\theta}_f(k_\tau, \omega) g^{-1} W(M) dM \right]^\theta \qquad (70)$$

where $\eta'_f(k_\tau, \omega)$ is the frequency dependence of the dynamic viscosity of an elementary fraction. The effect of surrounding macromolecules of different mass on relaxation time of the molecule under study is accounted for by the coefficient k_τ at deformation frequency [54, 80–82]. The coefficient k_τ is defined by the expression

$$k_\tau = \frac{M \eta_0}{\bar{M}_w \eta_{0f}}$$

which, after accounting for Eqs. 50 and 66, assumes the form

$$k_\tau = \left(\frac{\bar{M}_w}{M} \right)^{\theta-1} \left(\frac{\langle g \rangle_w}{g} \right)^\theta f^\theta(u) \qquad (71)$$

It was shown by Ferry [54] that the exponent at $\bar{M}_w/M$ may differ from $\theta - 1$. We therefore replace $\theta - 1$ by β in Eq. 71 and find β from experiment. The value of β remains constant for polymers of the same homologic series:

$$k = \left(\frac{\bar{M}_w}{M} \right)^\beta \left(\frac{\langle g \rangle_w}{g} \right)^{\beta+1} f^{\beta+1}(u) \qquad (72)$$

The relaxation frequency spectrum is found from

$$N(s) = \frac{2}{\pi s} \operatorname{Re}[se^{\pm i\pi/2} \eta'(se^{\pm i\pi/2})] \qquad (73)$$

from which we obtain

$$N(s) = \frac{2}{\pi s} \operatorname{Re} \left\{ se^{\pm i\pi/2} \left[\langle g \rangle_w f(u) \int_0^\infty \eta'^{1/\theta}(k_\tau, se^{\pm i\pi/2}) W(M) dM \right]^\theta \right\} \qquad (74)$$

Here $\eta'_f(k_{\tau_1}, se^{\pm i\pi/2})$ is the frequency dependence of the dynamic viscosity of an elementary fraction, in which the circular frequency ω is replaced by $se^{\pm i\pi/2}$.

Formation of Viscoelastic Functions from Molecular Structure Parameters

For an analytical description of viscoelastic properties it is sufficient to describe one of the viscoelastic functions, since they are all interconnected. It is simplest to describe the function $\eta'(\omega)$. For this it is necessary to take an equation which would be sufficiently simple for practical use and sufficiently flexible for describing the specific effects of molecular parameters on the $\eta'(\omega)$ dependence. The coefficients of the expression must have definite physical meaning. These requirements are met by the equation

$$\eta' = \eta_\infty + \frac{\eta_0 - \eta_\infty}{1 + (\tau\omega)^\alpha} \tag{75}$$

which has been successfully applied for describing $\eta'(\omega)$ for LDPE [78]. For polymer melts we usually have $\eta_0 \gg \eta_\infty$. If the value of η' at the highest studied frequency exceeds η_∞ very strongly, Eq. 75 transforms into a three-constant equation

$$\eta' = \frac{\eta_0}{1 + (\omega\tau_r)^\alpha} \tag{76}$$

For polydisperse polymers dynamic viscosity can be represented by this dependence within a large range of frequencies [78, 96]. For characteristic relaxation time of a fraction we may use the relation $\tau_r = k_1(gM)^\theta$.

After introducing coefficient k_τ into the relaxation time in Eq. 76, we obtain an expression for frequency dependence of the elementary fraction:

$$\eta'_f = k(gM)^\theta \{1 + k_1 \omega g^{\theta-\beta+1} \langle g \rangle_w^{\beta+1} M^{\theta-\beta} \bar{M}_w^\beta f^{\beta+1}(u)]^{\alpha_0}\}^{-1} \tag{77}$$

If the frequency function η'_f is given in the form of Eq. 77, then the expression for calculating $\eta'(\omega)$ assumes [96] the form

$$\eta' = k \left\{ f(u)\langle g \rangle_w \int_0^\infty \frac{MW(M)dM}{\{1 + [k_1 \omega g^{\theta-\beta+1}\langle g \rangle_w^{\beta+1} M^{\theta-\beta} \bar{M}_w^\beta f^{\beta+1}(u)\alpha_0\}^{1/\theta}} \right\}^\theta \tag{78}$$

The constants k, k_1, θ, and α_0 are found from dependences of rheologic parameters on the molecular mass of the monomolecular polymers. Frequency dependences of dynamic viscosity obtained from Eq. 78 are presented in Figure 12.

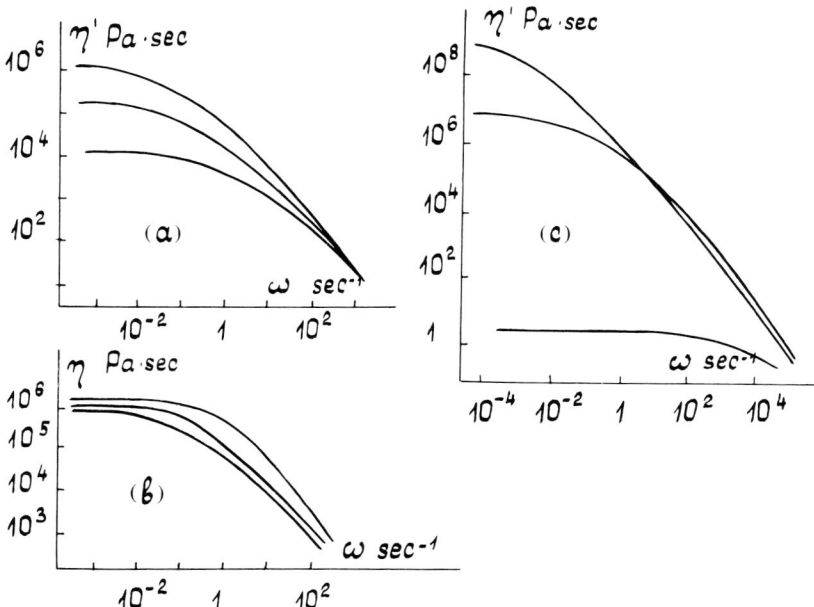

Figure 12 Effect of molecular parameters on frequency dependence of dynamic viscosity. (From Ref. 96.)

Dependence of Viscoelasticity Parameters of Melt on Molecular Structure

Frequency dependences of dynamic viscosity, as calculated from Eq. 78 for test samples of different molecular structure, are in fair agreement with known concepts on the effect of these parameters on the functions of viscoelasticity. It has been established that Eq. 78 satisfactorily agrees with experimental results (Fig. 13).

If we apply the approximation (Eq. 76) for a polydisperse polymer, the viscoelastic relaxation spectrum is determined by three parameters:

$$\eta_0 = \lim_{\omega \to 0} \eta'(\omega)$$

$$\tau_r = \frac{1}{\omega_r}; \qquad 2\eta'(\omega_r) = \eta_0$$

$$\alpha = -\frac{d(\log \eta')}{d(\log \omega)}; \qquad \omega \gg \omega_r$$

for which one obtains from Eq. 78 the following expressions:

$$\bar{M}_w^{-1} \int_0^\infty \frac{MW(M)dM}{\{1 + [k_1\tau_r^{-1}g^{\theta-\beta-1}\langle g\rangle_w^{\beta+1}M^{\theta-\beta}\bar{M}_w^\beta f^{\beta+1}(u)]^{\alpha_0}\}^{1/\theta}} = \left(\frac{1}{2}\right)^{1/\theta} \qquad (79)$$

$$\alpha = \alpha_0\omega \frac{\alpha_0 \int_0^\infty \dfrac{g^{(\theta-\beta-1)\alpha_0}\langle g\rangle_w^{(\beta+1)\alpha_0}M^{1+\alpha_0(\theta-\beta)}\bar{M}_w^{\alpha_0\beta}f^{\alpha_0(\beta+1)}(u)W(M)dM}{\{1 + [k_1\omega g^{\theta-\beta-1}\langle g\rangle_w^{\beta+1}M^{\theta-\beta}\bar{M}_w^\beta f^{\beta+1}(u)]^{\alpha_0}\}^{1+1/\theta}}}{\int_0^\infty \dfrac{MW(M)dM}{\{1 + [k_1\omega g^{\theta-\beta-1}\langle g\rangle_w^{\beta+1}M^{\theta-\beta}\bar{M}_w^\beta f^{\beta+1}(u)]^{\alpha_0}\}^{1/\theta}}} \qquad (80)$$

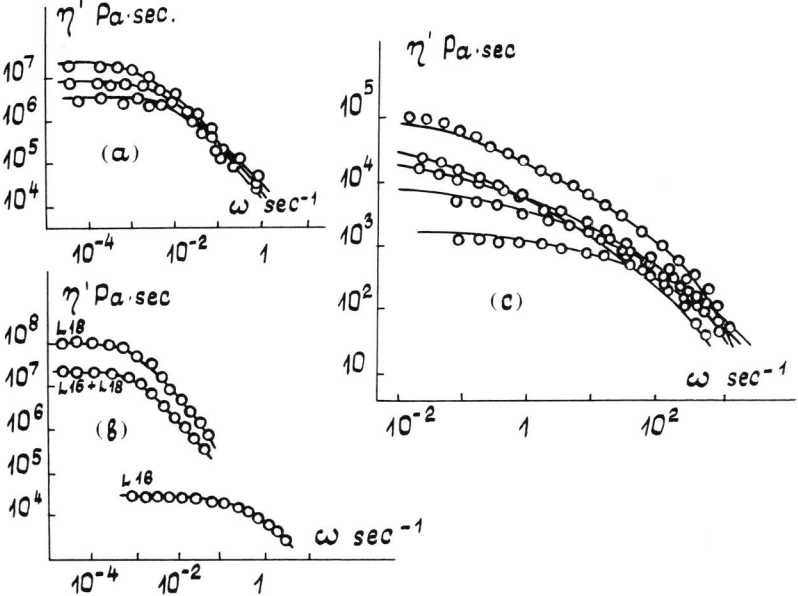

Figure 13 Dynamic viscosity of polystyrene (*a*), (*b*) and LDPE (*c*). Solid lines calculated using Eq. 78; dots, experimental data. (From Ref. 96.)

Coefficient α is calculated for $\omega \gg 1/\tau_r$. Expressions 50, 79, and 80 make it possible to analyze the dependences of η_0, τ_r, and α on molecular structure parameters (Fig. 14).

The relaxation spectrum of a polymer is interconnected with the frequency dependence of the components of the complex modulus. For the loss modulus the dependence is of the form

$$N(s) = \frac{2}{\pi s} \text{Re}[G''(se^{\pm i\pi/2})]_{s=\omega} \tag{81}$$

If the frequency dependence of dynamic viscosity is assumed in the form of Eq. 76, then we obtain from Eq. 81 for the relaxation spectrum

$$N(s) = \frac{2}{\pi} \eta_0 \sin\frac{\alpha\pi}{2} \left[(\tau_r s)^\alpha + 2\cos\frac{\alpha\pi}{2} + (\tau_r s)^{-\alpha} \right]^{-1} \tag{82}$$

The loss modulus is determined according to Eq. 39, and if the relaxation spectrum is given by Eq. 82, it assumes the form

$$G'(\omega) = \omega^2 \frac{\eta_0}{\tau_r} \psi_\omega \tag{83}$$

where ψ_ω is the deformation frequency function

$$\psi_\omega = 2\sin\frac{\alpha\pi}{2} \int_0^\infty \left[(z + \omega^2\tau^2)\left(z^{-\alpha} + 2\cos\frac{\alpha\pi}{2} + z^\alpha\right) \right]^{-1} dz \tag{84}$$

and $z = s\tau_r$ is a dimensionless relaxation frequency. In reduced coordinates $v = \omega\tau_r$ the dependence (Eq. 83) becomes transformed into

$$G'(v) = v^2 \frac{\eta_0}{\tau_r} \psi_v \tag{85}$$

where ψ_v is obtained from Eq. 84 by replacing $\omega\tau$ by v.

For linear polymers η_0 and τ_r show the same dependence on molecular mass, while ψ_v is a function of coefficient α only. In $G'-v$ coordinates the loss modulus therefore depends only on polydispersity and not on molecular mass. As a result, the $G'(\omega\tau)$ curves for linear polymer melts of differing MMD width do not coincide. For branched polymer melts, as may be seen in Figure 14, η_0 and τ_r are not similar functions of mass average molecular mass and long-chain branching; hence $G'(v)$ is only scarcely dependent on these molecular parameters. A similar picture is obtained for the modulus of loss: from Eq. 76 we have from G'':

$$G'' = \frac{\omega\eta_0}{1 + (\omega\tau_r)^\alpha} \tag{86}$$

Changing over to $G''-v$ coordinates, we obtain from Eq. 86

$$G''(v) = \frac{\eta_0}{\tau_r} \cdot \frac{v}{1 + v^\alpha} \tag{87}$$

From Eq. 87 we obtain the same conclusion for $G''(v)$ dependence on molecular parameters as for $G'(v)$.

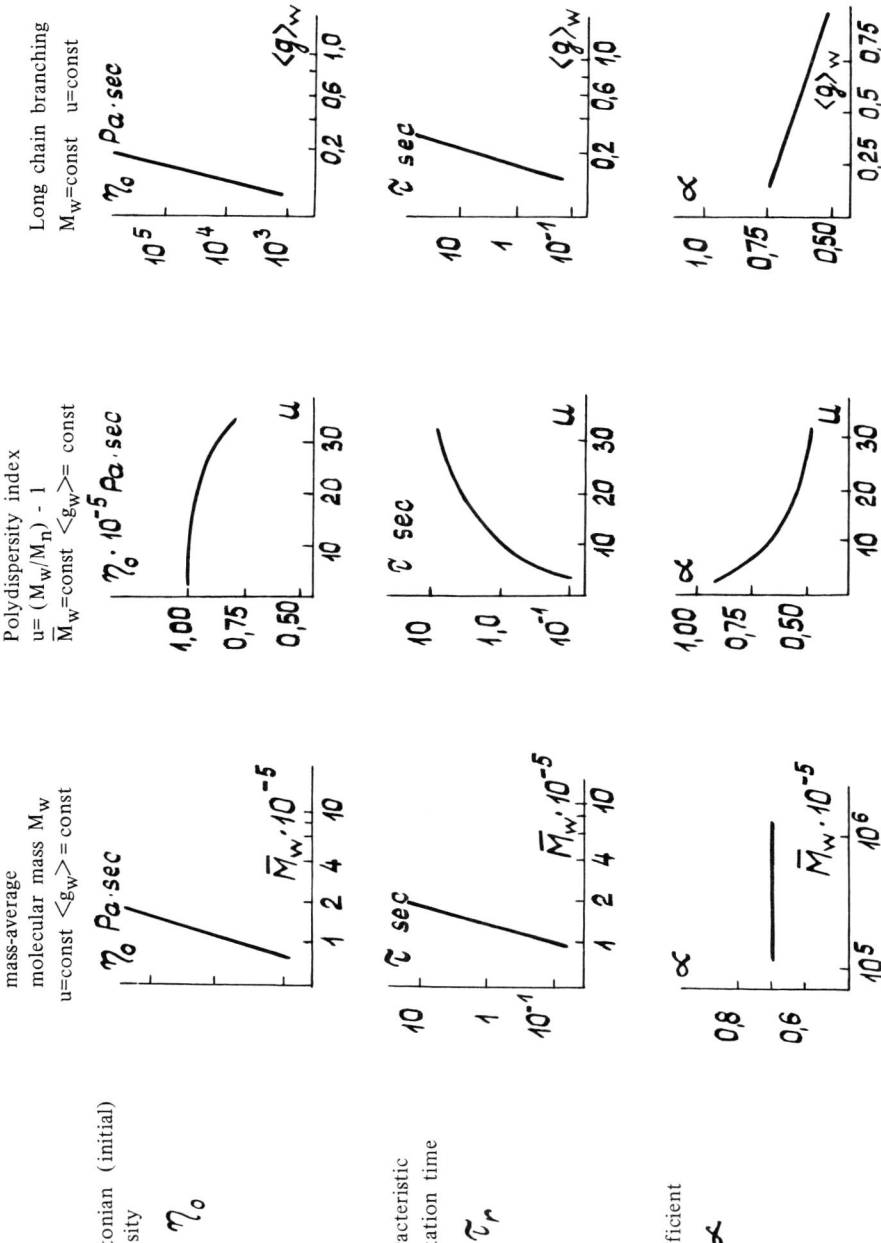

Figure 14 Effect of molecular parameters on viscoelasticity parameters.

Application of Analogies Between Viscosity and Viscoelasticity Parameters in the Study of Rheological and Molecular Parameter Dependences

A number of useful approximations of rheological parameter dependences on molecular characteristics may be obtained from the analogies

$$\eta'(\omega) = \eta(\dot\gamma) \tag{88a}$$

$$\zeta'(\omega) = \zeta(\dot\gamma) \tag{88b}$$

if $\omega = \dot\gamma$.

The first normal stress difference $\sigma_{11} - \sigma_{22} = 2\zeta\dot\gamma^2$ may be expressed through frequency functions of linear viscoelasticity by means of analogy. The coefficient of normal stresses at linear periodic deformation is determined by the dependence $\zeta' = G'/2\omega^2$. Hence by analogy with Eq. 88b we have

$$\sigma_{11} - \sigma_{22} = 2G'(\omega) \tag{89}$$

We may thus gain information on the effect of molecular structure parameters on the first normal stress difference by considering the previously discussed effect of molecular parameters on $G'(\omega)$ and juxtaposing $\sigma_{11} - \sigma_{22}$ and G' if $\dot\gamma = \omega$.

High-elastic deformations, which are accumulated by the polymer melt at shear flow, equal, according to the Lodge formula,

$$e = \frac{(\sigma_{11} - \sigma_{22})}{2\sigma_{12}} \tag{90}$$

From analogy with Eq. 88a we obtain $\eta'\omega = \eta\dot\gamma$ or $G'' = \sigma_{12}$ at $\omega = \dot\gamma$.

Taking notice of Eq. 89 and replacing σ_{12} by G'' in Eq. 90, we obtain

$$e = \frac{G'}{G''} = \cot \delta \tag{91}$$

at $\omega = \dot\gamma$. By using the dependence in Eq. 86, G may be represented in the form of Eq. 87. Approximating dynamic viscosity by means of Eq. 76, the storage modulus in G'–v coordinates is given by Eq. 85; hence, for reversible deformation, we have

$$e = v(1 + v^\alpha)\psi_v \tag{92}$$

As can be seen from (92), the effect of molecular parameters in highly elastic deformations at shear flow is determined, using dimensionless coordinates $v = \omega\tau_r = \dot\gamma\tau_r$, by the previously mentioned parameter ψ_v only, i.e., it basically depends on the width of molecular mass distribution.

High elastic deformation at stationary flow within the region of low rates of shear are proportional to the reduced velocity and equal $e \approx v\psi_v$, and at high rates of shear $e \approx v^{1+\alpha}\psi_v$. It follows that in natural coordinates e–v the effect of characteristic relaxation time and hence of molecular mass and branching on e is enhanced with increase in $\dot\gamma$. Similar effects follow also from experimental results (Fig. 15).

It may be seen from Eq. 42 that initial compliance is determined by the expression

$$I_0 = \frac{\zeta_0}{\eta_0^2} \tag{93}$$

Since we have

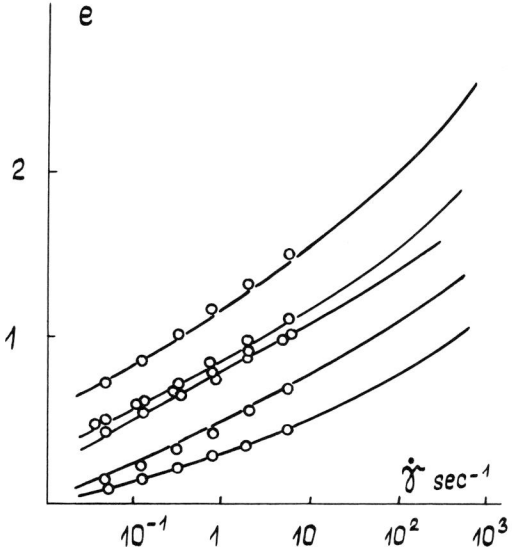

Figure 15 Dependence of highly elastic deformation on shear rate for LDPE of varying molecular structure. Molecular parameters are given in Eq. 78.

$$\zeta = \frac{\sigma_{11} - \sigma_{22}}{2\dot{\gamma}^2} \tag{94}$$

which, considering Eq. 89 transforms into $\dot{\gamma} = \omega$ at $\zeta = G'/\omega^2$, we obtain for Eq. 93

$$I_0 = \lim_{\omega \to 0} \frac{G'(\omega)}{\eta_0^2 \omega^2} \tag{95}$$

If we express $G'(\omega)$ according to Eq. 85, we get

$$I_0 = \frac{\tau_r}{\eta_0} \cdot \psi(0) \tag{96}$$

where

$$\psi_0 = \lim_{\omega \to 0} \psi_\omega.$$

It follows from Eq. 96 that for linear polymers, for which $\eta_0 \sim \tau_r$, initial compliance does not depend on molecular mass and $\psi(0)$ depends insignificantly on polydisperseness; hence J_0 depends on polydisperseness to only a small extent.

APPROXIMATED DEPENDENCES FOR TAKING INTO ACCOUNT RELATIVE CHANGES IN MOLECULAR STRUCTURE IN TECHNOLOGICAL PROCESSES

Linear Polymers

In the course of processing polymers we frequently observe mechanochemical reactions which depend on technological parameters. As a result the average molecular mass and the MMD width undergo changes. To ascertain changes in molecular parameters after processing or for the purpose of systematic monitoring it is necessary to determine them very quickly. Since physicochemical methods are time-consuming it is more viable to judge changes in molecular parameters from changes in rheological characteristics. Dependence (Eq. 50) for linear polymers ($\langle g \rangle_w = 1$) assumes the form $\eta_0 = k[\bar{M}_w f(u)]^\theta$; hence the relative change in viscosity may be expressed as

$$\frac{\eta_{02}}{\eta_{01}} = \left(\frac{\bar{M}_{w2}}{\bar{M}_{w1}}\right)^\beta \exp[k_n(u_2 - u_1)] \tag{97}$$

For polyethylene the value of 4.38×10^{-3} has been given for k_n [78]. However, the value of k depends on the type of MMD and, as shown by Briedis and Alkine [99], lies between 10^{-3} and 10^{-2}. Supposing that k has its maximum value and considering that the change in MMD width is such in the processing that u changes not more than twice its value, we find that the highest value of the factor $\exp[k(u_2 - u_1)]$ does not exceed 1.03–1.05 (depending on the initial width of MMD), which is less than the accuracy of determining η_{02}/η_{01}. In the calculation of M_w from changes in μ neglect of polydispersity produced errors not exceeding 1.2%. Hence the factor $\exp[k(u_2 - u_1)]$ may be omitted in Eq. 97, and the relative change in molecular mass at $\theta = 3.4$ can be determined according to Eq. 96 as follows:

$$\frac{\bar{M}_{w2}}{\bar{M}_{w1}} = \left(\frac{\eta_{02}}{\eta_{01}}\right)^{0.294} \tag{98}$$

For a linear polymer η_0 and τ_r are equally dependent on M_w; hence digression from the ratio $\eta_{02}/\eta_{01} = \tau_{r2}/\tau_{r1}$ indicates change in polydispersity.

The following value has been obtained [96] for the characteristic relaxation time of a linear polydisperse polymer:

$$\tau_r = k_r[\bar{M}_w(u + 1)^{1/2}]^{3.4} \tag{99}$$

From Eqs. 98 and 99 we determine the relative change in the polydispersity index:

$$\left(\frac{u_2 + 1}{u_1 + 1}\right) = \left(\frac{\tau_{r2}\eta_{01}}{\tau_{r1}\eta_{02}}\right)^{0.588} \tag{100}$$

It may be seen from Eqs. 98 and 100 that the change in mass average molecular mass and in polydispersity index can be found from changes in the basic characteristics of viscoelasticity η_0 and τ_r.

Branched Polymers

If cleavage of branched macromolecules takes place during processing, then one may consider that, in the case of a sufficiently large degree of branching, the ratio between the average number of knots before and after processing equals that between the mass average molecular mass before and after treatment. In other words, changes in branching in this case take place in agreement with the Mendelson–Drott hypothesis [98]. For mass average averaging we have

$$\ell = \frac{\langle n \rangle_w}{\bar{M}_w} = \text{const} \tag{101}$$

The change in long-chain branching at processing is described by the ratio

$$\frac{\langle n\rangle_{w2}}{\langle n\rangle_{w1}} = \frac{\tilde{M}_{w2}}{\tilde{M}_{w1}} \tag{102}$$

The change in Newtonian viscosity of a branched polymer, as found from Eq. 50, equals

$$\left(\frac{\eta_{02}}{\eta_{01}}\right)^{1/\theta} = \frac{\langle g\rangle_{w2}\tilde{M}_{w2}}{\langle g\rangle_{w1}\tilde{M}_{w1}} \tag{103}$$

In Eq. 103, as well as in the case of a linear polymer, the effect of polydispersity is neglected. The dependence (Eq. 103), considering the ratio in Eq. 102, transforms into an expression determining only the characteristics of long-chain branching, i.e., $\langle g\rangle_w$ and $\langle n\rangle_w$:

$$\left(\frac{\eta_{02}}{\eta_{01}}\right)^{1/\theta} = \frac{\langle g\rangle_{w2}\langle n\rangle_{w2}}{\langle g\rangle_{w1}\langle n\rangle_{w1}} \tag{104}$$

The conformational parameter of a branched molecule $\langle g\rangle_w$ is functionally linked with the branching index $\langle n\rangle_w$; hence it becomes possible to find both these values from changes in viscosity η_0. One mostly uses the Zimm–Stockmayer formula for this purpose:

$$\langle g\rangle_w = \frac{G}{\langle n\rangle_w}\left[\frac{1}{2}\left(\frac{2+\langle n\rangle_w}{\langle n\rangle_w}\right)^{1/2} \ln\frac{(2+\langle n\rangle_w)^{1/2}+\langle n\rangle_w^{1/2}}{(2+\langle n\rangle_w)^{1/2}-\langle n\rangle_w^{1/2}} - 1\right] \tag{105}$$

If $\langle n\rangle_{w1}$ is known, one finds $\langle g\rangle_{w1}$, which makes it possible to calculate the product $\langle g\rangle_{w2}\langle n\rangle_{w2}$. Since the dependence $\langle g\rangle_w = F(\langle n\rangle_w)$ is transformable into $\langle g\rangle_w\langle n\rangle_w = \langle n\rangle_w F(\langle n\rangle_w)$, we solve the latter expression numerically to find $\langle n\rangle_w$ and hence $\langle g\rangle_w$.

For determining the relative change in MMD width it is necessary to know the change in characteristic relaxation time. This can be found for a branched polymer from the dependence (Eq. 96)

$$\tau_r = k_1 \langle g\rangle_w^{\alpha_\tau} \tilde{M}_w^{\beta_\tau} F(u) \tag{106}$$

where k_1, α_τ, and β_τ are constants of the polymer of the given type of MMD, and $F(u)$ is the polydispersity function.

With the aid of Eqs. 50 and 106 we form a parameter which is independent of molecular mass:

$$\frac{\tau^{\theta/\beta_\tau}}{\eta_0} = \frac{k_1^{\theta/\beta_\tau}}{k} \cdot \langle g\rangle_w^{(\alpha_\tau - \beta_\tau)\theta/\beta_\tau} F^{\theta/\beta_\tau}(u) f^{-\theta}(u) \tag{107}$$

hence the following characteristic is singled out:

$$\varphi(u) = \frac{k_1^{\theta/\beta_\tau}}{k} \cdot f^{\theta/\beta_\tau}(u) \cdot f^{-\theta}(u) \tag{108}$$

which depends on the polydispersity index only.

The ratio between the characteristics $\varphi(u)$ before and after processing equals

$$\frac{\varphi(u_2)}{\varphi(u_1)} = \frac{\eta_{01}}{\eta_{02}}\left[\frac{\tau_2}{\tau_1}\left(\frac{\langle g\rangle_{w2}}{\langle g\rangle_{w1}}\right)^{\beta_\tau - \alpha_\tau}\right]^{\theta/\beta_\tau} \tag{109}$$

The function $\varphi(u)$ is, in first approximation, linearly dependent on the polymolecularity index (Eq. 100) (Fig. 16), i.e., we have $\varphi(u) = A_x + k_x u$. For distributions of sufficient width $k_x u \gg A$, hence

$$\frac{u_2}{u_1} = \frac{\eta_{01}}{\eta_{02}} \left[\frac{\tau_2}{\tau_1} \left(\frac{\langle g \rangle_{w2}}{\langle g \rangle_{w1}} \right)^{\beta_\tau - \alpha_\tau} \right]^{\theta/\beta_\tau} \quad (110)$$

for results obtained in Eq. 78, we have $\alpha_\tau = \theta + 2$, $\beta_\tau = \theta + 1$. In this case Eq. 110 is simplified and assumes the following form:

$$\frac{u_2}{u_1} = \frac{\eta_{01}}{\eta_{02}} \left[\frac{\tau_{r2}}{\tau_{r1}} \cdot \frac{\langle g \rangle_{w2}}{\langle g \rangle_{w1}} \right]^{\theta/(\theta+1)} \quad (111)$$

At definitive relative η_0 and τ_r values the relative change of the basic molecular parameters $\bar{M}_w$, u, and $\langle g \rangle_w$ or ($\langle n \rangle_w$) may be found from Eqs. 102, 104, and 111. For this purpose it is necessary to know the index characterizing the degree of dependence of Newtonian viscosity of branched polymer melts on mass average molecular mass, as well as the initial value of the branching index $\langle n \rangle_{w1}$, the corresponding value $\langle g \rangle_{w1}$. Considering that the dependence $\langle g \rangle_w = F(\langle n \rangle_w)$ does not change its nature within a narrow range of change of $\langle n \rangle_w$, it is sufficient to know the nominal characteristics of branching of the material of the given type, in order to determine the relative changes in long-chain branching.

Determination of Viscoelasticity Parameters η_0 and τ_r

The most convenient and shortest way of determining the parameters η_0 and τ_r lies in the determination of the resistance of a melt to linear periodic deformation at various frequencies. Briedis and colleagues [96, 101] proposed a method that makes use of the approximation of Eq. 76. If periodic deformation is effected at three frequencies characterized by the ratios

$$\frac{\omega_3}{\omega_2} = \frac{\omega_2}{\omega_1} = m$$

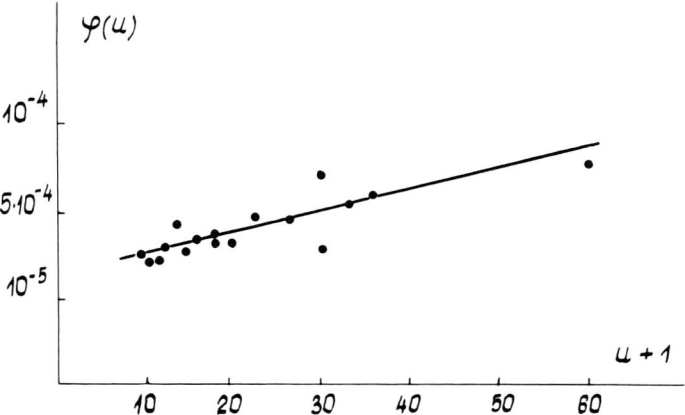

Figure 16 Dependence of the polydispersity parameter $\varphi(u)$ of the polydispersity index. (From Ref. 100.)

then we obtain from the calculated data three values of dynamic viscosity:

$$\eta_0 = \eta_2' \frac{2\eta_1'\eta_3' - \eta_2'\eta_3' - \eta_1'\eta_2'}{\eta_2'\eta_3' - \eta_2'^2} \tag{112}$$

$$\tau_r = \frac{1}{\omega_3}\left[\left(\frac{\eta_0}{\eta_3'}\right) - 1\right]^{\log m/\log[(\eta_0/\eta_3')-1]/[(\eta_0/\eta_2')-1]} \tag{113}$$

The frequencies must conform with the following conditions:

$\omega_1 \tau_r \ll 1$

$\omega_2 \tau_r \approx 1$

$\omega_3 \tau_r \gg 1$

The frequency ratio m must not be below 10, and ω_3 must not lie outside the region of frequencies which are beyond the high-elasticity plateau.

REFERENCES

1. L. Terentyeva and V. P. Budtov, *Vysokomol. Soedin.*, A2(3): 702 (1978).
2. G. V. Schulz, *Z. Phys. Chem.*, B47: 155 (1940).
3. G. V. Schulz, *Z. Phys. Chem.*, B43: 25 (1939).
4. L. H. Tung, *J. Polym. Sci.*, 20: 495 (1956).
5. J. H. Green, *Chem. Ind. (London)*, 1959: 924.
6. W. D. Lansing and E. O. Kraemer, *J. Am. Chem. Soc.*, 57: 1369 (1935).
7. H. Wesslaw, *Macromol. Chem.*, 20: 111 (1956).
8. H. Volker and F. Luig, *Angew. Macromol. Chem.*, 12: 43 (1970).
9. J. Waterson and H. Elias, *J. Macromol. Sci.*, 15: 459 (1971).
10. P. J. Flory, *Principles of Polymer Chemistry*, Cornell University Press, Ithaca, N.Y. (1953).
11. B. Gross, *Mathematical Structure of the Theories of Viscoelasticity*, Paris (1974).
12. L. A. Faitelson and M. G. Tsiprin, *Mekh. Polim.*, 3: 515 (1968).
13. T. G. Fox, *J. Polym. Sci.*, C3: 33 (1965).
14. G. S. Berry and T. G. Fox, *Adv. Polym. Sci.*, 5(3): 261 (1968).
15. F. Bueche, *J. Chem. Phys.*, 20: 1959 (1952).
16. F. Bueche, *J. Chem. Phys.*, 25: 599 (1965).
17. G. V. Vinogradov and A. Ya. Malkin, *Rheology of Polymers*, Mir, Moscow (1980).
18. V. A. Kargin and G. L. Slonimsky, *Dokl. Akad. Nauk. SSSR*, 64: 239 (1948).
19. V. A. Kargin and G. L. Slonimsky, *Zh. Fiz., Chim.*, 23: 563 (1949).
20. M. W. Volkenstein and O. B. Ptitsin, *Dokl. Akad. Nauk. SSSR*, 78(4): 657 (1951).
21. Yu. Ya. Gottlieb and M. V. Volkenstein, *Zh. Tekh. Fiz.*, 23: II 1936 (1948).
22. P. Rouse, *J. Chem. Phys.*, 21: 1273 (1953).
23. F. Bueche, *J. Chem. Phys.*, 22: 603 (1954).
24. B. Zimm, *J. Chem. Phys.*, 24: 269 (1956).
25. B. Zimm, G. Roe, and L. Epstein, *J. Chem. Phys.*, 24: 279 (1956).
26. J. G. Kirkwood and J. Riseman, *J. Chem. Phys.*, 16: 565 (1948).
27. N. K. Blinova, S. I. Sergienko, A. Ya. Malkin, Yu. G. Yanovsky, and G. V. Vinogradov, *Mekh. Polim.* 1: 132 (1973).
28. N. W. Tschoegl, *J. Chem. Phys.*, 39: 149 (1963).
29. A. V. Tobolsky and J. J. Aklonis, *J. Phys. Chem.*, 68: 1970 (1964).
30. A. V. Tobolsky, *J. Polym. Sci.*, C3: 157 (1965).
31. E. J. Bueche, *J. Appl. Phys.*, 26: 738 (1955).
32. L. B. Blizard, *J. Appl. Phys.*, 22: 730 (1951).

33. R. S. Marvin and H. J. Oser, *J. Res. Nat. Bur. Stand.*, *B66*: 171 (1962).
34. H. J. Oser and R. S. Marvin, *J. Res. Nat. Bur. Stand.*, *B67*: 87 (1963).
35. V. N. Pokrovsky, *Teoriya vyazkouprugovo povedeniya kontsentrirovanih polimerov pri malih tchastotah vozdeistviya.* Institut Fiz. Khim. Akad. Nauk. SSSR, Moscow (1970). (preprint)
36. A. S. Lodge and Y. Wu, *Rheol. Acta, 10*: 539 (1971).
37. A. S. Lodge, *Rheol. Acta, 5*: 222 (1966).
39. W. N. Greasley, *J. Chem. Phys., 54*: 5143 (1971).
40. J. D. Ferry, R. T. Landel, and M. C. Viliams, *J. Appl. Phys., 26*: 359 (1955).
41. P. De Gennes, *J. Chem. Phys., 55*: 572 (1971).
42. P. De Gennes, *Macromolecules, 9*: 587 (1976).
43. S. Edvards and J. Grant, *J. Phys. A. Math. Nucl. Gen., 6*: 1169 (1973).
44. W. M. Prest, R. Porter, and J. O. Reilly, *J. Appl. Polym. Sci., 14*: 2697 (1970).
45. A. Ya. Malkin, *Rheol. Acta, 7*: 335 (1968).
46. J. R. G. Treloar, *Physics of Rubber Elasticity,* Clarendon Press, Oxford (1949).
47. G. V. Vinogradov, E. A. Dzyura, A. Ya. Malkin, and V. A. Grechanovskii, *J. Polym. Sci. A-2, 9*: 1153 (1971).
48. A. Ya. Malkin, G. Zh. Zhangereva, M. P. Zabugina, and G. V. Vinogradov, *Vysokomol. Soedin., 18*: 572 (1976).
49. N. Neomoto, M. Moriwaki, H. Odani, and H. Kurata, *Macromolecules, 4*: 215 (1971).
50. D. Thomas, *Polym. Eng. Sci., 11*: 305 (1971).
51. G. V. Vinogradov, A. Ya. Malkin, Yu. G. Yanowskii, E. K. Borisenkova, B. V. Yarlikov, and G. V. Berezhnaya, *J. Polym. Sci. A-2, 10*: 1061 (1972).
52. S. Onogi, T. Masuda, and K. Kitagova, *Macromolecules, 3*: 109 (1970).
53. J. L. den Otter, *Rheol. Acta, 8*: 355 (1969).
54. J. D. Ferry, *Viscoelastic Properties of Polymers,* 2nd ed., New York (1970).
55. R. Sabia, *J. Appl. Polym. Sci., 7*: 347 (1963).
56. G. Palma, G. Pezzin, and S. Zaramella, *J. Polym. Sci., 33*: 23 (1971).
57. G. Lokati and L. Gargani, *J. Polym. Sci., Polym. Lett. Ed., 11*: 95 (1973).
58. V. P. Budtov and E. K. Vinogradov, *Vysokomol. Soedin., B16*: 250 (1974).
59. S. Saeda and T. Suzuki, *Kobunshi Kagaku,* 30(342): 598 (1973).
60. R. A. Mendelson, W. A. Bowles, and F. L. Finger, *J. Polym. Sci. A-2, 8*: 105 (1970).
61. S. Saeda, J. Yotsuyanagi, and K. Yamaguchi, *J. Appl. Polym. Sci., 15*: 277 (1971).
62. R. Schroff and M. Shida, *J. Polym. Sci. A-2, 8*: 1917 (1970).
63. J. Miltz and A. Ram, *Polym. Eng. Sci., 13*: 273 (1973).
64. W. Greasly and L. Segal, *AIChE J., 16*: 261 (1970).
65. R. A. Mendelson, W. A. Bowles, and F. L. Finger, *J. Polym. Sci. A-2, 8*: 127 (1970).
66. V. A. Grechanovsky, I. Ya. Poddubniy, and L. A. Nedoinov, *Vysokomol. Soedin., A-14*: 2267 (1972).
67. V. P. Budtov and E. K. Vinogradov, *Mekh. Polim., 4*: 753 (1972).
68. O. F. Bezrukov, V. P. Budtov, and V. P. Fokanov, in *Yadernomagnitniy resonans,* Leningrad, p. 83 (1971).
69. W. W. Graesly and E. S. Shinbach, *J. Polym. Sci. Polym. Phys. Ed., 12*: 2047 (1974).
70. B. H. Zimm and W. H. Stockmayer, *J. Chem. Phys., 17*: 1301 (1949).
71. R. Chartoff and B. Maxwell, *J. Polym. Sci. A-2, 8*: 445 (1970).
72. E. Schröder, *Plaste Kautsch.,* 20(4): 241 (1970).
73. E. Schröder and G. Winkler, *Plaste Kautsch., 21*: 269 (1974).
74. J. Miltz and A. Ram, *Polymer, 12*: 685 (1971).
75. V. P. Popov, *Zh. Prikl. Spektrosk., 15*: 1113 (1971).
76. I. P. Briedis and L. A. Faitelson, *Mekh. Polim., 1*: 120 (1976).
77. G. Labaig, P. Monge, and J. Bednarick, *J. Polym., 14*: 384 (1973).

78. E. Brauer, I. P. Briedis, V. I. Buchgalter, L. L. Sulzhenko, L. A. Faitelson, and P. Fidler, *Mekh. Polim.*, *2*: 283 (1977).
79. R. Schroff and M. Shida, *Polym. Eng. Sci.*, *11*: 200 (1971).
80. K. Ninomiya, *J. Colloid. Sci.*, *14*: 49 (1959).
81. K. Ninomiya, *J. Colloid. Sci.*, *17*: 759 (1962).
82. K. Ninomiya and J. D. Ferry, *J. Colloid. Sci.*, *18*: 421 (1963).
83. W. W. Greassley, *J. Chem. Phys.*, *54*: 5143 (1971).
84. N. Nakajima, Proceedings of the 5th International Congress of Rheology, Tokyo-Baltimore-Mancheten, Vol. 4, p. 295 (1970).
85. H. Booij, *Rheol. Acta*, *5*: 222 (1966).
86. H. Booij, *Rheol. Acta*, *5*: 215 (1966).
87. R. J. Tanner, J. M. Simons, *Chem. Eng. Sci.*, *22*: 1803 (1967).
88. A. I. Leonov, *Zh. Prikl. Mekh. Tech. Fiz.*, *4*: 78 (1964).
89. B. H. Bersted, *J. Appl. Polym. Sci.*, *19*: 2167 (1975).
90. B. H. Bersted, *J. Appl. Polym. Sci.*, *20*: 2705 (1976).
91. S. Pedersen and A. Ram, *Polym. Eng. Sci.*, *18*: 990 (1978).
92. S. Middleman, *The Flow of High Polymers*, Interscience, New York (1968).
93. C. H. Bamford, W. G. Barb, A. D. Jenkins, and P. G. Onyon, *The Kinetics of Vinyl Polymerization by Radical Mechanisms*, Butterworth, London (1958).
94. A. A. Berlin, S. A. Volfson, and N. S. Yenikolopov, *Kinetika Polimerizatsionnih Protsessov*, Khimiya, Moscow (1978).
95. S. Ya. Frenkel, *Vvedeniye v Statisticheskuyu Teoriyu Polimerizatsii*, Nauka, Moscow-Leningrad (1965).
96. I. P. Briedis, *Rheol. Acta*, *24*: 357 (1985).
97. R. Kuhn, H. Krömer, and G. Rossmalth, *Angew. Macromol. Chem.*, *40/41*: 316 (1976).
98. E. E. Drott and R. A. Mendelson, *J. Polym. Sci. A-2*, *8*: 1361 (1970).
99. I. P. Briedis and V. I. Alksne, *Mekh. Kompozitnih Mater.*, *2*: 291 (1981).
100. I. P. Briedis and V. V. Leitland, *Mekh. Kompozitnih Mater.*, *1*: 119 (1987).
101. I. P. Briedis, A. M. Savelyev, and V. V. Vasilenko, *Mekh. Kompozitnih Mater.*, *5*: 777 (1980).

17

Thermally Stimulated Depolarization Techniques for Studying Polymer Relaxation

Alexander L. Kovarskii
Institute of Chemical Physics
Academy of Sciences of the USSR
Moscow, USSR

INTRODUCTION	643
PRINCIPLES OF THE TECHNIQUE	645
Monorelaxational Process	646
Depolarization of Systems with a Distribution in Dynamic Parameters	648
Effects of Displacement of Charged Species	650
EXPERIMENTAL	651
Preparation of Samples	651
Polarization Conditions	651
Heating Regimes	652
The Technique of Identifying the Origin of Peaks	654
Instrumentation	655
THE RESULTS OF STUDYING POLYMERIC SYSTEMS	657
Polymer Thermograms	657
Studies Under High Pressure	662
Analysis of Distributions in Dynamic Parameters	666
TSD and Mechanisms of Relaxational Processes	672
REFERENCES	674

INTRODUCTION

The technique of thermostimulated depolarization (TSD)—also called thermostimulated discharge, thermostimulated currents (TSC), or electretothermal analysis—is used to study relaxation and electrophysical phenomena in dielectrics and semiconductors. The technique is based on the analysis of weak currents which appear in samples prepolarized in external electrical field, due to heat motion of polar or charged particles.

The emergence of the TSD technique is closely connected with investigations of the electret state of substances. Electrets are dielectrics which, being polarized in the external field, are able to retain a charge for a long period of time. They may be regarded as analogs of constant magnets [1–5]. When an electret is formed in an electrical field the so-called homo- and heterocharges are generated in samples. A homocharge, i.e., a charge on the surface of a sample, which coincides in sign with the polarity of the adjacent electrode, is generated through injections of outside charged particles into the sample. A heterocharge, which has a sign opposite to the electrode sign, is due to an alignment of the sample's dipoles or to a displacement of charged particles of additives. Heterocharges are important in studying relaxation processes with the help of the TSD technique. To get rid of undesirable homocharges one makes use of tightly fitted (spray-coated) electrodes.

The lifetime of an electret state depends on temperature. The discharge of electrets is accelerated at the temperatures of physical (relaxational) and phase transitions in a substance, which are accompanied by unfreezing of the motion of polar and charged particles. Thus heating an electret leads to its discharge, i.e., generation of a current whose intensity reaches maximal values in the range of transitions. The TSD technique is based on the analysis of temperature dependences of such currents.

One example of such temperature dependence, a *thermogram,* is seen in Figure 1. Maxima are observed in the area of the polymer's transition from glassy into high-elastic state (T_g), at $T < T_g$ due to unfreezing of small-scale molecular motions and at $T > T_g$ due to electroconductivity. The TSD thermograms are clearly seen to be similar to the temperature dependences of the loss factor, obtained through the technique of dielectrical relaxation (DR) in an alternating field. Both techniques are related and enter into the arsenal of relaxation spectroscopy. The differences between them lie primarily in frequency ranges. The TSD technique permits investigating relaxations in the range of infralow frequencies (10^{-1}–10^{-4} Hz) where application of DR has some difficulties. An important advance of the TSD technique is a greater resolution in studying relaxation processes with close dynamic parameters. The greatest differences in the frequencies of such processes are observed precisely in the area of low frequencies. The TSD technique makes it possible to analyze processes differing in activation energies by 10% and in frequency factors by five times [3–4]. As compared to DR, the TSD technique is more sensitive, being very simple and productive.

It should be noted that TSD has much in common with the techniques of thermoactivation spectroscopy (thermostimulated conductivity, polarization, luminescence) and is often used in combination with them.

The TSD theory was developed in the 1960s by scientists of various countries. System-

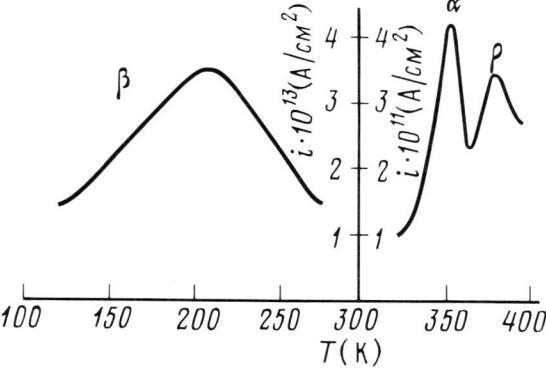

Figure 1 PVC thermogram. $T_p = 383$ K, $t_p = 25$ min, $F = 10$ kV cm^{-1}, $b = 4$ K min^{-1}. (From Ref. 51.)

atic studies involving TSD were started by Bucci and Fieschi [6, 7] on ionic crystals. In the early 1970s the technique was applied to study polymers. A notable contribution to the development of this field was made by the fundamental work of Turnhout [3, 4], Vanderschueren [8–11], Lushcheikin [5], and others [12–20].

It was found that apart from determining physical and phase transitions temperatures the TSD technique can be successfully used to determine the frequencies and activation parameters of relaxation processes, analyze the distribution of segmental motions in dynamic parameters, and to study physical and chemical processes which induce changes in electrical properties and molecular dynamics of polymers. There is one more area of application of TSD, which is beyond the scope of this chapter: the use of TSD to determine electrophysical parameters of polymer systems such as concentration of traps and their energetic parameters [4, 5, 21].

PRINCIPLES OF THE TECHNIQUE

The TSD technique is based on the theory of polarization of dielectrics and the kinetic regularities of their thermal degradation [1–5]. Polarization of a dielectric in an external electrical field is composed of several components which can be divided into two groups. The first comprises the components immediately following the electric field, caused by a displacement of electron orbitals and nuclei of atoms. The second group comprises the components caused by diffusional motion of particles, mainly by orientational motion of constant dipoles (dipole polarization) and by macrodisplacement of charged particles: ions and free electrons (space-charge polarization) (Fig. 2). With polar polymers the main role is played by dipole orientational polarization; with nonpolar polymers space-charge polarization is the major factor. In two-phase systems, including amorphous crystalline polymers, the interfacial polarization can be significant.

The polarization P generated in a sample by the external field is proportional to the dipole concentration n with the moment μ and the field intensity F. Provided $\mu F \ll kT$, in accordance with the Debye equation,

$$P = \frac{n\mu^2 F}{3kT} \tag{1}$$

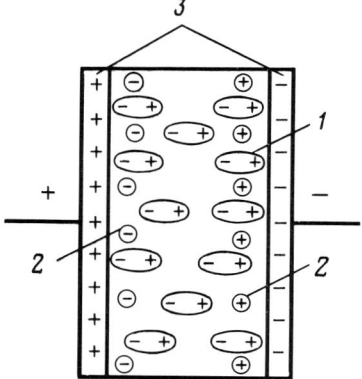

Figure 2 Schematic of metallized electret in an external electric field: (1) dipoles, (2) charged particles, (3) electrodes.

On increasing the intensity of the polarizing field up to the values corresponding to the condition of strong fields ($\mu F > kT$), the polarization reaches its maximal value equal to $n\mu$. Thus a constant electrical field was used to polarize dipoles or charged particles in a sample. The TSD technique involves freezing this polarization and subsequent heating of the system, most often through a linear time–temperature program (Fig. 3). The thermal destruction of polarization produces a current in the external circuit. The theoretical dependence of residual polarization and TSD current on time and temperature depends on whether the system consists of species with a single relaxation time or is described by a set of relaxation times. With polymers it is the latter case that is always realized. To analyze the results, however, one should consider both cases. This section demonstrates how to analyze the case of dipole polarization and follows a similar approach for other types of electret charges.

Monorelaxational Process

Assuming that the dipoles of a system are characterized by a single relaxation time, the rate of depolarization at a constant temperature and the TSD current are described by a simple kinetic equation

$$i(t) = \frac{dP(t)}{dt} = -\frac{P(t)}{\tau} \tag{2}$$

where $P(t)$ is the residual polarization at the moment of time t. By integrating Eq. 2 we obtain

$$P(t) = P_0 \exp\left(-\frac{t}{\tau}\right) \tag{3}$$

Here P_0 is the initial polarization calculated from Eq. 1. When the relaxation time does not remain constant in the course of an experiment the depolarization rate is described as

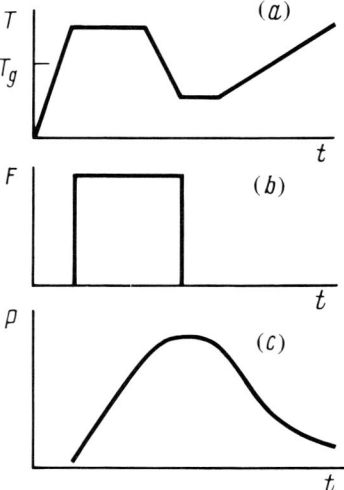

Figure 3 Variation of temperature T, field intensity F, and polarization P in TSD experiments.

$$P(t) = P_0 \exp\left(-\int_0^t \frac{dt}{\tau}\right) \quad (4)$$

The density of current generated in the external circuit is determined from Eqs. 2 and 4:

$$i(t) = -\left(\frac{P_0}{\tau}\right) \exp\left(-\int_0^t \frac{dt}{\tau}\right) \quad (5)$$

Since the TSD experiments are conducted through a temperature program, Eq. 5 can be rewritten as a function of temperature only. Provided the temperature is changed linearly with time, $T = T_0 + bt$, we obtain

$$i(T) = -\left(\frac{P_0}{\tau}\right) \exp\left(-b^{-1} \int_{T_0}^T \frac{dT}{\tau}\right) \quad (6)$$

where b is the rate of linear heating.

Having differentiated Eq. 6, one can find the condition under which the TSD current reaches a maximum:

$$\frac{d\tau(T)}{dT} = -b^{-1} \quad (7)$$

For further calculations the type of temperature dependence of the relaxation time must be specified. To this end one normally uses the Arrhenius and WLF equations:

$$\tau = \tau_0 \exp\left(\frac{E}{kT}\right) \quad (8)$$

where E is the activation energy and

$$\frac{\tau}{\tau_g} = \exp \frac{C_1(T - T_g)}{C_2 + T - T_g} \quad (9)$$

where $C_1 = 40$ and $C_2 = 52$ K. When the temperature dependence τ is described by Eq. 8 the expression for the TSD current maximum assumes the form

$$\tau_m = \frac{kT_m^2}{bE} \quad (10)$$

For the temperature dependence of Eq. 9 the condition for the maximum can be written as

$$\tau_m = \frac{(C_2 + T_m - T_g)^2}{C_1 C_2 b} \quad (11)$$

There are several ways of calculating relaxation times and activation energies from experimental TSD peaks.

Relaxation time can be determined from Eq. 2. The P value is determined by integrating the current, i.e., from the area under the TSD curve. Then

$$\tau(T) = b^{-1} \int_T^{T_\infty} \frac{i\, dT}{i} \quad (12)$$

Having obtained in this manner the values of τ at different temperatures, one can estimate the activation energy with respect to Eq. 8. The activation energy can also be determined from the initial section of a TSD peak.

Combining Eqs. 6 and 8 we obtain

$$i(T) = -\left(\frac{P_0}{\tau_0}\right)\exp\left[-\frac{E}{kT} - \frac{1}{b\tau_0}\int_{T_0}^{T}\exp\left(-\frac{E}{kT}\right)dT\right] \quad (13)$$

Having differentiated with respect to inverse temperature and assuming that at the initial stage of depolarization the second term in the brackets (Eq. 13) is equal to zero, we have

$$\ln i = \text{const} - \frac{E}{kT} \quad (14)$$

Thus the activation energy can be estimated from the slope angle of the dependence $\ln i$ on $1/T$.

Another way to determine E is to register TSD peaks at different heating rates. According to Eq. 10

$$E = k\ln\frac{b_1 T_{m2}^2/b_2 T_{m1}^2}{T_{m2}^2 - T_{m1}^2} \quad (15)$$

Moreover, E can be estimated from the halfwidth of peak ΔT using the expression obtained from Eq. 13:

$$\Delta T \approx \frac{2.5 kT_m^2}{E} \quad (16)$$

The obtained E values can be readily used to estimate τ_m from Eq. 10.

Depolarization of Systems with a Distribution in Dynamic Parameters

A feature of polymeric electrets is a wide distribution in frequencies and activation energies of molecular motion. These distributions manifest themselves primarily in a sharp widening of TSD peaks. Equations 14 and 12 become inapplicable to calculate E and τ. In the first case the slope angle of the initial section of the TSD curve is determined by the width of distribution in τ and E; in the second case the widening of peaks leads to lower values of currents at a given temperature and consequently to elevated values of τ. Equation 6 yields too low values of E. Thus only Eqs. 10 and 15 are valid for calculating the most probable values of τ and E, since normally the distribution function maximum coincides with that of the TSD peak [15].

At quasi-continuous distribution in relaxation times, the temperature dependence of TSD current is described as

$$i(T) = -P_0\int_0^{\infty}\frac{F(\tau)}{\tau}\exp\left[-\frac{1}{b\tau_0}\int_{T_0}^{T}\exp\left(-\frac{E}{kT}\right)dT\right]d\tau \quad (17)$$

where $F(\tau)$ is the distribution function in relaxation times. One can assume that the distribution in relaxation times is caused by the distributions in activation energies or frequency factor. In this case we have

$$i(T) = -P_0\int_0^{\infty}F(E)\tau_0^{-1}\exp\left[-\frac{E}{kT} - \frac{1}{b\tau_0}\int_{T_0}^{T}\exp\left(-\frac{E}{kT}\right)dT\right]d\tau$$

$$i(T) = -P_0\exp\left(-\frac{E}{kT}\right)\int_0^{\infty}\frac{F(\tau_0)}{\tau_0}\exp\left[-\frac{1}{b\tau_0}\int_{T_0}^{T}\exp\left(-\frac{E}{kT}\right)dT\right]d\tau \quad (18)$$

The molecular dynamics in polymers is normally characterized by distributions in all parameters τ, τ_0, and E, which complicates considerably the procedure of data treatment. There are three main approaches to solving this problem: the fitting method, the transformation method, and the fractional techniques of conducting experiments. Their main features are given below.

In the fitting method the analytical expressions of the known distribution functions [22–24] are substituted into Eqs. 17 and 18 and agreement between experimental and calculated TSD curves is achieved through varying the width and asymmetry parameter of these functions with the help of computers. One variant of this approach is the analysis of kinetic curves for isothermal charge drop. These curves are treated using equations which have been proposed to describe the so-called polychromatic kinetics, such as the Williams–Watts equation [25]:

$$\frac{P(t)}{P_0} = \exp(-t\nu_0)^\lambda \qquad (19)$$

One more equation has been proposed to suit the same ends [26]:

$$\frac{P(t)}{P_0} = (1 + \lambda\nu_0 t)^{1/\lambda} \qquad (20)$$

where $P(t)$ and P_0 are the values of polarization at the moment of time t and at the beginning of the process, λ is the parameter characterizing the shape of kinetic curve, ν_0 is the average value of relaxation times of dipoles, which determines the initial depolarization rate:

$$\nu_0 = \frac{d[P(t)/P_0]}{dt}\bigg|_{t=0} = \int_\nu \nu F(\nu) d\nu$$

The analysis of isothermal depolarization data makes it possible to estimate not only the width of distributions but their temperature dependence as well.

In the transformation method correlations among parameters T, E, and τ are used to transform a TSD curve into the distribution curve in all these parameters. Within the framework of this approach each point on the TSD curve is assumed to represent the maximum of the depolarization current for a small group of dipoles with the same values of τ, E, and τ_0. In this case the condition for maximum (Eq. 10) is valid for every point of the TSD curve:

$$\frac{T_i^2}{E_i} = \frac{\tau_i b}{k} \qquad (21)$$

A combination of Eq. 21 with the Arrhenius equation yields

$$\frac{T_i^2}{E_i} = \frac{\tau_{0i} b}{k} \exp\frac{E_i}{kT_i} \qquad (22)$$

This relationship was used by Pfister and Abkovitz [20] to obtain distributions in activation energies and frequency factors under the assumption that a distribution exists only for one of these parameters, the other being constant for all species (the so-called one-dimensional distribution). The relative number of species necessary to construct the distribution function is determined from Eq. 1:

$$n_i \sim \frac{i_i T_i^3}{bF} \qquad (23)$$

The most probable value of E is determined from Eq. 10. It has been shown [27] that when using Eq. 10 one can assume that $E/T^2 = $ const. In this case there is no need to use the approximation of one-dimensional distribution and one can easily plot distributions in all dynamic parameters independently. The value E/T^2 is determined for each polymer from the peaks recorded at different heating rates. The relative number of species for this particular case is determined by the correlation $n_i \sim i_i T_i$. As compared to the current the temperature is changed much less drastically, which is why the dependence of i on T is evidently the distribution of the polymer's glass transition temperatures. Distributions in activation energies are obtained through constructing dependencies of iT on E. Equation 10 is used to calculate relaxation times. On the basis of calculated values for E_i and τ_i one calculates τ_{0i}.

Fractional techniques (FTSD) are based on pulse (short-term) polarization or depolarization. A TSD curve is divided into individual peaks or into their initial sections corresponding to small groups of dipoles. The obtained data are treated using the equations of the monorelaxational process (Eqs. 2–12) [21, 28]. The FTSD technique is considered later.

Effects of Displacement of Charged Species

The processes of dipole-orientational depolarization considered above play an important role in polar polymers and their analysis provides the theoretical and experimental bases of the TSD technique for polymers. In some cases, however, an investigator comes across phenomena caused by displacement of charged species. The currents of charge displacement are primarily observed in nonpolar and weakly polar polymers, two-phase systems, and in polar systems at $T > T_g$. Let us briefly consider the origin of these phenomena.

A distinguishing characteristic of polymer systems is a relatively high concentration of low molecular weight additives, ions included. In addition, a polymer system always contains free electrons trapped during its production, processing, storage, use, and polarization.

When polymeric electrets are heated above T_g thermograms show the current maximum which is due to unfreezing of the motion of ions displaced during polarization (space-charge relaxation). The activation energy of this process coincides with the value of E for electrical conductivity, which reveals the identity of both processes. The intensity of the peak caused by electrical conductivity (ρ peak) strongly depends on polarization conditions, presence of additives, and γ irradiation. Identification techniques for these peaks are given later ("Experimental" section).

With two-phase systems one may observe additional peaks caused by polarization at the interface (Maxwell–Wagner effect). The interface boundaries accumulate charges because of differences in dielectric constants and conductivity of the phases.

The general relationship describing the TSD current caused by displacement of ions and electrons is as follows [4, 21, 29]:

$$i(T) = \frac{nq^2\delta^2\omega F}{6kTL} \exp\left[-\frac{E}{kT} - \int_{T_0}^{T} \frac{2\omega}{b} \exp\left(-\frac{E}{kT}\right) dT\right] \quad (24)$$

Where n is the concentration of weakly bound ions or electrons, q is the charge, δ is the depth of traps' location, F is the width of the barrier, ω is the frequency of species oscillations in the equilibrium state, and L is the sample thickness.

On deriving Eq. 24 it was assumed that the decay of charges occurs monomolecularly and faster than their retrapping, and that the work of transferring an ion (electron) in the field is much lower than the energy of heat motion ($qF\delta/2L \ll kT$). Relationships 24 and 13 are similar and thus to calculate the activation parameters of a discharge one can make use of the approaches developed for data analysis on dipole-orientational polarization. Expressions to calculate the number of carriers and the capture depth are

$$n = \frac{2PL}{q\delta^2} \tag{25}$$

and

$$\delta = \frac{2PL}{\sigma_{ef} + P} \tag{26}$$

Here σ_{ef} is the initial surface density of a charge. The charge quantity accumulated in a sample during polarization is described as

$$Q = \frac{\epsilon\epsilon_0 SFt}{x_0}\left[t + \frac{t_f L}{x_0}\exp\left(\frac{E_d}{kT}\right)\right]^{-1} \tag{27}$$

where t_f is the lifetime of charge carriers in the sample, E_d is the diffusion activation energy, S is the area of electrodes, ϵ is dielectrical permittivity, ϵ_0 is an electrical constant, and x_0 is the initial thickness of the charge localization region.

EXPERIMENTAL

Preparation of Samples

Samples are prepared in the form of films or disks. These should not be too thin to avoid surface effects or too thick to avoid temperature gradients. The optimal range of thicknesses is 100–500 μm. There should be no polar or charged additives in the samples. Molding is preferable to film casting from solution. In the latter case the samples should be dried in vacuo at elevating temperature. The surface should be cleaned with great care, or the TSD peaks may be distorted and additional peaks may appear on a thermogram.

The electrically conductive coating on both sides of a sample is obtained either by vacuum spraying of an oxidation-resistant metal (aluminum, gold) or by coating with a thin metallic foil. The first technique was shown to provide better resolution.

It is recommended that prior to polarization the samples be treated thermally in a short-circuit regime to eliminate parasite charges.

Polarization Conditions

Optimization is required for such parameters as polarization temperature T_p, exposition t_p, and strength of the DC electric field F.

The exposure at a chosen temperature should satisfy the condition $t_p > \tau$ at which the equilibrium polarization is established corresponding to Eq. 1. One can estimate τ from Eqs. 8 and 9 assuming that $\tau_g \simeq 10^2$ sec (the E values for different polymers are listed in Table 3). The best polarization conditions for polymers are supposed to be $T_p = T_g + (30–50°)$ and $t_p = 15–30$ min. When only low-temperature β peaks are registered, with most polymers it is possible to carry out polarization at room temperature during the same period of time. With polar polymers the intensity of peaks is proportional to the strength of the polarizing field at $F < 1$ MV cm^{-1} (Eq. 1). The sensitivity of the technique is thus dependent on F. At large F, however, the probability of breakdown and injection of charged species drastically increases. The most widely used range is 5–50 kV cm^{-1}.

Heating and cooling of samples in a DC electric field is not the only way of obtaining polymeric electrets. It has been found that dipole polarization also results from exposing dipoles to magnetic field (magnetoelectrets) [30] and mechanical stress molding and orientation (mechanoelectrets) [31].

Heating Regimes

The main regime of conducting the TSD experiments is the linear heating program. The heating rate is very important, determining as it does the characteristic frequency of the technique, the intensity of TSD peaks, and their position on the temperature scale in accordance with Eq. 15. As b is increased the maximum of TSD current also increases, i.e., the sensitivity of the technique is better. At the same time the resolution worsens. The best rate for a linear heating is 1–4° min^{-1}. In those cases when Eq. 15 is used to determine the activation energy, the heating rate is increased to 20° min^{-1}.

To analyze distributions one uses pulse or fractional techniques (FTSD) [9, 21, 28], which allow polarization or depolarization of small groups of dipoles with close relaxation times. Thus, for instance, Hino [19] polarized samples by short pulses over 20 sec while the temperature was increased by 10° in each interval between the pulses. The depolarized samples yielded a series of fractional peaks which were analyzed using the equation of monorelaxational process to obtain the distribution function in τ.

Another variant of FTSD is the regime of "sawtooth" heating, which consists of alternating cycles of heating and cooling [21]. This regime is characterized by the following parameters: the effective heating rate $b' = T_g\alpha$ (Fig. 4), the depth of temperature oscillation ΔT, the fraction step δT, and the number of fractions n. When studying polymers the best values for these parameters are $\Delta T \simeq 20°$, $\delta T = 3–5°$, $b' \simeq 20°$ min^{-1}.

As a result of such sawtooth heating, a TSD curve is divided into n components with each component being caused by depolarization of a small group of dipoles with close frequencies. Through the slope angle of temperature dependencies of fractional currents in ln i–$1/T$ coordinates one can estimate the activation energy of rotational motion for a

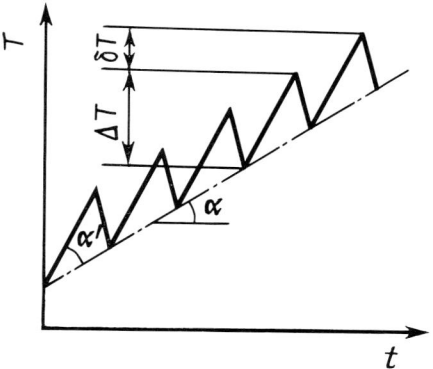

Figure 4 Temperature variation pattern in time at fractional heating.

given group of dipoles. The relative number of species in groups is proportional to the charge isolated by the given fraction and estimated from the area under the curve.

The envelope of FTSD current curves coincides with the TSD curve obtained at the rate of linear heating $b = b'$. The sawtooth heating allows one to find in each experiment 10–15 points comprising the distribution function (Fig. 5).

Thermocleaning, a technique similar to the one considered above, is used to separate two overlapping peaks [32]. Its principle is clear from Figure 6. A temperature step introduced into the linear heating regime makes it possible to suppress the low-temperature peak, isolating the high-temperature peak.

The intensity of peaks, position of maxima, and relaxation times are affected not only by polarization conditions and heating regimes. Since the glass transition process is linked with structural rearrangement and is a kinetic process, the TSD peak parameters (T_m, E, τ) may depend on a cooling rate of a sample after its polarization. There are two cases: (a) when the cooling rate is comparable to that of structural relaxation (annealing) and (b) when it is greater (quenching). Experimental data [3, 4] indicate that in the latter case, i.e., with quenching of samples, one observes less intensive and wider TSD peaks with the current maximum at lower temperature than is the case of annealing. It is known that

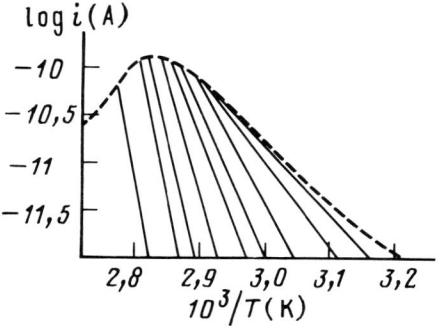

Figure 5 Temperature dependencies of thermostimulated depolarization currents in the Arrhenius coordinates, obtained for PVC at fractional heating (solid lines). Thermostimulated depolarization curve obtained in the regime of linear heating at the rate of 2° min^{-1} (dashed line). (From Ref. 77.)

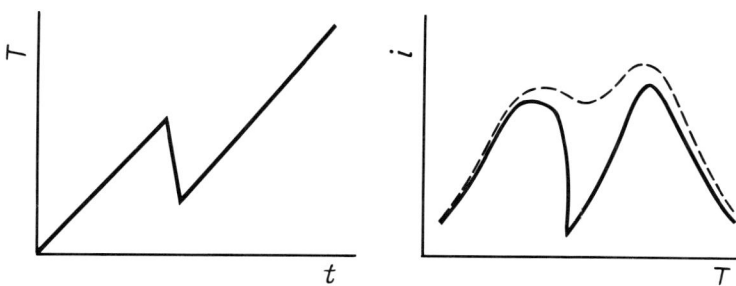

Figure 6 Thermograms obtained in the regimes of linear heating (left-hand curve) and partial thermal purification (right-hand curves).

varying the cooling rate within narrow limits does not lead to a considerable change in T_m. In TSD experiments one normally works in the 5–20° min^{-1} range of cooling rates.

The Technique of Identifying the Origin of Peaks

In studying polymers, particularly weakly polar ones, it is often necessary to establish what kind of polarization, dipole or charge, caused TSD peaks. This task is complicated by the fact that both processes are described by similar equations (Eqs. 13 and 24). The main approach to solve the problem is to vary polarization conditions: the field strength F, time t_p, and temperature T_p [21]. Figure 7 shows the dependence of the maxima current on F, t_p, and T_p. It is seen that for the two polarization mechanisms the dependences $i = f(F)$ and $i = f(T_p)$ are different. In dipole polarization the dependence of i on F shows a linear section corresponding to the condition of weak fields $\mu F \ll kT$ and a plateau in the area of strong fields. In the case of space-charge polarization following from Eq. 27 there is no such plateau, the dependence on field being either parabolic (at short polarization times $t_p \ll t_f L/x_0$) or linear ($t_p \gg t_f L/x_0$, where t_f is the transit time of charge carrier and x_0 is the thickness of its localization area). The dependence of i on T_p in the case of dipole polarization varies depending on polarization time being longer or shorter than relaxation time. In the former case there is an exponential peak growth with temperature; in the latter, a parabolic decrease occurs. At large polarization times in space-charge polarization the peak intensity is no longer dependent on temperature. To establish the polarization mechanism in this case one has to vary all three parameters of the polarization regime.

One can also identify the polarization mechanism by exposing a sample to the action of various external agents, e.g., γ irradiation. Polymer irradiation was found to result in a sharp decrease of peaks of space-charge origin due to neutralization of charges [4]. ρ peaks can be revealed by creating an elevated concentration of charged species in a polymer, introducing ionogenic additives, or charging samples by electron beam, corona discharge, etc.

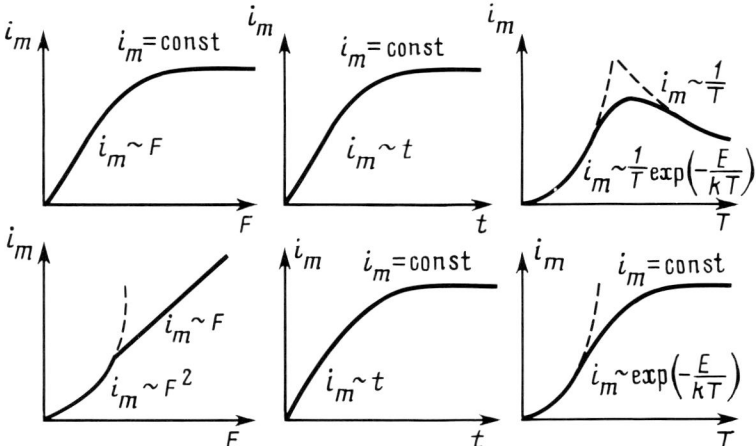

Figure 7 Dependences of maximal TSD current on polarization conditions in case of dipole (upper row) and charge (bottom row) polarization. (From Ref. 20.)

Instrumentation

A simple unit to study relaxational processes in polymers with the use of the TSD technique is comprised of a temperature-controlled chamber for samples, a power source, an electrometer, and a temperature programmer.

Samples were polarized using a stabilized DC voltage source. The electrometer's sensitivity with respect to current should not be less than 10^{-15} A. The temperature regime in the chamber is controlled by a proportional-type programmer. The temperature dependence of the TSD current is recorded by a two-coordinate plotter. The second recorder is needed to control the linearity of the temperature program. The temperature is measured by a thermocouple. More sophisticated units possess a system of data recording on a magnetic band for storing and processing.

The measuring chamber must satisfy the following requirements:

1. It must work in a wide temperature range (when studying polymers the range is $100 < T < 500$ K).
2. It must have low time lag (drift) and provide a fast switchover to the heating or cooling regime.
3. It must ensure uniform sample heating and good linearity of temperature program.
4. It must be able to prevent thermooxidative degradation of samples.

Figure 8 shows a chamber that meets these conditions. A quartz Dewar flask (3) is placed in a grounded metallic body (1) on Teflon gaskets (2). Nitrogen vapors are supplied through the Dewar flask's bottom neck (4) from an electrical evaporator placed in a vessel with liquid nitrogen.

Racks (6 and 7) supporting electrodes (8 and 9), connecting thick copper wires (10) for a resistivity thermometer, and a heating element are all fixed on quartz plate (11) located in the lid of the chamber (5). The rack of measuring electrode 6 is surrounded by protecting metallic tube (12). Isolating tube 13 is made of quartz. The electrometer or power supply is connected to socket 14. Heating elements and the resistance thermometer are located in two cells (16).

To register temperatures one uses thermocouple 15, whose junction is located in a grounded electrode (8). Electrodes 8 and 9 (25 mm in diameter) are made of stainless steel.

The chamber's design ensures the following parameters: the range of working temperatures from 90 to 770 K; the maximal cooling rate (in nitrogen vapors) $20°$ min^{-1}; linear, stepwise, and sawtooth heating regimes. Polymer oxidation under polarization is impossible.

Figure 9 is a schematic of a chamber for registering TSD currents at high pressure. The chamber consists of a highly resistant steel body (1) with row steel jacket (3) and two screw caps (2). The upper cap is fitted with a cone electric input (4), the bottom one with six inputs. Polycarbonate cones (5) are provided to insulate electric inputs. The screw cap sealing (6) is a set of copper, Teflon, and oil-resistant rubber rings. The working liquid, transformer oil, is supplied through orifice 7.

The chamber contains a measuring cell consisting of two electrodes (9) with heating elements (10). A uniform heating of a sample (11) is ensured by the same heating power on both electrodes. The heating elements are connected in series. Connected to the frame, a (grounded) bottom electrode has a built-in thermocouple (12) and a platinum resistance thermometer. The upper measuring electrode is isolated from heating elements and shield-

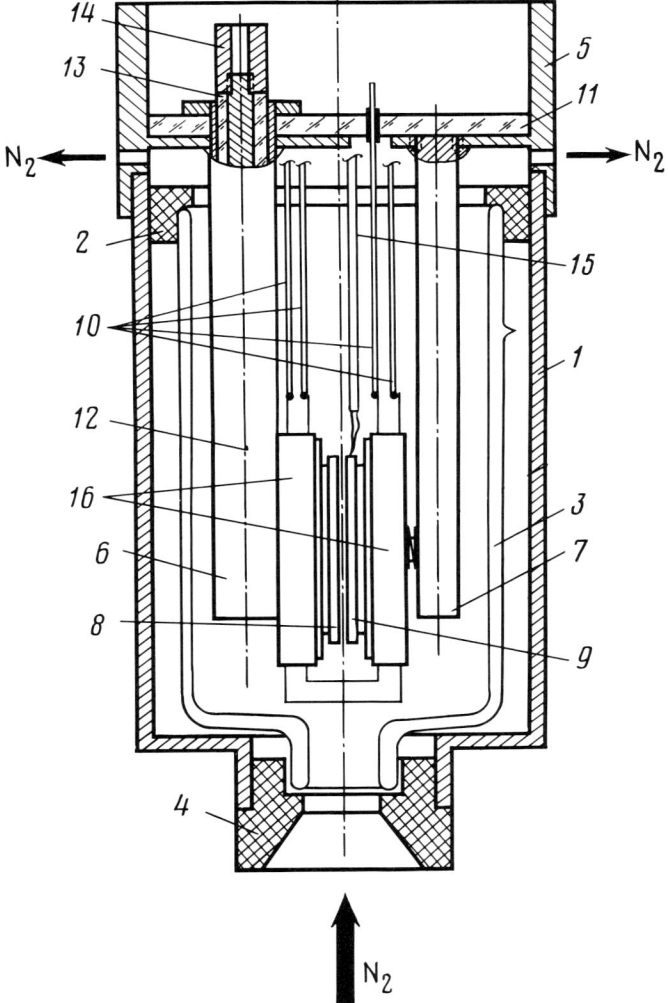

Figure 8 TSD current measuring chamber.

ed to prevent interference. The upper screw cap's electric input is for the measuring signal output and for the input of polarizing voltage. The bottom cap's inputs are to supply voltage to the heating element for the resistance thermometer and thermocouple. At low temperatures the system works at external cooling of the chamber by dry ice. The chamber is linked to a liquid compressor by a steel capillary (8). The pressure is measured by an external manometer. Transformer oil is the working liquid. The chamber allows registration of TSD currents at 400 MPa pressure and in the 233–450 K temperature range.

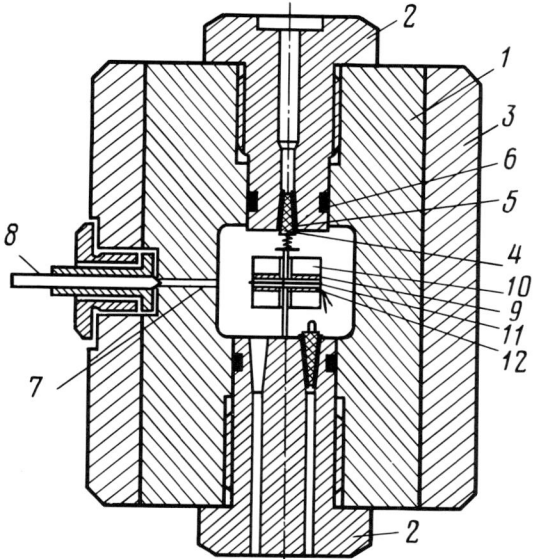

Figure 9 Chamber to register TSD currents at high pressures.

THE RESULTS OF STUDYING POLYMERIC SYSTEMS

Polymer Thermograms

The temperature dependences of TSD currents recorded over a wide temperature range (100–400 K) show three clear-cut maxima typical of practically all polymeric systems (Fig. 1). Two of them, near T_g (α peak) and at $T < T_g$ (β peak) are due to the motion of macromolecules' kinetic elements, whereas the high-temperature maximum at $T > T_g$ (ρ peak) is due to the unfreezing of translational mobility of foreign ions, i.e., with electric conductivity of a polymer. With polar polymers the maximum density values at the maximum of α peak are 10^{-10}–10^{-11} A cm^{-2}; with weakly polar polymers values are from 10^{-12}–10^{-13} A cm^{-2} (at the intensity of polarizing field 10–20 kV cm^{-1}) (Fig. 10). The i value at the maxima of β peaks is normally one to two orders of magnitude lower than with α peaks. The intensity of a ρ peak strongly depends on polarization conditions, electrode material, and additives. The peak intensity decreases as the degree of the polymer's crystallinity is increased, because there is a reduction in the number of relaxators in amorphous regions.

The width of peaks depends on the polymer, changing from 10 to 30 K (at half-height) with α process (T_g) and from 30 to 100 K with β process. The main factor determining the width of peaks is the width of relaxation time spectrum. (This will be seen in the section on "Analysis of Distributions in Dynamic Parameters.") Let us note that in systems with a narrow distribution in τ, such as supercooled liquids, the width of α peak at half-height is no higher than 3–5 K [33]. The charge isolated in α processes is from 4 to 15 times larger than that in β processes (Table 1).

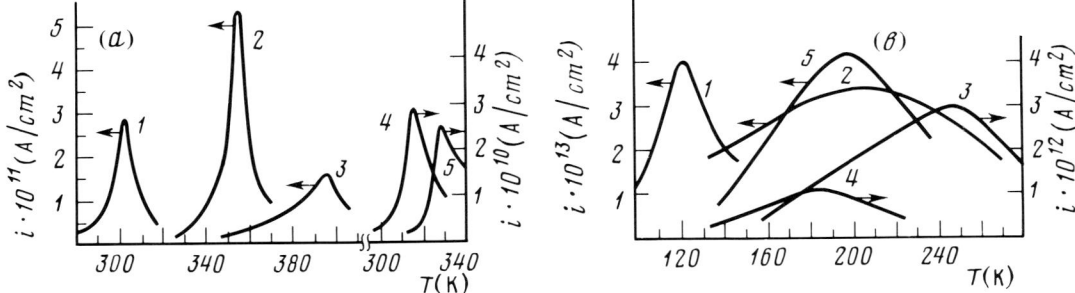

Figure 10 TSD peaks of (a) α processes and (b) β processes in (1) PVA, (2) PVC, (3) PMMA, (4) N6, and (5) PTFCE. Polarizing voltage 10 kV cm^{-1}; heating rate 4 K min^{-1}. (From Refs. 51, 71, and 72.)

Table 1 Charges Q Released in α and β Processes of Some Polymers ($F = 10$ kV cm^{-1})

Polymer	Q_α (nC cm^{-2})	Q_β (nC cm^{-2})
PMMA	5.3	0.3
PVC	5.4	0.4
PTFCE	0.8	0.2
PETP	1.4	0.1
PS	0.32	0.08

Source: Data from Refs. 27, 34, and 35.

The causes of the differences in Q_α and Q_β values are considered in "TSD and Mechanisms of Relaxational Processes," below.

Let us now consider the temperature ranges of α and β peaks (Table 2). Normally the temperature of an α peak maximum is in agreement with T_g (according to dilatometry data [3–5]) and is not significantly different from T_g obtained through the technique of dielectrical relaxation at low frequencies (1–100 Hz). It should be noted that the characteristic frequency of the TSD technique is determined by the expression $f = bE/kT_m$ (Eq. 10) depending on T_m, and E lies in the 10^{-1}–10^{-4} Hz range (at $b = 1$–$20°$ min^{-1}). This is considerably lower than the bottom frequency limit in the traditional DR technique. Since the apparent activation energy of the α process is very large, the differences in T_g obtained at various frequencies are small (Fig. 11). On the other hand, with β processes, which are characterized by relatively low E, the temperatures of the maxima are strongly dependent on the frequency of the technique. Thus the $T_\beta/T_\alpha(T_g)$ ratio obtained through the DR technique at 100 Hz is for many polymeric systems equal to 0.75 [43], while on the basis of TSD data ($b = 1$–$4°$ min^{-1}) it reaches ~0.6 (Fig. 12). It is precisely these two values that characterize the resolution of the techniques, which is considerably better with TSD. The maxima of β peaks on TSD thermograms are observed at temperature from 130 to 180° below T_g. Even larger differences are observed for α and γ processes. The data analysis (Fig. 12) shows that γ peaks observed at the temperature dependences of the

Table 2 Temperatures of α and β Peaks Maxima in Polar Polymers

Polymer	T_α (K)	T_β (K)	Reference
Polyvinyl chloride (PVC)	340–360	188–205	5, 11, 27, 36
Polytrifluorochloroethylene (PTFCE)	328	198	27
Polymethyl methacrylate (PMMA)	378–396	242–248	3–5, 8, 11, 15, 16, 27, 34
Polyvinylacetate (PVA)	303	123–133	11, 27, 34
Polycarbonate (PC)	413–433	193	5, 37, 38
Polystyrene (PS)	360–378	260	5, 35
Nylon-6 (N6)	313–316	173–181	5, 27, 34
Nylon-66 (N66)	313	165	5, 17
Polyethylene terephthalate (PETP)	357	168–203	5, 37, 39
Polyvinyl chloride (PVC)	348	253	40

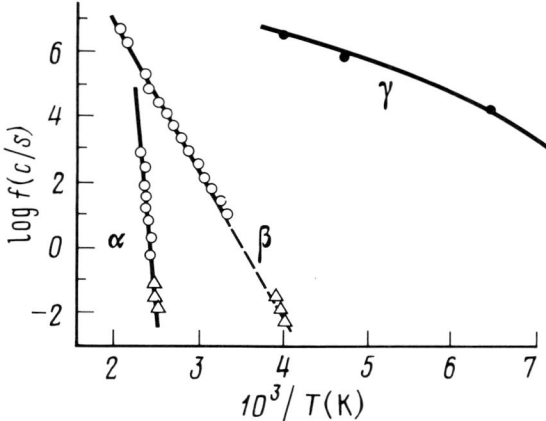

Figure 11 Arrhenius dependencies of frequencies of α, β, and γ processes in PMMA on the basis of data of dielectrical (open circles) and mechanical (closed circles) relaxations and TSD (triangles). (Data on dielectrical and mechanical relaxations from Refs. 41 and 42.)

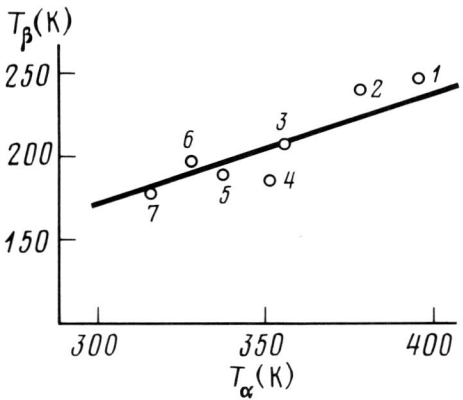

Figure 12 Correlation between T_α and T_β in polymers: (1) PMMA, (2) PS, (3) PVC, (4) PETP, (5) N66, (6) PTFCE, and (7) N6.

dielectric loss factor can be detected using the TSD technique only at very low temperature ($T < 100$ K). For this reason one can hardly identify the additional maxima on β peaks observed in some work on TSD with γ processes studied by DR. One can be absolutely sure that at least in some cases the additional maxima are caused by admixtures. At the same time, reproducible and unambiguous intricate thermograms were obtained for a number of homopolymers. Among these are "twin" α peaks in polycarbonate with maxima at 293 K and 418–423 K [5] and PETP (338 and 357 K) [5, 37]. With PMMA one also observes unusually wide or even split off TSD peaks [3, 8, 16].

The origin of twin peaks is normally explained by the existence of dense and loose packings in amorphous areas. The wide relaxation time spectra in PMMA (discussed later) support this conclusion. The splitting of β peaks in PETP into two maxima [37] has been explained by the existence of gauche- and transisomers. Thus owning to the technique's high resolution it is possible to separate the peaks of overlapping relaxational processes. It was shown that the technique of high pressures allows even better resolution [44].

Increasing the heating rate results in a shift of maxima toward the area of higher temperatures, which agrees with Eq. 10. Due to high activation energy the shift with α peaks is much less pronounced than that with β peaks. The E values calculated from the shift of maxima (Eq. 15) make up 200–500 kJ mol^{-1} with α processes and 15–100 kJ mol^{-1} with β processes, being close to the activation energies determined by the technique of dielectrical relaxation (Table 3). The application of Eqs. 14, 8, and 12 to estimate E leads to incorrect, underestimated values due to the widening of TSD peaks caused by a set of relaxation times. Thus, for instance, the E values for the α process in PVC calculated from Eqs. 14, 8, 12, and 15 are equal to 176, 184, and 422 kJ mol^{-1}, respectively.

The relaxation times in the maxima of α peaks lie in the 40–200 sec range, with β peaks in the 100–500 sec range ($b = 1–5°$ min^{-1}).

In the case of nonpolar polymers the TSD peaks can be caused both by the release of

Table 3 Comparison of Activation Energies Determined by TSD and DR Techniques

Polymer	Process type	E_{TSD}[a] (kJ mol^{-1})	E_{DR} (kJ mol^{-1})
PVA	α	182	185–291
	β	17	20–46
PVC	α	422	294–525
	β	40	40–84
PMMA	α	355	420
	β	67	88
PTFCE	α	205	281
	β	40	71

[a]Calculated from Eq. 15.
Source: TSD data from Refs. 27 and 34; DR data from Refs. 41–43.

charge carrier at certain temperatures and by the presence of polar oxygen-containing groups formed during storage and processing. The intensity of both dipole and charge peaks is not strong, the current densities being two to three orders of magnitude lower than with polar polymers [45–50]. As with dipole-orientational processes the charge release normally occurs in relaxation transition ranges. Thus three maxima are observed with high-density polyethylene (PE): at 135, 240, and 300 K (Fig. 13) [51]. The peak at 240 K coincides with T_g determined by other techniques, such as thermoluminescence [50]. The peak at 135 K can be identified with the β process. In fact T_β/T_α in this polymer is 0.56, close to the average value of 0.6 for most polymers (Fig. 12). With low-density PE both peaks are shifted toward the low-temperature range. Both polymers exhibit peaks at ~300 K, which seem to be caused by the release of charges from traps located at the interface surface (the Maxwell–Wagner effect). A narrow (2–3 K) and intensive peak is observed at the melting point of PE, which can be explained by the existence of deep traps [51].

Additives and admixtures play an important role in the TSD technique. The additives' particles can perform several functions. They can plasticize polymers, cause shifts of the main peak's maxima, or cause additional peaks of dipole (polar additives) or charge origin (charged species). Moreover, some additives can serve as effective electron traps, which enhances the probability of the emergence of new maxima on thermograms. Figure 14 shows a thermogram of PS with 3% wt of polar molecules of a dye, paranitropara-dimethylaminoazobenzene ($\mu = 8$ D). The relaxation of the additive causes the emergence of two clear-cut maxima at 261 and 166 K. These peaks have dipole origin, indicating that α and β processes may be caused by only one type of kinetic units (see "TSD and Mechanisms of Relaxational Processes"). In addition, there is a peak above 270 K which seems to be connected with the relaxation of charges localized on dye molecules and released upon unfreezing their motion.

Introduction of water, methanol [3, 4, 10], and some other additives [52, 53, 59] into polymers was found to cause a drastic change in the intensity and position of the ρ peak.

On analyzing thermograms it is useful to know how the thermal prehistory of a sample affects the TSD maxima. Various annealing (physical aging) and quenching regimes have been used repeatedly by different authors to prepare electret [4, 12, 13]. Some typical

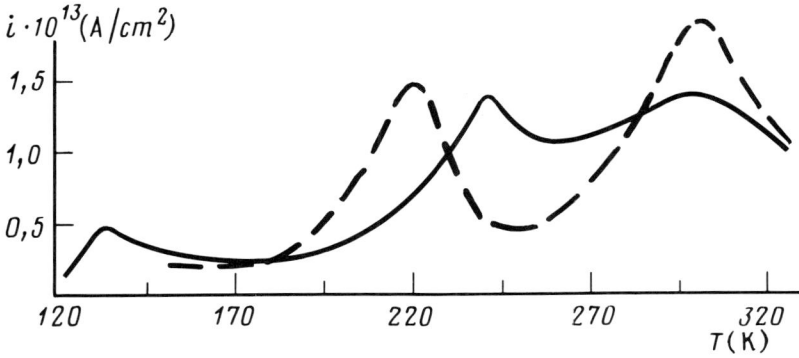

Figure 13 Thermogram for high-density polyethylene (solid line) and low-density polyethylene (dashed line). $T_p = 333$ K, $F = 20$ kV cm^{-1}, $b = 2°$ min^{-1}.

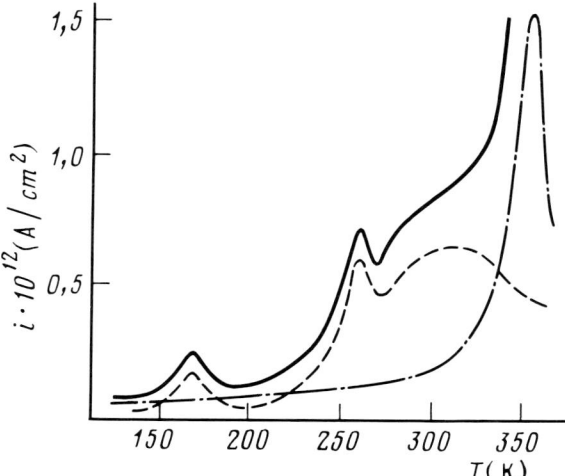

Figure 14 Thermogram for PS containing PNPDMAAB (solid line). Dashed line represents relaxation of the additive; dash-dot line represents relaxation of pure PS. (From Ref. 51.)

results are shown in Figure 15. In most cases thermal treatment has but a slight effect on the β peak, while α peaks undergo conspicuous changes. Quenching reduces the intensity and widens the peak; annealing brings about its narrowing and growth. The area under the peaks (charge) does not practically change. The ratio of charges released in α and β processes also remains constant. The reason for the changes lies apparently in the fact that in quenching a considerable number of large fluctuations are frozen. These fluctuations are loosely packed regions where unfreezing of segmental motions proceeds at lower temperatures. This leads to a widening of the distribution function in relaxation times.

Low-amplitude β processes do not require significant fluctuations of free volume and are less sensitive to thermal treatment.

The TSD technique has found wide application in the study of physical and chemical processes in polymers. The procedures adopted to this end are basically the same as those used in other relaxational processes: the control over the changes in transition temperatures, frequencies, and activation parameters of segmental motion. Crosslinking [54–56], crystallization [10, 37, 57], orientation [3, 58], aging [3, 10, 60–62], and other processes have been studied. Some works have been devoted to application of the technique to investigate binary systems: filled and plasticized polymers, mixtures, and copolymers [3, 10, 35, 44, 48, 52, 53, 59, 63–70].

Studies Under High Pressure

Hydrostatic pressure has been widely used recently as a technique for measuring changes in the free volume and estimating its role in molecular dynamics [71]. An important relaxation parameter determined in such experiments is the activation volume V^*, the minimal fluctuation of free volume near the kinetic unit, which is required for its motion. The activation volume is one of the few experimental parameters that allow determination

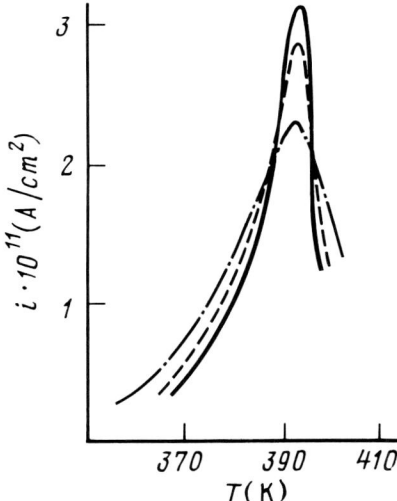

Figure 15 TSD peaks for PMMA recorded after cooling samples at different rate: 9 K min^{-1} (solid line), 20 K min^{-1} (dashed line), 460 K min^{-1} (dash-dot line). Heating rate is the same for all curves (6.5 K min^{-1}). (From Ref. 51.)

of sizes of kinetic units in a polymer. The baric dependence of relaxation times is determined by the Frenkel–Eiring equation:

$$\tau(p) = \tau(p \to 0) \exp\frac{pV^*}{kT} \tag{28}$$

where $\tau(p \to 0)$ are the values of relaxation times at normal pressure.

On the basis of experimental data obtained under uniform compression one can estimate the activation energy at constant volume E_V, which is the potential barrier of motion. The relationship between E_V and activation energy at constant pressure (activation enthalpy) E_p is

$$E_p = E_V + P_t V^* \tag{29}$$

where $P_t = T\alpha K$ is thermal pressure, α is the coefficient of thermal expansion, and K is the bulk modulus.

The $P_t V^*$ value is the work of formation of the activation volume. The E_V/E_p ratio is usually used as a criterion to estimate the barrier and free volume contributions into the energetics of molecular motion.

Compression is accompanied by a reduction in the polymer's free volume, which leads to decreased molecular mobility, increased T_g, and a shift of TSD peaks toward higher temperatures (Fig. 16).

The shape, width, and intensity of peaks shows only slight changes, at least up to pressures of 400–500 MPa.

The shift of the maximum of TSD α peaks with pressure is in a good agreement with the shift of the polymer's glass transition temperature dT_g/dp (Table 4).

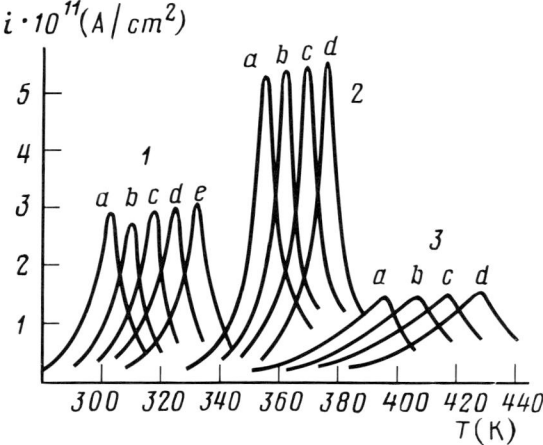

Figure 16 TSD peaks for (1) PVA, (2) PVC, (3) PMMA, recorded at different pressures: a, 0.1 MPa; b, 50MPa; c, 100MPa; d, 150 MPa; e, 200 MPa. $b = 4$ K min^{-1}. (From Refs. 27 and 72.)

Table 4 Activation Volumes of Relaxational Processes on the Basis of TSD and DR Data

Polymer	Process type	$\frac{dT}{dp}$ (K MPa^{-1})	E_p (kJ mol^{-1})	V^* (TSD) (cm^3 mol^{-1})	V^* (DR) (cm^3 mol^{-1})
PVA	α	0.15	182	186	139
PVC	α	0.14	422	342	321
PTFCE	α	0.12	205	164	
N6	α	0.10	150	93	
PMMA	α	0.22	355	398	21
	β	0.04	67	22	

Source: TSD data from Refs. 27 and 72; DR data from Refs. 73 and 74.

In the first work devoted to studying TSD curves for nylon-6 at high pressure [75] a linear temperature program was substituted by a smooth decrease in pressure from high to normal at a constant temperature. The equation $\tau(p) = P(p)/i(p)$, similar to Eq. 12, was used to determine relaxation times. The activation volume of the α process calculated from the baric dependence of τ at 293 K was 30.4 cm^3 mol^{-1}, which is considerably lower than the V^* values for relaxational processes in the area of T_g. These are, according to data collected through different techniques, 100–500 cm^3 mol^{-1}. The difference clearly is due to neglected distributions in all these parameters and application of equations for a monorelaxational process to calculate relaxation times at different T and p. This conclusion is supported by the underestimated V^* values obtained by Sharma and Jain [44], who used the same technique for data treatment of TSD in PVC and PVA copolymers. The activation volume of the α process measured at 350 K in the range of pressures up to 200

MPa equals 26.3 cm³ mol⁻¹, while with a dielectrical relaxation technique V^* reaches 321 cm³ mol⁻¹ in PVC and 139 cm³ mol⁻¹ in PVA [73].

A satisfactory correlation of TSD data with the results of other techniques is provided by the following procedure for analyzing thermograms. It follows from Eqs. 10 and 28 that

$$\tau(p) = \frac{kT[T_m(p)]^2}{b[E(p \to 0) + pV^*]} \qquad (30)$$

Provided that the shift in maxima of TSD peaks at external pressure is due to an increase in the activation energy of the process by pV^* at a constant relaxation time at the maximum τ_m, we can write

$$\frac{[T_m(p)]^2}{E(p \to 0) + pV^*} = \frac{[T_m(p \to 0)]^2}{E(p \to 0)}$$

and

$$\frac{[T_m(p)]^2}{[T_m(p \to 0)]^2} - 1 = \frac{pV^*}{E(p \to 0)} \qquad (31)$$

It follows from Eq. 31 that V^* can be determined easily from the slope angle of $[T_m(p)]^2/[T_m(p \to 0)]^2 - 1$ dependences on p. Figure 17 shows these dependences for several polymers. Equation 15 was used to determine the activation energies. The linearity of dependences seen in Figure 17 confirms the validity of Eq. 31. The average values of activation volumes are in agreement with V^* values obtained through the dielectrical relaxation technique (Table 4). Using Eqs. 30 and 31 one can also obtain the distribution in V^* (discussed in the next section).

Another way to determine V^* is to analyze isothermal depolarization at different pressures using Eqs. 19 and 20. This technique also allows determination of the activation

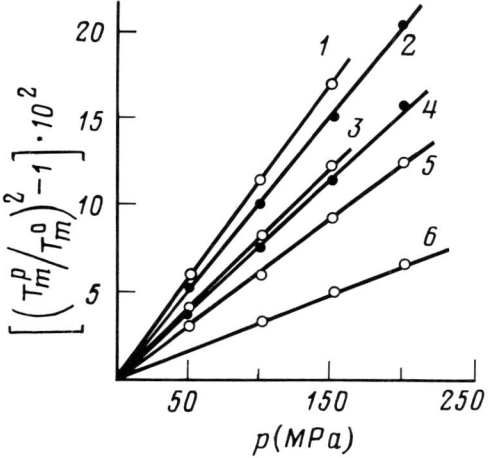

Figure 17 Dependence of $[T_m(p)/T_m(p \to 0)]^2 - 1$ on pressure for α process in (1) PMMA, (2) PVA, (3) PVC, (4) PTFCE, (5) N6, and β process in (6) PMMA. (From Refs. 27 and 72.)

energy at a constant volume. To this end isothermal depolarization is registered at various p and T satisfying the condition $V = $ const in agreement with the pVT diagram of a polymer.

Isothermal depolarization data were studied by Kovarskii [51], using Eq. 20 for *PVA* in the 300–320 K temperature range. It was found that $V^* = 140$ cm^3 mol^{-1}, $E_p = 161$ kJ mol^{-1}, and $E_V = 102$ kJ mol^{-1}. The differences between E_p and V^* values obtained through the conventional TSD technique and the procedure just described do not exceed 20%. The E_V/E_p ratio is 0.63, which is considerably lower than with β processes in polymers (0.8–0.9 [71]). This indicates that the energy of free volume formation makes a substantial contribution into the energetics of the α process, while the main contribution to β processes is made by the barrier of rotational motions.

Analysis of Distributions in Dynamic Parameters

Unlike other physical methods of studying polymers (dielectrical and mechanical relaxation in alternating fields, NMR, photon correlation spectroscopy, etc.) the TSD technique has at its disposal several approaches to obtain distribution functions in dynamic parameters. Apart from the traditional approach of fitting parameters of known distributions and experimental data, the fractional technique and the transformation method have been developed, as discussed earlier. Therefore, one can compare distributions obtained via different techniques and estimate the validity of information obtained.

Initial information on the distribution width in dynamic parameters can be obtained directly from TSD peaks. The width of a peak, assuming that the process is monorelaxational, is described by Eq. 16. If we use average values of activation energies of relaxational processes (Table 3), calculations yield the width of 5–10° for α peaks and 10–20° for β peaks. The width of TSD peaks for polymers is 40–60° for α processes and 60–120° for β process, i.e., it is much greater than the theory of monorelaxational processes suggests. Thus TSD peaks yield information on the distribution of relaxation transition temperatures; with the α process it is practically the distribution in glass transition temperatures [21].

Analyses of distributions in relaxation times have been carried out for PETP [19], PVDF [20], PMMA [12, 15, 76], PC [5, 9], PVC, PVA, PCTFE, and N6 [27, 34, 51]. These distributions were obtained through fractional [19], transformation [12, 15, 20, 27, 34], and fitting [9, 51, 76] techniques.

The method of direct transformation of TSD peaks into spectra of relaxation times is the most appealing because of its simplicity and productivity. It should be noted that the level of approximations which determines the errors in calculations of $F(\tau)$ is nearly the same for all three methods.

The main parameters characterizing distributions are shape, width, and asymmetry. The shape of distributions in τ has been analyzed by several workers [9, 20, 27, 34, 51, 76]. An exponential distribution in τ in the area of T_g, close to symmetrical, was obtained through the transformation technique with PVDF [20]. The distribution in τ of α processes in PMMA was shown to be described by the Havriliak–Negami function [76] and that of β processes by the Fuoss–Kirkwood function [22]. Analysis of distributions in τ, E, and τ_0 of the α process in five polymers [27, 34] revealed that they can be described easily with exponential equations of the following type:

$$F(X) = \begin{cases} C \exp\dfrac{X - X_m}{A(X)} & \text{at } X < X_m \\ C \exp\dfrac{X_m - X}{B(X)} & \text{at } X > X_m \end{cases} \quad (32)$$

where X are E values, $\ln \tau$ or $\ln \tau_0$ of corresponding distributions, X_m the value of these parameters at the distribution function maximum, $A(X)$ and $B(X)$ half-widths of distributions at the $1/e$ height at $X < X_m$ and $X > X_m$, respectively. Comparison of plots with Eqs. 32 and that of Havriliak–Negami (Fig. 18) shows that they can be used successfully to describe $F(\tau)$ for α processes in polymers. Another advantage of Eqs. 32, apart from the simplicity of analysis, is a clear physical meaning of the width (width at $1/e$ height) and asymmetry ($\gamma = A/B$) parameters. As compared to Gauss, Fuoss–Kirkwood, and Cole–Cole distributions, Eqs. 32 are noted for sharper peaks and steeper "wings."

Figure 19 shows distributions in τ in α processes for several polymers. Table 5 lists the parameters. The widest distribution is observed with α process in PMMA. Let us note that double maxima of TSD current were observed in a number of works [3, 16, 18], which confirms the wide distribution in τ in the polymer. No differentiation of distribution widths is observed with respect to the polymers' crystallinity. The parameter sensitive to crystallinity is the distribution asymmetry. The branch of high values of τ distributions is steeper with crystalline than with amorphous polymers, which are characterized by close to symmetrical distributions. Evidently this is related to the existence in the amorphous regions of crystalline polymers of a considerable share of molecular packings with the density close to that of crystalline regions.

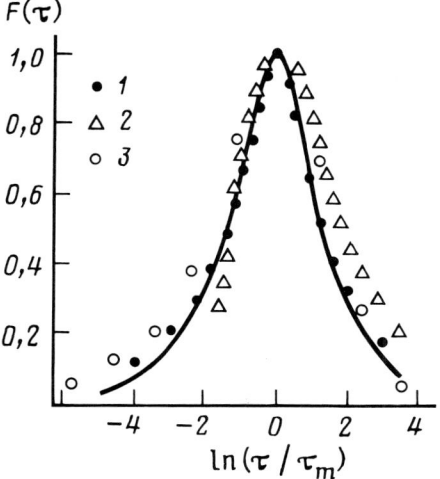

Figure 18 Distributions in relaxation times of α process in PVA at 303 K, obtained through (1) TSD technique, (2) ITP, and (3) DR technique. Solid line represents the Havriliak–Negami function with $\alpha = 0.315$ and $\beta = 0.98$. [Data for (1) and (2) from Ref. 51; data for (3) from Ref. 24.]

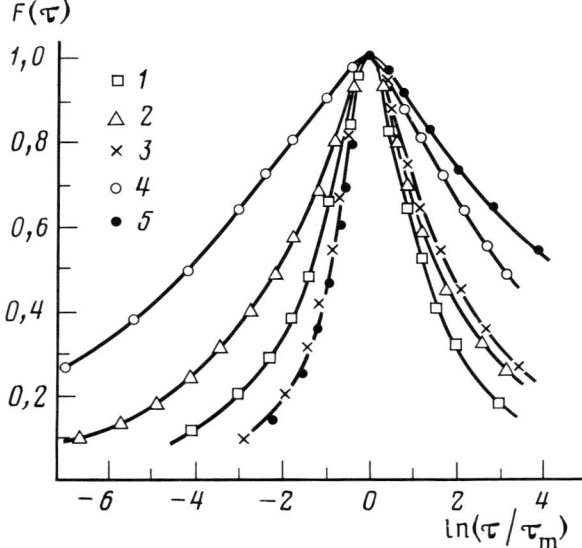

Figure 19 Distributions in relaxation times at temperatures of α peaks maximum in (1) PVA, (2) PVC, (3) N6, (4) PMMA, and (5) PTFCE. (From Ref. 27.)

The correctness of distributions obtained through the transformation of TSD peaks is supported by the results of their comparison with the data of dielectrical relaxation in alternating fields (Fig. 18). There is a satisfactory agreement between the distribution parameters obtained through TSD and ITD techniques.

In earlier work on TSD there was an active argument concerning the reason for distribution in τ: Is it caused by the frequency factor or activation energy distributions? Results obtained for a large number of polymers through the FTSD technique suggest that there is a set of activation energies with β process [51]. A similar distribution in E was shown [77] to occur in α processes (Fig. 20). There is also a distribution in the frequency factor in these processes. Thus in analyzing TSD data one must take into account distributions in all dynamic parameters.

Table 5 lists the distribution parameters in activation energies of α processes [27, 34]. The procedure of determining $F(E)$ was discussed earlier. Analysis shows that the $F(E)$ have the same shape and asymmetry as $F(\tau)$. The E values are purely empirical with no physical meaning, because the α process in a wide temperature range is not described by the Arrhenius equation. At the same time a narrow temperature range of extrapolations $(T_g \pm 20°)$ used in such cases makes it possible to apply the Arrhenius equation. It should be noted also that fractional techniques, normally based on Eq. 14, include the activation energy. In addition, the activation parameters and their distributions are useful in comparing characteristics of α and β processes. The equations of the free volume model are known to be inapplicable for β processes (Eqs. 9 and 11). An example of the use of the WLF equation to analyze distributions in α processes is provided in Ref. 20.

Distributions in relaxation times can also be obtained on the basis of data on isothermal

Table 5 Parameters of Distribution Functions in Relaxation Times τ, Energies E, and Activation Volumes V^* of the α Process: The Most Probable Values τ_m, E_m, V_m^*, the Width at $1/e$ Height Δ, and Assymetry γ

	Polymer				
Parameters	PVC	PVA	PTFCE	PMMA	N6
T_m (K)	356	303	328	396	316
$\ln \tau_m$ (sec)	3.6	4.1	4.2	4.0	4.4
$\Delta \ln \tau$ (sec)	5.4	3.4	7.0	9.5	3.0
$\gamma (\ln \tau)$	1.1	1.1	0.18	1.3	0.48
E_m (kJ mol^{-1})	422	182	205	355	150
ΔE (kJ mol^{-1})	31	17	38	65	20
$\gamma (E)$	1.1	1.1	0.19	1.3	0.54
V_m^* (cm^3 mol^{-1})	342	186	162	398	93
ΔV^* (cm^3 mol^{-1})	22	8.5	24	29	10
$\gamma (V^*)$	1.1	1.1	0.2	1.3	0.5

Source: Data from Refs. 27, 34, and 72.

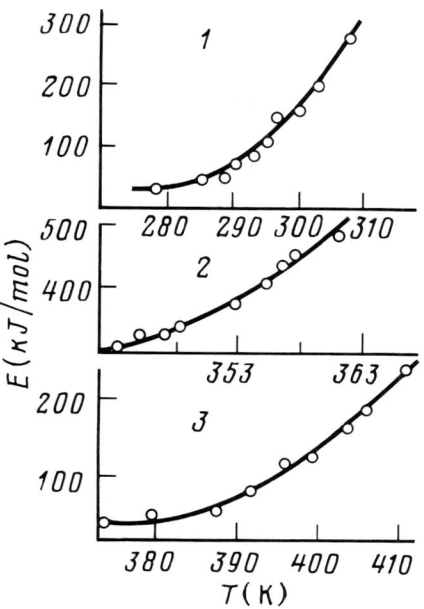

Figure 20 Dependence of activation energy of α process in (1) PVA, (2) PVC, and (3) PMMA on maximal fraction temperature. (From Ref. 77.)

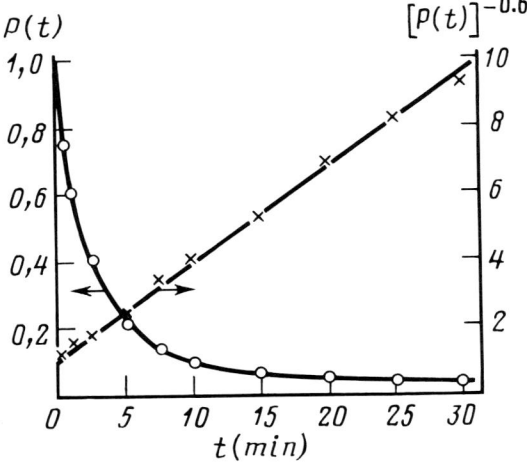

Figure 21 Depolarization kinetics of PVA at 303 K in normal coordinates and in coordinates of Eq. 20 with $1/\lambda = -0.6$. (From Ref. 51.)

depolarization kinetics. An analysis following Eq. 20 has been carried out [51]. Figure 21 shows a kinetic depolarization curve in PVA at 303 K. It is seen to be linearized in $[P(t)]^{-0.6}$–t coordinates. ν_0 equals 8.10^{-3} sec^{-1}. Use of the inverse Laplace transformation of Eq. 20 allows one to obtain the distribution function seen in Figure 18. The correlation between the results of both techniques (TSD and ITD) is considered to be reasonable.

Distributions in τ for β processes obtained through the same transformation technique as the data in Table 5 are different from the respective distributions of α processes. The former are wider and steeper (Fig. 22) and they cannot be described by Eqs. 32, but Cole–Cole and Fuoss–Kirkwood functions can describe them. With PMMA the parameter β in the Cole–Cole distribution, calculated from TSD data, is 0.33 (248 K). According to data of dielectrical relaxation in an alternating field, β with PMMA decreases from 0.5 to 0.42 as the temperature is decreased from 413 to 323 K [78]. Taking into account the differences in temperature, the agreement of data obtained through these techniques can be considered satisfactory.

The $F(\tau)$ width with β processes is 1.2–2 times greater than with α processes. However, the temperatures at which the distributions are determined are substantially different; consequently, to compare them correctly one has to reduce them to the same temperature. Since the temperature dependences of β processes normally obey the Arrhenius equation, the latter can be used to extrapolate the width $\Delta \ln \tau$ of distribution in the form

$$(\Delta \ln \tau)_{T_1} - (\Delta \ln \tau)_{T_2} = \frac{\Delta E(T_1 - T_2)}{k} \tag{33}$$

where ΔE is the width of the distribution in E.

Calculated on the basis of Eq. 33, the distribution width of the β process $(\Delta \ln \tau)_{1/e}$ at T_g is 8.6 for PMMA, i.e., it is close to the distribution width of the α process (9.5). Note that the width of distributions in activation energies for α and β processes is also close despite huge differences in the most probable E values.

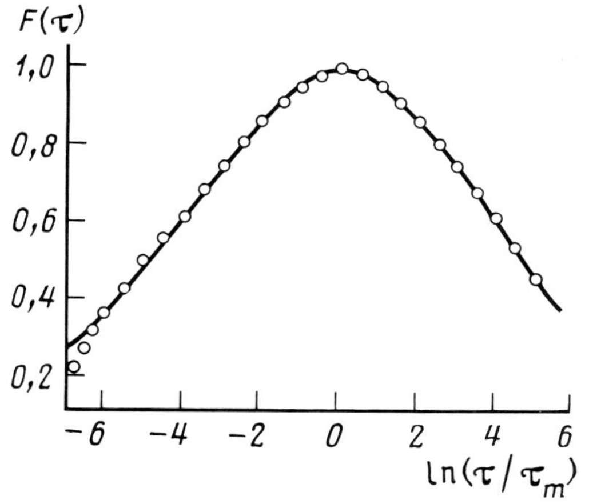

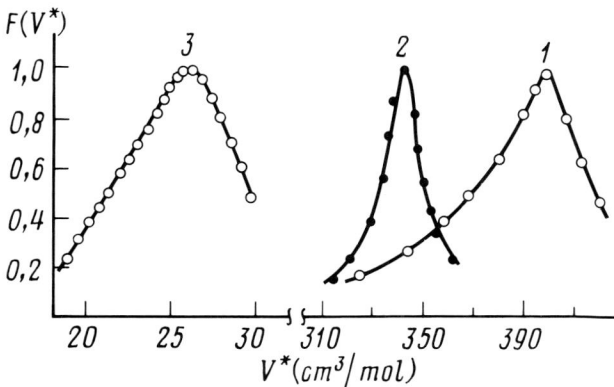

Figure 22 Distribution in relaxation times of β process in PMMA at 248 K (open circles). Solid line represents the Cole–Cole distribution with β = 0.33. (From Ref. 51.)

Plots of TSD curves over a wide range of hydrostatic pressures allows one to obtain distributions in activation volumes (Fig. 23). To this end use was made of Eq. 31 [27, 72], which can be written as

$$\frac{[T_i(p)]^2}{[T_i(p \to 0)]^2} - i = \frac{pV_i^*}{E_i(p \to 0)}$$

where $T_i(p)$ and $T_i(p \to 0)$ are the abscissas of a given point at TSD curves obtained at one and the same y coordinate and E_i comes from the distribution in activation energies. The relative number of species is determined from a relationship $n_i \sim i_i T_i$ (see p. 650).

The width of $F(V^*)$ is only slightly different for α and β processes, changing with

Figure 23 Distributions in activation volumes of α relaxation in (1) PMMA, (2) PVA, and β relaxation in (3) PMMA. (From Refs. 27 and 72.)

different polymers from 15 to 67 cm³ mol⁻¹. On the basis of the correlation between the activation volumes of molecular motion and compressibility [71], one can assume that the distribution in V^* reflects the set of compressibility coefficients of amorphous regions, which in its turn is the result of structural and dynamic inhomogeneity of these regions. That is why the main features (shape, width, and asymmetry) are repeated in distributions in all dynamic parameters for one and the same polymer.

TSD and Mechanisms of Relaxational Processes

The mechanisms of the main α and β processes are widely discussed and reviewed [79–82]. Recent experimental data do not support the idea that prevailed for a long time, that α and β processes are caused by motions of kinetic elements of a different scale (large backbone segments and side groups), related to the specificity of the glassy state. These processes were found to be typical of amorphous substances both organic and inorganic, polymeric and low molecular weight.

Two relaxational processes are observed in vitrifiable liquids which consist of rigid molecules with no internal rotation [80]. Introduction of polar molecules into nonpolar (or slightly polar) polymer brings about not one but two relaxation maxima on TSD thermograms (Fig. 14). These data strongly suggest that α and β processes are typical of the system with a single type of kinetic unit.

By using the DSC technique to study the dependence of activation energy of α and β processes on the chain length of polymer, it was found that both processes involve the same kinetic unit commensurable with the statistical Kuhn segment (3–12 monomer units) [81, 82]. In this way the existence of two processes differing in frequency and activation parameters is due to the difference in the motion mechanism of one and the same segment. Owing to its specificity the TSD technique allows one to determine these differences.

One of these features is the irreversible nature of electret discharge. This means that when there is complete discharge as a result of relaxation transition at temperature T_1 upon heating an electret with only one type of dipole, then all the relaxation transitions at $T > T_1$ will not be seen on TSD thermograms. On the other hand, when studying polymers by

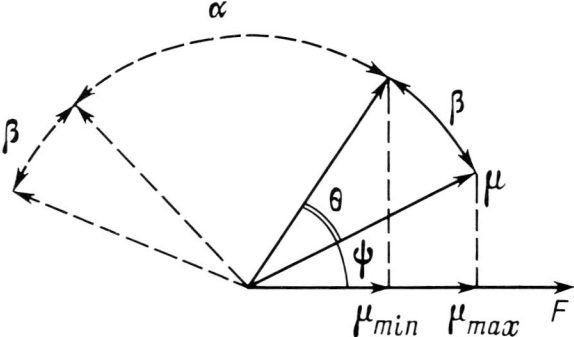

Figure 24 Reorientation of dipole μ in α and β processes. ψ is the angle of dipole alignment in the field F. θ is the apical angle of the rotation cone in the β process. μ_{max} is projection of μ vector on the direction of external field at switched on F; μ_{min} is the minimal value of projection μ at depolarization caused by the β process. (From Ref. 84.)

dielectrical relaxation in alternating fields, all types of molecular motion of a given dipole differing in frequencies will be active.

The appearance of two maxima on TSD curves at $T < T_g$ and T_g, observed in almost all polymers, indicates that depolarization in the low-temperature process has not been complete. The only reason for this effect in systems with a single type of dipole or with a set of kinetically equivalent dipoles is that in the low-temperature process the dipole rotation is restricted so that the orientation can change only within a cone with the apex angle θ (Fig. 24).

To get out from the area covered by the critical angle (α process) the vector of dipole moment has to overcome a high-potential barrier, which is unlikely at $T < T_g$ [83, 84]. The existence of such low-amplitude rotational motion of segments in a glassy state of a polymer results from a small free volume and is supported by low experimental values of E and V^*. A direct method to analyze rotational angles is double electron–electron resonance (ELDOR), which helped to establish that the rotational angles of species in liquids and polymers at $T < T_g$ are 0.1–0.4 rad (6–23°) [85, 86].

One can easily show that the apex angle of the rotational cone determines the ratio of charges released in α and β processes. It follows from Figure 24 that the charge released in the β process is equal to

$$Q_\beta = \frac{n(\mu_{max} - \mu_{min})}{S} = n\mu \left[\overline{\cos \psi} - \overline{\cos \left(\frac{\psi + \theta}{2} \right)} \right] \tag{34}$$

where n is the number of dipoles, μ_{max}, μ_{min} are the projections of the dipole moment vector μ on the direction of the polarizing field at the beginning and at the end of depolarization, ψ is the angle of dipole alignment, determined by polarization conditions, and S is the sample's area.

The total charge released in α and β processes is

$$Q_\alpha + Q_\beta = \frac{n\mu_{max}}{S} = n\mu \times \frac{\overline{\cos \psi}}{S} \tag{35}$$

It follows from Eqs. 34 and 35 that

$$\frac{Q_\alpha}{Q_\beta} = \frac{\overline{\cos (\psi + \theta/2)}}{\overline{\cos \psi} - \overline{\cos (\psi + \theta/2)}} \tag{36}$$

Thus the charge ratio, i.e., the ratio of the areas under the TSD peaks of α and β processes, depends on both the polarization conditions (ψ angle) and the rotational motion's amplitude of segments at $T < T_g$. The validity of Eq. 35 has been experimentally confirmed [84] by the correlation between activation volumes and charges (Fig. 25).

The meaning of this dependence is as follows. Both activation volumes and charges released depend on one and the same parameter, the amplitude of rotational segmental motions. Calculations made on the basis of Eq. 36 indicate that at $\psi > 65°$ (angles typical of $F = 10$–50 kV cm^{-1}) experimental Q_α/Q_β values are matched by angles θ in the 5–20° range. These θ values are in a good agreement with those obtained through the ELDOR technique [85, 86].

This, the main difference in mechanisms of α and β processes, is linked with the amplitude of segmental motion. β relaxation processes are characterized by low-amplitude rotational motion, which requires relatively low energies and activation volumes. α processes involve intensive structural rearrangements which accompany unfreezing of

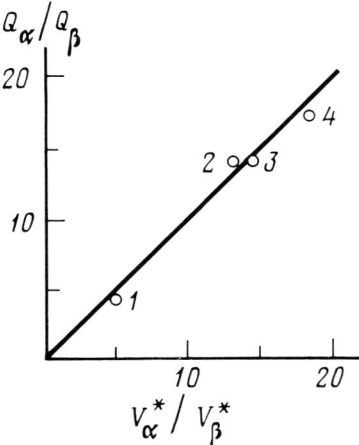

Figure 25 Dependence of the ratio of charges released at depolarization in α and β processes in (1) PTFCE, (2) PVC, (3) PETP, and (4) PMMA on the ratio of activation volumes. (From Ref. 84.)

large-amplitude rotational and translational motions of segments. Both processes appear to involve the same kinetic units. These can be estimated from numerous data on activation volumes of the rotation of low molecular weight species in polymers, obtained with the help of spin probe ESR spectroscopy [87]. The ratio of activation volume and van der Waals species volume V^*/V_w was found to be only slightly dependent on size and polymer nature, being 0.1 at $T < T_g$ (β process). Using V^* values listed in Table 4 one can easily obtain the segments' volumes, $V_w = 120-350$ cm^3 mol^{-1}, which comprises 5–10 monomer units. The same values of V_w for segments are obtained from activation volumes of the α process where $V^*/V_w \approx 1$ [51, 71].

REFERENCES

1. B. Gross, *Charge Storage in Solid Dielectrics*, Elsevier, Amsterdam (1964).
2. A. N. Gubkin, *Electrets*, Nauka, Moscow (1978).
3. J. van Turnhout, *Thermally Stimulated Discharge of Polymer Electrets*, Elsevier, Amsterdam (1975).
4. J. van Turnhout, in *Electrets* (G. M. Sessler, ed.), Springer-Verlag, Berlin (1980).
5. G. A. Lushcheikin, *Polymer Electrets*, Khimia, Moscow (1976; 2nd ed., 1984).
6. C. Bucci and R. Fieschi, *Phys. Rev. Lett.*, *12*: 16 (1964).
7. C. Bucci and R. Fieschi, *Phys. Rev.*, *148*: 816 (1966).
8. J. Vanderschueren, *J. Polym. Sci. Polym. Lett.*, *10*: 543 (1972).
9. J. Vanderschueren, *Appl. Phys. Lett.*, *25*: 270 (1974).
10. J. Vanderschueren, Ph.D. dissertation, Liége (1974).
11. J. Vanderschueren, *J. Polym. Sci. Polym. Phys. Ed.*, *15*: 873 (1977).
12. Ch. Solunov, T. Vassilev, and I. Iovtchev, *Physics*, *11*(4): 87 (1973).
13. Ch. Solunov and T. Vassilev, *J. Polym. Sci. Polym. Phys. Ed.*, *12*: 1273 (1974).
14. Ch. Ponevski and Ch. Solunov, *J. Polym. Sci. Polym. Phys. Ed.*, *13*: 1467 (1975).
15. Ch. A. Solunov and Ch. S. Ponevsky, *J. Polym. Sci. Polym. Phys. Ed.*, *15*: 969 (1977).
16. M. Randak and V. Adamek, *Plaste Kautsch.*, *19*: 905 (1972).
17. D. Chatain, C. Lacabanne, and M. Maitrot, *Phys. Stat. Sol.*, *A13*: 303 (1972).

18. C. Lacabanne and D. Chatain, *J. Polym. Sci. Polym. Phys. Ed.*, *11*: 2315 (1973).
19. T. Hino, *Jpn. J. Appl. Phys.*, *12*: 611 (1973).
20. G. Pfister and M. A. Abkovitz, *J. Appl. Phys.*, *45*: 1001 (1974).
21. J. A. Gorochovatskij, *The Fundamentals of Thermodepolarization Analysis*, Science, Moscow (1981).
22. R. M. Fuoss and J. G. Kirkwood, *J. Am. Chem. Soc.*, *63*: 385 (1941).
23. K. S. Cole and R. H. Cole, *J. Chem. Phys.*, *9*: 341 (1941).
24. S. Havriliak and S. Negami, *Polymer*, *8*: 161 (1967).
25. G. Williams and D. C. Watts, *Trans. Faraday Soc.*, *66*: 80 (1970).
26. V. A. Kutyrkin, I. R. Mardaleishvili, O. N. Karpukhin, and V. M. Anisimov, *Kinet. Catal. (USSR)*, *25*: 1310 (1984).
27. A. L. Kovarskii, S. A. Mansimov, and A. L. Buchachenko, *Polymer*, *27*: 1014 (1986).
28. R. A. Creswell and M. M. Perlman, *J. Appl. Phys.*, *41*: 2365 (1970).
29. M. M. Perlman, *J. Electrochem. Soc.*, *119*: 892 (1971).
30. M. L. Khara and C. S. Bhatnagar, *Ind. J. Pure Appl. Phys.*, *7*: 497 (1969).
31. V. E. Gul', G. A. Lushcheikin, and B. A. Dogadkin, *Dokl. AN USSR*, *149*: 302 (1963).
32. A. Servini and A. K. Jonscher, *Thin Solid Films*, *3*: 341 (1969).
33. M. Davies, P. J. Hains, and G. Williams, *J. Chem. Soc. Faraday Trans. II*, *69*: 1785 (1973).
34. S. A. Mansimov, M. K. Kerimov, H. B. Gezalov, and A. L. Kovarskii, *Vysokomol. Soedin. USSR*, *28A*: 1996 (1986).
35. A. L. Kovarskii and V. N. Saprygin, *Polymer*, *23*: 974 (1982).
36. J. Tornhout, *Polym. J.*, *2*: 173 (1971).
37. G. A. Lushcheikin and L. I. Voitenshtock, *Vysokomol. Soedin. USSR*, *16*: 1364 (1974).
38. J. Aoki and J. O. Brittain, *J. Polym. Sci. Polym. Phys. Ed.*, *14*: 1297 (1976).
39. J. Belana, P. Colomer, M. Pujal, and S. Montserrat, *J. Macromol. Sci. Phys.*, *B23*: 467 (1984).
40. A. R. McGhie, *Polymer*, *13*: 371 (1972).
41. N. G. McCrum, B. E. Read, and G. Williams, *Inelastic and Dielectric Effects in Polymeric Solids*, Wiley, London (1967).
42. D. W. McCall, in *Relaxation in Solid Polymers, Molecular Dynamics and Structure of Solids* (R. S. Carter and J. J. Rush, eds.), National Bureau of Standards Special Publication 301, Washington (1969).
43. R. F. Boyer, *Encyclopedia of Polymer Science and Technology*, Supplement N 2, Wiley, New York, pp. 745–839 (1977).
44. J. K. N. Sharma and K. K. Jain, *High Temp. High Press.*, *14*: 293 (1982).
45. P. Fischer and P. Röhl, *J. Polym. Sci. Polym. Phys. Ed.*, *14*: 531 (1976).
46. M. Pineri, P. Berticat, and E. Marchal, *J. Polym. Sci. Polym. Phys. Ed.*, *14*: 1325 (1976).
47. D. Chatain, C. Lacabanne, and J. C. Monpagens, *Macromol. Chem.*, *178*: 583 (1977).
48. M. Goel, V. Gupta, and P. Pillai, *Polym. Bull.* *7*: 103 (1982).
49. J.-P. Reboul and A. Toureille, *J. Polym. Sci. Polym. Phys. Ed.*, *22*: 21 (1984).
50. J. H. Ranicar and R. S. Fleming, *J. Polym. Sci. Polym. Phys. Ed.*, *10*: 1979 (1972).
51. A. L. Kovarskii, Ph.D. dissertation, Moscow (1988).
52. V. N. Saprygin and A. L. Kovarskii, *Vysokomol. Soedin. USSR*, *26B*: 364 (1984).
53. P. C. Mehendru, K. Jane, and P. Mehendru, *J. Phys.*, *D9*: 83 (1976).
54. P. Eyerer, *J. Appl. Polym. Sci.*, *16*: 2461 (1972).
55. D. L. Shelley and S. F. Huber, Annual Reports, Conference on Electrical Insulation and Dielectric Phenomena, National Academy of Science–National Research Council, Washington, p. 100 (1975).
56. P. K. Pillai, P. K. Nair, and R. Nath, *Polymer*, *17*: 921 (1976).
57. T. Hashimoto, M. Shiraki, and T. Sakai, *J. Polym. Sci. Polym. Phys. Ed.*, *13*: 2401 (1975).
58. R. J. Comstock, S. I. Stupp, and S. H. Carr, *J. Macromol. Sci.*, *B13*: 101 (1977).
59. S. K. Shrivastava, J. D. Ranade, and A. P. Shrivastava, *Phys. Lett.*, *69A*: 465 (1979).

60. Y. Aoki and J. O. Brittain, *J. Polym. Sci. Polym. Phys. Ed.*, *15*: 199 (1977).
61. M. Kryszewski, H. Kasica, J. Patora, and J. Piotrowski, *J. Polym. Sci.*, *C30*: 243 (1970).
62. T. Mizutani and M. Yeda, *J. Phys.*, *D11*: 185 (1978).
63. T. Tanaka, S. Hayashi, and K. Shibayama, *J. Appl. Phys.*, *48*: 3478 (1977).
64. G. A. Baum, *J. Appl. Polym. Sci.*, *17*: 2855 (1973).
65. P. K. C. Pillai, T. C. Goel, and S. F. Xavier, *Eur. Polym. J.*, *15*: 1149 (1980).
66. B. C. Simionescu, E. Neagu, M. Leanca, and C. I. Simionescu, *Polym. Bull.*, *8*: 15 (1982).
67. J. Vanderschueren, A. Janssens, M. Landing, and J. Niezette, *Polymer*, *23*: 395 (1982).
68. P. K. C. Pillai, B. K. Gupta, and M. Goel, *J. Polym. Sci. Polym. Phys. Ed.*, *19*: 1461 (1981).
69. H. Kaoru, A. Masao, S. Isao, and T. Toshiaki, *Kobunshi Ronbunshu*, *40*: 317 (1983).
70. J. X. Wen and T. Takamatsu, *Kobunshi Ronbunshu*, *40*: 135 (1983).
71. A. L. Kovarskii, *Vysokomol. Soedin. USSR*, *28A*: 1347 (1986).
72. S. A. Mansimov, A. L. Kovarskii, and M. K. Kerimov, *Vysokomol. Soedin. USSR*, *28A*: 2000 (1986).
73. S. Saito, H. Sasabe, T. Nakajima, and K. Yada, *J. Polym. Sci. A-2*, *6*: 1297 (1968).
74. H. Sasabe and S. Saito, *J. Polym. Sci. A-2*, *6*: 1401 (1968).
75. B. Ai, P. Destruel, H. T. Giam, and R. Loussier, *Phys. Rev. Lett.*, *34*: 84 (1975).
76. M. Kryszewski, M. Zielinski, and S. Sapieha, *Polymer*, *17*: 212 (1976).
77. A. L. Kovarskii and V. N. Saprygin, *Vysokomol. Soedin. USSR*, *26A*: 1949 (1984).
78. Y. Ishida and K. Yamafuji, *Koll. Z.Z. Polym.*, *177*: 97 (1961).
79. G. Williams, *Adv. Polym. Sci.*, *33*: 60 (1979).
80. G. P. Iohari and M. Goldstein, *J. Chem. Phys.*, *55*: 4245 (1971).
81. V. A. Bershtein, V. M. Egorov, Yu. Ya. Emelyanov, and V. A. Stepanov, *Polym. Bull.*, *9*: 98 (1983).
82. V. A. Bershtein, V. M. Egorov, and Yu. Ya. Emelyanov, *Vysokomol. Soedin. USSR*, *27A*: 2451 (1985).
83. J. K. Moscicki, G. Williams, and S. M. Aharoni, *Polymer*, *22*: 1361 (1981).
84. A. L. Kovarskii, S. A. Mansimov, and A. L. Buchachenko, *Dokl. AS USSR*, *291*: 1142 (1986).
85. V. A. Benderskii, S. T. Kirillov, and N. P. Piven, "Molecular Motions as Studied by Double Electron-Electron Resonance (ELDOR), Magnetic Resonance and Related Phenomena," Proceedings of the 20th Congress Ampere, August 21–26, Tallin, p. 21 (1978).
86. V. A. Benderskii and N. P. Piven, *J. Phys. Chem. USSR*, *59*: 1329 (1985).
87. A. M. Wasserman and A. L. Kovarskii, *Spin Labels and Probes in Chemical Physics of Polymers*, Nauka, Moscow (1986).

18

Structural Characterization of Styrene–Butadiene Copolymers by Ozonolysis—GPC and HPLC Methods

Yasuyuki Tanaka
Tokyo University of Agriculture and Technology
Koganei, Tokyo, Japan

INTRODUCTION	678
PREPARATION AND PACKING OF GPC AND HPLC GELS	679
DETERMINATION OF SEQUENCE DISTRIBUTION IN SBR	680
Ozonolysis of SBR	680
Sequence Distribution of Anionically Polymerized SBR	682
Sequence Distribution of Commercial SBR	686
S–B–S-Type Triblock Copolymers	690
SEQUENCE STRUCTURE OF STYRENE AND 1,2-BUTADIENE UNITS	699
Configurational Sequences of Styrene Units	699
Arrangement of Styrene and 1,2-Butadiene Units	700
ANALYSIS OF SEQUENCE DISTRIBUTION IN CURED SBR	704
DETERMINATION OF CHEMICAL COMPOSITION DISTRIBUTION IN SBR	707
REFERENCES	709

New GPC and HPLC methods were proposed for the determination of the sequence distribution of styrene units, arrangement of styrene and 1,2-butadiene units, and chemical composition distribution in styrene–butadiene copolymers. High-resolution polystyrene gel for GPC and polyacrylonitrile gel for HPLC were prepared for the measurement of ozonolysis products from copolymer and chemical composition distribution of SBR respectively. Ozonolysis conditions were established to cleave all the double bonds of butadiene units in copolymer. The high-resolution GPC of the ozonolysis products showed peaks derived from a styrene monad to 12 or longer sequences and sequences consisting of styrene and 1,2 units flanked by 1,4-butadiene units as well as a peak from block styrene sequences. The sequence distribution determined from the ozonolysis–GPC was in good agreement with that calculated by copolymerization parameters for radical and anionic copolymers. The method was applied to the characterization of commercial SBR samples and triblock copolymers. Tapered-block structures and block styrene sequences were clearly identified by a combination of the GPC analysis of original and ozonolysis products. The method was applicable to the analysis of cured SBR by ozonolysis of finely powdered samples in suspension. The tacticity of styrene units and arrangement of styrene and 1,2 units were characterized by HPLC separation followed by ^{13}C NMR analysis of GPC fractions. The probability of addition of styrene and 1,2 units was discussed.

The chemical composition distribution of SBR was determined by HPLC measurement using polyacrylonitrile gel as stationary phase with gradient elution of hexane/chloroform. The observed distribution was almost independent of the molecular weight of SBR. The HPLC method was proved to provide quantitative information about the blending of copolymers.

INTRODUCTION

Styrene–butadiene rubber (SBR) is commercially produced by emulsion polymerization or anionic polymerization initiated with alkyllithium, which is characterized by the variety of the sequence distribution of styrene units, ranging from almost random copolymers to S–B–S-type triblock copolymers. The sequence distribution can be controlled by the modification of the polymerization conditions, e.g., addition of a polar compound as a randomizing agent, control of the conversion, the controlled charge of styrene and butadiene, or use of a coupling agent to make the block sequences. The approximate distribution of styrene units can be predicted from the copolymerization parameters. In practice, however, it seems very difficult to determine the sequence distribution in the case of partial block and triblock copolymer. The occurrence of 1,2 addition of butadiene units may further complicate the situation.

The significance of the sequence distribution has long been recognized as a dominant factor in characterizing mechanical and thermal properties of SBR and styrene–butadiene block copolymers. The sequence distribution of styrene units in SBR has been studied using NMR and chemical degradation–gas chromatographic methods. Mochel [1] tried to determine the amount of short styrene sequences by the curve resolution of phenyl proton signals. The limitations of the method were shown by Tanaka et al. [2] using model compounds. Katrizky and Weiss [3, 4] and Segre et al. [5] assigned the ^{13}C NMR signals of SBR and measured the diad and triad sequence distributions of styrene and butadiene units. Similarly, Randall [6] determined the sequence distribution from the ^{13}C NMR spectrum of the hydrogenated SBR. Gas chromatographic analysis of the metathesis products of SBR has been applied to the determination of sequence distribution [7, 8]. However, these methods provide information only on the average sequence length or distribution of short sequences such as diad or triad. As to S–B–S triblock copolymer, Minoura and Hatanaka [9] and Leblanc [10] analyzed the distribution of block styrene sequences by gel permeation chromatography (GPC) measurement of oxidative degradation products from block copolymers. However, the analysis of the sequence distribution by these methods has not been enough to characterize the relationships among polymerization conditions, sequence structures, and physical properties.

A new method for the characterization of the sequence distribution of styrene and 1,2-butadiene units in SBR uses high-resolution GPC analysis of ozonolysis products. The ozonolysis–GPC method was proved to provide direct information about the sequence distribution of styrene units from short to block sequences and also on the arrangement of styrene and 1,2-butadiene sequences.

Copolymers usually have a distribution of chemical composition as well as molecular weight distribution. The determination of chemical composition distribution is another fundamental item in the characterization of copolymers. Thus far, chemical composition distribution has been analyzed by cross-fractionation and thin-layer chromatography. However, the former method requires troublesome procedures using two sets of solvents and nonsolvents. Although the latter method is rapid and convenient, sometimes it does not provide quantitative information; even a homopolymer or a copolymer having a narrow chemical composition distribution tends to show a trail with long tailing. Recently, high-performance liquid chromatography (HPLC) has been applied to the determination of the chemical composition distribution of styrene–methyl methacrylate copolymer and styrene–acrylonitrile copolymer [11–13]. In these measurements the use of silica gel as a stationary phase restricted the application of the HPLC method.

A new HPLC method to characterize the chemical composition distribution of SBR and styrene–methyl methacrylate copolymer uses a crosslinked acrylonitrile gel as stationary phase. The separation was found to be essentially independent of the molecular weight distribution of copolymer.

This chapter describes the structural characterization of SBR, styrene–butadiene block copolymers, and styrene–isoprene block copolymers using GPC and HPLC methods. First we consider the procedure of preparing polystyrene and polyacrylonitrile gels, because these GPC and HPLC methods make use of high-resolution analytical and preparative columns packed with these gels. Then we review the analysis of sequence distribution and arrangement of styrene and 1,2-butadiene units in these copolymers by the ozonolysis–GPC method. Finally, we deal with the determination of chemical composition distribution of SBR by HPLC.

PREPARATION AND PACKING OF GPC AND HPLC GELS

Polystyrene gel was prepared by suspension polymerization in aqueous medium using polyvinyl alcohol as a stabilizer. Table 1 gives the details of the polymerization conditions and evaluation of the packed columns [14, 15]. The suspension containing particles averaging 6–8 μm was obtained by using a high-speed homogenizer at about 20,000 rpm at 0°C and was polymerized at 80°C with slow stirring. After the resulting gel was washed several times with hot water followed by acetone, small particles were removed by decantation in acetone.

Packing of the GPC columns was carried out with an ordinary slurry method using the apparatus illustrated in Figure 1. The flow rate during the packing was increased stepwise

Table 1 Polymerization Conditions for Polystyrene Gel[a]

Run no.	Water (ml)	Styrene (g)	DVB[b] (g)	Diluent (g)	PVA (g)	Particle size (μm)	NTP[c]
1	700	38.4	34.5	Toluene 70	3.5	8.6	22,300
2	700	32.5	40.3	Toluene 70	3.5	7.5	24,700

[a]Initiator: 2,2'-azobis-(2,4-dimethyl valeronitrile) 2g; polymerization: 80°C, 10 hr.
[b]Divinylbenzene (purity 56%).
[c]Number of theoretical plates for 7.5 mm i.d. × 500 mm column.

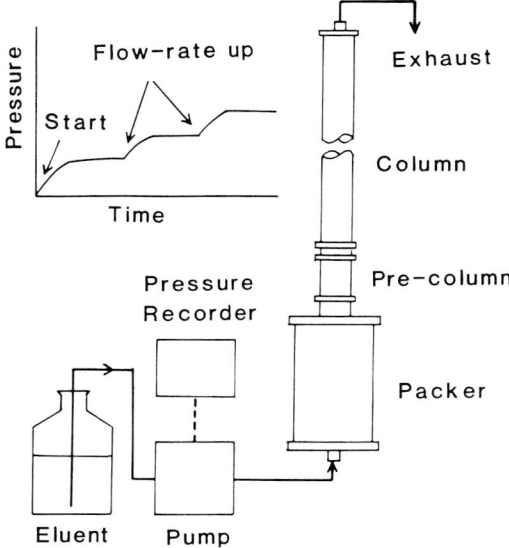

Figure 1 Packing of high-resolution GPC column.

to attain a constant pressure for each step and controlled around 80–100 kg cm^{-2} at the final stage. The number of theoretical plates was found to be 29,000–32,000 for a preparative GPC column (21.2 mm i.d. × 600 mm) using chloroform as an eluent. The apparatus was set upside down in the case of packing of polystyrene gel or polyacrylonitrile gel with other eluents.

In a similar way polyacrylonitrile gel was prepared by copolymerization of acrylonitrile with ethylene dimethacrylate using isoamyl alcohol as a diluent [16]. The method is given in Table 2.

DETERMINATION OF SEQUENCE DISTRIBUTION IN SBR

Ozonolysis of SBR

Ozonolysis of SBR [17, 18] was carried out by blowing 1.0–1.5% ozonated oxygen into 0.2–0.3% (w/v) SBR solution in methylene chloride through a glass filter at −20 to −30°C until the uptake of ozone ceased; this was detected by a color change of a

Table 2 Polymerization Conditions for Polyacrylonitrile Gel[a]

Water[b] (ml)	Acrylonitrile (ml)	Ethylene–dimethacrylate (ml)	Isoamyl alcohol (ml)	Particle size (μm)
625	50	25	50	2.5–5.0

[a]Initiator: 2,2′-azobis-(2,4-dimethyl valeronitrile) 1.9 g; polymerization: 80°C, 10 hr.
[b]Solution containing NaCl (8 wt %) and polyvinyl alcohol (0.6 wt %).

potassium iodide aqueous solution attached to the outlet of the reaction vessel. The ozonide was decomposed to alcohols by dropping lithium aluminum hydride in ethyl ether into the reaction mixture at 0°C and raising the temperature to reflux the solution for 30 min followed by the addition of a small amount of water. GPC measurements of ozonolysis products were made with an UV detector at 254 nm by using an analytical column (7.5 mm i.d. × 600 mm) or three preparative columns (21.2 mm i.d. × 600 mm) in series packed with polystyrene gel having an exclusion limit of 3×10^3. Chloroform was used as an eluent. The GPC fractions were analyzed by ^{1}H and ^{13}C NMR using a JEOL FX-200.

The degradation of all the butadiene double bonds was confirmed by ^{1}H NMR analysis of the ozonolysis products from cis-1,4-polybutadiene (cis-1,4, 96%; trans-1,4, 2%; 1,2, 2%) and anionically polymerized polybutadiene (cis-1,4, 42%; trans-1,4, 47%; 1,2, 11%) under these ozonolysis conditions. It was confirmed that the styrene units were intact under these ozonolysis conditions using a model mixture of polystyrene and polybutadiene; it showed only a single GPC peak corresponding to that of the original polystyrene, which will be shown later. It was also confirmed that 1,4-butanediol and 3-hydroxymethyl-1,6-hexanediol derived from 1,4-1,4 and 1,4-1,2-1,4 sequences, respectively, were not detected due to their low extinction coefficients in UV detector and very low solubility in chloroform.

As a model copolymer, styrene and butadiene were copolymerized with benzoil peroxide as an initiator and terminated at 4.3% conversion. The sample was analyzed by the ozonolysis–GPC method using high-resolution GPC columns having the exclusion limit of 3×10^3 as seen in Figure 2. By FD-MS, ^{1}H NMR, and ^{13}C NMR measurements of the

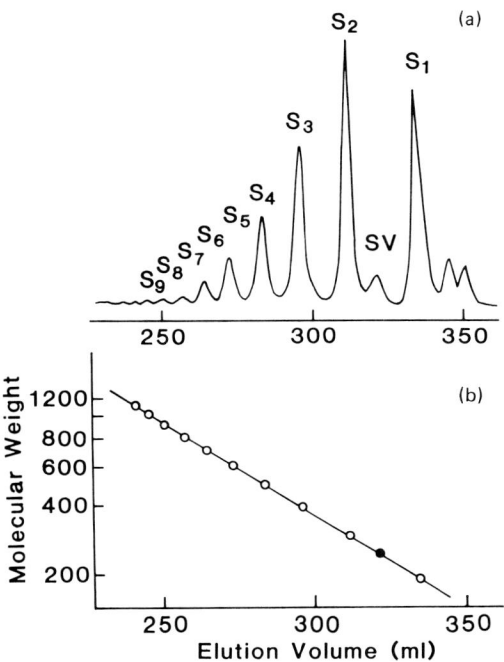

Figure 2 High-resolution ozonolysis–GPC curves for SBR initiated with benzoil peroxide and GPC calibration.

Table 3 Extinction Coefficients of Styrene Oligomers

n	$E_{254}{}^a$ (mol L^{-1} cm^{-1})	E_{254}/n
1	242	242
2	570	285
3	811	270
4	975	244
5	1391	278
6	1471	245
Average		261

^{a}Extinction coefficient at 254 nm for styrene oligomers:
$$CH_3CH_2(CH_2CH)_nCH_2CH_2CH_3$$
$$|$$
$$Ph$$

collected fractions, the peaks at 333, 320, 310, and 295 ml were assigned to the products derived from the 1,4-St-1,4, 1,4-St-1,2-1,4 (and 1,4-1,2-St-1,4), 1,4-(St)$_2$-1,4, and 1,4-(St)$_3$-1,4 sequences as given by the general formula (Eq. 1), respectively, which were abbreviated as S$_1$, SV (VS), S$_2$, and S$_3$.

$$HOCH_2CH_2\text{—} \text{—}(CH_2CH)_m\text{—} \text{—}(CH_2CH)_n\text{—} \text{—}CH_2CH_2OH \qquad (1)$$
$$| \qquad\qquad |$$
$$CH_2OH \qquad C_6H_5$$

A plot of log MW vs. elution volume gave a straight line for S$_1$ to S$_{10}$ and SV (VS).

The extinction coefficients at 254 nm were determined for the model compounds corresponding to S$_1$ to S$_6$. The extinction coefficient per styrene unit showed practically the same value as those given in Table 3. This finding indicates the integrated intensity of each GPC peak can be used for the analysis of the sequence distribution of styrene units.

Sequence Distribution of Anionically Polymerized SBR

Styrene and butadiene were polymerized by a molar ratio of 1:4.1 in toluene with n-butyllithium [18]. The conversion and content of styrene units in the copolymer increased gradually up to about 75% conversion, where almost all of the butadiene was consumed, and afterward both increased rapidly as seen in Figure 3. After the consumption of butadiene, the residual styrene is expected to be incorporated into the copolymer chains to make block styrene sequences. A similar composition–conversion relationship was reported for the copolymerization of styrene and butadiene with alkyllithium in nonpolar solvents [19–21].

The sequence distribution of the samples collected at different conversions was determined by using the ozonolysis–GPC method. Figure 4 shows the ozonolysis–GPC curves of these samples by using two analytical columns in series, with exclusion limits of 3×10^3 and 3×10^4. The molecular weight distribution of the block styrene peak expressed by $\overline{M}_w/\overline{M}_n$ was found to be 1.6–1.7 for each.

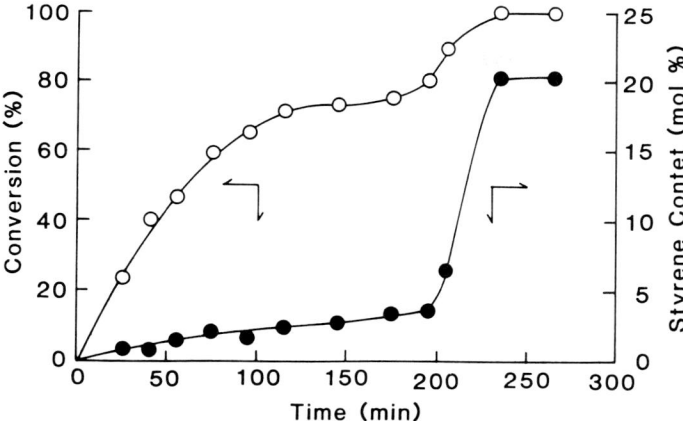

Figure 3 Conversion and copolymer composition curves of copolymerization with n-butyllithium.

These findings clearly show that the styrene units predominantly exist as isolated sequences at the initial stage of polymerization. The diad and triad styrene sequences increased with increasing conversion up to 75%. At the final stage of polymerization, block styrene sequences were formed in addition to the styrene monad-to-triad sequences. The number average length of the block styrene sequences increased from 43 to 65 with increasing conversion from 89.3 to 100%, whereas the distribution of short styrene sequences remained unaltered. This demonstrates that partially blocked copolymers are synthesized after the consumption of butadiene as predicted from the composition–conversion relationship.

Figure 5 shows the high-resolution GPC curves of the ozonolysis products of SBR at conversions of 75.6 and 89.3%. It is remarkable that the peaks due to the short styrene sequences overlap that of the block styrene sequences at 89.3% conversion. The peak due to block sequences reduced the extent of overlap with increasing conversion from 89.3 to 100%. These findings demonstrate that a small amount of butadiene still remains even at the final stage of polymerization to make an isolated butadiene unit in long styrene sequences.

The monomer reactivity ratios of $r_{St} = 0.16$ and $r_{Bd} = 13$ were calculated for copolymers initiated with n-butyllithium using the integrated equation of Mayo and Lewis. As to the radical SBR seen in Figure 3, the monomer reactivity ratios of $r_{St} = 0.94$ and $r_{Bd} = 1.2$ were determined using the Fineman–Ross equation. The sequence distribution of the styrene units calculated from the copolymerization parameters was compared with that determined by the ozonolysis–GPC measurement as shown in Figure 6. Here both 1,4-$(St)_n$-1,4 and 1,4-$(St)_n$-1,2-1,4 sequences are counted as the $(St)_n$ sequence, although the S_nV sequence is statistically anticipated to contain VS_n and S_mVS_{n-m} sequences. The composition of S_n sequences will be discussed later. The observed sequence distribution for the radical copolymer at low conversion showed good agreement with that calculated from monomer reactivity ratios. Only a slight difference was observed between the observed and calculated distribution for the copolymer initiated with n-butyllithium and terminated at 75.6% conversion. This may be due to the error inherent in the determina-

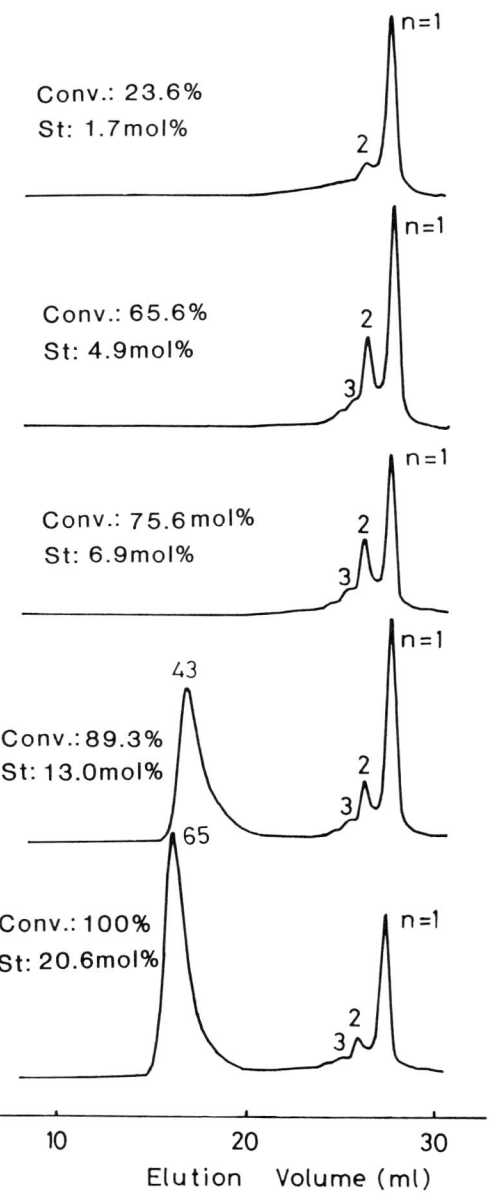

Figure 4 Ozonolysis–GPC curves of copolymer samples at various conversions.

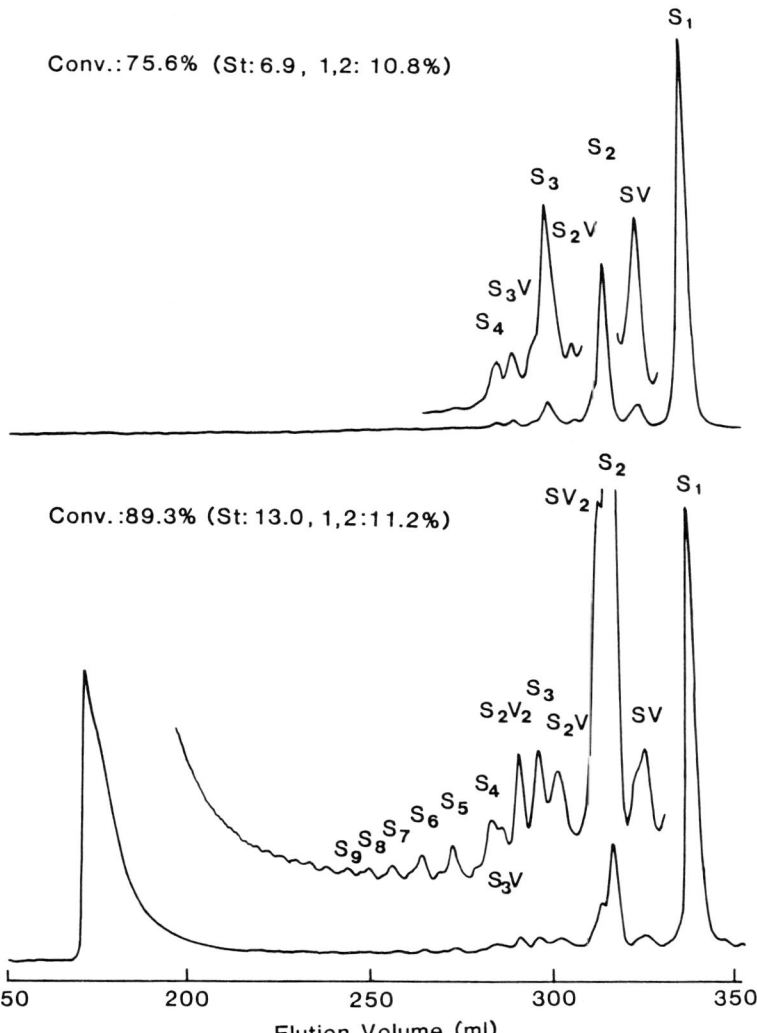

Figure 5 High-resolution ozonolysis–GPC curves of copolymer samples at 75.6% and 89.3% conversions.

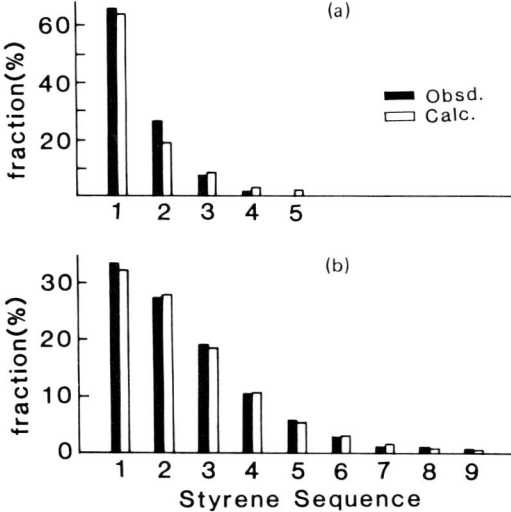

Figure 6 Sequence distribution of styrene units in (a) anionically polymerized sample at conversion 76.5% and (b) copolymer sample initiated with benzoil peroxide at conversion 4.3% (styrene, 42.6%; 1,2, 11.2%; and 1,4, 46.2%).

tion of monomer reactivity ratios by the integrated equation. These fairly good agreements demonstrate the validity of the ozonolysis–GPC method to characterize the sequence distribution of styrene units in SBR.

Sequence Distribution of Commercial SBR

Commercially produced random and partially blocked SBR were analyzed using the ozonolysis–GPC method [22]. The microstructures of the samples were analyzed by ^{1}H NMR as listed in Table 4.

Table 4 Microstructure of Commercial SBR Samples

	Mol %		
Sample	Styrene unit	1,2-Butadiene unit	1,4-Butadiene unit
SBR-1006 (N)	14.2	14.5	71.2
SBR-1502 (J)	13.8	13.5	72.7
Tufdene-2000R (A)	15.1	10.4	74.6
Solprene-1204 (A)	13.5	27.4	59.1
Tufdene-2003 (A)	15.1	9.8	75.1
Solprene-1205 (A)	14.0	13.1	72.9
Kraton-1101 (S)	18.9	7.7	73.4
Solprene-411 (A)	15.4	10.0	74.6

Key: N, Nippon Zeon Co. Ltd.; J, Japan Synthetic Rubber Co. Ltd.; A, Asahi Chemical Industry Co. Ltd.; S, Shell Chemical Co. Ltd.

Random SBR With Emulsion Polymerization

The ozonolysis–GPC curves of SBR-1006 and SBR-1502 are seen in Figures 7 and 8, respectively, together with the sequence distribution of styrene units therefrom. Both samples showed only short styrene sequences from S_1 to S_4V. The number average sequence lengths of styrene units were found to be 1.22 and 1.17 for SBR-1006 and SBR-1502, respectively.

Random SBR With Anionic Polymerization

Figure 9 shows the ozonolysis–GPC and the sequence distribution of styrene units for Tufdene-2000R. The same information for Solprene-1204 is presented in Figure 10. Both samples exhibited a broad peak corresponding to the long styrene sequences around the exclusion limit of the GPC; its relative intensity is 8% for Tufdene-2000R and 19% for Solprene-1204. The number average lengths of the styrene sequences are calculated to be 1.26 and 1.36, respectively, for Tufdene-2000R and Solprene-1204.

It is reported that SBR having a random sequence distribution can be prepared by the addition of a polar compound as a randomizing agent [23], by programmed addition of monomers, or by continuous polymerization [24]. The use of amine, ether, or alkali metal *tert*-butoxide as a randomizing agent is always accompanied by the increase of 1,2-

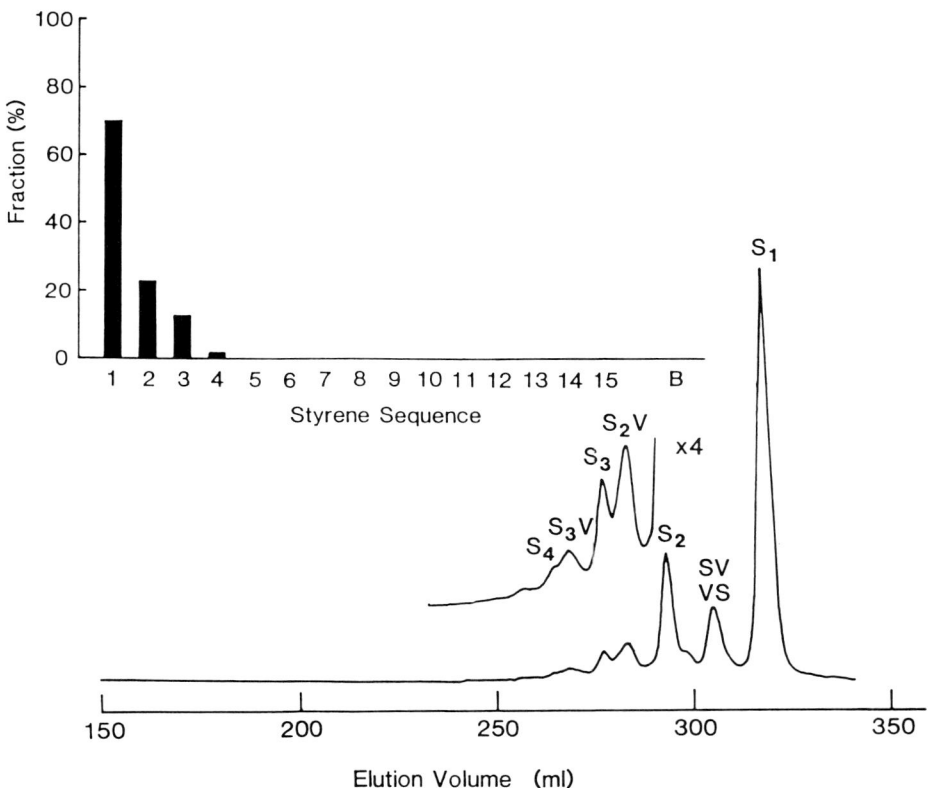

Figure 7 High-resolution ozonolysis–GPC curve for SBR-1006 and sequence distribution of styrene units.

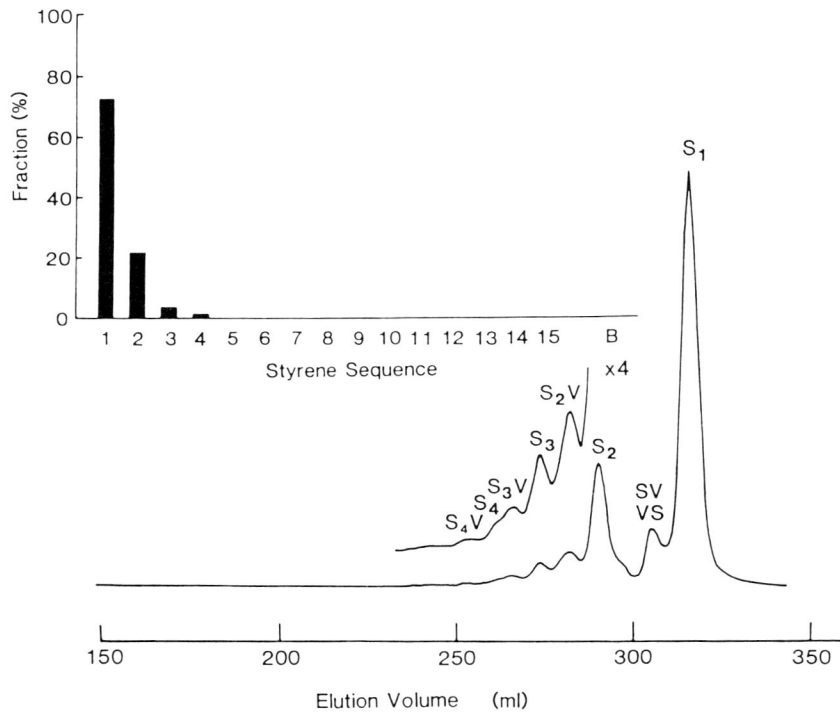

Figure 8 High-resolution ozonolysis–GPC curve for SBR-1502 and sequence distribution of styrene units.

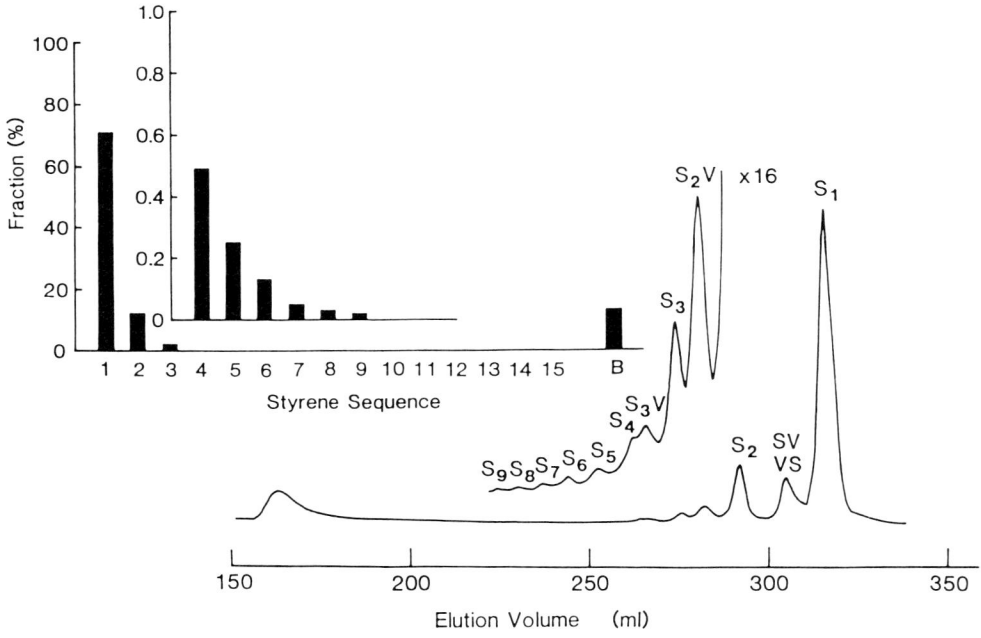

Figure 9 High-resolution ozonolysis–GPC curve for anionically polymerized random SBR (Tufdene-2000R) and sequence distribution of styrene units.

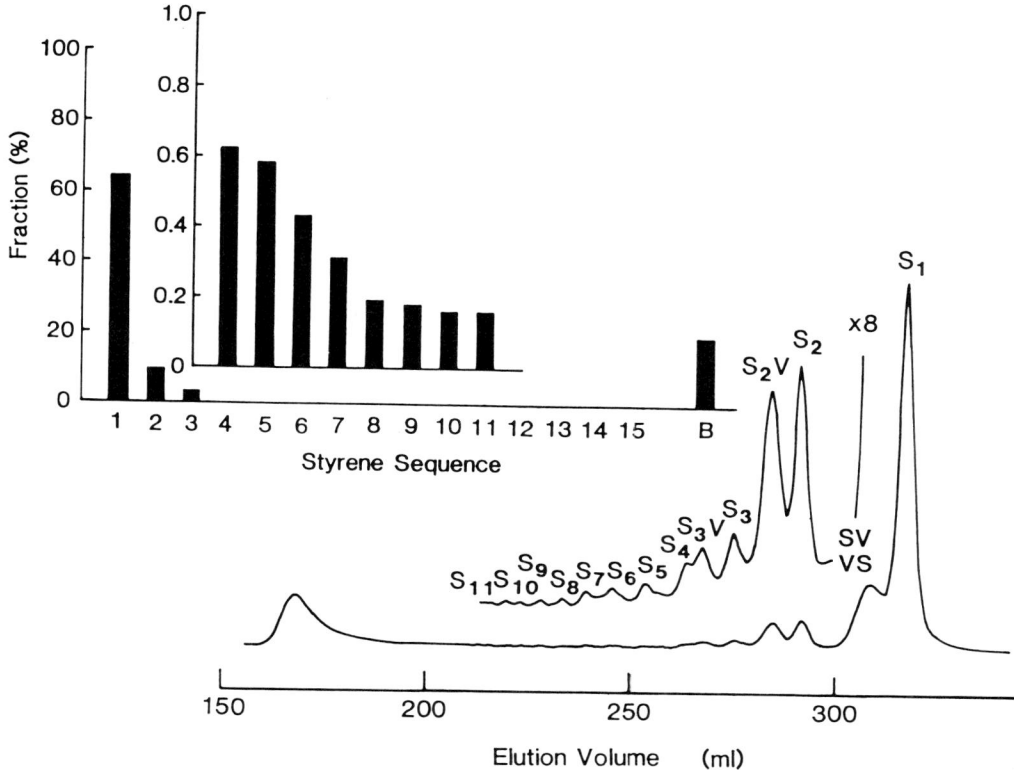

Figure 10 High-resolution ozonolysis–GPC curve for anionically polymerized random SBR (Solprene-1204) and sequence distribution of styrene units.

butadiene units from 15 to 80% depending on the structure and amount of the randomizer [25]. The content of 1,2-butadiene units in Solprene-1204 is about three times that of Tufdene-2000R, as can be seen by the relative intensity of SV, S_2V, and S_3V peaks in Figure 10. This means that Solprene-1204 is prepared using a relatively polar randomizing agent or in a polar solvent. In a usual anionic copolymerization by a molar ratio of styrene to butadiene of 15:85 in a nonpolar solvent, about 57–80% of styrene units are expected to exist in block sequences longer than 12 styrene units on the assumption of 100% conversion and monomer reactivity ratios of $r_{St} = 0.025$ and $r_{Bd} = 15$ [26], $r_{St} = 0.1$ and $r_{Bd} = 12.5$ [21], $r_{St} = 0.16$ and $r_{Bd} = 13$ [18], or $r_{St} = 0.04$ and $r_{Bd} = 26$ [27]. The presence of relatively small amounts of the long styrene sequences in Tufdene-2000R suggests that the sequence distribution is controlled by conversion or by a programmed addition of monomers. The findings mentioned indicate that both samples are random SBR having small amounts of long styrene sequences.

Partially Blocked SBR With Anionic Polymerization

Figures 11 and 12 show the ozonolysis–GPC of partially blocked SBR samples, Tufdene-2003 and Solprene-1205, together with the sequence distribution of styrene units obtained from the GPC curves. The content of the block styrene sequences is 55% for Tufdene-

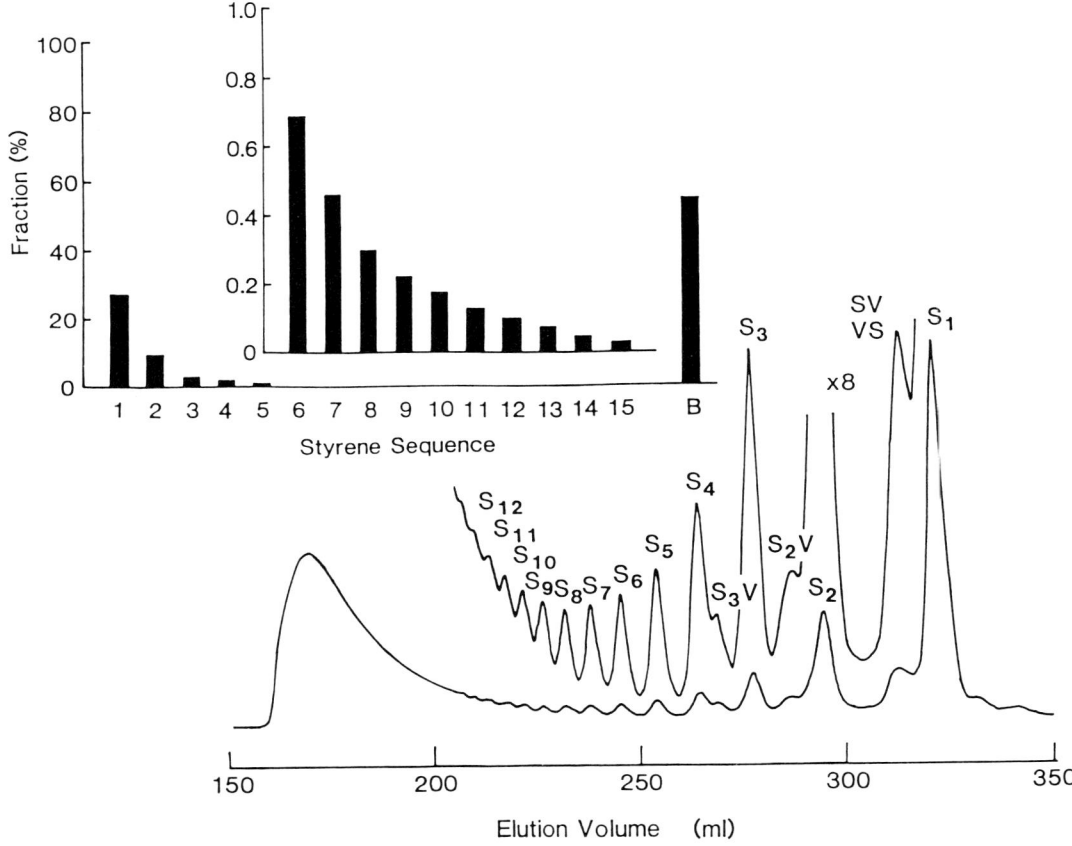

Figure 11 High-resolution ozonolysis–GPC curve for partially blocked SBR (Tufdene-2003) and sequence distribution of styrene units.

2003 and 49% for Solprene-1205. A long tailing of the block styrene peak was observed for Tufdene-2003, while the block styrene peak in Solprene-1205 is isolated from the tailing of short styrene peaks. This difference may be ascribed to the difference of polymerization conditions such as controlled addition of the monomers or polarity of the solvent.

S–B–S-Type Triblock Copolymers

It is well known that block copolymers exhibit characteristic properties reflecting the block structure of the styrene units [28]. For example, S–B–S-type triblock copolymers show rubberlike properties similar to conventional vulcanized elastomers at ambient temperatures without vulcanization, and they can be molded just as a normal thermoplastic resin can at elevated temperature. The molecular weight and molecular weight distribution of the block styrene sequences are regarded as the most important factors governing thermoplastic and other physical properties [29]. There are three typical routes for synthesizing S–B–S triblock copolymers by using organolithium initiators [30]:

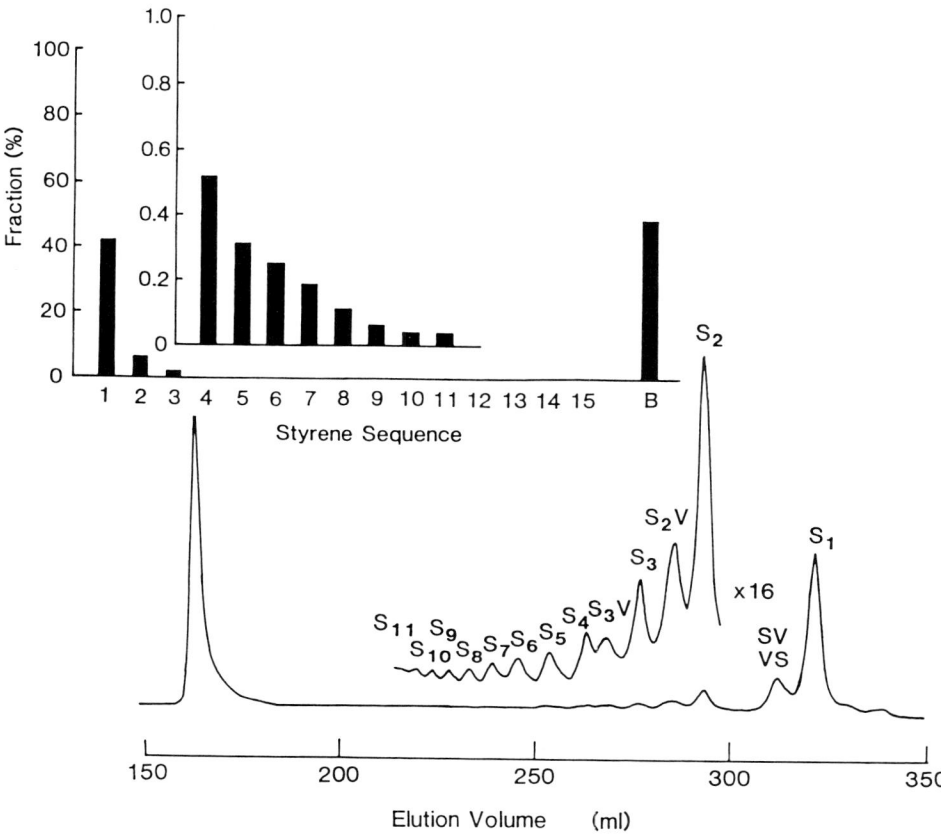

Figure 12 High-resolution ozonolysis–GPC curve for partially blocked SBR (Solprene-1205) and sequence distribution of styrene units.

1. The use of a monolithium initiator and a three-stage copolymerization of S, B, and S.
2. The use of a dilithio initiator and a two-step copolymerization of S and B.
3. The use of a monolithium initiator and a two-stage copolymerization followed by the coupling of S–B sequences. Branched or so-called star-shaped copolymers are synthesized by the use of multifunctional coupling agents.

GPC measurement on copolymer itself proved to be a powerful tool to analyze the purity of S–B–S triblock copolymers [31, 32]. However, there is no direct method for the characterization of the detailed structure of block and so-called tapered-block sequences. The ozonolysis–GPC method was applied to the analysis of the block styrene sequences in commercially obtained S–B–S and S–I–S copolymers. The microstructures of the samples are listed in Table 5. Linear and star-shaped triblock copolymers are distinguished by a combination of GPC measurements of the original copolymer and ozonolysis products.

Table 5 Microstructure of Commercial Triblock Copolymers

Sample	Microstructure (mol%)		
	Styrene	1,4 Unit	1,2 (3,4) Unit
S–B–S			
KX-65 (S)	16.8	75.1	8.1
Solprene-411 (A)	15.4	74.6	10.1
Clearen 530L (D)	73.0	18.0	9.0
S–I–S			
Kraton-1107 (S)	10.0	84.2	5.8
TR-1112 (S)	10.3	83.8	5.9

Key: S, Shell Chemical Co. Ltd.; A, Asahi Chemical Industry Co., Ltd.; D, Denki Kagaku Kogyo Co., Ltd.

Structure of KX-65. Figure 13 shows GPC curves of KX-65, a typical S–B–S triblock copolymer. A small isolated peak in the original sample disappeared by soxhlet extraction with acetone. The extracted fraction was found by ^{1}H NMR analysis to be block SBR containing 95.1% styrene units. GPC measurements of the extracted fraction and the ozonolysis product of the fraction using two columns with exclusion limits of 5×10^4 in series showed a molecular weight of $10._3 \times 10^3$ ($\overline{M}_w/\overline{M}_n = 1.04$) and $9.9_6 \times 10^3$ ($\overline{M}_w/\overline{M}_n = 1.05$), respectively. Thus the values of the ozonized and original samples are

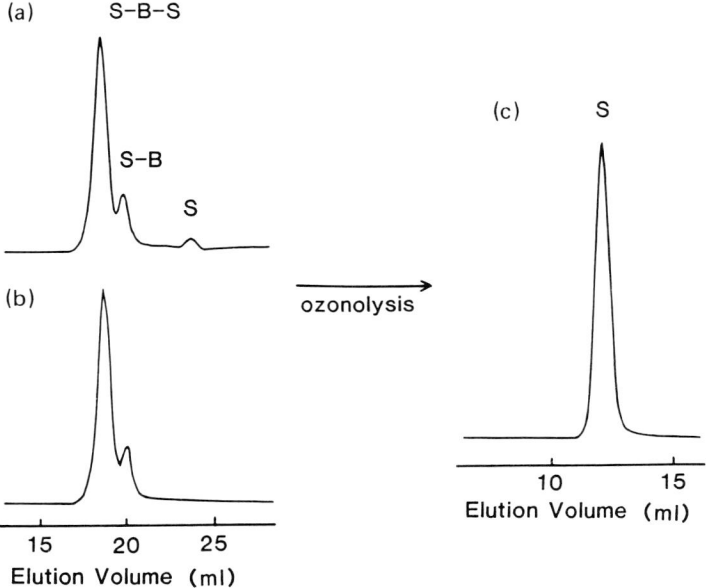

Figure 13 GPC curves for triblock SBR (KX-65): (*a*) original sample; (*b*) after soxhlet extraction; and (*c*) ozonolysis products.

in good agreement, indicating that this extracted fraction is a styrene block copolymer having a trace of butadiene units at the terminal.

The molecular weight of the shoulder peak in Figure 13a was estimated to be 47×10^3, which is approximately equal to half that of the main peak of 97×10^3. These values are expected to be slightly larger than the true values because these molecular weights were estimated by using the calibration curve obtained from standard polystyrenes. However, the difference of the uncorrected and corrected values, which were calculated according to the method of Chang [33], was found to be less than 10%, i.e., within the experimental error. Therefore, the molecular weights of the block copolymers were estimated without any corrections. The molecular weight of the block styrene sequence corresponding to S was found to be 12×10^3 from the ozonolysis–GPC curve shown in Figure 13c. The main and shoulder fractions showed the same chemical composition consisting of 16.8 mol % styrene units by ^{1}H NMR analysis. The molecular weight of the S–B diblock copolymer containing 16.8 mol % styrene units was calculated to be 43×10^3 by considering the molecular weight of the S sequence. This value is in good agreement with that observed for the shoulder peak in Figure 13a. These facts clearly demonstrate that the main and shoulder peaks correspond to the S–B–S and S–B copolymer synthesized by a coupling reaction of the S–B sequence with a bifunctional coupling agent.

Figure 14a shows the high-resolution GPC curve of the ozonolysis products from

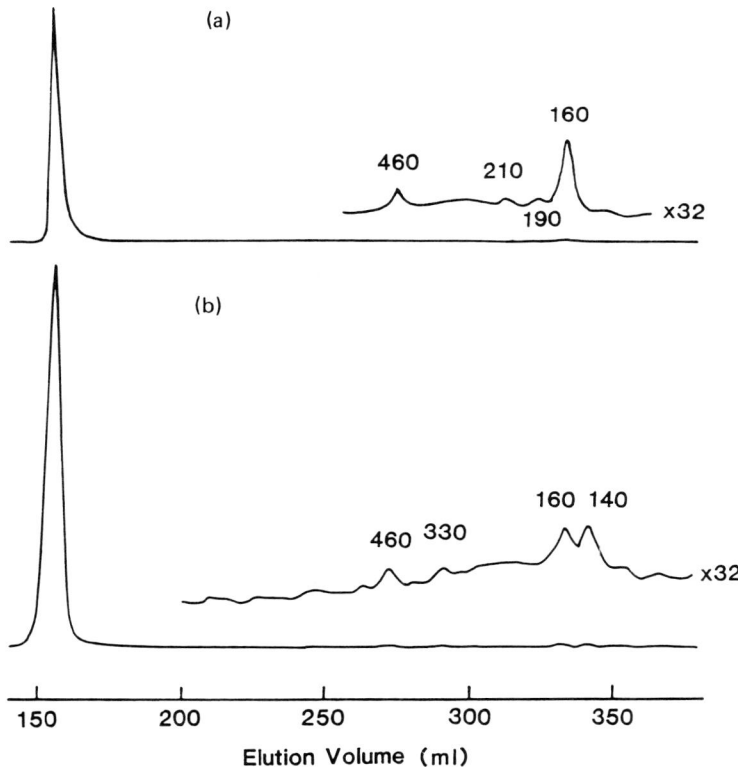

Figure 14 High-resolution ozonolysis–GPC curves (a) for KX-65 and (b) for a model mixture of polybutadiene and polystyrene.

KX-65. The block styrene sequence was observed as a single peak. The small peaks were assigned to the compounds having molecular weight of 460–160 on the basis of the calibration curve for styrene oligomers. Similar peaks were detected in the ozonolysis–GPC of a model mixture composed of 30 mol % styrene and anionically polymerized polybutadiene as seen in Figure 14b. This indicates that these small peaks are derived from impurities, initiator residues, and/or sequences of 1,2-butadiene units. Therefore, it can be concluded that this sample is almost pure S–B–S triblock copolymers containing small amounts of S–B and S sequences.

Structure of Solprene-411. Solprene-411 showed a small GPC peak with a molecular weight of 25×10^3, which disappeared by soxhlet extraction as shown in Figures 15a and b. This S fraction appeared by ^{1}H NMR measurement to be block SBR containing 96.4 mol % of styrene units. The molecular weights of the S fraction and the ozonolysis product from the fraction were 23×10^3 ($\overline{M}_w/\overline{M}_n = 1.26$) and 20×10^3 ($\overline{M}_w/\overline{M}_n = 1.28$), respectively. This indicates that the fraction is an S–B diblock copolymer having molecular weights of 20×10^3 and 3×10^3 for S and B, respectively. The molecular weight of the block styrene sequence agreed very closely with 20×10^3 determined by ozonolysis–GPC of the sample after soxhlet extraction as shown in Figure 15c.

The molecular weight of the main peak in Figure 15a was found to be 340×10^3, which is about four times that of the shoulder peak, which was estimated to be 90×10^3. The styrene contents in the main and shoulder peaks were found to be 15.3 and 15.9 mol %, respectively, by ^{1}H NMR analysis of the collected fractions. These findings

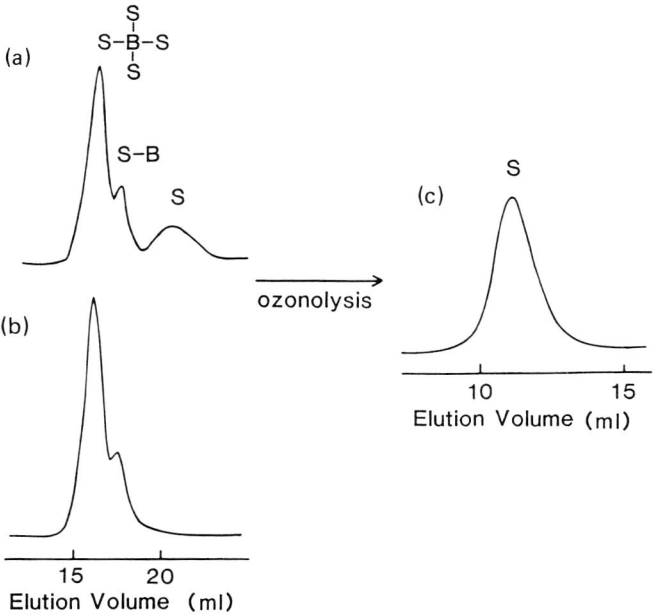

Figure 15 GPC curves for triblock SBR (Solprene-411): (a) original sample; (b) after soxhlet extraction; and (c) ozonolysis products.

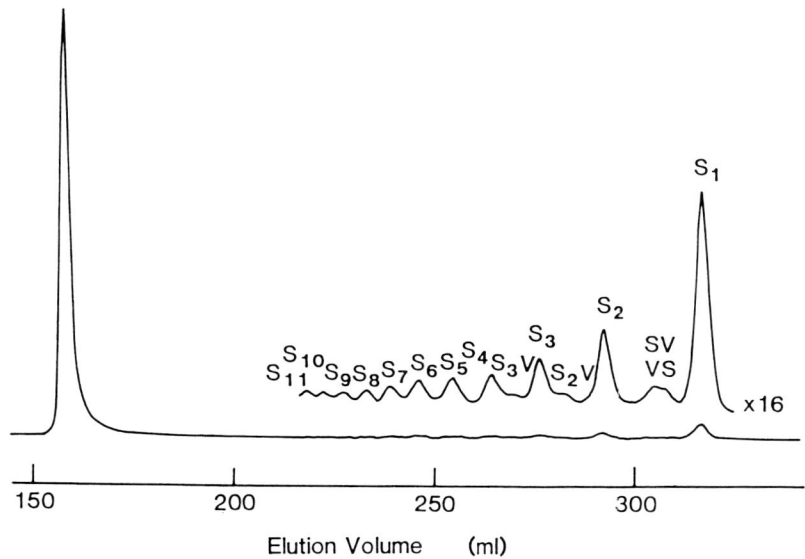

Figure 16 High-resolution ozonolysis–GPC curve for Solprene-411.

demonstrate that the main peak is a typical four-branched or star-shaped block copolymer prepared by the coupling reaction of the S–B diblock sequences.

The high-resolution GPC measurement of the ozonolysis products indicates that the sample is a typical tapered block copolymer containing about 15% of short styrene sequences as seen in Figure 16.

Structure of Clearen 530-L. Figure 17 shows the GPC and ozonolysis–GPC curves for Clearen 530-L, a typical high-styrene thermoplastic elastomer. The GPC of the ozonolysis products showed a bimodal distribution of block styrene sequences, designated as S_m and S_n, corresponding to molecular weights of 60×10^3 and 20×10^3 as shown in Figure 17c. The presence of short styrene sequences amounting to 23% was revealed in the high-resolution GPC curves as seen in Figure 17d, where block sequences were observed as a single peak due to the exclusion limit of 3×10^3 of the GPC columns. The GPC of the sample after soxhlet extraction showed a relatively broad main peak, as in Figure 17a. By recycling measurement, the main peak corresponding to the molecular weight of 134×10^3 exhibited two shoulder peaks, as indicated by arrows in Figure 17b. The molecular weight of a small shoulder peak in Figure 17a is estimated to be 60×10^3, which corresponds to that of the S_m sequence.

The chemical composition distribution was examined according to the HPLC method using polyacrylonitrile gel as stationary phase with gradient elution of hexane/chloroform [16]. The detail of the method is described later. The sample after soxhlet extraction showed two main peaks and two small peaks, as seen in Figure 18. The styrene content of the fractions A, C, and D was found to be 51.0, 72.0, and 82.8%, respectively, by ^{1}H NMR analysis of the collected fractions. These findings suggest that the sample is a mixture of two S–B–S-type copolymers accompanied with S–B copolymers, i.e., S_m–B–S_m + S_m–B and S_n–B–S_n + S_n–B. From the chemical composition distribution it seems

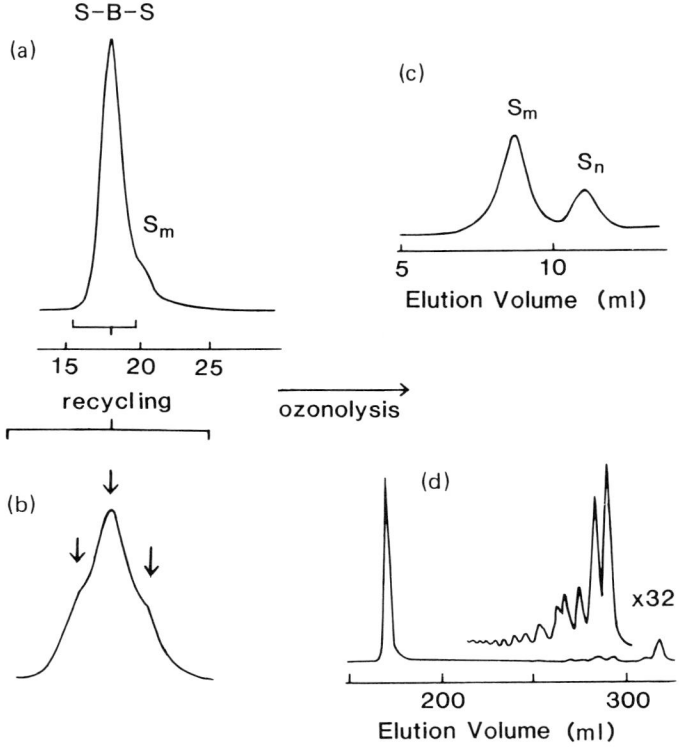

Figure 17 GPC curves for Clearen 530L: (*a*) after soxhlet extraction; (*b*) recycling measurement of main peak; (*c*) ozonolysis products; and (*d*) high-resolution measurement of ozonolysis products.

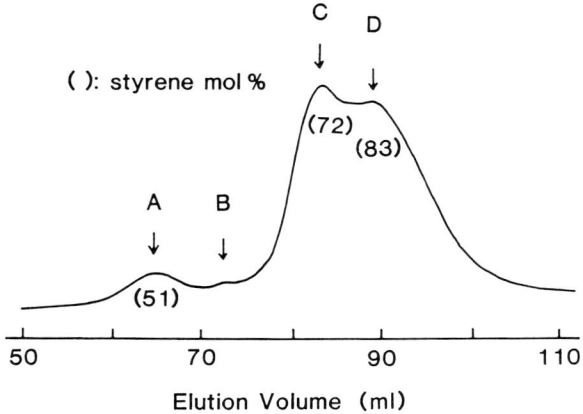

Figure 18 HPLC separation of Clearen 530L by chemical composition using polyacrylonitrile gel as stationary phase.

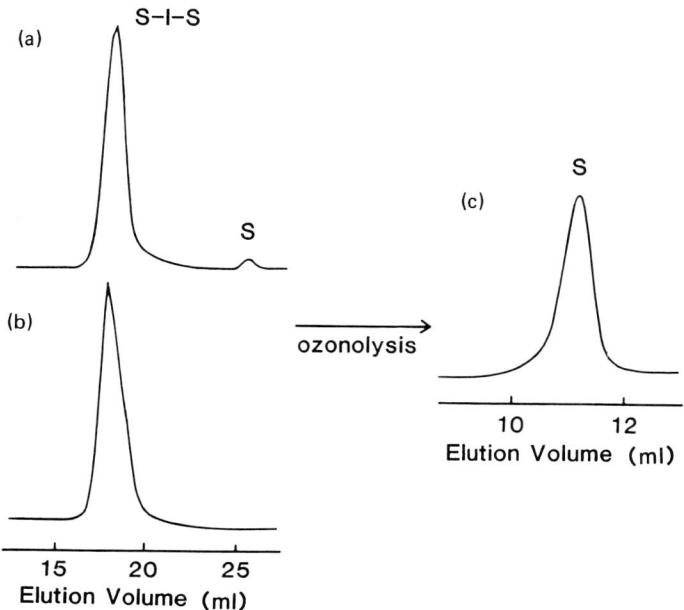

Figure 19 GPC curves for S–I–S triblock copolymer (Kraton-1107): (a) original sample; (b) after soxhlet extraction; and (c) ozonolysis products.

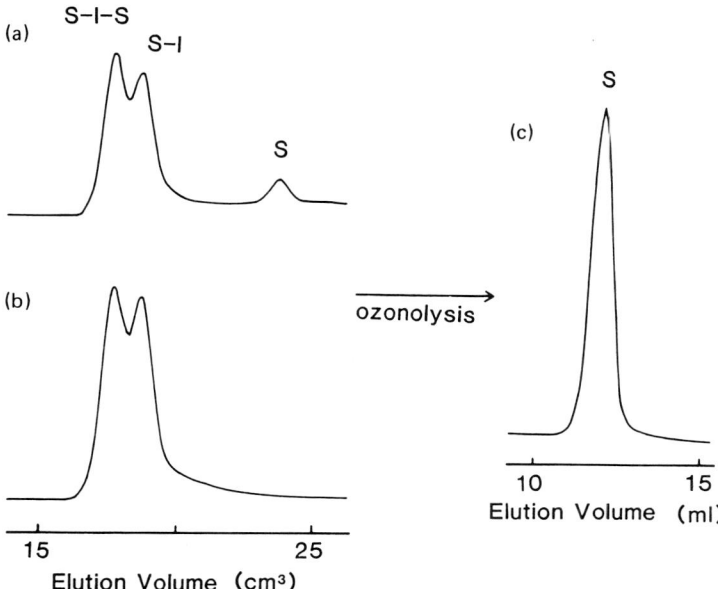

Figure 20 GPC curves for S–I–S triblock copolymer (TR-1112): (a) original sample; (b) after soxhlet extraction; (c) ozonolysis products.

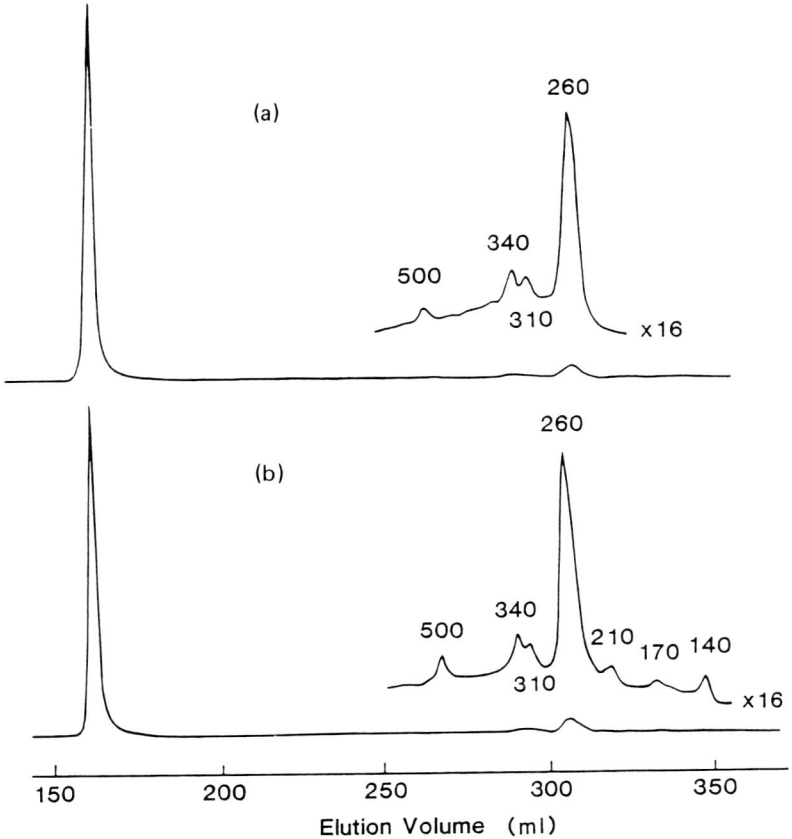

Figure 21 High-resolution ozonolysis–GPC curves (*a*) for Kraton-1107 and (*b*) for TR-1112.

less likely that the samples composed of S_n–B–S_n, S_m–B–S_n, and S_m–B–S_m copolymers accompanied those with S_m–B and S_n–B copolymers.

Structure of S–I–S Triblock Copolymers. Figures 19 and 20 show the GPC curves of Kraton-1107 and TR-1112, respectively. Each sample contained a small amount of polystyrene designated as S corresponding to 9×10^3 and 16×10^3 molecular weight. These values are in accord with the molecular weight of the block styrene sequences observed by ozonolysis–GPC. The main peak in Kraton-1107 corresponded to a molecular weight of 186×10^3. It was established as S–I–S copolymer by considering its chemical composition. On the other hand, TR-1112 showed two equivalent peaks corresponding to molecular weights of 93×10^3 and 210×10^3. The first was considered to be S–I copolymer and the second was believed to be S–I–S block copolymer.

Both samples showed small peaks amounting to 6% in Kraton-1107 and 10% in TR-1112 in the high-resolution GPC of the ozonolysis products as shown in Figure 21. These peaks do not correspond to short styrene sequences; they were more likely due to small amounts of additives and/or sequences containing 3,4-isoprene units.

SEQUENCE STRUCTURE OF STYRENE AND 1,2-BUTADIENE UNITS [34]

Configurational Sequences of Styrene Units

The ozonolysis–GPC method makes use of high-resolution GPC columns packed with polystyrene gel. The gel was also applicable to the HPLC separation of styrene dimer to pentamer by tacticity using isopropyl ether as an eluent [35, 36]. The assignment of ^{13}C NMR signals in polystyrene was successfully carried out by using these diastereomers of styrene dimer to pentamer as model compounds [37, 38]. The configurational sequence of styrene units in SBR is characterized by a combination of GPC and HPLC separations of the ozonolysis products followed by NMR analysis of the collected fractions.

The ozonolysis products derived from styrene sequences flanked by 1,4-butadiene units such as 1,4-(styrene)-1,4, 1,4-(styrene)$_2$-1,4, and 1,4-(styrene)$_3$-1,4 were fractionated by GPC (cf. Fig. 12). The collected fractions S_2 and S_3 were further separated into diastereomers by HPLC by using a preparative column packed with polystyrene gel using methanol as an eluent as shown in Figure 22. The HPLC fractions were collected and were subjected to ^{1}H and ^{13}C NMR analyses. The fraction, which eluted earlier in S_2, exhibited methine proton signals centered at 2.22 and 2.39 ppm; and the fraction, which eluted later, at 2.45 and 2.67 ppm. They are assigned to the isomers in racemic (r) and meso (m) configuration, respectively, by comparison with 4,6-diphenylnonane [35]. Similarly, the three fractions in S_3 were estimated to be the diastereomers in the rr, $mr + rm$, and mm configurations, respectively. These assignments were further confirmed by the carbon signals characteristic of configurational sequences in S_2 and S_3 [37]. The aliphatic carbon signals in S_2 were assigned by using two-dimensional NMR measurements (C-H COSY) and DEPT (distortionless enhancement by polarization transfer) measurements. The assignments for S_2 and S_1 are listed in Table 6.

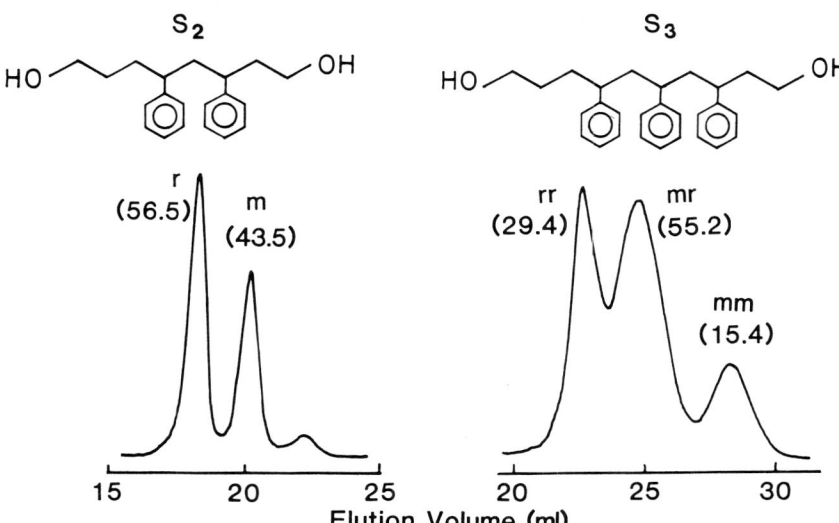

Figure 22 HPLC separation of S_2 and S_3 fractions using polystyrene gel with methanol as eluent.

Table 6 Assignment of C-13 NMR Signals in S_1 and S_2.

Sequence		Chemical shift (ppm from TMS)							
		C-1	C-2	C-3	C-4	C-5	C-6	C-7	C-8
S_1		60.25	39.35	41.62	32.44	30.20	62.17		
S_2	r	60.85	40.32	39.87	43.65	42.98	33.68	30.55	62.64
	m	60.81	38.94	39.72	44.19	42.75	31.76	30.37	62.64

Key:

S_1 1   2   3 4   5   6
    HOCH₂CH₂CHCH₂CH₂CH₂OH
                |
                Ph

S_2 1   2   3 4   5 6   7   8
    HOCH₂CH₂CHCH₂CHCH₂CH₂OH
                | |
                Ph   Ph

On the basis of the assignments of the HPLC peaks the probability of racemic addition (P_r) of styrene units was calculated to be 0.56 and 0.58 for radical and commercially obtained anionic SBR samples (Solprene-1205), respectively. The observed intensity ratios in S_3 are in good agreement with the values obtained by assuming Bernoullian statistics for both samples. These P_r values also agreed very closely with those determined by ^{13}C NMR for radical and anionic polystyrenes [38, 39]. This indicates that the configurational sequence in styrene–butadiene copolymer is controlled in a way similar to styrene homopolymer.

Arrangement of Styrene and 1,2-Butadiene Units

The GPC fraction indicated by SV (VS) was assigned to the sequence consisting of styrene and 1,2-butadiene units flanked by 1,4-butadiene units, which contains two isomers differing in the order of addition of both units as given in Eqs. 2 and 3:

SV   HOCH₂CH₂—CHCH₂—CHCH₂—CH₂CH₂OH   (2)
                     | |
                   C₆H₅   CH₂OH

VS   HOCH₂CH₂—CHCH₂—CHCH₂—CH₂CH₂OH   (3)
                     | |
                   CH₂OH   C₆H₅

The SV sequence is formed by the addition of a 1,2 unit to a 1,4-styrene sequence (Eq. 2); in the SV sequence, the styrene is added to a 1,4-1,2 chain (Eq. 3). The VS + SV fraction isolated by GPC was further separated by recycling on HPLC. After recycling five times, a distinct small peak indicated by D1 was isolated; after shaving the small peak, the residual fraction was separated into two discernible peaks, D2 and D3, by recycling 12 times as indicated in Figure 23.

Aliphatic carbon signals in these fractions are seen in Figure 24. Fraction D3 exhibited 18 signals, while both D1 and D2 fractions showed only 9 signals. The observed chemical shifts of fractions D1 and D2 are in fair agreement with those calculated for VS according to an empirical method. The characteristic signals at 33.39, 33.47, 34.97 ppm in fraction D3 were assigned satisfactorily to the carbon atoms in SV. Fractions D1 and D2 are

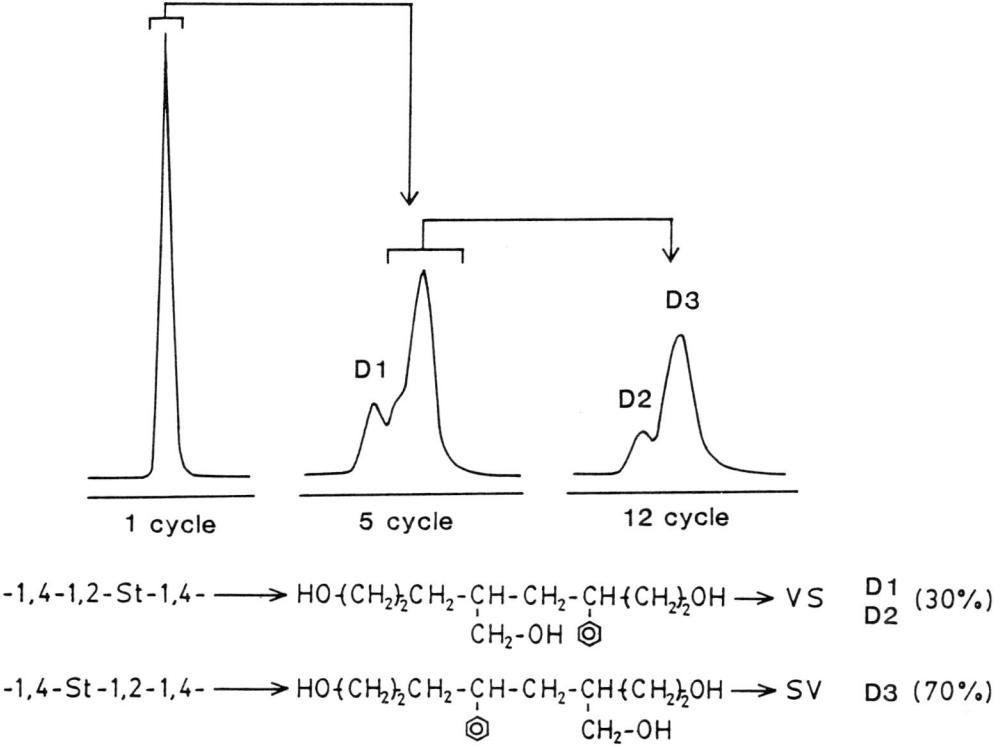

Figure 23 HPLC separation of SV + VS fraction using polystyrene gel with methanol as eluent.

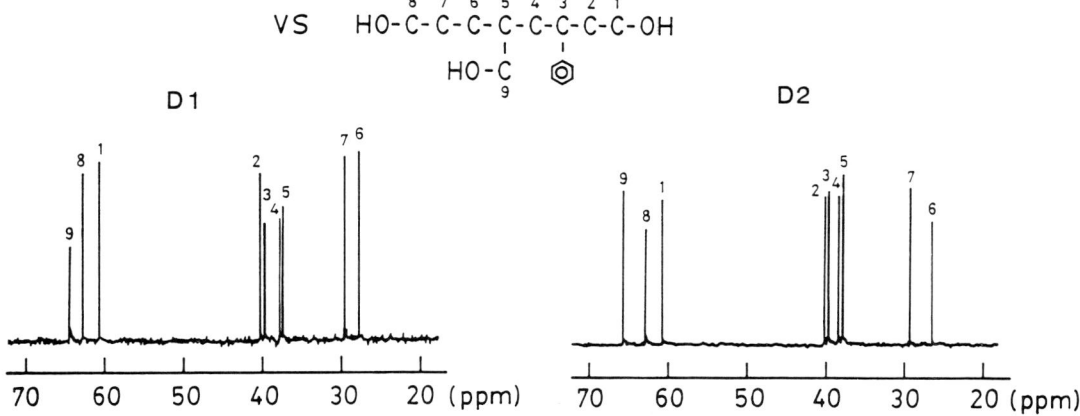

Figure 24 Carbon-13 NMR spectra of HPLC fractions corresponding to VS and SV.

believed to be separated by cotacticity of the VS sequence. The spectrum of fraction D3 is presumed to be made up by the superposition of two diastereomers in SV. The signals corresponding to the C-3 and C-5 methine carbon atoms were assigned by DEPT measurement. The tentative assignments of the aliphatic carbon signals are indicated in Figure 24.

The probability of the formation of styrene–styrene, styrene-1,2, and styrene-1,4 sequences flanked by 1,4 units can be determined from the relative intensities of S_2, SV, and S_1, respectively. The ratio of observed to calculated values of these fractions was found to be 0.35 for S_2 and SV and 1.2 for S_1. Calculated values were obtained by assuming Bernoullian statistics. This finding indicates the predominance of the addition of a 1,4 unit to a terminal styrene unit.

The S_2V fraction collected by GPC was further separated into four fractions by recycling four times using HPLC as seen in Figure 25. It is expected that S_2 is composed of three isomeric sequences—SSV, VSS, and SVS—reflecting the arrangement of styrene and 1,2 units. The configurations of the four collected fractions, T1 to T4, were assigned by ^{1}H NMR, ^{13}C NMR, and FT–IR using S_1, S_2, SV, and VS as model compounds.

The splitting pattern of the phenyl proton signal in the fractions T3 and T4 is similar to that of S_1 and S_2, while fractions T1 and T2 showed a striking resemblance to S_3 as seen in Figure 26. Thus fractions T1 and T2 may be assigned to an SSV or VSS sequence showing splitting similar to S_3 due to the shielding effect of the V unit.

Fractions T1 and T2 showed infrared bands characteristic of S_2 at 1280 and 1380 cm^{-1}, while the bands of fractions T3 and T4 in this region have a strong resemblance to those of SV, as seen in Figure 27. The ^{13}C NMR spectra of fractions T1 and T2 indicate the presence of two isomers in each fraction. By comparison of the chemical shifts of the diastereomers in S_2 and S_3 as well as in SV and VS, fraction T1 was assigned to the structures corresponding to SSV and VSS with a racemic configuration and fraction T2 to SSV and VSS structures with a meso configuration with respect to the styrene units. Fractions T3 and T4 were estimated to be SVS which separated into two fractions by cotacticity.

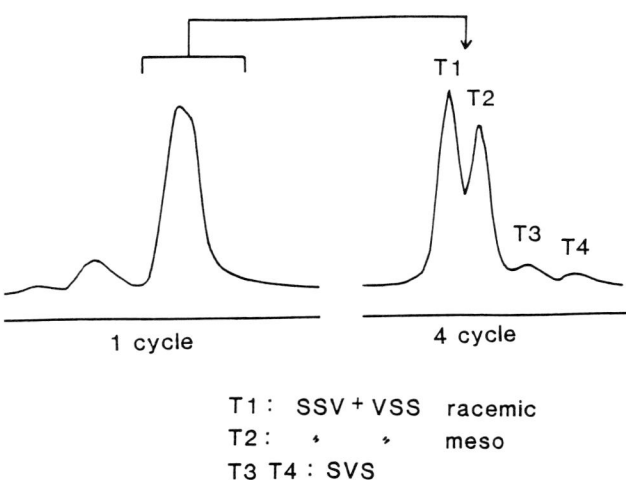

Figure 25 HPLC separation of S_2V fraction using polystyrene gel with methanol as eluent.

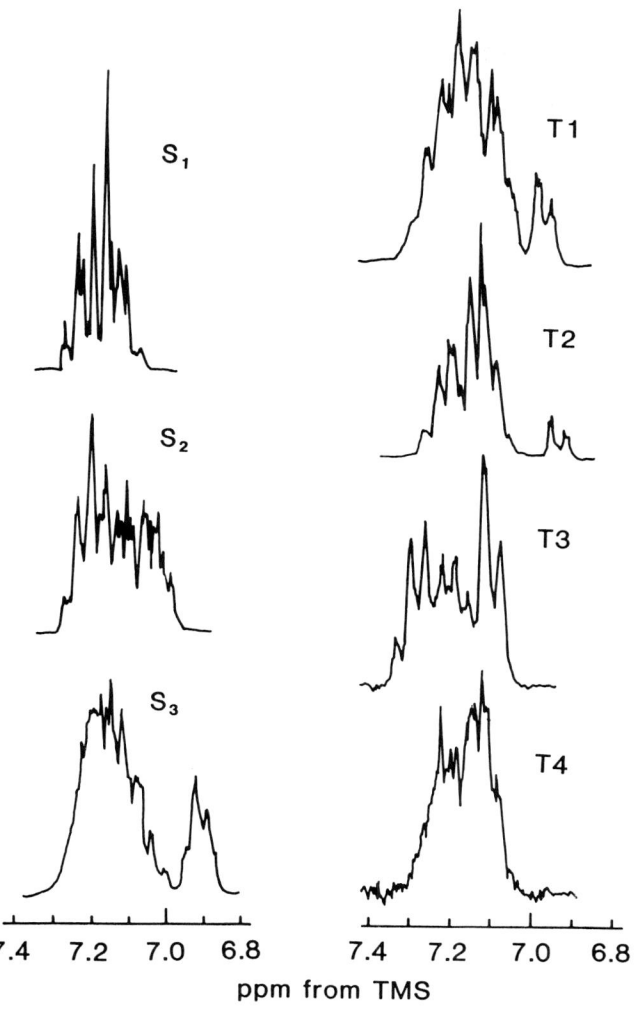

Figure 26 Comparison of phenyl proton signals of S_1, S_2, and S_3 fractions with those of T1 to T4 fractions in S_2V.

The intensity ratio between the SSV + VSS and SVS sequences was found to be 7:1 for anionic SBR. This finding indicates that the probability of the addition of a 1,2 unit to a terminal styrene is less than that of styrene to a 1,2 unit, taking account of the fact that the amount of 1,2 unit is about 1.8 times the styrene unit in this sample. This agrees well with the finding that VS is about one-fourth of SV. These facts suggest that the 1,2 unit in S_n sequences is arranged at the terminal as S_nV and not as an isolated sequence such as $S_{n-m}VS_m$. In the case of anionically polymerized SBR samples containing about 10% of 1,2 units, small shoulder peaks were observed for each S_n peak, which can be assigned to S_nV on the basis of the GPC calibration curves (cf. Figs. 5, 10, and 11).

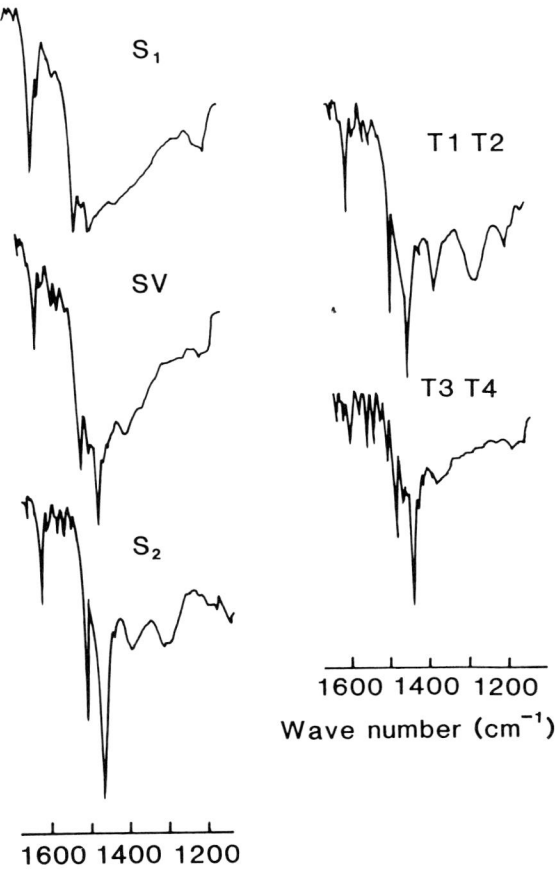

Figure 27 Comparison of FT–IR spectra of S_1, S_2, SV, and S_2V fractions with those of T1 + T2 and T3 + T4 fractions in S_2V.

ANALYSIS OF SEQUENCE DISTRIBUTION IN CURED SBR

The structural analysis of cured rubbers has been carried out by using mainly infrared spectroscopy and gas chromatographic analysis of chemical or thermal degradation products. Carbon-13 NMR analysis of thermally solubilized fraction was proposed for the characterization of cured rubbers [40]. However, it is difficult to obtain quantitative information on the sequence distribution of copolymers. The ozonolysis–GPC method is expected to be a powerful tool to analyze the sequence structure of cured SBR when the ozonolysis reaction proceeds quantitatively, as in the case of uncured SBR [41]. The cured SBR samples with pure gum stocks and with carbon black filler were powdered into fine particles (25–400 μm in diameter) by grinding after they were steeped in methylene chloride for 24 hr. The powdered samples were extracted with acetone in a soxhlet extractor for 20 hr and steeped in methylene chloride for 24 hr before ozonolysis. Ozonization was carried out in suspension of the powdered sample in methylene chloride and reductive degradation with lithium aluminum hydride in a way similar to uncured SBR.

The ozonolysis products soluble in methylene chloride–ethyl ether increased with increasing amounts of ozone and attained a maximum equivalent to those from the original SBR when blown through 100–150% ozone against the double bonds of butadiene units in cured SBR with pure gum stocks, which were estimated from the microstructure and the rubber content in the sample. The fractions of S_1, S_2, S_3, and S_n sequences determined by GPC were plotted against the ratio of ozone to the double bonds as seen in Figure 28. Here, S_n is the long styrene sequences showing a peak around the exclusion limit of the GPC column. It was observed that an instantaneous uptake of ozone ceased when 3.7% of ozone was introduced into the reaction vessel. Proton NMR analysis of the collected S_n fraction detected 77–47% of residual butadiene units for ozonolysis products obtained by the addition of 3.7–80% ozone, which decreased almost 0% by blowing through 100–150% of ozone. These findings indicate that a quantitative ozonization occurs only on the surface of the swollen rubber particles at the initial stage of ozonolysis and proceeds gradually into the inner part, resulting in complete ozonolysis of the double bonds by introducing excess ozone.

It is thought that the ozonolysis of styrene units progresses by adding excess amounts of ozone to SBR. In the case of polystyrene, it was found that only 10% of polystyrene was recovered in methylene chloride–ethyl ether solution and the number average molecular weight decreased from 50×10^3 to 14×10^3 by blowing through 400% ozone against styrene units. In the case of SBR, the amount of ozone is exactly adjusted to the amount equal to the double bonds in butadiene units. It was confirmed that the styrene units in polystyrene were intact under these ozonolysis conditions due to instantaneous

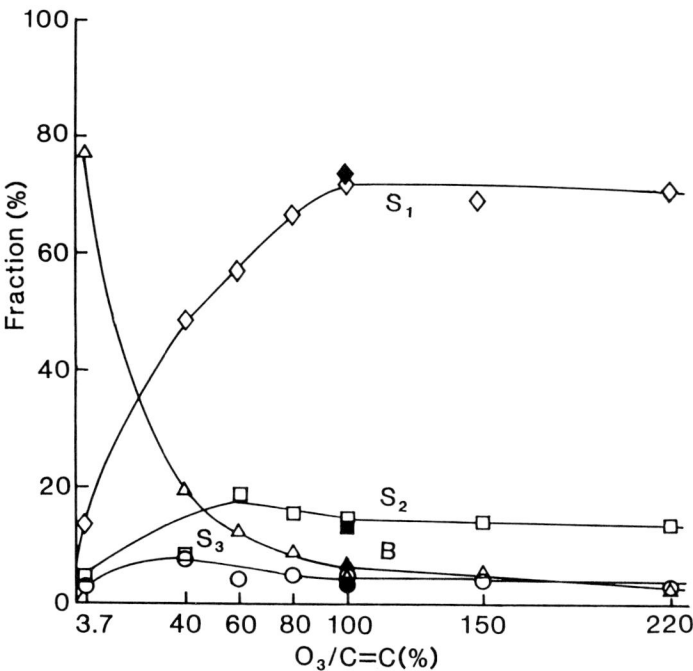

Figure 28 Amounts of S_1, S_2, S_3, and S_n fractions against ratio of ozone to butadiene units in cured SBR; closed symbols are those from uncured SBR.

consumption of ozone by double bonds. However, the amount of the long styrene sequences decreased by adding an excess amount of ozone, which indicates the progress of decomposition of styrene sequences in the case of cured SBR. Accordingly, 100–130% ozone is anticipated to be an appropriate condition for the sample; this depends on the degree of swelling and on particle size of the powdered sample.

Figure 29 shows the GPC curves of the ozonolysis products from cured SBR samples with pure gum stocks and with carbon black filler by blowing through 130% ozone against the double bonds in butadiene units together with that of the uncured sample. The short styrene sequences form S_1 to S_{10} and long styrene sequences were clearly observed for the cured samples. However, the resolution of GPC was lower than that of uncured sample and additional peaks were observed for both samples, which may be derived from by-products of the ozonolysis reaction or from the crosslinked chains. The sequence distributions of the cured samples are in fairly good agreement with that of the original sample, as shown in Figure 30.

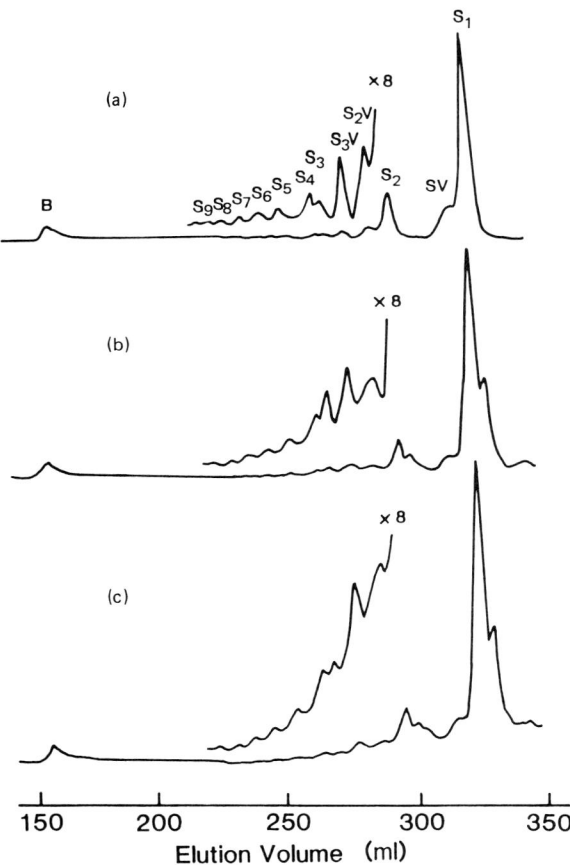

Figure 29 High-resolution ozonolysis–GPC curves for Tufdene-2000R: (a) original sample; (b) cured sample with pure gum stocks; and (c) cured sample with carbon black filler.

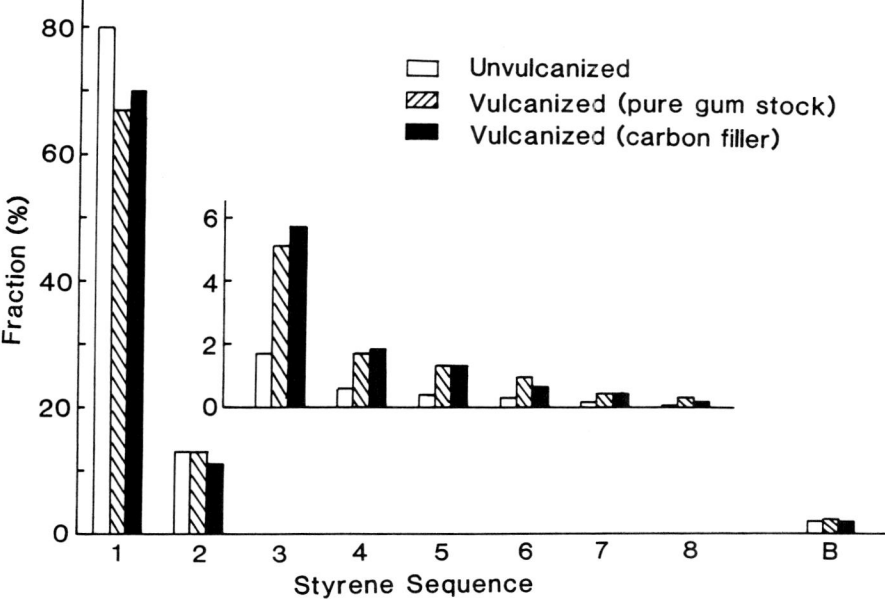

Figure 30 Sequence distribution of styrene units in Tufdene-2000R.

DETERMINATION OF CHEMICAL COMPOSITION DISTRIBUTION IN SBR

The determination of chemical composition distribution is important for the evaluation of the physical properties of SBR, especially to check the blending of SBR having different chemical composition. Tanaka and colleagues proposed a new HPLC method for the measurement of chemical composition distribution in SBR and styrene–methyl methacrylate copolymer using crosslinked acrylonitrile gel as a stationary phase [16, 42].

HPLC measurement was carried out for a mixture of polystyrene and six model SBR samples containing 11, 21, 33, 41, 58, and 82 mol % of styrene units, which were prepared by bulk copolymerization using benzoil peroxide as initiator adjusting the conversion to less than 6%. A complete separation was observed for these samples by using a gradient elution starting from a less polar eluent, i.e., hexane to chloroform ratio 90 : 10 to 55 : 45 as shown in Figure 31. The standard deviation of the elution volume was less than 1.0 ml, which corresponded to 1.4 mol % of styrene units. It was confirmed that the area of each peak was proportional to the styrene weight percent in each sample. These findings demonstrate that the HPLC can be converted into the chemical composition distribution.

Copolymers usually have a distribution of chemical composition as well as molecular weight distribution. At lower conversion, where the change of monomer composition is negligible, a narrow chemical composition distribution is expected only due to an instantaneous spread in composition. The effect of the molecular weight on the elution volume was examined for three SBR samples obtained by fractionation with GPC. The elution volume of the peak of these three fractions slightly increased from 48.1 to 49.1 ml

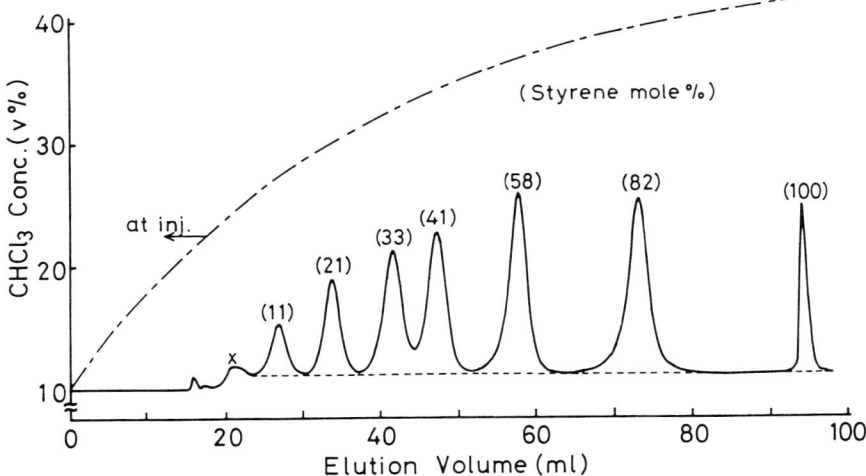

Figure 31 HPLC separation of model mixture of SBR samples and polystyrene.

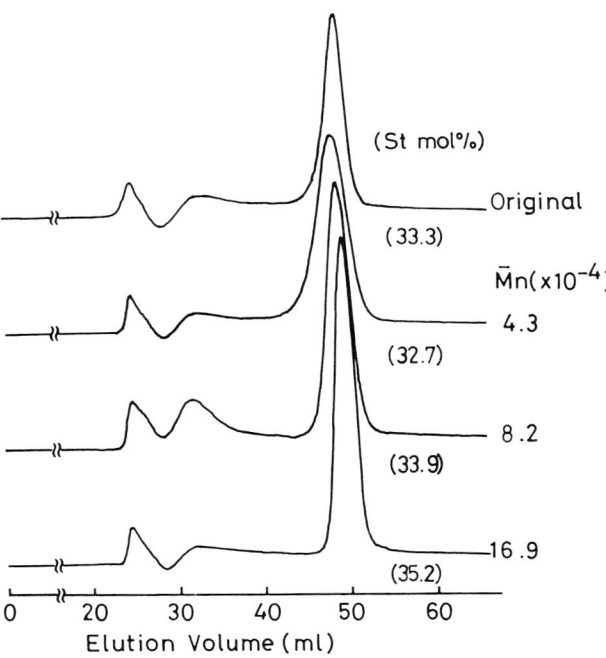

Figure 32 HPLC separation of original and fractionated SBR samples.

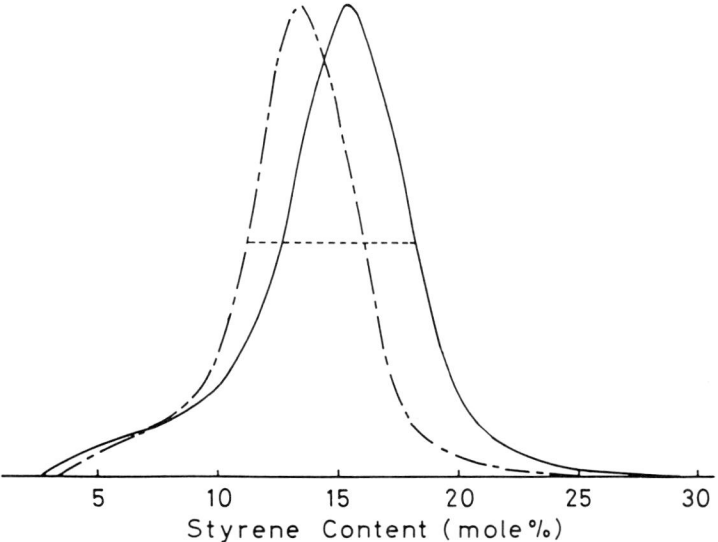

Figure 33 Chemical composition distribution of emulsion SBR samples; solid curve represents SBR-1502 and dashed curve is SBR-1006.

with the increase of molecular weight, as seen in Figure 32. The drift is within the experimental error. Therefore, it is concluded that the SBR samples were separated mainly according to chemical composition, practically independent of molecular weight.

The chemical composition distribution was determined for commercially obtained emulsion SBR samples as shown in Figure 33. The average content of styrene units determined by HPLC is in good agreement with that from ^{1}H NMR measurement for both samples. It is observed that cold rubber (#1502) has the distribution narrower than that of hot rubber (#1006); the half-width of the peak is 4.5 mol % styrene units for cold rubber and 6.0 mol % for hot rubber. This HPLC method is also applicable to partially blocked SBR and triblock SBR (cf. Fig. 18), although the elution volume is dependent on the sequence length of styrene units as well as the chemical composition.

REFERENCES

1. V. D. Mochel, *Rubber Chem. Technol.*, 40: 1200 (1967).
2. Y. Tanaka, H. Sato, K. Saito, and K. Miyashita, *Rubber Chem. Technol.*, 54: 685 (1981).
3. A. R. Katrizky and D. E. Weiss, *J. Chem. Soc. Perkin Trans.*, 2: 21 (1975).
4. A. R. Katrizky and D. E. Weiss, *J. Chem. Soc. Perkin Trans.*, 2: 27 (1975).
5. A. L. Segre, M. Delfini, F. Conti, and A. Boicelli, *Polymer*, 16: 338 (1975).
6. J. C. Randall, *J. Polym. Sci. Polym. Phys. Ed.*, 15: 1451 (1977).
7. E. Thorn-Csányi, C. Hennemann, and H. Perner, *Makro Mainz Prep.*, 1: 381 (1979).
8. L. Michajolv and H. J. Harwood, *J. Am. Chem. Soc. Div. Org. Coatings Plast. Prep.*, 30(2): 129 (1970).
9. Y. Minoura and Y. Hatanaka, *Nippon Gomu Kyokaishi*, 43: 838 (1970).
10. J. L. Leblanc, *Trib. CEBEDEAU*, 28(378): 231 (1975).
11. M. Danielewicz and M. Kubín, *J. Appl. Polym. Sci.*, 26: 951 (1981).

12. G. Glöckner, H. Kroschwitz, and Ch. Meissner, *Acta Polym., 33*: 614 (1982).
13. G. Glöckner, Jo H. M. van den Berg, N. L. J. Meijerink, T. G. Scholte, and R. Koningsveld, *Macromolecules, 17*: 962 (1984).
14. Y. Tanaka, J. Takeda, and K. Noguchi, U.S. Pat. 4,338,404 (1982).
15. Y. Tanaka, J. Takeda, and K. Noguchi, Jpn. Pat. 58-41464 (1983).
16. H. Sato, H. Takeuchi, S. Suzuki, and Y. Tanaka, *Makromol. Chem. Rapid Commun., 5*: 719 (1984).
17. Y. Tanaka, H. Sato, and Y. Nakafutami, *Polymer, 22*: 1721 (1981).
18. Y. Tanaka, H. Sato, Y. Nakafutami, and Y. Kashiwazaki, *Macromolecules, 16*: 1925 (1983).
19. A. A. Korotokov and N. N. Chesnokova, *Vysokomol. Soedin., 2*: 365 (1960).
20. J. Kuntz, *J. Polym. Sci., 54*: 569 (1961).
21. Y. L. Spirin, A. A. Arest-Yakubovich, D. K. Polyakov, A. R. Gantmakher, and S. S. Medvedev, *J. Polym. Sci., 58*: 1181 (1962).
22. Y. Tanaka, H. Sato, and J. Adachi, *Rubber Chem. Technol., 59*: 16 (1986).
23. H. L. Hsieh and W. H. Glaze, *Rubber Chem. Technol., 43*: 22 (1970).
24. T. C. Bouton and S. Futamura, *Rubber Age, 106*(3): 33 (1974).
25. T. A. Antkowiak, A. E. Oberster, A. F. Halasa, and D. P. Tate, *J. Polym. Sci. A-1, 10*: 1319 (1972).
26. V. D. Mochel and B. L. Johnson, *Rubber Chem. Technol., 43*: 1138 (1979).
27. A. F. Johnson and D. J. Worsfold, *Makromol. Chem., 85*: 273 (1965).
28. Y. Tanaka, H. Sato, and J. Adachi, *Rubber Chem. Technol., 60*: 25 (1987).
29. P. Guyot, J. C. Favier, M. Fontanille, and P. Sigwalt, *Polymer, 23*: 73 (1982).
30. A. F. Halasa, D. N. Schulz, D. P. Tate, and V. D. Mochel, in *Advances in Organometallic Chemistry,* Vol. 18, Academic Press, New York, p. 55 (1980).
31. R. D. Mate and M. R. Ambler, *Sep. Sci., 6*: 139 (1971).
32. P. Dreyfuss, J. L. Fetters, and D. R. Hansen, *Rubber Chem. Technol., 53*: 728 (1980).
33. F. S. Chang, *J. Chromatog., 55*: 67 (1971).
34. Y. Tanaka, Y. Nakafutami, Y. Kashiwazaki, J. Adachi, and K. Tadokoro, *Rubber Chem. Technol., 60*: 207 (1987).
35. Y. Tanaka, H. Sato, and K. Miyashita, *Makromol. Chem. Rapid Commun., 1*: 551 (1980).
36. H. Sato, K. Saito, K. Miyashita, and Y. Tanaka, *Makromol. Chem., 182*: 2259 (1981).
37. H. Sato and Y. Tanaka, *Macromolecules, 17*: 1964 (1984).
38. H. Sato, Y. Tanaka, and K. Hatada, *J. Polym. Sci. Polym. Phys. Ed., 21*: 1667 (1983).
39. H. Sato and Y. Tanaka, in *NMR and Macromolecules* (J. C. Randall, ed.) American Chemical Society, Washington, p. 181 (1984).
40. D. D. Werstler, *Rubber Chem. Technol., 53*: 1191 (1980).
41. Y. Tanaka, K. Nunogaki and J. Adachi, *Rubber Chem. Technol., 61*: 36 (1988).
42. H. Sato, H. Takeuchi, and Y. Tanaka, *Macromolecules, 19*: 2613 (1986).

19
Gelation of Polymer Solutions

Jean-Michel Guenet
Institute Charles Sadron (CNRS)
Strasbourg, France

INTRODUCTION	711
FORMATION	713
Gels from Stereoregular Polymers	713
Gels from Atactic Polymers	718
THERMODYNAMICS	721
MOLECULAR STRUCTURE	730
MORPHOLOGY	741
RHEOLOGY	749
SWELLING BEHAVIOR	756
CONCLUDING REMARKS	759
REFERENCES	760

INTRODUCTION

The physical gelation of polymer solutions leads to the formation of macroscopic networks called thermoreversible gels or physical gels. These terms embody numerous systems, some of which, such as gelatin [1], are widely known. Actually, gelation occurs not only with macromolecules but also with small molecules such as V_2O_5 [2] or, as recently discovered, with steroid derivatives [3]. The scope of this chapter, however, is restricted to gels of synthetic polymers.

Before going further, it seems necessary to define the concept of a gel. The generally accepted definition for a gel is a permanent tridimensional network swollen by a solvent and possessing the properties of a solid. In this respect, covalently crosslinked macromolecules, such as vulcanized rubber, fit the definition. Yet establishing a precise definition of a physical gel is about as easy as, say, passing a camel through the eye of a needle. As emphasized in the 1920s by D. Jordan Lloyd, "The colloidal condition, the gel, is one which is easier to recognize than to define" [4]. This statement seems to hold true today. The only property shared by all the known thermoreversible gels is, as indicated by their name, the reversibility with temperature of the fusion–formation process. If one accepts a possible chemical degradation of the polymer, the heating and cooling cycles can be repeated at length without any alteration of the gel characteristics. Also, the term "physical gel" emphasizes their origin in the formation of physical links, which, as will be seen later, are not necessarily of a crystalline nature. If one attempts to characterize a physical gel through its mechanical properties, particularly by seeking a solidlike behavior, one fails. As a matter of fact, some physical gels can also exhibit properties close to those of a liquid, although there is no doubt that the gellike aspect and consistency do not result from high viscosity. While there is no absolute criterion, the difference between a very viscous solution and a gel is recognized by simple observation.

In the author's opinion, thermoreversible gelation can be defined as follows: It can be recognized that thermoreversible gelation of a polymer solution has occurred when this solution no longer behaves as it should. This definition is purposely vague in order to encompass all kinds of thermoreversible gels that are currently known. For example, the gelation of isotactic polystyrene is recognized since cooling the solution to low temperatures should have produced a whitish "paté" due to the growth of classical crystals and not the transparent jelly that is actually seen. The reader will discover throughout this review that this definition is relevant.

When did mankind discover thermoreversible gelation? This question is difficult to answer. Yet, as gelatin is simply produced by cooking bones in boiling water, this phenomenon may be as old as the human conquest of fire. Scientifically speaking, von Nägeli pioneered, in the second half of the nineteenth century, the systematic investigations of these systems [5]. He studied natural objects such as cell walls and starch grains and postulated a discontinuous granular structure of very small crystalline particles. As early as 1899, Hardy [6] realized that gelation could also be caused by liquid–liquid phase separation frozen-in at its early stage by crystallization, a concept which is now "revisited." As will be seen, these early ideas, some of which are more than a century old, are impressively modern. Curiously enough, for some years physical gels were not investigated as much as they deserved. This long-lasting lack of interest might have originated because the problem seemed too difficult to tackle, mainly because of the lack of powerful enough techniques, or it might have been thought that the problem was too simple, a common prejudice in science, and did not deserve further consideration as far as basic research was concerned. From time to time, the formation of such systems would be reported, but they would generally be regarded as undesirable by-products formed during the preparation of classical crystals. However, while physical gelation was thought to be a property of water-soluble systems, its occurrence with synthetic polymers in apolar solvents was more and more frequently observed. It is difficult to trace this discovery since it may have been mentioned incidentally in a paper devoted to anything else. However, the mechanical properties of polyvinyl chloride (PVC) gels were determined as early as 1954 [7]. This investigation was certainly dictated by the industrial potential of this polymer.

Keller's group in Bristol should be credited for the renewal of interest in the physical gels of synthetic polymers. Their investigation of both the thermal properties and the short-range structure of isotactic polystyrene (iPS) gels [8] gave unexpected results, which changed an apathetic audience into an eager one. Whereas it had been thought that iPS thermoreversible gels were due to fringed micellar crystallization in which the chain adopts the usual threefold helical form (3_1 helix), Keller's group not only showed that the gel has a much lower melting point than 3_1 crystals, but also that the reflections of this form from x-ray diffraction were totally absent. Instead, a reflection at 5.1 Å was observed, the meaning of which is still controversial as will be seen in this chapter.

Science is often comparable to a tree in springtime, when many flowers blossom at the same time. So it was with physical gels. Particularly worth mentioning is the making of ultra-high-modulus polyethylene fibers by stretching physical gels of this polymer [9]. Moduli of about 100 GPa can be reached, an achievement as good if not better than Kevlar. Also, the discovery of the physical gelation of atactic polystyrene [10] (aPS), which was unexpected since this polymer cannot crystallize, poses a challenging problem of fundamental science.

This chapter surveys the current knowledge of physical gels from synthetic polymers. Some results and concepts are quite new and cannot be taken for granted as definitive. Accordingly, this review is only an assessment of the state of the art of physical gels. It is hoped that this survey is as comprehensive as possible. However, the reader must realize that some publications may be unintentionally overlooked, and as a result unfortunately not quoted here.

FORMATION

Physical gelation of polymer solutions occurs in a wide range of polymer concentrations (from 1% or less to 40% or more). The formation process of a thermoreversible gel is often correlated to the ease with which the polymeric component crystallizes. If the polymer can crystallize under the form of chain-folded crystals, then a special treatment, such as a rapid quench to low temperatures,† is in principle required to bypass this type of crystallization. Otherwise, if the polymer is noncrystallizable or poorly crystallizable, the final product after cooling to room temperature will be a gel provided that the appropriate solvent is chosen.

Instead of listing all the systems known with their formation process, which might be very cumbersome, the most representative thermoreversible gels are described. Unless new gelation processes have been discovered since this was written, those detailed here should cover what is currently known.

Although any rigid and definite classification is premature at present, it seems that two major types of gels can be schematically distinguished: gels obtained from solutions of stereoregular polymers and gels produced from solutions of atactic polymers.

Gels from Stereoregular Polymers

The gels prepared from isotactic polystyrene (iPS), polyethylene (PE), poly-4-methyl pentene-1 (P4MP1), and syndiotactic PMMA have been widely studied and exhibit some peculiar features.

Isotactic polystyrene is a slowly crystallizing polymer. This allows the crystallization and the gelation phenomena to be easily differentiated. Figure 1a displays the visual aspect of a solution once turned into a gel and Figure 1b shows the same solution where chain-folded crystals have been allowed to grow. The pioneering work of Lemstra and Challa [11] has permitted definition of the conditions to form a transparent gel as in Figure 1a. For instance, they showed that in *trans*-decalin chain-folded crystals growth predominates at high temperatures (between 40 and $\sim 100°C$) while below 20°C gelation occurs. Between 20 and 40°C, they observed that chain-folded growth still takes place but with much slower kinetics. This turns out to be a most unusual situation since, in principle, the

---

†Here it is understood to mean temperatures lower than room temperature (usually from 0°C down to the solvent melting temperature).

Figure 1 Visual aspect of a 10% iPS solution in *trans*-decalin (*a*) turned into a gel and (*b*) after chain-folded crystals have been allowed to grow at +50°C.

higher the undercooling, the faster the crystallization. Apparently the usual crystallization gives way suddenly to gelation, as noticed by Girolamo et al. [8]. The temperature of 20°C turns out to be close to the θ-temperature of polystyrene in decalin [12] (actually θ ≃ 18°C in *trans*-decalin). Accordingly, Wellinghoff et al. [13] suggested that gelation might simply result from the occurrence of a liquid–liquid phase separation† frozen-in at its early stage by crystallization (the same concept described some 80 years earlier by

---

†Below a miscibility gap, two domains are to be considered. One is located between the binodal and the spinodal curves where the system is metastable and will undergo liquid–liquid phase separation by nucleation and growth. The other is below the spinodal curve, where the system is characterized by unphysical properties so that it is unstable. As a result, the phase separation mechanism is different and proceeds via diffusion. This latter mechanism is designated as spinodal decomposition and is said to create a network structure at its early stage.

Hardy). In addition, the liquid–liquid phase separation is supposed to proceed via spinodal decomposition mainly because this mechanism is said to give off a network structure [14].

Recent investigations of gel formation by differential scanning calorimetry (DSC) [15] (see Fig. 2) have allowed the determination of the gelation temperature (T_{gel}) at zero-cooling rate to be achieved in various solvents (Fig. 3). Whereas cooling very slowly (i.e., as close as possible to the zero-cooling rate) would only yield chain-folded crystals, this type of experiment allows an upper "critical" gelation temperature to be estimated. It is indeed noticed that above this extrapolated temperature, no gelation will take place. Conversely, if the system is quickly quenched to below this temperature, even a few tenths of a degree below, only gelation will take place.

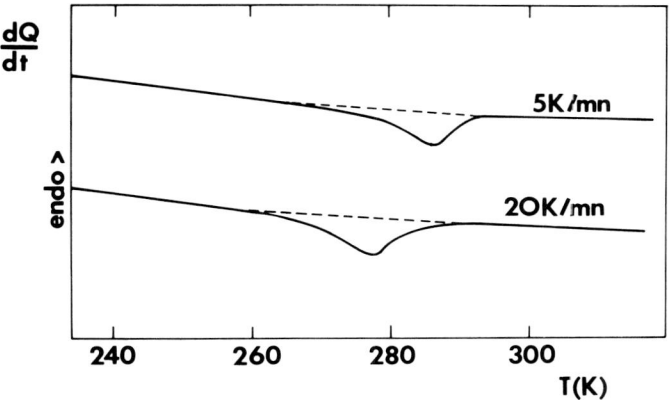

Figure 2 Typical differential scanning calorimetry (DSC) thermograms showing the formation exotherm observed on cooling a 15% iPS solution in *cis*-decalin.

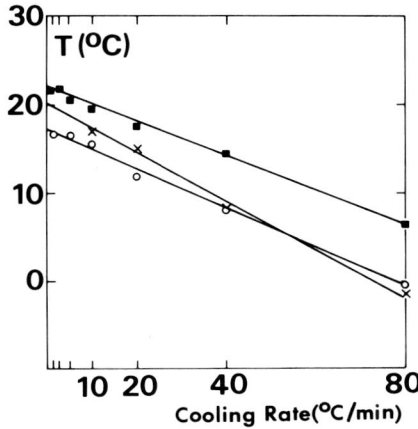

Figure 3 Plot of the onset of the gel formation exotherm as a function of the cooling rate. ■ = *trans*-decalin; ○ = *cis*-decalin; × = 1-chlorodecane. (Data from Ref. 15.)

These DSC investigations also show that T_{gel} is virtually independent of the polymer molecular weight. If the phenomenon were directly linked to the occurrence of liquid–liquid phase separation, T_{gel} should have varied as the critical demixing temperature T_c, that is, there should have been a substantial lowering with a decrease in the polymer molecular weight. Accordingly, while there is little doubt that liquid–liquid phase separation participates in gel formation at low temperatures and, as will be seen in this chapter, certainly influences gel morphology, this mechanism is not compulsory.

Results reported in Figure 3 indicate that T_{gel} for 15% solutions does not vary drastically with the solvent type. Yet at higher polymer concentrations the discrepancy is marked. Figure 4 is a gel formation phase diagram for *cis*-decalin and *trans*-decalin. In *cis*-decalin the phase diagram exhibits a maximum, whereas it does not in *trans*-decalin. Also, when the gel formation enthalpy, ΔH_{gel}, is plotted against polymer concentration for either solvent (Fig. 5), three facts emerge: (a) ΔH_{gel} reaches a maximum in both solvents, (b) ΔH_{gel} does not depend on molecular weight, and (c) the position of the maximum is solvent dependent ($C \simeq 30\%$ in *cis*-decalin and $C \simeq 40\%$ in *trans*-decalin). These types of results are found for molecular compounds and may thus indicate that the solvent plays a role in the gelation process. Such a conclusion had already been proposed by Sundararajan et al. [16] on the basis of x-ray diffraction investigations. The gel structure will be described further in the following section.

Polyethylene is a rapidly crystallizing polymer. Its crystallization kinetics impede the formation of metastable solutions at room temperature. However, it can form a gel [9]. This property apparently depends on the mechanical treatment applied while the solution was above the crystal melting temperature [17]. If the solution is stirred then cooled to

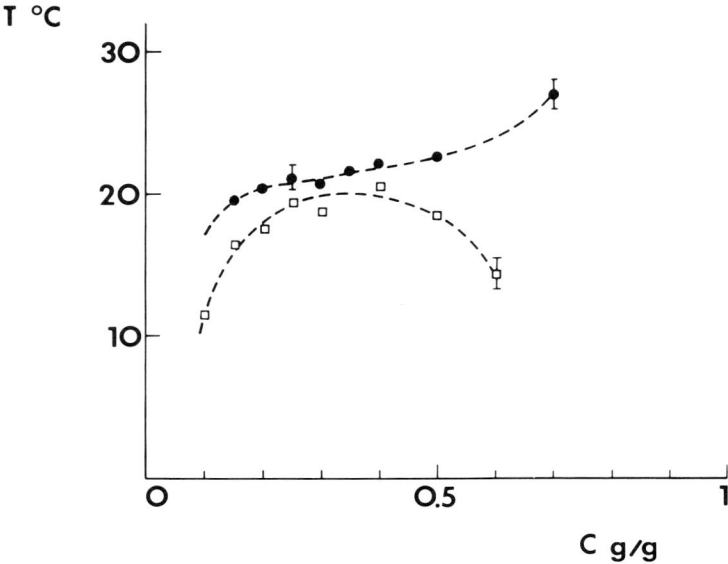

Figure 4 Gel formation temperature–concentration phase diagram. The values of the gelation temperature, T_{gel}, were determined at a cooling rate of 5°C min^{-1}; results are close to those obtained by extrapolation to zero cooling rate. □ = gels in *cis*-decalin; ● = gels in *trans*-decalin. (Data from Ref. 15.)

room temperature, the gel will set while in the quiescent state. Conversely, if the solution is not stirred but directly quenched to room temperature, a turbid but fluid suspension will be obtained, resulting from the formation of classical chain-folded crystals.

The current explanation of this behavior is that the stirring at high temperature promotes the building of rigid, rodlike chain portions, which, once the solution attains room temperature, will associate together without undergoing a folding process [17].

Making use of the propensity of polyethylene to form physical gels instead of flat crystals under specific conditions, Smith and Lemstra [9] were able to produce high-modulus polyethylene fibers (Young's modulus $\simeq$ 90 GPa and tensile strength $\simeq$ 3 GPa). Further studies by Nahr et al. cast some light on the gel formation conditions.

The gel formation of poly-4-methyl pentene-1 is quite peculiar. Charlet et al. [18] report that it depends strongly on the thermal history of the bulk polymer sample. Also, unlike polystyrene, the gelation phenomenon takes place at high temperature while the

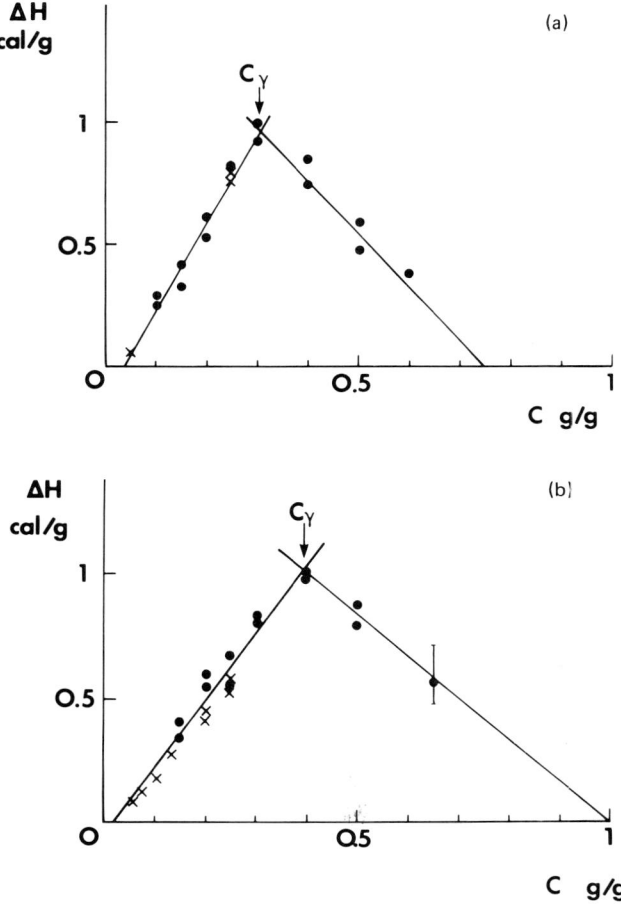

Figure 5 Variation of the gel formation enthalpy (area of the gel formation exotherm) as a function of polymer concentration: (*a*) gels in *cis*-decalin where ● = MW = 7 × 10⁴ and × = MW = 3.2 × 10⁵; (*b*) gels in *trans*-decalin, with symbols defined as in (*a*). (Data from Ref. 15.)

growth of chain-folded crystals occurs at lower temperatures. If the polymer is used as received (nascent state: directly recovered without melting after polymerization), then the solution turns into a gel above the dissolution temperature of chain-folded crystals, T_d, as long as it stays below the solation temperature, which can be higher than the boiling point in cyclic parafinic solvents. Conversely, for bulk polymer samples slowly recrystallized from the melt, their solutions remain clear at T_d but form gels on isothermal annealing at a temperature lower than T_d and higher than the crystallization temperature of chain-folded crystals. However, the gels formed with either system must eventually contain the same structures, since experiments carried out after gel drying do not show any difference. From careful studies of the heat of gel formation, Phuong-Nguyen and Delmas [19] conclude that the physical junctions consist of cohesive associations of regularly spiraled helices, stabilized by the solvent. Although the route followed to form the gel is different from iPS, the solvent seems to play a very similar role.

Although a large amount of information on the gelation of syndiotactic polymethyl methacrylate (s-PMMA) is available [20–24], the mechanism of gel formation is not totally elucidated. The final structure of the solutions depends strongly on the solvent, and the thermal history is reported to have some influence [23]. The participation of the solvent in the helix stabilization as well as in the formation of intermolecular associations has been suggested [25]. According to Berghmans et al. [26], the thermoreversible gelation of s-PMMA in o-xylene proceeds from a two-step mechanism: a rapid conformational change of the chain followed by a slower molecular association. It is noteworthy that a formation exotherm as in iPS can be observed in these systems.

Gels from Atactic Polymers

Solutions from polyvinyl chloride (PVC) and atactic polystyrene (aPS) do not form gels under the same conditions.

Atactic polyvinyl chloride† (PVC) produces physical gels in a large variety of solvents [27–30]. Only solutions in very good solvents such as tetrahydrofuran (THF) or cyclohexanone do not form a gel. Yet if one uses PVC synthesized at very low temperatures ($-40°C$ for instance), which possesses a larger content of syndiotactic sequences, gels are also obtained in very good solvents.

Gelation is quasi-instantaneous when the solution is rapidly cooled to room temperature [27]. No formation exotherm can be seen by DSC, which raises some questions as to the gel formation mechanism. The gelation temperature T_{gel} is a monotonically increasing function of temperature [29] and thus is similar to the liquid line observed in crystallizing systems. Two gelation mechanisms are currently considered. The first (i.e., the older) involves the crystallization of the syndiotactic sequences [28, 31, 32]; the second requires the creation of hydrogen bonds between the PVC chains [34]. The former mechanism implies cooperativeness and therefore should be represented by a first-order thermodynamic transition, whereas the latter does not and thus must be of second order or higher. Currently neither model can account for all the experimental results. The crystallization mechanism is in apparent conflict with the absence of a formation exotherm. The hydrogen-bonding mechanism conflicts with the existence of small electric dipole moments in the pregels, which can exist only if ordered structures are present.

---

†NMR characterization gives the following values for the triads: atactic PVC (synthesized at $+50°C$), hetero = 0.5, iso = 0.18, syndio = 0.32; PVC synthesized at $-40°C$, hetero = 0.51, iso = 0.09, syndio = 0.4.

The gel formation of polyvinyl alcohol in water is to some extent similar to that of PVC [35].

The physical gelation of atactic polystyrene (aPS), although unexpected, is now receiving increasing attention [10, 36–39]. Unlike PVC, no major inconsistencies have arisen as far as thermal properties are concerned. Methodical investigations of this phenomenon by Tan et al. [10] have revealed that gelation occurs well below room temperature and only in certain solvents. The most interesting system by far is carbon disulfide, (CS_2)–aPS, in which the gelation temperature approaches room temperature for high concentrations.

The determination of the gelation temperature has been carried out by several experimental methods. Originally, Tan et al. [10] employed the ball-drop method (Fig. 6), which consists of taking the gel temperature as the point where a small metallic sphere cannot move any longer. However, the conclusions drawn from this technique are questionable since the effects of chain entanglements, which may lead to a pseudo-gel behavior, are not correctly evaluated. Rheological experiments, with either a classical rheometer [36] or a sphere rheometer [37], have nevertheless confirmed Tan et al.'s findings. Worth emphasizing is that the measurements performed with the sphere rheometer were carried out in a sealed tube, which eliminates possible artifacts arising from solvent loss by evaporation (CS_2 is a very volatile solvent).

The gelation temperature can also be obtained by thermal analysis (DSC). As a matter of fact, unlike PVC, a gel formation exotherm can be observed in the aPS–CS_2 system [39]. As can be seen in Figure 7, this transition is distinct from the glass transition and is therefore not linked to any phenomenon associated with T_g. As with iPS, the temperature–polymer concentration gelation diagram exhibits a maximum (Fig. 8), which suggests that the gelation of aPS in carbon disulfide proceeds from the formation of a kind of stoichiometric compound. The origin of this "compound" has not yet been elucidated.

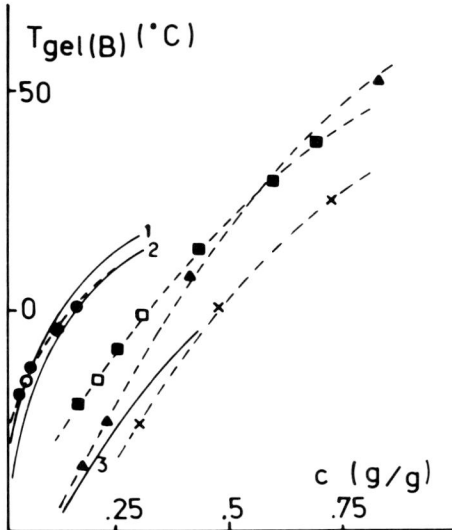

Figure 6 Determination of the gelation temperature T_{gel} by the ball drop method (closed symbols) and the sphere rheometer (open symbols) for atactic polystyrene in carbon disulfide. (Results from Gan et al., Ref. 38.) Full lines represent the results of Tan et al. [10] where 1 is for MW = 2×10^6, 2 for MW = 10^6, and 3 for MW = 4×10^3. (From Ref. 38.)

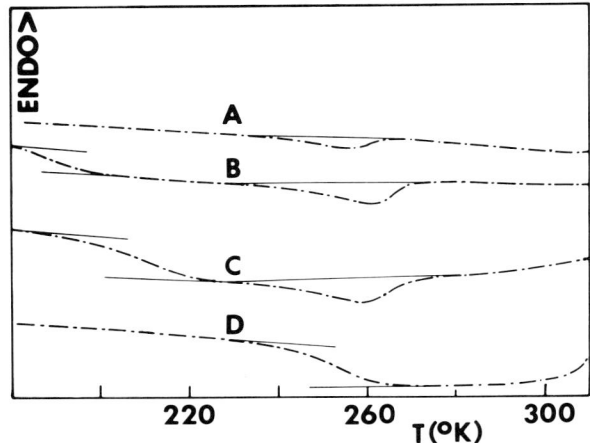

Figure 7 Typical DSC thermograms obtained on cooling at $-20°C$ min^{-1} atactic polystyrene/CS$_2$ samples of different concentrations (in w/w): A = 30.3%, B = 47.8%, C = 58.8%, and D = 74.7%. MW = 1.8×10^5. (From Ref. 39.)

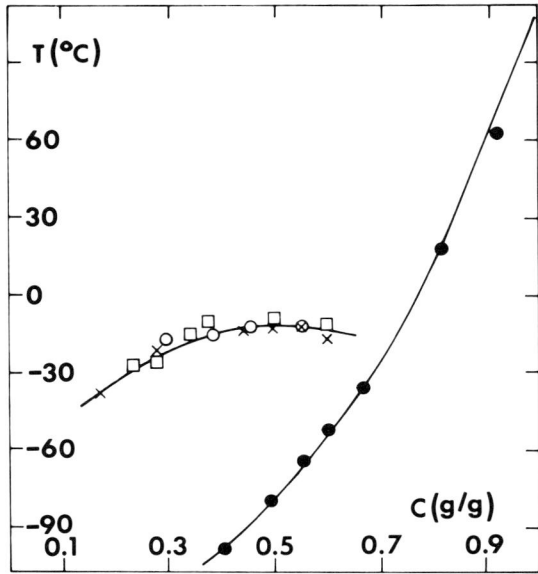

Figure 8 Gel formation temperature–concentration phase diagram of atactic polystyrene in carbon disulfide for different molecular weights: ×, MW = 1.8×10^5; □, MW = 1.7×10^6; and ○, MW = 5×10^6. Closed circles correspond to the glass transition temperature. Cooling rate: 20°C min^{-1}. (From Ref. 39.)

THERMODYNAMICS

This section details the thermal behavior of gels while heating. From this behavior, and particularly from the melting or transition temperature, the temperature–concentration phase diagram can be established. This is a useful basis for drawing some conclusions as to the "thermodynamic nature" of the gel.

Thus far only a few systems have been methodically investigated. They are discussed in this section, but they are not expected to be representative of all the thermoreversible gels. As will be seen, in some systems the thermodynamics is unusual, and similar behavior may be discovered later in other gels.

In this respect, the melting behavior of isotactic polystyrene is complex [8, 15, 40], depending strongly on the solvent used as well as on the thermal treatment. When it comes to assessing the effect of the solvent type [40], two cases are apparent:

1. The original gel molecular structure† does not transform into the 3_1 structure prior to the gel melting. This situation is encountered in *cis-* and *trans-*decalins.
2. The original gel structure transforms into the 3_1 form prior to gel melting (which accordingly becomes a metastable melting process). This situation occurs in 1-chlorodecane and 1-chlorododecane, for example [40].

A transformation like the second is illustrated through two types of DSC experiments. The first consists of scanning the gel at different heating rates. In Figure 9a, two endotherms are seen: a low-melting endotherm associated with the gel melting and a high-melting endotherm corresponding to the fusion of 3_1 crystals. For a high heating rate the low-melting endotherm predominates, whereas the inverse situation is seen for a slow heating rate. This indicates that the presence of the 3_1 crystals arises from the transformation of the original gel structure. Such a statement is supported by the second type of DSC experiment, which consists of scanning a gel annealed below the gel melting temperature (determined from the high heating rates). The DSC thermogram in this case reveals only the high-melting endotherm (Fig. 9b).

Whether the gel transforms into the 3_1 form prior to its metastable melting temperature or not represents a first step in a tentative classification. Gels in either form of decalin, once methodically studied, unveil a lower level of classification: either the gel is due to the formation of a congruently-melting compound (*cis-*decalin) or to a compound characterized by a singular point (*trans-*decalin) [15]. This is illustrated by a measure of the gel melting enthalpy as a function of the heating rate. For a system at thermodynamic equilibrium, the melting enthalpy should be independent of the heating rate. As can be seen in Figure 10, in both solvents the gel melting enthalpy is heating rate dependant. Yet, the extrapolation to zero heating rate gives finite values which are in close agreement with those found on cooling. The heating rate effect is not yet explained but may originate from a superheating phenomenon. A plot of ΔH_{gel} as a function of polymer concentration also reveals a maximum located in the vicinity $C \simeq 28\%$ in *cis-*decalin. In *trans-*decalin, the melting is spread over a larger temperature domain (Figure 10b) which renders more difficult a study of the melting enthalpy as a function of the heating rate.

---

†As will be seen in the section devoted to the gel molecular structure, the case of iPS is not definitively settled. As a result, it seems wiser to use a vague term rather than use the terminology of either model. Therefore, "original" here means the structure of the nascent gel.

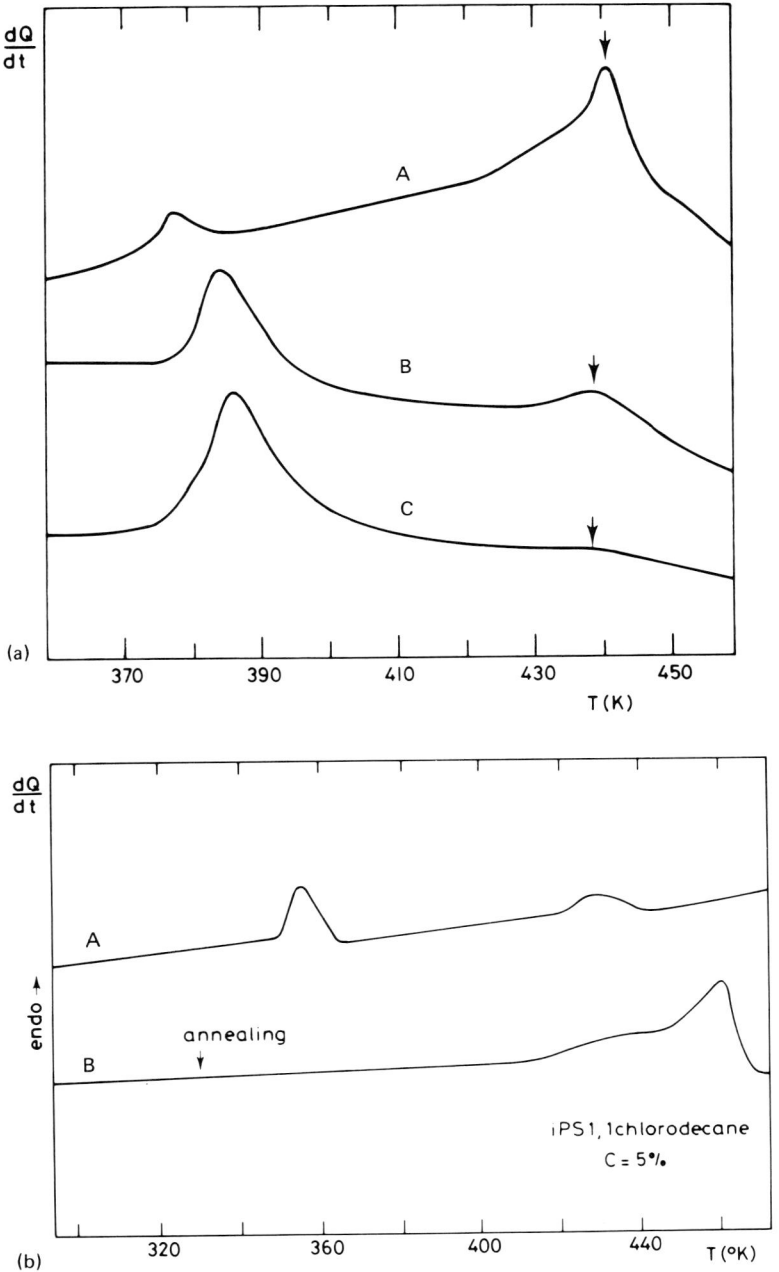

Figure 9 (*a*) Thermograms obtained from differential scanning calorimetry (DSC). Heating rates: A, 20°C min^{-1}; B, 40°C min^{-1}; C, 80°C min^{-1}. Isotactic polystyrene MW = 3.2 × 10^5 in 1-chlorododecane (C_{pol} = 5%). Gels prepared through a rapid quench from 180 to −5°C. (*b*) Thermogram obtained on the nascent gel (A) and the gel once annealed at the temperature indicated by the arrow (B). After annealing the low-melting endotherm has vanished to the detriment of the high-melting endotherm. Gels in 1-chlorodecane C_{pol} = 5% prepared by a quench to −20°C. (From Ref. 40.)

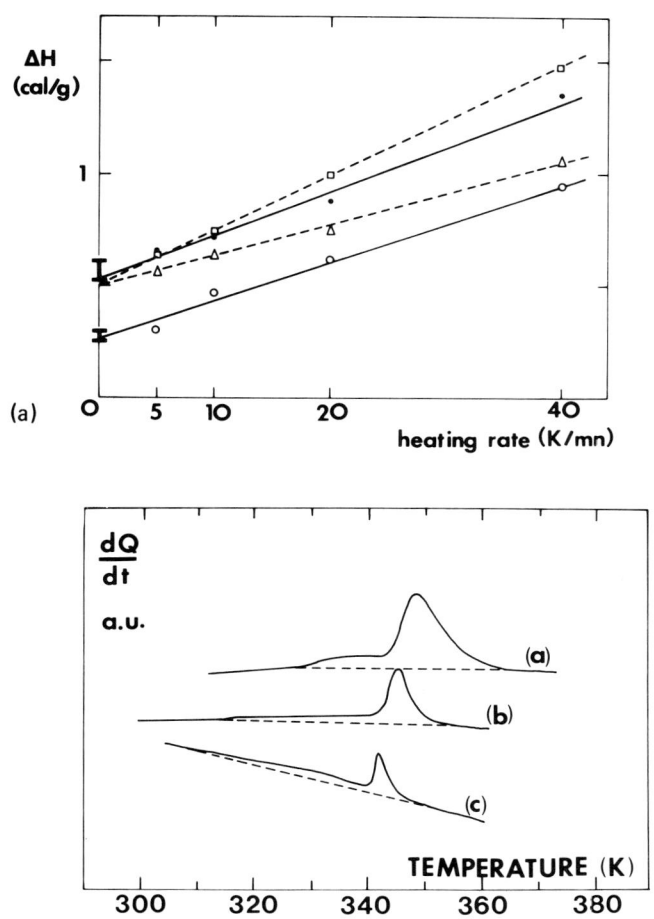

Figure 10 iPS gels. (*a*) Gel melting enthalpy as a function of the heating rate in *cis*-decalin. Concentrations as indicated. Bars on the ordinate = values measured on cooling (○, 10%; ▲, 20%; △, 40%, MW = 3.2 × 10^5; □, 40%, MW = 7.4 × 10^4). (*b*) Typical endotherm gel melting in *trans*-decalin gels for different heating rate (from top to bottom, 40°C/mn, 20°C/mn, 10°C/mn), C = 20%.

The phase diagram of gels in *cis*-decalin and *trans*-decalin [15] can be established without fundamental problems (Fig. 11; here a monotectic transition is observed at T_M, the meaning of which will be discussed in the paragraph dealing with the multiblock copolymer). The first fact that emerges from these diagrams is their striking resemblance to those determined on cooling. In addition, these diagrams emphasize the difference of the gel nature with the solvent type. These diagrams are typical of what is obtained with a congruently melting compound (*cis*-decalin) and compound exhibiting a singular point (*trans*-decalin) [41].

Also worth mentioning are the experiments performed on the crystallization behavior of the solvent in the gels. These experiments consist of determining by extrapolation the concentration at which $\Delta H_{\text{solvent}} = 0$ [42]. From this concentration one can determine a

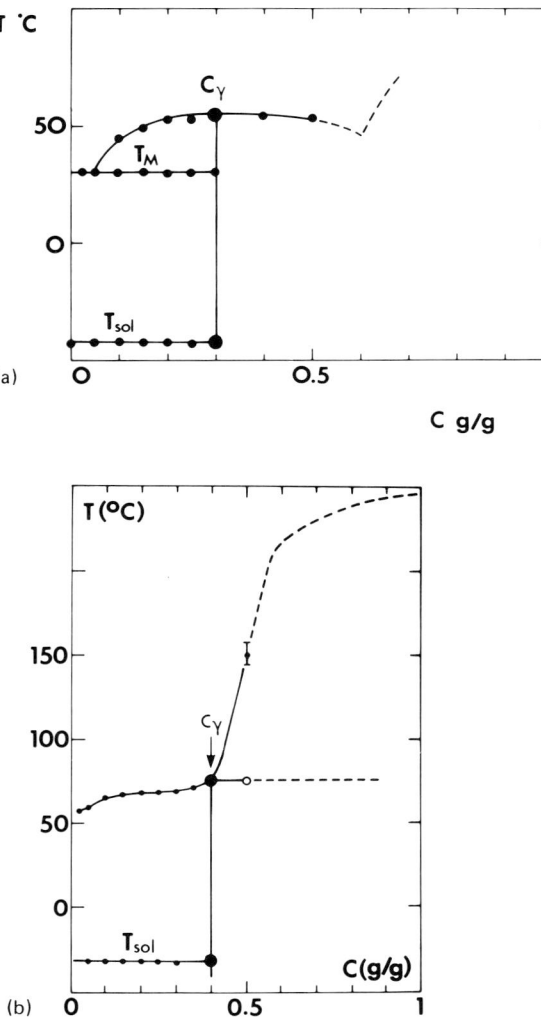

Figure 11 Temperature–concentration phase diagram for iPS–decalin gels. T_M represents the temperature at which a monotectic transition occurs. This invariant is due to the gel formation mechanism, which involves a liquid–liquid phase separation. T_{sol} = solvent melting temperature. The transition temperatures were obtained from an extrapolation of values determined from heating rates ranging from 5 to 40°C min^{-1}. (a) = iPS-*cis*-decalin, (b) = iPS-*trans*-decalin

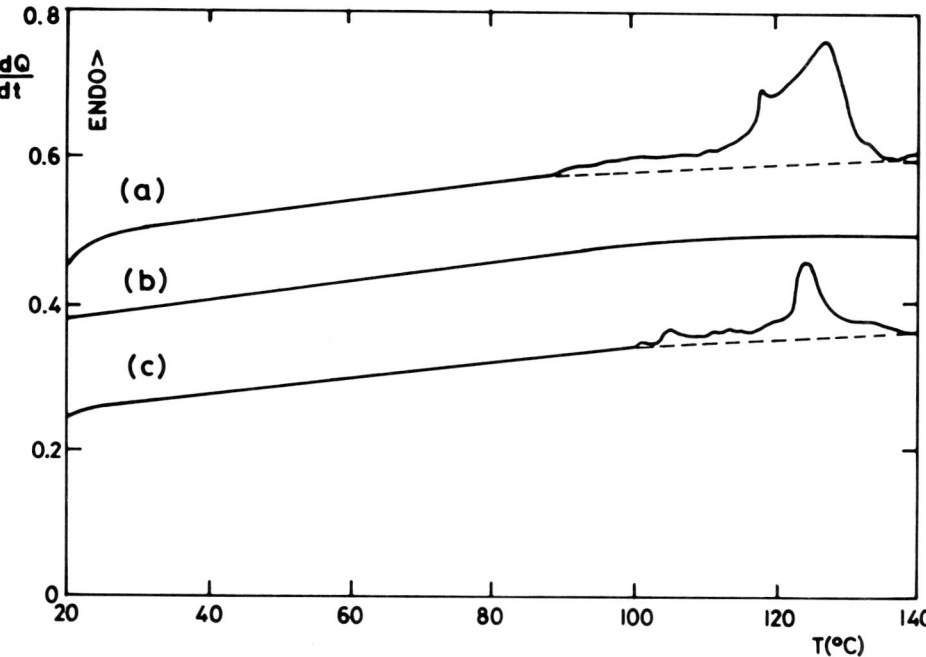

Figure 12 DSC thermograms for a PVC–diethyl malonate gel, $C_{pol} = 17.5\%$ (w/w). Heating rate = 20°C min^{-1}. (a) First run on the gel, aged 30 min at 20°C; (b) second run after melting in the hermetically sealed DSC pan; and (c) third run after recovering the gel from the DSC pan, cutting it into small pieces, and putting it into a new pan.

number $\bar{\alpha}$ which stands for the average number of solvent molecules "linked"† per monomer unit. The following values have been found for this parameter: $\bar{\alpha} \simeq 1.89$ in *cis*-decalin, $\bar{\alpha} \simeq 1.15$ in *trans*-decalin, $\bar{\alpha} \simeq 0.69$ in 1-chlorodecane, and $\bar{\alpha} \simeq 0.5$ in chlorododecane. One immediately perceives that the solvents in which the gel structure does not transform into the 3_1 form correspond to $\bar{\alpha} > 1$ (i.e., all the monomers are "screened" on average) and the case of $\bar{\alpha} \simeq 1$ corresponds to the "compound" with a singular point. Interestingly, the "degree of solvation" decreases markedly in 1-chlorodecane and 1-chlorododecane once the gel has been transformed into the 3_1 form. Since crystals of this form are anhydrous, these results are consistent with original gel structures being solvated. From these experiments, it has been concluded that the solvent plays an active role as opposed to the role of a mere diluent [42].

The thermal properties of polyvinyl chloride gels are reminiscent of those observed for iPS. The DSC thermograms seen in Figure 12 are typical examples of the behavior of freshly prepared gels [43]. The DSC trace obtained on the first run (curve a) exhibits a

---

†Whatever the system, there are always some solvent molecules that do not crystallize. This is often referred to as the depletion layer. This phenomenon does not necessarily mean that these uncrystallized molecules are trapped and immobilized by the polymer but simply that they undergo a special interaction with the polymer, preventing them from crystallization. These molecules are pejoratively designated "molecules linked to the polymer."

pronounced endotherm whose shape is not easily reproducible but whose position does correspond to the gel melting. Rescanning this sample, once molten and cooled down to room temperature, gives the thermogram in curve b. No endotherm is detectable whatsoever, as reported earlier by Yang and Geil [34]. If the gel of curve b is recovered then cut off into small pieces and transferred to another DSC pan, one obtains the thermogram of curve c, in which the endotherm has reappeared. From these experiments, it is concluded that the melting enthalpy measured at the gel melting point is overestimated if not totally artifactual. As a matter of fact, the variation of ΔH_{gel} with the heating rate extrapolates to zero for zero heating rate. Unlike iPS, it is not thought that the links are solvated since the same behavior is found independent of the solvent. There is currently no satisfactory explanation for the absence of a gel melting endotherm. Even the hydrogen-bonding model does not seem to be appropriate since hydrogen bonds are not expected to persist at such melting temperatures (100°C and above). As a matter of fact, hydrogels such as agarose gels, where hydrogen bonds are said to ensure gel cohesion, melt at lower temperatures. Is this phenomenon a challenging problem or does there exist a straightforward explanation? This remains an open question at the moment.

Aged PVC gels also exhibit unexpected thermal behavior. Thermograms drawn in Figure 13 [43] illustrate what is observed after aging. Curve a represents the thermogram recorded on a gel which was aged in the preparation test tube (i.e., *not* melted in the DSC pan beforehand). As above, the high melting endotherm is present, as is an endotherm at lower temperature. Curiously, the latter is still observed after subsequent aging on a gel melted in the DSC pan beforehand (Fig. 13, curve b). It has been suggested that this low-melting endotherm could be associated with the melting of structures characterized by a lower order [44, 45]. Yet one is faced with the paradoxical situation where the poorly ordered structures exhibit a reversible endotherm (at least after subsequent aging), where-

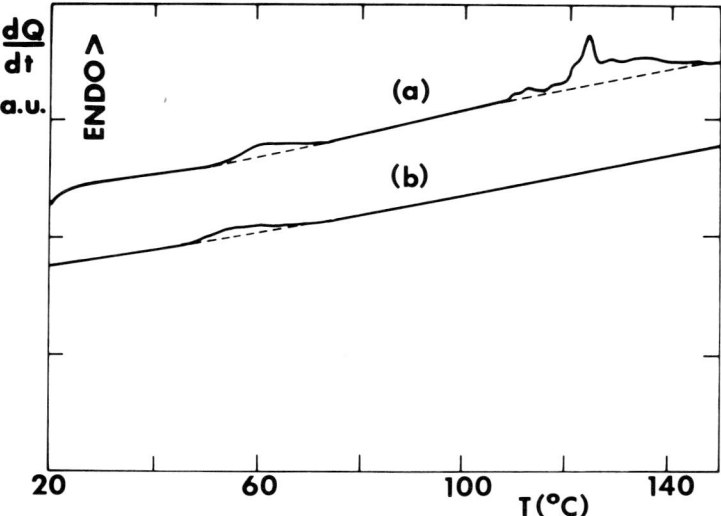

Figure 13 Effect of aging on PVC gels: DSC thermogram for a PVC–diethyl malonate gel, $C_{pol} = 17.5\%$ (w/w), heating rate = 20°C min^{-1}. (*a*) Gel aged at 20°C in a test tube for 24 hr then scanned in the DSC; (*b*) Gel once molten in the DSC pan and annealed in the pan for 24 hr at 20°C.

as the supposedly highly ordered structures do not. It is worth mentioning that the lower order structures are believed to be responsible for the evolution of PVC gels' mechanical properties with time [45].

No strange thermal properties are found with atactic polystyrene, which may be regarded to some extent as being paradoxical since physical gelation was absolutely not expected to occur with this uncrystallizable polymer. Figure 14 displays typical DSC thermograms obtained on heating [39]. As emphasized in the preceding section, the gel melting is distinct from the glass transition temperature. Also, it does not correspond to the overshoot frequently observed together with the T_g jump. The temperature–concentration phase diagram established from these experiments (Fig. 15) clearly indicates that there is no significant molecular weight effect [39]. In addition, this diagram is strikingly reminiscent of what was obtained for iPS in cis-decalin. This suggests, as already hinted by the formation phase diagram (Fig. 11), that gelation of aPS originates in the formation of a stoichiometric polymer + solvent "compound" [38, 39]. In spite of the experimental scatter, the variation of the gel melting enthalpy does exhibit a maximum near $C_{pol} \simeq 50\%$ (Fig. 16), giving further credence to the existence of a "compound." It is not yet known whether regular sequences are involved in the physical links and if they are, which sequences (iso or syndio). Recent experiments performed with epimerized polystyrenes (this allows the tacticity to be varied continuously from 100% iso to atactic polystyrene) suggest, however, that essentially the longest syndiotactic sequences participate in the physical links [46].

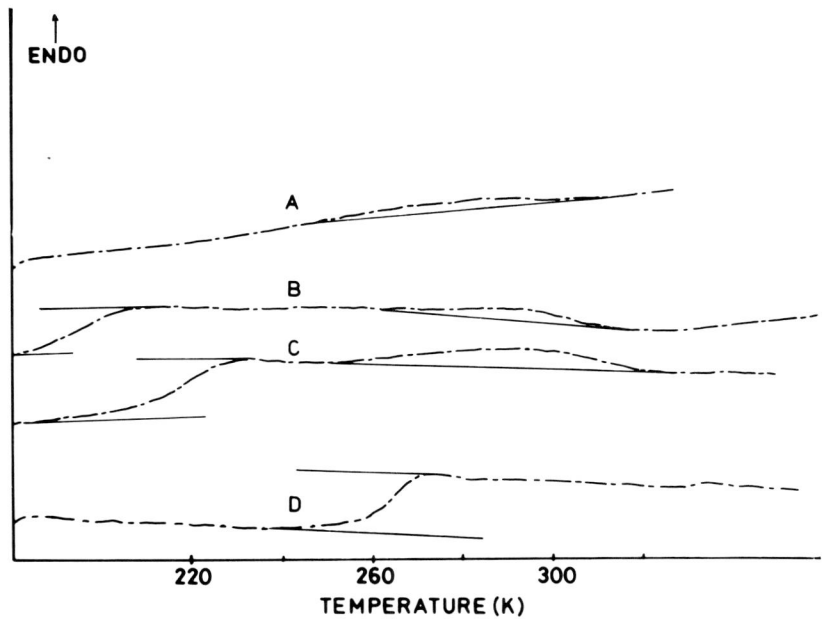

Figure 14 Typical DSC thermograms of aPS–CS_2 gels (MW = 1.8×10^5); heating rate 20°C min^{-1}. A = C_{aPS} = 30.3%; B = C_{aPS} = 47.8%; C = C_{aPS} = 58.8%; and D = C_{aPS} = 74.7% (concentrations w/w). (From Ref. 39.)

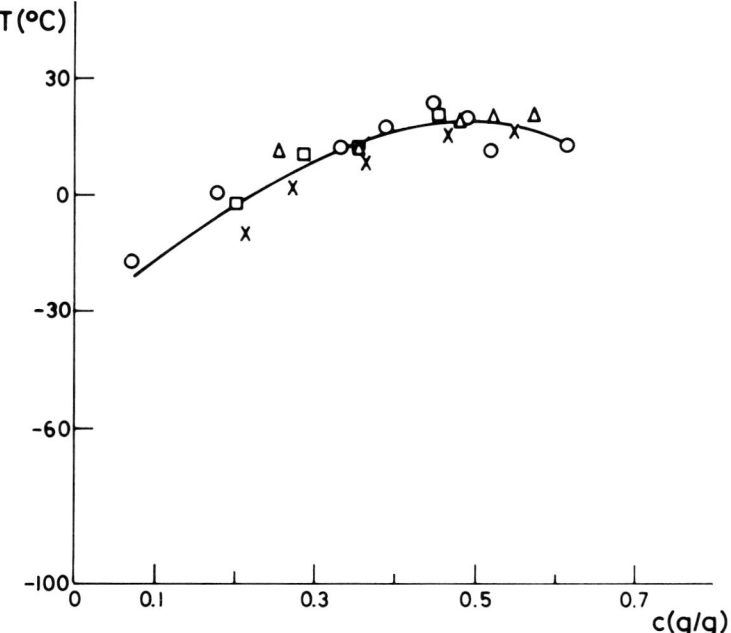

Figure 15 Temperature–concentration phase diagram for aPS–CS$_2$ gels prepared from different molecular weights: ○, MW = 1.8×10^5; △, MW = 1.7×10^6; X, MW = 6×10^3; □, MW = 5.6×10^6. The curve is only a guideline for the eyes. It points out, however, the absence of any molecular weight effect. (From Ref. 39.)

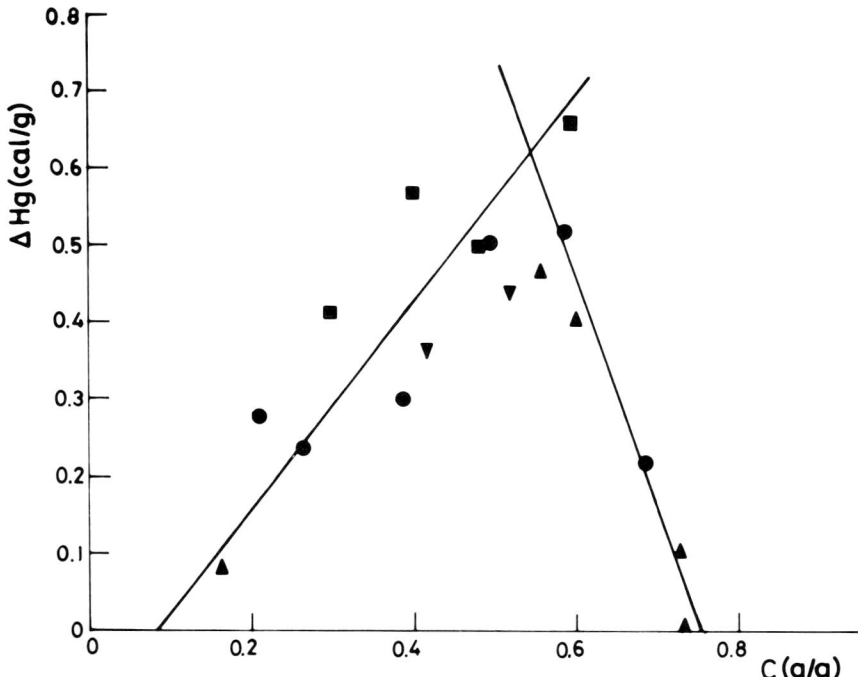

Figure 16 Variation of the gel melting enthalpy as a function of polymer concentration for aPS–CS$_2$ gels. ●, MW = 1.7×10^6; □, MW = 5.6×10^6; ▲, MW = 1.8×10^5; and ▼, MW = 6×10^3. (From Ref. 39.)

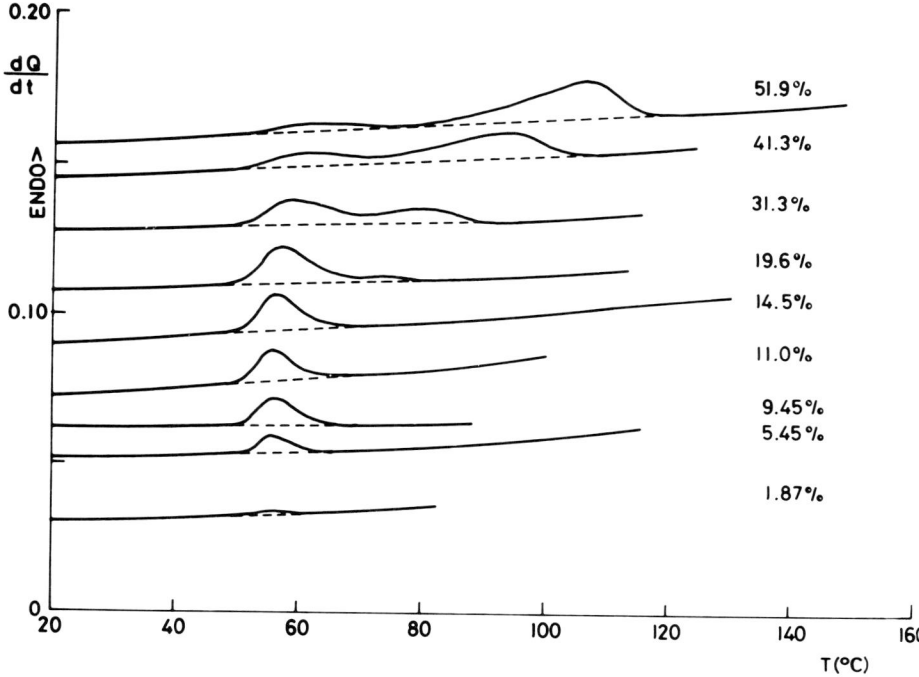

Figure 17 Gels from the PDMS–COSP multiblock copolymer in *trans*-decalin: DSC thermograms recorded on gels prepared through a rapid quench from 180 to −18°C, then annealed 1 day at −18°C and 7 days at 30°C. This procedure allows the thermogram resolution to be improved. Concentrations in w/w indicated. Heating rate 20°C min^{-1}. (From Ref. 48.)

Another type of gel, prepared from a multiblock copolymer, has recently been investigated. This copolymer is made up of alternating blocks of polydimethyl siloxane and of a crystalline polymer† designated as COSP (80% PDMS to 20% COSP w/w; $M_{w(PDMS)} = 17,000$, $M_{w(COSP)} = 4000$ [47]).

Typical DSC thermograms are given in Figure 17 for a wide range of copolymer concentrations that give the phase diagram of Figure 18 (copolymer in *trans*-decalin) [48]. These thermograms reveal several important features. Below a concentration C_M only the low-melting endotherm exists; correspondingly, above C_M two endotherms are visible, the low-melting being sharper than the high-melting one. The low-melting endotherm lies always at the same temperature independent of the copolymer concentration. At the highest concentrations, the low-melting endotherm vanishes. The invariance of the low endotherm temperature T_M with concentration has been interpreted by He et al. [48] as resulting from a monotectic transition. This transition is observed when a liquid–liquid phase separation occurs prior to crystallization, which is accordingly the gel formation mechanism. Interestingly, due to the supercooling effect, the monotectic line lies well above the miscibility gap.‡ Beyond a given concentration ($C \simeq 60\%$), the system forms a solid solution.

---

†This crystallizable polymer (COSP) is poly[1-(dimethyl)silyl,4-(dimethylethylene)silyl benzene].
‡When a solution is quenched within a miscibility gap to always the same temperature, then the concentration of the polymer-rich phase formed under these conditions remains unchanged independent of the initial solution

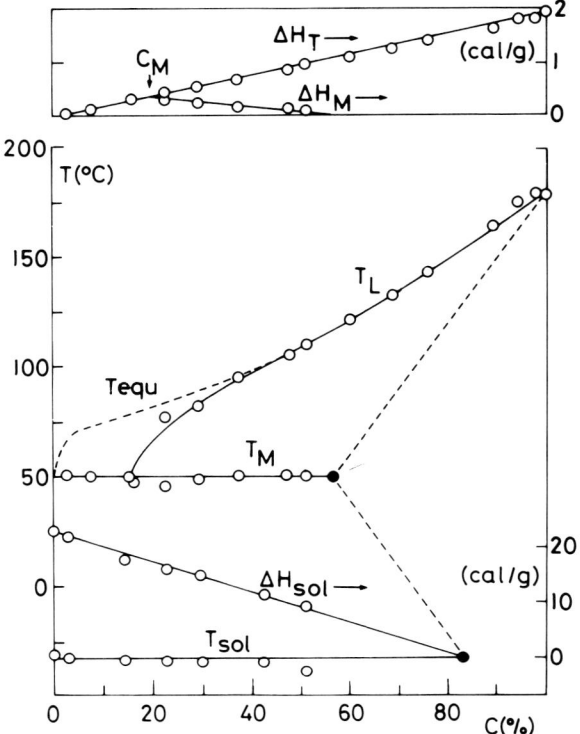

Figure 18 Temperature–concentration phase diagram for PDMS–COSP copolymer–*trans*-decalin gels prepared as indicated in Figure 17. Tamman's diagrams are also given (ΔH vs. C_{pol}). (From Ref. 48.)

The temperature–concentration phase diagram (Fig. 18) was found to be of further use particularly to account for the gel's mechanical properties with temperature.

Similar behavior has been reported with random copolymers, but no systematic investigation as a function of concentration has been carried out [49–51].

MOLECULAR STRUCTURE

In this section the gel molecular structure ranging from a few angstroms to a few hundred angstroms is examined. The structure at a larger scale will be presented in the next section, which is devoted to gel morphology.

Experimental investigations into these gel structures have been carried out with several

---

concentration. As a result, crystals formed from this phase will always have the same melting temperature, hence the existence of the monotectic line. In the case of polymers, the temperature of crystallization (when this phenomenon takes place within the miscibility gap provided that the system has reached this gap unaltered) will be different from the melting temperature (which depends on the crystal's characteristics, such as its thickness). As a result, the monotectic transition can be located well above the miscibility gap.

techniques: x-ray and neutron diffraction, elastic and quasi-elastic light scattering, neutron scattering, and infrared spectroscopy.

The gel molecular structure in isotactic polystyrene is still under discussion. As outlined in the introduction, Girolamo et al. [8] showed by x-ray diffraction on a dried (actually still containing solvent) and stretched gel that at 5.1 Å a meridional reflection could be observed which does not originate from the diffraction by 3_1 crystals. The first interpretation given of that unexpected reflection was to postulate the existence of chain defects such as head-to-head or tail-to-tail arrangements, which, on crystallizing together, would form the physical links of the network [8, 52]. However, ^{13}C NMR experiments indicate that, if such defects exist, they total under 1%. Allegedly, such a low content cannot account for the gel rigidity. In a later paper, Atkins et al. [53] realized that a conformation so far unobserved yet theoretically predicted, arising from quasi tt arrangements, could account for the 5.1 Å reflection provided that the resulting 12-fold helix possess a six-screw symmetry (the step of this helix is then a dimer). While this type of helical form was consistent with the existence of the 5.1 Å reflection, Atkins et al. noticed that the experimental intensity did not fit the theoretical [53]. Later, Sundararajan et al. [16] proposed that this helical form might be stabilized by solvent molecules. Such a suggestion was appealing since this type of stabilization is not common in the field of polymers yet is known in other areas of science. The 12-fold helical form received further support from experiments performed on dried and highly stretched gels [54]. Their diffraction pattern display unmistakable, if very weak, layer lines at 30.6 Å, which correspond exactly to the pitch of the 12_1 helix.

However, the main criticism that can be raised against these experiments is that they are representative of the dried and stretched state but not of the nascent gel state. In addition, there is no unquestionable proof that the 12-fold helical form owes its existence to the gelation phenomenon or that gelation is a consequence of the growth of 12_1 helices.

Making use of the unlikeliness of a coherent cross section between hydrogen and deuterium, a series of experiments were performed by Guenet [42] to try to shed some light on the gel structure in the nascent state. If the polymer is deuterated and the solvent hydrogenated, the diffraction pattern is mainly representative of the polymer. Accordingly, the polymer structure can be investigated without drying the gel. The results obtained by Guenet were rather unexpected and can be summarized as follows:

1. In a gel prepared from deuterated polystyrene and deuterated *cis*-decalin (under these conditions there is no difference between neutrons and x-rays), the diffraction pattern is similar to that of a liquid, the strongest reflection corresponding to 5.1 Å (actually ~5.3 Å) (Fig. 19).
2. When the gel is composed of deuterated polystyrene and hydrogenated *cis*-decalin, the 5.1 Å reflection is virtually absent (Fig. 20). Accordingly, the 5.1 Å reflection does not seem to characterize the polymer.
3. Finally, the spectrum of liquid *cis*-decalin is virtually identical to that of the gel and particularly exhibits the reflection at 5.3 Å (Fig. 21).

From these results it can be concluded that the 5.3 Å reflection certainly arises from the solvent only, hence the probable absence of any 12_1 form with a six-screw symmetry in nascent gels. Correspondingly, the liquidlike diffraction pattern of the gel suggests that crystallization is far from being the driving force governing gelation. Similar conclusions have been drawn by Aharoni et al. concerning P4MP1 [55].

The fact that the 5.3 Å reflection is of liquid order origin leads one to reconsider the

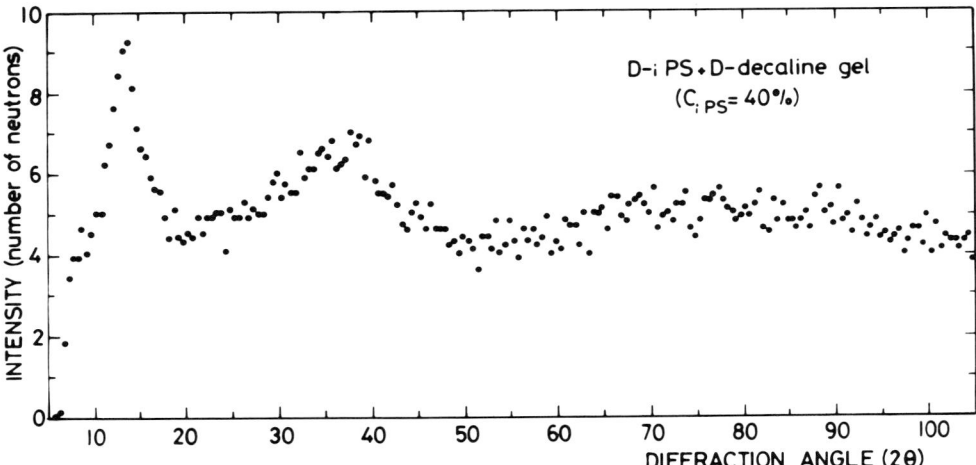

Figure 19 Neutron diffraction pattern of a gel prepared with perdeuterated isotactic polystyrene (C_8D_8) and perdeuterated *cis*-decalin ($C_{10}D_{18}$). Gel obtained by a quench to 0°C, concentration $C_{pol} = 40\%$ (w/w). (From Ref. 42.)

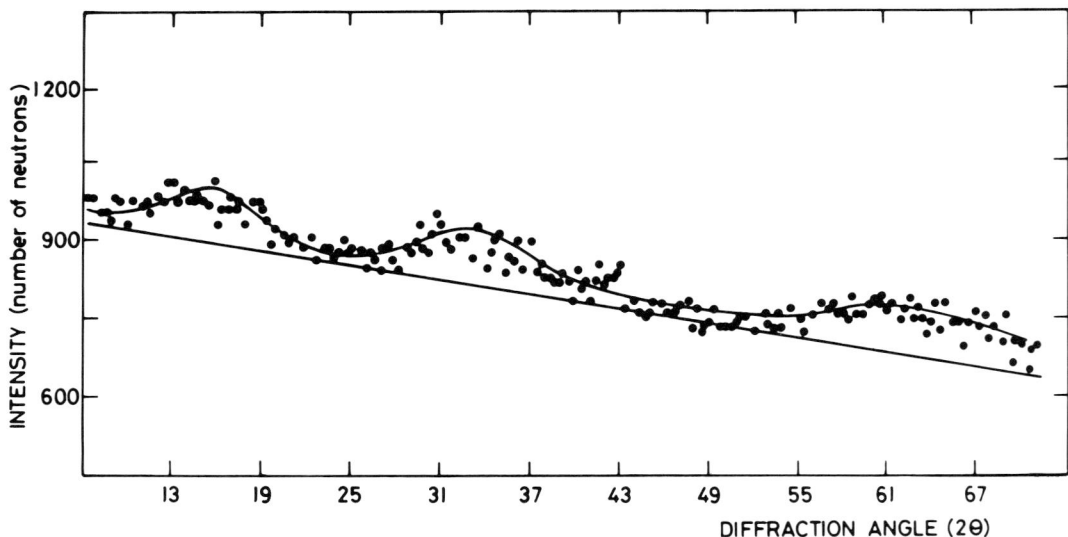

Figure 20 Neutron diffraction pattern of a gel prepared with perdeuterated isotactic polystyrene (C_8D_8) and protonated *cis*-decalin ($C_{10}H_{18}$). Gel obtained by quenching to 0°C, concentration $C_{pol} = 20\%$ (w/w). (From Ref. 42.)

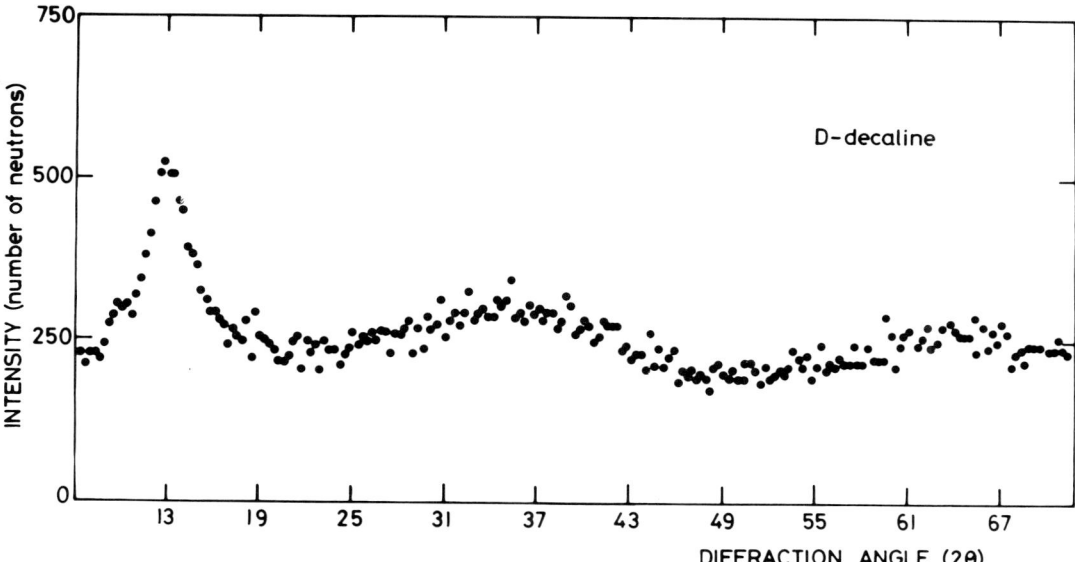

Figure 21 Neutron diffraction pattern of perdeuterated *cis*-decalin ($C_{10}D_{18}$) at 20°C. (From Ref. 42.)

distance it characterizes. As a matter of fact the strongest reflection in isotropic liquids usually corresponds to the distance d between first neighbors, which is given through the Ehrenfest relation for spherically averaged systems [56]:

$$1.23\lambda = 2d \sin \frac{\theta}{2} \tag{1}$$

From this relation one ends up with $d = 6.3–6.5$ Å, values which turn out to be very close to the pitch of the threefold helical form (6.65 Å). Apparently the distance between solvent molecules in the liquid state has something to do with the gelation phenomenon. This distance is directly related to the solvent molecular size, that is, its molar volume. In this respect, it was noticed that a plot of the solvent molar volume as a function of the parameter α, previously defined, exhibits two regimes, with a crossover at $\alpha = 1$ (Fig. 22). According to Guenet [42], the distance $d \simeq 6.5$ Å may hint that the chain actually adopts a near-threefold helical form whose cavities, created by the benzyl rings (see Fig. 23), play a role in the gelation process. If the solvent molecules have a smaller size than the average cavity's size, "stable" gels with "highly solvated structures" are obtained (stable in the sense that they do not transform into the 3_1 form prior to gel melting). Conversely, if the solvent molecules are characterized by a larger size, "unstable" gels with "poorly solvated structures" are produced. From all these results and considerations, Guenet [42] proposed the ladderlike model. This model is reminiscent of nematic liquid–crystalline polymer. As in nematic systems, rodlike chain portions are considered, but in addition, in the ladderlike model the chain–chain interaction is thought to be "mediated" by the solvent. Recent NMR experiments have shown that the solvent molecules possess virtually the same mobility in the gels and in the liquid state. This result

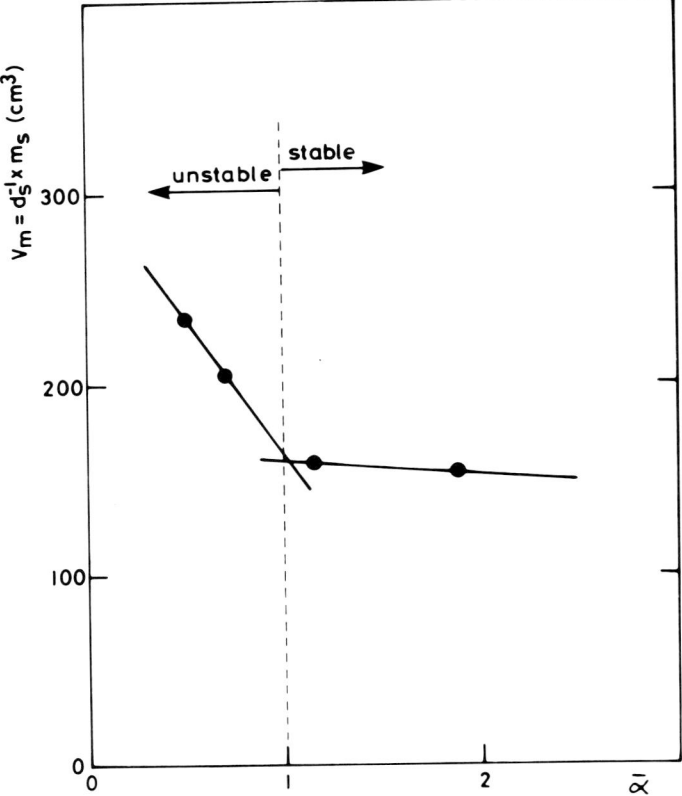

Figure 22 Plot of the molar volume as a function of α (number of solvent molecules "bound" per monomer unit) for solvents used in isotactic polystyrene gels. From left to right: 1-chlorododecane, 1-chlorodecane, *trans*-decalin, and *cis*-decalin. (From Ref. 42.)

emphasizes that the system is quite peculiar and has not unveiled all its secrets. We might be dealing with some kind of "plastic compound" by analogy with plastic crystals where the molecule can be nearly as mobile as in the liquid state.

Neutron scattering allows the chain trajectory to be elucidated by labeling a small percentage of the chains (~1%). Recent experiments carried out on iPS gels have revealed that the chain possesses a wormlike conformation ($R_g \simeq M^{0.5}$ where R_g is the radius of gyration) with a persistence length b of about 80 Å [57, 58] (see Fig. 24). These results show that the transformation of the solution into a gel does not alter the chain global behavior but enhances the persistence length (in the bulk or in θ solvent, $b \simeq 20$ Å). Eventually, the chain conformation in the gel is about halfway between that in the bulk amorphous state [59] and that in the semicrystalline state [60]. It also turns out to be the conformation observed for nematic liquid crystalline polymers [61, 62]. This latter fact partly supports the ladderlike model. It must be noted, however, that these results are also consistent with the existence of two turns of 12_1 helix. Yet the wormlike trajectory is not what would be expected from crystallization.

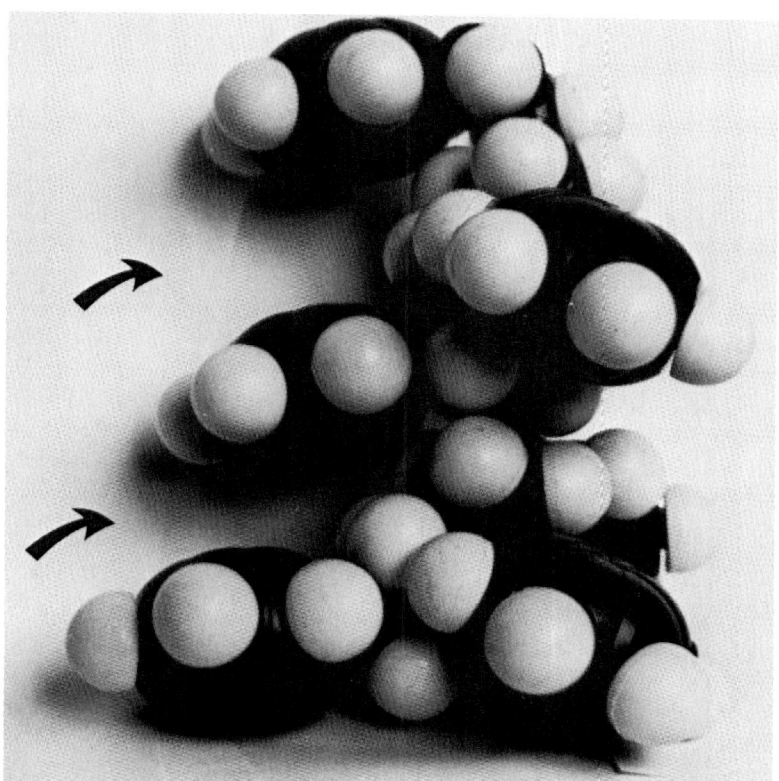

Figure 23 Representation by means of Ealing models of a portion of isotactic polystyrene chain which adopts a near-3, helical conformation. Arrows indicate the cavities created by the positioning of the benzene rings under this conformation. Such cavities are large enough to receive quite easily solvent molecules of the size of *cis*-decalin.

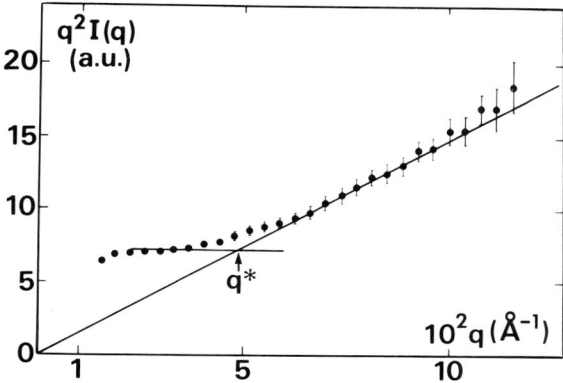

Figure 24 Typical Kratky plot [$q^2I(q)$ vs. q where $I(q)$ is the intensity scattered by the deuterium-labeled chains and $q = 4\pi/\lambda \times \sin \theta/2$] for a 15% PS–*cis*-decalin gel containing 1% of deuterated chains (MW = 2.5×10^5) prepared through a quench to 0°C. This plot displays two regimes: below q^* a regime where $I(q) \simeq 2/q^2R^2$ (R^2 is the mean square radius of gyration) and above q^* a regime where $I(q) \simeq \pi/qb$ (b is the statistical unit length). The first regime [$I(q) \simeq q^{-2}$] is characteristic of a gaussian coil while the second [$I(q) \simeq q^{-1}$] is due to a rodlike conformation. If q^* is taken as being the intersect between the asymptotes of both regimes, then it reads $q^* = 12/\pi b$. (From Ref. 57.)

According to Guenet [42], these results suggest that the gel state may be the third state for this kind of polymer, the first two states being the crystalline and the amorphous states.

No x-ray diffraction data are available so far with atactic polystyrene. As a result, it is not yet known whether the physical links are constituted of a special type of helix or resemble those of iPS.

The gel structure at larger scale has been investigated by elastic light scattering. The principle of this study is based on an observation made by Guenet et al. [63]. These authors noticed that gelation and the enhanced low angle scattering (ELAS) usually occur in the same solvents and correspondingly that the ELAS is absent in solvents wherein gelation is said to occur (incidentally, these results were the first to show the absence of the ELAS in some good solvents). The ELAS phenomenon, which takes place in moderately concentrated solutions, has been known since the early 1960s [64–66] but had failed to be predicted theoretically. In principle, if the solution were homogeneous, the intensity scattered from it [67] should read

$$I(q) \simeq (q^2 + \xi^2)^{-1} \tag{2}$$

where q is the scattering vector [$q = (4 \times \pi/\lambda) \times \sin \theta/2$, with λ the wavelength and θ the scattering angle] and ξ the screening length as defined by Edwards [68]. A plotting of $I^{-1}(q)$ vs. q^2 should accordingly give a straight line. Instead, a strong downturn is experimentally observed at the smallest angles [63–66], which is generally characteristic of the presence of aggregates (see Fig. 25). Yet how can aggregates be formed with an uncrystallizable polymer and, over and above, in very good solvents? These paradoxes

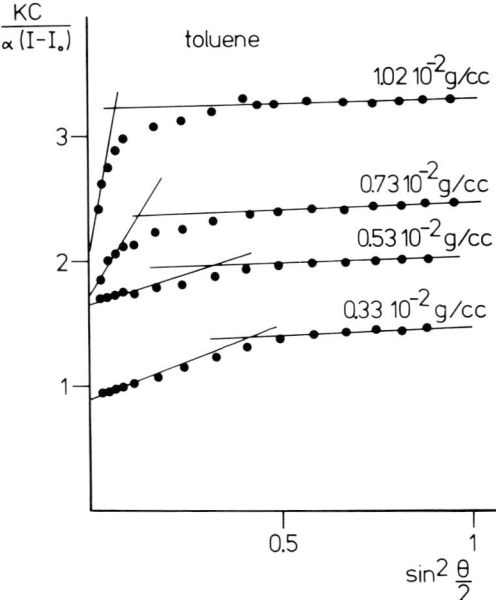

Figure 25 Variation of $I^{-1}(q)$ vs. q^2 for moderately concentrated solutions of atactic polystyrene in toluene (MW = 2.95×10^6 MW/MN = 1.06). Concentrations as indicated. (From Ref. 63.)

have long puzzled the scientists involved in the study of polymer solutions. In light of Tan et al.'s findings [10], Guenet et al. [63] and later Gan et al. [38, 39] proposed and gave support to gelation and ELAS being in fact different manifestations of the same phenomenon, i.e., the formation of a polymer + solvent stoichiometric compound (with the suitable sequences). Obviously, the formation of a polymer + solvent complex can be promoted when the polymer–solvent interaction is favorable. This condition is certainly met with good solvents, which consistently accounts for the observations of ELAS in good solvents.

The correlation between these phenomena is illustrated by experiments which show the ELAS temperature dependence [38]. As can be seen from Figure 26, the ELAS vanishes at a temperature higher than room temperature ($T \simeq 45°C$ in THF and $T \simeq 43°C$ in p-dioxane) but reappears on cooling of the solution. In addition, the intensities scattered by the solution before heating and after cooling back to room temperature are virtually identical. This proves the reversibility of the phenomenon and consequently invalidates

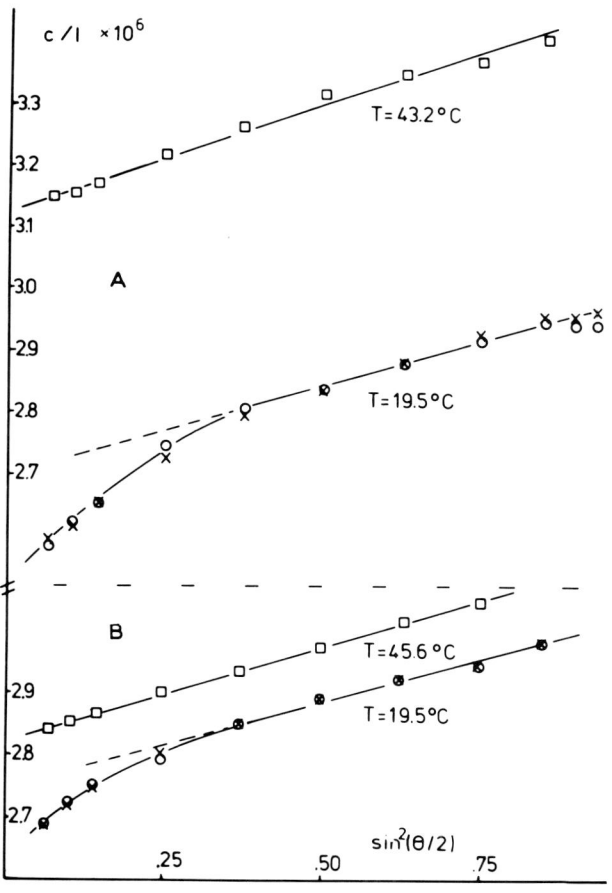

Figure 26 Variation of $I^{-1}(q)$ vs. q^2 for moderately concentrated solutions of atactic polystyrene in para-dioxane (A) $C = 5.7 \times 10^{-3}$ g cm^{-3} and in THF (B) $C = 5.6 \times 10^{-3}$ g cm^{-3} at different temperatures as indicated. MW $= 1.7 \times 10^6$. (From Ref. 38.)

previous assumptions which considered dissolution to be hampered by reptation, or snake-like movement [69].

Once the analogy between gelation and ELAS is accepted (some authors disagree [70]), the large-scale gel structure can be inferred from elastic light scattering studies at room temperature on moderately concentrated solutions. (In fact one deals to some extent with pregels.) The analysis of the experimental curves at low angle can be achieved by means of the Debye-Bueche theory later modified by Koberstein et al. [71]. This theory allows an average length a between physical crosslinks to be determined. This length a is of the order of the chain dimension for the polymer concentration where the ELAS is first detected. Then a varies as C^{-1}. The analysis of the intensity at larger angle, which varies linearly, can be performed with Eq. 1 and provides one with a value of the screening length ξ. Despite the presence of aggregates, it is found that the screening length varies, as theoretically predicted, as follows [72]:

$$\xi \simeq C^{-3/4}$$

This shows that gelation only slightly perturbs the usual solution "geometry" and accordingly suggests that the gel structure is rather of the fringed micellar type (see Fig. 27). A morphological study of the gel state is required but does not appear at the moment to be easy or straightforward.

Recent investigations by elastic [73] and quasi-elastic [74] light scattering on polyvinyl chloride pregels have unveiled unexpected types of behavior. A pregel is understood to mean a system that is formed below the critical gel concentration (usually of about 2% in PVC, this concentration being molecular weight-dependent). As a result, while aggregates resembling small gel particles are produced, they are not connected together. If the aggregates possess a fringed micellar structure, it is expected that the scattered intensity would be very similar to that observed in aPS, that is, exhibiting the ELAS phenomenon. Such a situation is not obtained in solvents such as diethyl malonate where, in fact, the reverse effect is seen [73]. The inverse of the scattered intensity displays a strong upturn

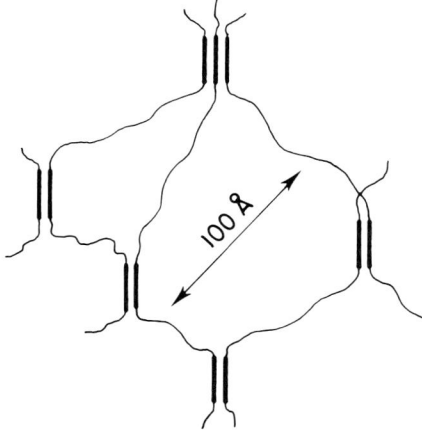

Figure 27 Representation of a possible gel structure often referred to as the fringed micelle model. As indicated, in this model the typical distance between physical links is of the order of 100 Å.

(Fig. 28). Correspondingly, the intensity exhibits a maximum which characterizes a distance of about 500 Å. The interpretation that should be proposed to account for this effect is not yet clear. However, Mutin and Guenet [73], after a thorough review of all the cases liable to produce such an intensity pattern, including possible chemical degradation, came to the conclusion that this upturn arises genuinely from the pregel structure. Obviously, aggregates formed via hydrogen bonding would not be supposed to give rise to such an intensity pattern. It seems that the minimum condition is the presence of crystals or at least of very ordered regions. Optical and electric birefringence experiments confirm the existence of such regions in the aggregates [74]. However, as already stressed, the fringed micellar structure of Figure 27 cannot account for the above observations. Consequently, an alternative interpretation must be sought. As will be seen in the section devoted to morphology, a "fibrillar-type" model may be a good candidate. So far these light-scattering experiments can only eliminate models but lack the additional information to enable one to propose another structure.

Similar anomalies have been found by inelastic light scattering [74]. From the so-called autocorrelation function (time correlation as opposed to space correlation), the particle diffusion coefficient can be obtained. A plot of the diffusion coefficient as a function of the squared scattering vector reveals an upturn at the smallest values (Fig. 29). This unusual variation is simply due to the effect of the static scattered intensity on the diffusion coefficient determination. From further investigations as a function of temperature [75], it is concluded that two types of links are present (designated as strong and weak). The strong links are supposed to be formed with the more regular syndiotactic

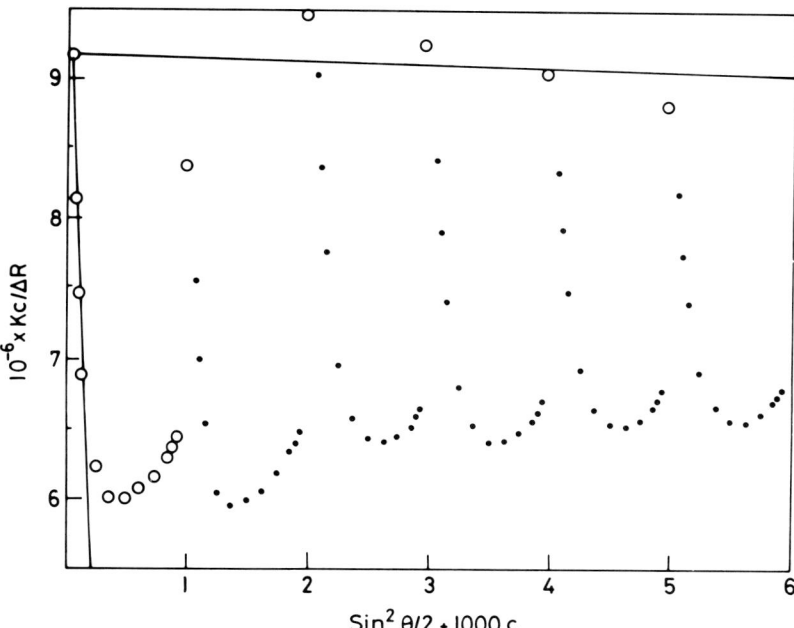

Figure 28 Typical Zimm plot $[C \times I^{-1}(q)$ vs. $q^2 + kC]$ for PVC (MW $\simeq 1.2 \times 10^5$) in diethyl malonate. Starting concentration $C = 5 \times 10^{-3}$ g cm^{-3}. (From Ref. 73.)

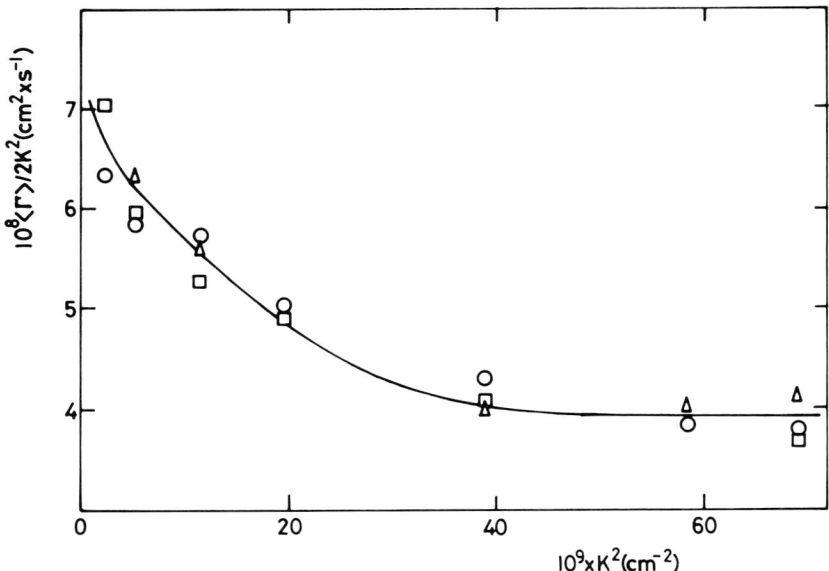

Figure 29 Variation of the diffusion coefficient D ($D = \Gamma/2K^2$) as a function of the squared scattering vector K^2 (K is used in quasi-elastic light scattering in lieu of q) for PVC in diethyl malonate. ○, $C = 5 \times 10^{-3}$ g cm^{-3}; □, the starting solution diluted to $C = 3 \times 10^{-3}$ g cm^{-3}; and △, diluted to $C = 10^{-3}$ g cm^{-3}.

sequences, whereas the weak links are thought to originate from the sequences possessing a lower order.

X-ray studies on "dried" and stretched gels have indeed revealed the characteristic reflections of syndiotactic crystals [33]. In addition, another reflection at 5.1 Å was observed and supposed to arise from other types of crystals [33], that is, chain-folded crystals as opposed to micellar crystals. Guenet [42] has, however, pointed out that this reflection may also be due to the solvent (dimethyl phthalate was used, which is hard to get rid of and has virtually the same molar volume as decalin).

The structures involved in syndiotactic polymethyl methacrylate have been investigated essentially by infrared spectroscopy and x-ray diffraction on oriented swollen films [76]. From the infrared observations, it is inferred that there exist sequences with regular conformation. The intramolecular order is maintained after removal of the solvent. From their recent investigations on swollen films, Kusuyama et al. [76] conclude that the chain adopts a 37 $d,1$-repeating unit's helical form, which might also be found in gels. In addition, it is strongly suspected that the solvent stabilizes the helix and also participates in the intermolecular associations.

Concerning polyethylene [9] or the multiblock copolymer [48], the gels seem to be of a crystalline nature.

To briefly summarize this section, it appears that some systems undergo gelation with the help of the solvent while others form gels by mere crystallization.

MORPHOLOGY

The determination of gel morphology is certainly the most delicate part of gel investigation. Since the solvent is the major component in a gel, the system must be dried and thin slice obtained by cross-sectioning prior to examination in the electron microscope. As emphasized in the preceding section, the extent to which the drying process affects the morphology is not known. In addition, the ultramicrotome section may either alter the original morphology or create artifactual structures. One way of avoiding both the drying process and the microtome sectioning consists of investigating gel samples by optical microscopy. However, while some observations can be made, in most cases the resolution is too low to draw unquestionable conclusions. For some systems, the use of both techniques enables one to get a fairly good idea of the actual structure (i.e., in the wet state).

Studies of isotactic polystyrene gels have been carried out by means of both techniques [40, 77]. Optical microscope pictures obtained via phase contrast [40] reveal a "salt and pepper" structure, which is more visible once the gel has been transformed into the 3_1 helical form (Fig. 30). The electron microscope preparations were obtained through a particular property of the gel once 3_1 transformation has taken place. Under this helical form, the gel is no longer soluble in toluene (the gel was soluble when in the nascent state [40]). However, some kind of "weathering" takes place which gives gel particles of different sizes. Some particles happen to be thin enough to be observed in the electron microscope. Figure 31 is a typical picture for a gel prepared in 1-chlorododecane and transformed into the threefold helical form [40]. As can be seen, this picture reveals an array of threads, which is nothing less than the network. Also, some dark spots are seen which are 3_1 chain-folded crystals that grew from the dilute phase independently of the network. It is inferred that the threads are responsible for the salt and pepper aspect observed by optical microscopy.

Apparently the gelation phenomenon tends to promote the formation of fiberlike objects. There is nevertheless still a doubt as to whether the 3_1 transformation has simply frozen-in fiberlike structures already existing in the nascent gel or has also altered the original morphology. Investigations carried out with gel of iPS in nitrobenzene using a more sophisticated preparation method [77] show a kind of Swiss cheese structure, which suggests that the fiberlike objects are not necessarily present in the original gel structure.

An interesting case is observed in diethyl malonate [40]. In this solvent the 3_1 transformation occurs spontaneously at room temperature. Guenet et al. [40] have shown, by means of the toluene extraction method, that three types of objects can be distinguished (Fig. 32):

1. Chain-folded crystals.
2. Fiberlike objects such as those seen with 1-chlorododecane gels.
3. Small shish kebabs.

According to Guenet et al. [40], the existence of these objects can be accounted for by the simultaneous occurrence of the 3_1 transformation and of the crystallization of chains located in the dilute phase. Once under the threefold helical form, the fiberlike objects can act as nuclei for the crystallization of the remaining chains, hence the formation of shish kebabs. As soon as the dilute solution does not contain free polymer chains any longer, the 3_1 fiberlike objects to come from 3_1 transformation will remain bare.

The fiberlike morphology has also been observed in various systems such as nylon

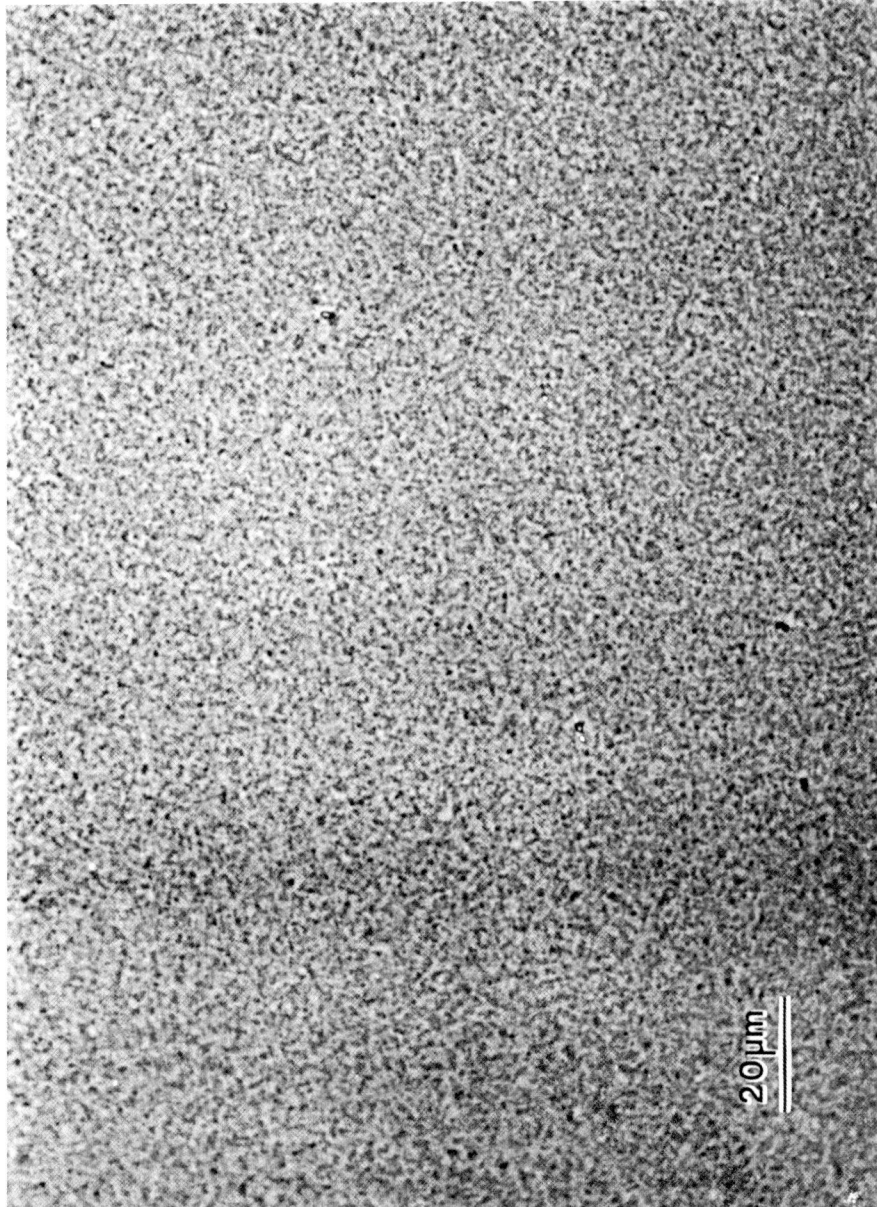

Figure 30 Phase contrast microscopy. Gel from diethyl malonate prepared between glass slides ($C = 10\%$ in w/w quenching temperature 0°C). Observation after subsequent annealing to transform the original structure into the threefold helical form. (From Ref. 40.)

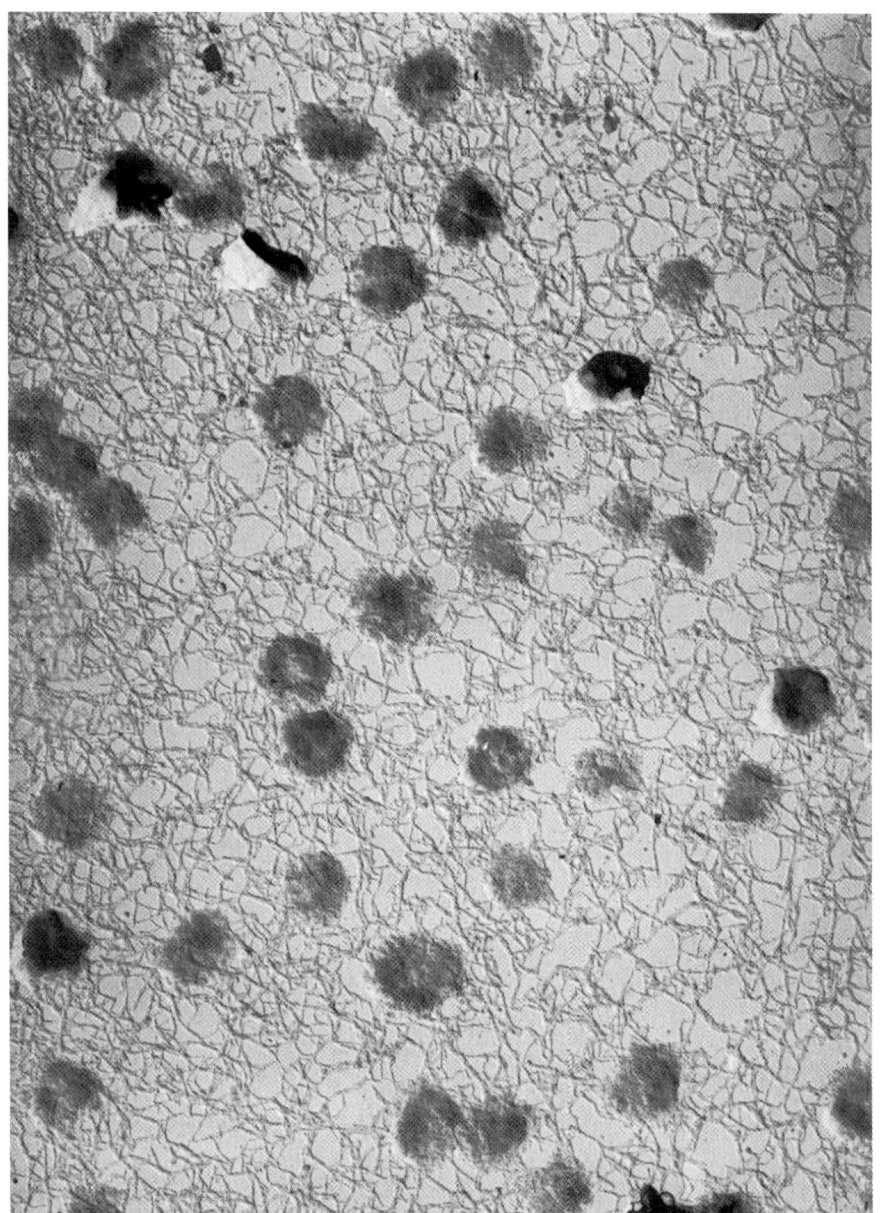

Figure 31 Electron micrograph of a thin gel particle obtained from a 7% (w/w) gel in 1-chloro-dodecane (quenching temperature −5°C, aged a week at 20°C then annealed at 72°C and finally toluene extracted). (From Ref. 40.)

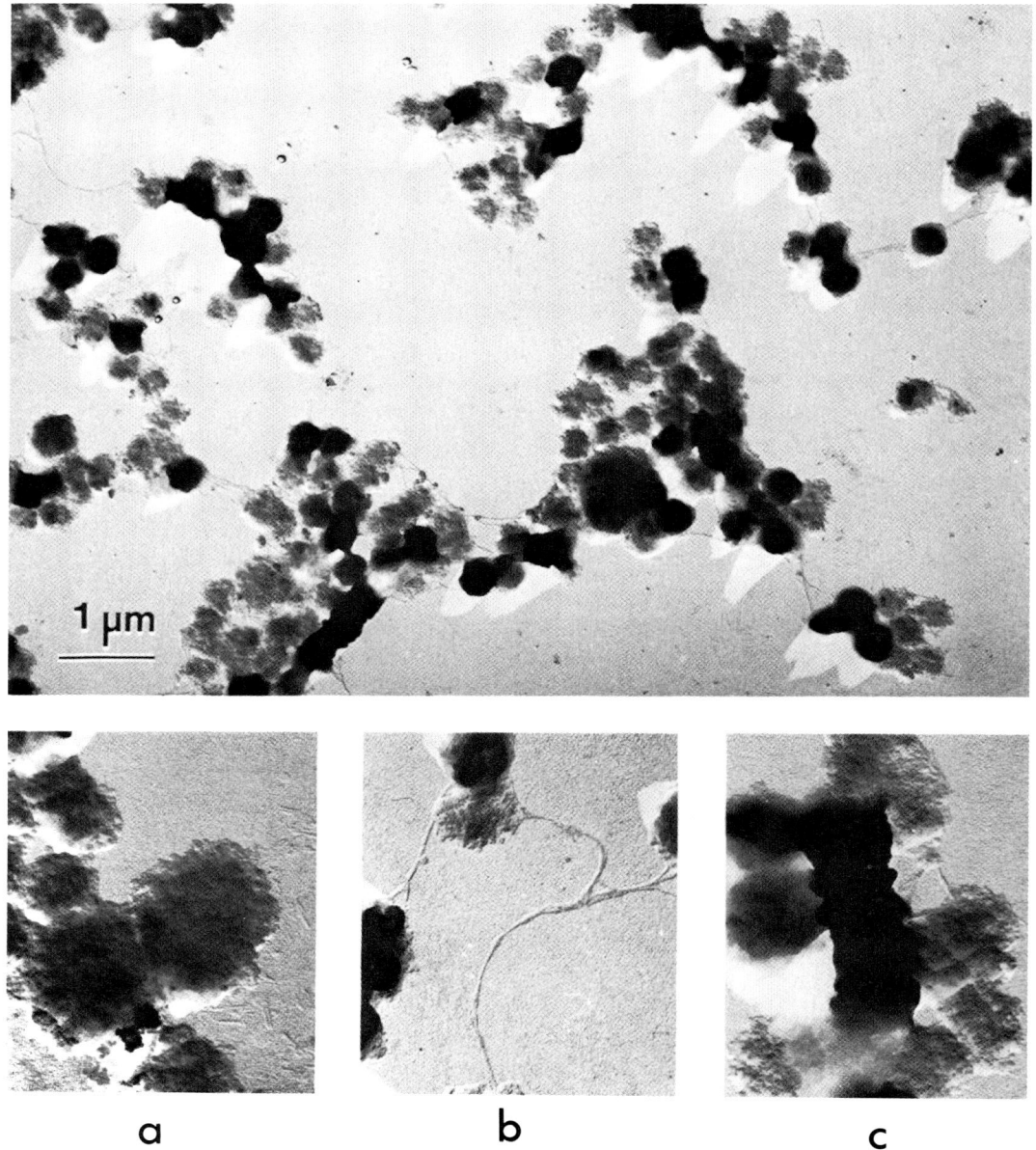

Figure 32 Electron micrograph of the material toluene-extracted from a 7% (w/w) gel in diethyl malonate aged two weeks at room temperature. Enlarged areas show (*a*) flat crystals (chain-folded crystals), (*b*) threads, and (*c*) shish kebobs. (From Ref. 40.)

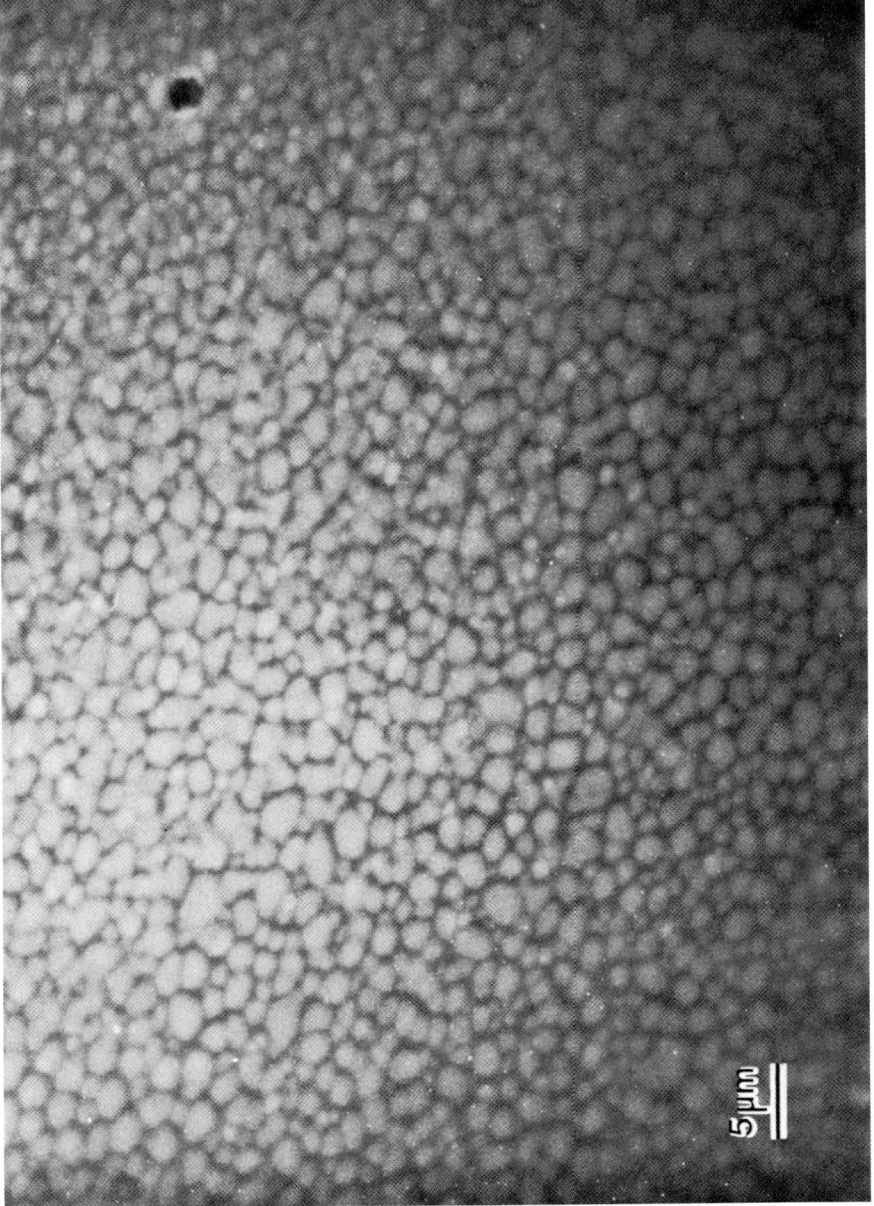

Figure 33 Phase contrast microscopy. PVC–benzyl alcohol gel (12.5% w/w) formed between glass slides through a quench to room temperature.

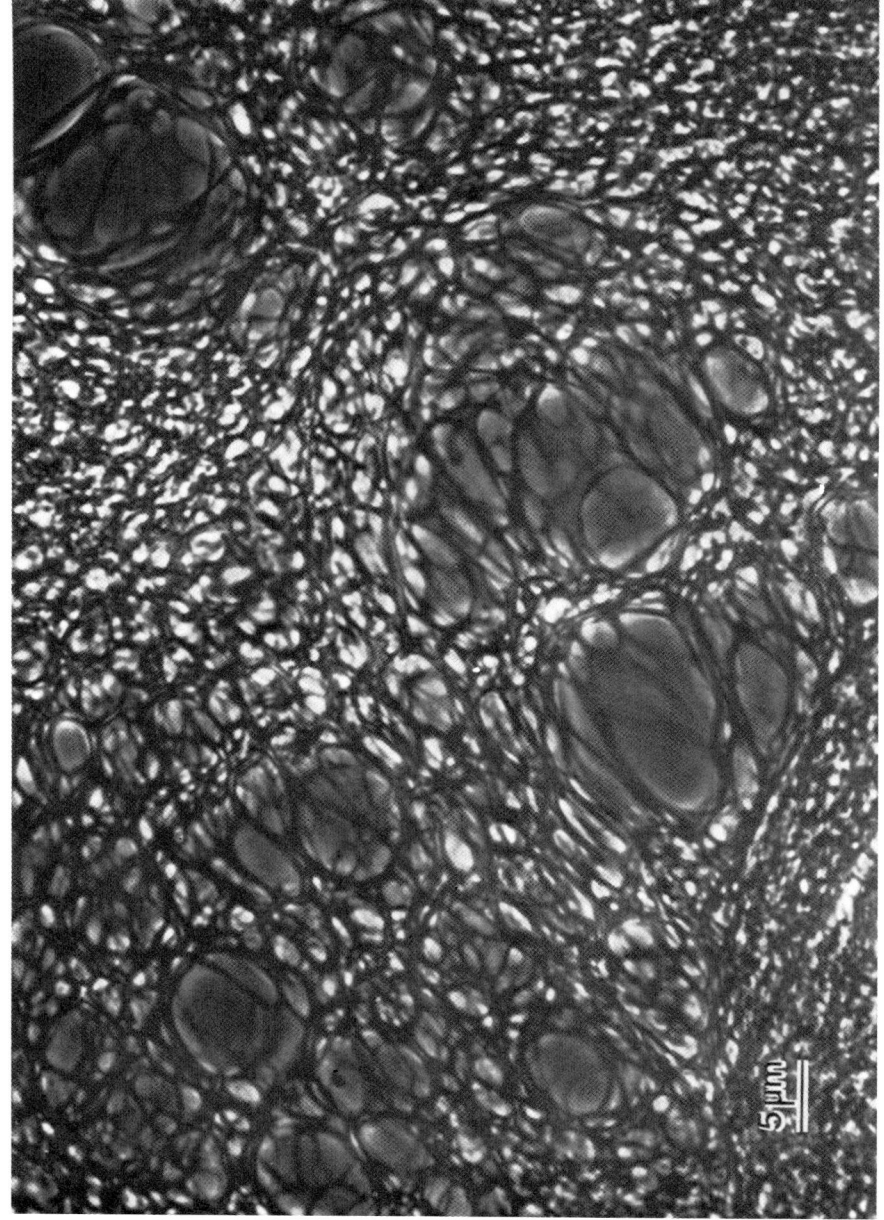

Figure 34 Phase contrast microscopy. PVC–poor solvent (cyclohexanol/hexanol 60:40) gel ($C = 12.5\%$ w/w) formed between glass slides by a quench to room temperature.

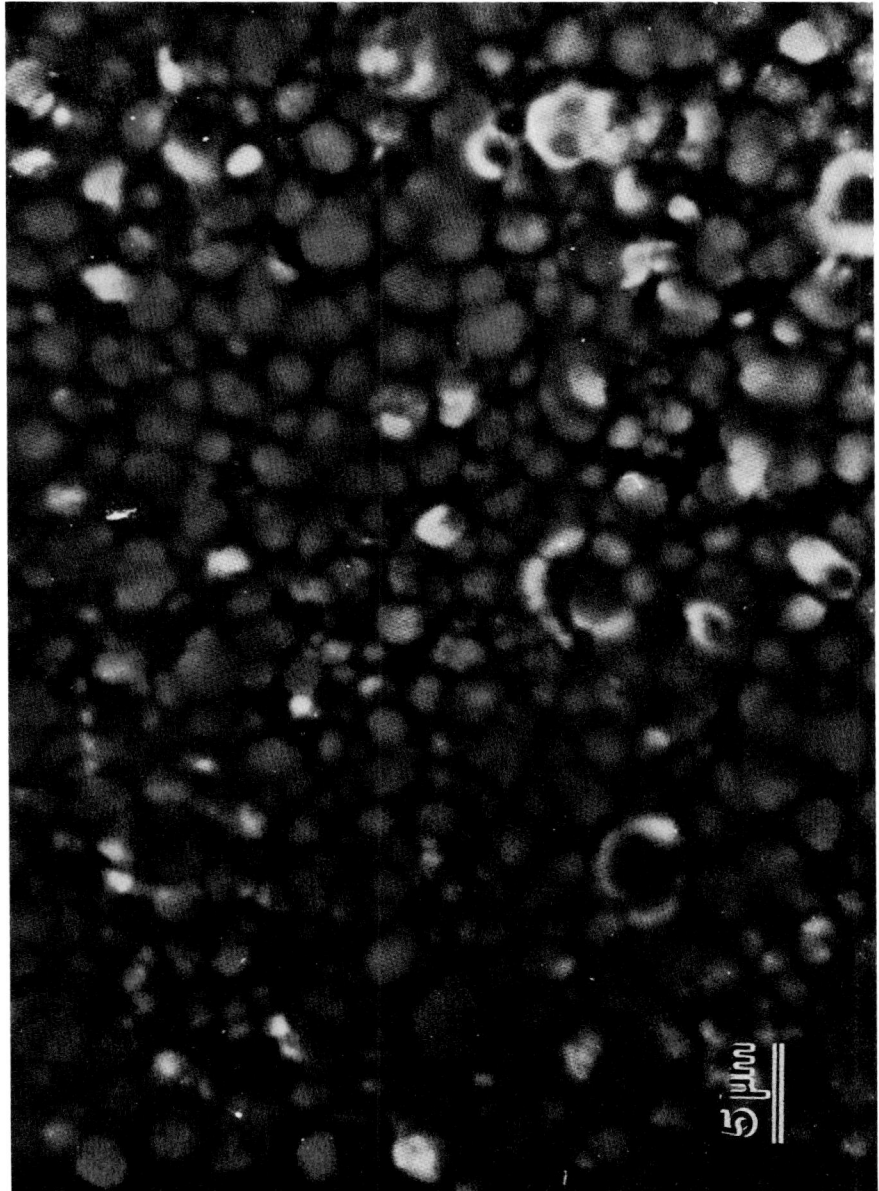

Figure 35 Phase contrast microscopy. Gel from PDMS–COSP copolymer in 1-phenyl dodecane. $C = 10\%$ (w/w) quenched between glass slides to room temperature.

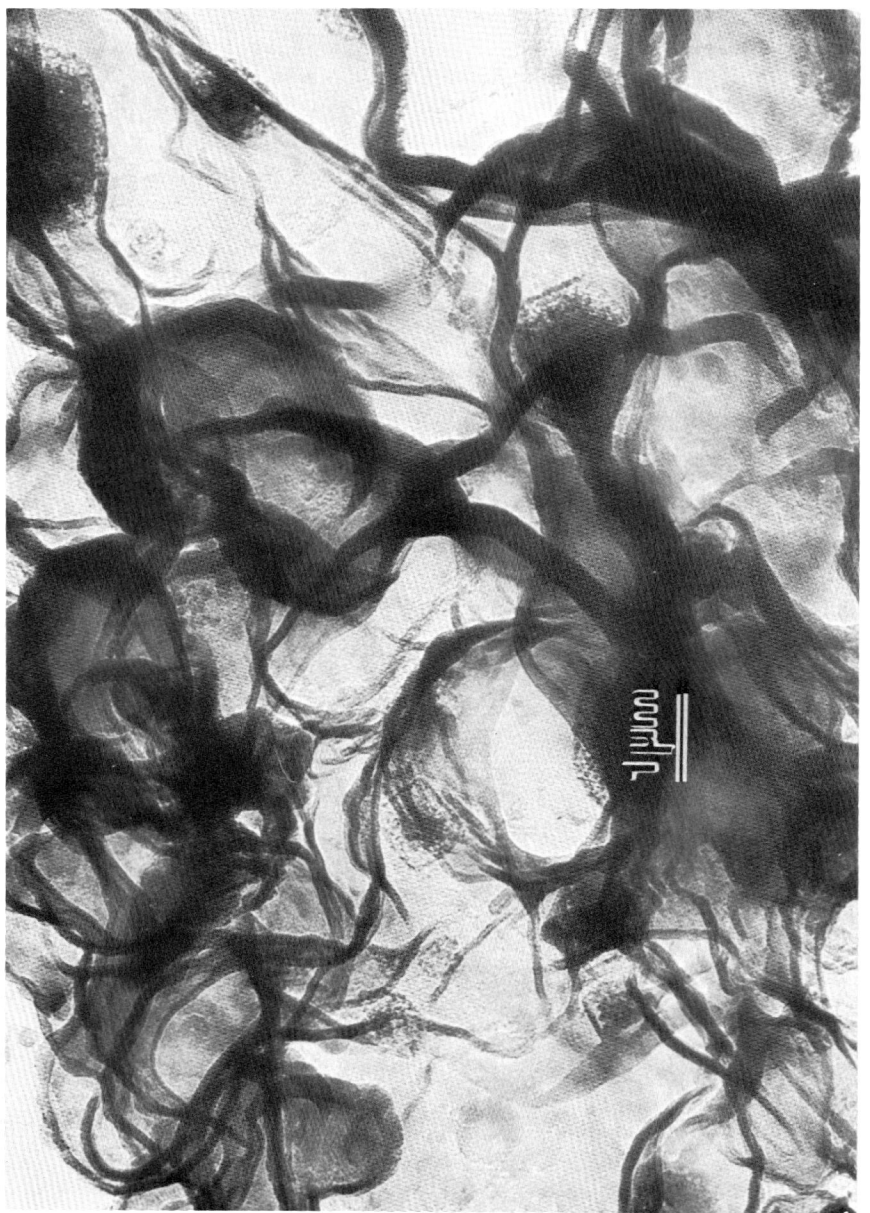

Figure 36 Electron micrograph taken from a microtome-sectioned gel-dried sample (PVC–benzyl alcohol, $C = 12.5\%$ w/w). This picture reveals features that are nearly identical to those observed by optical microscopy.

[78], poly-L-benzyl glutamate [79], and also in gels prepared from small molecules [80], which suggests a kind of "universal" morphology. In the case of polymers, it is usually thought that such a morphology results from spinodal decomposition [79]. Worth mentioning is a similar morphology observed in polyethylene gels prepared by stirring [81, 82].

Gels of polyvinyl chloride [43] or PDMS–COSP copolymer [83], when prepared by a quench into a miscibility gap, exhibit a similar structure. Interestingly, however, the mesh size is larger than in the other systems so that optical microscopy is powerful enough to see the network quite distinctly. Figures 33 and 34 are pictures of PVC gels formed at room temperature in benzyl alcohol ($\theta \simeq 150°C$ in this solvent) and in a blend of solvents (solvent + precipitant) [43]. Figure 35 portrays the morphology of a PDMS–COSP copolymer prepared in 1-phenyloctane [83]. Concerning PVC gels in benzyl alcohol, electron micrographs of ultramicrotome sectioned samples reveal the same features (Fig. 36). In solvents that are not known to be θ solvents for PVC, such as bromobenzene, Yang and Geil [34] found a similar morphology characterized, however, by a smaller mesh size (0.1 μm). The formation of fiberlike objects may accordingly not be restricted to systems formed below a miscibility gap. In the case of PVC, it may arise from the impossibility for this polymer to crystallize under the form of chain-folded crystals. The highest crystal stability is then achieved by predominantly unidirectional growth. As mentioned in the section on gel molecular structure, the fiberlike morphology may explain the particular intensity pattern given off by PVC pregels.

RHEOLOGY

The simplest way to investigate the gel mechanical behavior consists of determining the reduced stress on compression (often referred to as the compression modulus E). From theoretical considerations, which in particular consider the sample's incompressibility, it can be shown that in most cases the compression modulus reads

$$\sigma_R \quad \text{or} \quad E = \sigma/\left(\lambda - \left(\frac{1}{\lambda^2}\right)\right) \tag{3}$$

where σ is the true stress and λ the strain ($\lambda = l/l_0$ where l_0 is the initial sample's height and l its height after deformation).

Concerning chemically crosslinked gels, numerous theories have been developed which derive the variation of E as a function of different parameters such as the polymer concentration. In the latter case, scaling laws are expected [84]:

$$E \sim C^v \tag{4}$$

where the exponent v depends among other things on the chain statistics.

Other types of investigations can also be carried out such as the time-relaxation behavior once a displacement has been imposed onto the sample or the storage and loss moduli G' and G'' by means of a rheometer.

Numerous studies have been performed on "biological" gels. As mentioned in the introduction, Walter [7] seems to be the first researcher who systematically investigated the mechanical properties of a physical gel obtained from a synthetic polymer: polyvinyl chloride. The results were in accordance with a fringed micellar structure. In particular, Walter found a power law variation for the modulus as a function of concentration ($E \simeq$

C^3) and a variation of approximately kT as a function of temperature. This latter result is predicted for gaussian elastic networks.

Recent measurements by Mutin and Guenet [43] have shown that the value of 3 for the exponent is found in a large variety of solvents and especially in diethyl malonate. As in this solvent the structure is not thought to resemble a fringed micelle, this poses the query as to whether a power law variation is an unquestionable criterion to anticipate for the molecular structure. In addition, should the gel possess this structure, an exponent of 2 should be found.

The time-relaxation properties are slightly dependent on the solvent type. For instance, in a double logarithmic plot of σ_R vs. t, the slope $m = d \log \sigma_R/d \log t$ is slightly larger in diethyl malonate than in diethyl adipate (see Fig. 37) and also increases with longer times [43]. Such a discrepancy has been explained by Mutin and Guenet [43] as being the result of the presence of a larger amount of "weak" links in diethyl malonate than in diethyl adipate. These additional links account for the higher modulus in diethyl malonate than in diethyl adipate. Moreover, these weak links are more mobile, a property which may stem from the fact that they are solvated. The same authors have also shown that the appearance of the weak links is certainly responsible for the increase of modulus with time, something that was already suspected in plasticized PVC.

Concerning the PDMS–COSP multiblock copolymer, He et al. [48] have investigated the behavior of the compression modulus as a function of both concentration and temperature. For as-prepared gels, they have found that E varies with a power law:

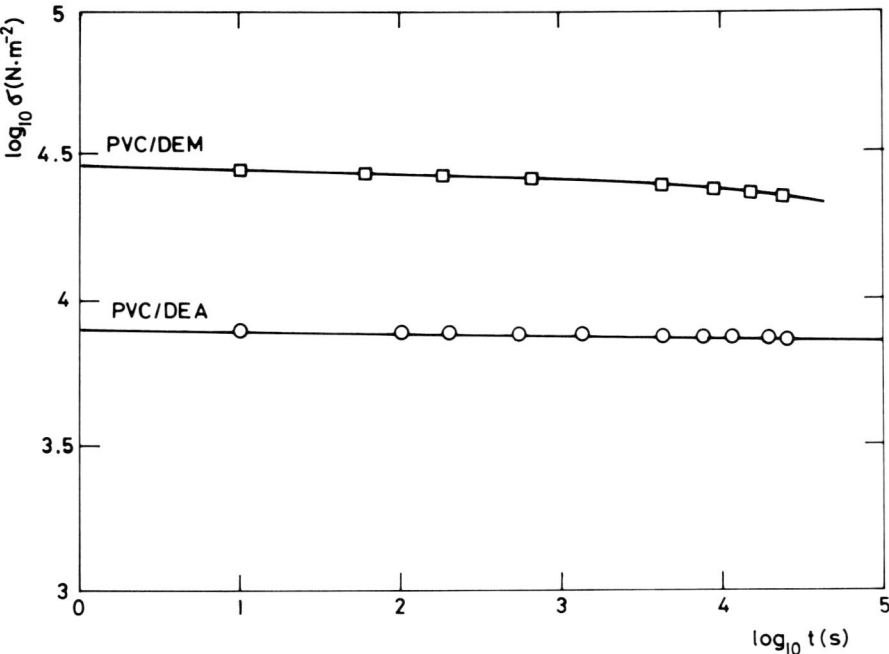

Figure 37 Plot of the stress relaxation at constant strain ($\lambda = 0.8$). $C_{PVC} = 17.5\%$ (w/w) in both solvents; □ = PVC–diethyl malonate; ○ = PVC–diethyl adipate.

$$E \simeq C^{4.5 \pm 0.2} \tag{5}$$

The value of this exponent is quite high and unpredicted by any theories.

The variation of E as a function of temperature is interesting since it can be phenomenologically accounted for through the knowledge of the temperature–concentration phase diagram. In view of the experimental results, He et al. [48] suggested a mechanism of gel partial melting occurring at T_M (see section on thermodynamics). In this respect, the modulus will depend on the amount of polymer-rich phase since this phase ensures the network character. As this amount is temperature dependent above T_M, the ratio $E(T)/E(20)$ where $E(T)$ = modulus at T and $E(20)$ = modulus at 20°C is given in first approximation by

$$\frac{E(T)}{E(20)} = \frac{C_{\text{prep}} - C_{M1}(T)}{C_{\alpha s}(T) - C_{M1}(T)} \times \frac{C_\alpha}{C_{\text{prep}}} \tag{6}$$

where C_{prep} is the gel concentration, C_α the solid solution concentration at T_M, and $C_{M1}(T)$ and $C_{\alpha s}(T)$ are the values of the liquid curve and of the solid curve as a function of temperature.

Through the use of Eq. 6 and the simplified phase diagram of Figure 38, He et al. [48] have been able to reproduce the variation of E with temperature (Fig. 39).

Worth mentioning also is that in time-relaxation experiments a slope of $m \simeq 0.01$–0.03 was found, a result comparable to the slope measured in PVC.

Gels of isotactic polystyrene possess rheological properties that contrast with those detailed above [85]. First of all, the time-relaxation behavior indicates considerable

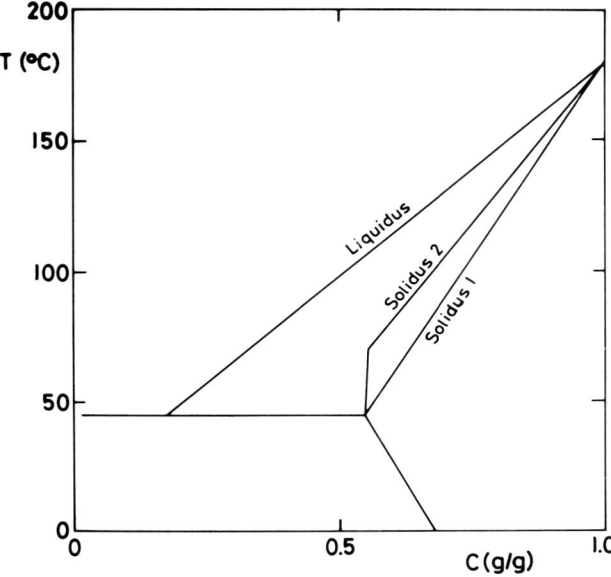

Figure 38 Simplified phase diagram for PDMS-COSP–*trans*-decalin gels after the phase diagram of Figure 18. Here two solid curves are considered since there is no way to determine them experimentally. (From Ref. 48.)

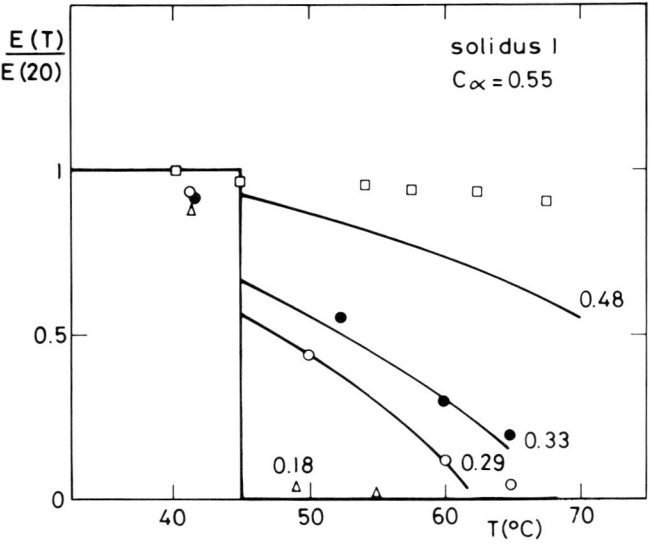

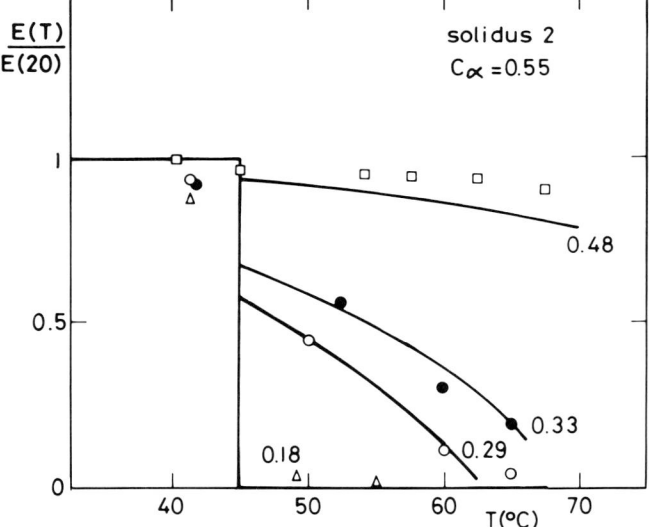

Figure 39 Theoretical variations (heavy lines) of the compression modulus as a function of temperature using Eq. 6 and experimental results (concentrations in w/w as indicated). The best agreement is achieved with solid curve 2. (From Ref. 48.)

mobility as ascertained by the value of m which is close to $m \simeq 0.1$ in *cis*-decalin and approaches $m \simeq 0.15$ in *trans*-decalin and 1-chlorodecane (see Fig. 40, for instance). One is no longer dealing with systems that can be regarded as having a solidlike behavior. In addition, it is tempting to associate the high value of m with the fact that these gels are considered noncrystalline since the "crystalline" gels (PVC and the PDMS–COSP multiblock copolymer) exhibit far lower m values.

The variation of the compression modulus (in fact the isochronal reduced stress at $t = 120$ sec) as a function of polymer concentration does not follow a simple power law [85]. Instead, a variation displaying "features" is seen. These are particularly marked in *cis*-decalin (Fig. 41). These features are less pronounced in *trans*-decalin and 1-chlorodecane [86], yet as can be seen in Figure 42 the behavior shows a strong departure from linearity in a double logarithmic scale. Worth noting is the almost constant value of E over a concentration domain ranging from $C \simeq 0.2$–0.35 g/g in *cis*-decalin. The origin of these features is currently unknown. Yet it is likely that both the gel morphology and the molecular structure (magnitude of the persistence length) depend on the polymer concentration, which implies that the description of the system by means of power laws proves improper.

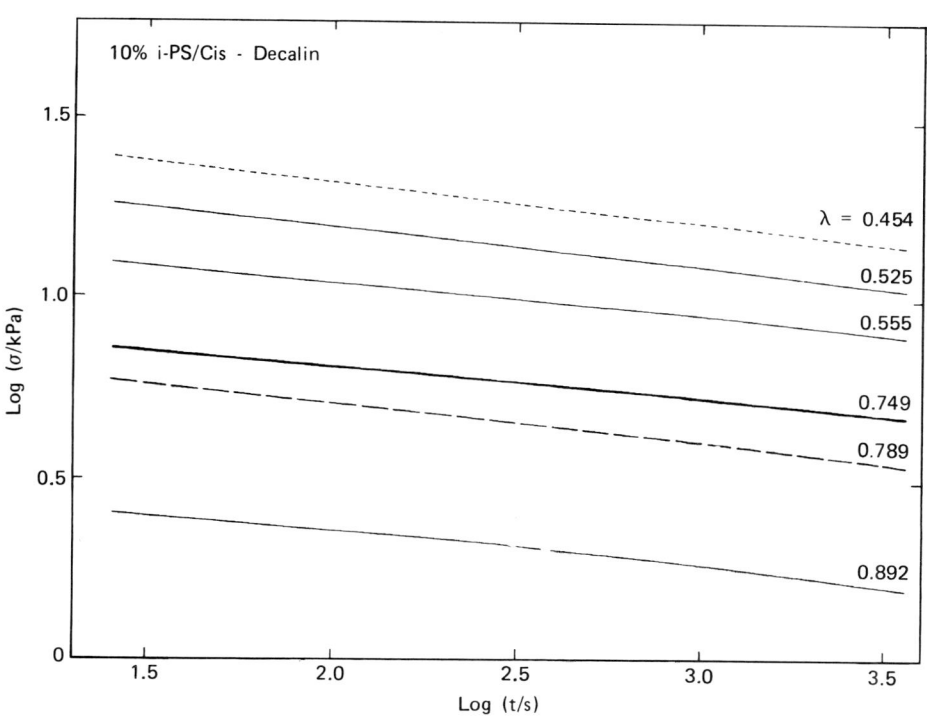

Figure 40. Compression stress relaxation responses for 10% iPS–*cis*-decalin gels at different deformations, as indicated. Differing lines refer to different molecular weights (short dashed line, MW = 5.24×10^5, MW/MN = 2; long dashed line, MW = 6.06×10^5, MW/MN = 2; light solid line and bold solid line correspond to MW = 1.44×10^6, the latter standing for an experiment performed with the sample immersed in the preparation solvent. (From Ref. 85.)

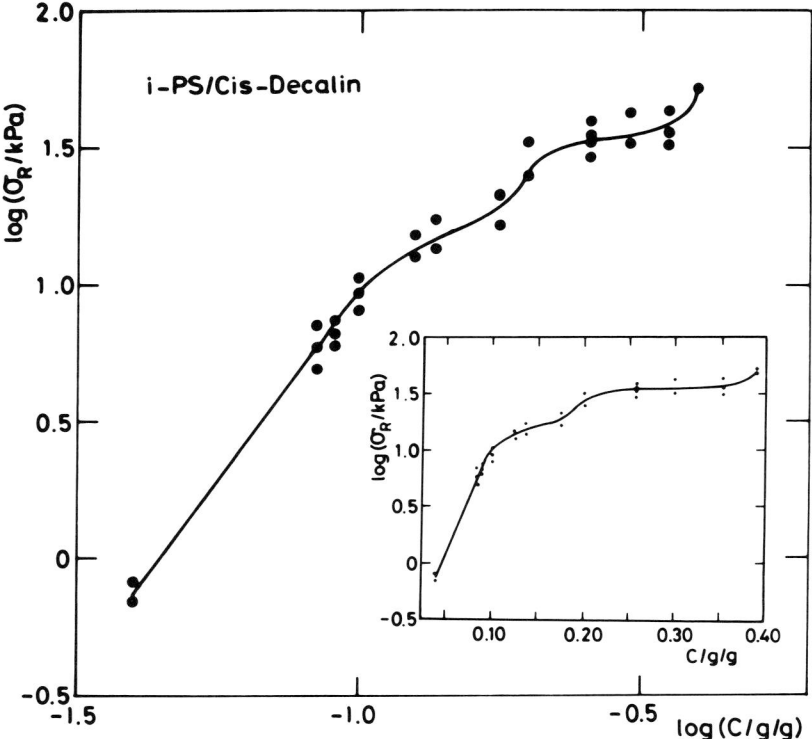

Figure 41 Variation of the compression modulus as a function of polymer concentration in iPS–cis-decalin gels (log–log representation). The inset, where the logarithm of the modulus is plotted as a function of concentration, emphasizes that the "features" are not due to the mode of representation. (From Ref. 85.)

McKenna and Guenet [87] have also noticed that the compression modulus is dependent on the quenching temperature. For instance, this parameter is three time larger for samples prepared at $+15°C$ than for those prepared at $-20°C$ or below. It is suspected that the persistence length may play an important role.

Berghmans et al. [26] recently reported on unusual behavior of syndiotactic polymethyl methacrylate in orthoxylene. By means of a rheometer, they determined the evolution of G' as a function of both time and temperature. The most striking result is certainly the apparent reversibility with temperature of the gelation process. Figure 43 portrays these evolutions. Upon cooling to $63°C$ the moduli start to grow after a short induction period. A decrease of temperature to $58°C$ accelerates the gel formation and an equilibrium is attained. By increasing the temperature again to $63°C$, the modulus retrieves the variation as a function of time it had when first quenched to that temperature. An additional change to $23°C$ and back to $63°C$ shows that there is virtually no hysteresis effect.

Concerning atactic polystyrene, only a few investigations have been carried out. From the studies of Clark et al. [36] by means of a classical rheometer, or that of Gan et al. [37]

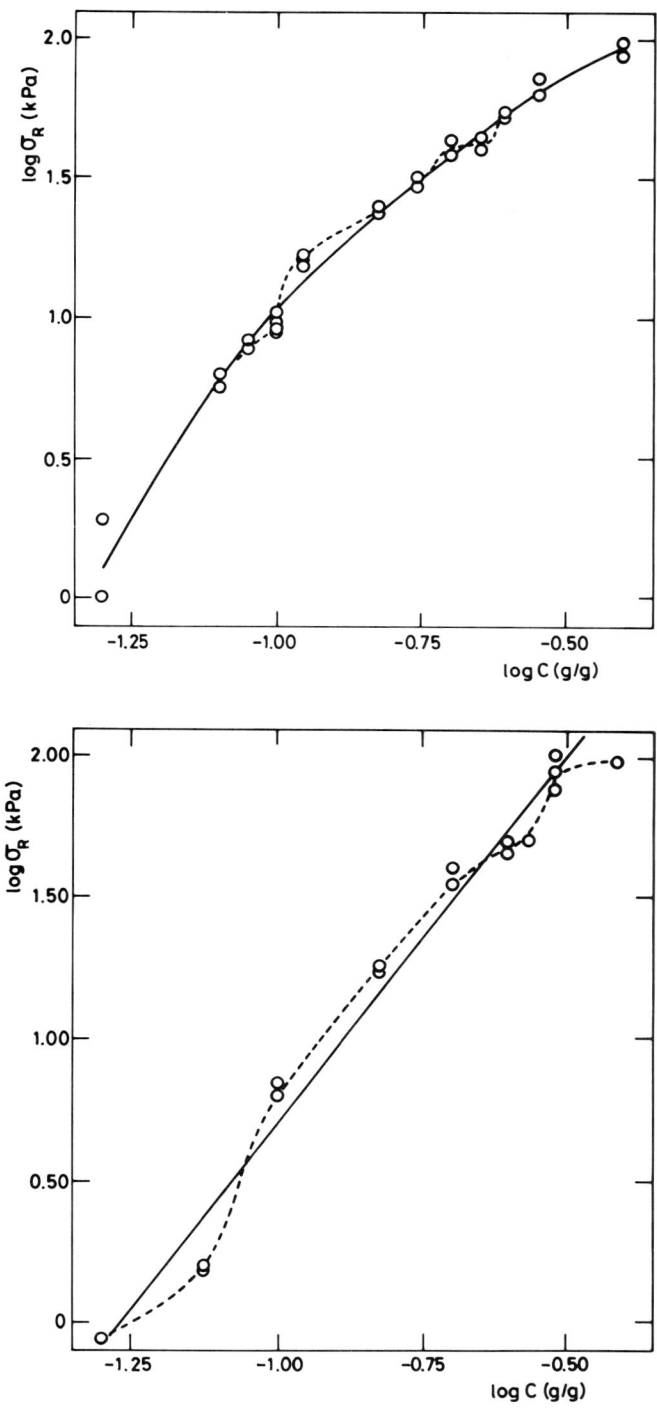

Figure 42 Variations of the compression modulus as a function of concentration for (a) iPS–trans-decalin gels in a double logarithmic scale and (b) iPS/1-chlorodecane gels. (From Ref. 84.)

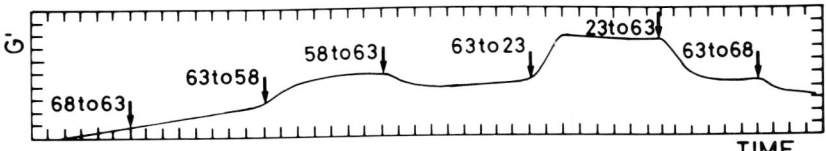

Figure 43 Variation as a function of time and of the thermal treatment of G' in s-PMMA–orthoxylene gels. (Data from Ref. 26.)

with a sphere rheometer, it can be concluded that aPS networks in CS_2 are weak gels. Moduli of about 2000–6000 dynes cm^{-2} are measured. Once the functionality has been derived from the classical rubber elasticity, these moduli give approximately two to three trifunctional crosslinks per chain.

Finally, although the measurements were not carried out on the gels but on the fibers spun from the gels, polyethylene and polypropylene are noteworthy. As emphasized in the introduction, polyethylene fibers reach Young's modulus of about 100 GPa [9]. For polypropylene, Peguy and Manley [88] obtained 36 GPa, a value close to the theoretical one ($\simeq 41$ GPa).

SWELLING BEHAVIOR

While the swelling properties constitute an important and well-documented area of covalent gels, only a small number of similar investigations have been achieved on physical gels. Here, two systems which are familiar to the author are described.

An extensive study of the swelling behavior of the PDMS–COSP multiblock copolymer was carried out by He et al. [48]. Since the PDMS block is amorphous at room temperature, it is expected that the gels swell once immersed in an excess of preparation solvent. Such is the case (here copolymer + *trans*-decalin). In Figure 44 the values of the equilibrium swelling ratio G^∞ are plotted as a function of polymer concentration C_{prep} [$G^\infty = (P/P_o)t \to \infty$ where P_o and P are the original gel sample weight and the gel sample weight after swelling respectively]. For a covalent gel, the C^* theorem [89] states that the system should swell up to the overlap concentration C^* of the chains linking the knots. As a result, G^∞ reads

$$G^\infty = \frac{C_{\text{prep}}}{C^*} \tag{7}$$

This relation ignores the network defects such as trapped entanglements, yet it has been shown to pertain in covalent gels in a large range of concentration [90]. Should the physical gels be characterized by a fringed micellar structure such as that seen in Figure 27, one would expect them to behave as or nearly as a covalent gel. The slope of Eq. 7 should be close to unity (taking $C^* \simeq 0.04$ g cm^{-3} [91] for the PDMS sequence of MW 17,000), which is not the case since 0.16 is found instead. According to He et al. [48] such a discrepancy is too large to be attributed to defects only. Another structure ought to be considered. As detailed in the previous sections, gel morphology is rather of fiberlike nature and the model seen in Figure 45 may be appropriate to describe it. Basing their analysis on this model, He et al. [48] consider that the amorphous part of the polymer-rich phase, which by virtue of the preparation method always has the same concentration, should as a result always swell to the same concentration. Accordingly, the swelling ratio

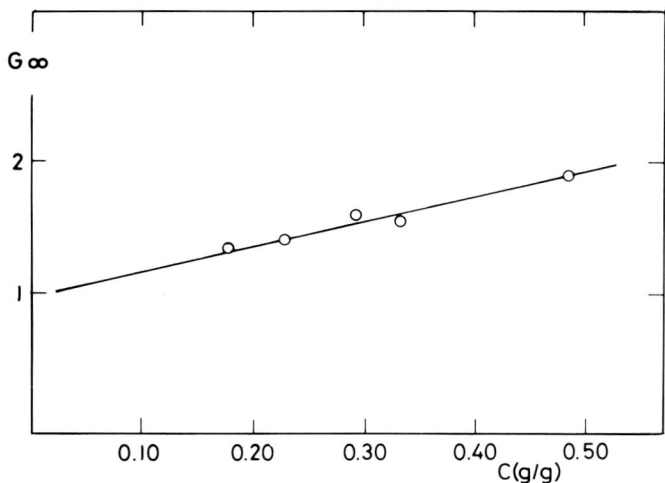

Figure 44 Variation of the equilibrium swelling into G^∞ as a function of the preparation concentration C_{prep} for PDMS-COSP–*trans*-decalin gels. (From Ref. 48.)

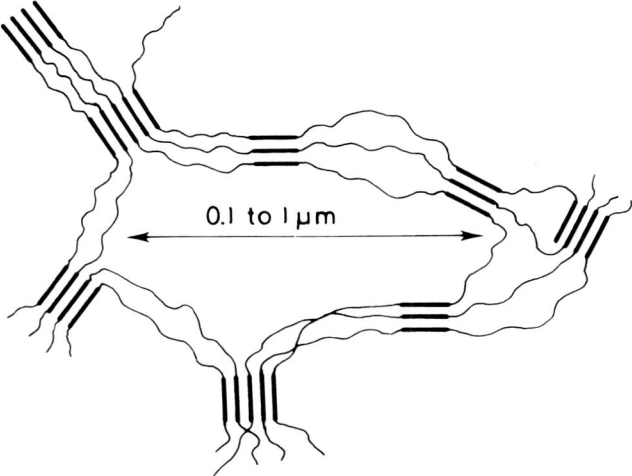

Figure 45 Fiberlike model. Here, unlike the fringed micelle model, the characteristic distance between effective knots is about 0.1–1 µm.

is proportional to the amount of polymer-rich phase, X_α, and can be straightforwardly derived:

$$G^\infty = 1 + \left(\frac{C_\alpha}{C_\gamma} - 1\right) \times X_\alpha \tag{8}$$

where C_α and C_γ are the solid solution concentration and the concentration reached by the polymer-rich phase after swelling respectively (X_α can be approximated to $X_\alpha \simeq C_{prep}/C_\alpha$).

In Figure 46 the experimental results are replotted as a function of C_{prep}/C_α. From the slope He et al. [48] found $C_\gamma \simeq 0.27$, a value which seems reasonable. The model of Figure 45 (fiberlike model) seems to be a more appropriate type of model than the fringed micellar one to account for the swelling behavior.

The swelling behavior of polyvinyl chloride is linked to the solvent type as well as to the chain microstructure (more or less stereoregular samples).

Mutin and Guenet [43] have studied the swelling behavior as a function of time in conjunction with the gel compression modulus. Their results are drawn in Figure 47, which contains the information on the modulus, the evolution of gel concentration with time, the solvent type, and the PVC type (for the sake of simplicity PVC + 50 and PVC-40 are used to designate the highly atactic sample and the more syndiotactic sample respectively). In principle, all the solvents used exhibit similar quality toward PVC. In some solvents such as diethyl adipate (DEA) or bromonaphthalene (BN) the swelling is quite important while the modulus remains virtually constant. In other solvents such as diethyl succinate (DES) or diethyl oxalate (DEO) the gel first swells, then slightly deswells, the net result being a small swelling. Finally, in diethyl malonate (DEM) the gels slightly deswell while those prepared from PVC-40 deswell considerably in bromonaphthalene. Meanwhile the modulus increases, which indicates that additional links are being formed. The similar behavior obtained in DEM for the PVC + 50 and in BN for the PVC-40 suggests that the solvent may play a role in the formation of these links, which, incidentally, are thought to be the weak links seen in light-scattering experiments. The exact effect of the solvent is not known. Either it forms solvated structures with the less regular chain portions, which facilitates their "crystallization," or it promotes the formation of "anhydrous" structure by solvating less particular monomer arrangements along the chain.

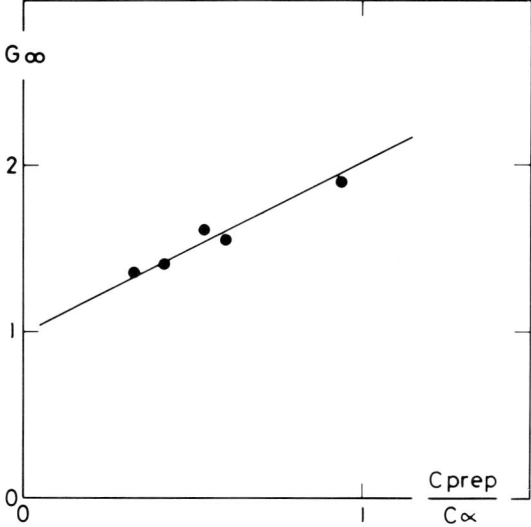

Figure 46 Plot of the equilibrium swelling ratio G^∞ as a function of C_{prep}/C_α. (From Ref. 48.)

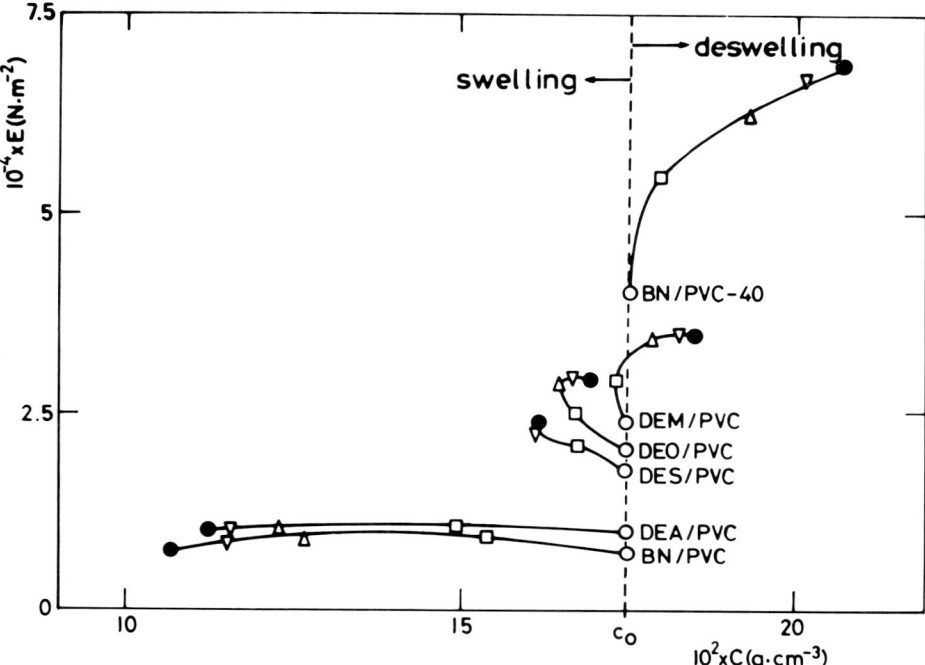

Figure 47 Variation of the compression modulus as a function of both the concentration and aging time for gels of PVC in various solvents. Original concentration in all the gels $C_0 = 17.5\%$ w/w. The samples were aged in an excess of preparation solvent. PVC + 50 stands for PVC synthesized at +50°C while PVC-40 is for PVC synthesized at −40°C. The PVC-40 possesses a higher syndiotactic tryad content. Solvents as indicated (BN, bromonaphthalene; DEA, diethyl adipate; DES, diethyl succinate; DEO, diethyl oxalate; DEM, diethyl malonate). ○, 1 hr aging; □, 1 day aging; △, 7 days aging; ▽, 14 days aging; and ●, 21 days aging.

CONCLUDING REMARKS

It is hoped that this review of current knowledge in the area of thermoreversible gels will assist both the scientist, whose interest in the domain is steadily growing, and the layman who simply desires to be informed of general principles and current developments.

It is quite possible that, after reading this review, the fictional sleuth Sherlock Holmes would have said: "You have erred perhaps in attempting to put color and life into each of your statements instead of confining yourself to the task of placing upon record that severe reasoning from cause to effect which is really the only notable feature about the thing."

Yet it is the hope of the author that color and life (and I hope this chapter contains color and life) have helped emphasize the "hot" problems in this domain. Particularly, the molecular structure in PVC, iPS, and aPS deserves further investigation. Similarly, the role of the solvent, which has been all too often regarded as a mere diluent, must also be studied. Another challenging task seems to be the development of theories that can account for the existing experimental results as well as predict further types of behavior.

Evidently, as thermoreversible gels and covalent gels are poles apart, especially when the morphology and structure are considered, theories proposed for the latter are most of the time inappropriate for the former.

Finally, what are the future trends, and more precisely the applications, in this field? Two of them are currently known: (a) polyethylene fibers of high Young's modulus and (b) PVC fibers (Rhône–Poulenc), which are spun from a PVC gel (PVC/ketone/CS_2). Since physical gels tend to possess a morphology which is not obtained under other conditions, they may be the starting point for making materials with unique properties.

REFERENCES

1. See, for instance, P. H. von Hippel, in *Treatise on Collagen* (G. N. Ramachandran, ed.), Academic Press, New York, chap. 6 (1967); A. G. Ward and A. Courts, eds., *The Science and Technology of Gelatin*, Academic Press, New York (1977).
2. See, for instance, A. Frey-Wyssling and K. Mühlethaler, *Vierteljahrsschr. Naturforsch. Ges. Zürich*, 89: 214 (1944).
3. See, for instance, P. Terech, R. Ramasseul, and F. Volino, *J. Colloid Int. Sci.*, 91: 280 (1983).
4. D. J. Lloyd, in *Colloid Chemistry*, Vol. 1 (J. Alexander, ed.), New York (1926).
5. von Nägeli, in *Pflanzenphysiologische Untersuchungen*, Zürich (1858).
6. W. B. Hardy, *J. Physiol.*, 24:288 (1899); *Proc. Roy. Soc.*, 66: 95 (1900).
7. A. T. Walter, *J. Polym. Sci.*, 13: 207 (1954).
8. M. Girolamo, A. Keller, K. Miyasaka, and N. Overbergh, *J. Polym. Sci. Polym. Phys. Ed.*, 14: 39 (1976).
9. P. J. Lemstra and P. Smith, *J. Mater. Sci.*, 15: 505 (1980).
10. H. M. Tan, A. Hiltner, H. Moet, and E. Baer, *Macromolecules*, 16: 28 (1983).
11. P. J. Lemstra and G. Challa, *J. Polym. Sci. Polym. Phys. Ed.*, 13: 1809 (1975).
12. G. C. Berry, *J. Chem. Phys.*, 44: 4450 (1966).
13. S. J. Wellinghoff, J. Shaw, and H. Baer, *Macromolecules*, 12: 032 (1979).
14. J. W. Cahn, *J. Am. Ceram. Soc.*, 52: 118 (1969).
15. J. M. Guenet and G. B. McKenna, to appear in *Macromolecules* (1988).
16. P. R. Sundararajan, N. J. Tyrer, and T. L. Bluhm, *Macromolecules*, 15: 286 (1982).
17. K. A. Nahr, P. J. Barham, and A. Keller, *Macromolecules*, 15: 464 (1982).
18. G. Charlet, H. Phuong-Nguyen, and G. Delmas, *Macromolecules*, 17: 1200 (1984).
19. H. Phuong-Nguyen and G. Delmas, *Macromolecules*, 18: 1235 (1985).
20. K. Koennecke and G. Rehage, *Colloid Polym. Sci.*, 259: 1062 (1981).
21. J. Spevacek, *J. Polym. Sci. Polym. Phys. Ed.*, 16: 523 (1978).
22. J. Spevacek, B. Schneider, M. Bohdanecky, and A. Sikora, *J. Polym. Sci. Polym. Phys. Ed.*, 20: 1623 (1982).
23. J. Spevacek and B. Schneider, *Makromol. Chem.*, 176: 3409 (1975).
24. J. Spevacek and B. Schneider, *Polymer*, 19: 63 (1978).
25. J. Spevacek et al., *J. Polym. Sci. Polym. Phys. Ed.*, 22: 617 (1984).
26. H. Berghmans et al., *Polymer*, 28: 97 (1987).
27. W. Aiken, T. Alfrey, Jr., A. Janssen, and H. Mark, *J. Polym. Sci.*, 2: 178 (1947).
28. T. Alfrey, Jr., N. Wiederhorn, R. S. Stein, and A. Tobolsky, *Ind. Eng. Chem.*, 41: 701 (1949).
29. M. A. Harrison, P. H. Morgan, and G. S. Park, *Eur. Polym. J.*, 8: 1361 (1972).
30. S. A. Leharne and G. S. Park, *Eur. Polym. J.*, 21: 383 (1985).
31. A. Takahashi, T. Nakamura, and I. Kagawa, *Polym. J.*, 3: 207 (1972).
32. H. C. Haas and R. L. McDonald, *J. Polym. Sci. Polym. Chem. Ed.*, 11: 1133 (1973).
33. P. J. Lemstra, A. Keller, and M. Cudby, *J. Polym. Sci. Polym. Phys. Ed.*, 16: 1507 (1978).

34. Y. C. Yang and P. H. Geil, *J. Macromol. Sci., B22*(3): 463 (1983).
35. A. Takahashi and S. Hiramitsu, *Polym. J., 6*: 103 (1974).
36. J. Clark, S. T. Wellinghoff, and W. G. Miller, *Am. Chem. Soc. Div. Polym. Chem. Polym. Prep., 24*(2): 86 (1983).
37. Y. S. Gan, J. François, J. M. Guenet, B. Gauthier-Manuel, and C. Allain, *Makromol. Chem. Rapid Commun., 6*: 225 (1985).
38. Y. S. Gan, J. François, and J. M. Guenet, *Macromolecules, 19*: 173 (1986).
39. J. François, Y. S. Gan, and J. M. Guenet, *Macromolecules, 19*: 2755 (1986).
40. J. M. Guenet, B. Lotz, and J. C. Wittmann, *Macromolecules, 18*: 420 (1985).
41. A. Reisman, in *Phase Equilibria,* Academic Press, New York (1970).
42. J. M. Guenet, *Macromolecules, 19*: 1961 (1986).
43. P. H. Mutin, thesis, Strasbourg (1986); P. H. Mutin and J. M. Guenet (to be submitted).
44. J. A. Juijn, A. Gisolf, and W. A. de Jong, *Kolloid-Z.Z. Polym., 194*: 5 (1973).
45. A. Dorrestijn, P. J. Lemstra, and H. Berghmans, *Polym. Commun., 24*: 226 (1983).
46. Y. S. Gan, thesis, Strasbourg (1986).
47. C. Prud'homme, thesis, Strasbourg (1980); X. He et al., *Polym. Bull., 17*: 45 (1987).
48. X. He, J. Herz, and J. M. Guenet, *Macromolecules, 20*: 2003 (1987).
49. A. Takahashi, *Polym. J., 4*: 379 (1973).
50. H. Berghmans, F. Govaerts, and N. Overbergh, *J. Polym. Sci. Polym. Phys. Ed., 17*: 1251 (1979).
51. R. C. Domszy, R. Alamo, C. O. Edwards, and L. Mandelkern, *Macromolecules, 19*: 310 (1986).
52. E. D. T. Atkins et al., *J. Polym. Sci. Polym. Phys. Ed., 25*: 211 (1977).
53. E. D. T. Atkins et al., *J. Polym. Sci. Polym. Phys. Ed., 18*: 71 (1980).
54. E. D. T. Atkins et al., *Polymer, 22*: 1161 (1981).
55. S. M. Aharoni, G. Charlet, and G. Delmas, *Macromolecules, 14*: 1390 (1981).
56. See, for instance, A. Guinier, in *Théorie et technique de la radiocristallographie,* Dunod, Paris (1955).
57. J. M. Guenet, *Phys. Rev. Lett., 58*: 1532 (1987).
58. J. M. Guenet, *Macromolecules, 20*: 2874 (1987).
59. J. M. Guenet, C. Picot, and H. Benoit, *Macromolecules, 12*: 86 (1979).
60. J. M. Guenet, *Macromolecules, 13*: 387 (1980).
61. B. Millaud, A. Thierry, and A. Skoulios, *J. Phys.* (*Paris*), *39*: 1109 (1978).
62. B. Millaud and C. Strazielle, *Polymer, 20*: 563 (1979).
63. J. M. Guenet, N. F. F. Willmott, and P. A. Ellsmore, *Polym. Commun., 24*: 230 (1983).
64. A. J. Hyde and R. B. Taylor, *Makromol. Chem., 62*: 204 (1963).
65. A. A. Tager, V. M. Andreeva, and E. M. Evsina, *Vysokomol. Soedin., 6*: 1901 (1964).
66. H. Benoit and C. Picot, *Pure Appl. Chem., 12*: 545 (1966).
67. H. Benoit and M. Benmouna, *Polymer, 25*: 1059 (1984).
68. S. F. Edwards, *Proc. Phys. Soc. London, 88*: 265 (1966).
69. T. L. Yu and A. M. Jamieson, *Macromolecules, 13*: 1590 (1980).
70. B. L. Hager, G. C. Berry, and H. H. Tsai, *J. Polym. Sci. Polym. Phys. Ed., 25*: 387 (1987).
71. J. T. Koberstein, C. Picot, and H. Benoit, *Polymer, 26*: 641 (1985).
72. M. Daoud et al., *Macromolecules, 8*: 804 (1975).
73. P. H. Mutin and J. M. Guenet, *Polymer, 27*: 1098 (1986).
74. S. J. Candau et al., *Polymer, 28*: 1334 (1987).
75. P. H. Mutin et al., *Polymer, 29*: 31 (1988).
76. Kusuyama et al., *Polym. Commun., 14*: 119 (1983).
77. J. H. Aubert and R. L. Clough, *Polymer 26*: 2047 (1985).
78. J. E. Stamhuis and A. J. Pennings, *Br. Polym. J., 10*: 221 (1978).
79. K. Tohyama and W. G. Miller, *Nature London, 289*: 813 (1981).
80. See, for instance, R. H. Wade et al., *J. Colloid Int. Sci., 114*: 442 (1986).

81. A. R. Postema, W. Hoogsteen, A. J. Pennings, *Polym. Commun.* 28: 148 (1987).
82. D. J. Dijkstra and A. J. Pennings, *Polym. Bull.*, 17: 507 (1987).
83. X. He, thesis, Strasbourg (1986).
84. P. G. de Gennes, in *Scaling Concepts in Polymer Physics*, Cornell University Press, Ithaca, N.Y. (1979).
85. J. M. Guenet and G. B. McKenna, *J. Polym. Sci. Polym. Phys. Ed.*, 24: 2499 (1986).
86. G. B. McKenna and J. M. Guenet, *J. Polym. Sci. Polym. Phys. Ed.*, 26: 267 (1988).
87. G. B. McKenna and J. M. Guenet, to appear in *Polym. Commun.* (1988).
88. A. Peguy and R. St. John Manley, *Polym. Commun.*, 25: 39 (1984).
89. P. G. de Gennes, *Macromolecules*, 9: 587 (1976).
90. J. P. Munch et al., *J. Phys. (Paris)*, 58: 971 (1977).
91. M. Beltzung, J. Herz, and C. Picot, *Macromolecules*, 16: 2 (1983).

Index

A

AB type monomers, 183
ABA type block copolymers, 143, 153, 154
Absorbance, 497
Absorption intensity, 497
Absorption of electromagnetic radiation, 496
Acenapthylene, 288
Acid chloride route, 178
Acrylates, 3
Acrylic acid, 288
Acrylic polyethylene, 539
Acrylonitrile, 3, 267
Acrylonitrile butadiene styrene, 539
Activation energy, 252, 253, 648, 652, 663, 665
Activation enthalpy, 663
Activation equilibrium, 50
Activation volume, 662, 673
Addition law to the relaxation spectrum, 629
Addition polymerization, 67, 424
Addition reaction of methyl radical, 252
Adherent films, 285
Adhesives, 76, 169
Agglomeration of primary PVC particles, 356
Aging, 726
Aldehydes, 288
Alignment of molecular chain axes, 520

Aliphatic polyester, 376
Alkali metal, 16
Alkene polymerization, 111, 115, 124
Alkene polymers, 119
Alkyl groups, 115
Alkyllithium initiators, 15
Aluminizing, 547
Aluminum, 651
Aluminum alkoxides, 113
Amidation reactions, 151
Amines, 15
Amino aromatic compounds, 284
Amino derivatives, 329
Ammonium laurate, 362
Ammonium salts, 360
Amorphous ethylene propylene elastomers, 129
Amorphous polymers, 125, 540, 667
Amorphous regions, 657
Anion exchange resin, 478
Anionic block copolymerization, 153, 155
Anionic chain copolymerizations, 25
Anionic chain polymerizations, 15
Anionic initiators, 2
Anionic polymerization, 14, 16, 153
Anionic polymerization of epoxides, 20
Anionic SBR, 700, 703
Anisotropic crystalline materials, 542
Anisotropic light, 523

Anisotropic transmission properties, 530
Annealing, 578, 584, 653
Anodic peak potentials of monomers, 274
Anodic voltammograms, 276
Appearance of plastics, 538
ARB polycondensation, 243
ARB polymerization, 379
Aromatic additives, 259
Aromatic dicarboxylic acids, 183
Aromatic hydrocarbons, 106, 259
Aromatic monomers, 13
Aromatic polyarylates, 177
Aromatic polyesters, 95, 177, 179
Aromatic polyphosphonates, 191
Aromatic polysulfonates, 185
Aromatic rings, 193
Arrhenius equation, 649, 659, 670
ASTM methods, 540
ASTM terminology, 540
Atactic polymers, 713
Atactic polyolefins, 119
Atactic polystyrene, 754
Atomic absorption, 471, 472
Atomic absorption spectrometer, 472
Atomic absorption spectroscopy, 471
Atomic composition, 484
Autocorrelation function, 357
Automotive applications, 169
Average chain length, 223
Average molecular mass, 607
Average molecular weight in radical polymerization, 408
Average number of branches, 396
Averaging of viscosities, 628
Avogadro's number, 241, 366
Azo catalysts, 146
Azoacyl peroxide, 148
Azobiscyano pentanoic acid, 142
Azobiscyanovaleric acid, 152
Azooxocarbenium initiator, 164
Azoperoxidic compounds, 136, 147
Azoperoxidic initiators, 147

B

Backbone bonds, 240
Backbone structure of polystyrene derivatives, 321
Backmixing, 231
Batch chain polymerization, 444
Batch isothermal chain addition polymerization, 446
Batch polycarbonate polymerization process, 93
Batch polymerization, 71, 89
Batch polymerization reactors, 71
Batch reactors, 76, 214, 413
Benzene, 284
Bertrand lens, 516
Biaxial indicatrices, 509, 522
Biaxial ordering, 523
Biaxial polymeric systems, 511
Biaxial properties of polymer films, 511
Biaxially deformed polymers, 531
Bifunctional reactant, 34
Bifunctional step growth polymerization, 391
Bimodal distribution, 451
Bimodal particle size distributions, 350
Bimolecular ionization, 11
Bimolecular rate constants, 252
Bimolecular reaction, 37
Bimolecular termination, 5, 9
Binary copolymerization, 26
Birefringence, 513, 519, 520, 541, 578, 583
Birefringence measurement, 581
Birefringent polymer liquid crystals, 519
Birefringent properties, 583
Birefringent strain pattern, 583
Block copolymerization, 136
Block copolyamides, 151
Block copolymer, 85, 147
Block copolymer formation, 165
Block copolymer micrographs, 166
Block copolymer morphology, 167
Block copolymer structures, 134
Block copolymer synthesis, 156
Block copolymerization matrix, 137
Block copolymerization of lactams, 155
Block copolymerization via the anionic mechanism, 153
Block copolymerization via the step growth procedure, 148
Block copolymers, 133–160
Block copolymers by cationic polymerization, 156

Block copolymers by coupling reactions, 159
Block copolymers of propylene, 115
Block polymers, 133
Block styrene sequence, 682, 683, 689, 690, 694
Boltzmann constant, 36, 240
Bolza problem, 436
Bond angle restrictions, 241
Bond dissociation, 403
Bond dissociation energy, 310
Bond fission, 310
Branch point, 385
Branched alkyl group, 114
Branched polymers, 636
Branching, 391, 401, 606, 620
Branching index, 628
Breadth of the MWD, 442, 451
Brittle fracture, 586
Bueche theory, 614
Bulk density of the resin, 359
Bulk phase polymerization, 85
Bulk polymerization, 70, 96, 254
Bulk styrene polymerization, 71
Butadiene, 115, 681, 699, 705
Butadiene polymerization, 117
Butadiene units, 700
Butyl lithium, 153

C

Cable insulation, 349
Calorimetric measurement, 357
Carbenium ion, 158
Carbon black, 567
Carbon tetrachloride, 330
Carbonium ion, 12, 14, 22
Carboxyl terminated polymers, 152
Carboxylic acids, 283
Catalyst deashing, 80
Catalyst decay, 30
Catalyst for ethylene polymerization, 108
Catalyst for propylene polymerization, 108
Cation, 2
Cation donor, 10
Cation exchange resin, 479
Cationic copolymerization, 25, 274

Cationic electroinitiation of styrene, 283
Cationic initiators, 11, 249
Cationic polymerization, 9, 10, 12, 156, 257, 289
Cationic polymerization of styrene, 273, 282
Cationic polymerization of trioxane, 285
Cationogen, 10
Centrifugal forces, 482
Chain copolymerization, 23
Chain extension in polymer melts, 528
Chain fold surfaces, 525
Chain folded crystals, 718, 741
Chain growth polymerization, 213, 375, 376
Chain initiating species, 3
Chain length, 381
Chain polymerization, 1, 4, 437
Chain reactions, 431
Chain statistics, 240
Chain stiffness, 185
Chain structure of various polyethylenes, 78
Chain transfer, 12, 156, 165
Chain transfer agent, 83, 353
Chain transfer coefficient, 355
Chain transfer reaction, 136, 254, 410
Chain transfer to monomer, 354
Chain transfer to organic halides, 158
Characteristic of polyarylates, 185
Characteristic relaxation time, 637
Characterization of block copolymers, 133, 165
Characterize branching, 620
Characterizing molecular mass, 609
Charge transfer, 259
Charged species, 650
Chemical crosslinks, 169
Chemical composition distribution, 695
Chemical composition distribution of SBR, 678, 707
Chemical degradation, 711
Chlorinated hydrocarbons, 179
Chlorine cleavage, 330
Chlorine radicals, 336
Cholesteric liquid crystalline polymer, 515
Chromatography, 473, 479
Chromic acid, 545

Circularly polarized light, 523
Classification of chromatographic technique, 473
Classification of Ziegler Natta catalyst, 105
Closed loop analysis, 461
Coates graph, 56
Coatings, 169
Cold crystallization, 493, 495
Cole-Cole distribution, 670
Collision frequency, 377
Collision rate, 36
Colloidal condition, 711
Coloration, 308, 312
Commercial development of polyarylates, 204
Commercial process utilizing Ziegler Natta catalysts, 126
Commercial SBR, 677
Composition distribution of SBR, 679
Compression modulus, 749, 753
Compression stress relaxation responses, 753
Compressive stress, 583
Concept of noninferiority, 439
Condensation reaction, 139
Condensing reagents, 184
Configurational sequences of styrene units, 699
Conjugated dienes, 115
Conoscopic image, 516
Constant current electrolysis, 281
Constant potential electrolysis, 281, 282
Continuous flow stirred tank reactors, 221
Continuous irradiation, 311
Continuous melt polycarbonate polymerization process, 94
Continuous polymerization, 76
Continuous polymerization reactors, 99
Continuous reactors, 73
Continuous stirred bed reactor, 83, 87
Continuous stirred tank reactors, 81, 214, 432
Contour plots for tubular reactors, 224
Contrast, 589
Control of polymerization reactors, 98
Control the molecular weight, 214
Conveying, 69

Cooling cycles, 711
Copolymer, 586, 592, 681, 729
Copolymer composition, 99, 298, 683
Copolymer sequence distribution, 70
Copolymerization, 23, 113, 257, 265, 538, 683
Copolymerization equation, 25
Copolymerization of acrylonitrile, 680
Copolymerization of ethylene, 250
Copolymerization of ethylene and propylene, 129
Copolymerization of monomers, 27
Copolymerization of vinyl ethers, 257
Copolymerization reactors, 454
Copolymers, 678, 707
Copolymers of ethylene, 115
Corotating extruder, 95
COSP polymer, 749
Cotacticity, 702
Coupling reactions, 165
Coupling reactions of polymer anion, 160
Coupling reactions of polymer cations, 164
Covalent bond formation, 9
Critical angles, 542
Critical polymerization potential, 300
Cross propagation, 24
Crosslinked acrylonitrile, 679
Crosslinked polymers, 540
Crosslinked thermoset materials, 540
Crosslinking, 135, 308, 430, 496, 538
Crosslinking formation, 335, 336
Crosslinking reaction, 312, 317
Cryogenic microtomy, 538
Crystal melting temperature, 716
Crystal morphology, 567
Crystalline alkene polymers, 119
Crystalline cured EPDM rubbers, 592
Crystalline isotactic polymer, 125
Crystalline melting points, 495
Crystalline polyarylates, 185, 199
Crystalline polymers, 119, 125
Crystalline polyolefins, 119
Crystalline regions, 667
Crystallinity, 556, 581
Crystallization, 352, 488, 492, 493, 712, 714, 734, 740, 741, 758
Crystallizing, 718

Cured rubbers, 704
Cyclic amides, 153
Cyclic dimers, 221, 240
Cyclic esters, 153
Cyclic ethers, 19
Cyclic monomers, 19, 156, 211
Cyclic oligomers, 214, 240
Cyclic oxides, 153
Cyclic polymerization of divinyl acetals, 262
Cyclic polymerization of divinyl ethers, 260
Cyclic voltammetry, 276, 278, 279
Cyclic voltammogram of isoprene, 291
Cyclization reactions, 242
Cyclization reactions in nylon-6 polymerization, 240
Cyclohexane, 166
Cyclohexylene dinitrilotetraacetic acid, 481

D

Debye equation, 645
Decomposition of this polyazo initiator, 146
Deformation history, 567
Deformation of macromolecules, 611
Deformation rates, 611
Degradation, 308
Degree of branching, 430
Degree of chain alignment, 528
Degree of solvation, 725
Dehydrochlorination process, 349, 357
Dehydrogenation, 485
Depolarization, 643, 646, 648, 652
Depolarization current, 649
Depolarization kinetics, 670
Determinations, 167
Devolatilizing extruder, 92
Dialkyl fumarates, 253
Diamine aromatic radicals, 146
Diazo compounds, 146
Diazotation, 146
Diblock copolymer, 133, 168
Dicarboxylic acid chloride, 149
Dicarboxylic acids, 139, 177, 182, 184
Dichloromethane, 178
Dielectric constant, 177

Diene polymerization with Ziegler Natta catalysts, 118
Diethylaluminum chloride, 158
Differential rate equations, 54
Differential scanning calorimetry, 487, 715
Diffusion, 182, 228, 272, 308, 411
Diffusion coefficient, 277
Diffusion controlled copolymerization, 29
Diffusion controlled reactions, 433
Diffusivity coefficients, 336
Difunctional anionic initiator, 155
Difunctional initiators, 153
Dilithio compounds, 153
Dilithio initiators, 154
Diluent recovery, 82
Diols, 177
Diphenols, 95
Dipole orientational depolarization, 650
Dipole orientational polarization, 651
Dipole polarization, 646
Dipole rotation, 673
Direct cationic initiation, 276
Direct electron transfer initiation, 273
Direct polycondensation of bisphenol, 184
Direct polycondensations, 184
Direct polyesterification, 183
Direct reader ICP, 477
Dispersion, 592, 610
Dispersion polymerization, 70
Dissolution process, 351
Distribution function, 648
Distributions in activation volumes, 671
Distributions in relaxation times, 667, 671
Divinyloxy compounds, 265
Domain shapes, 167
Domains in block copolymers, 135
Double bond formation, 325
DSC measurements, 490, 672, 716
DCS thermogram, 491, 494, 720, 721, 729
DSC thermograms for a PVC diethyl malonate gel, 725
Dupont solution polymerization process, 81
Dynamic parameters, 644, 666

Dynamic viscosity, 630
Dynamic viscosity of polystyrene, 631
Dynamics of polystyrene, 311

E

Ealing models, 735
Economization of PVC production, 368
Effective heating rate, 652
Elas phenomenon, 738
Elastic materials, 547
Elastic networks, 750
Elastic properties, 616
Elastic thermoplastics, 540
Elasticity parameters of the melt, 615
Elastomers, 103, 540
Electrochemical activation, 272
Electrochemical behavior, 278
Electrochemical cationic polymerization, 272
Electrochemical initiation, 274
Electrochemical organic synthesis, 272
Electrochemical oxidation of monomeric heteroatomic compounds, 284
Electrochemical reaction, 272
Electrochemical redox behavior, 281
Electrochemically initiated polymerization of methoxystyrenes, 283
Electroinitiated cationic copolymerization of indene, 295
Electroinitiated cationic copolymerizations, 293
Electroinitiated cationic initiation, 275
Electroinitiated cationic polymerization, 271, 273, 274
Electroinitiated polymerization, 11, 271, 281, 282, 287, 289, 291, 295
Electroinitiated polymerization of isocyanate, 290
Electrolysis, 281
Electrolysis of isoprene, 292
Electromagnetic field, 482
Electromagnetic radiation, 496
Electron beam etching, 563
Electron beam interaction, 600
Electron donation, 330
Electron spectroscopy, 308
Electronic transitions, 497

Electrophoretic measurements, 357
Electropolymerization, 302
Elemental analysis, 501
Emulsifiers, 360
Emulsion polymerization, 70, 76, 77, 88, 360, 362, 363, 378
Emulsion polymerization of isoprene, 147
Emulsion polymerization of VCM, 360, 370
Emulsion processes, 352
Emulsion SBR samples, 709
Encapsulation of catalyst sites, 30
Energy absorption, 488
Energy dispersive spectrometer, 500
Energy exposure, 545
Energy of activation, 37
Energy transfer effects, 228
Engineering plastics, 177, 207, 400
Engineering thermoplastics, 572
Entanglements, 614
Enthalpy, 488, 721
Entropy, 240
Epoxy, 544
Epoxy curing reaction, 544
Epoxy resins, 544
Equilibrium conversion, 398
Equilibrium swelling, 758
ESR spectra for propagating radicals, 255, 256
ESR spectroscopy, 137
Esterification, 35, 42, 89, 149
Etch delineated morphology, 552
Etchant temperature, 558
Etched polypropylene, 560
Etched surface morphology for polychlorotrifluoroethylene, 556
Etched surface structure, 552
Etched topography, 549
Etching process, 549
Ethers, 285
Ethylene, 3
Ethylene dichloride, 9
Ethylene homopolymer, 128
Ethylene polymerization, 80
Ethylene polymerization reactor technology, 77
Ethylene propylene block copolymer, 87

Ethylene propylene elastomers, 103, 109, 126
Euler Lagrange equations, 445
Exclusion chromatography, 474
Exothermic polycondensation, 88
Exothermic polymerization reactions, 98
Exposure time, 545
Extended chain morphologies, 494
Extended chain orientation, 203
Extinction between crossed polars, 512
Extruder polymerization processes, 97
Extruder polymerization reactor, 96
Extruder profiles, 349

F

Fabrication process, 511
Failure analysis, 538, 541
Faraday's law, 272
Fatigue, 598
Fatigue properties, 598
Fiber, 589
Fiberglass, 600
Film coating, 288
Film formation, 289
Film properties, 150
Finishing operations, 213
First order catalyst deactivation, 445
First order decay, 446
First order kinetics, 50
Flame atomization, 473
Flame resistant, 302
Flame retardant characteristics, 191
Flameless methods, 472
Flash evaporation, 128
Flash tank, 80
Flexible PVC, 349
Flexible spacer units, 195
Flexible tubing, 349
Flexural modulus, 199, 203
Flooring, 349
Flory-Schulz distributions, 60
Flow chart for nylon 6 manufacture, 244
Fluidized bed, 82
Fluidized bed reactor, 82, 88, 127
Fluorescence microscopy, 543
Fluorescing properties of polycarbonate, 574
Fluorine liberation, 324

Footwear, 169
Form birefringence, 509, 531
Formation exotherm, 718
Formation of a block copolymer, 145
Formation of epoxy resins, 383
Formation of styrene-styrene, 702
Fourier transform infrared spectrometry, 497
Fourier transform mathematical techniques, 498
Fractional precipitation, 166
Fractionation data for polypropylene, 125
Fractionation techniques, 125
Fractography, 564
Fracture mode, 567
Fracture markings, 568
Fracture mode characteristics, 564
Fracture sites, 600
Fracture surface roughness, 567
Free energy, 2
Free ions, 17, 18
Free radical, 1, 2
Free radical formation, 336
Free radical polymerization, 70, 160, 352, 411
Free radical polymerization reactor, 457
Free volume, 673
Frequency dependences of viscoelasticity parameters, 616, 625
Frequency factor, 36
Frequency function, 630
Frequency spectra of relaxation, 607
Friction coefficient, 614
Friction welding, 581
FSTD technique, 668
Fuoss Kirkwood function, 666, 670
Fusion formation process, 711

G

Gas chromatograms, 326
Gas chromatograms of neat isopropylbenzene, 324
Gas chromatography, 322, 473, 474
Gas phase olefin polymerization, 83
Gas phase polymerization, 82, 85, 87, 127

Gas phase processes, 83, 126
Gas phase reactors, 127
Gaussian distribution function, 52, 53
GC/MS instrument, 484
Gel, 677, 711, 717, 750
Gel cohesion, 726
Gel effect, 353, 446, 447, 448
Gel effect in radical polymerization, 410
Gel formation, 715, 719
Gel formation enthalpy, 716, 717
Gel formation temperature concentration phase diagram, 720
Gel melting, 721
Gel melting enthalpy, 727
Gel molecular structure, 730
Gel morphology, 716, 741
Gel permeation chromatography, 167, 254, 474
Gel point, 384, 423
Gel point conversion, 399
Gel state, 738
Gel structure, 721, 738
Gelation, 384, 711, 715, 718, 737, 738
Gelation of syndiotactic polymethyl methacrylate, 718
Gelation phenomena, 713, 717, 733, 741
Gelation process, 716
Gelation temperature, 719
Gels, 713, 729, 733
Gels from atactic polymers, 718
Gels from stereoregular polymers, 713
Gels of isotactic polystyrene, 751
Gibbs free energy, 419
Glass fiber, 549, 552
Glass fiber orientation, 552
Glass reinforced polyesters, 195
Glass state, 423
Glass transition, 185, 494, 495, 539
Glass transition temperature, 88, 135, 496, 663
Glass transition temperature of polycarbonate, 578
Glass transition temperatures, 185
Glass transition temperatures of aromatic polyphosphonates, 192
Glassy segments, 135
Gold, 651
GPC analysis, 677

GPC calibration curves, 703
GPC curves for triblock, 692
GPC measurement, 691
GPC methods, 678, 679
GPC peak, 681
Graft copolymer, 135
Graph theory, 55
Graph polymer, 76
Greassley network theory, 624
Grignard reaction, 160
Group transfer polymerization, 1
Growth polymerization, 88

H

Halogen carbon bonds of halogenoisopropylbenzenes, 330
Halogen cleavage, 320
Halogen containing copolymer resist, 338
Halogen containing isopropyl benzene derivatives, 327
Halogen free substituent effect on photolysis, 324
Halogen substituent effect on photolysis, 324
Halogenated olefins, 3
Hamiltonian, 450
Hammett constants, 333, 334
Harmonic time function, 615
Heat capacity, 490
Heat exchangers, 228
Heat of fusion, 492
Heat of polymerization, 82
Heat of reaction, 87, 231, 252
Heat removal, 69, 76
Heat retention, 556
Heat sensitive initiators, 403
Heat treated fibers of thermotropic liquids, 199
Heat treated nylon, 66, 552
Heat treatment, 203
Heat treatment of polyester fibers, 196
Heating regimes, 652
Heats of polymerization, 69
Heptane, 104, 125
Heteroatomic molecules, 284
Heterogeneous catalysts, 105

Heterogeneous catalytic reactions, 435
Heterogeneous Ziegler Natta catalysts, 111
Hexafunctional polymerization gelation, 403
Hexagonal spherulite, 560
Hexane, 104
High density polyethylene, 77, 128
High elastic deformations, 634
High elasticity state, 616
High energy radiation, 403
High magnification transmission electron micrograph, 560
High molecular weight polyarylates, 181
High molecular weight polymer formation, 283
High pressure liquid chromatography, 474
High pressure technology, 78
High resolution GPC, 698
High resolution GPC curves, 683, 693
High resolution ozonolysis, 687, 688, 695
High resolution ozonolysis GPC curves, 681, 685, 706
High temperature polycondensation, 183
High temperature solution polycondensation, 181
High temperature solution polymerization, 130
Higher molecular weight species, 385
High-polarity solvents, 9
High-pressure ethylene polymerization, 76
High-resolution microscopy, 574
High-temperature polycondensation methods, 179
High radiation beams, 319
Hips polymerization, 75
Hips process, 75
Homogeneous catalysts, 105, 129
Homopolymer separation, 166
Homopolymer blend, 165
Homopolymer effect on stress strain behavior, 166
Homopolymer formation, 148, 153, 156
Homopolymerization, 25, 114, 253
Homopropagation, 24

Hot water stripping, 129
HPLC gels, 679
HPLC measurement, 678, 707
HPLC methods, 677, 679, 695, 709
HPLC separation, 678, 699, 702 708
Huckel molecular orbital, 274
Hybrid processes, 361
Hydrogen abstraction, 11, 319, 333
Hydrogen abstraction by chlorine radicals, 332
Hydrogen chloride acceptors, 181
Hydrogen flluoride benzene solution, 284
Hydrogen peroxide, 7
Hydrogenation, 485
Hydrolization of polyester, 167
Hydrolysis, 265
Hydrolytic polymerization, 212
Hydrolytic stability, 150
Hydroxy derivatives, 328
Hydroxyl terminated polybutadienes, 145
Hydroxyl terminated polystyrene, 161
Hyperfine splitting constants of propagating radicals, 257
Hysteresis, 754

I

Ideal polymerization, 445, 626
Ideal reactor, 227
Image analysis, 556
Image enhancement, 593
Impact strength, 195
Indane, 289
Indirect cationic initiation, 275
Induction period, 50
Inductively coupled plasma, 476
Inductively coupled plasma atomic emission spectroscopy, 476
Industrial polymerization, 70
Infrared spectrometry, 496, 731, 740
Inherent viscosity of polyarylate, 179
Inhibition constant, 9
Inhomogeneity of Ziegler Natta catalysts, 124
Inifer technique, 159
Initial elasticity, 619

Initiation, 57, 403, 407
Initiation mechanism, 275
Initiator efficiency, 6, 404
Initiator species, 1
Initiators, 403
Injection molded copolyesters, 195
Injection molded plastics, 196, 201
Injection molded polypropylene, 560
Injection molding, 193, 203
Intensity ratio, 703
Interfacial polycondensation, 178, 180
Interfacial polymerization, 82, 88, 89
Interferometers, 546
Internal morphology, 552
Internal viscosity, 613
Inter-esterification process, 95
Intramolecular rearrangements of a-radicals, 336
Intrinsic viscosities of electroinitiated polymerization, 287
Inverse emulsion polymerization, 70
Iodine derivatives, 109
Ion chromatography, 478
Ion chromatography flow scheme, 478
Ion exchange chromatography, 474
Ion selective electrodes, 480
Ionic chain polymerization, 9
Ionic initiators, 20
Ionic polymerization, 9, 10, 70
Ionic ring opening polymerizations, 20
Ionization interference, 473
Ionizing radiation, 12
Ionogenic additives, 654
IPS gels, 723, 734
Irreversible electron transfer, 278
Irreversible kinetics schemes, 460
Irreversible polymerization, 423
Isochromes, 581
Isoclines, 581
Isocyanates, 290
Isoprene, 115, 291
Isopropylbenzene, 324
Isotactic, 119
Isotactic polyolefins, 120, 121
Isotactic polypropylene, 85, 123
Isotactic polystyrene, 121, 712, 713, 735
Isotactic polystyrene gels, 741
Isothermal batch reactor, 215

Isothermal crystallization, 493
Isothermal depolarization, 649, 665
Isothermal polymerization of nylon, 6, 214
ITD techniques, 668

J

Jet flow geometries, 528

K

Kargin-Slonimsky-rouse model, 614
Kargin-Slonimsky-rouse theory, 613
Kinetic data, 242
Kinetic equation, 57, 646
Kinetic model, 446
Kinetic model for reversible non linear step growth polymerization, 384
Kinetic relations, 5
Kinetic units, 663
Kinetics, 2
Kinetics of group transfer polymerization, 47
Kinetics of polymerization, 289
Knife frosting, 547
Kraton thermoplastic elastomer, 135

L

Lactamates of alkali metals, 213
Lambert-Beer law, 8
Laminar flow, 224
LaPlace transforms, 58
Laser energy, 563
Laser etching, 563
Law of selective miscibility, 533
LDPE melts, 612
Leather substitutes, 349
Legendre Clebsch condition, 462
Levorotatory copolymer, 259
Lewis acids, 10, 22
Light absorption, 547
Light optical microscopy, 544
Linear chains, 377, 378, 423
Linear periodic deformation, 634
Linear polyesters, 177
Linear polymers, 158, 605, 632, 635

Linear step growth polymerization, 385
Linear thermal expansion, 193
Linear viscoelasticity, 611
Liquid chromatography, 679
Liquid crystalline polymer textures, 532
Liquid crystallinity, 200, 533
Liquid liquid chromatography, 474
Liquid phase etching of polypropylene, 558
Liquid propylene, 87
Liquid slurry polymerization, 80, 85
Liquid solid chromatography, 474
Lithium aluminum hydride, 704
Lithographic fields, 338
Living polymerization, 57, 431
Local chain axis, 523
Locus of polymerization, 355
Long chain branches, 356
Long chain branches in PVC, 356
Long-chain branching, 619, 628, 638
Loop reactors, 80, 129
Low density polyethylene, 78, 129
Low frequency periodic deformation, 619
Low molecular weight model compounds, 308, 320
Low molecular weight polymer, 310
Low molecular weight products, 272
Low pressure ethylene polymerization process, 79
Low pressure IDPE technology, 78
Low temperature solution polycondensation, 180
Low voltage scanning capability, 591
Luminescence, 644

M

Macroazoinitiators, 136, 143
Macrobisperoxides, 136, 141
Macrodisplacement, 645
Macroradicals, 136
Magnifications, 545
Maltese cross contrast from spherulites, 525
Mass polymerization, 356
Mass polymerization of VCM, 357
Mass spectrometer analyzer, 484

Mass spectrometry, 322, 481
Mass transfer of monomer, 30
Mass transfer resistances, 235
Material mixing, 69
Matrix dilution techniques, 501
Maxwell Wagner effect, 650, 660
Maxwell's equation, 508
Mayo Lewis equation, 24, 25
Mean molecular weights, 607
Measurement of birefringence, 513
Mechanical goods, 169
Mechanical properties, 104, 730
Mechanical straining, 600
Mechanism of gel partial melting, 751
Mechanism of radical polymerization, 413
Mechanism of relaxational processes, 672
Mechanochemical synthesis, 138
Melt polycondensation, 89
Melt polymerization, 88, 92, 95
Melt processing, 195
Melt spinning, 193
Melting, 539
Melting behavior of isotactic polystyrene, 721
Melting peak of PTFE, 491
Melting point for nylon, 6
Melting points, 185
Melting temperatures, 723
Melting thermograms, 492
Melt/solid phase polycondensation, 181
Mendelson-Drott hypothesis, 636
Mercury light, 543
Metal counterion, 16
Metal shadowing, 546
Metallography, 544
Metallurgy, 586
Methyl methacrylates, 152, 249
Methyl vinyl ether, 253
Micellar crystallization, 712
Michelson interferometer, 497, 498
Microcracks, 572
Micrometry, 584
Micromixing, 223
Microphase separation, 135, 167
Microscope hardware, 516
Microscopic surface structure, 546

774 / Index

Microscopy techniques, 537, 544
Microsuspension, 361
Microtomed nylon-6 structures, 549
Microtomy, 547
Minimum time problem, 444
Mixing, 586
Mixing patterns, 433
MMD parameter determination, 610
Modeling, 436
Modification of molecular weight distribution, 451
Modified plate and frame filter press, 71
Modulus of elasticity, 598
Modulus temperature relationship, 167
Molding, 578
Molding shrinkage, 195, 498
Molecular dynamics, 645, 649, 662
molecular mass, 616
Molecular mass distribution, 606, 634
Molecular motion, 663
Molecular ordering in polymers, 505
Molecular parameters, 605
Molecular reactivity ratios, 250
Molecular size, 37
Molecular structure, 272, 623, 504, 730
Molecular weight, 10, 31, 37, 70, 165, 167, 254, 457, 693, 694
Molecular weight control, 113, 350
Molecular weight distribution, 1, 51, 71, 96, 156, 272, 376, 384, 429
Molecular weight of the polyarylates, 182
Molecular weight of the polymer, 412
Monochromatic light source, 546
Monodisperse polymers, 612
Monofunctional initiators, 153
Monolithium initiator, 691
Monomer conversion, 221
Monomer reactivity ratios, 683, 689
Monomers, 9, 67, 69, 375, 538
Monomolecular polymer, 630
Monorelaxational process, 646, 650
Monorelaxational processes, 666
Morphologies of plastics, 542
Morphology, 540–545, 739, 741, 760
Morphology of polymer blends, 540
Multiblock copolymer, 133, 145, 165, 740, 750

Multicomponent analysis, 504
Multifunctional cationic initiators, 158
Multifunctional monomers, 423
Multifunctional polymerization, 385, 395, 423
Multiobjective decision analysis, 439, 440
Multiple melting peaks, 492
Multiple steady states, 100
Multiplicity of catalyst species, 30
MWD properties, 442

N

Narrow molecular mass distributions, 612
Nernst equation, 480
Network copolymers, 454
Network model, 614
Neumann's principle, 509
Neutron activation analysis, 485
Newtonian viscosity, 609, 612, 615, 617, 618
Nitrobenzene, 8
NMR analysis, 694, 705
NMR experiments, 731, 733
NMR spectroscopic technique, 111
NMR spectroscopy, 266
NMR techniques, 354
Nomarski differential interference contrast, 543
Noncrystalline transparent solid polymers, 581
Nonequilibrium phase diagram, 723
Nonionic surfactants, 169, 359
Nonpolar polymers, 660
Non-plug flow, 223
Normal stress coefficient, 615
Normal stress difference, 615
Notched toughness, 185
Nuclear magnetic resonance, 30
Nuclear magnetic resonance spectrometry, 486
Nucleating effects, 494
Nucleophiles, 15
Nucleophilic catalysis, 48, 49
Number average molecular weight, 236, 447, 457

Nylon-6, 244, 545, 549, 552
Nylon-6 manufacture, 243
Nylon-6,6 polymerization, 211
Nylon salt, 89, 243
Nylons, 211
Nylon-12 from lauryl lactum, 245
Nylon-6 polymerization, 221
Nylon-6 reactors, 230
Nylon-6 with additives, 549
Nylon-7, 245
Nylon-8, 245
Nylon-9, 245
N-vinylcarbazole, 288

O

Oil resistant coatings for textiles, 245
Olefin monomers, 77
Oligomers, 90, 95, 385, 453
Oligomeric peroxy compounds, 139
Oligomerization, 266
Oligoperoxide copolymerization, 141
Oligoperoxides, 136, 138
Oligoperoxides used in block polymerization, 140
Optical anisotropy in polymers, 506
Optical crystallographers, 506
Optical diffraction pattern, 516
Optical hot-stage microscopy, 583
Optical indicatrix, 507, 508, 511
Optical light microscopy, 541
Optical methods, 166
Optical microscope, 542, 546, 593
Optical microscope heating stage, 574
Optical microscopy, 505, 545, 563, 741
Optical pleochroism, 523
Optical technique, 574
Optically active copolymers, 260
Optimal control theory, 435
Optimal regulation, 436
Optimization of polymerization processes, 429, 442
Optimization of polymerization reactors, 452
Optimization of transient operation, 459
Optimization of tubular reactors, 236
Ordering in polymers, 524
Organic disulfides, 136

Organic glasses, 540
Organic materials, 540
Organoaluminum compounds, 104, 113
Organolithium initiators, 690
Organometallic compounds, 15, 20, 105, 109, 259
Organometallic derivatives, 10
Orientation birefringence, 509, 520
Orthogonal directions, 508
Oxidation, 308
Oxidation reduction reactions, 7
Oxidation resistant metal, 651
Oxide acetylene flame, 472
Oxyheterocyclic monomers, 284
Ozonolysis, 705
Ozonolysis and HPLC methods, 677
Ozonolysis conditions, 705
Ozonolysis GPC method, 682
Ozonolysis of SBR, 680
Ozonolysis of styrene, 705
Ozonolysis products, 691, 698, 706
Ozonolysis products of SBR, 683
Ozonolysis–GPC, 689, 693
Ozonolysis–GPC curves of copolymer, 684
Ozonolysis–GPC curves of SBR 1006, 687

P

Packing of GPC columns, 679
Parabolic velocity profile, 224
Paramorphoses, 533
Partially blocked sbr with anionic polymerization, 689
Particle formation at low conversion, 356
Particle nucleation, 357
Particle number density, 357
Particle size distribution, 76, 350
Partition chromatography, 474
Passivation effect, 295
Pentane, 9
Peracetic acid, 147
Perbenzoic acid, 147
Perchloric acid, 11
Periodic deformation, 615
Peroxides, 6

Peroxidized polymers, 141
Peroxy groups, 148
Perturbational Huckel molecular orbital considerations, 274
PET, 195
PET manufacture, 89
PET polymers, 89
PET reactors, 381
Phase arrangements, 581
Phase behavior, 69
Phase contrast, 542
Phase contrast microscopy, 543
Phase delineation, 574
Phase diagram, 716, 751
Phase inversion point, 76
Phase plane analysis, 462
Phase segregation in block copolymers, 135
Phenol acetate, 182
Phenol silyl ether route, 183
Phenols, 284
Phenyl ester, 181
Phillips Petroleum company, 128
Phonograph records, 349
Phosphonyl radicals, 251
Photochemical initiation, 7
Photochemical reactions of polystyrene derivatives, 309
Photocoloration, 312
Photocoloration of polystyrene, 339
Photocoloration of polystyrene derivatives, 319, 334, 335
Photocoloration problem, 325
Photoinduced phenyl radicals, 321
Photoinitiated polymerizations, 7
Photolysis, 313, 319
Photolysis for polystyrene, 308, 311
Photolytic initiation, 272
Photomultiplier, 472
Photon correlation spectroscopy, 357
Photooxidation, 317
Photooxidation of polystyrene, 317
Physical gelation, 711, 713, 717
Physical ranging, 661
Pinner synthesis, 145
Plasma sprayed coatings, 206
Plasticized PVC, 348, 750
Plasticizer, 70, 350
Plasticizer absorption, 437

Plug flow velocity profiles, 228
POB copolyester, 195
Pointcare sphere construction, 518
Poisoning, 113
Poisson distribution, 52
Polar orientation, 521
Polar polymers, 657
Polar solvents, 151, 154
Polarization, 644, 646, 651
Polarization conditions, 651, 654, 673
Polarization state of light, 513
Polarization times, 654
Polarized azimuth, 519
Polarized light, 542, 572, 589
Polarized light microscopy, 542
Polyacrylates, 120
Polyacrylonitrile gel, 695
Polyamication reaction, 44
Polyamide, 66, 88, 90
Polyamide formation, 45
Polyarylate polymerization process, 95
Polyarylates, 95, 179, 180, 184, 201, 204, 207
Polyazoester synthesis, 145, 181
Polyazoesters, 145
Polyazomide formula, 145
Polyazomides, 145
Polybutadiene, 103, 134, 149
Polybutylene terephthalate, 539
Polycarbonate polybutyleneterephthalate, 578
Polycarbonate polydimethylsiloxane block copolymer, 150
Polycarbonates, 88
Polycarbonate-polybutyleneterephthalate photoelasticity, 581
Polychlorotrifluoroethylene, 539, 556
Polychromatic kinetics, 649
Polychromatic light, 546
Polycondensation, 67, 87, 88, 182
Polycondensation equilibrium, 90
Polycondensation of ARB monomers, 453
Polycondensation of diacid chlorides, 181
Polycondensation of dicarboxylic acid chlorides, 181
Polycondensation reactions, 87, 89
Polycyclic hydrocarbons, 285

Polydiene polyester block copolymers, 155
Polydisperse polymers, 616
Polydispersed index, 381
Polydispersed long-chain species, 380
Polydispersity, 442, 606, 632
Polydispersity index, 221, 222, 230, 376, 398, 423, 620
Polyester, 142
Polyesterification, 36, 37, 42
Polyesterification of adipic acid, 39–41
Polyesterification reaction, 42
Polyesters, 88
Polyether blocks, 167
Polyetheramide block copolymer, 151
Polyethlene melts, 129
Polyethylene, 77, 78, 511, 556, 717
Polyethylene fibers, 756
Polyisoprene, 103
Polyisoprene blocks, 147
Polymer chain axes, 511
Polymer chain structure, 78
Polymer characteristics, 539
Polymer chemistry, 538
Polymer drying steps, 82
Polymer films, 531
Polymer growth, 2
Polymer linkage, 88
Polymer molecular weight, 285, 292
Polymer phases, 546
Polymer physics, 538
Polymer science, 538
Polymer separation, 82
Polymer solvent interactions, 412
Polymer thermograms, 657
Polymer washing, 88
Polymeric azo initiators, 147
Polymeric electrets, 650
Polymeric polymers, 620
Polymerization, 6, 71, 538, 684
Polymerization by electrochemical technique, 272
Polymerization catalysts, 125
Polymerization chemistry of alkenes, 104
Polymerization conditions for polystyrene gel, 679
Polymerization in a CSTR, 460
Polymerization in screw reactors, 95

Polymerization initiators, 139
Polymerization kinetic systems, 55
Polymerization kinetics, 1, 8, 67, 430
Polymerization mechanism of Ziegler Natta catalysts, 109
Polymerization models, 451
Polymerization of allylic compounds, 253
Polymerization of caprolactum, 212
Polymerization of dienes, 115
Polymerization of hexafunctional monomer, 391
Polymerization of isoprene, 116
Polymerization of nylon-6, 213
Polymerization of olefins, 77
Polymerization of propylene, 111
Polymerization of styrene, 12, 18, 19, 142, 273, 441
Polymerization of tetrahydrofuran, 285
Polymerization of vinyl chloride, 347
Polymerization potential, 297
Polymerization rates, 1, 8, 13, 22, 368
Polymerization reaction engineering, 70
Polymerization reactions, 51, 114
Polymerization reactor configuration, 70
Polymerization reactor dynamics, 460
Polymerization reactors, 68, 70, 126, 460
Polymerization temperature, 352, 353, 369
Polymerization zone, 626
Polymerography, 538–542
Polymethacrylates, 120
Polymethyl methacrylate, 70, 598
Polymolecularity, 637
Polynuclear aromatics, 497
Polyolefin helixes, 121
Polyolefins, 121
Polyoxyethylene, 155
Polyoxyethylene glycol, 143
Polypropylene, 125, 539, 545
Polypropylene fractionations, 125
Polypropylene process comparison, 86
Polypropylene spherulite, 560
Polypropylene spherulite morphology, 560
Polystyrene gel, 699
Polystyrene, 70, 120–149, 160, 166, 539, 540, 714, 718

Polystyrene derivative polymer films, 336
Polystyrene derivatives, 310
Polystyrene film, 308
Polystyrene gel, 702
Polysulfones, 150
Polytetrafluoroethylene, 206
Polyurethanes, 45, 88, 583
Polyvinyl acetate, 70
Polyvinyl chloride, 120, 712, 718
Polyvinylcyclohexane, 120
Pontryagin's minimum principle, 462
Porosity requirements, 350
Postpolymerization, 283
Power law, 617, 753
Preparation of block copolymers, 152
Preparation of PVC by emulsion, 348
Prepolymers, 73
Primary photochemical reaction, 310
Primary PVC particles, 356
Primary radical formation, 328
Primary radicals, 5, 6
Probability of ionic reactions, 46
Probability of propagation, 414, 423
Production of polyamides, 452
Production yields for halogen containing isopropyl benzenes, 327
Program cooled crystallization, 493
Propagation, 4, 15, 54, 57, 407
Propagation rate constant, 413
Propagation rates, 28
Propagation reactions, 26, 122, 404, 414, 445
Properties of olefins, 122
Properties of thermotropic liquid crystalline polyarylates, 198
Propylene oxides, 19
Propylene polymerization reactors, 84
Propylene polymerization, 84, 85, 125
Propylene reactivity, 114
Protogen, 10
Proton donor, 10
Protonic acids, 10
Pseudo homogeneous catalysts, 105
PVC gels, 726, 749, 760
PVC grafted stabilizer, 360
PVC grains, 363
PVC industry, 348

PVC latex, 350
PVC lattices, 350
PVC pregels, 749
PVC properties, 350
PVC resins, 350
PVC thermogram, 644
Pyridine, 184

Q

Quality aspects, 350
Quality control, 433, 541
Quantitative microscopy, 541
Quantum yields, 314
Quantum yields of deexcitation, 314
Quasi cleavage, 586
Quaterpolymers, 260
Quenching regimes, 661

R

Radial refractive index, 527
Radiation induced changes in polymer films, 317
Radiation induced chemical reactions, 319
Radiation induced reactions, 317, 330
Radiation induced reactions of monomer units, 308
Radiation induced reactions of polystyrene derivatives, 307
Radiationless deactivation of polystyrene derivatives, 314
Radical balance, 365
Radical cationdication, 275
Radical cations, 14
Radical chain polymerization, 3, 5
Radical copolymerization of isobutyl vinyl ether, 260
Radical copolymerization of PVC, 266
Radical copolymerizations of cis-propenyl vinyl ether, 267
Radical polymerization, 9, 13, 249, 403, 407, 411, 445
Radical polymerization of cis-propenyl vinyl ether, 262
Radical polymerization of divinyl ethers, 260

Radical polymerication of methyl methacrylate, 417
Radical polymerization of MMA, 259
Radical polymerization of vinyl ethers, 249, 253
Radical polymerizations of divinyl formal divinyl acetal, 263
Radical reactivity of vinyl ethers, 250, 251
Radical termination, 136
Radiofrequency coupling, 476
Radiolysis, 313
Raman spectrum, 496
Random SBR with anionic polymerization, 687
Random sequence distribution, 687
Randomizing agent, 687
Raoult's law, 231, 379
Rate constant for esterification, 35
Rate constants, 4, 5, 27, 36, 236, 252, 384
Rate of copolymerization, 28
Rate of generation of polymer radicals, 408
Rate of polymerization, 5, 243, 285
Rate of propagation, 5, 17
Rate of the reaction, 272
Ray velocity, 514
Reaction between linear chains, 378
Reaction mechanism, 1
Reaction media in polymerization processes, 431
Reaction rate, 450
Reaction rate of the a-radicals, 311
Reaction temperature, 260
Reactivity of tert butyl vinyl, 252
Reactivity ratios, 24, 126, 292, 293, 300, 302
Reactivity ratios for radical polymerization, 258
Reactor, 67
Reactor cooling, 358
Reactor design, 30, 430
Reactor dynamics, 433
Reactor fouling, 76
Reactor models, 227
Reactor performance, 223
Reactor staging, 76

Reactors for polyamide polymerization, 89
Reactors for polycarbonates, 92
Reactors for polyethylene terephthalate, 89
Real tubular reactors, 223
Recrystallization, 545
Redox initiation system, 369
Redox systems, 360
Reducing the high melting point, 201
Reflectance, 541
Reflected light, 574
Reflux condenser, 228, 356
Reflux condenser cooling, 358
Refractive index, 135, 507, 510, 541, 581
Refractive index changes, 583
Refractive index gradients, 516
Relative intensity, 689
Relative reactivities of olefins, 114
Relative reactivities of various monomers, 250
Relaxation, 600, 661
Relaxation frequencies, 615
Relaxation function, 615
Relaxation parameters, 662
Relaxation time spectrum, 624
Relaxation times, 193, 613, 647–664
Relaxation transition temperatures, 666
Relaxational processes, 661, 664
Reorientation of diple, 672
Residence time, 381
Residence time distribution, 76, 226
Residence time distribution on chain length distribution, 432
Residence times in cascades, 222
Resinography, 538
Resonance structures, 46
Reversible linear step growth polymerization, 377
Reversible radical polymerization, 419
Rheological experiments, 719
Rheological parameters, 620
Rheology, 586, 749
Rheometer, 749, 754
Ring opening polymerization, 19, 20, 22
Roughness, 567
Rubber elasticity, 756

Rubber latex with polystyrene, 593
Rubber vulcanizates, 153, 169

S

Sample preparation, 543
SBR, 707
Scaling laws, 749
Scanning electron microscope, 567, 591, 600
Scanning ICP, 477
Schulz distribution, 609
Screw reactors, 96
Sealents, 169
Secondary emulsifiers, 359
Secondary photochemical reactions, 312
Secondary suspension stabilizer, 358, 359
Seed polymerization, 361
Segmental diffusion, 411
Selectivity of ionic polymerization, 9
Self reinforced polyesters, 195
Self termination, 31
Self-propagation, 24
Sem imaging techniques, 592
Semibatch copolymerization, 442
Semibatch operation, 455
Semicrystalline, 491
Sensitivity to radiation, 339
Sequence distribution in SBR, 680
Sequence distribution of anionically polymerized SBR, 682
Sequence distribution of commercial SBR, 686
Sequence distribution of styrene, 678, 686
Shearing flow, 611
Sintering techniques, 206
Skin free s PVC, 360
Slurry polymerization, 128, 129
Smectic liquid crystals, 512
Sodium dodecyl sulfate, 362
Sodium hydroxide production, 348
Sodium naphthalene, 153
Sodium naphthalene difunctional initiators, 153
Softening point, 153
Solid catalyzed coordination polymerization, 100

Solid phase process, 182
Solid state polymerization, 88, 245
Soluble transition metal compounds, 107
Solution chromatography, 474
Solution polycondensation, 88, 145
Solution polymerization, 70, 76, 81, 88, 349
Solution processes, 129
Solvent extraction, 166
Sonic intensity, 138
Sources of birefringence, 509
Soxhlet apparatus, 166
Specific heat, 490
Spectrometry, 472
Spherulite crystallography, 527
Spherulites, 524, 526, 549, 556, 560
Spherulitic microstructure, 527
Spin density, 257
Spray drying, 361
Stabilizing forces, 357
Star block copolymers, 133, 159
Star branched polymer, 163
Static reactors, 71
Stationary flow, 612, 615, 634
Steady state flow, 612
Step growth polymerization, 67, 87, 148, 161, 377, 381, 385, 403
Step growth polymerization of bifunctional monomers, 423
Step polymerization, 1, 4, 34, 42, 60
Step reactions, 431, 452
Stereomicroscopy, 541, 542, 598
Stereoregular polymerizations, 376
Stereoregular polymers, 262
Stereoregularity, 272, 430
Stereoscopic image, 600
Stereoscopical microscope, 542
Stereospecific polymerization reactions, 119
Steric hindrances, 241
Stirling approximation, 52
Stirred tank reactor, 356
Strain, 600
Stress birefringence, 578, 581
Stress in glass or polymer, 541
Stress relaxation, 578
Stress rupture resistance, 185
Stress strain diagram, 168
Stretched gels, 740

Structural characterization of styrene butadiene copolymers, 677
Structure of clearen, 530–531, 695
Structure of KX-65, 692
Structure of propagating radical, 255
Structure of solprene, 411, 694
Structure property relationships in polyarylates, 185
Structures of functional group species, 392
Styrene, 3, 282, 681, 699
Styrene butadiene copolymer, 700
Styrene methyl methacrylate copolymer, 679
Styrene polymerization, 70
Styrene-rubber morphology, 76
Substitution on aromatic rings, 197
Supercooled liquids, 657
Surface activity, 30
Surface chemistry, 545
Surface reflectance, 545
Surface relief, 546
Surface roughness, 546, 598
Surface texture, 549
Surface topography, 552, 560
Surface morphology, 546
Suspension polymerization, 70, 76, 128, 355–365, 369
Suspension polymerization of VCM using redox initiation, 371
Suspension polymerization rate curves, 363
Suspension process, 128
Suspension stabilizers, 359, 362
Swelling, 757, 758
Swelling behavior, 756
Swelling behavior of polyvinyl chloride, 758
Swelling of PVC, 351
Swelling of PVC at low pressures, 351
Swelling ratio, 756
Syncatalyst system, 10
Syndiotactic, 119
Syndiotactic polydiene, 123
Syndiotactic polyolefin, 124
Syndiotactic polypropylene, 119
Syndiotacticity, 352
Synthesis of aromatic polyesters, 178
Synthesis of block copolymers, 135, 155
Synthesis of oligoperoxides, 139
Synthesis of polypropylene, 128
Synthesize multiblock copolymers, 148
Synthetic polyamides, 211
Synthetic rubbers, 103
S-B-S triblock copolymers, 690, 697, 698

T

Tapered blocks, 154
Telechelic polydienes, 142
Telechelics, 152
Temperature concentration phase diagram, 727, 728
Temperature control, 455
Temperature effect in radical polymerization, 418
Temperatures concentration phase diagram, 724
Tensile load, 583
Terephthalic acids, 184, 204
Termination, 4, 407
Termination of polymer radicals, 405
Termination rate, 6
Termination rate constant, 29
Termination reaction, 28
Tetrahydrofurane, 156
Tetramethyl silane, 486
Textile industry, 540
TGA spectrum of calcium oxalate, 488
Thermal analysis, 487
Thermal analysis curve, 488
Thermal gravimetric analysis, 487
Thermal homolysis, 6
Thermal initiation, 448
Thermal initiation of the polyazoester, 145
Thermal polymerization of styrene, 447
Thermal properties, 185
Thermal properties of polyvinyl chloride, 725
Thermal treatments, 494
Thermodynamic principle, 493
Thermodynamics, 721
Thermograms, 644, 653
Thermograms for high-density polyethylene, 661
Thermoplastic, 169

Thermoplastic elastomer, 134, 143, 146, 150
Thermoplastics, 135, 539, 540
Thermoreversible gelation, 712
Thermosetting plastics, 540
Thermostimulated conductivity, 644
Thermostimulated depolarization currents, 653
Thermostimulated discharge, 643
Thermotropic liquid crystalline aromatic polyarylates, 193
Thermotropic liquid crystalline copolyarylates, 200, 201
Thermotropic liquid crystalline polyarylates, 191
Thermotropic liquid crystallinity, 195, 203
Thin film casting, 538
Thin film deposition, 546
Third order reaction, 44
Three-stage copolymerization, 691
Time conversion curves, 50
Time resolved spectroscopy, 330
Titanium, 104
Toluene, 104
Tower-type reactor, 127
Tracer studies, 226
Transesterification, 151
Transfer agent, 409
Transformation technique, 670
Transition metal, 104
Transition metal carbon bonds, 110
Transition metal carbonyls, 137
Transition metal compound, 32, 104, 106
Transition metal olefin complex, 33
Transition metallic catalysts, 84
Transition state rate theory, 36
Transitional motions, 674
Transmission electron microscopy, 547, 591
Transmittance, 497
Transparent coating, 541
Transparent films, 166
Transport phenomena, 433
Triblock copolymers, 133, 151, 677, 678, 690, 691
Triblock SBR solprene, 411, 694

Trihydroxy terminated polyisobutylene, 152
Truncation error, 379
TSD current, 652–655
TSD current measuring chamber, 656
TSD currents, 656
TSD experiments, 652
TSD peak parameters, 653
TSD peaks, 652, 660, 666
TSD technique, 643, 658, 663, 672
TSD theory, 644
TSD thermograms, 644, 658, 672
Tubular ethylene polymerization reactor, 76
Tubular reactor, 626
Turbidimetric titration, 166
Turbidity titrator, 166
Two phase polycondensation, 179
Two phase polymerization processes, 353
Two phase systems, 650
Type of chain polymerization, 3
Type of polymerization reactor, 444

U

Ultrasonic degradation, 138
Ultrasonic waves, 138
Ultrasound, 298
Ultraviolet stability, 185
Ultraviolet stabilizer, 185
Uncured SBR, 704
Unfilled nylon-6, 547
Uniaxial indicatrix, 521
Uniaxially deformed conventional polymer, 528
Uniaxially sheared films of liquid crystalline polymers, 529
Union Carbide Company, 127
Unsaturated monomers, 3
Urethane formation, 152
Uses of PVC resins, 349
UV light, 136
UV radiation, 137

V

Vanadium, 104

Vanadium based polymerization catalysts, 112
Vanadium catalyzed redox initiator system, 370
Vapor liquid equilibrium, 229, 235, 245, 379
VCM activity, 368
VCM production, 348
VCM soluble electrolytes, 357
VCM soluble quaternary ammonium salts, 358
VCM suspension, 356
Vectra, 201
Velocity profile, 223, 224
Vinyl acetate, 76
Vinyl acetate emulsion, 136
Vinyl chloride polymerization, 352
Vinyl ethers, 3, 249
Vinyl monomers, 137, 145, 148, 249, 257
Vinyl polymerization, 14
Vinylcyclohexane, 115
Viscoelastic functions, 629
Viscoelastic properties, 165, 611, 613, 628
Viscoelasticity parameters, 625, 631, 634
Viscosity of epoxy resins, 544
VK column, 228, 229, 231
VK tube, 231
Volatilization, 42
Voltammetry, 276
Voltammograms, 281
Volume relaxation effects, 351

W

Water immiscible organic liquids, 179
Water solubility of VCM, 361
Water soluble initiators, 360
Wavelength dispersive XFR spectrometer, 500
Weight average molecular weights, 384, 447
Weld interface, 578
Welding, 578
Wiped film reactor, 89

X

XRF excitation, 499
Xylene, 476
Xylene etch, 549
X-ray diffraction, 524, 716, 736, 740
X-ray fluorescence instrumentation, 499
X-ray fluorescence spectrometry, 498

Y

Young's modulus, 756, 760

Z

Zeiss Contron image processor, 597
Ziegler catalyst, 164
Ziegler Natta catalysis, 29, 103, 110, 113–129
Ziegler Natta polymerization, 29–33

About the Editor

NICHOLAS P. CHEREMISINOFF heads the product development group in the Polymers Technology Division of Exxon Chemical Company. He is responsible for directing the research and development of specialty elastomers for the consumer and various industry segment markets, and conducts research on polymer rheology/processing and development of quality control instrumentation for polymer manufacturing. Dr. Cheremisinoff has had extensive experience in the chemical and allied industries, with particular interests in the design and scale-up of multiphase reactors. The author/co-author of over thirty-five books, he received his B.S., M.S., and Ph.D. degrees in chemical engineering from Clarkson College of Technology.